Calculus

for JEE Main & Advanced

2000+
Problems with Solutions...

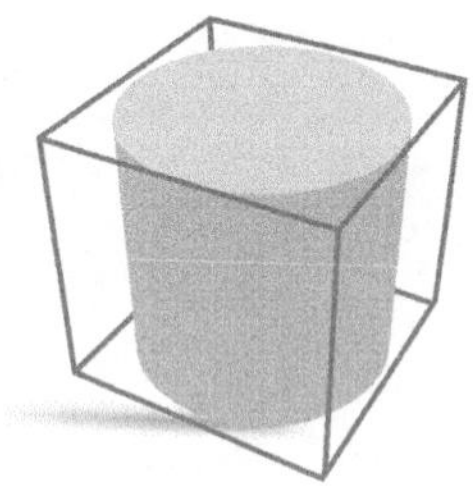

Unproblem JEE

Calculus

for JEE Main & Advanced

2000+

Problems with Solutions...

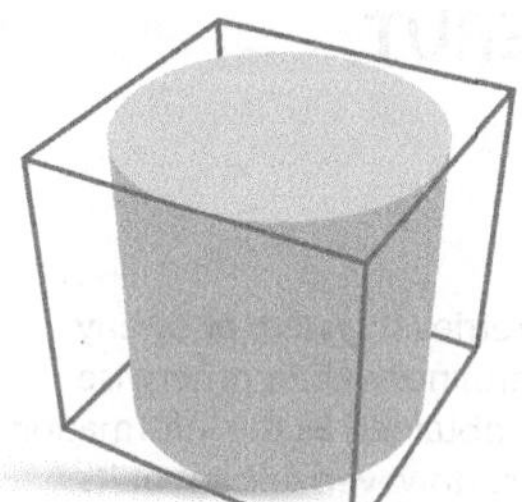

Vinod Kumar Pareek
B.Tech

arihant
ARIHANT PRAKASHAN (Series), MEERUT

arihant

ARIHANT PRAKASHAN (Series), MEERUT

Administrative & Production Offices

Regd. Office
'Ramchhaya' 4577/15, Agarwal Road, Darya Ganj, New Delhi -110002
Tele: 011- 47630600, 43518550

Head Office
Kalindi, TP Nagar, Meerut (UP) - 250002
Tel: 0121-7156203, 7156204

Sales & Support Offices
Agra, Ahmedabad, Bengaluru, Bareilly, Chennai, Delhi, Guwahati, Hyderabad, Jaipur, Jhansi, Kolkata, Lucknow, Nagpur & Pune.

ISBN : 978-93-26199-95-7

PRICE : ₹525.00

PO No : TXT-XX-XXXXXXX-X-XX

Published by Arihant Publications (India) Ltd.

For further information about the books published by Arihant, log on to www.arihantbooks.com or e-mail at info@arihantbooks.com

Follow us on

Preface

Unproblem JEE Series is an attempt to enrich and enliven the Mathematical Problem Skills in JEE & other engineering entrances aspirants. Most of the Problem Books available in the segment are not substantially capable to serve the purpose for which an engineering aspirant goes for the problem books. They are just a compilation of problems packed in different chapters without any logical arrangement & purposeful grading. Thus, they fail to provide the readers with the approach required to solve complex & structured problems generally asked in JEE.

The Second book of Unproblem JEE Series, **'Problems in Calculus'** is an effort to detoxify the fear of calculus that many students have by giving them the fact that Calculus does not require any special knowledge rather it is just one additional step beyond algebra & trigonometry. Through the problems given in the book, students will develop the strengths towards calculus through without having any additional burden to them.

Problems in each chapter are arranged to promote problem-solving ability and make the students understand that concepts & techniques are logically connected. In this way, students will become comfortable with concepts & will know how and when to use them while solving a problem.

The chapters in the book are divided into three levels and each level is a complete treatment of the chapter but at different levels of difficulty.

- Each chapter is divided into three major heads; **Level 1** 'Starter Level', **Level 2** 'JEE Main Level' & **Level 3** 'JEE Advanced Level'.
- **Level 1 'Starter Level'** consists of preliminary questions that test basic concepts. After completing this level student will have a solid understanding of fundamentals of the chapter necessary to solve the problems.
- **Level 2 'JEE Main Level'** has single choice correct questions and numerical value questions aimed especially for JEE Main.
- **Level 3 'JEE Advanced Level'** consists of various types of questions; MCQ with one or more than One Correct Answer, Comprehension Based Questions, Match the Columns and Numerical Value problems as per JEE Advanced Pattern.
- The problems given in JEE Advanced Level are made to be conceptually challenging resulting in practical and relevant practice for JEE Advanced.
- Detailed solutions to all the problems have been made available online just to help the students in case they stuck somewhere while solving some questions.

At the end, I can say confidently that if you rigorously practice the problems given in this book then surely you will get an edge in Mathematics and will increase your chance of getting good JEE Score.

However, I have taken the utmost care while writing this book but if any error or whatsoever has skipped my attention then please bring it to my notice so that it can be rectified. I would also like to appeal all my fellow Mathematics Teachers to have a look at the book and share their valuable suggestions for its further improvement.

Vinod Kumar Pareek

B.Tech

Contents

CHAPTER

01

FUNCTIONS

LEVEL 01 \ Starter Level

1.1 Domain and Range of Functions

1 Find the domain of each of the following functions

(i) $f(x)=\dfrac{x-1}{x+2}$ (ii) $f(x)=\dfrac{1}{x^2-3x+2}$

(iii) $f(x)=\dfrac{x^2+4}{x^2+2}$ (iv) $f(x)=\sqrt{4-x}$

(v) $f(x)=\dfrac{1}{\sqrt{4-x}}$ (vi) $f(x)=\sqrt{x-1}-\sqrt{7-x}$

(vii) $y=\dfrac{1}{\sqrt[3]{2x-5}}$

2 Find the domain of the following functions

(i) $f(x)=\sqrt{|x|-2}$ (ii) $f(x)=\sqrt{|x|-2}+\sqrt{4-|x|}$

(iii) $f(x)=\sqrt{\dfrac{|x|-1}{5-|x|}}$

3 Find the range and domain of the following functions

(i) $f(x)=\text{sgn}\,x^2$ (ii) $f(x)=\text{sgn}(x^2+5)$

(iii) $f(x)=\text{sgn}(-x^2+4x-7)$

4 Find the domain of the following functions

(i) $f(x)=\dfrac{1}{[x+2]}$ (ii) $f(x)=\dfrac{1}{\sqrt{x-[x]}}$

5 Find the domain and range of following functions

(i) $y=\dfrac{1}{\{x\}}$ (ii) $y=\log_{10}\{x\}$

(iii) $y=[\sin\{x\}]$, where [·] greatest integer function.

6 Find the domain of inverse trigonometric functions

(i) $f(x) = \sin^{-1}(3x + 4)$

(ii) $f(x) = \sin^{-1}(\log_3 x)$

(iii) $f(x) = \sin^{-1}\left(\frac{3}{4 + 2\sin x}\right)$

(iv) $f(x) = \sin^{-1}\left(\frac{3 - 2x}{5}\right) + \sqrt{3 - x}$

(v) $f(x) = \cos^{-1}(2x^2 - 3)$

7 Find the domain of the following functions

(i) $f(x) = \log_2(x^2 - 4x + 3)$

(ii) $f(x) = \frac{\log(3 - x)}{x^2 - 5x + 4}$

(iii) $f(x) = \frac{1}{\log(9 - x^2)}$

(iv) $f(x) = \log_{1/2}\left(\frac{2 - x}{x - 4}\right)$

8 Find the range of the following functions

(i) $f(x) = \frac{x}{1 + x^2}$

(ii) $f(x) = \frac{2 + x^2}{x^2}$

(iii) $f(x) = \frac{x^2 - 3x + 2}{x^2 + x - 6}$

(iv) $f(x) = \sqrt{4 - x^2}$

(v) $f(x) = \frac{1}{2 - \cos 5x}$

(vi) $f(x) = 4^x - 2^x + 1$

(vii) $f(x) = \cot^{-1}(2x - x^2)$

1.2 Types of Functions

9 Identify that the following functions is one-one or many one and into or onto.

(i) $f : R \to R, f(x) = |x|$

(ii) $f : [0, \infty) \to [0, 1], f(x) = \frac{x}{1 + x}$

(iii) If the function $f(x)$ and $g(x)$ are defined on $R \to R$ such that

$$f(x) = \begin{cases} 0, & x \in \text{rational} \\ x, & x \notin \text{rational} \end{cases} \text{ and } g(x) = \begin{cases} 0, & x \in \text{irrational} \\ x, & x \notin \text{irrational} \end{cases}$$, then check the nature of $(f - g)(x)$

(iv) $f : N \to Z, f(n) = \begin{cases} \frac{n-1}{2}, & \text{when } n \text{ is odd} \\ -\frac{n}{2}, & \text{when } n \text{ is even} \end{cases}$

10 If $A = \{1, 2, 3\}$ and $B = \{a, b, c, d\}$ for $f : A \to B$, then find the total number of

(i) mappings

(ii) one-one mappings

(iii) many-one mappings

(iv) onto mappings

(v) into mappings

(vi) constant mappings

11 If $A = \{1, 2, 3, 4\}$ and $B = \{a, b, c\}$ for $f : A \to B$, then find the total number of

(i) mappings

(ii) one-one mappings

(iii) many-one mappings

(iv) onto mappings

(v) into mappings

(vi) constant mappings

1.3 Even and Odd Functions

12 Determine the nature of each of the following functions

(i) $f(x) = x\sin(x^2 + 3)$ (ii) $f(x) = \sin x + \cos x$

(iii) $f(x) = \log\left(\frac{2-x}{2+x}\right)$ (iv) $f(x) = x\sin^2 x + \tan(x^5)$

(v) $f(x) = \sqrt{x^2 - 2x + 4} - \sqrt{x^2 + 2x + 4}$

1.4 Periodic Functions

13 Find the period of the following functions

(i) $f(x) = \sin 4x$ (ii) $f(x) = 5\operatorname{cosec}(3x - 5) + 7$

(iii) $f(x) = \frac{a^x}{a^{[x]}}$, where [·] represents the greatest integer function.

(iv) $f(x) = |\sin x| + \cos x$ (v) $f(x) = \frac{|\sin x - \cos x|}{|\sin x| + |\cos x|}$

(vi) $f(x) = |\sin 3x| - \tan\left(\frac{6x}{5}\right)$

1.5 Identical Functions

14 Check whether the following functions are identical or not.

(i) $f(x) = \ln x^2$ and $g(x) = 2\ln x$

(ii) $f(x) = \operatorname{cosec} x$ and $g(x) = \frac{1}{\sin x}$

(iii) $f(x) = \operatorname{sgn}(x^2 + 1)$ and $g(x) = \sin^2 x + \cos^2 x$

(iv) $f(x) = \log_e x$ and $g(x) = \frac{1}{\log_x e}$

1.6 Composition of Functions

15 Let $f : \{2,3,4,5\} \to \{3, 4, 5, 9\}$ and $g : \{3, 4, 5, 9\} \to \{2, 11, 15\}$ be a function defined as $f(2) = 3$, $f(3) = 4$, $f(4) = f(5) = 5$, $g(3) = g(4) = 7$ and $g(5) = g(9) = 11$, then find *gof*.

16 Let two real functions f and g are defined as $f : R \to R$ and $g : R \to R$, $f(x) = x^2 - 2$ and $g : R \to R$, $g(x) = x - 5$, then find *fog* and *gof*.

17 Find *fog* and *gof* for the functions $f(x) = \sin x$ and $g(x) = \sqrt{x + 5}$.

18 Find *fog* and *gof* for the functions $f(x) = \cos x$ and $g(x) = |x|$.

1.7 Inverse of Functions

19 Find the inverse of the following functions

(i) $f(x) = x^2 - x + 1$

(ii) $f(x) = \frac{x-1}{x-2}$

(iii) $f(x) = e^{5x+13}$

(iv) $f(x) = 2(3)^x + 1$

(v) $f : [-1, 1] \to [-1, 1]$, $f(x) = x^2 \operatorname{sgn}(x)$

1.8 Functional Equations

20 If $f(x)+2f(1-x)=x^2+2\ \forall\ x \in R$, then find $f(x)$.

21 If $f(x)+f\left(\frac{x-1}{x}\right)=1+x\ \forall\ x \in R$, then find $f(x)$.

22 If $f(x+y)=f(x)+f(y)-xy-1\ \forall\ x, y \in N$ and $f(1)=1$, then find the value of $\sum_{r=1}^{5} f(r)$.

23 If $f\left(\sqrt{\frac{x-1}{x+1}}\right)=3x$, what is the value of $f(3)$?

24 The function $f: R \to R$ satisfies the condition $mf(x-1)+nf(-x)=2|x|+1$. If $f(-2)=5$ and $f(1)=1$, then find the value of $m+n$.

LEVEL 02 \ JEE Main Level

Single Choice Correct Questions

1.1 Domain and Range of Functions

1 The domain of the function $f(x)=\sqrt{1-\sqrt{1-\sqrt{1-x^2}}}$ is

(a) $(-\infty, 1)$
(b) $(-1, \infty)$
(c) $[0, 1]$
(d) $[-1, 1]$

2 The domain of the function $f(x)=\sqrt{(x-4)-2\sqrt{x-5}}-\sqrt{x-4+2\sqrt{x-5}}$ is

(a) $(5, \infty)$
(b) $(4, \infty)$
(c) $[5, \infty)$
(d) $[4, \infty)$

3 The domain of the function $f(x)=\sqrt{\frac{2-|x|}{3-|x|}}$ is

(a) $(-\infty, \infty)-[-2, 2]$
(b) $(-\infty, \infty)-[-3, 3]$
(c) $[-2, 2] \cup (-\infty, -3) \cup (3, \infty)$
(d) None of these

4 The domain of the function $f(x)=\frac{1}{\sqrt{[x]^2-[x]-6}}$ is

(a) $(-\infty, -2) \cup [4, \infty)$
(b) $(-\infty, -2] \cup [4, \infty)$
(c) $(-\infty, -2) \cup (4, \infty)$
(d) None of these

5 Let $[x]$ denotes the greatest integer $\le x$, where $x \in R$. If the domain of real valued function $f(x)=\sqrt{\frac{|[x]|-2}{|[x]|-3}}$ is $(-\infty, a) \cup [b, c) \cup [4, \infty)$, $a<b<c$, then the value of $a+b+c$ is **[JEE Main 2021]**

(a) 8
(b) 1
(c) -2
(d) -3

6 The domain of $f(x)=\frac{1}{\sqrt{|[|x|-1]|-5}}$ is, where $[\cdot]$ represents the greatest integer function

(a) $x \in R-\{7\}$
(b) $x \in (-\infty, -5] \cup [5, \infty)$
(c) $x \in (-\infty, -7] \cup [7, \infty)$
(d) $x \in R-\{5\}$

7 The domain of function $f(x)=\dfrac{1}{\sqrt{\{\sin x\}+\{\sin(\pi+x)\}}}$, where $\{\cdot\}$ denotes the fractional part function, is

(a) $x \in R$

(b) $x \in R - \dfrac{n\pi}{2}, n \in I$

(c) $x \in R - \left(n\pi + \dfrac{\pi}{4}\right), n \in I$

(d) None of these

8 The domain of the function $f(x)=\dfrac{3}{x}+2^{\sin^{-1}x}+\dfrac{1}{\sqrt{x-2}}$ is

(a) $[-1, 1]$

(b) $R-\{0\}$

(c) $[-1, 0] \cup (0, 1)$

(d) ϕ

9 The domain of definition of the function $f(x)=\cos^{-1}[2-3x^2]$, where $[\cdot]$ denotes the greatest integer function is

(a) $(-1, 1)$

(b) $[-1, 1]$

(c) $(-1, 1)-\{0\}$

(d) $[-1, 1]-\{0\}$

10 The real valued function $f(x)=\dfrac{\operatorname{cosec}^{-1}x}{\sqrt{x-[x]}}$, where $[x]$ denotes the greatest integer less than or equal to x, is defined for all x belonging to **[JEE Main 2021]**

(a) all real except integers

(b) all non-integers except the interval $[-1, 1]$

(c) all integers except 0, -1 and 1

(d) all real except the interval $[-1, 1]$

11 The domain of the function $f(x)=\sqrt{\sec^{-1}\left(\dfrac{2-|x|}{4}\right)}$ is

(a) $(-\infty, -6)$

(b) $(-\infty, -6) \cup (6, \infty)$

(c) $(-\infty, -6] \cup [6, \infty)$

(d) $(-\infty, -2] \cup [2, \infty)$

12 The domain of the function $f(x)=\sin^{-1}\left(\dfrac{3x^2+x-1}{(x-1)^2}\right)+\cos^{-1}\left(\dfrac{x-1}{x+1}\right)$ is **[JEE Main 2021]**

(a) $\left[0, \dfrac{1}{4}\right]$

(b) $[-2, 0] \cup \left[\dfrac{1}{4}, \dfrac{1}{2}\right]$

(c) $\left[\dfrac{1}{4}, \dfrac{1}{2}\right] \cup \{0\}$

(d) $\left[0, \dfrac{1}{2}\right]$

13 The domain of the function $\operatorname{cosec}^{-1}\left(\dfrac{1+x}{x}\right)$ is **[JEE Main 2021]**

(a) $\left(-1, \dfrac{-1}{2}\right] \cup (0, \infty)$

(b) $\left[-\dfrac{1}{2}, 0\right) \cup [1, \infty)$

(c) $\left(-\dfrac{1}{2}, \infty\right) - \{0\}$

(d) $\left[-\dfrac{1}{2}, \infty\right) - \{0\}$

14 The domain of definition of the function $f(x)=\sqrt[4]{\dfrac{1-3^x}{5^{-x}-5}}$ is

(a) $(-\infty, -1) \cup [0, \infty)$

(b) $(-\infty, -1] \cup [0, \infty)$

(c) $(-\infty, -1) \cup (0, \infty)$

(d) None of these

15 The domain of the definition of the function $f(x) = \log_4[\log_5\{\log_3(18x - x^2 - 77)\}]$ is

(a) (8, 10) (b) [8, 10] (c) $(-\infty, 8)$ (d) $[10, \infty)$

16 The domain of definition of the function $f(x) = \log_{10}\{1 - \log_{10}(x^2 - 5x + 16)\}$ is

(a) (2, 3) (b) [2, 3] (c) $(3, \infty)$ (d) $(-\infty, 2)$

17 The domain of the definition of the function $f(x) = \dfrac{1}{4 - x^2} + \log_{10}(x^3 - x)$ is **[JEE Main 2019]**

(a) $(-1, 0) \cup (1, 2) \cup (3, \infty)$ (b) $(-2, -1) \cup (-1, 0) \cup (2, \infty)$
(c) $(-1, 0) \cup (1, 2) \cup (2, \infty)$ (d) $(1, 2) \cup (2, \infty)$

18 The domain of the function $f(x) = \sin^{-1}\left\{\log_2\left(\dfrac{1}{2}x^2\right)\right\}$ is

(a) $[-2, -1) \cup [1, 2]$ (b) $(-2, -1) \cup [1, 2]$
(c) $[-2, -1] \cup [1, 2]$ (d) $(-2, -1) \cup (1, 2)$

19 Let $f(x) = \max\{\sin t : 0 \le t \le x\}$, $g(x) = \min\{\sin t, 0 \le t \le x\}$ and $h(x) = [f(x) - g(x)]$, where [] is the greatest integer function, then range of $h(x)$ is

(a) {0, 1} (b) {1, 2}
(c) {0, 1, 2} (d) $\{-3, -2, -1, 0, 1, 2, 3\}$

20 The range of the function $f(x) = \dfrac{e^x - e^{|x|}}{e^x + e^{|x|}}$ is

(a) [0, 1) (b) $(1, \infty)$ (c) $(-1, 0]$ (d) $(-\infty, 0)$

21 Let $f : (1, 3) \to R$ be a function defined by $f(x) = \dfrac{x[x]}{1 + x^2}$, where $[x]$ denotes the greatest integer $\le x$. Then, the range of f is **[JEE Main 2020]**

(a) $\left(\dfrac{2}{5}, \dfrac{3}{5}\right] \cup \left(\dfrac{3}{4}, \dfrac{4}{5}\right)$ (b) $\left(\dfrac{2}{5}, \dfrac{4}{5}\right]$
(c) $\left(\dfrac{3}{5}, \dfrac{4}{5}\right)$ (d) $\left(\dfrac{2}{5}, \dfrac{1}{2}\right) \cup \left(\dfrac{3}{5}, \dfrac{4}{5}\right]$

22 The range of the function $f(x) = \ln\left\{\dfrac{1}{\sqrt{[\cos x] - [\sin x]}}\right\}$, $[x] \le x$ is

(a) $\left(0, \dfrac{\pi}{4}\right)$ (b) $\left\{\dfrac{\pi}{2}\right\}$ (c) {0} (d) None of these

23 The range of the function $f(x) = \log_2(\sqrt{(x-4)} + \sqrt{6 - x})$ is

(a) $\left[0, \dfrac{1}{2}\right]$ (b) $\left[\dfrac{1}{2}, 1\right]$ (c) [0, 1] (d) (0, 1)

24 Let $f: R \to R$ be defined by $f(x) = \dfrac{x}{1 + x^2}$, $x \in R$. Then, the range of f is **[JEE Main 2019]**

(a) $\left[-\dfrac{1}{2}, \dfrac{1}{2}\right]$ (b) $(-1, 1) - \{0\}$ (c) $R - \left[-\dfrac{1}{2}, \dfrac{1}{2}\right]$ (d) $R - [-1, 1]$

25 The range of the function $f(x) = \log_e(3x^2 - 4x + 5)$ is

(a) $\left[-\infty, \log_e \dfrac{11}{3}\right)$ (b) $\left[\log_e \dfrac{11}{3}, \infty\right)$
(c) $\left[-\log_e \dfrac{11}{3}, \log_e \dfrac{11}{3}\right]$ (d) None of these

1.2 Types of Functions

26 The function $f:[0, 2] \to [1, 4]$ is defined by $f(x) = x^3 - 5x^2 + 7x + 1$ is

(a) one-one and onto (b) onto but not one-one
(c) one-one but not onto (d) neither one-one nor onto

27 Let $f: R \to R$ be defined by $f(x) = \frac{x-a}{x-b}$, where $a \neq b$. Then, f is

(a) injective but not surjective (b) surjective but not injective
(c) bijective (d) None of these

28 Let x denotes the total number of one-one functions from a set A with 3 elements to a set B with 5 elements and y denotes the total number of one-one functions from the set A to the set $A \times B$. Then, **[JEE Main 2021]**

(a) $2y = 91x$ (b) $2y = 273x$ (c) $y = 91x$ (d) $y = 273x$

29 If $f(x) = \frac{\sin([x]\pi)}{x^2 + x + 1}$, where $[\cdot]$ denotes the greatest integer function, then

(a) f is one-one (b) f is not one-one and non-constant
(c) f is a constant function (d) None of these

30 Let a function $f:(0, \infty) \longrightarrow (0, \infty)$ be defined by $f(x) = \left|1 - \frac{1}{x}\right|$. Then, f is **[JEE Main 2019]**

(a) injective only (b) both injective as well as surjective
(c) not injective but it is surjective (d) neither injective nor surjective

31 Let $f: R \to R$ be defined as $f(x) = 3^{-|x|} - 3^x + \text{sgn}(e^{-x}) + 2$ (where $\text{sgn}\, x$ denotes signum function of x). Then, which of the following is correct?

(a) f is injective but not surjective (b) f is surjective but not injective
(c) f is injective as well as surjective (d) f is neither injective nor surjective

32 The function $f: R \to \left[-\frac{1}{2}, \frac{1}{2}\right]$ defined as $f(x) = \frac{x}{1+x^2}$ is **[JEE Main 2017]**

(a) invertible (b) injective but not surjective
(c) surjective but not injective (d) neither injective nor surjective

33 Let a function f is defined from $R \to R$ as $f(x) = \begin{cases} m - x & , x \leq 1 \\ 2mx + 1 & , x > 1 \end{cases}$. If the function is onto on R, then the range of m, is

(a) $[-2, \infty)$ (b) $[-2, 0)$ (c) $\{-2\}$ (d) $(0, \infty)$

34 If $f:(-1, 1) \to B$ be a function defined by $f(x) = \tan^{-1}\left(\frac{2x}{1-x^2}\right)$, then f is both one-one and onto, when B is the interval

(a) $\left[-\frac{\pi}{2}, \frac{\pi}{2}\right]$ (b) $\left(-\frac{\pi}{2}, \frac{\pi}{2}\right)$ (c) $\left[0, \frac{\pi}{2}\right]$ (d) $(0, \pi/2)$

35 The number of functions from $f: \{a, b, c, d, e, f\} \to \{p, q, r\}$ is

(a) 6^3 (b) 3^6 (c) $3!$ (d) $\frac{6!}{3!}$

36 Set A consists of 6 different elements and set B consists of 4 different elements. The number of mapping which can be defined from the set $A \to B$ which are surjective is

(a) 256 (b) 432 (c) 840 (d) 1560

37 The number of functions f from $\{1, 2, 3, \dots, 20\}$ onto $\{1, 2, 3, \dots, 20\}$ such that $f(k)$ is a multiple of 3, whenever k is a multiple of 4, is

(a) $(15)! \times 6!$ (b) $5^6 \times 15$ (c) $5! \times 6!$ (d) $6^5 \times (15)!$

38 The number of bijection that can be defined from $A : \{1, 2, 8, 9\}$ to $B : \{3, 4, 5, 10\}$ is

(a) 4^4 (b) 4^2 (c) 24 (d) 18

1.3 Even and Odd Functions

39 A function $f(x)$ is given by $f(x) = \begin{cases} x^2 \sin\frac{\pi x}{2}, & |x| < 1 \\ x|x|, & |x| \geq 1 \end{cases}$ is

(a) an even function (b) an odd function
(c) a periodic function (d) None of these

40 If $f(x) = \dfrac{x(\sin x + \tan x)}{\left[\dfrac{x+\pi}{\pi}\right] - 1/2}$ where $x \neq n\pi$ and [] denotes the greatest integer function, then

(a) $f(x)$ is an odd function (b) $f(x)$ is an even function
(c) $f(x)$ is neither even nor odd function (d) None of these

41 If $g : [-2, 2] \to R$, where $f(x) = x^3 + \tan x + \left[\dfrac{x^2+1}{p}\right]$ is an odd function, then the value of parameter p, where $[\cdot]$ denotes the greatest integer function is

(a) $-5 < p < 5$ (b) $p < 5$ (c) $p > 5$ (d) None of these

42 If $f(x) = \dfrac{x}{e^x - 1} + \dfrac{x}{2} + 1$, then $f(x)$ is

(a) an even function (b) odd function
(c) neither even nor odd function (d) None of these

43 Let $f : R \to R$ such that $f(x+y) + f(x-y) = 2f(x)f(y)\ \forall\ x, y \in R$ and $f(0) \neq 0$, then $f(x)$ is an

(a) even function (b) odd function
(c) periodic function (d) None of these

1.4 Periodic Functions

44 Let $f(x) = a^x$, $(a > 0)$ be written as $f(x) = f_1(x) + f_2(x)$, where $f_1(x)$ is an even function and $f_2(x)$ is an odd function. Then, $f_1(x+y) + f_1(x-y)$ equals **[JEE Main 2019]**

(a) $2f_1(x+y) \cdot f_2(x-y)$ (b) $2f_1(x+y) \cdot f_1(x-y)$
(c) $2f_1(x) \cdot f_2(y)$ (d) $2f_1(x) \cdot f_1(y)$

45 The fundamental period of $f(x) = 4\cos^4\left(\dfrac{x-\pi}{4\pi^2}\right) - 2\cos\left(\dfrac{x-\pi}{2\pi^2}\right)$ is equal to

(a) π^3 (b) $4\pi^2$ (c) $3\pi^2$ (d) $2\pi^3$

46 Let $[x]$ denotes the greatest integer function. If the function $f(x) = \tan(\sqrt{[n]}\, x)$ has period $\dfrac{\pi}{3}$, then the values of λ is

(a) [3, 4] (b) [9, 10] (c) (9, 10) (d) [9, 10)

47 The period of the function $f(x) = ||\sin x| - |\cos x||$, is

(a) $\dfrac{\pi}{2}$ (b) π (c) 2π (d) None of these

1.5 Identical Functions

48 Which of the following functions is/are identical to $|x-2|$?

(a) $f(x)=\sqrt{x^2-4x+4}$
(b) $g(x)=|x|-|2|$
(c) $h(x)=\dfrac{|x-2|^2}{|x-2|}$
(d) $p(x)=\left|\dfrac{x^2-x-2}{x+1}\right|$

49 Which part of following are identical?

(a) $\sin^{-1}(\sin x)$ and $\sin(\sin^{-1}x)$
(b) $\log_e e^x$ and $e^{\log_e x}$
(c) $\log_e x^2$ and $2\log_e x$
(d) $x|x|$ and $x^2\,\text{sgn}x$

50 If the functions are defined as $f(x)=\sqrt{x}$ and $g(x)=\sqrt{1-x}$, then what is the common domain of the following functions?

$f+g,\ f-g,\ f/g,\ g/f,\ g-f$, where $(f\pm g)(x)=f(x)\pm g(x)$, $(f/g)(x)=\dfrac{f(x)}{g(x)}$ **[JEE Main 2021]**

(a) $0\le x\le 1$
(b) $0\le x<1$
(c) $0<x<1$
(d) $0<x\le 1$

51 The value of x for which $f(x)=\cos x$ and $g(x)=\dfrac{1}{\sqrt{1+\tan^2 x}}$ are identical is

(a) $x\in R$
(b) $x\in\bigcup_{n\in Z}\left(-\dfrac{\pi}{2}+2n\pi,\dfrac{\pi}{2}+2n\pi\right)$
(c) $x\in\bigcup_{n\in Z}(2n\pi,2n\pi+\pi)$
(d) None of these

1.6 Composition of Functions

52 Let $f:R\left\{\dfrac{\alpha}{6}\right\}\to R$ be defined by $f(x)=\dfrac{5x+3}{6x-\alpha}$. Then, the value of α for which $(fof)(x)=x$, for all $x\in R-\left\{\dfrac{\alpha}{6}\right\}$ is **[JEE Main 2021]**

(a) 5
(b) 6
(c) 8
(d) no such α exists

53 If $f(x)=\dfrac{1}{x}$ and $g(x)=\dfrac{1}{x^2}$ and $h(x)=x^2$, then

(a) $fog(x)=x^2,\ x\ne 0$ and $h(g(x))=\dfrac{1}{x^2}$
(b) $h(g(x))=\dfrac{1}{x^2},\ x\ne 0$ and $fog(x)=x^2$
(c) $fog(x)=x^2,\ x\ne 0$ and $hog(x)=(g(x))^2,\ x\ne 0$
(d) None of the above

54 For $x\in\left(0,\dfrac{3}{2}\right)$, let $f(x)=\sqrt{x}$, $g(x)=\tan x$ and $h(x)=\dfrac{1-x^2}{1+x^2}$. If $\phi(x)=((hof)og)(x)$, then $\phi\left(\dfrac{\pi}{3}\right)$ is equal to **[JEE Main 2019]**

(a) $\tan\dfrac{\pi}{12}$
(b) $\tan\dfrac{11\pi}{12}$
(c) $\tan\dfrac{7\pi}{12}$
(d) $\tan\dfrac{5\pi}{12}$

55 If $g(x) = x^2 + x - 2$ and $\frac{1}{2} gof(x) = 2x^2 - 5x + 2$, then which is not a possible $f(x)$?

(a) $2x - 3$ (b) $-2x + 2$

(c) $x - 3$ (d) None of these

56 If $g(x) = x^2 + x - 1$ and $(gof)(x) = 4x^2 - 10x + 5$, then $f\left(\frac{5}{4}\right)$ is equal to **[JEE Main 2020]**

(a) $-\frac{1}{2}$ (b) $\frac{3}{2}$ (c) $-\frac{3}{2}$ (d) $\frac{1}{2}$

57 Let $f(x) = \frac{kx}{x+1}$, $x \neq -1$, then the value of k is, $f(f(x)) = x$

(a) 1 (b) -1 (c) 2 (d) 0

58 For $x \in R - \{0, 1\}$, let $f_1(x) = \frac{1}{x}$, $f_2(x) = 1 - x$ and $f_3(x) = \frac{1}{1-x}$ be three given functions. If a function, $J(x)$ satisfies $(f_2 o J o f_1)(x) = f_3(x)$, then $J(x)$ is equal to **[JEE Main 2019]**

(a) $f_2(x)$ (b) $f_3(x)$

(c) $f_1(x)$ (d) $\frac{1}{x} f_3(x)$

59 If $f(x) = \frac{1-x}{1+x}$; $x \neq 0$, then $f[f(x)] + f\left[f\left(\frac{1}{x}\right)\right]$

(a) < 2 (b) ≥ 2 (c) $= 2$ (d) None of these

60 Let $f, g : N \to N$, such that $f(n+1) = f(n) + f(1) \; \forall \; n \in N$ and g be any arbitrary function. Which of the following statements is not true? **[JEE Main 2021]**

(a) if fog is one-one, then g is one-one (b) if f is onto, then $f(n) = n \; \forall \; n \in N$

(c) f is one-one (d) if g is onto, then fog is one-one

1.7 Inverse of Functions

61 Let $f : \left[-\frac{\pi}{3}, \frac{2\pi}{3}\right] \to [0, 4]$ be a function defined by $f(x) = \sqrt{3} \sin x + \cos x + 2$, then $f^{-1}(x)$ equals

(a) $\frac{2\pi}{3} - \cos^{-1}\left(\frac{x-2}{2}\right)$ (b) $\sin^{-1}\left(\frac{x-2}{2}\right) + \frac{\pi}{3}$

(c) $\sin^{-1}\left(\frac{x-2}{2}\right)$ (d) $\sin^{-1}\left(\frac{x+2}{2}\right) + \frac{\pi}{6}$

62 If the function $f : [1, \infty] \to [1, \infty)$ is defined by $f(x) = 2^{x(x-1)}$, then $f^{-1}(x)$ is

(a) $\left(\frac{1}{2}\right)^{x(x-1)}$ (b) $\frac{1}{2}(1 + \sqrt{1 + 4\log_2 x})$

(c) $\frac{1}{2}(1 - \sqrt{1 + 4\log_2 x})$ (d) not defined

63 The inverse function of $f(x) = \frac{8^{2x} - 8^{-2x}}{8^{2x} + 8^{-2x}}$, $x \in (-1, 1)$ is **[JEE Main 2020]**

(a) $\frac{1}{4}(\log_8 e)\log_e\left(\frac{1-x}{1+x}\right)$ (b) $\frac{1}{4}\log_e\left(\frac{1-x}{1+x}\right)$

(c) $\frac{1}{4}(\log_8 e)\log_e\left(\frac{1+x}{1-x}\right)$ (d) $\frac{1}{4}\log_e\left(\frac{1+x}{1-x}\right)$

64 The inverse of $f(x)=\dfrac{e^{3x}-e^{-3x}}{e^{3x}+e^{-3x}}$ is

(a) $\dfrac{1}{6}\log\left(\dfrac{1+x}{1-x}\right)$ (b) $\dfrac{1}{6}\log\left(\dfrac{x}{1-x}\right)$ (c) $\dfrac{1}{6}\log\left(\dfrac{1+x}{1-x}\right)$ (d) $\dfrac{1}{6}\log\left(\dfrac{1-x}{1+x}\right)$

65 The inverse of $y=5^{\log x}$ is **[JEE Main 2017]**

(a) $x=5^{\log y}$ (b) $x=y^{\frac{1}{\log 5}}$

(c) $x=5^{\frac{1}{\log y}}$ (d) $x=y^{\log 5}$

66 Let $a>1$ be a real number and $f(x)=\log_a x^2$ for $x>0$. If f^{-1} is the inverse function of f and b and c are real numbers, then $f^{-1}(b+c)$ is equal to

(a) $f^{-1}(b)\cdot f^{-1}(c)$ (b) $f^{-1}(b)+f^{-1}(c)$

(c) $\dfrac{1}{f(b+c)}$ (d) $\dfrac{1}{f^{-1}(b)+f^{-1}(c)}$

67 Let $f:R-\{3\}\to R-\{1\}$ be defined by $f(x)=\dfrac{x-2}{x-3}$. Let $g:R\to R$ be given as $g(x)=2x-3$.

Then, the sum of all the values of x for which $f^{-1}(x)+g^{-1}(x)=\dfrac{13}{2}$ is equal to **[JEE Main 2021]**

(a) 7 (b) 2 (c) 5 (d) 3

68 Suppose $f(x)=(x+1)^2$ for $x\geq -1$. If $g(x)$ is the function whose graph is the reflection of the graph of $f(x)$ with respect to the line $y=x$, then $g(x)$ equals

(a) $-\sqrt{x}-1, x\geq 0$ (b) $\dfrac{1}{(x+1)^2}, x>-1$

(c) $\sqrt{x+1}, x\geq -1$ (d) $\sqrt{x}-1, x\geq 0$

69 Consider function $f:A\to B$ and $g:B\to C$, $A, B, C\subseteq R$ such that $(gof)^{-1}$ exists, then **[JEE Main 2021]**

(a) f and g both are one-one (b) f and g both are onto

(c) f is one-one and g is onto (d) f is onto and g is one-one

70 If $f(x)=\sin x+\cos x$, $g(x)=x^2-1$, then $g(f(x))$ is invertible in domain

(a) $\left[-\dfrac{\pi}{2},0\right]$ (b) $\left[-\dfrac{\pi}{4},\dfrac{\pi}{4}\right]$ (c) $\left[-\dfrac{\pi}{2},\pi\right]$ (d) $(0,\pi)$

71 If $f:R\to R$, $g:R\to R$ defined as $f(x)=3x-4$ and $g(x)=2+3x$, then $(g^{-1}of^{-1})$ is

(a) 1 (b) $\dfrac{1}{2}$ (c) $\dfrac{1}{3}$ (d) $\dfrac{1}{5}$

72 Let $f:R\to R$, $g:R\to R$ be two functions given by $f(x)=2x-3$, $g(x)=x^3+5$, then $(fog)^{-1}(x)$ is equal to

(a) $\left(\dfrac{x+7}{2}\right)^{1/3}$ (b) $\left(x-\dfrac{7}{2}\right)^{1/3}$ (c) $\left(\dfrac{x-2}{7}\right)^{1/3}$ (d) $\left(\dfrac{x-7}{2}\right)^{1/3}$

1.8 Functional Equations

73 Let $f(x)=ax^7+bx^3+cx-5$ where a, b and c are constants. If $f(-7)=7$, then $f(7)$ equals

(a) -17 (b) -7 (c) 17 (d) -3

74 Let f be a real valued function of real and positive argument such that

$f(x)+3xf\left(\frac{1}{x}\right)=2(x+1)$ for all real $x>0$

Then, the value of $f(10099)$ is

(a) 550 (b) 505 (c) 5050 (d) 10010

75 A function $f(x)$ is given by $f(x)=\frac{5^x}{5^x+5}$, then the sum of the series

$$f\left(\frac{1}{20}\right)+f\left(\frac{2}{20}\right)+f\left(\frac{3}{20}\right)+\ldots+f\left(\frac{39}{20}\right) \text{ is equal to}$$ [JEE Main 2021]

(a) $\frac{29}{2}$ (b) $\frac{49}{2}$ (c) $\frac{39}{2}$ (d) $\frac{19}{2}$

76 If $f(x)+2f\left(\frac{1}{x}\right)=3x, x\neq 0$ and $S=\{x\in R: f(x)=f(-x)\}$, then S

(a) is an empty set (b) contains exactly one element

(c) contains exactly two elements (d) contains more than two elements

77 Let $f: N\to N$ is a function such that $f(m+n)=f(m)+f(n)$ for every $m, n\in N$. If $f(6)=18$, then $f(2)\cdot f(3)$ is equal to [JEE Main 2021]

(a) 6 (b) 54 (c) 18 (d) 36

78 A real valued function $f(x)$ satisfies the functional equation $f(x-y)=f(x)\cdot f(y)-f(a-x)\,f(a+y)$ where a is a given constant and $f(0)=1$, $f(2a-x)$ is equal to

(a) $f(a)+f(a-x)$ (b) $f(-x)$ (c) $-f(x)$ (d) $f(x)$

79 Let $a, b, c\in R$. If $f(x)=ax^2+bx+c$ be such that $a+b+c=3$ and $f(x+y)=f(x)+f(y)+xy$, $\forall\, x, y\in R$, then $\sum_{n=1}^{10} f(n)$ is equal to [JEE Main 2017]

(a) 330 (b) 165 (c) 190 (d) 255

Numerical Value Type Questions

80 The value of $\left[\frac{3}{4}\right]+\left[\frac{3}{4}+\frac{1}{100}\right]+\left[\frac{3}{4}+\frac{2}{100}\right]+\ldots+\left[\frac{3}{4}+\frac{99}{100}\right]$ is ....

81 If $[x]$ is the greatest integer less than or equal to x, then $\sum_{n=8}^{100}\left[\frac{(-1)^n n}{2}\right]$ is equal to .... [JEE main 2021]

82 If x and y satisfy the equation $y=2[x]+3$ and $y=3[x-2]$, where $[\cdot]$ denotes the greatest integer function, then $[x+y]$ is ....

83 If the function $f: R\to A$ given by $f(x)=\frac{x^2}{x^2+1}$ is surjection, then $A\in[k, m)$, then the value of $k+m$ is ....

84 Let $A: \{0, 1, 2, 3, 4, 5, 6, 7\}$, then the number of bijective functions $f: A\to A$ such that $f(1)+f(2)=3-f(3)$ is equal to ....

85 Let $A=\{a, b, c\}$ and $B=\{1, 2, 3, 4\}$. Then, the number of elements in the set $C=\{f: A\to B \mid 2\in f(A) \text{ and } f \text{ is not one-one}\}$ is .... [JEE Main 2020]

86 The function $f(x)=\dfrac{x+1}{x^3+1}$ can be written as the sum of an even function $g(x)$ and an odd function $f(x)$, then the value of $g(0)$ is ....

87 The function $f(x)=\min\{x-[x], -x-[-x]\}$ is a periodic function with period ....

88 Let $f: R \to R$ be a continuous and differential function such that $(f(x^2+1))^{\sqrt{x}}=5$ for $\forall x \in (0, \infty)$.

Then, the value of $\left(f\left(\dfrac{16+y^2}{y^2}\right)\right)^{\frac{4}{\sqrt{y}}}$ for $y \in (0, \infty)$ is equal to ....

89 Let $f: R-\{0\} \to R$ be any function such that $f(x)+2f\left(\dfrac{1}{x}\right)=3x$. The sum of the values of x for which $f(x)=1$ is ....

90 If $a+\alpha=1, b+\beta=2$ and $af(x)+\alpha f\left(\dfrac{1}{x}\right)=bx+\dfrac{\beta}{x}, x \neq 0$, then the value of expression

$\dfrac{f(x)+f\left(\dfrac{1}{x}\right)}{x+\dfrac{1}{x}}$ is .... **[JEE Main 2021]**

91 A non-zero function $f(x)$ is symmetrical about the line $y=x$, then the value of λ such that $f^2(x)=(f^{-1}(x))^2-\lambda xf(x)\ f^{-1}(x)+3x^2f(x)\ \forall x \in R^f$ is ....

92 Let $f(x)$ and $g(x)$ be functions which take integer arguments.

Let $f(x+y)=f(x)+g(y)+8$ for all integers x and y. Let $f(x)=x$ for all negative integers x and let $g(8)=17$. The value of $f(0)$ is ....

LEVEL 03 \ JEE Advanced Level

Single Choice Correct Questions

1 The complete set of values of a such that $\dfrac{x^2-x}{1-ax}$ all real values is

(a) $[0, \infty)$ (b) $[1, \infty)$ (c) $(-\infty, 1]$ (d) R

2 Let $f`: D \to R$ be defined as $f(x)=\dfrac{x^2+2x+\alpha}{x^2+4x+3\alpha}$, where D and R denotes the domain of f and the set of all real numbers, respectively. If f is surjective mapping, then the range of α is

(a) $0 \le \alpha \le 1$ (b) $0 < \alpha \le 1$

(c) $0 \le \alpha < 1$ (d) $0 < \alpha < 1$

3 The domain of the function

$f(x)=\sqrt{2\{x\}^2-3\{x\}+1}, x \in [-1, 1]$, where {} denotes the fractional part function, is

(a) $\left(-1, -\dfrac{1}{2}\right) \cup (0, 1/2)$ (b) $[-1, -1/2] \cup [0, 1/2]$

(c) $(-1, -1/2) \cup \left[0, \dfrac{1}{2}\right] \cup \{1\}$ (d) $[-1, -1/2] \cup [0, 1/2] \cup \{1\}$

4 The domain of the derivative of the function $f(x)=\begin{cases}\tan^{-1}x & \text{if } |x|\le 1\\ \frac{1}{2}(|x|-1) & \text{if } |x|>1\end{cases}$ is

(a) $R-\{0\}$ (b) $R-\{1\}$ (c) $R-\{-1\}$ (d) $R-\{-1,1\}$

5 If the function $f: R\rightarrow R$ is defined by $f(x)=|x|(x-\sin x)$, then which of the following statements is true?

(a) f is one-one but not onto (b) f is onto, but not one-one
(c) f is both one-one and onto (d) f is neither one-one nor onto

6 The function $f: R\rightarrow R$ defined as $f(x)=\frac{1}{2}\ln(\sqrt{\sqrt{x^2+1}+x}+\sqrt{\sqrt{x^2+1}-x})^2$ is

(a) one-one and onto both (b) one-one but not onto
(c) onto but not one-one (d) neither one-one nor onto

7 If $f(x)=\tan^2\left(\frac{\pi x}{n^2-5n+8}\right)+\cot(n+m)\pi x$, $n\in N$, $m\in Q$, is a periodic function with 2 as its fundamental period, then m can not belong to

(a) $(-\infty,-2)\cup(-1,\infty)$ (b) $(-\infty,-3)\cup(-2,\infty)$
(c) $(-3,-2)\cup(-2,-1)$ (d) $(-3,-5/2)\cup(-5/2,-2)$

8 For $x\in R$, $x\neq 0, 1$, let $f_0(x)=\frac{1}{1-x}$ and $f_{n+1}(x)=f_0(f_n(x))$, $n=0,1,2,\ldots$, then the value of $f_{100}(3)+f_1\left(\frac{2}{3}\right)+f_2\left(\frac{3}{2}\right)$ is equal to

(a) $\frac{4}{3}$ (b) $\frac{1}{3}$ (c) $\frac{5}{3}$ (d) $\frac{8}{3}$

9 Let $f(x)=4x(1-x)$, $0\le x\le 1$. The number of solution of $f(f(f(x)))=\frac{x}{3}$ is

(a) 2 (b) 4 (c) 8 (d) 16

10 The number of solutions of $2\cos x=|\sin x|$, $0\le x\le 4\pi$, is

(a) 0 (b) 2 (c) 4 (d) infinite

11 Let f be a real valued periodic function defined for all real numbers such that for some fixed $a>0$, $f(x+a)=\frac{1}{2}+\sqrt{f(x)-\{f(x)\}^2}$ for all x. Then, period of $f(x)$ is

(a) a (b) $2a$ (c) $3a$ (d) $4a$

More than One Correct

12 Which of the following pairs of function is/are identical?

(a) $f(x)=\tan(\tan^{-1}x)$ and $g(x)=\cot(\cot^{-1}x)$
(b) $f(x)=\text{sgn}(x)$ and $g(x)=\text{sgn}(\text{sgn}(x))$
(c) $f(x)=\cot^2 x\cdot\cos^2 x$ and $g(x)=\cot^2 x-\cos^2 x$
(d) $f(x)=e^{\ln\sec^{-1}x}$ and $g(x)=\sec^{-1}x$

13 Let $f:\left(-\frac{\pi}{2},\frac{\pi}{2}\right)\rightarrow R$ be given by $f(x)=\{\log(\sec x+\tan x)\}^3$. Then,

(a) $f(x)$ is an odd function (b) $f(x)$ is a one-one function
(c) $f(x)$ is an onto function (d) $f(x)$ is an even function

14 Let $f(x)=\sin\left(\frac{\pi}{6}\sin\left(\frac{\pi}{2}\sin x\right)\right)$ for all $x \in R$ and $g(x)=\frac{\pi}{2}\sin x$ for all $x \in R$. Let $fog(x)$ denote $f(g(x))$ and $gof(x)$ denote $g(f(x))$. Then, which of the following is (are) true?

(a) Range of f is $\left[-\frac{1}{2}, \frac{1}{2}\right]$

(b) Range of fog is $\left[-\frac{1}{2}, \frac{1}{2}\right]$

(c) $\lim_{x \to 0} \frac{f(x)}{g(x)} = \frac{\pi}{6}$

(d) There is an $x \in R$ such that $(gof)(x) = 1$

15 If $f_1(x)=x$, $f_2(x)=1-x$, $f_3(x)=\frac{1}{x}$, $f_4(x)=\frac{1}{1-x}$, $f_5(x)=\frac{x}{x-1}$ and $f_6(x)=\frac{x-1}{x}$. Suppose that $f_6(f_m(x))=f_4(x)$ and $f_n(f_4(x))=f_3(x)$, then

(a) $m=5$ (b) $n=5$ (c) $m=6$ (d) $n=6$

16 Let $f(x):\left[\frac{2\pi}{3}, \frac{5\pi}{3}\right] \to [0, 4]$ be a function defined by $f(x)=\sqrt{3}\sin x - \cos x + 2$,then

(a) $f^{-1}(1)=\frac{4\pi}{3}$ (b) $f'(1)=\pi$

(c) $f^{-1}(2)=\frac{5\pi}{6}$ (d) $f^{-1}(2)=\frac{7\pi}{6}$

17 Let $f: R \to R$ such that $f(x+y)+f(x-y)=2f(x)\cdot f(y)$ for all $x, y \in R$. Then,

(a) $f(x)$ is an even function, if $f(0) \neq 0$
(b) $f(x)$ is an odd function, if $f(0) \neq 0$
(c) $f(x)$ is an even function, if $f(0) = 0$
(d) $f(x)$ is an odd function, if $f(0) = 0$

18 The set of all values of x satisfying $\{x\}=x[x]$, where $[x]$ represents greatest integer function and $\{x\}$ represents fractional part of x, is

(a) 0 (b) $-\frac{1}{2}$ (c) $-1<x<1$ (d) R

19 Let $f(x)+f(y)=f(x\sqrt{1-y^2}+y\sqrt{1-x^2})$ [$f(x)$ is not identically zero]. Then,

(a) $f(4x^3-3x)+3f(x)=0$ (b) $f(4x^3-3x)=3f(x)$

(c) $f(2x\sqrt{1-x^2})+2f(x)=0$ (d) $f(2x\sqrt{1-x^2})=2f(x)$

20 Consider real valued function satisfying $2f(\sin x)+f(\cos x)=x$, then

(a) domain of $f(x)$ is R (b) domain of $f(x)$ is $[-1, 1]$

(c) range of $f(x) \cup \left[-\frac{2\pi}{3}, \frac{\pi}{3}\right]$ (d) range of $f(x)$ is R

21 Let a function $f(x)$, $x \neq 0$ be such that $f(x)+f\left(\frac{1}{x}\right)=f(x)\cdot f\left(\frac{1}{x}\right)$, then $f(x)$ can be

(a) $1-x^{2013}$ (b) $\sqrt{|x|}+1$

(c) $\frac{\pi}{2\tan^{-1}|x|}$ (d) $\frac{2}{1+k\ln|x|}$

Linked Comprehension Type

Paragraph I Directions (Q.Nos. 22-24)

Let $f(x) = f_1(x) - 2f_2(x)$, where

$$f_1(x) = \begin{cases} \min\{x^2, |x|\} & , |x| \le 1 \\ \max\{x^2, |x|\} & , |x| > 1 \end{cases} \text{ and } f_2(x) = \begin{cases} \min\{x^2, |x|\} & , |x| > 1 \\ \max\{x^2, |x|\} & , |x| \le 1 \end{cases}$$

$$\text{Let } g(x) = \begin{cases} \min\{f(t) : -3 \le t \le x, & -3 \le x < 0\} \\ \max\{f(t) : 0 \le t \le x, & 0 \le x \le 3\} \end{cases}$$

22 For $-3 \le x \le -1$, the range of $g(x)$ is

(a) $[-1, 3]$ (b) $[-1, -15]$
(c) $[-1, 9]$ (d) None of these

23 For $x \in (-1, 0)$, $f(x) + g(x)$ is

(a) $x^2 - 2x + 1$ (b) $x^2 + 2x - 1$ (c) $x^2 + 2x + 1$ (d) $x^2 - 2x - 1$

24 The graph of $y = g(x)$ in its domain is broken at

(a) 1 point (b) 2 points
(c) 3 points (d) None of these

Matrix Match Type Questions

25 Let $E_1 = \{(x \in R : x \ne 1)$ and $\dfrac{x}{x-1} > 0\}$ and $E_2 = \left\{x \in E_1 : \sin^{-1}\left(\log_e\left(\dfrac{x}{x-1}\right)\right) \text{ is a real number}\right\}$.

(here, the inverse trigonometric function $\sin^{-1} x$ assumes value in $\left[-\dfrac{\pi}{2}, \dfrac{\pi}{2}\right]$)

Let $f : E_1 \to R$ be the function defined by $f(x) = \log\left(\dfrac{x}{x-1}\right)$ and $g : E_2 \to R$ be a function defined by $g(x) = \sin^{-1}\left(\log\left(\dfrac{x}{x-1}\right)\right)$.

Match the following

	Column I		Column II
(i)	The range of f	(A)	$\left(-\infty, \dfrac{1}{1-e}\right] \cup \left[\dfrac{e}{1-e}, \infty\right)$
(ii)	The range of g contains	(B)	$(0, 1)$
(iii)	The domain of f contains	(C)	$\left[-\dfrac{1}{2}, \dfrac{1}{2}\right]$
(iv)	The domain of g is	(D)	$(-\infty, 0) \cup (0, \infty)$
		(E)	$\left(-\infty, \dfrac{e}{e-1}\right]$
		(F)	$(-\infty, 0) \cup \left(\dfrac{1}{2}, \dfrac{e}{e-1}\right]$

[JEE Advanced 2018]

26 Column I contains two functions $f(x)$ and $g(x)$ and Column II contains values of x for which both the functions are identical.

Match the following

	Column I		Column II
(i)	$f(x)=\tan^{-1}\left(\frac{2x}{1-x^2}\right)$ and $g(x)=2\tan^{-1}x$	(A)	$x\in\{-1,1\}$
(ii)	$f(x)=\sin^{-1}(\sin x)$ and $g(x)=\sin(\sin^{-1}x)$	(B)	$x\in(0,1)$
(iii)	$f(x)=\log_{x^2}25$ and $g(x)=\log_x 5$	(C)	$x\in(-1,1)$
(iv)	$f(x)=\sin^{-1}x+\cos^{-1}x$ and $g(x)=\sec^{-1}x+\text{cosec}^{-1}x$	(D)	$x\in[-1,1]$

27 Match the following

	Column I		Column II
(i)	If $f(x)=\begin{cases} x+1 & , x<0 \\ x^2-1 & , x\geq 0 \end{cases}$, then $f(f(x))$ for $-1\leq x\leq 0$ is	(A)	$\frac{x-3}{2}$
(ii)	If $f\left(\frac{2\tan x}{1+\tan^2 x}\right)=\frac{(\cos 2x+1)(\sec^2 x+2\tan x)}{2}$, then $f(x)$ is	(B)	x^2+2x
(iii)	$f(x+y+1)=(\sqrt{f(x)}+\sqrt{f(y)})^2\ \forall\ x,y\in R$ and $f(0)=1$, then $f(x)$ is	(C)	$x+1$
(iv)	If $4<x<5$ and $f(x)=\left[\frac{x}{4}\right]+2x+2$, where [] denotes the greatest integer function, then $f^{-1}(x)$ is	(D)	$(x+1)^2$

28 Match the following

	Column I (Function)		Column II (Period)
(i)	$f(x)=e^{\cos^4 \pi x+x-[x]+\cos^2 \pi x}$	(A)	1/3
(ii)	$f(x)=\cos 2\pi\{2x\}+\sin 2\pi\{2x\}$	(B)	1/4
(iii)	$f(x)=\sin 3\pi\{x\}+\tan \pi[x]$	(C)	1/2
(iv)	$f(x)=3x-[3x+a]-b$, where $a,b\in R^+$	(D)	1

where $\{x\}$ is the fractional part of x and [] is the greatest integer function.

Numerical Value Type Questions

Question Stem Directions (Q.Nos. 29-31) The answer to each question is a **NUMERICAL VALUE.**

Let $P(x)$ be a quadratic polynomial with real coefficients such that for all real x the relation $2[1+P(x)] = P(x-1) + P(x+1)$ holds. If $P(0) = 8$ and $P(2) = 32$, then

29 Sum of all the coefficients of $P(x)$ is ....

30 If the range of $P(x)$ is $[m, \infty)$, then the value of m is ....

31 If α, β are the zero of the polynomial $p(x)$, then $\frac{\alpha}{\beta}+\frac{\beta}{\alpha}$ is equal to ....

Question Stem Directions (Q.Nos. 32-34) The answer to each question is a **NUMERICAL VALUE.**

If $(f(x))^2 \times f\left(\frac{1-x}{1+x}\right) = 64x, x \neq 0, 1$, then

32 If $f(x) = (g(x))^{1/3}$, then $g(1/2)$ is equal to ....

33 If $m\,f\left(\frac{1}{2}\right) = n\,f\left(\frac{1}{3}\right)$, then the value of n^m is (where m and n are prime numbers) ....

34 The domain of $f(x)$ is $R - \{k_1, k_2\}$, then $k_1 + k_2$ is equal to ....

35 The number of integral value of x for which the function $\sqrt{\sin x + \cos x} + \sqrt{7x - x^2 - 6}$ is defined is ....

36 The set of real values of x satisfying the equality $\left[\frac{3}{x}\right] + \left[\frac{4}{x}\right] = 5$ (where [] denotes the greatest integer function) belongs to the interval $\left(a, \frac{b}{c}\right]$ where $a, b, c \in N$ and $\frac{b}{c}$ is in its lowest form. Then, the value of $a + b + c + abc$ is ....

37 The number of integral solutions of the equation $\{x+1\} + 2x = 4[x+1] - 6$ is ....

38 Let x be a set with exactly 5 elements and y be a set with exactly 7 elements, if α the number of one-one function from x to y an β is the number of onto function from y to x, then the value of $\frac{\beta - \alpha}{5!}$ is ....

39 The number of integral value of n for which the period of the function $\cos(nx)\sin\left(\frac{5x}{n}\right)$ is 3π is ....

40 Let $f(x) = \frac{x}{1+x}$ and let $g(x) = \frac{rx}{1-x}$. Also, let s be the set of all real number r such that $f(g(x)) = g(f(x))$ for infinitely many real number x. The number of elements in sets is ....

41 If a polynomial function $f(x)$ satisfies $f(f(f(x)) = 8x + 21$, then the sum of all coefficients of $f(x)$ is ....

42 Let $g(x)$ is a polynomial function satisfying $g(x)\,g(y) = g(x) + g(y) + g(xy) - 2$ for all $x, y \in R$. If $g(2) = 5$, then $g(5)$ is ....

ANSWER KEY

Level 1

1. (i) $R - \{2\}$ (ii) $R - \{1, 2\}$ (iii) R (iv) $(-\infty, 4]$
   (v) $(-\infty, 4)$ (vi) $[1, 7]$ (vii) $R - \left\{\frac{5}{2}\right\}$
2. (i) $(-\infty, -2] \cup [2, \infty]$ (ii) $[-4, -2] \cup [2, 4]$ (iii) $(-5, -1] \cup [1, 5)$
3. (i) Domain = R, Range = $\{0, 1\}$
   (ii) Domain = R, Range = $\{1\}$
   (iii) Range = $\{-1\}$, Domain = R
4. (i) $R - [-2, -1)$ (ii) $R - I$
5. (i) Domain = $R - I$, Range = $(1, \infty)$
   (ii) Domain = $R - I$, Range = $(-\infty, 0)$
   (iii) Domain = R, Range = $\{0\}$
6. (i) $\left[-\frac{5}{3}, -1\right]$ (ii) $\left[\frac{1}{3}, 3\right]$ (iii) $\left[-\frac{\pi}{2}, -\frac{\pi}{6}\right]$ (iv) $[-1, 3]$ (v) $[-\sqrt{2}, -1] \cup [1, \sqrt{2}]$
7. (i) $(-\infty, 1) \cup (3, \infty)$ (ii) $(-\infty, 3) - \{1\}$ (iii) $(-3, 3) - \{-2\sqrt{2}, 2\sqrt{2}\}$ (iv) $(2, 4)$
8. (i) $\left[-\frac{1}{2}, \frac{1}{2}\right]$ (ii) $(1, \infty)$ (iii) $R - 1, \frac{1}{5}$ (iv) $[0, 2]$
   (v) $\left[\frac{1}{3}, 1\right]$ (vi) $\left[\frac{3}{4}, \infty\right)$ (vii) $\left[\frac{\pi}{4}, \pi\right)$
9. (i) many one, into (ii) one-one, onto (iii) one-one, onto (iv) one-one, onto
10. (i) 64 (ii) 24 (iii) 40
    (iv) 0 (v) 64 (vi) 4
11. (i) 81 (ii) 0 (iii) 81 (iv) 36 (v) 45 (vi) 3
12. (i) odd function (ii) neither even nor odd function (iii) odd function
    (iv) odd function (v) odd function
13. (i) $\frac{\pi}{2}$ (ii) $\frac{2\pi}{3}$ (iii) 1
    (iv) $\frac{\pi}{2}$ (v) π (vi) $\frac{5\pi}{3}$
14. (i) not identical (ii) identical (iii) identical (iv) not identical
15. $gof = \{(2, 7), (3, 7), (4, 11), (5, 11)\}$
16. $(fog)(x) = x^2 - 10x + 23$
    $gof(x) = x^2 - 7$
17. $fog(x) = \sin(\sqrt{x + 5})$
    $gof(x) = \sqrt{\sin x + 5}$
18. $fog\,(x) = \cos|x|$
    $gof(x) = |\cos x|$
19. (i) $\frac{1}{2} + \sqrt{x - \frac{3}{4}}$ (ii) $\frac{2x - 1}{x - 1}$ (iii) $\frac{\log x - 13}{5}$
    (iv) $\log_3\left(\frac{x - 1}{3}\right)$ (v) $\begin{cases} \sqrt{x}, & 0 \le x \le 1 \\ -\sqrt{-x}, & -1 \le x \le 0 \end{cases}$
20. $\frac{(x-2)^2}{3}$ 21. $\frac{1 + x^2 - x^3}{2x(1 - x)}$ 22. -15 23. $\frac{-15}{4}$ 24. $\frac{4}{3}$

Level 2

1. (d)	**2.** (c)	**3.** (c)	**4.** (b)	**5.** (c)	**6.** (c)	**7.** (b)
8. (d)	**9.** (d)	**10.** (b)	**11.** (c)	**12.** (c)	**13.** (d)	**14.** (a)
15. (a)	**16.** (a)	**17.** (c)	**18.** (c)	**19.** (c)	**20.** (c)	**21.** (d)
22. (c)	**23.** (b)	**24.** (a)	**25.** (b)	**26.** (b)	**27.** (a)	**28.** (a)
29. (c)	**30.** (d)	**31.** (d)	**32.** (c)	**33.** (b)	**34.** (b)	**35.** (b)
36. (d)	**37.** (a)	**38.** (c)	**39.** (b)	**40.** (a)	**41.** (c)	**42.** (a)
43. (a)	**44.** (d)	**45.** (d)	**46.** (d)	**47.** (a)	**48.** (a)	**49.** (d)
50. (c)	**51.** (b)	**52.** (a)	**53.** (c)	**54.** (b)	**55.** (c)	**56.** (a)
57. (b)	**58.** (b)	**59.** (b)	**60.** (d)	**61.** (a)	**62.** (b)	**63.** (c)
64. (c)	**65.** (b)	**66.** (a)	**67.** (c)	**68.** (d)	**69.** (c)	**70.** (b)
71. (c)	**72.** (d)	**73.** (a)	**74.** (c)	**75.** (c)	**76.** (c)	**77.** (b)
78. (c)	**79.** (a)	**80.** (75)	**81.** (4)	**82.** (30)	**83.** (1)	**84.** (720)
85. (19)	**86.** (1)	**87.** (1)	**88.** (25)	**89.** (–1)	**90.** (2)	**91.** (3)
92. (17)						

Level 3

1. (b)	**2.** (d)	**3.** (d)	**4.** (c)	**5.** (c)	**6.** (d)	**7.** (c)
8. (c)	**9.** (c)	**10.** (c)	**11.** (b)	**12.** (a, b, c)	**13.** (a, b, c)	**14.** (a, b, c)
15. (b, c)	**16.** (a, d)	**17.** (a, d)	**18.** (a, b)	**19.** (a, d)	**20.** (b, c)	**21.** (a, b, c, d)
22. (a)	**23.** (b)	**24.** (a)				

25. (i)–D, (ii)–B, (iii)–A, (iv)–A

26. (i)–B, C, (ii)–A, B, C, D (iii)–B, (iv)–A

27. (i)–B, (ii)–C, (iii)–D, (iv)–A

28. (i)–D, (ii)–C, (iii)–D, (iv)–A

29. (19)	**30.** (–17)	**31.** (10.5)	**32.** (48)	**33.** (9)	**34.** (0)	**35.** (3)
36. (20)	**37.** (1)	**38.** (199)	**39.** (8)	**40.** (2)	**41.** (5)	**42.** (26)

SOLUTIONS

Level I

1. (i) We have, $f(x)=\dfrac{x-1}{x+2}$

$f(x)$ is defined if $x+2\neq 0$.

Thus, domain $=R-\{2\}$.

(ii) Here, $y=\dfrac{1}{x^2-3x+2}$

$=\dfrac{1}{(x-1)(x-2)}$ is defined,

if $(x-1)(x-2)\neq 0$

$\Rightarrow \quad x\neq 1,2$

Thus, domain $=R-\{1, 2\}$

(iii) Here, $f(x)=\dfrac{x^2+4}{x^2+2}$ is defined everywhere as $x^2+2>0$ for all x.

Hence, domain $=R$

(iv) Here, $f(x)=\sqrt{4-x}$ is defined if $4-x\geq 0$

$\Rightarrow \quad x\leq 4$

Hence, domain of $f(x)=(-\infty, 4]$

(v) Here, $f(x)=y=\dfrac{1}{\sqrt{4-x}}$ is defined if

$4-x>0 \qquad [\because 4-x\neq 0]$

$\Rightarrow \quad x<4$

Hence, domain of $f(x)=(-\infty, 4)$

(vi) $f(x)=\sqrt{x-1}-\sqrt{7-x}$ is defined, if

$x-1\geq 0$ and $7-x\geq 0$

$x\geq 1$ and $x\leq 7$

Hence, domain of $f(x)=[1,7]$

(vii) Here, $y=\dfrac{1}{\sqrt[3]{2x-5}}$ is defined, if

$2x-5\neq 0$

$\Rightarrow \quad x\neq \dfrac{5}{2}$

Domain $=R-\left\{\dfrac{5}{2}\right\}$

2. (i) Here, $f(x)=\sqrt{|x|-2}$ is defined, if

$|x|-2\geq 0$

$|x|\geq 2$

$\Rightarrow \quad x\geq 2$ and $x\leq -2$

Hence, domain $=(-\infty,-2]\cup[2,\infty)$

(ii) Given, $f(x)=\sqrt{|x|-2}+\sqrt{4-|x|}$

$f(x)$ is defined, if

$|x|-2\geq 0$ and $4-|x|\geq 0$

$\Rightarrow \quad |x|\geq 2$ and $|x|\leq 4$

$\Rightarrow \quad x\geq 2, x\leq -2$ and $-4\leq x\leq 4$

$\Rightarrow \quad x\in[-4,-2]\cup[2,4]$

Domain $=[-4,-2]\cup[2,4]$

(iii) We have,

$$f(x)=\sqrt{\frac{|x|-1}{5-|x|}}$$

$f(x)$ is defined, if $\dfrac{|x|-1}{5-|x|}\geq 0$ and $5-|x|\neq 0$

$\Rightarrow \quad \dfrac{|x|-1}{|x|-5}\leq 0$ and $|x|\neq 5$

$\Rightarrow \quad 1\leq |x|\leq 5$ and $|x|\neq 5$

Domain $=(-5,-1]\cup[1,5)$

3. (i) We have,

$$f(x)=\text{sgn}\,x^2$$

We know that $x^2\geq 0\ \forall x\in R$

$$\therefore \quad \text{sgn}(x^2)=\begin{cases}0, & x^2=0\\ 1, & x^2\neq 0\end{cases}=\begin{cases}0, & x=0\\ 1, & x\neq 0\end{cases}$$

Hence, Domain $=R$

Range $=\{0,1\}$

(ii) $f(x)=\text{sgn}(x^2+5)$

We know that $x^2\geq 0$

So, $\quad x^2+5\geq 0, \forall x\in R$

So, $\quad f(x)=\text{sgn}(x^2+5)=1$

Hence, domain $=R$

Range $=\{1\}$

(iii) We have,

$$f(x)=\text{sgn}(-x^2+4x-7)$$
$$=\text{sgn}(-x^2+4x-4-3)$$
$$=\text{sgn}(-(x-2)^2-3)$$

We know that

$(x-2)^2\geq 0\Rightarrow -(x-2)^2\leq 0\Rightarrow -(x-2)^2-3\leq -3$

So, i.e. $(x-2)^2-3<0$

Hence, $\text{sgn}(-(x-2)^2-3)=-1$

Range $=\{-1\}$

4. (i) We have, $f(x)=\dfrac{1}{[x+2]}$

Now, $[x+2]=0$

$\Rightarrow \quad 0\leq (x+2)<1$

$\Rightarrow \quad -2\leq x<-1$

Thus, domain $=R-[-2,-1)$

(ii) We have, $f(x)=\frac{1}{\sqrt{x-[x]}}$

Denominator = 0 gives $x-[x]=0$

$\Rightarrow \quad [x]=x$

$\Rightarrow \quad x\in I$

Also, $\quad x-[x]>0$

$\Rightarrow \quad x>[x]$

$\Rightarrow \quad x\in R$

Hence, domain $=R-I$

5. (i) For Domain $\{x\}\neq 0$

$x\in R-I$

Range $\quad y=\frac{1}{\{x\}}$

We know that

$0\le\{x\}<1$

$\Rightarrow \quad 1<\frac{1}{\{x\}}<\infty$

Range $=(1,\infty)$

(ii) $y=\log_{10}\{x\}$

Domain $\{x\}>0$

$\Rightarrow \quad x\in R-I$

Range $\quad y=\log_{10}\{x\}$

$\{x\}\in(0,1)$

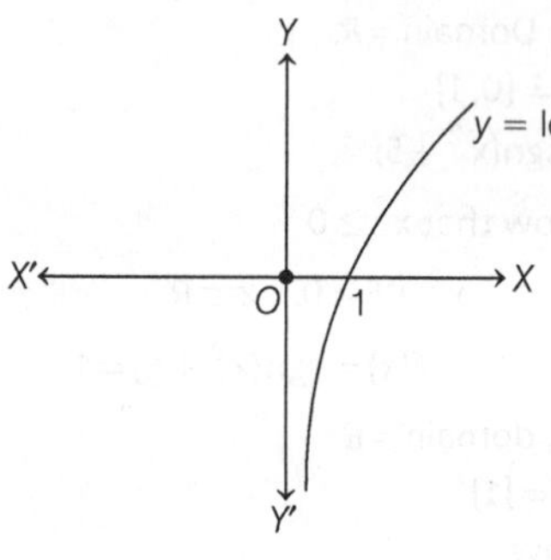

Then, range $=(-\infty,0)$

(iii) $y=[\sin\{x\}]$

Domain $=R$

Range

We know that $0\le\{x\}<1$

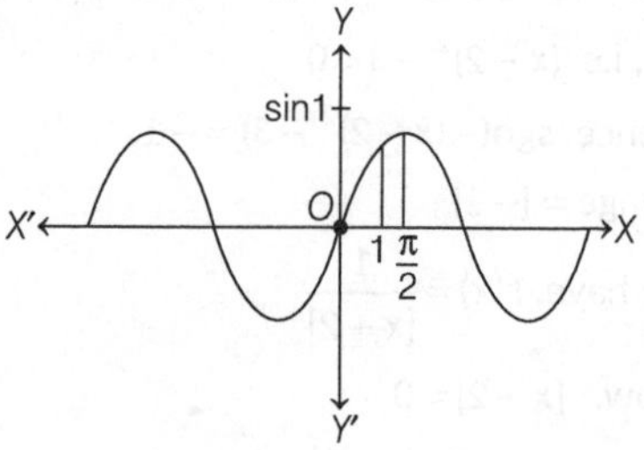

$0\le\sin\{x\}<1$

$\therefore \quad [\sin\{x\}]=0$

Range $=\{0\}$

6. (i) $f(x)=\sin^{-1}(3x+4)$

f is defined, if $-1\le 3x+4\le 1$

$\Rightarrow \quad -5\le 3x\le -3 \Rightarrow -\frac{5}{3}\le x\le -1$

$\Rightarrow \quad x\in\left[-\frac{5}{3},-1\right]$

Hence, Domain $=\left[\frac{-5}{3},-1\right]$

(ii) The given function $f(x)=\sin^{-1}(\log_3 x)$ is defined, if

$-1\le\log_3 x\le 1$

$\Rightarrow \quad 3^{-1}\le x\le 3^1 \Rightarrow \frac{1}{3}\le x\le 3$

Domain $=\left[\frac{1}{3},3\right]$

(iii) The given function is defined, if

$-1\le\frac{3}{4+2\sin x}\le 1$

$\Rightarrow \quad -1\le\frac{4+2\sin x}{3}\le 1$

$\Rightarrow \quad -3\le 4+2\sin x\le 3$

$\Rightarrow \quad -7\le 2\sin x\le -1$

$\Rightarrow \quad -\frac{7}{2}\le\sin x\le -\frac{1}{2}$

But $\quad -1\le\sin x\le 1$

$\Rightarrow \quad -1\le\sin x\le -1/2 \Rightarrow \frac{-\pi}{2}\le x\le -\pi/6$

Thus, domain $=\left[\frac{-\pi}{2},\frac{-\pi}{6}\right]$

(iv) We have, $f(x)=\sin^{-1}\left(\frac{3-2x}{5}\right)+\sqrt{3-x}$, f is defined if

$-1\le\frac{3-2x}{5}\le 1$ and $3-x\ge 0$

$\Rightarrow \quad -5\le 3-2x\le 5$ and $x\le 3$

$\Rightarrow \quad -8\le -2x\le 2$ and $x\le 3$

$\Rightarrow \quad -2\le 2x\le 8$ and $x\le 3$

$\Rightarrow \quad -1\le x\le 4$ and $x\le 3$

Hence, domain is $[-1,3]$

(v) We have, $f(x)=\cos^{-1}(2x^2-3)$

We know that for $\cos^{-1}x$, x lies in the range of $[-1,1]$.

$\therefore \quad -1\le 2x^2-3\le 1$

$\Rightarrow \quad 2\le 2x^2\le 4$

$\Rightarrow \quad 1\le x^2\le 2$

$\Rightarrow \quad x^2\ge 1$ and $x^2\le 2$

$\Rightarrow \quad x^2-1\ge 0$ and $x^2-2\le 0$

$\Rightarrow \quad x\in(-\infty,-1]\cup[1,\infty)$ and $x\in[-\sqrt{2},\sqrt{2}]$

Hence, domain is $[-\sqrt{2},-1]\cup[1,\sqrt{2}]$

7. (i) We have, $f(x)=\log_2(x^2-4x+3)$

$f(x)$ is defined, if

$$x^2-4x+3>0$$

$$\Rightarrow \quad (x-1)(x-3)>0$$

$$\Rightarrow \quad x<1 \text{ or } x>3$$

Hence, domain $=(-\infty,1)\cup(3,\infty)$

(ii) We have, $f(x)=\dfrac{\log(3-x)}{x^2-5x+4}$

Here, the function $\log(3-x)$ is defined, if

$$3-x>0$$

$$\Rightarrow \quad x<3$$

$$\Rightarrow \quad x\in(-\infty,3)$$

Also, $x^2-5x+4=0$ gives $x=1,4$

Hence, domain $=(-\infty,3)-\{1\}$

(iii) We have, $f(x)=\dfrac{1}{\log(9-x^2)}$

The function f is defined, if

$$9-x^2>0$$

$$\Rightarrow \quad x^2-9<0$$

$$\Rightarrow \quad (x-3)(x+3)<0$$

$$\Rightarrow \quad x\in(-3,3)$$

Also, $9-x^2\neq 1$ and $9-x^2\neq 0$

$$\Rightarrow \quad x^2\neq 8 \text{ and } x^2\neq 9$$

$$\Rightarrow \quad x^2\neq \pm 2\sqrt{2} \text{ and } x^2\neq \pm 3$$

Hence, domain is $(-3,3)-\{-2\sqrt{2},2\sqrt{2}\}$

(iv) We have, $f(x)=\log_{1/2}\left(\dfrac{2-x}{x-4}\right)$

$f(x)$ is defined if

$$\frac{2-x}{x-4}>0$$

$$\Rightarrow \quad \frac{x-2}{x-4}<0 \Rightarrow x\in(2,4)$$

Hence, domain is $(2,4)$

8. (i) Clearly, $f(x)$ is defined for all $x\in R$ so domain of $f=R$

Now, let $\quad y=\dfrac{x}{1+x^2}$

$$\Rightarrow \quad x^2y+y=x \Rightarrow x^2y-x+y=0$$

$$\Rightarrow \quad x=\frac{1\pm\sqrt{1-4y^2}}{2y}$$

For x to be real, $1-4y^2\geq 0$ and $y\neq 0$

$$-\frac{1}{2}\leq y\leq \frac{1}{2} \text{ and } y\neq 0$$

Also, $y=0$ for $x=0$

Hence, range of $f(x)$ is $\left[\dfrac{-1}{2},\dfrac{1}{2}\right]$

(ii) We observe that $f(x)$ is defined for all $x\in R-\{0\}$

So, domain $f=R-\{0\}$

Let $y=\dfrac{2+x^2}{x^2} \Rightarrow x=\pm\sqrt{\dfrac{2}{y-1}}$

For x to be real, we must have $y-1>0$

$$\Rightarrow \quad y>1 \Rightarrow y\in(1,\infty)$$

Hence, range of $f=(1,\infty)$

(iii) $f(x)=\dfrac{x^2-3x+2}{x^2+x-6}$

$f(x)$ is defined for $x^2+x-6\neq 0$

i.e. $\quad x\neq -3,2$

$\therefore$ Domain $=R-\{-3,2\}$

Let $\quad y=\dfrac{x^2-3x+2}{x^2+x-6}=\dfrac{x-1}{x+3}$

$$\Rightarrow \quad x=\frac{3y+1}{1-y}$$

Clearly, f is real for $1-y\neq 0$

i.e. $\quad y\neq 1$

Also, when $x=2$, then $y=\dfrac{1}{5}$

Hence, Range of $f=R-1,\dfrac{1}{5}$

(iv) We have, $f(x)=\sqrt{4-x^2}$

For Domain

$$4-x^2\geq 0$$

$$\Rightarrow \quad x^2-4\leq 0$$

$$\Rightarrow \quad x\in[-2,2]$$

So domain $=[-2,2]$

Now, let $y=\sqrt{4-x^2}$

$$\Rightarrow \quad x=\sqrt{4-y^2}$$

For x to be real,

$$4-y^2\geq 0$$

$$y^2-4\leq 0$$

$$y\in[-2,2]$$

But $y=\sqrt{4-x^2}\geq 0$

Hence, range of $f=[0,2]$

(v) We have, $f(x)=\dfrac{1}{2-\cos 5x}$

$$-1\leq \cos 5x\leq 1 \Rightarrow -1\leq -\cos 5x\leq 1$$

$$\Rightarrow \quad 1\leq 2-\cos 5x\leq 3$$

$$\Rightarrow \quad \frac{1}{3}\leq \frac{1}{2-\cos 5x}\leq 1$$

$$\Rightarrow \quad \frac{1}{3}\leq f(x)\leq 1$$

$\therefore$ Range $=\left[\dfrac{1}{3},1\right]$

(vi) Given that $f(x)=4^x-2^x+1=y$ (let)

$$y=(2^x)^2-2^x+1$$

$$y=\left(2^x-\frac{1}{2}\right)^2+\frac{3}{4}\geq\frac{3}{4}$$

$$\therefore \quad y\geq\frac{3}{4}$$

Hence, range $=\left[\frac{3}{4},\infty\right)$

(vii) $f(x)=\cot^{-1}(2x-x^2)$

$\because \quad y\in(0,\pi)$ ... (i)

and $\cot y=2x-x^2$

$\Rightarrow \quad (x-1)^2=1-\cot y\geq 0$

$\Rightarrow \quad \cot y\leq 1$

$\Rightarrow \quad y\geq\frac{\pi}{4}$ ... (ii)

From Eqs. (i) and (ii),

$$\frac{\pi}{4}\leq y<\pi$$

Hence, range $=\left[\frac{\pi}{4},\pi\right)$

9. (i) We have, $f(x)=|x|$

Graph of $f(x)$ is as shown

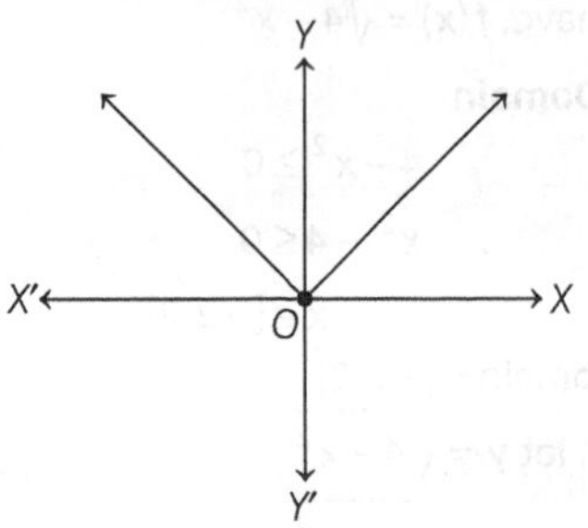

From the graph observe that $f(x)$ is many one function.

We observe that range of f is $[0,\infty]$.

Hence, $f(x)$ is into function.

(ii) Given that $f:[0,\infty)\rightarrow[0,1]$

$$f(x)=\frac{x}{1+x}$$

Let $\quad f(x_1)=f(x_2)$

$$\Rightarrow \quad \frac{x_1}{1+x_1}=\frac{x_2}{1+x_2}$$

$\Rightarrow \quad x_1x_2+x_1=x_1x_2+x_2 \Rightarrow x_1=x_2$

Thus $f(x)$ is one-one.

Let $\quad y=\frac{x}{1+x}$

$\Rightarrow \quad y+yx=x \Rightarrow x=\frac{y}{1-y}$

As, $x\geq 0, \frac{y}{1-y}\geq 0$

$\Rightarrow \quad \frac{y}{y-1}\leq 0$

$\Rightarrow \quad 0\leq y<1$

or range of f is $[0, 1)$.

Therefore, f is onto.

(iii) We have, $(f-g)R\rightarrow R$

$$(f-g)(x)=\begin{cases}-x, & \text{if } x\in\text{rational}\\ x, & \text{if } x\in\text{irrational}\end{cases}$$

Clearly, $(f-g)(x)$ is one-one and onto.

(iv) When n is even,

let $f(2m_1)=f(2m_2)$

$\Rightarrow \quad -\frac{2m_1}{2}=-\frac{2m_2}{2}\Rightarrow m_1=m_2$

When n is odd, let

$$f(2m_1+1)=f(2m_2+1)$$

$$\Rightarrow \quad \frac{2m_1+1-1}{2}=\frac{2m_2+1-1}{2}$$

$\Rightarrow \quad m_1=m_2$

Therefore, $f(x)$ is one-one.

Also, when n is even

$$\frac{-n}{2}=\frac{-2m}{2}=-m$$

When n is odd, $\frac{n-1}{2}=\frac{2m+1-1}{2}=m$

Hence, range of function is Z.

Therefore, function is onto.

10. Here, $m=3$ and $n=4\,(n>m)$

(i) Total number of mappings $=n^m=4^3=64$

(ii) Total number of one-one mappings $={}^nP_m={}^4P_3=4\cdot3\cdot2=24$

(iii) Total number of many-one mappings $=n^m-{}^nP_m=64-24=40$

(iv) Total number of onto mappings $=0$ (here $n>m$)

(v) Total number of into mappings $=n^m=4^3=64$ (here $n>m$)

(vi) Total number of constant mappings $=n=4$

11. Here, $m=4$ and $n=3\,(n<m)$

(i) Total number of mappings $=n^m=3^4=81$

(ii) Total number of one-one mappings $=0$ $\quad[\because n<m]$

(iii) Total number of many-one mappings $=n^m=3^4=81$ $\quad[\because n<m]$

(iv) Total number of onto mappings

$$=\sum_{r=0}^{2}(-1)^r\cdot{}^3C_r\cdot(3-r)^4$$

$$={}^3C_0(3-0)^4-{}^3C_1\cdot(3-1)^4+{}^3C_2\cdot(3-2)^4$$

$$=81-48+3=36$$

(v) Total number of into mappings

$$= \sum_{r=1}^{2} (-1)^{r-1} \cdot {}^3C_r \cdot (3-r)^4$$

$$= {}^3C_1 \cdot (3-1)^4 - {}^3C_2 \cdot (3-2)^4 = 48 - 3 = 45$$

(vi) Total number of constant mappings $= n = 3$

12. (i) Given, $f(x) = x\sin(x^2+3)$

$$\Rightarrow \quad f(-x) = (-x)\sin((-x)^2+3)$$

$$= -x\sin(x^2+3) = -f(x)$$

$\Rightarrow f(x)$ is an odd function

(ii) $f(-x) = \sin(-x) + \cos(-x)$

$$= -\sin x + \cos x$$

$f(-x)$ is neither equal to $f(x)$ nor $-f(x)$ so, $f(x)$ is neither even nor odd function.

(iii) $f(x) = \log\left(\dfrac{2-x}{2+x}\right)$

$$f(-x) = \log\left(\frac{2+x}{2-x}\right) = \log\left(\frac{2-x}{2+x}\right)^{-1}$$

$$= -\log\left(\frac{2-x}{2+x}\right) = -f(x)$$

$\Rightarrow f(x)$ is an odd function.

(iv) $f(x) = x\sin^2 x + \tan(x^5)$

$$f(-x) = (-x)\sin^2(-x) + \tan((-x)^5)$$

$$= -x\sin^2 x + \tan(-x^5)$$

$$= -x\sin^2 x - \tan(x^5)$$

$$= -(x\sin^2 x + \tan x^5)$$

$f(-x) = -f(x)$

So, $f(x)$ is an odd function.

(v) $f(x) = \sqrt{x^2-2x+4} - \sqrt{x^2+2x+4}$

$$f(-x) = \sqrt{x^2+2x+4} - \sqrt{x^2-2x+4}$$

$$= -(\sqrt{x^2-2x+4} - \sqrt{x^2+2x+4})$$

$f(-x) = -f(x)$

So, $f(x)$ is an odd function.

13. (i) $f(x) = \sin 4x$

Period of $\sin x$ is 2π.

So, period of $\sin 4x = \dfrac{2\pi}{4} = \dfrac{\pi}{2}$

(ii) $f(x) = 5\operatorname{cosec}(3x-5)+7$

Period of $5\operatorname{cosec}(3x-5)+7$ is $\dfrac{2\pi}{3}$.

(iii) $f(x) = \dfrac{a^x}{a^{[x]}} = a^{x-[x]} = a^{\{x\}}$

where $\{\cdot\}$ is fractional part function period of $\{\cdot\}$ is 1.

So, period of $f(x)$ is 1.

(iv) $f(x) = |\sin x| + |\cos x|$

Period of $|\sin x|$ and $|\cos x|$ is π.

So, we consider period of $f(x)$ to be π.

But $f\left(\dfrac{\pi}{2}+x\right) = \left|\sin\left(\dfrac{\pi}{2}+x\right)\right| + \left|\cos\left(\dfrac{\pi}{2}+x\right)\right|$

$$= |\cos x| + |-\sin x|$$

$$= |\cos x| + |\sin x|$$

$$f\left(\frac{\pi}{2}+x\right) = f(x)$$

So, period of $f(x) = \dfrac{\pi}{2}$

(v) We have, $|\sin x - \cos x| = \sqrt{2}\left|\sin\left(x - \dfrac{\pi}{4}\right)\right|$, which is the period π.

Period of $|\sin x| + |\cos x|$ is $\dfrac{\pi}{2}$.

Therefore, period of $f(x)$ is LCM of $\left(\pi, \dfrac{\pi}{2}\right)$.

$\therefore$ Period of $f(x) = \pi$

(vi) $f(x) = |\sin 3x| - \tan\left(\dfrac{6x}{5}\right)$

Period of $|\sin 3x| = \dfrac{\pi}{3}$

and period of $\tan\left(\dfrac{6x}{5}\right) = \dfrac{\pi}{6/5} = \dfrac{5\pi}{6}$

Therefore, period of $f(x) = $ LCM of $\left(\dfrac{\pi}{3}, \dfrac{5\pi}{6}\right)$

$$= \frac{\text{LCM of } (\pi, 5\pi)}{\text{HCF of } (3,6)} = \frac{5\pi}{3}$$

14. (i) We have, $f(x) = \ln x^2$

Domain of $f = R - \{0\}$

Now, $g(x) = 2\ln x$

$D_g : x \in (0, \infty)$

Domain is different. The functions are not identical.

(ii) We have, $f(x) = \operatorname{cosec} x$

$D_f : R - \{n\pi\}, n \in Z$

$R_f : (-\infty, -1] \cup [1, \infty)$

and $g(x) = \dfrac{1}{\sin x}$

Domain $= R - \{n\pi\}, n \in Z \{\because \sin x = 0 \Rightarrow x = n\pi\}$

$R_f = (-\infty, -1] \cup [1, \infty)$

So, f and g are identical functions.

(iii) We have, $f(x) = \operatorname{sgn}(x^2+1)$ and

Domain $= R$

and $(x^2+1) > 0$

so, $\operatorname{sgn}(x^2+1) = 1$

Range $= \{1\}$

Again, $g(x)=\sin^2 x+\cos^2 x$

Domain $=R$

and $g(x)=1$

Range $=\{1\}$

f and g are identical.

(iv) $f(x)=\log_e x$

$D_f: x\in(0,\infty)$

But $g(x): \dfrac{1}{\log_x e}$

$D_g: x\in(0,\infty)-\{1\}$

So, f and g are not identical.

15. We have,

$gof(2)=g(f(2))=g(3)=7$

$gof(3)=g(f(3))=g(4)=7$

$gof(4)=g(f(4))=g(5)=11$

$gof(5)=g(f(5))=g(5)=11$

So, $gof=\{(2,7),(3,7),(4,11),(5,11)\}$

16. We have, $f(x)=x^2-2,\ g(x)=x-5$

$$(fog)(x)=f(g(x))=f(x-5)$$
$$=(x-5)^2-2$$
$$=x^2-10x+25-2=x^2-10x+23$$
$$gof(x)=g(f(x))$$
$$=g(x^2-2)$$
$$=(x^2-2)-5=x^2-7$$

17. We have, $f(x)=\sin x$ and $g(x)=\sqrt{x+5}$

$$fog(x)=f(g(x))=f(\sqrt{x+5})=\sin(\sqrt{x+5})$$

and $gof(x)=g(f(x))=g(\sin x)=\sqrt{\sin x+5}$

18. We have, $f(x)=\cos x$ and $g(x)=|x|$

$$fog(x)=f(g(x))=f(|x|)=\cos|x|$$

and $gof(x)=g(f(x))=g(\cos x)=|\cos x|$

19. (i) We have, $f(x)=x^2-x+1$

$$\Rightarrow\quad y=\left(x-\frac{1}{2}\right)^2+\frac{3}{4}$$
$$\Rightarrow\quad \left(x-\frac{1}{2}\right)^2=y-\frac{3}{4}$$
$$\Rightarrow\quad x-1/2=\sqrt{y-\frac{3}{4}}$$
$$\Rightarrow\quad x=\frac{1}{2}+\sqrt{y-\frac{3}{4}}$$
$$\Rightarrow\quad f^{-1}(y)=1/2+\sqrt{y-\frac{3}{4}}$$

Hence, $f^{-1}=\dfrac{1}{2}+\sqrt{x-\dfrac{3}{4}}$

(ii) We have, (let) $y=f(x)=\dfrac{x-1}{x-2}$

$$\Rightarrow\quad xy-2y=x-1$$
$$\Rightarrow\quad x(y-1)=2y-1$$
$$x=\frac{2y-1}{y-1}$$
$$f^{-1}(y)=\frac{2y-1}{y-1}$$
$$\therefore\quad f^{-1}(x)=\frac{2x-1}{x-1}$$

(iii) We have, $f(x)=e^{5x+13}=y$ (let)

$$\Rightarrow\quad 5x+13=\log y$$
$$\Rightarrow\quad 5x=\log y-13$$
$$\Rightarrow\quad x=\frac{\log y-13}{5}$$
$$\Rightarrow\quad f^{-1}(y)=\frac{\log y-13}{5}$$

Hence, $f^{-1}(x)=\dfrac{\log x-13}{5}$

(iv) We have, $f(y)=2(3)^x+1$

$$\Rightarrow\quad 2\cdot 3^x=y-1\Rightarrow 3^x=\frac{y-1}{2}$$
$$\Rightarrow\quad x=\log_3\left(\frac{y-1}{2}\right)$$
$$\therefore\quad f^{-1}(x)=\log_3\left(\frac{x-1}{3}\right)$$

(v) $f(x)=x^2\operatorname{sgn}(x)=\begin{cases}x^2, & x>0\\ 0, & x=0\\ -x^2, & x<1\end{cases}$

$$\therefore\quad f(x)=\begin{cases}x^2, & 0\le x\le 1\\ -x^2, & -1\le x<0\end{cases}$$

Now, $y=x^2, 0\le x\le 1$

Since $y\in[0,1]$, we have $x=\sqrt{y}$

For $y=-x^2, -1\le x\le 0$

Since $y\in[-1,0]$, we have $x=-\sqrt{-y}$

Thus, $f^{-1}(x)=\begin{cases}\sqrt{x}, & 0\le x\le 1\\ -\sqrt{-x}, & -1\le x\le 0\end{cases}$

20. We have, $f(x)+2f(1-x)=x^2+2$ ...(i)

x replace by $1-x$, we get

$$f(1-x)+2f(x)=(1-x^2)+2 \quad \text{...(ii)}$$

From Eqs. (i) and (ii), we get

$$f(x)+2[(1-x^2)+2-2f(x)]=x^2+2$$
$$\Rightarrow\quad f(x)-4f(x)=x^2+2-2(1-x)^2-4$$
$$\Rightarrow\quad -3f(x)=x^2-2+4x-2x^2-2$$

$$\Rightarrow \quad -3f(x) = -x^2 + 4x - 4$$

$$\Rightarrow \quad f(x) = \frac{x^2 - 4x + 4}{3} = \frac{(x-2)^2}{3}$$

21. We have,

$$f(x) + f\left(\frac{x-1}{x}\right) = 1 + x \quad \text{...(i)}$$

x replace by $\frac{x-1}{x}$, we get

$$f\left(1 - \frac{1}{x}\right) + f\left(\frac{1}{1-x}\right) = 1 + 1 - \frac{1}{x} \quad \text{...(ii)}$$

and replacing x by $\frac{1}{1-x}$ in Eq (i), we get

$$f\left(\frac{1}{1-x}\right) + f(x) = 1 + \frac{1}{1-x} \quad \text{...(iii)}$$

Now, adding Eq. (i) and Eq. (iii) and subtracting Eq. (ii), we get

$$2f(x) = 1 + x + 1 + \frac{1}{1-x} - 2 + \frac{1}{x}$$

$$\Rightarrow \quad 2f(x) = x + \frac{1}{1-x} + \frac{1}{x}$$

$$\Rightarrow \quad 2f(x) = \frac{x^2(1-x) + x + (1-x)}{x(1-x)}$$

$$= \frac{1 + x^2 - x^3}{x(1-x)}$$

$$\Rightarrow \quad f(x) = \frac{1 + x^2 - x^3}{2x(1-x)}$$

22. We have, $f(x+y) = f(x) + f(y) - xy - 1$

Put $x = y = 1$, we get

$$f(2) = f(1) + f(1) - 2 = 1 + 1 - 2 = 0 \quad [\because f(1) = 1]$$

For $x = 2$, $y = 1$

$$\therefore \quad f(3) = f(2) + f(1) - 2 - 1 = 0 + 1 - 3 = -2 \quad [\because f(2) = 0]$$

For $x = y = 2$

$$f(4) = f(2) + f(2) - 4 - 1 = 0 - 5 = -5 \quad [\because f(2) = 0]$$

and $x = 3$, $y = 2$

$$f(5) = f(3) + f(2) - 6 - 1$$

$$= -2 + 0 - 7 = -9 \quad [\because f(2) = 0, f(3) = -2]$$

$$\therefore \quad \sum_{r=1}^{5} f(r) = f(1) + f(2) + f(3) + f(4) + f(5)$$

$$= 1 + 0 - 2 - 5 - 9 = -15$$

23. Put $\sqrt{\frac{x-1}{x+1}} = 3$

$$\Rightarrow \quad \frac{x-1}{x+1} = 9 \Rightarrow 9x + 9 = x - 1$$

$$\Rightarrow \quad 8x = -10 \Rightarrow x = -\frac{5}{4}$$

$$\Rightarrow \quad f(3) = 3(-5/4) = -\frac{15}{4}$$

24. We have, $mf(x-1) + nf(-x) = 2|x| + 1$

Put $x = -1$ ...(i)

$$mf(-2) + nf(1) = 3$$

Again, put $x = 2$

$$mf(1) + nf(-2) + 5 \quad \text{...(ii)}$$

Adding Eqs. (i) and (ii), we get

$$(m+n)(f(-2) + f(1)) = 8$$

$$\Rightarrow \quad (m+n)[5+1] = 8$$

$$\Rightarrow \quad (m+n) = \frac{8}{6}$$

$$m + n = \frac{4}{3}$$

Level 2

1. (d) Clearly, $f(x)$ is defined, if

$$1 - \sqrt{1 - \sqrt{1 - x^2}} \geq 0, 1 - \sqrt{1 - x^2} \geq 0 \text{ and } 1 - x^2 \geq 0$$

$$\Rightarrow 1 \geq 1 - \sqrt{1 - x^2}, 1 \geq 1 - x^2 \text{ and } x^2 - 1 \leq 0$$

$$\Rightarrow \quad \sqrt{1 - x^2} \geq 0, x^2 \geq 0 \text{ and } x^2 - 1 \leq 0$$

$$\Rightarrow \quad x^2 - 1 \leq 0$$

$$\Rightarrow \quad -1 \leq x \leq 1$$

$$\Rightarrow \quad x \in [-1, 1]$$

2. (c) We have,

$$f(x) = \sqrt{(x-4) - 2\sqrt{x-5}} - \sqrt{x - 4 + 2\sqrt{x-5}}$$

$$= \sqrt{x - 5 - 2\sqrt{x-5} + 1} - \sqrt{x - 5 + 2\sqrt{x-5} + 1}$$

$$= \sqrt{(\sqrt{x-5})^2 - 2\sqrt{x-5} + (1)^2}$$

$$- \sqrt{(\sqrt{x-5})^2 + 2\sqrt{x-5} + (1)^2}$$

$$= \sqrt{(\sqrt{x-5} - 1)^2} - \sqrt{(\sqrt{x-5} + 1)^2}$$

$$= |\sqrt{x-5} - 1| - |\sqrt{x-5} + 1|$$

So, $x \geq 5$

Hence, domain $= [5, \infty)$

3. (c) $f(x)$ is defined for

$$\frac{2 - |x|}{3 - |x|} \geq 0$$

$$\Rightarrow \quad \frac{|x| - 2}{|x| - 3} \geq 0$$

$$\Rightarrow \quad |x| \leq 2 \text{ or } |x| > 3$$

$$\Rightarrow \quad x \in [-2, 2] \text{ or } x \in (-\infty, -3) \cup (3, \infty)$$

$$\Rightarrow \quad x \in [-2, 2] \cup (-\infty, -3) \cup (3, \infty)$$

4. (b) $f(x)$ is defined for $[x]^2 - [x] - 6 > 0$

$$\Rightarrow \quad \{[x] - 3\}\{[x] + 2\} > 0$$

$$\Rightarrow \quad [x] < -2 \text{ or } [x] > 3$$

$$\Rightarrow \quad x \in (-\infty, -2] \cup [4, \infty)$$

5. (c) For domain $\dfrac{|[x]|-2}{|[x]|-3}\geq 0$

Case I When $|[x]|-2\geq 0$ and $|[x]|-3>0$

$\therefore \quad x\in(-\infty,-3)\cup[4,\infty)$ ... (i)

Case II When $|[x]|-2\leq 0$ and $|[x]|-3<0$

$\therefore \quad x\in[-2,3]$ ... (ii)

From (i) and (ii), we get

The domain of the function

$$=(-\infty,-3)\cup[-2,3)\cup[4,\infty)$$

$\therefore \quad a=-3, b=-2, c=3$

$\therefore \quad a+b+c=-3-2+3=-2$

6. (c) $f(x)=\dfrac{1}{\sqrt{|[|x|-1]|-5}}$ is defined.

Therefore, $\quad |[|x|-1]|-5>0$

$\Rightarrow \quad |[|x|-1]|>5$

$\Rightarrow \quad [|x|-1]<-5$ or $[|x|-1]>5$

$\Rightarrow \quad |x|-1<-5$ or $|x|-1\geq 6$

$\Rightarrow \quad |x|<-4$ or $|x|\geq 7$

$\Rightarrow \quad |x|\geq 7$

$x\in(-\infty,-7]\cup[7,\infty)$

7. (b) We have, $f(x)=\dfrac{1}{\sqrt{\{\sin x\}+\{\sin(\pi+x)\}}}$

$$=\frac{1}{\{\sin x\}+\{-\sin x\}}$$

Now, $\{\sin x\}+\{-\sin x\}$

$$=\begin{cases}0, & \sin x \text{ is an integer}\\ 1, & \sin x \text{ is not an integer}\end{cases}$$

For $f(x)$ to get defined $\{\sin x\}+\{-\sin x\}\neq 0$

$\Rightarrow \quad \sin x\neq$ integer

$\Rightarrow \quad \sin x\neq \pm 1, 0$

$\Rightarrow \quad \sin x\neq \dfrac{n\pi}{2}, n\in I$

8. (d) We have,

$$f(x)=\frac{3}{x}+2^{\sin^{-1}x}+\frac{1}{\sqrt{x-2}}$$

Let $\quad f(x)=p(x)+q(x)+r(x)$,

where, $\quad p(x)=\dfrac{3}{x}, q(x)=2^{\sin^{-1}x}$

and $\quad r(x)=\dfrac{1}{\sqrt{x-2}}$

Clearly, domain of $p(x)=R-\{0\}$

domain of $q(x)=[-1,1]$

domain of $r(x)=(2,\infty)$

$\therefore$ Domain of $f=(R-\{0\})\cap[-1,1]\cap(2,\infty)$

$=\phi$ (null set)

9. (d) We have, $f(x)=\cos^{-1}[2-3x^2]$

$f(x)$ is defined, when

$-1\leq[2-3x^2]\leq 1$

$\Rightarrow \quad -1\leq 2-3x^2<2$

$\Rightarrow \quad -3\leq -3x^2<0$

$\Rightarrow \quad 1\geq x^2>0$

$\Rightarrow \quad 0<x^2\leq 1$

$\therefore \quad x\in[-1,1]-\{0\}$

Hence, domain is $[-1,1]-\{0\}$.

10. (b) Domain of $\operatorname{cosec}^{-1}x$ is $(-\infty,-1]\cup[1,\infty)$ and $x-[x]>0$

$\Rightarrow \quad \{x\}>0$

$\Rightarrow \quad x\neq Z$ [Z is integer]

$\therefore$ Required domain $=(-\infty,-1]\cup[1,\infty)-Z$

11. (c) We have, $f(x)=\sqrt{\sec^{-1}\left(\dfrac{2-|x|}{4}\right)}$

$\sec^{-1}$ function always take positive values which are $[0,\pi]-\left\{\dfrac{\pi}{2}\right\}$.

Hence, the given function is defined if $\dfrac{2-|x|}{4}\leq -1$

or $\dfrac{2-|x|}{4}\geq 1 \Rightarrow |x|\geq 6$ or $|x|\leq -2$

i.e. $\quad x\in(-\infty,-6]\cup[6,\infty)$

12. (c) We have,

$$f(x)=\sin^{-1}\left(\frac{3x^2+x-1}{(x-1)^2}\right)+\cos^{-1}\left(\frac{x-1}{x+1}\right)$$

So, $\quad -1\leq\dfrac{x-1}{x+1}\leq 1\Rightarrow 0\leq x<\infty$ ... (i)

and $\quad -1\leq\dfrac{3x^2+x-1}{(x-1)^2}\leq 1$

$\Rightarrow \quad x\in\left[\dfrac{1}{4},\dfrac{1}{2}\right]\cup\{0\}$ ... (ii)

From Eqs. (i) and (ii),

$$\text{Domain}=\left[\frac{1}{4},\frac{1}{2}\right]\cup\{0\}$$

13. (d) We know that the domain of $\operatorname{cosec}^{-1}x$ is $(-\infty,-1]\cup[1,\infty)$.

So, for $\operatorname{cosec}^{-1}\left(\dfrac{1+x}{x}\right)$

$$\frac{1+x}{x}\in(-\infty,-1]\cup[1,\infty)$$

$\Rightarrow \quad \dfrac{1}{x}\in(-\infty,-2)\cup[0,\infty)$

$\Rightarrow \quad x\in\left[\dfrac{-1}{2},0\right)\cup(0,\infty)$

$\Rightarrow \quad x\in\left[\dfrac{-1}{2},\infty\right)-\{0\}$

14. (a) We have, $f(x)=\sqrt[4]{\dfrac{1-3^x}{5^{-x}-5}}$

$f(x)$ is defined, when $\dfrac{1-3^x}{5^{-x}-5}\geq 0$

or $\qquad \dfrac{3^x-1}{5^{-x}-5}\leq 0$

Case I If $3^x-1\geq 0$ and $5^{-x}-5<0$

$\Rightarrow \qquad 3^x\geq 1$ and $5^{-x}<5^1$

$\Rightarrow \qquad 3^x\geq 3^0$ and $5^{-x}<5^1$

$\therefore \qquad x\geq 0$ and $-x<1$

$\Rightarrow \qquad x\geq 0$ and $x>-1$

Hence, $x\in[0,\infty)$.

Case II If $3^x-1\leq 0$ and $5^{-x}-5>0$

$\Rightarrow \qquad 3^x\leq 1$ and $5^{-x}>5$

$\Rightarrow \qquad 3^x\leq 3^0$ and $5^{-x}>5^1$

$\therefore \qquad x\leq 0$ and $-x>1$

$\Rightarrow \qquad x\leq 0$ and $x<-1$

Hence, $x\in(-\infty,-1)$.

Combining both cases, we get

$$x\in(-\infty,-1)\cup[0,\infty)$$

Hence, domain is $(-\infty,-1)\cup[0,\infty)$.

15. (a) Since, $\log x$ is defined for $x>0$, therefore

$f(x)=\log_4[\log_5\{\log_3(18x-x^2-77)\}]$ exists, if

$\log_5\{(\log_3(18x-x^2-77)\}>0$

$\Rightarrow \qquad \log_3(18x-x^2-77)>5^0$

$\Rightarrow \qquad 18x-x^2-77>3$

$\Rightarrow \qquad x^2-18x+80<0$

$\Rightarrow \qquad (x-8)(x-10)<0$

$\Rightarrow \qquad x\in(8,10)$

Hence, domain of given function is (8, 10).

16. (a) We have,

$f(x)=\log_{10}\{1-\log_{10}(x^2-5x+16)\}$

$f(x)$ is defined when

$1-\log_{10}(x^2-5x+16)>0$ and $x^2-5x+16>0$

$\Rightarrow \qquad \log_{10}(x^2-5x+16)<1$ and $x^2-5x+16>0$

$\Rightarrow \qquad x^2-5x+16<10^1$ and $\left(x-\dfrac{5}{2}\right)^2+\dfrac{39}{4}>0$

$\Rightarrow \qquad x^2-5x+6<0$ and $\left(x-\dfrac{5}{2}\right)^2+\dfrac{39}{4}>0$

$\Rightarrow \qquad (x-2)(x-3)<0$ and $x\in R$

$\Rightarrow \qquad x\in(2,3)$ and $x\in R$

$\Rightarrow \qquad x\in(2,3)$

Hence, domain is $(2,3)$.

17. (c) Given, function $f(x)=\dfrac{1}{4-x^2}+\log_{10}(x^3-x)$

For domain of $f(x)$

$4-x^2\neq 0 \Rightarrow x\neq\pm 2 \qquad$ ...(i)

and $\qquad x^3-x>0$

$\Rightarrow x(x-1)(x+1)>0$

From Wavy curve method,

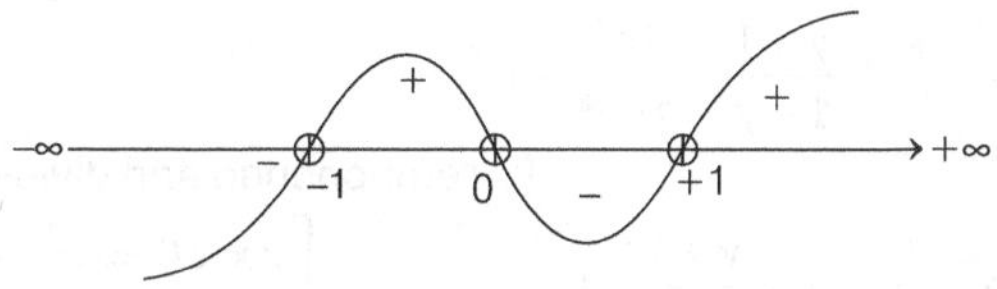

$x\in(-1,0)\cup(1,\infty) \qquad$ ...(ii)

From Eqs. (i) and (ii), we get the domain of $f(x)$ as $(-1,0)\cup(1,2)\cup(2,\infty)$.

18. (c) For $f(x)$ to be defined, we must have

$$-1\leq\log_2\left(\frac{1}{2}x^2\right)\leq 1$$

$\Rightarrow \qquad 2^{-1}\leq\dfrac{1}{2}x^2\leq 2^1 \qquad$ [$\because$ the base $=2>1$]

$\Rightarrow \qquad 1\leq x^2\leq 4 \qquad$ ... (i)

Now, $1\leq x^2\Rightarrow x^2-1\geq 0$ i.e. $(x-1)(x+1)\geq 0$

$\Rightarrow \qquad x\leq -1$ or $x\geq 1 \qquad$ ... (ii)

Also, $x^2\leq 4\Rightarrow x^2-4\leq 0$ i.e. $(x-2)(x+2)\leq 0$

$\Rightarrow \qquad -2\leq x\leq 2 \qquad$ ...(iii)

From Eqs. (ii) and (iii), we get the domain of f

$$=\{(-\infty,-1]\cup[1,\infty)\}\cap[-2,2]$$
$$=[-2,-1]\cup[1,2]$$

19. (c) We have,

$$f(x)=\begin{cases}\sin x, & 0\leq x<\pi/2\\ 1, & \dfrac{\pi}{2}\leq x\leq 2\pi\end{cases}$$

and $\qquad g(x)=\begin{cases}0, & 0\leq x\leq\pi\\ \sin x, & \pi<x<\dfrac{3\pi}{2}\\ -1, & \dfrac{3\pi}{2}\leq x\leq 2\pi\end{cases}$

Then, $\qquad h(x)=\begin{cases}0, & 0\leq x<\pi/2\\ 1, & \pi/2\leq x<\dfrac{3\pi}{2}\\ 2, & x\geq\dfrac{3}{2}\end{cases}$

Hence, the range of $h(x)$ is $\{0,1,2\}$.

20. (c) We have,

$$f(x)=\frac{e^x-e^{|x|}}{e^x+e^{|x|}}$$

$$=\begin{cases}0 & , \quad x\ge 0\\ \dfrac{e^x-e^{-x}}{e^x+e^{-x}} & , \quad x<0\end{cases}$$

Range for $x \ge 0$ is $\{0\}$ ...(i)

and let $y=f(x)=\dfrac{e^x-e^{-x}}{e^x+e^{-x}}$ for $x<0$

$$\Rightarrow \quad \frac{y+1}{1-y}=\frac{2e^x}{2e^{-x}}=e^{2x}$$

[$\because$ componendo and dividendo]

$$\Rightarrow \quad 0<\frac{y+1}{1-y}<1 \qquad \left[\begin{array}{l}\because x<0 \Rightarrow e^{2x}<1\\ \therefore 0<e^{2x}<1\end{array}\right]$$

$$\Rightarrow \quad 0>\frac{y+1}{y-1}>-1$$

$$\Rightarrow \quad \frac{y+1}{y-1}<0 \text{ and } \frac{y+1}{y-1}>-1$$

$$\Rightarrow \quad \frac{y+1}{y-1}<0 \text{ and } \frac{2y}{y+1}>0$$

$$\Rightarrow \quad y\in(-1,1) \text{ and } y\in(-\infty,0)\cup(1,\infty)$$

Hence, $y\in(-1,0)$ ...(ii)

From Eqs. (i) and (ii),

Range $=(-1,0]$

21. (d) The given function $f:(1,3)\to R$, defined by

$$f(x)=\frac{x[x]}{1+x^2}$$

$$=\begin{cases}\dfrac{x(1)}{1+x^2}, & x\in(1,2)\\ \dfrac{x(2)}{1+x^2}, & x\in[2,3)\end{cases}$$

$$\Rightarrow \quad f(x)=\begin{cases}\dfrac{x}{1+x^2}, & x\in(1,2)\\ \dfrac{2x}{1+x^2}, & x\in[2,3)\end{cases}$$

is a decreasing function, so

$$\lim_{h\to 0} f(3-h)\Rightarrow \frac{2(3)}{1+9}=\frac{3}{5}$$

$$f(2)=\frac{4}{5}$$

Similarly, $f(x\to 2^-)$ tends to $\frac{2}{5}$ and $f(x\to 1^+)$ tends to $\frac{1}{2}$.

So, range of the given function f is $\left(\frac{2}{5},\frac{1}{2}\right)\cup\left(\frac{3}{5},\frac{4}{5}\right]$.

22. (c) Since $f(x)$ is defined, when

$$[\cos x]>[\sin x],$$

which is possible only when

$$[\cos x]=1 \text{ and } [\sin x]=0$$

$$\therefore \quad f(x)=\ln\{1\}=0$$

Hence, range $=\{0\}$.

23. (b) Given, $f(x)=\log_2(\sqrt{x-4}+\sqrt{6-x})$

Let $y=f(x)=\log_2(\sqrt{x-4}+\sqrt{6-x})$

$$\Rightarrow \quad 2^y=\sqrt{x-4}+\sqrt{6-x}$$

On squaring, we get

$$2^{2y}=(x-4)+(6-x)+2\sqrt{x-4}\sqrt{6-x}$$

$$\Rightarrow \quad 2^{2y}-2=2\sqrt{(x-4)(6-x)}$$

$$\Rightarrow \quad 2^{2y-1}-1=\sqrt{(x-4)(6-x)}\ge 0$$

$$\therefore \quad 2^{2y-1}-1\ge 0 \qquad \text{... (i)}$$

Also, $(2^{2y-1}-1)^2=-x^2+10x-24$

$$\Rightarrow \quad -(2^{2y-1}-1)^2=x^2-10x+24$$

$$=(x-5)^2-1$$

$$\Rightarrow \quad 1-(2^{2y-1}-1)^2=(x-5)^2\ge 0$$

$$\Rightarrow \quad (2^{2y-1}-1)^2\le 1$$

$$\Rightarrow \quad -1\le 2^{2y-1}-1\le 1 \qquad \text{...(ii)}$$

From Eqs. (i) and (ii),

$$0\le 2^{2y-1}-1\le 1$$

$$\Rightarrow \quad 1\le 2^{2y-1}\le 2$$

$$\Rightarrow \quad 2^0\le 2^{2y-1}\le 2^1$$

$$\Rightarrow \quad 0\le 2y-1\le 1$$

$$\Rightarrow \quad 1\le 2y\le 2$$

$$\therefore \quad \frac{1}{2}\le y\le 1$$

Hence, range of $f(x)=\left[\frac{1}{2},1\right]$

24. (a) We have, $f(x)=\dfrac{x}{1+x^2}, x\in R$

$f(x)$ is an odd function and maximum occur at $x=1$

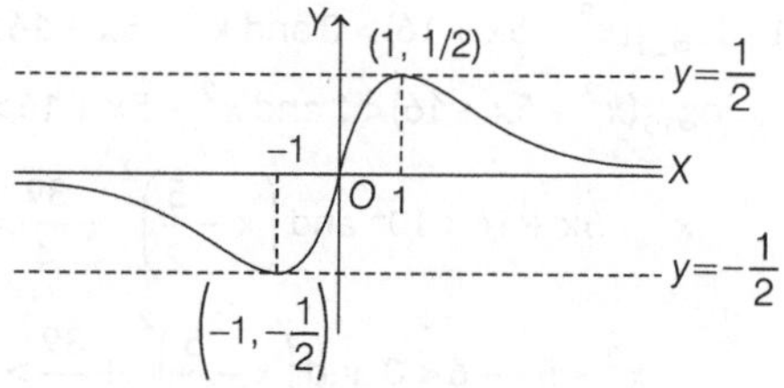

From the graph, it is clear that range of $f(x)$ is $\left[-\frac{1}{2},\frac{1}{2}\right]$.

25. (b) $f(x)$ is defined, if $3x^2 - 4x + 5 > 0$

$$\Rightarrow \quad 3\left[x^2 - \frac{4}{3}x + \frac{5}{3}\right] > 0$$

$$\Rightarrow \quad 3\left[\left(x - \frac{2}{3}\right)^2 + \frac{11}{9}\right] > 0,$$

which is true for all real x.

$\therefore$ Domain of $(f) = (-\infty, \infty)$.

Let $\quad y = \log_e(3x^2 - 4x + 5)$

$$\Rightarrow \quad e^y = 3x^2 - 4x - 5$$

$$\Rightarrow \quad 3x^2 - 4x + (5 - e^y) = 0$$

For x to be real,

$$16 - 12(5 - e^y) \geq 0$$

$$\Rightarrow \quad 12e^y \geq 44$$

$$\Rightarrow \quad e^y \geq \frac{11}{3}$$

$$\Rightarrow \quad y \geq \log_e \frac{11}{3}$$

Range of $f = \left[\log_e \frac{11}{3}, \infty\right)$

26. (b) We have, $f(x) = x^3 - 5x^2 + 7x + 1$

$$f'(x) = 3x^2 - 10x + 7$$

$$= (3x - 7)(x - 1)$$

Now, $f'(x) > 0$

$$\therefore \quad x \in (-\infty, 1) \cup \left(\frac{7}{3}, \infty\right)$$

So, $f(x)$ is increasing in [0, 1] and decreasing in [1, 2].

So, $f(x)$ is many one function.

Now, $\quad f(0) = 1$

$f(1) = 4$

$f(2) = 3$

$\therefore$ Range of f is [1, 4].

Thus, $f(x)$ is onto.

27. (a) Let $x, y \in R$ be such that

$$f(x) = f(y)$$

$$\Rightarrow \quad \frac{x - a}{x - b} = \frac{y - a}{y - b}$$

$$\Rightarrow \quad (x - a)(y - b) = (x - b)(y - a)$$

$$\Rightarrow \quad xy - bx - ay + ab = xy - ax - by + ab$$

$$\Rightarrow \quad (a - b)x = (a - b)y$$

$$\Rightarrow \quad x = y$$

$\therefore$ f is injective.

Let $y \in R$. Then,

$$f(x) = y$$

$$\Rightarrow \quad \frac{x - a}{x - b} = y$$

$$\Rightarrow \quad x - a = xy - by$$

$$\Rightarrow \quad x = \frac{a - by}{1 - y}$$

Clearly, $x \notin R$ for $y = 1$

$\therefore$ f is not surjective.

28. (a) $x = \{f : A \to B, f \text{ is one-one}\}$

$y = \{g : A \to A \times B, g \text{ is one-one}\}$

Number of elements in $A = 3$ i.e. $|A| = 3$

Similarly, $|B| = 5$

Then, $|A \times B| = |A| \times |B| = 3 \times 5 = 15$

Now, number of one-one function from A to B will be

$${}^5P_3 = \frac{5!}{(5-3)!} = \frac{5!}{2!} = 5 \times 4 \times 3 = 60$$

$$\therefore \quad x = 60$$

Now, number of one-one function from A to $A \times B$ will be

$$= {}^{15}P_3 = \frac{15!}{(15-3)!} = \frac{15!}{12!}$$

$$= 15 \times 14 \times 13 = 2730$$

$$\therefore \quad y = 2730$$

Thus, $2 \times (2730) = 91 \times (60)$

$$2y = 91x$$

29. (c) We have, $f(x) = \dfrac{\sin([x]\pi)}{x^2 + x + 1}$

We have given, $[x]\pi$, we know that greatest integer x value always be an integer.

So, we can write

$$[x]\pi = n\pi, n \in Z$$

$$\Rightarrow \quad f(x) = \frac{\sin(nx\pi)}{x^2 + x + 1}$$

$$\therefore \quad f(x) = 0 \qquad [\because \sin(n\pi) \text{ is always zero}]$$

$\therefore$ $f(x)$ is a constant function.

30. (d) We have, $f(x) = \dfrac{|x - 1|}{x} = \begin{cases} -\dfrac{(x-1)}{x}, & \text{if } 0 < x \leq 1 \\ \dfrac{x - 1}{x}, & \text{if } x > 1 \end{cases}$

$$= \begin{cases} \dfrac{1}{x} - 1, & \text{if } 0 < x \leq 1 \\ 1 - \dfrac{1}{x}, & \text{if } x > 1 \end{cases}$$

Now, let us draw the graph of $y = f(x)$

Note that when $x \to 0$, then $f(x) \to \infty$, when $x = 1$, then $f(x) = 0$ and when $x \to \infty$, then $f(x) \to 1$

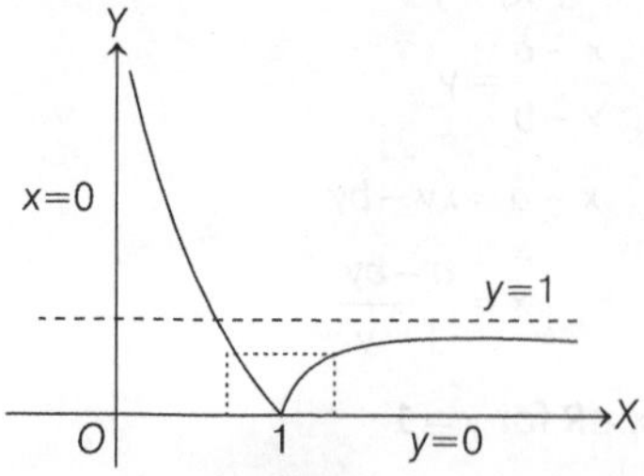

Clearly, $f(x)$ is not injective because, if $f(x) < 1$, then f is many-one, as shown in figure.

Also, $f(x)$ is not surjective because range of $f(x)$ is $[0, \infty[$ and but in problem codomain is $(0, \infty)$, which is incorrect.

$\therefore f(x)$ is neither injective nor surjective

31. (d) We have, $f(x) = \begin{cases} 3 & , x \le 0 \\ 3^{-x} - 3^x + 3 & , x > 0 \end{cases}$

(as $\operatorname{sgn}(e^{-x}) = 1, \forall x \in R$)

Clearly, $f(x)$ is many one function, for $x > 0$, $f(x)$ is decreasing, hence range of $f(x)$ for $x > 0$ is $(-\infty, 3)$.

$\therefore$ Range of $f(x)$ is $(-\infty, 3]$, which is subset of R.

So, function is not on to.

Hence, f is neither injective nor surjective.

32. (c) We have, $f(x) = \dfrac{x}{1+x^2}$

$$\therefore \quad f\left(\frac{1}{x}\right) = \frac{\frac{1}{x}}{1 + \frac{1}{x^2}} = \frac{x}{1+x^2} = f(x)$$

$$\therefore \quad f\left(\frac{1}{2}\right) = f(2) \text{ or } f\left(\frac{1}{3}\right) = f(3) \text{ and so on.}$$

So, $f(x)$ is many-one function.

Again, let $\quad y = f(x) \Rightarrow y = \dfrac{x}{1+x^2}$

$\Rightarrow \quad y + x^2 y = x \Rightarrow yx^2 - x + y = 0$

As, $\quad x \in R$

$\therefore \quad (-1)^2 - 4(y)(y) \ge 0 \Rightarrow 1 - 4y^2 \ge 0$

$\Rightarrow \quad y \in \left[\dfrac{-1}{2}, \dfrac{1}{2}\right]$

$\therefore$ Range = Codomain

$= \left[\dfrac{-1}{2}, \dfrac{1}{2}\right]$

So, $f(x)$ is surjective.

Hence, $f(x)$ is surjective but not injective.

33. (b) We have, $f(x) = \begin{cases} m - x & , x \le 1 \\ 2mx + 1 & , x > 1 \end{cases}$

$f(x)$ is onto, then

Range of f = Codomain of f

For $x \le 1$ $f(x) = m - x$, slope of f is negative.

At $x = 1$, $f(1) = m - 1$

For $x > 1$ f tends to $2m + 1$, m must be negative.

For $f(x)$ is a onto function.

$2m + 1 \ge m - 1$

and slope $m \ge -2$

Then, $m \ge -2$ and $m < 0$

So, range of m is $[-2, 0)$.

34. (b) $f(x) = \tan^{-1}\left(\dfrac{2x}{1-x^2}\right)$

Put $x = \tan\theta = \tan^{-1}\left(\dfrac{2\tan\theta}{1 - \tan^2\theta}\right)$

$f(x) = 2\theta$

Then, f is one-one and onto.

$-1 < \tan\theta < 1$

$\Rightarrow \quad \dfrac{-\pi}{4} < \theta < \dfrac{\pi}{4}$

$\Rightarrow \quad \dfrac{-\pi}{2} < 2\theta < \dfrac{\pi}{2}$

$\therefore \quad B$ is $\left(-\dfrac{\pi}{2}, \dfrac{\pi}{2}\right)$.

35. (b) The number of functions from $f : \{a_1, a_2 \ldots a_m\} \to \{b_1, b_2 \ldots b_n\}$ is $= n^m$

So, the number of function from $f : \{a, b, c, d, e, f\} \to \{p, q, r\}$ is $= 3^6$.

36. (d) Set A consist of 6 elements and B consist of 4 different elements, then

Number of surjective mapping

$= 4^6 - {}^4C_1(3)^6 + {}^4C_2(2)^6 - {}^4C_3(1)^6$

$= 4096 - 4(729) + 6(64) - 4$

$= 1560$

37. (a) According to given information, we have, if $k \in \{4, 8, 12, 16, 20\}$

Then, $f(k) \in \{3, 6, 9, 12, 15, 18\}$

$[\because$ codomain $(f) = \{1, 2, 3, \ldots, 20\}]$

Now, we need to assign the value of $f(k)$ for $k \in \{4, 8, 12, 16, 20\}$ this can be done in ${}^6C_5 \cdot 5!$ ways $= 6 \cdot 5! = 6!$ and remaining 15 element can be associated by $15!$ ways.

$\therefore$ Total number of onto functions $= 15!6!$

38. (c) There are 4 inputs {1, 2, 8, 9} and 4 outputs {3, 4, 5, 10}. Hence, function will be bijective if and only if each output is connected with only one input. Therefore, number of bijective function is $4! = 24$

39. (b) We have, $f(x) = \begin{cases} x^2 \sin\dfrac{\pi x}{2}, & |x| < 1 \\ x|x|, & |x| \geq 1 \end{cases}$

$$\Rightarrow \quad f(x) = \begin{cases} -x^2, & x \leq -1 \\ x^2 \sin\dfrac{\pi x}{2}, & -1 < x < 1 \\ x^2, & x \geq 1 \end{cases}$$

For all $x \in (-1, 1)$

$$f(-x) = (-x)^2 \sin\left(\frac{-\pi x}{2}\right) = -x^2 \sin\frac{\pi x}{2} = -f(x)$$

Let $x \in [1, \infty)$, then $x = 1 + k, k \geq 0$

Clearly, $-x = -1 - k \in (-\infty, -1]$

Now, $f(-x) = f(-1-k) = -(1+k)^2$

and $f(x) = (1+k)^2$

$\therefore f(-x) = -f(x)$ for all $x \in (-\infty, -1] \cup [1, \infty)$.

Thus, $f(-x) = -f(x)$ for all $x \in R$.

Hence, $f(x)$ is an odd function.

40. (a) $f(x) = \dfrac{x(\sin x + \tan x)}{\left[\dfrac{x+\pi}{\pi}\right] - 1/2} = \dfrac{x(\sin x + \tan x)}{\left[\dfrac{x}{\pi}\right] + 1 - 1/2}$

$$f(x) = \frac{x(\sin x + \tan x)}{\left[\dfrac{x}{\pi}\right] + 1/2}$$

Now, $f(-x) = \dfrac{(-x)(\sin(-x) + \tan(-x))}{\left[-\dfrac{x}{\pi}\right] + 1/2}$

$$f(-x) = \frac{x(\sin x + \tan x)}{-1 - \left[\dfrac{x}{\pi}\right] + 1/2}$$

$$= -\left(\frac{x(\sin x + \tan x)}{\left[\dfrac{x}{\pi}\right] + 1/2}\right)$$

$f(-x) = -f(x)$

Hence, $f(x)$ is an odd function.

41. (c) We have, $g(x) = x^3 + \tan x + \left[\dfrac{x^2+1}{p}\right]$

On $g(-x) = (-x)^3 + \tan(-x) + \left[\dfrac{(-x)^2+1}{p}\right]$

$$g(-x) = -x^3 - \tan x + \left[\frac{x^2+1}{p}\right]$$

Since, $g(x)$ is an odd function.

$$\therefore \quad g(x) + g(-x) = 0$$

$$\Rightarrow \quad 2\left[\frac{x^2+1}{p}\right] = 0$$

$$\Rightarrow \quad \left[\frac{x^2+1}{p}\right] = 0$$

$$\Rightarrow \quad 0 \leq \frac{x^2+1}{p} < 1$$

Now, $x \in [-2, 2]$

$$\therefore \quad x^2 + 1 \in [1, 5]$$

$$\therefore \quad p > 5$$

42. (a) Since, $f(x) = \dfrac{x}{e^x - 1} + \dfrac{x}{2} + 1$ ... (i)

$$\therefore \quad f(-x) = \frac{-x}{e^{-x} - 1} - \frac{x}{2} + 1$$

$$= \frac{xe^x}{e^x - 1} - \frac{x}{2} + 1 \quad \text{... (ii)}$$

Subtracting Eq. (ii) from Eq. (i), we get

$$f(x) - f(-x) = \frac{x}{e^x - 1}(1 - e^x) + x = -x + x$$

$$f(x) - f(-x) = 0 \Rightarrow f(-x) = f(x)$$

Hence, $f(x)$ is an even function.

43. (a) Putting x, y equal to zero in the given relation

$$2f(0) = 2f(0) \cdot f(0)$$

$$\Rightarrow \quad f(0)[f(0) - 1] = 0$$

$\Rightarrow \quad f(0) = 0$ or $f(0) = 1$ but given that $f(0) \neq 0$.

$$\therefore \quad f(0) = 1$$

Again, putting $x = 0$ in the given relation

$$f(y) + f(-y) = 2f(0)f(y)$$

$$\Rightarrow \quad f(y) + f(-y) = 2f(y)$$

$$\Rightarrow \quad f(-y) = f(y)$$

$\therefore f(x)$ is an even function.

44. (d) Given, function $f(x) = a^x, a > 0$ is written as sum of an even and odd functions $f_1(x)$ and $f_2(x)$, respectively.

Clearly, $f_1(x) = \dfrac{a^x + a^{-x}}{2}$ and $f_2(x) = \dfrac{a^x - a^{-x}}{2}$

So, $f_1(x+y) + f_1(x-y)$

$$= \frac{1}{2}[a^{x+y} + a^{-(x+y)}] + \frac{1}{2}[a^{x-y} + a^{-(x-y)}]$$

$$= \frac{1}{2}\left[a^x a^y + \frac{1}{a^x a^y} + \frac{a^x}{a^y} + \frac{a^y}{a^x}\right]$$

$$= \frac{1}{2}\left[a^x\left(a^y + \frac{1}{a^y}\right) + \frac{1}{a^x}\left(\frac{1}{a^y} + a^y\right)\right]$$

$$=\frac{1}{2}\left(a^x+\frac{1}{a^x}\right)\left(a^y+\frac{1}{a^y}\right)$$

$$=2\left(\frac{a^x+a^{-x}}{2}\right)\left(\frac{a^y+a^{-y}}{2}\right)=2f_1(x)\cdot f_1(y)$$

45. (d) We have, $f(x)=4\cos^4\left(\frac{x-\pi}{4\pi^2}\right)-2\cos\left(\frac{x-\pi}{2\pi^2}\right)$

$$=\left\{2\cos^2\left(\frac{x-\pi}{4\pi^2}\right)\right\}^2-2\cos\left(\frac{x-\pi}{2\pi^2}\right)$$

$$=\left[1+\cos\left(\frac{x-\pi}{2\pi^2}\right)\right]^2-2\cos\left(\frac{x-\pi}{2\pi^2}\right)$$

$$=1+\cos^2\left(\frac{x-\pi}{2\pi^2}\right)$$

Clearly, period of $f(x)=\frac{\pi}{\frac{1}{2\pi^2}}=2\pi^3$

46. (d) Clearly, $f(x)=\tan(\sqrt{[n]}x$ has period $\frac{\pi}{\sqrt{[n]}}$

but given that period is $\frac{\pi}{3}$.

Therefore, $\frac{\pi}{\sqrt{[n]}}=\frac{\pi}{3}\Rightarrow[n]=9\Rightarrow n\in[9,10)$

47. (a) We have,

$$f\left(\frac{\pi}{2}+x\right)=\left|\left|\sin\left(\frac{\pi}{2}+x\right)\right|-\left|\cos\left(\frac{\pi}{2}+x\right)\right|\right|$$

$$f\left(\frac{\pi}{2}+x\right)=||\cos x|-|\sin x||$$

$$\Rightarrow f\left(\frac{\pi}{2}+x\right)=||\sin x|-|\cos x||$$

$$\Rightarrow f\left(\frac{\pi}{2}+x\right)=f(x)\text{ for all }x\in R$$

Hence, $f(x)$ is periodic with period $\frac{\pi}{2}$.

48. (a) $f(x)=\sqrt{x^2-4x+4}=\sqrt{(x-2)^2}=|x-2|$

$$g(x)=|x|-|2|=|x|-2=\begin{cases}-x-2, & x<0\\ x-2, & x>0\end{cases}$$

Thus, $g(x)$ is not same as $|x-2|$

$$h(x)=\frac{|x-2|^2}{|x-2|}=|x-2|,\ x\neq 2$$

This is not same as $|x-2|$ as $h(x)$ has domain $R-\{2\}$.

$$p(x)=\left|\frac{x^2-x-2}{x+1}\right|=\left|\frac{(x-2)(x+1)}{(x+1)}\right|=|x-2|,\ x\neq -1$$

Thus, $p(x)$ is not same as $|x-2|$ as $p(x)$ has domain $R-\{-1\}$.

49. (d) Here,

(a) $\sin^{-1}(\sin x)$ is defined for $x\in\left[-\frac{\pi}{2},\pi/2\right]$

While $\sin(\sin^{-1}x)$ is defined for $x\in[-1,1]$.

So, $\sin^{-1}(\sin x)$ and $\sin(\sin^{-1}x)$ is not identical.

(b) $\log_e e^x$ is defined for $x>0$

and $e^{\log_e x}$ is defined for $x>0$.

So, $\log_e e^x$ and $e^{\log_e x}$ is not identical.

(c) $\log_e x^2$ is defined for all $x\in R-\{0\}$ while $2\log x$ is defined for $x>0$.

$\therefore \log_e x^2$ and $2\log x$ is not identical.

(d) Let $f(x)=x|x|$ and $g(x)=x^2\operatorname{sgn}x$

$\therefore D_f=R$ and $D_g=R$

$\therefore D_f=D_g$

and $f(x)=x|x|$ and $g(x)=x^2\frac{|x|}{x}=x|x|$

$\Rightarrow R_f=R_g$

Hence, $x|x|$ and $x^2\operatorname{sgn}x$ is identical.

50. (c) Given, $f(x)=\sqrt{x}$ and $g(x)=\sqrt{1-x}$

$\therefore$ Domain of $f(x)=D_1$ is $x\geq 0$

i.e. $D_1: x\in(0,\infty)$

and domain of $g(x)=D_2$ is $1-x\geq 0$

$\Rightarrow x\leq 1$

i.e. $D_2: x\in(-\infty,1]$

As, we know that the domain of $f+g, f-g, g-f$ will be $D_1\cap D_2$ as well as the domain for f/g is $D_1\cap D_2$ except all those value(s) of x, such that $g(x)=0$.

Similarly, for $\frac{g}{f}$ is $D_1\cap D_2$ but $f(x)\neq 0$.

Hence, common domain for $(f+g),(f-g),\left(\frac{f}{g}\right),\left(\frac{g}{f}\right)$ and $(g-f)$ will be $0<x<1$.

51. (b) Given that

$f(x)=\cos x$ has domain R and range $[-1,1]$

and $g(x)=\frac{1}{\sqrt{1+\tan^2 x}}=\frac{1}{\sqrt{\sec^2 x}}=|\cos x|$

has domain $R-\left\{(2n+1)\frac{\pi}{2}\right\}, n\in Z$ as $\tan x$ is not defined for $x=(2n+1)\frac{\pi}{2}, n\in Z$

Also, range of $|\cos x|$ is [0, 1].

Hence, $f(x)$ and $g(x)$ are identical in first and fourth quadrant

$$x\in\bigcup_{n\in Z}\left(-\frac{\pi}{2}+2n\pi,\frac{\pi}{2}+2n\pi\right)$$

52. (a) Let $y = f(x) = \dfrac{5x+3}{6x-\alpha}$ ... (i)

$\Rightarrow \quad 5x + 3 = 6xy - \alpha y$

$\Rightarrow \quad x(6y-5) = \alpha y + 3$

$\Rightarrow \quad x = \dfrac{\alpha y + 3}{6y-5}$

$\Rightarrow \quad f^{-1}(y) = \dfrac{\alpha y + 3}{6y-5}$

So, $\quad f^{-1}(x) = \dfrac{\alpha x + 3}{6x-5}$ ... (ii)

Now, $(fof)\, x = x$

$\Rightarrow \quad f(x) = f^{-1}(x)$

$$\frac{5x+3}{6x-\alpha} = \frac{\alpha x+3}{6x-5}$$

By comparing, $\alpha = 5$

53. (c) We have,

$$f(x) = \frac{1}{x},\ g(x) = \frac{1}{x^2} \text{ and } h(x) = x^2$$

$f(g(x)) = x^2,\ x \neq 0$

$h(g(x)) = \dfrac{1}{x^4} = (g(x))^2,\ x \neq 0$

54. (b) Given, for $x \in (0, 3/2)$, functions

$f(x) = \sqrt{x}$ ... (i)

$g(x) = \tan x$ ... (ii)

and $\quad h(x) = \dfrac{1-x^2}{1+x^2}$ ... (iii)

Also given, $\phi(x) = ((hof)og)(x) = (hof)(g(x))$

$= h(f(g(x))) = h(f(\tan x))$

$= h(\sqrt{\tan x}) = \dfrac{1-(\sqrt{\tan x})^2}{1+(\sqrt{\tan x})^2}$

$= \dfrac{1-\tan x}{1+\tan x} = \tan\left(\dfrac{\pi}{4} - x\right)$

Now, $\quad \phi\left(\dfrac{\pi}{3}\right) = \tan\left(\dfrac{\pi}{4} - \dfrac{\pi}{3}\right)$

$= \tan\left(\dfrac{3\pi - 4\pi}{12}\right)$

$= \tan\left(-\dfrac{\pi}{12}\right)$

$= -\tan\left(\dfrac{\pi}{12}\right)$

$= \tan\left(\pi - \dfrac{\pi}{12}\right)$

$= \tan\left(\dfrac{11\pi}{12}\right)$

55. (c) We have, $\dfrac{1}{2} gof(x) = 2x^2 - 5x + 2$

$\Rightarrow g(f(x)) = 4x^2 - 10x + 4$

$\therefore \quad [\{f(x)\}^2 + \{f(x)\} - 2] = 4x^2 - 10x + 4$

$\Rightarrow (f(x))^2 + f(x) - (4x^2 - 10x + 6) = 0$

$\therefore \quad f(x) = \dfrac{-1 \pm \sqrt{1 + 4(4x^2 - 10x + 6)}}{2}$

$= \dfrac{-1 \pm \sqrt{(16x^2 - 40x + 25)}}{2}$

$= \dfrac{-1 \pm (4x-5)}{2}$

$= 2x - 3 \text{ or } -2x + 2$

56. (a) Since, the $g(x) = x^2 + x - 1$ and $gof(x)$ $= 4x^2 - 10x + 5$ both are quadratic equations, therefore $f(x)$ must be a linear equation.

Now, let $f(x) = ax + b$,

So, $gof(x) = 4x^2 - 10x + 5$

$\Rightarrow g(ax + b) = 4x^2 - 10x + 5$

$\Rightarrow (ax+b)^2 + (ax+b) - 1 = 4x^2 - 10x + 5$

$[\because g(x) = x^2 + x - 1]$

$\Rightarrow a^2x^2 + a(2b+1)x + (b^2 + b - 1) = 4x^2 - 10x + 5$

On comparing the coefficients of terms, we get

$a^2 = 4$ ...(i)

$\Rightarrow \quad a(2b+1) = -10$ ...(ii)

$\Rightarrow \quad b^2 + b - 1 = 5$ ...(iii)

From Eq. (i), we have $a = \pm 2$

From Eq. (iii), we have $b = -3, 2$

and the equation is getting satisfied by $a = -2$ and $b = 2$, so $f(x) = -2x + 2$

$\therefore \quad f\left(\dfrac{5}{4}\right) = -2\left(\dfrac{5}{4}\right) + 2 = -\dfrac{5}{2} + 2 = -\dfrac{1}{2}$

57. (b) We have, $f(x) = \dfrac{kx}{x+1}, x \neq -1$

Now, $f(f(x)) = x$

$\Rightarrow \quad \dfrac{k\left(\dfrac{kx}{x+1}\right)}{\dfrac{kx}{x+1} + 1} = x$

$\Rightarrow \quad \dfrac{k^2x}{(k+1)x+1} = x$

$\Rightarrow (k+1)x^2 + (1-k^2)x = 0$

Now, it is true for all real x

$\Rightarrow \quad k + 1 = 0 \text{ and } 1 - k^2 = 0$

$\Rightarrow \quad k = -1$

58. (b) We have,

$$f_1(x)=\frac{1}{x},\ f_2(x)=1-x \text{ and } f_3(x)=\frac{1}{1-x}$$

Also, we have

$$(f_2 \circ J \circ f_1)(x)=f_3(x)$$

$$\Rightarrow \quad f_2((J \circ f_1)(x))=f_3(x)$$

$$\Rightarrow \quad f_2(J(f_1(x)))=f_3(x)$$

$$\Rightarrow \quad 1-J(f_1(x))=\frac{1}{1-x}$$

$$[\because f_2(x)=1-x \text{ and } f_3(x)=\frac{1}{1-x}]$$

$$\Rightarrow \quad 1-J\left(\frac{1}{x}\right)=\frac{1}{1-x} \qquad [\because f_1(x)=\frac{1}{x}]$$

$$\Rightarrow \quad J\left(\frac{1}{x}\right)=1-\frac{1}{1-x}=\frac{1-x-1}{1-x}=\frac{-x}{1-x}$$

Now, put $\frac{1}{x}=X$, then

$$J(X)=\frac{\frac{-1}{X}}{1-\frac{1}{X}}=\frac{-1}{X-1}=\frac{1}{1-X}$$

$$\Rightarrow \quad J(X)=f_3(X) \text{ or } J(x)=f_3(x)$$

59. (b) $f[f(x)]+f\left[f\left(\frac{1}{x}\right)\right]$

$$=f\left(\frac{1-x}{1+x}\right)+f\left(\frac{1-\frac{1}{x}}{1+\frac{1}{x}}\right), x\neq 0$$

$$=\frac{1-\frac{1-x}{1+x}}{1+\frac{1-x}{1+x}}+\frac{1-\frac{x-1}{x+1}}{1+\frac{x-1}{x+1}}$$

$$=x+\frac{1}{x}=\left(\sqrt{x}-\frac{1}{\sqrt{x}}\right)^2+2\geq 2$$

60. (d) Given, $f(n+1)=f(n)+f(1), \forall\ n\in N$

$$\Rightarrow \quad f(n+1)-f(n)=f(1)$$

It is an AP with common difference $=f(1)$

Also, general term $=T_n=f(1)+(n-1)f(1)=nf(1)$

$$\Rightarrow \quad f(n)=nf(1)$$

Clearly, $f(n)$ is one-one.

For *fog* to be one-one, *g* must be one-one.

For *f* to be onto, $f(n)$ should take all the values of natural numbers.

As, $f(x)$ is increasing, $f(1)=1$

$$\Rightarrow \quad f(n)=n$$

If *g* is many-one, then *fog* is many one.

So, if *g* is onto, then *fog* is one-one.

61. (a) We have,

$$y=\sqrt{3}\sin x-\cos x+2$$

$$y=2\left(\frac{\sqrt{3}}{2}\sin x-\frac{1}{2}\cos x\right)+2$$

$$\Rightarrow \quad \frac{y}{2}-1=\frac{\sqrt{3}}{2}\sin x-\frac{1}{2}\cos x$$

$$\Rightarrow \quad \frac{y-2}{2}=-\cos\left(\frac{\pi}{3}+x\right)$$

$$\Rightarrow \quad \frac{y-2}{2}=\cos\left(\pi-\left(\frac{\pi}{3}+x\right)\right)$$

$$\Rightarrow \quad \cos^{-1}\left(\frac{y-2}{2}\right)=\frac{2\pi}{3}-x$$

$$\Rightarrow \quad x=\frac{2\pi}{3}-\cos^{-1}\left(\frac{y-2}{2}\right)$$

Therefore, $f^{-1}(x)=\frac{2\pi}{3}-\cos^{-1}\left(\frac{x-2}{2}\right)$

62. (b) *x* is positive by definition

$$y=2^{x(x-1)}$$

$$\Rightarrow \quad x^2-x-\log_2 y=0$$

$$x=\frac{1}{2}(1\pm\sqrt{1+4\log_2 y})$$

Since, $x\geq 1$

So, $x=\frac{1}{2}(1+\sqrt{1+4\log_2 y})$

$$\Rightarrow \quad f^{-1}(y)=\frac{1}{2}(1+\sqrt{1+4\log_2 y})$$

$$\Rightarrow \quad f^{-1}(x)=\frac{1}{2}(1+\sqrt{1+4\log_2 x})$$

63. (c) Given function,

$$f(x)=\frac{8^{2x}-8^{-2x}}{8^{2x}+8^{-2x}}, x\in(-1,1)=\frac{8^{4x}-1}{8^{4x}+1}=y \text{ (let)}$$

On applying componendo and dividendo rule, we get

$$8^{4x}=\frac{1+y}{1-y}$$

On applying logarithm having base '8' both sides, we get

$$4x=\log_8\left(\frac{1+y}{1-y}\right)$$

$$\Rightarrow \quad x=\frac{1}{4}\log_8\left(\frac{1+y}{1-y}\right)=\frac{1}{4}(\log_8 e)\log_e\left(\frac{1+y}{1-y}\right)$$

[by base change property of logarithm $\log_a b=\log_a e\cdot\log_e b$]

By interchanging the variables *x* and *y*, we get the inverse function of $f(x)$ and it is

$$f^{-1}(x)=\frac{1}{4}(\log_8 e)\log_e\left(\frac{1+x}{1-x}\right).$$

64. (c) We have,

$$y = \frac{e^{3x} - e^{-3x}}{e^{3x} + e^{-3x}}$$

$$y = \frac{e^{6x} - 1}{e^{6x} + 1}$$

$$\Rightarrow \quad \frac{y+1}{y-1} = \frac{e^{6x} - 1 + e^{6x} + 1}{e^{6x} - 1 - e^{6x} - 1}$$

$$\Rightarrow \quad \frac{y+1}{y-1} = \frac{2e^{6x}}{-2}$$

$$\Rightarrow \quad \frac{1+y}{1-y} = e^{6x}$$

Taking log on both sides, we get

$$\log\left(\frac{1+y}{1-y}\right) = 6x$$

$$\Rightarrow \quad x = \frac{1}{6}\log\left(\frac{1+y}{1-y}\right)$$

Hence, $f^{-1} = \frac{1}{6}\log\left(\frac{1+x}{1-x}\right)$

65. (b) We have, $y = 5^{\log x}$

$$\Rightarrow \quad \log y = \log x \cdot \log 5$$

$$\Rightarrow \quad \log x = \frac{\log y}{\log 5}$$

$$\Rightarrow \quad \log x = \log_5 y$$

$$\Rightarrow \quad x = e^{\log_5 y}$$

$$\Rightarrow \quad x = y^{\log_5 e}$$

$$\Rightarrow \quad x = y^{\frac{1}{\log 5}}$$

66. (a) We have, $f(x) = \log_a x^2$

$$y = \log_a x^2$$

$$\Rightarrow \quad y = 2\log_a x \Rightarrow x = a^{y/2}$$

So, $f^{-1}(x) = a^{x/2}$

Now, $f^{-1}(b+c) = a^{\frac{b+c}{2}} = a^{b/2} \cdot a^{c/2}$

$$= f^{-1}(b) \cdot f^{-1}(c)$$

67. (c) Given, $f(x) = \frac{x-2}{x-3}$

$$g(x) = 2x - 3$$

Let $\quad y = f(x) = \frac{x-2}{x-3}$

$$\Rightarrow \quad xy - 3y = x - 2 \Rightarrow xy - x = 3y - 2$$

$$\Rightarrow \quad x(y-1) = 3y - 2$$

$$\Rightarrow \quad x = \frac{3y-2}{y-1}$$

$$\Rightarrow \quad f^{-1}(y) = \frac{3y-2}{y-1} \Rightarrow f^{-1}(x) = \frac{3x-2}{x-1}$$

Similarly, $g^{-1}(x) = \frac{x+3}{2}$

Given, $f^{-1}(x) + g^{-1}(x) = \frac{13}{2}$

$$\Rightarrow \quad \frac{3x-2}{x-1} + \frac{x+3}{2} = \frac{13}{2}$$

$$\Rightarrow \quad x^2 + 8x - 7 = 13(x-1)$$

$$\Rightarrow \quad x^2 - 5x + 6 = 0$$

$$\Rightarrow \quad (x-2)(x-3) = 0$$

$$\Rightarrow \quad x = 2, 3$$

$\therefore$ Sum $= 2 + 3 = 5$

68. (d) We have, $f(x) = (x+1)^2$

$$\Rightarrow \quad y = (x+1)^2, (x \geq -1)$$

Its reflection $g(x)$ in the line $y = x$ is given by

$$x = (y+1)^2$$

$$\Rightarrow \quad y + 1 = \pm\sqrt{x}$$

$$\Rightarrow \quad y = \sqrt{x} - 1 \; (x \geq 0)$$

So, $\quad g(x) = \sqrt{x} - 1$

69. (c) $\therefore (gof)^{-1}$ exists $\Rightarrow gof$ is bijective

$\Rightarrow f$ must be one-one and g must be onto.

70. (b) We have, $f(x) = \sin x + \cos x$ and $g(x) = x^2 - 1$

$$g(f(x)) = g(\sin x + \cos x)$$

$$= (\sin x + \cos x)^2 - 1$$

$$g(f(x)) = \sin^2 x + \cos^2 x + 2\sin x \cos x - 1$$

$$= 1 + \sin 2x - 1$$

$$g(f(x)) = \sin 2x$$

Let $y = \sin 2x$ is invertible.

$$\therefore \quad x = \frac{1}{2}\sin^{-1} y$$

$$= \frac{1}{2}\left[-\frac{\pi}{2}, \frac{\pi}{2}\right] = \left[-\frac{\pi}{4}, \frac{\pi}{4}\right]$$

71. (c) We have, $f(x) = 3x - 4$

$$\Rightarrow \quad y = 3x - 4$$

$$\Rightarrow \quad x = \frac{y+4}{3}$$

$$\Rightarrow \quad f^{-1}(x) = \left(\frac{x+4}{3}\right) \text{ and } g(x) = 2 + 3x$$

$$\Rightarrow \quad y = 2 + 3x$$

$$\Rightarrow \quad x = \left(\frac{y-2}{3}\right)$$

$$\Rightarrow \quad g^{-1}(x) = \left(\frac{x-2}{3}\right)$$

Now, $g^{-1}of^{-1}(x) = g^{-1}(f^{-1}(x))$

$$= g^{-1}\left(\frac{x+4}{3}\right) = \left(\frac{\frac{x+4}{3} - 2}{3}\right) = \frac{x-2}{9}$$

$\therefore \quad g^{-1}of^{-1}(5) = \frac{5-2}{9} = \frac{3}{9} = \frac{1}{3}$

72. (d) We have,

$f(x) = 2x - 3$ and $g(x) = x^3 + 5$ are bijection and hence f^{-1} and g^{-1} both exists.

Now, $y = 2x - 3$

$\Rightarrow \quad x = \frac{y+3}{2}$

$\therefore \quad f^{-1}(y) = \frac{y+3}{2}$

$\therefore \quad f^{-1}(x) = \frac{x+3}{2}$ ... (i)

and $y = x^3 + 5 \Rightarrow x = (y-5)^{1/3}$

$\Rightarrow \quad g^{-1}(y) = (y-5)^{1/3}$

$\therefore \quad g^{-1}(x) = (x-5)^{1/3}$ ... (ii)

$\therefore \quad (fog)^{-1} = (g^{-1}of^{-1})(x)$

$= g^{-1}[f^{-1}(x)]$

$= g^{-1}\left[\frac{x+3}{2}\right]$ [from Eqs. (i) and (ii)]

$= \left[\frac{x+3}{2} - 5\right]^{1/3}$

$= \left(\frac{x-7}{2}\right)^{1/3}$

73. (a) Given that $f(x) = ax^7 + bx^3 + cx - 5$

At $x = 7$,

$f(7) = a(7)^7 + b(7)^3 + c(7) - 5$ ... (i)

and when $x = -7$

$f(-7) = a(-7)^7 + b(-7)^3 + c(-7) - 5$ ... (ii)

Adding Eqs. (i) and (ii), we get

$f(7) + f(-7) = -10$

$\Rightarrow \quad f(7) + 7 = -10 \Rightarrow f(7) = -17$

74. (c) We have, $f(x) + 3xf\left(\frac{1}{x}\right) = 2(x+1)$

$f(x) + 3xf\left(\frac{1}{x}\right) = 2x + 2$ ... (i)

Put $x \to \frac{1}{x}$,

$f\left(\frac{1}{x}\right) + \frac{3}{x}f(x) = \frac{2}{x} + 2$

Multiplying by $3x$,

$3xf\left(\frac{1}{x}\right) + 9f(x) = 6 + 6x$ ... (ii)

Subtracting Eq. (ii) from Eq. (i),

$8f(x) = 4x + 4$

$f(x) = \frac{x+1}{2}$

Now, $f(10099) = \frac{10099+1}{2} = \frac{10100}{2} = 5050$

75. (c) Given, $f(x) = \frac{5^x}{5^x + 5}$, then $f(2-x) = \frac{5^{2-x}}{5^{2-x} + 5}$

$= \frac{5}{5^x + 5}$

This gives, $f(x) + f(2-x) = \frac{5^x + 5}{5^x + 5} = 1$

$\Rightarrow \quad f\left(\frac{1}{20}\right) + f\left(2 - \frac{1}{20}\right) = f\left(\frac{1}{20}\right) + f\left(\frac{39}{20}\right) = 1$

Similarly,

$f\left(\frac{2}{20}\right) + f\left(\frac{38}{20}\right) = 1$ and so on,

$\therefore \quad f\left(\frac{1}{20}\right) + f\left(\frac{2}{20}\right) + \ldots + f\left(\frac{38}{20}\right) + f\left(\frac{39}{20}\right)$

$= 1 + 1 + \ldots + 1 + f\left(\frac{20}{20}\right)$

$= 19 + f(1) = 19 + \frac{1}{2} = \frac{39}{2}$

76. (c) We have,

$f(x) + 2f\left(\frac{1}{x}\right) = 3x$ ... (i)

Replacing x by $\frac{1}{x}$

$f\left(\frac{1}{x}\right) + 2f(x) = \frac{3}{x}$ ... (ii)

Solving Eqs. (i) and (ii), we get

$f(x) = \left(\frac{2}{x} - x\right)$

$\therefore \quad f(x) = f(-x)$

$\Rightarrow \quad \left(\frac{2}{x} - x\right) = \left(-\frac{2}{x} + x\right)$

$\Rightarrow \quad \frac{4}{x} - 2x = 0$

$\Rightarrow \quad 2x^2 - 4 = 0$

$\Rightarrow \quad x = \pm\sqrt{2}$

Hence, $S = \{x \in R : f(x) = f(-x)\} = \{-\sqrt{2}, \sqrt{2}\}$

77. (b) We have, $f(m+n)=f(m)+f(n)$

Put $m=n=1$

$$f(2)=2f(1) \quad \ldots \text{(i)}$$

Put $m=2, n=1$

$$f(3)=f(2)+f(1)$$

From Eq. (i),

$$f(3)=2f(1)+f(1)$$
$$f(3)=3f(1) \quad \ldots \text{(ii)}$$

Put $m=3, n=3$

$$f(6)=f(3)+f(3)$$

From Eq. (ii),

$$18=6f(1)$$
$$f(1)=3$$

So, $\quad f(2)=6, f(3)=9$

Hence, $f(2)\cdot f(3)=6\times 9=54$

78. (c) We have, $f(x-y)=f(x)\cdot f(y)-f(a-x)f(a+y)$

Putting $x=y=0$, we get

$$f(0)=f(0)f(0)-f(a)f(a)$$
$$\Rightarrow \quad [f(a)]^2=1^2-1=0$$
$$\Rightarrow \quad f(a)=0$$

Put $x=0, y=a$, we get

$$f(-a)=f(a)-f(a)f(2a)$$
$$f(-a)=0$$

Again, putting $x=a, y=a$, we get

$$f(0)=[f(a)]^2-f(0)f(2a)$$
$$\Rightarrow \quad 1=0-f(2a)$$
$$\Rightarrow \quad f(2a)=-1$$

Now putting $x=2a, y=x$, we get

$$f(2a-x)=f(2a)f(x)-f(-a)f(a+x)$$
$$=(-1)f(x)-(0)f(a+x)$$
$$f(2a-x)=-f(x)$$

79. (a) We have,

$$f(x)=ax^2+bx+c$$

Now, $f(x+y)=f(x)+f(y)+xy$

Put $y=0 \Rightarrow f(x)=f(x)+f(0)+0$

$$\Rightarrow \quad f(0)=0 \Rightarrow c=0$$

Again, put $y=-x$

$$\therefore \quad f(0)=f(x)+f(-x)-x^2$$
$$\Rightarrow \quad 0=ax^2+bx+ax^2-bx-x^2$$
$$\Rightarrow \quad 2ax^2-x^2=0$$
$$\Rightarrow \quad a=\frac{1}{2}$$

Also, $\quad a+b+c=3$

$$\Rightarrow \quad \frac{1}{2}+b+0=3 \Rightarrow b=\frac{5}{2}$$

$$\therefore \quad f(x)=\frac{x^2+5x}{2}$$

Now, $\quad f(n)=\frac{n^2+5n}{2}=\frac{1}{2}n^2+\frac{5}{2}n$

$$\therefore \quad \sum_{n=1}^{10} f(n)=\frac{1}{2}\sum_{n=1}^{10} n^2+\frac{5}{2}\sum_{n=1}^{10} n$$
$$=\frac{1}{2}\cdot\frac{10\times 11\times 21}{6}+\frac{5}{2}\times\frac{10\times 11}{2}$$
$$=\frac{385}{2}+\frac{275}{2}=\frac{660}{2}=330$$

80. (75) We have,

$$\left[\frac{3}{4}\right]+\left[\frac{3}{4}+\frac{1}{100}\right]+\left[\frac{3}{4}+\frac{2}{100}\right]+\ldots+\left[\frac{3}{4}+\frac{99}{100}\right]$$
$$=[0.75]+[0.75+0.01]+[0.75+0.02]+\ldots+$$
$$[0.75+0.25]+[0.75+0.26]+\ldots+[0.75+0.99]$$
$$=0+0\ldots+0+\underbrace{1+1+\ldots+1}_{75 \text{ times}}=75$$

81. (4) We have, $\sum_{n=8}^{100}\left[\frac{(-1)^n n}{2}\right]$

$$=[4]+[-4.5]+[5]+[-5.5]+\ldots+[-49.5]+[50]$$
$$=4-5+5-6+6+\ldots-50+50=4$$

82. (30) We have,

$$y=2[x]+3=3[x-2]$$
$$\Rightarrow \quad 2[x]+3=3[x]-6$$
$$\Rightarrow \quad [x]=9$$

and $\quad y=2[x]+3=2\times 9+3=21$

$$\therefore \quad [x+y]=[x+21]=[x]+21=9+21=30$$

83. (1) Given, $f: R \to A$ is defined by

$$f(x)=\frac{x^2}{x^2+1}$$

Since, f is surjection.

So, $R(f)=$ codomain A

Let $\quad y=\frac{x^2}{x^2+1}$

$$\Rightarrow \quad y=x^2(1-y)$$
$$\Rightarrow \quad x=\sqrt{\frac{y}{1-y}}$$
$$\Rightarrow \quad \frac{y}{1-y}\ge 0 \Rightarrow y\in[0,1)$$

So, the range of f is $[0,1)$.

Given, that f is surjection.

So, A is $[0,1)$.

By comparing $k=0$ and $m=1$

$\Rightarrow k+m=0+1=1$

84. (720) We have, $f(1)+f(2)=3-f(3)$

$\Rightarrow \quad f(1)+f(2)+f(3)=3$

The only possibility is $0+1+2=3$

$\Rightarrow$ Elements 1, 2, 3 in the domain can be mapped with 0, 1, 2 only.

So, number of bijective functions

$=3!\times 5! = 6\times 120 = 720$

85. (19) The desired functions will contain either one element or two elements in its codomain and '2' will be the element of the each codomain.

So, the possible codomain are {2}, {1, 2,}, {2, 3} or {2, 4}

$\therefore$ Total number of functions

$=1+3\times(2^3-2)=1+(3\times 6)=19$

86. (1) $g(x)=\dfrac{f(x)+f(-x)}{2}=\dfrac{1}{2}\left[\dfrac{x+1}{x^3+1}-\dfrac{1-x}{1-x^3}\right]$

$=\dfrac{1}{2}\left[\dfrac{1}{x^2-x+1}-\dfrac{1}{1+x+x^2}\right]$

$=\dfrac{1}{2}\left[\dfrac{2(x^2+1)}{(x^2+1)^2-x^2}\right]=\dfrac{x^2+1}{x^4+x^2+1}$

$\therefore \quad g(0)=1$

87. (1) We have, $f(x)=\min\{x-[x],-x-[-x]\}$

The graph of $f(x)$ is

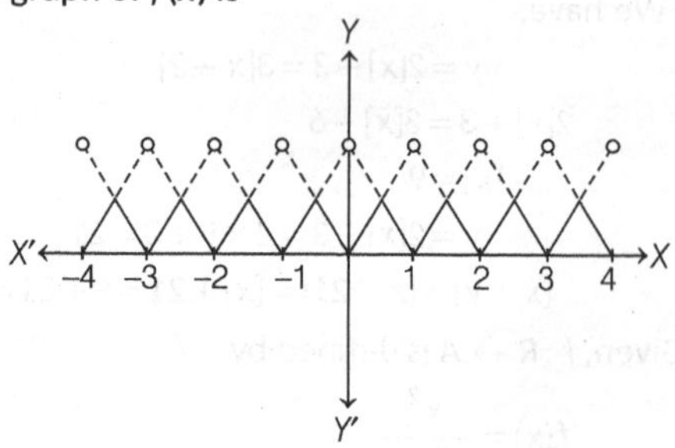

Clearly, it is periodic function with period 1.

88. (25) We have, $\left(f\left(\dfrac{16+y^2}{y^2}\right)\right)^{\frac{4}{\sqrt{y}}}=\left(f\left(\dfrac{16}{y^2}+1\right)\right)^{\frac{4}{\sqrt{y}}}$

$=\left(f\left(\left(\dfrac{4}{y}\right)^2+1\right)\right)^{\frac{4}{\sqrt{y}}}=\left[\left(f\left(\left(\dfrac{4}{y}\right)^2+1\right)\right)^{\frac{2}{\sqrt{y}}}\right]^2$

$=5^2=25$

89. (−1) Given, $f(x)+2f\left(\dfrac{1}{x}\right)=3x$ ...(i)

Replacing x by $\dfrac{1}{x}$ in Eq. (i), then

$f\left(\dfrac{1}{x}\right)+2f(x)=\dfrac{3}{x}$

$\Rightarrow 4(x)+2f\left(\dfrac{1}{x}\right)=\dfrac{6}{x}$ ...(ii)

Subtracting Eq. (i) from Eq. (ii), we get

$3f(x)=\dfrac{6}{x}-3x$

$\Rightarrow \quad f(x)=\dfrac{2}{x}-x=1$ [given]

$\Rightarrow \quad 2-x^2=x$

$\Rightarrow \quad x^2+x-2=0$

$\Rightarrow (x+2)(x-1)=0 \Rightarrow x=-2,1$

Hence, sum of the values of $x=-2+1=-1$

90. (2) Given, $a+\alpha=1$

$b+\beta=2$

$\because \quad a\cdot f(x)+\alpha\cdot f\left(\dfrac{1}{x}\right)=bx+\dfrac{\beta}{x}$ ...(i)

Replacing x by $\dfrac{1}{x}$,

$af\left(\dfrac{1}{x}\right)+\alpha f(x)=\dfrac{b}{x}+\beta x$ ...(ii)

Adding Eqs. (i) and (ii), we get

$(a+\alpha)\left[f(x)+f\left(\dfrac{1}{x}\right)\right]=\left(x+\dfrac{1}{x}\right)(b+\beta)$

$\Rightarrow \quad \dfrac{f(x)+f\left(\dfrac{1}{x}\right)}{x+\dfrac{1}{x}}=\dfrac{b+\beta}{a+\alpha}=\dfrac{2}{1}=2$

91. (3) Given, $f(x)\neq 0$

$\because f(x)$ is symmetrical about $y=x$, then

$f^{-1}(x)=f(x)=y=x$

$\Rightarrow f^2(x)=(f^{-1}(x))^2-\lambda x\, f(x)\cdot f^{-1}(x)+3x^2 f(x)$

$\Rightarrow \quad x^2=x^2-\lambda x\cdot x\cdot x+3x^2\cdot x$

$\Rightarrow \quad \lambda x^3=3x^3$

$\therefore \quad \lambda=3, x\neq 0$

92. (17) We have, $f(x+y)=f(x)+g(y)+8$

If $x=-8$ and $y=8$

$f(0)=f(-8)+g(8)+8$

$f(0)=f(-8)+17+8$ ... (i)

Now, $f(x)=x$ for all negative integer x.

$\therefore \quad f(-8)=-8$

Put in Eq. (i),

$f(0)=-8+17+8=17$

Level 3

1. (b) Let $y=\frac{x^2-x}{1-ax}$

$\Rightarrow x^2-x=y-axy \Rightarrow x^2+x(ay-1)-y=0$

Since, x is real, we get

$(ay-1)^2+4y\geq 0$

$\Rightarrow a^2y^2+2y(2-a)+1\geq 0, \forall\, y\in R$

As $a^2>0$

$4(2-a)^2-4a^2\leq 0$

$\Rightarrow 4-4a\leq 0$

$\Rightarrow a\in[1,\infty)$

2. (d) Let $y=\frac{x^2+2x+\alpha}{x^2+4x+3\alpha}$ ... (i)

$\Rightarrow (y-1)x^2+(4y-2)x+(3ay-\alpha)=0$

If range of Eq. (i) is R, then roots of this equation must be real number for all $y\in R$.

$\Rightarrow (4y-2)^2-4(y-1)(3\alpha y-\alpha)\geq 0, \forall\, y\in R$

$\Rightarrow (4-3\alpha)y^2-4(1-\alpha)y+(1-\alpha)\geq 0, \forall\, y\in R$

This is true, $\forall\, y\in R$, if $4-3\alpha>0$

$\Rightarrow \alpha<\frac{4}{3}$

and $16(1-\alpha)^2-4(4-3\alpha)(1-\alpha)\leq 0 \Rightarrow \alpha(\alpha-1)\leq 0$

[if $x^2+2x+\alpha=0$ and $x^2+4x+3\alpha$ have a common root, then $\alpha=0,1$. So, range of f is not equal to r]

Hence, $0<\alpha<1$.

3. (d) We have, $f(x)=\sqrt{2\{x\}^2-3\{x\}+1}$

$f(x)$ is defined for $2\{x\}^2-3\{x\}+1\geq 0$

$\Rightarrow (2\{x\}-1)(\{x\}-1)\geq 0$

$\therefore \{x\}\leq\frac{1}{2}$ or $\{x\}\geq 1$ [not possible]

$\Rightarrow \{x\}\leq 1/2$

Now, graph of $y=\{x\}$ and $y=\frac{1}{2}$

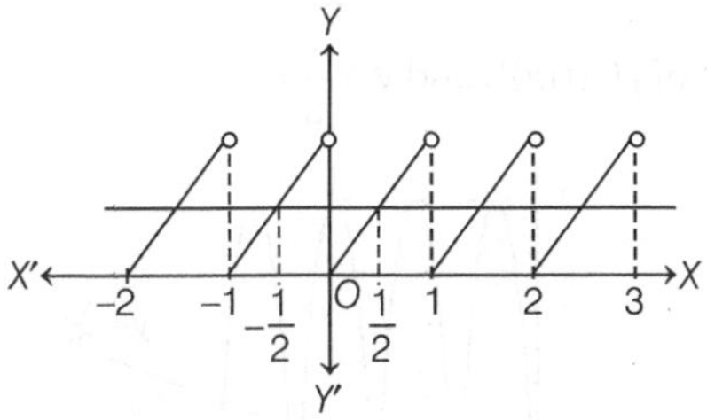

From graph domain

$=[-1,-1/2]\cup[0,1/2]\cup\{1\}$

4. (c) Given, $f(x)=\begin{cases}\tan^{-1}x, & |x|\leq 1\\ \frac{1}{2}(|x|-1), & |x|>1\end{cases}$

$f(x)=\begin{cases}\tan^{-1}x, & -1<x<1\\ \frac{1}{2}(x-1) & x>1\\ \frac{1}{2}(-x-1) & x<-1\end{cases}$

$f'(x)=\begin{cases}\frac{1}{1+x^2}, & -1<x<1\\ \frac{1}{2}, & x>1\\ -\frac{1}{2}, & x<-1\end{cases}$

The change points of $f(x)$ are 1 and -1.

RHD at $x=1$ is $\frac{1}{2}$ and LHD is $\frac{1}{1+1^2}=\frac{1}{2}$

Hence, $f'(x)$ at 1 is $\frac{1}{2}$

Now, consider $x=-1$

RHD at $x=-1$ is $\frac{1}{1+(-1)^2}=\frac{1}{2}$ and LHD is $-\frac{1}{2}$.

Hence, $f'(x)$ is not defined at $x=-1$

$\therefore f'(x)$ is defined for all points except -1.

5. (c) We have,

$f(x)=|x|(x-\sin x)$

$f(-x)=-(|x|(x-\sin x))$

$f(-x)=-f(x)$

$\Rightarrow f(x)$ is an odd function.

Now, $f(x)=\begin{cases}x^2-x\sin x & x\geq 0\\ -x^2+x\sin x & x<0\end{cases}$

It is continuous and differentiable, so

$f'(x)=2x-\sin x-x\cos x, x\geq 0$

$=(x-\sin x)+x(1-\cos x)\geq 0$

$f'(x)=-2x+x\cos x+\sin x, x<0$

$=-(x-\sin x)-x(1-\cos x)<0$

So, $f(x)$ is an increasing function, $\forall\, x\in R$

Hence, $f(x)$ is one-one and range of $f=R$

Hence, $f(x)$ is onto also.

6. (d) Given,

$f(x)=\frac{1}{2}\ln(\sqrt{\sqrt{x^2+1}+x}+\sqrt{\sqrt{x^2+1}-x})^2$

Now, putting $x\to(-x)$

$f(x)=\frac{1}{2}\ln(\sqrt{\sqrt{(-x)^2+1}-x}+\sqrt{\sqrt{(-x)^2+1}+x})^2$

$$=\frac{1}{2}\ln\left(\sqrt{\sqrt{x^2+1}-x}+\sqrt{\sqrt{x^2+1}+x}\right)^2$$

$f(-x)=f(x)$

Therefore, $f(x)$ is an even function.

So, $f(x)$ is many one function.

Now, we know that $\sqrt{x^2+1}>0,\ x\in R$

and $\sqrt{\sqrt{x^2+1}+x}+\sqrt{\sqrt{x^2+1}-x}>1$

$\therefore\ f(x)>0$

Hence, range of codomain of f.

Hence, f is not onto function.

7. (c) Period of $\tan^2\left(\frac{\pi x}{n^2-5n+8}\right)$

$$=\frac{\pi}{\pi/(n^2-5n+8)}=(n^2-5n+8)$$

and period of $\cot(n+m)\pi x=\frac{\pi}{(n+m)\pi}=\frac{1}{(n+m)}$

$\therefore$ Period of $f(x)=$

$$\text{LCM}\left(n^2-5n+8,\frac{1}{n+m}\right)=2\ \text{(given)}$$

Case I : $n^2-5n+8=1$ or $n^2-5n+8=2$

$\Rightarrow\quad n^2-5n+7=0$ or $n^2-5n+6=0$

$\therefore\quad n=\frac{5\pm\sqrt{3}i}{2}$ or $n=2,3$ (where, $i=\sqrt{-1}$)

But $\quad n\in N$

$\therefore\quad n=2,3$

Case II : $\left(\frac{1}{n+m}\right)\lambda=2,\ \lambda\in I$

$$\Rightarrow\quad \frac{1}{n+m}\not>1$$

For $n=2$, $\frac{1}{2+m}\not>1\ \Rightarrow\frac{1}{2+m}-1\not>0$

$$\Rightarrow\quad\frac{-1-m}{2+m}\not>0\quad\Rightarrow\quad\frac{m+1}{m+2}\not<0$$

$\therefore\ m\notin(-2,-1)$ ... (i)

and for $n=3$, $\frac{1}{3+m}\not>0\Rightarrow\frac{1}{3+m}-1\not>0$

$$\Rightarrow\quad\frac{-2-m}{3+m}\not>0\Rightarrow\frac{m+2}{m+3}\not<0$$

$\therefore\quad m\notin(-3,-2)$ ... (ii)

Combining Eqs. (i) and (ii), we get

$$m\notin(-3,-2)\cup(-2,-1)$$

8. (c) We have, $f_0(x)=\frac{1}{1-x}$

and $f_{n+1}(x)=f_0(f_n(x))$

$\therefore\quad f_1(x)=f_0(f_0(x))=f_0\left(\frac{1}{1-x}\right)$

$$=\frac{1}{1-\frac{1}{1-x}}=\frac{x-1}{x}$$

and $f_2(x)=f_0(f_1(x))=f_0\left(\frac{x-1}{x}\right)=\frac{1}{1-\frac{x-1}{x}}=x$

Thus, $f_3(x)=f_0(f_2(x))=f_0(x)$

$f_4(x)=f_0(f_3(x))=f_1(x)$

$f_5(x)=f_0(f_4(x))=f_2(x)$

$f_6(x)=f_0(f_5(x))=f_3(x)$

In general $f_{3n}(x)=f_3(x)$

$f_{3n+1}(x)=f_1(x)$

$f_{3n+2}f(x)=f_2(x)$

$\therefore\quad f_{100}(x)=f_1(x)=\frac{x-1}{x}$

Hence, $f_{100}(3)+f_1\left(\frac{2}{3}\right)+f_2\left(\frac{3}{2}\right)$

$$=f_1(3)+f_1\left(\frac{2}{3}\right)+f_2\left(\frac{3}{2}\right)$$

$$=\frac{3-1}{3}+\frac{\frac{2}{3}-1}{2/3}+\frac{3}{2}=\frac{2}{3}-\frac{1}{2}+\frac{3}{2}=\frac{5}{3}$$

9. (c) We have, $f(x)=4x(1-x),\ 0\le x\le 1$

$$f(f(x))=4(4x(1-x)(1-4x(1-x))$$
$$=4(4x(1-x)(1-4x+4x^2)$$
$$=16x(1-x)(2x-1)^2$$

$f(f(f(x)))=16(4x(1-x)(2x-1)^2(8x-8x^2-1)$

Now, graph of $f(x)$

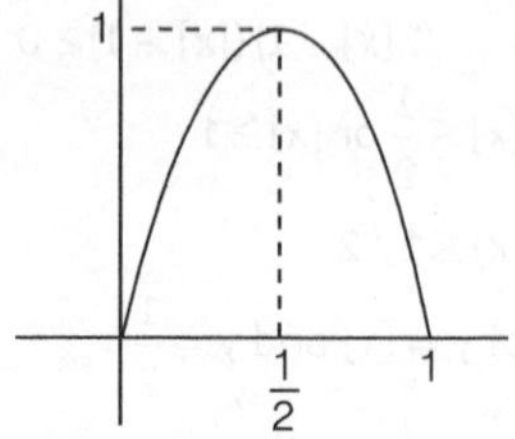

Graph of $f(f(f(x)))$ and $y=\frac{x}{3}$ is

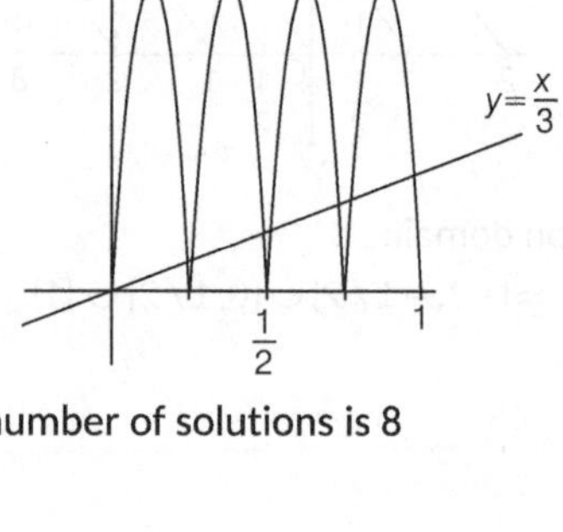

Hence, number of solutions is 8

10. (c) Given, equation $2\cos x = |\sin x|, 0 \le x \le 4\pi$

The graph of $y = 2\cos x$ and $y = |\sin x|$ as shown.

Their points of intersection represents the solution of the given equation.

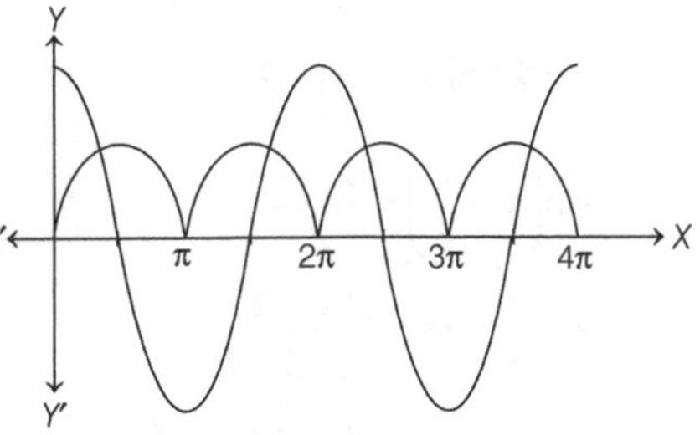

We find that graphs intersects at four points. Hence, the equation has four solutions.

11. (b) We have, $f(x+a) = \frac{1}{2} + \sqrt{f(x) - \{f(x)\}^2}$

$$\therefore \quad f(x+a+a) = \frac{1}{2} + \sqrt{f(x+a) - \{f(x+a)\}^2}$$

$$\Rightarrow f(x+2a) = \frac{1}{2} + \sqrt{f(x+a)\{1 - f(x+a)\}}$$

$$\Rightarrow f(x+2a) = \frac{1}{2} + \sqrt{\left\{\frac{1}{2} + \sqrt{f(x) - \{f(x)\}^2}\right\}\left\{\frac{1}{2} - \sqrt{f(x) - \{f(x)\}^2}\right\}}$$

$$\Rightarrow f(x+2a) = \frac{1}{2} + \sqrt{\frac{1}{4} - \{f(x) - \{f(x)\}^2\}}$$

$$\Rightarrow f(x+2a) = \frac{1}{2} + \sqrt{\frac{1}{4} - f(x) + (f(x))^2}$$

$$\Rightarrow f(x+2a) = \frac{1}{2} + \sqrt{\left\{f(x) - \frac{1}{2}\right\}^2}$$

$$= \frac{1}{2} + \left|f(x) - \frac{1}{2}\right|$$

$$f(x+2a) = \frac{1}{2} + f(x) - \frac{1}{2} \qquad [\because f(x) \ge 1/2]$$

$\Rightarrow f(x+2a) = f(x)$ for all x

Hence, $f(x)$ is a periodic function with period $2a$.

12. (a, b, c) $f(x) = \tan(\tan^{-1} x) = x, \forall x \in R$

and $g(x) = \cot(\cot^{-1} x) = x, \forall\, x \in R$

Hence, this pair is identical.

$$\text{Now, } f(x) = \text{sgn}(x) = \begin{cases} 1, & x > 0 \\ 0, & x = 1 \\ -1, & x < 1 \end{cases}$$

Domain $= R$ and Range $= \{-1, 0, 1\}$

$$\text{and } g(x) = \text{sgn}(\text{sgn}(x) = \begin{cases} 1, & \text{sgn}(x) > 0 \\ 0, & \text{sgn}(x) = 0 \\ -1, & \text{sgn}(x) < 0 \end{cases}$$

$$= \begin{cases} 1, & x > 0 \\ 0, & x = 0 \\ -1, & x < 0 \end{cases}$$

Domain $= R$ and Range $= \{-1, 0, 1\}$.

Hence, this pair is identical.

Again, $f(x) = \cot^2 x \cos^2 x = \frac{\cos^2 x}{\sin^2 x} \cdot \cos^2 x$

Domain $= R - n\pi, n \in Z$

and range $= [0, \infty)$

$$g(x) = \cot^2 x - \cos^2 x$$

$$= \cos^2 x\left(\frac{1}{\sin^2 x} - 1\right)$$

$$= \frac{\cos^2 x(1 - \sin^2 x)}{\sin^2 x}$$

$$= \frac{\cos^2 x \cdot \cos^2 x}{\sin^2 x} = f(x)$$

Domain $= R - n\pi$ and range $= [0, \infty)$

Hence, this pair is identical.

Again, $f(x) = e^{\ln \sec^{-1} x}$

$f(x)$ is defined, when

$$\sec^{-1} x > 0 \text{ and } (x \le -1 \text{ or } x \ge 1)$$

$$\Rightarrow \quad x > 1 \text{ and } x \in (-\infty, -1] \cup [1, \infty)$$

So, domain $= [1, \infty)$

whereas, $g(x) = \sec^{-1} x$ has domain $[-\infty, -1] \cup [1, \infty)$.

Hence, this pair is not identical.

13. (a, b, c) We have,

$$f(x) = \{\log(\sec x + \tan x)\}^3, x \in \left(-\frac{\pi}{2}, \frac{\pi}{2}\right)$$

$$\therefore \quad f(-x) = \{\log(\sec x - \tan x)\}^3$$

$$= \left\{\log\left(\frac{1}{\sec x + \tan x}\right)\right\}^3$$

$$= \{-\log(\sec x + \tan x)\}^3$$

$$= -f(x)$$

So, $f(x)$ is an odd function.

Let $g(x) = \sec x + \tan x, x \in \left(-\frac{\pi}{2}, \frac{\pi}{2}\right)$.

Then, $g'(x) = \sec x(\sec x + \tan x)\ 0$ for all $x \in \left(-\frac{\pi}{2}, \frac{\pi}{2}\right)$.

$\Rightarrow g(x)$ is strictly increasing an $\left(-\frac{\pi}{2}, \frac{\pi}{2}\right)$.

Also, $\phi(x) = \log_e x$ is increasing on R.

Therefore, $(\phi o g)(x) = \phi(g(x)) = f(x)$ is increasing on $\left(-\frac{\pi}{2}, \frac{\pi}{2}\right)$.

Hence, $f(x)$ is increasing on $\left(-\frac{\pi}{2}, \frac{\pi}{2}\right)$.

Consequently, $f(x)$ is one-one on $\left(-\frac{\pi}{2}, \frac{\pi}{2}\right)$.

Clearly, $g(x) \in (0, \infty)$ for all $x \in \left(-\frac{\pi}{2}, \frac{\pi}{2}\right)$

So, $f(x) = \log(g(x)) \in (-\infty, \infty)$.

Hence, $f(x)$ is an onto function.

14. (a, b, c) We have, $g(x) = \frac{\pi}{2}\sin x$

$$\therefore \quad g(g(x)) = g\left(\frac{\pi}{2}\sin x\right) = \frac{\pi}{2}\sin\left(\frac{\pi}{2}\sin x\right)$$

Thus, $f(x) = \sin\left(\frac{1}{3}g(g(x))\right)$

We observe that $-\frac{\pi}{2} \le \frac{\pi}{2}\sin x \le \frac{\pi}{2}$ for all $x \in R$

$\Rightarrow$ Range $g(x) = \left[-\frac{\pi}{2}, \frac{\pi}{2}\right]$

$$\Rightarrow \quad g(x) \in \left[-\frac{\pi}{2}, \frac{\pi}{2}\right]$$

$$\Rightarrow \quad g(g(x)) \in \left[-\frac{\pi}{2}, \frac{\pi}{2}\right]$$

$$\Rightarrow \quad \frac{1}{3}g(g(x)) \in \left[-\frac{\pi}{6}, \frac{\pi}{6}\right]$$

15. (b, c) $f_6(f_m(x)) = f_4(x)$

$$\Rightarrow \quad \frac{f_m(x) - 1}{f_m(x)} = \frac{1}{1-x}$$

$$\Rightarrow \quad 1 - \frac{1}{f_m(x)} = \frac{1}{1-x}$$

$$\Rightarrow \quad \frac{1}{f_m(x)} = 1 - \frac{1}{1-x} = \frac{-x}{1-x}$$

$$\Rightarrow \quad f_m(x) = \frac{1-x}{-x} = \frac{x-1}{x} = f_6(x)$$

So, $m = 6$ and $f_n(f_4(x)) = f_3(x)$

$$\Rightarrow \quad f_n\left(\frac{1}{1-x}\right) = \frac{1}{x}$$

Replacing $\frac{1}{1-x}$ by x, then x by $\left(1 - \frac{1}{x}\right)$

$$f_n(x) = \frac{1}{\left(1 - \frac{1}{x}\right)}$$

$$= \frac{x}{x-1}$$

$$f_n(x) = f_5(x)$$

$$\Rightarrow \quad n = 5$$

16. (a, d) We have, $y = \sqrt{3}\sin x - \cos x + 2,\ x \in \left[\frac{2\pi}{3}, \frac{5\pi}{3}\right]$

$$y = 2\left[\frac{\sqrt{3}}{2}\sin x - \frac{1}{2}\cos x\right] + 2$$

$$y = 2\sin\left(x - \frac{\pi}{6}\right) + 2$$

$$\Rightarrow \quad \sin\left(x - \frac{\pi}{6}\right) = \frac{y-2}{2}$$

$$\Rightarrow \quad x - \frac{\pi}{6} = \sin^{-1}\left(\frac{y-2}{2}\right)$$

$$\Rightarrow \quad x = \sin^{-1}\left(\frac{y-2}{2}\right) + \frac{\pi}{6}$$

So, $\quad f^{-1}(x) = \sin^{-1}\left(\frac{x-2}{2}\right) + \frac{\pi}{6}$

Now, $f^{-1}(1) = \sin^{-1}\left(\frac{1-2}{2}\right) + \frac{\pi}{6}$

$$= \sin^{-1}\left(-\frac{1}{2}\right) + \frac{\pi}{6} = \frac{7\pi}{6} + \frac{\pi}{6} = \frac{4\pi}{3}$$

$$f^{-1}(2) = \sin^{-1}(0) + \frac{\pi}{6} = \pi + \frac{\pi}{6} = \frac{7\pi}{6}$$

17. (a, d) We have,

$f(x+y) + f(x-y) = 2f(x) \cdot f(y)\ \forall\ x, y \in R$ ... (i)

$\Rightarrow \quad 2f(0) = 2(f(0))^2$ [replacing x, y by zero]

$\Rightarrow 2f(0)(f(0) - 1) = 0$

Case I When $f(0) = 0$

Replacing x by 0 in Eq. (i), we get

$$f(y) + f(-y) = 2f(0)f(y),\ \forall\ y \in R$$

$$\Rightarrow \quad f(-y) = -f(y) \text{ for all } y \in R$$

$\Rightarrow \quad f(x)$ is an odd function.

Case II When $f(0) = 1$

Replacing x by 0 in Eq. (i), we get

$$f(y) + f(-y) = 2f(0)f(y)$$

$$\Rightarrow \quad f(-y) = f(y),\ \forall\ y \in R$$

$\Rightarrow \quad f(x)$ is an even function.

18. (a, b) We have, $\{x\} = x[x]$

$$\{x\} = x(x - \{x\})$$

$$\{x\} = x^2 - x\{x\}$$

$$\Rightarrow \quad \{x\} = \frac{x^2}{1+x}$$

Now, $\quad 0 \le \{x\} < 1$

$$\Rightarrow \quad \frac{x^2}{1+x} \ge 0 \Rightarrow 1 + x > 0$$

$$\Rightarrow \quad x > -1 \Rightarrow \frac{x^2}{1+x} < 1$$

$$\Rightarrow \quad x^2 < 1 + x$$

$$\Rightarrow \quad x^2 - x - 1 < 0$$

$$\Rightarrow \quad \frac{1-\sqrt{5}}{2} < x < \frac{1+\sqrt{5}}{2}$$

Given, equation can be written as $x - [x] = x[x]$

$$\frac{x}{1+x} = [x] = \lambda$$

$$x = \frac{\lambda}{1-\lambda}$$

$$\frac{1-\sqrt{5}}{2} < \frac{\lambda}{1-\lambda} < \frac{1+\sqrt{5}}{2}$$

$$\Rightarrow \quad \frac{-1-\sqrt{5}}{2} < \lambda < \frac{\sqrt{5}-1}{2}$$

$$\lambda = 0 \Rightarrow x = 0$$

$$\lambda = -1 \Rightarrow x = -\frac{1}{2}$$

19. (a, d) Given, $f(x) + f(y) = (x\sqrt{1-y^2} + y\sqrt{1-x^2})$ ... (i)

Replace y by x. Then, $2f(x) = f(2x\sqrt{1-x^2})$

$3f(x) = f(x) + 2f(x) = f(x) + f(2x\sqrt{1-x^2})$

$= f[x\sqrt{1-4x^2(1-x^2)} + 2x\sqrt{1-x^2}\ \sqrt{1-x^2}]$

$= f[x\sqrt{(2x^2-1)^2} + 2x(1-x^2)]$

$= f[x|2x^2-1| + 2x - 2x^3]$

$= f(2x^3 - x + 2x - 2x^3)$ or $f(x - 2x^3 + 2x - 2x^3)$

$= f(x)$ or $f(3x - 4x^3)$

$\therefore f(x) = 0$ or $3f(x) = f(3x - 4x^3)$

Consider, $3f(x) = f(3x - 4x^3)$

Replacing x by $-x$, we get

$3f(-x) = f(4x^3 - 3x)$ ... (ii)

Also, from Eq. (i), $f(x) + f(-x) = f(0)$

Putting $x = y = 0$ in Eq. (i), we have $f(0) = 0$

or $\quad f(x) + f(-x) = 0$

Thus, $f(x)$ is an odd function.

Now, from Eq. (ii), $-3f(x) = f(4x^3 - 3x)$

or $\quad f(4x^3 - 3x) + 3f(x) = 0$

20. (b, c) Given, $\quad 2f(\sin x) + f(\cos x) = x$ ...(i)

Replacing x by $\left(\frac{\pi}{2} - x\right)$, we get

$$2f(\cos x) + f(\sin x) = \frac{\pi}{2} - x$$

From Eqs. (i) and (ii), we get

$$3f(\sin x) = 3x - \frac{\pi}{2}$$

$$\Rightarrow \quad f(\sin x) = x - \frac{\pi}{6} \Rightarrow f(x) = \sin^{-1} x - \frac{\pi}{6}$$

$f(x)$ has the domain $[-1, 1]$.

Also, $\sin^{-1} x \in \left[-\frac{\pi}{2}, \frac{\pi}{2}\right] \Rightarrow \sin^{-1} x - \frac{\pi}{6} \in \left[-\frac{2\pi}{3}, \frac{\pi}{3}\right]$

21. (a, b, c, d) $\left(f\left(\frac{1}{x}\right) - 1\right) = \frac{1}{(f(x)-1)}$ i.e. $\left(f\left(\frac{1}{x}\right) - 1\right)$ is reciprocal of $(f(x) - 1)$.

(a) $f(x) = 1 - x^{2013}$

$$f(x) - 1 = -x^{2013} \text{ and } f\left(\frac{1}{x}\right) - 1 = \frac{-1}{x^{2013}}$$

(b) $f(x) = \sqrt{x} + 1$

$$f(x) - 1 = \sqrt{x} \text{ and } f\left(\frac{1}{x}\right) - 1 = \frac{1}{\sqrt{x}}$$

Now, for $f(x) = \dfrac{\frac{\pi}{2}}{\tan^{-1}|x|}$

$$f(x) - 1 = \frac{\cot^{-1}|x|}{\tan^{-1}|x|} \text{ and } f\left(\frac{1}{x}\right) - 1 = \frac{\tan^{-1}|x|}{\cot^{-1}|x|}$$

Also, for $f(x) = \dfrac{2}{1 + k\ln|x|}$

$$f(x) - 1 = \frac{1 - k\ln|x|}{1 + k\ln|x|} \text{ and } f\left(\frac{1}{x}\right) - 1 = \frac{1 + k\ln|x|}{1 - k\ln|x|}$$

Sol. (Q.Nos. 22-24)

$f_1(x) = x^2$ and $f_2(x) = |x|$

or $\quad f(x) = f_1(x) - 2f_2(x) = x^2 - 2|x|$

Graph of $f(x)$

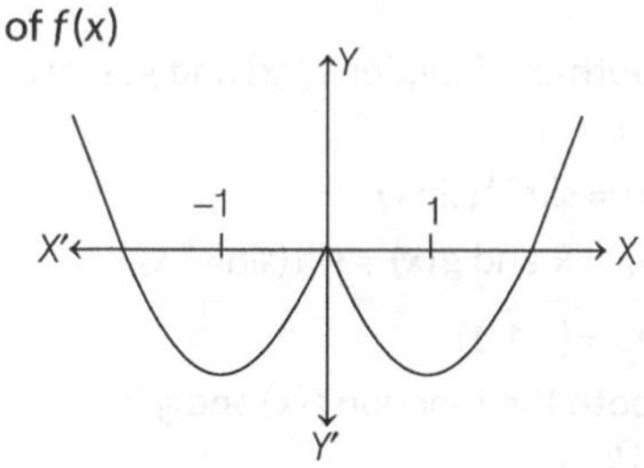

$$g(x) = \begin{cases} f(x) & , -3 \le x < -1 \\ -1 & , -1 \le x < 0 \\ 0 & , 0 \le x \le 2 \\ f(x) & , 2 < x \le 3 \end{cases}$$

$$= \begin{cases} x^2 + 2x & , -3 \le x < -1 \\ -1 & , -1 \le x < 0 \\ 0 & , 0 \le x \le 2 \\ x^2 - 2x & , 2 < x \le 3 \end{cases}$$

22. (a) The range of $g(x)$ for $[-3, -1]$ is $[-1, 3]$.

23. (b) For $x \in (-1, 0)$, $f(x) + g(x) = x^2 + 2x - 1$.

24. (a) Clearly, the graph is broken at $x = 0$.
Hence, graph is broken at 1 point.

25. (i) → (D), (ii) → (B), (iii) → (A), (iv) → (A)

$E_1 : \dfrac{x}{x-1} > 0 \Rightarrow x \in (-\infty, 0) \cup (1, \infty)$

Also, $E_2 : -1 \le \log\left(\dfrac{x}{x-1}\right) \le 1$

$\Rightarrow \quad \dfrac{1}{e} \le \dfrac{x}{x-1} \le e \Rightarrow \dfrac{1}{e} \le 1 + \dfrac{1}{x-1} \le e$

$\Rightarrow \quad \dfrac{1}{e} - 1 \le \dfrac{1}{x-1} \le e-1$

$\Rightarrow \quad (x-1) \in \left(-\infty, \dfrac{e}{1-e}\right] \cup \left[\dfrac{1}{e-1}, \infty\right)$

$\Rightarrow \quad x \in \left(-\infty, \dfrac{1}{1-e}\right] \cup \left[\dfrac{e}{e-1}, \infty\right)$

Now, $\dfrac{x}{x-1} \in (0, \infty) - \{1\}, \forall\, x \in E_1$

$\Rightarrow \quad \log\left(\dfrac{x}{x-1}\right) \in (-\infty, \infty) - \{0\}$

$\Rightarrow \quad \sin^{-1}\left(\log\left(\dfrac{x}{x-1}\right)\right) \in \left[-\dfrac{\pi}{2}, \dfrac{\pi}{2}\right] - \{0\}$

$\left\{\text{considering } \log\left(\dfrac{x}{x-1}\right) \in [-1, 1] - \{0\}\right\}$

26. (i) → (B, C), (ii) → (A, B, C, D), (iii) → (B), (iv) → (A)

(i) $\because \tan^{-1}\left(\dfrac{2x}{1-x^2}\right) = \begin{cases} 2\tan^{-1}x, & -1 < x < 1 \\ \pi + 2\tan^{-1}x, & x < -1 \\ -\pi + 2\tan^{-1}x, & x > 1 \end{cases}$

Hence, both the functions $f(x)$ and $g(x)$ are identical for $x \in (-1, 1)$.

(ii) $\because f(x) = \sin^{-1}(\sin x)$

$\therefore \quad D_f = R$ and $g(x) = \sin(\sin^{-1} x)$

$\therefore \quad D_g = [-1, 1]$

Hence, both the function $f(x)$ and $g(x)$ are identical for $x \in [-1, 1]$.

(iii) $\because f(x) = \log_{x^2} 25$

$f(x)$ is defined when $x^2 > 0$ and $\ne 1$

$\therefore\ D_f = R \sim \{-1, 1\}$ and $g(x) = \log_x 5$

$g(x)$ is defined when $x > 0$ and $\ne 1$

$\therefore\ D_g = (0, \infty) \sim \{1\}$

Hence, both the functions $f(x)$ and $g(x)$ are identical for $x \in (0, \infty) \sim \{1\}$.

(iv) $\because f(x) = \sin^{-1}x + \cos^{-1}x = \dfrac{\pi}{2}$ for $x \in [-1, 1]$

and $g(x) = \sec^{-1}x + \text{cosec}^{-1}x = \dfrac{\pi}{2}$ for $x \in (-\infty, -1] \cup [1, \infty)$.

Hence, both the functions are identical only, if $x = \{-1, 1\}$.

27. (i) → (B), (ii) → (C), (iii) → (D), (iv) → (A)

(i) We have, $f(x) = \begin{cases} x+1, & x < 0 \\ x^2+1, & x \ge 0 \end{cases}$

$(fof)x = \begin{cases} 2x+1 & x < -1 \\ (x+1)^2 - 1 & -1 < x < 0 \end{cases}$

$= \begin{cases} 2x+1, & x < -1 \\ x^2 + 2x & -1 \le x \le 0 \end{cases}$

Hence, $f(f(x)) = x^2 + 2x, -1 \le x \le 0$

(ii) If $f\left(\dfrac{2\tan x}{1+\tan^2 x}\right) = \dfrac{(\cos 2x + 1)(\sec^2 x + 2\tan x)}{2}$

$= \dfrac{(2\cos^2 x - 1 + 1)(\sec^2 x + 2\tan x)}{2}$

$= \left(\dfrac{\sec^2 x + 2\tan x}{\sec^2 x}\right)$

$= 1 + \dfrac{2\tan x}{1+\tan^2 x}$

Therefore, $f(x) = 1 + x$

(iii) $f(x+y+1) = (\sqrt{f(x)} + \sqrt{f(y)})^2$ ...(i)

$\therefore \quad f(0) = 1 = (1+0)^2$ ...(ii)

Substituting $x = y = 0$

$f(1) = (1+1)^2 = 4$ ...(iii)

On substituting $x = 0, y = 1$ in Eq. (i),

$f(2) = (1+2)^2 = 9$ ...(iv)

and substituting $x = 1, y = 1$ in Eq. (i),

$f(3) = (2+2)^2 = (1+3)^2 = 16$ ...(v)

From Eqs. (ii), (iii), (iv) and (v), we get

$f(x) = (1+x)^2$

(iv) We have, $4 < x < 5$

$1 < \dfrac{x}{4} < \dfrac{5}{4}$

$\Rightarrow \quad \left[\dfrac{x}{4}\right] = 1$

So, $f(x) = 1 + 2x + 2 = 2x + 3$

Now, let $y = 2x + 3 \Rightarrow x = \dfrac{y-3}{2}$

Therefore, $f^{-1}(x) = \dfrac{x-3}{2}$

28. (i) → (D), (ii) → (C), (iii) → (D), (iv) → (A)

(i) $f(x) = e^{\cos^4 \pi x + x - [x] + \cos^2 \pi x}$

Since, have period 1 and $x - [x] = \{x\}$ has period 1.

Then, the period of $f(x) = 1$

(ii) $f(x) = \cos 2\pi\{2x\} + \sin 2\pi\{2x\}$

The $f(x)$ of $\{2x\}$ is 1/2, then the period of $f(x) = 1/2$

(iii) $f(x)=\sin 3\pi\{x\}+\tan\pi[x]$

Clearly, $\tan\pi[x]=0, \forall\, x\in R$ and the period of $\sin 3\pi\{x\}$ is equal to 1.

Then, period of $f(x)$ is 1.

(iv) $f(x)=3x-[3x+a]-b$

$=3x+a-[3x+a]-(a+b)$

$=\{3x+a\}-(a+b)$

Thus, period of $f(x)=1/3$

Sol. (Q.Nos. 29-31)

29. (19) **30.** (−17)

Let $P(x)=ax^2+bx+c$ and given that

$$2(1+P(x))=P(x-1)+P(x+1)$$

$\Rightarrow 2(1+ax^2+bx+c)$

$$=a(x-1)^2+b(x-1)+c+a(x+1)^2+b(x+1)+c$$

$\Rightarrow 2+2ax^2+2bx+2c=2ax^2+2bx+2a+2c$

$\Rightarrow \quad 2a=2$

$\therefore \quad a=1$

Also, $P(0)=8 \Rightarrow c=8$ and $P(2)=32$

$\Rightarrow 4a+2b+c=32 \Rightarrow 4+2b+8=32 \Rightarrow b=10$

Hence, $P(x)=x^2+10x+8$

Sum of coefficients of $P(x)=1+10+8=19$

and $\quad P(x)=x^2+10x+8=(x+5)^2-25+8$

$P(x)=(x+5)^2-17$

$P(x)+17=(x+5)^2$

$P(x)+17\geq 0$

$P(x)\geq -17$ or $P(x)\in[-17,\infty)$

$\therefore \quad m=-17$

31. (10.5) We have, $P(x)=x^2+10x+8$

α,β are the zeroes of the $P(x)$

$\therefore \quad \alpha+\beta=-10$ and $\alpha\beta=8$

Now, $\dfrac{\alpha}{\beta}+\dfrac{\beta}{\alpha}=\dfrac{\alpha^2+\beta^2}{\alpha\beta}=\dfrac{(\alpha+\beta)^2-2\alpha\beta}{\alpha\beta}$

$$=\frac{(-10)^2-2(8)}{8}=\frac{100-16}{8}=\frac{84}{8}=\frac{21}{2}=10.5$$

Sol. (Q.Nos. 32-34)

$\because (f(x))^2\times f\left(\dfrac{1-x}{1+x}\right)=64x \quad$ ... (i)

Replacing x by $\left(\dfrac{1-x}{1+x}\right)$ in Eq. (i), then

$$\left(f\left(\frac{1-x}{1+x}\right)\right)^2\times f(x)=64\left(\frac{1-x}{1+x}\right) \quad \text{... (ii)}$$

From Eq. (i),

$$(f(x))^4\times\left(f\left(\frac{1-x}{1+x}\right)\right)^2=(64)^2x^2 \quad \text{... (iii)}$$

Now, dividing Eq. (iii) by Eq. (ii), then

$(f(x))^3=64x^2\left(\dfrac{1+x}{1-x}\right)$ or $f(x)=\left(64x^2\left(\dfrac{1+x}{1-x}\right)\right)^{1/3}$

32. (48) Here, $g(x)=64x^2\left(\dfrac{1+x}{1-x}\right)$

$$\therefore \quad g\left(\frac{1}{2}\right)=64\left(\frac{1}{2}\right)^2\left(\frac{1+1/2}{1-1/2}\right)=16(3)=48$$

33. (9) $\because \quad f(x)=\left(64x^2\left(\dfrac{1+x}{1-x}\right)\right)^{1/3}$

$$\therefore \quad f\left(\frac{1}{2}\right)=\left(64\left(\frac{1}{4}\right)\left(\frac{1+1/2}{1-1/2}\right)\right)^{1/3}=(48)^{1/3} \quad \text{... (iv)}$$

and $\quad f\left(\dfrac{1}{3}\right)=\left(64\left(\dfrac{1}{9}\right)\left(\dfrac{1+1/3}{1-1/3}\right)\right)^{1/3}$

$$=\left(\frac{64}{9}\times 2\right)^{1/3}=\left(\frac{128}{9}\right)^{1/3} \quad \text{... (v)}$$

From Eqs. (iv) and (v), we get

$$\frac{f\left(\frac{1}{2}\right)}{f\left(\frac{1}{3}\right)}=\left(\frac{48\times 9}{128}\right)^{1/3}=\left(\frac{27}{8}\right)^{1/3}=\frac{3}{2}$$

$\Rightarrow \quad 2f\left(\dfrac{1}{2}\right)=3f\left(\dfrac{1}{3}\right)$

$\therefore \quad m=2$ and $n=3$

$\Rightarrow \quad n^m=3^2=9$

34. (0) $\because f(x)=\left(64x^2\left(\dfrac{1+x}{1-x}\right)\right)^{1/3}$

$f(x)$ is to be defined, when $x\neq -1, 1$. $\quad [\because x\neq 1$ given$]$

Hence, $D_f=R\sim\{-1,1\}$.

$\Rightarrow \quad k_1+k_2$

$=-1+1=0$

35. (3) $f(x)=\sqrt{\sin x+\cos x}+\sqrt{7x-x^2-6}$

$$=\sqrt{\sqrt{2}\sin\left(x+\frac{\pi}{4}\right)}+\sqrt{(x-6)(1-x)}$$

Now, $f(x)$ is defined, if $\sin\left(x+\dfrac{\pi}{4}\right)\geq 0$

and $\quad (x-6)(1-x)\geq 0$

i.e., $0\leq x+\dfrac{\pi}{4}\leq\pi$ or $2\pi\leq x+\dfrac{\pi}{4}\leq 3\pi$ and $1\leq x\leq 6$

i.e., $-\dfrac{\pi}{4}\leq x\leq\dfrac{3\pi}{4}$ or $\dfrac{7\pi}{4}\leq x\leq\dfrac{11\pi}{4}$ and $1\leq x\leq 6$

or $x\in\left[1,\dfrac{3\pi}{4}\right]\cup\left[\dfrac{7\pi}{4},6\right]$

Integers values of x are $x=1, 2$ and 6.

So, number of integral values is 3.

36. (20) For real x, $\left[\frac{3}{x}\right]+\left[\frac{4}{x}\right]=5$ holds obviously $x>1$, otherwise this will not hold.

For $x>1$ the only possibility to hold above equality that is

$$\left[\frac{3}{x}\right]=2 \text{ and } \left[\frac{4}{x}\right]=3$$

$$\Rightarrow \quad 2\le\frac{3}{x}<3 \text{ and } 3\le\frac{4}{x}<4$$

$$\Rightarrow \quad 1<x\le\frac{3}{2} \text{ and } 1<x\le\frac{4}{3}$$

The smallest interval is (1, 4/3].

$\therefore \quad a=1, b=4, c=3$

$\therefore \quad a+b+c+abc=8+12=20$

37. (1) We know that $\{x\}=x-[x]$

$$\therefore \quad \{x+1\}+2x=4[x+1]-6$$

$$\Rightarrow \quad (x+1)-[x+1]+2x=4[x+1]-6$$

$$\Rightarrow \quad 3x+1=5[x+1]-6$$

$$\Rightarrow \quad 3x+1=5[[x]+1]-6$$

$$\Rightarrow \quad 3x=5[x]-2 \quad \text{... (i)}$$

$$\Rightarrow \quad 3([x]+\{x\})=5[x]-2$$

$$\Rightarrow \quad 3\{x\}=2[x]-2 \quad \text{... (ii)}$$

Now, $0\le\{x\}<1$

$$\Rightarrow \quad 0\le 3\{x\}<3$$

$$\Rightarrow \quad 0\le 2[x]-2<3$$

$$\Rightarrow \quad 2\le 2[x]<5$$

$$\Rightarrow \quad 1\le[x]<\frac{5}{2}$$

$$\Rightarrow \quad [x]=1,2$$

From Eq. (i), if $[x]=1$, then $3x=5-2\Rightarrow x=1$,

if $[x]=2$, then $3x=5\times2-2\Rightarrow x=8/3$

Hence, $x=1$ is the only integral solution of given equation.

38. (119) We have,

$$n(x)=5 \text{ and } n(y)=7$$

$\alpha=$ number of one-one function $={}^7C_5\times5!$

β is the number of onto functions in y to x

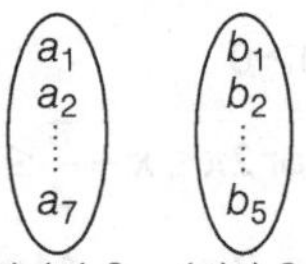

1,1,1,1,3   1,1,1,2,2

$$\beta=\frac{7!}{3!4!}\times5!+\frac{7!}{(2!)^3 3!}\times5!=({}^7C_3+3.{}^7C_3)5!$$

$$=4\times{}^7C_3\times5!$$

Hence, $\frac{\beta-\alpha}{5!}=4\times{}^7C_3-{}^7C_5=4\times35-21=119$

39. (8) Let $f(x)=\cos nx\sin\left(\frac{5x}{n}\right)$, $f(x)$ be periodic, then $f(x+\lambda)=f(x)$, where λ is period.

or $\cos(nx+n\lambda)\sin\left(\frac{5x+5\lambda}{n}\right)=\cos nx\sin\left(\frac{5x}{n}\right)$

At $x=0$, $\cos n\lambda\sin\left(\frac{5\lambda}{n}\right)=0$

If $\cos n\lambda=0$, then $n\lambda=r\pi+\frac{\pi}{2}, r\in Z$

$$\Rightarrow \quad n(3\pi)=r\pi+\frac{\pi}{2} \quad [\because \lambda=3\pi]$$

$$\Rightarrow \quad 3^{n-r}=\frac{1}{2} \quad \text{[which is not possible]}$$

Now, let $\sin\left(\frac{5\lambda}{n}\right)=0$, then

$$\frac{5\lambda}{n}=p\pi\,(p\in Z)$$

$$\Rightarrow \quad \frac{5(3\pi)}{n}=p\pi \quad [\because \lambda=3\pi]$$

$$\Rightarrow \quad n=\frac{15}{p}$$

For $p=\pm1, \pm3, \pm5, \pm15$, we have respectively

$n=\pm15, \pm5, \pm3, \pm1$ $\quad (\because n\in Z)$

Hence, the number of integral values is 8.

40. (2) We have,

$$f(x)=\frac{x}{1+x} \text{ and } g(x)=\frac{rx}{1-x}$$

$$(fog)(x)=\frac{\frac{rx}{1-x}}{1+\frac{rx}{1-x}}=\frac{rx}{1-x+rx} \quad \text{...(i)}$$

and $$g(f(x))=\frac{r\left(\frac{x}{1+x}\right)}{1-\frac{x}{1+x}}=rx \quad \text{...(ii)}$$

Now, $f(g(x))=g(f(X))$

$$\Rightarrow \quad \frac{rx}{1-x+rx}=rx$$

$$\Rightarrow \quad rx\left[\frac{1}{1-x+rx}-1\right]=0$$

$$\Rightarrow \quad rx\left[\frac{1}{1+x(r-1)}-1\right]=0$$

For infinite many real number

$r=0$ and $r-1=0$

$r=0, 1$

The number of elements in set S is 2.

41. (5) We have,

$$f(f(f(x))) = 8x + 21$$

$f(x)$ must be a linear function, therefore

$$f(x) = px + q$$

$$\therefore \quad f(f(x)) = p(f(x)) + q$$
$$= p(px + q) + q$$
$$= p^2x + pq + q$$

$$\therefore f(f(f(x))) = p(p^2x + pq + q) + q$$

$$\therefore f(f(f(x))) = p^3x + p^2q + pq + q$$

According to the question,

$$p^3x + p^2q + pq + q = 8x + 21$$

$$\therefore \quad p^3 = 8$$

$$\Rightarrow \quad p = 2$$

and $\quad p^2q + pq + q = 21$

$$\Rightarrow \quad q = 3$$

So, $f(x) = 2x + 3$

Hence, sum of all coefficients $= 2 + 3 = 5$

42. (26) We have,

$$g(x)\,g(y) = g(x) + g(y) + g(xy) - 2 \quad \text{... (i)}$$

Putting $\quad x = 2$ and $y = 1$

$$g(2)\,g(1) = g(2) + g(1) + g(2) - 2$$

$$\Rightarrow \quad 5g(1) = 5 + g(1) + 5 - 2$$

$$\Rightarrow \quad g(1) = 2$$

Putting $y = \dfrac{1}{x}$ in Eq. (i), we get

$$g(x)\cdot g\left(\frac{1}{x}\right) = g(x) + g\left(\frac{1}{x}\right) + g(1) - 2$$

$$\Rightarrow g(x)\cdot g\left(\frac{1}{x}\right) = g(x) + g\left(\frac{1}{x}\right) + 2 - 2 \qquad [\because g(1) = 2]$$

$$\Rightarrow g(x)\cdot g\left(\frac{1}{x}\right) = g(x) + g\left(\frac{1}{x}\right)$$

[$\because$ if $f(x)$ is a polynomial function of degree n satisfying $f(x) + f\left(\dfrac{1}{x}\right) = f(x)\cdot f\left(\dfrac{1}{x}\right)$, then $f(x) = x^n \pm 1$]

$$\Rightarrow \quad g(x) = x^n + 1$$

Now, $\quad g(2) = 5$

$$\Rightarrow \quad 2^n + 1 = 5$$

$$\Rightarrow \quad n = 2$$

$$\therefore \quad g(x) = x^2 + 1$$

Hence, $g(5) = 5^2 + 1 = 26$

CHAPTER

02

LIMITS

LEVEL 01 \ Starter Level

2.1 Limit of Algebraic Function

1 Evaluate the limit of following algebraic functions.

(i) $\lim_{x\to 3}\frac{4x+3}{x+2}$

(ii) $\lim_{x\to 2}\frac{x^2-4}{x-2}$

(iii) $\lim_{x\to \frac{-1}{3}}\frac{27x^3+1}{3x+1}$

(iv) $\lim_{x\to -2}\frac{x^2-x-6}{x^2-2x-8}$

(v) $\lim_{x\to 2}\frac{x^4-16}{x^2-3x+2}$

(vi) $\lim_{x\to \sqrt{3}}\frac{x^2-3}{3x^2-2\sqrt{3}x-3}$

(vii) $\lim_{x\to 1}\left(\frac{1}{x^2+x-2}-\frac{x}{x^3-1}\right)$

(viii) $\lim_{x\to 0}\frac{(a+x)^2-a^2}{x}$

(ix) $\lim_{x\to a}(x^2-a^2)\left(\frac{1}{x+a}+\frac{1}{x-a}\right)$

(x) $\lim_{x\to 2}\left[\frac{1}{x-2}-\frac{2(2x-3)}{x^3-3x^2+2x}\right]$

(xi) $\lim_{x\to 1}\frac{(\sqrt{5x-4}-\sqrt{x})}{(x^2-1)}$

(xii) $\lim_{x\to 2}\frac{4-\sqrt{5x+6}}{3-\sqrt{4x+1}}$

(xiii) $\lim_{x\to 1}\frac{\sqrt[5]{x^2}-2\sqrt[5]{x}+1}{4(x-1)^2}$

(xiv) $\lim_{x\to 3}\frac{x^5-243}{x^3-27}$

2 Evaluate the limits of following functions.

(i) $\lim_{x\to \infty}\frac{x^2-4x+3}{2x^2-x+4}$

(ii) $\lim_{x\to \infty}\sqrt{x^2+5x}-x$

(iii) $\lim_{x\to \infty}\frac{\sqrt{x^2+a^2}-\sqrt{x^2+b^2}}{\sqrt{x^2+c^2}-\sqrt{x^2+d^2}}$

(iv) $\lim_{n\to \infty}\frac{1+2+3+\ldots+n}{n^2}$

(v) $\lim_{n\to\infty} \frac{1^3+2^3+3^3+\ldots+n^3}{(n-1)^4}$

(vi) $\lim_{n\to\infty} \frac{n!+(n-1)!}{n!-(n-1)!}$

(vii) $\lim_{x\to-\infty} (\sqrt{x^2-4x}+x)$

2.2 Limit of Trigonometric Functions

3 Evaluate the limits of following trigonometric functions.

(i) $\lim_{x\to 0} \frac{\sin x - 2\sin 3x + \sin 5x}{x}$

(ii) $\lim_{x\to 0} \frac{\sin 2x + 4x}{2x + \sin 4x}$

(iii) $\lim_{x\to 0} \frac{1-\cos 4x}{1-\cos 6x}$

(iv) $\lim_{x\to a} \frac{\cos x - \cos a}{x-a}$

(v) $\lim_{x\to \frac{\pi}{2}} \frac{\sqrt{2}-\sqrt{1+\sin x}}{\sqrt{2}\cos^2 x}$

(vi) $\lim_{x\to \frac{\pi}{6}} \frac{2\sin^2 x + \sin x - 1}{2\sin^2 x - 3\sin x + 1}$

(vii) $\lim_{x\to\infty} \frac{\sin\left(\frac{a}{b^x}\right)}{\sin\left(\frac{b}{a^x}\right)}$, where $a<b$

(viii) $\lim_{x\to 3} \frac{x^2-4x+3}{x^2-9-\sin(x-3)}$

(ix) $\lim_{x\to\pi} \frac{1-\sin x+\cos x}{1+\sin x+\cos x}$

(x) $\lim_{x\to 2} \frac{(\cos\alpha)^x+(\sin\alpha)^x-1}{x-2}$

2.3 Limit of Exponential and Logarithmic Functions

4 Find the limits of following exponential and logarithmic functions.

(i) $\lim_{x\to 0} \frac{e^{2+x}-e^2}{x}$

(ii) $\lim_{x\to 0} \left(\frac{e^{ax}-e^{bx}}{x}\right)$

(iii) $\lim_{x\to 0} \left(\frac{2^{2+x}-4}{x}\right)$

(iv) $\lim_{x\to 0} \frac{\log(1+3x)}{3^x-1}$

(v) $\lim_{x\to 0} \frac{\log_3(1+2x)}{x}$

(vi) $\lim_{x\to 0} \frac{a^x+b^x-c^x-d^x}{x}$

(vii) $\lim_{x\to a} \frac{\log x - \log a}{x-a}$

(viii) $\lim_{x\to 0} \frac{\log(a+x)-\log(a-x)}{x}$

(ix) $\lim_{x\to \frac{\pi}{2}} \frac{a^{\cot x}-a^{\cos x}}{\cot x - \cos x}$

(x) $\lim_{x\to 0} \left(\frac{e^{x^2}-\cos x}{\sin^2 x}\right)$

2.4 Limit in the Form $[f(x)]^{g(x)}$

5 Evaluate the limits of the following functions.

(i) $\lim_{x\to\infty} \left(\frac{2x-4}{2x+2}\right)^{\frac{3x+2}{3}}$

(ii) $\lim_{x\to 0} (\cos x + \sin x)^{\frac{1}{x}}$

(iii) $\lim_{x\to 0} \left(\frac{\sin x}{x}\right)^{\frac{\sin x}{x-\sin x}}$

(iv) $\lim_{x\to 0} \left(\frac{a^x+b^x}{2}\right)^{\frac{3}{x}}$

(v) $\lim_{x\to \frac{7}{2}} (2x^2-9x+8)^{\cot(2x-7)}$

2.5 Limit using Sandwich and Leibnitz's Formula

6 Evaluate the limit of the following function

(i) $\lim_{x\to 0} \dfrac{\int_0^{x^2} \cos(t^2)\,dt}{x\sin x}$

(ii) $\lim_{x\to 0} \dfrac{\int_0^{x} \cos t\,dt}{x}$

(iii) $\lim_{n\to\infty} \dfrac{[n\sqrt{2}]}{n}$, [ ] is denotes greatest integer function

(iv) $\lim_{n\to\infty} \left(\dfrac{1}{1+n^2}+\dfrac{3}{2+n^2}+\dfrac{5}{3+n^2}+\dots+\dfrac{2n-1}{n+n^2}\right)$

LEVEL 02 \ JEE Main Level

Single Choice Correct Questions

2.1 Limit of Algebraic Function

1 $\lim_{x\to 3^+} \left\{\dfrac{[x]^2}{2}-\left[\dfrac{x}{5}\right]^2\right\}$ is equal to

(a) 0 (b) $\dfrac{5}{2}$ (c) $\dfrac{9}{2}$ (d) 4

2 $\lim_{x\to 1} \dfrac{\left(\sum_{k=1}^{50} x^k\right)-50}{x-1}$ is equal to

(a) 0 (b) 1225 (c) 1275 (d) 1

3 $\lim_{x\to a} \dfrac{(a+2x)^{\frac{1}{3}}-(3x)^{\frac{1}{3}}}{(3a+x)^{\frac{1}{3}}-(4x)^{\frac{1}{3}}}$ is equal to **[JEE Main 2020]**

(a) $\left(\dfrac{2}{9}\right)^{\frac{4}{3}}$ (b) $\left(\dfrac{2}{3}\right)^{\frac{4}{3}}$ (c) $\left(\dfrac{2}{3}\right)\left(\dfrac{2}{9}\right)^{\frac{1}{3}}$ (d) $\left(\dfrac{2}{9}\right)\left(\dfrac{2}{3}\right)^{\frac{1}{3}}$

4 $\lim_{x\to\infty} \left[\sqrt{x+\sqrt{x+\sqrt{x}}}-\sqrt{x}\right]$ is equal to

(a) 0 (b) $\dfrac{1}{2}$ (c) $\log 2$ (d) e^4

5 $\lim_{x\to 0} \dfrac{\sqrt{a^2-ax+x^2}-\sqrt{a^2+ax+x^2}}{\sqrt{a+x}-\sqrt{a-x}}$ equals

(a) a (b) $-a^{\frac{1}{2}}$ (c) $a^{\frac{1}{2}}$ (d) $-a^{\frac{3}{2}}$

6 The value of $\lim_{x\to 2} \dfrac{\sqrt{x+7}-3\sqrt{2x-3}}{\sqrt[3]{x+6}-2\sqrt[3]{3x-5}}$ is

(a) $\dfrac{34}{23}$ (b) $\dfrac{54}{25}$ (c) $\dfrac{33}{23}$ (d) $\dfrac{34}{33}$

7 $\lim_{x\to 1}\left[\left(\frac{4}{x^2-x^{-1}}-\frac{x^2-3x+1}{1-x^3}\right)^{-1}+3\frac{x^4-1}{x^3-x^{-1}}\right]$ is equal to

(a) 3 (b) 2 (c) 4 (d) $\frac{28}{3}$

8 The value of $\lim_{x\to-\infty}\sqrt{x^2-2x}-ax-b=0$, then

(a) $a=1$ and $b=0$
(b) $a=-1$ and $b=1$
(c) $a=-1$ and $b=0$
(d) $a=-1$ and $b=-2$

9 If $\lim_{x\to\infty}\{\sqrt{ax^4+bx^3+cx^2+dx+2}-\sqrt{ax^4+3x^3-4x^2+2x+e}\}$ is finite, then the value of b is

(a) 1 (b) 2 (c) 3 (d) 4

10 For a certain value of k, $\lim_{x\to-\infty}[(x^5+7x^4-5)^k-x]$ is finite and non-zero. The value of k and the value of limit is

(a) 0, 1 (b) $\frac{1}{5},\frac{7}{5}$ (c) $\frac{-1}{5},\frac{-7}{5}$ (d) None of these

11 Let $\alpha(a)$ and $\beta(a)$ be the roots of the equation $(\sqrt[3]{1+a}-1)x^2+(\sqrt{1+a}-1)x+(\sqrt[6]{1+a}-1)=0$, where $a>-1$. Then, $\lim_{a\to 0^+}\alpha(a)$ and $\lim_{a\to 0^+}\beta(a)$ are

(a) $\frac{-5}{2}$ and 1 (b) $\frac{-1}{2}$ and -1 (c) $\frac{-7}{2}$ and 2 (d) $\frac{-9}{2}$ and 3

12 For each $t\in R$, let $[t]$ be the greatest integer less than or equal to t. Then,
$\lim_{x\to 0^+}x\left(\left[\frac{1}{x}\right]+\left[\frac{2}{x}\right]+\left[\frac{3}{x}\right]+\ldots+\left[\frac{15}{x}\right]\right)$ is equal to

(a) 15
(b) 120
(c) 0
(d) Does not exist

13 Let $[t]$ denotes the greatest integer $\le t$. If for some $\lambda\in R-\{0,1\}$, $\lim_{x\to 0}\left|\frac{1-x+|x|}{\lambda-x+[x]}\right|=L$, then L is

[JEE Main 2020]

(a) 1 (b) 2 (c) $\frac{1}{2}$ (d) 0

14 If $\lim_{n\to\infty}\frac{n\cdot 5^n}{n(x-3)^n+n\cdot 5^{n+1}-5^n}=\frac{1}{5}$, where $n\in N$, then the number of integers in the range of x is

(a) 3 (b) 4 (c) 9 (d) 11

15 Let $f(x)=3$ and $f'(x)=3$, then $\lim_{x\to 3}\frac{xf(3)-3f(x)}{x-3}$ equals

(a) 3 (b) -3 (c) 6 (d) -6

16 Let $f(x)=5-|x-2|$ and $g(x)=|x+1|\in R$. If $f(x)$ attains maximum value at α and $g(x)$ attains minimum value at β, then $\lim_{x\to-\alpha\beta}\frac{(x-1)(x^2-5x+6)}{(x^2-6x+8)}$

[JEE Main 2019]

(a) $\frac{1}{2}$ (b) $\frac{-3}{2}$ (c) $\frac{-1}{2}$ (d) $\frac{3}{2}$

2.2 Limit of Trigonometric Function

17 $\lim_{x\to 0}\frac{x^a \sin^b x}{\sin(x^c)}$, where $a, b, c \in R-\{0\}$ exists and has non-zero value. Then,

(a) $a+b=c$ (b) $a+c=b$ (c) $b+c=a$ (d) None of these

18 $\lim_{x\to 0}\frac{(1-\cos 4x)(2-\cos x)}{x\tan 2x}$ is equal to

(a) 1 (b) $\frac{1}{4}$ (c) 2 (d) 4

19 $\lim_{x\to 1}\frac{\sqrt{1-\cos 2(x-1)}}{x-1}$ is equal to

(a) $\sqrt{2}$ (b) $-\sqrt{2}$ (c) 0 (d) Does not exist

20 $\lim_{n\to\infty}\cos(\pi\sqrt{n^2+n})$, where $n\in Z$, is equal to

(a) 0 (b) 1 (c) -1 (d) $\frac{1}{2}$

21 $\lim_{x\to 0}\frac{\cos\left(\frac{\pi}{2}\cos^2 x\right)}{x^2}$ is equal to

(a) 0 (b) $\frac{\pi}{4}$ (c) $\frac{\pi}{2}$ (d) π

22 $\lim_{x\to 0}\frac{\tan^2 x-x^2}{x^2\tan^2 x}$ is equal to

(a) $\frac{1}{3}$ (b) $\frac{2}{3}$ (c) 1 (d) $\frac{-1}{3}$

23 The integer n, for which $\lim_{x\to 0}\frac{(\cos x-1)(\cos x-e^x)}{x^n}$ is a finite non-zero number, is

(a) 1 (b) 2 (c) 3 (d) 4

24 If $\lim_{x\to\infty}\frac{x^3\sin\frac{1}{x}+2x+3}{x^2-2x+1}=l$, then the value of $\sqrt{l\sqrt{l\sqrt{l\ldots\infty}}}$ is

(a) 0 (b) 1 (c) 2 (d) None of these

25 Let α and β be the distinct roots of $ax^2+bx+c=0$, then $\lim_{x\to\alpha}\frac{1-\cos(ax^2+bx+c)}{(x-\alpha)^2}$ is equal to

(a) $\frac{a^2}{2}(\alpha-\beta)^2$ (b) 0 (c) $-\frac{a^2}{2}(\alpha-\beta)^2$ (d) $\frac{1}{2}(\alpha-\beta)^2$

26 $\lim_{x\to 0}\frac{1-\cos x\sqrt{\cos 2x}}{x^2}$ is equal to

(a) $\frac{1}{2}$ (b) 1 (c) $\frac{3}{2}$ (d) 0

27 $\lim_{x\to 0}\frac{\sqrt[4]{1+\sin x}-\sqrt[4]{1-\sin x}}{x}$ is equal to

(a) $\frac{1}{2}$ (b) 0 (c) -1 (d) 1

28 If $\lim_{x\to 0} \dfrac{\sin 2x - a\sin x}{\left(\dfrac{x}{3}\right)^3} = L$ exists finitely, then the value of $\left[\dfrac{L}{a}\right]$, where [ ] is greatest integer function.

(a) 13 (b) 14 (c) −13 (d) −14

29 If $\lim_{x\to 0}\left(\left[\dfrac{\sin^{-1}x}{x}\right]+\left[\dfrac{2^3\sin^{-1}2x}{x}\right]+\left[\dfrac{3^3\sin^{-1}3x}{x}\right]+\ldots+\left[\dfrac{n^3\sin^{-1}nx}{x}\right]\right)=100$, then value of n is, where $[\cdot]$ denotes G.I.F.

(a) 2 (b) 3 (c) 4 (d) 5

30 $\lim_{x\to 0^-}\sin^{-1}[\tan x] = l$, then $\{l\}$ is equal to, where [ ] is the greatest integer function and { } is the fractional part function.

(a) 0 (b) $1-\dfrac{\pi}{2}$ (c) $\dfrac{\pi}{2}-1$ (d) $2-\dfrac{\pi}{2}$

31 $\lim_{x\to 1^+}\dfrac{\sqrt{\pi}-\sqrt{\cos^{-1}x}}{\sqrt{x+1}}$ is equal to

(a) $\dfrac{1}{\pi}$ (b) $\dfrac{1}{\sqrt{\pi}}$ (c) $\dfrac{1}{\sqrt{2\pi}}$ (d) $\sqrt{\dfrac{2}{\pi}}$

32 If $\{x\}$ denotes the fractional part of x, then $\lim_{x\to 0}\dfrac{\{x\}}{\sin\{x\}}$ is

(a) 0 (b) 1 (c) −1 (d) Does not exist

33 The value of $\lim_{x\to 0}\left(\left[100\dfrac{\sin^{-1}x}{x}\right]+\left[100\dfrac{\tan^{-1}x}{x}\right]\right)$ is

(a) 198 (b) 199 (c) 200 (d) None of these

34 $\lim_{x\to\infty}\dfrac{\cot^{-1}(x^{-a}\log_a x)}{\sec^{-1}(a^x\log_x a)}$, $a>1$ is equal to

(a) 2 (b) $\log_a 2$ (c) 0 (d) 1

35 $\lim_{x\to\frac{\pi}{2}}\dfrac{[x/2]}{\log(\sin x)}$ is equal to, where [ ] is greatest integer function

(a) Does not exist (b) 1 (c) 0 (d) −1

36 Let $P=\lim_{n\to\infty}\left(\cos\dfrac{x}{2}\cdot\cos\dfrac{x}{4}\cdot\cos\dfrac{x}{8}\ldots\cos\left(\dfrac{x}{2^n}\right)\right)$, then the value of $\lim_{x\to 0}P$ is

(a) 0 (b) 1 (c) 2 (d) None of these

37 If $f(x)$ be a cubic polynomial and $\lim_{x\to 0}\dfrac{\sin^2 x}{f(x)}=\dfrac{1}{2}$, then $f(1)$ can not be equal to

(a) 0 (b) 1 (c) 2 (d) −3

38 $\lim_{x\to 0^-}\dfrac{\tan^{-1}(x+[x])}{[x]-x}$ is equal to, where $[x]$ is the greatest integer function

(a) $\dfrac{\pi}{2}$ (b) $\dfrac{\pi}{4}$ (c) π (d) 0

39 If $f(x)=\begin{cases}\dfrac{[x]^2+\sin[x]}{[x]}, & [x]\neq 0\\ 0, & [x]=0\end{cases}$, where $[x]$ denotes the greatest integer function, then $\lim_{x\to 0} f(x)$ is

(a) 0 (b) 1 (c) −1 (d) Does not exist

40 Let $f(x)=\begin{cases}\dfrac{\tan^2\{x\}}{x^2-[x]^2}, & x>0\\ 1, & x=0\\ \sqrt{\{x\}\cot\{x\}}, & x<0\end{cases}$, where $[x]$ is greatest integer function and $\{x\}$ is the fractional part of x, then

(a) $\lim_{x\to 0^+} f(x)=1$ (b) $\lim_{x\to 0^-} f(x)=2$ (c) $\lim_{x\to 0^+} f(x)=2$ (d) None of these

41 $\lim_{x\to 0}\dfrac{1+2\sin x-e^x-\ln(1+x)}{x^3}$ is equal to

(a) $\dfrac{2}{3}$ (b) $\dfrac{1}{3}$ (c) $\dfrac{1}{6}$ (d) $-\dfrac{1}{6}$

42 The value of $\lim_{x\to\infty} x^2\left(\sqrt{\left(1-\cos\dfrac{1}{x}\right)}\sqrt{\left(1-\cos\dfrac{1}{x}\right)}\sqrt{\left(1-\cos\dfrac{1}{x}\right)}\ldots\infty\right)$ is

(a) $\dfrac{1}{4}$ (b) $\dfrac{1}{2}$ (c) 1 (d) 0

43 The value of $\lim_{h\to 0} 2\left\{\dfrac{\sqrt{3}\sin\left(\dfrac{\pi}{6}+h\right)-\cos\left(\dfrac{\pi}{6}+h\right)}{\sqrt{3}h(\sqrt{3}\cosh-\sinh)}\right\}$ is

[JEE Main 2021]

(a) $\dfrac{4}{3}$ (b) $\dfrac{2}{\sqrt{3}}$ (c) $\dfrac{3}{4}$ (d) $\dfrac{2}{3}$

2.3 Limit of Exponential and Logarithmic Functions

44 $\lim_{x\to 0}\dfrac{a^{\tan x}-a^{\sin x}}{\tan x-\sin x}$, $a>0$ is equal to

(a) $\log a$ (b) $\dfrac{1}{2}\log a$ (c) $-\log a$ (d) $2\log a$

45 $\lim_{x\to 0}\dfrac{\log(a+x)-\log a}{x}+k\lim_{x\to e}\dfrac{\log x-1}{x-e}=1$, then

(a) $k=e\left(1-\dfrac{1}{a}\right)$ (b) $k=e(1+a)$ (c) $k=e(2-a)$ (d) None of these

46 $\lim_{x\to 2\pi}\dfrac{1-(\sec x)^{\sec x}}{\ln(\sec x)}$ is equal to

(a) 1 (b) −1 (c) 0 (d) None of these

47 $\lim_{x\to 0}\dfrac{\log(1+x+x^2)+\log(1-x+x^2)}{\sec x-\cos x}$ is equal to

(a) 2 (b) 0 (c) 1 (d) None of these

48 $\lim_{x\to\infty} \dfrac{e^x\left(\left(2^{x^n}\right)^{\frac{1}{e^x}}-\left(3^{x^n}\right)^{\frac{1}{e^x}}\right)}{x^n}$, $n \in N$ is equal to

(a) 0 (b) 1 (c) $\ln\frac{2}{3}$ (d) $\ln\frac{3}{2}$

49 $\lim_{x\to 1} \sec\frac{\pi}{2^x}\log_e x$ is equal to

(a) $\frac{2}{\pi}$ (b) $\frac{2}{\pi\log 2}$ (c) $\frac{-2}{\pi}\log 2$ (d) None of these

50 If $\lim_{x\to 0} \dfrac{\ln\cot\left(\frac{\pi}{4}-\beta x\right)}{\tan\alpha x} = 1$, then $\frac{\alpha}{\beta}$ is equal to

(a) 1 (b) 2 (c) 0 (d) 3

51 $\lim_{x\to 0} \dfrac{x\left(e^{\frac{(\sqrt{1+x^2+x^4}-1)}{x}}-1\right)}{\sqrt{1+x^2+x^4}-1}$ is equal to **[JEE Main 2020]**

(a) $\sqrt{e}$ (b) 1 (c) 0 (d) Does not exist

52 If $f(x)=\begin{cases} \dfrac{\left(e^{(x+3)\ln 27}\right)^{\frac{x}{27}}-9}{3^x-27} & ; \ x<3 \\ \lambda\left(\dfrac{1-\cos(x-3)}{(x-3)\tan(x-3)}\right) & ; \ x>3 \end{cases}$. If $\lim_{x\to 3} f(x)$ exists, then λ is equal to

(a) $\frac{2}{9}$ (b) $\frac{9}{2}$ (c) $\frac{2}{3}$ (d) None of these

53 The value of $\lim_{x\to 0^+} \dfrac{e^{(x^x-1)}-x^x}{[(x^2)^x-1]^2}$ is equal to

(a) 1 (b) $\frac{1}{8}$ (c) $\frac{3}{2}$ (d) $\frac{1}{4}$

54 If a and b are positive numbers and $\lim_{x\to 0} \dfrac{(1+a^2)+8e^{1/x}}{1+(2+b+b^2)e^{1/x}} = 2$, then the value of a^2+b^2 is

(a) 1 (b) 2 (c) 5 (d) 13

2.4 Limits in the Form of $\{f(x)\}^{g(x)}$

55 $\lim_{x\to\infty} \left(\dfrac{x+c}{x-c}\right)^x = 16$, then the value of c is

(a) log 2 (b) log 3 (c) log 4 (d) 0

56 $\lim_{x\to\infty} \left(\cos\dfrac{\pi}{x}\right)^x$ is equal to

(a) 1 (b) e^2 (c) 0 (d) e^{π}

57 The value of $\lim_{y \to \infty} \left(\frac{(a_1^{1/y} + a_2^{1/y} + a_3^{1/y} + \dots a_n^{1/y})^y}{n} \right)$, where $a_1, a_2, a_3, \dots, a_n > 0$, is

(a) $\frac{1}{n} \sum_{i=1}^{n} a_i$ (b) $\prod_{i=1}^{n} (a_i)^{\frac{1}{n}}$ (c) $n \sum_{i=1}^{n} a_i$ (d) $\prod_{i=1}^{n} (a_i)^n$

58 $\lim_{x \to a} \left[2 - \frac{a}{x} \right]^{\tan \frac{\pi x}{2a}}$ is equal to

(a) $e^{\frac{2}{\pi}}$ (b) $e^{\frac{\pi}{2}}$ (c) $e^{\frac{-2}{\pi}}$ (d) $e^{\frac{-\pi}{2}}$

59 $\lim_{x \to 1} \left(\tan \left(\frac{\pi}{4} + \log x \right) \right)^{\frac{1}{\log x}}$ is equal to

(a) e (b) e^{-1} (c) e^2 (d) e^{-2}

60 Let $f : R \to R$ be such that $f(1) = 3$ and $f'(1) = 6$. Then, $\lim_{x \to 0} \left\{ \frac{f(1+x)}{f(1)} \right\}^{\frac{1}{x}}$ equals

(a) e^3 (b) $e^{\frac{1}{2}}$ (c) 1 (d) e^2

61 If α and β be the roots of $ax^2 + bx + c = 0$, then $\lim_{x \to a} (1 + ax^2 + bx + c)^{\frac{1}{x-a}}$ is

(a) $a(\alpha - \beta)$ (b) $\ln | a(\alpha - \beta) |$ (c) $e^{a(\alpha-\beta)}$ (d) $e^{a|\alpha-\beta|}$

2.5 Limits Using Sandwich Theorem Series and Leibnitz's Formula

62 Let $f : R \to (0, \infty)$ be such that $f(x) + \frac{e^{x+x^2}}{f(x)} \le e^x + e^{x^2} \ \forall x > 0$, then $\lim_{x \to 1} f(x)$ is

(a) 1 (b) $\frac{1}{e}$ (c) $2e$ (d) e

63 The value of $\lim_{n \to \infty} \left(\frac{1}{\sqrt{n^2}} + \frac{1}{\sqrt{n^2+1}} + \frac{1}{\sqrt{n^2+2}} + \dots \frac{1}{\sqrt{n^2+2n}} \right)$ is

(a) 1 (b) 2 (c) 3 (d) $\frac{1}{2}$

64 If f is a positive function such that $f(x+T) = f(x) \, (T > 0) \ \forall x \in R$, then

$\lim_{n \to \infty} n \left(\frac{f(x+T) + 2f(x+2T) + \dots + nf(x+nT)}{f(x+T) + 4f(x+4T) + \dots + n^2 f(x+n^2 T)} \right)$ equals

(a) 2 (b) $\frac{2}{3}$ (c) $\frac{3}{2}$ (d) None of these

65 The value of $\lim_{n \to \infty} \prod_{n=0}^{n} \left(1 + \frac{1}{2^{2^n}} \right)$ is

(a) 1 (b) 2 (c) 4 (d) None of these

66 The value of $\lim_{x \to \infty} \dfrac{\left(\int_0^{x+y} e^{t^2} dt\right)^2}{\int_0^{x+y} e^{2t^2} dt}$ is equal to

(a) 0 (b) 1 (c) −1 (d) None of these

67 The value of $\lim_{x \to 0} \left\{\dfrac{1}{x^5}\left(\int_0^x e^{-t^2} dt\right) - \dfrac{1}{x^4} + \dfrac{1}{3x^2}\right\}$ is

(a) $\dfrac{1}{6}$ (b) $-\dfrac{1}{6}$ (c) $-\dfrac{1}{10}$ (d) $\dfrac{1}{10}$

68 $\lim_{x \to 0^+} \dfrac{\int_0^{x^2} (\sin\sqrt{t})\, dt}{x^3}$ is equal to

[JEE Main 2021]

(a) $\dfrac{2}{3}$ (b) $\dfrac{3}{2}$ (c) 0 (d) $\dfrac{1}{15}$

Numerical Value Type Questions

69 $\lim_{x \to \infty} \dfrac{\sqrt{n^3 - 2n^2 + 1} + \sqrt[3]{n^4 + 1}}{\sqrt[4]{n^6 + 6n^5 + 2} - \sqrt[5]{n^7 + 3n^3 + 1}}$ equals ......

70 If P, Q and R are positive real numbers such that $\lim_{x \to \infty} (\sqrt{Px^2 + Qx} - Rx) = 3$, then $\dfrac{Q}{R}$ equals to ......

71 Let $\lim_{x \to 1} \dfrac{x^a - ax + a - 1}{(x-1)^2} = f(a)$. Then, the value of $f(4)$ is ....

72 Let f be a differential function such that $f'(2) = \dfrac{1}{4}$, then the value of $\lim_{x \to 0}\left(\dfrac{f(2+3x^4) - f(2-5x^4)}{x^4}\right)$ is ....

73 If $\lim_{x \to 0} \dfrac{\sin 3x + (1 - \cos 2x)}{x + P\tan x} = 1$, then the value of P is ......

74 If $L = \lim_{x \to \frac{\pi}{4}} \dfrac{(1 - \tan x)(1 - \sin 2x)}{(1 + \tan x)(\pi - 4x)^3}$, then the value of $\dfrac{1}{L}$ is equal to ......

75 $\sum_{k=1}^{n} \sin^{-1} a_k = \dfrac{n\pi}{2}$ for any $n \in N$ and $p = \prod_{k=1}^{n} (a_k)^k$ and $\lim_{x \to p} \dfrac{x^{1/3} - (3-2x)^{1/4}}{x^2 - x} = k$, then the value of $6(k + p)$ is ....

76 If $\lim_{x \to 0} \dfrac{a \sin x - \sin 2x}{\tan^3 x} = L$ is finite, then the value of $a + L$ is ....

77 If $\lim_{x \to 0} \dfrac{\{(a-n)nx - \tan x\} \sin x}{x^2} = 0$, where n is a non-zero real number, then $a = f(n)$, then value of $f(1)$ is ....

78 If $\lim_{x \to 0} \dfrac{ax - (e^{4x} - 1)}{ax(e^{4x} - 1)}$ exists and is equal to b, then the value of $a - 2b$ is .... [JEE Main 2021]

79 Let $\lim_{x \to 0} \dfrac{\sin 4x}{\tan\left(\dfrac{x}{k}\right)} = L_1$ and $\lim_{x \to 0} \dfrac{e^{4x} - 1}{x} = L_2$ and the value of $L_1 L_2$ is 16. The value of k is ....

80 Let p and q be two positive integer greater than 1. If $\lim_{x\to 0} \frac{e^{\cos(x^q)} - e}{x^p} = \frac{-e}{2}$, then the value of $\frac{q}{p}$ is ....

81 Let $L = \lim_{x\to 0} \frac{e^{x\sin(2007)x} - 1}{x\ln(1+x)}$ and $M = \lim_{x\to\infty} x\{2x - (\sqrt[3]{x^3 + x^2 + 1}) - \sqrt[3]{x^3 - x^2 + 1}\}$, then the value of $\frac{LM}{223}$ is ....

82 The value of $\lim_{x\to 0} \frac{(140)^x - (35)^x - (28)^x - (20)^x + 7^x + 5^x + 4^x - 1}{x\sin^2 x} = P\ln 2 \cdot \ln Q \cdot \ln 7$, then $P + Q$ is equal to ....

83 If $\lim_{x\to\infty}\left[\cos\left(2\pi\left(\frac{x}{1+x}\right)^a\right)\right]^{x^2} = e^{-ka^2\pi^2}$, $a \in R$, then value of k is ....

84 If $\lim_{x\to 1}(1 + ax + bx^2)^{\frac{c}{x-1}} = e^5$, then the value of $\frac{a^2c}{b}$ is ....

85 If $\lim_{x\to 0}(a\tan bx^2 - 3\cos^2 x + 4)^{\frac{1}{3x^2}} = e^{\frac{7}{3}}$, $a, b \in R$, then ab is equal to ....

86 Let $L_1 = \lim_{x\to\infty}\int_0^x \frac{(1+\cos t)^2}{x}dt$ and $L_2 = \lim_{x\to 0}\int_0^x \frac{(1+\cos t)^2}{x}dt$, then the value of $4(L_1 + L_2)$ is ....

87 Let $f : R \to R$ be a positive increasing function with $\lim_{x\to\infty}\frac{f(3x)}{f(x)} = 1$. Then, $\lim_{x\to\infty}\frac{f(2x)}{f(x)}$ is equal to ....

88 If $I = \lim_{n\to\infty}\sum_{r=2}^{n}\left\{(r+1)\sin\frac{\pi}{r+1} - r\sin\frac{\pi}{r}\right\}$, then the value of $[I]$, where [] denotes the greatest integer function, is ....

LEVEL 03 \ JEE Advanced Level

Single Choice Correct Questions

1 If $\lim_{n\to\infty}\left(\frac{2n^3 + 3n^2 + 5}{n(n+1)} - an + b\right) = 3$, then

(a) $a = 2$ and $b = 2$ (b) $a = 2$ and $b = -2$ (c) $a = 1$ and $b = 1$ (d) $a = -2$ and $b = 1$

2 If $f(a) = a^2$, $g(a) = b^2$ and $f'(a) = 3g'(a)$, then $\lim_{x\to a}\frac{\sqrt{f(x)} - a}{\sqrt{g(x)} - b}$, where $a, b \in R^+$, is

(a) $\frac{b^2}{a^2}$ (b) $\frac{3b}{a}$ (c) $\frac{2b}{a}$ (d) None of these

3 Let $P(x)$ be a real polynomial of degree 3 having extreme values at $x = 1$ and $x = 3$, if $\lim_{x\to 0}\frac{P(x)}{x} = 2$, then $P(-3)$ is

(a) 4 (b) -8 (c) -16 (d) 0

4 The value of $\lim_{x \to \pi} \dfrac{2^{\cot x} + 3^{\cot x} - 5^{1+\cot x} + 2}{(2^{\cot x})^2 + (9^{\cot x})^{1/2} - 5^{\cot x} + 1}$ is

(a) 5 (b) 2 (c) −2 (d) Does not exist

5 The value of $\lim_{x \to 0^+} \left(\dfrac{\cot^{-1}(\ln \{x\})}{\pi} \right)^{-\ln\{x\}}$ is equal to, where $\{x\}$ denotes fractional part function

(a) $e^{\frac{1}{\pi}}$ (b) $e^{\frac{-1}{\pi}}$ (c) $e^{\frac{-2}{\pi}}$ (d) $e^{\frac{2}{\pi}}$

6 For each $t \in R$, let $[t]$ be the greatest integer less than or equal to t.

Then, $\lim_{x \to 1^+} \dfrac{(1-|x| + \sin|1-x|) \sin\left(\frac{\pi}{2}[1-x]\right)}{|1-x|\,[1-x]}$ is equal to

(a) 0 (b) −1 (c) 1 (d) Does not exist

7 $\lim_{x \to 0} \dfrac{Ae^x - B\cos x + Ce^{-x}}{x \sin x} = 2$, then the value of $A + B + C$ is

(a) 4 (b) 2 (c) 0 (d) − 4

8 If $\lim_{n \to \infty} \dfrac{4^n}{(4 - \sqrt{3} + 2\sin\theta)^{n+2}}$ exists and is equal to p, $(p \neq 0)$, where $\theta \in \left(0, \frac{\pi}{2}\right)$, then value of $\left(\dfrac{p + \cos\theta}{p}\right)$ is equal to

(a) 7 (b) 8 (c) 9 (d) 10

9 $f(x)$ is the function such that $\lim_{x \to 0} \dfrac{f(x)}{x} = 1$, if $\lim_{x \to 0} \dfrac{x(1 + a\cos x) - b\sin x}{(f(x))^3} = \dfrac{5}{6}$, then

(a) $a = -1$ and $b = -2$ (b) $a = -2$ and $b = -1$
(c) $a = 1$ and $b = -2$ (d) $a = -1$ and $b = +2$

10 If $f(n+1) = \dfrac{1}{2}\left(f(n) + \dfrac{9}{f(n)}\right)$, $n \in N$ and $f(n) > 0\ \forall\, n \in N$, then $\lim_{n \to \infty} f(n)$ is equal to

(a) 3 (b) −3 (c) $\dfrac{1}{2}$ (d) None of these

11 If $\lim_{x \to 2^-} \dfrac{k \cdot e^{\frac{1}{|x+2|}} - 1}{2 - e^{\frac{1}{|x+2|}}} = \lim_{x \to -2^+} \sin\left(\dfrac{x^4 - 16}{x^5 + 32}\right)$, then k is equal to

(a) 1 (b) 2 (c) $\sin\dfrac{2}{5}$ (d) $\sin\dfrac{3}{5}$

12 $\lim_{x \to \infty} \log_{x-1} x \cdot \log_x (x+1) \cdot \log_{(x+1)} (x+2) \ldots \log_{(x^5-1)} x^5$ is equal to

(a) 1 (b) 0 (c) 5 (d) 10

13 Let $a = \min[x^2 + 2x + 3, x \in R]$ and $b = \lim_{x \to 0} \dfrac{\sin x \cos x}{e^x - e^{-x}}$. Then, the value of $\sum_{r=0}^{n} a^r b^{n-r}$ is

(a) $\dfrac{2^{n+1} + 1}{3 \cdot 2^n}$ (b) $\dfrac{2^{n+1} - 1}{3 \cdot 2^n}$ (c) $\dfrac{2^n - 1}{3 \cdot 2^n}$ (d) $\dfrac{4^{n+1} - 1}{3 \cdot 2^n}$

14 At the end points A and B of a fixed segment of length L. Lines are drawn meeting in C and making angles θ and 2θ, respectively with the given segment. Let D be the foot of the altitude CD and x represents the length AD, then the value of x as θ tends to zero i.e. $\lim_{\theta \to 0} x$ is

(a) $\frac{L}{3}$ (b) $\frac{2L}{3}$ (c) L (d) $\frac{L}{2}$

15 Let $f''(x)$ be continuous at $x=0$ and $f''(0)=4$, then value of $\lim_{x \to 0} \frac{2f(x)-3f(2x)+f(4x)}{x^2}$ is

(a) 11 (b) 2 (c) 12 (d) 10

More than One Correct Options

16 Let $k>0$ and $\lambda = \lim_{x \to 0} \frac{k(1-4\sqrt{K^2-x^2})}{x^2\sqrt{K^2-x^2}}$ be finite, then

(a) $k=\frac{1}{2}$ (b) $k=\frac{1}{4}$ (c) $\lambda=8$ (d) $\lambda=4$

17 Which of the following limits vanishes?

(a) $\lim_{x \to \infty} x^{\frac{1}{4}} \sin \frac{1}{\sqrt{x}}$

(b) $\lim_{x \to \frac{\pi}{2}} (1-\sin x)\tan x$

(c) $\lim_{x \to \infty} \frac{3x^2+5}{x^2-2x+5} \cdot \text{Sgn}(x)$

(d) $\lim_{x \to 3} \frac{[x]^2-9}{x^2-9}$, where [ ] denotes greatest integer function.

18 Which of the following is/are true?

(a) If $\lim_{x \to a} \{f(x)+g(x)\}$ exists, then both $\lim_{x \to a} f(x)$ and $\lim_{x \to a} g(x)$ exist.

(b) If $\lim_{x \to a} f(x)$ and $\lim_{x \to a} g(x)$ exists, then $\lim_{x \to a} \{f(x)+g(x)\}$ exists.

(c) $\lim_{x \to a} \{f(x)\cdot g(x)\}$ exists, then both $\lim_{x \to a} f(x)$ and $\lim_{x \to b} g(x)$ exist.

(d) If $\lim_{x \to a} f(x)$ and $\lim_{x \to a} g(x)$ exist, then $\lim_{x \to a} f(x)\cdot g(x)$ exists.

19 Let $f: R \to R$ be a function, we say that f has

Property I If $\lim_{h \to 0} \frac{f(h)-f(0)}{\sqrt{|h|}}$ exist and finite and

Property II If $\lim_{h \to 0} \frac{f(h)-f(0)}{h^2}$ exist and is finite, then which of the following is/are correct?

(a) $f(x)=x^{2/3}$ has property I

(b) $f(x)=\sin x$ has property II

(c) $f(x)=|x|$ has property I

(d) $f(x)=x|x|$ has property II

20 For a function $f(x)=\frac{\ln(\{\sin x\}\{\cos x\}+1)}{\{\sin x\}\{\cos x\}}$, where {} denotes fractional part function, then

(a) $f(0^-)=f\left(\frac{\pi^+}{2}\right)$

(b) $f(0^+)=f\left(\frac{\pi}{2}\right)$

(c) $\lim_{x \to 0} f(x)=1$

(d) $\lim_{x \to \frac{\pi}{2}} f(x)=1$

21 Let $f(x)=\dfrac{1-x\,(1+|1-x|)}{|1-x|}\cos\left(\dfrac{1}{1-x}\right)$ for $x \neq 1$, then

(a) $\lim_{x\to 1^+} f(x)=0$ (b) $\lim_{x\to 1^-} f(x)$ does not exist

(c) $\lim_{x\to 1^-} f(x)=0$ (d) $\lim_{x\to 1^+} f(x)$ does not exist

22 If $\lim_{x\to 0}\left(\dfrac{a^x+b^x+c^x}{3}\right)^{k/x}$, $(a,b\,c,k>0)$

(a) equals 1, if $k=1$ (b) equals abc, if $k=3$

(c) equal abc, if $k=1$ (d) equal $(a^2b^2c^2)^{1/3}$, if $k=2$

23 If $\lim_{x\to 0}\dfrac{1-\cos x}{e^{ax}-bx-1}=1$, then

(a) $a=1$ and $b=1$ (b) $a=2$ and $b=0$ (c) $a=-2$ and $b=0$ (d) $a=-1$ and $b=-1$

24 If $m, n \in N$, $\lim_{x\to 0}\dfrac{\sin x^n}{(\sin x)^m}$ is

(a) 0, if $n>m$ (b) 1, if $n=m$ (c) ∞, if $n<m$ (d) $\dfrac{n}{m}$, if $n<m$

25 If $f(x)=\lim_{n\to\infty}\{n(x^{1/n}-1)\}$ for $x>0$, then which of the following is/are true?

(a) $f\left(\dfrac{1}{x}\right)$ (b) $f\left(\dfrac{1}{x}\right)=\dfrac{1}{f(x)}$ (c) $f\left(\dfrac{1}{x}\right)=-f(x)$ (d) $f(xy)=f(x)+f(y)$

26 If $\alpha,\beta\in\left(-\dfrac{\pi}{2},0\right)$ such that $(\sin\alpha+\sin\beta)+\dfrac{\sin\alpha}{\sin\beta}=0$ and $(\sin\alpha+\sin\beta)\dfrac{\sin\alpha}{\sin\beta}=-1$ and

$\lambda=\lim_{n\to\infty}\dfrac{1+(2\sin\alpha)^{2n}}{(2\sin\beta)^{2n}}$, then

(a) $\alpha=-\dfrac{\pi}{6}$ (b) $\alpha=-\dfrac{\pi}{3}$ (c) $\lambda=2$ (d) $\lambda=1$

27 If $f(x)=\left(\dfrac{|x|}{|x|+2}\right)^{-x}$, then

(a) $\lim_{x\to-\infty} f(x)=0$ (b) $\lim_{x\to 0} f(x)=1$ (c) $\lim_{x\to\infty} f(x)=e^2$ (d) $\lim_{x\to-\infty} f(x)=e^{-2}$

28 Let $f(x)=\lim_{m\to\infty}\left\{\lim_{n\to\infty}\{\cos^{2m}(n!\,\pi x)\}\right\}$, where $x\in R$, then

(a) $f(x)=1$, if x is rational (b) $f(x)=0$, if x is irrational

(c) $f(x)=0$, if x is rational (d) $f(x)=1$, if x is irrational

29 Consider the function $f(x)=\left(\dfrac{ax+1}{bx+2}\right)^x$, where $a^2+b^2\neq 0$, then $\lim_{x\to\infty} f(x)$

(a) exists for all values of a and b (b) is zero for $0<a<b$

(c) is non-existent for $a>b>0$ (d) is $e^{-\frac{1}{a}}$ or $e^{-\frac{1}{b}}$ for $a=b$

Linked Comprehension Type

Paragraph I Directions (Q.Nos. 30-32)

Consider two functions $f(x)=\lim_{n\to\infty}\left(\cos\frac{x}{\sqrt{n}}\right)^n$ and $g(x)=-x^{4b}$, where $b=\lim_{x\to\infty}(\sqrt{x^2+x+1}-\sqrt{x^2+1})$. Then,

30 $f(x)$ is

(a) e^{-x^2} (b) $e^{-\frac{x^2}{2}}$ (c) e^x (d) $e^{\frac{x^2}{2}}$

31 $g(x)$ is

(a) $-x^2$ (b) x^2 (c) x^4 (d) $-x^4$

32 Number of solutions of $f(x)+g(x)=0$ is

(a) 0 (b) 1 (c) 2 (d) 4

Paragraph II Directions (Q.Nos. 33-35)

If $\lim_{x\to 0}\frac{axe^x-b\log(1+x)+cxe^{-x}}{x^2\sin x}=2$, then answer the following questions.

33 Which is/are correct ?

(a) $a=3$ (b) $b=12$ (c) $c=10$ (d) $b=4$

34 Roots of equation $ax^2+bx+c=0$ are

(a) real and unequal (b) imaginary

(c) rational (d) irrational

35 Domain of $f(x)=\sqrt{\frac{ax^2-b}{c}}$ is

(a) $[-2, 2]$ (b) $(-\infty,-2]\cup[2,\infty)$

(c) $[-3, 3]$ (d) None of these

Matrix Match Type Questions

36 Match the following

Column I		Column II
$\lim_{x\to 0}f(x)=i$, where (i) $f(x)=\frac{\tan[e^2]x^2-\tan[-e^2]x^2}{\sin^2 x}$, where $[x]$ is the greatest integer function	(A)	$\frac{1}{2}$
(ii) $f(x)=\frac{\left[\frac{5}{2}+\tan x+\tan^2 x\right]-\left[\frac{5}{2}\right]}{\tan x}$	(B)	2
(iii) $f(x)=\frac{\sqrt[3]{1+x^2}-\sqrt[4]{1-2x}}{x+x^2}$	(C)	15
(iv) $f(x)=\lim_{x\to 0}\frac{1-\cos x+2\sin x-\sin^3 x-x^2+3x^4}{\tan^3 x-6\tan^2 x+x-5x^3}$	(D)	0

37 Match the following

Column I	Column II
(i) If $\lim_{x\to 0} \dfrac{\int_0^x \left(\dfrac{t^2}{\sqrt{b+t}}\right)dt}{ax-\sin x}=1, (b\neq 0)$, then (a,b) lies on the line	(A) $x-y+3=0$
(ii) If $\lim_{x\to\infty} \sqrt{x^4+x^2-1}-ax^2+b=0, a>0$, then (a,b) lies on the line	(B) $3x+4y-5=0$
(iii) If $\lim_{x\to\infty} \sqrt{x^2+x+1}-(ax+b)=0, a>0$, then (a,b) lies on the line	(C) $x+2y+3=0$
(iv) $\lim_{x\to\infty} \dfrac{(1+a^3)+8e^{1/x}}{2+(1-b^3)e^{1/x}}=2$, then (a^3,b^3) lies on the line	(D) $x+6y+2=0$

38 Match the following

Column I	Column II
(i) $\lim_{x\to-\infty} \dfrac{3^{\sin x}+2x+1}{\sin x-\sqrt{x^2+1}}$	(A) 0
(ii) $\lim_{m\to\infty}\lim_{n\to\infty} \sum_{k=1}^{m} \dfrac{\sqrt[n]{(k-1)^n+k^n}}{m^2}$	(B) $\dfrac{1}{2}$
(iii) $\lim_{n\to\infty} \left(\dfrac{n}{n^2-4}+\dfrac{4^n(-1)^n}{2^n-1}\right)^{-1}$	(C) -1
(iv) $\lim_{x\to-\infty} \left\{x+\sqrt{x^2+2x\cos\dfrac{1}{\lvert x\rvert}}\right\}$	(D) 2

39 Match the following

Column I	Column II
(i) If $L=\lim_{x\to\frac{\pi}{2}^+} \dfrac{\cos[\tan^{-1}(\tan x)]}{x-\dfrac{\pi}{2}}$, then L is	(A) -2
(ii) $\lim_{n\to\infty}\left[\tan\left(\sum_{r=1}^{n}\tan^{-1}\left(\dfrac{4}{4r^2+3}\right)\right)\right]$ is equal to	(B) -1
(iii) If $k=\lim_{x\to-\infty} \dfrac{(3x^4+2x^2)\sin\left(\dfrac{1}{x}\right)+\lvert x\rvert^3+5}{\lvert x\rvert^3+\lvert x\rvert^2+\lvert x\rvert+1}$, then k is equal to	(C) 2
(iv) If $f(x)=\lim_{t\to 0}\left[\dfrac{2x}{\pi}\tan^{-1}\left(\dfrac{x}{t^2}\right)\right]$, then $f(-1)$ is	(D) 1

Numerical Value Type Questions

Question Stem Directions (Q.Nos. 40-42) The answer to each question is a **NUMERICAL VALUE.** For problems,

Let $f(x)=\dfrac{\tan^{-1}(1-\{x\})\cos^{-1}(1-\{x\})\sin^{-1}(1-\{x\})}{\sqrt{2\{x\}(1-\{x\})^2}}$, $\{x\}$ denotes the fractional part of x, then

40 If $P=\lim\limits_{x\to 0^+} f(x)$, then $\cos\left(\dfrac{2P}{\pi}\right)$ is equal to ....

41 If $Q=\lim\limits_{x\to 0^-} f(x)$, then the value of $\sin^{-1}\sqrt[3]{2Q}$ is ....

42 $\dfrac{Q^2}{P}$ is equal to ....

Question Stem Directions (Q.Nos. 43-45) The answer to each question is a **NUMERICAL VALUE.**

Consider the limit $\lim\limits_{x\to 0}\dfrac{1}{x^3}\left(\dfrac{1}{\sqrt{1+x}}-\dfrac{(1+ax)}{(1+bx)}\right)$ exists finite and has the value equal to l, then the value of

43 a is ....

44 $a+b$ is ....

45 $\left|\dfrac{b}{l}\right|$ is ....

46 Let $f(x)=3x^{10}-7x^8+5x^6-21x^3+3x^2-7$, then the value of $265\left(\lim\limits_{h\to 0}\dfrac{h^4+3h^2}{[f(1-h)-f(1)]\sin 5h}\right)$ is ....

47 If α and β, $(\alpha<\beta)$ are the roots of the equation

$$\lim_{t\to\infty}\cos^{-1}\left[\sin\left\{\tan^{-1}\left(\frac{\sqrt{tx}}{\sqrt{tx^2-3tx+t+1-x}}\right)\right\}\right]=\frac{\pi}{6},$$ then the value of $(8^\alpha+2^\beta-\alpha\beta)$ is ....

48 Let $g(x)=\dfrac{(x-1)^n}{\log\cos^m(x-1)}$, $0<x<2$, m and n are integers, $m\neq 0$, $n>0$ and let P be the left hand derivative of $|x-1|$ at $x=1$. If $\lim\limits_{x\to 1^+} g(x)=P$, then $m+n$ is ....

49 Let $f''(x)$ be continuous at $x=0$. If $\lim\limits_{x\to 0}\dfrac{af(x)-3f(2x)+bf(4x)}{\sin^2 x}$ exists and $f(0)\neq 0$ and $f'(0)\neq 0$, then the value of $a+3b$ is ....

50 Let $L=\lim\limits_{x\to\infty}\left(x\log x+2x\log\sin\left(\dfrac{1}{\sqrt{x}}\right)\right)$, then the value of $\left(-\dfrac{2}{L}\right)$ is ....

51 Let $f(x)=\dfrac{\left(\dfrac{\pi}{2}-\cos^{-1}(1-\{x\}^2)\right)[\cos^{-1}(1-\{x\})]^2}{2(\{x\}-\{x\}^3)}$. If $f(0^+)=p$ and $f(0^-)=q$, then the value of $\left(\dfrac{p\pi}{q}\right)$ ($\{k\}$ denotes the fractional part of k) is ....

52 If α and β are two distinct real roots of the equation $ax^3+x-1-a=0$ $(a\neq -1, 0)$, none of which is equal to unity, then the value of $\lim\limits_{x\to\frac{1}{\alpha}}\dfrac{(1+a)x^3-x^2-a}{(e^{1-\alpha x}-1)(x-1)}$ is $\dfrac{al}{\alpha}(k\alpha-\beta)$.

Then, the value of kl is ......

53 If the value of $\lim_{x\to 0}\dfrac{(1+3x+2x^2)^{\frac{1}{x}}-(1+3x-2x^2)^{\frac{1}{x}}}{x}=ke^3$, then the value of $4k$ is ....

54 Let $S=\dfrac{1}{2}\tan\dfrac{\pi}{4}+\dfrac{1}{2^2}\tan\dfrac{\pi}{8}+\dfrac{1}{2^3}\tan\dfrac{\pi}{16}+\ldots\ldots+\dfrac{1}{2^n}\tan\left(\dfrac{\pi}{2^{n+1}}\right)$. If $\lim_{n\to\infty} S=L$, then the value of $(100\pi L)$ is ....

55 The sequence $(a_{n-1}), n\in N$ is an arithmetical progression and d is its common difference. If $\lim_{n\to\infty}\left(1-\dfrac{d^2}{a_1^2}\right)\left(1-\dfrac{d^2}{a_2^2}\right)\ldots\ldots\left(1-\dfrac{d^2}{a_n^2}\right)$ converges to $\dfrac{1}{4}$ and $a_1=8$, then the value of d is ....

ANSWER KEY

Level 1

1. (i) 3 (ii) 4 (iii) 3 (iv) $\frac{5}{6}$ (v) 32 (vi) $\frac{1}{2}$ (vii) $-\frac{1}{9}$ (viii) $2a$ (ix) $2a$ (x) $-\frac{1}{2}$ (xi) 1 (xii) $\frac{15}{16}$ (xiii) 0.01 (xiv) 15

2. (i) $\frac{1}{2}$ (ii) $\frac{5}{2}$ (iii) $\frac{a^2-b^2}{c^2-d^2}$ (iv) $\frac{1}{2}$ (v) $\frac{1}{4}$ (vi) 1 (vii) 2

3. (i) 0 (ii) 1 (iii) $\frac{4}{9}$ (iv) $-\sin a$ (v) $\frac{1}{8}$ (vi) -3 (vii) 0 (viii) $\frac{2}{5}$ (ix) -1 (x) $\cos^2\alpha\ln\cos\alpha+\sin^2\alpha\ln\sin\alpha$

4. (i) e^2 (ii) $a-b$ (iii) $4\ln 2$ (iv) $\frac{3}{\log 3}$ (v) $2\log_3 e$ (vi) $\ln\left(\frac{ab}{cd}\right)$ (vii) $\frac{1}{a}$ (viii) $\frac{2}{a}$ (ix) $\log a$ (x) $\frac{3}{2}$

5. (i) e^{-3} (ii) e (iii) $\frac{1}{e}$ (iv) $(ab)^{3/2}$ (v) $e^{5/2}$

6. (i) 1 (ii) 1 (iii) $\sqrt{2}$ (iv) 1

Level 2

1. (c)	**2.** (c)	**3.** (c)	**4.** (b)	**5.** (b)	**6.** (a)	**7.** (a)
8. (b)	**9.** (c)	**10.** (b)	**11.** (b)	**12.** (b)	**13.** (b)	**14.** (c)
15. (d)	**16.** (a)	**17.** (a)	**18.** (d)	**19.** (d)	**20.** (a)	**21.** (c)
22. (b)	**23.** (c)	**24.** (b)	**25.** (a)	**26.** (c)	**27.** (a)	**28.** (d)
29. (d)	**30.** (d)	**31.** (c)	**32.** (d)	**33.** (b)	**34.** (d)	**35.** (c)
36. (b)	**37.** (c)	**38.** (b)	**39.** (d)	**40.** (a)	**41.** (d)	**42.** (b)
43. (a)	**44.** (a)	**45.** (a)	**46.** (b)	**47.** (c)	**48.** (c)	**49.** (b)
50. (b)	**51.** (b)	**52.** (c)	**53.** (b)	**54.** (b)	**55.** (c)	**56.** (a)
57. (b)	**58.** (c)	**59.** (c)	**60.** (d)	**61.** (c)	**62.** (d)	**63.** (b)
64. (c)	**65.** (b)	**66.** (a)	**67.** (d)	**68.** (a)	**69.** (1)	**70.** (6)
71. (6)	**72.** (2)	**73.** (2)	**74.** (32)	**75.** (11)	**76.** (3)	**77.** (2)
78. (5)	**79.** (1)	**80.** (0.5)	**81.** (2)	**82.** (7)	**83.** (2)	**84.** (5)
85. (4)	**86.** (22)	**87.** (1)	**88.** (1)			

Level 3

1. (a)	**2.** (b)	**3.** (c)	**4.** (d)	**5.** (b)	**6.** (a)	**7.** (a)
8. (c)	**9.** (b)	**10.** (a)	**11.** (c)	**12.** (c)	**13.** (d)	**14.** (b)
15. (c)	**16.** (b, c)	**17.** (a, b)	**18.** (b, d)	**19.** (a, c)	**20.** (a, b)	**21.** (c, d)
22. (b, d)	**23.** (a, d)	**24.** (a, b, c)	**25.** (c, d)	**26.** (a, c)	**27.** (b, d)	**28.** (a, b)
29. (b, c, d)	**30.** (b)	**31.** (a)	**32.** (c)	**33.** (a, b)	**34.** (a, c)	**35.** (b)

36. (i)–C, (ii)–D, (iii)–A, (iv)–B

37. (i)–A, (ii)–D, (iii)–B, (iv)–C

38. (i)–D, (ii)–B, (iii)–A, (iv)–C

39. (i)–D, (ii)–C, (iii)–A, (iv)–B

40. (0.7)	**41.** (1)	**42.** (1)	**43.** (0.25)	**44.** (1)	**45.** (24)	**46.** (3)
47. (9)	**48.** (4)	**49.** (5)	**50.** (6)	**51.** (4)	**52.** (1)	**53.** (16)
54. (200)	**55.** (6)					

SOLUTIONS

Level 1

1. (i) $\lim_{x\to 3}\frac{4x+3}{x+2}=\frac{4(3)+3}{3+2}=\frac{15}{5}=3$

(ii) $\lim_{x\to 2}\frac{x^2-4}{x-2}=\lim_{x\to 2}\frac{(x+2)(x-2)}{(x-2)}=2+2=4$

(iii) $\lim_{x\to\frac{-1}{3}}\frac{27x^3+1}{3x+1}=\lim_{x\to\frac{-1}{3}}\frac{(3x)^3+(1)^3}{3x+1}$

$$=\lim_{x\to\frac{-1}{3}}\frac{(3x+1)(9x^2+1-3x)}{3x+1}$$

$$=9\left(\frac{-1}{3}\right)^2+1-3\left(\frac{-1}{3}\right)$$

$$=1+1+1=3$$

(iv) $\lim_{x\to -2}\frac{x^2-x-6}{x^2-2x-8}=\lim_{x\to -2}\frac{x^2-3x+2x-6}{x^2-4x+2x-8}$

$$=\lim_{x\to -2}\frac{(x-3)(x+2)}{(x-4)(x+2)}$$

$$=\frac{-2-3}{-2-4}=\frac{5}{6}$$

(v) $\lim_{x\to 2}\frac{x^4-16}{x^2-3x+2}=\lim_{x\to 2}\frac{(x^2)^2-(4)^2}{x^2-2x-x+2}$

$$=\lim_{x\to 2}\frac{(x^2+4)(x^2-4)}{(x-2)(x-1)}$$

$$=\lim_{x\to 2}\frac{(x^2+4)(x-2)(x+2)}{(x-2)(x-1)}$$

$$=\frac{8\times 4}{1}=32$$

(vi) $\lim_{x\to\sqrt{3}}\frac{x^2-3}{3x^2-2\sqrt{3}x-3}$

$$=\lim_{x\to\sqrt{3}}\frac{(x+\sqrt{3})(x-\sqrt{3})}{3x^2-3\sqrt{3}x+\sqrt{3}x-3}$$

$$=\lim_{x\to\sqrt{3}}\frac{(x+\sqrt{3})(x-\sqrt{3})}{3x(x-\sqrt{3})+\sqrt{3}(x-\sqrt{3})}$$

$$=\lim_{x\to\sqrt{3}}\frac{(x+\sqrt{3})(x-\sqrt{3})}{(3x+\sqrt{3})(x-\sqrt{3})}$$

$$=\frac{2\sqrt{3}}{4\sqrt{3}}=\frac{1}{2}$$

(vii) $\lim_{x\to 1}\left(\frac{1}{x^2+x-2}-\frac{x}{(x-1)(x^2+x+1)}\right)$

$$=\lim_{x\to 1}\left(\frac{(x-1)(x^2+x+1)-x(x^2+x-2)}{(x-1)(x^2+x+1)(x^2+x-2)}\right)$$

$$=\lim_{x\to 1}\left(\frac{x^3-1-x^3-x^2+2x}{(x-1)(x^2+x+1)(x^2+x-2)}\right)$$

$$=\lim_{x\to 1}-\left(\frac{x^2-2x+1}{(x-1)(x^2+x+1)(x^2+x-2)}\right)$$

$$=\lim_{x\to 1}\frac{-(x-1)^2}{(x-1)(x^2+x+1)(x+2)(x-1)}$$

$$=\frac{-1}{(1+1+1)(1+2)}=\frac{-1}{9}$$

(viii) $\lim_{x\to 0}\frac{(a+x)^2-a^2}{x}=\lim_{x\to 0}\frac{(a+x+a)(a+x-a)}{x}$

$$=\lim_{x\to 0}\frac{x(2a+x)}{x}=2a$$

(ix) $\lim_{x\to a}(x^2-a^2)\left(\frac{1}{x+a}+\frac{1}{x-a}\right)$

$$=\lim_{x\to a}(x^2-a^2)\left(\frac{x-a+x+a}{(x+a)(x-a)}\right)$$

$$=\lim_{x\to a}\frac{(x^2-a^2)(2x)}{x^2-a^2}=2a$$

(x) $\lim_{x\to 2}\left[\frac{1}{x-2}-\frac{2(2x-3)}{x^3-3x^2+2x}\right]$

$$=\lim_{x\to 2}\left[\frac{1}{x-2}-\frac{2(2x-3)}{x(x^2-3x+2)}\right]$$

$$=\lim_{x\to 2}\left[\frac{1}{x-2}-\frac{2(2x-3)}{x(x-1)(x-2)}\right]$$

$$=\lim_{x\to 2}\left[\frac{x(x-1)-2(2x-3)}{x(x-1)(x-2)}\right]$$

$$=\lim_{x\to 2}\left[\frac{x^2-5x+6}{x(x-1)(x-2)}\right]=\lim_{x\to 2}\left[\frac{(x-2)(x-3)}{x(x-1)(x-2)}\right]$$

$$=\frac{-1}{2(1)}=\frac{-1}{2}$$

(xi) $\lim_{x\to 1}\frac{(\sqrt{5x-4}-\sqrt{x})}{(x^2-1)}$

$$=\lim_{x\to 1}\frac{\sqrt{5x-4}-\sqrt{x}}{(x+1)(x-1)}\times\frac{(\sqrt{5x-4}+\sqrt{x})}{(\sqrt{5x-4}+\sqrt{x})}$$

$$=\lim_{x\to 1}\frac{(5x-4)-(x)}{(x+1)(x-1)(\sqrt{5x-4}+\sqrt{x})}$$

$$=\lim_{x\to 1}\frac{4x-4}{(x+1)(x-1)(\sqrt{5x-4}+\sqrt{x})}$$

$$=\lim_{x\to 1}\frac{4(x-1)}{(x+1)(x-1)(\sqrt{5x-4}+\sqrt{x})}=\frac{4}{2(2)}=1$$

(xii) $\lim_{x\to 2}\frac{4-\sqrt{5x+6}}{3-\sqrt{4x+1}}\times\frac{3+\sqrt{4x+1}}{3+\sqrt{4x+1}}\times\frac{4+\sqrt{5x+6}}{4+\sqrt{5x+6}}$

$$= \lim_{x\to 2}\frac{16-(5x+6)}{9-(4x+1)}\times\left(\frac{3+\sqrt{4x+1}}{4+\sqrt{5x+6}}\right)$$

$$= \lim_{x\to 2}\frac{10-5x}{8-4x}\times\left(\frac{3+\sqrt{4x+1}}{4+\sqrt{5x+6}}\right)$$

$$= \lim_{x\to 2}\frac{5(2-x)}{4(2-x)}\times\left(\frac{3+\sqrt{4x+1}}{4+\sqrt{5x+6}}\right)$$

$$=\frac{5}{4}\times\left(\frac{3+3}{4+4}\right)=\frac{5}{4}\times\frac{6}{8}=\frac{15}{16}$$

(xiii) Let $\sqrt[5]{x}=y$. When $x\to 1\Rightarrow y\to 1$.

$$\lim_{y\to 1}\frac{y^2-2y+1}{4(y^5-1)^2}$$

$$= \lim_{y\to 1}\frac{(y-1)^2}{4[(y-1)(y^4+y^3+y^2+y+1)]^2}$$

$$=\frac{1}{4(1+1+1+1+1)^2}$$

$$=\frac{1}{4\times 25}=\frac{1}{100}=0.01$$

(xiv) $\lim_{x\to 3}\frac{x^5-243}{x^3-27}=\lim_{x\to 3}\frac{x^5-(3)^5}{x^3-(3)^3}$

$$= \lim_{x\to 3}\frac{\frac{x^5-(3)^5}{x-3}}{\frac{x^3-3^3}{x-3}}$$

$$=\frac{5(3)^{5-1}}{3(3)^{3-1}} \qquad \left[\because \lim_{x\to a}\frac{x^n-a^n}{x-a}=na^{n-1}\right]$$

$$=\frac{5\times 3^4}{3\times 3^2}$$

$$=5\times 3=15$$

2. (i) $\lim_{x\to\infty}\frac{x^2-4x+3}{2x^2-x+4}=\lim_{x\to\infty}\frac{x^2\left(1-\frac{4}{x}+\frac{3}{x^2}\right)}{x^2\left(2-\frac{1}{x}+\frac{4}{x^2}\right)}$

$$=\frac{1}{2} \qquad \left[\text{when } x\to\infty,\text{ then } \frac{1}{x}\to 0\right]$$

(ii) $\lim_{x\to\infty}\sqrt{x^2+5x}-x$

$$= \lim_{x\to\infty}\sqrt{x^2+5x}-x\times\frac{\sqrt{x^2+5x}+x}{\sqrt{x^2+5x}+x}$$

$$= \lim_{x\to\infty}\frac{(x^2+5x)-x^2}{\sqrt{x^2+5x}+x}=\lim_{x\to\infty}\frac{5x}{x\left(\sqrt{1+\frac{5}{x}}+1\right)}$$

$$=\frac{5}{\sqrt{1}+1}=\frac{5}{2}$$

(iii) $\lim_{x\to\infty}\frac{\sqrt{x^2+a^2}-\sqrt{x^2+b^2}}{\sqrt{x^2+c^2}-\sqrt{x^2+d^2}}\times\frac{\sqrt{x^2+a^2}+\sqrt{x^2+b^2}}{\sqrt{x^2+a^2}+\sqrt{x^2+b^2}}\times\frac{\sqrt{x^2+c^2}+\sqrt{x^2+d^2}}{\sqrt{x^2+c^2}+\sqrt{x^2+d^2}}$

$$= \lim_{x\to\infty}\frac{(x^2+a^2)-(x^2+b^2)}{(x^2+c^2)-(x^2+d^2)}\times\frac{\sqrt{x^2+c^2}+\sqrt{x^2+d^2}}{\sqrt{x^2+a^2}+\sqrt{x^2+b^2}}$$

$$= \lim_{x\to\infty}\frac{a^2-b^2}{c^2-d^2}\times\frac{x\left(\sqrt{1+\frac{c^2}{x^2}}+\sqrt{1+\frac{d^2}{x^2}}\right)}{x\left(\sqrt{1+\frac{a^2}{x^2}}+\sqrt{1+\frac{b^2}{x^2}}\right)}$$

$$=\frac{a^2-b^2}{c^2-d^2}\left(\frac{1+1}{1+1}\right)=\frac{a^2-b^2}{c^2-d^2}$$

(iv) $\lim_{n\to\infty}\frac{1+2+3+\ldots+n}{n^2}=\lim_{n\to\infty}\frac{n\left(\frac{n+1}{2}\right)}{n^2}$

$$= \lim_{n\to\infty}\frac{n^2\left(1+\frac{1}{n}\right)}{2n^2}=\frac{1}{2}$$

(v) $\lim_{n\to\infty}\frac{1^3+2^3+3^3+\ldots+n^3}{(n-1)^4}=\lim_{n\to\infty}\frac{\left(\frac{n(n+1)}{2}\right)^2}{(n-1)^4}$

$$= \lim_{n\to\infty}\frac{n^2(n+1)^2}{4(n-1)^4}$$

$$= \lim_{n\to\infty}\frac{n^2\cdot n^2\left(1+\frac{1}{n}\right)^2}{4\cdot n^4\left(1-\frac{1}{n}\right)^4}=\frac{1}{4}$$

(vi) $\lim_{n\to\infty}\frac{n!+(n-1)!}{n!-(n-1)!}=\lim_{n\to\infty}\frac{(n-1)!(n+1)}{(n-1)!(n-1)}$

$$= \lim_{n\to\infty}\frac{n\left(1+\frac{1}{n}\right)}{n\left(1-\frac{1}{n}\right)}=1$$

(vii) Let $y=-x$, when $x\to-\infty\Rightarrow y\to\infty$

$$= \lim_{y\to\infty}(\sqrt{y^2+4y}-y)$$

$$= \lim_{y\to\infty}\sqrt{y^2+4y}-y\times\frac{\sqrt{y^2+4y}+y}{\sqrt{y^2+4y}+y}$$

$$= \lim_{y\to\infty}\frac{(y^2+4y)-y^2}{\sqrt{y^2+4y}+y}$$

$$= \lim_{y\to\infty}\frac{4y}{y\left(\sqrt{1+\frac{4}{y}}+1\right)}=\frac{4}{1+1}=\frac{4}{2}=2$$

3. (i) $\lim_{x\to 0}\frac{\sin x-2\sin 3x+\sin 5x}{x}$

$$=\lim_{x\to 0}\left(\frac{\sin x}{x}-2\frac{\sin 3x}{x}+\frac{\sin 5x}{x}\right)$$

$$=\lim_{x\to 0}\left(\frac{\sin x}{x}-2\frac{\sin 3x}{3x}\times 3+\frac{\sin 5x}{5x}\times 5\right)$$

$$=1-6+5=0$$

(ii) $\lim_{x\to 0}\frac{\sin 2x+4x}{2x+\sin 4x}=\lim_{x\to 0}\frac{\frac{\sin 2x}{2x}\times 2x+4x}{2x+\frac{\sin 4x}{4x}\times 4x}$

$$=\lim_{x\to 0}\frac{x\left(\frac{\sin 2x}{2x}\times 2+4\right)}{x\left(2+\frac{\sin 4x}{4x}\times 4\right)}$$

$$=\frac{2+4}{2+4}=\frac{6}{6}=1$$

(iii) $\lim_{x\to 0}\frac{1-\cos 4x}{1-\cos 6x}=\lim_{x\to 0}\frac{2\sin^2 2x}{2\sin^2 3x}$

$$=\lim_{x\to 0}\frac{2\left(\frac{\sin 2x}{2x}\times 2x\right)^2}{2\left(\frac{\sin 3x}{3x}\times 3x\right)^2}=\lim_{x\to 0}\frac{\left(\frac{\sin 2x}{2x}\right)^2\times 4x^2}{\left(\frac{\sin 3x}{3x}\right)^2\times 9x^2}=\frac{4}{9}$$

(iv) $\lim_{x\to a}\frac{\cos x-\cos a}{x-a}$

$$=\lim_{x\to a}\frac{-2\sin\left(\frac{x+a}{2}\right)\sin\left(\frac{x-a}{2}\right)}{x-a}$$

$$=\lim_{x\to a}\frac{-\sin\left(\frac{x+a}{2}\right)\sin\left(\frac{x-a}{2}\right)}{\left(\frac{x-a}{2}\right)}$$

$$=-\sin\left(\frac{2a}{2}\right)\cdot(1)=-\sin a$$

(v) $\lim_{x\to\frac{\pi}{2}}\frac{\sqrt{2}-\sqrt{1+\sin x}}{\sqrt{2}\cos^2 x}$

$$=\lim_{x\to\frac{\pi}{2}}\frac{\sqrt{2}-\sqrt{1+\sin x}}{\sqrt{2}\cos^2 x}\times\frac{\sqrt{2}+\sqrt{1+\sin x}}{\sqrt{2}+\sqrt{1+\sin x}}$$

$$=\lim_{x\to\frac{\pi}{2}}\frac{2-(1+\sin x)}{\sqrt{2}\cos^2 x(\sqrt{2}+\sqrt{1+\sin x})}$$

$$=\lim_{x\to\frac{\pi}{2}}\frac{1-\sin x}{(\sqrt{2}\cos^2 x)(\sqrt{2}+\sqrt{1+\sin x})}$$

Put $x=\frac{\pi}{2}+h$

$$=\lim_{h\to 0}\frac{1-\sin\left(\frac{\pi}{2}+h\right)}{\sqrt{2}\cos^2\left(\frac{\pi}{2}+h\right)\left(\sqrt{2}+\sqrt{1+\sin\left(\frac{\pi}{2}+h\right)}\right)}$$

$$=\lim_{h\to 0}\frac{1-\cos h}{\sqrt{2}\sin^2 h(\sqrt{2}+\sqrt{1+\cos h})}$$

$$=\lim_{h\to 0}\frac{2\sin^2\frac{h}{2}}{\sqrt{2}\sin^2 h(\sqrt{2}+\sqrt{1+\cos h})}$$

$$=\lim_{h\to 0}\frac{2\frac{\sin^2\frac{h}{2}}{h^2/4}\times\frac{h^2}{4}}{\sqrt{2}\frac{\sin^2 h}{h^2}\times h^2(\sqrt{2}+\sqrt{1+\cos h})}$$

$$=\frac{1}{2\sqrt{2}(\sqrt{2}+\sqrt{1+1})}$$

$$=\frac{1}{2\sqrt{2}(\sqrt{2}+\sqrt{2})}$$

$$=\frac{1}{2\sqrt{2}}\cdot\frac{1}{2\sqrt{2}}=\frac{1}{8}$$

(vi) $\lim_{x\to\frac{\pi}{6}}\frac{2\sin^2 x+\sin x-1}{2\sin^2 x-3\sin x+1}$

$$=\lim_{x\to\frac{\pi}{6}}\frac{2\sin^2 x+2\sin x-\sin x-1}{2\sin^2 x-2\sin x-\sin x+1}$$

$$=\lim_{x\to\frac{\pi}{6}}\frac{2\sin x(\sin x+1)-1(\sin x+1)}{2\sin x(\sin x-1)-1(\sin x-1)}$$

$$=\lim_{x\to\frac{\pi}{6}}\frac{(2\sin x-1)(\sin x+1)}{(2\sin x-1)(\sin x-1)}=\frac{\frac{1}{2}+1}{\frac{1}{2}-1}=\frac{\frac{3}{2}}{\frac{-1}{2}}=-3$$

(vii) $\lim_{x\to\infty}\frac{\sin\left(\frac{a}{b^x}\right)}{\left(\frac{a}{b^x}\right)}\times\left(\frac{a}{b^x}\right)\times\frac{\left(\frac{b}{a^x}\right)}{\sin\left(\frac{b}{a^x}\right)\times\left(\frac{b}{a^x}\right)}$

$$=\frac{a}{b}\lim_{x\to\infty}\left(\frac{a}{b}\right)^x=\frac{a}{b}(0)$$

$$\left[\text{when } a<b,\left(\frac{a}{b}\right)<1,\text{ so }\left(\frac{a}{b}\right)^\infty=0\right]$$

$$=0$$

(viii) $\lim_{x\to 3}\frac{x^2-4x+3}{x^2-9-\sin(x-3)}$

$$=\lim_{x\to 3}\frac{x^2-3x-x+3}{(x+3)(x-3)-\sin(x-3)}$$

$$=\lim_{x\to 3}\frac{(x-3)(x-1)}{(x+3)(x-3)-\sin(x-3)}$$

$$=\lim_{x\to 3}\frac{(x-3)(x-1)}{(x-3)\left[(x+3)-\frac{\sin(x-3)}{x-3}\right]}$$

$$=\frac{3-1}{(3+3)-1}=\frac{2}{5}$$

(ix) $\lim_{x\to\pi}\dfrac{1-\sin x+\cos x}{1+\sin x+\cos x}$

By L-Hospital rule,

$$=\lim_{x\to\pi}\frac{-\cos x-\sin x}{\cos x-\sin x}=\frac{-(-1)-0}{(-1)-0}=-1$$

(x) $\lim_{x\to 2}\dfrac{(\cos\alpha)^x+(\sin\alpha)^x-1}{x-2}$

$$=\lim_{x\to 2}\frac{(\cos\alpha)^x+(\sin\alpha)^x-\sin^2\alpha-\cos^2\alpha}{x-2}$$

$$=\lim_{x\to 2}\frac{\cos^2\alpha\{(\cos\alpha)^{x-2}-1\}+\sin^2\alpha\{(\sin\alpha)^{x-2}-1\}}{(x-2)}$$

$$=\lim_{x\to 2}\left[\cos^2\alpha\left(\frac{(\cos\alpha)^{x-2}-1}{x-2}\right)+\sin^2\alpha\left(\frac{(\sin\alpha)^{x-2}-1}{x-2}\right)\right]$$

$$=\cos^2\alpha\ln\cos\alpha+\sin^2\alpha\ln\sin\alpha$$

4. (i) $\lim_{x\to 0}\dfrac{e^{2+x}-e^2}{x}=\lim_{x\to 0}\dfrac{e^2\cdot e^x-e^2}{x}$

$$=\lim_{x\to 0}\frac{e^2(e^x-1)}{x}=e^2$$

(ii) $\lim_{x\to 0}\left(\dfrac{e^{ax}-e^{bx}}{x}\right)=\lim_{x\to 0}\left(\dfrac{e^{ax}-1-e^{bx}+1}{x}\right)$

$$=\lim_{x\to 0}\left(\frac{e^{ax}-1}{x}-\frac{e^{bx}-1}{x}\right)$$

$$=\lim_{x\to 0}\left(\frac{e^{ax}-1}{ax}\times a-\frac{e^{bx}-1}{bx}\times b\right)=a-b$$

(iii) $\lim_{x\to 0}\left(\dfrac{2^{2+x}-4}{x}\right)=\lim_{x\to 0}\left\{4\left(\dfrac{2^x-1}{x}\right)\right\}$

$$=4\ln 2\qquad\left[\because\lim_{x\to 0}\frac{a^x-1}{x}=\ln a\right]$$

(iv) $\lim_{x\to 0}\dfrac{\log(1+3x)}{3^x-1}=\lim_{x\to 0}\dfrac{\log(1+3x)}{3x}\cdot\dfrac{3x}{3^x-1}$

$$=3\cdot\frac{1}{\log 3}=\frac{3}{\log 3}$$

(v) $\lim_{x\to 0}\dfrac{\log_3(1+2x)}{x}$

$$=\lim_{x\to 0}\frac{\log_e(1+2x)}{\log_e 3\cdot x}\times\frac{2}{2}\qquad\left[\because\log_a b=\frac{\log_e b}{\log_e a}\right]$$

$$=\frac{2}{\log_e 3}=2\log_3 e\qquad\left[\because\log_a b=\frac{1}{\log_b a}\right]$$

(vi) $\lim_{x\to 0}\dfrac{a^x+b^x-c^x-d^x}{x}$

$$=\lim_{x\to 0}\frac{a^x-1+b^x-1-c^x+1-d^x+1}{x}$$

$$=\lim_{x\to 0}\frac{a^x-1}{x}+\lim_{x\to 0}\frac{b^x-1}{x}-\lim_{x\to 0}\frac{c^x-1}{x}-\lim_{x\to 0}\frac{d^x-1}{x}$$

$$=\ln a+\ln b-\ln c-\ln d=\ln\left(\frac{ab}{cd}\right)$$

(vii) $\lim_{x\to a}\dfrac{\log x-\log a}{x-a}=\lim_{x\to a}\dfrac{\log\left(\frac{x}{a}\right)}{a\left(\frac{x}{a}-1\right)}$

$$=\lim_{x\to a}\frac{\log\left\{1+\left(\frac{x}{a}-1\right)\right\}}{a\left(\frac{x}{a}-1\right)}=\frac{1}{a}$$

(viii) $\lim_{x\to 0}\dfrac{\log\left(\frac{a+x}{a-x}\right)}{x}=\lim_{x\to 0}\dfrac{\log\left\{1+\left(\frac{a+x}{a-x}-1\right)\right\}}{x}$

$$=\lim_{x\to 0}\frac{\log\left(1+\frac{2x}{a-x}\right)}{x}$$

$$=\lim_{x\to 0}\frac{\log\left(1+\frac{2x}{a-x}\right)}{\frac{2x}{a-x}}\times\frac{2}{a-x}=\frac{2}{a}$$

(ix) $\lim_{x\to\frac{\pi}{2}}\dfrac{a^{\cos x}(a^{\cot x-\cos x}-1)}{\cot x-\cos x}=1\times\log a=\log a$

(x) $\lim_{x\to 0}\left(\dfrac{e^{x^2}-\cos x}{\sin^2 x}\right)=\lim_{x\to 0}\left(\dfrac{e^{x^2}-1+1-\cos x}{\sin^2 x}\right)$

$$=\lim_{x\to 0}\left(\frac{e^{x^2}-1}{\sin^2 x}+\frac{1-\cos x}{1-\cos^2 x}\right)$$

$$=\lim_{x\to 0}\left(\frac{\frac{e^{x^2}-1}{x^2}}{\left(\frac{\sin^2 x}{x^2}\right)}+\frac{1-\cos x}{(1+\cos x)(1-\cos x)}\right)$$

$$=1+\frac{1}{(1+1)}=1+\frac{1}{2}=\frac{3}{2}$$

5. (i) $\lim_{x\to\infty}\left(\dfrac{2x-4}{2x+2}\right)^{\frac{3x+2}{3}}$ $[1^\infty$ form$]$

$$=e^{\lim_{x\to\infty}\left(\frac{2x-4}{2x+2}-1\right)\left(\frac{3x+2}{3}\right)}$$

$$=e^{\lim_{x\to\infty}\left(\frac{2x-4-2x-2}{2x+2}\right)\left(\frac{3x+2}{3}\right)}$$

$$=e^{\lim_{x\to\infty}\left(\frac{-6}{2x+2}\times\frac{3x+2}{3}\right)}=e^{\lim_{x\to\infty}\left(\frac{-6x-4}{2x+2}\right)}=e^{-3}$$

(ii) $\lim_{x\to 0}(\cos x+\sin x)^{\frac{1}{x}}$

$$= e^{\lim_{x\to 0}(\cos x+\sin x-1)\frac{1}{x}}$$

$$= e^{\lim_{x\to 0}\frac{\{\sin x-(1-\cos x)\}}{x}}$$

$$= e^{\lim_{x\to 0}\left(\frac{\sin x}{x}-\frac{2\sin^2 x/2}{x}\right)}$$

$$= e^{(1-0)} = e$$

(iii) $\lim_{x\to 0}\left(\frac{\sin x}{x}\right)^{\frac{\sin x}{x-\sin x}}$

$$= e^{\lim_{x\to 0}\left(\frac{\sin x}{x}-1\right)\left(\frac{\sin x}{x-\sin x}\right)} = e^{\lim_{x\to 0}\left(\frac{\sin x-x}{x}\right)\cdot\frac{\sin x}{(x-\sin x)}}$$

$$= e^{\lim_{x\to 0}\left(-\frac{\sin x}{x}\right)} = e^{-1} = \frac{1}{e}$$

(iv) $\lim_{x\to 0}\left(\frac{a^x+b^x}{2}\right)^{\frac{3}{x}} = e^{\lim_{x\to 0}\left(\frac{a^x+b^x}{2}-1\right)\frac{3}{x}}$

$$= e^{\frac{3}{2}\lim_{x\to 0}\left(\frac{a^x-1+b^x-1}{x}\right)}$$

$$= e^{\frac{3}{2}\lim_{x\to 0}\left(\frac{a^x-1}{x}+\frac{b^x-1}{x}\right)} = e^{\frac{3}{2}(\ln a+\ln b)}$$

$$= e^{\frac{3}{2}(\ln ab)} = (ab)^{\frac{3}{2}}$$

(v) $e^{\lim_{x\to 7/2}(2x^2-9x+8-1)\cot(2x-7)}$

$$= e^{\lim_{x\to 7/2}\frac{(2x^2-9x+7)}{\tan(2x-7)}} = e^{\lim_{x\to 7/2}\frac{(2x-7)(x-1)}{\tan(2x-7)}}$$

$$= e^{\left(\frac{7}{2}-1\right)} = e^{\frac{5}{2}}$$

6 (i) $\lim_{x\to 0}\frac{\int_0^{x^2}\cos t^2\,dt}{x\sin x}$

$$= \lim_{x\to 0}\frac{\int_0^{x^2}\cos t^2\,dt}{x^2\frac{\sin x}{x}}$$

$$= \lim_{x\to 0}\frac{\int_0^{x^2}\cos t^2\,dt}{x^2}\times\lim_{x\to 0}\frac{x}{\sin x}$$

$$\lim_{x\to 0}\frac{\int_0^{x^2}\cos t^2\,dt}{x^2}\times 1$$

Applying L'hospital rule, we get

$$\lim_{x\to 0}\frac{\cos x^4\cdot 2x}{2x}$$

$$\Rightarrow \quad \cos 0 = 1$$

(ii) $\lim_{x\to 0}\frac{\int_0^{x}\cos t\,dt}{x}$

Using L'hospital rule, we get

$$\lim_{x\to 0}\frac{\cos x}{1} = 1$$

(iii) $\lim_{n\to\infty}\frac{[n\sqrt{2}]}{n}$

We know, $\quad x-1\le[x]<x$

$\because \quad n\sqrt{2}-1\le[n\sqrt{2}]<n\sqrt{2}$

$$\Rightarrow \quad \frac{n\sqrt{2}-1}{n}\le\frac{[n\sqrt{2}]}{n}<\frac{n\sqrt{2}}{n}$$

$$\Rightarrow \quad \sqrt{2}-\frac{1}{n}\le\frac{[n\sqrt{2}]}{n}<\sqrt{2}$$

By Sandwich theorem,

$$\lim_{n\to\infty}\left(\sqrt{2}-\frac{1}{n}\right)\le\lim_{n\to\infty}\frac{[n\sqrt{2}]}{n}<\lim_{n\to\infty}\sqrt{2}$$

$$\Rightarrow \quad \sqrt{2}\le\lim_{n\to\infty}\frac{[n\sqrt{2}]}{n}<\sqrt{2}$$

$$\Rightarrow \quad \lim_{x\to\infty}\frac{[n\sqrt{2}]}{2}=\sqrt{2}$$

(iv) $\lim_{n\to\infty}\left(\frac{1}{1+n^2}+\frac{3}{2+n^2}+\frac{5}{3+n^2}+\ldots+\frac{2n-1}{n+n^2}\right)$

$$\because \frac{1}{n+n^2}+\frac{3}{n+n^2}+\ldots+\frac{2n-1}{n+n^2}<\frac{1}{1+n^2}+\frac{3}{2+n^2}+\ldots+\frac{2n-1}{n+n^2}+\ldots\frac{2n-1}{1+n^2}$$

$$\Rightarrow \frac{1+3+5+\ldots+2n-1}{n+n^2}<\frac{1}{1+n^2}+\frac{3}{2+n^2}+\ldots+\frac{2n-1}{n+n^2}<\frac{1}{1+n^2}+\frac{3}{1+n^2}+\frac{5}{1+n^2}+\ldots+\frac{2n-1}{1+n^2}$$

$$\Rightarrow \frac{n^2}{n+n^2}<\frac{1}{1+n^2}+\frac{3}{2+n^2}+\ldots+\frac{2n-1}{n+n^2}<\frac{n^2}{1+n^2}$$

$\therefore \quad \lim_{n\to\infty}\frac{n^2}{n+n^2}=1$ and $\lim_{n\to\infty}\frac{n^2}{1+n^2}=1$

$$\because \lim_{n\to\infty}\frac{1}{1+n^2}+\frac{3}{2+n^2}+\ldots+\frac{2n-1}{n+n^2}=1$$

Level 2

1. (c) $\lim_{x\to 3^+}\left\{\frac{[x]^2}{2}-\left[\frac{x}{5}\right]^2\right\}$

$$= \lim_{x\to 3^+}\frac{[x]^2}{2}-\lim_{x\to 3^+}\left[\frac{x}{5}\right]^2=\frac{9}{2}-0=\frac{9}{2}$$

2. (c) We have, $\lim_{x\to 1}\dfrac{\left(\sum_{k=1}^{50}x^k\right)-50}{x-1}$

$$= \lim_{x\to 1}\frac{(x+x^2+\ldots+x^{50})-50}{x-1}$$

$$= \lim_{x\to 1}\left\{\frac{x-1}{x-1}+\frac{x^2-1}{x-1}+\ldots+\frac{x^{50}-1}{x-1}\right\}$$

$$=1+2+3+\ldots+50$$

$$=\frac{50(51)}{2}=1275$$

3. (c) We have, $\lim_{x\to a}\dfrac{(a+2x)^{\frac{1}{3}}-(3x)^{\frac{1}{3}}}{(3a+x)^{\frac{1}{3}}-(4x)^{\frac{1}{3}}}$

$$= \lim_{x\to a}\frac{\dfrac{(a+2x)-(3x)}{(a+2x)^{\frac{2}{3}}+(3x)^{\frac{2}{3}}+(a+2x)^{\frac{1}{3}}(3x)^{\frac{1}{3}}}}{\dfrac{(3a+x)-(4x)}{(3a+x)^{\frac{2}{3}}+(4x)^{\frac{2}{3}}+(3a+x)^{\frac{1}{3}}(4x)^{\frac{1}{3}}}}$$

$$\left[\begin{array}{l}\because a^3-b^3=(a-b)(a^2+b^2+ab)\\ \Rightarrow (a-b)=\dfrac{a^3-b^3}{a^2+b^2+ab}\end{array}\right]$$

$$= \lim_{x\to a}\frac{\dfrac{(a-x)}{(a+2x)^{\frac{2}{3}}+(3x)^{\frac{2}{3}}+(a+2x)^{\frac{1}{3}}(3x)^{\frac{1}{3}}}}{\dfrac{3(a-x)}{(3a+x)^{\frac{2}{3}}+(4x)^{\frac{2}{3}}+(3a+x)^{\frac{1}{3}}(4x)^{\frac{1}{3}}}}$$

$$=\frac{\dfrac{1}{\dfrac{3(3a)^{\frac{2}{3}}}{3}}}{3(4a)^{\frac{2}{3}}}=\frac{(4)^{\frac{2}{3}}}{3(3)^{\frac{2}{3}}}=\frac{2^{\frac{4}{3}}}{3\cdot(3)^{\frac{2}{3}}}=\left(\frac{2}{3}\right)\left(\frac{2}{9}\right)^{\frac{1}{3}}$$

4. (b) We have, $\lim_{x\to\infty}\left[\sqrt{x+\sqrt{x+\sqrt{x}}}-\sqrt{x}\right]$

$$= \lim_{x\to\infty}\frac{x+\sqrt{x+\sqrt{x}}-x}{\sqrt{x+\sqrt{x+\sqrt{x}}}+\sqrt{x}}$$

$$= \lim_{x\to\infty}\frac{\sqrt{x+\sqrt{x}}}{\sqrt{x+\sqrt{x+\sqrt{x}}}+\sqrt{x}}$$

$$= \lim_{x\to\infty}\frac{\sqrt{x}(\sqrt{1+x^{-1/2}})}{\sqrt{x}\left[\left(\sqrt{1+\sqrt{x^{-1}+x^{-3/2}}}\right)+1\right]}$$

$$=\frac{1}{1+1}=\frac{1}{2}$$

5. (b) We have, $\lim_{x\to 0}\dfrac{\sqrt{a^2-ax+x^2}-\sqrt{a^2+ax+x^2}}{\sqrt{a+x}-\sqrt{a-x}}$

$$= \lim_{x\to 0}\frac{\sqrt{a^2-ax+x^2}-\sqrt{a^2+ax+x^2}}{\sqrt{a+x}-\sqrt{a-x}}\times\frac{\sqrt{a^2-ax+x^2}+\sqrt{a^2+ax+x^2}}{\sqrt{a^2-ax+x^2}+\sqrt{a^2+ax+x^2}}\times\frac{\sqrt{a+x}+\sqrt{a-x}}{\sqrt{a+x}+\sqrt{a-x}}$$

$$= \lim_{x\to 0}\frac{(a^2-ax+x^2)-(a^2+ax+x^2)}{(a+x)-(a-x)}\cdot\frac{\sqrt{a+x}+\sqrt{a-x}}{\sqrt{a^2-ax+x^2}+\sqrt{a^2+ax+x^2}}$$

$$= \lim_{x\to 0}\frac{-2ax}{2x}\times\left(\frac{\sqrt{a+x}+\sqrt{a-x}}{\sqrt{a^2-ax+x^2}+\sqrt{a^2+ax+x^2}}\right)$$

$$=\frac{-a(2\sqrt{a})}{2a}=-\sqrt{a}$$

6. (a) We have, $\lim_{x\to 2}\dfrac{\sqrt{x+7}-3\sqrt{2x-3}}{\sqrt[3]{x+6}-2\sqrt[3]{3x-5}}$

Let $x-2=t$, if $x\to 2\Rightarrow t\to 0$

$$= \lim_{t\to 0}\frac{(t+9)^{1/2}-3(2t+1)^{1/2}}{(t+8)^{1/3}-2(3t+1)^{1/3}}$$

$$= \lim_{t\to 0}\frac{3\left(1+\dfrac{t}{9}\right)^{\frac{1}{2}}-3(1+2t)^{\frac{1}{2}}}{2\left(1+\dfrac{t}{8}\right)^{\frac{1}{3}}-2(1+3t)^{\frac{1}{3}}}$$

$$= \lim_{t\to 0}\frac{3\left[\left(1+\dfrac{t}{18}\right)-(1+t)\right]}{2\left[\left(1+\dfrac{t}{24}\right)-(1+t)\right]}\qquad [\because (1+x)^n=1+nx]$$

$$=\frac{3}{2}\lim_{t\to 0}\frac{t\left(\dfrac{1}{18}-1\right)}{t\left(\dfrac{1}{24}-1\right)}=\frac{3\left(\dfrac{-17}{18}\right)}{2\left(\dfrac{-23}{24}\right)}=\frac{34}{23}$$

7. (a) We have,

$$\lim_{x\to 1}\left[\left(\frac{4}{x^2-x^{-1}}-\frac{x^2-3x+1}{1-x^3}\right)^{-1}+3\frac{x^4-1}{x^3-x^{-1}}\right]$$

$$= \lim_{x\to 1}\left[\left(\frac{4x}{x^3-1}+\frac{x^2-3x+1}{x^3-1}\right)^{-1}+\frac{3x(x^4-1)}{x^4-1}\right]$$

$$= \lim_{x\to 1}\left[\left(\frac{x^2+x+1}{x^3-1}\right)^{-1}+3x\right]$$

$$= \lim_{x\to 1}\left[\left(\frac{x^2+x+1}{(x-1)(x^2+x+1)}\right)^{-1}+3x\right]$$

$$= \lim_{x\to 1}(x-1+3x)\ =3$$

8. (b) We have, $\lim_{x\to -\infty} \sqrt{x^2-2x} - ax - b = 0$

$$\Rightarrow \lim_{x\to \infty} \sqrt{x^2+2x} + ax - b = 0$$

$$\left[\because \lim_{x\to -\infty} f(x) = \lim_{x\to \infty} f(-x)\right]$$

$$\Rightarrow \lim_{x\to \infty} x\left(1+\frac{2}{x}\right)^{\frac{1}{2}} + ax - b = 0$$

$$\Rightarrow \lim_{x\to \infty} x\left(1+\frac{2}{2x}\right) + ax - b = 0 \quad \left[\begin{array}{l}\text{if } x\to 0,\\ (1+x)^n = 1+nx\end{array}\right]$$

$$\Rightarrow \lim_{x\to \infty} x + 1 + ax - b = 0$$

$$\Rightarrow \lim_{x\to \infty} x(1+a) + 1 - b = 0$$

For finite limit, $1+a=0 \Rightarrow a=-1$

and $1-b=0 \Rightarrow b=1$

9. (c) We have,

$$\lim_{x\to \infty} \{\sqrt{ax^4+bx^3+cx^2+dx+2} - \sqrt{ax^4+3x^3-4x^2+2x+e}\}$$

$$= \lim_{x\to \infty} \frac{(ax^4+bx^3+cx^2+dx+2)-(ax^4+3x^3-4x^2+2x+e)}{\sqrt{ax^4+bx^3+cx^2+dx+2}+\sqrt{ax^4+3x^3-4x^2+2x+e}}$$

$$= \lim_{x\to \infty} \frac{(b-3)x^3+(c+4)x^2+(d-2)x+(2-e)}{\sqrt{ax^4+bx^3+cx^2+dx+2}+\sqrt{ax^4+3x^3-4x^2+2x+e}}$$

The degree of numerator is 3 and the degree of denominator is 2. Therefore, for the limit to be finite, we must have,

$$b-3=0 \Rightarrow b=3$$

10. (b) We have, $\lim_{x\to -\infty} [(x^5+7x^4-5)^k - x]$

$$\Rightarrow \lim_{x\to -\infty} \left[x^{5k}\left(1+\frac{7}{x}-\frac{5}{x^5}\right)^k - x\right]$$

$$\Rightarrow \lim_{x\to -\infty} x\left[x^{5k-1}\left(1+\frac{7}{x}-\frac{5}{x^5}\right)^k - 1\right]$$

For a finite value, $5k-1=0$

$$k=\frac{1}{5}$$

Now, $\lim_{x\to -\infty} x\left[(1+X)^{\frac{1}{5}} - 1\right]$, where $X = \frac{7}{x} - \frac{5}{x^5}$

$$\lim_{x\to -\infty} x\left[1+\frac{1}{5}X+\ldots-1\right] = \lim_{x\to -\infty} x\left[\frac{1}{5}\left(\frac{7}{x}-\frac{5}{x^5}\right)\right]$$

$$\Rightarrow \lim_{x\to -\infty} \frac{1}{5}\left[7-\frac{5}{x^4}\right] = \frac{7}{5}$$

11. (b) Let $1+a=t^6$

$$(t^2-1)x^2+(t^3-1)x+(t-1)=0$$

$$\Rightarrow (t+1)x^2+(t^2+t+1)x+1=0$$

When $a\to 0$, then $t\to 1$

$$\Rightarrow 2x^2+3x+1=0$$

$$\Rightarrow 2x^2+2x+x+1=0$$

$$\Rightarrow (2x+1)(x+1)=0$$

So, $x=\frac{-1}{2}$ and $x=-1$

Hence, $\lim_{a\to 0^+} \alpha(a) = \frac{-1}{2}$ and $\lim_{a\to 0^+} \beta(a) = -1$

12. (b) We have,

$$\lim_{x\to 0^+} x\left(\left[\frac{1}{x}\right]+\left[\frac{2}{x}\right]+\left[\frac{3}{x}\right]+\ldots+\left[\frac{15}{x}\right]\right)$$

$$\Rightarrow \lim_{x\to 0^+} x\left(\frac{1}{x}-\left\{\frac{1}{x}\right\}+\frac{2}{x}-\left\{\frac{2}{x}\right\}+\ldots+\frac{15}{x}-\left\{\frac{15}{x}\right\}\right)$$

$$[\because [x] = x - \{x\}]$$

$$\lim_{x\to 0^+} x\left[\left(\frac{1}{x}+\frac{2}{x}+\frac{3}{x}+\ldots+\frac{15}{x}\right) - \left(\left\{\frac{1}{x}\right\}+\left\{\frac{2}{x}\right\}+\ldots+\left\{\frac{15}{x}\right\}\right)\right]$$

$$\lim_{x\to 0^+} \left[x\left(\frac{1+2+3+\ldots+15}{x}\right) - x\left(\left\{\frac{1}{x}\right\}+\left\{\frac{2}{x}\right\}+\ldots+\frac{15}{x}\right)\right]$$

$$=(1+2+3+\ldots+15)-0$$

$$\left[\because \left\{\frac{1}{x}\right\} \text{is fraction part function, so } 0 \le \left\{\frac{1}{x}\right\} < 1\right]$$

$$=\frac{15(15+1)}{2}=120$$

13. (b) If $\lim_{x\to 0} \left|\frac{1-x+|x|}{\lambda - x + [x]}\right| = L$ exists, then LHL = RHL

$$\text{RHL} = \lim_{h\to 0} \left|\frac{1-(0+h)+|0+h|}{\lambda-(0+h)+[0+h]}\right| = \lim_{h\to 0} \left|\frac{1-h+h}{\lambda-h+0}\right| = \frac{1}{|\lambda|}$$

$$\text{LHL} = \lim_{h\to 0} \left|\frac{1-(0-h)+|0-h|}{\lambda-(0-h)+[0-h]}\right|$$

$$= \lim_{h\to 0} \left|\frac{1+h+h}{\lambda+h-1}\right| = \frac{1}{|\lambda-1|}$$

So, LHL = RHL

$$\frac{1}{|\lambda|} = \frac{1}{|\lambda-1|} \Rightarrow \lambda = \frac{1}{2}$$

Then, $\lim_{x\to 0} \left|\frac{1-x+|x|}{\lambda-x+[x]}\right| = 2$

14. (c) We have, $\lim_{n\to \infty} \frac{n\cdot 5^n}{n(x-3)^n + n\cdot 5^{n+1} - 5^n} = \frac{1}{5}$

$$\Rightarrow \lim_{n\to \infty} \frac{n\cdot 5^n}{n\cdot 5^n\left[\left(\frac{x-3}{5}\right)^n + 5 - \frac{1}{n}\right]} = \frac{1}{5}$$

$$\Rightarrow \quad \lim_{n\to\infty} \frac{1}{\left(\frac{x-3}{5}\right)^n + 5 - \frac{1}{n}} = \frac{1}{5}$$

This will be possible only when

$$\Rightarrow \quad -1 < \frac{x-3}{5} < 1$$

$$\Rightarrow \quad -5 < x-3 < 5$$

$$\Rightarrow \quad -2 < x < 8$$

Hence, the possible values of x are

−1, 0, 1, 2, 3, 4, 5, 6 and 7.

15. (d) We have, $\lim_{x\to 3} \frac{xf(3) - 3f(x)}{x-3}$

$$= \lim_{x\to 3} \frac{xf(3) - 3f(3) + 3f(3) - 3f(x)}{x-3}$$

$$= \lim_{x\to 3} \frac{-3f(x) + 3f(3) + xf(3) - 3f(3)}{x-3}$$

$$= \lim_{x\to 3} -3\left(\frac{f(x) - f(3)}{x-3}\right) + \lim_{x\to 3} \frac{f(3)(x-3)}{x-3}$$

$$= -3 \lim_{x\to 3}\left(\frac{f(x) - f(3)}{x-3}\right) + f(3)$$

$$= -3f'(x) + f(3)$$

$$= -3\times 3 + 3 = -9 + 3 = -6$$

16. (a) We have, $f(x) = 5 - |x-2|$

Here, $f(x)$ attains maximum value at $x = 2$.

So, $\alpha = 2$ and $g(x) = |x+1|$

$g(x)$ attains minimum value at $x = -1$

So, $\beta = -1$

So, $\alpha\beta = 2(-1) = -2$

So, $\lim_{x\to -\alpha\beta} \frac{(x-1)(x^2-5x+6)}{(x^2-6x+8)} = \lim_{x\to 2} \frac{(x-1)(x-2)(x-3)}{(x-4)(x-2)}$

$$= \frac{(2-1)(2-3)}{(2-4)} = \frac{-1}{-2} = \frac{1}{2}$$

17. (a) $\lim_{x\to 0} \frac{x^a \sin^b x}{\sin(x^c)} = \lim_{x\to 0} x^a \left(\frac{\sin x}{x}\right)^b \times \left(\frac{x^c}{\sin x^c}\right) \cdot x^{b-c}$

$$= \lim_{x\to 0} x^{a+b-c}$$

The limit will have non-zero value of $a+b-c = 0$

$$\Rightarrow \quad a + b = c$$

18. (d) $\lim_{x\to 0} \frac{(2\sin^2 2x)(2-\cos x)}{2x^2\left(\frac{\tan 2x}{2x}\right)}$

$$= \lim_{x\to 0} \frac{\left(2\times\frac{\sin 2x}{2x}\right)^2 (2-\cos x)}{\left(\frac{\tan 2x}{2x}\right)}$$

$$= 4\times 1 = 4$$

19. (d) $\lim_{x\to 1} \frac{\sqrt{1-\cos 2(x-1)}}{x-1} = \lim_{x\to 1} \frac{\sqrt{2}\,|\sin(x-1)|}{x-1}$

Now, $\lim_{x\to 1^+} \frac{\sqrt{2}\,|\sin(x-1)|}{x-1} = \lim_{x\to 1^+} \frac{\sqrt{2}\sin(x-1)}{x-1} = \sqrt{2}$

and $\lim_{x\to 1^-} \frac{\sqrt{2}\,|\sin(x-1)|}{x-1} = \lim_{x\to 1^-} \frac{-\sqrt{2}\sin(x-1)}{x-1} = -\sqrt{2}$

So, $\lim_{x\to 1^+} f(x) \neq \lim_{x\to 1^-} f(x)$

Limit does not exists.

20. (a) $\lim_{n\to\infty} \cos(\pi\sqrt{n^2+n}) = \lim_{n\to\infty} (-1)^n \cos(n\pi - \pi\sqrt{n^2+n})$

$$= \lim_{n\to\infty} (-1)^n \cos\{\pi(n - \sqrt{n^2+n})\}$$

$$= \lim_{n\to\infty} (-1)^n \cos\left(\frac{-n\pi}{n+\sqrt{n^2+n}}\right)$$

$$= \lim_{n\to\infty} (-1)^n \cos\left(\frac{\pi}{1+\sqrt{1+\frac{1}{n}}}\right)$$

$$= (-1)^n \cos\frac{\pi}{2} = 0$$

21. (c) $\lim_{x\to 0} \frac{\cos\left(\frac{\pi}{2}\cos^2 x\right)}{x^2} = \lim_{x\to 0} \frac{\cos\left(\frac{\pi}{2}(1-\sin^2 x)\right)}{x^2}$

$$= \lim_{x\to 0} \frac{\cos\left(\frac{\pi}{2} - \frac{\pi}{2}\sin^2 x\right)}{x^2}$$

$$= \lim_{x\to 0} \frac{\sin\left(\frac{\pi}{2}\sin^2 x\right)}{\frac{\pi}{2}\sin^2 x} \times \frac{\pi}{2} \times \frac{\sin^2 x}{x^2}$$

$$= \frac{\pi}{2}$$

22. (b) $\lim_{x\to 0} \frac{\tan^2 x - x^2}{x^2\tan^2 x} = \lim_{x\to 0} \frac{(\tan x - x)(\tan x + x)}{x^4\left(\frac{\tan^2 x}{x^2}\right)}$

$$= \lim_{x\to 0} \left(\frac{\tan x - x}{x^3}\right) \cdot \frac{\tan x + x}{x\left(\frac{\tan^2 x}{x^2}\right)}$$

$$= \lim_{x\to 0} \left(\frac{x + \frac{x^3}{3} + \frac{2x^5}{15} + \ldots - x}{x^3}\right) \cdot \frac{1}{\left(\frac{\tan^2 x}{x^2}\right)} \left(\frac{\tan x}{x} + 1\right)$$

$$= \frac{1}{3} \cdot (1)(1+1) = \frac{2}{3}$$

23. (c) We have, $\lim_{x\to 0}\frac{(\cos x-1)(\cos x-e^x)}{x^n}$

$$=\lim_{x\to 0}-\frac{(1-\cos x)(\cos x-e^x)}{x^2\cdot x^{n-2}}$$

$$=\lim_{x\to 0}\frac{-(\cos x-e^x)}{2x^{n-2}}\qquad\left[\because \lim_{x\to 0}\frac{1-\cos x}{x^2}=\frac{1}{2}\right]$$

Applying L-Hospital rule,

$$\lim_{x\to 0}\frac{-(-\sin x-e^x)}{2(n-2)x^{n-3}}=\lim_{x\to 0}\frac{\sin x+e^x}{2(n-2)x^{n-3}}$$

The limit is finite, if $n-3=0\Rightarrow n=3$

24. (b) We have, $\lim_{x\to\infty}\left(\frac{x^3\sin\frac{1}{x}+2x+3}{x^2-2x+1}\right)=l$

$$\Rightarrow\quad \lim_{x\to\infty}\frac{\left(\frac{\sin 1/x}{1/x}\times x^2+2x+3\right)}{x^2-2x+1}=l$$

$$\Rightarrow\quad \lim_{x\to\infty}\left(\frac{\left(\frac{\sin 1/x}{1/x}\right)+\frac{2}{x}+\frac{3}{x^2}}{1-\frac{2}{x}+\frac{1}{x^2}}\right)=l$$

$$\Rightarrow\quad l=1$$

Hence, $\sqrt{l\sqrt{l\sqrt{l\ldots\infty}}}=\sqrt{l\sqrt{l\sqrt{l\ldots\infty}}}=1$

25. (a) $\lim_{x\to\alpha}\frac{1-\cos a(x-\alpha)(x-\beta)}{(x-\alpha)^2}$

$$=\lim_{x\to\alpha}\frac{2\sin^2\left(\frac{a}{2}(x-\alpha)(x-\beta)\right)}{(x-\alpha)^2}$$

$$=\lim_{x\to\alpha}\frac{2\sin^2\left(\frac{a}{2}(x-\alpha)(x-\beta)\right)}{\left(\frac{a}{2}\right)^2(x-\alpha)^2(x-\beta)^2}\times(x-\beta)^2\times\left(\frac{a}{2}\right)^2$$

$$=\frac{a^2}{2}(\alpha-\beta)^2$$

26. (c) $\lim_{x\to 0}\frac{1-\cos x\sqrt{\cos 2x}}{x^2}\times\frac{1+\cos x\sqrt{\cos 2x}}{1+\cos x\sqrt{\cos 2x}}$

$$=\lim_{x\to 0}\frac{1-\cos^2 x\cos 2x}{x^2[1+\cos x\sqrt{\cos 2x}]}$$

$$=\lim_{x\to 0}\frac{\sin^2 x+\cos^2 x-\cos^2 x\cos 2x}{x^2[1+\cos x\sqrt{\cos 2x}]}$$

$$=\lim_{x\to 0}\frac{\sin^2 x+\cos^2 x(1-\cos 2x)}{x^2[1+\cos x\sqrt{\cos 2x}]}$$

$$=\lim_{x\to 0}\frac{\sin^2 x+\cos^2 x(2\sin^2 x)}{x^2(1+\cos x\sqrt{\cos 2x})}$$

$$=\lim_{x\to 0}\frac{\sin^2 x(1+2\cos^2 x)}{x^2(1+\cos x\sqrt{\cos 2x})}=\frac{3}{1+1}=\frac{3}{2}$$

27. (a) $\lim_{x\to 0}\frac{(1+\sin x)^{\frac{1}{4}}-(1-\sin x)^{\frac{1}{4}}}{x}$

$$=\lim_{x\to 0}\frac{[(1+\sin x)^{1/4}-(1-\sin x)^{1/4}][(1+\sin x)^{1/4}+(1-\sin x)^{1/4}][1+\sin x]^{1/2}+(1-\sin x)^{1/2}}{x[(1+\sin x)^{1/4}+(1-\sin x)^{1/4}][(1+\sin x)^{1/2}+(1-\sin x)^{1/2}]}$$

$$=\lim_{x\to 0}\frac{(1+\sin x)-(1-\sin x)}{x[(1+\sin x)^{1/4}+(1-\sin x)^{1/4}][(1+\sin x)^{1/2}+(1-\sin x)^{1/2}]}$$

$$=\lim_{x\to 0}\frac{2\sin x}{x[(1+\sin x)^{1/4}+(1-\sin x)^{1/4}][(1+\sin x)^{1/2}+(1-\sin x)^{1/2}]}$$

$$=\frac{2}{2\times 2}=\frac{1}{2}$$

28. (d) $\lim_{x\to 0}\frac{\sin 2x-a\sin x}{\frac{x^3}{27}}=\lim_{x\to 0}\cdot\frac{2\sin x\cos x-a\sin x}{x^3}\times 27$

$$=\lim_{x\to 0}\left(\frac{\sin x}{x}\right)\frac{(2\cos x-a)}{x^2}\times 27$$

This limit to exist finitely, it must be $\left(\frac{0}{0}\right)$ form.

$$2\cos(0)-a=0\Rightarrow 2-a=0$$

$$a=2$$

So, $\quad L=\lim_{x\to 0}\left(\frac{\sin x}{x}\right)2\left(\frac{\cos x-1}{x^2}\right)\times 27$

$$L=1\times 2\times\left(\frac{-1}{2}\right)\times 27=-27$$

So, $\quad \left[\frac{L}{a}\right]=\left[\frac{-27}{2}\right]=[-13.5]=-14$

29. (d) In vicinity of $x=0, |\sin^{-1}x|>|x|$

$$\Rightarrow\frac{\sin^{-1}x}{x}>1,\text{ in vicinity of } x=0$$

$$\therefore\lim_{x\to 0}\left(\left[\frac{\sin^{-1}x}{x}\right]+\left[\frac{2^3\sin^{-1}2x}{2x}\right]+\ldots\left[\frac{n^3\sin^{-1}nx}{x}\right]\right)=100$$

$$\Rightarrow\quad 1^3+2^3+3^3+\ldots+n^3=100$$

$$\Rightarrow\quad\left(\frac{n(n+1)}{2}\right)^2=100\Rightarrow\frac{n(n+1)}{2}=10$$

$$\Rightarrow\quad n(n+1)=20\Rightarrow n=5$$

30. (d) As, $x\to 0^-, \tan x\to 0^-\Rightarrow[\tan x]=-1$

$$\Rightarrow\quad \sin^{-1}(-1)=-\frac{\pi}{2}$$

$$\therefore\quad \lim_{x\to 0^-}\sin^{-1}[\tan x]=-\frac{\pi}{2}$$

Hence, $\{l\}=\left\{-\frac{\pi}{2}\right\}=\left\{2-\frac{\pi}{2}-2\right\}=2-\frac{\pi}{2}$

31. (c) $\lim_{x\to 1^+} \frac{\sqrt{\pi}-\sqrt{\cos^{-1}x}}{\sqrt{x+1}}$

Put $\cos^{-1}x=\theta \Rightarrow x=\cos\theta$

$$\therefore \lim_{\theta\to\pi}\frac{\sqrt{\pi}-\sqrt{\theta}}{\sqrt{1+\cos\theta}}=\lim_{\theta\to\pi}\frac{\pi-\theta}{\sqrt{1+\cos\theta}}\left(\frac{1}{\sqrt{\pi}+\sqrt{\theta}}\right)$$

$$=\lim_{\theta\to\pi}\frac{\pi-\theta}{\sqrt{2\cos^2\frac{\theta}{2}}}\lim_{\theta\to\pi}\left(\frac{1}{\sqrt{\pi}+\sqrt{\theta}}\right)$$

$$=\lim_{k\to 0}\frac{-k}{\sqrt{2\cos^2\left(\frac{\pi}{2}+\frac{k}{2}\right)}\cdot 2\sqrt{\pi}} \quad [\text{putting } \theta=\pi+k]$$

$$=\lim_{k\to 0}\frac{-k}{2\sqrt{2\pi}\left|\cos\left(\frac{\pi}{2}+\frac{k}{2}\right)\right|}$$

$$=\lim_{k\to 0}\frac{k}{2\sqrt{2\pi}\left(\sin\frac{k}{2}\right)} \quad \left[\left|\cos\left(\frac{\pi}{2}+\frac{k}{2}\right)\right|=-\sin\frac{k}{2}\right]$$

$$=\lim_{k\to 0}\frac{\frac{k}{2}}{\sqrt{2\pi}\left(\sin\frac{k}{2}\right)}=\frac{1}{\sqrt{2\pi}}$$

32. (d) We have, $\lim_{x\to 0^-}\frac{\{x\}}{\sin\{x\}}=\lim_{x\to 0^-}\frac{x-[x]}{\sin(x-[x])}$

$$=\lim_{x\to 0^-}\frac{x+1}{\sin(x+1)} \quad [\text{when } x\to 0^-, [x]=-1]$$

$$=\frac{1}{\sin 1}=\text{cosec } 1$$

and $\lim_{x\to 0^+}\frac{\{x\}}{\sin\{x\}}=\lim_{x\to 0^+}\frac{x-[x]}{\sin(x-(x))}$

$$=\lim_{x\to 0^+}\frac{x}{\sin x} \quad [\text{when } x\to 0^+, [x]=0]$$

$$=1$$

So, $\lim_{x\to 0}\frac{\{x\}}{\sin\{x\}}$ does not exist.

33. (b) We know that

$$\lim_{x\to 0}\frac{\sin^{-1}x}{x}=1 \quad [\text{but a value is more than 1}]$$

and $$\lim_{x\to 0}\frac{\tan^{-1}x}{x}=1 \quad [\text{but a value is less than 1}]$$

So, $\lim_{x\to 0}\left(\left[100\frac{\sin^{-1}x}{x}\right]+\left[100\frac{\tan^{-1}x}{x}\right]\right)=100+99=199$

34. (d) $\lim_{x\to\infty}\frac{\cot^{-1}(x^{-a}\log_a x)}{\sec^{-1}(a^x\log_x a)}=\lim_{x\to\infty}\frac{\cot^{-1}\left(\frac{\log_a x}{x^a}\right)}{\sec^{-1}\left(\frac{a^x}{\log_a x}\right)}$

As, $x\to\infty, \left(\frac{\log_a x}{x}\right)\to 0$ and $x\to\infty, \left(\frac{a^x}{\log_a x}\right)\to\infty$

$$=\frac{\cot^{-1}(0)}{\sec^{-1}(\infty)}$$

$$=\frac{\pi/2}{\pi/2}=1$$

35. (c) We have,

$$\text{LHL}=\lim_{x\to\frac{\pi}{2}^-}\frac{\left[\frac{x}{2}\right]}{\log(\sin x)}=\lim_{h\to 0}\frac{\left[\frac{\pi}{4}-\frac{h}{2}\right]}{\log\sin\left(\frac{\pi}{2}-h\right)}$$

$$=\lim_{h\to 0}\frac{\left[\frac{\pi}{4}-\frac{h}{2}\right]}{\log_e\cosh}=\lim_{h\to 0}\frac{0}{\log_e\cosh}=0$$

$$\text{RHL}=\lim_{x\to\frac{\pi}{2}^+}\frac{\left[\frac{x}{2}\right]}{\log\sin x}=\lim_{h\to 0}\frac{\left[\frac{\pi}{4}+\frac{h}{2}\right]}{\log\left\{\sin\left(\frac{\pi}{2}+h\right)\right\}}$$

$$=\lim_{h\to 0}\frac{\left[\frac{\pi}{4}+\frac{h}{2}\right]}{\log\cosh}=\lim_{h\to 0}\frac{0}{\log\cosh}=0$$

Hence, $\lim_{x\to\frac{\pi}{2}}\frac{\left[\frac{x}{2}\right]}{\log(\sin x)}=0$

36. (b) Let $A=\cos\left(\frac{x}{2}\right)\cdot\cos\left(\frac{x}{2^2}\right)\cdot\cos\left(\frac{x}{2^3}\right)\ldots\cos\left(\frac{x}{2^n}\right)$

$$\Rightarrow A\sin\left(\frac{x}{2^n}\right)=\cos\left(\frac{x}{2}\right)\cdot\cos\left(\frac{x}{2^2}\right)\ldots\cos\left(\frac{x}{2^n}\right)\sin\left(\frac{x}{2^n}\right)$$

$$=\frac{1}{2}\cos\frac{x}{2}\cdot\cos\left(\frac{x}{2^2}\right)\ldots\cos\left(\frac{x}{2^{n-1}}\right)\cdot\sin\left(\frac{x}{2^{n-1}}\right)$$

$$=\frac{1}{2^2}\cos x\cos\left(\frac{x}{2^2}\right)\ldots\cos\left(\frac{x}{2^{n-2}}\right)\sin\left(\frac{x}{2^{n-2}}\right)$$

$$\Rightarrow A\sin\left(\frac{x}{2^n}\right)=\frac{1}{2^n}\sin x \Rightarrow A=\frac{\frac{1}{2^n}\sin x}{\sin\left(\frac{x}{2^n}\right)}$$

Now, $$P=\lim_{n\to\infty}\frac{\frac{1}{2^n}\sin x}{\sin\left(\frac{x}{2^n}\right)}$$

$$P=\lim_{n\to\infty}\frac{\frac{\sin x}{x}}{\frac{\sin\left(\frac{x}{2^n}\right)}{\left(\frac{x}{2^n}\right)}}=\frac{\sin x}{x}$$

Now, $$\lim_{x\to 0}P=\lim_{x\to 0}\frac{\sin x}{x}=1$$

37. (c) We have, $\lim_{x\to 0}\frac{\sin^2 x}{f(x)}=\frac{1}{2}$

$$\Rightarrow \quad f(x)=x^2(ax+b), a\neq 0$$

$$\Rightarrow \quad \lim_{x\to 0}\frac{\sin^2 x}{x^2(ax+b)}=\frac{1}{2}\Rightarrow \frac{1}{b}=\frac{1}{2}\Rightarrow b=2$$

So $\quad f(x)=x^2(ax+2), a\neq 0$

Now, $f(1)=(a+2)$ but $a\neq 0$

So, $\quad f(1)\neq 2$

38. (b) $\lim_{x\to 0^-}\frac{\tan^{-1}(x+[x])}{[x]-x}=\lim_{h\to 0}\frac{\tan^{-1}(-h+[-h])}{[-h]+h}$

$$=\lim_{h\to 0}\frac{\tan^{-1}(-1-h)}{-1+h}$$

$$=\tan^{-1}(1)=\frac{\pi}{4}$$

39. (d) We have,

$$\lim_{x\to 0^-}[x]=-1 \text{ and } \lim_{x\to 0^+}[x]=0$$

$$\therefore \quad \lim_{x\to 0^-}f(x)=\lim_{x\to 0^-}\frac{(-1)^2+\sin(-1)}{(-1)}=-1+\sin 1$$

and $\quad \lim_{x\to 0^+}f(x)=0 \qquad [\because f(x)=0 \text{ for } 0\le x<1]$

So, $\lim_{x\to 0}f(x)$ does not exist.

40. (a) $\text{RHL}=\lim_{x\to 0^+}f(x)=\lim_{h\to 0}\frac{\tan^2\{h\}}{h^2-[h]^2}$

$$=\lim_{x\to 0}\frac{\tan^2 h}{h^2}=1$$

$$\text{LHL}=\lim_{h\to 0}\sqrt{\{-h\}\cot\{-h\}}=\lim_{h\to 0}\sqrt{(1-h)\cot(1-h)}$$

$$=\sqrt{\cot 1}$$

41. (d) $\lim_{x\to 0}\frac{1+2\sin x-e^x-\ln(1+x)}{x^3}$

$$=\lim_{x\to 0}\frac{1+2\left(x-\frac{x^3}{3!}+\frac{x^5}{5!}+\dots\right)-\left(1+x+\frac{x^2}{2!}+\frac{x^3}{3!}\right)-\left(x-\frac{x^2}{2}+\frac{x^3}{3}\right)}{x^3}$$

$$=\lim_{x\to 0}\frac{x^3\left(\frac{2}{3!}-\frac{1}{3!}-\frac{1}{3}\right)+\text{higher power of } x}{x^3}$$

$$=\frac{1}{3}-\frac{1}{6}-\frac{1}{3}=\frac{-1}{6}$$

42. (b) Put $\frac{1}{x}=y$, if $x\to\infty$, then $y\to 0$

$$\lim_{y\to 0}\frac{\sqrt{(1-\cos y)\sqrt{(1-\cos y)\sqrt{(1-\cos y)\dots\infty}}}}{y^2}$$

$$=\lim_{y\to 0}\frac{(1-\cos y)^{\frac{1}{2}+\frac{1}{2^2}+\frac{1}{2^3}+\dots\infty}}{y^2}=\lim_{y\to 0}\frac{(1-\cos y)}{y^2}$$

$$\left[\because \frac{1}{2}+\frac{1}{2^2}+\frac{1}{2^3}+\dots=\frac{\frac{1}{2}}{1-\frac{1}{2}}=1\right]$$

$$=\lim_{y\to 0}\frac{(1-\cos y)(1+\cos y)}{y^2(1+\cos y)}$$

$$=\lim_{y\to 0}\frac{\sin^2 y}{y^2(1+\cos y)}=\frac{1}{1+1}=\frac{1}{2}$$

43. (a) $\lim_{h\to 0}2\left\{\frac{\sqrt{3}\sin\left(\frac{\pi}{6}+h\right)-\cos\left(\frac{\pi}{6}+h\right)}{\sqrt{3}h(\sqrt{3}\cos h-\sin h)}\right\}$

$$=\lim_{h\to 0}2\left(\frac{\sqrt{3}\left(\frac{1}{2}\cos h+\frac{\sqrt{3}}{2}\sin h\right)-\left(\frac{\sqrt{3}}{2}\cos h-\frac{1}{2}\sin h\right)}{\sqrt{3}h(\sqrt{3}\cos h-\sin h)}\right)$$

$$=\lim_{h\to 0}4\left(\frac{\sin h}{h}\right)\cdot\frac{1}{\sqrt{3}}\left(\frac{1}{\sqrt{3}\cos h-\sin h}\right)$$

$$=4\cdot 1\cdot\frac{1}{\sqrt{3}}\left(\frac{1}{\sqrt{3}}\right)=\frac{4}{3}$$

44. (a) We have, $\lim_{x\to 0}\frac{a^{\tan x}-a^{\sin x}}{\tan x-\sin x}$

$$=\lim_{x\to 0}\frac{\left(\frac{a^{\tan x}-1}{\tan x}\right)\times\tan x-\left(\frac{a^{\sin x}-1}{\sin x}\right)\sin x}{\tan x-\sin x}$$

$$=\lim_{x\to 0}\log a\cdot\frac{\tan x-\sin x}{\tan x-\sin x}=\log a$$

45. (a) We have,

$$\lim_{x\to 0}\frac{\log(a+x)-\log a}{x}+k\lim_{x\to e}\frac{\log x-1}{x-e}=1$$

Applying L-Hospital rule,

$$\lim_{x\to 0}\frac{\frac{1}{a+x}-0}{1}+k\lim_{x\to e}\frac{\frac{1}{x}}{1}=1$$

$$\Rightarrow \quad \frac{1}{a}+k\cdot\frac{1}{e}=1\Rightarrow\frac{k}{e}=1-\frac{1}{a}\Rightarrow k=e\left(1-\frac{1}{a}\right)$$

46. (b) $\lim_{x\to 2\pi}\frac{1-(\sec x)^{\sec x}}{\ln(\sec x)}$

Let $\sec x=t$

When $x\to 2\pi$, then $t\to 1$

$$\lim_{t\to 1}\frac{1-t^t}{\ln(t)}$$

Applying L-Hospital rule,

$$\lim_{t\to 1}\frac{-t^t(1+\ln t)}{\dfrac{1}{t}}=-1$$

47. (c) We have, $\lim\limits_{x\to 0}\dfrac{\log(1+x+x^2)+\log(1-x+x^2)}{\sec x-\cos x}$

$$=\lim_{x\to 0}\frac{\log[(1+x^2)^2-x^2]}{1-\cos^2 x}\times\cos x$$

$$=\lim_{x\to 0}\frac{\log[1+(x^2+x^4)]}{\sin^2 x}\times\cos x$$

$$=\lim_{x\to 0}\frac{\log[1+(x^2+x^4)]}{(x^2+x^4)}\times\frac{(x^2+x^4)\cos x}{\sin^2 x}$$

$$=\lim_{x\to 0}\frac{\log[1+(x^2+x^4)]}{(x^2+x^4)}\cdot\frac{x^2}{\sin^2 x}\cdot(1+x^2)\cos x=1$$

48. (c) We have,

$$\lim_{x\to\infty}\frac{e^x\left(2^{\left(\frac{x^n}{e^x}\right)}-3^{\left(\frac{x^n}{e^x}\right)}\right)}{x^n}=\lim_{x\to\infty}\frac{2^{\left(\frac{x^n}{e^x}\right)}-3^{\left(\frac{x^n}{e^x}\right)}}{\left(\dfrac{x^n}{e^x}\right)}$$

Now, we find $\lim\limits_{x\to\infty}\dfrac{x^n}{e^x}$

Using L-Hospital rule, $\lim\limits_{x\to\infty}\dfrac{nx^{n-1}}{e^x}$

Applying L-Hospital rule n times,

$$=\lim_{x\to\infty}\frac{n!}{e^x}=0$$

So, $\lim\limits_{x\to\infty}\dfrac{2^{\left(\frac{x^n}{e^x}\right)}-3^{\left(\frac{x^n}{e^x}\right)}}{\left(\dfrac{x^n}{e^x}\right)}$ $\left[\dfrac{0}{0}\text{ form}\right]$

$$=\lim_{x\to\infty}\left(\frac{2^{\left(\frac{x^n}{e^x}\right)}-1}{\dfrac{x^n}{e^x}}-\frac{3^{\left(\frac{x^n}{e^x}\right)}-1}{\left(\dfrac{x^n}{e^x}\right)}\right)=\ln 2-\ln 3=\ln\frac{2}{3}$$

49. (b) We have, $\lim\limits_{x\to 1}\sec\dfrac{\pi}{2^x}\log_e x$

$$=\lim_{x\to 1}\frac{\log x}{\cos\left(\dfrac{\pi}{2^x}\right)}=\lim_{x\to 1}\frac{\log[1+(x-1)]}{\sin\left(\dfrac{\pi}{2}-\dfrac{\pi}{2^x}\right)}$$

$$=\lim_{x\to 1}\frac{\dfrac{\log[1+(x-1)]}{x-1}}{\dfrac{\sin\left(\dfrac{\pi}{2}-\dfrac{\pi}{2^x}\right)}{\left(\dfrac{\pi}{2}-\dfrac{\pi}{2^x}\right)}}\cdot\lim_{x\to 1}\frac{x-1}{\left(\dfrac{\pi}{2}-\dfrac{\pi}{2^x}\right)}$$

$$=\lim_{x\to 1}\frac{x-1}{\pi\left(\dfrac{2^{x-1}-1}{2^x}\right)}=\lim_{x\to 1}\frac{2^x}{\pi}\left(\frac{x-1}{2^{x-1}-1}\right)=\frac{2}{\pi\log 2}$$

50. (b) $\lim\limits_{x\to 0}\dfrac{\ln\cot\left(\dfrac{\pi}{4}-\beta x\right)}{\tan\alpha x}=1$

$$\Rightarrow\quad\lim_{x\to 0}-\frac{\ln\tan\left(\dfrac{\pi}{4}-\beta x\right)}{\tan\alpha x}=1$$

$$\Rightarrow\quad-\lim_{x\to 0}\frac{\ln\left[\left(\dfrac{1-\tan\beta x}{1+\tan\beta x}-1\right)+1\right]}{\tan\alpha x}=1$$

$$\Rightarrow\quad-\lim_{x\to 0}\frac{\ln\left(\dfrac{-2\tan\beta x}{1+\tan\beta x}+1\right)}{\tan\alpha x}=1$$

$$\Rightarrow\quad-\lim_{x\to 0}\frac{\ln\left(\dfrac{-2\tan\beta x}{1+\tan\beta x}+1\right)}{\left(\dfrac{-2\tan\beta x}{1+\tan\beta x}\right)}\times\frac{\left(\dfrac{-2\tan\beta x}{1+\tan\beta x}\right)}{\dfrac{\tan\alpha x}{\alpha x}\times\alpha x}=1$$

$$\Rightarrow\quad-\lim_{x\to 0}\frac{-2\dfrac{\tan\beta x}{\beta x}\times\beta x}{(1+\tan\beta x)\cdot\alpha x}=-\left(\frac{-2\beta}{\alpha}\right)=1$$

$$\Rightarrow\quad\frac{2\beta}{\alpha}=1$$

$$\Rightarrow\quad\frac{\alpha}{\beta}=2$$

51. (b) $\lim\limits_{x\to 0}\dfrac{x\left(e^{\frac{(\sqrt{1+x^2+x^4}-1)}{x}}-1\right)}{\sqrt{1+x^2+x^4}-1}$

$$=\lim_{x\to 0}\frac{e^{(\sqrt{1+x^2+x^4}-1)/x}-1}{\left(\dfrac{\sqrt{1+x^2+x^4}-1}{x}\right)}$$

$$=1\qquad\left[\because\lim_{x\to 0}\frac{e^x-1}{x}=1\right]$$

52. (c) $\lim\limits_{x\to 3}f(x)$ exists

$$\Rightarrow\quad\lim_{x\to 3^-}f(x)=\lim_{x\to 3^+}f(x)$$

$$=\lim_{x\to 3^-}\frac{(27)^{\frac{(x+3)x}{27}}-9}{3^x-27}$$

$$=\lim_{x\to 3^+}\lambda\frac{1-\cos(x-3)}{(x-3)^2}$$

$$\Rightarrow \lim_{x\to 3^-} \frac{3^2\left(3^{\left(\frac{x^2+3x}{9}-2\right)}-1\right)}{3^3(3^{x-3}-1)} = \frac{\lambda}{2}$$

$$\Rightarrow \lim_{x\to 3} \frac{1}{3}\frac{x^2+3x-18}{9(x-3)} = \frac{\lambda}{2}$$

$$\Rightarrow \frac{1}{27}\cdot 9 = \frac{\lambda}{2}$$

$$\Rightarrow \lambda = \frac{2}{3}$$

53. (b) We know that $\lim_{x\to 0^+} x^x = 1$

$$\therefore \lim_{x\to 0^+} (x^x - 1) = 0$$

Now, $\lim_{x\to 0^+} \frac{e^{(x^x-1)} - x^x}{[(x^2)^x - 1]^2}$

$$= \lim_{x\to 0^+} \frac{\left(1 + (x^x-1) + \frac{(x^x-1)^2}{2!} + \frac{(x^x-1)^3}{3!}\ldots\right) - x^x}{(x^x-1)^2(x^x+1)^2}$$

$$= \lim_{x\to 0^+} \frac{\frac{(x^x-1)^2}{2!} + \frac{(x^x-1)^3}{3!} + \ldots}{(x^x-1)^2(x^x+1)^2}$$

$$= \lim_{x\to 0^+} \frac{(x^x-1)^2\left[\frac{1}{2!} + \frac{(x^x-1)}{3!} + \ldots\right]}{(x^x-1)^2(x^x+1)^2}$$

$$= \frac{1}{2}\cdot\frac{1}{2^2} = \frac{1}{2}\times\frac{1}{4} = \frac{1}{8}$$

54. (b) RHL $= \lim_{x\to 0^+} \frac{(1+a^2)+8e^{1/x}}{1+(2+b+b^2)e^{1/x}} = 2$

$$\Rightarrow \lim_{x\to 0^+} \frac{(1+a^2)e^{-1/x}+8}{e^{-1/x}+(2+b+b^2)} = 2$$

$$\Rightarrow \frac{8}{2+b+b^2} = 2$$

$$\Rightarrow 2+b+b^2 = 4$$

$$\Rightarrow b^2+b-2 = 0$$

$$\Rightarrow b^2+2b-b-2 = 0$$

$$\Rightarrow (b+2)(b-1) = 0$$

$$\Rightarrow b = 1$$

Now, LHL $\lim_{x\to 0^-} \frac{(1+a^2)+8e^{\frac{1}{x}}}{1+(2+b+b^2)e^{\frac{1}{x}}} = 2$

$$\Rightarrow 1+a^2 = 2$$

$$\Rightarrow a^2 = 1 \Rightarrow a = 1$$

Now, $a^2+b^2 = 1+1 = 2$

55. (c) $\lim_{x\to\infty}\left(\frac{x+c}{x-c}\right)^x = 16$

$$\Rightarrow e^{\lim_{x\to\infty}\left(\frac{x+c}{x-c}-1\right)x} = 16$$

$$\Rightarrow e^{\lim_{x\to\infty}\left(\frac{x+c-x+c}{x-c}\right)x} = 16$$

$$\Rightarrow e^{\lim_{x\to\infty}\left(\frac{2cx}{x-c}\right)} = 16 \Rightarrow e^{2c} = 16$$

Taking log both sides

$$2c = \log 16$$

$$c = \frac{1}{2}\log 16$$

$$c = \log 4$$

56. (a) Let $L = \lim_{x\to\infty}\left(\cos\frac{\pi}{x}\right)^x$

$$\log L = \lim_{x\to\infty} x\log\left(\cos\frac{\pi}{x}\right)$$

$$\Rightarrow \log L = \lim_{x\to\infty} \frac{\log\left(\cos\frac{\pi}{x}\right)}{\frac{1}{x}} \qquad \left[\frac{0}{0}\text{form}\right]$$

$$\Rightarrow \log L = \lim_{x\to\infty} \frac{-\frac{\pi}{x^2}\tan\left(\frac{\pi}{x}\right)}{\frac{-1}{x^2}}$$

[using L-Hospital rule]

$$= \lim_{x\to\infty} +\pi\tan\frac{\pi}{x}$$

$$\log L = +\pi(0) \Rightarrow \log L = 0$$

$$L = e^0 = 1$$

57. (b) Let $\frac{1}{y} = x$, if $y\to\infty$, then $x\to 0$

$$= \lim_{x\to 0}\left(\frac{a_1^{\,x}+a_2^{\,x}+a_3^{\,x}+\ldots+a_n^{\,x}}{n}\right)^{\frac{1}{x}}$$

$$= e^{\lim_{x\to 0}\left(\frac{a_1^{\,x}+a_2^{\,x}+a_3^{\,x}+\ldots+a_n^{\,x}}{n}-1\right)^{\frac{1}{x}}}$$

$$= e^{\lim_{x\to 0}\left(\frac{a_1^{\,x}+a_2^{\,x}+\ldots a_n^{\,x}-n}{nx}\right)}$$

$$= e^{\lim_{x\to 0}\frac{1}{n}\left[\left(\frac{a_1^{\,x}-1}{x}+\frac{a_2^{\,x}-1}{x}+\ldots+\frac{a_n^{\,x}-1}{x}\right)\right]}$$

$$= e^{\frac{1}{n}[\log a_1+\log a_2+\log a_3+\ldots+\log a_n]}$$

$$= e^{\frac{1}{n}[\log(a_1\cdot a_2\cdot a_3\ldots a_n)]}$$

$$= e^{\log(a_1\cdot a_2\cdot a_3\ldots a_n)^{\frac{1}{n}}} = (a_1\cdot a_2\cdot a_3\ldots a_n)^{\frac{1}{n}}$$

$$= \prod_{i=1}^{n}(a_i)^{\frac{1}{n}}$$

58. (c) $\lim_{x\to a}\left[2-\frac{a}{x}\right]^{\tan\frac{\pi x}{2a}} = \lim_{x\to a}\left[1+\left(1-\frac{a}{x}\right)\right]^{\tan\frac{\pi x}{2a}}$

$$= e^{\lim_{x\to a}\left(1-\frac{a}{x}\right)\tan\frac{\pi x}{2a}}$$

$$= e^{\lim_{x\to a}\frac{\left(1-\frac{a}{x}\right)}{\cot\frac{\pi x}{2a}}} = e^{\lim_{x\to a}\frac{\left(1-\frac{a}{x}\right)}{\tan\left(\frac{\pi}{2}-\frac{\pi x}{2a}\right)}}$$

$$= e^{\lim_{x\to a}\frac{\frac{-a}{x}\left(1-\frac{x}{a}\right)}{\tan\frac{\pi}{2}\left(1-\frac{x}{a}\right)}\times\frac{\frac{\pi}{2}}{\frac{\pi}{2}}}$$

$$= e^{\frac{-a}{a}\times\frac{1}{\pi/2}} = e^{\frac{-2}{\pi}}$$

59. (c) We have, $\lim_{x\to 1}\left\{\tan\left(\frac{\pi}{4}+\log x\right)\right\}^{\frac{1}{\log x}}$ [1^∞ form]

$$= \lim_{x\to 1}\left(\frac{1+\tan(\log x)}{1-\tan(\log x)}\right)^{\frac{1}{\log x}}$$

$$= e^{\lim_{x\to 1}\left(\frac{1+\tan(\log x)}{1-\tan(\log x)}-1\right)\cdot\frac{1}{\log x}}$$

$$= e^{\lim_{x\to 1}\left(\frac{2\tan(\log x)}{1-\tan(\log x)}\right)\cdot\frac{1}{\log x}}$$

$$= e^2$$

60. (d) We have, $\lim_{x\to 0}\left\{\frac{f(1+x)}{f(1)}\right\}^{\frac{1}{x}}$

$$= \lim_{x\to 0}\left\{1+\frac{f(1+x)-f(1)}{f(1)}\right\}^{\frac{1}{x}}$$

$$= e^{\lim_{x\to 0}\frac{f(1+x)-f(1)}{xf(1)}}$$

$$= e^{\frac{1}{f(1)}\lim_{x\to 0}\frac{f(x+1)-f(1)}{x}}$$

$$= e^{\frac{f'(1)}{f(1)}} = e^{\frac{6}{3}} = e^2$$

61. (c) α and β are the roots of the equation,

$$ax^2+bx+c=0$$

$\Rightarrow$ $a\alpha^2+b\alpha+c=0$ ...(i)

and $a\beta^2+b\beta+c=0$ ...(ii)

Now, $\lim_{x\to\alpha}(1+ax^2+bx+c)^{\frac{1}{x-\alpha}} = e^{\lim_{x\to\alpha}\left(\frac{ax^2+bx+c}{x-\alpha}\right)}$

$$= e^{\lim_{x\to\alpha}\frac{2ax+b}{1}}$$

[by L-Hospital rule}

From Eqs. (i) and (ii),

$$a(\alpha^2-\beta^2)+b(\alpha-\beta)=0$$

$\Rightarrow$ $(\alpha-\beta)[a(\alpha+\beta)+b]=0$

$\because$ $\alpha\neq\beta$, so $a\alpha+a\beta+b=0$

$$a\alpha+b=-a\beta$$

$$= e^{2a\alpha+b} = e^{a\alpha+(a\alpha+b)}$$

$$= e^{a\alpha-a\beta} = e^{a(\alpha-\beta)}$$

62. (d) We have, $f(x)+\frac{e^{x+x^2}}{f(x)}\le e^x+e^{x^2}$

$\Rightarrow$ $[f(x)]^2-(e^x+e^{x^2})f(x)+e^{x+x^2}\le 0$

$\Rightarrow [f(x)]^2-(e^x+e^{x^2})f(x)+e^x\cdot e^{x^2}\le 0$

$\Rightarrow$ $[f(x)-e^x][f(x)-e^{x^2}]\le 0$

So, $e^{x^2}\le f(x)\le e^x\ \forall x\in(0,1)$

and $e^x\le f(x)\le e^{x^2}\ \forall x\in(1,\infty)$

$\therefore$ $\lim_{x\to 1}e^x = \lim_{x\to 1}e^{x^2} = e$

Hence, by Sandwich theorem

$$\lim_{x\to 1}f(x)=e$$

63. (b) We have, $\frac{1}{\sqrt{n^2+2n}}\le\frac{1}{\sqrt{n^2}}\le\frac{1}{\sqrt{n^2}}$

$$\frac{1}{\sqrt{n^2+2n}}\le\frac{1}{\sqrt{n^2+1}}\le\frac{1}{\sqrt{n^2}}$$

$$\vdots$$

$$\frac{1}{\sqrt{n^2+2n}}\le\frac{1}{\sqrt{n^2+2n}}\le\frac{1}{n^2}$$

$\Rightarrow$ $\frac{2n}{\sqrt{n^2+2n}}\le S_n\le\frac{2n}{\sqrt{n^2}}$ [adding vertically]

But $\lim_{n\to\infty}\frac{2n}{\sqrt{n^2+2n}}=2$

and $\lim_{n\to\infty}\frac{2n}{\sqrt{n^2}}=2$

$\therefore$ By Sandwich theorem,

$$\lim_{n\to\infty}S_n=2$$

64. (c) Given, $f(x+T)=f(x)$

So, $f(x+T)=f(x+2T)=f(x+3T)$

$$=\ldots f(x+nT)$$

$$=f(x)$$

Now, $\lim_{n\to\infty}\frac{nf(x)[1+2+\ldots+n]}{f(x)(1+4+\ldots+n^2)}$

$$= \lim_{n\to\infty}\frac{n(1+2+\ldots+n)}{(1^2+2^2+\ldots+n^2)}$$

$$= \lim_{n\to\infty}\frac{n\left(\frac{n(n+1)}{2}\right)}{\frac{n(n+1)(2n+1)}{6}}=\frac{3}{2}$$

65. (b) We have, $\lim_{n\to\infty}\prod_{n=0}^{n}\left(1+\frac{1}{2^{2^n}}\right)$

Multiplying and dividing by $\left(1-\frac{1}{2}\right)$,

$$=\lim_{n\to\infty}\frac{\left(1-\frac{1}{2}\right)\left(1+\frac{1}{2}\right)\left(1+\frac{1}{2^2}\right)\ldots\left(1+\frac{1}{2^{2^n}}\right)}{1-\frac{1}{2}}$$

$$=\lim_{n\to\infty}\frac{1-\left(\frac{1}{2^{2^n}}\right)^2}{1-\frac{1}{2}}=\frac{1}{\frac{1}{2}}=2$$

66. (a) We have, $\lim_{x\to\infty}\frac{\left(\int_0^{x+y}e^{t^2}dt\right)^2}{\int_0^{x+y}e^{2t^2}dt}$

$$=\lim_{x\to\infty}\frac{2\left(\int_0^{x+y}e^{t^2}dt\right)\cdot e^{(x+y)^2}}{e^{2(x+y)^2}}$$

$$=\lim_{x\to\infty}\frac{2\int_0^{x+y}e^{t^2}dt}{e^{(x+y)^2}}$$

$$=\lim_{x\to\infty}\frac{2\cdot e^{(x+y)^2}\cdot 1}{e^{(x+y)^2}\cdot 2(x+y)}$$

$$=\lim_{x\to\infty}\frac{1}{x+y}=0$$

67. (d) Let $\lim_{x\to 0}\frac{3\int_0^x e^{-t^2}dt-3x+x^3}{3x^5}$ $\left[\frac{0}{0}\text{ form}\right]$

$$=\lim_{x\to 0}\frac{3\frac{d}{dx}\int_0^x e^{-t^2}dt-3+3x^2}{15x^4}$$

[applying L-Hospital rule]

By applying Newtons Leibinitz's formula,

$$=\lim_{x\to 0}\frac{3e^{-x^2}-3+3x^2}{15x^4}\quad\left[\frac{0}{0}\text{ form}\right]$$

Applying L-Hospital rule

$$=\lim_{x\to 0}\frac{-3e^{-x^2}(-2x)+6x}{60x^3}=\lim_{x\to 0}\frac{-6x(e^{-x^2}-1)}{60x^3}$$

$$=\lim_{x\to 0}\frac{1}{10}\left(\frac{(e^{-x^2}-1)}{-x^2}\right)=\frac{1}{10}$$

68. (a) We have, $\lim_{x\to 0^+}\frac{\int_0^{x^2}(\sin\sqrt{t})\,dt}{x^3}$

Using L-Hospital rule,

$$=\lim_{x\to 0^+}\frac{2x\cdot\sin\sqrt{x^2}-0}{3x^2}$$

$$=\lim_{x\to 0^+}\frac{2x\cdot\sin|x|}{3x^2}$$

$$=\lim_{x\to 0^+}\frac{2\sin x}{3x}=\frac{2}{3}$$

69. (1) We have, $\lim_{n\to\infty}\frac{\sqrt{n^3-2n^2+1}+\sqrt[3]{n^4+1}}{\sqrt[4]{n^6+6n^5+2}-\sqrt[5]{n^7+3n^3+1}}$

$$=\lim_{n\to\infty}\frac{n^{3/2}\sqrt{1-\frac{2}{n}+\frac{1}{n^3}}+n^{4/3}\sqrt[3]{1+\frac{1}{n^4}}}{n^{6/4}\sqrt[4]{1+\frac{6}{n}+\frac{2}{n^6}}-n^{7/5}\sqrt[5]{1+\frac{3}{n^4}+\frac{1}{n^7}}}$$

$$=\lim_{n\to\infty}\frac{n^{3/2}\left(\sqrt{1-\frac{2}{n}+\frac{1}{n^3}}+n^{-1/6}\sqrt[3]{1+\frac{1}{n^4}}\right)}{n^{3/2}\left(\sqrt[4]{1+\frac{6}{n}+\frac{2}{n^6}}-n^{-1/10}\sqrt[5]{1+\frac{3}{n^4}+\frac{1}{n^7}}\right)}$$

$$=\frac{1+0}{1+0}=1$$

70. (6) We have, $\lim_{x\to\infty}(\sqrt{Px^2+Qx}-Rx)=3$

$$\Rightarrow\quad\lim_{x\to\infty}\frac{Px^2+Qx-R^2x^2}{\sqrt{Px^2+Qx}+Rx}=3$$

$$\Rightarrow\quad\lim_{x\to\infty}\frac{x^2(P-R^2)+Qx}{x\left(\sqrt{P+\frac{Q}{x}}+R\right)}=3$$

So, $P-R^2=0$ and $\frac{Q}{\sqrt{P}+R}=3$

$$P=R^2\qquad\ldots\text{(i)}$$

From Eq. (i),

$$\frac{Q}{R+R}=3\Rightarrow\frac{Q}{2R}=3\Rightarrow\frac{Q}{R}=6$$

71. (6) We have,

$$f(a)=\lim_{x\to 1}\frac{x^a-ax+a-1}{(x-1)^2}$$

$$=\lim_{h\to 0}\frac{(1+h)^a-a(1+h)+a-1}{h^2}$$

$$=\lim_{h\to 0}\frac{\left(1+ah+\frac{a(a-1)}{2!}h^2+\ldots\right)-a-ah+a-1}{h^2}$$

$$f(a)=\frac{a(a-1)}{2}$$

$$\therefore\quad f(4)=\frac{4(4-1)}{2}=6$$

72. (2) $\lim_{x\to 0}\dfrac{f(2+3x^4)-f(2-5x^4)}{x^4}$

By applying L-Hospital rule,

$$=\lim_{x\to 0}\frac{f'(2+3x^4)\cdot 12x^3-f'(2-5x^4)(-20x^3)}{4x^3}$$

$$=\lim_{x\to 0}\frac{x^3\{12f'(2+3x^4)+20f'(2-5x^4)\}}{4x^3}$$

$$=\frac{12f'(2)+20f'(2)}{4}$$

$$=\frac{12\times\frac{1}{4}+20\times\frac{1}{4}}{4}=\frac{8}{4}=2 \qquad \left[\because f'(2)=\frac{1}{4}\text{(given)}\right]$$

73. (2) $\lim_{x\to 0}\dfrac{\sin 3x+(1-\cos 2x)}{x+P\tan x}=1$

$$\Rightarrow \lim_{x\to 0}\frac{3\dfrac{\sin 3x}{3x}+\left(\dfrac{1-\cos 2x}{x}\right)}{1+P\left(\dfrac{\tan x}{x}\right)}=1$$

$$\Rightarrow \frac{3+0}{1+P}=1\Rightarrow 3=1+P\Rightarrow P=2$$

74. (32) We have,

$$\lim_{x\to\frac{\pi}{4}}\frac{(1-\tan x)(1-\sin 2x)}{(1+\tan x)(\pi-4x)^3}=\lim_{x\to\frac{\pi}{4}}\frac{\tan\left(\frac{\pi}{4}-x\right)(1-\sin 2x)}{(\pi-4x)(\pi-4x)^2}$$

$$=\lim_{x\to\frac{\pi}{4}}\frac{\tan\left(\frac{\pi}{4}-x\right)(1-\sin 2x)}{4\left(\frac{\pi}{4}-x\right)(\pi-4x)^2}$$

$$=\frac{1}{4}\lim_{x\to\frac{\pi}{4}}\frac{1-\sin 2x}{(\pi-4x)^2}$$

Applying L-Hospital rule,

$$=\frac{1}{4}\lim_{x\to\frac{\pi}{4}}\frac{-2\cos 2x}{2(\pi-4x)(-4)}=\frac{1}{16}\lim_{x\to\frac{\pi}{4}}\frac{\cos 2x}{(\pi-4x)}$$

Again applying L-Hospital rule,

$$=\frac{1}{16}\lim_{x\to\frac{\pi}{4}}\frac{-2\sin 2x}{(-4)}=\frac{1}{32}$$

Now, $\dfrac{1}{L}=32$

75. (11) $\sum_{k=1}^{n}\sin^{-1}a_k=\dfrac{n\pi}{2}$

$$\Rightarrow \sin^{-1}a_1+\sin^{-1}a_2+\ldots+\sin^{-1}a_n=\frac{n\pi}{2}$$

$$\Rightarrow \sin^{-1}a_1=\sin^{-1}a_2=\sin^{-1}a_3\ldots=\frac{\pi}{2}$$

$$\left[\because \frac{-\pi}{2}\le\sin^{-1}x\le\frac{\pi}{2}\right]$$

$$\Rightarrow a_1=a_2=a_3\ldots=a_n=1$$

So, $P=\prod_{k=1}^{n}(1)^k=1$

Now, $\lim_{x\to 1}\dfrac{x^{1/3}-(3-2x)^{1/4}}{x^2-x}$

$$=\lim_{h\to 0}\frac{(1+h)^{1/3}-(1-2h)^{1/4}}{(1+h)^2-(1+h)} \qquad [\text{put } x=1+h]$$

$$=\lim_{h\to 0}\frac{\left(1+\frac{1}{3}h\right)-\left(1-\frac{1}{2}h\right)}{(1+2h)-(1+h)}$$

$$=\lim_{h\to 0}\frac{h\left(\frac{1}{3}+\frac{1}{2}\right)}{h}=\frac{1}{3}+\frac{1}{2}=\frac{5}{6}$$

So, $6(k+p)=6\left(\dfrac{5}{6}+1\right)=6\left(\dfrac{11}{6}\right)=11$

76. (3) We have, $\lim_{x\to 0}\dfrac{a\sin x-\sin 2x}{\tan^3 x}$

Applying L-Hospital rule,

$$=\lim_{x\to 0}\frac{a\cos x-2\cos 2x}{3\tan^2 x\sec^2 x}$$

Here, denominator is zero. So, for finite limit numerator must be zero, when $x\to 0$.

So, $a-2=0\Rightarrow a=2$

Now, $L=\lim_{x\to 0}\dfrac{2\sin x-\sin 2x}{\tan^3 x}$

$$=\lim_{x\to 0}\frac{2\sin x-2\sin x\cos x}{\tan^3 x}$$

$$=\lim_{x\to 0}\frac{2\sin x(1-\cos x)}{\tan^3 x}$$

$$=\lim_{x\to 0}\frac{2\sin x\left(2\sin^2\frac{x}{2}\right)}{\tan^3 x}$$

$$=\lim_{x\to 0}\frac{4\left(\dfrac{\sin x}{x}\right)\left(\dfrac{\sin\frac{x}{2}}{x/2}\right)^2\cdot\dfrac{1}{4}}{\left(\dfrac{\tan x}{x}\right)^3}=1$$

So, $a+L=2+1=3$

77. (2) $\lim_{x\to 0}\dfrac{\{(a-n)nx-\tan x\}\sin nx}{x^2}=0$

$$\Rightarrow \lim_{x\to 0}\left(\frac{(a-n)nx(\cos x)-\sin x}{x\cos x}\cdot\frac{\sin nx}{nx}\cdot n\right)=0$$

$$\Rightarrow n\lim_{x\to 0}\left\{\frac{(a-n)(nx)\left(1-\frac{x^2}{2!}+\frac{x^4}{4!}-\ldots\right)-\left(x-\frac{x^3}{3!}+\ldots\right)}{x\left(1-\frac{x^2}{2!}+\frac{x^4}{4!}-\ldots\right)}\right\}=0$$

$$\Rightarrow \lim_{x\to 0} \frac{\left[(a-n)n\left(1-\frac{x^2}{2!}+\frac{x^4}{4!}\right)-\left(1-\frac{x^2}{3!}+\ldots\right)\right]}{\left(1-\frac{x^2}{2!}+\frac{x^4}{4!}+\ldots\right)}=0$$

$$\Rightarrow \lim_{x\to 0} \frac{[(a-n)(n)-1]-\left[\frac{(a-n)n}{2}-\frac{1}{6}\right]x^2+\ldots}{\left(1-\frac{x^2}{2!}+\frac{x^4}{4!}\ldots\right)}=0$$

$$\Rightarrow \quad (a-n)n-1=0$$

$$\Rightarrow \quad (a-n)n=1 \Rightarrow a-n=\frac{1}{n}$$

$$\Rightarrow \quad a=n+\frac{1}{n}=f(n)$$

$$f(1)=1+\frac{1}{1}=2$$

78. (5) $$\lim_{x\to 0} \frac{ax-(e^{4x}-1)}{ax(e^{4x}-1)}=b$$

$$\Rightarrow \lim_{x\to 0} \frac{ax-\left(1+(4x)+\frac{(4x)^2}{2!}+\ldots-1\right)}{ax\left(1+4x+\frac{(4x)^2}{2!}+\ldots-1\right)}=b$$

$$\Rightarrow \quad \lim_{x\to 0} \frac{(a-4)x-\frac{(4x)^2}{2!}+\ldots}{4ax^2+\frac{16\cdot a}{2!}x^3+\ldots}=b$$

Limit exists, if $a-4=0$

$a=4$

and $\frac{-\frac{16}{2}}{4a}=b \Rightarrow \frac{-8}{4a}=b \Rightarrow \frac{-1}{2}=b \qquad \{a=4\}$

Now, $a-2b=4-2\left(\frac{-1}{2}\right)=4+1=5$

79. (1) $$L_1=\lim_{x\to 0} \frac{\sin 4x}{\tan\left(\frac{x}{k}\right)}=\lim_{x\to 0} \frac{\frac{\sin 4x}{4x}\times 4x}{\frac{\tan\left(\frac{x}{k}\right)}{\left(\frac{x}{k}\right)}\times\left(\frac{x}{k}\right)}=4k$$

and $$L_2=\lim_{x\to 0} \frac{e^{4x}-1}{4x}\times 4=4$$

Now, $L_1L_2=16$

$4\cdot(4k)=16 \Rightarrow k=1$

80. (0.5) We have, $p\geq 2$ and $q\geq 2$

$$\lim_{x\to 0} \frac{e^{\cos(x^q)}-e}{x^p}=\frac{-e}{2}$$

$$\Rightarrow e\lim_{x\to 0} \frac{e^{\cos(x^q)-1}}{\{\cos(x^q)-1\}}\times\left(\frac{\cos(x^q)-1}{(x^q)^2}\right)\frac{x^{2q}}{x^p}=\frac{-e}{2}$$

$$\Rightarrow e\times 1\times \lim_{x\to 0} \frac{-2\sin^2\frac{x^q}{2}}{x^{2q}}\times \lim_{x\to 0} x^{2q-p}=\frac{-e}{2}$$

$$\Rightarrow e\times 1\times\left(-\frac{1}{2}\right)\times \lim_{x\to 0} x^{2q-p}=\frac{-e}{2}$$

$\therefore \lim_{x\to 0} x^{2q-p}$ must be equal to 1.

$$\Rightarrow \quad 2q-p=0 \quad \text{or} \quad \frac{q}{p}=\frac{1}{2}=0.5$$

81. (2) $$L=\lim_{x\to 0} \frac{e^{x\sin(2007)x}-1}{x\sin(2007x)}\cdot\frac{\sin(2007)x}{\ln(1+x)}$$

$$=\lim_{x\to 0} \frac{\sin(2007)x}{(2007)x}\cdot\frac{(2007)x}{\ln(1+x)}=2007$$

Now, $$M=\lim_{x\to\infty} -x[(x^3+x^2+1)^{1/3}+(x^3-x^2+1)^{1/3}-2x]$$

$$=\lim_{x\to\infty} -x\left[x\left(1+\frac{1}{x}+\frac{1}{x^3}\right)^{\frac{1}{3}}+x\left(1-\frac{1}{x}+\frac{1}{x^3}\right)^{\frac{1}{3}}-2x\right]$$

Put $\frac{1}{t}=x$

$$=\lim_{t\to 0} \frac{-[(1+t+t^3)^{1/3}+(1-t+t^3)^{1/3}-2]}{t^2}$$

$$=\lim_{t\to 0} -\frac{1}{t^2}\left[1+\frac{1}{3}(t+t^3)-(t+t^3)^2\ldots+1+\frac{1}{3}(-t+t^3)-\frac{1}{9}(-t+t^3)^2-2\right]$$

$$=\lim_{t\to 0} \frac{\frac{2}{9}t^2}{t^2}=\frac{2}{9}$$

Hence, $\frac{LM}{223}=\frac{2}{9}\times\frac{2007}{223}=2$

82. (7) $$\lim_{x\to 0} \frac{(4\cdot 5\cdot 7)^x-(7\cdot 5)^x-(4\cdot 7)^x-(4\cdot 5)^x+7^x+5^x+4^x-1}{x\sin^2 x}$$

$$\Rightarrow \lim_{x\to 0} \frac{(4^x-1)(5^x-1)(7^x-1)}{x\frac{\sin^2 x}{x^2}\times x^2}$$

$[\because abc-ab-bc-ca+a+b+c-1=(a-1)(b-1)(c-1)]$

$$=\lim_{x\to 0}\left(\frac{4^x-1}{x}\right)\cdot\left(\frac{5^x-1}{x}\right)\cdot\left(\frac{7^x-1}{x}\right)\cdot\left(\frac{x^2}{\sin^2 x}\right)$$

$$=\ln 4\cdot\ln 5\cdot\ln 7=\ln 2^2\cdot\ln 5\cdot\ln 7=2\ln 2\cdot\ln 5\cdot\ln 7$$

By comparing, $P=2$ and $Q=5$

Hence, $P+Q=2+5=7$

83. (2) We have, $\lim_{x\to\infty}\left[\cos\left(2\pi\left(\frac{x}{1+x}\right)^a\right)\right]^{x^2}$

$$= e^{\lim_{x\to\infty}\left[\cos\left(2\pi\left(\frac{x}{1+x}\right)^a\right)-1\right]x^2} \quad \text{...(i)}$$

Let $\quad L = \lim_{x\to\infty}\left[\cos\left(2\pi\left(\frac{x}{1+x}\right)^a\right)-1\right]x^2$

Let $\quad \frac{x}{1+x} = t \Rightarrow x = \frac{t}{1-t}$

When $x\to\infty$, then $t\to 1$

$$\Rightarrow \quad L = \lim_{t\to 1}\left[\cos 2\pi t^a - 1\right]\left(\frac{t}{1-t}\right)^2$$

$$= \lim_{t\to 1}\left[\frac{\cos(2\pi - 2\pi t^a)-1}{\{\pi(1-t^a)\}^2}\right]\left(\frac{t}{1-t}\right)^2 \cdot \pi^2(1-t^a)^2$$

$$= \lim_{t\to 1}\frac{-2\sin^2(\pi - \pi t^a)}{\{\pi(1-t^a)\}^2}\cdot t^2\cdot\pi^2\left(\frac{1-t^a}{1-t}\right)^2$$

$$= -2(1)\pi^2 a^2 = -2\pi^2 a^2$$

So, given limit is $e^{-2\pi^2 a^2}$.

On comparing with $e^{-ka^2\pi^2}$

$$k = 2$$

84. (5) $\lim_{x\to 1}(1+ax+bx^2)^{\frac{c}{x-1}} = e^5$

$$\Rightarrow \quad e^{\lim_{x\to 1}(ax+bx^2)\frac{c}{x-1}} = e^5$$

$$\Rightarrow \quad \lim_{x\to 1}\frac{c(ax+bx^2)}{x-1} = 5$$

$$\Rightarrow \quad \lim_{h\to 0}\frac{c\{a(h+1)+b(h+1)^2\}}{(h+1)-1} = 5$$

$$\Rightarrow \lim_{h\to 0}\frac{(ac+b)+h(ac+2b)+bh^2}{h} = 5$$

$\Rightarrow \quad ac+b=0$ and $ac+2b=5$

On solving $b=5$ and $ac=-5$

Also, the form must be 1^∞ for which $a+b=0$

i.e. $\quad a=-5$ and $c=1$

Now, $\quad \frac{a^2c}{b} = \frac{(-5)^2\times 1}{5} = 5$

85. (4) We have,

$$\lim_{x\to 0}(a\tan bx^2 - 3\cos^2 x + 4)^{\frac{1}{3x^2}} = e^{\frac{7}{3}}$$

$$\Rightarrow \ e^{\lim_{x\to 0}\frac{1}{3x^2}(a\tan bx^2 - 3\cos^2 x + 3)} = e^{\frac{7}{3}}$$

$$\Rightarrow \ e^{\lim_{x\to 0}\frac{a\tan bx^2}{3x^2\times b}\times b + \frac{3(1-\cos^2 x)}{3x^2}} = e^{\frac{7}{3}}$$

$$\Rightarrow \quad e^{\left(\frac{ab}{3}+1\right)} = e^{\frac{7}{3}}$$

$$\Rightarrow \quad \frac{ab}{3}+1 = \frac{7}{3} \Rightarrow \frac{ab}{3} = \frac{7}{3}-1$$

$$\Rightarrow \quad \frac{ab}{3} = \frac{4}{3} \Rightarrow ab = 4$$

86. (22) $L_1 = \lim_{x\to\infty}\int_0^x \frac{(1+\cos t)^2}{x}dt = \lim_{x\to\infty}\frac{1}{x}\int_0^x(1+\cos t)^2\,dt$

$$= \lim_{x\to\infty}\frac{1}{x}\int_0^x(1+2\cos t+\cos^2 t)\,dt$$

$$= \lim_{x\to\infty}\frac{1}{x}\int_0^x\left\{1+2\cos t+\left(\frac{1+\cos 2t}{2}\right)\right\}dt$$

$$= \lim_{x\to\infty}\frac{1}{x}\int_0^x\left(\frac{3}{2}+2\cos t+\frac{\cos 2t}{2}\right)dt$$

$$= \lim_{x\to\infty}\frac{1}{x}\left(\frac{3}{2}x+2\sin x+\frac{\sin 2x}{4}\right) = \frac{3}{2}$$

and $L_2 = \lim_{x\to 0}\frac{1}{x}\int_0^x(1+\cos t)^2\,dt = \lim_{x\to 0}(1+\cos x)^2 = 4$

Now, $\quad 4(L_1+L_2) = 4\left(\frac{3}{2}+4\right) = 4\times\frac{11}{2} = 22$

87. (1) $f(x)$ is a positive increasing function

$$\Rightarrow \quad 0 < f(x) < f(2x) < f(3x)$$

$$\Rightarrow \quad 0 < \frac{f(x)}{f(x)} < \frac{f(2x)}{f(x)} < \frac{f(3x)}{f(x)}$$

$$\Rightarrow \quad \lim_{x\to\infty}\frac{f(x)}{f(x)} \le \lim_{x\to\infty}\frac{f(2x)}{f(x)} \le \lim_{x\to\infty}\frac{f(3x)}{f(x)}$$

$$\Rightarrow \quad 1 \le \lim_{x\to\infty}\frac{f(2x)}{f(x)} \le 1$$

By Sandwich theorem,

$$\lim_{x\to\infty}\frac{f(2x)}{f(x)} = 1$$

88. (1) $I = \lim_{n\to\infty}\sum_{r=2}^{n}\left\{(r+1)\sin\frac{\pi}{r+1} - r\sin\frac{\pi}{r}\right\}$

$$= \lim_{n\to\infty}\left(3\sin\frac{\pi}{3} - 2\sin\frac{\pi}{2} + 4\sin\frac{\pi}{4} - 3\sin\frac{\pi}{3} + 5\sin\frac{\pi}{5} - 4\sin\frac{\pi}{4} + \ldots + n\sin\frac{\pi}{n} - (n-1)\sin\frac{\pi}{n-1} + (n+1)\sin\frac{\pi}{n+1} - n\sin\frac{\pi}{n}\right)$$

$$I = \lim_{n\to\infty}\left((n+1)\sin\frac{\pi}{n+1} - 2\sin\frac{\pi}{2}\right)$$

$$I = \lim_{n\to\infty}\left[(n+1)\frac{\sin\left(\frac{\pi}{n+1}\right)}{\left(\frac{\pi}{n+1}\right)}\times\left(\frac{\pi}{n+1}\right) - 2\right]$$

$I = \pi - 2$

So, $[I] = 1$

Level 3

1. (a) $\lim_{n\to\infty}\left(\dfrac{2n^3+3n^2+5}{n^2+n}-an+b\right)=3$

$\Rightarrow \lim_{n\to\infty}\left(\dfrac{2n^3+3n^2+5-an^3-an^2+bn^2+bn}{n^2+n}\right)=3$

$\Rightarrow \lim_{n\to\infty}\left(\dfrac{n^3(2-a)+n^2(3-a+b)+bn+5}{n^2+n}\right)=3$

For a finite limit,

$2-a=0\Rightarrow a=2$

and $3-a+b=3$

$\Rightarrow 3-2+b=3 \Rightarrow b=2$

2. (b) $\lim_{x\to a}\dfrac{\sqrt{f(x)}-a}{\sqrt{g(x)}-b}=\lim_{x\to a}\dfrac{f(x)-a^2}{g(x)-b^2}\times\dfrac{\sqrt{g(x)}+b}{\sqrt{f(x)}+a}$

$=\lim_{x\to a}\dfrac{f(x)-f(a)}{g(x)-g(a)}\times\dfrac{\sqrt{g(x)}+b}{\sqrt{f(x)}+a}$

$=\lim_{x\to a}\dfrac{f'(a)}{g'(a)}\times\dfrac{\sqrt{g(x)}+b}{\sqrt{f(x)}+a}$

$=\dfrac{f'(a)}{g'(a)}\times\dfrac{\sqrt{g(a)}+b}{\sqrt{f(a)}+b}$

$=3\dfrac{(2b)}{(2a)}=\dfrac{3b}{a}$

3. (c) Let $P(x)=ax^3+bx^2+cx+d$

Then, $P'(x)=3ax^2+2bx+c$

Now, $\lim_{x\to 0}\dfrac{P(x)}{x}=2$

$\Rightarrow \lim_{x\to 0}\left(\dfrac{ax^3+bx^2+cx+d}{x}\right)=2$

$\Rightarrow \lim_{x\to 0}\left(ax^2+bx+c+\dfrac{d}{x}\right)=2$

$\Rightarrow d=0$ and $c=2$

$\therefore P'(x)=3ax^2+2bx+2$

$P(x)$ has extreme values at $x=1$ and $x=3$.

$\therefore P'(1)=3a+2b+2=0$

and $P'(3)=27a+6b+2=0$

On solving $a=\dfrac{2}{9}$ and $b=\dfrac{-4}{3}$

So, $P(x)=\dfrac{2}{9}x^3-\dfrac{4}{3}x^2+2x$

Now, $P(-3)=\dfrac{2}{9}(-3)^3-\dfrac{4}{3}(-3)^2+2(3)=-16$

4. (d) At $x\to\pi^+$, $\cot x\to\infty$

$\Rightarrow 2^{\cot x}, 3^{\cot x}, 5^{\cot x}\to\infty$

and $x\to\pi^-$, $\cot x\to-\infty$

$\Rightarrow 2^{\cot x}, 3^{\cot x}, 5^{\cot x}\to 0$

Now, RHL $=\lim_{x\to\pi^+}\dfrac{2^{\cot x}+3^{\cot x}-5^{1+\cot x}+2}{(2^{\cot x})^2+(9^{\cot x})^{1/2}-5^{\cot x}+1}$

$=\lim_{x\to\pi^+}\dfrac{2^{\cot x}+3^{\cot x}-5\cdot5^{\cot x}+2}{4^{\cot x}+3^{\cot x}-5^{\cot x}+1}$

$=\lim_{x\to\pi^+}\dfrac{\left(\frac{2}{5}\right)^{\cot x}+\left(\frac{3}{5}\right)^{\cot x}-5+\frac{2}{5^{\cot x}}}{\left(\frac{4}{5}\right)^{\cot x}+\left(\frac{3}{5}\right)^{\cot x}-1+\frac{1}{5^{\cot x}}}=\dfrac{-5}{-1}=5$

and LHL $=\lim_{x\to\pi^-}\dfrac{2^{\cot x}+3^{\cot x}-5^{1+\cot x}+2}{(2^{\cot x})^2+(9^{\cot x})^{1/2}-5^{\cot x}+1}$

$=\dfrac{2}{1}=2$

Hence, LHL$\neq$ RHL,

So, limits does not exist.

5. (b) $\lim_{x\to 0^+}\left(\dfrac{\cot^{-1}(\ln\{x\})}{\pi}\right)^{-\ln\{x\}}$

$=e^{\lim_{x\to 0^+}\left(\frac{\cot^{-1}(\ln\{x\})}{\pi}-1\right)(-\ln\{x\})}$

$=e^{\lim_{x\to 0^+}\left(\frac{\cot^{-1}(\ln x)-\pi}{\pi}\right)(-\ln x)}$ $[\because x\to 0^+, \{x\}=x]$

$=e^{\lim_{x\to 0^+}\left(\frac{\pi-\cot^{-1}(\ln x)}{\pi}\right)\ln(x)}$

$=e^{\lim_{x\to 0^+}\left(\frac{\cot^{-1}(-\ln x)}{\pi}\right)\ln(x)}$

$=e^{\lim_{x\to 0^+}\left(\frac{-\tan^{-1}\left(-\frac{1}{\ln x}\right)}{\left(\frac{-1}{\ln x}\right)\pi}\right)}=e^{\frac{-1}{\pi}}$

6. (a) We have, $\lim_{x\to 1^+}\dfrac{(1-|x|+\sin|1-x|)\sin\left(\frac{\pi}{2}[1-x]\right)}{|1-x|[1-x]}$

$\left[\begin{array}{l}\text{if } x\to 1^+, |1-x|=-(1-x)\\ \text{and } [1-x]=-1\end{array}\right]$

$=\lim_{x\to 1^+}\dfrac{\{1-x+\sin(-(1-x))\}\sin\left(\frac{\pi}{2}(-1)\right)}{-(1-x)(-1)}$

$=\lim_{x\to 1^+}\dfrac{\{1-x-\sin(1-x)\}\sin\left(-\frac{\pi}{2}\right)}{(1-x)}$

$=\lim_{x\to 1^+}\dfrac{(x-1)+\sin(1-x)}{1-x}$

$=\lim_{x\to 1^+}\left(\dfrac{x-1}{1-x}+\dfrac{\sin(1-x)}{1-x}\right)$

$=-1+1=0$

7. (a) $\lim_{x\to 0} \dfrac{Ae^x - B\cos x + Ce^{-x}}{x\sin x}$

$$= \lim_{x\to 0} \frac{A\left(1+x+\dfrac{x^2}{2!}+\ldots\right) - B\left(1-\dfrac{x^2}{2!}+\ldots\right) + C\left(1-x+\dfrac{x^2}{2!}+\ldots\right)}{x\left(x-\dfrac{x^3}{3!}+\ldots\right)}$$

$$= \lim_{x\to 0} \frac{(A-B+C)+(A-C)x+\left(\dfrac{A+B+C}{2}\right)x^2+\ldots}{x^2\left(1-\dfrac{x^2}{3!}+\ldots\right)}$$

For existance of limit, constant term and coefficient of x must be zero.

So, $A - B + C = 0$ ...(i)

and $A - C = 0$ ...(ii)

and $\dfrac{A+B+C}{2} = 2$

$\Rightarrow$ $A + B + C = 4$ ...(iii)

On solving, we get

$A = 1, B = 2$ and $C = 1$

$A + B + C = 1 + 2 + 1 = 4$

8. (c) We have,

$$p = \lim_{n\to\infty}\left(\frac{4}{(4-\sqrt{3}+2\sin\theta)}\right)^n \cdot \left(\frac{1}{(4-\sqrt{3}+2\sin\theta)^2}\right)$$

$$= \lim_{n\to\infty}\left(\frac{2}{2+\sin\theta-\dfrac{\sqrt{3}}{2}}\right)\left(\frac{1}{(4-)\sqrt{3}+2\sin\theta)^2}\right)$$

If $\theta\in\left(0,\dfrac{\pi}{3}\right) \Rightarrow \sin\theta\in\left(0,\dfrac{\sqrt{3}}{2}\right) \Rightarrow p$ does not exist

If $\theta\in\left(\dfrac{\pi}{3},\dfrac{\pi}{2}\right) \Rightarrow \sin\theta\in\left(\dfrac{\sqrt{3}}{2},1\right) \Rightarrow p = 0$

$\therefore$ For existence of limit $\theta = \dfrac{\pi}{3}$

So, $p = \dfrac{1}{4^2} = \dfrac{1}{16}$

Hence, $\dfrac{p+\cos\theta}{p} = 1 + \dfrac{\cos\theta}{p} = 1 + \dfrac{\cos\dfrac{\pi}{3}}{\left(\dfrac{1}{16}\right)} = 1 + 8 = 9$

9. (b) Given, $\lim_{x\to 0}\dfrac{f(x)}{x} = 1$

$\therefore$ $\lim_{x\to 0}\dfrac{x(1+a\cos x)-b\sin x}{x^3\cdot\left(\dfrac{f(x)}{x}\right)^3} = \dfrac{5}{6}$

$$\Rightarrow \lim_{x\to 0}\frac{(1+a\cos x)-\dfrac{b}{x}\sin x}{x^2} = \frac{5}{6}$$

$$\Rightarrow \lim_{x\to 0}\frac{\left(1+a\left(1-\dfrac{x^2}{2!}+\dfrac{x^4}{4!}+\ldots\right)-\dfrac{b}{x}\left[x-\dfrac{x^3}{3!}+\dfrac{x^5}{5!}-\ldots\right]\right)}{x^2} = \frac{5}{6}$$

For finite limit, $1 + a - b = 0 \Rightarrow b = a + 1$ ...(i)

and $\dfrac{-a}{2!}+\dfrac{b}{3!}=\dfrac{5}{6} \Rightarrow \dfrac{-a}{2}+\dfrac{b}{6}=\dfrac{5}{6}$

$\Rightarrow$ $-3a + b = 5 \Rightarrow -3a + a + 1 = 5$

$\Rightarrow$ $-2a = 4 \Rightarrow a = -2$

From Eq. (i), $b = -2 + 1 = -1$

10. (a) As, $n\to\infty$ $\lim_{n\to\infty} f(n) = \lim_{n\to\infty} f(n+1) = k$ (say)

We have, $f(n+1) = \dfrac{1}{2}\left(f(n)+\dfrac{9}{f(n)}\right)$

$\Rightarrow$ $\lim_{n\to\infty} f(n+1) = \lim_{n\to\infty}\dfrac{1}{2}\left(f(n)+\dfrac{9}{f(n)}\right)$

$\Rightarrow$ $k = \dfrac{1}{2}\left(k+\dfrac{9}{k}\right)$

$\Rightarrow$ $2k = \dfrac{k^2+9}{k}$

$\Rightarrow$ $k^2 = 9 \Rightarrow k = \pm 3$

$\therefore$ $\lim_{n\to\infty} f(n) = 3$ $[\because f(n) > 0]$

11. (c) $\lim_{x\to 2^-}\dfrac{ke^{\frac{1}{|x+2|}}-1}{2-e^{\frac{1}{|x+2|}}} = \lim_{x\to 2^-}\dfrac{k-e^{\frac{-1}{|x+2|}}}{2\cdot e^{\frac{-1}{x+2}}-1} = -k$

and $\lim_{x\to -2^+}\sin\left(\dfrac{x^4-16}{x^5+32}\right) = \lim_{x\to -2}\sin\left[\dfrac{\dfrac{x^4-(-2)^4}{x-(-2)}}{\dfrac{x^5-(-2)^5}{x-(-2)}}\right]$

$= \sin\left(-\dfrac{2}{5}\right) = -\sin\dfrac{2}{5}$

So, $-k = -\sin\dfrac{2}{5} \Rightarrow k = \sin\dfrac{2}{5}$

12. (c) $\lim_{x\to\infty}\log_{x-1}x\cdot\log_x(x+1)\cdot\log_{(x+1)}(x+2)\ldots\log_{(x^5-1)}x^5$

$$= \lim_{x\to\infty}\frac{\log x}{\log(x-1)}\cdot\frac{\log(x+1)}{\log x}\cdot\frac{\log(x+2)}{\log(x+1)}\cdots\frac{\log x^5}{\log(x^5-1)}$$

$$= \lim_{x\to\infty}\frac{\log x^5}{\log(x-1)} = \lim_{x\to\infty}\frac{5\log x}{\log(x-1)}$$

Using L-Hospital rule,

$$= \lim_{x\to\infty} \frac{5\cdot\frac{1}{x}}{\frac{1}{x-1}} = \lim_{x\to\infty}\frac{5(x-1)}{x} = 5$$

13. (d) We have,

$$a = \min[x^2+2x+3, x\in R]$$

$$\Rightarrow \quad a = \frac{-D}{4a} = \frac{-(4-12)}{4} = 2$$

and $$b = \lim_{x\to 0}\frac{\sin x\cos x}{e^x - e^{-x}} = \lim_{x\to 0}\frac{\sin x\cos x}{x\cdot e^{-x}\left(\frac{e^{2x}-1}{x}\right)}$$

$$= \lim_{x\to 0}\frac{\sin x\cos x}{2xe^{-x}\left(\frac{e^{2x}-1}{2x}\right)}$$

$$= \frac{1}{2}\left[\because \lim_{x\to 0}\frac{\sin x}{x} = 1 \text{ and } \lim_{x\to 0}\frac{e^x-1}{x} = 1\right]$$

Now, $$\sum_{r=0}^{n} a^r b^{n-r} = \sum_{r=0}^{n}(2)^r\left(\frac{1}{2}\right)^{n-r}$$

$$= \left(\frac{1}{2}\right)^n + 2\left(\frac{1}{2}\right)^{n-1} + 2^2\left(\frac{1}{2}\right)^{n-2} + 2^3\left(\frac{1}{2}\right)^{n-3} + \ldots$$

$$= \frac{1+4+4^2+\ldots 4^n}{2^n}$$

$$= \frac{1(4^{n+1}-1)}{2^n(4-1)} = \frac{4^{n+1}-1}{3\cdot 2^n}$$

14. (b) In ΔABC, $\tan\theta = \frac{CD}{AD} = \frac{CD}{x} \Rightarrow CD = x\tan\theta$ ...(i)

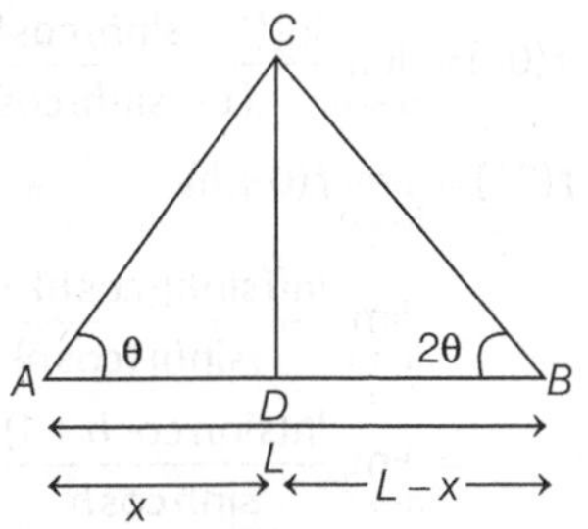

and $$\tan 2\theta = \frac{CD}{BD} = \frac{CD}{L-x}$$

$$\Rightarrow \quad CD = (L-x)\tan 2\theta$$

$$\Rightarrow \quad x\tan\theta = (L-x)\tan 2\theta \quad \text{[from Eq. (i)]}$$

$$\Rightarrow \quad x\tan\theta = (L-x)\left(\frac{2\tan\theta}{1-\tan^2\theta}\right)$$

$$\Rightarrow \quad x = \frac{2(L-x)}{1-\tan^2\theta} \Rightarrow x = \frac{2L}{3-\tan^2\theta}$$

$$\therefore \quad \lim_{\theta\to 0}\frac{2L}{3-\tan^2\theta} \Rightarrow \frac{2L}{3}$$

15. (c) $$\lim_{x\to 0}\frac{2f(x)-3f(2x)+f(4x)}{x^2}$$

Applying L-Hospital rule

$$= \lim_{x\to 0}\frac{2f'(x)-6f'(2x)+4f'(4x)}{2x}$$

$$= \lim_{x\to 0}\frac{2f''(x)-12f''(2x)+16f''(4x)}{2}$$

$$= \frac{6f''(0)}{2} = 3\times 4 = 12$$

16. (b,c) It is given that

$$\lambda = \lim_{x\to 0}\frac{k(1-4\sqrt{k^2-x^2})}{x^2\sqrt{k^2-x^2}} \text{ exist and finite}$$

$$\Rightarrow \quad \lim_{x\to 0} k(1-4\sqrt{k^2-x^2}) = 0$$

$$\Rightarrow \quad 1-4k = 0 \Rightarrow k = \frac{1}{4}$$

[$\because k > 0$, given]

$$\therefore \quad \lambda = \lim_{x\to 0}\frac{1-\sqrt{1-16x^2}}{x^2\sqrt{1-16x^2}}$$

$$\lambda = \lim_{x\to 0}\frac{1-(1-16x^2)}{x^2\sqrt{1-16x^2}\times(1+\sqrt{1-16x^2})}$$

$$\lambda = \frac{16}{1\times 2} = 8$$

17. (a, b) (a) $$\lim_{x\to\infty} x^{\frac{1}{4}}\sin\left(\frac{1}{\sqrt{x}}\right) = \lim_{x\to\infty} x^{\frac{1}{4}}\left(\frac{\sin\frac{1}{\sqrt{x}}}{\frac{1}{\sqrt{x}}}\right)\times\frac{1}{\sqrt{x}}$$

$$= \lim_{x\to\infty}\left(\frac{\sin\left(\frac{1}{\sqrt{x}}\right)}{\frac{1}{\sqrt{x}}}\right)\times\frac{1}{x^{\frac{1}{4}}} = 0$$

(b) $$\lim_{x\to\frac{\pi}{2}}(1-\sin x)\tan x$$

Let $x = \frac{\pi}{2}+h$

$$= \lim_{h\to 0}(1-\cos h)(-\cot h)$$

$$= \lim_{h\to 0}\frac{(1-\cos h)}{h^2}\cdot\frac{-h^2}{\tan h}$$

$$= \lim_{h\to 0}\frac{1-\cos h}{h^2}\cdot\frac{h}{\tan h}(-h) = \frac{1}{2}\cdot 1\cdot(0) = 0$$

(c) $$\lim_{x\to\infty}\frac{3x^2+5}{x^2-2x+5}\operatorname{sgn}(x) = \lim_{x\to\infty}\frac{x^2\left(3+\frac{5}{x^2}\right)}{x^2\left(1-\frac{2}{x}+\frac{5}{x^2}\right)}\cdot 1$$

$$= 3 \neq 0$$

(d) $\lim_{x\to 3} \frac{[x]^2-9}{x^2-9}$

$$\lim_{x\to 3^+} \frac{[x]^2-9}{x^2-9} = \lim_{x\to 3^+} \frac{9-9}{x^2-9} = 0$$

$$\lim_{x\to 3^-} \frac{[x]^2-9}{x^2-9} = \lim_{x\to 3^-} \frac{4-9}{x^2-9} = \infty$$

Hence, limit does not exist.

18. (b, d) (a) Let $f(x) = \frac{\sqrt{2+x}}{x}$ and $g(x) = \frac{-\sqrt{2-x}}{x}$, then

$$= \lim_{x\to 0}\left[\frac{\sqrt{2+x}}{x} + \frac{(-\sqrt{2-x})}{x}\right]$$

$$= \lim_{x\to 0}\left[\frac{\sqrt{2+x}-\sqrt{2-x}}{x}\right]$$

$$= \lim_{x\to 0} \frac{2+x-2+x}{x(\sqrt{2+x}+\sqrt{2-x})}$$

$$= \lim_{x\to 0} \frac{2}{(\sqrt{2+x}+\sqrt{2-x})} = \frac{1}{\sqrt{2}}$$

$\therefore \lim_{x\to 0} \{f(x)+g(x)\}$ exists but neither

$\lim_{x\to 0} \frac{\sqrt{2+x}}{x}$ nor $\lim_{x\to 0} \frac{-\sqrt{2-x}}{x}$ exists.

$\therefore$ Option (a) is false.

(b) $\lim_{x\to a} f(x) = l_1$ and $\lim_{x\to a} g(x) = l_2$

$\Rightarrow \lim_{x\to a} [f(x)+g(x)] = l_1 + l_2$

$\Rightarrow \lim_{x\to a} [f(x)+g(x)]$ exists

$\therefore$ Option (b) is true.

(c) Let $f(x) = \frac{|x|}{x}$ and $g(x) = |x|$

Then, $\lim_{x\to 0^-} \frac{|x|}{x} = -1$

and $\lim_{x\to 0^+} \frac{|x|}{x} = 1$

$\lim_{x\to 0} f(x)$ does not exist.

and $\lim_{x\to 0} |x| = 0$

$\Rightarrow \lim_{x\to 0} g(x)$ exists.

But $\lim_{x\to 0} f(x)\cdot g(x) = \lim_{x\to 0} \frac{|x|^2}{x} = \lim_{x\to 0} \frac{x^2}{x} = 0$

$\lim_{x\to 0} f(x)\cdot g(x)$ exists.

Thus, option (c) is false.

(d) $\lim_{x\to a} f(x) = l_1$ and $\lim_{x\to a} g(x) = l_2$

Then, $\lim_{x\to a} f(x)\cdot g(x) = l_1 \cdot l_2$

$\therefore$ Option (d) is true.

19. (a, c) (a) $f(x) = x^{2/3} = \lim_{h\to 0} \frac{h^{2/3}-0}{\sqrt{|h|}}$

$$= \lim_{h\to 0} \frac{|h|^{2/3}}{|h|^{1/2}} \qquad [\because h^{2/3} \text{ is positive}]$$

$$= \lim_{h\to 0} |h|^{1/6} = 0$$

So, $f(x) = x^{2/3}$ has property I is true.

(b) $f(x) = \sin x$

$$\lim_{h\to 0} \frac{\sin h - 0}{h^2} = \lim_{h\to 0} \frac{\sin h}{h\times h} = \lim_{h\to 0} \frac{1}{h}$$

So, limit does not exist.

(c) $f(x) = |x| = \lim_{h\to 0} \frac{|h|-0}{\sqrt{|h|}}$

$$= \lim_{h\to 0} \frac{|h|}{|h|^{1/2}} = \lim_{h\to 0} |h|^{1/2} = 0$$

So, $f(x) = |x|$ has property I.

(d) $f(x) = x|x|$

$$\lim_{h\to 0} \frac{h|h|-0}{h^2} = \lim_{h\to 0} \frac{|h|}{h}$$

So, limit does not exist.

20. (a, b) $\because f(0^-) = \lim_{h\to 0} f(0-h) = \lim_{h\to 0} \frac{\ln(\{-\sin h\}\{\cos h\}+1)}{\{-\sin h\}\{\cos h\}}$

$\because \quad h > 0,\ \sin h > 0 \Rightarrow -\sin h < 0$

So, $\quad \{-\sin h\} = 1 - \sin h$

$\therefore \quad f(0^-) = \lim_{h\to 0} \frac{\ln[(1-\sin h)\cos h + 1]}{(1-\sin h)\cos h} = \ln 2 \quad \text{...(i)}$

and $\quad f(0^+) = \lim_{h\to 0} f(0+h)$

$$= \lim_{h\to 0} \frac{\ln(\{\sin h\}\{\cos h\}+1)}{\{\sin h\}\{\cos h\}}$$

$$= \lim_{h\to 0} \frac{\ln(\sin h \cos h + 1)}{\sin h \cos h} = 1 \quad \text{...(ii)}$$

Now, $f\left(\frac{\pi^+}{2}\right) = \lim_{h\to 0} f\left(\frac{\pi}{2}+h\right)$

$$= \lim_{h\to 0} \frac{\ln\left(\left\{\sin\left(\frac{\pi}{2}+h\right)\right\}\left\{\cos\left(\frac{\pi}{2}+h\right)\right\}+1\right)}{\left\{\sin\left(\frac{\pi}{2}+h\right)\right\}\left\{\cos\left(\frac{\pi}{2}+h\right)\right\}}$$

$$= \lim_{h\to 0} \frac{\ln(\{\cos h\}\{-\sin h\}+1)}{\{\cos h\}\{-\sin h\}}$$

$= f(0^-)$ [from Eq. (i)]

and $f\left(\frac{\pi^-}{2}\right) = \lim_{h\to 0} f\left(\frac{\pi}{2}-h\right)$

$$= \lim_{h\to 0} \frac{\ln\left(\left\{\sin\frac{\pi}{2}-h\right\}\left\{\cos\frac{\pi}{2}-h\right\}+1\right)}{\left\{\sin\frac{\pi}{2}-h\right\}\left\{\cos\frac{\pi}{2}-h\right\}}$$

$$= \lim_{h\to 0} \frac{\ln(\{\cos h\}\{\sin h\}+1)}{\{\cos h\}\{\sin h\}}$$

$= f(0^+)$ [from Eq. (ii)]

Now, $f(0^-) \neq f(0^+)$, so limit doesn't exist at $x = 0$

and $f\left(\frac{\pi^+}{2}\right) \neq f\left(\frac{\pi^-}{2}\right)$ So, limit doesn't exists at $x = \frac{\pi}{2}$.

21. (c, d) We have,

$$f(x) = \frac{1 - x(1+|1-x|)}{|1-x|}\cos\left(\frac{1}{1-x}\right)$$

$$\lim_{x\to 1^+} f(x) = \lim_{h\to 0} \frac{1-(1+h)(1+h)}{h}\cos\frac{1}{h}$$

$$= \lim_{h\to 0} \frac{-h^2-2h}{h}\cos\frac{1}{h}$$

$$= \lim_{h\to 0} (-h-2)\cos\frac{1}{h}$$

Thus, $\lim_{x\to 1^+} f(x)$ does not exist. $\left[\because \cos\frac{1}{h} = \cos\infty \in [-1,1]\right]$

and $\lim_{x\to 1^-} f(x) = \lim_{h\to 0} \frac{1-(1-h)(1+h)}{h}\cos\frac{1}{h}$

$$= \lim_{h\to 0} \frac{1-(1-h^2)}{h}\cos\frac{1}{h}$$

$$= \lim_{h\to 0} \frac{h^2}{h}\cos\frac{1}{h} = \lim_{h\to 0} h\cos\left(\frac{1}{h}\right) = 0$$

So, $\lim_{x\to 1^-} f(x) = 0$

22. (b, d) Let $L = \lim_{x\to 0}\left(\frac{a^x+b^x+c^x}{3}\right)^{\frac{k}{x}}, a,b,c,k > 0$

$$= e^{\lim_{x\to 0}\left(\frac{a^x+b^x+c^x}{3}-1\right)\left(\frac{k}{x}\right)}$$

$$= e^{\lim_{x\to 0}\left[\frac{a^x-1}{x}+\frac{b^x-1}{x}+\frac{c^x-1}{x}\right]\frac{k}{3}}$$

$$= e^{\frac{k}{3}[\ln a + \ln b + \ln c]}$$

$$= e^{\frac{k}{3}[\ln(abc)]} = (abc)^{\frac{k}{3}}$$

$$\Rightarrow \quad L = \begin{cases} abc, & \text{if } k = 3 \\ (a^2b^2c^2)^{1/3}, & \text{if } k = 2 \\ (a\,b\,c)^{1/3}, & \text{if } k = 1 \end{cases}$$

23. (a, d) We have, $\lim_{x\to 0}\frac{1-\cos x}{e^{ax}-bx-1} = 1$

$$\Rightarrow \lim_{x\to 0} \frac{\left(\frac{2\sin^2\frac{x}{2}}{\left(\frac{x}{2}\right)^2}\right)\cdot\left(\frac{x}{2}\right)^2}{\left(1+\frac{ax}{1!}+\frac{a^2x^2}{2!}+\ldots\right)-bx-1} = 1$$

$$\Rightarrow \lim_{x\to 0} \frac{x^2}{x(a-b)+\frac{a^2x^2}{2}+\ldots} = 2$$

$$\Rightarrow \lim_{x\to 0} \frac{x}{(a-b)+\frac{a^2x}{2}+\ldots} = 2$$

If limit exists, then

$$a - b = 0 \Rightarrow a = b$$

and $\frac{1}{\frac{a^2}{2}} = 2 \Rightarrow a^2 = 1 \Rightarrow a = \pm 1$

So, if $a = 1, b = 1$ and if $a = -1, b = -1$

24. (a, b, c) $\lim_{x\to 0}\frac{\sin x^n}{(\sin x)^m} = \lim_{x\to 0}\frac{\frac{\sin x^n}{x^n}\times x^n}{\left(\frac{\sin x}{x}\right)^m \times x^m}$

$$= \lim_{x\to 0} x^{n-m}$$

If $m = n$, then $\lim_{x\to 0} x^{\circ} = 1$

If $n > m$, then $\lim_{x\to 0}(x)^{(\text{positive integer})} = 0$

If $m > n$, then $\lim_{x\to 0}(x)^{(\text{negative integer})} = \infty$

25. (c, d) $f(x) = \lim_{n\to\infty} n(x^{1/n}-1) = \lim_{n\to\infty}\left(\frac{x^{1/n}-1}{1/n}\right)$

Let $\frac{1}{n} = y$

$$= \lim_{y\to 0}\left(\frac{x^y-1}{y}\right)$$

$$f(x) = \log x$$

Now, $f\left(\frac{1}{x}\right) = \log\left(\frac{1}{x}\right)$

$$= \log 1 - \log x$$
$$= -\log x = -f(x)$$

and $f(xy) = \log(xy)$

$$= \log x + \log y$$
$$= f(x) + f(y)$$

26. (a, c) Given, $(\sin\alpha+\sin\beta)+\dfrac{\sin\alpha}{\sin\beta}=0$ ...(i)

$\Rightarrow \quad (\sin\alpha+\sin\beta)=\dfrac{-\sin\alpha}{\sin\beta}$

and $\quad (\sin\alpha+\sin\beta)\dfrac{\sin\alpha}{\sin\beta}=-1$

$\Rightarrow \quad \dfrac{\sin^2\alpha}{\sin^2\beta}=1 \Rightarrow \sin\alpha=\sin\beta$

Substituting in Eq. (i),

$$2\sin\alpha+1=0$$

$$\sin\alpha=\frac{-1}{2} \Rightarrow \alpha=\frac{-\pi}{6}$$

So, $\quad \lambda=\lim_{n\to\infty}\dfrac{1+(-1)^{2n}}{(-1)^{2n}}=2$

27. (b, d) $\lim_{x\to-\infty} f(x)=\lim_{x\to-\infty}\left(\dfrac{|x|}{|x|+2}\right)^{-x}=\lim_{x\to-\infty}\left(\dfrac{-x}{-x+2}\right)^{-x}$

Let $y=-x$

$$=\lim_{y\to\infty}\left(\frac{y}{y+2}\right)^{y}=e^{\lim_{y\to\infty}\left(\frac{-2}{y+2}\right)\cdot y}=e^{-2}$$

and $\quad \lim_{x\to\infty} f(x)=\lim_{x\to\infty}\left(\dfrac{x}{x+2}\right)^{-x}$

$$=e^{\lim_{x\to\infty}\left(\frac{x}{x+2}-1\right)(-x)}$$

$$=e^{\lim_{x\to\infty}\left(\frac{+2x}{x+2}\right)}=e^{2}$$

and $\quad \lim_{x\to0} f(x)=\lim_{x\to0}\left(\dfrac{|x|}{|x|+2}\right)^{-x}$

$\text{LHL}=\lim_{x\to0^-}\left(\dfrac{|x|}{|x|+2}\right)^{-x}=\lim_{h\to0}\left(\dfrac{h}{h+2}\right)^{h}=l_1$ (let)

$\Rightarrow \quad \ln l_1=\lim_{h\to0} h\ln\left(\dfrac{h}{h+2}\right)$

$$=\lim_{h\to0}\left(\frac{h+2}{h}\right)\left(\frac{h+2-h}{(h+2)^2}\right)\left(\frac{-h^2}{1}\right)$$

$$=\lim_{h\to0}\frac{-2h}{(h+2)}=0$$

$\Rightarrow \quad \ln l_1=0 \Rightarrow l_1=e^0=1$

and $\text{RHL}=\lim_{x\to0^+}\left(\dfrac{x}{x+2}\right)^{-x}=\lim_{x\to0^+}\left(\dfrac{x+2}{x}\right)^{x}=l_2$ (let)

$\Rightarrow \quad \ln l_2=\lim_{x\to0^+} x\ln\left(\dfrac{x+2}{x}\right)$

$$=\lim_{x\to0^+}\frac{\ln\left(\frac{x+2}{x}\right)}{\frac{1}{x}}$$

$$=\lim_{x\to0^+}\left(\frac{x}{x+2}\right)\times\left(\frac{-2}{x^2}\right)\left(\frac{-x^2}{1}\right)$$

$\Rightarrow \quad \ln l_2=0$

$\Rightarrow \quad l_2=e^0=1$

So $\quad \lim_{x\to0} f(x)=1$

28. (a, b) We have, $f(x)=\lim_{m\to\infty}\left\{\lim_{n\to\infty}[\cos^{2m}(n!\,\pi x)]\right\}, x\in R$

Case I When x is rational.

Let $x=\dfrac{p}{q}$, where $p,q,\in z, q\neq0$

When $n\to\infty$, then $n!$ would contain q as one of its factor.

$\Rightarrow n!\pi x$ is multiple of π.

So, $\quad \cos^2(n!\pi x)=1$

$\therefore \quad \lim_{m\to\infty}\left\{\lim_{n\to\infty}[\cos^{2m}(n!\pi x)]\right\}=1$

Case II When x is irrational, then $n!\pi x$ is not multiple of π.

So, $\cos(n!\pi x)$ be between -1 to 1.

So, $\cos^2(n!\,\pi x)$ be between 0 to 1.

$\Rightarrow \lim_{m\to\infty}\left[\lim_{n\to\infty}[\cos^{2m}(n!\pi x)]\right]=0$

29. (b, c, d) We have,

$$\lim_{x\to\infty}\left(\frac{ax+1}{bx+2}\right)^{x}=\lim_{x\to\infty}\left(\frac{a+\frac{1}{x}}{b+\frac{2}{x}}\right)^{x}=\left(\frac{a}{b}\right)^{\infty}$$

If $0<\dfrac{a}{b}<1$, then $\left(\dfrac{a}{b}\right)^{\infty}=0$

$\therefore \quad \lim_{x\to\infty} f(x)=0$, if $0<a<b$

If $\quad \dfrac{a}{b}>1$, then $\left(\dfrac{a}{b}\right)^{\infty}=\infty$

$\lim_{x\to\infty} f(x)$ is non existent, if $a>b>0$

If $\quad a=b$

$$\lim_{x\to\infty} f(x)=\lim_{x\to\infty}\left(\frac{ax+1}{ax+2}\right)^{x}$$

$$=e^{\lim_{x\to\infty}\left(\frac{ax+1}{ax+2}-1\right)x}$$

$$=e^{\lim_{x\to\infty}\frac{-x}{ax+2}}=e^{-\frac{1}{a}}$$

or $\quad =e^{-\frac{1}{b}}$ $\quad [\because a=b]$

Sol. (Q.Nos. 30-32)

30. (b) **31.** (a) **32.** (c)

$$f(x)=\lim_{n\to\infty}\left(\cos\frac{x}{\sqrt{n}}\right)^n = e^{\lim_{n\to\infty}\left(\cos\frac{x}{\sqrt{n}}-1\right)\cdot n}$$

Substituting $n=\frac{1}{t^2}$, we get

$$f(x)=e^{\lim_{t\to 0}\left(\frac{\cos tx-1}{t^2}\right)}$$

$$=e^{\lim_{t\to 0}-x^2\left(\frac{2\sin^2\left(\frac{tx}{2}\right)}{4(t^2x^2/4)}\right)}=e^{-\frac{1}{2}x^2}$$

Now, $\quad g(x)=-x^{4b}$ [given]

where, $\quad b=\lim_{x\to\infty}(\sqrt{x^2+x+1}-\sqrt{x^2+1})$

$$=\lim_{x\to\infty}\left(\frac{x^2+x+1-x^2-1}{\sqrt{x^2+x+1}+\sqrt{x^2+1}}\right)=\frac{1}{2}$$

$\therefore \quad g(x)=-x^{4(1/2)}=-x^2$

$\therefore f(x)+g(x)=0 \quad \Rightarrow f(x)=-g(x)\Rightarrow e^{-\frac{x^2}{2}}=x^2$

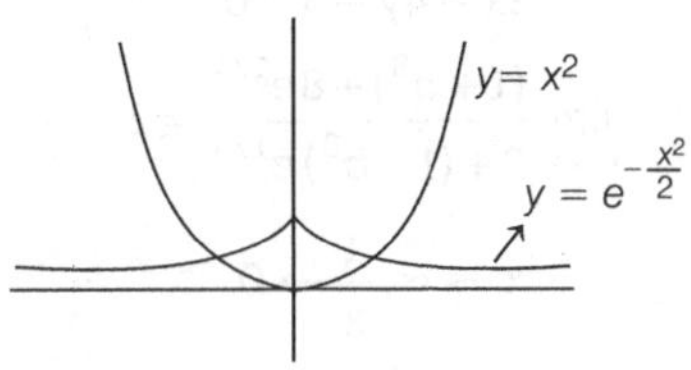

The graph of $f(x)$ and $-g(x)$ intersect each other at two points.

So, number of solution $=2$

Sol. (Q.Nos. 33-35)

33. (a, b) **34.** (a, c) **35.** (b)

$$\lim_{x\to 0}\frac{axe^x-b\log(1+x)+cxe^{-x}}{x^2\sin x}=2$$

Applying L-Hospital rule,

$$\lim_{x\to 0}\frac{ax\,e^x+ae^x-\frac{b}{(1+x)}+ce^{-x}-cxe^{-x}}{x^2\cos x+2x\sin x}=2$$

$\Rightarrow \quad a-b+c=0 \quad$ ...(i)

Again, $$\lim_{x\to 0}\frac{2ae^x+axe^x+\frac{b}{(1+x)^2}-2ce^{-x}+cxe^{-x}}{-x^2\sin x+2x\cos x+2\sin x+2x\cos x}=2$$

$\Rightarrow \quad 2a+b-2c=0 \quad$ ... (ii)

$$\Rightarrow \lim_{x\to 0}\frac{2ae^x+ae^x+axe^x-\frac{2b}{(1+x)^3}+3ce^{-x}-cxe^{-x}}{(-x^2\cos x-2x\sin x+6\cos x-4x\sin x)}=2$$

$\Rightarrow \quad \frac{3a-2b+3c}{6}=2$

$\Rightarrow \quad 3a-2b+3c=12 \quad$ ... (iii)

Eq. (i) + Eq. (ii) gives,

$$3a-c=0$$

Eq. (ii) $\times$ 2 + Eq. (iii) gives,

$$7a-c=12$$

$\Rightarrow \quad a=3, c=9$ and $b=12$

$$ax^2+bx+c=0$$

$\Rightarrow \quad 3x^2+12x+9=0$

$\Rightarrow \quad x^2+4x+3=0$

$\Rightarrow \quad (x+1)(x+3)=0$

$\Rightarrow x=-1\,,-3$, which are real unequal and rational.

$$f(x)=\sqrt{\frac{ax^2-b}{c}}\Rightarrow\frac{ax^2-b}{c}\geq 0$$

$\Rightarrow \quad \frac{3x^2-12}{9}\geq 0 \Rightarrow x^2-4\geq 0$

$\Rightarrow \quad x\in(-\infty,-2]\cup[2,\infty)$.

36. (i) $\to$ (C), (ii) $\to$ (D), (iii) $\to$ (A), (iv) $\to$ (B)

(i) $f(x)=\frac{\tan[e^2]x^2-\tan[-e^2]x^2}{\sin^2 x}$

$$=\frac{\tan 7x^2-\tan(-8)x^2}{\sin^2 x}=\frac{\tan 7x^2+\tan 8x^2}{\sin^2 x}$$

$$\lim_{x\to 0}f(x)=\frac{7x^2+8x^2}{x^2}=15$$

(ii) $f(x)=\frac{\left[\frac{5}{2}+\tan x+\tan^2 x\right]-\left[\frac{5}{2}\right]}{\tan x}$

$$\because \frac{5}{2}+\tan x+\tan^2 x=\tan^2 x+\tan x+\frac{1}{4}+\frac{9}{4}$$

$$=\left(\tan x+\frac{1}{2}\right)^2+\frac{9}{4}\to\frac{5}{2}\text{ as }x\to 0$$

$$=\lim_{x\to 0}\frac{\left[\frac{5}{2}+\tan x+\tan^2 x\right]-\left[\frac{5}{2}\right]}{\tan x}$$

$$=\lim_{x\to 0}\frac{2-2}{\tan x}=0$$

(iii) $\lim_{x\to 0}f(x)=\lim_{x\to 0}\frac{(1+x^2)^{1/3}-(1-2x)^{1/4}}{x(1+x)}$

$$=\lim_{x\to 0}\frac{\left(1+\frac{1}{3}x^2+\left(\frac{1}{3}\right)\left(\frac{-2}{3}\right)\frac{x^4}{2!}+\ldots\right)-\left(1-\frac{1}{4}(2x)+\frac{1}{4}\left(-\frac{3}{4}\right)\left(\frac{4x^2}{2!}\right)-\ldots\right)}{x(1+x)}$$

$$=\lim_{x\to 0}\left[\frac{\frac{1}{2}x+\frac{17}{24}x^2+\ldots}{x(1+x)}\right]=\frac{1}{2}$$

(iv) $\lim_{x\to 0} \dfrac{1-\cos x+2\sin x-\sin^3 x-x^2+3x^4}{\tan^3 x-6\sin^2 x+x-5x^3}$

$$= \lim_{x\to 0} \frac{2\left(\frac{x}{2}\right)^2+2x-x^3-x^2+3x^4}{x^3-6x^2+x-5x^3}=2$$

[when $x \to 0$, then $\sin x = x$]

37. (i) → (A), (ii) → (D), (iii) → (B), (iv) → (C)

(i) $\lim_{x\to 0} \dfrac{\int_0^x \frac{t^2}{\sqrt{b+t}}dt}{ax-\sin x}=1$

By L- Hospital rule,

$$= \lim_{x\to 0} \frac{\frac{x^2}{\sqrt{b+x}}(1)-0}{a-\cos x}=1$$

$$= \lim_{x\to 0} \frac{x^2}{(\sqrt{b+x})(a-\cos x)}=1$$

$\Rightarrow \quad \sqrt{b}(a-1)=0$ [for limit to be exist]

$\Rightarrow \quad a=1$ [$\because b\neq 0$]

$$\Rightarrow \lim_{x\to 0} \frac{x^2}{(\sqrt{b+x})(1-\cos x)}=1$$

$$\Rightarrow \lim_{x\to 0} \frac{x^2}{\sqrt{b+x}\left(2\sin^2\frac{x}{2}\right)}$$

$$\Rightarrow \lim_{x\to 0} \frac{4(x/2)^2}{2\sqrt{b+x}\sin^2\frac{x}{2}} \Rightarrow \frac{1}{2\sqrt{b}}(4)(1)^2=1$$

$\Rightarrow \quad \sqrt{b}=2 \Rightarrow b=4$

$\therefore \quad (a,b)\equiv(1,4)$

which satisfies the line $x-y+3=0$

(ii) $\lim_{x\to\infty} \sqrt{x^4+x^2-1}-ax^2+b=0$

$$\Rightarrow \lim_{x\to\infty} \frac{x^4+x^2-1-(ax^2-b)^2}{\sqrt{x^4+x^2-1}+(ax^2-b)}=0$$

$$\Rightarrow \lim_{x\to\infty} \frac{x^4(1-a^2)+x^2(1+2ab)+(-1-b^2)}{\sqrt{x^4+x^2-1}+(ax^2-b)}=0$$

$\Rightarrow \quad (1-a^2)=0 \Rightarrow a=\pm 1$

but $\quad a>0 \Rightarrow a=1$

$$\Rightarrow \lim_{x\to\infty} \frac{(1+2b)-\left(\frac{1+b^2}{x^2}\right)}{\sqrt{1+\frac{1}{x^2}-\frac{1}{x^4}}+\left(1-\frac{b}{x^2}\right)}=0$$

$$\Rightarrow \frac{1+2b}{2}=0 \Rightarrow b=-\frac{1}{2}$$

$(a,b)\equiv\left(1,-\frac{1}{2}\right)$, which lies on the line $x+6y+2=0$

(iii) $\lim_{x\to\infty} \sqrt{x^2+x+1}-(ax+b)=0$

$$\Rightarrow \lim_{x\to\infty} \frac{(x^2+x+1)-(ax+b)^2}{\sqrt{x^2+x+1}+(ax+b)}=0$$

$$\Rightarrow \lim_{x\to\infty} \frac{x^2(1-a^2)+x(1-2ab)+(1-b^2)}{\sqrt{x^2+x+1}+(ax+b)}=0$$

$\Rightarrow \quad a^2-1=0$

$a=\pm 1,$

but $\quad a>0$

So, $\quad a=1$

$$\Rightarrow \lim_{x\to\infty} \frac{x(1-2b)+(1-b^2)}{\sqrt{x^2+x+1}+(x+b)}=0$$

$$\Rightarrow \lim_{x\to\infty} \frac{(1-2b)+\left(\frac{1-b^2}{x}\right)}{\sqrt{1+\frac{1}{x}+\frac{1}{x^2}}+\left(1+\frac{b}{x}\right)}=0$$

$$\Rightarrow \frac{1-2b}{2}=0 \Rightarrow b=\frac{1}{2}$$

$\therefore (a,b)=(1, 1/2)$, which has on the line

$3x+4y-5=0$

(iv) $\lim_{x\to\infty} \dfrac{(1+a^3)+8e^{1/x}}{2+(1-b^3)e^{1/x}}=2$

As, $\quad x\to\infty, \frac{1}{x}\to 0$

$\Rightarrow \quad e^{1/x}\to e^0=1$

$$\Rightarrow \frac{(1+a^3)+8}{2+(1-b^3)}=2$$

$$\Rightarrow \frac{a^3+9}{3-b^3}=2$$

$\Rightarrow \quad a^3+2b^3+3=0$

$\Rightarrow (a^3,b^3)$ satisfy the line $x+2y+3=0$

38. (i) → (D), (ii) → (B), (iii) → (A), (iv) → (C)

(i) $\lim_{x\to-\infty} \dfrac{3^{\sin x}+2x+1}{\sin x-\sqrt{x^2+1}} = \lim_{x\to-\infty} \dfrac{\frac{3^{\sin x}}{x}+2+\frac{1}{\sin x}}{\frac{\sin x}{x}-\frac{1}{x}\sqrt{x^2+1}}$

$$= \lim_{x\to-\infty} \frac{\frac{3^{\sin x}}{x}+2+\frac{1}{\sin x}}{\frac{\sin x}{x}-\frac{1}{x}|x|\sqrt{1+\frac{1}{x^2}}}$$

$$= \lim_{x\to-\infty} \frac{\frac{3^{\sin x}}{x}+2+\frac{1}{\sin x}}{\frac{\sin x}{x}+\sqrt{1+\frac{1}{x^2}}}=2$$

(ii) $\lim_{m\to\infty}\lim_{n\to\infty}\sum_{k=1}^{m}\frac{\sqrt[n]{(k-1)^n+k^n}}{m^2}$

$$=\lim_{m\to\infty}\lim_{n\to\infty}\left(\frac{1+\sqrt[n]{1^n+2^n}+\sqrt[n]{2^n+3^n}+\ldots+\sqrt[n]{(m-1)^n+m^n}}{m^2}\right)$$

$$=\lim_{m\to\infty}\lim_{n\to\infty}\left(\frac{1+2\cdot n\sqrt{1+\left(\frac{1}{2}\right)^n}+3\cdot n\sqrt{1+\left(\frac{2}{3}\right)^n}+\ldots+m\cdot n\sqrt{1+\left(\frac{m-1}{m}\right)^n}}{m^2}\right)$$

$$=\lim_{m\to\infty}\left(\frac{1+2+3+\ldots..+m}{m^2}\right)$$

$$=\lim_{m\to\infty}\frac{m(m+1)}{2m^2}=\frac{1}{2}$$

(iii) $\lim_{n\to\infty}\left[\frac{n}{n^2-4}+\frac{4^n(-1)^n}{2^n-1}\right]^{-1}$

$$=\lim_{n\to\infty}\left[\frac{\frac{1}{n}}{1-\frac{4}{n^2}}+\frac{4^n(-1)^n}{2^n-1}\right]^{-1}$$

$$=\lim_{n\to\infty}\left(\frac{1}{\frac{1/n}{1-4/n^2}+\frac{4^n(-1)^n}{2^n-1}}\right)$$

$$=\lim_{n\to\infty}\left(\frac{1}{\frac{4^n(-1)^n}{2^n-1}}\right)=\lim_{n\to\infty}\frac{1}{\left(\frac{4^n-1+1}{(2^n-1)}\right)(-1)^n}$$

$$=\lim_{n\to\infty}\frac{1}{\left(2^n+1+\frac{1}{2^n-1}\right)(-1)^n}=\frac{1}{\infty}=0$$

(iv) $\lim_{x\to-\infty}\left\{x+\sqrt{x^2+2x\cos\frac{1}{|x|}}\right\}$

$$=\lim_{x\to-\infty}\left\{x+\sqrt{x^2+2x\cos\frac{1}{x}}\right\}$$

$$\left[\because x\to-\infty\Rightarrow|x|=-x \quad \cos\left(\frac{1}{|x|}\right)=\cos\left(-\frac{1}{x}\right)=\cos\frac{1}{x}\right]$$

$$=\lim_{x\to-\infty}\left(\frac{x^2-\left(x^2+2x\cos\frac{1}{x}\right)}{x-\sqrt{x^2+2x\cos\frac{1}{x}}}\right)$$

$$=\lim_{x\to-\infty}\left(\frac{-2x\cos\frac{1}{x}}{x-|x|\sqrt{1+\frac{2}{x}\cos\frac{1}{x}}}\right)$$

$$=\lim_{x\to-\infty}\left(\frac{-2x\cos\frac{1}{x}}{x+x\sqrt{1+\frac{2}{x}\cos\frac{1}{x}}}\right)$$

$$=\frac{-2}{2}=-1$$

39. (i)$\to$(D), (ii)$\to$(C), (iii)$\to$(A), (iv)$\to$(B)

(i) We have,

$$\lim_{x\to\frac{\pi}{2}^+}\frac{\cos[\tan^{-1}(\tan x)]}{x-\frac{\pi}{2}} \qquad \left[\frac{0}{0}\text{form}\right]$$

$$=\lim_{x\to\frac{\pi}{2}^+}\frac{\cos(x-\pi)}{x-\frac{\pi}{2}}$$

$\left[x\to\frac{\pi^+}{2}, x \text{ is in II quadrant, so} \tan^{-1}(\tan x)=x-\pi\right]$

$$=\lim_{x\to\frac{\pi}{2}^+}\frac{\cos(\pi-x)}{x-\frac{\pi}{2}}$$

$$=\lim_{x\to\frac{\pi}{2}^+}\frac{-\cos x}{x-\frac{\pi}{2}} \qquad \left[\frac{0}{0}\text{form}\right]$$

Applying in L-Hospital rule,

$$=\lim_{x\to\frac{\pi}{2}^+}\frac{-(-\sin x)}{1}=\sin\frac{\pi}{2}=1$$

(ii) We have,

$$\sum_{r=1}^{n}\tan^{-1}\left(\frac{4}{4r^2+3}\right)=\sum_{r=1}^{n}\tan^{-1}\left(\frac{1}{r^2+\frac{3}{4}}\right)$$

$$=\sum_{r=1}^{n}\tan^{-1}\left(\frac{1}{1+r^2-\frac{1}{4}}\right)$$

$$=\sum_{r=1}^{n}\tan^{-1}\left(\frac{\left(r+\frac{1}{2}\right)-\left(r-\frac{1}{2}\right)}{1+\left(r+\frac{1}{2}\right)\left(r-\frac{1}{2}\right)}\right)$$

$$=\sum_{r=1}^{n}\left[\tan^{-1}\left(r+\frac{1}{2}\right)-\tan^{-1}\left(r-\frac{1}{2}\right)\right]$$

$$=\left(\tan^{-1}\frac{3}{2}-\tan^{-1}\frac{1}{2}\right)+\left(\tan^{-1}\frac{5}{2}-\tan^{-1}\frac{3}{2}\right)+\left(\tan^{-1}\frac{7}{2}-\tan^{-1}\frac{5}{2}\right)+\ldots+\left(+\tan^{-1}\left(n+\frac{1}{2}\right)-\tan^{-1}\left(n-\frac{1}{2}\right)\right)$$

$$=\tan^{-1}\left(n+\frac{1}{2}\right)-\tan^{-1}\frac{1}{2}$$

Now, $\lim_{n\to\infty}\tan\left[\tan^{-1}\left(n+\frac{1}{2}\right)-\tan^{-1}\left(\frac{1}{2}\right)\right]$

$$=\lim_{n\to\infty}\frac{\tan\left[\tan^{-1}\left(n+\frac{1}{2}\right)\right]-\tan\left(\tan^{-1}\frac{1}{2}\right)}{1+\tan\left[\tan^{-1}\left(n+\frac{1}{2}\right)\right]\tan\left(\tan^{-1}\frac{1}{2}\right)}$$

$$=\lim_{n\to\infty}\frac{n+\frac{1}{2}-\frac{1}{2}}{1+\left(n+\frac{1}{2}\right)\left(\frac{1}{2}\right)}=\lim_{n\to\infty}\frac{n}{1+\frac{n}{2}+\frac{1}{4}}$$

$$=\lim_{n\to\infty}\frac{1}{\frac{1}{n}+\frac{1}{2}+\frac{1}{4n}}=\frac{1}{1/2}=2$$

(iii) We have, $\lim_{x\to-\infty}\frac{(3x^4+2x^2)\sin\left(\frac{1}{x}\right)+|x|^3+5}{|x|^3+|x|^2+|x|+1}$

$$=\lim_{x\to-\infty}\frac{(3x^3+2x)\left(\frac{\sin\frac{1}{x}}{1/x}\right)+(-x)^3+5}{(-x)^3+(-x)^2+(-x)+1}$$

$\left[\text{if } x\to-\infty, \text{ then } |x|=-x\right]$

$$=\lim_{x\to-\infty}\frac{2x^3+2x+5}{-x^3+x^2-x+1}$$

$$=\lim_{x\to-\infty}\frac{2+\frac{2}{x^2}+\frac{5}{x^3}}{-1+\frac{1}{x}-\frac{1}{x^2}+\frac{1}{x^3}}=-2$$

(iv) We have, $f(x)=\lim_{t\to0}\left[\frac{2x}{\pi}\tan^{-1}\left(\frac{x}{t^2}\right)\right]$

when $t\to0$, $\frac{1}{t}\to\infty$ and $\frac{1}{t^2}\to\infty$.

$$=\frac{2x}{\pi}\tan^{-1}(\infty)=\frac{2x}{\pi}\times\frac{\pi}{2}$$

$f(x)=x$

So, $f(-1)=-1$

Sol. (Q.Nos. 40-42)

40. (0.7) **41.** (1) **42.** (1)

$$f(x)=\frac{\tan^{-1}(1-\{x\})\cos^{-1}(1-\{x\})\sin^{-1}(1-\{x\})}{\sqrt{2\{x\}}(1-\{x\})^2}$$

$$P=\lim_{x\to0^+}f(x)$$

$$=\lim_{x\to0^+}\frac{\tan^{-1}(1-x)\cos^{-1}(1-x)\sin^{-1}(1-x)}{\sqrt{2x}(1-x)^2}$$

$[\because x\to0^+\Rightarrow\{x\}=x]$

$$=\lim_{x\to0^+}\frac{\tan^{-1}(1-x)}{1-x}\cdot\frac{\sin^{-1}(1-x)}{1-x}\cdot\frac{\cos^{-1}(1-x)}{\sqrt{2}x}$$

$$=\frac{\pi}{4}\cdot\frac{\pi}{2}\cdot\lim_{x\to0^+}\frac{\cos^{-1}(1-x)}{\sqrt{2}\sqrt{x}}$$

Let $\cos^{-1}(1-x)=\theta$

$x\to0^+,(1-x)\in(0,1)$

$\Rightarrow\quad \theta\in\left(0,\frac{\pi}{2}\right)\Rightarrow\sin\theta>0$

$$\frac{\pi^2}{8}\lim_{\theta\to0^+}\frac{\theta}{\sqrt{2}\sqrt{1-\cos\theta}}=\frac{\pi^2}{8\sqrt{2}}\lim_{\theta\to0^+}\frac{\theta}{\sqrt{2\sin^2\frac{\theta}{2}}}$$

$$=\frac{\pi^2}{8\sqrt{2}}\lim_{\theta\to0^+}\frac{\theta}{\sqrt{2}\sin\frac{\theta}{2}}\left[\because\sin\frac{\theta}{2}>0\right]$$

$$=\frac{\pi^2}{16}\lim_{\theta\to0^+}\frac{2\left(\frac{\theta}{2}\right)}{\sin\frac{\theta}{2}}\Rightarrow P=\frac{\pi^2}{8}$$

So, $\cos\left(\frac{2P}{\pi}\right)=\cos\left(\frac{2}{\pi}\times\frac{\pi^2}{8}\right)=\cos\left(\frac{\pi}{4}\right)=\frac{1}{\sqrt{2}}=0.7$

Now, $Q=\lim_{x\to0^-}f(x)$

$$=\lim_{x\to0^-}\frac{\tan^{-1}(1-\{x\})\cos^{-1}(1-\{x\})\sin^{-1}(1-\{x\})}{\sqrt{2\{x\}}(1-\{x\})^2}$$

$[\because x\to0^-\Rightarrow\{x\}=1+x]$

$$=\lim_{x\to0^-}\frac{\tan^{-1}(1-(1+x))\cos^{-1}(1-(1+x))\sin^{-1}(1-(1+x))}{\sqrt{2(1+x)}(1-(1+x))^2}$$

$$=\lim_{x\to0^-}\frac{\tan^{-1}(-x)\cos^{-1}(-x)\sin^{-1}(-x)}{\sqrt{2}\sqrt{1+x}(-x)^2}$$

Replacing x by $-x$ and limit $x \to 0^+$

$$= \lim_{x\to 0^+} \frac{\tan^{-1}x}{x}\cdot\frac{\sin^{-1}x}{x}\cdot\frac{\cos^{-1}x}{\sqrt{2}\sqrt{1-x}} = 1\cdot 1\cdot\frac{\pi/2}{\sqrt{2}(1)}$$

$$Q = \frac{\pi}{2\sqrt{2}}$$

So, $\sin^{-1}\sqrt{2}Q = \sin^{-1}\sqrt{2}\times\frac{\pi}{2\sqrt{2}} = \sin\frac{\pi}{2} = 1$

Clearly, $P = \frac{\pi^2}{8}$ and $Q = \frac{\pi}{2\sqrt{2}}$

$$Q^2 = \left(\frac{\pi}{2\sqrt{2}}\right)^2 = \frac{\pi^2}{8} = P \Rightarrow \frac{Q^2}{P} = 1$$

Sol. (Q.Nos. 43-45)

43. (0.25) **44.** (1) **45.** (24)

We have, $\lim_{x\to 0}\frac{1}{x^3}\left(\frac{1}{\sqrt{1+x}} - \frac{(1+ax)}{(1+bx)}\right)$

$$= \lim_{x\to 0}\frac{(1+bx)-(1+ax)\sqrt{1+x}}{x^3}$$

$$= \lim_{x\to 0}\frac{(1+bx)-(1+ax)\left(1+\frac{x}{2}-\frac{x^2}{8}+\frac{x^3}{16}\right)}{x^3}$$

$$= \lim_{x\to 0}\frac{bx - \frac{x}{2}+\frac{x^2}{8}-\frac{x^3}{16}-ax-\frac{ax^2}{2}+\frac{ax^3}{8}}{x^3}$$

$$= \lim_{x\to 0}\frac{x\left(b-\frac{1}{2}-a\right)+x^2\left(\frac{1}{8}-\frac{a}{2}\right)+x^3\left(\frac{a}{8}-\frac{1}{16}\right)}{x^3}$$

Coefficients of x and x^2 is equal to zero.

So, $\frac{1}{8} = \frac{a}{2} \Rightarrow a = \frac{1}{4} = 0.25$

and $b - \frac{1}{2} - a = 0 \Rightarrow b - \frac{1}{2} - \frac{1}{4} = 0 \Rightarrow b = \frac{3}{4}$

So, $l = \frac{a}{8} - \frac{1}{16}$

$$= \frac{1}{4\times 8} - \frac{1}{16} = \frac{1}{32} - \frac{1}{16} = \frac{1-2}{32} = \frac{-1}{32}$$

Here, $a + b = \frac{1}{4} + \frac{3}{4} = 1$ and $\left|\frac{b}{l}\right| = \left|\frac{3/4}{-1/32}\right| = |-24| = 24$

46. (3) $\lim_{h\to 0}\frac{h^4+3h^2}{[f(1-h)-f(1)]\sin 5h}$

$$= \lim_{h\to 0}\frac{h^2(h^2+3)}{[f(1-h)-f(1)]\sin 5h}$$

$$= \lim_{h\to 0}\frac{h^2+3}{(-1)\left(\frac{f(1-h)-f(1)}{-h}\right)\cdot\frac{\sin 5h}{h}\times\frac{5}{5}}$$

$$= \frac{-3}{f'(1)\cdot 5} \quad \text{... (i)}$$

Now, $f(x) = 3x^{10} - 7x^8 + 5x^6 - 21x^3 + 3x^2 - 7$

$$f'(x) = 30x^9 - 56x^7 + 30x^5 - 63x^2 + 6x$$

$$f'(1) = 30 - 56 + 30 - 63 + 6 = -53$$

So, $265\left(\lim_{h\to 0}\frac{h^4+3h^2}{[f(1-h)-f(1)]\sin 5h}\right)$

$$= 265\times\left(\frac{-3}{(-53)\times 5}\right) = 3$$

47. (9) $\lim_{t\to\infty}\frac{\sqrt{tx}}{\sqrt{tx^2-3tx+t+1}-x} = \tan\left[\sin^{-1}\left(\cos\frac{\pi}{6}\right)\right]$

$$= \lim_{t\to\infty}\frac{\sqrt{t}(\sqrt{x})}{\sqrt{t}\left(\sqrt{x^2-3x+1+\frac{1}{t}}-\frac{x}{t}\right)} = \sqrt{3}$$

$\Rightarrow \quad \frac{\sqrt{x}}{\sqrt{x^2-3x+1}} = \sqrt{3} \Rightarrow x = 3x^2 - 9x + 3$

$\Rightarrow \quad 3x^2 - 10x + 3 = 0$

$\Rightarrow \quad (3x-1)(x-3) = 0 \Rightarrow x = \frac{1}{3}, 3$

So, $\alpha = \frac{1}{3}$ and $\beta = 3$ [given, $\alpha < \beta$]

Hence, $8^\alpha + 2^\beta - \alpha\beta = 8^{1/3} + 2^3 - \frac{1}{3}\times 3 = 2 + 8 - 1 = 9$

48. (4) P be the left hand derivative of $|x-1|$.

So, for $x < 1$ $\quad |x-1| = -(x-1)$

$$P = -1$$

Now, $\lim_{x\to 1^+}\frac{(x-1)^n}{\log\cos^m(x-1)} = -1$

$\Rightarrow \quad \lim_{h\to 0}\frac{h^n}{\log\cos^m h} = -1$

$\Rightarrow \quad \lim_{h\to 0}\frac{h^n}{m\log\cos h} = -1$

Applying L-Hospital rule,

$\Rightarrow \quad \lim_{h\to 0}\frac{nh^{n-1}\cos h}{mx(-\sin h)} = -1$

$\Rightarrow \quad \lim_{h\to 0}\frac{nh^{n-1}\cos h}{m\left(-\frac{\sin h}{h}\times h\right)} = -1$

$\Rightarrow \quad \lim_{h\to 0}\frac{nh^{n-2}\cos h}{-m\left(\frac{\sin h}{h}\right)} = -1$

So, for limit existence

$$n - 2 = 0 \Rightarrow n = 2$$

Then, $\frac{n}{-m} = -1 \Rightarrow n = m$

So, $m = 2$

Hence, $m + n = 2 + 2 = 4$

49. (5) We have, $L = \lim_{x\to 0} \dfrac{af(x) - 3f(2x) + bf(4x)}{\dfrac{\sin^2 x}{x^2} \times x^2}$

For the limit to exist, we have

$$af(0) - 3f(0) + bf(0) = 0$$

$$a + b = 3 \quad \text{...(i)} \quad [\because f(0) \neq 0]$$

or $\quad L = \lim_{x\to 0} \dfrac{af'(x) - 6f'(2x) + 4bf'(x)}{2x}$

For the limit to exist,

$$af'(0) - 6f'(0) + 4bf'(0) = 0 \quad [\because f'(0) \neq 0, \text{ given}]$$

$$a + 4b = 6 \quad \text{... (ii)}$$

From Eqs. (i) and (ii),

$$3b = 3$$

$\Rightarrow \quad b = 1$ and $a = 2$

Hence, $\quad a + 3b = 2 + 3 = 5$

50. (6) Here, $L = \lim_{x\to\infty} x\log x + 2x\log\sin\left(\dfrac{1}{\sqrt{x}}\right)$

$$= \lim_{\theta\to 0} \frac{1}{\theta^2}\log\left(\frac{1}{\theta^2}\right) + \frac{2}{\theta^2}\log(\sin\theta) \quad \left[\text{put } x = \frac{1}{\theta^2}\right]$$

$$= \lim_{\theta\to 0} \frac{2\log(\sin\theta) - 2\log\theta}{\theta^2}$$

$$= \lim_{\theta\to 0} \frac{2\left[\cot\theta - \dfrac{1}{\theta}\right]}{2\theta} \quad \text{[using L-Hospital rule]}$$

$$= \lim_{\theta\to 0} \frac{2[\theta\cot\theta - 1]}{2\theta^2} = \lim_{\theta\to 0}\left[\frac{\theta - \tan\theta}{\theta^2\tan\theta}\right]$$

$$= \lim_{\theta\to 0} \frac{\theta - \tan\theta}{\theta^3}\cdot\frac{\theta}{\tan\theta} = \lim_{\theta\to 0}\frac{\theta - \tan\theta}{\theta^3}$$

$$L = \lim_{\theta\to 0} \frac{\theta - \left(\theta + \dfrac{\theta^3}{3} + ...\right)}{\theta^3} = -\frac{1}{3}$$

So, $\quad -\dfrac{2}{L} = 6$

51. (4) $p = f(0^+) = \lim_{x\to 0^+} f(x)$

$$[\text{if } x = 0 + h, h \to 0, \{x\} = \{0 + h\} = h]$$

$$= \lim_{h\to 0} \frac{\sin^{-1}(1 - h^2)[\cos^{-1}(1 - h)]^2}{2(h - h^3)}$$

$$= \lim_{h\to 0} \frac{\sin^{-1}(1 - h^2)}{2(1 - h^2)} \times \frac{[\cos^{-1}(1 - h)]^2}{h}$$

$$= \lim_{h\to 0} \frac{\sin^{-1}(1 - h^2)}{2(1 - h^2)} \times \lim_{h\to 0}\frac{[\cos^{-1}(1 - h)]^2}{h}$$

$$= \frac{\sin^{-1}(1)}{2}\lim_{h\to 0}\frac{2\cos^{-1}(1 - h)\dfrac{-1}{\sqrt{1 - (1 - h)^2}}(0 - 1)}{1}$$

$$= \frac{\pi}{4}\lim_{h\to 0}\frac{2\cos^{-1}(1 - h)}{\sqrt{1 - (1 - h)^2}}$$

Applying L-Hospital rule,

$$= \frac{\pi}{4}\lim_{h\to 0}\frac{2\times\dfrac{-1}{\sqrt{1 - (1 - h)^2}}\times(0 - 1)}{\dfrac{1}{2\sqrt{1 - (1 - h)^2}}\times[0 - 2(1 - h)(0 - 1)]}$$

$$= \frac{\pi}{2}\lim_{h\to 0}\frac{1}{\dfrac{2(1 - h)}{2}} = \frac{\pi}{2}\times 1 = \frac{\pi}{2}$$

$$q = f(0^-) = \lim_{h\to 0^-} f(x)$$

$$[x \to 0^- \Rightarrow x = 0 - h, h \to 0 \{x\} = \{0 - h\} = 1 - h]$$

$$= \lim_{h\to 0}\frac{\sin^{-1}[1 - (1 - h)^2]\cdot[\cos^{-1}(1 - (1 - h))]^2}{2[(1 - h) - (1 - h)^3]}$$

$$= \lim_{h\to 0}\frac{\sin^{-1}(1 - 1 - h^2 + 2h)(\cos^{-1}h)^2}{2(1 - h)[1 - (1 - h)^2]}$$

$$= \lim_{h\to 0}\frac{\sin^{-1}(2h - h^2)(\cos^{-1}h)^2}{2(1 - h)(1 - 1 - h^2 + 2h)}$$

$$= \lim_{h\to 0}\frac{\sin^{-1}(2h - h^2)}{(2h - h^2)}\times\frac{(\cos^{-1}h)^2}{2(1 - h^2)}$$

$$q = \frac{1\times\left(\dfrac{\pi}{2}\right)^2}{2} = \frac{\pi^2}{8}$$

Now, $\quad \dfrac{p\pi}{q} = \dfrac{\left(\dfrac{\pi}{2}\right)\pi}{\dfrac{\pi^2}{8}} = \dfrac{8}{2} = 4$

52. (1) We have, $\quad ax^3 + x - 1 - a = 0$

$\Rightarrow \quad a(x^3 - 1) + (x - 1) = 0$

$\Rightarrow \quad (x - 1)\{ax^2 + ax + (a + 1)\} = 0$

It is given that α and β are two distinct roots of $ax^3 + x - 1 - a = 0$.

Therefore, α and β are roots of $ax^2 + ax + (a + 1) = 0$ ...(i)

Hence, $\dfrac{1}{\alpha}$ and $\dfrac{1}{\beta}$ are roots of $(a + 1)x^2 + ax + a = 0$

$$\therefore (a + 1)\left(x - \frac{1}{\alpha}\right)\left(x - \frac{1}{\beta}\right) = (a + 1)x^2 + ax + 1$$

Now, $\lim_{x\to\frac{1}{\alpha}} \dfrac{(1 + a)x^3 - x^2 - a}{(e^{1-\alpha x} - 1)(x - 1)}$

$$= \lim_{x\to\frac{1}{\alpha}} \frac{a(x^3 - 1) + x^2(x - 1)}{(e^{1-\alpha x} - 1)(x - 1)}$$

$$= \lim_{x\to\frac{1}{\alpha}} \frac{a(x^2 + x + 1) + x^2}{\dfrac{(e^{1-\alpha x} - 1)}{(1 - \alpha x)}\times(1 - \alpha x)}$$

$$=-\frac{1}{\alpha}\lim_{x\to\frac{1}{\alpha}}\frac{(a+1)x^2+ax+a}{\left(x-\frac{1}{\alpha}\right)}$$

$$=-\frac{1}{\alpha}\lim_{x\to\frac{1}{\alpha}}\frac{(a+1)\left(x-\frac{1}{\alpha}\right)\left(x-\frac{1}{\beta}\right)}{\left(x-\frac{1}{\alpha}\right)}$$

$$=\frac{-(a+1)}{\alpha}\lim_{x\to\frac{1}{\alpha}}\left(x-\frac{1}{\beta}\right)$$

$$=-\frac{(a+1)}{\alpha}\left(\frac{1}{\alpha}-\frac{1}{\beta}\right)$$

$$=\frac{(a+1)}{\alpha^2\beta}(\alpha-\beta)$$

$\therefore\quad \frac{al}{\alpha}(k\alpha-\beta)=\frac{a+1}{\alpha^2\beta}(\alpha-\beta)$

$\Rightarrow\quad \frac{al}{\alpha}(k\alpha-\beta)=\frac{(a+1)a}{\alpha(a+1)}(\alpha-\beta)$

$\left[\because \text{from Eq.(i), product of roots is } \alpha\beta=\frac{a+1}{a}\right]$

$\Rightarrow\quad l(k\alpha-\beta)=\alpha-\beta$

$\Rightarrow\quad l=k=1$

So, $\quad k\cdot l=1$

53. (16) $\lim_{x\to 0}\frac{(1+3x+2x^2)^{\frac{1}{x}}-(1+3x-2x^2)^{\frac{1}{x}}}{x}$

$$=\lim_{x\to 0}\frac{[(1+3x+2x^2)^{\frac{1}{x}}-e^3]-[(1+3x-2x^2)^{\frac{1}{x}}-e^3]}{x}$$

$$=\lim_{x\to 0}\left[\frac{(1+3x+2x^2)^{\frac{1}{x}}-e^3}{x}\right]-\lim_{x\to 0}\left[\frac{(1+3x-2x^2)^{\frac{1}{x}}-e^3}{x}\right]$$

$=L_1-L_2$

$$L_1=\lim_{x\to 0}\left[\frac{(1+3x+2x^2)^{\frac{1}{x}}-e^3}{x}\right]$$

$$=\lim_{x\to 0}\left[\frac{e^{\frac{\ln(1+3x+2x^2)}{x}}-e^3}{x}\right]$$

$$=\lim_{x\to 0}\left[\frac{e^{\frac{\ln[1+x(3+2x)]}{x}}-e^3}{x}\right]$$

$$=\lim_{x\to 0}\frac{e^{(3+2x)-\frac{x(3+2x)^2}{2}\cdots\cdots}-e^3}{x}$$

$$=\lim_{x\to 0}e^3\left[\frac{e^{2x-\frac{x(3+2x)^2}{2}}-1}{x}\right]$$

$$=e^3\left[2-\frac{9}{2}\right]=-\frac{5}{2}e^3$$

and $L_2=\lim_{x\to 0}\frac{(1+3x-2x^2)^{\frac{1}{x}}-e^3}{x}$

$$=\lim_{x\to 0}\frac{e^{\frac{\ln(1+3x-2x^2)}{x}}-e^3}{x}$$

$$=\lim_{x\to 0}\frac{e^{3-2x-\frac{x}{2}(3-2x)^2}-e^3}{x}$$

$$=\lim_{x\to 0}e^3\left(\frac{e^{-2x-\frac{x}{2}(3-2x)^2}-1}{x}\right)$$

$$=e^3\left(-2-\frac{9}{2}\right)=\frac{-13}{2}e^3$$

So, $\quad L=L_1-L_2$

$$=-\frac{5}{2}e^3-\left(-\frac{13}{2}e^3\right)$$

$L=4e^3$

So, $k=4$

Now, $4k=16$

54. (200) $S=\sum_{n=1}^{n}\frac{1}{2^n}\tan\frac{x}{2^{n-1}},\ x=\frac{\pi}{4}$

Since, $\quad \tan x=\cot x-2\cot 2x$

$\therefore\quad \frac{1}{2}\tan x=\frac{1}{2}\cot x-\cot 2x$

$$\frac{1}{2^2}\tan\frac{x}{2}=\frac{1}{2^2}\cot\frac{x}{2}-\frac{1}{2}\cot x$$

$$\frac{1}{2^n}\tan\frac{x}{2^{n-1}}=\frac{1}{2^n}\cot\left(\frac{x}{2^{n-1}}\right)-\frac{1}{2^{n-1}}\cot\left(\frac{x}{2^{n-2}}\right)$$

$\therefore\quad S=\left(\frac{\cot\left(\frac{x}{2^{n-1}}\right)}{2^n}-\cot 2x\right)$

$$S=\left(\frac{1}{2^n}\cot\left(\frac{x}{2^{n-1}}\right)-\cot 2x\right)$$

Now, $\lim_{n\to\infty} = S = \lim_{n\to\infty}\left(\frac{1}{2}\cdot\frac{x}{2^{n-1}}\cdot\frac{1}{x}\cot\left(\frac{x}{2^{n-1}}\right) - \cot 2x\right)$

$$= \lim_{n\to\infty}\left(\frac{1}{2x}\frac{\frac{x}{2^{n-1}}}{\tan\left(\frac{x}{2^{n-1}}\right)} - \cot 2x\right) = \frac{1}{2x} - \cot 2x$$

Put $x = \frac{\pi}{4}$

$\therefore \quad S = \frac{2}{\pi}$, so $(100\pi)L = 100\pi\left(\frac{2}{\pi}\right) = 200$

55. (6) We have,

$$\left(1-\frac{d^2}{a_1^2}\right)\left(1-\frac{d^2}{a_2^2}\right)\ldots\ldots\left(1-\frac{d^2}{a_n^2}\right)$$

$$= \prod_{i=1}^{n}\left(1-\frac{d^2}{a_i^2}\right)$$

$$= \prod_{i=1}^{n}\left(\frac{a_i^2 - d^2}{a_i^2}\right)$$

$$= \prod_{i=1}^{n}\frac{(a_i - d)\cdot(a_i + d)}{a_i^2}$$

$$= \prod_{i=1}^{n}\frac{a_{i-1}}{a_i}\cdot\frac{a_{i+1}}{a_i}$$

$[a_i - d = a_{i-1},\ a_i + d = a_{i+1}]$

$$= \left(\frac{a_0}{a_1}\cdot\frac{a_1}{a_2}\ldots\ldots\frac{a_{n-1}}{a_n}\right)\left(\frac{a_2}{a_1}\cdot\frac{a_3}{a_2}\ldots\ldots\frac{a_{n+1}}{a_n}\right)$$

$$= \left(\frac{a_0}{a_n}\right)\left(\frac{a_{n+1}}{a_1}\right)$$

$$= \left(\frac{a_1 - d}{a_n}\right)\left(\frac{a_n + d}{a_1}\right)$$

$$= \left(\frac{a_1 - d}{a_1}\right)\left(\frac{a_n + d}{a_n}\right)$$

$$= \left(1 - \frac{d}{a_1}\right)\left(1 + \frac{d}{a_n}\right)$$

Now, $\lim_{n\to\infty}\prod_{i=1}^{n}\left(1-\frac{d^2}{a_i^2}\right)$

$$= \lim_{n\to\infty}\left(1-\frac{d}{a_1}\right)\left(1+\frac{d}{a+(n-1)d}\right)$$

$$= 1 - \frac{d}{a_1}$$

Now, $\quad 1 - \frac{d}{a_1} = \frac{1}{4} \qquad [a_1 = 8]$

$$1 - \frac{d}{8} = \frac{1}{4}$$

$\Rightarrow \quad \frac{d}{8} = \frac{3}{4}$

$$d = 6$$

CHAPTER

03

CONTINUITY AND DIFFERENTIABILITY

LEVEL 01 \ Starter Level

3.1 Continuity and Discontinuity

1 Discuss the continuity of the following functions at the indicated point(s).

(i) $f(x) = 2x - |x|$ at $x = 0$

(ii) $f(x) = \begin{cases} \dfrac{|x-a|}{x-a}, & x \neq a \\ 1, & x = a \end{cases}$ at $x = a$

(iii) $f(x) = \begin{cases} (x-a)\sin\left(\dfrac{1}{x-a}\right), & x \neq a \\ 0, & x = a \end{cases}$ at $x = a$

(iv) $f(x) = [x]$, where $[x]$ denotes the greatest integer function, at $x = k$, $k \in Z$.

(v) $f(x) = \begin{cases} \dfrac{\sin 4x}{\tan 3x}, & x < 0 \\ \dfrac{4}{3}, & x = 0 \\ \dfrac{\log(1+4x)}{e^{3x}-1}, & x > 0 \end{cases}$ at $x = 0$

(vi) $f(x) = \begin{cases} |1-4x^2|, & 0 \le x < 1 \\ [x^2 - 2x], & 1 \le x < 2 \end{cases}$ at $x = 1$

2 Find the value of constant k, so that the given function is continuous at the indicated point.

(i) $f(x) = \begin{cases} \dfrac{1-\cos 2kx}{x^2}, & x \neq 0 \\ 8, & x = 0 \end{cases}$ at $x = 0$

(ii) $f(x) = \begin{cases} \dfrac{x^3 + x^2 - 16x + 20}{(x-2)^2}, & x \neq 2 \\ k, & x = 2 \end{cases}$ at $x = 2$

(iii) $f(x) = \begin{cases} \dfrac{\sqrt{1+kx} - \sqrt{1-kx}}{x}, & -1 \le x < 0 \\ 2x^2 + 3x - 2, & 0 \le x < 1 \end{cases}$ at $x = 0$

(iv) $f(x)=\begin{cases}\dfrac{\cos 3x-\cos x}{x^2}, & x\neq 0\\ k, & x=0\end{cases}$ at $x=0$ (v) $f(x)=\begin{cases}\dfrac{1-\sqrt{2}\sin x}{\pi-4x}, & x\neq\dfrac{\pi}{4}\\ k, & x=\dfrac{\pi}{4}\end{cases}$ at $x=\dfrac{\pi}{4}$

(vi) $f(x)=\begin{cases}(\sin x)^{\frac{1}{\pi-2x}}, & x\neq\dfrac{\pi}{2}\\ k, & x=\dfrac{\pi}{2}\end{cases}$

3 Find a relation between a and b, so that the function $f(x)=\begin{cases}4x^3-2x^2+3x+1, & x\geq 1\\ ax+b, & x<1\end{cases}$ is continuous at $x=1$.

4 The function $f(x)$ is defined as $f(x)=\begin{cases}x^2+ax+b, & 0\leq x<2\\ 3x+2, & 2\leq x\leq 4\\ 2ax+5b, & 4<x\leq 8\end{cases}$

If f is continuous on $[0, 8]$, find the values of a and b.

5 If $f(x)=\dfrac{1-\sin x}{(\pi-2x)^2}$, when $x\neq\dfrac{\pi}{2}$ and $f\left(\dfrac{\pi}{2}\right)=k$.

$f(x)$ will be continuous function at $x=\dfrac{\pi}{2}$, the value of k is equal to

6 The value of $f(0)$ for the function $f(x)=\dfrac{2-\sqrt{x+4}}{\sin 2x}$, $x\neq 0$ is continuous at $x=0$, is

3.2 Differentiability

7 Find the points of non-differentiability of following functions.

(i) $f(x)=|x|+|x-1|$ (ii) $f(x)=|\log|x||, x\neq 0$

(iii) $f(x)=\sin^{-1}(\sin x)$ (iv) $f(x)=\begin{cases}x^3, & x^2<1\\ x, & x^2\geq 1\end{cases}$

3.3 Continuity and Differentiability

8 Discuss the continuity and differentiability of the function.

(i) $f(x)=\begin{cases}x[x], & 0\leq x<2\\ (x-1)[x], & 2\leq x\leq 3\end{cases}$ at $x=2$ (ii) $f(x)=\begin{cases}|x-3|, & x\geq 1\\ \dfrac{x^2}{4}-\dfrac{3x}{2}+\dfrac{13}{4}, & x<1\end{cases}$ at $x=1, 3$

(iii) $f(x)=\begin{cases}-x, & x<0\\ x^2, & 0\leq x<1\\ x^3-2x+2, & x\geq 1\end{cases}$ at $x=0$ and $x=1$

(iv) $f(x)=\begin{cases}\dfrac{x}{1+e^{\frac{1}{x}}}, & x\neq 0\\ 0, & x=0\end{cases}$ at $x=0$ (v) $f(x)=x^3\,\text{sgn}(x)$ at $x=0$

9 If the derivative of the function $f(x)=\begin{cases}ax^2+b, & x<-1\\ bx^2+ax+4, & x\geq -1\end{cases}$ is everywhere continuous, then find a and b.

LEVEL 02 \ JEE Main Level

Single Choice Correct Questions

3.1 Continuity and Discontinuity

1 If $f(x)=\begin{cases} px^2-q, & 0\le x<1 \\ 3, & x=1 \\ x+2, & 1<x\le 2 \end{cases}$, then the value of (p, q) for which $f(x)$ can not be continuous at $x=1$, is

(a) (4, 1) (b) (3, 0) (c) (8, 5) (d) (6, 2)

2 If the function $f(x)=\begin{cases} \dfrac{e^{-4x}+\ln(1+4x)+a}{x^3}, & x\ne 0 \\ b, & x=0 \end{cases}$ is continuous at $x=0$, then values of a and b are

(a) $a=1$ and $b=\dfrac{64}{3}$ (b) $a=-1$ and $b=\dfrac{32}{3}$

(c) $a=-1$ and $b=\dfrac{64}{3}$ (d) $a=1$ and $b=\dfrac{32}{3}$

3 If $f(x)=\begin{cases} \log(\sec^2 x)^{\cot^2 x}, & x\ne 0 \\ k, & x=0 \end{cases}$ is continuous at $x=0$, then k is equal to

(a) e^{-1} (b) 1 (c) e (d) 0

4 If $f(x)=\dfrac{p+q^{\frac{1}{x}}}{r+s^{\frac{1}{x}}}$, $s<1, q<1, r\ne 0, f(0)=1$ is right continuous at $x=0$, then

(a) $p=0$ (b) $p=q$ (c) $p=r$ (d) None of these

5 If $f(x)=\dfrac{x\left(e^{\frac{1}{x}}-e^{-\frac{1}{x}}\right)}{e^{\frac{1}{x}}+e^{-\frac{1}{x}}}$, $x\ne 0$ is continuous at $x=0$, then $f(0)$ is equal to

(a) 1 (b) 2 (c) 3 (d) 0

6 If $f(x)=\begin{cases} a^2\cos^2 x+b^2\sin^2 x, & x\le 0 \\ e^{ax+b}, & x>0 \end{cases}$ is continuous at $x=0$, then

(a) $2\log|a|=b$ (b) $2\log|b|=e$ (c) $\log a=2\log|b|$ (d) $a=b$

7 If $f(x)=\dfrac{2-(256-7x)^{\frac{1}{8}}}{(5x+32)^{\frac{1}{5}}-2}$, $(x\ne 0)$, then for f to be continuous everywhere, $f(0)$ is equal to

(a) -1 (b) 1 (c) $\dfrac{1}{2^6}$ (d) $\dfrac{7}{2^6}$

8 Let $f(x)=\cos 2x\cdot\cot\left(\dfrac{\pi}{4}-x\right)$. If $f(x)$ is continuous at $x=\dfrac{\pi}{4}$, then the value of $f\left(\dfrac{\pi}{4}\right)$ is

(a) 2 (b) -2 (b) $-\dfrac{1}{2}$ (d) $\dfrac{1}{2}$

9 If $f(x)=\begin{cases}\dfrac{1-\sin x}{(\pi-2x)^2}\cdot\dfrac{\log \sin x}{\log(1+\pi^2-4\pi x+4x^2)}, & x\neq\dfrac{\pi}{2}\\ k, & x=\dfrac{\pi}{2}\end{cases}$ is continuous at $x=\dfrac{\pi}{2}$, then k is equal to

(a) $-\dfrac{1}{16}$ (b) $-\dfrac{1}{32}$ (c) $-\dfrac{1}{64}$ (d) $-\dfrac{1}{28}$

10 If $[x]$ denotes the greatest integer not exceeding x and if the function f is defined by

$$f(x)=\begin{cases}\dfrac{a+2\cos x}{x^2}, & x<0\\ b\tan\dfrac{\pi}{[x+4]}, & x\geq 0\end{cases}$$ is continuous at

$x=0$, then the order pair (a, b) is

(a) $(-2, 1)$ (b) $(-2, -1)$ (c) $(-1, \sqrt{3})$ (d) $(-2, -\sqrt{3})$

11 The function $f(x)=\begin{cases}x^2-ax+3, & x \text{ is rational}\\ 2-x, & x \text{ is irrational}\end{cases}$ is continuous at exactly two points, then the possible values of a is

(a) $(2, \infty)$ (b) $(-\infty, 3)$ (c) $(-\infty, -1)\cup(3, \infty)$ (d) R

12 Find the values of a and b so that the function $f(x)=\begin{cases}\dfrac{x^2}{a}, & 0\leq x<1\\ a, & 1\leq x<\sqrt{2}\\ \dfrac{2b^2-4b}{x^2}, & \sqrt{2}\leq x<\infty\end{cases}$ is continuous for $0\leq x<\infty$. Then, the most suitable values of a and b are

(a) $a=1$ and $b=1$ (b) $a=-1$ and $b=1+\sqrt{2}$ (c) $a=-1$ and $b=1$ (d) None of these

13 If $f(x)=[\tan x]+\sqrt{\tan x-[\tan x]}$, $0\leq x<\dfrac{\pi}{2}$, where $[\cdot]$ denotes the greatest integer function, then

(a) $f(x)$ is continuous in $\left[0, \dfrac{\pi}{2}\right)$ (b) $f(x)$ is not continuous at $x=0$

(c) $f(x)$ is continuous at $x=0, \dfrac{\pi}{4}$ (d) $f(x)$ has infinite point of discontinuity

14 The function $f(x)=\begin{cases}[x]+\sqrt{x-[x]}, & x\geq 0\\ \sin x, & x<0\end{cases}$ is

(a) continuous only for all non-negative integers
(b) continuous only for all positive integers
(c) discontinuous only for the negative integers
(d) continuous for all real numbers

15 Let f be function defined by $f(x)=\begin{cases}\dfrac{x^2-1}{x^2-2|x-1|-1}, & \text{if } x\neq\\ \dfrac{1}{2}, & \text{if } x=1\end{cases}$

(a) The function is continuous at all values of x
(b) The function is continuous only for $x>1$
(c) The function is continuous at $x=1$
(d) The function is not continuous at $x=1$

16 If $f(x)=\dfrac{\ln(e^{x^2}+2\sqrt{x})}{\tan(\sqrt{x})}$ is continuous at $x=0$, then $f(0)$ must be equal to

(a) 0 (b) 1 (c) e^2 (d) 2

17 If $f(x)=\text{sgn}(\sin^2 x-\sin x-1)$ has four points of discontinuity for $x\in(0, n\pi)$, $n\in N$, then

(a) the minimum value of n is 5 (b) the maximum value of n is 6

(c) there is exactly two possible values of n (d) None of these

18 If $f(x)=\begin{cases}\dfrac{\lambda|x^2-5x+6|}{\mu(5x-6-x^2)}; & x<2\\ \mu; & x=2\\ e^{\frac{\tan(x-2)}{x-[x]}}; & x>2\end{cases}$ is continuous at $x=2$, then the sum of λ and μ is

(where [] is greatest integer function) [JEE Main 2021]

(a) $e+e^2$ (b) $e-e^2$ (c) e^2-e (d) e^2+e+1

19 Let $f(x)=\begin{cases}3\sin x+a^2-10a+30, & x\notin Q\\ 4\cos x, & x\in Q\end{cases}$ which one of the following statements is correct?

(a) $f(x)$ is continuous for all x, when $a=5$

(b) $f(x)$ must be continuous for all x, when $a=5$

(c) $f(x)$ is continuous for all $x=2n\pi-\tan^{-1}\dfrac{3}{4}$, $n\in Z$, when $a=5$

(d) $f(x)$ is continuous for all $x=2n\pi-\tan^{-1}\dfrac{4}{3}$, $n\in Z$, when $a=5$

20 Let $f:[-1,3]\to R$ be defined as $f(x)=\begin{cases}|x|+[x] & -1\le x<1\\ x+|x| & 1\le x<2\\ |x|+[x] & 2\le x\le 3\end{cases}$ [JEE Main 2019]

where [ ] denotes the greatest integer less than or equal type. Then, f is discontinuous at

(a) four or more points (b) only one point (c) only two points (d) only three points

21 Let $[t]$ denotes the greatest integer $\le t$ and $\lim\limits_{x\to 0} x\left[\dfrac{4}{x}\right]=A$. Then, the function $f(x)=[x^2]\sin\pi x$ is discontinuous, when x is equal to [JEE Main 2020]

(a) $\sqrt{A+1}$ (b) $\sqrt{A+21}$ (c) $\sqrt{A}$ (d) $\sqrt{A+5}$

22 Let $f(x)=\begin{cases}\dfrac{x}{2}-1, & 0\le x<1\\ \dfrac{1}{2}, & 1\le x\le 2\end{cases}$ and $g(x)=(2x+1)(x-k)+3$, $x\ge 0$, then $g(f(x))$ is continuous at $x=1$, if k is equal to

(a) 0 (b) $-\dfrac{1}{2}$ (c) $\dfrac{1}{2}$ (d) 1

23 If $\alpha, \beta(\alpha<\beta)$ are the points of discontinuity by the function $f(f(f(x)))$, where $f(x)=\dfrac{1}{1-x}$, then the set of values of a for which the points (α, β) and (a, a^2) lies on the same side of line $x+2y-3=0$, is

(a) $\left(-\dfrac{3}{2}, 1\right)$ (b) $\left[-\dfrac{3}{2}, 1\right]$ (c) $[1, \infty)$ (d) $\left(-\infty, -\dfrac{3}{2}\right]$

24 If $f(x) = [x] + [-x]$, where $[x]$ is greatest integer function, then $f(x)$ has

(a) isolated point discontinuity
(b) missing point discontinuity
(c) jump type discontinuity
(d) asymptotic discontinuity

25 If $f(x) = \begin{cases} \sin^{-1}\{\sin x\}, & x < \frac{\pi}{2} \\ \pi[x] - 1, & x \geq \frac{\pi}{2} \end{cases}$, where $\{\cdot\}$ denotes fractional part of x, then at $x = \frac{\pi}{2}$ $f(x)$ has

(a) missing point discontinuity
(b) jump type discontinuity
(c) oscillating type discontinuity
(d) None of these

26 If $f(x) = \begin{cases} x[x]^2 \log_{1+x} 2, & \text{for } -1 < x < 0 \\ \dfrac{\ln(e^{x^2} + 2\sqrt{\{x\}})}{\tan \sqrt{x}}, & \text{for } 0 < x < 1 \end{cases}$, then (where [ ] and {} are greatest integer function and fractional part function)

(a) if $f(0) = \ln 2 \Rightarrow f$ is continuous at $x = 0$
(b) f has an irremovable discontinuity at $x = 0$
(c) f has removable type discontinuity at $x = 0$
(d) None of the above

27 Let $f(x) = \begin{cases} x + q, & \text{if } x < 0 \\ |x - 1|, & \text{if } x \geq 0 \end{cases}$ and $g(x) = \begin{cases} x + 1, & x < 0 \\ (x-1)^2 + b, & x \geq 0 \end{cases}$

where a and b are non-negative numbers. If $gof(x)$ is continuous for all real x. Then, the values of a and b are

(a) $a = 1$ and $b = 2$
(b) $a = 2$ and $b = 3$
(c) $a = 2$ and $b = 0$
(d) $a = 1$ and $b = 0$

28 If $y = \dfrac{1}{t^2 + t - 2}$, where $t = \dfrac{1}{x - 1}$, then the number of points, where $f(x)$ is discontinuous, is

(a) 1 (b) 2 (c) 3 (d) 4

3.2 Differentiability

29 Let $f(x) = \begin{cases} a\sqrt{x+1}, & 0 < x < 3 \\ bx + 2, & 3 \leq x < 5 \end{cases}$, if $f(x)$ is differentiable at $x = 3$, then the values of a and b are

(a) $a = \frac{2}{5}$ and $b = \frac{8}{5}$
(b) $a = \frac{8}{5}$ and $b = \frac{2}{5}$
(c) $a = -\frac{8}{5}$ and $b = -\frac{2}{5}$
(d) $a = +\frac{8}{5}$ and $b = -\frac{2}{5}$

30 For $x \in R$, $f(x) = |\log 2 - \sin x|$ and $g(x) = f(f(x))$, then **[JEE Main 2016]**

(a) $g'(0) = \cos(\log 2)$
(b) $g'(0) = -\cos(\log 2)$
(c) g is differentiable at $x = 0$ and $g'(0) = -\sin(\log 2)$
(d) g is non-differentiable at $x = 0$

31 For what triplets the real numbers (a, b, c) with $a \neq 0$, the function

$$f(x) = \begin{cases} x, & x \leq 1 \\ ax^2 + bx + c, & \text{otherwise} \end{cases}$$

is differentiable for all real x?

(a) $\{(a, 1-2a, a), a \in R, a \neq 0\}$
(b) $\{(a, 1-2a, c), a, c \in R, a \neq 0\}$
(c) $\{(a, b, c), a, b, c \in R, a + b + c = 1\}$
(d) $\{(a, 1-2a, 0), a \in R, a \neq 0\}$

32 The number of points, where the function $f(x) = \text{Max.}\{a - x, a + x, b\} - \infty < x < \infty, 0 < a < b$ cannot be differentiable, is

(a) 1 (b) 2 (c) 3 (d) None of these

33 If $f(x)$ is differentiable everywhere, then

(a) $|f|$ is differentiable everywhere (b) $|f|^2$ is differentiable everywhere

(c) $f|f|$ is non-differentiable at some point (d) $f + |f|$ is differentiable everywhere

34 S be the set in which function defined $f(x) = \sin|x| - |x| + 2(x - \pi)\cos|x|$ is not differentiable, then number of elements in S is

(a) 0 (b) 1 (c) 2 (d) 4

35 Number of points in $[0, \pi]$, where $f(x) = \left[\dfrac{x \tan x}{\sin x + \cos x}\right]$ is non-differentiable, (where [ ] denotes the greatest integer function is

(a) 0 (b) 4 (c) 9 (d) None of these

36 If $f(x) = \begin{cases} 3x + 5, & x \le 1 \\ x^2 + px + q, & x > 1 \end{cases}$, then the values of p and q, if $f(x)$ is differentiable at $x = 1$, is

(a) $p = 0$ and $q = 7$ (b) $p = 3$ and $q = 4$ (c) $p = -1$ and $q = 8$ (d) $p = 1$ and $q = 6$

37 Let $f(x) = \begin{cases} (x-2)^2 \sin \dfrac{1}{(x-2)} - |x|, & x \ne 2 \\ -2, & x = 2 \end{cases}$ be a real valued function. Then, the number of points, where $f(x)$ is non-differentiable, is

(a) 1 (b) 2 (c) 3 (d) 4

38 If $f(x) = \begin{cases} x^p \sin\left(\dfrac{1}{x}\right), & x \ne 0 \\ 0, & x = 0 \end{cases}$ is continuous but non-differentiable at $x = 0$, then

(a) $p \in (-1, 1)$ (b) $p \in (0, 2]$ (c) $p \in (0, 1]$ (d) $p \in [1, 2)$

3.3 Continuity and Differentiability

39 If a function $f(x)$ defined by

$$f(x) = \begin{cases} ae^x + be^{-x}, & -1 \le x < 1 \\ cx^2, & 1 \le x \le 3 \\ ax^2 + 2cx, & 3 < x \le 4 \end{cases}$$

be continuous for some $a, b, c \in R$ and $f'(0) + f'(2) = e$, then the value of a is

(a) $\dfrac{1}{e^2 - 3e + 13}$ (b) $\dfrac{e}{e^2 - 3e - 13}$ (c) $\dfrac{e}{e^2 + 3e + 13}$ (d) $\dfrac{e}{e^2 - 3e + 13}$

40 The function $f(x) = [x^2] + [-x]^2$, where [ ] denotes the greatest integer function, is

(a) continuous and derivable at $x = 2$ (b) neither continuous nor derivable at $x = 2$

(c) continuous but not derivable at $x = 2$ (d) None of these

41 If $f(x) = \begin{cases} \dfrac{a^{2[x] + \{x\}} - 1}{2[x] + \{x\}}; & x \ne 0 \\ \log a; & x = 0 \end{cases}$, where [.] and {} denotes greatest integer and fractional part function, respectively, then

(a) $f(x)$ is continuous at $x = 0$ (b) $f(x)$ is discontinuous at $x = 0$

(c) $f(x)$ is continuous for all $x \in R$ (d) $f(x)$ is differentiable at $x = 0$

42 Let $f : R \to R$ be any function, define $g : R \to R$ by $g(x) = |f(x)| \,\forall\, x$, then

(a) g is continuous, if f is not continuous
(b) g is not continuous, if f is not continuous
(c) g is continuous, if f is continuous
(d) g is continuous, if f is differentiable

43 If $y = \sin^{-1}|\sin x|$, then

(a) y is discontinuous at $x = \frac{n\pi}{2}$

(b) y is non-differentiable at $x = \frac{n\pi}{2}$

(c) y is continuous and differentiable at $x = \frac{n\pi}{2}, n \in Z$

(d) None of the above

44 If $f(x) = \begin{cases} -x - \frac{\pi}{2}, & x \le -\frac{\pi}{2} \\ -\cos x, & -\frac{\pi}{2} < x \le 0 \\ x - 1, & 0 < x \le 1 \\ \ln x, & x > 1 \end{cases}$, then which of the following is not correct?

(a) $f(x)$ is continuous at $x = -\frac{\pi}{2}$
(b) $f(x)$ is non-differentiable at $x = 0$
(c) $f(x)$ is differentiable at $x = 1, -\frac{3}{2}$
(d) $f(x)$ is discontinuous at $x = 0$

45 If $2x + 4|y| = 7y$, then y as a function of x is

(a) continuous at $x = 0$
(b) differentiable at $x = 0$
(c) $\frac{dy}{dx} = \frac{2}{11} \,\forall\, x \in R$
(d) None of these

46 Number of points, where the function $f(x) = \text{Max} \cdot \{\text{sgn}(x), -\sqrt{9 - x^2}, x^3\}$ is continuous but non-differentiable, is

(a) 3 (b) 4 (c) 5 (d) 6

47 If $f(x) = \begin{cases} \int_0^x (5 + |1 - t|)dt, & x > 2 \\ 5x + 1, & x \le 2 \end{cases}$, then **[JEE Main 2021]**

(a) $f(x)$ is not continuous at $x = 2$
(b) $f(x)$ is everywhere differentiable
(c) $f(x)$ is continuous but non-differentiable at $x = 2$
(d) $f(x)$ is non-differentiable at $x = 1$

48 If the function $f(x) = \left[\frac{(x-5)^3}{A}\right] \sin(x - 5) + a\cos(x - 2)$,

where [ ] denotes the greatest integer function, is continuous and differentiable in (7, 9), then

(a) $A \in [8, 64]$ (b) $A \in (0, 8)$ (c) $A \in (64, \infty)$ (d) $A \in [8, 16]$

49 Let $g(x)$ be a polynomial of degree one and $f(x)$ be defined by

$$f(x) = \begin{cases} g(x), & x \le 0 \\ |x|^{\sin x}, & x > 0 \end{cases}$$

If $f(x)$ is continuous satisfying $f'(1) = f(-1)$, then $g(x)$ is equal to

(a) $(1 + \sin 1)x + 1$
(b) $(1 - \sin 1)x + 1$
(c) $(1 - \sin 1)x - 1$
(d) $(1 + \sin 1)x - 1$

50 Let $S(x)=\int_{x^2}^{x^3} \ln t\, dt\ (x>0)$ and $H\ x)=\frac{S'(x)}{x}$. Then, $H(x)$ is

(a) continuous but not derivable in its domain
(b) derivable and continuous in its domain
(c) neither derivable nor continuous in its domain
(d) derivable but not continuous in its domain

Numerical Value Type Questions

51 The value of a and b for which the function

$$f(x)=\begin{cases} \frac{\sin(a+1)x+\sin x}{x}, & x<0 \\ b, & x=0 \\ \frac{\sqrt{x+x^2}-\sqrt{x}}{x^{\frac{3}{2}}}, & x>0 \end{cases}$$

is continuous for all x in R, are, then the value of $b-2a$ is .....

52 Let $f(x)=\frac{\tan[e^2]x^2-\tan[-e^2]x^2}{\sin^2 x}$, $x\neq 0$. Then, the value of $f(0)$, so that f is a continuous function, is ......

53 $f(x)=\begin{cases} \frac{8^x-4^x-2^x+1}{x^2}, & x>0 \\ e^x \sin x+\pi x+\lambda \ln 4, & x\leq 0 \end{cases}$ is continuous at $x=0$, then the value of e^{λ} is ......

54 Let $f(x)=\frac{\sin 3x+A\sin 2x+B\sin x}{x^5}$, $x\neq 0$. If $f(x)$ is continuous at $x=0$, then $f(0)$ is equal to ....

55 If $f(x)=\begin{cases} (1+|\sin x|)^{\frac{a}{|\sin x|}}, & -\frac{\pi}{4}<x<0 \\ b, & x=0 \\ e^{\frac{\cot 4x}{\cot 2x}}, & 0<x<\frac{\pi}{4} \end{cases}$ is continuous, then the value of $(a+\ln b)$ is ......

[JEE Main 2021]

56 If $f(x)=\frac{(2^x-1)^3}{\sin(x(\log 2)\log(1+x^2\log 4)}$ is continuous everywhere, then the integral part of $f(0)$ is ....

57 The number of points of discontinuity of $f(x)=[3x]^2-\{3x\}^2$ (where $[\cdot]$ denotes the greatest integer function and $\{\cdot\}$ denotes the fractional part of x) in the interval $(-2, 2)$ is ......

58 Let f be a continuous function on R such that $f\left(\frac{1}{2^n}\right)=(\sin e^n)e^{-n^2}+\frac{5n^2}{n^2+3}$. Then, the value of $f(0)$ is ....

59 If function g and h are defined as

$$g(x)=\begin{cases} x^2+1, & x\in Q \\ px^2, & x\notin Q \end{cases} \text{ and } h(x)=\begin{cases} px, & x\in Q \\ 2x+q, & x\notin Q \end{cases}$$

If $(g+h)(x)$ is continuous at $x=1$ and $x=3$, then the value of $3p+q$ is ....

60 Let $f(x)=\sin^{-1}\left(\frac{2x}{1+x^2}\right)$. If p is the number of discontinuity points and q is the number of non-differentiable points of $f(x)$. Then, the value of $(p+q+2)$ is ....

61 If the function $f(x)=\begin{cases} x^2 e^{2(x-1)}; & 0\le x\le 1 \\ a\ \text{sgn}(x+1)\cos(2x-2)+bx^2; & 1<x\le 2 \end{cases}$ is differentiable at $x=1$, then the value of $a+2b$ is ....

62 $f(x)=|\sin x|+[\cos x]$, $x\in[0, 2\pi]$, where [ ] denotes the greatest integer function. Total number of points, where $f(x)$ is non-differentiable, is ....

63 Let $f(x)=\begin{cases} -1, & -2\le x<0 \\ x^2-1, & 0\le x\le 2 \end{cases}$ and $g(x)=|f(x)|+f|x|$.

The number of points in interval $(-2, 2)$, where f is continuous but non-differentiable is ......

64 If $f(x)=|x|-1$ and $g(x)=|x|$, then the number of points, where gof is continuous but non-differentiable is ....

65 Let $f(x)$ is differentiable for all x and $\lim_{h\to 0}\frac{1}{h}f(3+h)=7$, then $f'(3)$ is equal to ....

LEVEL 03 \ JEE Advanced Level

Single Choice Correct Questions

1 The function $f(x)=\frac{4-x^2}{|4x-x^3|}$ is

(a) discontinuous at only one point
(b) discontinuous at exactly two points
(c) discontinuous at exactly three points
(d) None of these

2 Let $f(x)=\lim_{n\to\infty}\left(\frac{x^{2n-1}+ax^2+bx}{x^{2n}+1}\right)$. If $f(x)$ is continuous for all $x\in R$, then

(a) $a=1$ and $b=1$
(b) $a=0$ and $b=1$
(c) $a=1$ and $b=0$
(d) $a=-1$ and $b=1$

3 If $f(x)=\text{sgn}\ (\cos 2x-2\sin x+3)$, then $f(x)$

(a) is continuous over its domain
(b) has a missing point discontinuity
(c) has isolated point discontinuity
(d) has irremovable discontinuity

4 Consider, $f(x)=\left[\frac{2(\sin x-\sin^3 x)+|\sin x-\sin^3 x|}{2(\sin x-\sin^3 x)-|\sin x-\sin^3 x|}\right]$, $x\neq\frac{\pi}{2}$ for $x\in(0, \pi)$, $f\left(\frac{\pi}{2}\right)=3$, where [ ] denotes the greatest integer function, then

(a) f is continuous and differentiable at $x=\frac{\pi}{2}$

(b) f is continuous but non-differentiable at $x=\frac{\pi}{2}$

(c) f is neither continuous non-differentiable at $x=\frac{\pi}{2}$

(d) None of the above

5 If $f(x)=\text{Min}\ \{|x|, 2-|x-1|, |x-1|\}$, then the number of points where the function $f(x)$ is non-differentiable, is

(a) 5
(b) 6
(c) 7
(d) None of these

6 The left hand derivative of $f(x) = [x] \cdot \sin \pi x$ at $x = k$, k is an integer, is

(a) $(-1)^k (k-1)\pi$ (b) $(-1)^{k-1}(k-1)\pi$ (c) $(-1)^k k\pi$ (d) $(-1)^{k-1} k\pi$

7 Let $[x]$ be the greatest integer less than or equal to x. Then, $f(x) = x\cos\{\pi(x+[x]\}$ is continuous at

[JEE (Adv.) 2017]

(a) $x = -1$ (b) $x = 0$ (c) $x = 2$ (d) $x = 1$

8 Let $f(x)$ be defined in $[-2, 2]$ by

$$f(x) = \begin{cases} \text{Max}(\sqrt{4-x^2}, \sqrt{1+x^2}, & -2 \le x \le 0 \\ \text{Min}(\sqrt{4-x^2}, \sqrt{1+x^2}), & 0 < x \le 2 \end{cases}, \text{ then } f(x)$$

(a) is continuous at all points (b) is not continuous at more than one point

(c) is non-differentiable only at one point (d) is non-differentiable at more than one point

9 If $f(x) = \begin{cases} \dfrac{1}{x} - \dfrac{2}{e^{2x}-1}, & x \ne 0 \\ 1, & x = 0 \end{cases}$, then

(a) $f(x)$ is differentiable at $x = 0$

(b) $f(x)$ is non-differentiable at $x = 0$

(c) $f'(0) = \dfrac{1}{3}$

(d) $f(x)$ is continuous but non-differentiable at $x = 0$

10 Let $g(x)$ be a cubic polynomial and $f(x) = g(x)|(x-1)(x-2)(x-3)^2(x-4)^3|$ is differentiable for $x \in R$ and $f'(3) + f'''(4) = 0$, then the value of $g(5)$ is equal

(a) 12 (b) 15

(c) 16 (d) None of these

11 A function f is satisfies the relation $f(x) = f'(x) + f''(x) + \dots \infty$ terms, where $f(x)$ is differentiable indefinitely. If $f'(-1) = 1$, then $f(-1)$ is equal to

(a) 0 (b) 1 (c) 2 (d) 4

12 Let f be a function defined by $2f(\sin x) + f(\cos x) = x \;\forall\; x \in R$, then set of points, where f is non-differentiable, is

(a) set of all natural numbers (b) set of all natural numbers

(c) $\{0, 1, -1\}$ (d) $\{1, -1\}$

More than One Correct

13 The function $f(x) = \sqrt{1-\sqrt{1-x^2}}$

(a) has its domain $-1 \le x \le 1$

(b) has finite one sided derivative at the point $x = \pm 1$

(c) is continuous and differentiable at $x = 0$

(d) is continuous but non-differentiable at $x = 0$

14 Which of the following has/have removable discontinuity at the indicated point?

(a) $f(x) = \dfrac{x-7}{|x-7|}$ at $x = 7$ (b) $f(x) = \dfrac{3-\sqrt{x}}{9-x}$ at $x = 9$

(c) $f(x) = \dfrac{1}{\ln|x|}$ at $x = 1$ (d) $f(x) = \dfrac{x^2-1}{x^3-1}$ at $x = 1$

15 Consider, the function $f(x)=\begin{cases}\dfrac{x}{1+|x|}, & |x|\geq 1\\ \dfrac{x}{1-|x|}, & |x|<1\end{cases}$, then

(a) $f(x)$ is differentiable at $x=0$
(b) $f(x)$ is continuous at $x=1$
(c) $f(x)$ is continuous at $x=-1$
(d) $f(x)$ is neither continuous nor differentiable at $x=-1$

16 Let $[x]$ denotes the greatest integer function. If $f(x)=[x \sin \pi x]$, then $f(x)$ is

(a) continuous at $x=0$
(b) continuous in $(-1, 0)$
(c) differentiable in $(-1, 1)$
(d) differentiable at $x=1$

17 Which of the following function(s) is/are continuous on $(0, \pi)$?

(a) $\tan x$
(b) $\int_0^x t \sin \frac{1}{t}\, dt$
(c) $\begin{cases}1, & 0<x\leq \dfrac{3\pi}{4}\\ 2\sin\left(\dfrac{2x}{9}\right), & \dfrac{3\pi}{4}<x\leq \pi\end{cases}$
(d) $\begin{cases}x\sin x, & 0<x\leq \dfrac{\pi}{2}\\ \dfrac{\pi}{2}\sin(\pi+x), & \dfrac{\pi}{2}<x<\pi\end{cases}$

18 Given, $f(x)=\begin{cases}\left[3-\left[\cot^{-1}\left(\dfrac{2x^3-3}{x^2}\right)\right]\right], & x>0\\ \{x^2\}\cos\left(e^{\frac{1}{x}}\right), & x<0\end{cases}$, where $\{x\}$ is fractional part function and [ ] denotes the greatest integer function, then which of the following statement does not hold good?

(a) $f(0^-)=0$
(b) $f(0^+)=3$
(c) $f(0)=0 \Rightarrow$ Continuity of f at $x=0$
(d) Irremovable discontinuity of f at $x=0$

19 Which of the following function(s), not defined at $x=0$ has/have removable discontinuity at $x=0$?

(a) $f(x)=\dfrac{1}{1+2^{\cot x}}$
(b) $f(x)=\cos\left(\dfrac{|\sin x|}{x}\right)$
(c) $f(x)=x\sin\left(\dfrac{\pi}{x}\right)$
(d) $f(x)=e^{\frac{1}{x}}$

20 If $f(x)=\begin{cases}|x|-3, & x<a\\ |x-2|+a, & x\geq a\end{cases}$ and $g(x)=\begin{cases}2-|x|, & x<2\\ \text{sgn}(x)-b, & x\geq 2\end{cases}$ and $h(x)=f(x)+g(x)$ is discontinuous at exactly one point, then which of the following values of a and b are possible?

(a) $a=-3$ and $b=0$
(b) $a=2$ and $b=1$
(c) $a=2$ and $b=0$
(d) $a=-3$ and $b=2$

21 Consider the function $f(x)=\begin{cases}\dfrac{1-a^x+xa^x\ln a}{a^x\cdot x^2}; & x<0\\ \dfrac{2^x a^x-x\ln 2-x\ln a-1}{x^2}; & x>0\end{cases}$, if $f(x)$ is continuous at $x=0$, then

(a) $a=2^{-\frac{1}{2}}$
(b) $a=2^{\frac{1}{2}}$
(c) $f(0)=\dfrac{1}{8}(\ln 2)^2$
(d) $f(0)=-\dfrac{1}{8}(\ln 2)^2$

22 Let $f(x) = x^3 - x^2 + x + 1$ and $g(x) = \begin{cases} \text{Max. } f(t), & 0 \le t \le x \text{ for } 0 \le x \le 1 \\ 3 - x, & x \le 2 \end{cases}$, then

(a) $g(x)$ is continuous for all $x \in [0, 2]$ (b) $g(x)$ is discontinuous at $x = 1$
(c) $g(x)$ is non-differentiable at $x = 1$ (d) $g(x)$ is differentiable for all $x \in (0, 2)$

23 Which of the following function is thrice differentiable at $x = 0$?

(a) $f(x) = |x| \sin^3 x$ (b) $f(x) = x|\tan^3 x|$ (c) $f(x) = x^3|x|$ (d) $f(x) = |x^3|$

24 Let $[x]$ be the greatest integer function. Then, at which of the following points the function $f(x) = x \cos\{\pi(x + [x])\}$ is discontinuous? **[JEE Advanced 2017]**

(a) $x = -1$ (b) $x = 0$ (c) $x = 2$ (d) $x = 1$

25 Let $g : R \to R$ be a differentiable function with $g(0) = 0$, $g'(0) = 0$ and $g'(1) \ne 0$. Let $f(x) = \begin{cases} \dfrac{x}{|x|}, & x \ne 0 \\ 0, & x = 0 \end{cases}$ and $h(x) = e^{|x|}\ \forall\, x \in R$. Let $(foh)(x)$ denotes $f\{h(x)\}$ and $(hof)(x)$ denotes $h\{f(x)\}$. Then, which of the following is (are) true?

(a) f is differentiable at $x = 0$ (b) h is differentiable at $x = 0$
(c) foh is differentiable at $x = 0$ (d) hof is differentiable at $x = 0$

26 If $f(x) = \lim_{t \to \infty} \dfrac{|a + \sin \pi x|^t - 1}{|a + \sin \pi x|^t + 1}$, $x \in (0, 4)$, then

(a) if $a = 1$, then $f(x)$ has 3 points of discontinuity.
(b) if $a = 3$, then $f(x)$ has no points of discontinuity
(c) if $a = 0.5$, then $f(x)$ has 6 points of discontinuity
(d) if $a = 0$, then $f(x)$ has 4 points of discontinuity

27 The function $f(x) = \sin x \cdot \text{sgn } x$ is

(a) discontinuous no where (b) an even function
(c) an odd function (d) differentiable for all x

28 Given, $f(x) = \begin{cases} \dfrac{\log_a (a|[x] + [-x]|)^x \, a^{\left(\frac{2}{\frac{[x]+[-x]}{|x|}}\right) - 5}}{3 + a^{\frac{1}{|x|}}}, & |x| \ne 0, a > 1 \\ 0, & x = 0 \end{cases}$

where [ ] represents the integral part function, then

(a) f is continuous at $x = 0$ (b) f is discontinuous at $x = 0$
(c) f is differentiable at $x = 0$ (d) f is non-differentiable at $x = 0$

29 The function $f(x) = \begin{cases} (1 - \text{sgn } x) \text{ sgn } x, & x \le 1 \\ \dfrac{1}{1 - e^{x/(1-x)}}, & x > 1 \end{cases}$, discontinuous at

(a) $x = 0$ (b) $x = 2$ (c) $x = 1$ (d) $x = 3$

30 A function f is defined as $f(x) = \begin{cases} 1, & -\infty < x < 0 \\ 1 + |\sin x|, & 0 \le x < \dfrac{\pi}{2} \\ 2 + \left(x - \dfrac{\pi}{2}\right), & \dfrac{\pi}{2} \le x < \infty \end{cases}$, then

(a) $f(x)$ is differentiable at $x = 0$

(b) $f(x)$ is non-differentiable at $x = 0$

(c) $f(x)$ is differentiable at $x = \frac{\pi}{2}$

(d) $f(x)$ is non-differentiable at $x = \frac{\pi}{2}$

Linked Comprehension Type

Paragraph I Directions (Q.Nos. 31-33)

If $|c| \le \frac{1}{2}$ and $f(x)$ is differentiable function at $x = 0$ given by

$$f(x) = \begin{cases} b \sin^{-1}\left(\frac{c+x}{2}\right), & -\frac{1}{2} < x < 0 \\ \frac{1}{2}, & x = 0, \\ \frac{e^{\frac{ax}{2}} - 1}{x}, & 0 < x < \frac{1}{2} \end{cases} \text{ then}$$

31 The value of a is

(a) $\frac{1}{2}$ (b) $\frac{3}{2}$ (c) 2 (d) 1

32 Which of the following is true?

(a) $16b^2 = 4 - c^2$

(b) $16b^2 = 1 - 4c^2$

(c) $64b^2 = 4 - c^2$

(d) None of these

33 The value of $f'(0)$ is

(a) $\frac{1}{2}$ (b) $\frac{1}{4}$ (c) $\frac{1}{8}$ (d) 0

Paragraph II Directions (Q.Nos. 34-36)

A function $f(x)$ is said to have jump discontinuity at a point $x = a$, if both of the limits LHL and RHL exist and finite at $x = a$ but not equal and $f(a)$ may be equal to either of these limits. The value of | LHL – RHL | is known as jump of discontinuity.

34 If $f(x) = [\sin x] - [\cos x]$, then the jump of discontinuity of $f(x)$ at $x = \frac{\pi}{2}$ where $[\cdot]$ denotes G.I.F. is

(a) $+\frac{1}{2}$ (b) 1 (c) 2 (d) None of these

35 Number of jump discontinuities in $y = f(x) \cdot g(x)$,

where $f(x) = \begin{cases} x+1, & x \ge 0 \\ x^2, & x < 0 \end{cases}$ and $g(x) = \begin{cases} \sin x, & x < 0 \\ 2x^2, & x \ge 0 \end{cases}$ is

(a) 1 (b) 2 (c) 3 (d) None of these

36 If $f(x) = \begin{cases} \cos^{-1}\{\cot x\}, & x < \frac{\pi}{2} \\ \pi[x] - 1, & x \ge \frac{\pi}{2} \end{cases}$, then jump of discontinuity of $f(x)$ at $x = 1$ where $[\cdot]$ and $\{\cdot\}$ denotes G.I.F. and F.P.F. respectively, is

(a) $-\frac{\pi}{2} + 1$ (b) $+1 + \frac{\pi}{2}$ (c) $\frac{\pi}{2} - 1$ (d) None of these

Paragraph III Directions (Q.Nos. 37-39)

The function $y = f(x)$ is defined by the parametric equations

$$x = 2t - |t - 1| \text{ and } y = 2t^2 + t|t|$$

On the basis of above information, answer the following questions

37 The value of $f(2)$ is

(a) 3 (b) 6 (c) 9 (d) 0

38 $f(x)$ consumes the value $\frac{1}{2}$ at x equal to

(a) $\frac{\sqrt{2}-1}{\sqrt{2}}$ (b) $\frac{\sqrt{3}-\sqrt{2}}{\sqrt{2}}$ (c) $\frac{\sqrt{2}+1}{2}$ (d) $\frac{\sqrt{3}+1}{2}$

39 The number of values of x at which f is discontinuous, is

(a) 0 (b) 1 (c) 2 (d) 4

Matrix Match Type Questions

40 Match the following.

	Column I		Column II
(i)	$f(x) = \frac{1}{x-1}$	(A)	Removable discontinuity
(ii)	$f(x) = \frac{x^3 - x}{x^2 - 1}$	(B)	Non-removable discontinuity
(iii)	$f(x) = \frac{\|x-1\|}{(x-1)}$	(C)	Jump of discontinuity
(iv)	$f(x) = \sin\left(\frac{1}{x-1}\right)$	(D)	Discontinuity due to vertical asymptotes
		(E)	Missing point discontinuity
		(F)	Oscillating discontinuity

41 Match the following.

	Column I		Column II
(i)	$f(x) = (x+1)\|x-1\|$ is	(A)	continuous in $(-1, 1)$
(ii)	$f(x) = \text{Min}(\|x\|, 1-\|x\|)$ is	(B)	differentiable in $(-1, 1)$
(iii)	$f(x) = \|\sin^{-1} x\|$ is	(C)	differentiable in $(0, 1)$
(iv)	$f(x) = x\|x\|$ is	(D)	non-differentiable in $(-1, 1)$

42 Match the following.

	Column I		Column II
(i)	$f: [0, 2] \to [0, 8]$ is continuous, then the number of roots of equation $f(x) - x^3 = 0$ is	(A)	exactly two
(ii)	If $f(x)$ is derivable, then between the two consecutive roots of $f(x) = 0$, the number of roots of $f(x) = 0$ is	(B)	atleast one
(iii)	The number of values of x at which $\|2 - \|x-2\|\|$ is non-differentiable	(C)	atmost one
(iv)	If $f(x) = \begin{cases} x^2, & \text{if } x \text{ is rational} \\ 5x - 6, & \text{if } x \text{ is irrational} \end{cases}$, then the number of values of x at which $f(x)$ is continuous	(D)	exactly three

43 Let $f : R \to R$, $f_2 : \left(-\frac{\pi}{2}, \frac{\pi}{2}\right) \to R$, $f_3 : \left(-1, e^{\frac{\pi}{2}} - 2\right) \to R$ and $f_4 : R \to R$ be functions defined by

(i) $f_1(x) = \sin\left(\sqrt{1 - e^{-x^2}}\right)$

(ii) $f_2(x) = \begin{cases} \dfrac{|\sin x|}{\tan^{-1} x}, & \text{if } x \neq 0 \\ 1, & \text{if } x = 0 \end{cases}$

(iii) $f_3(x) = [\sin\{\log(x + 2)\}]$, where for $t \in R$, $[t]$ denotes the greatest integer function

(iv) $f_4(x) = \begin{cases} x^2 \sin \dfrac{1}{x}, & x \neq 0 \\ 0, & x = 0 \end{cases}$

Match the following.

	Column I		Column II
(i)	The function f_1 is	(A)	not continuous at $x = 0$
(ii)	The function f_2 is	(B)	continuous at $x = 0$ and non-differentiable at $x = 0$
(iii)	The function f_3 is	(C)	differentiable at $x = 0$ and its derivative is not continuous at $x = 0$
(iv)	The function f_4 is	(D)	differentiable at $x = 0$ and its derivative is continuous at $x = 0$

44 Match the following.

	Column I		Column II
(i)	If $f(x) = \begin{cases} \dfrac{a + 3\cos x}{x^2}, & x < 0 \\ b\tan\left(\dfrac{\pi}{[x+3]}\right), & x \geq 0 \end{cases}$, is continuous at $x = 0$, where [ ] denotes the greatest integer function, then $[a - 2b]$ is	(A)	1
(ii)	If $f(x) = \text{sgn}(x^2 - ax + 1)$ has exactly one point of discontinuity, then the value of $\lvert a \rvert$ can be	(B)	0
(iii)	If $f(x) = \begin{cases} p + \dfrac{\sin[x]}{x}, & x > 0 \\ 2; & x = 0 \\ q + \left[\dfrac{\sin x - x}{x^3}\right], & x < 0 \end{cases}$, where [ ] denotes the greatest integer function, is continuous at $x = 0$, then $\lvert p - q \rvert$ is equal to	(C)	2
(iv)	Let $f(x) = \begin{cases} \dfrac{a\lvert x^2 - x - 2 \rvert}{2 + x - x^2}, & x < 2 \\ b, & x = 2 \\ \dfrac{x - [x]}{x - 2}, & x > 2 \end{cases}$, where [ ] denotes greatest integer function, is continuous at $x = 2$, then the value of $(1 - a)^2 + (1 - b)^2$ is	(D)	-2

Numerical Value Type Questions

Question Stem Directions (Q.Nos. 45-47) The answer to each question is a **NUMERICAL VALUE.**

If $f(x)=\begin{cases}[x], & -2\le x\le -\frac{1}{2}\\ 2x^2-1, & -\frac{1}{2}<x\le 2\end{cases}$ and $g(x)=f(|x|)+|f(x)|$, where $[x]$ represents the greatest integer function.

45 The number of points, where $|f(x)|$ is non-differentiable, is ....

46 The number of points, where $g(x)$ is non-differentiable, is ....

47 The number of points, where $g(x)$ is discontinuous, is ....

Question Stem Directions (Q.Nos. 48-50) The answer to each question is a **NUMERICAL VALUE.**

Let the equations $x^3+2x^2+px+q=0$ and $x^3+x^2+px+r=0$ have two common roots and the third root of each equation represented by α and β, respectively.

If $f(x)=\begin{cases}e^{x\log_{1+x}|\alpha+\beta|}, & -1<x<0\\ a, & x=0\\ b\cdot\dfrac{\ln(e^{x^2}+\alpha\beta\sqrt{x})}{\sin\sqrt{x}}, & 0<x<1\end{cases}$

is continuous at $x=0$

48 The value of $\alpha^2+\beta^2$ is

49 The value of $\lim\limits_{x\to 0^-} f(x)$ is

50 The value of $2(a+b)$ is

51 Let $f(x)=\begin{cases}ax(x-1)+b; & x<1\\ x+2; & 1\le x\le 3\\ px^2+qx+2; & x>3\end{cases}$ is continuous for all except $x=1$ but $|f(x)|$ is differentiable everywhere and $f'(x)$ is continuous at $x=3$ and $|a+b+p+q|=\frac{k}{18}$, then the value of k ....

52 If m is the number of non-differentiable points of $f(x)=\dfrac{1}{\log|x^2-4|}$ and the value of n for which $\lim\limits_{x\to 0}\dfrac{x^n-\sin x^n}{x-\sin^n x}$ has a non-zero finite value, then the value of $m+n$ is ....

53 Let $f(x)=\begin{cases}\sqrt{4x^2-12x+9}\cdot\{x\}, & x\ge 1\\ \cos\left(\frac{\pi}{2}(|x|-\{x\})\right), & x<1\end{cases}$. The number of points in $x\in[-1,2]$, where $f(x)$ is non-differential, is .... (where $\{\cdot\}$ is fractional part of x)

54 The number of points of non differentiability of the function $f(x)=|||x-1|-1|-1|$ are ....

55 Let $f : R \to R$ and $g : R \to R$ be respectively, given by $f(x) = |x| + 1$ and $g(x) = x^2 + 1$. Define $h : R \to R$ by

$$h(x) = \begin{cases} \text{Max } \{f(x), g(x)\}, & x \le 0 \\ \text{Min } \{f(x), g(x)\}, & x > 0 \end{cases}$$

Then, number of points which $h(x)$ is non-differentiable is ....

56 If $P(x)$ be a polynomial satisfying $P(x) - 2P'(x) = 3x^3 - 27x^2 + 38x + 1$. If function

$$f(x) = \begin{cases} \dfrac{P''(x) + 18}{6}, & x \neq \dfrac{\pi}{2} \\ \sin^{-1}(ab) + \cos^{-1}(a + b - 3ab), & x = \dfrac{\pi}{2} \end{cases}$$

is continuous at $x = \dfrac{\pi}{2}$, then $(a + b)$ is equal to ....

57 Let $f(x) = [[x] + \{x^2\}] + \{[x^2] + \{x\}\}$, then number of points, where $|f(x)|$ is non-derivable in $[-3, 3]$ is equal to .... {where [ ] denotes the greatest integer function and {} denotes fractional part function}.

58 Let $f : R \to R$ and f be a differentiable function such that $f(x + 2y) = f(x) + 4f(y) + 2y(2x - 1)\ \forall\ x, y \in R$ and $f'(0) = 1$, then $f(3) + f'(3)$ is equal to ....

59 If $g(x) = \lim\limits_{m \to \infty} \dfrac{x^m f(x) + h(x) + 3}{2x^m + 4x + 1}$, when $x \neq 1$ and $g(1) = e^3$ such that $f(x)$, $g(x)$ and $h(x)$ are continuous function at $x = 1$ and $f(1) - h(1) = a\{b - g(1)\}$, then $a + b$ is ......

60 Let $f(x)$ is defined as

$$f(x) = \begin{cases} (\cos x - \sin x)^{\text{cosec } x}, & -\dfrac{\pi}{2} < x < 0 \\ a, & x = 0 \\ \dfrac{e^{\frac{1}{x}} + e^{\frac{2}{x}} + e^{\frac{3}{x}}}{ae^{\frac{2}{x}} + be^{\frac{3}{x}}}, & 0 < x < \dfrac{\pi}{2} \end{cases}$$

If $f(x)$ is continuous at $x = 0$, then ab is ......

61 Let $f(x)$ be a differentiable function on R such that $2f(x + y) + f(x - y) = 3f(x) + 3f(y) + 2xy$ $\forall\ x, y \in R$. If $f'(0) = 0$, then the value of $f(2)$ is ......

ANSWER KEY

Level 1

1. (i) continuous (ii) discontinuous (iii) continuous
   (iv) discountinuous (v) continuous (vi) discontinuous
2. (i) ± 2 (ii) 7 (iii) -2
   (iv) -4 (v) $\frac{1}{4}$ (vi) 1
3. $a+b=6$
4. $a=3, b=-2$
5. $\frac{1}{8}$
6. $-\frac{1}{8}$
7. (i) $x=0, x=1$ (ii) $x=\pm 1$ (iii) $x=(2n+1)\frac{\pi}{2}, x \in Z$ (iv) $x=\pm 1$
8. (i) Continuous but non-differentiable
   (ii) Continuous at $x=1$ and $x=3$
   differentiable at $x=1$ and non-differentiable at $x=3$
   (iii) Continuous at $x=0$ and $x=1$
   non-differentiable at $x=0$ and $x=1$
   (iv) Continuous but non-differentiable at $x=0$
   (v) Continuous and differentiable
9. $a=2, b=3$

Level 2

1. (d)	**2.** (b)	**3.** (b)	**4.** (c)	**5.** (d)	**6.** (a)	**7.** (d)
8. (a)	**9.** (c)	**10.** (b)	**11.** (c)	**12.** (c)	**13.** (a)	**14.** (d)
15. (d)	**16.** (d)	**17.** (c)	**18.** (b)	**19.** (c)	**20.** (d)	**21.** (a)
22. (c)	**23.** (a)	**24.** (a)	**25.** (b)	**26.** (b)	**27.** (d)	**28.** (c)
29. (b)	**30.** (a)	**31.** (a)	**32.** (b)	**33.** (b)	**34.** (a)	**35.** (d)
36. (d)	**37.** (a)	**38.** (c)	**39.** (d)	**40.** (b)	**41.** (b)	**42.** (c)
43. (b)	**44.** (d)	**45.** (a)	**46.** (c)	**47.** (c)	**48.** (c)	**49.** (b)
50. (b)	**51.** (3.5)	**52.** (15)	**53.** (2)	**54.** (1)	**55.** (1)	**56.** (0)
57. (10)	**58.** (5)	**59.** (2)	**60.** (4)	**61.** (3)	**62.** (5)	**63.** (1)
64. (3)	**65.** (7)					

Level 3

1. (c)	**2.** (b)	**3.** (b)	**4.** (a)	**5.** (a)	**6.** (a)	**7.** (b)
8. (d)	**9.** (a)	**10.** (a)	**11.** (c)	**12.** (d)	**13.** (a, d)	**14.** (b, d)
15. (a, d)	**16.** (a, b, c, d)	**17.** (b, c)	**18.** (b, d)	**19.** (b, c)	**20.** (a, b, c)	**21.** (a, c)
22. (a, c)	**23.** (a, b, c)	**24.** (a, c, d)	**25.** (a, d)	**26.** (a, b, d)	**27.** (a, b)	**28.** (a, c)
29. (a, c)	**30.** (b, c)	**31.** (d)	**32.** (c)	**33.** (c)	**34.** (b)	**35.** (d)
36. (c)	**37.** (a)	**38.** (b)	**39.** (a)			

40. (i)–B, D, (ii)–A, E (iii)–B, C, (iv)–B, F
41. (i)–A, B (ii)–A, D, (iii)–A, C, D, (iv)–A, B, C
42. (i)–B, (ii)–C, (iii)–D, (iv)–A
43. (i)–B, (ii)–A, (iii)–D, (iv)–C
44. (i)–D, (ii)–C, (iii)–A, (iv)–B

45. (3)	**46.** (3)	**47.** (2)	**48.** (5)	**49.** (3)	**50.** (9)	**51.** (54)
52. (7)	**53.** (3)	**54.** (5)	**55.** (3)	**56.** (2)	**57.** (1)	**58.** (19)
59. (4)	**60.** (1)	**61.** (4)				

SOLUTIONS

Level 1

1. (i) Given, $f(x)=\begin{cases}2x-(-x), & x<0\\ 2x-(x), & x\geq 0\end{cases}$

$=\begin{cases}2x+x, & x<0\\ x, & x\geq 0\end{cases}$

$=\begin{cases}3x & x<0\\ x, & x\geq 0\end{cases}$

$$\text{LHL}=\lim_{x\to 0^-} f(x)=\lim_{x\to 0^-}(3x)=0$$

$$\text{RHL}=\lim_{x\to 0^+} f(x)=\lim_{x\to 0^+}(x)=0$$

$f(0)=0$

$\therefore$ LHL = RHL = $f(0)$

So, $f(x)$ is continuous at $x=0$.

(ii) Given, $f(x)=\begin{cases}\dfrac{(x-a)}{x-a}=1, & x>a\\ 1, & x=a\\ \dfrac{-(x-a)}{x-a}=-1, & x<a\end{cases}$

$$\text{LHL}=\lim_{x\to a^-} f(x)=-1$$

$$\text{RHL}=\lim_{x\to a^+} f(x)=1$$

$f(a)=a$

LHL $\neq$ RHL $=f(a)$

So, $f(x)$ is discontinuous at $x=a$.

(iii) $$\text{LHL}=\lim_{x\to a^-} f(x)$$

$$=\lim_{h\to 0} f(a-h)$$

$$=\lim_{h\to 0}(a-h-a)\sin\left(\frac{1}{a-h-a}\right)$$

$$=\lim_{h\to 0}(-h)\sin\left(-\frac{1}{h}\right)$$

$$=\lim_{h\to 0} h\sin\frac{1}{h}$$

$$=0$$

$$\text{RHL}=\lim_{x\to a^+} f(x)$$

$$=\lim_{h\to 0} f(a+h)$$

$$=\lim_{h\to 0}(a+h-a)\sin\left(\frac{1}{a+h-a}\right)$$

$$=\lim_{h\to 0} h\sin\left(\frac{1}{h}\right)$$

$$=0$$

$f(0)=0$

So, $f(x)$ is continuous at $x=0$.

(iv) Given, $f(x)=\begin{cases}k-1, & k-1\leq x<k\\ k, & k\leq x<k+1\end{cases}$

$$\text{LHL}=\lim_{x\to k^-} f(x)=k-1$$

and $$\text{RHL}=\lim_{x\to k^+} f(x)=k$$

LHL $\neq$ RHL

So, $f(x)$ is discontinuous, when $x=k\in Z$.

(v) $$\text{LHL}=\lim_{x\to 0^-} f(x)=\lim_{h\to 0} f(0-h)$$

$$=\lim_{h\to 0}\frac{\sin 4(-h)}{\tan 3(-h)}=\lim_{h\to 0}\frac{\dfrac{\sin 4h}{4h}\times 4h}{\dfrac{\tan 3h}{3h}\times 3h}=\frac{4}{3}$$

$$\text{RHL}=\lim_{x\to 0^+} f(x)$$

$$=\lim_{h\to 0} f(0+h)$$

$$=\lim_{h\to 0}\frac{\log(1+4h)}{e^{3h}-1}=\lim_{h\to 0}\frac{\dfrac{\log(1+4h)}{4h}\times 4}{\dfrac{e^{3h}-1}{3h}\times 3}$$

$$=\frac{4}{3}$$

$$f(0)=\frac{4}{3}$$

So, $f(x)$ is continuous at $x=0$.

(vi) We have,

$$|1-4x^2|=\begin{cases}1-4x^2, & 0\leq x<\dfrac{1}{2}\\ 4x^2-1, & \dfrac{1}{2}\leq x<1\end{cases}$$

If $1\leq x<2$, then $0\leq x-1<1$

$\therefore$ $[x^2-2x]=[(x-1)^2-1)]$

$=[(x-1)^2]-1=0-1=-1$

$\therefore$ $$f(x)=\begin{cases}1-4x^2, & 0\leq x<\dfrac{1}{2}\\ 4x^2-1, & \dfrac{1}{2}\leq x<1\\ -1, & 1\leq x<2\end{cases}$$

At, $x=1$

$$\text{LHL}=\lim_{x\to 1^-} f(x)=\lim_{x\to 1^-}(4x^2-1)$$

$$=4-1=3$$

$$\text{RHL}=\lim_{x\to 1^+} f(x)=-1$$

LHL $\neq$ RHL

So, $f(x)$ is discontinuous at $x=1$.

2. (i) $f(x)$ is continuous at $x=0$

So, $\lim_{x \to 0} f(x) = f(0)$

$$\Rightarrow \lim_{x \to 0} \frac{1-\cos 2kx}{x^2} = 8$$

$$\Rightarrow \lim_{x \to 0} \frac{2\sin^2 kx}{k^2x^2} k^2 = 8$$

$$\Rightarrow 2k^2 = 8$$

$$k^2 = 4$$

$$k = \pm 2$$

(ii) $f(x)$ is continuous at $x=2$

So, $\lim_{x \to 2} f(x) = f(2)$

$$\Rightarrow \lim_{x \to 2} \frac{x^3+x^2-16x+20}{(x-2)^2} = k$$

$$\Rightarrow \lim_{x \to 2} \frac{(x-2)^2(x+5)}{(x-2)^2} = k$$

$$\Rightarrow k = 7$$

(iii) $f(x)$ is continuous at $x=0$

$\lim_{x \to 0} f(x) = f(0)$

$$= \lim_{x \to 0} \frac{\sqrt{1+kx}-\sqrt{1-kx}}{x} \times \frac{\sqrt{1+kx}+\sqrt{1-kx}}{\sqrt{1+kx}+\sqrt{1-kx}}$$

$$= \lim_{x \to 0} \frac{(1+kx)-(1-kx)}{x(\sqrt{1+kx}+\sqrt{1-kx})}$$

$$= \lim_{x \to 0} \frac{2kx}{x(\sqrt{1+kx}+\sqrt{1-kx})}$$

$$= \frac{2k}{2} = k$$

$f(0) = -2$

So, $k=-2$

(iv) If $f(x)$ is continuous at $x=0$, then

$\lim_{x \to 0} f(x) = f(0)$

$$\Rightarrow \lim_{x \to 0} \frac{\cos 3x - \cos x}{x^2} = k$$

$$\Rightarrow \lim_{x \to 0} -\frac{2\sin 2x \sin x}{x^2} = k$$

$$\Rightarrow -4 \lim_{x \to 0} \left(\frac{\sin 2x}{2x}\right)\left(\frac{\sin x}{x}\right) = k$$

$$k = -4$$

(v) Given, $f(x)$ is continuous at $x=\frac{\pi}{4}$

So, $\lim_{x \to \frac{\pi}{4}} f(x) = f\left(\frac{\pi}{4}\right)$

$$\Rightarrow \lim_{x \to \frac{\pi}{4}} \frac{1-\sqrt{2}\sin x}{\pi - 4x} = k$$

$$\Rightarrow \sqrt{2} \lim_{x \to \pi/4} \frac{\frac{1}{\sqrt{2}} - \sin x}{\pi - 4x} = k$$

$$\Rightarrow \frac{\sqrt{2}}{4} \lim_{x \to \frac{\pi}{4}} \frac{\sin\frac{\pi}{4} - \sin x}{\frac{\pi}{4} - x} = k$$

$$\Rightarrow \frac{1}{4\sqrt{2}} \lim_{x \to \frac{\pi}{4}} \frac{2\sin\left(\frac{\pi}{8}-\frac{x}{2}\right)\cos\left(\frac{\pi}{8}+\frac{x}{2}\right)}{\left(\frac{\pi}{8}-\frac{x}{2}\right)} = k$$

$$\Rightarrow \frac{1}{2\sqrt{2}} \cos\frac{\pi}{4} = k \Rightarrow k = \frac{1}{4}$$

(vi) Given, $f(x)$ is continuous at $x=\frac{\pi}{2}$

So, $f\left(\frac{\pi}{2}\right) = \lim_{x \to \frac{\pi}{2}} (\sin x)^{\frac{1}{\pi-2x}}$

$$= e^{\lim_{x \to \frac{\pi}{2}} \frac{(\sin x - 1)}{\pi - 2x}}$$

$$= e^{\lim_{x \to \frac{\pi}{2}} \frac{\cos x}{-2}} \quad \text{(using LHR)}$$

$$k = e^0 = 1$$

3. Given, $f(x)$ is continuous at

$x=1$

So, LHL = RHL = $f(1)$

$$\text{RHL} = \lim_{x \to 1^+} (4x^3 - 2x^2 + 3x + 1)$$

$$= 4 - 2 + 3 + 1 = 6$$

$$\text{LHL} = \lim_{x \to 1^-} (ax + b)$$

$$= a + b$$

So, $a+b=6$

4. Given, $f(x)$ is continuous on $[0,8]$.

So, $f(x)$ is continuous at

$x=2$ and $x=4$.

So, at $x=2$

LHL = RHL = $f(2)$

$$\Rightarrow 4 + 2a + b = 8$$

$$\Rightarrow 2a + b = 4 \quad \text{...(i)}$$

At $x=4$,

$$8a + 5b = 14 \quad \text{...(ii)}$$

From Eqs. (i) and (ii),

$$8a + 5(4-2a) = 14$$

$$\Rightarrow 8a + 20 - 10a = 14$$

$$\Rightarrow 2a = 6$$

$$\Rightarrow a = 3$$

From Eq. (i), $b = -2$

5. $f(x)$ is continuous at $x=\frac{\pi}{2}$, if

$$\lim_{x\to\frac{\pi}{2}} f(x)=f\left(\frac{\pi}{2}\right)$$

$$\Rightarrow \qquad K=\lim_{x\to\frac{\pi}{2}} \frac{1-\sin x}{(\pi-2x)^2}$$

$$=\lim_{h\to 0}\frac{1-\sin\left(\frac{\pi}{2}-h\right)}{(\pi-\pi+2h)^2}$$

$$=\lim_{h\to 0}\frac{1-\cos h}{4h^2}=\frac{1}{4}\times\frac{1}{2}$$

$$k=\frac{1}{8}$$

6. $f(x)$ is continuous at $x=0$

So, $\lim_{x\to 0} f(x)=f(0)$

$$\Rightarrow \qquad f(0)=\lim_{x\to 0}\frac{2-\sqrt{x+4}}{\frac{\sin 2x}{2x}\times 2x}\times\frac{2+\sqrt{x+4}}{2+\sqrt{x+4}}$$

$$=\lim_{x\to 0}\frac{4-(x+4)}{2x(2+\sqrt{x+4})}$$

$$=\lim_{x\to 0}\frac{-x}{2x(2+\sqrt{x+4})}$$

$$=\frac{-1}{2(2+2)}=-\frac{1}{8}$$

7. (i) $$f(x)=\begin{cases}-x-x+1, & x<0\\ x-x+1, & 0\le x<1\\ x+x-1, & x\ge 1\end{cases}$$

$$f(x)=\begin{cases}-2x+1, & x<0\\ 1, & 0\le x<1\\ 2x-1, & x\ge 1\end{cases}$$

$$f'(x)=\begin{cases}-2, & x<0\\ 0, & 0<x<1\\ 2, & x>1\end{cases}$$

$f(x)$ is non-differentiable at $x=0$ and $x=1$.

(ii) We have, $f(x)=|\log|x||, x\neq 0$

Using graphical transformation on $y=\log x$, we can get $y=|\log|x||$ as follow

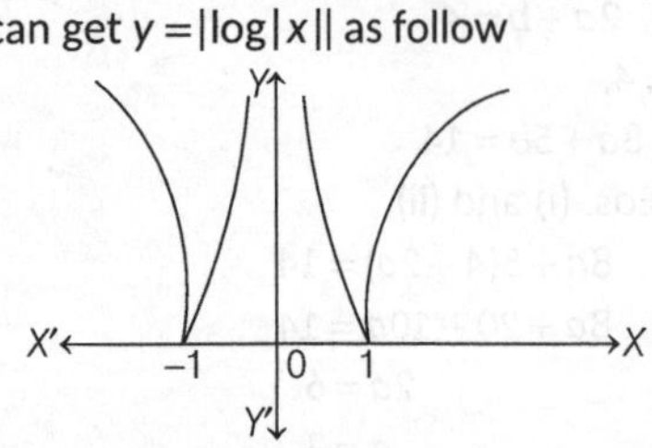

From the graph of the function $f(x)$, $f(x)$ is non-differentiable at $x=\pm 1$.

(iii) The graph of $f(x)$ is

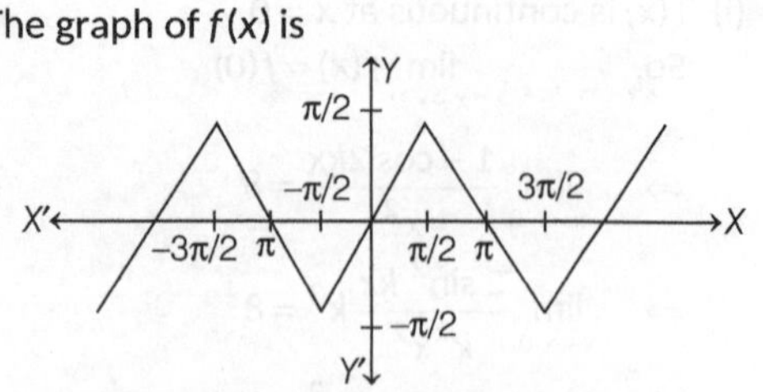

From the graph of the function, $f(x)$ is non-differentiable at $x=(2n+1)\frac{\pi}{2}, n\in Z$.

(iv) The graph of $f(x)$

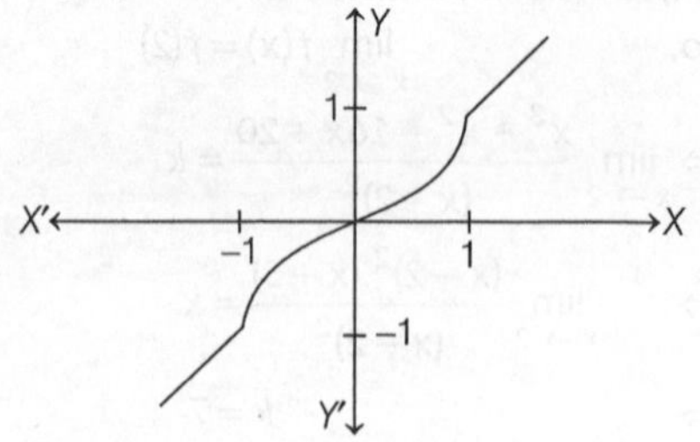

Thus, $f(x)$ is non-differentiable at $x=\pm 1$.

8. (i) At $x=2$,

$$\text{LHL}=\lim_{x\to 2^-} f(x)=\lim_{x\to 2^-} x[x]=2\times 1=2$$

$$\text{RHL}=\lim_{x\to 2^+} f(x)=\lim_{x\to 2^+}(x-1)[x]=(2-1)(2)=2$$

$$f(2)=(2-1)[2]=2$$

$\therefore$ $f(x)$ is continuous at $x=2$.

$$\text{Now, LHD}=\lim_{h\to 0}\frac{f(2-h)-f(2)}{-h}$$

$$=\lim_{h\to 0}\frac{(2-h)[2-h]-2}{-h}$$

$$=\lim_{h\to 0}\frac{(2-h)(1)-2}{-h}=1$$

$$\text{RHD}=\lim_{h\to 0}\frac{f(2+h)-f(2)}{h}$$

$$\therefore \lim_{h\to 0}\frac{(2+h-1)[2+h]-2}{h}=2$$

$$\text{LHD}\neq\text{RHD}$$

So, $f(x)$ is non-differentiable at $x=2$.

(ii) Given, $$f(x)=\begin{cases}\frac{x^2}{4}-\frac{3x}{2}+\frac{13}{4}, & x<1\\ 3-x, & 1\le x\le 3\\ x-3, & x\ge 3\end{cases}$$

Clearly, it can be seen that $f(x)$ is continuous at $x=1$ and $x=3$.

$$f'(x)=\begin{cases}\frac{2x}{4}-\frac{3}{2}, & x<1\\ -1, & 1<x<3\\ 1, & x>3\end{cases}$$

At $x=1$

$$f'(1^-)=\frac{1}{2}-\frac{3}{2}=-1$$

$$f'(1^+)=-1$$

So, $f(x)$ is differentiable at $x=1$ but $f(x)$ is non-differentiable at $x=3$.

(iii) At $x=0$,

LHL $=0$

RHL $=0$

$\therefore$ $f(x)$ is continuous at $x=0$

At $x=1$,

$$\text{LHL}=\lim_{x\to 1^-} f(x)=\lim_{x\to 1}(1)^2=1$$

$$\text{RHL}=\lim_{x\to 1^+} f(x)=\lim_{x\to 1^+}(x^3-2x+2)=1$$

$f(1)=1$

LHL = RHL = $f(1)$

So, $f(x)$ is continuous at $x=1$.

$$\text{Now, } f'(x)=\begin{cases}-1, & x<0\\ 2x, & 0<x<1\\ 3x^2-2, & x>1\end{cases}$$

At $x=0$,

$$f'(0^-)=-1$$

$$f'(0^+)=2(0)=0$$

$f(x)$ is non-differentiable at $x=0$

and $f'(1^-)=2(1)=2$

$$f'(1^+)=3(1)-2=1$$

$f(x)$ is non-differentiable at $x=1$.

(iv) Given, $f(0+h)=\lim_{h\to 0}\dfrac{0+h}{1+e^{\frac{1}{0+h}}}=\lim_{h\to 0}\dfrac{h}{1+e^{\frac{1}{h}}}=0$

$$f(0-h)=\lim_{h\to 0}\frac{0-h}{1+e^{-\frac{1}{h}}}=0$$

$$f(0+h)=f(0-h)=f(0)$$

So, $f(x)$ is continuous at $x=0$.

$$\text{Now, RHD}=\lim_{h\to 0}\frac{\dfrac{h}{1+e^{\frac{1}{h}}}-0}{h}=\lim_{h\to 0}\frac{1}{1+e^{\frac{1}{h}}}=0$$

$$\text{LHD}=\lim_{h\to 0}\frac{\dfrac{-h}{1+e^{-\frac{1}{h}}}-0}{-h}=\lim_{h\to 0}\frac{1}{1+e^{-\frac{1}{h}}}=1$$

LHD $\neq$ RHD

So, $f(x)$ is non-differentiable at $x=0$.

(v) Given, $\text{sgn}(x)=\begin{cases}1, & x>0\\ 0, & x=0\\ -1 & x<0\end{cases}$

$$\text{So, } f(x)=x^3\,\text{sgn}(x)=\begin{cases}x^3, & x>0\\ 0, & x=0\\ -x^3, & x<0\end{cases}$$

Clearly, at $x=0$

$$\text{LHL}=\text{RHL}=f(0)=0$$

So, $f(x)$ is continuous at $x=0$.

$$\text{and}\quad f'(x)=\begin{cases}3x^2, & x>0\\ -3x^2, & x<0\end{cases}$$

$$f'(0^-)=0$$

$$f'(0^+)=0$$

So, $f(x)$ is differentiable at $x=0$.

9. We have,

$$f(x)=\begin{cases}ax^2+b, & x<-1\\ bx^2+ax+4, & x\geq -1\end{cases}$$

$$\text{So,}\quad f'(x)=\begin{cases}2ax, & x<-1\\ 2bx+a, & x>-1\end{cases}$$

Since, $f(x)$ is differentiable, it must be continuous at $x=-1$

$\therefore$ LHL $=a+b$

RHL $=b-a+4$

$f(-1)=b-a+4$

So, $b-a+4=a+b$

$\Rightarrow$ $2a=4$

$a=2$

Again, $f'(x)$ is continuous.

$\therefore$ $-2b+a=-2a$ [putting $a=2$]

$\Rightarrow$ $-2b+2=-4$

$\Rightarrow$ $-2b=-6$

$\Rightarrow$ $b=3$

Level 2

1. (d) We observe that $\lim_{x\to 1^+} f(x)=f(1)$.

So, $f(x)$ is right continuous at $x=1$. Thus, $f(x)$ will not be continuous at $x=1$. Only, if

$$\lim_{x\to 1^-} f(x)\neq f(1)$$

$$\Rightarrow\quad \lim_{x\to 1^-}(px^2-q)\neq 3 \Rightarrow p-q\neq 3$$

Clearly, option (d) satisfies this condition.

2. (b) $f(x)$ is continuous at $x=0$, so

$$b=\lim_{x\to 0}\frac{e^{-4x}+\log(1+4x)+a}{x^3}$$

$$b=\lim_{x\to 0}\frac{\left(1+(-4x)+\frac{(-4x)^2}{2!}+\frac{(-4x)^3}{3!}+\ldots\right)+\left(4x-\frac{16x^2}{2}+\frac{64x^3}{3}+\ldots\right)+a}{x^3}$$

$$=\lim_{x\to 0}\frac{(1+a)+\left(\frac{-64}{6}+\frac{64}{3}\right)x^3+\ldots}{x^3}$$

So, $1+a=0$ and $-\frac{64}{6}+\frac{64}{3}=b$

$$a=-1\Rightarrow\frac{-64+128}{6}=b$$

$$\frac{64}{6}=b\Rightarrow b=\frac{32}{3}$$

3. (b) Since, f is continuous at $x=0$.

$\therefore\ f(0)=\lim_{x\to 0}\log(\sec^2 x)^{\cot^2 x}$

Therefore, we have

$$k=\lim_{x\to 0}\cot^2 x\cdot\log(\sec^2 x)$$

$$=\lim_{x\to 0}\cot^2 x\cdot\log(1+\tan^2 x)$$

$$=\lim_{x\to 0}\frac{\log(1+\tan^2 x)}{\tan^2 x}$$

$$=1\qquad\left[\because\lim_{x\to 0}\frac{\log(1+x)}{x}=1\right]$$

4. (c) If f is right continuous, so

$$\lim_{x\to 0^+}f(x)=f(0)=1$$

$$\Rightarrow\quad\lim_{x\to 0^+}\frac{p+q^{\frac{1}{x}}}{r+s^{\frac{1}{x}}}=1$$

As, $x\to 0^+;\ \frac{1}{x}\to\infty$

Given, $s<1\Rightarrow s=\frac{1}{a},\ a>1$

Similarly, $q=\frac{1}{b},\ b>1$

$$\Rightarrow\quad\lim_{x\to 0^+}\frac{p+\left(\frac{1}{b}\right)^{\frac{1}{x}}}{r+\left(\frac{1}{a}\right)^{\frac{1}{x}}}=1=\frac{p+\left(\frac{1}{b^{\infty}}\right)}{r+\left(\frac{1}{a^{\infty}}\right)}=1$$

$$\Rightarrow\quad\frac{p}{r}=1$$

$$\Rightarrow\quad p=r$$

5. (d) Given, $f(x)$ is continuous at $x=0$

So, LHL = RHL = $f(0)$

$$\Rightarrow\quad\lim_{x\to 0}\frac{x\left(e^{\frac{1}{x}}-e^{-\frac{1}{x}}\right)}{e^{\frac{1}{x}}+e^{-\frac{1}{x}}}$$

$$=\lim_{x\to 0}\frac{x\cdot e^{\frac{1}{x}}\left(1-e^{-\frac{2}{x}}\right)}{e^{\frac{1}{x}}\left(1+e^{-\frac{2}{x}}\right)}$$

$$\left[\text{when } x\to 0,\ e^{-\frac{2}{x}}=e^{-\infty}=0\right]$$

$$=\frac{0(1-0)}{(1+0)}$$

$$f(0)=0$$

6. (a) Since, $f(x)$ is continuous at $x=0$.

$\therefore$ LHL = RHL

$$\lim_{x\to 0^-}f(x)=\lim_{x\to 0^+}f(x)$$

$$\Rightarrow\lim_{x\to 0^-}a^2\cos^2 x+b^2\sin^2 x=\lim_{x\to 0^+}e^{ax+b}$$

$$\Rightarrow\quad a^2=e^b$$

$$\Rightarrow\quad 2\log|a|=b$$

[taking log on both sides]

7. (d) For f to be continuous everywhere

So, $f(0)=\lim_{x\to 0}\frac{2-(256-7x)^{\frac{1}{8}}}{(5x+32)^{\frac{1}{5}}-2}$

$$=\lim_{x\to 0}\frac{1-\left(1-\frac{7x}{256}\right)^{\frac{1}{8}}}{\left(1+\frac{5x}{32}\right)^{\frac{1}{5}}-1}$$

$$=\lim_{x\to 0}\frac{1-\left(1-\frac{1}{8}\cdot\frac{7x}{256}+\ldots\right)}{\left[1+\frac{1}{5}\cdot\frac{5x}{32}+\ldots\right]-1}$$

$$=\lim_{x\to 0}\frac{\left(\frac{1}{8}\cdot\frac{7x}{256}+\text{other terms higher power of } x\right)}{\left(\frac{x}{32}+\text{other terms higher power of } x\right)}$$

$$=\frac{\frac{1}{8}\cdot\frac{7}{256}}{\frac{1}{32}}=\frac{7}{64}$$

$$=\frac{7}{2^6}$$

8. (a) We have, $f(x) = \cos 2x \cdot \cot\left(\frac{\pi}{4} - x\right)$

$f(x)$ is continuous at $x = \frac{\pi}{4}$

$$\Rightarrow \lim_{x \to \frac{\pi}{4}} f(x) = f(\pi/4)$$

$$f(\pi/4) = \lim_{x \to \frac{\pi}{4}} \frac{\cos 2x}{\tan\left(\frac{\pi}{4} - x\right)}$$

Applying L-Hospital rule,

$$= \lim_{x \to \frac{\pi}{4}} \frac{-2\sin 2x}{-\sec^2\left(\frac{\pi}{4} - x\right)}$$

$$= 2$$

9. (c) $f(x)$ is continuous at $x = \frac{\pi}{2}$

So, $\lim_{x \to \frac{\pi}{2}} = f\left(\frac{\pi}{2}\right)$

$$\Rightarrow k = \lim_{x \to \frac{\pi}{2}} \frac{1 - \sin x}{(\pi - 2x)^2} \cdot \frac{\log(\sin x)}{\log(1 + \pi^2 - 4\pi x + 4x^2)}$$

$$= \lim_{h \to 0} \frac{1 - \cos h}{4h^2} \cdot \frac{\log(\cos h)}{\log(1 + 4h^2)} \quad \left[\because x = \frac{\pi}{2} - h\right]$$

$$= -\lim_{h \to 0} \left(\frac{1 - \cos h}{4h^2}\right)^2 \cdot \frac{\log(1 + \cos h - 1)}{\cos h - 1} \cdot \frac{4h^2}{\log(1 + 4h^2)}$$

$$= -\lim_{h \to 0} \left(\frac{\sin^2 \frac{h}{2}}{2h^2}\right)^2 \frac{\log\{1 + (\cos h - 1)\}}{(\cos h - 1)} \times \frac{4h^2}{\log(1 + 4h^2)}$$

$$= -\frac{1}{64} \lim_{h \to 0} \left(\frac{\sin \frac{h}{2}}{\frac{h}{2}}\right)^4 \frac{\log\{1 + (\cos h - 1)\}}{(\cos h - 1)} \cdot \frac{4h^2}{\log(1 + 4h^2)}$$

$$k = -\frac{1}{64}$$

10. (b) Since, the function is continuous at $x = 0$

So, $f(0^-) = f(0^+)$

$$\Rightarrow \lim_{x \to 0^-} \frac{a + 2\cos x}{x^2} = \lim_{x \to 0^+} b \tan \frac{\pi}{[x + 4]}$$

For a finite limit, the numerator has to be 0 for the limit to be of the form $\frac{0}{0}$.

So, $a + 2 = 0 \Rightarrow a = -2$

$$\Rightarrow \lim_{x \to 0^-} \frac{-2 + 2\cos x}{x^2} = \lim_{x \to 0^+} b \tan \frac{\pi}{[x + 4]}$$

Using L-Hospital rule,

$$\Rightarrow \lim_{x \to 0^-} \frac{-2\sin x}{2x} = b \tan \frac{\pi}{4}$$

$$-1 = b$$

11. (c) $f(x)$ is continuous at any point, then

$$x^2 - ax + 3 = 2 - x$$

$$\Rightarrow x^2 + x(1 - a) + 1 = 0$$

For this quadratic, to have two solutions

$$b^2 - 4ac > 0$$

$$\Rightarrow b^2 > 4ac$$

$$\Rightarrow (1 - a)^2 > 4 \times 1 \times 1$$

$$\Rightarrow (1 - a)^2 - 4 > 0$$

$$\Rightarrow (1 - a + 2)(1 - a - 2) > 0$$

$$\Rightarrow (3 - a)(-a - 1) > 0$$

$$\Rightarrow (a - 3)(a + 1) > 0$$

$$\Rightarrow a \in (-\infty, -1) \cup (3, \infty)$$

12. (c) Since, f is continuous for $0 \le x < \infty$, f is continuous at $x = 1$

So, $\lim_{x \to 1^-} \frac{x^2}{a} = a$

$$\therefore a = \pm 1$$

Since, f is continuous at $x = \sqrt{2}$.

$$\lim_{x \to \sqrt{2}} \frac{2b^2 - 4b}{x^2} = a$$

$$\therefore \frac{2b^2 - 4b}{2} = a$$

$$\Rightarrow 2b^2 - 4b = 2a \Rightarrow b^2 - 2b = a$$

When $a = 1, b^2 - 2b = 1$

$$\Rightarrow b = 1 \pm \sqrt{2}$$

When $a = -1, b^2 - 2b = -1$

$$\Rightarrow b^2 - 2b + 1 = 0$$

$$\Rightarrow (b - 1)^2 = 0 \Rightarrow b = 1$$

13. (a) We have,

$$f(x) = [\tan x] + \sqrt{\tan x - [\tan x]}$$
$$= [t] + \sqrt{t - [t]}$$

Let $t = \tan x$

Clearly, $0 \le t < \infty$ at $0 \le x < \frac{\pi}{2}$

Possible points of discontinuity may be at which $t \in N$.

Let $t = k \in N$

$$\text{LHL} = \lim_{t \to k^-} [t] + \sqrt{t - [t]}$$

$$= \lim_{h \to 0} [k - h] + \sqrt{(k - h) - [k - h]}$$

$$= \lim_{h\to 0} (k-1+\sqrt{k-h-k+1})$$

$$= k$$

$$\text{RHL} = \lim_{t\to k^+} [t]+\sqrt{t-[t]}$$

$$= \lim_{h\to 0} [k+h]+\sqrt{(k+h)-[k+h]}$$

$$= \lim_{h\to 0} k+\sqrt{k+h-k}$$

$$= k$$

$\therefore$ The function is continuous at $t = k \in N$.

Thus, function is continuous for all $\left[0, \frac{\pi}{2}\right)$.

14. (d) We have, $f(x) = \begin{cases} [x]+\sqrt{x-[x]}, & x \geq 0 \\ \sin x, & x < 0 \end{cases}$

$$= \begin{cases} [x]+\sqrt{\{x\}}, & x \geq 0 \\ \sin x, & x < 0 \end{cases}$$

Clearly, $f(x)$ is continuous for $x < 0$.
Now, let $x = k \in N$, then

$$\lim_{x\to k^-} f(x) = (k-1)+\sqrt{k-(k-1)} = k$$

$$\lim_{x\to k^+} f(x) = k+\sqrt{k-k} = k$$

and $\quad f(k) = k$

$\Rightarrow \quad f(x)$ is continuous at $x = k \in N$.

Now, let $x = a$, $x > 0$, $x \notin N$.

$$f(a^-) = k+\sqrt{a-k}$$

$$f(a^+) = k+\sqrt{a-k}$$

and $\quad f(a) = k+\sqrt{a-k}$

$\Rightarrow f(x)$ is continuous at $x = a$.

$\Rightarrow$ Hence, $f(x)$ is continuous for all $x \in R$.

15. (d) We have, $f(x) = \begin{cases} \dfrac{x^2-1}{x^2-2|x-1|-1}, & \text{if } x \neq 1 \\ \dfrac{1}{2}, & \text{if } x = 1 \end{cases}$

$$f(1^-) = \lim_{x\to 1^-}\left(\frac{x^2-1}{x^2-2(-x+1)-1}\right) = \lim_{x\to 1^-}\left(\frac{x^2-1}{x^2+2x-3}\right)$$

$$= \lim_{x\to 1^-}\left(\frac{(x+1)(x-1)}{(x+3)(x-1)}\right)$$

$$= \lim_{x\to 1^-}\left(\frac{x+1}{x+3}\right) = \frac{2}{4} = \frac{1}{2}$$

$$\text{and } f(1^+) = \lim_{x\to 1^+}\left(\frac{x^2-1}{x^2-2(x-1)-1}\right)$$

$$= \lim_{x\to 1^+}\left(\frac{x^2-1}{x^2-2x+1}\right) = \lim_{x\to 1^+}\left(\frac{x+1}{x-1}\right) = \infty$$

and $f(1) = \dfrac{1}{2}$

Hence, $f(x)$ is discontinuous for $x = 1$.

16. (d) Given, $f(x)$ is continuous at $x = 0$

$$f(0) = \lim_{x\to 0} \frac{\ln(e^{x^2}+2\sqrt{x})}{\dfrac{\tan\sqrt{x}}{\sqrt{x}} \times \sqrt{x}}$$

$$= \lim_{x\to 0} \frac{\ln(e^{x^2}+2\sqrt{x})}{\sqrt{x}}$$

$$= \lim_{x\to 0} \ln(e^{x^2}+2\sqrt{x})^{\frac{1}{\sqrt{x}}}$$

$$= \lim_{x\to 0} (e^{x^2}+2\sqrt{x}-1)\frac{1}{\sqrt{x}}$$

$$= \lim_{x\to 0}\left(\frac{e^{x^2}-1}{\sqrt{x}}+2\right)$$

$$= 2$$

17. (c) We have, $f(x) = \text{sgn}(\sin^2 x - \sin x - 1)$

$$= \text{sgn}\left[\left(\sin x - \frac{1}{2}\right)^2 - \frac{5}{4}\right]$$

For $f(x)$ to be discontinuous,

$$\left(\sin x - \frac{1}{2}\right)^2 = \frac{5}{4}$$

So, $\quad \sin x - \dfrac{1}{2} = \pm\dfrac{\sqrt{5}}{2}$

$$\sin x = \frac{1\pm\sqrt{5}}{2}$$

$$\Rightarrow \quad \sin x = \frac{1-\sqrt{5}}{2} \qquad [-1 \leq \sin x \leq 1]$$

There will be two values of x between π and 2π for which $\sin x = \dfrac{1-\sqrt{5}}{2}$

$\therefore$ For 4 points of discontinuity, n can take the value 4 units.

Hence, there is exactly two possible values of n.

18. (b) We have,

$$f(x) = \begin{cases} \dfrac{\lambda|x^2-5x+6|}{\mu(5x-6-x^2)}; & x < 2 \\ \mu; & x = 2 \\ e^{\frac{\tan(x-2)}{x-[x]}}; & x > 2 \end{cases}$$

At $x = 2$

$f(2) = \mu$

$$\text{RHL} = f(2+h) = \lim_{h\to 0} e^{\frac{\tan h}{2+h-[2+h]}}$$

$$= \lim_{h \to 0} e^{\frac{\tan h}{2+h-2}}$$

$$= \lim_{h \to 0} e^{\frac{\tan h}{h}}$$

$$= e$$

$$\text{LHL} = f(2-h) = \lim_{h \to 0} \frac{\lambda|(-h)(-1-h)|}{-\mu(-h)(-1-h)}$$

$$\left[f(x) = \frac{\lambda|(x-2)(x-3)|}{-\mu(x-2)(x-3)}, x<2\right]$$

$$= \lim_{h \to 0} \frac{-\lambda \times h(1+h)}{\mu \times h(1+h)} = -\frac{\lambda}{\mu}$$

Function $f(x)$ is continuous at $x = 2$

$\therefore$ LHL = RHL = $f(2)$

$$-\frac{\lambda}{\mu} = e = \mu$$

So, $\mu = e$

and $-\frac{\lambda}{\mu} = e$

$$-\lambda = \mu e$$

$$-\lambda = e^2$$

$$\lambda = -e^2$$

Now, $\lambda + \mu = -e^2 + e$

$$= e - e^2$$

19. (c) If $f(x)$ is continuous for all x, then $3 \sin x + a^2 - 10a + 30 = 4 \cos x \,\forall\, x$.

$\Rightarrow \quad (a-5)^2 + 5 = 4 \cos x - 3 \sin x$

$\Rightarrow \quad (a-5)^2 + 5 = 5 \cos(x + \theta)$,

where $\theta = \tan^{-1} \frac{3}{4} \forall x$.

We observe that LHS ≥ 5 and RHS ≤ 5.

So, both are equal, when $a = 5$ and $\cos(x + \theta) = 1$

$\Rightarrow \quad a = 5$

and $\quad x = 2n\pi - \tan^{-1} \frac{3}{4}$

20. (d) We have,

$$f(x) = \begin{cases} -x-1, & -1 \leq x < 0 \\ x+0, & 0 \leq x < 1 \\ x+x, & 1 \leq x < 2 \\ x+2, & 2 \leq x < 3 \\ x+3, & x = 3 \end{cases}$$

$$f(x) = \begin{cases} -x-1, & -1 \leq x < 0 \\ x, & 0 \leq x < 1 \\ 2x, & 1 \leq x < 2 \\ x+2, & 2 \leq x < 3 \\ x+3, & x = 3 \end{cases}$$

At $x = 0$,

LHL $= -1$ and RHL $= 0$

At $x = 1$,

LHL $= 1$ and RHL $= 2$

At $x = 2$,

LHL $= 2(2) = 4$ and RHL $= 2 + 2 = 4$

Here, LHL = RHL, so function is continuous at $x = 2$.

At $x = 3$,

LHL $= 3 + 2 = 5$

RHL $= 3 + 3 = 6$

LHL $\neq$ RHL

So, function is discontinuous at $x = 0$, 1 and 3.

21. (a) We have, $A = \lim_{x \to 0} x\left[\frac{4}{x}\right]$

$$= \lim_{x \to 0} x\left(\frac{4}{x} - \left\{\frac{4}{x}\right\}\right)$$

$$= \lim_{x \to 0} \left(4 - x\left\{\frac{4}{x}\right\}\right)$$

$$A = 4$$

Now, check when

(a) $x = \sqrt{A+1} \Rightarrow x = \sqrt{5}$

(b) $x = \sqrt{A+21} \Rightarrow x = 5$

(c) $x = \sqrt{A} \Rightarrow x = 2$

(d) $x = \sqrt{A+5} \Rightarrow x = 3$

$f(x) = [x^2] \sin \pi x$ will be discontinuous at non-integral multiple of π. So, $f(x)$ is discontinuous at $x = \sqrt{A+1}$.

22. (c) $g\{f(x)\} = \begin{cases} g\left(\frac{x}{2} - 1\right), & 0 \leq x < 1 \\ g\left(\frac{1}{2}\right), & 1 \leq x \leq 2 \end{cases}$

$$= \begin{cases} \frac{(x-1)(x-2-2k)+6}{2}, & 0 \leq x < 1 \\ 4-2k, & 1 \leq x \leq 2 \end{cases}$$

$$\lim_{x \to 1^-} g\{f(x)\} = 3$$

$$g\{f(1)\} = 4 - 2k$$

and $\lim_{x \to 1^+} g\{f(1)\} = 4 - 2k$

$f\{g(x)\}$ is continuous at $x = 1$

So, $4 - 2k = 3$

$$2k = 1$$

$$k = \frac{1}{2}$$

23. (a) We have, $f(x) = \frac{1}{1-x}, x \neq 1$

Now, $\quad f\{f(x)\} = f\left(\frac{1}{1-x}\right) = \frac{1}{1 - \frac{1}{1-x}} = \frac{x-1}{x}$

$f\{f(x)\}$ is defined for all $x \neq 0, 1$

Now, $f(f(f(x))) = f\left(\frac{x-1}{x}\right) = \frac{1}{1-\frac{x-1}{x}} = x \ \forall\ x \neq 0, 1$

Thus, the set of points of discontinuity is $\{0,1\}$.

$\therefore \qquad \alpha = 0$ and $\beta = 1$

It is given that $(\alpha, \beta) = (0, 1)$ and (a, a^2) lie on the same side of line $x + 2y - 3 = 0$.

So, $(0+2-3)(a+2a^2-3) > 0$

$$2a^2 + a - 3 < 0$$

$$(2a+3)(a-1) < 0$$

$$\Rightarrow \quad -\frac{3}{2} < a < 1$$

$$\Rightarrow \quad a \in \left(-\frac{3}{2}, 1\right)$$

24. (a) We have,

$$f(x) = [x] + [-x]$$

At $x = 1$,

$$\text{LHL} = \lim_{x \to 1^-} ([x] + [-x])$$
$$= \lim_{h \to 0} ([1-h] + [-(1-h)])$$
$$= \lim_{h \to 0} [1-h] + [h-1]$$
$$= 0 - 1 = -1$$
$$\text{RHL} = \lim_{x \to 1^+} ([x] + [-x])$$
$$= \lim_{h \to 0} ([1+h] + [-(1+h)])$$
$$= 1 - 2 = -1$$

and $f(1) = [1] + [-1] = 0$

Thus, LHL = RHL $\neq f(1)$

So, $f(x)$ has isolated point discontinuity.

25. (b) $\text{LHL} = \lim_{x \to \frac{\pi}{2}^-} \sin^{-1}\{\sin x\}$

$$= \lim_{x \to \frac{\pi}{2}^-} \sin^{-1}(\sin x - [\sin x]),$$

where $[\cdot]$ denotes G.I.F

$$= \lim_{x \to \frac{\pi}{2}^-} \sin^{-1}(\sin x - 0)$$
$$= \lim_{x \to \frac{\pi}{2}} \sin^{-1}(\sin x)$$
$$= \frac{\pi}{2}$$

$$\text{RHL} = \lim_{x \to \frac{\pi}{2}^+} (\pi[x] - 1) = \pi - 1$$

Thus, LHL $\neq$ RHL, $f(x)$ has jump type discontinuity.

26. (b) $\text{RHL} = \lim_{x \to 0^+} \frac{\ln(e^{x^2} + 2\sqrt{\{x\}})}{\tan\sqrt{x}}$

$$= \lim_{x \to 0^+} \frac{\ln(e^{x^2} + 2\sqrt{x} + 1 - 1)}{\sqrt{x}\left(\frac{\tan\sqrt{x}}{\sqrt{x}}\right)(e^{x^2} + 2\sqrt{x} - 1)} \times (e^{x^2} + 2\sqrt{x} - 1)$$

$$= 2$$

$$\text{LHL} = \lim_{x \to 0^-} \frac{x[x]^2 \ln 2}{\ln(x+1)} = \ln 2$$

Thus, LHL $\neq$ RHL

Hence, f has irremovable discontinuity at $x = 0$.

27. (d) Since $gof(x)$ is continuous.

Then, $f(0) = \lim_{x \to 0^-} f(x) = \lim_{h \to 0} f(0-h)$

$$\Rightarrow \quad 1 = \lim_{h \to 0} (0 - h + a)$$
$$\Rightarrow \quad 1 = a$$

and $g(0) = \lim_{x \to 0^-} g(x) = \lim_{h \to 0} f(0-h)$

$$1 + b = \lim_{h \to 0} (0 - h + 1)$$
$$\Rightarrow \quad 1 + b = 1$$
$$\Rightarrow \quad b = 0$$

28. (c) We have,

$t = \frac{1}{x-1}$ is discontinuous at $x = 1$

Also, $y = \frac{1}{t^2 + t - 2}$

$y = \frac{1}{(t+2)(t-1)}$ is discontinuous at $t = -2$ and $t = 1$

When $t = -2$, then $\frac{1}{x-1} = -2 \Rightarrow x = \frac{1}{2}$

When $t = 1$, then $\frac{1}{x-1} = 1 \Rightarrow x = 2$

So, $f(x)$ is discontinuous at three points $x = 1, \frac{1}{2}$ and 2.

29. (b) $f(x)$ is differentiable at $x = 3$.

So, $f(x)$ must be continuous at $x = 3$

$$\therefore \quad \lim_{x \to 3^-} f(x) = \lim_{x \to 3^+} f(x) = f(3)$$

$\Rightarrow \quad a\sqrt{3+1} = 3b + 2 \Rightarrow 2a = 3b + 2$ ...(i)

Now, $f'(x) = \begin{cases} \frac{a}{2\sqrt{x+1}}, & 0 < x < 3 \\ b, & 3 < x < 5 \end{cases}$

$f(x)$ is differentiable at $x = 3$.

$\therefore \quad f'(3^-) = f'(3^+)$

$$\Rightarrow \quad \frac{a}{2\sqrt{3+1}} = b \Rightarrow a = 4b \quad \text{...(ii)}$$

Substituting value in Eq. (i),

$$\Rightarrow \quad 8b = 3b + 2$$

$$5b = 2 \Rightarrow b = \frac{2}{5}$$

So, $a = 4\left(\frac{2}{5}\right)$

$$a = \frac{8}{5}$$

30. (a) We have, $f(x) = |\log 2 - \sin x|$

$\therefore\ g(x) = f\{f(x)\} = |\log 2 - \sin|\log 2 - \sin x||$,

In the neighbourhood of $x = 0$.

$g(x) = \log 2 - \sin(\log 2 - \sin x)$

Hence, $g(x)$ is differentiable at $x = 0$, as it is sum and composite of differentiable function.

$g'(x) = \cos(\log 2 - \sin x) \cos x$

$\therefore\ g'(0) = \cos(\log 2)$

31. (a) We have,

$$f(x) = \begin{cases} x, & x \le 1 \\ ax^2 + bx + c, & \text{otherwise} \end{cases}$$

Function $f(x)$ is differentiable for all $x \in R$.

So, it must be continuous at $x = 1$

So, $f(1^-) = \lim_{x \to 1^-} f(x) = 1$

$f(1^+) = \lim_{x \to 1^+} f(x) = a + b + c$

$f(1) = 1$

$\therefore\ f(1^-) = f(1^+) = f(1)$

$\Rightarrow \quad a + b + c = 1 \quad$ ...(i)

Also, $f'(x) = \begin{cases} 1, & x \le 1 \\ 2ax + b, & \text{otherwise} \end{cases}$

LHD $= \lim_{x \to 1^-} f'(x) = 1$

RHD $= \lim_{x \to 1^+} f'(x) = 2a + b$

$f(x)$ is differentiable at $x = 1$.

So, $\quad 2a + b = 1 \quad$ ...(ii)

From Eqs. (i) and (ii),

$a + b + c = 1$

$\Rightarrow a + (1 - 2a) + c = 1$

$\Rightarrow \quad a = c$

$\therefore \quad a = a, b = 1 - 2a$ and $c = a$

So, $\{(a, 1 - 2a, a);\ a \in R,\ a \ne 0\}$ is the triplet.

32. (b) Given, function is

$f(x) = \text{Max}\{a - x, a + x, b\}, -\infty < x < \infty, 0 < a < b$

So, graph of $f(x)$

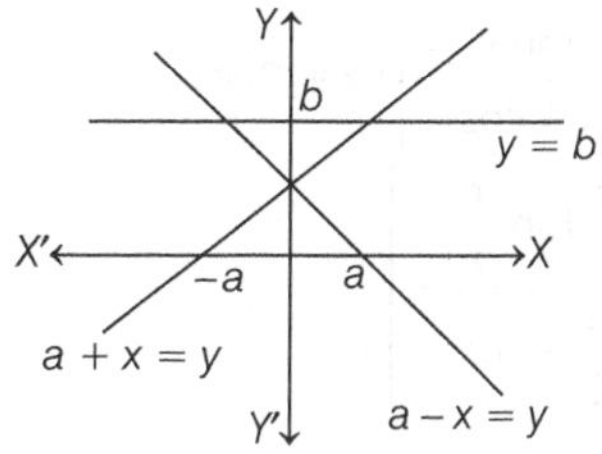

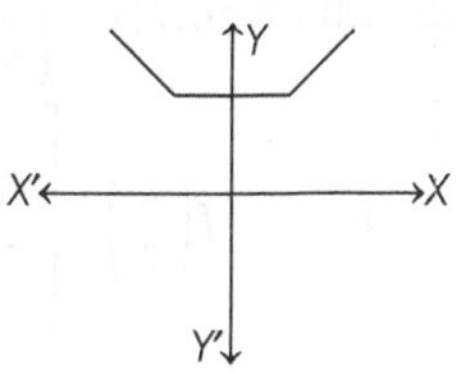

We know that a function cannot be differentiable at sharp edges.

$\therefore$ The function $f(x)$ is non-differentiable at 2 points.

33. (b) (a) If $f(x)$ is differentiable everywhere, then $|f|$ is differentiable everywhere.

Let $f(x) = x$.

$\Rightarrow |f(x)| = |x|$, which is non-differentiable at $x = 0$.

$\therefore\ |f|$ is non-differentiable everywhere.

(b) $|f|^2$ is differentiable everywhere.

Let $f(x) = x$

$\Rightarrow |f(x)|^2 = x^2$

Again $f(x) = e^x$

$|f(x)| = e^{2x}$

Differentiable everywhere

$\therefore\ |f|^2$ is differentiable everywhere.

(c) $f|f|$ is non-differentiable at some point

$f(x) = e^x$

$\Rightarrow |f(x)| = e^x$

So, $f|f| = e^{2x}$, differentiable everywhere.

So, this option is not correct.

(d) $f + |f|$ is differentiable everywhere.

Let $f(x) = x, |f| = |x|$

$f + |f| = \begin{cases} 0, & x < 0 \\ 2x, & x \ge 0 \end{cases}$, which is non-differentiable at $x = 0$.

34. (a)Given, $f(x) = \sin|x| + 2(x - \pi)\cos|x| - |x|$

$$\Rightarrow f(x) = \begin{cases} \sin x - x + 2(x - \pi)\cos x, & x \ge 0 \\ -\sin x + x + 2(x - \pi)\cos x, & x < 0 \end{cases}$$

$\therefore f(x)$ is continuous for all $x \in R$.

Now, $f'(x) = \begin{cases} \cos x - 1 + 2\cos x - 2\sin x(x - \pi), & x \ge 0 \\ -\cos x + 1 + 2\cos x - 2\sin x(x - \pi), & x < 0 \end{cases}$

$\therefore\ f(x)$ is differentiable at $x > 0$ and $x < 0$.

Now, $f'(0^-) = -1 + 1 + 2 - 0 = 2$

and $\ f'(0^+) = 1 - 1 + 2 - 0 = 2$

$\Rightarrow \quad f'(0^-) = f'(0^+)$

$\Rightarrow f'(x)$ is differentiable for all $x \in R$.

Hence, S is an empty set ϕ.

$\therefore$ Number of elements in $S = 0$ (number of element)

35. (d) Given, $f(x)=\left[\dfrac{x\tan x}{\sin x+\cos x}\right]; x\in[0,\pi]$

At $\quad f\left(\dfrac{\pi}{2}\right)=\left[\dfrac{\frac{\pi}{2}\tan\frac{\pi}{2}}{\sin\frac{\pi}{2}+\cos\frac{\pi}{2}}\right]$

$f\left(\dfrac{\pi}{2}\right)=\infty$

and $\quad f(0)=\left[\dfrac{0\tan 0}{\sin 0+\cos 0}\right]=0$

$\Rightarrow f(x)$ is continuous in $\left[0,\dfrac{\pi}{2}\right]$.

But, if [ ] denotes greatest integer function, then $f(x)$ will break infinity many times.

$\therefore$ $f(x)$ is non-differentiable at infinite times in $[0,\pi]$.

36. (d) We have,

$$f(x)=\begin{cases}3x+5, & x\le 1\\ x^2+px+q, & x>1\end{cases}$$

$$\therefore \quad f'(x)=\begin{cases}3 & x<1\\ 2x+p, & x>1\end{cases}$$

$f(x)$ is differentiable at $x=1$

Then, it must be continuous at $x=1$

So, $\lim\limits_{x\to 1^+} f(x)=\lim\limits_{x\to 1^-} f(x)$

$\Rightarrow \quad 1+p+q=3+5$

$\Rightarrow \quad p+q=7$

Also, $f'(1^+)=f'(1^-)$

$\Rightarrow \quad 2+p=3$

$p=1$

and $\quad p+q=7$

$1+q=7$

$q=6$

37. (a) Clearly, $(x-2)^2\sin\dfrac{1}{(x-2)}$ is differentiable at $x=0$ but $|x|$ is not.

So, $f(x)$ is non-differentiable at $x=0$.

Now, for differentiability at $x=2$

$$\lim_{x\to 2}\left[(x-2)^2\sin\left(\frac{1}{x-2}\right)-|x|\right]$$

$=0-2$

$=-2$

$f(2)=-2$

So, $f(x)$ is continuous at $x=2$.

Now, $f'(2^+)=\lim\limits_{h\to 0}\dfrac{f(2+h)-f(2)}{h}$

$$=\lim_{h\to 0}\frac{h^2\sin\frac{1}{h}-|2+h|-(-2)}{h}$$

$$=\lim_{h\to 0}\frac{h^2\sin\frac{1}{h}-2-h+2}{h}$$

$$=\lim_{h\to 0}\frac{h^2\sin\frac{1}{h}-h}{h}$$

$=-1$

Similarly $f'(2^-)=-1$

So, $f(x)$ is differentiable at $x=2$.

38. (c) Given, $f(x)$ is continuous at $x=0$.

So, $\lim\limits_{x\to 0} f(x)$

$=\lim\limits_{x\to 0} x^p\sin\dfrac{1}{x}$

$=(0)^p$ [finite value between -1 to 1]

$=0$, if $p>1$

Now, $f'(0)=\lim\limits_{h\to 0}\dfrac{f(h)-f(0)}{h}$

$$=\lim_{h\to 0}\frac{h^a\sin\left(\frac{1}{h}\right)}{h}$$

$$=\lim_{h\to 0}h^{a-1}\sin\left(\frac{1}{h}\right)$$

This limit will not exist, if $a-1\le 0\Rightarrow a\le 1$

So, $f(x)$ is continuous but non-differentiable at $x=0$, when $p\in(0,1]$.

39. (d) Since, function is continuous at $x=1$ and 3.

$\therefore \quad f(1^-)=f(1^+)$

$\Rightarrow \quad ae+be^{-1}=c \quad$ ...(i)

and $\quad f(3)=f(3^+)$

$\Rightarrow \quad 9c=9a+6c\Rightarrow c=3a \quad$ ...(ii)

From Eqs. (i) and (ii), we get

$b=ae(3-e) \quad$ ...(iii)

$$\text{So, } f'(x)=\begin{cases}ae^x-be^{-x}, & -1<x<1\\ 2cx, & 1<x<3\\ 2ax+2c, & 3<x<4\end{cases}$$

$f'(0)=a-b$ and $f'(2)=4c$

Given, $f'(0)+f'(2)=e$

$\Rightarrow \quad a-b+4c=e \quad$ ...(iv)

From Eqs. (i), (ii), (iii) and (iv), we get

$\Rightarrow \quad a-3ae+ae^2+12a=e$

$\Rightarrow \quad 13a-3ae+ae^2=e$

$\Rightarrow \quad a=\dfrac{e}{e^2-3e+13}$

40. (b) We have,

$$f(x)=[x^2]+[-x]^2$$

$$\lim_{x\to 2^-} f(x)=\lim_{h\to 0} f(2-h)=\lim_{h\to 0}[(2-h)^2]+[-2+h]^2$$

$$\Rightarrow \quad \lim_{x\to 2^-} f(x)=3+(-2)^2=7$$

and $\lim_{x\to 2^+} f(x)=\lim_{h\to 0} f(2+h)$

$$=\lim_{h\to 0}[(2+h)^2]+[-2-h]^2$$

$$=4+(-3)^2=4+9=13$$

Clearly, $\lim_{x\to 2^-} f(x)\neq \lim_{x\to 2^+} f(x)$

So, $f(x)$ is discontinuous at $x=2$.

Consequently, it is non-differentiable at $x=2$.

41. (b) Given, $f(x)=\begin{cases}\dfrac{a^{2[x]+\{x\}}-1}{2[x]+\{x\}}, & x\neq 0\\ \log a, & x=0\end{cases}$

$$=\begin{cases}\dfrac{a^{[x]+x}-1}{[x]+x}, & x\neq 0\\ \log a, & x=0\end{cases} \qquad (\because [x]+\{x\}=x)$$

To check the continuity of $f(x)$ at $x=0$.

$$f(0^+)=f(0^-)=f(0)$$

$$f(0^+)=\lim_{h\to 0}\frac{a^{[h]+h}-1}{[h]+h}$$

$$=\lim_{h\to 0}\frac{a^h-1}{h}=\log a$$

$$f(0^-)=\lim_{h\to 0}\frac{a^{[-h]-h}-1}{[-h]-h}$$

$$=\lim_{h\to 0}\frac{a^{-1-h}-1}{-1-h}$$

$$=\frac{a^{-1}-1}{-1}$$

$$=1-a^{-1}$$

$$f(0^-)=1-\frac{1}{a}$$

$$f(0^+)\neq f(0^-)$$

So, $f(x)$ is discontinuous at $x=0$

42. (c) From the property of modulus function it is clear that modulus function is continuous along the entire number line.

Thus, the function defined by $g(x)=|f(x)|$ is continuous, if $f(x)$ is continuous function.

43. (b) Using graphical transformation, the graph of $|\sin x|$ function has period π

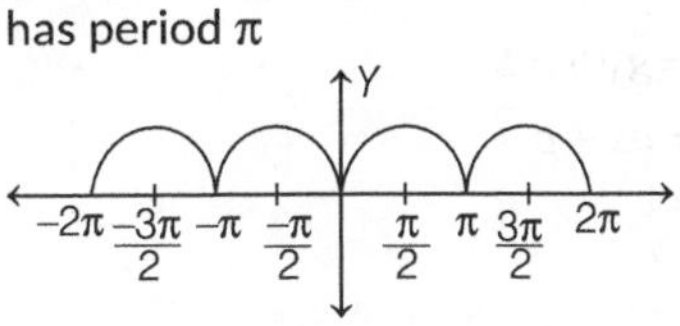

So, graph of $\sin^{-1}|\sin x|$

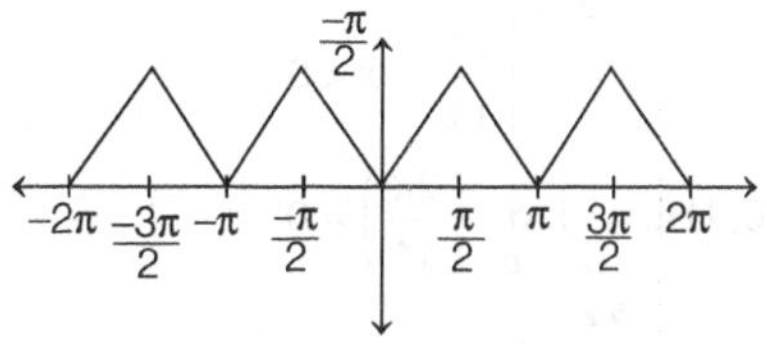

From the graph of the function $f(x)$ is continuous for all x and $f(x)$ is non-differentiable at $x=\dfrac{n\pi}{2}, n\in Z$.

44. (d) For $x=-\dfrac{\pi}{2}$,

$$\lim_{x\to -\frac{\pi}{2}^-} f(x)=\lim_{x\to -\frac{\pi}{2}^-}\left(-x-\frac{\pi}{2}\right)=0$$

$$\lim_{x\to -\frac{\pi}{2}^+} f(x)=\lim_{x\to -\frac{\pi}{2}^+}(-\cos x)=-\cos\left(-\frac{\pi}{2}\right)=0$$

and $f\left(-\dfrac{\pi}{2}\right)=-\left(-\dfrac{\pi}{2}\right)-\dfrac{\pi}{2}=0$

So, function is continuous at $x=-\dfrac{\pi}{2}$

Now, at $x=0$

$$\lim_{x\to 0^-} f(x)=\lim_{x\to 0^-}(-\cos x)=-1$$

$$\lim_{x\to -0^+} f(x)=\lim_{x\to 0^+}(x-1)=-1$$

and $\qquad f(0)=-\cos 0=-1$

$\therefore$ $f(x)$ is continuous at $x=0$.

Now, $f'(x)=\begin{cases}-1, & x<-\dfrac{\pi}{2}\\ \sin x, & -\dfrac{\pi}{2}<x<0\\ 1, & 0<x<1\\ \dfrac{1}{x}, & x>1\end{cases}$

Clearly, $f(x)$ is non-differentiable at $x=0$.

It is differentiable at $x=1, -\dfrac{3}{2}$

45. (a) We have,

$$2x+4|y|=7y$$

If $\qquad y\geq 0$

$$2x+4y=7y$$

$$\Rightarrow \quad 3y=2x$$

$$\Rightarrow \quad y=\frac{2x}{3}$$

If $\qquad y<0$

$$2x-4y=7y$$

$$\Rightarrow \quad 11y=2x$$

$$y=\frac{2x}{11}$$

$$\Rightarrow \qquad y=\begin{cases}\dfrac{2x}{3}, & x\geq 0\\ \dfrac{2x}{11}, & x<0\end{cases}$$

At $x=0$, LHL $=\lim\limits_{x\to 0^-}\left(\dfrac{2x}{11}\right)=0$

RHL $=\lim\limits_{x\to 0^+}\dfrac{2x}{3}=0$

So, function is continuous at $x=0$.

Now, $\dfrac{dy}{dx}=\begin{cases}\dfrac{2}{3}, & x\geq 0\\ \dfrac{2}{11}, & x<0\end{cases}$

y is non-differentiable at $x=0$.

46. (c) Let $g(x)=-\sqrt{9-x^2}$

$g(x)$ is defined for $x\in[-3,3]$.

So, $f(x)$ is defined on $[-3,3]$.

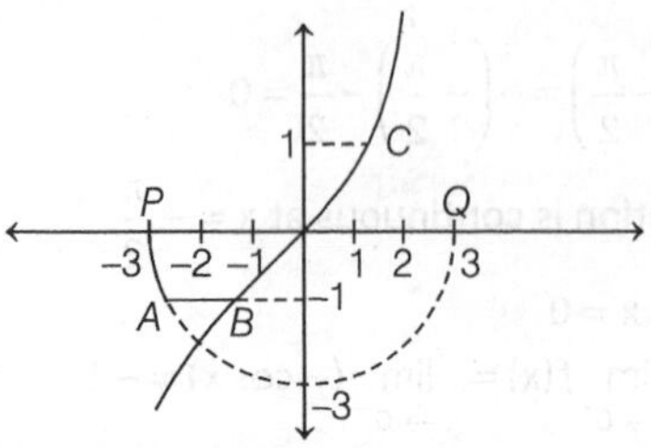

It is clear that the graph $f(x)$ is continuous but non-differentiable at A, B and C.

At P, right hand derivative is $-\infty$ and at Q, left hand derivative is ∞.

So, $f(x)$ is non-differentiable at P and Q.

Hence, $f(x)$ is non-derivatives at 5 points.

47. (c) $f(x)=\begin{cases}\int_0^x (5+|1-t|)\,dt; & x>2\\ 5x+1 & x\leq 2\end{cases}$

At $x=2$,

LHL $=\lim\limits_{x\to 2^-}(5x+1)=11$

RHL $=\lim\limits_{x\to 2^+}\int_0^x (5+|1-t|)dt$

$=\int_0^1 (5+(1-t))dt+\int_1^2 (5-(1-t))dt$

$=\int_0^1 (6-t)dt+\int_1^2 (4+t)dt$

$=\left[6t-\dfrac{t^2}{2}\right]_0^1+\left[4t+\dfrac{t^2}{2}\right]_1^2$

$=\left(6-\dfrac{1}{2}\right)+\left(8+2-4-\dfrac{1}{2}\right)=\dfrac{11}{2}+\dfrac{11}{2}=11$

$f(2)=11$

So, $f(x)$ is continuous at $x=2$.

Now, LHD at $x=2$ is $\dfrac{d}{dx}(5x+1)|_{x=2}=5$

RHD at $x=2$ is $\dfrac{d}{dx}\int_0^x (5+|1-t|)\,dt\Big|_{x=2}=6$

LHD $\neq$ RHD

So, function is non-differentiable at $x=2$.

48. (c) $f(x)=\left[\dfrac{(x-5)^3}{A}\right]\sin(x-5)+a\cos(x-2)$

$[x]$ will be continuous and differentiable, if $\dfrac{(x-5)^3}{A}$ lies between two consecutive integers.

Given, $A>0$ and $(x-5)^3>0$ for given interval, the two integers have to be non-negative.

Let $n\leq\dfrac{(x-5)^3}{A}<n+1$

Let us consider two cases (i) $n=0$ and (ii) $n>0$.

Case I If $n=0$

$$0\leq\frac{(x-5)^3}{A}<1$$

Here, $\qquad 0\leq\left(\dfrac{x-5}{A}\right)^3\Rightarrow A\geq 0$

and if $\dfrac{(x-5)^3}{A}<1\Rightarrow A>(x-5)^3$

For interval $(7,9)$,

$A>(9-5)^3\Rightarrow A>64$

$\therefore \qquad A\in(64,\infty)$

Case II If $n>0$

$$\frac{1}{n}>\frac{A}{(x-5)^3}>\frac{1}{n+1}$$

The inequality holds true for all values of x is given domain.

So, $A>\dfrac{(9-5)^3}{n+1}$ and $A<\dfrac{(7-5)^3}{n}$

Hence, $\dfrac{64}{n+1}<\dfrac{8}{n}\Rightarrow n<\dfrac{1}{7}$

Since, n is an integer and greater than 0.

This is not possible.

49. (b) $f(0^+)=\lim\limits_{x\to 0^+}|x|^{\sin x}$

$=e^{\lim\limits_{x\to 0}\sin x\log|x|}$

$=e^{\lim\limits_{x\to 0}\frac{\log x}{\text{cosec } x}}=e^0=1$

$f(0^-)=g(0)=1$

Let $g(x)=ax+b$

Then, $b=1$

$\Rightarrow g(x) = ax + 1$

For $x > 0$,

$$f'(x) = e^{\sin x \ln|x|}\left[\cos x \ln|x| + \frac{\sin x}{x}\right]$$

$$f'(1) = 1[0 + \sin 1] = \sin 1$$

$$f(-1) = -a + 1 \text{ or } a = 1 - \sin 1$$

or $g(x) = (1 - \sin 1)x + 1$

50. (b) We have,

$$S(x) = \int_{x^2}^{x^3} \ln t\, dt$$

$$S'(x) = \ln(x^3)\cdot 3x^2 - \ln(x^2)(2x)$$

$$= 9x^2 \ln x - 4x \ln x$$

$$= (9x - 4)x \ln x$$

$$\therefore H(x) = \frac{S'(x)}{x} = (9x - 4)\ln x$$

which is clearly continuous and differentiable in its domain.

51. (3.5) At $x = 0$, $f(0) = b$

For $x < 0$,

$$\text{LHL} = \lim_{x\to 0^-} f(0-h)$$

$$= \lim_{h\to 0^-} \frac{\sin(a+1)(-h) + \sin(-h)}{-h}$$

$$= \lim_{h\to 0^-} \frac{-\{\sin(a+1)h + \sin h\}}{-h}$$

$$= \lim_{h\to 0^-}\left(\frac{\sin(a+1)h}{(a+1)h}\times(a+1) + \frac{\sin h}{h}\right)$$

$$= a + 1 + 1 = a + 2$$

$$\text{RHL} = \lim_{h\to 0^+} f(0+h)$$

$$= \lim_{h\to 0^+} \frac{(h+h^2)^{\frac{1}{2}} - h^{\frac{1}{2}}}{h^{\frac{3}{2}}}$$

$$= \lim_{h\to 0} \frac{(1+h)^{\frac{1}{2}} - 1}{h}$$

$$= \lim_{h\to 0} \frac{\left(1 + \frac{1}{2}h - 1\right)}{h} \quad [\because (1+x)^n = 1 + nx, \text{ if } x \to 0]$$

$$= \frac{1}{2}$$

$f(x)$ is continuous at $x = 0$

$\therefore$ LHL = RHL = $f(0)$

$$\Rightarrow a + 2 = \frac{1}{2} = b$$

$$\Rightarrow \quad a = -\frac{3}{2} \text{ and } b = \frac{1}{2}$$

$$\text{Hence, } b - 2a = \frac{1}{2} - \left(2\times -\frac{3}{2}\right) = \frac{1}{2} + 3 = \frac{7}{2} = 3.5$$

52. (15) f is a continuous function, so

$$f(0) = \lim_{x\to 0} \frac{\tan[e^2]x^2 - \tan[-e^2]x^2}{\sin^2 x}$$

$$= \lim_{x\to 0} \frac{\tan[7.38]x^2 - \tan[-7.38]x^2}{\sin^2 x}$$

$$[\because e = 2.718 \therefore e^2 = 7.38]$$

$$= \lim_{x\to 0} \frac{\tan 7x^2 - \tan(-8)x^2}{\sin^2 x}$$

$$= \lim_{x\to 0} \frac{\frac{\tan 7x^2}{7x^2}\times 7 + \frac{\tan 8x^2}{8x^2}\times 8}{\frac{\sin^2 x}{x^2}}$$

$$= 7 + 8 = 15$$

53. (2) $\because f(x)$ is continuous at $x = 0$

So, $\lim_{x\to 0^+} f(0+h) = f(0)$

$$\Rightarrow \lim_{h\to 0^+} f(0+h) = \lambda \ln 4$$

$$\Rightarrow \quad \lim_{h\to 0}\left(\frac{8^h - 4^h - 2^h + 1}{h^2}\right) = \lambda \ln 4$$

$$\Rightarrow \lim_{h\to 0}\left(\frac{4^h(2^h - 1) - 1(2^h - 1)}{h^2}\right) = \lambda \ln 4$$

$$\Rightarrow \quad \lim_{h\to 0} \frac{(2^h - 1)(4^h - 1)}{h^2} = \lambda \ln 4$$

$$\Rightarrow \quad \ln 2 \ln 4 = \lambda \ln 4$$

So, $\lambda = \ln 2 \Rightarrow e^\lambda = 2$

54. (1) $f(x) = \dfrac{\sin 3x + A\sin 2x + B\sin x}{x^5}$

$$= \lim_{x\to 0} \frac{\left\{3x - \frac{(3x)^3}{3!} + \frac{(3x)^5}{5!} + \ldots\right\} + A\left\{2x - \frac{(2x)^3}{3!} + \frac{(2x)^5}{5!} - \ldots\right\} + B\left\{x - \frac{x^3}{3!} + \frac{x^5}{5!} + \ldots\right\}}{x^5}$$

$$= \lim_{x\to 0} \frac{(3 + 2A + B)x}{x^5} - \frac{x^3\left(\frac{9}{2} + \frac{4A}{3} + \frac{B}{6}\right)}{x^5} + \frac{x^5}{x^5}\left(\frac{3^5}{5!} + \frac{32}{5!} + \frac{B}{5!}\right)$$

For existence of limit,

$$3 + 2A + B = 0 \text{ and } \frac{9}{2} + \frac{4A}{3} + \frac{B}{6} = 0$$

$$\Rightarrow \quad 2A + B + 3 = 0 \text{ and } 27 + 8A + B = 0$$

$$\Rightarrow \quad 6A + 24 = 0 \Rightarrow A = -4 \text{ and } B = 5$$

$$f(0)=\frac{3^5}{5!}+\frac{32A}{5!}+\frac{B}{5!}$$
$$=\frac{3^5}{5!}+\frac{2^7}{5!}+\frac{5}{5!}$$
$$=\frac{243-128+5}{120}=\frac{120}{120}=1$$

55. (1) At $x=0$,

$f(0)=b$

$$\text{RHL}=f(0+h)=\lim_{h\to 0} e^{\frac{\cot 4h}{\cot 2h}}$$
$$=\lim_{h\to 0} e^{\frac{\tan 2h}{\tan 4h}}$$
$$=\lim_{h\to 0} e^{\frac{\frac{\tan 2h}{2h}\times 2h}{\frac{\tan 4h}{4h}\times 4h}}$$
$$=e^{\frac{1}{2}}$$

$$\text{LHL}=f(0-h)=\lim_{h\to 0}(1+|\sin h|)^{\frac{a}{|\sin h|}}$$
$$=\lim_{h\to 0}(1+\sin h)^{\frac{a}{\sin h}}$$
$$=e^{\lim\limits_{h\to 0}(1+\sin h-1)\cdot\frac{a}{\sin h}}$$
$$=e^a$$

$\because$ $f(x)$ is continuous.

So, LHL = RHL = $f(0)$

$$e^a=e^{\frac{1}{2}}=b \Rightarrow e^a=e^{\frac{1}{2}}$$

So, $a=\frac{1}{2}$ and $e^{\frac{1}{2}}=b$

$$\ln b=\ln e^{\frac{1}{2}}$$
$$\ln b=\frac{1}{2}$$

Hence, $a+\ln b$

$$=\frac{1}{2}+\frac{1}{2}=1$$

56. (0) We have,

$$\lim_{x\to 0} f(x)=\lim_{x\to 0}\frac{\left(\frac{2^x-1}{x}\right)^3}{\log 2\left(\frac{\sin(x\log 2)}{x\log 2}\right)\frac{\log(1+x^2\log 4)}{x^2\log 4}\cdot\log 4}$$
$$=\frac{(\log 2)^3}{(\log 2)(2\log 2)}=\frac{1}{2}\log 2$$

Therefore, $f(0)$ is defined to be $\frac{1}{2}\log 2$.

So, $[f(0)]=0$

57. (10) We have,

$$f(x)=[3x]^2-\{3x\}^2$$
$$=([3x]+\{3x\})([3x]-\{3x\})$$
$$=3x([3x]-\{3x\})$$
$$=3x(3x-\{3x\}-\{3x\})$$
$$=3x(3x-2\{3x\})$$

We know that $\{3x\}$ is discontinuous at integers.

Now, $-2<x<2$

$\Rightarrow -6<3x<6$

$\therefore$ Integers $=\{-5,-4,-3,-2,-1,0,1,2,3,4,5\}$

But $f(x)=3x(3x-2\{3x\})$ is continuous at $x=0$.

$\therefore$ Total number of points of discontinuity are 10.

58. (5) As, f is continuous, so

$f(0)=\lim\limits_{x\to 0} f(x)=\lim\limits_{n\to\infty} f(x_n)$ for any sequence x_n

such that $\lim\limits_{n\to\infty} x_n=0$.

Thus,

$$f(0)=\lim_{n\to\infty} f\left(\frac{1}{2^n}\right)=\lim_{n\to\infty}\left((\sin e^n)e^{-n^2}+\frac{5n^2}{n^2+3}\right)$$
$$=\lim_{n\to\infty}\left((\sin e^n)e^{-n^2}+\frac{5}{1+\frac{3}{n^2}}\right)$$
$$=0+5=5$$

59. (2) We have,

$$(g+h)x=g(x)+h(x)$$
$$=\begin{cases}x^2+1+px, & x\in Q\\ px^2+2x+q, & x\notin Q\end{cases}$$

If $(g+h)(x)$ is continuous at $x=1$ and $x=3$.

$\Rightarrow$ 1 and 3 are roots of equation

$$x^2+1+px=px^2+2x+q$$
$$\Rightarrow x^2(1-p)+x(p-2)+1-q=0$$

Now, $\frac{2-p}{1-p}=4$ and $\frac{1-q}{1-p}=3$

$\Rightarrow$ $p=\frac{2}{3}$ and $q=0$

$\Rightarrow$ $3p+q=3\times\frac{2}{3}+0=2$

60. (4) We have, $f(x)=\sin^{-1}\left(\frac{2x}{1+x^2}\right)$

Clearly, $f(x)$ is non-differentiable at $x=-1$ and $x=1$.

But it is continuous everywhere. So, $p=0$ and $q=2$

Hence, $p+q+2=0+2+2=4$

61. (3) LHL $= \lim_{x\to 1^-} f(x) = \lim_{x\to 1^-} x^2 e^{2(x-1)} = 1$

Also, $f(1)=1$

RHL $= \lim_{x\to 1^+} f(x) = \lim_{x\to 1^+} a\,\text{sgn}(x+1)\cos 2(x-1) + bx^2$

$= a+b$

For continuity $a+b=1$

Also, LHD $= \lim_{h\to 0} \dfrac{(1-h)^2 e^{-2h}-1}{-h}$

$= \lim_{h\to 0} 2e^{-2h} - he^{-2h} + \left(\dfrac{e^{-2h}-1}{-h}\right)$

$= 2+0+2=4$

and RHD $= \lim_{h\to 0} \dfrac{a\,\text{sgn}(2+h)\cos 2h + b(1+h)^2 - 1}{h}$

$= \lim_{h\to 0} \dfrac{a\cos 2h + b + bh^2 + 2bh - (a+b)}{h}$ $[\because a+b=1]$

$= \lim_{h\to 0} a\left(\dfrac{\cos 2h - 1}{h}\right) + bh + 2b = 2b$

$f(x)$ is differentiable at $x=1$, if $2b=4 \Rightarrow b=2$

Now, $a+b=1 \Rightarrow a=-1$

So, $a+2b=-1+2(2)=3$

62. (5) We have, $f(x)=|\sin x|+[\cos x]$, $x\in[0,2\pi]$

$$f(x)=\begin{cases} 1, & x=0 \\ \sin x, & x\in\left(0,\dfrac{\pi}{2}\right] \\ \sin x-1, & x\in\left(\dfrac{\pi}{2},\pi\right] \\ -\sin x, & x\in\left(\pi,\dfrac{3\pi}{2}\right) \\ -\sin x, & x\in\left[\dfrac{3\pi}{2},2\pi\right) \\ 1, & x=2\pi \end{cases}$$

$f(x)$ is discontinuous at $x=0, \dfrac{\pi}{2}, \dfrac{3\pi}{2}$ and 2π

$$\text{Now, } f'(x)=\begin{cases} \cos x, & x\in\left(0,\dfrac{\pi}{2}\right) \\ \cos x, & x\in\left(\dfrac{\pi}{2},\pi\right) \\ -\cos x, & x\in\left(\pi,\dfrac{3\pi}{2}\right) \\ -\cos x, & x\in\left(\dfrac{3\pi}{2},2\pi\right) \end{cases}$$

$\Rightarrow f(x)$ is non-differentiable at $x=0, \dfrac{\pi}{2}, \pi, \dfrac{3\pi}{2}$ and 2π

i.e. 5 points.

63. (1) We have,

$$f(x)=\begin{cases} -1, & -2\le x<0 \\ x^2-1, & 0\le x<2 \end{cases}$$

$g(x)=|f(x)|+f|x|$

$$\text{Now, } |f(x)|=\begin{cases} 1, & -2\le x<0 \\ 1-x^2, & 0\le x\le 1 \\ x^2-1, & 1<x\le 2 \end{cases}$$

$f(|x|)=\{x^2-1,\ -2\le x\le 2$

$$g(x)=\begin{cases} x^2, & -2\le x<0 \\ 0, & 0\le x\le 1 \\ 2x^2-2, & 1<x\le 2 \end{cases}$$

$g(x)$ is a continuous function.

$$\text{Now, } g'(x)=\begin{cases} 2x, & -2<x<0 \\ 0, & 0<x<1 \\ 4x, & 1<x<2 \end{cases}$$

At $x=1$

LHD $=0$ and RHD $=4$

LHD $\neq$ RHD

So, $g(x)$ is non-differentiable at $x=0$.

64. (3) $f(x)=|x|-1$ and $g(x)=|x|$

Then, $gof = g\{f(x)\}$

$g(|x|-1)=||x|-1|$

$$\text{So,}\quad gof=\begin{cases} |x-1|, & x\ge 0 \\ |x+1|, & x<0 \end{cases}$$

$$=\begin{cases} x-1, & x\ge 1 \\ 1-x, & 0\le x<1 \\ x+1, & -1\le x\le 0 \\ -x-1, & x<-1 \end{cases}$$

Graph of gof is

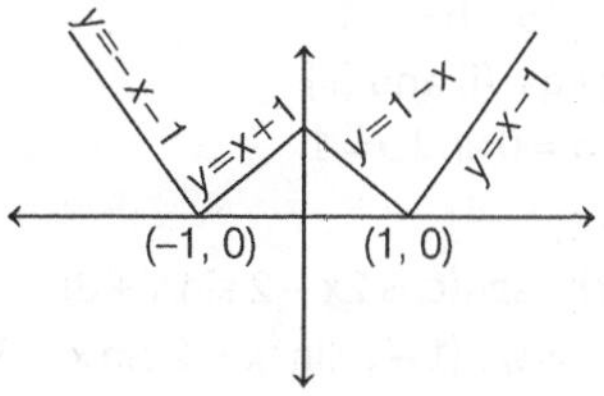

Clearly, gof is continuous but non-differentiable at three points -1, 0 and 1.

65. (7) Given that $f(x)$ is differentiable for all x.

So, it is everywhere continuous.

So, $f(3)$ is a finite quantity.

Now, $f'(3)=\lim_{h\to 0}\dfrac{f(3+h)-f(3)}{h}$

$\Rightarrow\quad f'(3)=\lim_{h\to 0}\dfrac{1}{h}f(3+h)-\lim_{h\to 0}\dfrac{f(3)}{h}$

$\Rightarrow\quad f'(3)=7-\lim_{h\to 0}\dfrac{f(3)}{h} \Rightarrow \lim_{h\to 0}\dfrac{f(3)}{h}=7-f'(3)$

RHS is a finite quantity, as $f'(3)$ exists.

So, LHS must be a finite, it is only possible, when $f(3)=0$

$\therefore\quad 0=7-f'(3)\Rightarrow f'(3)=7$

Level 3

1. (c) We have,

$$f(x)=\frac{4-x^2}{|4x-x^3|}=\frac{4-x^2}{|x||4-x^2|}$$

$f(x)$ is not defined at $x=0$ and ± 2

$$f(x)=\begin{cases}\dfrac{1}{|x|}, & \text{for } 4-x^2>0\\ \dfrac{-1}{|x|}, & \text{for } 4-x^2<0\end{cases}$$

$$=\begin{cases}-\dfrac{1}{x}, & \text{for } -2<x<0 \text{ or } x>2\\ \dfrac{1}{2}, & 0<x<2 \text{ or } -\infty<x<-2\end{cases}$$

$\Rightarrow f(x)$ is discontinuous at exactly three points $x=-2$, 0 and 2.

2. (b) We have, $f(x)=\lim\limits_{n\to\infty}\left(\dfrac{x^{2n-1}+ax^2+bx}{x^{2n}+1}\right)$

$$=\begin{cases}ax^2+bx, & -1<x<1\\ \dfrac{a-b-1}{2}, & x=-1\\ \dfrac{a+b+1}{2}, & x=1\\ \dfrac{1}{x}, & x>1 \text{ or } x<-1\end{cases}$$

For continuity at $x=1$,

$$\frac{a+b+1}{2}=a+b$$

$\Rightarrow \quad a+b=1$ ...(i)

For continuity at $x=-1$,

$a-b=-1$ ...(ii)

On solving Eqs. (i) and (ii)

$a=0$ and $b=1$

3. (b) We have,

$$f(x)=\text{sgn}(\cos 2x-2\sin x+3)$$
$$=\text{sgn}(1-2\sin^2 x-2\sin x+3)$$
$$=\text{sgn}(4-2\sin^2 x-2\sin x)$$
$$=\text{sgn}[(\sin x+2)(2-2\sin x)]$$

$$f(x)=\begin{cases}1, & x\neq\dfrac{\pi}{2}\\ 0, & x=\dfrac{\pi}{2}\end{cases}$$

So, $f(x)$ has a missing point discontinuity.

4. (a) Given, $f(x)=\left[\dfrac{2(\sin x-\sin^3 x)+|\sin x-\sin^3 x|}{2(\sin x-\sin^3 x)-|\sin x-\sin^3 x|}\right]$

$\therefore$ In neighbourhood of $\dfrac{\pi}{2}$, $\sin x>\sin^3 x$

$$\Rightarrow \quad f(x)=\left[\frac{3(\sin x-\sin^3 x)}{(\sin x-\sin^3 x)}\right]=3$$

$$\Rightarrow \quad f\left(\frac{\pi^-}{2}\right)=f\left(\frac{\pi^+}{2}\right)=f\left(\frac{\pi}{2}\right)=3$$

$\Rightarrow f(x)$ is continuous at $x=\dfrac{\pi}{2}$

$\Rightarrow \quad f'(x)=0\ \forall\ x\in(0,\pi)$

$\Rightarrow f(x)$ is continuous and differentiable at $x=\dfrac{\pi}{2}$.

5. (a) $f(x)=\text{Min}\{|x|, 2-|x-1|, |x-1|\}$

Draw the graph of $y=|x|$, $y=2-|x-1|$ and $y=|x-1|$

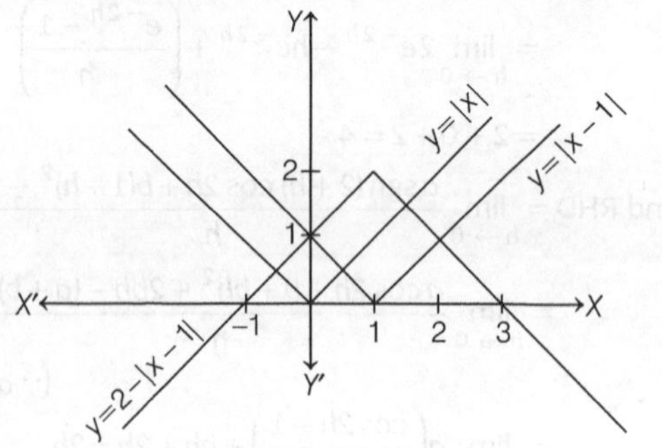

Graph of $y=\text{Min}\{|x|, |x-2|, 2-|x-1|\}$

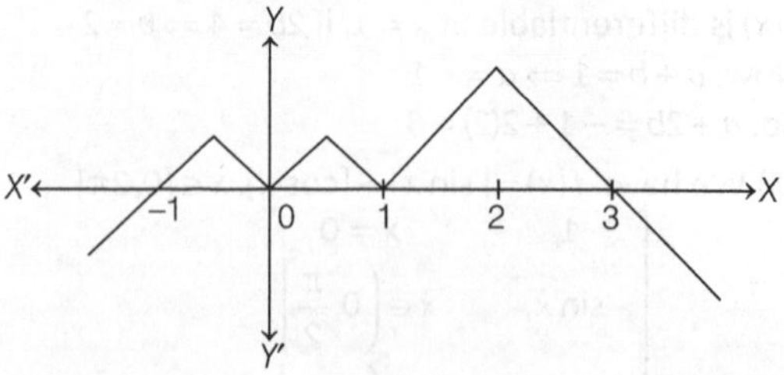

Clearly, function is non-differentiable at 5 points.

6. (a) We have,

$$f'(k^-)=\lim_{h\to 0}\frac{f(k-h)-f(k)}{-h}$$
$$=\lim_{h\to 0}\frac{[k-h]\sin\{\pi(k-h)\}}{-h}$$
$$=\lim_{h\to 0}\frac{(k-1)(-1)^{k-1}\sin(h\pi)}{-h}$$
$$=\lim_{h\to 0}(k-1)(-1)^k\left(\frac{\sin h\pi}{\pi h}\right)\pi$$
$$=(-1)^k(k-1)\pi$$

7. (b) We have, $f(x)=x\cos\{\pi(x+[x])\}$

$$=x\cos(\pi[x]+\pi x)$$

$$f(x)=\begin{cases}x\cos\pi x; & -2\le x<-1\\ -x\cos\pi x; & -1\le x<0\\ x\cos\pi x; & 0\le x<1\\ -x\cos\pi x; & 1\le x<2\\ x\cos\pi x; & 2\le x<3\end{cases}$$

At $\quad x=-1$,

$$\lim_{x\to -1^-}f(x)=\lim_{x\to -1^-}x\cos\pi x=-\cos(-\pi)=1$$

$$\lim_{x\to -1^+} f(x) = \lim_{x\to -1^+} -x\cos\pi x = \cos(-\pi) = -1$$

So, $f(x)$ is discontinuous at $x=-1$

At $\quad x=0$

$$\lim_{x\to 0^-} f(x) = \lim_{x\to 0^-} -x\cos\pi x = 0$$

$$\lim_{x\to 0^+} f(x) = \lim_{x\to 0^+} x\cos\pi x = 0 \text{ and } f(0)=0$$

So, $f(x)$ is continuous at $x=0$

At $\quad x=1$,

$$\lim_{x\to 1^-} f(x) = \lim_{x\to 1^-} x\cos\pi x = \cos\pi = -1$$

and $\lim_{x\to 1^+} f(x) = \lim_{x\to 1^+} -x\cos\pi x = -\cos\pi = 1$

$\therefore \quad \lim_{x\to 1^-} f(x) \neq \lim_{x\to 1^+} f(x)$

So, $f(x)$ is discontinuous at $x=1$

At $\quad x=2$,

$$\lim_{x\to 2^-} f(x) = \lim_{x\to 2^-} -x\cos\pi x = -2\cos 2\pi = -2$$

$$\lim_{x\to 2^+} f(x) = \lim_{x\to 2^+} x\cos\pi x = 2\cos 2\pi = 2$$

$\therefore \quad \lim_{x\to 2^-} f(x) \neq \lim_{x\to 2^+} f(x)$

So, $f(x)$ is discontinuous at $x=2$.

8. (d) We have,

$$f(x) = \begin{cases} \text{Max}(\sqrt{4-x^2}, \sqrt{1+x^2}), & -2 \le x \le 0 \\ \text{Min}(\sqrt{4-x^2}, \sqrt{1+x^2}), & 0 < x \le 2 \end{cases}$$

the graph of $f(x)$ is

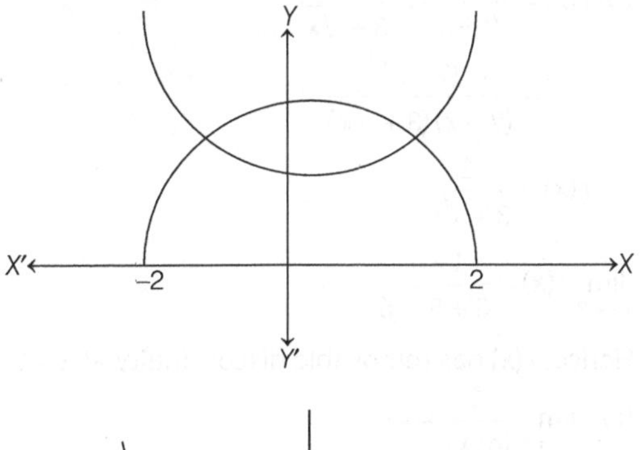

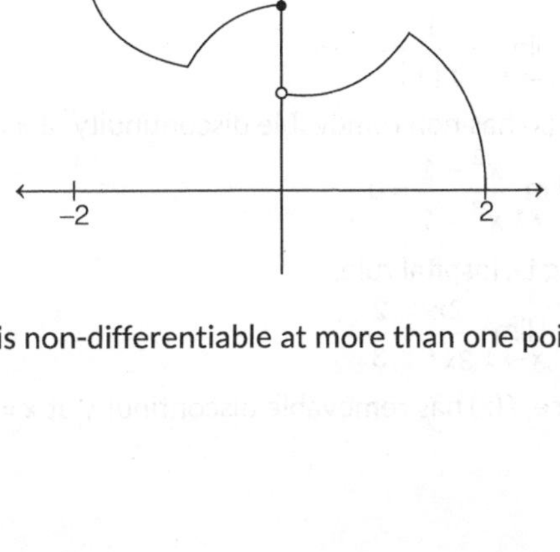

$f(x)$ is non-differentiable at more than one point.

9. (a) We have,

$$\lim_{x\to 0} f(x) = \lim_{x\to 0}\left(\frac{1}{x} - \frac{2}{e^{2x}-1}\right)$$

$$= \lim_{x\to 0} \frac{e^{2x}-1-2x}{x(e^{2x}-1)}$$

$$= \lim_{x\to 0} \frac{\left(1+2x+\frac{(2x)^2}{2!}+\frac{(2x)^3}{3!}+\dots\right)-1-2x}{x\left(1+2x+\frac{(2x)^2}{2!}+\dots-1\right)}$$

$$= \lim_{x\to 0} \left(\frac{\frac{(2x)^2}{2!}+\frac{(2x)^3}{3!}+\dots}{x\left(2x+\frac{(2x)^2}{2!}+\dots\right)}\right)$$

$$= \frac{\frac{2^2}{2!}}{2} = 1 = f(0)$$

So, $f(x)$ is continuous at $x=0$.

Now, LHD at $x=0$,

$$= \lim_{h\to 0} \frac{f(0-h)-f(0)}{-h}$$

$$= \lim_{h\to 0} \frac{f(-h)-1}{-h}$$

$$= \lim_{h\to 0} \left(\frac{\left(-\frac{1}{h} - \frac{2}{e^{-2h}-1} - 1\right)}{-h}\right)$$

$$= \lim_{h\to 0} \frac{e^{-2h}-1+2h+h(e^{-2h}-1)}{h^2(e^{-2h}-1)}$$

$$= \lim_{h\to 0} \frac{\left\{1-2h+\frac{(2h)^2}{2!}-\frac{(2h)^3}{3!}+\dots\right\} + h\left\{-2h+\frac{(2h)^2}{2!}-\frac{(2h)^3}{3!}+\dots\right\} - 1 + 2h}{h^2\left\{-2h+\frac{(2h)^2}{2!}-\frac{(2h)^3}{3!}+\dots\right\}}$$

$$= \lim_{h\to 0} \frac{\left(-\frac{8}{3!}+\frac{4}{2!}\right)h^3+\dots}{h^3\left(-2+\frac{4h}{2!}+\dots\right)} = -\frac{1}{3}$$

Similarly, RHD at $x=0 = -\frac{1}{3}$

$\therefore$ LHD = RHD

So, $f(x)$ is differentiable at $x=0$.

10. (a) $g(x)$ be a cubic polynomial.

So, let $g(x) = x^3 + bx^2 + cx + d$

Now, $f(x) = (x^3 + bx^2 + cx + d)|x-1||x-2||(x-3)^2|$
$|x-4|(x-4)^2$

$f(x)$ is differentiable for all real x. If $(x-1)$ and $(x-2)$ are factors of $g(x)$.

Clearly $f'(3) = 0$.

$\therefore f'''(4) = 0$ for which $(x-4)$ must be a factor of $g(x)$.

$\therefore \quad g(x) = x^3 + bx^2 + cx + d = (x-1)(x-2)(x-4)$

So, $g(5) = 4 \times 3 \times 1 = 12$

11. (c) We have,

$$f(x) = f'(x) + f''(x) + \ldots \infty \text{ terms} \quad \ldots(i)$$

Differentiating both sides

$$f'(x) = f''(x) + f'''(x) + \ldots \infty \text{ terms} \quad \ldots(ii)$$

We will subtract Eq. (ii) from Eq. (i), we get

$$f(x) - f'(x) = \{f'(x) + f''(x) + f'''(x) + \ldots\} - \{f''(x) + f'''(x) + \ldots\}$$

$\Rightarrow \quad f(x) - f'(x) = f'(x) \Rightarrow f(x) = 2f'(x)$

Now, at $x = -1$

$f(-1) = 2f'(-1)$

$\Rightarrow \quad f(-1) = 2(1) \quad$ [given, $f'(-1) = 1$]

$\Rightarrow \quad f(-1) = 2$

12. (d) We have,

$$2f(\sin x) + f(\cos x) = x \quad \ldots(i)$$

$$\Rightarrow \quad f(\cos x) = x - 2f(\sin x) \quad \ldots(ii)$$

Putting $x \to \frac{\pi}{2} - x$ in Eq. (i),

$$2f(\cos x) + f(\sin x) = \frac{\pi}{2} - x$$

From Eq. (ii),

$$2[x - 2f(\sin x)] + f(\sin x) = \frac{\pi}{2} - x$$

$$\Rightarrow \quad -4f(\sin x) + f(\sin x) = \frac{\pi}{2} - 3x$$

$$\Rightarrow \quad -3f(\sin x) = \frac{\pi}{2} - 3x$$

$$\Rightarrow \quad f(\sin x) = x - \frac{\pi}{6}$$

Putting $x \to \sin^{-1} x$

$$\Rightarrow \quad f(x) = \sin^{-1} x - \frac{\pi}{6}$$

Now, $\quad f'(x) = \frac{1}{\sqrt{1-x^2}}$

Non-differentiable at $\{-1, 1\}$.

13. (a, d) Given, $f(x) = \sqrt{1 - \sqrt{1-x^2}}$

For domain $\sqrt{1-x^2} \le 1$; $x^2 \le 1$

$\Rightarrow \quad x \in [-1, 1]$

Now, $f(0^-) = \lim_{x \to 0^-} f(x) = \lim_{x \to 0^-} \sqrt{1 - \sqrt{1-x^2}} = 0$

$f(0^+) = \lim_{x \to 0^+} f(x) - \lim_{x \to 0^+} \sqrt{1 - \sqrt{1-x^2}} = 0$

$f(0) = 0$

$\Rightarrow f(x)$ is continuous at $x = 0$.

Now, $f'(0^-) = \lim_{h \to 0} \frac{f(-h) - f(0)}{-h} = \lim_{h \to 0} \frac{f(-h)}{-h}$

$$= \lim_{h \to 0} \frac{\sqrt{1 - \sqrt{1-h^2}}}{-h} \times \frac{\sqrt{1 + \sqrt{1-h^2}}}{\sqrt{1 + \sqrt{1-h^2}}}$$

$$= \lim_{h \to 0} \frac{h}{-h\sqrt{1 + \sqrt{1-h^2}}} = -\frac{1}{\sqrt{2}}$$

$$f'(0^+) = \lim_{h \to 0} \frac{f(0+h) - f(0)}{h}$$

$$= \lim_{h \to 0} \frac{\sqrt{1 - \sqrt{1-h^2}}}{h}$$

$$= \lim_{h \to 0^+} \frac{h}{h\sqrt{1 + \sqrt{1-h^2}}} = \frac{1}{\sqrt{2}}$$

$\Rightarrow f(x)$ is non-differentiable at $x = 0$.

14. (b, d) (a) $f(x) = \frac{x-7}{|x-7|}$

$$f(x) = \begin{cases} -1, & x < 7 \\ 1, & x > 7 \end{cases}$$

Hence, $f(x)$ has non-removable discontinuity at $x = 7$

(b) $f(x) = \frac{3 - \sqrt{x}}{9 - x} \times \frac{3 + \sqrt{x}}{3 + \sqrt{x}}$

$$= \frac{(9-x)}{(9-x)(3+\sqrt{x})}$$

$$f(x) = \frac{1}{3 + \sqrt{x}}$$

$$\lim_{x \to 9} f(x) = \frac{1}{3+3} = \frac{1}{6}$$

Hence, $f(x)$ has removable discontinuity at $x = 9$

(c) $\lim_{x \to 1^+} \frac{1}{\ln|x|} = \infty$

and $\lim_{x \to 1^-} \cdot \frac{1}{\ln|x|} = -\infty$

So, $f(x)$ has non removable discontinuity at $x = 1$

(d) $\lim_{x \to 1} \frac{x^2 - 1}{x^3 - 1} = 0$

Using L-Hospital rule,

$$\lim_{x \to 1} \frac{2x}{3x^2} = \frac{2}{3}$$

Hence, $f(x)$ has removable discontinuity at $x = 1$.

15. (a, d) We have, $f(x)=\begin{cases}\dfrac{x}{1+|x|}, & |x|\ge 1\\ \dfrac{x}{1-|x|}, & |x|<1\end{cases}$

Clearly, $f(-1)=f(1)=\dfrac{1}{2}$ and $f(0)=0$

$f(x)$ is differentiable every, where possibly except for $x=0, \pm 1$. So, have to check differentiability only at $x=0, \pm 1$.

At $x=0$

$$f'(0)=\frac{f(0+h)-f(0)}{h}=\lim_{h\to 0}\frac{\dfrac{h}{1-|h|}-0}{h}$$

$$=\lim_{h\to 0}\frac{h}{h(1-|h|)}=\lim_{h\to 0}\frac{1}{1-|h|}$$

$\Rightarrow \quad \text{LHD}=\lim_{h\to 0^-}\dfrac{1}{1+h}=1$

and $\quad \text{RHD}=\lim_{h\to 0^+}\dfrac{1}{1+h}=1$

$\therefore \quad f'(0^-)=f'(0^+)=1$

So, function is differentiable at $x=0$

Now, at $x=1$

$\lim_{x\to 1^-}f(x)=\lim_{x\to 1^-}\dfrac{x}{1-x}=\infty$

So, $f(x)$ is discontinuous at $x=1$.

At $x=-1$

$$\lim_{x\to(-1)^-}f(x)=\lim_{x\to(-1)^-}\frac{x}{1-|x|}$$

$$=\lim_{x\to(-1)^-}\frac{x}{1+x}$$

$$=\infty$$

So, $f(x)$ is discontinuous at $x=-1$

Hence, it is non-differentiable at $x=-1$.

16. (a, b, c, d) $f(x)=[x\sin\pi x]$ for $x\in(-1,1)$

$\sin\pi x\in[-1,1]$

$\Rightarrow \quad x\sin\pi x\in(-1,1)$

$\Rightarrow \quad [x\sin\pi x]\in\{-1,0\}$

$\Rightarrow \quad f(x)=0\ \forall\ x\in[-1,1]$

$\Rightarrow f(x)$ is a continuous and differentiable function for all $x\in[-1,1]$.

17. (b, c) (a) We know that $\tan x$ is discontinuous at $x=\dfrac{\pi}{2}$

(b) $\int_0^x t\sin\dfrac{1}{t}\,dt$

$\because\ t\sin\dfrac{1}{t}$ is continuous in $(0,\pi)$.

So, its integral is continuous.

(c) $f(x)=\begin{cases}1, & 0<x\le\dfrac{3\pi}{4}\\ 2\sin\left(\dfrac{2x}{9}\right), & \dfrac{3\pi}{4}<x\le\pi\end{cases}$

$f\left(\dfrac{3\pi^-}{4}\right)=1$

$f\left(\dfrac{3\pi^+}{4}\right)=2\sin\left(\dfrac{2}{9}\cdot\dfrac{3\pi}{4}\right)$

$=2\sin\dfrac{\pi}{6}=1$

Also, $f\left(\dfrac{3\pi}{4}\right)=1$

$\Rightarrow f(x)$ is continuous in $(0,\pi)$.

(d) $f(x)=\begin{cases}x\sin x, & 0<x\le\dfrac{\pi}{2}\\ \dfrac{\pi}{2}\sin(\pi+x), & \dfrac{\pi}{2}<x<\pi\end{cases}$

$\Rightarrow f\left(\dfrac{\pi^-}{2}\right)=\dfrac{\pi}{2}$

$f\left(\dfrac{\pi^+}{2}\right)=\dfrac{\pi}{2}\sin\left(\dfrac{3\pi}{2}\right)=-\dfrac{\pi}{2}$

$f\left(\dfrac{\pi^-}{2}\right)\ne f\left(\dfrac{\pi^+}{2}\right)$

$\therefore f(x)$ is discontinuous in $(0,\pi)$.

18. (b, d) $f(0^+)=\lim_{h\to 0}f(0+h)$

$$=\lim_{h\to 0}3-\left[\cot^{-1}\left(\frac{2h^3-3}{h^2}\right)\right]$$

$=3-[\cot^{-1}-(\infty)]$

$=3-[\pi-\cot^{-1}\infty]$

$=3-3$

$=0$

$f(0^-)=\lim_{h\to 0}f(0-h)$

$$=\lim_{h\to 0}\{h^2\}\cos\left(e^{-\frac{1}{h}}\right)$$

$=0\times\cos(e^{-\infty})$

$=0\times\cos(0)$

$=0\times 1=0$

$\because \quad f(0^+)=f(0^-)$

$=f(0)=0$

$\therefore f(x)$ is continuous at $x=0$.

19. (b, c)

(a) $f(0^-) = \lim_{h\to 0} f(0-h)$

$$= \lim_{h\to 0} \frac{1}{1+2^{\cot(-h)}}$$

$$= \lim_{h\to 0} \frac{1}{1+2^{-\cot h}}$$

$$= \frac{1}{1+2^{-\infty}} = \frac{1}{1+0} = 1$$

and $f(0^+) = \lim_{h\to 0} f(0+h)$

$$= \lim_{h\to 0} \frac{1}{1+2^{\cot h}}$$

$$= \frac{1}{1+2^{\infty}} = \frac{1}{\infty} = 0$$

$\because \quad f(0^-) \neq f(0^+)$

(b) $f(0^-) = \lim_{h\to 0} f(0-h)$

$$= \lim_{h\to 0} \cos\left(\frac{|\sin(-h)|}{0-h}\right)$$

$$= \lim_{h\to 0} \cos\left(\frac{\sin h}{-h}\right)$$

$$= \cos(-1)$$

$$= \cos 1$$

and $f(0^+) = \lim_{h\to 0} f(0+h)$

$$= \lim_{h\to 0} \cos\left(\frac{|\sin(0+h)|}{0+h}\right)$$

$$= \lim_{h\to 0} \cos\left(\frac{\sin h}{h}\right)$$

$$= \cos 1$$

$\therefore \quad f(0^-) = f(0^+) \neq f(0)$

(c) $f(0^-) = \lim_{h\to 0} f(0-h)$

$$= \lim_{h\to 0} (-h)\sin\left(\frac{\pi}{-h}\right)$$

$$= \lim_{h\to 0} h\sin\left(\frac{\pi}{h}\right)$$

$= 0 \times$ (lie between -1 to 1)

$= 0$

and $f(0^+) = \lim_{h\to 0} f(0+h)$

$$= \lim_{h\to 0} h\sin\left(\frac{\pi}{h}\right)$$

$= 0 \times$ (lie between -1 to 1)

$= 0$

$\therefore \quad f(0^-) = f(0)^+ \neq f(0)$

(d) $f(0^-) = \lim_{h\to 0} f(0-h)$

$$= \lim_{h\to 0} e^{-\frac{1}{h}}$$

$$= e^{-\infty} = 0$$

$f(0^+) = \lim_{h\to 0} f(0+h)$

$$= \lim_{h\to 0} e^{\left(\frac{1}{h}\right)}$$

$$= e^{\infty}$$

$$= \infty$$

$f(0^-) \neq f(0^+)$

20. (a, b, c) $f(x)$ is continuous for all x, if it is continuous at $x = 1$ for which

$$|1| - 3 = |1-2| + a \Rightarrow a = -3$$

and $g(x)$ is continuous for all x, if it is continuous at $x = 2$ for which

$$2 - |2| = \operatorname{sgn}(2) - b \Rightarrow 0 = 1 - b \Rightarrow b = 1$$

Thus, $h(x) = f(x) + g(x)$ is continuous for all x, if $a = -3$ and $b = 2$.

Hence, $h(x) = f(x) + g(x)$ is discontinuous exactly one point for options (a), (b) and (c).

21. (a, c) $f(0^-) = \lim_{h\to 0} f(0-h)$

$$= \lim_{h\to 0} \frac{1 - a^{-h} - ha^{-h}\ln a}{a^{-h}(-h)^2}$$

Let $\quad P = \lim_{h\to 0} \frac{a^h - 1 - h\ln a}{h^2}$

Putting $\quad h = 2t$,

$$= \lim_{t\to 0} \frac{a^{2t} - 1 - 2t\ln a}{4t^2}$$

$$= \lim_{t\to 0} \left(\frac{(a^t-1)^2}{4t^2} + 2\frac{(a^t - 1 - t\ln a)}{4t^2}\right)$$

$$= \frac{1}{4}\lim_{t\to 0}\left(\frac{a^t - 1}{t}\right)^2 + \frac{2}{4}\lim_{t\to 0} \frac{a^t - 1 - t\ln a}{t^2}$$

$$P = \frac{1}{4}[\ln(a)]^2 + \frac{1}{2}P$$

$$\frac{1}{2}P = \frac{1}{4}(\ln a)^2 \Rightarrow P = \frac{1}{2}(\ln a)^2$$

$f(0^+) = \lim_{h\to 0} f(0+h)$

$$= \lim_{h\to 0} \frac{2^h \cdot a^h - h\ln 2 - h\ln a - 1}{h^2}$$

$$= \lim_{h\to 0} \frac{(a^h - 1)(2^h - 1) + (a^h - h\ln a - 1) + (2^h - h\ln 2 - 1)}{h^2}$$

$$= \lim_{h \to 0}\left(\frac{a^h - 1}{h}\right)\left(\frac{2^h - 1}{h}\right)$$

$$+ \lim_{h \to 0}\left(\frac{a^h - h\ln a - 1}{h^2}\right) + \lim_{h \to 0}\left(\frac{2^h - h\ln 2 - 1}{h^2}\right)$$

$$= (\ln a)(\ln 2) + \frac{1}{2}(\ln a)^2 + \frac{1}{2}(\ln 2)^2$$

$f(x)$ is continuous at $x = 0$.

So, $\frac{1}{2}(\ln a)^2 = (\ln a)(\ln 2) + \frac{1}{2}(\ln a)^2 + \frac{1}{2}(\ln 2)^2$

$\Rightarrow \quad 0 = (\ln a)(\ln 2) + \frac{1}{2}(\ln 2)^2$

$\Rightarrow \quad \ln a + \frac{1}{2}\ln 2 = 0$

$\Rightarrow \quad \ln a = -\frac{1}{2}\ln 2$

$\Rightarrow \quad \ln a = \ln 2^{-\frac{1}{2}}$

$a = 2^{-\frac{1}{2}}$

and $\quad f(0) = \frac{1}{2}(\ln a)^2 = \frac{1}{2}\left(-\frac{1}{2}\ln 2\right)^2$

$= \frac{1}{8}(\ln 2)^2$

22. (a, c) We have,

$$f(x) = x^3 - x^2 + x + 1$$

$$f'(x) = 3x^2 - 2x + 1$$

$$= 3\left(x^2 - \frac{2}{3}x + \left(\frac{1}{3}\right)^2\right) - 3\left(\frac{1}{3}\right)^2 + 1$$

$$= 3\left(x - \frac{1}{3}\right)^2 + \frac{2}{3} > 0$$

So, $f(x)$ is an increasing function.

So, Max $f(t) = f(x)$, $0 \le t \le x$

Now, $g(x) = \begin{cases} x^3 - x^2 + x + 1, & 0 \le x \le 1 \\ 3 - x, & 1 < x \le 2 \end{cases}$

At $\quad x = 1$,

LHL $= \lim_{x \to 1^-} f(x) = 2$

RHL $= \lim_{x \to 1^+} f(x) = 2$

Function is continuous at $x = 1$

Now, $g'(x) = \begin{cases} 3x^2 - 2x + 1, & 0 < x < 1 \\ -1, & 1 < x < 2 \end{cases}$

LHD $= 3 - 2 + 1 = 2$

and $\quad$ RHD $= -1$

LHD $\ne$ RHD

So, $g(x)$ is non-differentiable at $x = 1$

Hence, $g(x)$ is continuous for all $x \in [0, 2]$ but $g(x)$ is non-differentiable at $x = 1$.

23. (a, b, c) Let $f(x) = |x^3|$

$$= \begin{cases} -x^3, & x < 0 \\ x^3, & x > 0 \end{cases}$$

$$f'''(x) = \begin{cases} -6, & x < 0 \\ 6, & x > 0 \end{cases}$$

Hence, $f'''(x)$ does not exist.

and $\quad f(x) = x^3|x| = \begin{cases} -x^4, & x < 0 \\ x^4, & x > 0 \end{cases}$

$$f'''(x) = \begin{cases} -24x, & x < 0 \\ 24x, & x > 0 \end{cases}$$

Hence, $f'''(0) = 0$ it exists.

Similarly, $f(x) = |x|\sin^3 x$ and $f(x) = x|\tan^3 x$,

Also, $f'''(x) = 0$ and it exists.

24. (a, c, d) $f(x) = x\cos\{\pi(x + [x])\}$

$f(x) = x\cos([x]\pi + \pi x)$

Let's check continuity at $x = n \in Z$.

$f(n^+) = n\cos(n\pi + \pi n)$

$= n\cos 2n\pi$

$= n$

$f(n^-) = n\cos\{(n-1)\pi + n\pi\}$

$= n\cos(2n\pi - \pi)$

$= n\cos(2n - 1)\pi$

$= n(-1)$

$= -n$

It is discontinuous all integer points except 0.

25. (a, d) We have,

$g(0) = 0$, $g'(0) = 0$ and $g'(1) \ne 0$

$$f(x) = \begin{cases} g(x), & x > 0 \\ -g(x), & x < 0 \\ 0, & x = 0 \end{cases}$$

$\therefore \quad f'(x) = \begin{cases} g'(x) \\ -g'(x) \end{cases}$

$f'(0^+) = g'(0^+) = 0$

$f'(0^-) = g'(0^-) = 0$

Hence, $f(x)$ is differentiable at $x = 0$

$$h(x) = e^{|x|} = \begin{cases} e^{-x}, & x < 0 \\ e^x, & x \ge 0 \end{cases}$$

$\therefore \quad h'(x) = \begin{cases} -e^x, & x < 0 \\ e^x, & x > 0 \end{cases}$

$h'(0^+) = e^0 = 1$

$h'(0^-) = -e^0 = -1$

So, $h(x)$ is non-differentiable at $x = 0$.

Now, $f\{h(x)\} = g(e^{|x|}) \forall x \in R$

$$= \begin{cases} g(e^x), & x \ge 0 \\ g(e^{-x}), & x < 0 \end{cases}$$

$$(f(h(x)))' = \begin{cases} e^x g'(e^x), & x>0 \\ -e^{-x} g'(e^{-x}), & x<0 \end{cases}$$

$$(f(h(0^+)))' = e^0 g'(e^0) = g'(1)$$

$$(f(h(0^-)))' = -e^0 g'(e^0) = -g'(1)$$

Since, $g'(1) \neq 0$, $f\{h(x)\}$ is non-differentiable at $x = 0$.

$$h\{f(x)\} = \begin{cases} e^{|f(x)|}, & x \neq 0 \\ 1, & x = 0 \end{cases} = \begin{cases} e^{|g(x)|}, & x \neq 0 \\ 1, & x = 0 \end{cases}$$

Given $g(0) = 0$ and $g'(0) = 0$

Hence, $x = 0$ is repeated root of $g(x) = 0$.

Therefore, $h\{f(x)\}$ is differentiable at $x = 0$.

26. (a, b, d) Given, function is discontinuous, when

$$a + \sin \pi x = \pm 1$$

(a) Now, if $a = 1$, then $\sin \pi x = 0$ or $x = 1, 2$ and 3 then $f(x)$ has 3 points of discontinuity.

(b) If $a = 3$, then $\sin \pi x = 2$ not possible so $f(x)$ has no point of discontinuity.

(c) If $a = 0.5$ then $\sin \pi x = 0.5$

Therefore, x has 4 values, 2 each for one cycle of period 2.

(d) If $a = 0$, then $\sin \pi x = \pm 1$ or $x = \frac{1}{2}, \frac{3}{2}, \frac{5}{2}$ and $\frac{7}{2}$

So, $f(x)$ has 4 point of discontinuity.

27. (a, b) We have, $f(x) = \text{sgn}\, x \sin x$

$$f(x) = \begin{cases} \sin x, & x \geq 0 \\ -\sin x, & x < 0 \end{cases}$$

Graph of $f(x)$ is as shown below

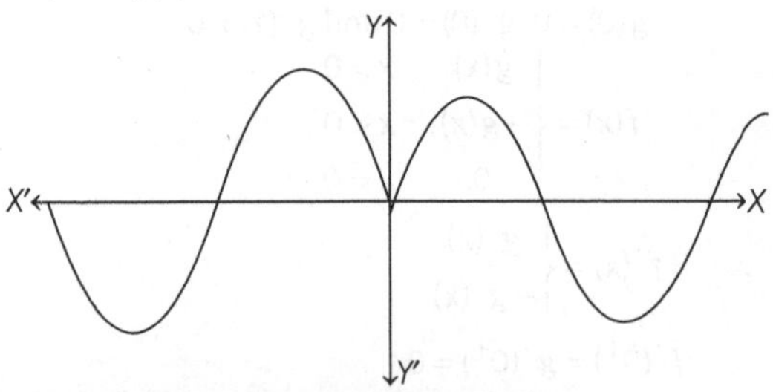

Clearly, $f(x)$ is discontinuous no where, but non-differentiable only at $x = 0$.

Also, $f(x)$ is an even function.

28. (a, c) $f'(0) = \lim_{h \to 0} \frac{f(0+h) - f(0)}{h}$

$$\text{LHD} = \lim_{h \to 0^-} \frac{\left\{\log_a (a|[h] + [-h]|)^h \cdot \dfrac{a^{\left(\frac{2}{\frac{[h]+[-h]}{|h|}} - 5\right)}}{3 + a^{\frac{1}{|h|}}}\right\} - 0}{h}$$

$$= \lim_{h \to 0^-} \frac{h \log_a (a|-1+0|) \dfrac{a^{\left(\frac{2}{\left(\frac{-1+0}{-h}\right)} - 5\right)}}{3 + a^{\frac{1}{h}}}}{h}$$

$$= \lim_{h \to 0^-} \{\log_a a\} \left(\frac{a^{2h-5}}{3 + a^{-\frac{1}{h}}}\right)$$

$$= \lim_{h \to 0^-} \frac{a^{2h-5}}{3 + a^{-\frac{1}{h}}} = 0$$

$$\text{RHD} = \lim_{h \to 0^+} \{\log_a (a|[h] + [-h]|^h\} \cdot \left(\frac{a^{\left(\frac{2}{\frac{[h]+[-h]}{|h|}} - 5\right)}}{3 + a^{\frac{1}{|h|}}}\right)$$

$$= \lim_{h \to 0^+} \frac{\{h \log_a (a|0-1|\} \left(\dfrac{a^{\left(\frac{2}{\frac{0-1}{h}} - 5\right)}}{3 + a^{\frac{1}{h}}}\right)}{h}$$

$$= \lim_{h \to 0^+} \{\log_a a\} \left(\frac{a^{-2h-5}}{3 + a^{\frac{1}{h}}}\right)$$

$$= \lim_{h \to 0} \frac{a^{-2h-5}}{3 + a^{-1/h}} = 0$$

$\therefore f'(0) = f'(0^+) = f'(0^-) = 0$

Hence, $f(x)$ is differentiable at $x = 0$ and also continuous at $x = 0$.

29. (a, c) We have,

$$f(x) = \begin{cases} (1 - \text{sgn}\, x)\, \text{sgn}\, x, & x \leq 1 \\ \dfrac{1}{1 - e^{\frac{x}{1-x}}}, & x > 1 \end{cases}$$

$$= \begin{cases} \{1 - (-1)\}(-1), & x < 0 \\ (1-1)(1), & 0 < x \leq 1 \\ \dfrac{1}{1 - e^{\frac{x}{1-x}}}, & x > 1 \end{cases}$$

$$= \begin{cases} -2, & x<0 \\ 0, & 0<x\le 1 \\ \dfrac{1}{1-e^{\frac{x}{1-x}}}, & x>1 \end{cases}$$

$f(x)$ is clearly discontinuous at $x = 0$.

Now, at $x = 1$

$\text{LHL} = f(1^-) = 0$

$$f(1^+) = \lim_{x\to 1^+} \frac{1}{1-e^{\frac{x}{1-x}}} = \lim_{h\to 0} \frac{1}{1-e^{\frac{1+h}{-h}}}$$

$$= \lim_{h\to 0} \frac{1}{\left(1-e^{-\frac{1}{h}}\cdot e^{-1}\right)} = 1$$

$\Rightarrow f(x)$ is discontinuous at $x = 1$

30. (b, c) Given,

$$f(x) = \begin{cases} 1, & -\infty < x < 0 \\ 1+|\sin x|, & 0 \le x < \dfrac{\pi}{2} \\ 2+\left(x-\dfrac{\pi}{2}\right), & \dfrac{\pi}{2} \le x < \infty \end{cases}$$

Clearly, $\lim_{x\to 0} f(x) = f(0) = 1$

So, $f(x)$ is continuous at $x = 0$.

$$f(0) = 1 \text{ and } f\left(\frac{\pi}{2}\right) = 2$$

At $x = 0$,

$$\text{LHD} = f'(0^-) = \lim_{h\to 0^-} \frac{f(0-h)-f(0)}{-h}$$

$$= \lim_{h\to 0} \frac{f(-h)-1}{-h}$$

$$= \lim_{h\to 0} \frac{1-1}{-h} = 0$$

$$\text{RHD} = f'(0^+) = \lim_{h\to 0^+} \frac{f(0+h)-f(0)}{h}$$

$$= \lim_{h\to 0} \frac{1+|\sin h|-1}{h}$$

$$= \lim_{h\to 0} \frac{|\sin h|}{h}$$

$$= \lim_{h\to 0} \frac{\sin h}{h} = 1$$

$\therefore \quad f'(0^+) \ne f'(0^-)$

Hence, $f(x)$ is non-differentiable at $x = 0$.

At $x = \dfrac{\pi}{2}$.

$$\text{LHL} = \lim_{h\to \frac{\pi}{2}^-} f(x) = 1+1 = 2$$

$$\text{RHL} = \lim_{h\to \frac{\pi}{2}^+} f(x) = 2+\frac{\pi}{2}-\frac{\pi}{2} = 2$$

$$f\left(\frac{\pi}{2}\right) = 2$$

So, $f(x)$ is continuous at $x = \dfrac{\pi}{2}$

$$\text{LHD} = f'\left(\frac{\pi^-}{2}\right) = \lim_{h\to 0^-} \frac{f\left(\frac{\pi}{2}-h\right)-f\left(\frac{\pi}{2}\right)}{-h}$$

$$= \lim_{h\to 0^-} \frac{1+\left|\sin\left(\frac{\pi}{2}-h\right)\right|-2}{-h}$$

$$= \lim_{h\to 0^-} \frac{|\cos h|-1}{-h} = \lim_{h\to 0^-} \frac{\cos h-1}{-h}$$

$$= \lim_{h\to 0} \frac{2\sin^2\frac{h}{2}}{h} = 0$$

$$\text{RHD} = f'\left(\frac{\pi^+}{2}\right) = \lim_{h\to 0^+} \frac{f\left(\frac{\pi}{2}+h\right)-f\left(\frac{\pi}{2}\right)}{h}$$

$$= \lim_{h\to 0} \frac{2+\left(h+\frac{\pi}{2}-\frac{\pi}{2}\right)^2-2}{h} = \lim_{h\to 0} \frac{h^2}{h} = 0$$

$$\therefore \quad f'\left(\frac{\pi^-}{2}\right) = f'\left(\frac{\pi^+}{2}\right)$$

So, $f(x)$ is differentiable at $x = \dfrac{\pi}{2}$.

Sol. (Q.Nos. 31-33)

31. $\to$ (d), **32.** $\to$ (c), **33.** $\to$ (c)

Given that

$$f(x) = \begin{cases} b\sin^{-1}\left(\dfrac{c+x}{2}\right), & -\dfrac{1}{2} < x < 0 \\ \dfrac{1}{2}, & x = 0 \\ \dfrac{e^{\frac{ax}{2}}-1}{x}, & 0 < x < \dfrac{1}{2} \end{cases}, \text{ where } |c| < \frac{1}{2}$$

$f(x)$ is differentiable at $x = 0$, then $f(x)$ will also be continuous at $x = 0$

$\Rightarrow \lim_{h\to 0} f(0+h) = f(0)$

$$\Rightarrow \lim_{h\to 0}\frac{e^{\frac{ah}{2}}-1}{h}=\frac{1}{2}$$

$$=\lim_{h\to 0}\frac{e^{\frac{ah}{2}}-1}{\frac{ah}{2}}\times\frac{a}{2}=\frac{1}{2}\Rightarrow a=1$$

Also, $Lf'(0)=Rf'(0)$

$$\Rightarrow \lim_{h\to 0}\frac{f(0-h)-f(0)}{-h}=\lim_{h\to 0}\frac{f(0+h)-f(0)}{h}$$

$$\Rightarrow \lim_{h\to 0}\frac{b\sin^{-1}\left(\frac{c-h}{2}\right)-\frac{1}{2}}{-h}=\lim_{h\to 0}\frac{\frac{e^{\frac{ah}{2}}-1}{h}-\frac{1}{2}}{h}$$

$$\Rightarrow \lim_{h\to 0}\frac{b\sin^{-1}\left(\frac{c-h}{2}\right)-\frac{1}{2}}{-h}=\lim_{h\to 0}\frac{2e^{\frac{ah}{2}}-2-h}{2h^2}$$

Applying L-Hospital rule,

$$\Rightarrow \lim_{h\to 0}\left(\frac{-\frac{1}{2}\frac{b}{\sqrt{1-\left(\frac{c-h}{2}\right)^2}}}{-1}\right)=\lim_{h\to 0}\frac{2e^{\frac{ah}{2}}\left(\frac{a}{2}\right)-1}{4h}$$

Putting $a=1$

$$=\lim_{h\to 0}\frac{e^{\frac{h}{2}}-1}{4h}=\lim_{h\to 0}\frac{e^{\frac{h}{2}}-1}{8\left(\frac{h}{2}\right)}$$

$$\therefore \quad \frac{b}{2\sqrt{1-\left(\frac{c^2}{4}\right)}}=\frac{1}{8}$$

$$\Rightarrow \quad 4b=\sqrt{1-\frac{c^2}{4}}$$

$$\Rightarrow \quad 16b^2=\frac{4-c^2}{4}$$

$$\Rightarrow \quad 64b^2=4-c^2$$

Clearly $f'(0)=\frac{1}{8}$

Sol. (Q.Nos. 34-36)

34. (b) $\lim_{x\to\frac{\pi}{2}^-}([\sin x]-[\cos x])$

$$=\lim_{h\to 0}\left(\left[\sin\left(\frac{\pi}{2}-h\right)\right]-\left[\cos\left(\frac{\pi}{2}-h\right)\right]\right)$$

$$=[1^-]-[0^+]$$

$$=0-0=0$$

and $\lim_{x\to\frac{\pi}{2}^+}([\sin x]-[\cos x])$

$$=\lim_{h\to 0}\left(\left[\sin\left(\frac{\pi}{2}+h\right)\right]-\left[\cos\left(\frac{\pi}{2}+h\right)\right]\right)$$

$$=[1^-]-[0^-]$$

$$=0-(-1)=1$$

Then, jump of discontinuity $=1-0=1$

35. (d) $h(x)=f(x)\cdot g(x)$

$$=\begin{cases}x^2\sin x, & x<0\\ 2x^2(x+1), & x\ge 0\end{cases}$$

$h(0^-)=0$

$h(0^+)=0$

$\Rightarrow h(x)$ is continuous for all $x\in R$.

$\Rightarrow$ There is no jump of discontinuity.

36. (c) We have, $f(x)=\begin{cases}\cos^{-1}\{\cot x\}, & x<\frac{\pi}{2}\\ \pi[x]-1, & x\ge\frac{\pi}{2}\end{cases}$

$$\text{LHL}=\lim_{x\to\frac{\pi}{2}^-}f(x)$$

$$=\lim_{x\to\frac{\pi}{2}^-}\cos^{-1}\{\cot x\}$$

$$=\lim_{x\to\frac{\pi}{2}^-}\cos^{-1}(\cot x-[\cot x])$$

$$=\lim_{h\to 0^-}\cos^{-1}(h-[h]) \qquad \{\because \text{putting } \cot x=h\}$$

$$=\lim_{h\to 0}\cos^{-1}(h)$$

$$=\frac{\pi}{2}$$

$$\text{RHL}=\lim_{x\to\frac{\pi}{2}^+}f(x)$$

$$=\lim_{x\to\frac{\pi}{2}^+}(\pi[x]-1)=\pi-1$$

Thus, jump of discontinuity $=(\pi-1)-\frac{\pi}{2}$

$$=\frac{\pi}{2}-1$$

Sol. (Q.Nos. 37-39)

First, we express f explicitly, which will help us in answering the questions.

Case I $t<0$. In such case for the given equations, we have

$x=2t-(1-t)=3t-1$

and $y=2t^2-t^2=t^2=\left(\frac{x+1}{3}\right)^2$

Also, $t < 0 \Rightarrow x < -1$.

Case II $0 \le t \le 1$. In this case, we have

$$x = 2t - (1 - t) = 3t - 1$$

and $$y = 2t^2 + t^2 = 3t^2 = 3\left(\frac{x+1}{3}\right)^2$$

Also, $0 \le t \le 1 \Rightarrow -1 \le x \le 2$.

Case III $t > 1$. In this case, we have

$x = 2t - (t - 1) = t + 1$

and $y = 2t^2 + t^2 = 3t^2 = 3(x-1)^2$

Also, $x > 2$. Therefore,

$$f(x) = \begin{cases} \frac{1}{9}(x+1)^2, & \text{for } x < -1 \\ \frac{1}{3}(x+1)^2, & \text{for } -1 \le x \le 2 \\ 3(x-1)^2, & \text{for } x > 2 \end{cases}$$

37. (a) We have, $f(2) = \frac{1}{3}(2+1)^2 = 3$

38. (b) Since, f is continuous on $[-1, 2]$ and

$$f(-1) = 0 < \frac{1}{2} < 3 = f(2)$$

By the intermediate value theorem for continuous functions, we have $f(x) = \frac{1}{2}$ for some $x \in [-1, 2]$.

Therefore, $\frac{1}{3}(x+1)^2 = \frac{1}{2} \Rightarrow x = \frac{\sqrt{3}-\sqrt{2}}{\sqrt{2}} \in (-1, 2)$

39. (a) Clearly, f is continuous at $x = -1$ and 2 and hence f is continuous for all x.

40. (i) $\to$ (B, D), (ii) $\to$ (A, E), (iii) $\to$ (B, C), (iv) $\to$ (B, F)

(i) We have, $f(x) = \frac{1}{x-1}$

Function is discontinuous at $x = 1$

So, $\lim_{x \to 1^+} f(x) = \lim_{x \to 1^+} \frac{1}{x-1} = \frac{1}{0} = \infty$

and $\lim_{x \to 1^-} f(x) = \lim_{x \to 1^-} \frac{1}{x-1} = -\infty$

$\therefore f(x)$ has discontinuity due to vertical asymptote, which is non-removable.

(ii) We have, $f(x) = \frac{x^3 - x}{x^2 - 1} = x, x \ne \pm 1$

$\therefore f(x)$ has missing point discontinuity, which is removable.

(iii) We have, $f(x) = \frac{|x-1|}{x-1} = \begin{cases} 1, & x > 1 \\ -1 & x < 1 \end{cases}$

$\therefore f(x)$ has jump discontinuity, which is non-removable.

(iv) We have, $f(x) = \sin\left(\frac{1}{x-1}\right)$

$\therefore \lim_{x \to 1^+} \sin\left(\frac{1}{x-1}\right) = \sin \infty =$ any value between -1 and 1. So, $f(x)$ has oscillating discontinuity, which is non-removable.

41. (i) $\to$ (A, B), (ii) $\to$ (A, D), (iii) $\to$ (A, C, D), (iv) $\to$ (A, B, C)

(i) We have, $f(x) = (x+1)|x-1|$

$$= \begin{cases} x^2 - 1, & x \ge 1 \\ 1 - x^2, & x < 1 \end{cases}$$

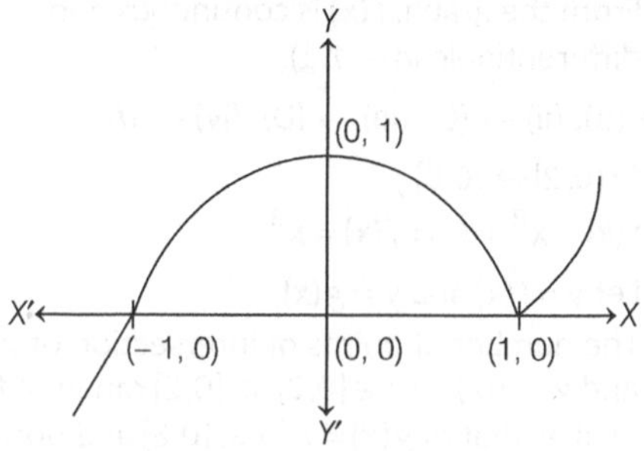

Clearly, by graph of $f(x)$,

$f(x)$ is continuous in $(-1, 1)$

and $f(x)$ is differentiable in $(-1, 1)$

(ii) We have, $f(x) = \text{Min}(|x|, 1 - |x|)$

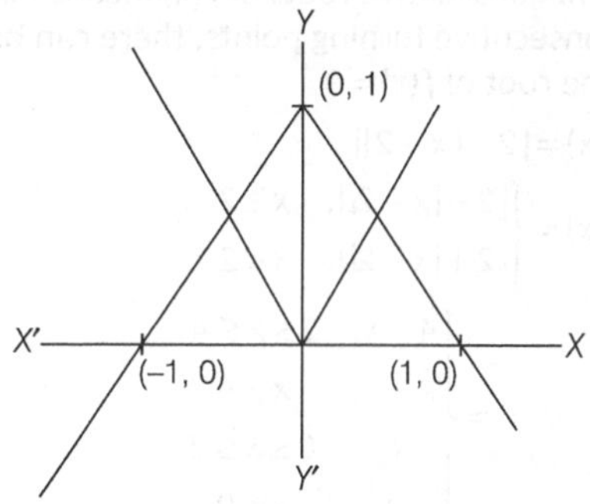

By graph of $f(x)$, $f(x)$ is continuous in $(-1, 1)$ and non-differentiable in $(-1, 1)$.

(iii) $f(x) = |\sin^{-1} x|$ is continuous

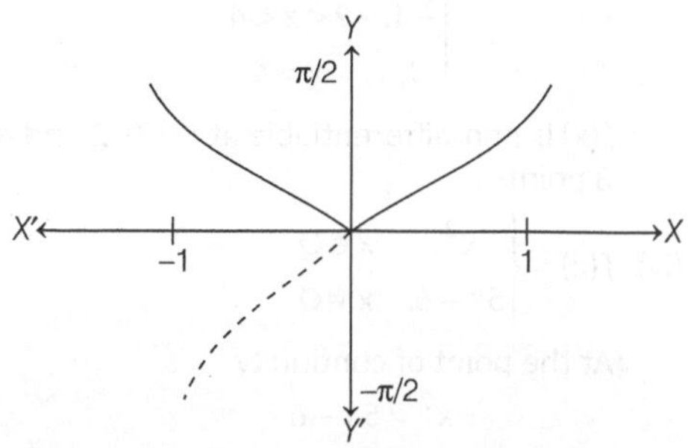

From the graph, $f(x)$ is non-differentiable at $x = 0$

$f(x)$ is differentiable in $(0, 1)$ but non-differentiable in $(-1, 1)$.

(iv) $y = x|x|$

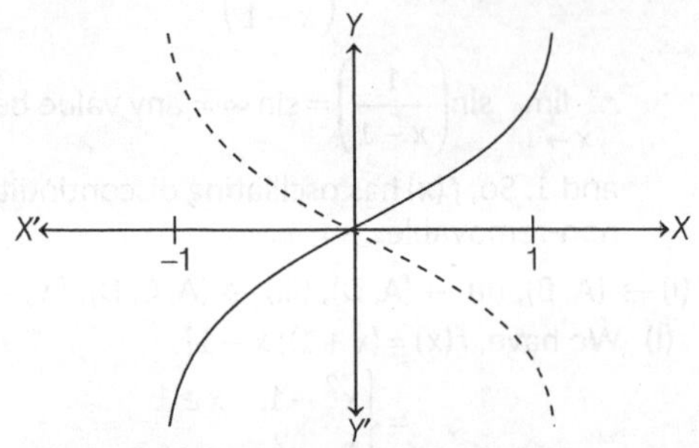

From the graph, $f(x)$ is continuous and differentiable in $(-1, 1)$.

42. (i) → (B), (ii) → (C), (iii) → (D), (iv) → (A)

(i) $f : [0, 2] \to [0, 8]$

$f(x) - x^3 = 0 \Rightarrow f(x) = x^3$

Let $y = f(x)$ and $y = g(x)$

The number of points of intersection of $y = f(x)$ and $y = g(x)$ for $x \in [0, 2]$. In $[0, 2]$ range of $f(x)$ as well as that of $g(x) = x^3$ i.e. $[0, 8]$ and both are continuous function.

Thus, atleast once $f(x)$ has to meet with the curve $g(x)$ i.e. $f(x) - x^3 = 0$ has alteast one root.

(ii) Two consecutive roots of $f(x)$ means between two consecutive turning points, there can be atmost one root of $f(x) = 0$

(iii) $f(x) = |2 - |x - 2||$

$$f(x) = \begin{cases} |2 - |x - 2||, & x \geq 2 \\ |2 + |x - 2||, & x < 2 \end{cases}$$

$$= \begin{cases} 4 - x, & 2 \leq x \leq 4 \\ x - 4, & x \geq 4 \\ x, & 0 \leq x \leq 2 \\ -x, & x < 0 \end{cases}$$

$$\text{So, } f'(x) = \begin{cases} -1, & x < 0 \\ 1, & 0 < x < 2 \\ -1, & 2 < x < 4 \\ 1, & x > 4 \end{cases}$$

$f(x)$ is non-differentiable at $x = 0$, 2 and 4 at exactly 3 points.

(iv) $f(x) = \begin{cases} x^2, & x \in Q \\ 5x - 6, & x \notin Q \end{cases}$

At the point of continuity,

$$x^2 = 5x - 6$$

$$\Rightarrow x^2 - 5x + 6 = 0$$

$x = 2$ and $x = 3$, then the number of values of x at which $f(x)$ is continuous, is 2.

43. (i) → (B), (ii) → (A), (iii) → (D), (iv) → (C)

(i) Clearly, $f_1(x)$ is continuous on R.

$$f_1'(0) = \lim_{h \to 0} \frac{\sin\sqrt{1 - e^{-h^2}} - 0}{h}$$

$$= \lim_{h \to 0} \frac{\sin\sqrt{1 - e^{-h^2}}}{\sqrt{1 - e^{-h^2}}} \times \sqrt{\frac{e^{-h^2} - 1}{-h^2}} \times \frac{|h|}{h}$$

$$= 1 \times 1 \times \lim_{h \to 0} \frac{|h|}{h}$$

This limit does not exist.

Hence, $f_1(x)$ is non-differentiable at $x = 0$.

(ii) $$\lim_{x \to 0} f_2(x) = \lim_{x \to 0} \frac{|\sin x|}{\tan^{-1} x}$$

$$= \lim_{x \to 0} \frac{|\sin x|}{|x|} \times \frac{x}{\tan^{-1} x} \cdot \frac{|x|}{x}$$

$$= 1 \times 1 \times \lim_{x \to 0} \frac{|x|}{x}$$

This limit does not exist.

So, $f_2(x)$ is not continuous at $x = 0$.

(iii) $f_3(x) = [\sin\{\log(x + 2)\}]$

When $x \to 0$, then $[\sin\{\log(x + 2)\}] \to 0$

Let us consider $[\sin\{\log(x + 2)\}] = 0$

$$\Rightarrow 0 \leq \sin\{\log(x + 2)\} < 1$$

$$\Rightarrow 0 \leq \log(x + 2) < \pi, \log(x + 2) \neq \frac{\pi}{2}$$

$$\Rightarrow 1 \leq x + 2 < e^{\pi}, x + 2 \neq e^{\frac{\pi}{2}}$$

$$\Rightarrow -1 \leq x < e^{\pi} - 2, x \neq e^{\frac{\pi}{2}} - 2$$

Thus, $f_3(x)$ is differentiable at $x = 0$ and derivative is continuous at $x = 0$.

(iv) $\lim_{x \to 0} f_4(x)$

$$= \lim_{x \to 0}\left(x^2 \sin\frac{1}{x}\right) = \lim_{x \to 0} x^2\left(\sin\frac{1}{x}\right)$$

$= 0 \times$ (value between -1 and 1)

$= 0$

Thus, $f_4(x)$ is continuous at $x = 0$

Also, $$f_4'(0) = \lim_{h \to 0} \frac{h^2 \sin\left(\frac{1}{h}\right) - 0}{h} = \lim_{h \to 0} h \sin\left(\frac{1}{h}\right) = 0$$

Thus, derivative of $f_4(x)$ exists at $x = 0$.

Also, $f_4'(x) = -\cos\frac{1}{x} + 2x \sin\frac{1}{x}, x \neq 0$.

44. (i) → (D), (ii) → (C), (iii) → (A), (iv) → (B)

(i) $$f(x) = \begin{cases} \dfrac{a + 3\cos x}{x^2}, & x < 0 \\ b\tan\left(\dfrac{\pi}{[x + 3]}\right), & x \geq 0 \end{cases}$$

$$f(0^-) = \lim_{x \to 0^-} f(x) = \lim_{x \to 0^-} \left(\frac{a + 3\cos x}{x^2} \right)$$

$\therefore$ Denominator tends to zero, numerator must also tends to zero. This implies $a = -3$

$$\therefore \text{LHL} = \lim_{x \to 0^-} \left(\frac{-3 + 3\cos x}{x^2} \right)$$

$$= \lim_{x \to 0^-} \frac{-3(1 - \cos x)}{x^2}$$

$$= \lim_{x \to 0^-} \frac{-3\left(2\sin^2 \frac{x}{2}\right)}{x^2}$$

$$= -\frac{3}{2}$$

$$\text{RHL} = f(0^+) = \lim_{x \to 0^+} f(x) = \lim_{x \to 0^+} b\tan\left(\frac{\pi}{[x+3]} \right)$$

$$= b\tan\left(\frac{\pi}{3} \right)$$

$$= b\sqrt{3}$$

Also, $f(0) = b\sqrt{3}$

$f(x)$ is continuous at $x = 0$.

So,

$$b\sqrt{3} = -\frac{3}{2}$$

$$b = -\frac{\sqrt{3}}{2}$$

$$\therefore [a - 2b] = [-3 + \sqrt{3}]$$

$$= [-3 + 1.732]$$

$$= [-1.268] = -2$$

(ii) If $f(x) = \text{sgn}(x^2 - ax + 1)$ is discontinuous at exactly one point, then

$x^2 - ax + 1 = 0$ must have only one real root.

$\therefore \; b^2 - 4ac = 0$

So, $(-a)^2 - 4 = 0$

$$a^2 = 4$$

$$a = \pm 2$$

$$\therefore \quad |a| = 2$$

(iii) At $x = 0$,

$$\text{LHL} = \lim_{x \to 0^-} f(x) = \lim_{x \to 0^-} q + \left[\frac{\sin x - x}{x^3} \right]$$

$$= \lim_{x \to 0^-} q + \left[\frac{\left[x - \frac{x^3}{3!} + \frac{x^5}{5!} - \dots \right] - x}{x^3} \right]$$

$$= \lim_{x \to 0^-} q + \left[-\frac{1}{3!} + \frac{x^2}{5!} - \dots \right]$$

$$= q + \left[-\frac{1}{6} \right]$$

$$= q - 1$$

$$\text{RHL} = \lim_{x \to 0^+} f(x) = \lim_{x \to 0^+} p + \frac{\sin[x]}{x}$$

$$= \lim_{x \to 0^+} \left(p + \frac{\sin 0}{x} \right) = p$$

Also, $f(0) = 2$

So, $q - 1 = 2$ and $p = 2$

$\Rightarrow \quad q = 3$ and $p = 2$

So, $|p - q| = |2 - 3| = |-1| = 1$

(iv) $$\lim_{x \to 2^-} f(x) = \lim_{x \to 2^-} \frac{a|(x-2)(x+1)|}{-(x-2)(x+1)} = a$$

$$\lim_{x \to 2^+} f(x) = \lim_{x \to 2^+} \frac{x - [x]}{x - 2} = \lim_{x \to 2^+} \frac{x-2}{x-2} = 1$$

Also, $f(2) = b$

So, $a = b = 1$

$\Rightarrow (1 - a^2) + (1 - b^2) = (1 - 1^2) + (1 - 1^2) = 0$

Sol. (Q.Nos. 45-47)

45. $\to$ (3), **46.** $\to$ (3), **47.** $\to$ (2)

$$f(x) = \begin{cases} [x], & -2 \le x \le \frac{1}{2} \\ 2x^2 - 1, & -\frac{1}{2} < x \le 2 \end{cases}$$

$$= \begin{cases} -2, & -2 \le x < -1 \\ -1, & -1 \le x \le -\frac{1}{2} \\ 2x^2 - 1, & -\frac{1}{2} < x \le 2 \end{cases}$$

$$|f(x)| = \begin{cases} 2, & -2 \le x < -1 \\ 1, & -1 \le x \le -\frac{1}{2} \\ |2x^2 - 1|, & -\frac{1}{2} < x \le 2 \end{cases}$$

$$= \begin{cases} 2, & -2x \le x < -1 \\ 1, & -1 \le x \le -\frac{1}{2} \\ 1 - 2x^2, & -\frac{1}{2} < x \le \frac{1}{\sqrt{2}} \\ 2x^2 - 1, & \frac{1}{\sqrt{2}} < x \le 2 \end{cases}$$

$$\text{and } f|x| = \begin{cases} -2, & -2 \le |x| < -1 \\ -1, & -1 \le |x| \le -\frac{1}{2} \\ 2|x|^2 - 1, & -\frac{1}{2} < |x| \le 2 \end{cases}$$

$f(x) = 2x^2 - 1, -2 \le x \le 2$

$$\therefore\ g(x)=f|x|+|f(x)|=\begin{cases}2x^2+1, & -2\le x<-1\\ 2x^2, & -1\le x\le -\frac{1}{2}\\ 0, & -\frac{1}{2}<x<\frac{1}{\sqrt{2}}\\ 4x^2-2, & \frac{1}{\sqrt{2}}\le x\le 2\end{cases}$$

Now, $g(-1^-)=\lim_{x\to -1^-}(2x^2+1)=3$

$g(-1^+)=\lim_{x\to -1^+}2x^2=2$

$g(x)$ is discontinuous at $x=-1$

$$g\left(-\frac{1}{2}^-\right)=\lim_{x\to -\frac{1}{2}^-}2x^2=\frac{1}{2}$$

$$g\left(-\frac{1}{2}^+\right)=\lim_{x\to -\frac{1}{2}^+}(0)=0$$

$g(x)$ is discontinuous at $x=-\frac{1}{2}$

Now, $g\left(\frac{1}{\sqrt{2}}^-\right)=0$

$$g\left(\frac{1}{\sqrt{2}}^+\right)=\lim_{x\to \frac{1}{\sqrt{2}}^+}(4x^2-2)=0$$

So, $g(x)$ is continuous at $x=0$.

Now, $g'\left(\frac{1}{\sqrt{2}}^-\right)=0$

$$g'\left(\frac{1}{\sqrt{2}}^+\right)=8\left(\frac{1}{\sqrt{2}}\right)=\frac{8}{\sqrt{2}}$$

Hence, $g(x)$ is non-differentiable at $x=\frac{1}{\sqrt{2}}$.

So, $|f(x)|$ is non-differentiable at $-1, -\frac{1}{2}$ and $\frac{1}{\sqrt{2}}$.

$g(x)$ is non-differentiable at $x=-1, -\frac{1}{2}$ and $\frac{1}{\sqrt{2}}$

and $g(x)$ is discontinuous at $x=-\frac{1}{2}$ and -1.

Sol. (Q.Nos. 48-50)

Let θ, ϕ and α are the roots of $x^3+2x^2+px+q=0$, then

$$\theta+\phi+\alpha=-2 \quad \text{...(i)}$$

$$\theta\phi+\phi\alpha+\theta\alpha=p \quad \text{...(ii)}$$

$$\theta\cdot\phi\cdot\alpha=-q \quad \text{...(iii)}$$

and let θ, ϕ and β are the roots of the equation

$$x^3+x^2+px+r=0$$

$$\therefore \quad \theta+\phi+\beta=-1 \quad \text{...(iv)}$$

$$\theta\phi+\phi\beta+\theta\beta=p \quad \text{...(v)}$$

$$\theta\cdot\phi\cdot\beta=-r \quad \text{...(vi)}$$

From Eqs. (i) and (iv), we get

$$\alpha-\beta=-1$$

From Eqs. (ii) and (v), we get

$$(\theta+\phi)(\alpha-\beta)=0\Rightarrow\theta+\phi=0$$

Now, from Eqs. (i) and (iv), we get

$$\alpha=-2 \text{ and } \beta=-1$$

48. (5) $\alpha^2+\beta^2=(-2)^2+(-1)^2=5$

49. (3) $\lim_{x\to 0^-}f(x)=\lim_{h\to 0}f(0-h)$

$$=\lim_{h\to 0}e^{-h\log_{1-h}|-3|}=\lim_{h\to 0}e^{-\frac{h\log 3}{\log(1-h)}}$$

$$=e^{\lim_{h\to 0}\left(\frac{-h}{\log(1-h)}\cdot\log 3\right)}=e^{\log 3}=3$$

50. (9) $\lim_{x\to 0^+}f(x)=\lim_{h\to 0}f(0+h)$

$$=\lim_{h\to 0}b\cdot\frac{\ln(e^{h^2}+2\sqrt{h})}{\sin\sqrt{h}}$$

$$=b\lim_{h\to 0}\cdot\frac{\ln e^{h^2}+\ln\left(1+\frac{2\sqrt{h}}{e^{h^2}}\right)}{\sin\sqrt{h}}$$

$$=b\cdot\lim_{h\to 0}\left(\frac{\frac{h^2}{\sqrt{h}}+\frac{\ln\left(1+\frac{2\sqrt{h}}{e^{h^2}}\right)}{\sqrt{h}}}{\frac{\sin\sqrt{h}}{\sqrt{h}}}\right)$$

$$=b\cdot\lim_{h\to 0}\left(\frac{h^{\frac{3}{2}}+\frac{\ln\left(1+\frac{2\sqrt{h}}{e^{h^2}}\right)\cdot\frac{2}{e^{h^2}}}{\frac{2\sqrt{h}}{e^{h^2}}}}{\frac{\sin\sqrt{h}}{\sqrt{h}}}\right)$$

$$=b\left(\frac{0+1\cdot\frac{2}{1}}{1}\right)=2b$$

Since, $f(x)$ is continuous at $x=0$, then

$$\lim_{x\to 0^-}f(x)=f(0)=\lim_{x\to 0^+}f(x)$$

$$\Rightarrow \quad 3=a=2b$$

$$\therefore \quad a=3 \text{ and } b=\frac{3}{2}$$

$$\Rightarrow \quad 2(a+b)=2\left(3+\frac{3}{2}\right)=9$$

51. (54) $f(x)$ is continuous at $x=3$

So, $\lim_{x\to 3^-} f(x) = \lim_{x\to 3^+} f(x)$

$\Rightarrow \quad 5 = 9p+3q+2$ ...(i)

$\Rightarrow \quad 3p+q=1$

Now, at $x=1$

$\lim_{x\to 1^+} f(x)=3$ and $\lim_{x\to 1^-} f(x)=b$

$f(x)$ is discontinuous at $x=1$

So, $\quad b\neq 3$ ...(ii)

Now, $f'(x)=\begin{cases}2ax-a, & x<1\\ 1, & 1<x<3\\ 2px+q, & x>3\end{cases}$

Now, $\lim_{x\to 3^+} f'(x)=6p+q$ and $\lim_{x\to 3^-} f'(x)=1$

$6p+q=1$ ...(iii)

From Eqs. (i) and (iii),

$p=0$ and $q=1$

Now, LHD for $|f(x)|$ at $x=1$,

$$\lim_{h\to 0}\frac{|-ab(h+1)+b|-3}{-h}=-a$$

So, $b=-3$ for LHD exist.

RHD for $|f(x)|$ at $x=1$ is 1.

So, $a=-1$

Now, $|a+b+p+q|=|-1-3+0+1|=3=\dfrac{k}{18}$

So, $k=54$

52. (7) We have,

$$f(x)=\frac{1}{\log|x^2-4|}$$

f is not defined, when

$(x^2-4)=0, |x^2-4|=1$

$(x^2-4)=0, (x^2-4)=\pm 1$

$\Rightarrow \quad x^2=4, x^2=4\pm 1$

$x=\pm 2, \pm\sqrt{5}$ and $\pm\sqrt{3}$

Thus, $m=6$

Also, $\lim_{x\to 0}\dfrac{x^n-\sin x^n}{x-\sin^n x}$

Its limit exists only, when $n=1$

Hence, the value of $m+n=7$

53. (3) Let

$$f(x)=\begin{cases}\sqrt{4x^2-12x+9}\cdot\{x\}, & x\geq 1\\ \cos\left(\dfrac{\pi}{2}(|x|-\{x\})\right), & x<1\end{cases}$$

$$=\begin{cases}|2x-3|\cdot\{x\} & x\geq 1\\ \cos\dfrac{\pi}{2}(|x|-\{x\}) & x<1\end{cases}$$

$$=\begin{cases}-(2x-3)(x-1), & 1\leq x\leq\dfrac{3}{2}\\ (2x-3)(x-1), & \dfrac{3}{2}\leq x\leq 2\\ \cos\left[\dfrac{\pi}{2}(-x-(x+1)\right], & -1\leq x<0\\ \cos\dfrac{\pi}{2}[x-(x)], & 0\leq x<1\end{cases}$$

$$=\begin{cases}-\sin\pi x, & -1\leq x<0\\ 1, & 0\leq x<1\\ -(2x-3)(x-1), & 1\leq x<\dfrac{3}{2}\\ (2x-3)(x-1), & \dfrac{3}{2}\leq x\leq 2\end{cases}$$

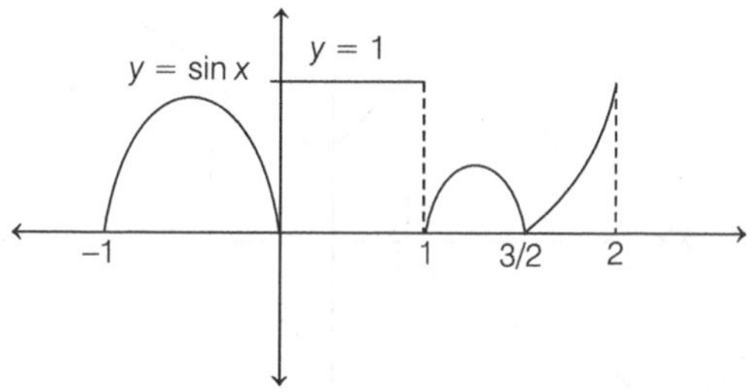

$\Rightarrow f(x)$ is non-differentiable at $x=0$, 1 and $\dfrac{3}{2}$

54. (5) We have, $f(x)=|||x-1|-1|-1|$

$$=\begin{cases}||x-2|-1, & x\geq 1\\ ||x|-1|, & x\leq 1\end{cases}$$

$$=\begin{cases}|-x-1|, & x<0\\ |x-1|, & 0\leq x<1\\ |-x+2-1|, & 1\leq x<2\\ |x-3|, & x\geq 2\end{cases}$$

$$=\begin{cases}-x-1, & x<-1\\ x+1, & -1\leq x<0\\ -x+1, & 0\leq x<1\\ x-1, & 1\leq x<2\\ -x+3, & 2\leq x<3\\ x-3, & x\geq 3\end{cases}$$

$\Rightarrow f'(x)$ is continuous for all $x\in R$.

$$\Rightarrow \quad f'(x)=\begin{cases}-1, & x<-1\\ 1, & -1<x<0\\ -1, & 0<x<1\\ 1, & 1<x<2\\ -1, & 2<x<3\\ 1, & x>3\end{cases}$$

$f(x)$ is non-differentiable at $x=-1$, 0, 1, 2 and 3.

So, $f(x)$ is non-differentiable at 5 points.

55. (3) We have,

$f(x) = |x| + 1$ and $g(x) = x^2 + 1$

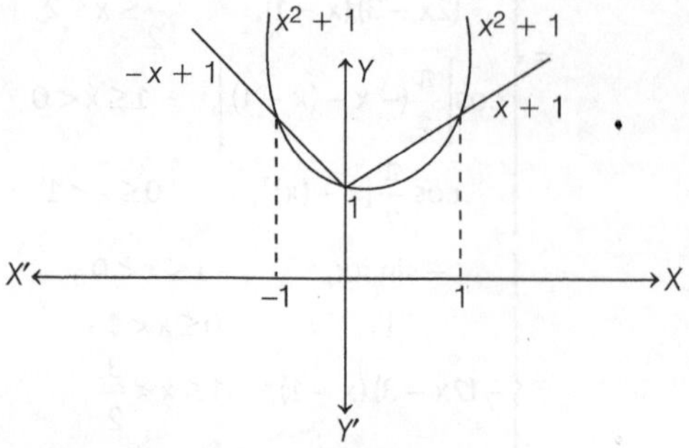

So, graph of $h(x)$ is

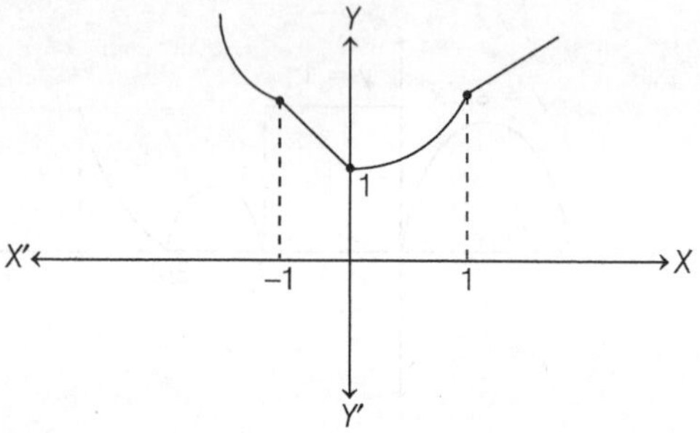

$f(x)$ is non-differentiable at $x = -1$, 0 and 1

So, $h(x)$ is non-differentiable at 3 points.

56. (2) We have,

$$P(x) = 3x^3 - 9x^2 + 2x + 5$$
$$P'(x) = 9x^2 - 18x + 2$$
$$P''(x) = 18x - 18$$

$$\Rightarrow \quad f(x) = \begin{cases} \dfrac{18x - 18 + 18}{6}, & x \neq \dfrac{\pi}{2} \\ \sin^{-1}(ab) + \cos^{-1}(a+b-3ab), & x = \dfrac{\pi}{2} \end{cases}$$

$$\Rightarrow \quad f(x) = \begin{cases} 3x, & x \neq \dfrac{\pi}{2} \\ \sin^{-1}(ab) + \cos^{-1}(a+b-3ab), & x = \dfrac{\pi}{2} \end{cases}$$

For continuity,

$$\sin^{-1}(ab) + \cos^{-1}(a+b-3ab) = \frac{3\pi}{2}$$

$\Rightarrow \quad ab = 1$ and $(a + b - 3ab) = -1 \Rightarrow a + b = 2$

57. (1) We have,

$$f(x) = [[x] + \{x\}] + \{[x^2] + \{x\}\}$$

$$= [x] + \{x\} \qquad \left\{ \begin{array}{l} \because [[x] + \{x\}] = [x] \\ \text{and } [x^2] + \{x\}\} = \{x\} \end{array} \right.$$

$$f(x) = x$$

So, $|f(x)| = |x|$

which is non-derivable at only one point at $x = 0$.

Hence, $|f(x)|$ is non-derivable at one point only.

58. (19) Putting $x = 0 = y \Rightarrow f(0) = 0$

$$f'(x) = \lim_{h \to 0} \frac{f(x+h) - f(x)}{h}$$

$$= \lim_{h \to 0} \frac{f(x) + 4f\left(\dfrac{h}{2}\right) + h(2x - 1) - f(x)}{h}$$

$$= \lim_{h \to 0} \frac{4f\left(\dfrac{h}{2}\right) - 4f(0) + h(2x - 1)}{h}$$

$$f'(x) = 2f'(0) + (2x - 1) \Rightarrow f'(x) = 2x + 1$$

$\Rightarrow \qquad f(x) = x^2 + x \qquad$ (given $f'(0) = 1$)

Now, $f(3) + f'(3) = 12 + 7 = 19$

59. (4) $\lim\limits_{x \to 1^+} g(x) = \lim\limits_{m \to \infty} \lim\limits_{x \to 1^+} \dfrac{x^m f(x) + h(x) + 3}{2x^m + 4x + 1}$

$$= \lim_{m \to \infty} \lim_{x \to 1^+} \left\{ \frac{f(x) + \dfrac{h(x) + 3}{x^m}}{2 + \dfrac{4x + 1}{x^m}} \right\}$$

$$g(1) = \frac{f(1)}{2} \Rightarrow f(1) = 2g(1) \qquad \text{...(i)}$$

Similarly, $g(1) = \lim\limits_{x \to 1^-} g(x) = \dfrac{h(1) + 3}{5}$

$\Rightarrow \qquad h(1) = 5g(1) - 3$

Now, $f(1) - h(1) = 2g(1) - \{5g(1) - 3\}$

$$= -3g(1) + 3$$
$$= 3[1 - g(1)]$$

By comparing $a = 3$

$$b = 1$$

So, $a + b = 3 + 1 = 4$

60. (1) Here, $f(x)$ is continuous at $x = 0$.

So, $f(0^-) = f(0^+) = f(0)$

$$\therefore \quad f(0^+) = \lim_{x \to 0^+} \frac{e^{1/x} + e^{2/x} + e^{3/x}}{ae^{2/x} + be^{3/x}}$$

$$\therefore \quad f(0^+) = \lim_{h \to 0} \frac{e^{\frac{3}{h}}\left(\dfrac{1}{e^{\frac{2}{h}}} + \dfrac{1}{e^{\frac{1}{h}}} + 1\right)}{e^{\frac{3}{h}}\left(\dfrac{a}{e^{\frac{1}{h}}} + b\right)}$$

$$= \frac{1}{b} \qquad \left[\lim_{h \to 0} \frac{1}{e^{\frac{1}{h}}} = 0\right]$$

and $f(0^-) = \lim\limits_{x \to 0^-} (\cos x - \sin x)^{\operatorname{cosec} x}$

$$= \lim_{h \to 0} (\cos h + \sin h)^{-\operatorname{cosec} h}$$

$$e^{\lim_{h\to 0}(\cos h+\sin h-1)\left(\frac{-1}{\sin h}\right)}$$

$$= e^{\lim_{h\to 0}\left(-2\sin^2\frac{h}{2}+2\sin\frac{h}{2}\cos\frac{h}{2}\right)\left(-\frac{1}{2\sin\frac{h}{2}\cos\frac{h}{2}}\right)}$$

$$= e^{\lim_{h\to 0}\frac{\sin\frac{h}{2}-\cos\frac{h}{2}}{\frac{h}{-}}} = e^{-1}$$

and $f(0)=a$

$\Rightarrow \quad a=e^{-1}=\frac{1}{b}$

$\Rightarrow \quad a=\frac{1}{e}$ and $b=e$

So, $ab=\frac{1}{e}\times e=1$

61. (4) We have,

$$2f(x+y)+f(x-y)=3f(x)+3f(y)+2xy$$

$\Rightarrow \quad 2f(x+y)-2f(x)=f(x)+3f(y)+2xy-f(x-y)$

Let $y=h$

$\Rightarrow \quad 2[f(x+h)-f(x)]=f(x)-f(x-h)+3f(h)+2xh$

Dividing by h on both sides,

$$\Rightarrow 2\left(\frac{f(x+h)-f(x)}{h}\right)=\frac{f(x)-f(x-h)}{h}+\left(\frac{3f(h)-0}{h}\right)+2x$$

$\Rightarrow \quad 2f'(x)=f'(x)+3f'(0)+2x$

$\Rightarrow \quad f'(x)=2x \qquad [\because f'(0)=0 \text{ (given)}]$

So, $f(x)=\frac{2x^2}{2}+C$

$f(x)=x^2+C \qquad [\because f(0)=0]$

$\Rightarrow \quad C=0$

So, $f(x)=x^2$

Then, $f(2)=4$

CHAPTER

04

METHODS OF DIFFERENTIATION

LEVEL 01 \ Starter Level

4.1 Fundamentals and Chain Rule

1 Differentiate each of the following with respect to x.

(i) $y = \frac{1}{x} + \frac{2}{x^2} + \frac{3}{x^3}$

(ii) $y = \sqrt{x} + x\sqrt{x} + x^2\sqrt{x}$

(iii) $y = \sin(4x+3)$

(iv) $y = \cot\sqrt{x}$

(v) $y = \sqrt{\tan x^2}$

(vi) $y = (2x^2 - 3x + 4)^6$

(vii) $y = \sqrt{x \sin x}$

(viii) $y = \sqrt{\cot\sqrt{x}}$

(ix) $y = \cos(\sin\sqrt{ax+b})$

(x) $y = \sqrt{\frac{a^2 - x^2}{a^2 + x^2}}$

(xi) $y = \log(x + \sqrt{x^2+1})$

4.2 Differentiation of Inverse Trigonometric Functions

2 Differentiate each of the following inverse trigonometric functions with respect to 'x'.

(i) $y = \sin^{-1}\frac{x}{a}$

(ii) $y = \log(\sin^{-1} x^4)$

(iii) $y = \tan^{-1}\left(\frac{\sin x}{1+\cos x}\right)$

(iv) $y = \tan^{-1}(\cot x) + \cot^{-1}(\tan x)$

(v) $y = \sin^{-1}(3x - 4x^3)$

(vi) $y = \tan^{-1}\left(\frac{1+x}{1-x}\right)$

(vii) If $y = \sin\left(2\tan^{-1}\sqrt{\frac{1-x}{1+x}}\right)$, then find $\frac{dy}{dx}$.

(viii) $y = \cos^{-1}\left(\frac{x - \sqrt{1-x^2}}{\sqrt{2}}\right)$

(ix) $y = \sin^{-1}\left(\frac{\sqrt{1+x} + \sqrt{1-x}}{2}\right)$

4.3 Differentiation of Implicit Functions

3 Differentiate each of the following with respect to 'x'.

(i) $x^{2/3} + y^{2/3} = a^{2/3}$

(ii) $xy^2 - x^2y - 5 = 0$

(iii) $x \sin 2y = y \cos 2x$

(iv) $y \log x = (x - y)$

(v) $\sec\left(\frac{x+y}{x-y}\right) = a$

(vi) $e^x + e^y = e^{x+y}$

(vii) $\sin(x + y) = \log(x + y)$

4.4 Differentiation of Logarithmic and Exponential Functions

4 Find $\frac{dy}{dx}$ for the following

(i) $y = x^y$

(ii) $y = (\cos x)^{\sin x}$

(iii) $y = \sqrt{(x-1)(x-2)(2x-3)}$

(iv) $x^y = y^x$

(v) $x^y = e^{x-y}$

(vi) $y = \left(x + \frac{1}{x}\right)^x$

4.5 Differentiation of Parametric Functions

5 Find $\frac{dy}{dx}$ for following parametric functions.

(i) $y = 2at$ and $x = at^2$

(ii) $x = a(t - \sin t)$ and $y = a(1 - \cos t)$

(iii) $y = 3t^2 + 2t - 1$ and $x = t^3 - 1$

(iv) $y = \sec 4t$ and $x = \tan^{-1} t$

(v) $x = e^{\cos 2t}$ and $y = e^{\sin 2t}$

(vi) $x = a \sin 2\theta (1 + \cos 2\theta)$ and $y = b \cos 2\theta (1 - \cos 2\theta)$

4.6 Differentiation of Infinite Series

6 Find $\frac{dy}{dx}$ for the following infinite series.

(i) $y = \sqrt{x + \sqrt{x + \sqrt{x + \ldots \infty}}}$

(ii) $y = \sqrt{\sin x + \sqrt{\sin x + \sqrt{\sin x + \ldots \infty}}}$

(iii) $y = x^{x^{x^{\cdots\infty}}}$

(iv) $y = e^{x + e^{x + e^{x + \ldots\infty}}}$

(v) $y = \cfrac{x}{1 + \cfrac{x}{2 + \cfrac{x}{1 + \cfrac{x}{2 + \ldots\infty}}}}$

(vi) $y = \frac{x(2+y)}{2+y+x}$

4.7 Differentiation of Second or Higher Order Derivatives

7 Find $\frac{d^2y}{dx^2}$ for the following

(i) $y = \frac{\log x}{x}$ (ii) $y = e^x \sin 3x$

(iii) $y = x^2 \cos x$ (iv) $y = \tan^{-1} x$

(v) $y = \log(\log x)$

4.8 Miscellaneous

8 If $f(x) = x|x|$, then find $\frac{dy}{dx}$.

9 If $f(x) = |x-1| + |x-3|$, then find $f'(2)$.

10 If $y = e^x \cos x$, then prove that $\frac{dy}{dx} = \sqrt{2}\, e^x \cos\left(x + \frac{\pi}{4}\right)$.

11 Let f be a function for which $f'(x) = \frac{1}{x^2+1}$. If $g(x) = f(3x-1)$. Find $g'(x)$.

12 If $\cos y = x \cos(a+y)$ with $\cos a \neq \pm 1$. Prove that $\frac{dx}{dy} = \frac{\sin a}{\cos^2(a+y)}$.

13 If $y = \frac{\sqrt{1-x^2}\,(2x+3)^{1/2}}{(x^2+2)^{2/3}}$, find $\frac{dy}{dx}$ at $x = 0$.

14 Differentiate $\sin x$ w.r.t. $\cos x$.

15 Differentiate x^x w.r.t. $x \log x$.

LEVEL 02 \ JEE Main Level

Single Choice Correct Questions

4.1 Basic Fundamentals and Chain Rule

1 If $y = \sin^n x \cos nx$, then $\frac{dy}{dx}$ is

(a) $n \sin^{n-1} x \cos(n+1)x$ (b) $n \sin^{n-1} x \sin(n+1)x$

(c) $n \sin^{n-1} x \cos(n-1)x$ (d) $n \sin^{n-1} x \cos nx$

2 If $y = |\cos x| + |\sin x|$, then $\frac{dy}{dx}$ at $x = \frac{2\pi}{3}$ is

(a) $\frac{1-\sqrt{3}}{2}$ (b) 0 (c) $\frac{1}{2}(\sqrt{3}-1)$ (d) None of these

3 If $f(x) = \sin\left\{\frac{\pi}{3}[x] - x^2\right\}$ for $2 < x < 3$ and $[x]$ is the greatest integer function, then $f'\left(\sqrt{\frac{\pi}{3}}\right)$ is

(a) $\sqrt{\frac{\pi}{3}}$ (b) $-\sqrt{\frac{\pi}{3}}$ (c) $-\sqrt{\pi}$ (d) None of these

4 If $x\sqrt{1+y}+y\sqrt{1+x}=0$, then $\frac{dy}{dx}$ is equal to

(a) $\frac{1}{(1+x)^2}$ (b) $-\frac{1}{(1+x)^2}$ (c) $\frac{1}{(1-x)^2}$ (d) $\frac{1}{(1+x)}$

5 If $f(x)=\frac{g(x)+g(-x)}{2}+\frac{2}{[h(x)+h(-x)]^{-1}}$, where g and h are differentiable function, then $f'(0)$ is

(a) 1 (b) $\frac{1}{2}$ (c) $\frac{3}{2}$ (d) 0

6 If $y=\frac{1}{1+x^{n-m}+x^{p-m}}+\frac{1}{1+x^{m-n}+x^{p-n}}+\frac{1}{1+x^{m-p}+x^{n-p}}$, then $\frac{dy}{dx}$ is equal to

(a) 1 (b) 0 (c) $m+n+p$ (d) $m-n+p$

7 If g is the inverse of f and $f'(x)=\frac{1}{1+x^5}$, then $g'(x)$ is equal to

(a) $1+x^5$ (b) $5x^4$ (c) $\frac{1}{1+[g(x)]^5}$ (d) $1+[g(x)]^5$

8 If $y=\log^n x$, where $\log^n$ mean log log log ...... (repeated n times), then $x\log x\log^2 x\log^3 x\ldots\log^{n-1} x\log^n x\frac{dy}{dx}$ is equal to

(a) $\log x$ (b) x (c) $\frac{1}{\log x}$ (d) $\log^n x$

4.2 Differentiation of Inverse Trigonometric Functions

9 If for $x\in\left(0,\frac{1}{4}\right)$ the derivative of $\tan^{-1}\left(\frac{6x\sqrt{x}}{1-9x^3}\right)$ is $\sqrt{x}\cdot g(x)$, then $g(x)$ equals **[JEE Main 2017]**

(a) $\frac{9}{1+9x^3}$ (b) $\frac{3x\sqrt{x}}{1-9x^3}$ (c) $\frac{3x}{1-9x^3}$ (d) $\frac{3}{1+9x^3}$

10 If $\cos^{-1}\left(\frac{x^2-y^2}{x^2+y^2}\right)=\log a$, then $\frac{dy}{dx}$ is

(a) $\frac{-x}{y}$ (b) $\frac{-y}{x}$ (c) $\frac{y}{x}$ (d) $\frac{x}{y}$

11 If $f(x)=\cot^{-1}\left(\frac{x^x-x^{-x}}{2}\right)$, then $f'(1)$ equals

(a) -1 (b) 1 (c) $\log_e 2$ (d) $-\log_e 2$

12 If $y=\tan^{-1}\frac{2^x}{1+2^{2x+1}}$, then $\frac{dy}{dx}$ at $x=0$ is

(a) 1 (b) 2 (c) $-\frac{1}{10}\ln 2$ (d) None of these

13 If $2y=\left(\cot^{-1}\left(\frac{\sqrt{3}\cos x+\sin x}{\cos x-\sqrt{3}\sin x}\right)\right)^2$, $x\in\left(0,\frac{\pi}{2}\right)$, then $\frac{dy}{dx}$ is equal to **[JEE Main 2019]**

(a) $\frac{\pi}{6}-x$ (b) $\frac{\pi}{3}-x$ (c) $x-\frac{\pi}{6}$ (d) $2x-\frac{\pi}{3}$

14 If $f(x)=\tan^{-1}\left(\frac{\sqrt{1+x^2}+\sqrt{1-x^2}}{\sqrt{1+x^2}-\sqrt{1-x^2}}\right)$, then $f'(x)$ equals

(a) $-\frac{x}{\sqrt{1-x^4}}$ (b) $\frac{x}{\sqrt{1-x^4}}$ (c) $\frac{-x^2}{\sqrt{1-x^4}}$ (d) $\frac{x^4}{\sqrt{1-x^4}}$

15 If $y=\tan^{-1}\frac{1}{1+x+x^2}+\tan^{-1}\frac{1}{x^2+3x+3}+\tan^{-1}\frac{1}{x^2+5x+7}+\ldots\ldots.n$ terms, then $y'(0)$ equals

(a) $\frac{1}{1+n^2}$ (b) $-\frac{n^2}{1+n^2}$ (c) $\frac{n}{1+n^2}$ (d) None of these

4.3 Differentiation of Implicit Functions

16 If $\sin y+e^{-x\cos y}=e^2$, then $\frac{dy}{dx}$ at $(1,\pi)$ is

(a) $\sin y$ (b) $-x\cos y$ (c) e (d) e^{-1}

17 If $\sin xy+\cos xy=0$, then $\frac{dy}{dx}$ is equal to

(a) $\frac{y}{x}$ (b) $-\frac{y}{x}$ (c) $-\frac{x}{y}$ (d) $\frac{x}{y}$

18 If $x\log_e(\log_e x)-x^2+y^2=4\ (y>0)$, then $\frac{dy}{dx}$ at $x=e$ is equal to **[JEE Main 2019]**

(a) $\frac{1+2e}{2\sqrt{4+e^2}}$ (b) $\frac{2e-1}{2\sqrt{4+e^2}}$ (c) $\frac{1+2e}{\sqrt{4+e^2}}$ (d) $\frac{e}{\sqrt{4+e^2}}$

4.4 Differentiation of Logarithmic and Exponential Function

19 If $y=x^{e^x}$, then $\frac{dy}{dx}$ is

(a) $y(\log x+e^x)$ (b) $y\log x\left(\frac{1}{2}+e^x\right)$

(c) $ye^x(\log x+x)$ (d) $ye^x\left(\log x+\frac{1}{x}\right)$

20 If $x^{2x}-2x^x\cot y-1=0$, then $y'(1)$ equals

(a) $-\log 2$ (b) -1 (c) 1 (d) $\log 2$

21 If $2^x+2^y=2^{x+y}$, then $\frac{dy}{dx}$ is equal to

(a) $\frac{2^x+2^y}{(2^x-2^y)}$ (b) $\frac{2^x+2^y}{1+2^{x+y}}$ (c) $2^{x-y}\cdot\frac{2^y-1}{1-2^x}$ (d) $\frac{2^{x+y}-2^x}{2^y}$

22 If $y=(\tan x)^{(\tan x)^{\tan x}}$, then find $\frac{dy}{dx}$ at $x=\frac{\pi}{4}$.

(a) 0 (b) 1 (c) -1 (d) 2

23 The differential coefficient of $f(\log_e x)$ w.r.t. x, where $f(x)=\log_e x$ is

(a) $\frac{x}{\log_e x}$ (b) $\frac{\log_e x}{x}$ (c) $\frac{1}{x\log_e x}$ (d) None of these

24 If $y = x^{(\ln x)^{\ln(\ln x)}}$, then $\frac{dy}{dx}$ is equal to

(a) $\frac{y}{x}[\ln x^{\ln x - 1} + 2 \ln x \ln(\ln x)]$

(b) $\frac{y}{x}[(\ln x^{\ln(\ln x)}) 2 \ln(\ln x) + 1]$

(c) $\frac{y}{x \ln x}[(\ln x)^2 + 2 \ln(\ln x)]$

(d) $\frac{y \cdot \ln y}{x \ln x}[2 \ln(\ln x) + 1]$

25 If $f(x) = |x|^{|\sin x|}$, then $f'\left(\frac{\pi}{4}\right)$ equals

(a) $\left(\frac{\pi}{4}\right)^{1/\sqrt{2}}\left(\frac{\sqrt{2}}{2}\ln\frac{4}{\pi} - \frac{2\sqrt{2}}{\pi}\right)$

(b) $\left(\frac{\pi}{4}\right)^{1/\sqrt{2}}\left(\frac{\sqrt{2}}{2}\ln\frac{4}{\pi} + \frac{2\sqrt{2}}{\pi}\right)$

(c) $\left(\frac{\pi}{4}\right)^{1/\sqrt{2}}\left(\frac{\sqrt{2}}{2}\ln\frac{4}{\pi} - \frac{2\sqrt{2}}{\pi}\right)$

(d) $\left(\frac{\pi}{4}\right)^{1/\sqrt{2}}\left(\frac{\sqrt{2}}{2}\ln\frac{\pi}{4} + \frac{2\sqrt{2}}{\pi}\right)$

26 For $x > 1$, if $(2x)^{2y} = 4e^{2x - 2y}$, then $(1 + \log_e 2x)\frac{dy}{dx}$ is equal to

(a) $\log_e 2x$ (b) $\frac{x \log 2x + \log 2}{x}$ (c) $x \log 2x$ (d) $\frac{x \log 2x - \log 2}{x}$

4.5 Differentiation of Parametric Functions

27 If $x = \operatorname{cosec}\theta - \sin\theta$ and $y = \operatorname{cosec}^n\theta - \sin^n\theta$, then $(x^2 + 4)\left(\frac{dy}{dx}\right)^2$ is equal to

(a) $n^2(y^2 + 4)$ (b) $n(y^2 + 4)$ (c) $n^3(y + 4)$ (d) None of these

28 If $x = a\cos^3\theta$ and $y = a\sin^3\theta$, then $\sqrt{1 + \left(\frac{dy}{dx}\right)^2}$ is equal to

(a) $\sec^2\theta$ (b) $\sec\theta$ (c) $|\tan\theta|$ (d) $|\sec\theta|$

29 If $r = \sqrt{x^2 + y^2 + z^2}$ and $x = 2\sin 3t$, $y = 2\cos 3t$ and $z = 8t$, then $\frac{dr}{dt}$ is equal to

(a) $\frac{32t}{\sqrt{1 + 16t^2}}$ (b) $\frac{16t}{\sqrt{1 + 16t^2}}$ (c) $\frac{t}{\sqrt{1 + 16t^2}}$ (d) $\frac{4t}{\sqrt{1 + 16t^2}}$

4.6 Differentiation of Infinite Series

30 If $y = a^{x^{a^{x^{a^{x \ldots \infty}}}}}$, then $\frac{dy}{dx}$ is equal to

(a) $\frac{x \log y}{1 - y \log x \log y}$

(b) $\frac{y^2 \log x}{x(1 - y \log x \log y)}$

(c) $\frac{y^2 \log y}{x(1 - y \log x \log y)}$

(d) None of these

31 If $y = \cfrac{x}{a + \cfrac{x}{b + \cfrac{x}{a + \ldots \infty}}}$, then $\frac{dy}{dx}$ is equal to

(a) $\frac{a}{ab + 2ay}$ (b) $\frac{1}{a + 2y}$ (c) $\frac{a}{ab + 2by}$ (d) $\frac{b}{ab + 2ay}$

32 Let $f(x) = x + \cfrac{1}{x + \cfrac{1}{x + \cfrac{1}{x + \ldots\ldots \infty}}}$, then the value of $f(100) \cdot f'(100)$ is

(a) 50 (b) 100 (c) 200 (d) None of these

4.7 Differentiation of Second or Higher Order Derivatives

33 If $e^y (x+1) = 1$, then

(a) $\frac{d^2y}{dx^2} = \frac{dy}{dx}$ (b) $\frac{d^2y}{dx^2} = \left(\frac{dy}{dx}\right)^2$ (c) $\frac{d^2y}{dx^2} = -\left(\frac{dy}{dx}\right)^2$ (d) None of these

34 $\frac{d^2x}{dy^2}$ is equal to

(a) $\left(\frac{d^2y}{dx^2}\right)^{-1}$ (b) $\left(\frac{d^2y}{dx^2}\right)^{-1}\left(\frac{dy}{dx}\right)^{-3}$ (c) $\left(\frac{d^2y}{dx^2}\right) \cdot \left(\frac{dy}{dx}\right)^{-2}$ (d) $-\left(\frac{d^2y}{dx^2}\right)\left(\frac{dy}{dx}\right)^{-3}$

35 If $x = 3 \tan t$ and $y = 3 \sec t$, then the value of $\frac{d^2y}{dx^2}$ at $t = \frac{\pi}{4}$, is **[JEE Main 2019]**

(a) $\frac{1}{6}$ (b) $\frac{1}{6\sqrt{2}}$ (c) $\frac{1}{3\sqrt{2}}$ (d) $\frac{3}{2\sqrt{2}}$

36 If $e^y + xy = e$, then ordered pair $\left(\frac{dy}{dx}, \frac{d^2y}{dx^2}\right)$ at $x = 0$, is **[JEE Main 2019]**

(a) $\left(\frac{1}{e}, \frac{-1}{e^2}\right)$ (b) $\left(\frac{-1}{e}, \frac{1}{e^2}\right)$ (c) $\left(\frac{1}{e}, \frac{1}{e^2}\right)$ (d) $\left(\frac{-1}{e}, \frac{-1}{e^2}\right)$

37 If a curve is represented parametrically by the equations $x = f(t)$ and $y = g(t)$, then $\left(\frac{d^2y}{dx^2}\right) \Big/ \left(\frac{d^2x}{dy^2}\right)$ is equal to (where $f'(t) \neq 0$ and $g'(t) \neq 0$)

(a) 1 (b) $\frac{g'(t)}{f'(t)}$ (c) $-\left(\frac{g'(t)}{f'(t)}\right)^2$ (d) $-\left(\frac{g'(t)}{f'(t)}\right)^3$

38 If $f(x) = x^n + 4$, then the value of $f(1) + \frac{f'(1)}{1!} + \frac{f''(1)}{2!} + \ldots\ldots + \frac{f^n(1)}{n!}$ is

(a) 2^{n-1} (b) $2^n + 4$

(c) $1 + \frac{1}{1!} + \frac{1}{2!} + \ldots + \frac{1}{n!}$ (d) None of these

39 If $y = 2\cos\theta - \cos 2\theta$ and $x = 2\sin\theta - \sin 2\theta$, $\theta \in [0, 2\pi)$, then the value of $\frac{d^2y}{dx^2}$ at $\theta = \pi$, is **[JEE Main 2020]**

(a) $\frac{-3}{4}$ (b) $\frac{3}{4}$ (c) $\frac{3}{8}$ (d) $\frac{3}{2}$

40 If $y = x \log\left(\frac{x}{a+bx}\right)$, then $x^3 \frac{d^2y}{dx^2}$ is equal to

(a) $x\frac{dy}{dx} - y$ (b) $\left(x\frac{dy}{dx} - y\right)^2$ (c) $y\frac{dy}{dx} - x$ (d) $\left(y\frac{dy}{dx} - x\right)^2$

41 If $y = \tan^{-1}\left(\dfrac{\ln \dfrac{e}{x^2}}{\ln ex^2}\right) + \tan^{-1}\left(\dfrac{3+2\ln x}{1-6\ln x}\right)$, then $\dfrac{d^2y}{dx^2}$ equals

(a) 0 (b) 1 (c) 2 (d) None of these

42 If $y = (A+Bx)e^{mx} + (m-1)^{-2}e^x$, then $\dfrac{d^2y}{dx^2} - 2m\dfrac{dy}{dx} + m^2y$ is equal to

(a) e^x (b) e^{mx} (c) e^{-m} (d) $e^{(1-m)x}$

43 If $x = \sin t$ and $y = \sin kt$, then value of $(1-x^2)y_2 - xy_1$ is

(a) $-k^2y$ (b) k^2y (c) ky^2 (d) $-ky^2$

44 If $\dfrac{d}{dx}[f(x)] = g(x)$ and $\dfrac{d}{dx}[g(x)] = f(x^2)$, then $\dfrac{d^2}{dx^2}[f(x^3)]$ can be expressed in the form $kx^af(x^b) + pxg(x^c)$, where $a, b, c, k, p \in N$, then the value of $(k+p+a+b+c)$ equals

(a) 26 (b) 27 (c) 28 (d) 30

45 Let $f(x) = \sin x$, $g(x) = x^2$ and $h(x) = \log x$, if $F(x) = h[f\{g(x)\}]$, then $\dfrac{d^2F}{dx^2}$ is equal to

(a) $2\cos^3 x$
(b) $2\cot x^2 - 4x^2\operatorname{cosec}^2 x^2$
(c) $2x\cot x^2$
(d) $-2\operatorname{cosec}^2 x$

46 Let $f(x)$ be a polynomial in x. Then, the second order derivative of $f(e^x)$ is

(a) $f''(e^x)\cdot e^x + f'(e^x)$
(b) $f'''(e^x)\cdot e^{2x} + f'(e^x)\cdot e^{2x}$
(c) $f''(e^x)\cdot e^{2x}$
(d) $f''(e^x)\cdot e^{2x} + f'(e^x)\cdot e^x$

47 If $h(x)$ is inverse of $g(x)$ and $g'(x) = \dfrac{1}{1+x^3}$, then $h''(x)$ is equal to

(a) $\dfrac{-3[h(x)]^2}{[1+\{h(x)\}^3]^3}$ (b) $\dfrac{-3[h(x)]^2}{[1+\{h(x)\}^3]^2}$ (c) $\dfrac{-3[h(x)]^2}{[1+\{h(x)\}^3]}$ (d) $3h^2(x)[1+h^3(x)]$

4.8 Miscellaneous

48 In a triangle, if the sides a and b are constant and the base angle A and B vary, then $\dfrac{dA}{dB}$ can be written as

(a) $\sqrt{\dfrac{a^2 - b^2\sin^2 A}{b^2 + a^2\sin^2 A}}$
(b) $\sqrt{\dfrac{a^2 - b^2\sin^2 A}{b^2 - a^2\sin^2 B}}$
(c) $\sqrt{\dfrac{a^2 + b^2\sin^2 A}{b^2 + a^2\sin^2 B}}$
(d) None of these

49 Let f be a differentiable function satisfying $f(x) + f(y) + f(z) + f(x)\cdot f(y)\cdot f(z) = 14\ \forall\ x, y, z \in R$, then

(a) $f'(x) < 0\ \forall\ x \in R$
(b) $f'(x) = 0\ \forall\ x \in R$
(c) $f'(x) > 0\ \forall\ x \in R$
(d) None of these

50 If $f(1) = 1$ and $f'(1) = 3$, then the derivative of $f[f\{f(x)\}] + [f(x)]^2$ at $x = 1$ is **[JEE Main 2019]**

(a) 15 (b) 12 (c) 9 (d) 33

51 If $f(x)=\begin{vmatrix} x^3 & x^4 & 3x^2 \\ 1 & -6 & 4 \\ p & p^2 & p^3 \end{vmatrix}$, where p is any constant, then $\frac{d^3}{dx^3}[f(x)]$ is

(a) proportional to x^3 (b) proportional to x
(c) proportional to x^2 (d) a constant

52 If $f(x)=\begin{vmatrix} x^2 & x^3 & x^4 \\ 2x & 3x^2 & 4x^3 \\ 2 & 6x & 12x^2 \end{vmatrix}$, then $f'(x)$ at $x=1$ is

(a) 24 (b) 12
(c) 0 (d) None of these

53 Let $f(x)=\begin{vmatrix} \cos x & x & 1 \\ 2\sin x & x^2 & 2x \\ \tan x & x & 1 \end{vmatrix}$. Then, $\lim\limits_{x\to 0}\frac{f'(x)}{x}$ is equal to

(a) 2 (b) −2 (c) −1 (d) 1

54 If $u=f(x^3)$, $v=g(x^2)$, $f'(x)=\cos x$ and $g'(x)=\sin x$, then $\frac{du}{dv}$ is

(a) $\frac{3}{2}x\cos x^3 \operatorname{cosec} x^2$ (b) $\frac{3}{2}x\sin x^3 \sec x^2$
(c) $\tan x$ (d) None of these

55 Differentiate $\sin^{-1}(4x\sqrt{1-4x^2})$ w.r.t. to $\sqrt{1-4x^2}$, if $x\in\left(-\frac{1}{2},-\frac{1}{2\sqrt{2}}\right)$

(a) $\frac{1}{x}$ (b) $-\frac{1}{x^{-2}}$
(c) $\frac{1}{x^2}$ (d) $-\frac{1}{x^2}$

56 Let $g(x)=\log_e[f(x)]$, where $f(x)$ is a twice differentiable positive function on $(0,\infty)$ such that $f(x+1)=x\,f(x)$. Then, for $N=1,2,3,\ldots$ $g''\left(N+\frac{1}{2}\right)-g''\left(\frac{1}{2}\right)$ is equal to

(a) $-4\left\{1+\frac{1}{9}+\frac{1}{25}+\ldots\ldots+\frac{1}{(2N-1)^2}\right\}$

(b) $4\left\{1+\frac{1}{9}+\frac{1}{25}+\ldots\ldots+\frac{1}{(2N-1)^2}\right\}$

(c) $-4\left\{1+\frac{1}{9}+\frac{1}{25}+\ldots\ldots+\frac{1}{(2N+1)^2}\right\}$

(d) $4\left\{1+\frac{1}{9}+\frac{1}{25}+\ldots\ldots+\frac{1}{(2N+1)^2}\right\}$

57 The derivative of $\log_{10}x$ w.r.t. x^3 is

(a) $\frac{1}{3x^3}\log_e 10$ (b) $\frac{3}{x^3}\log_{10}e$
(c) $\frac{1}{3x^3}\log_{10}e$ (d) None of these

Numerical Value Type Questions

58 If $y=\dfrac{5^{\log_{4\sqrt{5}}x}-9^{\log_{729}(x+1)^6}}{49^{4\log_{2401}x}-x-1}$ and $\dfrac{dy}{dx}=ax-b$, then the value of $a+b$ is ....

59 The value of x, $x\in(2,\infty)$, where $f(x)=\sqrt{x+\sqrt{8x-16}}+\sqrt{x-\sqrt{8x-16}}$ is not differentiable is ....

60 Let $f:R\to R$ be defined as $f(x)=x^3+3x^2+6x-5+4e^{2x}$ and $g(x)=f^{-1}(x)$, then $28g'(-1)$ is ....

61 If $g(x)=(ax^2+bx+c)\sin x+(dx^2+ex+f)\cos x$ is such that $g'(x)=x^2\sin x$. Then, the value of $a+b+c+d+e+f$ is ....

62 If $y=\sin^{-1}\ln\left(\dfrac{x^2}{2}\right)$ and the value of $(xy')^2\left(\left(\ln\dfrac{x^2}{2}\right)^2-1\right)$ is k, then $|k|$ is equal to

63 If $y=\sum\limits_{k=1}^{6}k\cos^{-1}\left\{\dfrac{3}{5}\cos kx-\dfrac{4}{5}\sin kx\right\}$, then $\dfrac{dy}{dx}$ at $x=0$ is .... **[JEE Main 2020]**

64 If $\log((e-1)e^{xy}+x^2)=x^2+y^2$, then $\dfrac{dy}{dx}$ at (1, 0) is ....

65 If $y\sqrt{x^2+1}=\log(\sqrt{x^2+1}-x)$, then $(x^2+1)\dfrac{dy}{dx}+xy+k=0$, then the value of k is ....

66 If $4xe^{xy}=y+5\sin^2x$, then the value of $y'(0)$ is ....

67 If $x^2+y^2=t-\dfrac{1}{t}$ and $x^4+y^4=t^2+\dfrac{1}{t^2}$, then the value of $x^my\dfrac{dy}{dx}-1=0$, then m is equal to ....

68 If $(\sin x)(\cos y)=\dfrac{1}{2}$, then the value of $\left|\dfrac{d^2y}{dx^2}\right|$ at $\left(-\dfrac{\pi}{4},\dfrac{\pi}{4}\right)$ is ....

69 If $y=y(x)$ is an implicit function of x such that $\log_e(x+y)=4xy$, then $\dfrac{d^2y}{dx^2}$ at $x=0$ is equal to **[JEE Main 2021]**

70 $P=a^2\cos^2\theta+b^2\sin^2\theta$ and if $P+\dfrac{d^2P}{d\theta^2}=2a^2+2b^2-kP$, then the value of k is ....

71 If $y=e^{-2x}\cos 3x$ and $\dfrac{d^2y}{dx^2}+a\dfrac{dy}{dx}+by=0$, then the value of $a+b$ is ....

72 Differential coefficient of $\sec^{-1}\left(\dfrac{1}{2x^2-1}\right)$ w.r.t. $\sqrt{1-x^2}$ at $x=\dfrac{1}{2}$ is ....

73 If $y(x)=\begin{vmatrix}\sin x & \cos x & \sin x+\cos x+1\\ 23 & 17 & 13\\ 1 & 1 & 1\end{vmatrix}$, $x\in R$, then $\dfrac{d^2y}{dx^2}+y$ is ....

74 Let $f:(-1,1)\to R$ be the differentiable function with $f(0)=-1$ and $f'(0)=1$. If $g(x)=[f\{2f(x)\}+2]^2$, then $|g'(0)|$ is ....

75 A function f satisfies the relation $f(x)=f'(x)+f''(x)+\ldots\ldots\infty$ terms, where $f(x)$ is differentiable indefinitely. If $f'(-1)=1$, then $f(-1)$ is equal to ....

76 The derivative of $\tan^{-1}\dfrac{\sqrt{1+x^2}-1}{x}$ w.r.t. $\tan^{-1}\dfrac{2x\sqrt{1-x^2}}{1-2x^2}$ at $x=\dfrac{1}{2}$ is ....

LEVEL 03 \ JEE Advanced Level

Single Choice Correct Questions

1 If y is a function of x and $\log(x+y)=3xy$, then the value of $y'(0)$ is

(a) 1 (b) −1 (c) 2 (d) 0

2 If $f(x)=(2x-3\pi)^5+\dfrac{4x}{3}$ and g is the inverse of f, then $g'(2\pi)$ is

(a) $\dfrac{4}{3}$ (b) $\dfrac{3}{4}$ (c) $\dfrac{30\pi^4+4}{3}$ (d) $\dfrac{3}{30\pi^4+4}$

3 Let $f(\alpha)=\cos\left[\cot^{-1}\left(\dfrac{\cos\alpha}{\sqrt{-\cos 2\alpha}}\right)\right]$, where $\dfrac{\pi}{4}<\alpha<\dfrac{\pi}{2}$, then the value of $\dfrac{d[f(\alpha)]}{d(\cot\alpha)}$ is

(a) 6 (b) 5 (c) 3 (d) 1

4 If $y=\dfrac{\cos 6x+6\cos 4x+15\cos 2x+10}{\cos 5x+5\cos 3x+10\cos x}$, then $\dfrac{dy}{dx}$ is equal to

(a) $2\sin x+\cos x$ (b) $-2\sin x$ (c) $\cos 2x$ (d) $\sin 2x$

5 If $y=\sqrt{(a-x)(x-b)}-(a-b)\tan^{-1}\sqrt{\dfrac{a-x}{x-b}}$, then $\dfrac{dy}{dx}$ is equal to

(a) 1 (b) $\sqrt{\dfrac{a-x}{x-b}}$

(c) $\sqrt{(a-x)(x-b)}$ (d) $\dfrac{1}{\sqrt{(a-x)(b-x)}}$

6 If $f(x)=|x-1|$ and $g(x)=f(f(f(x)))$, then for $x>2$, $g'(x)$ is equal to

(a) $1\ \forall\ x>2$ (b) 1 for $2<x<3$ (c) −1 for $2<x<3$ (d) None of these

7 If $e^x=\dfrac{\sqrt{1+t}-\sqrt{1-t}}{\sqrt{1+t}+\sqrt{1-t}}$ and $\tan\dfrac{y}{2}=\sqrt{\dfrac{1-t}{1+t}}$, then $\dfrac{dy}{dx}$ at $t=\dfrac{1}{2}$ is

(a) $-\dfrac{1}{2}$ (b) $\dfrac{1}{2}$ (c) 0 (d) None of these

8 If $f(x)=\sin^{-1}\{[3x+2]-\{3x+(x-\{2x\})\}\}$, $x\in\left(0,\dfrac{\pi}{12}\right)$ and $gof(x)=x\ \forall\ x\in\left(0,\dfrac{\pi}{12}\right)$, then $g'\left(\dfrac{\pi}{6}\right)$ is equal to [**Note** $\{y\}$ and $[y]$ denote fractional part function and greatest integer function, respectively]

(a) $\dfrac{\sqrt{3}}{8}$ (b) $\dfrac{-1}{4}$ (c) $\dfrac{1}{8}$ (d) $\dfrac{-\sqrt{3}}{4}$

9 Let $f(x)$ and $g(x)$ be two functions having finite non-zero third order derivatives $f'''(x)$ and $g'''(x)$ for all $x\in R$. If $f(x)\cdot g(x)=1$ for all $x\in R$, then $\dfrac{f'''}{f'}-\dfrac{g'''}{g'}$ is equal to

(a) $3\left(\dfrac{f''}{g}-\dfrac{g''}{f}\right)$ (b) $3\left(\dfrac{f''}{f}-\dfrac{g''}{g}\right)$ (c) $3\left(\dfrac{g''}{g}-\dfrac{f''}{g}\right)$ (d) $3\left(\dfrac{f''}{g}-\dfrac{g''}{f}\right)$

10 If a curve is given by $x = a\cos t + \frac{b}{2}\cos 2t$ and $y = a\sin t + \frac{b}{2}\sin 2t$, then the point for which $\frac{d^2y}{dx^2} = 0$ are given by

(a) $\sin t = \frac{2a^2 + b^2}{3ab}$ (b) $\cos t = -\left[\frac{a^2 + 2b^2}{3ab}\right]$ (c) $\tan t = \frac{a}{b}$ (d) None of these

11 If $y^2 = ax + bx + c$, then $y^3 \frac{d^2y}{dx^2}$ is

(a) a constant (b) a function of x only

(c) a function of y only (d) a function of x and y

More than One Correct Options

12 If $y = \log_7(\log_7 x)$, then $\frac{dy}{dx}$ is equal to

(a) $\frac{\log_7 e}{x\log_e x}$ (b) $\frac{1}{x\log_e x \cdot \log_e 7}$ (c) $\frac{1}{\log_e (7x)^x}$ (d) $\frac{\log_7 e \cdot \log_x e}{x}$

13 If $y = \frac{\sec x + \tan x}{(\sec x - \tan x)}$, then $\frac{dy}{dx}$ is

(a) $2\sec x(\sec x + \tan x)^2$ (b) $2\cos^3 x(1 + \sin x)^2$

(c) $2\sec x(\sec x - \tan x)^2$ (d) $2\sec^3 x(1 + \sin x)^2$

14 If $x = t + t^3$ and $y = t^2$, where $t \in R$, then for what value(s) of t is $\frac{dy}{dx} = \frac{1}{2}$?

(a) $\frac{1}{3}$ (b) 2 (c) 3 (d) 1

15 If $x = a\left[\cos t + \log\tan\frac{t}{2}\right]$ and $y = a\sin t$, then

(a) $\frac{dy}{dx} = 0$ (b) $\left(\frac{dy}{dx}\right)_{t=\frac{\pi}{2}} = \infty$ (c) $\left(\frac{d^2y}{dx^2}\right)_{t=\frac{\pi}{4}} = \frac{2\sqrt{2}}{a}$ (d) None of these

16 If $f_K(x) = \log_e[f_{K+1}(x)]$ for all $n \in N$ and $f_1(x) = x$, then $\frac{d}{dx}(f_n(x))$ is equal to

(a) $f_n(x)\frac{d}{dx}\{f_{n-1}(x)\}$ (b) $f_n(x)f_{n-1}(x)$

(c) $f_n(x) \cdot f_{n-1}(x) \ldots\ldots f_2(x) \cdot f_1(x)$ (d) None of these

17 Let $f(x) = x\sin\pi x$, $x > 0$, then for all natural number n, $f'(x)$ vanishes at

(a) a unique point in the interval $\left(n, n + \frac{1}{2}\right)$ (b) a unique point in the interval $\left(n + \frac{1}{2}, n + 1\right)$

(c) a unique point in the interval $(n, n + 1)$ (d) two points in the interval $(n, n + 1)$

18 If a function represented parametrically by the equation $x = \frac{1 + \log t}{t^2}$ and $y = \frac{3 + 2\log t}{t}$, then which statement is true.

(a) $y''(x - 2xy') = y$ (b) $yy' = 2x(y')^2 + 1$

(c) $xy' = 2y(y')^2 + 2$ (d) $y''(y - 4xy') = (y')^2$

19 Which of the following statements are correct?

(a) If $f(x) = |x-2|$, then $f[f(x)]' = 1$ for $x > 20$

(b) If $f(0) = a, f'(0) = b, g(0) = 0$ and $(fog)'(0) = c$, then $g'(0) = \frac{c}{b}$

(c) The differential coefficient of $f(\log x)$, where $f(x) = \log x$ is $\frac{1}{x \log x}$

(d) Differential coefficient of $2\tan^{-1} x$ with respect to $\sin^{-1} \frac{2x}{1+x^2}$ at $x = \frac{1}{2}$ is 1

20 Let $f(x) = 3x^2 + 4x\,g'(1) + g''(2)$ and $g(x) = 2x^2 + 3xf'(2) + f''(3)\ \forall x \in R$, then

(a) $f'(1) = 22 + 12f'(2)$ (b) $g'(2) = 44 + 12g'(1)$

(c) $f''(3) + g''(2) = 10$ (d) None of these

21 Let $f: R \to R, g: R \to R$ and $h: R \to R$ be differentiable function such that $f(x) = x^3 + 3x + 2$, $g[f(x)] = x$ and $h[g(g(x))] = x\ \forall x \in R$, then

(a) $g'(2) = \frac{1}{15}$ (b) $h'(1) = 666$ (c) $h(0) = 16$ (d) $h[g(3)] = 36$

Linked Comprehension Type

Paragraph I Directions (Q.Nos. 22-24)

Let x be the independent variable and y be a dependent variable defined as $y = f(x)$.

Assuming $y^{\frac{1}{m}} + y^{-\frac{1}{m}} = 2x$, where $m \in R - \{0\}$

22 The value of y is given by

(a) $(x - \sqrt{x^2-1})^m$ (b) $(x + \sqrt{x^2-1})^m$ (c) $(x + \sqrt{x^2+1})^m$ (d) $(x - \sqrt{x^2+1})^m$

23 $\frac{dy}{dx}$ is equal to

(a) $\frac{my}{\sqrt{x^2+1}}$ (b) $\frac{my}{\sqrt{x^2-1}}$ (c) $\frac{\sqrt{x^2-1}}{my}$ (d) $\frac{-\sqrt{x^2-1}}{my}$

24 $(x^2-1)\frac{d^2y}{dx^2} + x\frac{dy}{dx}$ is equal to

(a) my (b) m^2y (c) $\frac{m^2}{y}$ (d) my^2

Paragraph II Directions (Q.Nos. 25-27)

A curve is represented parametrically by the equations $x = f(t) = a^{\ln(b^t)}$ and $y = g(t) = b^{-\ln(a^t)}$, $a, b > 0$ and $a \neq 1, b \neq 1$ where $t \in R$.

25 Which of the following is not a correct expression for $\frac{dy}{dx}$?

(a) $-\frac{1}{f(t)}$ (b) $-g(t)$ (c) $\frac{-g(t)}{f(t)}$ (d) $\frac{f(t)}{g(t)}$

26 The value of $\frac{d^2y}{dx^2}$ at the point, where $f(t) = g(t)$ is

(a) 0 (b) $-\frac{1}{2}$ (c) 2 (d) 3

27 The value of $\frac{f(t)}{f'(t)} \cdot \frac{f''(-t)}{f'(-t)} + \frac{f(-t)}{f'(-t)} \cdot \frac{f''(t)}{f'(t)} \forall t \in R$ is equal to

(a) 2 (b) $\frac{1}{2}$ (c) −2 (d) 4

Paragraph III Directions (Q.Nos. 28-30)

Sometime a function can be written as the product of two functions such that it is easy to determine the *n*th derivatives of two functions separately. In such cases, the *n*th derivative of the product can be written by applying Leibnitz's theorem, which can be stated as below.

If *u* and *v* are functions of *x* having derivative of *n*th order and $y = f \cdot g$, then

$y_n = f_n g + {}^nC_1 f_{n-1} g_1 + {}^nC_2 f_{n-2} g_2 + \ldots\ldots + {}^nC_k f_{n-k} g_k \ldots\ldots {}^nC_n f g_n$, where suffixes denotes derivative w.r.t. '*x*'.

28 If $y = x^5 e^{2x}$, then $y_{10}(0)$ is equal to

(a) 2^{10} (b) 315×2^{10}

(c) 195×2^{10} (d) 315×2^{12}

29 If $y = x^2 \cos x$, then $y_7\left(\frac{\pi}{2}\right)$ is equal to

(a) $\frac{\pi^2}{2} - 72$ (b) $\frac{\pi^2}{4} - 56$

(c) 0 (d) $\frac{\pi^2}{4} + 56$

30 If $y = \frac{\sin^{-1} x}{\sqrt{1-x^2}}$, then $y_{n+2}(0)$ is equal to

(a) $(n+1)^2 y_n(0)$ (b) $n^2 y_n(0)$

(c) $(n-1)^2 y_n(0)$ (d) $(1-n)^2 y_n(0)$

Matrix Match Type Questions

31 Match the following

	Column I		Column II
(i)	Let $y(x) = \cos(3\cos^{-1} x)$, $x \in [-1, 1]$ and $x \pm \frac{\sqrt{3}}{2}$. Then, $\frac{1}{y(x)}\left\{(x^2-1)\frac{d^2y(x)}{dx^2} + x\frac{dy(x)}{dx}\right\}$ equals	(A)	10
(ii)	If $u = x^2 + y^2$ and $x = s + 3t$, $y = 2s - t$, then $\frac{d^2u}{ds^2}$ is	(B)	-2
(iii)	If $y = f(x)$ is given by $x = t^5 - 5t^3 - 20t + 7$ and $y = 4t^3 - 3t^2 - 18t + 3$, then $-5\frac{dy}{dx}$ at $t = 1$ is	(C)	1
(iv)	If $y = \sin^{-1}(\cos x)$, $x \in (\pi, 2\pi)$, then $\frac{dy}{dx}$ is	(D)	9

32 Match the following

	Column I		Column II
(i)	If $f(x)$ is a polynomial satisfying the relation $f(x)+f(2x)=5x^2-18$, then $f'(-1)$ equals	(A)	2
(ii)	Let $y=x^3-2$ and $x=3z^2+5$, then the value of $\frac{dy}{dz}$ at $z=0$ is	(B)	1
(iii)	Let $p(x)$ be a polynomial of degree 4 with $p(2)=-1$, $p'(2)=0$, $p''(2)=2$, $p'''(2)=-12$ and $p''''(2)=24$. Then, $p''(3)$ is	(C)	-2
(iv)	Let $y=f(x)$ and $x=\frac{1}{z}$. If $\frac{d^2y}{dx^2}=2\lambda z^3\frac{dy}{dz}+z^4\frac{d^2y}{dz^2}$, then value of λ is	(D)	0

33 Match the following

	Column I		Column II
(i)	The derivative of $\tan^{-1}\left(\frac{2x}{1-x^2}\right)$ w.r.t. $\sin^{-1}\left(\frac{2x}{1+x^2}\right)$ is	(A)	-2
(ii)	If $y=e^x(a\cos x+b\sin x)$, where a and b are constants and $y''-2y'+\lambda y=0$, then λ equals	(B)	4
(iii)	If $y=(\tan^{-1}x)^2$, then $(x^2+1)^2y''+2x(x^2+1)y'+\lambda=0$, then λ equals	(C)	2
(iv)	If $y=e^{\sqrt{x}}+e^{-\sqrt{x}}$ and $kx\frac{d^2y}{dx^2}+2\frac{dy}{dx}=y$, then the value of k is	(D)	1

34 Match the following

	Column I		Column II
(i)	Let $f(x)=x^n$, n being a non-negative integer. The number of values of n for which the equality $f'(a+b)=f'(a)+f'(b)$ is valid for $a, b>0$ is	(A)	$\frac{1}{\sqrt{2}}$
(ii)	The derivative of $f(\tan x)$ w.r.t. $g(\sec x)$ at $x=\frac{\pi}{4}$, where $f'(1)=2$ and $g'(\sqrt{2})=4$, is	(B)	-4
(iii)	Let $f(x)=\sin x$, $g(x)=2x$ and $h(x)=\cos x$, if $\phi(x)=[go(fh)](x)$, then $\phi''\left(\frac{\pi}{4}\right)$ is	(C)	2
(iv)	The function y is defined by the equation $xy-\log y=1$ satisfies $x(y\cdot y''+y'^2)-y''+k\cdot yy'=0$, the value of k is	(D)	3

35 Match the following.

	Column I		Column II
(i)	If $x^a y^b = (x+y)^{a+b}$, then $\frac{dy}{dx}$ is	(A)	$\frac{y}{x}$
(ii)	If $y = \cot^{-1}\left(\frac{\sqrt{1+\sin x}+\sqrt{1-\sin x}}{\sqrt{1+\sin x}-\sqrt{1-\sin x}}\right)$, when $x \in \left(0, \frac{\pi}{2}\right)$, then $\frac{dy}{dx}$ is	(B)	$\frac{x^2}{y^2}$
(iii)	If $\sqrt{1-x^6}+\sqrt{1-y^6} = a(x^3 - y^3)$ and $\frac{dy}{dx} = f(x, y)\sqrt{\frac{1-y^6}{1-x^6}}$, then $f(x, y)$ is	(C)	independent of x
(iv)	If $x = t - \frac{1}{t}$ and $y = t + \frac{1}{t}$, then $\frac{dy}{dx}$ is equal to	(D)	$\frac{x}{y}$

Numerical Value Type Questions

Question Stem Directions (Q.Nos. 36-38) The answer to each question is a **NUMERICAL VALUE.**

A non-zero polynomial with coefficient has property that $f(x) = f'(x) \cdot f''(x)$.

36 The degree of polynomial $f(x)$ is ....

37 The leading coefficient of $f(x)$ is ....

38 If $f(0) = 12$, then $\left(\frac{-1}{2}\right)$ times sum of the roots of the equation $f(x) = 0$ is ....

Question Stem Directions (Q.Nos. 39-41) The answer to each question is a **NUMERICAL VALUE.**

If $f(x)$ is a polynomial satisfying $f(x) = x^3 + x^2 f'(1) + x f''(2) + f'''(3)\ \forall\ x \in R$, then

39 The value of $f(1)$ is ....

(a) 6 (b) 8 (c) 4 (d) -2

40 $f''(2) - 2f'(1)$ is equal to ....

41 $f(0) + f(3)$ is equal to ....

42 If $f : R \to R$ is a differentiable function with $f(0) = 1$ and satisfying the equation $f(x+y) = f(x) \cdot f'(y) + f'(x) \cdot f(y)\ \forall\ x, y \in R$, then the value of $\log_e [f(4)]$ is ....

43 Let $g(x) = f(x) \sin x$, where $f(x)$ is twice differentiable function on $(-\infty, \infty)$ such that $f'(-\pi) = 1$. Then, the value of $|g''(-\pi)|$ is ....

44 If $(y - x^5)^2 = x(1+x^2)^2$, then the value of $\left(\frac{dy}{dx}\right)$ at (1, 3) is ....

45 If $y = \frac{x^2}{2} + \frac{1}{2} x\sqrt{x^2+1} + \ln\sqrt{x + \sqrt{x^2+1}}$, then $xy' + \ln y' = ky$, then value of k is ....

46 If $f\left(\frac{x+y}{3}\right) = \frac{2 + f(x) + f(y)}{3}$ for all real x and y and $f'(2) = 2$, then the value of $f(3)$ is ....

47 If $F(x)=\left[f\left(\frac{x}{2}\right)\right]^2+\left[g\left(\frac{x}{2}\right)\right]^2$, where $f''(x)=-f(x)$ and $g(x)=f'(x)$ and given that $F(5)=5$, then $F(10)$ is equal to ....

48 Prove that if $(a+bx)\,e^{y/x}=x$ and $x^k\frac{d^2y}{dx^2}=\left(x\frac{dy}{dx}-y\right)^2$. Then, the value of k is ....

49 Let $f(x)=\cos^{-1}(4x^3-3x)$. If $f'(0)=S$ and $f'\left(-\frac{4}{5}\right)=K$ and $f'\left(\frac{4}{5}\right)=G$, then the value of $2S-3K+2G$ is ....

50 If $y=\frac{5x}{\sqrt[3]{(1-x)^2}}+\cos^2(2x+1)$, then $\frac{dy}{dx}=\frac{A\,(3-x)}{3\,(1-x)^{5/3}}-C\sin(Bx+2)$, then $A+B+C$ is equal to ....

51 Let $P(x)$ be a polynomial of degree 4 such that $P(1)=P(3)=P(5)=P'(7)=0$. If the real number $x\neq 1$, 3 and 5 is such that $P(x)=0$ can be expressed as $x=\frac{p}{q}$, where p and q are coprimes, then $p+q$ is equal to ....

52 If $y=\frac{a+bx^{3/2}}{x^{5/4}}$ and $\frac{dy}{dx}$ vanishes, when $x=5$, then $\left(\frac{a}{b}\right)^2$ is equal to ....

53 Let $f(x)=\exp(x^3+x^2+x)$ for any real number x and let g be the inverse of f, then the value of $g'(e^3)=\frac{1}{ke^3}$, then k is equal to ....

54 If $f(x)=\sin\left[\cos^{-1}\left(\frac{1-2^{2x}}{1+2^{2x}}\right)\right]$ and its first derivative with respect to x is $-\frac{b}{a}\log 2$, when $x=1$, where a and b are integers, then the minimum value of $|a^2-b^2|$ is .... **[JEE Main 2021]**

55 If a curve is represented parametrically by the equations

$$x=\sin\left(t+\frac{7\pi}{12}\right)+\sin\left(t-\frac{\pi}{12}\right)+\sin\left(t+\frac{3\pi}{12}\right) \text{ and}$$

$$y=\cos\left(t+\frac{7\pi}{12}\right)+\cos\left(t-\frac{\pi}{12}\right)+\cos\left(t+\frac{3\pi}{12}\right)$$, then the value of $\frac{d}{dt}\left(\frac{x}{y}-\frac{y}{x}\right)$ at $t=\frac{\pi}{8}$ is ....

ANSWER KEY

Level 1

1. (i) $-\left(\frac{1}{x^2}+\frac{4}{x^3}+\frac{9}{x^4}\right)$ (ii) $\frac{1}{2\sqrt{x}}+\frac{3}{2}\sqrt{x}+\frac{5}{2}x\sqrt{x}$ (iii) $4\cos(4x+3)$

(iv) $-\frac{1}{2\sqrt{x}}\operatorname{cosec}^2\sqrt{x}$ (v) $\frac{x\sec^2 x^2}{\sqrt{\tan x^2}}$

(vi) $6(4x-3)(2x^2-3x+4)^5$ (vii) $\frac{x\cos x+\sin x}{2\sqrt{x\sin x}}$ (viii) $\frac{-\operatorname{cosec}^2 x}{4\sqrt{x\cot\sqrt{x}}}$

(ix) $\frac{-a}{2\sqrt{ax+b}}(\sin(\sin\sqrt{ax+b}))\cos\sqrt{ax+b}$ (x) $\frac{-2a^2x}{(a^2-x^2)^{1/2}(a^2+x^2)^{3/2}}$ (xi) $\frac{1}{\sqrt{x^2+1}}$

2. (i) $\frac{1}{\sqrt{a^2-x^2}}$ (ii) $\frac{4x^3}{\sin^{-1}x^4\sqrt{1-x^8}}$ (iii) $\frac{1}{2}$ (iv) -2

(v) $\frac{3}{\sqrt{1-x^2}}$ (vi) $\frac{1}{1+x^2}$ (vii) $\frac{-x}{\sqrt{1-x^2}}$ (viii) $\frac{-1}{\sqrt{1-x^2}}$ (ix) $\frac{-1}{2\sqrt{1-x^2}}$

3. (i) $\frac{-y^{1/3}}{x^{1/3}}$ (ii) $\frac{y^2-2xy}{x^2-2xy}$ (iii) $\frac{2y\sin 2x+\sin 2y}{\cos 2x-2x\cos 2y}$ (iv) $\frac{x-y}{x(1+\log x)}$

(v) $\frac{y}{x}$ (vi) $-e^{y-x}$ (vii) -1

4. (i) $\frac{y^2}{x(1-y\log x)}$ (ii) $(\cos x)^{\sin x}\left[\frac{-\sin^2 x}{\cos x}+\cos x\log\cos x\right]$

(iii) $\frac{1}{2}\sqrt{(x-1)(x-2)(2x-3)}\left[\frac{1}{x-1}+\frac{1}{x-2}+\frac{2}{2x-3}\right]$ (iv) $\frac{y(x\log y-y)}{x(y\log x-x)}$

(v) $\frac{\log x}{(1+\log x)^2}$ (vi) $\left(x+\frac{1}{x}\right)^x\left[\frac{x^2-1}{x^2+1}+\log\left(x+\frac{1}{x}\right)\right]$

5. (i) $\frac{1}{t}$ (ii) $\cot\frac{t}{2}$ (iii) $\frac{6t+2}{3t^2}$ (iv) $4(1+t^2)\sec 4t\tan 4t$

(v) $\frac{-y\log x}{x\log y}$ (vi) $\frac{b}{a}\tan\theta$

6. (i) $\frac{1}{2y-1}$ (ii) $\frac{\cos x}{2y-1}$ (iii) $\frac{y^2}{x(1-y\log x)}$ (iv) $\frac{y}{1-y}$

(v) $\frac{1}{y+1}$ (vi) $\frac{1}{y+1}$

7. (i) $\frac{-3+2\log x}{x^3}$ (ii) $2e^x(3\cos 3x-4\sin 3x)$ (iii) $(2-x^2)\cos x-4x\sin x$

(iv) $\frac{-2x}{(1+x^2)^2}$ (v) $\frac{-(1+\log x)}{(x\log x)^2}$

8. $2|x|$

9. 0

11. $\frac{3}{9x^2-6x+2}$

13. $\frac{1}{\sqrt[3]{4}\sqrt{3}}$

14. $-\cot x$

15. x^x

Level 2

1. (a) **2.** (c) **3.** (b) **4.** (b) **5.** (d) **6.** (b) **7.** (d)
8. (d) **9.** (a) **10.** (c) **11.** (a) **12.** (c) **13.** (c) **14.** (a)
15. (b) **16.** (c) **17.** (b) **18.** (b) **19.** (d) **20.** (b) **21.** (c)
22. (a) **23.** (c) **24.** (b) **25.** (d) **26.** (d) **27.** (a) **28.** (d)
29. (a) **30.** (c) **31.** (d) **32.** (b) **33.** (b) **34.** (d) **35.** (b)
36. (b) **37.** (d) **38.** (b) **39.** (c) **40.** (b) **41.** (a) **42.** (a)
43. (a) **44.** (c) **45.** (b) **46.** (d) **47.** (d) **48.** (b) **49.** (b)
50. (d) **51.** (b) **52.** (b) **53.** (b) **54.** (a) **55.** (a) **56.** (a)
57. (c) **58.** (1) **59.** (4) **60.** (2) **61.** (1) **62.** (4) **63.** (91)
64. (2) **65.** (1) **66.** (4) **67.** (3) **68.** (4) **69.** (40) **70.** (3)
71. (17) **72.** (4) **73.** (6) **74.** (4) **75.** (2) **76.** (0.1732)

Level 3

1. (c) **2.** (b) **3.** (d) **4.** (b) **5.** (b) **6.** (c) **7.** (a)
8. (d) **9.** (b) **10.** (b) **11.** (a) **12.** (a, b, d) **13.** (a, d) **14.** (a, d)
15. (b, c) **16.** (a, c) **17.** (b, c) **18.** (b, d) **19.** (a, b, c, d) **20.** (a, b, c) **21.** (b, c)
22. (b) **23.** (b) **24.** (b) **25.** (c) **26.** (c) **27.** (a) **28.** (d)
29. (b) **30.** (a)
31. (i)–D, (ii)–A, (iii)–B, (iv)–C
32. (i)–C, (ii)–D, (iii)–A, (iv)–B
33. (i)–D, (ii)–C, (iii)–A, (iv)–B
34. (i)–C, (ii)–A, (iii)–B, (iv)–D
35. (i)–A, (ii)–C, (iii)–B, (iv)–D
36. (3) **37.** (0.05) **38.** (9) **39.** (4) **40.** (12) **41.** (0) **42.** (2)
43. (2) **44.** (8) **45.** (2) **46.** (8) **47.** (5) **48.** (3) **49.** (11)
50. (11) **51.** (100) **52.** (5) **53.** (6) **54.** (481) **55.** (8)

SOLUTIONS

Level 1

1. (i) We have, $y = \frac{1}{x} + \frac{2}{x^2} + \frac{3}{x^3}$

$$y = x^{-1} + 2x^{-2} + 3x^{-3}$$

On differentiating w.r.t. x, we get

$$\frac{dy}{dx} = -x^{-2} - 4x^{-3} - 9x^{-4}$$

$$= -\left(\frac{1}{x^2} + \frac{4}{x^3} + \frac{9}{x^4}\right)$$

(ii) We have, $y = \sqrt{x} + x\sqrt{x} + x^2\sqrt{x}$

$$y = x^{1/2} + x^{3/2} + x^{5/2}$$

On differentiating w.r.t. x, we get

$$\frac{dy}{dx} = \frac{1}{2}x^{-1/2} + \frac{3}{2}x^{1/2} + \frac{5}{2}x^{3/2}$$

$$= \frac{1}{2\sqrt{x}} + \frac{3}{2}\sqrt{x} + \frac{5}{2}x\sqrt{x}$$

(iii) We have, $y = \sin(4x+3)$

On differentiating w.r.t. x, we get

$$\frac{dy}{dx} = \cos(4x+3) \times (4+0)$$

$$\frac{dy}{dx} = 4\cos(4x+3)$$

(iv) We have, $y = \cot\sqrt{x}$

On differentiating w.r.t. x, we get

$$\frac{dy}{dx} = -\operatorname{cosec}^2\sqrt{x} \times \frac{1}{2\sqrt{x}}$$

$$= -\frac{1}{2\sqrt{x}}\operatorname{cosec}^2\sqrt{x}$$

(v) We have, $y = \sqrt{\tan x^2}$

On differentiating w.r.t. x, we get

$$\frac{dy}{dx} = \frac{1}{2\sqrt{\tan x^2}} \times \sec^2 x^2 \times 2x = \frac{x\sec^2 x^2}{\sqrt{\tan x^2}}$$

(vi) We have, $y = (2x^2 - 3x + 4)^6$

On differentiating w.r.t. x, we get

$$\frac{dy}{dx} = 6(2x^2 - 3x + 4)^5 \cdot (4x - 3)$$

$$= 6(4x-3)(2x^2 - 3x + 4)^5$$

(vii) We have, $y = \sqrt{x\sin x}$

On differentiating w.r.t. x, we get

$$\frac{dy}{dx} = \frac{1}{2\sqrt{x\sin x}} \times [x(\cos x) + \sin x(1)]$$

$$\frac{dy}{dx} = \frac{x\cos x + \sin x}{2\sqrt{x\sin x}}$$

(viii) We have, $y = \sqrt{\cot\sqrt{x}}$

On differentiating w.r.t. x, we get

$$\frac{dy}{dx} = \frac{1}{2\sqrt{\cot\sqrt{x}}} \times (-\operatorname{cosec}^2\sqrt{x}) \cdot \frac{1}{2\sqrt{x}}$$

$$\frac{dy}{dx} = \frac{-\operatorname{cosec}^2\sqrt{x}}{4\sqrt{x\cot\sqrt{x}}}$$

(ix) We have, $y = \cos(\sin\sqrt{ax+b})$

On differentiating w.r.t. x, we get

$$\frac{dy}{dx} = -\sin(\sin\sqrt{ax+b}) \cdot \cos\sqrt{ax+b} \cdot \frac{1}{2\sqrt{ax+b}} \times a$$

$$= \frac{-a}{2\sqrt{ax+b}}(\sin(\sin\sqrt{ax+b})) \cdot \cos\sqrt{ax+b}$$

(x) We have, $y = \sqrt{\frac{a^2 - x^2}{a^2 + x^2}}$

On differentiating w.r.t. x, we get

$$\frac{dy}{dx} = \frac{1}{2\sqrt{\frac{a^2 - x^2}{a^2 + x^2}}} \times \left(\frac{(a^2 + x^2)(-2x) - (a^2 - x^2)(2x)}{(a^2 + x^2)^2}\right)$$

$$= \frac{\sqrt{a^2 + x^2}}{2\sqrt{a^2 - x^2}}(2x)\left[\frac{-a^2 - x^2 - a^2 + x^2}{(a^2 + x^2)^2}\right]$$

$$= \frac{\sqrt{a^2 + x^2}}{2\sqrt{a^2 - x^2}}(2x)\left[\frac{-2a^2}{(a^2 + x^2)^2}\right]$$

$$= \frac{-2a^2x}{\sqrt{a^2 - x^2}\,(a^2 + x^2)^{3/2}}$$

$$= \frac{-2a^2x}{(a^2 - x^2)^{1/2}(a^2 + x^2)^{3/2}}$$

(xi) We have, $y = \log(x + \sqrt{x^2 + 1})$

On differentiating w.r.t. x, we get

$$\frac{dy}{dx} = \frac{1}{x + \sqrt{x^2 + 1}} \times \left(1 + \frac{1}{2\sqrt{x^2 + 1}} \cdot 2x\right)$$

$$= \frac{1}{(x + \sqrt{x^2 + 1})}\left(\frac{\sqrt{x^2 + 1} + x}{\sqrt{x^2 + 1}}\right)$$

$$\frac{dy}{dx} = \frac{1}{\sqrt{x^2 + 1}}$$

2. (i) We have, $y=\sin^{-1}\frac{x}{a}$

On differentiating w.r.t. x, we get

$$\frac{dy}{dx}=\frac{1}{\sqrt{1-\left(\frac{x}{a}\right)^2}}\cdot\left(\frac{1}{a}\right)=\frac{1}{\sqrt{\frac{a^2-x^2}{a^2}}}\cdot\frac{1}{a}$$

$$\frac{dy}{dx}=\frac{1}{\sqrt{a^2-x^2}}$$

(ii) We have, $y=\log(\sin^{-1}x^4)$

On differentiating w.r.t. x, we get

$$\frac{dy}{dx}=\frac{1}{\sin^{-1}x^4}\cdot\frac{1}{\sqrt{1-(x^4)^2}}\cdot 4x^3$$

$$=\frac{4x^3}{\sin^{-1}x^4\cdot\sqrt{1-x^8}}$$

(iii) We have, $y=\tan^{-1}\left(\frac{\sin x}{1+\cos x}\right)$

$$y=\tan^{-1}\left(\frac{2\sin\frac{x}{2}\cos\frac{x}{2}}{2\cos^2\frac{x}{2}}\right)=\tan^{-1}\left(\tan\frac{x}{2}\right)$$

$$y=\frac{x}{2}$$

On differentiating w.r.t. x, we get

$$\frac{dy}{dx}=\frac{1}{2}$$

(iv) $y=\tan^{-1}(\cot x)+\cot^{-1}(\tan x)$

$$=\tan^{-1}\left(\tan\left(\frac{\pi}{2}-x\right)\right)+\cot^{-1}\left(\cot\left(\frac{\pi}{2}-x\right)\right)$$

$$=\left(\frac{\pi}{2}-x\right)+\left(\frac{\pi}{2}-x\right)$$

$$y=\pi-2x$$

On differentiating w.r.t. x, we get

$$\frac{dy}{dx}=-2$$

(v) We have, $y=\sin^{-1}(3x-4x^3)$

Put $x=\sin\theta$

$$=\sin^{-1}(3\sin\theta-4\sin^3\theta)$$

$$=\sin^{-1}(\sin 3\theta)=3\theta$$

$$y=3\sin^{-1}x$$

On differentiating w.r.t. x, we get

$$\frac{dy}{dx}=3\cdot\frac{1}{\sqrt{1-x^2}}$$

(vi) $y=\tan^{-1}\left(\frac{1+x}{1-x}\right)$

Put $x=\tan\theta$

$$=\tan^{-1}\left(\frac{1+\tan\theta}{1-\tan\theta}\right)$$

$$=\tan^{-1}\left(\frac{\tan\frac{\pi}{4}+\tan\theta}{1-\tan\frac{\pi}{4}\tan\theta}\right)$$

$$=\tan^{-1}\left(\tan\left(\frac{\pi}{4}+\theta\right)\right)$$

$$y=\frac{\pi}{4}+\theta\Rightarrow y=\frac{\pi}{4}+\tan^{-1}x$$

On differentiating w.r.t. x, we get

$$y=\frac{1}{1+x^2}$$

(vii) We have, $y=\sin\left[2\tan^{-1}\left(\sqrt{\frac{1-x}{1+x}}\right)\right]$

Put $x=\cos\theta$

$$=\sin\left[2\tan^{-1}\sqrt{\frac{1-\cos\theta}{1+\cos\theta}}\right]$$

$$=\sin\left[2\tan^{-1}\sqrt{\frac{2\sin^2\frac{\theta}{2}}{2\cos^2\frac{\theta}{2}}}\right]$$

$$=\sin\left[2\tan^{-1}\left(\tan\frac{\theta}{2}\right)\right]=\sin\left[2\cdot\frac{\theta}{2}\right]$$

$$=\sin\theta=\sin(\cos^{-1}x)$$

$$y=\sqrt{1-x^2}$$

On differentiating w.r.t. x, we get

$$\frac{dy}{dx}=\frac{1}{2\sqrt{1-x^2}}\times(-2x)$$

$$\frac{dy}{dx}=\frac{-x}{\sqrt{1-x^2}}$$

(viii) We have, $y=\cos^{-1}\left(\frac{x-\sqrt{1-x^2}}{\sqrt{2}}\right)$

Let $x=\sin\theta$

$$=\cos^{-1}\left(\frac{\sin\theta-\sqrt{1-\sin^2\theta}}{\sqrt{2}}\right)$$

$$=\cos^{-1}\left(\frac{\sin\theta-\cos\theta}{\sqrt{2}}\right)$$

$$=\cos^{-1}\left(\frac{1}{\sqrt{2}}\sin\theta-\frac{1}{\sqrt{2}}\cos\theta\right)$$

$$=\cos^{-1}\left(\sin\frac{\pi}{4}\sin\theta-\cos\frac{\pi}{4}\cos\theta\right)$$

$$=\cos^{-1}\left[-\left(\cos\frac{\pi}{4}\cos\theta-\sin\frac{\pi}{4}\sin\theta\right)\right]$$

$$= \cos^{-1}\left[-\left(\cos\left(\frac{\pi}{4}+\theta\right)\right)\right]$$

$[\because \cos^{-1}(-x) = \pi - \cos^{-1} x]$

$$= \pi - \cos^{-1}\cos\left(\frac{\pi}{4}+\theta\right)$$

$$= \pi - \left(\frac{\pi}{4}+\theta\right)$$

$$= \frac{3\pi}{4} - \theta$$

$$y = \frac{3\pi}{4} - \sin^{-1} x$$

On differentiating w.r.t. x, we get

$$\frac{dy}{dx} = \frac{-1}{\sqrt{1-x^2}}$$

(ix) We have, $y = \sin^{-1}\left(\frac{\sqrt{1+x}+\sqrt{1-x}}{2}\right)$

Put $x = \cos 2\theta$

$$= \sin^{-1}\left(\frac{\sqrt{1+\cos 2\theta}+\sqrt{1-\cos 2\theta}}{2}\right)$$

$$= \sin^{-1}\left(\frac{\sqrt{2\cos^2\theta}+\sqrt{2\sin^2\theta}}{2}\right)$$

$$= \sin^{-1}\left(\frac{\sqrt{2}\cos\theta+\sqrt{2}\sin\theta}{2}\right)$$

$$= \sin^{-1}\left(\frac{1}{\sqrt{2}}\cos\theta+\frac{1}{\sqrt{2}}\sin\theta\right)$$

$$= \sin^{-1}\left(\sin\frac{\pi}{4}\cos\theta+\cos\frac{\pi}{4}\sin\theta\right)$$

$$= \sin^{-1}\left(\sin\left(\frac{\pi}{4}+\theta\right)\right) = \frac{\pi}{4}+\theta$$

$$y = \frac{\pi}{4}+\frac{1}{2}\cos^{-1} x$$

On differentiating w.r.t. x, we get

$$\frac{dy}{dx} = \frac{1}{2}\left(\frac{-1}{\sqrt{1-x^2}}\right)$$

$$\frac{dy}{dx} = \frac{-1}{2\sqrt{1-x^2}}$$

3. (i) Given, $x^{2/3}+y^{2/3} = a^{2/3}$

On differentiating w.r.t. x, we get

$$\frac{2}{3}x^{-1/3}+\frac{2}{3}y^{-1/3}\frac{dy}{dx} = 0$$

$$\frac{2}{3y^{1/3}}\frac{dy}{dx} = -\frac{2}{3x^{1/3}}$$

$$\frac{dy}{dx} = -\frac{y^{1/3}}{x^{1/3}}$$

(ii) We have, $xy^2 - x^2y - 5 = 0$

On differentiating w.r.t. x, we get

$$x\left(2y\frac{dy}{dx}\right)+y^2(1)-\left[x^2\frac{dy}{dx}+y(2x)\right] = 0$$

$$\Rightarrow \quad 2xy\frac{dy}{dx}+y^2-x^2\frac{dy}{dx}-2xy = 0$$

$$\Rightarrow \quad \frac{dy}{dx}(2xy-x^2) = 2xy-y^2$$

$$\Rightarrow \quad \frac{dy}{dx} = \frac{2xy-y^2}{2xy-x^2}$$

$$= \frac{y^2-2xy}{x^2-2xy}$$

(iii) We have, $x\sin 2y = y\cos 2x$

On differentiating w.r.t. x, we get

$$x\left(2\cos 2y\frac{dy}{dx}\right)+\sin 2y(1)$$

$$= y(-2\sin 2x)+\cos 2x\left(\frac{dy}{dx}\right)$$

$$\Rightarrow \quad \frac{dy}{dx}(2x\cos 2y-\cos 2x) = -(2y\sin 2x+\sin 2y)$$

$$\Rightarrow \quad \frac{dy}{dx} = \frac{-(2y\sin 2x+\sin 2y)}{2x\cos 2y-\cos 2x}$$

$$\Rightarrow \quad \frac{dy}{dx} = \frac{2y\sin 2x+\sin 2y}{\cos 2x-2x\cos 2y}$$

(iv) Given, $y\log x = (x-y)$

On differentiating w.r.t. x, we get

$$y\left(\frac{1}{x}\right)+\log x\left(\frac{dy}{dx}\right) = 1-\frac{dy}{dx}$$

$$\Rightarrow \quad \frac{dy}{dx}(1+\log x) = 1-\frac{y}{x}.$$

$$\Rightarrow \quad \frac{dy}{dx} = \frac{x-y}{x(1+\log x)}$$

(v) We have, $\sec\left(\frac{x+y}{x-y}\right) = a$

$$\Rightarrow \quad \frac{x+y}{x-y} = \sec^{-1} a$$

$$\Rightarrow \quad \frac{x+y+x-y}{x+y-x+y} = \frac{\sec^{-1}a+1}{\sec^{-1}a-1}$$

$$\Rightarrow \quad \frac{x}{y} = \frac{\sec^{-1}a+1}{\sec^{-1}a-1}$$

$$\Rightarrow \quad \frac{y}{x} = \frac{\sec^{-1}a-1}{\sec^{-1}a+1} \quad \text{...(i)}$$

$$\Rightarrow \quad y = \left(\frac{\sec^{-1}a-1}{\sec^{-1}a+1}\right)x$$

$$\Rightarrow \quad \frac{dy}{dx} = \left(\frac{\sec^{-1}a-1}{\sec^{-1}a+1}\right)(1)$$

$$\Rightarrow \quad \frac{dy}{dx}=\frac{\sec^{-1}a-1}{\sec^{-1}a+1}$$

From Eqs. (i), we get

$$\frac{dy}{dx}=\frac{y}{x}$$

(vi) We have, $e^x+e^y=e^{x+y}$ ...(i)

On differentiating w.r.t. x, we get

$$e^x+e^y\frac{dy}{dx}=e^{x+y}\cdot\left(1+\frac{dy}{dx}\right)$$

$$\Rightarrow \quad e^x+e^y\frac{dy}{dx}=e^{x+y}+e^{x+y}\frac{dy}{dx}$$

$$\Rightarrow \quad \frac{dy}{dx}(e^y-e^{x+y})=e^{x+y}-e^x$$

$$\Rightarrow \quad \frac{dy}{dx}=\frac{e^{x+y}-e^x}{e^y-e^{x+y}}$$

From Eq. (i), we get

$$\frac{dy}{dx}=\frac{e^x+e^y-e^x}{e^y-e^x-e^y}=\frac{e^y}{-e^x}$$

$$\frac{dy}{dx}=-e^{y-x}$$

(vii) We have, $\sin(x+y)=\log(x+y)$

On differentiating w.r.t. x, we get

$$\cos(x+y)\left(1+\frac{dy}{dx}\right)=\frac{1}{x+y}\left(1+\frac{dy}{dx}\right)$$

$$\Rightarrow \cos(x+y)+\cos(x+y)\frac{dy}{dx}=\frac{1}{(x+y)}+\frac{1}{x+y}\frac{dy}{dx}$$

$$\Rightarrow \quad \frac{dy}{dx}\left(\cos(x+y)-\frac{1}{x+y}\right)=\frac{1}{x+y}-\cos(x+y)$$

$$\Rightarrow \quad \frac{dy}{dx}=-1$$

4. (i) We have, $y=x^y \Rightarrow \log y=y\log x$

On differentiating w.r.t. x, we get

$$\frac{1}{y}\frac{dy}{dx}=y\left(\frac{1}{x}\right)+\log x\left(\frac{dy}{dx}\right)$$

$$\Rightarrow \quad \frac{dy}{dx}\left(\frac{1}{y}-\log x\right)=\frac{y}{x}$$

$$\Rightarrow \quad \frac{dy}{dx}\left(\frac{1-y\log x}{y}\right)=\frac{y}{x}$$

$$\Rightarrow \quad \frac{dy}{dx}=\frac{y^2}{x(1-y\log x)}$$

(ii) We have, $y=(\cos x)^{\sin x}$

$\Rightarrow \quad \log y=\sin x\log(\cos x)$

On differentiating w.r.t. x, we get

$$\frac{1}{y}\frac{dy}{dx}=\sin x\left(\frac{1}{\cos x}\cdot(-\sin x)\right)+\log(\cos x)(\cos x)$$

$$\frac{1}{y}\frac{dy}{dx}=\frac{-\sin^2 x}{\cos x}+\cos x\log(\cos x)$$

$$\frac{dy}{dx}=(\cos x)^{\sin x}\left[\frac{-\sin^2 x}{\cos x}+\cos x\log\cos x\right]$$

(iii) $y=\sqrt{(x-1)(x-2)(2x-3)}$

$$\ln y=\frac{1}{2}[\ln(x-1)+\ln(x-2)+\ln(2x-3)]$$

On differentiating w.r.t. x, we get

$$\frac{1}{y}\frac{dy}{dx}=\frac{1}{2}\left[\frac{1}{x-1}+\frac{1}{x-2}+\frac{2}{2x-3}\right]$$

$$\frac{dy}{dx}=\frac{1}{2}\sqrt{(x-1)(x-2)(2x-3)}\left[\frac{1}{x-1}+\frac{1}{x-2}+\frac{2}{2x-3}\right]$$

(iv) We have, $x^y=y^x$

Taking log both sides, we get

$$y\log x=x\log y$$

$$y\cdot\frac{1}{x}+\log x\cdot\frac{dy}{dx}=x\cdot\frac{1}{y}\frac{dy}{dx}+\log y(1)$$

$$\frac{dy}{dx}\left(\log x-\frac{x}{y}\right)=\log y-\frac{y}{x}$$

$$\frac{dy}{dx}\left(\frac{y\log x-x}{y}\right)=\frac{x\log y-y}{x}$$

$$\frac{dy}{dx}=\frac{y(x\log y-y)}{x(y\log x-x)}$$

(v) We have, $x^y=e^{x-y}$

Taking log both sides, we get

$$y\log x=(x-y)\log e$$

$$y\log x=x-y \quad \text{...(i)}$$

$$\Rightarrow \quad y(\log x+1)=x$$

$$\Rightarrow \quad \frac{y}{x}=\frac{1}{1+\log x} \quad \text{...(ii)}$$

$$\Rightarrow \quad \frac{y}{x}+\log x\frac{dy}{dx}=1-\frac{dy}{dx}$$

$$\Rightarrow \quad \frac{dy}{dx}(\log x+1)=1-\frac{y}{x}$$

$$\Rightarrow \quad \frac{dy}{dx}=\frac{x-y}{x(1+\log x)}$$

Using Eqs. (i) and (ii), we get

$$\frac{dy}{dx}=\frac{y\log x}{x(1+\log x)}=\frac{\log x}{(1+\log x)^2}$$

(vi) We have, $y=\left(x+\frac{1}{x}\right)^x$

Taking log both sides, we get

$$\log y=x\log\left(x+\frac{1}{x}\right)$$

On differentiating w.r.t. x, we get

$$\frac{1}{y}\frac{dy}{dx}=x\left(\frac{1}{x+\frac{1}{x}}\right)\left(1-\frac{1}{x^2}\right)+\log\left(x+\frac{1}{x}\right)\cdot 1$$

$$=x\left(\frac{x}{x^2+1}\right)\left(\frac{x^2-1}{x^2}\right)+\log\left(x+\frac{1}{x}\right)$$

$$\frac{1}{y}\frac{dy}{dx}=\frac{x^2-1}{x^2+1}+\log\left(x+\frac{1}{x}\right)$$

$$\Rightarrow \quad \frac{dy}{dx}=y\left(\frac{x^2-1}{x^2+1}+\log\left(x+\frac{1}{x}\right)\right)$$

$$=\left(x+\frac{1}{x}\right)^x\left[\frac{x^2-1}{x^2+1}+\log\left(x+\frac{1}{x}\right)\right]$$

5. (i) Given, $y=2\,at$

On differentiating w.r.t. t, we get

$$\frac{dy}{dt}=2a \quad \text{...(i)}$$

and $x=at^2$

Again, on differentiating w.r.t. t, we get

$$\frac{dx}{dt}=2at \quad \text{...(ii)}$$

From Eqs. (i) and (ii),

$$\Rightarrow \quad \frac{dy}{dx}=\frac{\frac{dy}{dt}}{\frac{dx}{dt}}=\frac{2a}{2at} \Rightarrow \frac{dy}{dx}=\frac{1}{t}$$

(ii) Given, $x=a(t-\sin t)$

$$\frac{dx}{dt}=a(1-\cos t) \quad \text{...(i)}$$

and $y=a(1-\cos t)$

$$\frac{dy}{dt}=a(\sin t) \quad \text{...(ii)}$$

From Eqs. (i) and (ii),

$$\frac{dy}{dx}=\frac{\frac{dy}{dt}}{\frac{dx}{dt}}=\frac{a\sin t}{a(1-\cos t)}$$

$$=\frac{\sin t}{1-\cos t}=\frac{2\sin\frac{t}{2}\cos\frac{t}{2}}{2\sin^2\frac{t}{2}}$$

$$\frac{dy}{dx}=\cot\frac{t}{2}$$

(iii) Given, $y=3t^2+2t-1$

On differentiating w.r.t. t, we get

$$\frac{dy}{dt}=6t+2$$

and $x=t^3-1$

$$\frac{dx}{dt}=3t^2$$

$$\text{Now, } \frac{dy}{dx}=\frac{dy/dt}{dx/dt}=\frac{6t+2}{3t^2}$$

(iv) Given, $y=\sec 4t$

On differentiating w.r.t. t, we get

$$\frac{dy}{dt}=4\sec 4t\tan 4t$$

and $x=\tan^{-1}t$

$$\frac{dx}{dt}=\frac{1}{1+t^2}$$

$$\text{Now, } \frac{dy}{dx}=\frac{\frac{dy}{dt}}{\frac{dx}{dt}}=4(1+t^2)\sec 4t\tan 4t$$

(v) Given, $x=e^{\cos 2t}$

On differentiating w.r.t. t, we get

$$\frac{dx}{dt}=e^{\cos 2t}\cdot(-\sin 2t\cdot 2)$$

and $y=e^{\sin 2t}$

On differentiating w.r.t. t, we get

$$\frac{dy}{dt}=e^{\sin 2t}(\cos 2t\cdot 2)$$

$$\text{Now, } \frac{dy}{dx}=\frac{\frac{dy}{dt}}{\frac{dx}{dt}}=\frac{2\cos 2t\cdot e^{\sin 2t}}{-2\sin 2t\, e^{\cos 2t}}$$

$$=-\frac{y\log x}{x\log y} \quad \left[\begin{array}{l}\because x=e^{\cos 2t},\\ \log x=\cos 2t\\ \text{and } y=e^{\sin 2t},\\ \log y=\sin 2t\end{array}\right]$$

(vi) We have, $x=a\sin 2\theta(1+\cos 2\theta)$

$$=a\left(\sin 2\theta+\frac{1}{2}\cdot 2\sin 2\theta\cos 2\theta\right)$$

$$\Rightarrow \quad x=a\left(\sin 2\theta+\frac{1}{2}\sin 4\theta\right)$$

On differentiating w.r.t. θ, we get

$$\frac{dx}{d\theta}=a(2\cos 2\theta+2\cos 4\theta)$$

$$\frac{dx}{d\theta}=2a(\cos 2\theta+\cos 4\theta)$$

$$=2a\cdot 2\cos 3\theta\cos\theta$$

and $y=b\cos 2\theta(1-\cos 2\theta)$

$$=b(\cos 2\theta-\cos^2 2\theta)$$

$$=b\left(\cos 2\theta-\left(\frac{1+\cos 4\theta}{2}\right)\right)$$

$$y=b\left(\cos 2\theta-\frac{1}{2}-\frac{1}{2}\cos 4\theta\right)$$

On differentiating w.r.t. θ, we get

$$\frac{dy}{d\theta}=b(-2\sin 2\theta+2\sin 4\theta)$$
$$=2b(\sin 4\theta-\sin 2\theta)$$
$$=2b\cdot 2\cos 3\theta\sin\theta$$

So, $$\frac{dy}{dx}=\frac{2b\cdot 2\cos 3\theta\sin\theta}{2a\cdot 2\cos 3\theta\cos\theta}=\frac{b}{a}\tan\theta$$

6. (i) We have, $y=\sqrt{x+\sqrt{x+\sqrt{x+\ldots\infty}}}$

$$y=\sqrt{x+y}$$

On squaring both sides, we get

$$y^2=x+y$$
$$\Rightarrow\quad y^2-y=x$$

On differentiating w.r.t. x, we get

$$2y\frac{dy}{dx}-\frac{dy}{dx}=1$$
$$\Rightarrow\quad \frac{dy}{dx}(2y-1)=1$$
$$\Rightarrow\quad \frac{dy}{dx}=\frac{1}{2y-1}$$

(ii) We have, $y=\sqrt{\sin x+\sqrt{\sin x+\sqrt{\sin x+\ldots\infty}}}$

$$y=\sqrt{\sin x+y}$$

On squaring both sides, we get

$$y^2=\sin x+y$$
$$\Rightarrow\quad y^2-y=\sin x$$

On differentiating w.r.t. x, we get

$$2y\frac{dy}{dx}-\frac{dy}{dx}=\cos x$$
$$\Rightarrow\quad \frac{dy}{dx}(2y-1)=\cos x$$
$$\Rightarrow\quad \frac{dy}{dx}=\frac{\cos x}{2y-1}$$

(iii) We have, $y=x^{x^{x\ldots\infty}}$

$$\Rightarrow\quad y=x^y$$
$$\Rightarrow\quad \log y=y\log x$$

On differentiating w.r.t. x, we get

$$\frac{1}{y}\frac{dy}{dx}=y\cdot\frac{1}{x}+\log x\cdot\frac{dy}{dx}$$
$$\Rightarrow\quad \frac{dy}{dx}\left(\frac{1}{y}-\log x\right)=\frac{y}{x}$$
$$\Rightarrow\quad \frac{dy}{dx}\left(\frac{1-y\log x}{y}\right)=\frac{y}{x}$$
$$\Rightarrow\quad \frac{dy}{dx}=\frac{y^2}{x(1-y\log x)}$$

(iv) We have, $y=e^{x+e^{x+e^{x+\ldots\ldots\infty}}}$

$$\Rightarrow\quad y=e^{x+y}$$
$$\Rightarrow\quad \log y=x+y$$

On differentiating w.r.t. x, we get

$$\frac{1}{y}\frac{dy}{dx}=1+\frac{dy}{dx}$$
$$\Rightarrow\quad \frac{dy}{dx}\left(\frac{1}{y}-1\right)=1\Rightarrow\frac{dy}{dx}\left(\frac{1-y}{y}\right)=1$$
$$\Rightarrow\quad \frac{dy}{dx}=\frac{y}{1-y}$$

(v) We have, $y=\cfrac{x}{1+\cfrac{x}{2+\cfrac{x}{1+\cfrac{x}{2+\ldots\infty}}}}\Rightarrow y=\cfrac{x}{1+\cfrac{x}{2+y}}$

$$\Rightarrow y=\frac{x}{\frac{2+y+x}{2+y}}=\frac{x(2+y)}{2+y+x}$$
$$\Rightarrow 2y+y^2+yx=2x+xy$$
$$\Rightarrow y^2+2y=2x$$

and then proceed as in sol. (vi).

(vi) We have, $$y=\frac{x(2+y)}{2+y+x}$$

$$\Rightarrow\quad y(2+y+x)=x(2+y)$$
$$\Rightarrow\quad 2y+y^2+xy=2x+xy$$
$$\Rightarrow\quad y^2+2y=2x$$

On differentiating w.r.t. x, we get

$$2y\frac{dy}{dx}+2\frac{dy}{dx}=2$$
$$\Rightarrow\quad \frac{dy}{dx}(y+1)=1$$
$$\Rightarrow\quad \frac{dy}{dx}=\frac{1}{y+1}$$

7. (i) We have,

$$y=\frac{\log x}{x}$$
$$\Rightarrow\quad \frac{dy}{dx}=\frac{1}{x}\frac{d}{dx}(\log x)+\log x\frac{d}{dx}\left(\frac{1}{x}\right)$$
$$\Rightarrow\quad \frac{dy}{dx}=\frac{1}{x^2}-\frac{\log x}{x^2}$$
$$\Rightarrow\quad \frac{d^2y}{dx^2}=\frac{-2}{x^3}-\frac{1}{x^3}+\frac{2\log x}{x^3}=\frac{-3+2\log x}{x^3}$$

(ii) We have,

$$y=e^x\sin 3x$$
$$\Rightarrow\quad \frac{dy}{dx}=e^x(3\cos 3x)+e^x\sin 3x$$
$$=e^x(3\cos 3x+\sin 3x)$$

$$\Rightarrow \frac{d^2y}{dx^2} = e^x(3\cos 3x + \sin 3x) + e^x(-9\sin 3x + 3\cos 3x)$$

$$\Rightarrow \frac{d^2y}{dx^2} = e^x(6\cos 3x - 8\sin 3x)$$

$$= 2e^x(3\cos 3x - 4\sin 3x)$$

(iii) We have, $y = x^2 \cos x$

$$\frac{dy}{dx} = 2x\cos x - x^2 \sin x$$

$$\Rightarrow \frac{d^2y}{dx^2} = 2\cos x - 2x\sin x - 2x\sin x - x^2\cos x$$

$$\Rightarrow \frac{d^2y}{dx^2} = (2 - x^2)\cos x - 4x\sin x$$

(iv) We have, $y = \tan^{-1} x$

$$\Rightarrow \frac{dy}{dx} = \frac{1}{1+x^2}$$

$$\Rightarrow \frac{d^2y}{dx^2} = \frac{-2x}{(1+x^2)^2}$$

(v) We have, $y = \log(\log x)$

$$\frac{dy}{dx} = \frac{1}{x\log x}$$

$$\Rightarrow \frac{d^2y}{dx^2} = \frac{-1}{(x\log x)^2}\left(\frac{x}{x} + \log x\right)$$

$$\Rightarrow \frac{d^2y}{dx^2} = \frac{-(1+\log x)}{(x\log x)^2}$$

8. $y = x|x| = \begin{cases} x^2, & x \geq 0 \\ -x^2, & x < 0 \end{cases}$

$$\frac{dy}{dx} = \begin{cases} 2x, & x > 0 \\ -2x, & x < 0 \end{cases}$$

So, $\dfrac{dy}{dx} = 2|x|$

9. We have, $f(x) = |x-1| + |x-3|$

$$f(x) = \begin{cases} -(x-1)-(x-3); & x < 1 \\ (x-1)-(x-3); & 1 \leq x < 3 \\ (x-1)+(x-3); & x \geq 3 \end{cases}$$

$$= \begin{cases} -2x+4; & x < 1 \\ 2; & 1 \leq x < 3 \\ 2x-4; & x \geq 3 \end{cases}$$

Thus, $f'(2) = 0$

10. We have, $y = e^x \cos x$

On differentiating w.r.t. x, we get

$$\frac{dy}{dx} = e^x\cos x - e^x \sin x$$

$$= e^x(\cos x - \sin x)$$

$$= \sqrt{2}\, e^x\left(\frac{1}{\sqrt{2}}\cos x - \frac{1}{\sqrt{2}}\sin x\right)$$

$$= \sqrt{2}\, e^x\left(\cos x \cos\frac{\pi}{4} - \sin x \sin\frac{\pi}{4}\right)$$

$$\Rightarrow \frac{dy}{dx} = \sqrt{2}e^x \cos\left(x + \frac{\pi}{4}\right)$$ **Hence proved.**

11. We have, $f'(x) = \dfrac{1}{x^2+1}$

and $\quad g(x) = f(3x-1)$

On differentiating w.r.t. x, we get

$$g'(x) = f'(3x-1)\cdot 3$$

$$= 3f'(3x-1)$$

$$= 3\cdot\frac{1}{(3x-1)^2+1} = \frac{3}{9x^2-6x+2}$$

12. Given, $\cos y = x\cos(a+y)$

$$\Rightarrow x = \frac{\cos y}{\cos(a+y)}$$

On differentiating w.r.t. y, we get

$$\frac{dx}{dy} = \frac{-\cos(a+y)\sin y + \cos y \sin(a+y)}{\cos^2(a+y)}$$

$$\Rightarrow \frac{dx}{dy} = \frac{\sin(a+y-y)}{\cos^2(a+y)}$$

$$\Rightarrow \frac{dx}{dy} = \frac{\sin a}{\cos^2(a+y)}$$ **Hence proved.**

13. We have, $y = \dfrac{(1-x^2)^{1/2}(2x+3)^{1/2}}{(x^2+2)^{2/3}}$

Taking log both sides, we get

$$\log y = \frac{1}{2}\log(1-x^2) + \frac{1}{2}\log(2x+3) - \frac{2}{3}\log(x^2+2)$$

On differentiating w.r.t. x, we get

$$\frac{1}{y}\frac{dy}{dx} = \frac{1}{2}\left(\frac{1}{1-x^2}\right)(-2x) + \frac{1}{2}\left(\frac{1}{2x+3}\right)(2) - \frac{2}{3}\left(\frac{1}{x^2+2}\right)(2x)$$

$$\Rightarrow \frac{dy}{dx} = y\left[\frac{-x}{1-x^2} + \frac{1}{2x+3} - \frac{4x}{3(x^2+2)}\right]$$

At $x = 0$,

$$\left.\frac{dy}{dx}\right|_{x=0} = \frac{\sqrt{3}}{\sqrt[3]{4}} \times \frac{1}{3} = \frac{1}{\sqrt[3]{4}\sqrt{3}}$$

14. Let $u = \sin x$

$$\frac{du}{dx} = \cos x$$

and $v = \cos x$

$$\frac{dv}{dx} = -\sin x$$

Now, $\dfrac{du}{dv} = \dfrac{\frac{du}{dx}}{\frac{dv}{dx}} = \dfrac{\cos x}{-\sin x} = -\cot x$

15. Let $u = x^x$

$$\log u = x \log x$$

$$\frac{1}{u}\frac{du}{dx} = x \cdot \frac{1}{x} + \log x \cdot 1$$

$$\frac{du}{dx} = u(1 + \log x)$$

$$\frac{du}{dx} = x^x (1 + \log x)$$

and $v = x \log x$

$$\frac{dv}{dx} = x\left(\frac{1}{x}\right) + \log x \cdot 1 = 1 + \log x$$

Now, $$\frac{du}{dv} = \frac{\frac{du}{dx}}{\frac{dv}{dx}} = \frac{x^x (1 + \log x)}{(1 + \log x)}$$

$$\frac{du}{dv} = x^x$$

Level 2

1. (a) We have, $y = \sin^n x \cos nx$

On differentiating w.r.t. x, we get

$$\frac{dy}{dx} = n \cdot \sin^{n-1} x \cdot \cos x \cos nx + \sin^n x[-n \sin nx]$$

$$= n \sin^{n-1} x [\cos x \cos nx - \sin x \sin nx]$$

$$= n \sin^{n-1} x \cdot \cos(n+1)x$$

2. (c) In neighbourhood of $x = \frac{2\pi}{3}$,

$|\cos x| = -\cos x$ and $|\sin x| = \sin x$

So, $y = -\cos x + \sin x$

$\therefore$ $$\frac{dy}{dx} = \sin x + \cos x$$

At $x = \frac{2\pi}{3}$, $$\left.\frac{dy}{dx}\right|_{x=\frac{2\pi}{3}} = \sin\frac{2\pi}{3} + \cos\left(\frac{2\pi}{3}\right)$$

$$= \frac{\sqrt{3}}{2} - \frac{1}{2} = \frac{\sqrt{3}-1}{2}$$

3. (b) For $2 < x < 3$, we have $[x] = 2$

$\therefore$ $$f(x) = \sin\left(\frac{2\pi}{3} - x^2\right)$$

$\Rightarrow$ $$f'(x) = -2x \cos\left(\frac{2\pi}{3} - x^2\right)$$

$\Rightarrow$ $$f'\left(\sqrt{\frac{\pi}{3}}\right) = -2\sqrt{\frac{\pi}{3}} \cos\left(\frac{2\pi}{3} - \frac{\pi}{3}\right)$$

$$= -2\sqrt{\frac{\pi}{3}} \cos\left(\frac{\pi}{3}\right)$$

$$= -2\sqrt{\frac{\pi}{3}} \times \frac{1}{2} = -\sqrt{\frac{\pi}{3}}$$

4. (b) We have, $x\sqrt{1+y} = -y\sqrt{1+x}$

On squaring both sides, we get

$$x^2(1+y) = y^2(1+x)$$

$\Rightarrow$ $$x^2 - y^2 + x^2y - xy^2 = 0$$

$\Rightarrow$ $$(x+y)(x-y) + xy(x-y) = 0$$

$\Rightarrow$ $$(x-y)(x+y+xy) = 0$$

Either $x = y$ or $x + y + xy = 0$

$$x \neq y$$

$\Rightarrow$ $$x + y + xy = 0$$

$\Rightarrow$ $$y = \frac{-x}{1+x}$$

$\Rightarrow$ $$\frac{dy}{dx} = -\frac{1}{(1+x)^2}$$

5. (d) We have,

$$f(x) = \frac{g(x) + g(-x)}{2} + \frac{2}{[h(x) + h(-x)]^{-1}}$$

$$f(x) = \frac{g(x)}{2} + \frac{g(-x)}{2} + 2h(x) + 2h(-x)$$

On differentiating w.r.t. x, we get

$$f'(x) = \frac{g'(x)}{2} - \frac{g'(-x)}{2} + 2h'(x) - 2h'(-x)$$

$$f'(0) = 0$$

6. (b) $$y = \frac{1}{1 + x^{n-m} + x^{p-m}} + \frac{1}{1 + x^{m-n} + x^{p-n}} + \frac{1}{1 + x^{m-p} + x^{n-p}}$$

$$\Rightarrow y = \frac{1}{1 + \frac{x^n}{x^m} + \frac{x^p}{x^m}} + \frac{1}{1 + \frac{x^m}{x^n} + \frac{x^p}{x^n}} + \frac{1}{1 + \frac{x^m}{x^p} + \frac{x^n}{x^p}}$$

$$\Rightarrow y = \frac{x^m}{x^m + x^n + x^p} + \frac{x^n}{x^m + x^p + x^n} + \frac{x^p}{x^p + x^m + x^n}$$

$$\Rightarrow y = \frac{x^m + x^n + x^p}{x^m + x^n + x^p} = 1 \Rightarrow \frac{dy}{dx} = 0$$

7. (d) Since, g is inverse of f.

$$f[g(x)] = x$$

$\Rightarrow$ $$f'[g(x)] \cdot g'(x) = 1$$

$\Rightarrow$ $$g'(x) = 1 + [g(x)]^5 \quad \left[\because f'(x) = \frac{1}{1+x^5}\right]$$

8. (d) Given, $y = \log^n x$

$\Rightarrow$ $$\frac{dy}{dx} = \frac{1}{\log^{n-1} x \cdot \log^{n-2} x \ldots x}$$

$\Rightarrow$ $$\log^{n-1} x \cdot \log^{n-2}(x) \ldots. x \frac{dy}{dx} = 1$$

So, $\log^n x \cdot \log^{n-1} x \cdot \log^{n-2} x \ldots. x \frac{dy}{dx} = \log^n x$

9. (a) We have, $y = \tan^{-1}\left(\frac{6x\sqrt{x}}{1-9x^3}\right)$

$$= \tan^{-1}\left(\frac{2(3x^{3/2})}{1-(3x^{3/2})^2}\right)$$

$$\left[\because 2\tan^{-1}x = \tan^{-1}\left(\frac{2x}{1-x^2}\right)\right]$$

$$y = 2\tan^{-1}(3x^{3/2})$$

On differentiating w.r.t. x, we get

$$y' = 2\cdot\frac{1}{1+(3x^{3/2})^2}\cdot 3\cdot\frac{3}{2}x^{1/2}$$

$$y' = \frac{9\sqrt{x}}{1+9x^3}$$

But given that $y' = \sqrt{x}\cdot g(x)$

Hence, $g(x) = \frac{9}{1+9x^3}$.

10. (c) We have, $\cos^{-1}\left(\frac{x^2-y^2}{x^2+y^2}\right) = \log a$

$$\Rightarrow \quad \cos^{-1}\left(\frac{1-\left(\frac{y}{x}\right)^2}{1+\left(\frac{y}{x}\right)^2}\right) = \log a$$

Put $\frac{y}{x} = \tan\theta$

$$\Rightarrow \quad \cos^{-1}\left(\frac{1-\tan^2\theta}{1+\tan^2\theta}\right) = \log a$$

$$\Rightarrow \quad \cos^{-1}(\cos 2\theta) = \log a \Rightarrow 2\theta = \log a$$

$$\Rightarrow \quad 2\tan^{-1}\left(\frac{y}{x}\right) = \log a$$

$$\Rightarrow \quad \frac{y}{x} = \tan\left(\frac{\log a}{2}\right)$$

On differentiating both sides w.r.t. x, we get

$$\frac{x\frac{dy}{dx} - y\cdot 1}{x^2} = 0$$

$$\Rightarrow \quad x\frac{dy}{dx} - y = 0$$

$$\Rightarrow \quad \frac{dy}{dx} = \frac{y}{x}$$

11. (a) We have, $f(x) = \cot^{-1}\left(\frac{x^x - x^{-x}}{2}\right)$

$$\Rightarrow \quad f'(x) = -\frac{1}{1+\left(\frac{x^x-x^{-x}}{2}\right)^2}\cdot\frac{d}{dx}\left(\frac{x^x-x^{-x}}{2}\right)$$

$$\Rightarrow \quad f'(x) = \frac{-2}{4+(x^x-x^{-x})^2}\frac{d}{dx}(x^x - x^{-x})$$

$$= \frac{-2}{4+(x^x-x^{-x})^2}\frac{d}{dx}(e^{x\log x} - e^{-x\log x})$$

$$\Rightarrow \quad f'(x) = \frac{-2}{(x^x+x^{-x})^2} \times \left\{e^{x\log x}\frac{d}{dx}(x\log x) - e^{-x\log x}\frac{d}{dx}(-x\log x)\right\}$$

$$\Rightarrow f'(x) = \frac{-2}{(x^x+x^{-x})^2}[x^x(1+\log x) + x^{-x}(1+\log x)]$$

$$\Rightarrow f'(x) = \frac{-2(1+\log x)}{(x^x+x^{-x})^2}\cdot(x^x+x^{-x}) = \frac{-2(1+\log x)}{(x^x+x^{-x})}$$

$$\Rightarrow f'(1) = \frac{-2}{1+1} = -1$$

12. (c) $y = \tan^{-1}\left[\frac{2^x}{1+2(2^{2x})}\right]$

$$\Rightarrow \quad y = \tan^{-1}\left[\frac{2(2^x)-2^x}{1+2\cdot 2^x 2^x}\right]$$

$$= \tan^{-1}(2\cdot 2^x) - \tan^{-1}(2^x)$$

$$\Rightarrow \quad y = \tan^{-1}(2^{x+1}) - \tan^{-1}(2^x)$$

$$\Rightarrow \frac{dy}{dx} = \frac{1}{1+(2)^{(2x+2)}}\cdot 2^{x+1}\cdot\ln 2 - \frac{1}{1+(2^x)^2}\cdot 2^x\cdot\ln 2$$

$$\Rightarrow \left(\frac{dy}{dx}\right)_{x=0} = \frac{1}{5}(2)\ln 2 - \frac{1}{2}\ln 2$$

$$= \ln 2\left(\frac{2}{5}-\frac{1}{2}\right) = \ln 2\left(\frac{4-5}{10}\right) = -\frac{1}{10}\ln 2$$

13. (c) We have, $2y = \left(\cot^{-1}\left(\frac{\sqrt{3}\cos x + \sin x}{\cos x - \sqrt{3}\sin x}\right)\right)^2$

Numerator and denominator divided by 'cos x', we get

$$2y = \left(\cot^{-1}\left(\frac{\sqrt{3}+\tan x}{1-\sqrt{3}\tan x}\right)\right)^2$$

$$= \left[\cot^{-1}\left(\tan\left(\frac{\pi}{3}+x\right)\right)\right]^2$$

$$= \left[\frac{\pi}{2} - \tan^{-1}\left(\tan\left(\frac{\pi}{3}+x\right)\right)\right]^2$$

$$= \left[\frac{\pi}{2} - \left(\frac{\pi}{3}+x\right)\right]^2$$

$$\Rightarrow \quad 2y = \left(\frac{\pi}{6}-x\right)^2$$

On differentiating w.r.t. x, we get

$$2\frac{dy}{dx}=2\left(\frac{\pi}{6}-x\right)\cdot(0-1)$$

$$\Rightarrow \quad \frac{dy}{dx}=x-\frac{\pi}{6}$$

14. (a) We have, $y=\tan^{-1}\left(\frac{\sqrt{1+x^2}+\sqrt{1-x^2}}{\sqrt{1+x^2}-\sqrt{1-x^2}}\right)$

Clearly, $0\le x^2\le 1$

Let $\quad 2\theta=\cos^{-1}x^2$

$$\Rightarrow \quad \theta\in\left[0,\frac{\pi}{4}\right] \text{ and } \cos 2\theta = x^2$$

$$\Rightarrow \quad y=\tan^{-1}\left(\frac{\sqrt{1+\cos 2\theta}+\sqrt{1-\cos 2\theta}}{\sqrt{1+\cos 2\theta}-\sqrt{1-\cos 2\theta}}\right)$$

$$=\tan^{-1}\left(\frac{\cos\theta+\sin\theta}{\cos\theta-\sin\theta}\right)$$

$$=\tan^{-1}\left(\frac{1+\tan\theta}{1-\tan\theta}\right)$$

$$=\tan^{-1}\left(\tan\left(\frac{\pi}{4}+\theta\right)\right)=\frac{\pi}{4}+\theta$$

$$\Rightarrow \quad y=\frac{\pi}{4}+\frac{1}{2}\cos^{-1}x^2$$

On differentiating w.r.t. x, we get

$$\frac{dy}{dx}=-\frac{1}{2}\frac{1}{\sqrt{1-x^4}}\times(2x) \Rightarrow \frac{dy}{dx}=\frac{-x}{\sqrt{1-x^4}}$$

15. (b) $1+x+x^2=1+x(x+1)$

$$\Rightarrow x^2+3x+3=1+x^2+3x+2=1+(x+1)(x+2)$$

$$\Rightarrow x^2+5x+7=1+x^2+5x+6=1+(x+2)(x+3)$$

$$\therefore \; T_1=\tan^{-1}\frac{1}{1+x+x^2}=\tan^{-1}(x+1)-\tan^{-1}x$$

$$T_2=\tan^{-1}\left(\frac{1}{x^2+3x+3}\right)=\tan^{-1}(x+2)-\tan^{-1}(x+1)$$

$$T_3=\tan^{-1}\left(\frac{1}{x^2+5x+7}\right)=\tan^{-1}(x+3)-\tan^{-1}(x+2)$$

$$\vdots \qquad\qquad \vdots \qquad\qquad \vdots$$

$$T_n=\tan^{-1}(x+n)-\tan^{-1}(x+n-1)$$

$$\therefore \; y=S_n=\tan^{-1}(x+n)-\tan^{-1}x$$

$$\frac{dy}{dx}=\frac{1}{1+(x+n)^2}-\frac{1}{1+x^2}$$

At $x=0$,

$$\left(\frac{dy}{dx}\right)_{x=0}=\frac{1}{1+n^2}-1=\frac{1-1-n^2}{1+n^2}=\frac{-n^2}{1+n^2}$$

16. (c) We have, $\sin y+e^{-x\cos y}=e^2$

On differentiating w.r.t. x, we get

$$\cos y\frac{dy}{dx}+(e^{-x\cos y})\left(x\sin y\frac{dy}{dx}-\cos y\right)=0$$

At $(1,\pi)$,

$$\cos\pi\frac{dy}{dx}+e^{-\cos\pi}\left(\sin\pi\frac{dy}{dx}-\cos\pi\right)=0$$

$$\Rightarrow \quad -\frac{dy}{dx}+e(0+1)=0$$

$$\Rightarrow \quad -\frac{dy}{dx}=-e$$

$$\Rightarrow \quad \frac{dy}{dx}=e$$

17. (b) $\sin xy+\cos xy=0$

$$\Rightarrow \quad \sin xy=-\cos xy$$

On differentiating both sides w.r.t. x, we get

$$\cos(xy)\left(y+x\frac{dy}{dx}\right)=\sin(xy)\left(y+x\frac{dy}{dx}\right)$$

$$\Rightarrow \quad y+x\frac{dy}{dx}=0$$

$$\Rightarrow \quad \frac{dy}{dx}=-\frac{y}{x}$$

18. (b) We have, $x\log_e(\log_e x)-x^2+y^2=4$

On differentiating both sides w.r.t. x, we get

$$\log(\log x)+x\frac{1}{x\log x}-2x+2y\frac{dy}{dx}=0 \qquad \text{...(i)}$$

When $x=e, y=\sqrt{4+e^2}$, put these values in Eq. (i), we get

$$0+1-2e+2\sqrt{4+e^2}\frac{dy}{dx}=0$$

$$\Rightarrow \quad \frac{dy}{dx}=\frac{2e-1}{2\sqrt{4+e^2}}$$

19. (d) We have, $y=x^{e^x}\Rightarrow \log y=e^x\log x$

$$\therefore \quad \frac{1}{y}\frac{dy}{dx}=e^x\cdot\frac{1}{x}+\log x\cdot e^x$$

$$\Rightarrow \quad \frac{1}{y}\frac{dy}{dx}=e^x\left(\log x+\frac{1}{x}\right) \Rightarrow \frac{dy}{dx}=y\cdot e^x\left(\log x+\frac{1}{x}\right)$$

20. (b) Let $u=x^x$, then $\log u=x\log x$

$$\therefore \quad \frac{1}{u}\frac{du}{dx}=x\cdot\frac{1}{x}+\log x$$

$$\frac{du}{dx}=u(1+\log x)$$

$$\therefore \quad \frac{du}{dx}=x^x(1+\log x)$$

At $x=1$,

$$\frac{du}{dx}=1 \qquad \text{...(i)}$$

Now, $v=x^{2x}$

$$\frac{dv}{dx}=2 \text{ at } x=1 \quad ...(ii)$$

Also, when $x=1$, we have

$$1-2\cot y-1=0$$

$$\therefore \quad \cot y=0 \quad ...(iii)$$

Now, differentiating the given relation

$$v-2u\cot y-1=0$$

$$\frac{dv}{dx}-2\frac{du}{dx}\cot y+2u\,\text{cosec}^2 y\frac{dy}{dx}=0$$

At $x=1$,

$$2-2\cdot 1(0)+2u(1+\cot^2 y)\frac{dy}{dx}=0$$

From Eqs. (i), (ii) and (iii), we get

$$1+1(1+0)\frac{dy}{dx}=0$$

$$\Rightarrow \quad \frac{dy}{dx}=-1$$

21. (c) We have, $2^x+2^y=2^{x+y}$

$$\Rightarrow \quad 2^x\ln 2+2^y\ln 2\frac{dy}{dx}=2^{x+y}\ln 2\left(1+\frac{dy}{dx}\right).$$

$$\Rightarrow (2^{x+y}\ln 2-2^y\ln 2)\frac{dy}{dx}=2^x\ln 2-2^{x+y}\ln 2$$

$$\Rightarrow \quad \frac{dy}{dx}=\frac{2^x-2^{x+y}}{2^{x+y}-2^y}=\frac{2^x(1-2^y)}{2^y(2^x-1)}$$

$$\Rightarrow \quad \frac{dy}{dx}=2^{x-y}\left(\frac{1-2^y}{2^x-1}\right)$$

$$\Rightarrow \quad \frac{dy}{dx}=2^{x-y}\left(\frac{2^y-1}{1-2^x}\right)$$

22. (a) We have, $y=(\tan x)^{(\tan x)^{\tan x}}$

$$\log y=(\tan x)^{\tan x}\log\tan x$$

Again, taking log on both sides, we get

$$\log\log y=(\tan x\log\tan x)+\log(\log\tan x)$$

On differentiating w.r.t. x, we get

$$\frac{1}{y\log y}\frac{dy}{dx}=\log\tan x\sec^2 x+\tan x\cdot\frac{\sec^2 x}{\tan x}+\frac{\sec^2 x}{\tan x\cdot\log\tan x}$$

$$\frac{dy}{dx}=y\log y\cdot\sec^2 x\left[\log\tan x+1+\frac{1}{\tan x\log\tan x}\right]$$

At $x=\frac{\pi}{4}$,

$$y=1 \text{ and } \log y=0$$

So, $\left(\frac{dy}{dx}\right)_{x=\frac{\pi}{4}}=0$

23. (c) We have, $f(x)=\log_e x$

$$\Rightarrow \quad f(\log_e x)=\log_e(\log_e x)$$

On differentiating w.r.t. x, we get

$$\frac{d}{dx}f(\log_e x)=\frac{1}{\log_e x}\cdot\frac{1}{x}=\frac{1}{x\log_e x}$$

24. (b) We have, $y=x^{(\ln x)^{\ln(\ln x)}}$

$$\Rightarrow \quad \ln y=(\ln x)^{\ln(\ln x)}\cdot\ln x \quad ...(i)$$

$$\Rightarrow \quad \ln(\ln y)=\ln(\ln x)\cdot\ln(\ln x)+\ln(\ln x)$$

On differentiating w.r.t. x, we get

$$\frac{1}{\ln y}\cdot\frac{1}{y}\frac{dy}{dx}=\frac{2\ln(\ln x)}{\ln x}\cdot\frac{1}{x}+\frac{1}{x\ln x}=\frac{2\ln(\ln x)+1}{x\ln x}$$

$$\therefore \quad \frac{dy}{dx}=\frac{y}{x}\cdot\frac{\ln y}{\ln x}[2\ln(\ln x)+1]$$

On substituting the value of $\ln y$ from Eq. (i), we get

$$\frac{dy}{dx}=\frac{y}{x}(\ln x)^{\ln(\ln x)}[2\ln(\ln x)+1]$$

25. (d) We have, $f(x)=|x|^{|\sin x|};\ \frac{\pi}{4}\in\left(0,\frac{\pi}{2}\right)$

$$\Rightarrow \quad f(x)=x^{\sin x} \text{ for } x\in\left(0,\frac{\pi}{2}\right)$$

$$\Rightarrow \quad \ln f(x)=(\sin x)\ln x$$

$$\Rightarrow \quad \frac{1}{f(x)}\cdot f'(x)=\frac{\sin x}{x}+(\ln x)(\cos x)$$

$$\Rightarrow \quad f'(x)=f(x)\left[\frac{\sin x}{x}+\cos x\cdot\ln x\right]$$

$$\Rightarrow \quad f'(x)=(x)^{\sin x}\left[\frac{\sin x}{x}+\cos x\cdot\ln x\right]$$

$$\Rightarrow \quad f'\left(\frac{\pi}{4}\right)=\left(\frac{\pi}{4}\right)^{1/\sqrt{2}}\left[\frac{\frac{1}{\sqrt{2}}}{\left(\frac{\pi}{4}\right)}+\frac{1}{\sqrt{2}}\ln\frac{\pi}{4}\right]$$

$$=\left(\frac{\pi}{4}\right)^{1/\sqrt{2}}\left[\frac{2\sqrt{2}}{\pi}+\frac{\sqrt{2}}{2}\ln\frac{\pi}{4}\right]$$

26. (d) We have, $(2x)^{2y}=4e^{2x-2y}$

$$\Rightarrow \quad 2y\ln 2x=\ln 4+2x-2y$$

$$\Rightarrow \quad y=\frac{x+\ln 2}{1+\ln 2x}$$

$$\Rightarrow \quad y'=\frac{(1+\ln 2x)-(x+\ln 2)\frac{1}{x}}{(1+\ln 2x)^2}$$

$$\Rightarrow \quad y'(1+\ln 2x)^2=\left[\frac{x\ln 2x-\ln 2}{x}\right]$$

27. (a) As, $x = \operatorname{cosec}\theta - \sin\theta$

$$x^2 + 4 = (\operatorname{cosec}\theta - \sin\theta)^2 + 4 = (\operatorname{cosec}\theta + \sin\theta)^2$$

and $y^2 + 4 = (\operatorname{cosec}^n\theta - \sin^n\theta)^2 + 4$

$$= (\operatorname{cosec}^n\theta + \sin^n\theta)^2$$

Now, $\dfrac{dy}{dx} = \dfrac{dy/d\theta}{dx/d\theta}$

$$= \frac{n(\operatorname{cosec}^{n-1}\theta)(-\operatorname{cosec}\theta\cot\theta) - n\sin^{n-1}\theta\cdot\cos\theta}{-\operatorname{cosec}\theta\cot\theta - \cos\theta}$$

$$= \frac{n(\operatorname{cosec}^n\theta\cot\theta + \sin^{n-1}\theta\cdot\cos\theta)}{(\operatorname{cosec}\theta\cot\theta + \cos\theta)}$$

$$= \frac{n\cot\theta(\operatorname{cosec}^n\theta + \sin^n\theta)}{\cot\theta(\operatorname{cosec}\theta + \sin\theta)}$$

$$= \frac{n(\operatorname{cosec}^n\theta + \sin^n\theta)}{\operatorname{cosec}\theta + \sin\theta} = \frac{n\sqrt{y^2+4}}{\sqrt{x^2+4}}$$

On squaring both sides, we get

$$\left(\frac{dy}{dx}\right)^2 = \frac{n^2(y^2+4)}{(x^2+4)}$$

$$\Rightarrow (x^2+4)\left(\frac{dy}{dx}\right)^2 = n^2(y^2+4)$$

28. (d) We have, $x = a\cos^3\theta$ and $y = a\sin^3\theta$

On differentiating w.r.t. θ, we get

$$\frac{dx}{d\theta} = -3a\cos^2\theta(\sin\theta)$$

and $$\frac{dy}{d\theta} = 3a\sin^2\theta(\cos\theta)$$

Now, $\dfrac{dy}{dx} = \dfrac{3a\sin^2\theta\cos\theta}{-3a\cos^2\theta\sin\theta} = -\dfrac{\sin\theta}{\cos\theta}$

$$\frac{dy}{dx} = -\tan\theta$$

$$\therefore \quad 1 + \left(\frac{dy}{dx}\right)^2 = 1 + \tan^2\theta$$

$$\Rightarrow \quad 1 + \left(\frac{dy}{dx}\right)^2 = \sec^2\theta$$

$$\Rightarrow \quad \sqrt{1 + \left(\frac{dy}{dx}\right)^2} = |\sec\theta|$$

29. (a) We have, $x = 2\sin 3t, y = 2\cos 3t$ and $z = 8t$

$$r = \sqrt{x^2 + y^2 + z^2}$$

$$= \sqrt{4\sin^2 3t + 4\cos^2 3t + 64t^2}$$

$$r = \sqrt{4 + 64t^2} \Rightarrow r^2 = 4 + 64t^2$$

On differentiating w.r.t. t, we get

$$2r\frac{dr}{dt} = 2 \times 64 \times t$$

$$\Rightarrow \quad \frac{dr}{dt} = \frac{64t}{r}$$

$$= \frac{64t}{\sqrt{4+64t^2}} = \frac{64t}{2\sqrt{1+16t^2}} = \frac{32t}{\sqrt{1+16t^2}}$$

30. (c) We have, $\quad y = a^{x^{a^{x^{a^{x\ldots\infty}}}}}$

$$\Rightarrow \quad y = a^{x^y}$$

$$\Rightarrow \quad \log y = x^y \log a$$

$$\Rightarrow \quad \log\log y = y\log x + \log(\log a)$$

$$\Rightarrow \quad \frac{1}{\log y}\cdot\frac{1}{y}\frac{dy}{dx} = \frac{dy}{dx}\log x + y\cdot\frac{1}{x} + 0$$

$$\Rightarrow \frac{dy}{dx}\left(\frac{1}{y\log y} - \log x\right) = \frac{y}{x}$$

$$\Rightarrow \quad \frac{dy}{dx} = \frac{y^2\log y}{x(1 - y\log x\log y)}.$$

31. (d) We have, $y = \cfrac{x}{a + \cfrac{x}{b + \cfrac{x}{a + \ldots\infty}}}$

$$\Rightarrow \quad y = \cfrac{x}{a + \cfrac{x}{b+y}}$$

$$\Rightarrow \quad y = \frac{x(b+y)}{a(b+y)+x}$$

$$\Rightarrow \quad a(b+y)y + xy = x(b+y)$$

$$\Rightarrow \quad a(by' + 2yy') + y + xy' = b + y + xy'$$

$$\Rightarrow \quad y'(ab + 2ay) = b$$

$$\Rightarrow \quad y' = \frac{b}{ab + 2ay}$$

32. (b) We have, $f(x) = x + \cfrac{1}{x + \cfrac{1}{x + \cfrac{1}{x + \ldots\infty}}}$

$$\Rightarrow \quad y = x + \frac{1}{x+y} = \frac{x^2 + xy + 1}{x+y}$$

$$\Rightarrow \quad xy + y^2 = xy + x^2 + 1$$

$$\Rightarrow \quad y^2 = 1 + x^2 \Rightarrow y = \pm\sqrt{1+x^2}$$

$$\Rightarrow \quad y = \sqrt{1+x^2}$$

[$\because$ y cannot be negative, as x is positive]

$$\Rightarrow \quad \frac{dy}{dx} = \frac{2x}{2\sqrt{1+x^2}} = \frac{x}{\sqrt{1+x^2}}$$

$$\Rightarrow \quad \left.\frac{dy}{dx}\right|_{100} = \frac{100}{\sqrt{1+100^2}} \text{ and } y|_{100} = \sqrt{1+100^2}$$

$$\therefore \quad f(100)f'(100) = 100$$

33. (b) We have, $e^y (x+1) = 1$

On differentiating w.r.t. x, we get

$$e^y (1) + (x+1) \cdot e^y \cdot \frac{dy}{dx} = 0$$

$$\Rightarrow \quad (x+1) \cdot e^y \frac{dy}{dx} = -e^y$$

$$\Rightarrow \quad \frac{dy}{dx} = -\frac{1}{x+1}$$

Again, differentiating w.r.t. x, we get

$$\frac{d^2y}{dx^2} = \frac{1}{(x+1)^2}$$

Hence, $\frac{d^2y}{dx^2} = \left(\frac{dy}{dx}\right)^2$

34. (d) $\frac{dx}{dy} = \frac{1}{\left(\frac{dy}{dx}\right)}$

$$\Rightarrow \frac{d^2x}{dy^2} = \frac{d}{dy}\left(\frac{1}{\frac{dy}{dx}}\right) = \frac{d}{dx}\frac{1}{\left(\frac{dy}{dx}\right)} \cdot \frac{dx}{dy}$$

$$= -\frac{1}{\left(\frac{dy}{dx}\right)^2} \cdot \frac{d^2y}{dx^2} \cdot \frac{1}{\left(\frac{dy}{dx}\right)}$$

$$= -\frac{1}{\left(\frac{dy}{dx}\right)^3} \cdot \frac{d^2y}{dx^2}$$

$$= -\left(\frac{dy}{dx}\right)^{-3} \cdot \frac{d^2y}{dx^2}$$

35. (b) We have, $x = 3\tan t$ and $y = 3\sec t$

Clearly, $\frac{dy}{dx} = \frac{\frac{dy}{dt}}{\frac{dx}{dt}} = \frac{\frac{d}{dt}(3\sec t)}{\frac{d}{dt}(3\tan t)}$

$$= \frac{3\sec t \tan t}{3\sec^2 t} = \frac{\tan t}{\sec t} = \sin t$$

and $\frac{d^2y}{dx^2} = \frac{d}{dx}\left(\frac{dy}{dx}\right) = \frac{d}{dt}\left(\frac{dy}{dx}\right) \cdot \frac{dt}{dx}$

$$= \frac{\frac{d}{dt}\left(\frac{dy}{dx}\right)}{\frac{dx}{dt}}$$

$$= \frac{\frac{d}{dt}(\sin t)}{\frac{d}{dt}(3\tan t)}$$

$$= \frac{\cos t}{3\sec^2 t} = \frac{\cos^3 t}{3}$$

Now, $\frac{d^2y}{dx^2}\left(\text{at } t = \frac{\pi}{4}\right) = \frac{\cos^3 \frac{\pi}{4}}{3} = \frac{1}{3(2\sqrt{2})} = \frac{1}{6\sqrt{2}}$

36. (b) We have, $e^y + xy = e$

On differentiating w.r.t. x, we get

$$e^y \frac{dy}{dx} + y + x\frac{dy}{dx} = 0 \quad \text{...(i)}$$

Again, differentiating w.r.t. x, we get

$$e^y \frac{d^2y}{dx^2} + e^y\left(\frac{dy}{dx}\right)^2 + 2\frac{dy}{dx} + x\frac{d^2y}{dx^2} = 0 \quad \text{...(ii)}$$

On putting $x = 0$ in $e^y + xy = e$, we get

$$y = 1$$

Putting $x = 0$, $y = 1$ in Eq. (i), we get

$$e\frac{dy}{dx} + 1 = 0$$

$$\Rightarrow \quad \frac{dy}{dx} = -\frac{1}{e}$$

Putting $x = 0, y = 1$ and $\frac{dy}{dx} = -\frac{1}{e}$ in Eq. (ii), we get

$$\frac{d^2y}{dx^2} = \frac{1}{e^2}$$

37. (d) $\frac{dy}{dx} = \frac{\frac{dy}{dt}}{\frac{dx}{dt}} = \frac{g'(t)}{f'(t)}$

$$\Rightarrow \frac{d^2y}{dx^2} = \frac{d}{dt}\left(\frac{g'(t)}{f'(t)}\right) \cdot \frac{dt}{dx}$$

$$\Rightarrow \frac{d^2y}{dx^2} = \frac{f'(t)g''(t) - g'(t)f''(t)}{[f'(t)]^3}$$

and $\frac{dx}{dy} = \frac{\frac{dx}{dt}}{\frac{dy}{dt}} = \frac{f'(t)}{g'(t)}$

$$\Rightarrow \frac{d^2x}{dy^2} = \frac{d}{dt}\left(\frac{f'(t)}{g'(t)}\right) \cdot \frac{dt}{dy}$$

$$= \frac{g'(t)f''(t) - f'(t)g''(t)}{[g'(t)]^3}$$

$$\therefore \frac{\left(\frac{d^2y}{dx^2}\right)}{\left(\frac{d^2x}{dy^2}\right)} = -\left(\frac{g'(t)}{f'(t)}\right)^3$$

38. (b) $f(1)=5, f'(x)=nx^{n-1}$, so $f'(1)=n$

$f''(1)=n(n-1), \ldots$

$\vdots$

$f^n(1)=n(n-1)\ldots 2\cdot 1$

Thus, $f(1)+\dfrac{f'(1)}{1!}+\dfrac{f''(1)}{2!}+\ldots\dfrac{f^n(1)}{n!}$

$$=5+\frac{n}{1!}\frac{n+(n-1)}{2!}+\ldots+\frac{n!}{n!}$$

$$=4+(1+1)^n=4+2^n$$

39. (c) It is given that $x=2\sin\theta-\sin 2\theta$

and $y=2\cos\theta-\cos 2\theta, \theta\in[0,2\pi]$

$$\therefore \frac{dy}{dx}=\frac{\frac{dy}{d\theta}}{\frac{dx}{d\theta}}=\frac{-2\sin\theta+2\sin 2\theta}{2\cos\theta-2\cos 2\theta}=\frac{\sin 2\theta-\sin\theta}{\cos\theta-\cos 2\theta}$$

$$\therefore \frac{d^2y}{dx^2}=\frac{d}{dx}\left(\frac{dy}{dx}\right)=\frac{d}{d\theta}\left(\frac{dy}{dx}\right)\times\frac{d\theta}{dx}$$

$$=\frac{d}{d\theta}\left(\frac{\sin 2\theta-\sin\theta}{\cos\theta-\cos 2\theta}\right)\times\frac{1}{\frac{dx}{d\theta}}$$

$$=\frac{(\cos\theta-\cos 2\theta)(2\cos 2\theta-\cos\theta)-(\sin 2\theta-\sin\theta)(-\sin\theta+2\sin 2\theta)}{(\cos\theta-\cos 2\theta)^2}$$

$$\times\frac{1}{(2\cos\theta-2\cos 2\theta)}$$

$$\therefore \left.\frac{d^2y}{dx^2}\right|_{\theta=\pi}=\frac{(-1-1)(2+1)-(0-0)(-0+0)}{2(-1-1)^3}=\frac{-2\times 3}{-2\times 8}=\frac{3}{8}$$

40. (b) We have, $y=x\log\left(\dfrac{x}{a+bx}\right)$

$$\Rightarrow \frac{y}{x}=\log x-\log(a+bx)$$

On differentiating w.r.t. x, we get

$$\frac{x\frac{dy}{dx}-y}{x^2}=\frac{1}{x}-\frac{1}{a+bx}b$$

$$\Rightarrow x\frac{dy}{dx}-y=\frac{ax}{a+bx} \quad \ldots(i)$$

Again, differentiating both sides w.r.t. x, we get

$$x\frac{d^2y}{dx^2}+\frac{dy}{dx}-\frac{dy}{dx}=\frac{(a+bx)a-axb}{(a+bx)^2}$$

$$\Rightarrow x\frac{d^2y}{dx^2}=\frac{a^2}{(a+bx)^2}$$

$$\Rightarrow x^3\frac{d^2y}{dx^2}=\frac{a^2x^2}{(a+bx)^2}=\left(x\frac{dy}{dx}-y\right)^2$$

[from Eq. (i)]

41. (a) We have, $y=\tan^{-1}\left(\dfrac{\ln\frac{e}{x^2}}{\ln ex^2}\right)+\tan^{-1}\left(\dfrac{3+2\ln x}{1-6\ln x}\right)$

$$\Rightarrow y=\tan^{-1}\left(\frac{1-2\ln x}{1+2\ln x}\right)+\tan^{-1}\left(\frac{3+2\ln x}{1-6\ln x}\right)$$

Let $a=\dfrac{1-2\ln x}{1+2\ln x}$ and $b=\dfrac{3+2\ln x}{1-6\ln x}$

So, $\quad a+b=\dfrac{16(\ln x)^2+4}{(1+2\ln x)(1-6\ln x)}$

and $(1-ab)=\dfrac{-8(\ln x)^2-2}{(1+2\ln x)(1-6\ln x)}$

$$\Rightarrow \frac{a+b}{1-ab}=\frac{16(\ln x)^2+4}{-[8(\ln x)^2+2]}=-2$$

So, $y=\tan^{-1}a+\tan^{-1}b=\tan^{-1}\left(\dfrac{a+b}{1-ab}\right)$

$\Rightarrow y=\tan^{-1}(-2)$

Then, $\dfrac{dy}{dx}=0$ and $\dfrac{d^2y}{dx^2}=0$

42. (a) We have, $y=(A+Bx)e^{mx}+(m-1)^{-2}\cdot e^x$

$$y\cdot e^{-mx}=(A+Bx)+(m-1)^{-2}\cdot e^{(1-m)x}$$

$$\Rightarrow e^{-mx}\cdot y_1-my\,e^{-mx}=B-(m-1)^{-1}\cdot e^{-(m-1)x}$$

$$\Rightarrow e^{-mx}\cdot y_2-y_1e^{-mx}\cdot m-m[e^{-mx}\cdot y_1-ye^{-mx}\cdot m]=e^{-(m-1)x}$$

$$\Rightarrow e^{-mx}\cdot y_2-2my_1e^{-mx}+m^2y\cdot e^{-mx}=e^{-(m-1)x}$$

$$\Rightarrow y_2-2my_1+m^2y=e^x$$

43. (a) $\because x=\sin t$ and $y=\sin kt$

$$\Rightarrow \frac{dx}{dt}=\cos t \text{ and } \frac{dy}{dt}=k\cos kt$$

$$\therefore y_1=\frac{dy/dt}{dx/dt}=\frac{k\cos kt}{\cos t}$$

$\Rightarrow y_1\cos t=k\cos kt$

On squaring both sides, we have

$$y_1^2\cos^2 t=k^2\cos^2 kt$$

$$\Rightarrow y_1^2(1-\sin^2 t)=k^2(1-\sin^2 kt)$$

$$\Rightarrow y_1^2(1-x^2)=k^2(1-y^2)$$

On differentiating both sides w.r.t. x, we get

$$(1-x^2)(2y_1y_2)+y_1^2(-2x)=k^2(-2y_1y)$$

$$\Rightarrow (1-x^2)y_2-xy_1=-k^2y$$

44. (c) Given, $f'(x)=g(x)$

and $\quad g'(x)=f(x^2)=f''(x)$

Let $\quad y=f(x^3)$

$$\Rightarrow \quad \frac{dy}{dx} = f'(x^3)\cdot 3x^2$$

$$\Rightarrow \quad \frac{d^2y}{dx^2} = 3[3x^2x^2\cdot f''(x^3) + 2x\cdot f'(x^3)]$$

$$= 9x^4 f''(x^3) + 6xf'(x^3)$$

$$= 9x^4\cdot f(x^6) + 6xg(x^3)$$

$$= kx^a f(x^b) + pxg(x^c)$$

$\therefore \quad (k+p+a+b+c) = 9+6+4+6+3 = 28$

45. (b) We have, $f(x) = \sin x, g(x) = x^2$ and $h(x) = \log x$ and $F(x) = h[f\{g(x)\}]$

$$F(x) = \log \sin x^2$$

On differentiating both sides w.r.t. x, we get

$$\frac{dF(x)}{dx} = \frac{2x\cos(x^2)}{\sin x^2}$$

Again, differentiating both sides w.r.t. x, we get

$$\frac{d^2F(x)}{dx^2} = \frac{(2\cos x^2 - 4x^2 \sin x^2)\sin x^2 - (2x\cos x^2)^2}{\sin^2 x^2}$$

$$\Rightarrow \quad \frac{d^2F(x)}{dx^2} = 2\cot x^2 - 4x^2 \operatorname{cosec}^2 x^2$$

46. (d) Given, $f(x)$ be a polynomial in x.

On differentiating $f(e^x)$, we get

$$[f(e^x)]' = f'(e^x)\cdot e^x$$

Again, on differentiating, we get

$$[f(e^x)]'' = f'(e^x)\cdot e^x + f''(e^x)\cdot e^{2x}$$

Hence, second derivative of $f(e^x)$ is $f''(e^x)\cdot e^{2x} + f'(e^x)\cdot e^x$.

47. (d) $\because g[h(x)] = x \Rightarrow g'[h(x)]h'(x) = 1$

$$\Rightarrow \quad \frac{1}{1+[h(x)]^3}\cdot h'(x) = 1$$

$$\Rightarrow \quad h'(x) = 1 + [h(x)]^3$$

$$\Rightarrow \quad h''(x) = 3[h(x)]^2 h'(x)$$

$$\Rightarrow \quad h''(x) = 3h^2(x)[1 + h^3(x)]$$

48. (b) By sine formula,

$$\frac{a}{\sin A} = \frac{b}{\sin B}$$

$$\Rightarrow \quad \sin A = \frac{a\sin B}{b}$$

$$\Rightarrow \quad A = \sin^{-1}\left(\frac{a\sin B}{b}\right)$$

On differentiating both sides w.r.t. B, we get

$$\frac{dA}{dB} = \frac{1}{\sqrt{1-\left(\frac{a\sin B}{b}\right)^2}}\cdot\frac{a}{b}\cos B$$

$$= \frac{a\cos B}{\sqrt{b^2 - a^2\sin^2 B}} = \sqrt{\frac{a^2 - b^2\sin^2 A}{b^2 - a^2\sin^2 B}}$$

$$\left[\begin{aligned}&\because b\sin A = a\sin B\\ &\Rightarrow b^2\sin^2 A = a^2\sin^2 B\\ &\Rightarrow a^2 - b^2\sin^2 A = a^2\cos^2 B\end{aligned}\right]$$

49. (b) We have, $f(x) + f(y) + f(z) + f(x)\cdot f(y)\cdot f(z) = 14 \quad \forall\, x, y, z \in R$

Putting $x = y = z = 0$, we get

$$3f(0) + \{f(0)\}^3 = 14$$

$$\Rightarrow \quad [f(0)]^3 + 3f(0) - 14 = 0$$

$$\Rightarrow \quad f(0) = 2$$

Now, putting $x = y = z$, we get

$$3f(x) + [f(x)]^3 = 14\ \forall\, x \in R$$

On differentiating w.r.t. x, we get

$$3f'(x) + 3[f(x)]^2\cdot f'(x) = 0\ \forall\, x \in R$$

$$\Rightarrow \quad [\{f(x)\}^2 + 1]f'(x) = 0\ \forall\, x \in R$$

So, $f'(x) = 0\ \forall\, x \in R$

50. (d) We have, $f[f\{f(x)\}] + [f(x)]^2$

$$\frac{d}{dx}[[f\{f(f(x))\}] + [f(x)]^2]$$

$$= f'[f\{f(x)\}]\cdot f'[f(x)]\cdot f'(x) + 2f(x)\cdot f'(x)$$

Put $x = 1$

$$= f'[f\{f(1)\}] \times f'[f(1)]\cdot f'(1) + 2f(1)\cdot f'(1)$$

$$= f'[f(1)]\cdot f'(1)\cdot 3 + 2\cdot 1\cdot 3$$

$$= f'(1)\cdot 3\cdot 3 + 6$$

$$= 3\cdot 3\cdot 3 + 6$$

$$= 27 + 6 = 33$$

51. (b) We have, $f(x) = \begin{vmatrix} x^3 & x^4 & 3x^2 \\ 1 & -6 & 4 \\ p & p^2 & p^3 \end{vmatrix}$

$$f(x) = x^3(-6p^3 - 4p^2) - x^4(p^3 - 4p) + 3x^2(p^2 + 6p)$$

$$f(x) = -x^3(6p^3 + 4p^2) - x^4(p^3 - 4p) + 3x^2(p^2 + 6p)$$

Now, $f'(x) = -3x^2(6p^3 + 4p^2) - 4x^3(p^3 - 4p) + 6x(p^2 + 6p)$

$$f''(x) = -6x(6p^3 + 4p^2) - 12x^2(p^3 - 4p) + 6(p^2 + 6p)$$

$$f'''(x) = -6(6p^3 + 4p^2) - 24x(p^3 - 4p)$$

Hence, $\frac{d^3}{dx^3}[f(x)]$ is proportional to x.

52. (b) We have, $f(x) = \begin{vmatrix} x^2 & x^3 & x^4 \\ 2x & 3x^2 & 4x^3 \\ 2 & 6x & 12x^2 \end{vmatrix}$

On differentiating both sides w.r.t. x, we get

$$f'(x)=\begin{vmatrix} 2x & 3x^2 & 4x^3 \\ 2x & 3x^2 & 4x^3 \\ 2 & 6x & 12x^2 \end{vmatrix}+\begin{vmatrix} x^2 & x^3 & x^4 \\ 2 & 6x & 12x^2 \\ 2 & 6x & 12x^2 \end{vmatrix}$$

$$+\begin{vmatrix} x^2 & x^3 & x^4 \\ 2x & 3x^2 & 4x^3 \\ 0 & 6 & 24x \end{vmatrix}$$

$$=0+0+\begin{vmatrix} x^2 & x^3 & x^4 \\ 2x & 3x^2 & 4x^3 \\ 0 & 6 & 24x \end{vmatrix}$$

[expanding along R_3]

$\Rightarrow \quad f'(x)=24x^5-12x^5 \Rightarrow f'(x)=12x^5$

So, $\quad f'(1)=12(1)^5=12$

53. (b) We have, $f(x)=\begin{vmatrix} \cos x & x & 1 \\ 2\sin x & x^2 & 2x \\ \tan x & x & 1 \end{vmatrix}$

Operating $R_1 \to R_1 - R_3$

$\Rightarrow f(x)=\begin{vmatrix} \cos x-\tan x & 0 & 0 \\ 2\sin x & x^2 & 2x \\ \tan x & x & 1 \end{vmatrix}$

$\Rightarrow f(x)=(\cos x-\tan x)(x^2-2x^2)=x^2(\tan x-\cos x)$

$\Rightarrow f'(x)=2x(\tan x-\cos x)-x^2(\sec^2 x+\sin x)$

$\Rightarrow \dfrac{f'(x)}{x}=2\tan x-2\cos x-x\sec^2 x-x\sin x$

$\Rightarrow \lim\limits_{x\to 0}\left(\dfrac{f'(x)}{x}\right)=0-2=-2$

54. (a) $\dfrac{du}{dv}=\dfrac{\frac{du}{dx}}{\frac{dv}{dx}}=\dfrac{f'(x^3)\cdot 3x^2}{g'(x^2)\cdot 2x}$

$=\dfrac{\cos x^3\cdot 3x^2}{\sin x^2\cdot 2x}$

$=\dfrac{3}{2}x\cos x^3\cdot \operatorname{cosec} x^2$

55. (a) $y=\sin^{-1}(4x\sqrt{1-4x^2})$ and $v=\sqrt{1-4x^2}$

We know that,

$2\sin^{-1}x=\left\{-\pi-\sin^{-1}(2x\sqrt{1-x^2})\text{ for }\dfrac{-1}{\sqrt{2}}<x<-1\right.$

$\Rightarrow \quad y=\sin^{-1}(4x\sqrt{1-4x^2})=-\pi-2\sin^{-1}(2x)$

$\Rightarrow \quad \dfrac{dy}{dx}=\dfrac{-4}{\sqrt{1-4x^2}}$ and $\dfrac{dv}{dx}=\dfrac{-4x}{\sqrt{1-4x^2}}$

$\Rightarrow \quad \dfrac{dy}{dv}=\dfrac{1}{x}$

56. (a) $g(x+1)=\log[f(x+1)]=\log[xf(x)]$
$=\log x+\log f(x)=\log x+g(x)$

$\Rightarrow \quad g(x+1)-g(x)=\log x$

$\Rightarrow \quad g'(x+1)-g'(x)=\dfrac{1}{x}$

$\Rightarrow \quad g''(x+1)-g''(x)=-\dfrac{1}{x^2}$

$\Rightarrow \quad g''\left(1+\dfrac{1}{2}\right)-g''\left(\dfrac{1}{2}\right)=-4$

$\Rightarrow g''\left(2+\dfrac{1}{2}\right)-g''\left(1+\dfrac{1}{2}\right)=-\dfrac{4}{9}$

$\Rightarrow \quad g''\left(N+\dfrac{1}{2}\right)-g''\left(N-1+\dfrac{1}{2}\right)=\dfrac{-4}{(2N-1)^2}$

Suming up all terms

Hence,

$g''\left(N+\dfrac{1}{2}\right)-g''\left(\dfrac{1}{2}\right)=-4\left(1+\dfrac{1}{9}+\ldots\dfrac{1}{(2N-1)^2}\right)$

57. (c) Let $u=\log_{10}x \Rightarrow u=\dfrac{\log_e x}{\log_e 10}$

On differentiating both sides w.r.t. x, we get

$\dfrac{du}{dx}=\dfrac{1}{x\log_e 10}$

Now, let $\quad v=x^3$

$\Rightarrow \quad \dfrac{dv}{dx}=3x^2$

$\Rightarrow \quad \dfrac{du}{dv}=\dfrac{\frac{du}{dx}}{\frac{dv}{dx}}=\dfrac{\frac{1}{x\log_e 10}}{3x^2}$

$=\dfrac{1}{3x^3\log_e 10}=\dfrac{1}{3x^3}\log_{10}e$

58. (1) We have, $y=\dfrac{5^{\log_{4\sqrt{5}}x}-9^{\log_{729}(x+1)^6}}{49^{4\log_{2401}x}-x-1}$

$\Rightarrow \quad y=\dfrac{x^4-(x+1)^2}{x^2-x-1}$

$\Rightarrow \quad y=\dfrac{(x^2+x+1)[x^2-(x+1)]}{x^2-x-1}$

$\Rightarrow \quad y=\dfrac{(x^2+x+1)(x^2-x-1)}{x^2-x-1}$

$\Rightarrow \quad y=x^2+x+1$

On differentiating w.r.t. x, we get

$\dfrac{dy}{dx}=2x+1=ax-b \qquad$ (given)

$\Rightarrow \quad a=2$ and $b=-1$

So, $a+b=2+(-1)=1$

59. (4) We have, $f(x)=\sqrt{x+\sqrt{8x-16}}+\sqrt{x-\sqrt{8x-16}}$

Let $8x-16=t^2$

$$=\sqrt{\frac{t^2+16+8t}{8}}+\sqrt{\frac{t^2+16-8t}{8}}$$

$$=\frac{|t+4|}{2\sqrt{2}}+\frac{|t-4|}{2\sqrt{2}}$$

Hence, $f(x)$ is not differentiable at $x=4\in(2,\infty)$.

60. (2) We have, $g(x)=f^{-1}(x)$

$\Rightarrow \quad f[g(x)]=x$ ...(i)

$\therefore \quad f[g(-1)]=-1$

$\Rightarrow [g(-1)]^3+3[g(-1)]^2+6g(-1)-5+4e^{2g(-1)}=-1$

$\Rightarrow \quad [g(-1)+1]^3+3g(-1)-6+4e^{2g(-1)}=-1$

$\Rightarrow \quad [g(-1)+1]^3+3g(-1)+4e^{2g(-1)}=5$

It is clear that $g(-1)=0$ ...(ii)

On differentiating Eq. (i) w.r.t. x, we get

$f'[g(x)]\cdot g'(x)=1$

$\Rightarrow \quad f'[g(-1)]\cdot g'(-1)=1$

$\Rightarrow \quad f'(0)\cdot g'(-1)=1 \Rightarrow g'(-1)=\dfrac{1}{f'(0)}$

$$=\frac{1}{3(0)^2+6(0)+6-0+8e^0}$$

$$=\frac{1}{14}$$

$\Rightarrow \quad 28g'(-1)=28\times\dfrac{1}{14}=2$

61. (1) Given,

$g(x)=(ax^2+bx+c)\sin x+(dx^2+ex+f)\cos x$

$g'(x)=(2ax+b)\sin x+(ax^2+bx+c)\cos x$
$\quad +(2dx+e)\cos x-(dx^2+ex+f)\sin x$

$=\sin x\,(2ax+b-dx^2-ex-f)$
$\quad +\cos x\,(ax^2+bx+c+2dx+e)$

Given that, $g'(x)=x^2\sin x$

On comparing equations, we get

$a=0, b=2, c=0, d=-1, e=0$ and $f=0$

Hence, $a+b+c+d+e+f=1$.

62. (4) We have, $y=\sin^{-1}\left(\ln\left(\dfrac{x^2}{2}\right)\right)$

$\Rightarrow \quad \sin y=\ln\dfrac{x^2}{2}$

$\Rightarrow \quad \cos y\cdot y'=\dfrac{2}{x^2}\cdot\left(\dfrac{2x}{2}\right)$

$\Rightarrow \quad y'=\dfrac{2}{x\cos y}$

$\Rightarrow \quad (xy')^2=\dfrac{4}{\cos^2 y}$

$\Rightarrow (xy')^2\left\{\left[\ln\left(\dfrac{x^2}{2}\right)\right]^2-1\right\}=\dfrac{4}{\cos^2 y}\{\sin^2 y-1\}=-4$

$\Rightarrow \quad k=-4 \Rightarrow |k|=4$

63. (91)

Given, $y=\displaystyle\sum_{k=1}^{6}k\cos^{-1}\left\{\frac{3}{5}\cos(kx)-\frac{4}{5}\sin(kx)\right\}$

Let $\quad \dfrac{3}{5}=\cos\theta$ and $\dfrac{4}{5}=\sin\theta$

So, $\quad \dfrac{3}{5}\cos(kx)-\dfrac{4}{5}\sin(kx)=\cos(kx+\theta)$

where, $\tan\theta=4/3$

$\therefore \quad y=\displaystyle\sum_{k=1}^{6}k\cos^{-1}\cos(kx+\theta)=\sum_{k=1}^{6}(k^2x+k\theta)$

$$=x\left(\frac{6\times7\times13}{6}\right)+\left(\frac{6\times7}{2}\right)\theta$$

$\Rightarrow \quad y=91x+21\theta$

$\therefore \quad \left.\dfrac{dy}{dx}\right|_{x=0}=91$

64. (2) We have, $\log((e-1)e^{xy}+x^2)=x^2+y^2$

$\Rightarrow \quad (e-1)e^{xy}+x^2=e^{x^2+y^2}$

On differentiating both sides w.r.t. x, we get

$$(e-1)\cdot e^{xy}\cdot\left(x\frac{dy}{dx}+y\right)+2x=e^{x^2+y^2}\left(2x+2y\frac{dy}{dx}\right)$$

Putting $x=1$ and $y=0$, we get

$$(e-1)\frac{dy}{dx}+2=e(2+0)$$

$\Rightarrow \quad (e-1)\dfrac{dy}{dx}=2e-2$

$\Rightarrow \quad \dfrac{dy}{dx}=\dfrac{2(e-1)}{e-1} \Rightarrow \dfrac{dy}{dx}=2$

65. (1) We have, $y\sqrt{x^2+1}=\log(\sqrt{x^2+1}-x)$

On differentiating both sides w.r.t. x, we get

$$\frac{dy}{dx}\sqrt{x^2+1}+y\cdot\frac{1}{2\sqrt{x^2+1}}\cdot 2x$$

$$=\frac{1}{\sqrt{x^2+1}-x}\times\left\{\frac{1}{2}\frac{2x}{\sqrt{x^2+1}}-1\right\}$$

$\Rightarrow \quad (x^2+1)\dfrac{dy}{dx}+xy=\sqrt{x^2+1}\left(-\dfrac{1}{\sqrt{x^2+1}}\right)$

$\Rightarrow \quad (x^2+1)\dfrac{dy}{dx}+xy=-1$

$\Rightarrow (x^2+1)\dfrac{dy}{dx}+xy+1=0$

By comparing $(x^2+1)\frac{dy}{dx}+xy+k=0$, we get

$$k=1$$

66. (4) We have, $4xe^{xy}=y+5\sin^2 x$ ...(i)

Putting $x=0$ in above equation, we get $y=0$

$\therefore$ (0, 0) lies on the curve.

Now, differentiating Eq. (i) w.r.t. x, we get

$$4e^{xy}+4xe^{xy}\left(y+x\frac{dy}{dx}\right)=\frac{dy}{dx}+10\sin x\cos x \quad ...(ii)$$

Putting (0, 0) in Eq. (ii), we get

$$4+0=\left(\frac{dy}{dx}\right)_{(0,0)}+0$$

$$\Rightarrow \left(\frac{dy}{dx}\right)_{(0,0)}=4$$

67. (3) We have, $x^2+y^2=t-\frac{1}{t}$

On squaring both sides, we get

$$(x^2+y^2)^2=\left(t-\frac{1}{t}\right)^2$$

$$\Rightarrow x^4+y^4+2x^2y^2=t^2+\frac{1}{t^2}-2$$

$$\Rightarrow x^4+y^4+2x^2y^2=x^4+y^4-2 \quad \left[\text{given } x^4+y^4=t^2+\frac{1}{t^2}\right]$$

$$\Rightarrow 2x^2y^2=-2$$

$$\Rightarrow x^2y^2=-1$$

$$\Rightarrow y^2=-\frac{1}{x^2}$$

On differentiating w.r.t. x, we get

$$2y\frac{dy}{dx}=-(-2)x^{-3}$$

$$\Rightarrow y\frac{dy}{dx}=\frac{1}{x^3}$$

$$\Rightarrow x^3y\frac{dy}{dx}=1$$

$$\Rightarrow x^3y\frac{dy}{dx}-1=0$$

By comparing,

$$m=3$$

68. (4) We have,

$$\sin x\cos y=\frac{1}{2}$$

On differentiating w.r.t. x, we get

$$\cos x\cos y-\sin y\sin x\frac{dy}{dx}=0$$

$$\Rightarrow \frac{dy}{dx}=\cot x\cot y$$

$$\Rightarrow \frac{d^2y}{dx^2}=-\operatorname{cosec}^2 x\cdot\cot y-\operatorname{cosec}^2 y\cdot\cot x\cdot\frac{dy}{dx}$$

Now, $\left(\frac{dy}{dx}\right)_{\left(-\frac{\pi}{4},\frac{\pi}{4}\right)}=-1$

$$\Rightarrow \left(\frac{d^2y}{dx^2}\right)_{\left(-\frac{\pi}{4},\frac{\pi}{4}\right)}=-(2)(1)-2(1)(-1)(-1)$$

$$=-4$$

So, $\left|\left(\frac{d^2y}{dx^2}\right)_{\left(-\frac{\pi}{4},\frac{\pi}{4}\right)}\right|=4$

69. (40) We have, $\ln(x+y)=4xy$

$$\Rightarrow x+y=e^{4xy}$$

$$\Rightarrow 1+\frac{dy}{dx}=\left(4x\frac{dy}{dx}+4y\right)e^{4xy}$$

If $x=0$, then $y=1$

At (0, 1), $\frac{dy}{dx}=3$

$$\frac{d^2y}{dx^2}=e^{4xy}\left(4x\frac{dy}{dx}+4y\right)^2+e^{4xy}\left(4x\frac{d^2y}{dx^2}+\frac{4dy}{dx}+4\frac{dy}{dx}\right)$$

At $x=0$

$$\frac{d^2y}{dx^2}=16+24=40$$

70. (3) We have, $P=a^2\cos^2\theta+b^2\sin^2\theta$

$$\Rightarrow \frac{dP}{d\theta}=2a^2\cos\theta(-\sin\theta)+2b^2(\sin\theta\cos\theta)$$

$$=-a^2\sin 2\theta+b^2\sin 2\theta=\sin 2\theta(b^2-a^2)$$

$$\Rightarrow \frac{d^2P}{d\theta^2}=2(b^2-a^2)\cos 2\theta$$

Now, $P+\frac{d^2P}{d\theta^2}=a^2\cos^2\theta+b^2\sin^2\theta+2(b^2-a^2)\cos 2\theta$

$=a^2\cos^2\theta+b^2\sin^2\theta+2(b^2-a^2)(\cos^2\theta-\sin^2\theta)$

$=(2a^2\cos^2\theta+2a^2\sin^2\theta)-a^2\cos^2\theta+(2b^2\sin^2\theta+2b^2\cos^2\theta)-b^2\sin^2\theta-2b^2\sin^2\theta-2a^2\cos^2\theta$

$=2a^2+2b^2-3(a^2\cos^2\theta+b^2\sin^2\theta)$

$=2a^2+2b^2-3P$

By comparing with $2a^2+2b^2-kP$, we get

$$k=3$$

71. (17) We have, $y = e^{-2x}\cos 3x$

On differentiating both sides w.r.t. x, we get

$$y' = -2e^{-2x}\cos 3x - 3e^{-2x}\sin 3x$$
$$y' = -2y - 3e^{-2x}\sin 3x$$

Again, differentiating both sides w.r.t. x, we get

$$y'' = -2y' + 6e^{-2x}\sin 3x - 9e^{-2x}\cos 3x$$
$$y'' = -2y' + 2(-y' - 2y) - 9y$$
$$y'' = -4y' - 13y$$
$$y'' + 4y' + 13y = 0$$

Then, $a = 4$ and $b = 13$

$\therefore\ a + b = 4 + 13 = 17$

72. (4) Let $u = \sec^{-1}\left(\dfrac{1}{2x^2 - 1}\right)$

$\Rightarrow\quad u = \cos^{-1}(2x^2 - 1)$

On differentiating w.r.t. x, we get

$$\frac{du}{dx} = -\frac{4x}{\sqrt{1-(2x^2-1)^2}} = \frac{-4x}{\sqrt{1-4x^4-1+4x^2}}$$
$$= \frac{-4x}{\sqrt{4x^2-4x^4}} = \frac{-2x}{\sqrt{x^2-x^4}} = \frac{-2}{\sqrt{1-x^2}}$$

Now, let $v = \sqrt{1-x^2}$

$$\frac{dv}{dx} = \frac{1}{2\sqrt{1-x^2}} \times (-2x) = \frac{-x}{\sqrt{1-x^2}}$$

Now, $$\frac{du}{dv} = \frac{\dfrac{du}{dx}}{\dfrac{dv}{dx}} = \frac{\dfrac{-2}{\sqrt{1-x^2}}}{\dfrac{-x}{\sqrt{1-x^2}}} = \frac{2}{x}$$

At $x = \dfrac{1}{2}$, $\left.\dfrac{du}{dv}\right|_{x=\frac{1}{2}} = \dfrac{2}{1/2} = 4$

73. (6) We have, $y(x) = \begin{vmatrix} \sin x & \cos x & \sin x + \cos x + 1 \\ 23 & 17 & 13 \\ 1 & 1 & 1 \end{vmatrix}$

$$\Rightarrow\quad \frac{dy}{dx} = \begin{vmatrix} \cos x & -\sin x & \cos x - \sin x \\ 23 & 17 & 13 \\ 1 & 1 & 1 \end{vmatrix}$$

$$\Rightarrow\quad \frac{d^2y}{dx^2} = \begin{vmatrix} -\sin x & -\cos x & -\sin x - \cos x \\ 23 & 17 & 13 \\ 1 & 1 & 1 \end{vmatrix}$$

Now, $$\frac{d^2y}{dx^2} + y = \begin{vmatrix} -\sin x & -\cos x & -\sin x - \cos x \\ 23 & 17 & 13 \\ 1 & 1 & 1 \end{vmatrix} + \begin{vmatrix} \sin x & \cos x & \sin x + \cos x + 1 \\ 23 & 17 & 13 \\ 1 & 1 & 1 \end{vmatrix}$$

$$\Rightarrow\quad \frac{d^2y}{dx^2} + y = \begin{vmatrix} 0 & 0 & 1 \\ 23 & 17 & 13 \\ 1 & 1 & 1 \end{vmatrix}$$

$$\Rightarrow\quad \frac{d^2y}{dx^2} + y = 23 - 17 = 6$$

Alternate Method

Using $C_2 \to C_2 - C_1, C_3 \to C_3 - C_1$, we get

$$y = \begin{vmatrix} \sin x & \cos x - \sin x & \cos x + 1 \\ 23 & -6 & -10 \\ 1 & 0 & 0 \end{vmatrix}$$

$$= -10(\cos x - \sin x) + 6\cos x + 6 = 10\sin x - 4\cos x + 6$$

$$\Rightarrow\quad \frac{dy}{dx} = 10\cos x + 4\sin x$$

$$\Rightarrow\quad \frac{d^2y}{dx^2} = -10\sin x + 4\cos x = -y + 6 \Rightarrow \frac{d^2y}{dx^2} + y = 6$$

74. (4) Given, $f(0) = -1$ and $f'(0) = 1$

Also, given $g(x) = f[2\{f(x)\} + 2]^2$

Thus, $g'(x) = 2f\,[2f(x) + 2]f'[2\{f(x)\} + 2]\cdot 2f'(x)$

$$\Rightarrow\quad g'(0) = 2f[2f(0) + 2]f'[2\{f(0)\} + 2]\cdot 2f'(0)$$
$$= 2f(0)f'(0)\cdot 2f'(0)$$
$$= 2\cdot(-1)1\cdot 2(1)$$
$$\Rightarrow\quad g'(0) = -4$$

So, $|g'(0)| = 4$

75. (2) We have, $f(x) = f'(x) + f''(x) + \ldots \infty$ terms ...(i)

and $f'(-1) = 1$

$\Rightarrow\quad f'(x) = f''(x) + f'''(x) + f^{iv}(x) + \ldots$ ...(ii)

Now, we will be subtracting Eq. (ii) from Eq. (i), we get

$$f(x) - f'(x) = f'(x) \Rightarrow f(x) = 2f'(x)$$

Now, for $x = -1$

$$f(-1) = 2f'(-1) = 2(1)$$
$$\Rightarrow\quad f(-1) = 2$$

76. (0.1732) We have, $\tan^{-1}\left(\dfrac{\sqrt{1+x^2}-1}{x}\right)$

Put $x = \tan\theta$

$$\tan^{-1}\left(\frac{\sqrt{1+\tan^2\theta}-1}{\tan\theta}\right) = \tan^{-1}\left(\frac{\sec\theta - 1}{\tan\theta}\right)$$
$$= \tan^{-1}\left(\frac{1-\cos\theta}{\sin\theta}\right)$$
$$= \tan^{-1}\left(\tan\frac{\theta}{2}\right) = \frac{\theta}{2} = \frac{1}{2}\tan^{-1}x$$

and $\tan^{-1}\left(\dfrac{2x\sqrt{1-x^2}}{1-2x^2}\right)$

Put $x = \sin\theta$

$$\tan^{-1}\left(\frac{2\sin\theta\cos\theta}{1-2\sin^2\theta}\right)=\tan^{-1}\left(\frac{\sin 2\theta}{\cos 2\theta}\right)$$

$$=\tan^{-1}(\tan 2\theta)$$

$$=2\theta=2\sin^{-1}x$$

Now, $$\frac{\frac{d}{dx}\left(\frac{\tan^{-1}x}{2}\right)}{\frac{d}{dx}(2\sin^{-1}x)}=\frac{\frac{1}{2}\,\frac{1}{1+x^2}}{\frac{2}{\sqrt{1-x^2}}}=\frac{1}{4}\,\frac{\sqrt{1-x^2}}{1+x^2}$$

At $x=\frac{1}{2}$,

$$=\frac{\sqrt{3}}{4\cdot 2\left(\frac{5}{4}\right)}=\frac{\sqrt{3}}{10}=\frac{1.732}{10}=0.1732$$

Level 3

1. (c) We have, $\log(x+y)=3xy$

On differentiating both sides w.r.t. x, we get

$$\frac{1}{x+y}\left(1+\frac{dy}{dx}\right)=3\left(x\frac{dy}{dx}+y\right)$$

On putting $x=0$ and $y=1$

$$\frac{1}{0+1}\left(1+\frac{dy}{dx}\right)=3(0+1)$$

$$\Rightarrow \quad 1+\frac{dy}{dx}=3$$

$$\Rightarrow \quad \frac{dy}{dx}=2$$

2. (b) We know that $f(f^{-1}(x))=x$

$$f(g(x))=x \quad [\because f^{-1}(x)=g(x) \text{ given}]$$

On differentiating both sides w.r.t. x, we get

$$f'[g(x)]\cdot g'(x)=1$$

$$\Rightarrow \quad g'(x)=\frac{1}{f'[g(x)]}$$

At $x=2\pi$,

$$g'(2\pi)=\frac{1}{f'[g(2\pi)]} \quad \text{...(i)}$$

Now, $$f(x)=(2x-3\pi)^5+\frac{4x}{3} \quad \text{...(ii)}$$

At $x=\frac{3\pi}{2}$,

$$f\left(\frac{3\pi}{2}\right)=\frac{4}{3}\times\frac{3\pi}{2}=2\pi$$

On differentiating Eq. (ii) w.r.t. x, we get

$$f'(x)=5(2x-3\pi)^4\cdot 2+\frac{4}{3}$$

Now, from Eq. (i),

$$g'(2\pi)=\frac{1}{f'\left(\frac{3\pi}{2}\right)}=\frac{1}{5\left(2\times\frac{3\pi}{2}-3\pi\right)^4\times 2+\frac{4}{3}}=\frac{1}{4/3}$$

$$\Rightarrow \quad g'(2\pi)=\frac{3}{4}$$

Alternate Method

Given, $g^{-1}(x)=f(x)\Rightarrow g(f(x))=x$ ...(i)

On differentiating Eq. (i) w.r.t. x, we get

$$g'(f(x))f'(x)=1\Rightarrow g'(f(x))=\frac{1}{f'(x)}$$

$$\Rightarrow \quad g'(y)=\frac{1}{f'(x)} \quad [\text{let } g=f(x)]$$

Now, $y=2\pi$, where $x=\frac{3\pi}{2}$ $\left[\because f\left(\frac{3\pi}{2}\right)=2\pi\right]$

$$\therefore g'(2\pi)=\frac{1}{f'\left(\frac{3\pi}{2}\right)}=\frac{1}{5\left(2\times\frac{3\pi}{2}-3\pi\right)^4\times 2+\frac{4}{3}}$$

$$=\frac{1}{4/3}=\frac{3}{4}$$

3. (d) We have, $f(\alpha)=\cos\left[\cot^{-1}\left(\frac{\cos\alpha}{\sqrt{-\cos 2\alpha}}\right)\right]$

Let $\theta=\cot^{-1}\left(\frac{\cos\alpha}{\sqrt{-\cos 2\alpha}}\right)$

$$\cot\theta=\frac{\cos\alpha}{\sqrt{-\cos 2\alpha}}$$

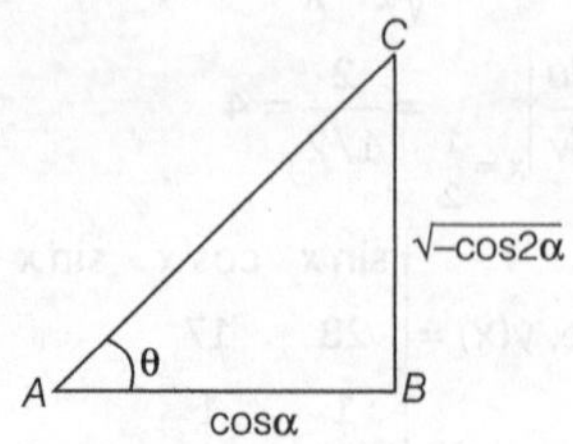

Now, $$AC=\sqrt{\cos^2\alpha-\cos 2\alpha}$$

$$=\sqrt{\cos^2\alpha-\cos^2\alpha+\sin^2\alpha}$$

$$=|\sin\alpha|$$

$$AC=\sin\alpha \quad \left[\because \sin\alpha>0 \text{ for } \alpha\in\left(\frac{\pi}{4},\frac{\pi}{2}\right)\right]$$

So, $$\theta=\cos^{-1}\left(\frac{\cos\alpha}{\sin\alpha}\right)$$

So, $$f(\alpha)=\cos\left[\cos^{-1}\left(\frac{\cos\alpha}{\sin\alpha}\right)\right]$$

$$f(\alpha)=\cot\alpha$$

$$\therefore \frac{df(\alpha)}{d(\cot\alpha)}=\frac{d(\cot\alpha)}{d(\cot\alpha)}=1$$

4. (b) We have,

$$y=\frac{\cos 6x+6\cos 4x+15\cos 2x+10}{\cos 5x+5\cos 3x+10\cos x}$$

$$=\frac{\cos 6x+(5+1)\cos 4x+(5+10)\cos 2x+10}{\cos 5x+5\cos 3x+10\cos x}$$

$$=\frac{(\cos 6x+\cos 4x)+5(\cos 4x+\cos 2x)+10(1+\cos 2x)}{\cos 5x+5\cos 3x+10\cos x}$$

$$=\frac{2\cos 5x\cos x+10\cos 3x\cos x+20\cos^2 x}{\cos 5x+5\cos 3x+10\cos x}$$

$$=\frac{2\cos x(\cos 5x+5\cos 3x+10\cos x)}{\cos 5x+5\cos 3x+10\cos x}$$

$$y=2\cos x$$

On differentiating both sides w.r.t. x, we get

$$\frac{dy}{dx}=2(-\sin x)=-2\sin x$$

5. (b) Let $x=a\cos^2\theta+b\sin^2\theta$

$$\Rightarrow \quad \frac{dx}{d\theta}=(b-a)\sin 2\theta$$

Here, $a-x=a-a\cos^2\theta-b\sin^2\theta=(a-b)\sin^2\theta$

and $\quad x-b=a\cos^2\theta+b\sin^2\theta-b=(a-b)\cos^2\theta$

Now, $y=\sqrt{(a-x)(x-b)}-(a-b)\tan^{-1}\sqrt{\dfrac{a-x}{x-b}}$

$$=(a-b)\sin\theta\cos\theta-(a-b)\tan^{-1}(\tan\theta)$$

$$=\frac{a-b}{2}\sin 2\theta-(a-b)\theta$$

On differentiating both sides w.r.t. θ, we get

$$\frac{dy}{d\theta}=(a-b)\cos 2\theta-(a-b)$$

Now, $\dfrac{dy}{dx}=\dfrac{(a-b)\cos 2\theta-(a-b)}{(b-a)\sin 2\theta}$

$$=\frac{1-\cos 2\theta}{\sin 2\theta}=\tan\theta=\sqrt{\frac{a-x}{x-b}}$$

6. (c) We have, $f(x)=|x-1|$

$\therefore \quad g(x)=f(f(f(x)))=f(f(|x-1|))=f(||x-1|-1|)$

$\Rightarrow \quad g(x)=|||x-1|-1|-1|$

$$\Rightarrow \quad g(x)=\begin{cases}||x-2|-1|, & x\geq 1\\ ||-x|-1|, & x<1\end{cases}$$

$$\Rightarrow \quad g(x)=\begin{cases}|x-3|, & x\geq 2\\ |2-x-1|, & 1\leq x<2\\ |x-1|, & 0\leq x<1\\ |-x-1|, & x<0\end{cases}$$

$$\Rightarrow \quad g(x)=\begin{cases}x-3, & x\geq 3\\ 3-x, & 2\leq x<3\\ x-1, & 1\leq x<2\\ 1-x, & 0\leq x<1\\ x+1, & -1\leq x<0\\ -x-1, & x<-1\end{cases}$$

$$\Rightarrow \quad g'(x)=\begin{cases}1, & x>3\\ -1, & 2<x<3\\ 1, & 1<x<2\\ -1, & 0<x<1\\ 1, & -1<x<0\\ -1, & x<-1\end{cases}$$

7. (a) We have, $e^x=\dfrac{\sqrt{1+t}-\sqrt{1-t}}{\sqrt{1+t}+\sqrt{1-t}}$

$$\Rightarrow e^x=\frac{1-\sqrt{\dfrac{1-t}{1+t}}}{1+\sqrt{\dfrac{1-t}{1+t}}}=\frac{1-\tan\dfrac{y}{2}}{1+\tan\dfrac{y}{2}}$$

$$\Rightarrow e^x=\tan\left(\frac{\pi}{4}-\frac{y}{2}\right)$$

On differentiating w.r.t. x, we get

$$e^x=\sec^2\left(\frac{\pi}{4}-\frac{y}{2}\right)\left(\frac{-1}{2}\right)\frac{dy}{dx}$$

$$\Rightarrow \frac{dy}{dx}=-2e^x\cos^2\left(\frac{\pi}{4}-\frac{y}{2}\right) \quad \text{...(i)}$$

When $t=\dfrac{1}{2}$,

$$\tan\frac{y}{2}=\frac{1}{\sqrt{3}} \Rightarrow \frac{y}{2}=\frac{\pi}{6}$$

$\therefore$ From Eq. (i),

$$\frac{dy}{dx}=-2[e^x]_{t=1/2}\cdot\cos^2\left(\frac{\pi}{4}-\frac{\pi}{6}\right)$$

$$=-2\left[\frac{\sqrt{\dfrac{3}{2}}-\sqrt{\dfrac{1}{2}}}{\sqrt{\dfrac{3}{2}}+\sqrt{\dfrac{1}{2}}}\right]\cdot\left[\frac{1}{\sqrt{2}}\cdot\frac{\sqrt{3}}{2}+\frac{1}{\sqrt{2}}\cdot\frac{1}{2}\right]^2$$

$$=-2\left[\frac{\sqrt{3}-1}{\sqrt{3}+1}\right]\cdot\left[\frac{\sqrt{3}+1}{2\sqrt{2}}\right]^2$$

$$=\frac{-2}{8}(3-1)=\frac{-1}{2}$$

8. (d) We have, $f(x)=\sin^{-1}\{[3x+2]-\{(3x+(x-\{2x\})\}\}$

$\because \quad x\in\left(0,\dfrac{\pi}{12}\right) \Rightarrow 2x\in\left(0,\dfrac{\pi}{6}\right)$

$\therefore$ $\{2x\}=2x$ and $3x\in\left(0,\dfrac{\pi}{4}\right)$

So, $[3x]=0$

$$f(x)=\sin^{-1}[2-(3x-x)]=\sin^{-1}[2-2x]$$

$\Rightarrow \quad f(x)=\sin^{-1}(1-2x)=y \qquad$ [as $[2-2x]=1-2x$]

$\Rightarrow 1-2x=\sin y$

$$\Rightarrow \quad x=\frac{1-\sin y}{2}=f^{-1}(y)$$

$$\therefore \quad f^{-1}(x) = \frac{1-\sin x}{2} = g(x)$$

$$\Rightarrow \quad g'(x) = \frac{-1}{2}\cos x \Rightarrow g'\left(\frac{\pi}{6}\right) = \frac{-1}{2}\left(\frac{\sqrt{3}}{2}\right) = -\frac{\sqrt{3}}{4}$$

9. (b) We have, $f(x) \cdot g(x) = 1$

On differentiating both sides w.r.t. x, we get

$$f'(g) + g'f = 0 \quad \text{...(i)}$$

Again, differentiating both sides w.r.t. x, we get

$$f''g + 2f'g' + f \cdot g'' = 0 \quad \text{...(ii)}$$

$$\Rightarrow \frac{f'''}{f'}(f'g) + \frac{g'''}{g'}(fg') + \frac{3f''}{f}(fg') + \frac{3g''}{g}(gf') = 0$$

$$\Rightarrow \left(\frac{f'''}{f'} + \frac{3g''}{g}\right)(f'g) = -\left(\frac{g'''}{g'} + \frac{3f''}{f}\right)(fg')$$

$$\Rightarrow -\left(\frac{f'''}{f'} - \frac{3g''}{g}\right) \cdot (fg') = -\left(\frac{g'''}{g'} + \frac{3f''}{g}\right) fg'$$

[using Eq. (i)]

$$\Rightarrow \quad \frac{f'''}{f'} + \frac{3g''}{g} = \frac{g'''}{g'} + \frac{3f''}{f}$$

$$\Rightarrow \quad \frac{f'''}{f'} - \frac{g'''}{g'} = \frac{3f''}{f} - \frac{3g''}{g} = 3\left(\frac{f''}{f} - \frac{g''}{g}\right)$$

10. (b) We have, $x = a\cos t + \frac{b}{2}\cos 2t$

$$\frac{dx}{dt} = -a\sin t - b\sin 2t \quad \text{...(i)}$$

and $\quad y = a\sin t + \frac{b}{2}\sin 2t$

$$\frac{dy}{dt} = a\cos t + b\cos 2t \quad \text{...(ii)}$$

$$\therefore \quad \frac{dy}{dx} = \frac{\frac{dy}{dt}}{\frac{dx}{dt}} = \frac{a\cos t + b\cos 2t}{-(a\sin t + b\sin 2t)}$$

$$\Rightarrow \quad \frac{dy}{dx} = \frac{-(a\cos t + b\cos 2t)}{(a\sin t + b\sin 2t)}$$

Now, $\frac{d^2y}{dx^2} = \frac{d}{dx}\left(\frac{dy}{dx}\right) = \frac{d}{dt}\left(\frac{dy}{dx}\right) \cdot \frac{dt}{dx}$

$$= -\frac{[(a\sin t + b\sin 2t)(-a\sin t - 2b\sin 2t) - (a\cos t + b\cos 2t)(a\cos t + 2b\cos 2t)]}{(a\sin t + b\sin 2t)^2} \times \frac{1}{-a\sin t - b\sin 2t}$$

$$= -\left[\frac{a^2\sin^2 t + 3ab\sin t\sin 2t + 2b^2\sin^2 2t + a^2\cos^2 t + 3ab\cos t\cos 2t + 2b^2\cos^2 2t}{(a\sin t + b\sin 2t)^3}\right]$$

$$= -\left[\frac{a^2(\sin^2 t + \cos^2 t) + 2b^2(\sin^2 2t + \cos^2 2t) + 3ab\cos(2t - t)}{(a\sin t + b\sin 2t)^3}\right]$$

$$\frac{d^2y}{dx^2} = -\left[\frac{a^2 + 2b^2 + 3ab\cos t}{(a\sin t + b\sin 2t)^3}\right]$$

Given, $\frac{d^2y}{dx^2} = 0$

$$\Rightarrow a^2 + 2b^2 + 3ab\cos t = 0 \Rightarrow \cos t = -\left(\frac{a^2 + 2b^2}{3ab}\right)$$

11. (a) Given, $y^2 = ax + bx + c$

On differentiating both sides w.r.t. x, we get

$$2y \cdot \frac{dy}{dx} = a + b \quad \text{...(i)}$$

Again, differentiating both sides w.r.t. x, we get

$$2\left(\frac{dy}{dx}\right)^2 + 2y\frac{d^2y}{dx^2} = 0$$

$$\Rightarrow \quad \left(\frac{dy}{dx}\right)^2 + y\left(\frac{d^2y}{dx^2}\right) = 0 \Rightarrow y\frac{d^2y}{dx^2} = -\left(\frac{dy}{dx}\right)^2$$

On multiplying both sides by y^2, we get

$$y^3\frac{d^2y}{dx^2} = -y^2\left(\frac{dy}{dx}\right)^2 \quad \text{[from Eq. (i)]}$$

From Eq. (i), we get

$$y^3\frac{d^2y}{dx^2} = -y^2\left(\frac{a+b}{2y}\right)^2 \Rightarrow y^3\frac{d^2y}{dx^2} = -\frac{(a+b)^2}{4}$$

So, $y^3\frac{d^2y}{dx^2}$ is a constant.

12. (a, b, d) We have, $y = \log_7(\log_7 x)$

$$\therefore \quad \frac{dy}{dx} = \frac{1}{\log_7 x} \cdot \log_7 e \cdot \frac{1}{x}\log_7 e = \frac{1}{x \cdot \log_e x \log_e 7}$$

$$\left[\because \log_e x = \frac{\log_a x}{\log_a e} \text{ and } \log_a b = \frac{1}{\log_b a}\right]$$

$$= \frac{\log_7 e}{x\log_e x} = \frac{\log_7 e \cdot \log_x e}{x}$$

13. (a, d) We have,

$$y = \frac{\sec x + \tan x}{(\sec x - \tan x)}$$

$$\Rightarrow \quad y = (\sec x + \tan x)^2$$

$$\Rightarrow \quad \frac{dy}{dx} = 2(\sec x\tan x + \sec^2 x)(\sec x + \tan x)$$

$$\Rightarrow \quad \frac{dy}{dx} = 2\sec x(\tan x + \sec x)(\sec x + \tan x)$$

$$\frac{dy}{dx} = 2\sec x(\sec x + \tan x)^2 = 2(1 + \sin x)^2 \cdot \sec^3 x$$

14. (a, d) We have, $x = t + t^3$

On differentiating both sides w.r.t. t, we get

$$\frac{dx}{dt} = 1 + 3t^2$$

and $\quad y = t^2$

$$\frac{dy}{dt} = 2t$$

Now, $\frac{dy}{dx} = \frac{\frac{dy}{dt}}{\frac{dx}{dt}} = \frac{2t}{1+3t^2} = \frac{1}{2}$ [given]

$$\Rightarrow \quad 3t^2 + 1 = 4t$$

$$\Rightarrow \quad 3t^2 - 4t + 1 = 0$$

$$\Rightarrow \quad 3t^2 - 3t - t + 1 = 0$$

$$\Rightarrow \quad 3t(t-1) - 1(t-1) = 0$$

$$\Rightarrow \quad (3t-1)(t-1) = 0$$

So, $3t - 1 = 0$ and $t - 1 = 0$

$$t = \frac{1}{3} \text{ and } t = 1$$

15. (b, c) We have, $x = a\left[\cos t + \log \tan \frac{t}{2}\right]$

On differentiating both sides w.r.t. t, we get

$$\frac{dx}{dt} = a\left[-\sin t + \frac{1}{\left(\tan \frac{t}{2}\right)} \cdot \sec^2\left(\frac{t}{2}\right) \cdot \frac{1}{2}\right]$$

$$= a\left[-\sin t + \frac{1}{2} \frac{\sec\left(\frac{t}{2}\right)}{\sin \frac{t}{2}}\right]$$

$$= a\left[-\sin t + \frac{1}{2 \sin \frac{t}{2} \cos \frac{t}{2}}\right]$$

$$= a\left[-\sin t + \frac{1}{\sin t}\right]$$

$$= a\left(\frac{1 - \sin^2 t}{\sin t}\right)$$

$$\Rightarrow \quad \frac{dx}{dt} = \frac{a \cos^2 t}{\sin t}$$

and $\quad y = a \sin t$

$$\Rightarrow \quad \frac{dy}{dt} = a \cos t$$

$$\Rightarrow \quad \frac{dy}{dx} = \frac{\frac{dy}{dt}}{\frac{dx}{dt}} = \frac{a \cos t}{\frac{a \cos^2 t}{\sin t}} = \frac{\sin t}{\cos t}$$

$$\Rightarrow \quad \frac{dy}{dx} = \tan t \quad \text{...(i)}$$

Now, $\left(\frac{dy}{dx}\right)_{t=\frac{\pi}{2}} = \tan \frac{\pi}{2} = \infty$

Again, differentiating Eq. (i), we get

$$\frac{d^2y}{dx^2} = \sec^2 t \cdot \frac{dt}{dx}$$

$$= \sec^2 t \cdot \left(\frac{\sin t}{a \cos^2 t}\right)$$

$$= \frac{\sin t \sec^4 t}{a}$$

$$\Rightarrow \quad \left(\frac{d^2y}{dx^2}\right)_{t=\frac{\pi}{4}} = \frac{\frac{1}{\sqrt{2}} \times (\sqrt{2})^4}{a} = \frac{2\sqrt{2}}{a}$$

16. (a, c) $f_k(x) = \log_e [f_{k+1}(x)], \forall\, n \in N$ and $f_1(x) = x$

$$\Rightarrow \quad f_{(k+1)}(x) = e^{f_k(x)}$$

$$\Rightarrow \quad f_2(x) = e^{f_1(x)} = e^x$$

$$\Rightarrow \quad f_3(x) = e^{f_2(x)} = e^{e^x}$$

$$\vdots$$

$$\Rightarrow \quad f_n(x) = e^{e^{e^{\cdot^{\cdot^{e^x}}}}} \quad \text{[e's are } (n-1) \text{ times]}$$

$$\Rightarrow \quad \frac{d}{dx} f_n(x) = f_n(x) f_{n-1}(x) \cdot f_{n-2}(x) \dots f_2(x) \cdot f_1(x)$$

$$= f_n(x) \cdot \frac{d}{dx} [f_{(n-1)}(x)]$$

17. (b, c) We have, $f(x) = x \sin \pi x$

$$f'(x) = \sin \pi x + \pi x \cos \pi x = 0$$

or $\quad \tan \pi x = -\pi x$

The graph of $y = \tan \pi x$ and $y = -\pi x$ is as shown in the figure.

Therefore,

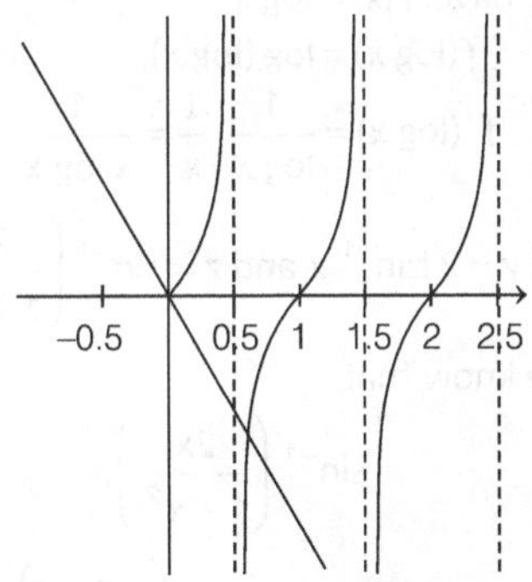

From the graph,

$$x \in \left(n + \frac{1}{2}, n+1\right) \text{ or } (n, n+1)$$

18. (b, d) We have, $x=\frac{1+\log t}{t^2}$ and $y=\frac{3+2\log t}{t}$

$$\frac{dy}{dx}=\frac{\frac{dy}{dt}}{\frac{dx}{dt}}$$

$$\Rightarrow \frac{dy}{dx}=\frac{\frac{t\left(\frac{2}{t}\right)-(3+2\log t)}{t^2}}{\frac{t^2\left(\frac{1}{t}\right)-(1+\log_e t)2t}{t^4}}=\left(\frac{-1-2\log t}{-1-2\log t}\right)t$$

$$\Rightarrow \frac{dy}{dx}=t$$

Eliminating $\log t$ from y, we get

$$y=\frac{1+2t^2x}{t}=\frac{1+2(y')^2x}{y'}$$

or $\quad yy'=1+2x(y')^2$

On differentiating both sides w.r.t. x, we get

$$y\cdot y''+(y')^2=4xy'\cdot y''+2(y')^2$$

$$\Rightarrow \quad y\cdot y''=4xy'\cdot y''+(y')^2$$

$$\Rightarrow \quad y''(y-4xy')=(y')^2$$

19. (a, b, c, d)

(a) We have, $f(x)=|x-2|$

$\Rightarrow \quad f[f(x)]=|f(x)-2|=||x-2|-2|$

$\therefore \quad f[f(x)]=|x-4|=x-4\ \forall\ x>20$

$\Rightarrow \quad f[f(x)]'=1\ \forall\ x>20$

(b) We have,

$f(0)=a, f'(0)=b, g(0)=0$ and $(fog)'(0)=c$ [given]

Now, $(fog)'(x)=f'[g(x)]\cdot g'(x)$

$\therefore \quad (fog)'(0)=f'[g(0)]\cdot g'(0)$

$c=f'(0)\cdot g'(0)$

$c=b\cdot g'(0)$

$g'(0)=\frac{c}{b}$

(c) We have, $f(x)=\log x$

$\Rightarrow \quad f(\log x)=\log(\log x)$

$\Rightarrow \quad f'(\log x)=\frac{1}{\log x}\cdot\frac{1}{x}=\frac{1}{x\log x}$

(d) Let $y=2\tan^{-1}x$ and $z=\sin^{-1}\left(\frac{2x}{1+x^2}\right)$

We know that

$$2\tan^{-1}x=\begin{cases}\sin^{-1}\left(\frac{2x}{1+x^2}\right), & \text{for } -1\le x\le 1\\ -\pi-\sin^{-1}\left(\frac{2x}{1+x^2}\right), & \text{for } x<-1\\ \pi-\sin^{-1}\left(\frac{2x}{1+x^2}\right), & \text{for } x>1\end{cases}$$

$$\Rightarrow \quad y=\begin{cases}z, & -1\le x\le 1\\ -\pi-z, & x<-1\\ \pi-z, & x>1\end{cases}$$

$$\Rightarrow \quad \frac{dy}{dz}=\begin{cases}1, & -1<x<1\\ -1, & x<-1\\ -1, & x>1\end{cases}$$

$$\Rightarrow \left(\frac{dy}{dz}\right)_{x=\frac{1}{2}}=1$$

20. (a, b, c) We have, $f(x)=3x^2+4xg'(1)+g''(2)$ ...(i)

$g(x)=2x^2+3xf'(2)+f''(3)$ ...(ii)

$\therefore \quad f'(x)=6x+4g'(1)$ ...(iii)

and $\quad g'(x)=4x+3f'(2)$ ...(iv)

$\Rightarrow \quad f''(x)=6.$ ...(v)

and $\quad g''(x)=4$ ...(vi)

Putting $x=1$ in Eq. (iii) and (iv), we get

$f'(1)=6+4g'(1)$ and $g'(1)=4+3f'(2)$

$=6+4[4+3f'(2)]$

$\Rightarrow \quad f'(1)=22+12f'(2)$

Putting $x=2$ in Eq. (iv), we get

$g'(2)=8+3f'(2)$

$\Rightarrow \quad g'(2)=8+3[12+4g'(1)]$

[putting $x=2$ in Eq. (iii) $f'(2)=12+4g'(1)$]

$g'(2)=44+12g'(1)$

Putting $x=3$ in Eq. (v) and $x=2$ in Eq. (vi), we get

$f''(3)=6$ and $g''(2)=4$

$\Rightarrow \quad f''(3)+g''(2)=6+4=10$

21. (b, c) We have, $f(x)=x^3+3x+2$

$\therefore \quad f(1)=6$

$\therefore$ In $g[f(x)]=x$ putting $x=1$

$g(6)=1$

Also, $g[f(x)]=x$

$\Rightarrow \quad g'[f(x)]\cdot f'(x)=1$

Putting $x=0$,

$\therefore \quad g'[f(0)]\cdot f'(0)=1$

$\therefore \quad g'(2)=\frac{1}{f'(0)}=\frac{1}{3}$

$f(3)=38$

$\therefore \quad g(38)=3$

$\therefore \quad h[g(3)]=h[g\{g(38)\}]=38$

$f(2)=16\Rightarrow g(16)=2$

$\therefore \quad h[g\{g(16)\}]=h[g(2)]=h(0)$

$\therefore \quad 16=h[g\{g(16)\}]=h(0)$

$f'(x)=3x^2+3$

$f'(6)=11$

$$f'(1)=6$$

$$\Rightarrow \quad g'(6)=\frac{1}{6}$$

$$h[g\{g(x)\}]=x$$

$$\Rightarrow \quad h'[g\{g(x)\}]\times g'[g(x)]\times g'(x)=1$$

Putting $x=236$,

$$h'[g\{g(236)\}]\times g'[g\{236)\}]\times g'(236)=1$$

$$\Rightarrow \quad h'[g(6)]\, g'(6)\times \frac{1}{f'(6)}=1$$

$$\Rightarrow \quad h'(1)=666$$

Sol. (Q.Nos. 22-24)

22. (b) **23.** (b) **24.** (b)

We have, $y^{1/m}+y^{-1/m}=2x$

$$\text{LHL}=y^{1/m}+y^{-1/m}$$

Let $y=(x+\sqrt{x^2-1})^m$

$$\text{So, } x+\sqrt{x^2-1}+\frac{1}{x+\sqrt{x^2-1}}\times\frac{x-\sqrt{x^2-1}}{x-\sqrt{x^2-1}}$$

$$=x+\sqrt{x^2-1}+\frac{x-\sqrt{x^2-1}}{x^2-(x^2-1)}$$

$$=x+\sqrt{x^2-1}+\frac{x-\sqrt{x^2-1}}{1}$$

$$=2x$$

So, $y=(x+\sqrt{x^2-1})^m$

Now, differentiating both sides w.r.t. x, we get

$$\frac{dy}{dx}=m\,(x+\sqrt{x^2-1})^{m-1}.\left(1+\frac{1}{2\sqrt{x^2-1}}2x\right)$$

$$=m\,(x+\sqrt{x^2-1})^{m-1}\left(\frac{x+\sqrt{x^2-1}}{\sqrt{x^2-1}}\right)$$

$$\frac{dy}{dx}=\frac{my}{\sqrt{x^2-1}} \quad \text{...(i)}$$

$$\text{and} \quad \sqrt{x^2-1}\,\frac{dy}{dx}=my$$

Again, differentiating both sides w.r.t. x, we get

$$\sqrt{x^2-1}\,\frac{d^2y}{dx^2}+\frac{dy}{dx}\left(\frac{1}{2\sqrt{x^2-1}}\cdot 2x\right)=m\,\frac{dy}{dx}$$

$$\Rightarrow \quad (x^2-1)\frac{d^2y}{dx^2}+x\,\frac{dy}{dx}=m\sqrt{x^2-1}\,\frac{dy}{dx}$$

From Eq. (i),

$$(x^2-1)\frac{d^2y}{dx^2}+x\,\frac{dy}{dx}=m^2y$$

Sol. (Q.Nos. 25-27)

We have,

$$x=f(t)=a^{\ln b^t}=a^{t\ln b} \quad \text{...(i)}$$

$$y=g(t)=b^{-\ln a^t}=(b^{\ln a})^{-t}=(a^{\ln b})^{-t}=a^{-t\ln b}$$

$$\therefore \quad y=g(t)=a^{\ln(b^{-t})}=f(-t) \quad \text{...(ii)}$$

From Eqs. (i) and (ii),

$$xy=1$$

25. (c) Now, $y=\dfrac{1}{x}$

$$\therefore \quad \frac{dy}{dx}=-\frac{1}{x^2}=-\frac{1}{f^2(t)}$$

Also, $xy=1$

$$\therefore \quad \frac{dy}{dx}=-\frac{1}{x^2}=-\frac{y^2}{1}$$

Also, $xy=1$

$$\Rightarrow \quad \frac{dy}{dx}=-\frac{y}{x}$$

$$=-\frac{g(t)}{f(t)}$$

26. (c) $f(t)=g(t)\Rightarrow f(t)=f(-t)\Rightarrow t=0$

[$\because f(t)$ is a one-one function]

At $t=0$, $x=y=1$

$$\because \quad xy=1$$

$$\frac{dy}{dx}=-\frac{1}{x^2}$$

$$\text{and} \quad \frac{d^2y}{dx^2}=\frac{2}{x^3}$$

At $x=1$, $\dfrac{d^2y}{dx^2}=2$

27. (a) $\because \quad xy=1$

$$f\cdot g=1$$

$$\therefore \quad fg'+gf'=0$$

$$\therefore \; fg''+g'f'+g'f'+gf''=0$$

$$\Rightarrow \quad fg''+gf''+2f'g'=0$$

$$\text{or} \quad \frac{f}{f'}\frac{g''}{g'}+\frac{gf''}{g'f'}=-2 \quad \text{...(iii)}$$

From Eq. (ii),

$$g(t)=f(-t)$$

$$\therefore \quad g'(t)=-f'(-t)$$

$$\text{and} \quad g''(t)=f''(-t)$$

Substituting in Eq. (iii), we get

$$\frac{f(t)}{f'(t)}\cdot\frac{f''(-t)}{-f'(-t)}+\frac{f(-t)}{-f'(-t)}\cdot\frac{f''(t)}{f'(t)}=-2$$

$$\Rightarrow \quad \frac{f(t)}{f'(t)}\cdot\frac{f''(-t)}{f'(-t)}+\frac{f(-t)}{f'(-t)}\cdot\frac{f''(t)}{f'(t)}=2$$

Sol. (Q.Nos. 28-30)

28. (d) Take $f = e^{2x}$ and $g = x^5$

Then,

$$y_{10} = (e^{2x})_{10} x^5 + {}^{10}C_1 (e^{2x})_9 5x^4 + {}^{10}C_2 (e^{2x})_8 20x^4 + {}^{10}C_3 (e^{2x})_7 80x^2 + {}^{10}C_4 (e^{2x})_6 (160x) + {}^{10}C_5 (e^{2x})_5 \cdot 160$$

Now, putting $x = 0$, we get

$$y_{10}(0) = 2^{12} \times 315$$

29. (b) Take $f = \cos x$ and $g = x^2$

Then, $y_7 = x^2 \cos\left(x + 7\frac{\pi}{2}\right) + {}^7C_1 2x \cos\left(x + 6\frac{\pi}{2}\right) + {}^8C_2 2\cos\left(x + \frac{5\pi}{2}\right)$

So, $y_7\left(\frac{\pi}{2}\right) = \left(\frac{\pi}{2}\right)^2 \cos 4\pi + 7 \cdot 2\left(\frac{\pi}{2}\right)\cos\left(\frac{7\pi}{2}\right) + \frac{8 \cdot 7}{2} \cdot 2 \cdot \cos 3\pi$

$\Rightarrow y_7\left(\frac{\pi}{2}\right) = \frac{\pi^2}{4} - 56$

30. (a) We have, $y = \frac{\sin^{-1} x}{\sqrt{1 - x^2}}$

$\Rightarrow \quad (1 - x^2) y^2 = (\sin^{-1} x)^2$

$\Rightarrow \quad -2xy^2 + (1 - x^2) 2yy_1 = \frac{2 \sin^{-1} x}{\sqrt{1 - x^2}} = 2y$

$\Rightarrow \quad -xy + (1 - x^2) \cdot y_1 = 1$

$\Rightarrow (1 - x^2) y_2 - 2xy_1 - xy_1 - y = 0$

$\Rightarrow \quad (1 - x^2) y_2 - 3xy_1 - y = 0$ ...(i)

Differentiating Eq. (i) w.r.t. 'x' we get

$(1 - x^2) y_3 - 2xy_2 - 3xy_2 - 3y_1 - y_1 = 0$

$(1 - x^2) y_3 - (2 + 3) xy_2 - 4y_1 = 0$ ...(ii)

Differentiating both sides of Eq. (ii) w.r.t. 'x' we get

$(1 - x^2) y_4 - (2.2 + 3) xy_3 - 9y_2 = 0$

Similarly,

$(1 - x^2) y_{n+2} - (2n + 3)\, xy_{n+1} - (n + 1)^2 y_n = 0$

Now, putting $x = 0$, we get

$$y_{n+2}(0) = (n + 1)^2 y_n(0)$$

31. (i) → (D), (ii) → (A), (iii) → (B), (iv) → (C)

(i) We have, $y = \cos(3\cos^{-1} x)$

$\Rightarrow \quad y' = \frac{3 \sin(3 \cos^{-1} x)}{\sqrt{1 - x^2}}$

$\Rightarrow \sqrt{1 - x^2}\, y' = 3 \sin(3 \cos^{-1} x)$

$\Rightarrow \frac{-x}{\sqrt{1 - x^2}} y' + \sqrt{1 - x^2}\, y'' = 3 \cos(3 \cos^{-1} x) \frac{-3}{\sqrt{1 - x^2}}$

$\Rightarrow \quad -xy' + (1 - x^2) y'' = -9y$

$\Rightarrow \quad \frac{1}{y}[(x^2 - 1) y'' + xy'] = 9$

(ii) We have,

$u = x^2 + y^2, x = s + 3t$ and $y = 2s - t$

$\therefore \quad \frac{dx}{ds} = 1$ and $\frac{dy}{ds} = 2$ ...(i)

and $\frac{d^2x}{ds^2} = 0, \frac{d^2y}{ds^2} = 0$

Now, $u = x^2 + y^2$

$\therefore \quad \frac{du}{ds} = 2x\frac{dx}{ds} + 2y\frac{dy}{ds}$

$\Rightarrow \frac{d^2u}{ds^2} = 2x\frac{d^2x}{ds^2} + 2\left(\frac{dx}{ds}\right)^2 + 2y\frac{d^2y}{ds^2} + 2\left(\frac{dy}{ds}\right)^2$

$= 2x(0) + 2(1)^2 + 2y(0) + 2(2)^2$

$= 2 + 8$

$= 10$

(iii) $\frac{dy}{dx} = \frac{dy/dt}{dx/dt}$

$= \frac{12t^2 - 6t - 18}{5t^4 - 15t^2 - 20}$

At $t = 1$,

$\frac{dy}{dx} = \frac{12 - 6 - 18}{5 - 15 - 20} = \frac{2}{5}$

or $\quad -5\left.\frac{dy}{dx}\right|_{t=1} = -5 \times \frac{2}{5} = -2$

(iv) We have, $y = \sin^{-1}(\cos x)$

$y = \frac{\pi}{2} - \cos^{-1}(\cos x) = \frac{\pi}{2} - (2\pi - x),$

$x \in (\pi, 2\pi)$

$= x - \frac{3\pi}{2}$

$\frac{dy}{dx} = 1$

32. (i) → (C), (ii) → (D), (iii) → (A), (iv) → (B)

(i) Let $f(x) = ax^2 + bx + c$

Then, $(ax^2 + bx + c) + (4ax^2 + 2bx + c) = 5x^2 - 18$

On solving, we get

$5a = 5 \Rightarrow a = 1$

$3b = 0 \Rightarrow b = 0$

$$2c = -18 \Rightarrow c = -9$$

Therefore, $f(x) = x^2 - 9$

So, $f'(x) = 2x$

Hence, $f'(-1) = 2(-1) = -2$

(ii) We have, $y = x^3 - 2$ and $x = 3z^2 + 5$

On differentiating, we get

$$\frac{dy}{dx} = 3x^2 \text{ and } \frac{dx}{dz} = 6z$$

So, $\frac{dy}{dz} = \frac{\frac{dy}{dx}}{\frac{dz}{dx}} = \frac{dy}{dx} \cdot \frac{dx}{dz} = (3x^2)(6z) = 3(3z^2 + 5)^2 \cdot 6z.$

Therefore, $\left(\frac{dy}{dz}\right)_{z=0} = 0.$

(iii) Let us take

$$p(x) = a(x-2)^4 + b(x-2)^3 + c(x-2)^2 + d(x-2) - 1$$

$\therefore\ p(2) = -1$

$0 = p'(2) = d$

$2 = p''(2) = 2c$ or $c = 1$

$-12 = p'''(2) = 6b$ or $b = -2$

$24 = p''''(2) = 24a$ or $a = 1$

Thus, $p''(x) = 12(x-2)^2 - 12(x-2) + 2$

or $p''(3) = 12 - 12(1) + 2 = 2$

(iv) We have, $\frac{dy}{dx} = f'(x)$ and $\frac{dx}{dz} = -\frac{1}{z^2}$

$\therefore\ \frac{dy}{dx} = \frac{dy}{dz} \cdot \frac{dz}{dx} = -\frac{dy}{dz} \cdot z^2$

$\Rightarrow \frac{d^2y}{dx^2} = \left[z^2 \frac{d^2y}{dz^2} + \frac{dy}{dz} \cdot 2z\right] z^2$

$\Rightarrow \frac{d^2y}{dx^2} = z^4 \cdot \frac{d^2y}{dz^2} + 2z^3 \frac{dy}{dz}$

$\Rightarrow \lambda = 1$

33. (i) $\rightarrow$ (D), (ii) $\rightarrow$ (C), (iii) $\rightarrow$ (A), (iv) $\rightarrow$ (B)

(i) Let $y = \tan^{-1}\left(\frac{2x}{1-x^2}\right)$

Putting $x = \tan\theta$

$$y = \tan^{-1}\left(\frac{2\tan\theta}{1-\tan^2\theta}\right) = \tan^{-1}(\tan 2\theta)$$

$\Rightarrow y = 2\theta$

So, $\frac{dy}{d\theta} = 2$

and $z = \sin^{-1}\left(\frac{2x}{1+x^2}\right)$

Putting $x = \tan\theta$

$$z = \sin^{-1}\left(\frac{2\tan\theta}{1+\tan^2\theta}\right)$$

$\Rightarrow z = \sin^{-1}(\sin 2\theta)$

$\Rightarrow z = 2\theta$

$\Rightarrow \frac{dz}{d\theta} = 2$

Now, $\frac{dy}{dz} = \frac{dy/d\theta}{dz/d\theta} = \frac{2}{2} = 1$

(ii) We have, $y = e^x(a\cos x + b\sin x)$ ...(i)

$\Rightarrow y' = e^x(-a\sin x + b\cos x) + e^x(a\cos x + b\sin x)$

$\Rightarrow y' = e^x(-a\sin x + b\cos x) + y$

$\Rightarrow y'' = e^x(-a\cos x - b\sin x) + e^x(-a\sin x + b\cos x) + y'$

$\Rightarrow y'' = -y + (y' - y) + y'$

$\Rightarrow y'' - 2y' + 2y = 0$

Hence, $\lambda = 2$

(iii) We have, $y = (\tan^{-1} x)^2$

$\Rightarrow \frac{dy}{dx} = 2\tan^{-1} x \cdot \frac{1}{1+x^2}$

$\Rightarrow (1+x^2)\frac{dy}{dx} = 2\tan^{-1} x$

$\Rightarrow (1+x^2)\frac{d^2y}{dx^2} + 2x\frac{dy}{dx} = \frac{2}{1+x^2}$

$\Rightarrow (1+x^2)^2\frac{d^2y}{dx^2} + 2x(1+x^2)\frac{dy}{dx} = 2$

$\Rightarrow (1+x^2)^2 y'' + 2x(1+x^2)y' - 2 = 0$

So, $\lambda = -2$

(iv) We have, $y = e^{\sqrt{x}} + e^{-\sqrt{x}}$

$\Rightarrow \frac{dy}{dx} = \frac{1}{2\sqrt{x}}(e^{\sqrt{x}} - e^{-\sqrt{x}})$

$\therefore\ 2\sqrt{x}\frac{dy}{dx} = e^{\sqrt{x}} - e^{-\sqrt{x}}$

$\Rightarrow 2\left[\sqrt{x}\frac{d^2y}{dx^2} + \frac{1}{2\sqrt{x}}\frac{dy}{dx}\right] = \frac{1}{2\sqrt{x}}(e^{\sqrt{x}} + e^{-\sqrt{x}})$

$\Rightarrow 2\left[\sqrt{x}\frac{d^2y}{dx^2} + \frac{1}{2\sqrt{x}}\frac{dy}{dx}\right] = \frac{1}{2\sqrt{x}} y$

On multiplying by $2\sqrt{x}$, we get

$$4x\frac{d^2y}{dx^2} + 2\frac{dy}{dx} = y$$

So, $k = 4$

34. (i) $\rightarrow$ (C), (ii) $\rightarrow$ (A), (iii) $\rightarrow$ (B), (iv) $\rightarrow$ (D)

(a) We have, $f(x) = x^n$

Therefore, $f'(x) = nx^{n-1}$

$\Rightarrow f'(a+b) = n(a+b)^{n-1}$

$\Rightarrow \quad f'(a) = na^{n-1}$

$\Rightarrow \quad f'(b) = nb^{n-1}$

Using the given condition,

$n(a+b)^{n-1} = na^{n-1} + nb^{n-1}$

$\Rightarrow \quad (a+b)^{n-1} = a^{n-1} + b^{n-1}$

The above equation will only true, when $n - 1 = 1 \Rightarrow n = 2$

(ii) Let $y = f(\tan x)$ and $z = g(\sec x)$

$$\frac{dy}{dx} = f'(\tan x) \cdot \sec^2 x$$

and $\dfrac{dz}{dx} = g'(\sec x) \cdot \sec x \tan x$

$$\therefore \quad \frac{dy}{dz} = \frac{f'(\tan x) \cdot \sec^2 x}{g'(\sec x) \cdot \sec x \tan x}$$

$$\left.\frac{dy}{dz}\right|_{x=\frac{\pi}{4}} = \frac{f'(1)(\sqrt{2})^2}{g'(\sqrt{2}) \cdot \sqrt{2}} = \frac{1}{\sqrt{2}}$$

(iii) $\phi(x) = [go(fh)](x)$

$= g[fh(x)] \quad [\because f \cdot h(x) = f(x) \cdot h(x) = \sin x \cdot \cos x]$

$= g(\sin x \cos x)$

$\phi(x) = 2 \sin x \cos x$

$\Rightarrow \quad \phi(x) = \sin 2x$

$\Rightarrow \quad \phi'(x) = 2 \cos 2x$

$\Rightarrow \phi''(x) = -4 \sin 2x$

Putting $x = \dfrac{\pi}{4}$

$$\phi''\left(\frac{\pi}{4}\right) = -4 \sin 2\left(\frac{\pi}{4}\right) = -4$$

(iv) Given, equation is $xy - \log y = 1$

$\Rightarrow \quad xy' + y(1) - \dfrac{1}{y} y' = 0$

$\Rightarrow \quad xyy' + y^2 - y' = 0$

Again, differentiating w.r.t. 'x' we get

$(xy - 1)\, y'' + y'(xy' + y \cdot 1) + 2yy' = 0$

$\Rightarrow \quad x(yy'' + y'^2) - y'' + 3yy' = 0$

So, $k = 3$

35. (i) → (A), (ii) → (C), (iii) → (B), (iv) → (D)

(i) We have, $x^a y^b = (x+y)^{a+b}$

Taking log both sides, we get

$a \ln x + b \ln y = (a+b) \ln (x+y)$

On differentiating both sides w.r.t. x, we get

$$\frac{a}{x} + \frac{b}{y}\frac{dy}{dx} = \frac{a+b}{x+y}\left(1 + \frac{dy}{dx}\right)$$

$$\Rightarrow \quad \frac{dy}{dx}\left(\frac{b}{y} - \frac{a+b}{x+y}\right) = \frac{a+b}{x+y} - \frac{a}{x}$$

$$\Rightarrow \quad \frac{dy}{dx}\left(\frac{bx + by - ay - by}{y(x+y)}\right) = \frac{ax + bx - ax - ay}{x(x+y)}$$

$$\Rightarrow \quad \frac{dy}{dx} = \frac{y(bx - ay)}{x(bx - ay)}$$

$$\Rightarrow \quad \frac{dy}{dx} = \frac{y}{x}$$

(ii) We have, $y = \cot^{-1}\left(\dfrac{\sqrt{1+\sin x} + \sqrt{1 - \sin x}}{\sqrt{1 + \sin x} - \sqrt{1 - \sin x}}\right)$

$$= \cot^{-1}\left(\frac{\sqrt{\left(\cos\frac{x}{2} + \sin\frac{x}{2}\right)^2} + \sqrt{\left(\cos\frac{x}{2} - \sin\frac{x}{2}\right)^2}}{\sqrt{\left(\cos\frac{x}{2} + \sin\frac{x}{2}\right)^2} - \sqrt{\left(\cos\frac{x}{2} - \sin\frac{x}{2}\right)^2}}\right)$$

$$= \cot^{-1}\left(\frac{\cos\frac{x}{2} + \sin\frac{x}{2} + \cos\frac{x}{2} - \sin\frac{x}{2}}{\cos\frac{x}{2} + \sin\frac{x}{2} - \cos\frac{x}{2} + \sin\frac{x}{2}}\right)$$

$$\left[\because \left|\cos\frac{x}{2} - \sin\frac{x}{2}\right| = \cos\frac{x}{2} - \sin\frac{x}{2} \text{ for } x \in \left(0, \frac{\pi}{2}\right)\right]$$

$$= \cot^{-1}\left(\frac{2\cos\frac{x}{2}}{2\sin\frac{x}{2}}\right) = \cot^{-1}\left(\cot\frac{x}{2}\right)$$

$\Rightarrow y = \dfrac{x}{2}$

On differentiating both sides w.r.t. x, we get

$$\frac{dy}{dx} = \frac{1}{2}$$

Hence, $\dfrac{dy}{dx}$ is independent of 'x'.

(iii) We have, $\sqrt{1 - x^6} + \sqrt{1 - y^6} = a(x^3 - y^3)$

Putting $x^3 = \sin\theta$ and $y^3 = \sin\phi$

The given equation reduces to

$\cos\theta + \cos\phi = a(\sin\theta - \sin\phi)$

$$\Rightarrow \quad 2\cos\left(\frac{\theta+\phi}{2}\right)\cos\left(\frac{\theta-\phi}{2}\right) = 2a\cos\left(\frac{\theta+\phi}{2}\right)\sin\left(\frac{\theta-\phi}{2}\right)$$

$$\Rightarrow \quad \cos\left(\frac{\theta-\phi}{2}\right) = a\sin\left(\frac{\theta-\phi}{2}\right)$$

$$\Rightarrow \quad \cot\left(\frac{\theta-\phi}{2}\right) = a$$

$$\Rightarrow \quad \frac{\theta-\phi}{2} = \cot^{-1} a$$

$$\Rightarrow \quad \theta - \phi = 2\cot^{-1} a$$

$\Rightarrow \sin^{-1}(x^3) - \sin^{-1}(y^3) = 2\cot^{-1} a$

On differentiating both sides w.r.t. x, we get

$$\frac{3x^2}{\sqrt{1-x^6}} - \frac{3y^2}{\sqrt{1-y^6}}\frac{dy}{dx} = 0$$

$$\Rightarrow \quad \frac{dy}{dx} = \frac{x^2\sqrt{1-y^6}}{y^2\sqrt{1-x^6}}$$

On comparing with $\frac{dy}{dx} = f(x, y)\sqrt{\frac{1-y^6}{1-x^6}}$, we get

$$f(x, y) = \frac{x^2}{y^2}$$

(iv) We have, $x = t - \frac{1}{t}$

$$\Rightarrow \quad \frac{dx}{dt} = 1 + \frac{1}{t^2}$$

and $\quad y = t + \frac{1}{t}$

$$\Rightarrow \quad \frac{dy}{dt} = 1 - \frac{1}{t^2}$$

$$\Rightarrow \quad \frac{dy}{dx} = \frac{\frac{dy}{dt}}{\frac{dx}{dt}} = \frac{1 - \frac{1}{t^2}}{1 + \frac{1}{t^2}} = \frac{t - \frac{1}{t}}{t + \frac{1}{t}}$$

$$\Rightarrow \quad \frac{dy}{dx} = \frac{x}{y}$$

Sol. (Q.Nos. 36-38)

Suppose, degree of $f(x) = n$, then degree of $f'(x) = n-1$ and degree of $f''(x) = n-2$

So, $n = (n-1) + (n-2)$

$\therefore \quad n = 3$

36. (3) Hence, degree of $f(x)$ is 3.

37. (0.05) Let $f(x) = ax^3 + bx^2 + cx + d$

From $f(x) = f'(x) \cdot f''(x)$, we have

$(ax^3 + bx^2 + cx + d) = (3ax^2 + 2bx + c)(6ax + 2b)$

Comparing coefficient of x^3, x^2, x and constant terms, we have

$$a = 18a^2 \Rightarrow a = \frac{1}{18} \qquad [\because a \neq 0]$$

$$b = 18ab \Rightarrow b = b$$

$$c = 4b^2 + 6ac \Rightarrow c = 6b^2$$

and $\quad d = 2bc \Rightarrow d = 12b^3$

$$\Rightarrow \quad f(x) = \frac{1}{18}x^3 + bx^2 + 6b^2x + 12b^3$$

Hence, leading coefficient of $f(x) = \frac{1}{18} = 0.05$

38. (9) $\because \quad f(0) = 12$

$\therefore \quad 0 + 0 + 0 + 12b^3 = 12$

$\therefore \quad b = 1$

$$\Rightarrow \quad f(x) = \frac{1}{18}x^3 + x^2 + 6x + 12$$

Now, $f(x) = 0$

$\therefore \quad x^3 + 18x^2 + 108x + 216 = 0$

Hence, sum of roots of $f(x) = 0$ is -18.

$$\therefore \quad \frac{-1}{2}(-18) = 9$$

Sol. (Q.Nos. 39-41)

We have,

$$f(x) = x^3 + x^2 f'(1) + x \cdot f''(2) + f'''(3) \qquad \text{...(i)}$$

$$\Rightarrow \quad f'(x) = 3x^2 + 2xf'(1) + f''(2) \qquad \text{...(ii)}$$

$$\Rightarrow f''(x) = 6x + 2f'(1) \qquad \text{...(iii)}$$

$$\Rightarrow f'''(x) = 6 \qquad \text{...(iv)}$$

Putting $x = 1, 2$ and 3 in Eqs. (ii), (iii) and (iv), respectively

$$f'(1) = 3 + 2f'(1) + f''(2)$$

$$f''(2) = 12 + 2f'(1)$$

and $\quad f''(3) = 6$

Solving these equations, we get

$$f'(1) = -5$$

$$f''(2) = 2$$

and $\quad f'''(3) = 6$

$$\therefore \quad f(x) = x^3 - 5x^2 + 2x + 6$$

$\Rightarrow f(0) = 6, f(1) = 4$ and $f(3) = -6$

39. (4) The value of $f(1) = 4$

40. (12) $f''(2) = 12 + 2f'(1)$

$f''(2) - 2f'(1) = 12$

41. (0) $f(0) + f(3) = 6 - 6 = 0$

42. (2) We have,

$$f(x+y) = f(x) \cdot f'(y) + f'(x) \cdot f(y) \qquad \text{...(i)}$$

Putting $y = 0$

$$f(x) = f(x) \cdot f'(0) + f'(x) \cdot f(0)$$

$$= f(x) f'(0) + f'(x) \qquad [\because f(0) = 1, \text{ given}] \text{...(ii)}$$

To find $f'(0)$, putting

$x = y = 0$ in Eq. (i),

$$\therefore \quad f(0) = 2f'(0)$$

$$\therefore \quad f'(0) = \frac{1}{2}$$

So, from Eq. (ii), we get

$$f'(x) = \frac{f(x)}{2}$$

$$\Rightarrow \quad \int \frac{f'(x)}{f(x)}dx = \int \frac{1}{2}dx$$

$$\Rightarrow \quad \log_e f(x) = \frac{x}{2} + C \qquad [\text{as } f(0) = 1]$$

$$\Rightarrow \quad \log f(x) = \frac{x}{2}$$

At $x = 4$, $\quad \log f(4) = \frac{4}{2} = 2$

43. (2) We have, $g(x)=f(x)\sin x$ ...(i)

On differentiating Eq. (i) w.r.t. x, we get

$$g'(x)=f(x)\cos x+f'(x)\sin x$$

Again, differentiating w.r.t. x, we get

$$g''(x)=f(x)(-\sin x)+f'(x)\cos x + f'(x)\cos x+f''(x)\sin x$$

$$\Rightarrow \quad g''(-\pi)=2f'(-\pi)\cos(-\pi)=2\times 1\times(-1)$$

$$\Rightarrow \quad |g''(-\pi)|=|-2|=2$$

44. (8) We have, $(y-x^5)^2=x(1+x^2)^2$

On differentiating both sides w.r.t. x, we get

$$2(y-x^5)\left(\frac{dy}{dx}-5x^4\right)=1(1+x^2)^2+x[2(1+x^2)(2x)]$$

On putting $x=1$ and $y=3$, we get

$$\frac{dy}{dx}=8$$

45. (2) We have, $y=\frac{x^2}{2}+\frac{1}{2}x\sqrt{x^2+1}+\ln\sqrt{x+\sqrt{x^2+1}}$

$$=\frac{x^2}{2}+\frac{1}{2}x\sqrt{x^2+1}+\frac{1}{2}\ln x+\sqrt{x^2+1}$$

$$\Rightarrow \quad y'=x+\frac{1}{2}\left\{\frac{x^2}{\sqrt{x^2+1}}+\sqrt{x^2+1}\right\}+\frac{1}{2\sqrt{x^2+1}}$$

$$=x+\frac{1}{2}\left\{\frac{x^2+1}{\sqrt{x^2+1}}\right\}+\frac{\sqrt{x^2+1}}{2}$$

$$\Rightarrow \quad 2y'=2x+\sqrt{x^2+1}+\sqrt{x^2+1}$$

$$\Rightarrow \quad 2y'=2x+2\sqrt{x^2+1}$$

$$\Rightarrow \quad y'=x+\sqrt{x^2+1}$$

Now, $xy'+\ln y'=x(x+\sqrt{x^2+1})+\ln(x+\sqrt{x^2+1})$

$$=2\left(\frac{x^2}{2}+\frac{x}{2}\sqrt{x^2+1}+\frac{1}{2}\ln(x+\sqrt{x^2+1})\right)$$

$$=2\left(\frac{x^2}{2}+\frac{x}{2}\sqrt{x^2+1}+\ln\sqrt{x+\sqrt{x^2+1}}\right)$$

$$=2y$$

So, $\quad k=2$

46. (8) We have, $f\left(\frac{x+y}{3}\right)=\frac{2+f(x)+f(y)}{3}$ ...(i)

Replacing x by $3x$ and y by 0,

$$f(x)=\frac{2+f(3x)+f(0)}{3}$$

$\Rightarrow \quad f(3x)-3f(x)+2=-f(0)$ ...(ii)

Putting $x=0$ and $y=0$, we get

$$f(0)=2$$

Now, $f'(x)=\lim_{h\to 0}\frac{f(x+h)-f(x)}{h}$

$$=\lim_{h\to 0}\frac{f\left(\frac{3x+3h}{3}\right)-f(x)}{h}$$

$$=\lim_{h\to 0}\frac{\frac{2+f(3x)+f(3h)}{3}-f(x)}{h}$$

$$=\lim_{h\to 0}\frac{f(3x)-3f(x)+f(3h)+2}{3h}$$

$$=\lim_{h\to 0}\frac{f(3h)-f(0)}{3h} \quad \text{[from Eq. (ii)]}$$

$$=f'(0)=c \quad \text{(say)}$$

$\Rightarrow \quad f'(x)=c$

At $x=2$, $\quad f'(2)=c=2$

$$f'(x)=2$$

On integrating both sides, we get

$$f(x)=2x+d$$

At $x=0$, $\quad f(0)=d$

$$d=2$$

So, $f(x)=2x+2$

Now, $f(3)=2\times 3+2=8$

47. (5) We have,

$$F(x)=\left[f\left(\frac{x}{2}\right)\right]^2+\left[g\left(\frac{x}{2}\right)\right]^2$$

$$\Rightarrow \quad F'(x)=2f\left(\frac{x}{2}\right)\cdot f'\left(\frac{x}{2}\right)\cdot\frac{1}{2}+2g\left(\frac{x}{2}\right)\cdot g'\left(\frac{x}{2}\right)\cdot\frac{1}{2}$$

$$\Rightarrow \quad F'(x)=f\left(\frac{x}{2}\right)\cdot f'\left(\frac{x}{2}\right)+g\left(\frac{x}{2}\right)\cdot g'\left(\frac{x}{2}\right)$$

Here, $g(x)=f'(x)$

On differentiating Eq. (i) w.r.t. x, we get

$$g'(x)=f''(x)$$

$$\Rightarrow \quad g'(x)=-f(x) \quad [\because f''(x)=-f(x)]$$

So, $\quad F'(x)=f\left(\frac{x}{2}\right)\cdot g\left(\frac{x}{2}\right)+g\left(\frac{x}{2}\right)\cdot\left[-f\left(\frac{x}{2}\right)\right]$

$\Rightarrow \quad F'(x)=0 \Rightarrow F(x)=c$

Now, $F(5)=5$

So, $\quad c=5$

$\therefore \quad F(x)=5$, then $F(10)=5$

48. (3) Given,

$$(a+bx)e^{y/x}=x \quad \text{...(i)}$$

and $\quad x^k\frac{d^2y}{dx^2}=\left(x\frac{dy}{dx}-y\right)^2$ ...(ii)

From Eq. (i), $\quad e^{y/x}=\frac{x}{a+bx}$

$$\Rightarrow \quad \frac{y}{x}=\ln\left(\frac{x}{a+bx}\right)$$

$$\Rightarrow \quad y = x\ln\left(\frac{x}{a+bx}\right)$$

$$\Rightarrow \quad \frac{dy}{dx} = \ln\left(\frac{x}{a+bx}\right) + x\left(\frac{a+bx}{x}\right)\left(\frac{a+bx-xb}{(a+bx)^2}\right)$$

$$\Rightarrow \quad \frac{dy}{dx} = \ln\left(\frac{x}{a+bx}\right) + \frac{a}{(a+bx)}$$

$$\Rightarrow \quad \frac{dy}{dx} = \frac{y}{x} + \frac{a}{a+bx} \quad \text{...(iii)}$$

$$\Rightarrow \quad x\frac{dy}{dx} - y = \frac{ax}{a+bx}$$

$$\Rightarrow \left(x\frac{dy}{dx} - y\right)^2 = \left(\frac{ax}{a+bx}\right)^2 \quad \text{...(iv)}$$

Also, from Eq. (iii), we get

$$\frac{d^2y}{dx^2} = y\left(\frac{-1}{x^2}\right) + \frac{1}{x}\cdot\frac{dy}{dx} + \frac{-ab}{(a+bx)^2}$$

$$\Rightarrow \quad \frac{d^2y}{dx^2} = \frac{-y}{x^2} + \frac{1}{x}\left(\frac{y}{x} + \frac{a}{a+bx}\right) - \frac{ab}{(a+bx)^2}$$

$$\Rightarrow \quad \frac{d^2y}{dx^2} = \frac{a}{(a+bx)x} - \frac{ab}{(a+bx)^2}$$

$$\Rightarrow \quad \frac{d^2y}{dx^2} = \frac{a^2+abx-abx}{x(a+bx)^2} = \frac{a^2}{x(a+bx)^2}$$

$$\Rightarrow x^3\frac{d^2y}{dx^2} = \frac{(ax)^2}{(a+bx)^2} = \left(x\frac{dy}{dx} - y\right)^2$$

$$\Rightarrow \quad k = 3$$

49. (11) We have, $f(x) = \cos^{-1}(4x^3 - 3x)$

$$= \begin{cases} 3\cos^{-1}x - 2\pi; & -1 \le x \le -\frac{1}{2} \\ 2\pi - 3\cos^{-1}x; & -\frac{1}{2} \le x \le \frac{1}{2} \\ 3\cos^{-1}x; & \frac{1}{2} \le x \le 1 \end{cases}$$

$$\therefore \quad f'(x) = \begin{cases} \frac{3}{\sqrt{1-x^2}}; & x \in \left(-\frac{1}{2}, \frac{1}{2}\right) \\ -\frac{3}{\sqrt{1-x^2}}; & x \in \left(-1, -\frac{1}{2}\right) \cup \left(\frac{1}{2}, 1\right) \end{cases}$$

Now, $S = f'(0) = 3$, $K = f'\left(-\frac{4}{5}\right) = -\frac{3}{\sqrt{1-\left(-\frac{4}{5}\right)^2}} = -5$

and $G = f'\left(\frac{4}{5}\right) = -\frac{3}{\sqrt{1-\left(\frac{4}{5}\right)^2}} = -5$

Hence, $2S - 3K + 2G = 2\times 3 - 3\times(-5) + 2(-5)$
$= 6 + 15 - 10 = 11$

50. (11) We have,

$$y = \frac{5x}{\sqrt[3]{(1-x)^2}} + \cos^2(2x+1)$$

$$\Rightarrow \quad y = 5x(1-x)^{-2/3} + \cos^2(2x+1)$$

$$\Rightarrow \quad \frac{dy}{dx} = 5\cdot 1(1-x)^{-2/3} + 5x\left(-\frac{2}{3}\right)(1-x)^{-\frac{5}{3}}\cdot(-1)$$
$$+ 2\cos(2x+1)\{-\sin(2x+1)\}\cdot 2$$

$$= 5(1-x)^{-5/3}\left[(1-x) + \frac{2}{3}(x)\right] - 2\sin(4x+2)$$

$$= 5(1-x)^{-5/3}\left(\frac{3-3x+2x}{3}\right) - 2\sin(4x+2)$$

$$\Rightarrow \quad \frac{dy}{dx} = \frac{5(3-x)}{3(1-x)^{-5/3}} - 2\sin(4x+2)$$

So, $A = 5, B = 4$ and $C = 2$

$\therefore\ A + B + C = 5 + 4 + 2 = 11$

51. (100) $P(x) = a(x-1)(x-3)(x-5)(x-b)$

$\Rightarrow\ P'(x) = a[(x-3)(x-5)(x-b) + (x-1)(x-5)(x-b)$
$+ (x-1)(x-3)(x-5) + (x-1)(x-3)(x-b)]$

$\Rightarrow\ P'(7) = a[4\times 2(7-b) + 6\times 2\times(7-b)$
$+ 6\times 4\times(7-b) + 6\times 4\times 2] = 0$

$\Rightarrow\ (7-b)(8+12+24) + 48 = 0$

$\Rightarrow\ (7-b)(44) = -48 \Rightarrow (b-7)44 = 48$

$\Rightarrow\ b - 7 = \frac{48}{44} \Rightarrow b = \frac{12}{11} + 7 \Rightarrow b = \frac{89}{11}$

So, $p = 89, q = 11$

Hence, $p + q = 89 + 11 = 100$

52. (5) We have, $y = \dfrac{a + bx^{3/2}}{x^{5/4}}$

Given, $\left.\dfrac{dy}{dx}\right|_{x=5} = 0$

On differentiating both sides w.r.t. x, we get

$$\frac{dy}{dx} = \left[\frac{x^{\frac{5}{4}}\frac{3}{2}bx^{\frac{1}{2}} - \frac{5}{4}(a+bx^{\frac{3}{2}})\cdot x^{\frac{1}{4}}}{x^{\frac{5}{2}}}\right]$$

Putting $x = 5$, we get

$$\frac{dy}{dx} = \left[\frac{5^{\frac{5}{4}}\left(\frac{3}{2}b5^{\frac{1}{2}}\right) - \frac{5}{4}(a+b\cdot 5^{\frac{3}{2}})\cdot 5^{\frac{1}{4}}}{5^{\frac{5}{2}}}\right]$$

Now, $\left.\dfrac{dy}{dx}\right|_{x=5} = 0$

$$\Rightarrow \frac{3}{2}b5^{1/2} - \frac{a}{4} - \frac{b}{4}5^{3/2} = 0 \Rightarrow \left(\frac{3}{2}b - \frac{5b}{4}\right)\cdot 5^{1/2} - \frac{a}{4} = 0$$

$$\Rightarrow \left(\frac{b}{4}\right)\sqrt{5} = \frac{a}{4} \Rightarrow \frac{a}{b} = \sqrt{5}$$

Hence, $\left(\frac{a}{b}\right)^2 = 5$

53. (6) Let $y = f(x) = e^{x^3 + x^2 + x}$

$$\therefore \quad \frac{dy}{dx} = f'(x) = e^{x^3 + x^2 + x}(3x^2 + 2x + 1)$$

$$\Rightarrow \quad g'(y) = \frac{dx}{dy} = \frac{1}{e^{x^3 + x^2 + x}(3x^2 + 2x + 1)} \quad \text{...(i)}$$

where $y = e^3$, then $e^3 = e^{x^3 + x^2 + x}$

$$\Rightarrow \quad 3 = x^3 + x^2 + x \Rightarrow x^3 + x^2 + x - 3 = 0$$

$$\Rightarrow \quad (x-1)(x^2 + 2x + 3) = 0$$

$$\therefore \quad x - 1 = 0 \qquad [\because x^2 + 2x + 3 \neq 0]$$

$$\Rightarrow \quad x = 1$$

Now, putting $y = e^3$ and $x = 1$ in Eq. (i), we get

$$g'(e^3) = \frac{1}{e^{1+1+1}(3+2+1)} = \frac{1}{6e^3}$$

So, $\quad k = 6$

54. (481) We have, $f(x) = \sin\left[\cos^{-1}\left(\frac{1-2^{2x}}{1+2^{2x}}\right)\right]$

Let $2^x = \tan\theta$

$$= \sin\left[\cos^{-1}\left(\frac{1-\tan^2\theta}{1+\tan^2\theta}\right)\right]$$

$$= \sin[\cos^{-1}(\cos 2\theta)] = \sin 2\theta$$

$$= \frac{2\tan\theta}{1+\tan^2\theta}$$

$$f(x) = \frac{2\cdot 2^x}{1+2^{2x}}$$

Now, $f'(x) = 2\left[\frac{(1+2^{2x})(2^x\log 2) - 2^x(2^{2x}\log 2\cdot 2)}{(1+2^{2x})^2}\right]$

So, $f'(1) = 2\left[\frac{5\cdot 2\log 2 - 2\cdot 4\cdot 2\log 2}{(1+4)^2}\right]$

$$= 2\left(\frac{10\log 2 - 16\log 2}{25}\right) = -\frac{12}{25}\log 2$$

Given that $f'(1) = \frac{-b}{a}\log 2$

So, $a = 25$ and $b = 12$

Then, $|a^2 - b^2| = |(25)^2 - (12)^2|$

$$= |625 - 144| = 481$$

55. (8) $\because \; x = \sin\left(t + \frac{7\pi}{12}\right) + \sin\left(t - \frac{\pi}{12}\right) + \sin\left(t + \frac{3\pi}{12}\right)$

$$= 2\sin\left(t + \frac{\pi}{4}\right)\cos\left(\frac{\pi}{3}\right) + \sin\left(t + \frac{3\pi}{12}\right)$$

$$= 2\sin\left(t + \frac{\pi}{4}\right)\times\frac{1}{2} + \sin\left(t + \frac{3\pi}{12}\right)$$

$$\Rightarrow \quad x = 2\sin\left(t + \frac{\pi}{4}\right)$$

and $\; y = \cos\left(t + \frac{7\pi}{12}\right) + \cos\left(t - \frac{\pi}{12}\right) + \cos\left(t + \frac{3\pi}{12}\right)$

$$= 2\cos\left(t + \frac{\pi}{4}\right)\times\cos\left(\frac{\pi}{3}\right) + \cos\left(t + \frac{\pi}{4}\right)$$

$$= 2\cos\left(t + \frac{\pi}{4}\right)\times\frac{1}{2} + \cos\left(t + \frac{\pi}{4}\right)$$

$$= 2\cos\left(t + \frac{\pi}{4}\right)$$

$$\Rightarrow \frac{x}{y} - \frac{y}{x} = \tan\left(t + \frac{\pi}{4}\right) - \cot\left(t + \frac{\pi}{4}\right)$$

$$= \frac{\sin^2\left(t + \frac{\pi}{4}\right) - \cos^2\left(t + \frac{\pi}{4}\right)}{\sin\left(t + \frac{\pi}{4}\right)\cos\left(t + \frac{\pi}{4}\right)}$$

$$= \frac{-2\left[\cos^2\left(t + \frac{\pi}{4}\right) - \sin^2\left(t + \frac{\pi}{4}\right)\right]}{2\sin\left(t + \frac{\pi}{4}\right)\cos\left(t + \frac{\pi}{4}\right)}$$

$$= -2\frac{\cos\left(2t + \frac{\pi}{2}\right)}{\sin\left(2t + \frac{\pi}{2}\right)} = -\frac{2\times(-\sin 2t)}{\cos 2t}$$

$$\Rightarrow \quad \left(\frac{x}{y} - \frac{y}{x}\right) = 2\tan 2t$$

$$\Rightarrow \quad \frac{d}{dt}\left(\frac{x}{y} - \frac{y}{x}\right) = 4\sec^2 2t$$

At $t = \frac{\pi}{8}$,

$$4\sec^2\left(\frac{\pi}{4}\right) = 4(\sqrt{2})^2 = 8$$

CHAPTER

05

ROLLE'S AND MEAN VALUE THEOREM

LEVEL 01 \ Starter Level

5.1 Rate of Change

1. Find the rate of change of the total surface area of a cylinder of radius r and height h, when the radius varies.
2. Find the rate of change of the volume of a sphere with respect to its diameter.
3. Find the rate of change of volume of cube with respect to its surface area, when edge of cube is 8 cm.
4. If the length of the rectangle is increasing at the rate of 2 cm/s and breadth is decreasing at the rate of 3 cm/s. At what rate is the
   (i) perimeter and (ii) area changes
   when length is 2 cm and breadth is 4 cm.
5. If the radius of circle is increasing at the rate of 2 cm/min. Find the rate of change in
   (i) circumference and (ii) area of circle
   when $r = 2$ cm.
6. If the side of a square is increasing at the rate of 4 cm/s. At what rate is the
   (i) perimeter and (ii) the area
   is increasing, when the side is 6 cm.
7. The radius of spherical soap bubble is increasing at the rate of 0.5 cm/s. Find the rate of change of its surface area, when radius is 7 cm.

8 Find an acute angle x, which increases twice as fast as its sine.

9 The bottom of a rectangular tank is 10 m by 6 m. Water is pumped into the tank at the rate of 30 m^3/min. Find the rate, at which the level of water in the tank is rising.

5.2 Rolle's and Lagrange's Mean Value Theorem

10 Discuss the applicability of Rolle's theorem for the following functions on the indicated interval.

(i) $f(x) = |x-1|$ on $[0, 2]$ (ii) $f(x) = (x-2)^{2/3}, x \in [1, 3]$

(iii) $f(x) = 2x^2 - 5x + 3$ on $[1, 3]$ (iv) $f(x) = \begin{cases} -4x+5, & 0 \le x \le 1 \\ 2x-3, & 1 < x \le 2 \end{cases}$

(v) $f(x) = [x]$ on $[2, 4]$

11 Verify Rolle's theorem for each of the following functions on the indicated interval.

(i) $f(x) = x^2 - 8x + 12$ on $[2, 6]$ (ii) $f(x) = \sin 2x$ on $\left[0, \frac{\pi}{2}\right]$

(iii) $f(x) = \sin x + \cos x$ on $\left[0, \frac{\pi}{2}\right]$ (iv) $f(x) = e^x \sin x$ on $[0, \pi]$

(v) $f(x) = e^{1-x^2}$ on $[-1, 1]$

12 Verify Lagrange's mean value theorem for the following functions on the indicated intervals.

(i) $f(x) = (x^2 - 1)$ on $[2, 3]$ (ii) $f(x) = \sqrt{25 - x^2}$ on $[-3, 4]$

(iii) $f(x) = x^3 - 5x^2 - 3x$ on $[1, 3]$ (iv) $f(x) = \frac{1}{4x-1}, 1 \le x \le 4$

13 Using Rolle's theorem. Find the points on the curve $y = 16 - x^2, x \in [-1, 1]$, where tangent is parallel to X-axis.

LEVEL 02 \ JEE Main Level

Single Choice Correct Questions

5.1 Rate of Change

1 A point on the parabola $y^2 = 8x$ at which the ordinate increases twice the rate of the abscissa, is

(a) (2, 2) (b) $\left(\frac{1}{2}, 2\right)$ (c) $\left(2, \frac{1}{2}\right)$ (d) $\left(\frac{1}{2}, \frac{1}{2}\right)$

2 Let $y = 4x - x^3$ and x is increasing at the rate of 2 units per second. How fast is the slope of the curve changing, when x = 3 units?

(a) 18 (b) 36 (c) 48 (d) 24

3 A particle moves along the curve $6y = x^3 + 2$. The number of points on the curve at which the y-coordinate changes eight times the x-coordinate, is

(a) 1 (b) 2 (c) 3 (d) 4

4 A small stone is dropped into a quiet lake and the waves move in a circle at the rate of 2.5 cm/s. At the instant when the radius of circular wave is 7.5 cm, the rate of change of the area of the circle is

(a) $\left(\frac{105}{2}\right)\pi$ (b) $\left(\frac{75}{2}\right)\pi$ (c) 51π (d) 52π

5 A ball is thrown vertically upwards and height x ft is reached by the ball in t s, is given by $x = 80t - 16t^2$. The ball reaches the maximum height in
(a) 2 s (b) 2.5 s (c) 3 s (d) 1.5 s

6 Distance travelled by a motor car in t sec after brakes are applied is s ft, where $s = 22t - 12t^2$. The distance travelled by the car before it stops, is
(a) 10.08 ft (b) 10 ft (c) 11 ft (d) 11.5 ft

7 The total revenue received from the sale of x units of a product is given by $R(x) = 3x^2 - 6y + 5$. Then, the marginal revenue when $x = 5$ units, is
(a) 30 (b) 24 (c) 36 (d) 40

8 Let the equation of a curve be $x = a(\theta + \sin\theta)$ and $y = a(1 - \cos\theta)$. If θ changes at a constant rate K, then the rate of change of slope of the tangent to the curve, when $\theta = \frac{\pi}{2}$, is
(a) K (b) $\frac{2K}{\sqrt{3}}$ (c) $\frac{K}{\sqrt{3}}$ (d) None of these

9 A balloon is pumped at the rate of a cm^3/s. The rate of increase of its surface area, when the radius is b cm, is
(a) $\frac{2a^2}{b^4}$ (b) $\frac{a}{2b}$ (c) $\frac{2a}{b}$ (d) None of these

10 Sand is draining from a conical filter, whose height and diameter are both 15 cm, into a cylindrical sand pot of diameter 15 cm. The rate at which sand drains from the filter into the pot is 100 cm^3/min. The rate in cm/min at which the level in the pot is rising at the instant, when the sand in the pot is 10 cm, is
(a) $\frac{9}{16\pi}$ (b) $\frac{25}{9\pi}$ (c) $\frac{5}{3\pi}$ (d) $\frac{16}{9\pi}$

11 A spherical iron ball of 10 cm radius is coated with a layer of ice of uniform thickness that melts at a rate of 50 cm^3/min. When the thickness of ice is 5 cm, then the rate (in cm/ min) at which of the thickness of ice decreases, is **[JEE Main 2020]**
(a) $\frac{5}{6\pi}$ (b) $\frac{1}{54\pi}$ (c) $\frac{1}{36\pi}$ (d) $\frac{1}{18\pi}$

12 The distance in seconds, described by a particle in time t, is given by $s = ae^t + \frac{b}{e^t}$. Then, the acceleration of the particle at time t is
(a) proportional to t (b) proportional to s
(c) s (d) constant

13 Two cyclist start from a junction of roads inclined at 60° with each other. Their velocities are 3 m/s and 4 m/s. Then, the rate at which two cyclists are separating, is
(a) $\sqrt{10}$ m/s (b) 1 m/s (c) $\sqrt{13}$ m/s (d) $\sqrt{5}$ m/s

14 Displacement s of a particle at time t expressed as $s = \frac{1}{2}t^3 - 24t$. The acceleration at the time, when the velocity vanishes, is
(a) 6 (b) 8 (c) 12 (d) 16

15 A particle is moving along the curve $x = at^2 + bt + c$. If $ac = b^2$, then the particle will be moving with uniform
(a) rotation (b) velocity
(c) acceleration (d) None of these

16 If $s = e^t$ (sint + cost) is the equation of motion of a moving particle, then acceleration at time t is given by

(a) $2e^t$ (cost + sint) (b) $2e^t$ (cost − sint)

(c) e^t (cost − sint) (d) e^t (cost + sint)

17 If the surface area of a cube is increasing at a rate of 3.6 cm^2/s, retaining its shape, then the rate of change of its volume (in cm^3/s), when the length of a side of the cube is 10 cm, is **[JEE Main 2020]**

(a) 18 (b) 10 (c) 9 (d) 20

18 If a car is moving such that the velocity acquired is proportional to the square root of the distance covered, then its acceleration is

(a) a constant (b) $\propto s^2$ (c) $\propto \frac{1}{s^2}$ (d) $\propto s$

19 If the rate of change in volume of spherical soap bubble is uniform, then the rate of change of surface area varies as

(a) square of radius

(b) square root of radius

(c) inversely proportional to radius

(d) cube of the radius

20 If at any instant t for a sphere, r denotes the radius, S denotes the surface area and V denotes the volume, then $\frac{dV}{dt}$ is

(a) $\frac{S}{2}\frac{dr}{dt}$ (b) $\frac{r}{2}\frac{dS}{dt}$ (c) $r\frac{dS}{dt}$ (d) $\frac{r^2}{2}\frac{dS}{dt}$

21 A water tank has the shape of an inverted right circular cone, whose semi-vertical angle is $\tan^{-1}(1/2)$. Water is poured into it at a constant rate of 5 m^3/min. Then, the rate (in m/min) at which the level of water is rising at the instant when the depth of water in the tank is 10 m, is **[JEE Main 2019]**

(a) $\frac{2}{\pi}$ (b) $\frac{1}{5\pi}$ (c) $\frac{1}{15\pi}$ (d) $\frac{1}{10\pi}$

22 The coordinate of the point on the ellipse $9x^2 + 4y^2 = 234$, where the ordinate decreases at the same rate at which the abscissa increases, are

(a) $\left(2\sqrt{2}, \frac{9\sqrt{2}}{2}\right)$ (b) $\left(-2\sqrt{2}, \frac{9\sqrt{2}}{2}\right)$

(c) $\left(\sqrt{2}, \frac{-9\sqrt{2}}{4}\right)$ (d) $(\sqrt{2}, -\sqrt{2})$

23 A variable triangle is inscribed in a circle of radius R. If the rate of change of a side is R times the rate of change of the opposite angle, then the angle is

(a) $\frac{\pi}{6}$ (b) $\frac{\pi}{3}$ (c) $\frac{\pi}{4}$ (d) $\frac{\pi}{2}$

24 The weight W of a certain stock of fish is given by $W = mw$, where m is the size of stock and w is the average weight of a fish. If m and w change with time t as $m = 2t^2 + 3$ and $w = t^2 - t + 2$. Then, the rate of change of w w.r.t. t, when $t = 1$, is

(a) 1 (b) 13 (c) 5 (d) 8

25 A man 2 m tall walks away from a lamp-post 6 m high at the rate of 4 km/h. The rate of increase of length of his shadow is

(a) 2 km/h (b) 3 km/h (c) 4 km/h (d) None of these

5.2 Rolle's and Lagrange's Mean Value Theorem

26 The function $f(x) = x(x+3)e^{-\frac{1}{2}x}$ satisfies the condition of Rolle's theorem in [-3, 0]. The value of c is

(a) 0 (b) -1 (c) -2 (d) -3

27 The value of c of Rolle's theorem for $f(x) = \log(x^2+2) - \log 3$ on [-1, 1] is

(a) 0 (b) 1 (c) -1 (d) Does not exist

28 Rolle's theorem holds for the function $f(x) = x^3 + mx^2 + nx$ on the interval [1, 2] and the value of $c = \frac{4}{3}$, then

(a) $m = 8, n = -5$ (b) $m = -5, n = 8$ (c) $m = 5, n = -8$ (d) $m = -5, n = -8$

29 Let $f(x) = \begin{cases} \dfrac{x^3 - 2x^2 - 5x + 6}{x - 1}, & \text{if } x \neq 1 \text{ and } x \in [-2, 3] \\ -6, & \text{if } x = 1 \end{cases}$.

Then, value of c in the Rolle's theorem for the function $f(x)$ on [-2, 3] is

(a) 1 (b) 2 (c) $-\frac{1}{2}$ (d) $\frac{1}{2}$

30 Let f be continuous on $[a, b]$, differentiable in (a, b) and $f(a) = f(b) = 0$. Further, suppose g is continuous on $[a, b]$ and differentiable in (a, b), then for the function $H(x) = f'(x) + f(x)g'(x)$. Which of the following statement is definitely true?

(a) $H(x) = 0$ for infinitely many x in (a, b) (b) $H(x) > 0$ for infinitely many x in (a, b)

(c) $H'(x) = 0$ for infinitely many x in (a, b) (d) $H(x) = 0$ for atleast one x in (a, b)

31 For all twice differentiable functions

$f : R \to R$, with $f(0) = f(1) = f'(0) = 0$ **[JEE Main 2020]**

(a) $f''(x) \neq 0$ at every point $x \in (0, 1)$ (b) $f''(x) = 0$ at every point $x \in (0, 1)$

(c) $f''(0) = 0$ (d) $f''(x) = 0$ at every point $x \in (0, 2)$

32 For the function $f(x) = x^3 - 6x^2 + ax + b$. If Rolle's theorem holds in [1, 3] with $c = 2 + \frac{1}{\sqrt{3}}$, then (a, b) is

(a) (11, 12) (b) (11, 11)

(c) (11, R) (d) (R, 0)

33 Let $f(x) = (x-4)(x-5)(x-6)(x-7)$, then

(a) $f'(x) = 0$ has four roots

(b) three roots of $f'(x) = 0$ lie in $(4, 5) \cup (5, 6) \cup (6, 7)$

(c) the equation $f'(x)$ has only two roots

(d) three roots of $f'(x) = 0$ lie in $(3, 4) \cup (4, 5) \cup (5, 6)$

34 If f and g are differentiable functions in [0, 1] satisfying $f(0) = g(1) = 2$, $g(0) = 0$ and $f(1) = 6$, then for some $c \in (0, 1)$

(a) $2f'(c) = g'(c)$ (b) $2f'(c) = 3g'(c)$

(c) $f'(c) = g'(c)$ (d) $f'(c) = 2g'(c)$

35 Let $f(x)$ be twice differentiable function for all real values of x and satisfies $f(1) = 1$, $f(2) = 4$ and $f(3) = 9$, then which of the following is definitely true?

(a) $f''(x) = 2, \forall x \in (1, 3)$ (b) $f''(x) = f'(x) = 5$ for some $x \in (2, 3)$

(c) $f''(x) = 3, \forall x \in (2, 3)$ (d) $f''(x) = 2$ for some $x \in (1, 3)$

36 The value of c in Rolle's theorem for $f(x) = \log\left[\dfrac{x^2 + ab}{x(a+b)}\right]$ in (a, b), where $a > 0$, is

(a) AM of a and b (b) GM of a and b (c) HM of a and b (d) $\dfrac{1}{a} + \dfrac{1}{b}$

37 If c is a point at which Rolle's theorem holds for the function, $f(x) = \log_e\left(\dfrac{x^2 + \alpha}{7x}\right)$ in the interval $[3, 4]$, where $\alpha \in R$, then $f''(c)$ is equal to **[JEE Main 2020]**

(a) $-\dfrac{1}{24}$ (b) $-\dfrac{1}{12}$ (c) $\dfrac{1}{12}$ (d) $\dfrac{\sqrt{3}}{7}$

38 If $f(x) = (x-a)(x-b)(x-c)$, where $a < b < c$ are real numbers, then the application of Rolle's theorem on f leads to

(a) $(a+b+c)(ab+bc+ca) = 3$ (b) $(a+b+c)^2 = 3(ab+bc+ca)$

(c) $(a+b+c)^2 > 3(ab+bc+ca)$ (d) $(a+b+c)^2 < 3(ab+bc+ca)$

39 $A(1, 0)$ and $B(e, 1)$ are points on the curve $y = \log_e x$. If P is a point on the curve at which the tangent to the curve is parallel to the chord AB, then the abscissa of P is

(a) $\dfrac{e-1}{2}$ (b) $e-1$ (c) $\dfrac{e+1}{\sqrt{2}}$ (d) $\dfrac{e-1}{\sqrt{2}}$

40 The value of c in the Lagrange's mean value theorem for the function $f(x) = x^3 - 4x^2 + 8x + 11$, when $x \in [0, 1]$, is **[JEE Main 2020]**

(a) $\dfrac{\sqrt{7}-2}{3}$ (b) $\dfrac{2}{3}$ (c) $\dfrac{4-\sqrt{5}}{3}$ (d) $\dfrac{4-\sqrt{7}}{3}$

41 If $a, b, c \in \left(0, \dfrac{\pi}{2}\right)$ and $a < c < b$, then $\dfrac{\cos a - \cos b}{\sin a - \sin b}$ may be equal to which of the following

(a) $\tan c$ (b) $\cot c$ (c) $-\tan c$ (d) $-\cot c$

42 If $0 < a < b < \dfrac{\pi}{2}$ and $f(a, b) = \dfrac{\tan b - \tan a}{b - a}$, then

(a) $f(a, b) \geq 2$ (b) $f(a, b) > 1$ (c) $f(a, b) \leq 1$ (d) None of these

43 Let a, b and c be real numbers and $a < b < c$. If f is continuous on $[a, c]$ and differentiable in (a, c). If $f'(x)$ is strictly increasing, then $(c-b) \cdot f(a) + (b-a) \cdot f(c)$ is

(a) greater than $(c-a) f(b)$ (b) less than $(c-a) f(b)$

(c) equal to $(c-a) f(b)$ (d) greater than $2(c-a) f(b)$

44 Suppose f is differentiable on R, $f(0) = 0$ and $1 \leq f'(x) \leq 2 \ \forall \ x \in R$, then

(a) $x \leq f(x) \leq 2x, \forall \ x \geq 0$ (b) $2x < f(x) \leq 3x, \forall \ x \geq 0$

(c) $3x < f(x) \leq 4x, \forall \ x > 0$ (d) $4x < f(x) \leq 5x, \forall \ x > 0$

45 A point on the curve $f(x) = \sqrt{x-2}$ in $[2, 3]$, when the tangent is parallel to the chord joining the end points, is

(a) $\left(\dfrac{9}{4}, \dfrac{1}{2}\right)$ (b) $\left(\dfrac{5}{2}, \dfrac{9}{4}\right)$ (c) $\left(\dfrac{5}{2}, \dfrac{1}{\sqrt{2}}\right)$ (d) None of these

46 Let $f(x)$ and $g(x)$ be defined and differentiable for $x \geq x_0$ and $f(x_0) = g(x_0)$, $f'(x) > g'(x)$ for $x > x_0$, then

(a) $f(x) < g(x)$ for some $x > x_0$ (b) $f(x) = g(x)$ for some $x > x_0$

(c) $f(x) > g(x)$ for some $x > x_0$ (d) None of these

47 Let $g(x) = [\log(1+x)]^{-1} - x^{-1}, x > 0$, then

(a) $1 < g(x) < 2$ (b) $-1 < g(x) < 0$ (c) $0 < g(x) < 1$ (d) None of these

48 Let f be any function continuous on $[a, b]$ and twice differentiable in (a, b). If for all $x \in (a, b)$, $f'(x) > 0$ and $f''(x) < 0$, then for any $c \in (a, b)$, $\frac{f(c) - f(a)}{f(b) - f(c)}$ is greater than **[JEE Main 2020]**

(a) $\frac{c-a}{b-c}$ (b) 1 (c) $\frac{b-c}{c-a}$ (d) $\frac{b+a}{b-a}$

49 Let the function $f : [-7, 0] \to R$ be continuous on $[-7, 0]$ and differentiable in $(-7, 0)$. If $f(-7) = -3$, $f'(x) \le 2$, $\forall x \in (-7, 0)$, then for all such function f, $f(-1) + f(0)$ lie in the interval

(a) $(-\infty, 20]$ (b) $[-3, 11]$ (c) $(-\infty, 11]$ (d) $[-6, 20]$ **[JEE Main 2020]**

Numerical Value Type Questions

50 A balloon which is always spherical is being inflated by pumping in gas at the rate of 1800 cm^3/s. When the radius of the balloon is 15 cm, the rate of the change of the radius is $\frac{K}{\pi}$ cm/s. Then, K is ....

51 The length of a rectangle is decreasing at the rate of 3 cm/s and width is increasing at the rate of 2 cm/s. When the length is 5 cm and width is 4 cm, the rate of change of area is ....

52 A stone is thrown vertically upwards and the height x ft reached by the stone in t s is given by $x = 100t - 40t^2$. The stone reached the maximum height in t s. Then, value of $4t$ is ....

53 An object is moving in the clockwise direction around the unit circle $x^2 + y^2 = 1$. As it passes through the point $\left(\frac{1}{2}, \frac{\sqrt{3}}{2}\right)$ its y-coordinate is decreasing at the rate of 3 units/s. The rate at which x-coordinate changes at this point is $K\sqrt{3}$, then K is ....

54 A kite is moving horizontally at a height of 151.2 m. If the speed of the kite is 10 m/s. When the kite is 250 m from the boy who is flying the kite, the height of the boy being 1.2 m, the rate of the string being let out is ....

55 Rolle's theorem is applicable to the function $f(x) = x^3 - 2x^2 - 4x + 7$ on the interval $[0, K]$. Then, the value of K is ....

56 The maximum possible integral value of $\frac{\beta - \alpha}{\tan^{-1}\beta - \tan^{-1}\alpha}$, where $0 < \alpha < \beta < \sqrt{3}$, is ....

57 If $f(x)$ is differentiable on $[a, b]$ such that $f(a) = 2$ and $f(b) = 6$, then there exists one c, $a < c < b$ such that $(b^3 - a^3) f'(c) = Kc^2$, then K is equal to ....

58 Let $f : [1, 3] \to [0, \infty)$ be continuous and differentiable function. If $[f(3) - f(1)][f^2(3) + f^2(1) + f(3) f(1)] = Kf^2(c) f'(c)$, where $c \in (1, 3)$, then the value of K is ....

59 Suppose that $f(0) = -3$ and $f'(x) \le 5$ for all values of x. Then, the largest value, which $f(2)$ can assume, is ....

60 Let $f(x)$ satisfy the requirement of Lagrange's mean value theorem in $[0, 3]$. If $f(0) = 0$ and $|f'(x)| \le \frac{1}{3}$, $\forall x \in [0, 3]$. Then, largest value of $|f(x)|$ is ....

61 Let f be a differentiable function for all $x \in R$. If $f(1) = -2$ and $f'(x) \ge 2$, $\forall x \in [1, 6]$ such that least value of $f(6)$ is m, then the value of $2(m+2)^3 + 17$ is ....

LEVEL 03 \ JEE Advanced Level

Single Choice Correct Questions

1 A water tank has shape of an inverted circular cone with circular top opening of radius 2 m and height 4 m. If water is being pumped into tank at rate of 2 m^3/min, then the rate at which water level is rising, when water is 3m deep, is

(a) $\frac{4}{9\pi}$ m/min (b) $\frac{2}{3\pi}$ m/min (c) $\frac{8}{3\pi}$ m/min (d) $\frac{8}{9\pi}$ m/min

2 A spherical ball 10 cm in radius is coated with a layer of ice of uniform thickness that melts at the rate of 50 cm^3/min. When the thickness of ice is 5 cm, then the rate at which the thickness of ice decreases, is (in cm/min)

(a) $\frac{1}{36\pi}$ (b) $\frac{1}{18\pi}$ (c) $\frac{1}{54\pi}$ (d) $\frac{5}{6\pi}$

3 The two equal sides of an isosceles triangle with fixed base b are decreasing at the rate of 3 cm/s. How fast is the area decreasing, when the two equal sides are equal to the base?

(a) $\sqrt{3}b$ (b) b (c) $\frac{b}{\sqrt{3}}$ (d) None of these

4 Sand is pouring from a pipe at the rate of 36 cm^3/s. The falling sand forms a cone on the ground in such a way that the height of the cone is always $\frac{1}{3}$ of the radius of the base. The rate at which the height of the sand cone is increasing, when the height is 4 cm, is

(a) $\frac{1}{2\pi}$ (b) $\frac{1}{4\pi}$ (c) $\frac{1}{8\pi}$ (d) None of these

5 A ladder 13 m long rests with one end against a vertical wall. The other on the floor, the lower end moves away from vertical wall at the rate of 2m/min. The rate at which the upper end falls, when its base is 12 m away from the wall.

(a) $\frac{3}{24}$ (b) $\frac{5}{24}$ (c) $\frac{24}{5}$ (d) None of these

6 A particle describes an ellipse, whose semi-axes are 4 m and 3 m with a constant speed 1 m/s. The velocity of the foot of the perpendicular from the particle on the major axis, when the particle is at a distance of 1 m from the major axis, is

(a) $\frac{2}{11}$ m/s (b) $\frac{11}{2}$ m/s (c) $\sqrt{\frac{2}{11}}$ m/s (d) None of these

7 If a, b and c are non-zero real numbers such that

$\int_0^1 (1+\cos^8 x)(ax^2+bx+c)\,dx = \int_0^2 (1+\cos^8 x)(ax^2+bx+c)\,dx = 0.$

Then, the roots of equation $ax^2+bx+c=0$

(a) Both roots lie between (0, 1)

(b) Both roots lie between (1, 2)

(c) One root lies between (0, 1) and other lies between (1, 2)

(d) None of the above

8 Let for the function $f(x) = \begin{cases} \cos^{-1} x, & -1 \le x \le 0 \\ mx + c, & 0 < x \le 1 \end{cases}$. Lagrange's mean value theorem is applicable on [−1, 1], then ordered pair (m, c) is

(a) $\left(1, -\frac{\pi}{2}\right)$ (b) $\left(1, \frac{\pi}{2}\right)$ (c) $\left(-1, -\frac{\pi}{2}\right)$ (d) $\left(-1, \frac{\pi}{2}\right)$

9 If $f(x)=\begin{cases} 3, & x=0 \\ -x^2+3x+a, & 0<x<1 \\ mx+b, & 1\le x\le 2 \end{cases}$ satisfies the hypothesis of the lagrange's mean value theorem, then the value of $(a+b)m$ is

(a) 2 (b) 3 (c) 4 (d) 7

10 f is continuous on $[0, 5]$ and differentiable in $(0, 5)$, further $f(0)=4$ and $f(5)=-1$, if $g(x)=\frac{f(x)}{x+1}$ for $0\le x\le 5$, then there exists $c\in(0, 5)$ such that $g'(c)$ is equal to

(a) $-\frac{4}{6}$ (b) $\frac{4}{6}$ (c) $\frac{5}{6}$ (d) $-\frac{5}{6}$

11 If the function $f(x)$ and $g(x)$ are continuous on $[a, b]$ and differentiable in (a, b), then in the interval (a, b) the equation $\begin{vmatrix} f'(x) & f(a) \\ g'(x) & g(a) \end{vmatrix} = \frac{1}{a-b}\begin{vmatrix} f(a) & f(b) \\ g(a) & g(b) \end{vmatrix}$

(a) has atleast one root (b) has exactly one root

(c) has atmost one root (d) no root

12 If $f(x)=(1-x)^{5/2}$ satisfies the relation $f(x)=f(0)+xf'(0)+\frac{x^2}{2}f''(\theta x)$, then as $x\to 1$, the value of θ is

(a) $\frac{4}{25}$ (b) $\frac{25}{4}$ (c) $\frac{25}{9}$ (d) $\frac{9}{25}$

13 If f is continuous on $[a, b]$ and differentiable in $(a, b)\,(ab>0)$, then there exists $c\in(a, b)$ such that $\frac{f(b)-f(a)}{\left(\frac{1}{b}\right)-\left(\frac{1}{a}\right)}$ is equal to

(a) $c^2f'(c)$ (b) $cf'(c)$ (c) $-cf'(c)$ (d) $-c^2f'(c)$

14 Let $f(x)=\frac{1}{x^2}$ and $g(x)=\frac{1}{x}$ on $[a, b]$, $0<a<b$. Let $\frac{f(b)-f(a)}{g(b)-g(a)}=\frac{f'(c)}{g'(c)}$ for some $a<c<b$. Then, c is

(a) AM of a and b (b) GM of a and b (c) HM of a and b (d) None of these

More than One Correct

15 A point moves in a fixed straight path $s=\sqrt{t}$, then

(a) acceleration $\propto v^3$ (b) acceleration is negative

(c) $v\propto\frac{1}{s}$ (d) None of these

16 Let $f(x)=\begin{bmatrix} x^{2\alpha+1}\ln x & ,x>0 \\ 0 & ,x=0 \end{bmatrix}$. If $f(x)$ satisfies Rolle's theorem in interval [0, 1], then α can be

(a) $-\frac{1}{2}$ (b) $-\frac{1}{3}$ (c) $-\frac{1}{4}$ (d) -1

17 Let $f(x)$ be a differentiable function and $f(a)=f(b)=0\,(a<b)$, then in the interval (a, b)

(a) $f(x)+f'(x)=0$ has atleast one root (b) $f(x)-f'(x)=0$ has atleast one root

(c) $f(x)\cdot f'(x)=0$ has atleast one real root (d) None of these

18 If $f(x) = x(x-2)(x-4)$, $1 \le x \le 4$, then the number satisfying the conditions of the mean value theorem is

(a) 1 (b) 3 (c) $\frac{5}{2}$ (d) $\frac{7}{2}$

19 In [0, 1], Lagrange's mean value theorem is applied to

(a) $f(x) = \begin{cases} \frac{1}{2} - x, & \text{if } x < \frac{1}{2} \\ \left(\frac{1}{2} - x\right)^2, & \text{if } x \ge \frac{1}{2} \end{cases}$ (b) $f(x) = \begin{cases} \frac{\sin x}{x}, & x \ne 0 \\ 1, & x = 0 \end{cases}$

(c) $f(x) = x|x|$ (d) $f(x) = |x|$

20 Let $f(x)$ be twice differentiable function such that $f''(x) > 0$ on [0, 2]. Then,

(a) $f(0) + f(2) = 2f(c)$ for atleast one $c \in (0, 2)$ (b) $f(0) + f(2) < 2f(1)$

(c) $f(0) + f(2) > 2f(1)$ (d) $2f(0) + f(2) > 3f\left(\frac{2}{3}\right)$

21 Let $f : [0, 8] \to R$ be differentiable function such that $f(0) = 0$, $f(4) = 1$ and $f(8) = 1$, then which one of the following holds good?

(a) There exists some $c_1 \in (0, 8)$, where $f'(c_1) = \frac{1}{4}$

(b) There exists some $c \in (0, 8)$, where $f'(c) = \frac{1}{12}$

(c) There exists $c_1, c_2 \in (0, 8)$, where $8f'(c_1) f(c_2) = 1$

(d) There exists some $\alpha, \beta \in (0, 2)$ such that $\int_0^8 f(t)\, dt = 3[\alpha^2 f(\alpha^3) + \beta^2 f(\beta^3)]$

22 Let $F(x) = [f(x)]^2 + [f'(x)]^2$, $F(0) = 6$, where $f(x)$ is thrice differentiable function such that $|f(x)| \le 1$, $\forall x \in [-1, 1]$, then choose the correct statement.

(a) There is atleast one point in each of the interval (−1, 0) and (0, 1), where $|f'(x)| \le 2$

(b) There is atleast one point in each of the interval (−1, 0) and (0, 1), where $F(x) \le 5$

(c) There is no point of local maxima of $F(x)$ in (−1, 1)

(d) For some $c \in (-1, 1)$, $F(c) \ge 6$, $F'(c) = 0$ and $F''(c) \le 0$

23 Let $f, g : [-1, 2] \to R$ be continuous functions which are twice differentiable on the interval (−1, 2). Let the values of f and g at the points −1, 0 and 2 be as given in the following table

	$x = -1$	$x = 0$	$x = 2$
$f(x)$	3	6	0
$g(x)$	0	1	−1

In each of the intervals (−1, 0) and (0, 2), the function $(f - 3g)''$ never vanishes. Then, the correct statement(s) is/are

(a) $f'(x) - 3g'(x) = 0$ has exactly three solutions in $(-1, 0) \cup (0, 2)$

(b) $f'(x) - 3g'(x) = 0$ has exactly one solution in (−1, 0)

(c) $f'(x) - 3g'(x) = 0$ has exactly one solution in (0, 2)

(d) $f'(x) - 3g'(x) = 0$ has exactly two solutions in (−1, 0) and exactly two solutions in (0, 2)

24 Let $f(x) = a_5 x^5 + a_4 x^4 + a_3 x^3 + a_2 x^2 + ax$, where a_i are real and $f(x) = 0$ has positive root α_0, then

(a) $f'(x) = 0$ has a root α_1 such that $0 < \alpha_1 < \alpha_0$ (b) $f'(x) = 0$ has atleast one real root

(c) $f''(x) = 0$ has atleast one real root (d) None of these

Linked Comprehension Type

Paragraph I Directions (Q.Nos. 25-27)

Suppose a, b and c be non-zero real numbers and $f(x)=(1+\cos^8 x)(ax^2+bx+c)$

25 Which of the following is not true?

(a) LMVT can be applied for $f(x)$

(b) Rolle's theorem can be applied for $f(x)$

(c) $f(x)=0$ has a solution between 0 and 1 if $c(a+b+c)<0$

(d) $f(x)=0$ has two solutions between 0 and 2 if $c(4a+2b+c)>0$

26 Suppose $\int_0^1 f(x)dx=0$, then $f(x)=0$ has

(a) only one solution between 0 and 1 (b) no solution between 0 and 1

(c) two solutions between 0 an 1 (d) None of these

27 If $f(x)=\int f(x)dx=\int_0^1 f(x)dx=\int_0^2 f(x)dx=0$, then which of the following can Rolle's theorem can be applied for $f(x)$

(a) [0, 1] (b) [0, 2] (c) [1, 2] (d) All of these

Paragraph II Directions (Q.Nos. 28-30)

If f is continuous on $[a, b]$, differentiable in (a, b) and $f(a)=f(b)$, then there exists $c\in(a, b)$ such that $f'(c)=0$.

28 If $f(a)=f(b)=0$, then the equation $f'(x)+\lambda f(x)=0$ has

(a) solution for all real λ (b) no solution for any real λ

(c) exactly one solution for all real λ (d) solution only for $\lambda=1$

29 If $\frac{a_0}{n+1}+\frac{a_1}{n}+\frac{a_2}{n-1}+\ldots+\frac{a_{n-1}}{2}+a_n=0$, where $a_0, a_1, a_2, \ldots, a_n$ are real, then the equation $a_0x^n+a_1x^{n-1}+a_2x^{n-2}+\ldots+a_n=0$ has a root in

(a) $(-\infty, -1)$ (b) $(-1, 0)$ (c) $(0, 1)$ (d) $(1, \infty)$

30 The number of values of c in Rolle's theorem for $f(x)=2x^3+x^2-4x-2$ in the interval $[-\sqrt{2}, \sqrt{2}]$ is

(a) 0 (b) 1 (c) 2 (d) 3

Matrix Match Type Questions

31 Match the following

	Column I		Column II
(i)	The rate of increase in the area of an equilateral triangle of side 30 cm, when each side is increasing at the rate of 0.1 cm/s, is	A.	5
(ii)	If the rate of decrease of $y=\frac{x^2}{2}-2x+5$ is thrice the rate of decrease of x, then x is equal to	B.	2
(iii)	The rate of change of volume of cube with respect to its surface area, when side is 8 cm, is	C.	$\frac{4}{3}$
(iv)	A particle moves along the curve $3y=2x^3+3$. If the rate of change of ordinate of a point is twice the rate of change of its abscissa, then the difference of the ordinates of the points is	D.	$\frac{3\sqrt{3}}{2}$

32 Match the following

	Column I		Column II
(i)	Let $f(x) = 2x(x-3)^n$ on [0, 3]. If $f(x)$ satisfies Rolle's theorem on [0, 3] for which $c = \frac{3}{4}$, then the value of n is	A.	$\frac{1}{2}$
(ii)	If $f(x) = 4^{\sin x}$, $x \in [0, \pi]$ satisfies Rolle's theorem, then the value of $\sin c$ is	B.	2
(iii)	The number of real roots of $x^4 - 4x - 2 = 0$ is	C.	1
(iv)	If $f(x) = x^{\alpha} \log x$ and $f(0) = 0$, then the value of $\alpha \in (0, 2)$ for which Rolle's theorem can be applied on [0, 1], is	D.	3

33 Match the following

	Column I		Column II
(i)	Value of c in LMVT for $f(x) = \sqrt{x^2 - 4}$ on the interval [2, 4] is	A.	4
(ii)	Value of c in LMVT for $f(x) = x^2 - 2x + 4$ on [1, 5] is	B.	3
(iii)	f is continuous on [2, 4] and differentiable in (2, 4). It is given that $f(2) = 5$ and $f(4) = 13$, then there exist, $c \in (2, 4)$ such that $f'(c)$ is equal to	C.	1
(iv)	If c is a value in Rolle's theorem for $f(x) = e^{-x} \sin x$ on the interval $[0, \pi]$, then $\tan c$ is equal to	D.	$\sqrt{6}$

34 Match the following

	Column I		Column II
(i)	The equation $x \log x = 3 - x$ has atleast one root in	A.	(0, 1)
(ii)	If $c = \sqrt{3}$ and $f(x) = x + \frac{1}{x}$, then interval x in which LMVT is applicable for $f(x)$, is	B.	(1, 3)
(iii)	If $c = \frac{1}{2}$ and $f(x) = 2x - x^2$, then interval of x in which LMVT is applicable for $f(x)$, is	C.	(0, 3)
(iv)	If $27a + 9b + 3c + d = 0$, then the equation $4ax^3 + 3bx^2 + 2cx + d = 0$ has atleast one root in	D.	(−1, 1)

Numerical Value Type Questions

Question Stem Directions (Q.Nos. 35-37) The answer to each question is a **NUMERICAL VALUE.**

A conical vessel 20 cm across the top and 15 cm deep is full of water. The vessel springs a leak at the bottom and losses water at 5 cm^3/s.

35 $\frac{k}{5\pi}$ cm^3/s is the rate at which water level is dropping at the instant, when the water is exactly half of the vessel height, then k is equal to ....

36 The amount of water (in cm^3), when the height of the water is 3 cm, is ....

37 The value of $\frac{d^2h}{dt^2}$, when the water is exactly 7.5 cm deep and $\frac{d^2V}{dt^2} = -\frac{4}{9}$ cm^3/s^2, is $\frac{1}{k}\left(\frac{-1}{k\pi} \frac{-2}{25\pi^2}\right)$, then k is equal to ....

38 The distance travelled by a particle moving in a straight line in time t is $s = \sqrt{at^2 + 2bt + c}$. Acceleration of the particle is proportional to s^{-P}, then the value of P is ....

39 Water is dropped at the rate of 2 m^3/s into a cone of semi-vertical angle 45°. If the rate at which periphery of water surface changes, when the height of the water in the cone is 2 m, is d, then the value of $5d$ is ....

40 l_1 and l_2 are the side length of two variable squares S_1 and S_2, respectively. If $l_1 = l_2 + l_2^3 + 6$, then rate of change of the area of S_2 with respect to rate of change of the area of S_1, when $l_2 = 1$ is equal to $\frac{1}{2^\lambda}$, then the value of λ is ....

41 The radius of the base of the cone is increasing at the rate of 3 cm/min and the altitude is decreasing at the rate of 4 cm/min. The rate of change of lateral surface area, when the radius is 7 cm and altitude is 24 cm, is $K\pi$ cm^2/s, the value of K ....

42 The radius of a right circular cylinder increases at a constant rate. Its altitude is a linear function of the radius and increases 3 times as fast as radius, when the radius is 1 cm and the altitude is 6 cm. When the radius is 6 cm, then the volume is increasing at the rate of 1 cm^3/s. When the radius is 36 cm, the volume is increasing at a rate of n cm^3/s, then the value of n ....

43 If $f(x)$ is twice differentiable such that $f(a) = 0, f(b) = 2, f(c) = -1, f(d) = 2$ and $f(e) = 0$, where $a < b < c < d < e$, then the minimum number of zeroes of $g(x) = [f'(x)]^2 + g'(x) f''(x)$ in the interval $[a, e]$, is ....

44 If the function $f(x) = x^3 - px^2 + qx$ is defined on [1, 3] satisfies the hypothesis of Rolle's theorem such that $c = \frac{5}{4}$, then value of $(p + 4q + 17)$ is ....

45 Let $f : [0, 4] \to R$ be a differentiable function and $\int_0^4 f(t)dt = K[\alpha f(\alpha^2) + \beta f(\beta^2)]$ for some $0 < \alpha$ and $\beta < 2$, then the value of K is ....

46 If the equation $x^3 - 3x + 1 = 0$ has three real roots α, β and γ such that $\alpha < \beta < \gamma$, then the value of $\{\alpha\} + \{\beta\} + \{\gamma\}$ is .... (where {} is fractional part function)

47 If f is continuous on [0, 2] and differentiable in (0, 2) such that $f(0) = 2, f(2) = 8$ and $f'(x) \le 3$ for all x in (0, 2), then the value of $f(1)$ is ....

48 Let $x^3 - 2Kx^2 - 4Kx + K^2 = 0$. If one root is less than 1 and the other is in (1, 4) and the third root is greater than 4 such that $K \in [m + \sqrt{n}, n(m + \sqrt{n})]$, then the value of $2(m + n - 1)^2$ is ....

ANSWER KEY

Level 1

1. $2\pi h + 4\pi r$
2. $\frac{1}{2}\pi D^2$
3. 2 cm
4. (i) – 2 cm/s (ii) 2 cm/s
5. (i) 4π cm/s (ii) 8π cm^2/min
6. (i) 16 cm/s (ii) 48 cm^2/s
7. 88 cm^2/s
8. $\frac{\pi}{3}$
9. 0.5 m/min
10. (i) Rolle's theorem is not applicable
    (ii) Rolle's theorem is not applicable
    (iii) Rolle's theorem is not applicable
    (iv) Rolle's theorem is not applicable
    (v) Rolle's theorem is not applicable
13. (0, 16)

Level 2

1. (b)	**2.** (b)	**3.** (b)	**4.** (b)	**5.** (b)	**6.** (a)	**7.** (b)
8. (a)	**9.** (c)	**10.** (d)	**11.** (d)	**12.** (c)	**13.** (c)	**14.** (c)
15. (c)	**16.** (b)	**17.** (c)	**18.** (a)	**19.** (c)	**20.** (b)	**21.** (b)
22. (a)	**23.** (b)	**24.** (b)	**25.** (a)	**26.** (c)	**27.** (a)	**28.** (b)
29. (d)	**30.** (d)	**31.** (b)	**32.** (c)	**33.** (b)	**34.** (d)	**35.** (d)
36. (b)	**37.** (c)	**38.** (c)	**39.** (b)	**40.** (d)	**41.** (c)	**42.** (b)
43. (a)	**44.** (a)	**45.** (a)	**46.** (c)	**47.** (c)	**48.** (a)	**49.** (a)
50. (2)	**51.** (2)	**52.** (5)	**53.** (3)	**54.** (8)	**55.** (2)	**56.** (3)
57. (12)	**58.** (6)	**59.** (7)	**60.** (1)	**61.** (2017)		

Level 3

1. (d)	**2.** (b)	**3.** (a)	**4.** (b)	**5.** (c)	**6.** (c)	**7.** (c)
8. (d)	**9.** (d)	**10.** (d)	**11.** (a)	**12.** (d)	**13.** (d)	**14.** (c)
15. (a, b, c)	**16.** (b, c)	**17.** (a, b, c)	**18.** (a, b)	**19.** (b, c, d)	**20.** (c, d)	**21.** (a, c, d)
22. (a, b, d)	**23.** (b, c)	**24.** (a, b, c)	**25.** (d)	**26.** (a)	**27.** (d)	**28.** (a)
29. (c)	**30.** (c)					

31. (i)–D, (ii)–A, (iii)–B, (iv)–C
32. (i)–D, (ii)–C, (iii)–B, (iv)–A,C
33. (i)–D, (ii)–B, (iii)–A, (iv)–C
34. (i)–B, (ii)–B, (iii)–A, (iv)–C

35. (1)	**36.** (12.56)	**37.** (7.5)	**38.** (3)	**39.** (5)	**40.** (5)	**41.** (54)
42. (33)	**43.** (6)	**44.** (59.2)	**45.** (2)	**46.** (1)	**47.** (5)	**48.** (200)

SOLUTIONS

Level 1

1. The surface area of cylinder,
$$S = 2\pi r(h + r)$$
$$\Rightarrow \quad S = 2\pi rh + 2\pi r^2$$
On differentiating w.r.t. r, we get
$$\frac{dS}{dr} = 2\pi h + 4\pi r$$

2. Let radius of sphere $= r$ and diameter $= D$
$$\therefore \quad r = \frac{D}{2}$$
Volume of sphere, $V = \frac{4}{3}\pi r^3$
$$\Rightarrow \quad V = \frac{4}{3}\pi\left(\frac{D}{2}\right)^3 = \frac{4}{3}\pi\frac{D^3}{8}$$
$$\Rightarrow \quad V = \frac{1}{6}\pi D^3$$
On differentiating w.r.t. D, we get
$$\frac{dV}{dD} = \frac{1}{6}\pi(3D^2)$$
$$= \frac{1}{2}\pi D^2$$

3. Let side of cube $= a$

Volume of cube, $V = a^3$
$$\therefore \quad \frac{dV}{da} = 3a^2 \quad \text{...(i)}$$
Surface area of cube, $S = 6a^2$
$$\therefore \quad \frac{dS}{da} = 12a \quad \text{...(ii)}$$
On dividing Eq. (i) by Eq. (ii), we get
$$\frac{dV}{dS} = \frac{\frac{dV}{da}}{\frac{dS}{da}} = \frac{3a^2}{12a} \Rightarrow \frac{dV}{dS} = \frac{a}{4}$$
At $a = 8$ cm,
$$\frac{dV}{dS} = \frac{8}{4} = 2 \text{ cm}$$

4. Let length of rectangle $= x$

and breadth of rectangle $= y$
$$\therefore \quad \frac{dx}{dt} = 2 \text{ cm/s}$$
$$\text{and} \quad \frac{dy}{dt} = -3 \text{ cm/s}$$
(i) Perimeter of rectangle $= 2(x + y)$
$$\Rightarrow \quad P = 2(x + y)$$
$$\Rightarrow \frac{dP}{dt} = 2\left(\frac{dx}{dt} + \frac{dy}{dt}\right) = 2(2 - 3) \Rightarrow \frac{dP}{dt} = -2 \text{ cm/s}$$
Hence, perimeter is decreasing at the rate of 2 cm/s.

(ii) Area of rectangle $= xy$
$$\Rightarrow \quad A = xy$$
On differentiating w.r.t. t, we get
$$\frac{dA}{dt} = x\frac{dy}{dt} + y\frac{dx}{dt}$$
$$= x(-3) + y(2) = -3x + 2y$$
At $x = 2$ cm and $y = 4$ cm,
$$\frac{dA}{dt} = -3(2) + 2(4) = 2 \text{ cm/s}$$
Hence, area is increasing at the rate of 2 cm/s.

5. Let radius of circle $= r$
$$\therefore \quad \frac{dr}{dt} = 2 \text{ cm/min} \quad \text{...(i)}$$
(i) Circumference of circle, $P = 2\pi r$
$$\Rightarrow \quad \frac{dP}{dt} = 2\pi\frac{dr}{dt} = 2\pi(2)$$
[from Eq. (i)]
$$\Rightarrow \quad \frac{dP}{dt} = 4\pi \text{ cm/min}$$
(ii) Area of circle, $A = \pi r^2$
$$\Rightarrow \quad \frac{dA}{dt} = \pi\left(2r\frac{dr}{dt}\right) = 2\pi r(2) \quad \text{[from Eq. (i)]}$$
$$\Rightarrow \quad \frac{dA}{dt} = 4\pi r$$
At $r = 2$ cm,
$$\frac{dA}{dt} = 8\pi \text{ cm}^2\text{/min}$$

6. Let the side of square $= a$
$$\therefore \quad \frac{da}{dt} = 4 \text{ cm/s} \quad \text{...(i)}$$
(i) Perimeter of square, $P = 4a$
$$\Rightarrow \quad \frac{dP}{dt} = 4\frac{da}{dt} = 4(4) \quad \text{[from Eq. (i)]}$$
$$\Rightarrow \quad \frac{dP}{dt} = 16 \text{ cm/s}$$
(ii) Area of square, $A = a^2$
$$\Rightarrow \quad \frac{dA}{dt} = 2a\frac{da}{dt} = 2a(4) \quad \text{[from Eq. (i)]}$$
$$\Rightarrow \quad \frac{dA}{dt} = 8a$$
At $a = 6$ cm,
$$\therefore \quad \frac{dA}{dt} = 8 \times 6 = 48 \text{ cm}^2\text{/s}$$

7. Let radius of spherical soap bubble $= r$

and $\frac{dr}{dt} = 0.5$ cm/s ...(i)

So, surface area, $S = 4\pi r^2$

$\Rightarrow \frac{dS}{dt} = 4\pi(2r)\frac{dr}{dt}$

$\Rightarrow \frac{dS}{dt} = 8\pi r(0.5)$ [from Eq. (i)]

$\Rightarrow \frac{dS}{dt} = 4\pi r$

At $r = 7$ cm, $\frac{dS}{dt} = 4\pi(7)$

$= 4\times\frac{22}{7}\times 7 = 88\ \text{cm}^2/\text{s}$

8. According to the question,

$\frac{dx}{dt} = 2\frac{d}{dx}(\sin x)$

$\Rightarrow \frac{dx}{dt} = 2\cos x\cdot\frac{dx}{dt}$

$\Rightarrow 2\cos x = 1$

$\Rightarrow \cos x = \frac{1}{2}$

$\Rightarrow x = \frac{\pi}{3}$

9. Let the volume of water tank is V.

$V = l\times b\times h$

$\Rightarrow V = 10\times 6\times h$

$\Rightarrow V = 60h$

$\Rightarrow \frac{dV}{dt} = 60\frac{dh}{dt}$

$\Rightarrow 30 = 60\frac{dh}{dt}$ $\left[\because \frac{dV}{dt} = 30\right]$

$\Rightarrow \frac{dh}{dt} = \frac{1}{2} = 0.5$ m/min

10. (i) We have,

$f(x) = |x-1| = \begin{cases} -x+1, & \text{when } 0 \le x < 1 \\ x-1, & 1 \le x \le 2 \end{cases}$

Since, a polynomial function is everywhere continuous and differentiable.

Therefore, $f(x)$ is continuous and differentiable $\forall\, x < 1$ as well as $x > 1$.

So, at $x = 1$, we have

$\lim_{x\to 1^-} f(x) = \lim_{x\to 1^-}(-x+1) = -1+1 = 0$

$\Rightarrow \lim_{x\to 1^+} f(x) = \lim_{x\to 1^+}(x-1) = 1-1 = 0$

$\Rightarrow f(1) = 0$

So, $f(x)$ is continuous on $[0, 2]$.

Now, LHD at $x = 1$,

$\lim_{h\to 0}\frac{f(1-h)-f(1)}{-h} = \lim_{h\to 0}\frac{[-(1-h)+1]-0}{-h}$

$= -1$

RHD at $x = 1$,

$\lim_{h\to 0}\frac{f(1+h)-f(1)}{h} = \lim_{h\to 0}\frac{(1+h-1)-0}{h}$

$= 1$

LHD $\ne$ RHD

So, $f(x)$ is non-differentiable at $x = 1$.

Hence, Rolle's theorem is not applicable in $[0, 2]$.

(ii) We have, $f(x) = (x-2)^{2/3}$

$\therefore f'(x) = \frac{2}{3}(x-2)^{-1/3}$

$\Rightarrow f'(x) = \frac{2}{3(x-2)^{1/3}}$

Clearly, $\lim_{x\to 2^+} f'(x) = \infty$

So, $f(x)$ is non-differentiable at $x = 2 \in (1, 3)$.

Hence, Rolle's theorem is not applicable to $f(x) = (x-2)^{2/3}$ on the interval $[1, 3]$.

(iii) We have, $f(x) = 2x^2 - 5x + 3$

$f(x)$ is a polynomial function, so $f(x)$ is everywhere continuous and differentiable.

$f(1) = 2(1)^2 - 5(1) + 3 = 0$

$f(3) = 2(3)^2 - 5(3) + 3 = 6$

$f(1) \ne f(3)$

So, Rolle's theorem is not applicable on $[1, 3]$.

(iv) We have, $f(x) = \begin{cases} -4x+5, & 0 \le x \le 1 \\ 2x-3, & 1 < x \le 2 \end{cases}$

At $x = 1$, LHL $= \lim_{x\to 1^-} f(x) = 1$

RHL $= \lim_{x\to 1^+} f(x) = -1$

LHL $\ne$ RHL

So, function is not continuous.

So, Rolle's theorem is not applicable.

(v) We have, $f(x) = [x]$

$f(x)$ is greatest integer function, which is discontinuous at every integer.

So, $f(x)$ is discontinuous on $x = 3$.

So, Rolle's theorem is not applicable on $f(x) = [x]$ in $[2, 4]$.

11. (i) We have, $f(x) = x^2 - 8x + 12$ is a polynomial function, which is continuous and differentiable everywhere.

Now, $f(2) = 2^2 - 8(2) + 12 = 16 - 16 = 0$

$f(6) = (6)^2 - 8(6) + 12 = 48 - 48 = 0$

$\Rightarrow f(2) = f(6)$

Now, $f'(x) = 2x - 8$

By Rolle's theorem,

$$f'(c) = 0$$

$\Rightarrow$ $2c - 8 = 0$

$\Rightarrow$ $c = 4 \in [2, 6]$

Hence, Rolle's theorem is verified.

(ii) We have, $f(x) = \sin 2x$

Since, $\sin x$ is continuous and differentiable everywhere.

$$f(0) = \sin 0 = 0$$

$$f\left(\frac{\pi}{2}\right) = \sin 2\left(\frac{\pi}{2}\right) = \sin \pi = 0$$

Now, $f'(x) = 2 \cos 2x$

By Rolle's theorem,

$$f'(c) = 0$$

$\Rightarrow$ $2 \cos 2c = 0$

$\Rightarrow$ $2c = \frac{\pi}{2}$

$\Rightarrow$ $c = \frac{\pi}{4} \in \left(0, \frac{\pi}{2}\right)$

Hence, Rolle's theorem is verified.

(iii) We have, $f(x) = \sin x + \cos x$

$f(x)$ is a trigonometric function, which is continuous and differentiable everywhere. So,

$$f(0) = \sin 0 + \cos 0 = 1$$

$$f\left(\frac{\pi}{2}\right) = \sin \frac{\pi}{2} + \cos \frac{\pi}{2} = 1$$

$\Rightarrow$ $f(0) = f\left(\frac{\pi}{2}\right)$

Now, $f'(x) = \cos x - \sin x$

By Rolle's theorem,

$$f'(c) = \cos c - \sin c = 0$$

$\Rightarrow$ $\tan c = 1$

$\Rightarrow$ $c = \frac{\pi}{4} \in \left(0, \frac{\pi}{2}\right)$

Hence, Rolle's theorem is verified.

(iv) We have, $f(x) = e^x \sin x$

$f(x)$ is product of exponential on trigonometric function, which is always continuous and differentiable.

So, $f(0) = e^0 \sin 0 = 0$

$$f(\pi) = e^{\pi} \sin \pi = 0$$

$\Rightarrow$ $f(0) = f(\pi)$

Now, $f'(x) = e^x \cos x + \sin x\, e^x$

$$= e^x (\sin x + \cos x)$$

Now, $f'(c) = 0$

$\Rightarrow$ $e^x (\sin x + \cos x) = 0$

$\Rightarrow$ $\sin x = -\cos x$

$\Rightarrow$ $\tan x = -1$

$\Rightarrow$ $x = \pi - \frac{\pi}{4}$

$\Rightarrow$ $x = \frac{3\pi}{4} \in (0, \pi)$

Hence, Rolle's theorem is verified.

(v) We have, $f(x) = e^{1-x^2}$

$f(x)$ is continuous and differentiable everywhere on $[-1, 1]$.

Now, $f(-1) = e^0 = 1$

$$f(1) = e^0 = 1$$

$\Rightarrow$ $f(-1) = f(1)$

Now, $f'(x) = e^{1-x^2}(-2x)$

$$= -2x \cdot e^{1-x^2}$$

By Rolle's theorem,

$$f'(c) = 0$$

$\Rightarrow$ $-2c \cdot e^{1-c^2} = 0$

$\Rightarrow$ $c = 0 \in (-1, 1)$

Hence, Rolle's theorem is verified.

12. (i) Given, $f(x) = x^2 - 1$ is a polynomial function, which is continuous and differentiable everywhere.

Now, $f'(x) = 2x$

Now, $f'(c) = \frac{f(b) - f(a)}{b - a}$

$\Rightarrow$ $2c = \frac{8 - 3}{3 - 2}$

$\Rightarrow$ $2c = 5 \Rightarrow c = \frac{5}{2} \in (2, 3)$

Hence, Lagrange's mean value theorem is verified.

(ii) Given, $f(x) = \sqrt{25 - x^2}$ is always continuous and differentiable.

$$f(-3) = \sqrt{25 - (-3)^2} = 4$$

$$f(4) = \sqrt{25 - (4)^2} = 3$$

Now, $f'(x) = \frac{1}{2\sqrt{25 - x^2}} \times (-2x)$

$$= \frac{-2x}{2\sqrt{25 - x^2}} = \frac{-x}{\sqrt{25 - x^2}}$$

Now, $f'(c) = \frac{f(b) - f(a)}{b - a}$

$\Rightarrow$ $\frac{-c}{\sqrt{25 - c^2}} = \frac{3 - 4}{4 + 3}$

$\Rightarrow$ $-7c = -\sqrt{25 - c^2}$

$\Rightarrow$ $49c^2 = 25 - c^2 \Rightarrow 50c^2 = 25$

$\Rightarrow$ $c^2 = \frac{25}{50} \Rightarrow c = \pm \frac{1}{\sqrt{2}} \in [-3, 4]$

Hence, Lagrange's mean value theorem is verified.

(iii) We have, $f(x)=x^3-5x^2-3x$

$f(x)$ is continuous and differentiable everywhere.

Now, $f(1)=(1)^3-5(1)^2-3(1)=-7$

$f(3)=(3)^3-5(3)^2-3(3)=-27$

$\Rightarrow \quad f'(x)=3x^2-10x-3$

Now, $\quad f'(c)=\dfrac{f(b)-f(a)}{b-a}$

$\Rightarrow 3c^2-10c-3=\dfrac{-27+7}{3-1}$

$\Rightarrow (3c-7)(c-1)=0$

$\Rightarrow \quad c=1,\dfrac{7}{3}$

Clearly, $c=\dfrac{7}{3}\in(1,3)$

Hence, Lagrange's mean value theorem is verified.

(iv) We have, $f(x)=\dfrac{1}{4x-1}$

$f(x)$ is continuous and differentiable in $x\in[1,4]$.

Now, $\quad f(1)=\dfrac{1}{4-1}=\dfrac{1}{3}$

$f(4)=\dfrac{1}{16-1}=\dfrac{1}{15}$

$\Rightarrow \quad f'(x)=\dfrac{-1}{(4x-1)^2}\cdot(4)$

Now, $\quad f'(c)=\dfrac{f(4)-f(1)}{4-1}$

$\Rightarrow \dfrac{-4}{(4c-1)^2}=\dfrac{\frac{1}{15}-\frac{1}{3}}{4-1}$

$\Rightarrow \dfrac{-4}{(4c-1)^2}=\dfrac{-4}{15(3)}$

$\Rightarrow \quad (4c-1)^2=45$

$\Rightarrow \quad 4c-1=\pm3\sqrt{5}$

$\Rightarrow \quad c=\dfrac{3\sqrt{5}+1}{4}\in(1,4)$

Hence, Lagrange's mean value theorem is verified.

13. We have, $\quad y=16-x^2$

y is continuous and differentiable on $[-1,1]$.

Now, at $x=-1, y=15$

At $x=1, y=15$

Now, $f'(x)=-2x$

By Rolle's theorem,

$f'(c)=0$

$\Rightarrow \quad -2c=0\Rightarrow c=0$

Now, when $x=0$, then, $y=16$

Hence, required point is (0, 16).

Level 2

1. (b) We have, $\quad y^2=8x$ ...(i)

On differentiating w.r.t. t, we get

$2y\dfrac{dy}{dt}=8\dfrac{dx}{dt}$

$\Rightarrow \quad 2y\left(2\dfrac{dx}{dt}\right)=8\dfrac{dx}{dt} \quad \left[\because \text{given}, \dfrac{dy}{dt}=2\dfrac{dx}{dt}\right]$

$\Rightarrow \quad 4y=8\Rightarrow y=2$

From Eq. (i), we get

$4=8x\Rightarrow x=\dfrac{1}{2}$

Hence, required point is $\left(\dfrac{1}{2},2\right)$.

2. (b) We have, $\quad y=4x-x^3$

On differentiating w.r.t. x, we get

$\dfrac{dy}{dx}=4-3x^2$

Let m be the slope, then

$m=4-3x^2$

On differentiating w.r.t. t, we get

$\dfrac{dm}{dt}=0-6x\dfrac{dx}{dt}$

Given that $\dfrac{dx}{dt}=2$ unit/sec, at $x=3$,

$\dfrac{dm}{dt}=-6(3)(2)=-36$

3. (b) We have, $6y=x^3+2$

On differentiating w.r.t. t, we get

$6\dfrac{dy}{dt}=3x^2\dfrac{dx}{dt}$

$\Rightarrow \quad 6\left(8\dfrac{dx}{dt}\right)=3x^2\dfrac{dx}{dt} \quad \left[\because \dfrac{dy}{dt}=8\dfrac{dx}{dt}\right]$

$\Rightarrow \quad 3x^2=48$

$\Rightarrow \quad x^2=16\Rightarrow x=\pm4$

If $x=4$, then $y=11$ and if $x=-4$, then $y=\dfrac{-31}{3}$

Therefore, the points on the curve are (4, 11) and $\left(-4,\dfrac{-31}{3}\right)$.

4. (b) Let r be the radius and A is the area of circle.

Then, $\quad A=\pi r^2$

$\Rightarrow \quad \dfrac{dA}{dt}=2\pi r\dfrac{dr}{dt}=2\pi r(2.5) \quad \left[\because \dfrac{dr}{dt}=2.5 \text{ cm/s}\right]$

$\Rightarrow \quad \dfrac{dA}{dt}=5\pi r$

At $r = 7.5 = \frac{15}{2}$,

$$\frac{dA}{dt} = 5\pi\left(\frac{15}{2}\right) = \left(\frac{75}{2}\right)\pi$$

5. (b) We have, $x = 80t - 16t^2$

At maximum height,

$$\frac{dx}{dt} = 0 \text{ i.e. } v = 0$$

$$\Rightarrow \quad 80 - 32t = 0$$

$$\Rightarrow \quad t = \frac{80}{32} = 2.5 \text{ s}$$

6. (a) We have, $s = 22t - 12t^2$

$$\Rightarrow \quad \frac{ds}{dt} = 22 - 24t$$

$$\Rightarrow \quad 22 - 24t = 0$$

$$\Rightarrow \quad t = \frac{22}{24} = \frac{11}{12}$$

So, distance travelled by car before it stops,

$$s = 22\left(\frac{11}{12}\right) - 12\left(\frac{11}{12}\right)^2$$

$$= \frac{11}{12}(22 - 11) = \frac{121}{12} = 10.08 \text{ ft}$$

7. (b) Given, $R(x) = 3x^2 - 6x + 5$

Marginal revenue is given by

$$\text{MR}(x) = \frac{dR}{dx} = 6x - 6,$$

When $x = 5$ units, then

$$\text{MR}(x) = 6 \times 5 - 6 = 24$$

$$\therefore \quad \text{MR}(5) = 24$$

8. (a) As, θ changes with constant rate.

$$\therefore \quad \frac{d\theta}{dt} = K$$

Given, $x = a(\theta + \sin\theta)$

$$\frac{dx}{d\theta} = a(1 + \cos\theta)$$

and $y = a(1 - \cos\theta)$

$$\frac{dy}{d\theta} = a\sin\theta$$

Then, $$\frac{dy}{dx} = m = \frac{a\sin\theta}{a(1 + \cos\theta)}$$

Rate of change of slope

$$\frac{dm}{dt} = \frac{1}{1 + \cos\theta} \cdot \frac{d\theta}{dt}$$

When $\theta = \frac{\pi}{2}$, then

$$\frac{dm}{dt} = K$$

9. (c) For spherical balloon,

$$V = \frac{4}{3}\pi r^3$$

$$\Rightarrow \quad \frac{dV}{dt} = 4\pi r^2 \frac{dr}{dt}$$

$$\Rightarrow \quad a = 4\pi r^2 \frac{dr}{dt}$$

At $r = b$ cm,

$$\Rightarrow \quad \frac{dr}{dt} = \frac{a}{4\pi b^2} \quad \text{...(i)}$$

Now, $S = 4\pi r^2$

$$\Rightarrow \quad \frac{dS}{dt} = 8\pi r \frac{dr}{dt}$$

From Eq. (i),

$$\frac{dS}{dt} = 8\pi r\left(\frac{a}{4\pi b^2}\right)$$

$$\Rightarrow \quad \frac{dS}{dt} = \frac{2ar}{b^2}$$

At $r = b$,

$$\frac{dS}{dt} = \frac{2a}{b} \text{ cm}^2/\text{s}$$

10. (d) For cylindrical pot,

$$V = \pi r^2 h$$

$$\Rightarrow \quad \frac{dV}{dt} = \pi r^2 \frac{dh}{dt} \quad [r = \text{constant}]$$

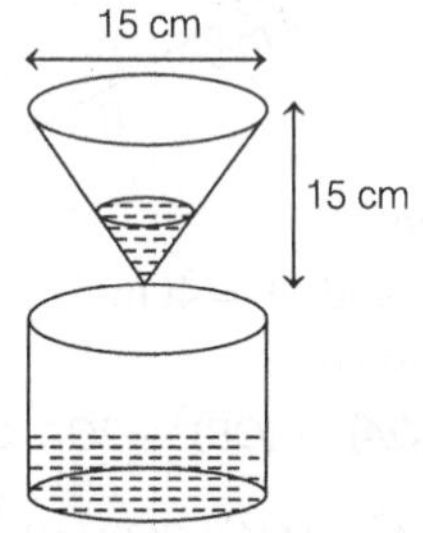

Hence, $$100 = \pi r^2 \frac{dh}{dt}$$

$$\Rightarrow \quad 100 = \pi \cdot \frac{225}{4}\frac{dh}{dt} \quad \left[\because r = \frac{15}{2} \text{ cm}\right]$$

$$\Rightarrow \quad \frac{dh}{dt} = \frac{400}{225\pi} = \frac{16}{9\pi} \text{ cm/min}$$

11. (d) It is given that a spherical iron ball of 10 cm radius is coated with a layer of ice of uniform thickness, let the thickness is x cm, then volume of the ball is

$$V = \frac{4}{3}\pi(10 + x)^3$$

On differentiating w.r.t. t, we get

$$\frac{dV}{dt} = 4\pi(10 + x)^2 \frac{dx}{dt}, \quad \text{...(i)}$$

where t is time in minute.

It is given, $\frac{dV}{dt} = -50 \text{ cm}^3/\text{min}$,

Now when x is 5 cm, then

$$-50 = 4\pi(10+5)^2 \frac{dx}{dt} \quad \text{[from Eq. (i)]}$$

$$\Rightarrow \quad \frac{dx}{dt} = -\frac{50}{4\pi(225)} = -\frac{1}{18\pi} \text{ cm/min}$$

Negative sign indicates the thickness of ice layer decreases with time.

12. (c) Given, $s = ae^t + \frac{b}{e^t}$

On differentiating w.r.t. t, we get

$$\frac{ds}{dt} = v = ae^t - \frac{b}{e^t}$$

Again, differentiating w.r.t. t, we get

$$\frac{d^2s}{dt^2} = \text{acceleration} = ae^t + \frac{b}{e^t}$$

$\therefore$ Acceleration $= s$

13. (c) Let both cyclists starts from O and reach A and B, respectively.

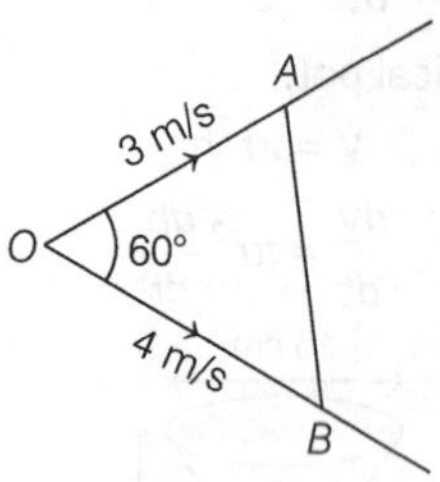

Let t be the time.

Thus, $OA = 3t$ m and $OB = 4t$ m

Now, using cosine rule,

$$(AB)^2 = (OA)^2 + (OB)^2 - 2OA \cdot OB \cos 60°$$

$$\Rightarrow \quad AB^2 = 9t^2 + 16t^2 - 2(3t)(4t)\frac{1}{2}$$

$$\Rightarrow \quad AB^2 = 13t^2$$

$$\Rightarrow \quad AB = \sqrt{13}t$$

On differentiating w.r.t. t, we get

$$\frac{d(AB)}{dt} = \sqrt{13} \text{ m/s}$$

14. (c) We have, $s = \frac{t^3}{2} - 24t$

Thus, velocity, $v = \frac{3}{2}t^2 - 24$

and acceleration, $a = \frac{dv}{dt} = \frac{d^2s}{dt^2} = 3t$

Velocity vanishes, when

$$\frac{3}{2}t^2 - 24 = 0$$

$$\Rightarrow \quad 3t^2 = 48$$

$$\Rightarrow \quad t^2 = 16$$

$$\Rightarrow \quad t = 4$$

Thus, acceleration, when $t = 4$

$$a = 3(4) = 12 \text{ units}$$

15. (c) We have, $x = at^2 + bt + c$

On differentiating w.r.t. t, we get

$$v = \frac{dx}{dt} = 2at + b$$

Again, differenatiating w.r.t. t, we get

$$\text{Acceleration} = \frac{d^2x}{dt^2} = 2a$$

Hence, the particle is moving with uniform acceleration.

16. (b) We have, $s = e^t(\sin t + \cos t)$

$$\Rightarrow \quad \frac{ds}{dt} = e^t(\sin t + \cos t) + e^t(\cos t - \sin t)$$

$$\Rightarrow \quad \frac{ds}{dt} = 2e^t(\cos t)$$

$$\text{Now,} \quad \frac{d^2s}{dt^2} = 2[e^t(\cos t) + e^t(-\sin t)]$$

$$= 2e^t(\cos t - \sin t)$$

Hence, acceleration at time t is $2e^t(\cos t - \sin t)$.

17. (c) Since, surface area of cube, $A = 6a^2 \text{ cm}^2$

It is given, $\frac{dA}{dt} = 3.6 \text{ cm}^2/\text{sec}$

$$\Rightarrow \quad 12a\frac{da}{dt} = 3.6 \text{ cm}^2/\text{sec} \quad \text{...(i)}$$

Now, as volume of cube, $V = a^3 \text{ cm}^3$

$$\therefore \quad \frac{dV}{dt} = 3a^2\frac{da}{dt} = 3a^2\frac{3.6}{12a} \quad \text{[from Eq. (i)]}$$

So, at $a = 10$ cm, $\frac{dV}{dt} = 0.9 \times 10 = 9 \text{ cm}^3/\text{sec}$

18. (a) Let v is the velocity of the car, when the distance covered is s.

Then, $\quad v \propto \sqrt{s} \quad$ [given]

$$\Rightarrow \quad v = K\sqrt{s} \Rightarrow \frac{dv}{ds} = \frac{K}{2\sqrt{s}}$$

$$\Rightarrow \quad v\frac{dv}{ds} = \frac{Kv}{2\sqrt{s}} = \frac{K^2}{2} = \text{constant}$$

Hence, the acceleration is constant.

19. (c) Let volume of spherical soap bubble is $V = \frac{4}{3}\pi r^3$, where r is radius.

$$\Rightarrow \quad \frac{dV}{dt} = 4\pi r^2\frac{dr}{dt} = C \text{ (constant)} \quad \text{...(i)}$$

and the surface area, $S = 4\pi r^2$

$$\frac{dS}{dt} = 8\pi r \frac{dr}{dt}$$

$$= 8\pi r \cdot \frac{C}{4\pi r^2} \quad \text{[from Eq. (i)]}$$

$$\Rightarrow \quad \frac{dS}{dt} = \frac{2C}{r}$$

$$\Rightarrow \quad \frac{dS}{dt} \propto \frac{1}{r}$$

So, the rate of change of surface area varies as inversely proportional to radius.

20. (b) Surface area of sphere, $S = 4\pi r^2$

$$\Rightarrow \quad \frac{dS}{dt} = 8\pi r \frac{dr}{dt} \quad \text{...(i)}$$

and the volume of sphere,

$$V = \frac{4}{3}\pi r^3$$

$$\Rightarrow \quad \frac{dV}{dt} = 4\pi r^2 \frac{dr}{dt}$$

$$= \frac{4\pi r^2}{8\pi r} \cdot \frac{dS}{dt} \quad \text{[from Eq. (i)]}$$

$$\Rightarrow \quad \frac{dV}{dt} = \frac{r}{2}\frac{dS}{dt}$$

21. (b) Given, semi-vertical angle of right circular cone

$$= \tan^{-1}\left(\frac{1}{2}\right)$$

Let $\alpha = \tan^{-1}\left(\frac{1}{2}\right)$

$$\Rightarrow \quad \tan\alpha = \frac{1}{2}$$

$$\Rightarrow \quad \frac{r}{h} = \frac{1}{2} \quad \left[\text{from figure, } \tan\alpha = \frac{r}{h}\right]$$

$$\Rightarrow \quad r = \frac{1}{2}h \quad \text{...(i)}$$

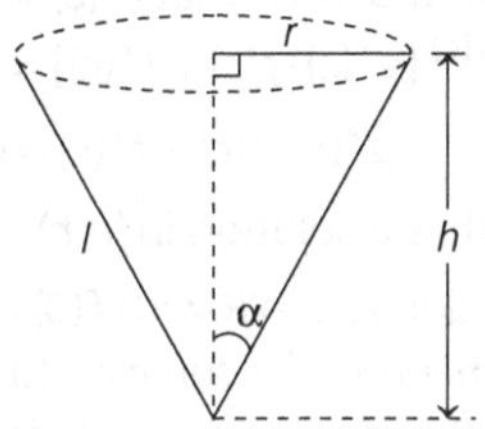

$\because$ Volume of cone is $(V) = \frac{1}{3}\pi r^2 h$

$$\therefore \quad V = \frac{1}{3}\pi\left(\frac{1}{2}h\right)^2 (h) = \frac{1}{12}\pi h^3 \quad \text{[from Eq. (i)]}$$

On differentiating both sides w.r.t. t, we get

$$\frac{dV}{dt} = \frac{1}{12}\pi(3h^2)\frac{dh}{dt}$$

$$\Rightarrow \quad \frac{dh}{dt} = \frac{4}{\pi h^2}\frac{dV}{dt}$$

$$\Rightarrow \quad \frac{dh}{dt} = \frac{4}{\pi h^2} \times 5 \quad \left[\because \text{given } \frac{dV}{dt} = 5 \text{ m}^3/\text{min}\right]$$

Now, at $h = 10$ m, the rate at which height of water level is rising $= \left.\frac{dh}{dt}\right|_{h=10}$

$$= \frac{4}{\pi(10)^2} \times 5 = \frac{1}{5\pi} \text{ m/min}$$

22. (a) We have, $9x^2 + 4y^2 = 234$

On differentiating w.r.t. t, we get

$$18x\frac{dx}{dt} + 8y\frac{dy}{dt} = 0$$

$$\Rightarrow \quad 18x\frac{dx}{dt} - 8y\frac{dx}{dt} = 0 \quad \left[\because -\frac{dy}{dt} = \frac{dx}{dt}\right]$$

$$\Rightarrow \quad 18x = 8y$$

$$\Rightarrow \quad y = \frac{9x}{4}$$

$$\therefore \quad 9x^2 + 4\left(\frac{9x}{4}\right)^2 = 234$$

$$\Rightarrow \quad 36x^2 + 81x^2 = 936$$

$$\Rightarrow \quad 117x^2 = 936$$

$$\Rightarrow \quad x^2 = 8 \Rightarrow x = \pm 2\sqrt{2}$$

Now, $y = \frac{9x}{4} \Rightarrow x = \pm 2\sqrt{2}$ and $y = \pm\frac{9\sqrt{2}}{2}$

$\therefore$ Coordinates are $\left(2\sqrt{2}, \frac{9\sqrt{2}}{2}\right), \left(-2\sqrt{2}, -\frac{9\sqrt{2}}{2}\right)$

23. (b) Let the side $BC = a$ and A be the opposite angle.

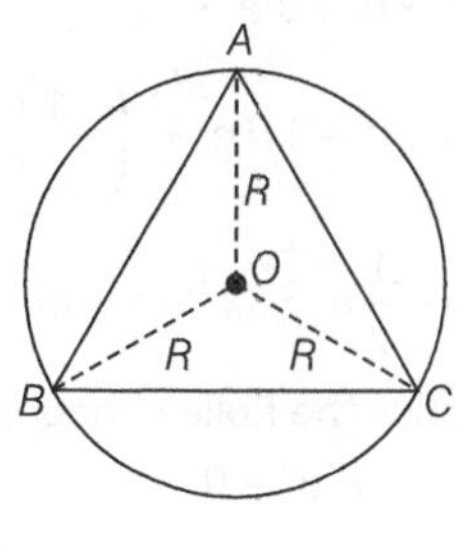

Then, $R = \frac{a}{2\sin A} \Rightarrow a = 2R\sin A$

On differentiating w.r.t. t, we get

$$\frac{da}{dt} = 2R\cos A\frac{dA}{dt}$$

$$\Rightarrow \quad R\frac{dA}{dt} = 2R\cos A\frac{dA}{dt} \quad \left[\because \frac{da}{dt} = R\frac{dA}{dt}\right]$$

$$\Rightarrow \quad \cos A = \frac{1}{2} \Rightarrow A = \frac{\pi}{3}$$

24. (b) We have,

$$W = mw,\ m = 2t^2 + 3 \text{ and } w = t^2 - t + 2$$

$$\therefore \quad \frac{dW}{dt} = \frac{dm}{dt} w + m \frac{dw}{t},\ \frac{dm}{dt} = 4t \text{ and } \frac{dw}{dt} = 2t - 1$$

At $t = 1$, we get

$$m = 5,\ w = 2,\ \frac{dm}{dt} = 4 \text{ and } \frac{dw}{dt} = 1$$

Hence, $\left(\frac{dW}{dt}\right)_{t=1} = 4 \times 2 + 5 \times 1 = 13$

25. (a) Given, $\frac{dx}{dt} = 4$ km/h

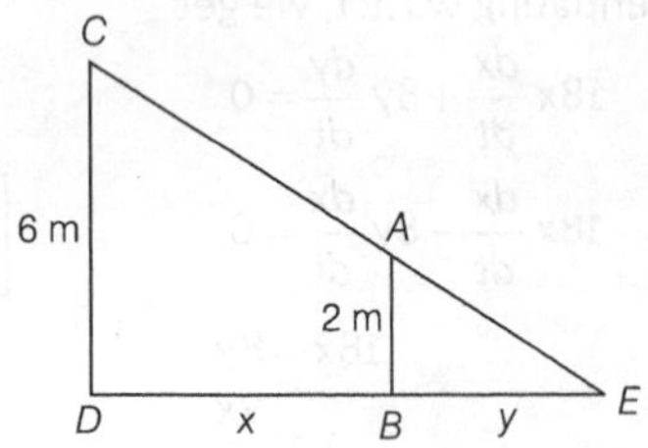

We have, $\frac{y}{x+y} = \frac{2}{6}$

$$\Rightarrow \quad \frac{y}{x+y} = \frac{1}{3}$$

$$\Rightarrow \quad 3y = x + y$$

$$\Rightarrow \quad x = 2y$$

On differentiating w.r.t. t, we get

$$\frac{dx}{dt} = 2\frac{dy}{dt}$$

$$\Rightarrow \quad \frac{dy}{dt} = \frac{4}{2} = 2 \text{ km/h}$$

26. (c) Given, $f(x) = x(x+3)e^{-\frac{1}{2}x}$

$$\therefore \quad f'(x) = (x^2 + 3x)e^{-\frac{1}{2}x}\left(-\frac{1}{2}\right) + (2x+3)e^{-\frac{1}{2}x}$$

$$= -\frac{1}{2}e^{-\frac{1}{2}x}(x^2 - x - 6)$$

Since, $f(x)$ satisfies the Rolle's theorem.

$$\therefore \quad f'(c) = 0$$

$$\Rightarrow \quad -\frac{1}{2}e^{-\frac{c}{2}}(c^2 - c - 6) = 0$$

$$\Rightarrow \quad c = 3, -2$$

But $c = 3 \notin [-3, 0]$, so $c = -2$

27. (a) Given, $f(x) = \log(x^2 + 2) - \log 3$

$$f(-1) = 0$$

and $\quad f(1) = 0$

$$\Rightarrow \quad f(-1) = f(1)$$

$$\Rightarrow \quad f'(x) = \frac{2x}{x^2 + 2}$$

By Rolle's theorem,

$$f'(c) = 0$$

$$\Rightarrow \quad \frac{2c}{c^2 + 2} = 0$$

$$\Rightarrow \quad c = 0 \in [-1, 1]$$

28. (b) We have, $\quad f(1) = f(2)$

$$\Rightarrow \quad 1 + m + n = 8 + 4m + 2n$$

$$\Rightarrow \quad 3m + n = -7 \qquad \text{...(i)}$$

and $\quad f'\left(\frac{4}{3}\right) = 0$

$$\Rightarrow \quad 3\left(\frac{4}{3}\right)^2 + 2m\left(\frac{4}{3}\right) + n = 0$$

$$\Rightarrow \quad 8m + 3n = -16 \qquad \text{...(ii)}$$

From Eqs. (i) and (ii), we get

$$m = -5 \text{ and } n = 8$$

29. (d) We have,

$$f(x) = \begin{cases} x^2 - x - 6, & \text{if } x \neq 1 \text{ and } x \in [-2, 3] \\ -6, & \text{if } x = 1 \end{cases}$$

Clearly, $\lim_{x \to 1} f(x) = -6 = f(1)$

$\Rightarrow$ f is continuous at $x = 1$

and hence f is continuous on $[-2, 3]$. f is also differentiable in $(-2, 3)$.

Further, $f(-2) = 4 + 2 - 6 = 0$

and $\quad f(3) = 9 - 3 - 6 = 0$

Thus, $\quad f(-2) = 0 = f(3)$.

Hence, $f'(c) = 0$ for some $c \in (-2, 3)$.

Therefore, $f'(c) = 0$ implies

$$2c - 1 = 0 \text{ or } c = \frac{1}{2} \in [-2, 3]$$

30. (d) Let $F(x) = e^{g(x)} f(x)$, $\forall\, x \in [a, b]$.

Since, $F(x)$ satisfies the conditions of Rolle's theorem, there exists $c \in (a, b)$ such that $F'(c) = 0$

$$\Rightarrow \quad e^{g(c)}[g'(c) \cdot f(c) + f'(c)] = 0$$

$$\Rightarrow \quad g'(c) \cdot f(c) + f'(c) = 0 \qquad [\because e^{g(c)} \neq 0]$$

Thus, $H(x) = 0$ has a solution in (a, b).

31. (b) Given function $f : R \to R$ with $f(0) = f(1) = f'(0) = 0$.

So, by Rolle's theorem, for some $c \in (0, 1)$ $f'(c) = 0$

and as $f'(0) = 0$ and function f is twice differential.

So, again for some $x \in (0, 1)$.

$$f''(x) = 0 \qquad \text{[by Rolle's theorem]}$$

32. (c) Given, $f(x) = x^3 - 6x^2 + ax + b$

Given, $(a, b) = (1, 3)$ and $c = 2 + \frac{1}{\sqrt{3}}$

$$f(1) = 1 - 6 + a + b = a + b - 5$$

and $\quad f(3) = 27 - 54 + 3a + b = 3a + b - 27$

According to Rolle's theorem,

$$a+b-5=3a+b-27$$

$$\Rightarrow \quad 2a=22$$

$$\Rightarrow \quad a=11$$

Now, $f'(x)=3x^2-12x+a$

According to Rolle's theorem,

$$f'(c)=0$$

$$\Rightarrow \quad 3c^2-12c+a=0$$

$$\Rightarrow \quad 3\left(2+\frac{1}{\sqrt{3}}\right)^2-12\left(2+\frac{1}{\sqrt{3}}\right)+a=0$$

$$\Rightarrow \quad a=11$$

Therefore, $(a,b)=(11,R)$

33. (b) Since, $f(4)=f(5)=f(6)=f(7)=0$

So, on applying Rolle's theorem to intervals [4, 5], [5, 6] and [6, 7], there exists $x_1\in(4,5)$, $x_2\in(5,6)$ and $x_3\in(6,7)$ such that $f'(x_1)=f'(x_2)=f'(x_3)=0$

Since, $f'(x)$ is a polynomial of degree three.

So, it cannot have four roots.

34. (d) Let $h(x)=f(x)-2g(x)$

As, $h(0)=h(1)=2$

Hence, using Rolle's theorem

$$h'(c)=0$$

$$\Rightarrow \quad f'(c)=2g'(c)$$

35. (d) Let $g(x)=f(x)-x^2$, we have

$$g(1)=0,\ g(2)=0 \text{ and } g(3)=0$$

$$[\because f(1)=1,\ f(2)=4 \text{ and } f(3)=9]$$

From Rolle's theorem on $g(x)$,

$g'(x)=0$ for atleast one $x\in(1,2)$

Let $g'(c_1)=0$, where $c_1\in(1,2)$

Similarly, $g(x)=0$ for atleast one $x\in(2,3)$

Let $g'(c_2)=0$, where $c_2\in(2,3)$.

Therefore, $g'(c_1)=g'(c_2)=0$

By Rolle's theorem, atleast one $x\in(c_1,c_2)$

such that $g''(x)=0$ or $f''(x)=2$ for some $x\in(1,3)$.

36. (b) Given, $f(x)=\log\left(\dfrac{x^2+ab}{x(a+b)}\right)$

Thus, $f(a)=f(b)$

Using Rolle's theorem,

$$f'(c)=0$$

So, $$f'(x)=\frac{x(a+b)}{x^2+ab}\left[\frac{x(a+b)\cdot 2x-(x^2+ab)(a+b)}{\{x(a+b)\}^2}\right]$$

$$\Rightarrow f'(x)=\frac{1}{x^2+ab}\left[\frac{(a+b)\{2x^2-(x^2+ab)\}}{x(a+b)}\right]$$

$$\Rightarrow f'(x)=\frac{x^2-ab}{x(x^2+ab)}$$

Therefore, $f'(c)=\dfrac{c^2-ab}{c(c^2+ab)}=0\Rightarrow c^2=ab \Rightarrow c=\sqrt{ab}$

Therefore, c is the GM of a and b.

37. (c) The given function, $f(x)=\log_e\left(\dfrac{x^2+\alpha}{7x}\right)$ holds the Rolle's theorem for the interval [3, 4], so

$$f(3)=f(4)$$

$$\Rightarrow \quad \log_e\left(\frac{9+\alpha}{21}\right)=\log_e\left(\frac{16+\alpha}{28}\right)$$

$$\Rightarrow \quad \frac{9+\alpha}{3}=\frac{16+\alpha}{4}$$

$$\Rightarrow \quad 36+4\alpha=48+3\alpha$$

$$\Rightarrow \quad \alpha=12 \quad \text{...(i)}$$

and $f'(c)=0$, for some $c\in(3,4)$

$$\Rightarrow \quad \left[\frac{7x}{x^2+\alpha}\times\frac{7x(2x+0)-(x^2+\alpha)7}{(7x)^2}\right]_{x=c}=0$$

$$\Rightarrow \quad \frac{c(2c)-(c^2+12)}{(c^2+12)c}=0 \quad [\because \alpha=12]$$

$$\Rightarrow \quad c^2-12=0;\ \because c\in(3,4)$$

$$\Rightarrow \quad c=\sqrt{12};\ \because c\in(3,4)$$

$$\therefore \quad f''(c=\sqrt{12})$$

$$=\frac{c(c^2+12)(2c)-(c^2-12)(3c^2+12)}{\{(c^2+12)c\}^2}$$

$$=\frac{(2\times12\times24)-(0\times48)}{(24)^2(12)}=\frac{1}{12}$$

38. (c) We have, $f(x)=(x-a)(x-b)(x-c)$

$$\Rightarrow \quad f(a)=f(b)=f(c)=0$$

$\Rightarrow$ a, b and c are three distinct roots of $f(x)=0$

So, by Rolle's theorem, $f'(x)$ has one real root in the interval (a,b) and other in the interval (b,c).

Thus, $f'(x)$ has two distinct real roots.

Now, $f(x)=(x-a)(x-b)(x-c)$

$$\Rightarrow \quad f(x)=x^3-x^2(a+b+c)+x(ab+bc+ca)-abc$$

$$\Rightarrow \quad f'(x)=3x^2-2x(a+b+c)+(ab+bc+ca)$$

As, $f'(x)$ has distinct roots.

$$\therefore \quad 4(a+b+c)^2-12(ab+bc+ca)>0$$

$$\Rightarrow \quad (a+b+c)^2>3(ab+bc+ca)$$

39. (b) Using Lagrange's mean value theorem for $f(x)=\log_e x$ on the interval [1, e], there exists $x_0\in(1,e)$ such that

$$f'(x_0)=\frac{f(e)-f(1)}{e-1}$$

$$\Rightarrow \quad \frac{1}{x_0}=\frac{1-0}{e-1}$$

$$\Rightarrow \quad x_0=e-1$$

40. (d) Given function, $f(x)=x^3-4x^2+8x+11$, when $x\in[0,1]$ is a continuous function in interval $x\in[0,1]$ and differentiable in interval $x\in(0,1)$, so according to Lagrange's mean value theorem for $x=c\in(0,1)$

$$f'(c)=\frac{f(1)-f(0)}{1-0}$$

$$\Rightarrow (3x^2-8x+8)_{x=c}=\frac{(1-4+8+11)-11}{1-0}$$

$$\Rightarrow \quad 3c^2-8c+8=5$$

$$\Rightarrow \quad 3c^2-8c+3=0$$

$$\Rightarrow \quad c=\frac{8-\sqrt{64-36}}{6} \qquad [\because c\in(0,1)]$$

$$=\frac{4-\sqrt{7}}{3}$$

41. (c) For the function, $f(x)=\cos x$

$$f'(c)=\frac{f(b)-f(a)}{b-a}$$

$$\Rightarrow \quad -\sin c=\frac{\cos b-\cos a}{b-a} \qquad \text{...(i)}$$

For the function, $f(x)=\sin x$

$$f'(c)=\frac{f(b)-f(a)}{b-a}$$

$$\Rightarrow \quad \cos c=\frac{\sin b-\sin a}{b-a} \qquad \text{...(ii)}$$

On dividing Eq. (i) by Eq. (ii), we get

$$\frac{\cos b-\cos a}{\sin b-\sin a}=\frac{-\sin c}{\cos c}$$

$$=-\tan c$$

42. (b) Consider, the function $f(x)=\tan x$ defined on $[a,b]$ such that $0<a<b<\frac{\pi}{2}$.

Applying Lagrange's mean value theorem, we get

$$f'(c)=\frac{f(b)-f(a)}{b-a} \text{ for some } c\in(a,b)$$

$$\Rightarrow \sec^2 c=\frac{\tan b-\tan a}{b-a}$$

$$\Rightarrow \quad f(a,b)=\sec^2 c$$

$$\Rightarrow \quad f(a,b)>1 \qquad \left[\because \sec^2 c>1, \text{as } c\in\left(0,\frac{\pi}{2}\right)\right]$$

43. (a) Using Lagrange's mean value theorem for f on $[a,b]$ and $[b,c]$, we have

$$\frac{f(b)-f(a)}{b-a}=f'(u)$$

and
$$\frac{f(c)-f(b)}{c-b}=f'(v)$$

For some $u\in(a,b)$ and $v\in(b,c)$, since $a<u<b<v<c$ and $f'(x)$ is strictly increasing, we have

$$\frac{f(b)-f(a)}{b-a}=f'(u)<f'(v)<\frac{f(c)-f(b)}{c-b}$$

$$[\because b-a>0 \text{ and } c-b>0]$$

Therefore,

$$\Rightarrow (c-b)[f(b)-f(a)]<(b-a)[f(c)-f(b)]$$

$$\Rightarrow (c-b)f(a)+(b-a)f(c)>(c-b+b-a)f(b)$$

$$>(c-a)f(b)$$

44. (a) Let $x>0$. Using Lagrange's mean value theorem on $[0, x]$, there exists $x_0\in(0,x)$ such that

$$\frac{f(x)-f(0)}{x-0}=f'(x)$$

$$\Rightarrow \quad f(x)=xf'(x_0)$$

So, $\quad 1\le f'(x)\le 2 \Rightarrow x\le xf'(x_0)\le 2x$

This is true for all $x>0$, when $x=0$, we have $f(x)=0$.

Thus, equity holds.

Hence, $x\le f(x)\le 2x, \forall\, x\ge 0$

45. (a) The given function, $f(x)=\sqrt{x-2}$ is continuous on [2, 3] and differentiable in (2, 3).

Therefore, by Lagrange's mean value theorem

$$f'(c)=\frac{f(b)-f(a)}{b-a}$$

$$\Rightarrow \quad f'(c)=\frac{f(3)-f(2)}{3-2}=\frac{\sqrt{3-2}-\sqrt{2-2}}{3-2}$$

$$\Rightarrow \quad f'(c)=1$$

$$\Rightarrow \quad \frac{1}{2\sqrt{c-2}}=1$$

$$\Rightarrow \quad 2\sqrt{c-2}=1$$

$$\Rightarrow \quad \sqrt{c-2}=\frac{1}{2}$$

$$\Rightarrow \quad c-2=\frac{1}{4}$$

$$\Rightarrow \quad c=\frac{9}{4}$$

So, $f(c)=\sqrt{\frac{9}{4}-2}=\frac{1}{2}$

$\therefore$ Point on the curve is $\left(\frac{9}{4},\frac{1}{2}\right)$.

46. (c) Consider a function $\phi(x)=f(x)\cdot g(x)$ on interval $[x_0, x]$.

By Lagrange's theorem,

$$\phi(x)-\phi(x_0)=\phi'(z)(x-x_0) \text{ for some } z\in(x_0,x)$$

Since, $\quad \phi(x_0)=0,\ \phi'(z)=f'(z)-g'(z)>0$

So, $\quad \phi(x)-\phi(x_0)=[f'(z)-g'(z)](x-x_0)$

$$\Rightarrow \quad \phi(x)=[f'(z)-g'(z)](x-x_0)>0$$

$$\Rightarrow \quad \phi(x)>0$$

Thus, $f(x)-g(x)>0$

$$\Rightarrow \quad f(x)>g(x)$$

47. (c) For $x > 0$, consider a function, $f(x) = \log(1+x)$ on the interval $[0, x]$.

Since, $f(x)$ is differentiable in $(0, x)$

By Lagrange's theorem, there is a $c \in (0, x)$ such that

$$\frac{\log(1+x) - \log(1+0)}{x-0} = f'(c) = \frac{1}{1+c}$$

$$\Rightarrow \quad \frac{\log(1+x)}{x} = \frac{1}{1+c} < 1$$

$$\Rightarrow \quad \log(1+x) < x \Rightarrow g(x) > 0$$

Also, $c \in (0, 1)$

$$\Rightarrow \quad c < x$$

$$\Rightarrow \quad \frac{1}{1+c} > \frac{1}{1+x}$$

$$\Rightarrow \quad \frac{\log(1+x)}{x} > \frac{1}{1+x}$$

$$\Rightarrow \quad \log(1+x) > \frac{x}{1+x} \Rightarrow g(x) < 1$$

48. (a) Let us use Lagrange's mean value theorem for $x \in [a, c]$,

$$\frac{f(c) - f(a)}{c - a} = f'(\alpha), \alpha \in (a, c)$$

Also, use Lagrange's mean value theorem for $x \in [c, b]$,

$$\frac{f(b) - f(c)}{b - c} = f'(\beta), \beta \in (c, b)$$

$\because \quad f''(x) < 0$

$\Rightarrow \quad f'(x)$ is decreasing.

$$\therefore \quad f'(\alpha) > f'(\beta)$$

$$\Rightarrow \quad \frac{f(c) - f(a)}{c - a} > \frac{f(b) - f(c)}{b - c}$$

$$\Rightarrow \quad \frac{f(c) - f(a)}{f(b) - f(c)} > \frac{c - a}{b - c}$$

49. (a) Let us use Lagrange's mean value theorem for $x \in [-7, -1]$,

$$\frac{f(-1) - f(-7)}{(-1+7)} \le 2$$

$$\Rightarrow \quad \frac{f(-1) + 3}{6} \le 2$$

$$\Rightarrow \quad f(-1) \le 9$$

Also, use Lagrange's mean value theorem for $x \in [-7, 0]$

$$\frac{f(0) - f(-7)}{0+7} \le 2$$

$$\Rightarrow \quad \frac{f(0) + 3}{7} \le 2$$

$$\Rightarrow \quad f(0) \le 11$$

$$\therefore \quad f(0) + f(-1) \le 20$$

So, $f(-1) + f(0)$ lie in the interval $(-\infty, 20]$.

50. (2) Let r be the radius and V be the volume.

Then, $\quad V = \frac{4}{3}\pi r^3$

$$\Rightarrow \quad \frac{dV}{dt} = 4\pi r^2 \frac{dr}{dt}$$

Now, $\quad \frac{dV}{dt} = 1800$ and $r = 15$ cm

$$\Rightarrow \quad 1800 = 4\pi(225)\frac{dr}{dt}$$

$$\Rightarrow \quad \frac{dr}{dt} = \frac{2}{\pi}$$

Hence, $\quad K = 2$

51. (2) Let x be the length and y be the width of the rectangle and A be the area.

So, $\quad \frac{dx}{dt} = -3$ cm/s

and $\quad \frac{dy}{dt} = 2$ cm/s

Now, $\quad A = xy$

$$\Rightarrow \quad \frac{dA}{dt} = x\frac{dy}{dt} + y\frac{dx}{dt}$$

When $x = 5$ and $y = 4$ cm, then

$$\frac{dA}{dt} = 5(2) + 4(-3) = 10 - 12$$

$$\Rightarrow \quad \frac{dA}{dt} = -2 \text{ cm/s}$$

Hence, rate of decrease in the area is 2 cm/s.

52. (5) We have, $x = 100t - 40t^2$

At maximum height,

$$\frac{dx}{dt} = 0 \text{ i.e. } v = 0$$

$$\frac{dx}{dt} = 100 - 80t = 0$$

$$t = \frac{100}{80} = \frac{5}{4} = 1.25 \text{ s}$$

$$\therefore \quad 4t = 4 \times 1.25 = 5$$

53. (3) We have, $x^2 + y^2 = 1$

On differentiating w.r.t. t, we get

$$2x\frac{dx}{dt} + 2y\frac{dy}{dt} = 0$$

At $\left(\frac{1}{2}, \frac{\sqrt{3}}{2}\right)$, $\quad \frac{dy}{dt} = -3$

So, $\quad 2 \cdot \frac{1}{2}\frac{dx}{dt} + 2 \cdot \frac{\sqrt{3}}{2}(-3) = 0$

$$\Rightarrow \quad \frac{dx}{dt} - 3\sqrt{3} = 0$$

$$\Rightarrow \quad \frac{dx}{dt} = 3\sqrt{3} \text{ units/s}$$

$$\therefore \quad K = 3$$

54. (8) Let the position of the kite at time t be at C.

$\therefore \quad BC = 151.2$ m

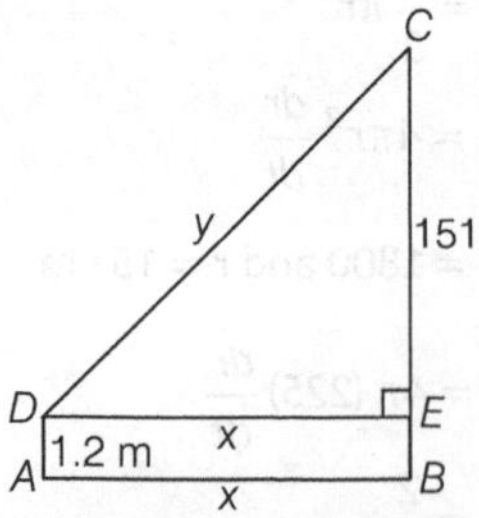

Let AD be the boy who is flying the kite.

Let $AB = x = DE$ and $DC = y$

$\therefore \quad CE = BC - BE = 151.2 - 1.2 = 150$ m

Therefore, from the right angled ΔCDE,

$$y^2 = x^2 + (150)^2$$

$$\therefore \quad 2y\frac{dy}{dt} = 2x\frac{dx}{dt} + 0$$

$$\Rightarrow \quad \frac{dy}{dt} = \frac{x}{y}\frac{dx}{dt}$$

$$\Rightarrow \quad \frac{dy}{dt} = \frac{10x}{y} \qquad \left[\because \frac{dx}{dt} = 10\,\text{m}/\text{s}\right]$$

$$\because \quad \frac{10\sqrt{y^2 - (150)^2}}{y}$$

When $y = 250$, then

$$\frac{dy}{dt} = \frac{10\sqrt{(250)^2 - (150)^2}}{250} = \frac{10 \times 200}{250} = 8\text{ m/s}$$

55. (2) Given, $f(x) = x^3 - 2x^2 - 4x + 7$

$$\Rightarrow \quad f'(x) = 3x^2 - 4x - 4$$

Since, Rolle's theorem is applicable on $f(x)$.

So, $\quad f'(K) = 0$

$$\Rightarrow \quad 3K^2 - 4K - 4 = 0 \Rightarrow (3K + 2)(K - 2) = 0$$

$$\Rightarrow \quad K = 2, \frac{-2}{3}$$

Hence, value of K is 2. $\quad$ [since, $k > 0$]

56. (3) Let $f(x) = \tan^{-1} x$, $0 < \alpha < \beta < \sqrt{3}$

Then, by using Lagrange's mean value theorem,

$$\frac{\tan^{-1}\beta - \tan^{-1}\alpha}{\beta - \alpha} = \frac{1}{1 + c^2},$$

where $0 < \alpha < c < \beta < \sqrt{3}$.

So, $\quad \frac{1}{4} < \frac{1}{1 + c^2} < \frac{1}{1}$

$$\Rightarrow \quad \frac{1}{4} < \frac{\tan^{-1}\beta - \tan^{-1}\alpha}{\beta - \alpha} < 1$$

$$\Rightarrow \quad 1 < \frac{\beta - \alpha}{\tan^{-1}\beta - \tan^{-1}\alpha} < 4$$

$\therefore$ Maximum possible integral value is 3.

57. (12) Let $g(x) = f(x) + Ax^3$ and choose A such that

$$g(a) = g(b)$$

Therefore, $A = \frac{-f(b) + f(a)}{b^3 - a^3} \Rightarrow A = \frac{-4}{b^3 - a^3}$

Since, $g(x)$ satisfies condition of Rolle's theorem.

$$g'(c) = 0 \text{ for some } c \in (a, b)$$

or $\quad f'(c) - 12\frac{c^2}{b^3 - a^3} = 0$

or $\quad (b^3 - a^3) f'(c) = 12c^2$

So, value of $K = 12$

58. (6) Let $F(x) = f^3(x)$

$F(x)$ is continuous and differentiable function on [1, 3].

$$\therefore \quad \frac{F(3) - F(1)}{3 - 1} = F'(c) \text{ for atleast one } c \in (1, 3)$$

$$\therefore \quad \frac{f^3(3) - f^3(1)}{2} = 3f^2(c) f'(c)$$

$$\Rightarrow \quad f^3(3) - f^3(1) = 6f^2(c) \cdot f'(c)$$

$$\Rightarrow \quad \{f(3) - f(1)\}\{f^2(3) + f^2(1) + f(3) \cdot f(1)\} = 6f^2(c) \cdot f'(c)$$

Hence, $K = 6$

59. (7) By Lagrange's mean value theorem in [0, 2],

$$\frac{f(2) - f(0)}{2 - 0} = f'(c), \text{ where } c \in (0, 2)$$

$$\Rightarrow \quad \frac{f(2) - (-3)}{2} = f'(c)$$

$$\Rightarrow \quad \frac{f(2) + 3}{2} \le 5 \qquad [f'(c) \le 5]$$

$$\Rightarrow \quad f(2) + 3 \le 10 \Rightarrow f(2) \le 7$$

So, largest value of $f(2)$ is 7.

60. (1) Let $x \in (0, 3)$. Since, $f(x)$ satisfies the requirements of Lagrange's mean value theorem in [0, 3]. So, it also satisfies in $[0, x]$ consequently, there exist $c \in (0, x)$ such that

$$f'(c) = \frac{f(x) - f(0)}{x - 0}$$

$$\Rightarrow \quad f'(c) = \frac{f(x)}{x}$$

$$\Rightarrow \quad \left|\frac{f(x)}{x}\right| = |f'(c)|$$

$$\Rightarrow \quad \left|\frac{f(x)}{x}\right| \le \frac{1}{3}$$

$$\Rightarrow \quad |f(x)| \le \frac{|x|}{3} \qquad [\because x > 0]$$

$$\Rightarrow \quad |f(x)| \le \frac{x}{3}$$

$$\Rightarrow \quad |f(x)| \le 1 \qquad [\because x \in (0, 3) \therefore |x| < 3]$$

Hence, largest value of $|f(x)| = 1$

61. (2017) Given, f is differentiable in (1, 6).

So, by Lagrange's mean value theorem,

$$\frac{f(6)-f(1)}{6-1}=f'(x)$$

$$\Rightarrow \quad \frac{f(6)-f(1)}{5}\geq 2$$

$$\Rightarrow \quad f(6)-f(1)\geq 10$$

$$\Rightarrow \quad f(6)\geq f(1)+10$$

$$\Rightarrow \quad f(6)\geq -2+10$$

$$\Rightarrow \quad f(6)\geq 8$$

Thus, $m=8$

Now, the value of $2(m+2)^3+17$

$$=2(8+2)^3+17$$

$$=2017$$

Level 3

1. (d) Given, for cone

$$R=2\text{ m}$$

$$H=4\text{ m}$$

and $$\frac{dV}{dt}=2\text{ m}^3/\text{min}$$

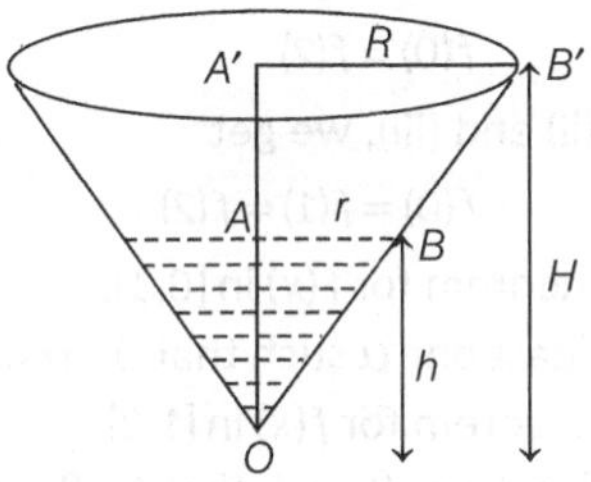

In $\Delta OA'B'$ and OAB,

$$\frac{r}{R}=\frac{h}{H}$$

$$\Rightarrow \quad r=\frac{R}{H}\cdot h$$

$$\Rightarrow \quad r=\frac{2}{4}h \Rightarrow r=\frac{1}{2}h$$

Now, $$V=\frac{1}{3}\pi r^2 h$$

$$\Rightarrow \quad V=\frac{1}{3}\pi\left(\frac{1}{2}h\right)^2 h$$

$$\Rightarrow \quad V=\frac{1}{12}\pi h^3$$

On differentiating w.r.t. t, we get

$$\frac{dV}{dt}=\frac{1}{4}\pi h^2\frac{dh}{dt}$$

$$\Rightarrow \quad \frac{dh}{dt}=\frac{8}{\pi h^2} \qquad \left[\because \frac{dV}{dt}=2\text{ m}^3/\text{min}\right]$$

$$\Rightarrow \quad \left.\frac{dh}{dt}\right|_{h=3}=\frac{8}{\pi(3)^2}$$

$$=\frac{8}{9\pi}\text{ m/min}$$

2. (b) Radius of spherical ball $=10$ cm

$$\frac{dV}{dt}=50\text{ cm}^3/\text{min}$$

Thickness of ice $=5$ cm $=T$ (let)

Total radius $=5+10=15$ cm

Volume $=\frac{4}{3}\pi(10+T)^3$

$$\frac{dV}{dt}=\frac{4}{3}\pi\left(3(10+T)^2\cdot\frac{dT}{dt}\right)$$

$$\Rightarrow \quad 50=4\pi(15)^2\frac{dT}{dt}$$

$$\Rightarrow \quad \frac{dT}{dt}=\frac{50}{4\pi\times 15\times 15} \Rightarrow \frac{dT}{dt}=\frac{1}{18\pi}\text{ cm/min}$$

3. (a) Let ABC is an isosceles triangle, where BC is the base of fixed length b.

Let the length of two equal sides of ΔABC be a.

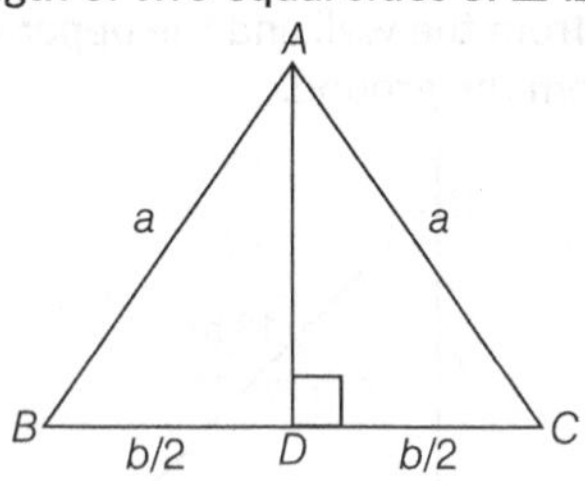

Draw $AD \perp BC$,

$$AD=\sqrt{a^2-\frac{b^2}{4}}$$

$\therefore$ Area of ΔABC, $A=\frac{1}{2}b\cdot\sqrt{a^2-\frac{b^2}{4}}$

On differentiating w.r.t. t, we get

$$\frac{dA}{dt}=\frac{1}{2}b\frac{2a}{2\sqrt{a^2-\frac{b^2}{4}}}\frac{da}{dt}$$

$$\Rightarrow \quad \frac{dA}{dt}=\frac{ab}{\sqrt{4a^2-b^2}}\frac{da}{dt}$$

Given, $\frac{da}{dt}=-3\text{ cm/s}$

$$\therefore \quad \frac{dA}{dt}=\frac{-3ab}{\sqrt{4a^2-b^2}}$$

When $a=b$

$$\frac{dA}{dt}=\frac{-3b^2}{\sqrt{3b^2}}=-\sqrt{3}b$$

Hence, the area of triangle is decreasing at the rate of $\sqrt{3}b\text{ cm}^2/\text{s}$.

4. (b) Let h be the height and r be the radius of the cone, then

$$h = \frac{r}{3} \quad \text{[given]}$$

$$\Rightarrow \quad r = 3h$$

$$\therefore \quad \text{Volume, } V = \frac{1}{3}\pi r^2 h$$

$$= \frac{1}{3}\pi(3h)^2 h$$

$$\Rightarrow \quad V = 3\pi h^3$$

$$\therefore \quad \frac{dV}{dt} = 9\pi h^2 \frac{dh}{dt}$$

$$\Rightarrow \quad 36 = 9\pi h^2 \frac{dh}{dt}$$

$$\Rightarrow \quad \frac{dh}{dt} = \frac{36}{9\pi h^2} = \frac{4}{\pi h^2}$$

When $h = 4$ cm, then

$$\frac{dh}{dt} = \frac{4}{\pi(4)^2} = \frac{1}{4\pi} \text{ cm/s}$$

5. (c) Let at time t the lower end P of ladder PQ be at a distance x from the wall and the upper end Q be at height y from the ground.

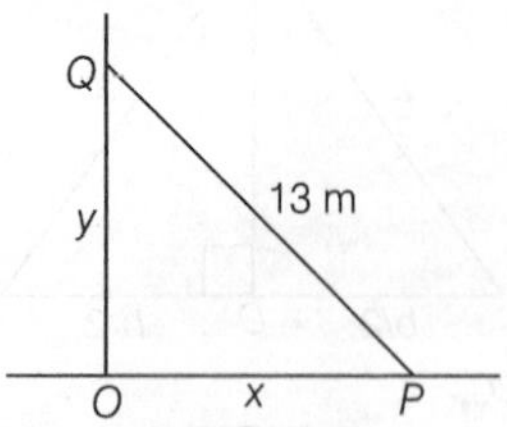

Then, $x^2 + y^2 = 13^2$

Putting $x = 12$, we get

$$y = 5$$

Now, $x^2 + y^2 = 13^2$

On differentiating w.r.t. t, we get

$$2x\frac{dx}{dt} + 2y\frac{dy}{dt} = 0$$

$$\Rightarrow \quad x\frac{dx}{dt} + y\frac{dy}{dt} = 0$$

When $x = 12, y = 5$ and $\frac{dx}{dt} = 2$, then

$$12 \times 2 + 5\frac{dy}{dt} = 0$$

$$\Rightarrow \quad \frac{dy}{dt} = \frac{-24}{5} \text{ m/min}$$

6. (c) Let $x = 4\cos\theta$ and $y = 3\sin\theta$

$$\frac{dx}{dt} = -4\sin\theta \cdot \frac{d\theta}{dt} \quad \text{...(i)}$$

$$\text{and} \quad \frac{dy}{dt} = 3\cos\theta\frac{d\theta}{dt} \quad \text{...(ii)}$$

Also, as $|v| = 1$

$$\Rightarrow \quad \left(\frac{dx}{dt}\right)^2 + \left(\frac{dy}{dt}\right)^2 = 1$$

$$\Rightarrow \quad \left(\frac{d\theta}{dt}\right)^2 = \frac{1}{16\sin^2\theta + 9\cos^2\theta} = \frac{1}{7\sin^2\theta + 9}$$

At $y = 1$, $\quad 1 = 3\sin\theta$

$$\Rightarrow \quad \sin\theta = \frac{1}{3}$$

$$\therefore \quad \frac{d\theta}{dt} = \sqrt{\frac{1}{\frac{7}{9} + 9}} = \frac{3}{2\sqrt{22}}$$

Putting in Eq. (i), we get

$$\frac{dx}{dt} = \frac{-4}{3}\left(\frac{3}{2\sqrt{22}}\right) = -\sqrt{\frac{2}{11}} \text{ m/s}$$

7. (c) Let $f(x) = \int (1 + \cos^8 x)(ax^2 + bx + c)\,dx$

$$\therefore \quad f'(x) = (1 + \cos^8 x)(ax^2 + bx + c) \quad \text{...(i)}$$

For given conditions,

$$f(1) - f(0) = 0$$

$$\Rightarrow \quad f(0) = f(1) \quad \text{...(ii)}$$

$$\text{and} \quad f(2) - f(0) = 0$$

$$\Rightarrow \quad f(0) = f(2) \quad \text{...(iii)}$$

From Eqs. (ii) and (iii), we get

$$f(0) = f(1) = f(2)$$

By Rolle's theorem for $f(x)$ in $[0, 1]$.

$f'(\alpha) = 0$ atleast one α such that $0 < \alpha < 1$

By Rolle's theorem for $f(x)$ in $[1, 2]$.

$f'(\beta) = 0$ atleast one β such that $1 < \beta < 2$

Now, from Eq. (i), $f'(\alpha) = 0$

$\Rightarrow (1 + \cos^8 \alpha)(a\alpha^2 + b\alpha + c) = 0$

$$\therefore \quad a\alpha^2 + b\alpha + c = 0 \quad [\because 1 + \cos^8 \alpha \neq 0]$$

i.e. α is the root of quadratic equation $ax^2 + bx + c = 0$

Similarly, β is a root of equation $ax^2 + bx + c = 0$.

But equation $ax^2 + bx + c = 0$ being a quadratic equation cannot have more than two roots.

Hence, equation $ax^2 + bx + c = 0$ has one root between 0 and 1 and other root between 1 and 2.

8. (d) For Lagrange's mean value theorem to be applicable, $f(x)$ is continuous in $[-1, 1]$ and differentiable in $(-1, 1)$.

Now, $f(0) = \cos^{-1}(0) = \frac{\pi}{2}$

$$\Rightarrow \quad f(0^-) = \frac{\pi}{2}$$

$$\Rightarrow \quad f(0^+) = m(0) + c = c$$

$$\text{So,} \quad c = \frac{\pi}{2}$$

Now, $\quad \text{LHD} = f'(x) = \frac{d}{dx}(\cos^{-1} x) = \frac{-1}{\sqrt{1-x^2}}$

$\Rightarrow \quad f'(0^-) = \frac{-1}{\sqrt{1-0}} = -1$

$\Rightarrow \quad f'(0^+) = m$

So, $\quad m = -1$

Hence, (m, c) will be equal to $\left(-1, \frac{\pi}{2}\right)$.

9. (d) According to Lagrange's mean value theorem, $f(x)$ is continuous.

$$f(0) = 3 = a$$

$\Rightarrow \quad f(1) = -1 + 3 + 3 = m + b$

So, $\quad m + b = 5 \quad$ ...(i)

and its derivative function is also continuous, then

$$f'(x) = \begin{cases} -2x + 3x, & 0 < x < 1 \\ m, & 1 \le x \le 2 \end{cases}$$

$\Rightarrow \quad f'(1) = -2 + 3 = m$

$\Rightarrow \quad m = 1$

From Eq. (i),

$$b = 4$$

Hence, $(a + b)\, m = (3 + 4)\, 1 = 7$

10. (d) We have, $0 \le x \le 5$

$\Rightarrow x + 1 \ne 0$. Obviously, g is continuous on $[0, 5]$ and differentiable in $(0, 5)$.

Hence, by Lagrange's mean value theorem, there exists $c \in (0, 5)$ such that

$$g'(c) = \frac{g(5) - g(0)}{5 - 0}$$

Now, $\quad g(5) = \frac{f(5)}{6} = \frac{-1}{6}$

$\Rightarrow \quad g(0) = \frac{f(0)}{1} = \frac{4}{1}$

Therefore, $g'(c) = \frac{-\frac{1}{6} - 4}{5} = \frac{-5}{6}$

11. (a) Consider, a function $\phi(x)$ is given by

$$\phi(x) = \begin{vmatrix} f(a) & f(x) \\ g(a) & g(x) \end{vmatrix}$$

Since, $f(x)$ and $g(x)$ are continuous and differentiable function.

Therefore, $\phi(x)$ is continuous on $[a, b]$ and differentiable in (a, b). Consequently by Lagrange's mean value theorem, there exist atleast one $c \in (a, b)$ such that

$$\phi'(c) = \frac{\phi(b) - \phi(a)}{b - a}$$

$$\Rightarrow \begin{vmatrix} f(a) & f'(c) \\ g(a) & g'(c) \end{vmatrix} = \frac{1}{b-a} \begin{vmatrix} f(a) & f(b) \\ g(a) & g(b) \end{vmatrix}$$

$$\Rightarrow \begin{vmatrix} f'(c) & f(a) \\ g'(c) & g(a) \end{vmatrix} = \frac{1}{a-b} \begin{vmatrix} f(a) & f(b) \\ g(a) & g(b) \end{vmatrix}$$

Hence, the equation $\begin{vmatrix} f'(x) & f(a) \\ g'(x) & g(a) \end{vmatrix} = \frac{1}{a-b} \begin{vmatrix} f(a) & f(b) \\ g(a) & g(b) \end{vmatrix}$ has atleast one root in (a, b).

12. (d) We have, $f(x) = (1 - x)^{5/2}$

$\Rightarrow \quad f'(x) = \frac{-5}{2}(1 - x)^{3/2}$

$\Rightarrow \quad f''(x) = \frac{15}{4}(1 - x)^{1/2}$

Also, $f(0) = 1, f'(0) = -\frac{5}{2}$

and $f''(\theta x) = \frac{15}{4}(1 - \theta x)^{1/2}$

Therefore, $(1 - x)^{5/2} = 1 - \frac{5}{2}x + \frac{x^2}{2}(1 - \theta x)^{1/2} \times \frac{15}{4}$

Taking limit $x \to 1$ on both sides, we get

$$0 = 1 - \frac{5}{2} + \frac{15}{8}(1 - \theta)^{1/2}$$

$\Rightarrow \quad (1 - \theta)^{1/2} = \frac{4}{5} \Rightarrow \theta = 1 - \frac{16}{25} = \frac{9}{25}$

13. (d) We have, $ab > 0 \Rightarrow 0 \notin (a, b)$.

Define, $F(x) = f\left(\frac{1}{x}\right), \forall\, x \in \left[\frac{1}{b}, \frac{1}{a}\right]$.

Obviously, F is continuous on $\left[\frac{1}{b}, \frac{1}{a}\right]$ and differentiable in $\left(\frac{1}{b}, \frac{1}{a}\right)$. Hence, by Lagrange's mean value theorem, there exists $d \in \left(\frac{1}{b}, \frac{1}{a}\right)$ such that

$$F'(d) = \frac{F\left(\frac{1}{a}\right) - F\left(\frac{1}{b}\right)}{\left(\frac{1}{a}\right) - \left(\frac{1}{b}\right)}$$

But $\quad F'(d) = f'\left(\frac{1}{d}\right)\left(\frac{-1}{d^2}\right) = -f'\left(\frac{1}{d}\right)\left(\frac{1}{d^2}\right)$

Therefore, $\frac{F\left(\frac{1}{a}\right) - F\left(\frac{1}{b}\right)}{\left(\frac{1}{a}\right) - \left(\frac{1}{b}\right)} = F'(d) = -f'\left(\frac{1}{d}\right)\left(\frac{1}{d^2}\right)$

$$\Rightarrow \frac{f(a) - f(b)}{\left(\frac{1}{a}\right) - \left(\frac{1}{b}\right)} = -f'\left(\frac{1}{d}\right)\left(\frac{1}{d^2}\right)$$

Now, put $\frac{1}{d} = c$

Therefore, $\frac{f(b) - f(a)}{\left(\frac{1}{b}\right) - \left(\frac{1}{a}\right)} = -c^2 f'(c)$

14. (c) We have, $f(x)=\frac{1}{x^2}$ and $g(x)=\frac{1}{x}$; $x\in[a,b]$; $0<a<b$.

By given condition, $\frac{\frac{1}{b^2}-\frac{1}{a^2}}{\frac{1}{b}-\frac{1}{a}}=\frac{\frac{-2}{c^3}}{\frac{-1}{c^2}}$; $a<c<b$

$$\Rightarrow \frac{a^2-b^2}{a^2b^2}\times\frac{ab}{a-b}=\frac{2}{c^3}\times\frac{c^2}{1}$$

$$\Rightarrow \frac{a+b}{ab}=\frac{2}{c}$$

$$\Rightarrow c=\frac{2ab}{a+b}$$

$\Rightarrow$ c is HM of a and b.

15. (a,b, c) We have, $s=\sqrt{t}$

$$\Rightarrow \frac{ds}{dt}=\frac{1}{2\sqrt{t}}=v$$

$$\Rightarrow \sqrt{t}=\frac{1}{2v} \quad \text{...(i)}$$

and $$a=\frac{d^2s}{dt^2}=-\frac{1}{4}t^{-3/2}$$

$$=-\frac{1}{4t\sqrt{t}}<0 \quad \text{...(ii)}$$

$$\therefore \quad a=-\frac{1}{4t\sqrt{t}}=\frac{-1}{4\left(\frac{1}{2v}\right)\left(\frac{1}{2v}\right)^2}=\frac{-8v^3}{4}=-2v^3$$

$$\Rightarrow a\propto v^3$$

$$\Rightarrow v=\frac{1}{2\sqrt{t}}$$

$$\Rightarrow v=\frac{1}{2s}$$

$$\Rightarrow v\propto\frac{1}{s}$$

16. (b, c) We have,

$$f(x)=\begin{cases}x^{2\alpha+1}\ln x, & x>0\\ 0, & x=0\end{cases}$$

If $f(x)$ satisfies Rolle's theorem, then $f(x)$ is continuous at $x=0$.

$$\lim_{x\to 0} f(x)=f(0)=0$$

$$\lim_{x\to 0} x^{2\alpha+1}\ln x=0$$

$$\lim_{x\to 0}\frac{\ln x}{x^{-(2\alpha+1)}}=0$$

Applying L-Hospital rule,

$$\lim_{x\to 0}\frac{\frac{1}{x}}{-(2\alpha+1)x^{-(2\alpha+1)-1}}=0$$

$$\Rightarrow -\frac{1}{2\alpha+1}\lim_{x\to 0}\frac{1}{x^{-(2\alpha+1)}}=0$$

Now, $2\alpha+1\neq 0$

$$\alpha\neq\frac{-1}{2}$$

Now, at $\alpha=-1$,

$$\lim_{x\to 0}\frac{1}{x}\neq 0$$

So, $\alpha\neq -1$

Hence, α can be $-\frac{1}{3}$ and $-\frac{1}{4}$.

17. (a, b, c) Consider, $g(x)=e^x f(x)$

Then, $g(a)=g(b)$

Using Rolle's theorem, there exists atleast one $c\in(a,b)$, for which $g'(c)=0$

or $e^c f(c)+e^c f'(c)=0$

or $f(c)+f'(c)=0$

Thus, $f(x)+f'(x)=0$ has atleast one root in (a,b).

Consider, $g(x)=e^{-x}f(x)$

Then, $g(a)=g(b)$

Using Rolle's theorem, there exists atleast one $c\in(a,b)$, for which

$$g'(c)=0$$

or $-e^{-c}f(c)+e^{-c}[f'(c)]=0$

or $-f(c)+f'(c)=0$

Thus, $f'(x)-f(x)=0$ has atleast one root in (a,b).

Now, consider $g(x)=[f(x)]^2$

Then, $g(a)=g(b)$

Using Rolle's theorem,

$$g'(c)=0$$

or $2f(c)\cdot f'(c)=0$

or $f(c)\cdot f'(c)=0$

Thus, $f'(x)\cdot f(x)=0$ has atleast one root in (a,b).

18. (a, b) We have,

$$f(x)=x(x-2)(x-4)$$

$$\Rightarrow f'(x)=(x-2)(x-4)+x(x-4)+x(x-2)$$

$$=3x^2-12x+8$$

Also, $f(1)=3$ and $f(4)=0$

Thus, $f'(x)=\frac{f(4)-f(1)}{4-1}=\frac{0-3}{3}$

$$\Rightarrow f'(x)=-1$$

$$\Rightarrow 3x^2-12x+8=-1$$

$$\Rightarrow 3x^2-12x+9=0$$

$$\Rightarrow x^2-4x+3=0\Rightarrow x^2-3x-x+3=0$$

$$\Rightarrow x(x-3)-1(x-3)=0$$

$$\Rightarrow (x-1)(x-3)=0\Rightarrow x=1,3$$

19. (b, c, d)

(a) Clearly, f is continuous on $[0, 1]$.

Now, $f'(x) = \begin{cases} -1 & \text{for } x < \frac{1}{2} \\ -2\left(\frac{1}{2} - x\right) & \text{for } x \geq \frac{1}{2} \end{cases}$

Clearly, $f'\left(\frac{1}{2} - 0\right) = -1$ and $f'\left(\frac{1}{2} + 0\right) = 0$

i.e. f is non-differentiable at $\frac{1}{2} \in (0, 1)$.

Therefore, Lagrange's mean value theorem is not applicable to $f(x)$ on $[0, 1]$.

(b) We have, $\lim_{x \to 0} \frac{\sin x}{x} = 1 = f(0)$

This implies, f is continuous on $[0, 1]$.

Also, $\frac{(\sin x)}{x}$ is differentiable for all $x \neq 0$ and hence Lagrange's mean value theorem is applicable.

(c) We have, $f(x) = x^2$ ($\because x \geq 0$) is continuous on $[0, 1]$ and differentiable on $(0, 1)$.

Hence, Lagrange's mean value theorem is applicable.

(d) We have, $f(x) = |x| = x$ ($\because x \geq 0$) and hence Lagrange's mean value theorem is applicable.

20. (c, d) We have, $f''(x) > 0, \forall x \in [0, 2]$

$\Rightarrow f'(x)$ is an increasing function.

Now, by Lagrange's mean value theorem.

$$f'(c_1) = \frac{f(1) - f(0)}{1 - 0}, c_1 \in (0, 1)$$

and $$f'(c_2) = \frac{f(2) - f(1)}{2 - 1}, c_2 \in (1, 2)$$

$$\Rightarrow \quad f'(c_1) < f'(c_2) \Rightarrow f(0) + f(2) > 2f(1)$$

Similarly, applying Lagrange's mean value theorem between $\left[0, \frac{2}{3}\right]$ and $\left[\frac{2}{3}, 2\right]$.

$$\frac{f(2) - f\left(\frac{2}{3}\right)}{\left(\frac{4}{3}\right)} > \frac{f\left(\frac{2}{3}\right) - f(0)}{\left(\frac{2}{3}\right)} \Rightarrow 2f(0) + f(2) > 3f\left(\frac{2}{3}\right)$$

21. (a, c, d) We have,

$f : [0, 8] \to R$ be differentiable function such that

$$f(0) = 0, f(4) = 1 \text{ and } f(8) = 1$$

By Lagrange's mean value theorem,

$$\frac{f(4) - f(0)}{4 - 0} = f'(c_1), c \in (0, 4)$$

$$f'(c_1) = \frac{1}{4}$$

Now, $\quad 8f'(c_1) \cdot f(c_2) = 1$

f is differentiable in $(0, 8)$ and continuous on $[0, 8]$

$$f'(c_1) = \frac{f(8) - f(0)}{8 - 0} = \frac{1}{8}, c_1 \in (0, 8)$$

So, $\quad 8 \cdot \frac{1}{8} \cdot f(c_2) = 1 \Rightarrow f(c_2) = 1$

So, $c_2 = 4$ or $c_2 = 8$

Hence, $c_2 \in (0, 8)$

(d) Let $t = x^3$

Let $g(x) = \int f(x^3)\, dx$ ...(i)

$$g(x) = \int_0^{x^3} f(t)\, dt \left[\begin{array}{l} t \in [0, 8], \\ \text{then } x \in [0, 2] \end{array}\right]$$

$g(x)$ is continuous on $[0, 2]$ and it is differentiable on $(0, 2)$.

Now, $$g'(\alpha) = \frac{g(2) - g(0)}{2 - 0} = \frac{\int_0^{2^3} f(t)\, dt - \int_0^{0^3} f(t)\, dt}{2}$$

$$g'(\alpha) = \frac{\int_0^8 f(t)\, dt}{2}$$

On differentiating Eq. (i), we get

$$g'(x) = 3x^2 f(x^3)$$

$$g'(\alpha) = 3\alpha^2 f(\alpha^3)$$

Similarly, $$g'(\beta) = \frac{\int_0^8 f(t)\, dt}{2}$$

$$g'(\beta) = 3\beta^2 f(\beta^3)$$

Now, $$g'(\alpha) + g'(\beta) = \int_0^8 f(t)\, dt$$

$$\Rightarrow 3\alpha^2 f(\alpha^3) + 3\beta^2 f(\beta^3) = \int_0^8 f(t)\, dt$$

$$= 3[\alpha^2 f(\alpha^3) + \beta^2 f(\beta^3)] = \int_0^8 f(t)\, dt$$

22. (a, b, d) We have,

$$F(x) = [f(x)]^2 + [f'(x)]^2, \; |f(x)| \leq 1 \Rightarrow f(x) \in [-1, 1]$$

By Lagrange's mean value theorem in $x \in (0, 1)$.

$$\Rightarrow \quad [f'(x)] = \frac{[f(1)] - [f(0)]}{1 - 0}, |f'(x)| \leq |f(1)| - |f(0)|$$

$|f'(x)| \leq 2$ for atleast one x in $(0, 1)$.

Similarly, $|f'(x)| \leq 2$ for atleast one x is $(-1, 0)$.

$$F(x) = [f(x)]^2 + [f'(x)]^2$$

For atleast one x in $(0, 1)$ and $(-1, 0)$.

$$|f'(x)| \leq 2 \text{ and } |f(x)| \leq 1$$

$$[f'(x)]^2 \leq 4 \text{ and } [f(x)]^2 \leq 1$$

$$\Rightarrow \quad [f'(x)]^2 + [f(x)]^2 \leq 5$$

$\Rightarrow F(x) \leq 5$ for atleast one x in $(-1, 0)$ and $(0, 1)$.

Now, $F(0) = 6$

$\Rightarrow$ There must exist a point of maxima in $(-1, 1)$.

[let c be the point of maxima]

23. (b, c) Let $h(x)=f(x)-3g(x)$

$$h(-1)=f(-1)-3g(-1)=3-0=3$$
$$h(0)=f(0)-3g(0)=6-3=3$$
$$h(2)=f(2)-3g(2)=0-(-3)=3$$

Thus, $h'(x)=0$ has atleast one root in $(-1,0)$ and atleast one root in $(0,2)$.

24. (a, b, c) Clearly, $f(0)=0$, so $f(x)=0$ has two real roots $0, \alpha_0 (>0)$.

Therefore, $f'(x)=0$ has a real root α, lying between 0 and α_0, so $0<\alpha_1<\alpha_0$.

Again, $f'(x)=0$ is fourth degree equation.

As imaginary roots occur in conjugate pairs $f'(x)$ will have another real root α_2.

Therefore, $f''(x)=0$ will have a real root lying between α_1 and α_2.

As, $f(x)=0$ is an equation of fifth degree, it will have atleast three real roots and so $f'(x)$ will have atleast two real roots.

Sol. (Q.Nos. 25-27)

25. (d) We have, $f(x)=(1+\cos^8 x)(ax^2+bx+c)$

Here, $f(x)$ is continuous and differentiable function in for $x \in R$.

$\therefore$ LMVT and Rolle's theorem is applied for $f(x)$.

Now, $f(0)=2c$ and $f(1)=(1+\cos^8 1)(a+b+c)$

$f(0)\cdot f(1)=2(1+\cos^8 1)c(a+b+c)<0 \quad [\because c(a+b+c)<0]$

$\therefore f(x)=0$ has a solution between 0 and 1

$f(2)=(1+\cos^8 2)(4a+2b+c)$

$f(0)\cdot f(2)=2(1+\cos^8 2)c(4a+2b+c)>0$

No solution between 0 and 2. $\quad [\because c(4a+2b+c)>0]$

26. (a) We have, $\int_0^1 (1+\cos^8 x)(ax^2+bx+c)dx=0$

Let $f(x)=\int_0^1 (1+\cos^8 x)(ax^2+bx+c)dx$

$f(1)-f(0)=0 \Rightarrow f(1)=f(0)$

$$f'(x)=f(x)$$

By Rolle's theorem for $f(x)$ in $[0,1]$.

$f'(c)=0=f(c)$ atleast one c such that $0<c<1$.

27. (d) We have, $f(x)=\int_0^1 f(x)dx=\int_0^2 f(x)dx=0$

From given conditions,

$$f(1)-f(0)=0 \Rightarrow f(1)=f(0)$$

and $\quad f(2)-f(0)=0 \Rightarrow f(0)=f(2)$

$\therefore f(0)=f(1)=f(2)$

Here, $f(x)$ is continuous and differentiable for all values of $x \in R$.

and $\quad f(0)=f(1)$

$$f(1)=f(2)$$
$$f(0)=f(2)$$

$\therefore$ Rolle's theorem is applied for the $[0,1]$, $[0,2]$ and $[1,2]$.

Sol. (Q.Nos. 28-30)

28. (a) Let $\phi(x)=e^{\lambda x}f(x)$, so that

ϕ is continuous on $[a,b]$, ϕ is differentiable in (a,b) and $\phi(a)=\phi(b)$ $\quad [\because f(a)=f(b)=0]$

Therefore, by Rolle's theorem,

$$\phi'(c)=0 \text{ for some } c\in(a,b)$$
$$e^{\lambda c}[\lambda f(c)+f'(c)]=0$$
$$\Rightarrow \quad f'(c)+\lambda f(c)=0$$

i.e. for each λ, there exist $c\in(a,b)$ such that

$$f'(c)+\lambda f(c)=0$$

i.e. $f'(x)+\lambda f(x)=0$ is solvable for all real values of λ.

29. (c) Let $f(x)=\dfrac{a_0x^{n+1}}{n+1}+\dfrac{a_1x^n}{n}+\ldots+\dfrac{a_{n-1}x^2}{2}+a_n x$

$f(x)$ is continuous and differentiable for all real x.

Also, $f(0)=0$ and $f(1)=0$

Therefore, by Rolle's theorem on $[0,1]$,

$f'(c)=0$ for some $c\in(0,1)$

i.e. $a_0c^n+a_1c^{n-1}+a_2c^{n-2}+\ldots+a_n=0,\ 0<c<1$,

i.e. $a_0x^n+a_1x^{n-1}+a_2x^{n-2}+\ldots+a_n=0$ has a root in $(0,1)$.

30. (c) We have, $f(x)=2x^3+x^2-4x-2$

Clearly, $f(-\sqrt{2})=f(\sqrt{2})=0$

Using Rolle's theorem,

$$6c^2+2c-4=0$$
$$\Rightarrow \quad (3c-2)(c+1)=0$$

So, $c=-1,\dfrac{2}{3}$. Both values belongs to $(-\sqrt{2},2)$.

31. (i) $\rightarrow$ (D), (ii) $\rightarrow$ (A), (iii) $\rightarrow$ (B), (iv) $\rightarrow$ (C),

(i) $\quad A=\dfrac{\sqrt{3}}{4}x^2$

$$\Rightarrow \frac{dA}{dt}=\frac{\sqrt{3}}{2}\left(x\frac{dx}{dt}\right)$$
$$=\frac{\sqrt{3}}{2}\times 30\times 0.1=\frac{3\sqrt{3}}{2}$$

(ii) We have, $y=\dfrac{x^2}{2}-2x+5$

$$\Rightarrow \quad \frac{dy}{dt}=(x-2)\frac{dx}{dt}$$
$$\Rightarrow \quad \frac{3\,dx}{dt}=(x-2)\frac{dx}{dt} \quad \left[\text{given that } \frac{dy}{dt}=3\frac{dx}{dt}\right]$$
$$\Rightarrow \quad x-2=3$$
$$\Rightarrow \quad x=5$$

(iii) Volume of cube, $V = x^3$

$$\Rightarrow \quad \frac{dV}{dx} = 3x^2$$

and $\quad S = 6x^2$

$$\Rightarrow \quad \frac{dS}{dx} = 12x$$

Now, $\dfrac{dV}{dS} = \dfrac{\frac{dV}{dx}}{\frac{dS}{dx}} = \dfrac{3x^2}{12x} = \dfrac{1}{4}x$

When $x = 8$ cm, then

$$\frac{dV}{dS} = \frac{8}{4} = 2$$

(iv) We have, $3y = 2x^3 + 3$

$$\Rightarrow \quad 3\frac{dy}{dt} = 6x^2\frac{dx}{dt}$$

$$\Rightarrow \quad 6\frac{dx}{dt} = 6x^2\frac{dx}{dt} \qquad \left[\frac{dy}{dt} = 2\frac{dx}{dt}\right]$$

$$\Rightarrow \quad x^2 = 1$$

$$\Rightarrow \quad x = \pm 1$$

If $x = 1$, then $y = \dfrac{5}{3}$

and if $x = -1$, then $y = \dfrac{1}{3}$

Hence, difference of the ordinates

$$= \frac{5}{3} - \frac{1}{3} = \frac{4}{3}$$

32. (i) → (D), (ii) → (C), (iii) → (B), (iv) → (A, C)

(i) We have, $f(x) = 2x(x-3)^n$

$$\Rightarrow \quad f'(x) = 2[nx(x-3)^{n-1} + (x-3)^n]$$

$$= 2(x-3)^n\left[\frac{nx}{x-3} + 1\right]$$

Since, Rolle's theorem holds for $c = \dfrac{3}{4}$

$$\Rightarrow \quad f'\left(\frac{3}{4}\right) = 0$$

$$\Rightarrow \quad \left(-\frac{9}{4}\right)^n\left[\frac{-n}{3} + 1\right] = 0$$

$$\Rightarrow \quad n = 3$$

(ii) We have, $f(x) = 4^{\sin x}$

∵ $f(x)$ satisfies the conditions of Rolle's theorem.

So, $\quad f'(c) = 0$

$$\Rightarrow \quad 4^{\sin c} \cdot \cos c \cdot \log 4 = 0$$

$$\Rightarrow \quad \cos c = 0$$

$$\Rightarrow \quad c = \frac{\pi}{2} \in \left(0, \frac{\pi}{2}\right)$$

Then, $\quad \sin c = \sin\dfrac{\pi}{2} = 1$

(iii) Let $f(x) = x^4 - 4x - 2$

Then, $f'(x) = 4x^3 - 4$

$$= 4(x^3 - 1)$$

$$= 4(x-1)(x-\omega)(x-\omega^2)$$

Thus, $f'(x) = 0$ has one real root and 2 imaginary roots.

So, $f(x) = 0$ has 2 real and 2 imaginary roots.

(iv) Given, $f(x) = x^\alpha \log x$

Since, Rolle's theorem is applicable on $[0, 1]$.

So, it must be continuous on [0, 1].

i.e. $\quad \lim_{x \to 0^+} f(x) = f(0) = 0$

Now, $\lim_{x \to 0^+} f(x) = \lim_{x \to 0^+} x^\alpha \cdot \log x$

$$= \lim_{x \to 0^+} \frac{\log x}{x^{-\alpha}}$$

$$= \lim_{x \to 0^+} \left(\frac{\frac{1}{x}}{-\alpha x^{-\alpha-1}}\right)$$

$$= \lim_{x \to 0^+} \left(\frac{1}{-\alpha x^{-\alpha}}\right) = \lim_{x \to 0^+} \left(\frac{x^\alpha}{-\alpha}\right)$$

It will be zero only, when $\alpha > 0$.

33. (i) → (D), (ii) → (B), (iii) → (A), (iv) → (C)

(i) Clearly, f is continuous on $[2, 4]$ and differentiable in $(2, 4)$.

Hence, by Lagrange's mean value theorem, there exists $c \in (2, 4)$ such that

$$f'(c) = \frac{f(4) - f(2)}{4 - 2}$$

$$\Rightarrow \quad \frac{2c}{2\sqrt{c^2 - 4}} = \frac{2\sqrt{3} - 0}{2}$$

$$\Rightarrow \quad c^2 = 3(c^2 - 4)$$

$$\Rightarrow \quad 2c^2 = 12 \Rightarrow c = \sqrt{6} \in (2, 4)$$

(ii) We have, $f(x) = x^2 - 2x + 4, x \in [1, 5]$, therefore

$$f(1) = 3 \text{ and } f(5) = 19$$

By Lagrange's mean value theorem, for some $c \in (1, 5)$

$$\frac{f(5) - f(1)}{5 - 1} = f'(c) = 2c - 2$$

$$\Rightarrow \quad 2c - 2 = 4$$

$$\Rightarrow \quad c = 3$$

(iii) We have,

$$f'(c) = \frac{f(4) - f(2)}{4 - 2} = \frac{13 - 5}{2}$$

$$\Rightarrow \quad f'(c) = 4$$

(iv) We have, $f(x) = e^{-x} \sin x$

f is continuous on $[0, \pi]$ and differentiable in $(0, \pi)$ and $f(0) = f(\pi) = 0$.

By Rolle's theorem, there exists $c \in (0, \pi)$ such that $f'(c) = 0$

$$\Rightarrow \quad e^{-c} (\cos c - \sin c) = 0$$

$$\Rightarrow \quad \cos c - \sin c = 0 \quad [\because e^{-C} \neq 0]$$

$$\Rightarrow \quad \tan c = 1$$

34. (i) → (B), (ii) → (B), (iii) → (A), (iv) → (C)

(i) Let $f'(x) = \log x - \dfrac{3}{x} + 1$

$$\Rightarrow \quad f(x) = (x-3)\log x + C$$

$$\Rightarrow \quad f(1) = f(3)$$

Hence, atleast one root lies in (1, 3).

(ii) Given, $\quad f(x) = x + \dfrac{1}{x}$

$$\Rightarrow \quad f'(x) = 1 - \frac{1}{x^2}$$

$$\Rightarrow \quad f'(\sqrt{3}) = 1 - \frac{1}{3} = \frac{2}{3}$$

$$\Rightarrow \quad \frac{f(b) - f(a)}{b - a} = f'(\sqrt{3}) = \frac{2}{3} \Rightarrow \frac{ab - 1}{ab} = \frac{2}{3}$$

Hence, required interval is (1, 3).

(iii) Given, $f(x) = 2x - x^2$

$$\Rightarrow \quad f'(x) = 2 - 2x$$

$$\Rightarrow \quad f'\left(\frac{1}{2}\right) = 2 - 2\left(\frac{1}{2}\right) = 1$$

Now, $\dfrac{f(b) - f(a)}{b - a} = f'\left(\dfrac{1}{2}\right)$

$$\Rightarrow \quad a + b = 1$$

So, required interval is (0, 1).

(iv) Given, $\quad f'(x) = 4ax^3 + 3bx^2 + 2cx + d$

$$\Rightarrow \quad f(x) = ax^4 + bx^3 + cx^2 + dx + e$$

$$\Rightarrow \quad f(0) = f(3) \Rightarrow 27a + 9b + 3c + d = 0$$

Hence, required interval is (0, 3).

Sol. (Q.Nos. 35-37)

35. (1) Volume of cone $= \dfrac{1}{3}\pi r^2 h = \dfrac{1}{3}\pi\left(\dfrac{2}{3}h\right)^2 \cdot h = \dfrac{4}{27}\pi h^3$

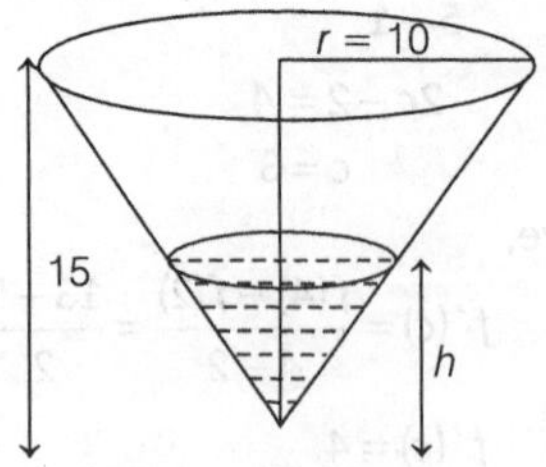

$$\Rightarrow \quad \frac{dV}{dt} = \frac{4}{9}\pi h^2 \frac{dh}{dt}$$

$$\Rightarrow \quad 5 = \frac{4}{9}\pi (7.5)^2 \frac{dh}{dt}$$

$$\Rightarrow \quad \frac{dh}{dt} = \frac{45}{4\pi(7.5)^2} = \frac{1}{5\pi} \text{ cm}^3/\text{s}$$

$$\Rightarrow \quad k = 1$$

36. (12.56) V (at $h = 3$) $= 4\pi$ cm^3 $= 12.56$ cm^3

37. (7.5) $\dfrac{dV}{dt} = \dfrac{4}{9}\pi h^2 \dfrac{dh}{dt}$

$$\Rightarrow \quad \frac{d^2V}{dt^2} = \frac{4}{9}\pi\left[2h\left(\frac{dh}{dt}\right)^2 + h^2 \frac{d^2h}{dt^2}\right]$$

$$\Rightarrow \quad -\frac{4}{9} = \frac{4}{9}\pi (7.5)\left[2\left(\frac{1}{5\pi}\right)^2 + 7.5\frac{d^2h}{dt^2}\right]$$

$$\Rightarrow \quad \frac{d^2h}{dt^2} = \frac{1}{7.5}\left(-\frac{1}{7.5\pi} - \frac{2}{25\pi^2}\right)$$

$$\Rightarrow \quad k = 7.5$$

38. (3) Given, $s = \sqrt{at^2 + 2bt + c}$

$$\Rightarrow \quad \frac{ds}{dt} = \frac{(at + b)}{\sqrt{at^2 + bt + c}} = \frac{at + b}{s}$$

Again, differentiating both sides w.r.t. t, we get

$$\frac{d^2s}{dt^2} = \frac{s(a) - (at + b)\dfrac{ds}{dt}}{s^2}$$

$$= \frac{as - (at + b)\left(\dfrac{at + b}{s}\right)}{s^2}$$

$$= \frac{as^2 - (at + b)^2}{s^3}$$

$$= \frac{a(at^2 + 2bt + c) - (a^2t^2 + b^2 + 2abt)}{s^3}$$

$$= \frac{(ac - b^2)}{s^3}$$

$\therefore$ Acceleration $\propto \dfrac{1}{s^3}$

$$\propto s^{-3}$$

Hence, $P = 3$

39. (5) We have, $\dfrac{dV}{dt} = 2\text{ m}^3/\text{s}$

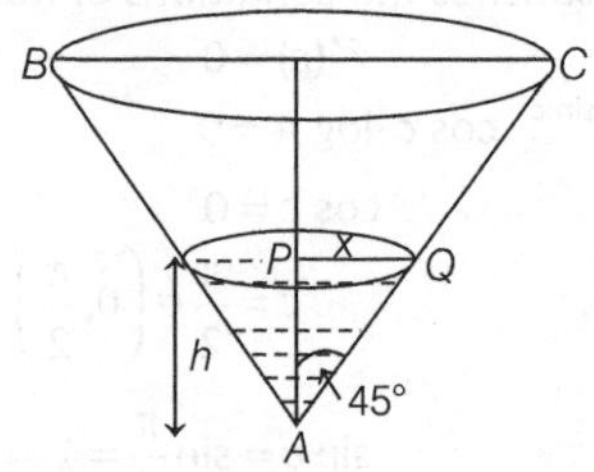

In ΔAPQ, $\dfrac{r}{h} = \tan 45°$

$\Rightarrow \quad r = h$

Now, volume of cone, $V = \dfrac{1}{3}\pi r^2 h$

$$\Rightarrow \quad V = \frac{1}{3}\pi r^3 \qquad [\because r = h]$$

$$\Rightarrow \quad \frac{dV}{dt} = \pi r^2 \frac{dr}{dt}$$

$$\Rightarrow \quad \frac{dr}{dt} = \frac{2}{\pi r^2} \qquad \left[\because \frac{dV}{dt} = 2\,m^3/s\right]$$

Now, perimeter, $P = 2\pi r$

$$\Rightarrow \quad \frac{dP}{dt} = \frac{d}{dt}(2\pi r) = 2\pi\left(\frac{2}{\pi r^2}\right) = \frac{4}{r^2}$$

When $h = 2$ m, then $r = 2$ m

Hence, $\dfrac{dP}{dt} = \dfrac{4}{4} = 1$ m/s

Hence, $5d = 5 \times 1 = 5$ m/s

40. (5) Let A_1 and A_2 be areas of squares S_1 and S_2 respectively, then

$A_1 = l_1^2$ and $A_2 = l_2^2$

$\dfrac{dA_1}{dl_1} = 2l_1$ and $\dfrac{dA_2}{dl_2} = 2l_2$

$$\frac{dA_2}{dA_1} = \frac{l_2}{l_1} \cdot \frac{dl_2}{dl_1} = \frac{l_2}{l_1(1 + 3l_2^2)}$$

When $l_2 = 1$, then $l_1 = 8$

$$\frac{dA_2}{dA_1} = \frac{1}{8(1+3)} = \frac{1}{32}$$

$$= \frac{1}{2^5} = \frac{1}{2^\lambda} \qquad \text{[given]}$$

Hence, $\lambda = 5$

41. (54) We know that $l^2 = r^2 + h^2$ ...(i)

When $r = 7$ cm, $h = 24$ cm, then

$$l^2 = 49 + 576$$

$$\Rightarrow \quad l = 25 \text{ cm}$$

Now, differentiating Eq. (i), w.r.t. t, we get

$$2l\frac{dl}{dt} = 2r\frac{dr}{dt} + 2h\frac{dh}{dt}$$

$$\Rightarrow \quad 25\frac{dl}{dt} = 7 \times 3 + 24(-4) \qquad \left[\begin{array}{l}\dfrac{dr}{dt} = 3\text{ cm/min} \\ \dfrac{dh}{dt} = -4\text{ cm/min}\end{array}\right]$$

$$\Rightarrow \quad 25\frac{dl}{dt} = -75 \Rightarrow \frac{dl}{dt} = -3\text{ cm/min}$$

Lateral surface area of cone, $S = \pi r l$

$$\frac{dS}{dt} = \pi\left(r\frac{dl}{dt} + l\frac{dr}{dt}\right)$$

$$\Rightarrow \quad \frac{dS}{dt} = 54\,\pi\text{ cm}^2/\text{min}$$

So, $\quad K = 54$

42. (33) We have, $\dfrac{dr}{dt} = K$ (constant), $\dfrac{dh}{dt} = 3K$

and $h = ar + C$

$$\frac{dh}{dt} = a\frac{dr}{dt} \Rightarrow 3K = a \cdot K \Rightarrow a = 3$$

So, $h = 3r + C$

When $h = 6, r = 1$, then

$$6 = 3 + C \Rightarrow C = 3$$

So, $h = 3r + 3$

Volume, $V = \pi r^2 h$

$$\Rightarrow \quad V = \pi r^2(3r + 3)$$

$$= 3\pi(r^3 + r^2)$$

$$\Rightarrow \quad \frac{dV}{dt} = 3\pi(3r^2 + 2r)\frac{dr}{dt} \qquad \text{...(i)}$$

When $r = 6$, $\dfrac{dV}{dt} = 1$, then

$$1 = 3\pi(108 + 12)\frac{dr}{dt} \Rightarrow \frac{dr}{dt} = \frac{1}{360\pi}$$

From Eq. (i),

When $r = 36$, then

$$\frac{dV}{dt} = 3\pi r(3r + 2)\frac{dr}{dt}$$

$$\Rightarrow \quad \frac{dV}{dt} = 3\pi \times 36(110)\frac{1}{360\,\pi}$$

$$\Rightarrow \quad \frac{dV}{dt} = 33\text{ cm}^3/\text{s}$$

43. (6) We have, $g(x) = [f'(x)]^2 + g'(x) \cdot f''(x)$

$$\Rightarrow \quad g(x) = \frac{d}{dx}[f(x) \cdot f'(x)]$$

By Rolle's theorem, between any two roots of a polynomial, there is a root of its derivative.

Now, $f(x) \cdot f'(x) = 0$

Either $f(x) = 0$ or $f'(x) = 0$

Thus, $f(x)$ has four zeroes at $x = a$ between b and c, between c and d and at e.

So, $f'(x)$ has atleast 3 zeroes.

Thus, $f(x) \cdot f'(x)$ has atleast 7 zeroes.

Therefore, $g(x)$ has atleast 6 zeroes.

Hence, the minimum number of zeroes of $g(x)$ is 6.

44. (59.2) Given, $f(x) = x^3 - px^2 + qx$

$$\Rightarrow \quad f(1) = f(3)$$

$$\Rightarrow \quad 1 - p + q = 27 - 9P + 3q \Rightarrow 4p - q = 13 \qquad \text{... (i)}$$

Also $f'(c) = 0$

$$\Rightarrow \quad 3c^2 - 2pc + q = 0$$

$$\Rightarrow 3\left(\frac{5}{4}\right)^2 - 2p\left(\frac{5}{4}\right) + q = 0$$

$$\Rightarrow \quad 40p - 16q = 75 \quad \text{... (ii)}$$

From Eqs. (i) and (ii),

$$p = \frac{133}{24} \text{ and } q = \frac{55}{6}$$

$$\therefore \; p + 4q + 17 = \frac{133}{24} + \frac{110}{3} + 17 = 59.2$$

45. (2) Replacing t by Z^2, we get

$$\int_0^4 f(t)\,dt = \int_0^2 2Zf(Z^2)\,dZ.$$

From Lagrange's mean value theorem,

$$\frac{\int_0^2 2Zf(Z^2)\,dZ - \int_0^0 2Zf(Z^2)\,dZ}{2-0} = 2\gamma f(\gamma^2),$$

where $\gamma \in (0, 2)$.

For $0 < \alpha < \gamma < \beta < 2$, using mean value theorem,

$$\int_0^2 2Zf(Z^2)\,dZ - 2[2\gamma f(\gamma^2)] = 2\left(\frac{2\alpha f(\alpha)^2 + 2\beta f(\beta^2)}{2}\right)$$

$$\Rightarrow \int_0^4 f(t)\,dt = 2\,[\alpha f(\alpha^2) + \beta f(\beta^2)] \text{ for } 0 < \alpha \text{ and } \beta < 2$$

Hence, $K = 2$

46. (1) Given, $f(x) = x^3 - 3x + 1 = 0$

$$\Rightarrow \quad f'(x) = 3x^2 - 3$$
$$= 3(x^2 - 1) = 3(x+1)(x-1)$$

Clearly, $f(x)$ is increasing in $(-\infty, -1) \cup (1, \infty)$ and decreasing in $(-1, 1)$.

Now, $f(-2) = -8 + 6 + 1 = -1$

$f(-1) = -1 + 3 + 1 = 3$

$f(0) = 1$

$f(1) = 1 - 3 + 1 = -1$

$f(2) = 8 - 6 + 1 = 3$

Now, $f(-2) \cdot f(-1) < 0$

Thus, there is a root lying in $(-2, -1)$.

So, $[\alpha] = -2$

Also, $f(0) \cdot f(1) < 0$

Thus, there is a root lying in $(0, 1)$.

So, $[\beta] = 0$

Again, $f(1) \cdot f(2) < 0$

Thus, there is a root lying in $(1, 2)$.

So, $[\gamma] = 1$

Since, α, β and γ are the roots of $x^3 - 3x + 1 = 0$

So, $\alpha + \beta + \gamma = 0$

Now, $\{\alpha\} + \{\beta\} + \{\gamma\} = \alpha - [\alpha] + \beta - [\beta] + \gamma - [\gamma]$

$= (\alpha + \beta + \gamma) - ([\alpha] + [\beta] + [\gamma])$

$= 0 - (-2 + 0 + 1)$

$= -(-1) = 1$

47. (5) When f is continuous on $[0, 1]$ and differentiable in $(0, 1)$.

So, by Lagrange's mean value theorem,

$$f'(x) = \frac{f(1) - f(0)}{1 - 0}$$

$$\Rightarrow \quad f(1) - f(0) = f'(x)$$

$$\Rightarrow \quad f(1) - 2 \le 3 \Rightarrow f(1) \le 5 \quad \text{...(i)}$$

When f is continuous on $[1, 2]$ and differentiable in $(1, 2)$.

So, by Lagrange's mean value theorem, we get

$$f'(x) = \frac{f(2) - f(1)}{2 - 1}$$

$$\Rightarrow \quad f(2) - f(1) = f'(x)$$

$$\Rightarrow \quad f(2) - f(1) \le 3$$

$$\Rightarrow \quad 8 - f(1) \le 3 \Rightarrow f(1) \ge 5 \quad \text{...(ii)}$$

From Eqs. (i) and (ii), we get

$$f(1) = 5$$

48. (200) Let $f(x) = x^3 - 2Kx^2 - 4Kx + K^2$,

where $f(0) = K^2 \ge 0$

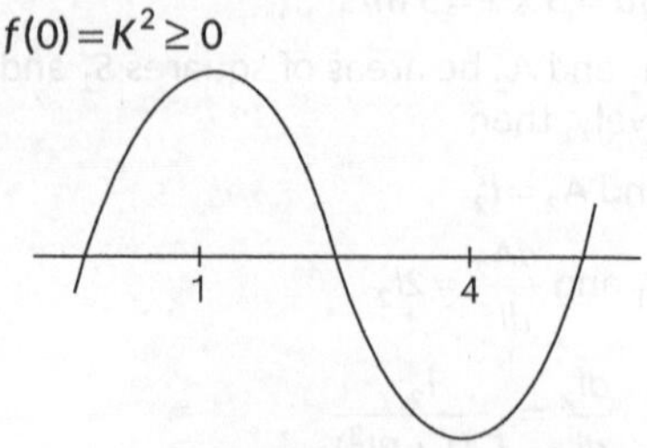

Clearly, $f(1) > 0$

$$\Rightarrow \quad K^2 - 6K + 1 > 0$$

$$\Rightarrow \quad (K-3)^2 - (2\sqrt{2})^2 > 0$$

$$\Rightarrow \quad (K-3)^2 > (2\sqrt{2})^2$$

$$\Rightarrow \quad K - 3 > 2\sqrt{2}, (K-3) < -2\sqrt{2}$$

$$\Rightarrow \quad K > (3 + 2\sqrt{2}) \text{ or } K < 3 - 2\sqrt{2} \quad \text{...(i)}$$

Also, $f(4) > 0$

$$\Rightarrow \quad 64 - 32k - 16k + K^2 > 0$$

$$\Rightarrow \quad K^2 - 48K + 64 > 0$$

$$\Rightarrow \quad (K-24)^2 - 512 > 0$$

$$\Rightarrow \quad (K-24)^2 - (16\sqrt{2})^2 > 0$$

$$\Rightarrow \quad (K-24)^2 > (16\sqrt{2})^2$$

$$\Rightarrow \quad K - 24 > 16\sqrt{2}$$

or $\quad K - 24 < -16\sqrt{2}$

$$\Rightarrow \quad K > (24 + 16\sqrt{2})$$

and $\quad K < (24 - 16\sqrt{2}) \quad \text{...(ii)}$

From Eqs. (i) and (ii), we get

$$3 + 2\sqrt{2} < K < 24 + 16\sqrt{2}$$

$$\Rightarrow \quad 3 + \sqrt{8} < K < 8(3 + \sqrt{8})$$

Thus, $m = 3$ and $n = 8$

Now, value of $2(m + n - 1)^2 = 2(3 + 8 - 1)^2 = 200$

CHAPTER

06

APPROXIMATION AND ERRORS, INCREASING AND DECREASING

LEVEL 01 \ Starter Level

6.1 Approximation and Errors

1 Using differentials, find the approximate values of following

(i) $\sqrt{49.02}$ (ii) $(66)^{\frac{1}{3}}$

(iii) $\log_e 4.04$, it being given that $\log_e 4 = 1.3863$.

(iv) $\cos 61°$, it being given that $\sin 60° = 0.86603$ and $1° = 0.01745$ rad.

2 Using differentials

(i) if $y = x^4 - 10$, if error in calculating the x is .01, then error in y, when $x = 3$.

(ii) if $y = \frac{1}{x^2}$, $x = 1$ and $\Delta x = -0.01$, then approximate error in y.

(iii) The radius of a circle measured as radius 7 cm with an error of 0.02 cm, then find the error in area.

(iv) Find the percentage error in calculating the surface area of a cubical base with an error of 1% is made in measuring the length of the cube.

3 Find the approximate value of $f(2.01)$, where $f(x) = 4x^2 + 5x + 2$.

4 Find the approximate change in the volume V of a cube of side x m, caused by increasing the side by 1%.

6.2 Increasing and Decreasing Functions

5 Check the nature of following differentiable functions

(i) $f(x) = e^x + \sin x, x \in R^+$.

(ii) $f(x) = x^3 - 3x^2 + 3x - 100$ on R.

(iii) $f(x) = \frac{1}{x} + 2, x \in R - \{0\}$.

(iv) $f(x) = \tan^{-1} x - x, x \in R$.

(v) $f(x) = \cos^{-1} x + \log\left(x + \frac{1}{x}\right)$.

(vi) $f(x) = \frac{4 \sin x}{2 + \cos x} - x, x \in \left[0, \frac{\pi}{2}\right]$.

6 For each of the following graph, comment whether $f(x)$ is increasing or decreasing or neither increasing nor decreasing at given point.

(i) 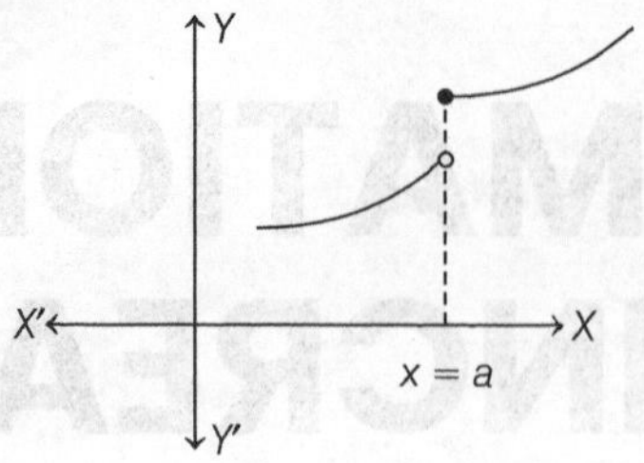

(ii) 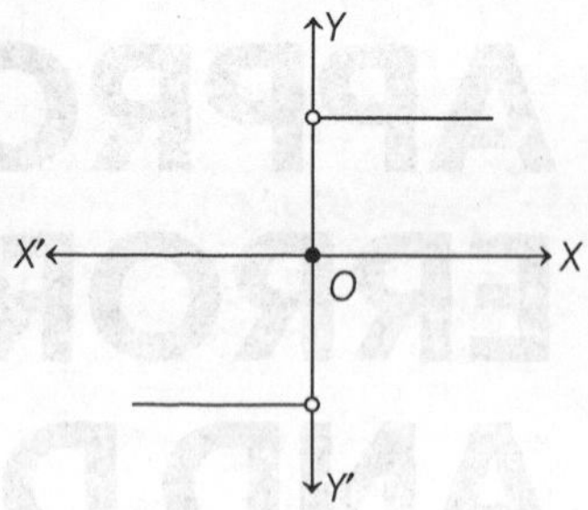

(iii) 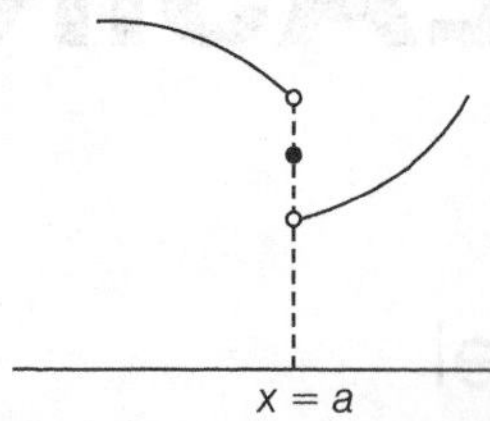

(iv) 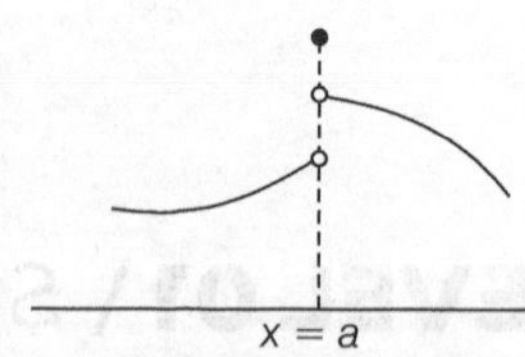

7 Find the intervals in which the following functions are increasing and decreasing.

(i) $f(x) = x^4 - 8x^3 + 22x^2 - 24x + 21$ (ii) $f(x) = \frac{x}{2} + \frac{2}{x}, x \neq 0$

(iii) $f(x) = (2^x - 1)(2^x - 2)^2$ (iv) $f(x) = \sin x + \cos x, x \in [0, 2\pi]$

(v) $f(x) = x^x, x > 0$

LEVEL 02 \ JEE Main Level

Single Choice Correct Questions

6.1 Approximation and Errors

1 If the radius of a sphere is measured as 9 cm with an error of 0.03 cm, then the approximate error in calculating its volume is

(a) $9.72\,\pi\text{ cm}^3$ (b) $7.92\,\pi\text{ cm}^3$ (c) $8.72\,\pi\text{ cm}^3$ (d) None of these

2 If an error of 1° is made in measuring the angle of a sector of radius 30 cm, then approximate error in its area is

(a) 450 cm^2 (b) $25\,\pi\text{ cm}^2$ (c) $2.5\,\pi\text{ cm}^2$ (d) None of these

3 While measuring the side of an equilateral triangle an error of $k\%$ is marked, then percentage error in its area is

(a) $k\%$ (b) $2k\%$ (c) $\frac{k}{2}\%$ (d) $3k\%$

4 If the ratio of base radius and height of a cone is 1 : 2 and percentage error in radius is 1%, then the error in volume is
(a) 1% (b) 2% (c) 3% (d) None of these

5 The height of cone increases by $k\%$, its semi-vertical angle remain same. Then, the approximate percentage increases in total surface area, is
(a) $k\%$ (b) $2k\%$ (c) $3k\%$ (d) $4k\%$

6 The time T of a complete oscillation of a simple pendulum of length l is given by the equation $T = 2\pi\sqrt{\frac{l}{g}}$, where g is constant. Then, the percentage error in T, when l is increased by 1%, is
(a) 1% (b) 0.5% (c) 2% (d) None of these

7 If the error committed in measuring the radius of the circle is 0.05%, then the corresponding error in calculating the area is
(a) 0.05% (b) 0.025% (c) 0.25% (d) 0.1%

8 The pressure P and the volume V of a gas are connected by the rotation $PV^{1.4} = \text{constant}$. Then, the percentage error in P corresponding to a decrease in $\frac{1}{2}\%$ in V, is
(a) 0.7% (b) 0.5% (c) 0.6% (d) None of these

9 If the radius of sphere is measured as 7 cm with an error of 0.02 cm, then the error in calculating the volume is
(a) 6.16 cm^3 (b) 18.48 cm^3 (c) 12.32 cm^3 (d) None of these

10 The focal length of mirror is given by $\frac{1}{v} - \frac{1}{u} = \frac{2}{f}$. If equal errors ($\alpha$) are made in measuring u and v, then the relative error in f is
(a) $\frac{2}{\alpha}$ (b) $\alpha\left(\frac{1}{u} + \frac{1}{v}\right)$ (c) $\alpha\left(\frac{1}{u} - \frac{1}{v}\right)$ (d) None of these

6.2 Increasing and Decreasing Functions

11 If $f : R \to R$, $f(x)$ is differentiable such that $f\{f(x)\} = k(x^5 + x)$, $(k \neq 0)$, then $f(x)$ is
(a) increasing (b) decreasing
(c) either increasing or decreasing (d) non-monotonic

12 Let $f(x) = 3\sin^4 x + 10\sin^3 x + 6\sin^2 x - 3$, $x \in \left[-\frac{\pi}{6}, \frac{\pi}{2}\right]$. Then, f is **[JEE Main 2021]**
(a) increasing in $\left(-\frac{\pi}{6}, \frac{\pi}{2}\right)$ (b) decreasing in $\left(0, \frac{\pi}{2}\right)$
(c) increasing in $\left(-\frac{\pi}{6}, 0\right)$ (d) decreasing in $\left(-\frac{\pi}{6}, 0\right)$

13 Which of the following statement is always true for differentiable function $f(x)$ and $g(x)$?
(a) If $f(x)$ is increasing, then $f'(x)$ is decreasing
(b) If $f(x)$ is increasing, then $\frac{1}{f(x)}$ is also increasing
(c) If f and g are positive functions and f is increasing and g is decreasing, then $\frac{f}{g}$ is decreasing function.
(d) If f and g are positive functions and f is decreasing and g is increasing, then $\frac{f}{g}$ is decreasing function.

14 Let $f : R \to R$ be defined as $f(x) = \begin{cases} -\frac{4}{3}x^3 + 2x^2 + 3x &, x > 0 \\ 3xe^x &, x \le 0 \end{cases}$. Then, f is increasing function in the interval **[JEE Main 2021]**

(a) $\left(-\frac{1}{2}, 2\right)$ (b) $(0, 2)$ (c) $\left(-1, \frac{3}{2}\right)$ (d) $(-3, -1)$

15 If $f(x) = kx^3 - 9x^2 + 9x + 3$ is monotonically increasing in every interval, then which one of the following is correct?

(a) $k < 3$ (b) $k \le 3$ (c) $k > 3$ (d) $k \ge 3$

16 The value of k in order that $f(x) = \sin x - \cos x - kx + 5$ decreases for all positive real values of x, is given by

(a) $k < 1$ (b) $k \ge 1$ (c) $k > \sqrt{2}$ (d) $k \le \sqrt{2}$

17 The function $f(x) = \frac{4x^3 - 3x^2}{6} - 2\sin x + (2x - 1)\cos x$ is **[JEE Main 2021]**

(a) increasing in $\left[\frac{1}{2}, \infty\right)$ (b) decreasing in $\left(-\infty, \frac{1}{2}\right)$

(c) increasing in $\left(-\infty, \frac{1}{2}\right)$ (d) decreases $\left[\frac{1}{2}, \infty\right)$

18 If $f(x) = \frac{x^2}{2 - 2\cos x}$ and $g(x) = \frac{x^2}{6x - 6\sin x}$, where $0 < x < 1$, then

(a) both f and g are increasing functions
(b) f is decreasing and g is increasing function
(c) f is increasing function and g is decreasing function
(d) both f and g are decreasing functions

19 If $P(x)$ is a polynomial function and $P'(x) > P(x)\ \forall\ x \ge 1$ and $P(1) = 0$, then

(a) $P(x) \ge 0, \forall\ x \ge 1$ (b) $P(x) < 0, \forall\ x \ge 1$
(c) $P(x) = 0, \forall\ x \ge 1$ (d) None of these

20 If $f'(x) = |x| - \{x\}$, where $\{x\}$ denotes the fractional part of x, then $f(x)$ is decreasing in

(a) $x < 0$ (b) $x > 1$ (c) $x < 1$ (d) $0 < x < 1$

21 The function $f(x) = \tan^{-1}(\sin x + \cos x)$ is an increasing function in

(a) $\left(0, \frac{\pi}{2}\right)$ (b) $\left(-\frac{\pi}{2}, \frac{\pi}{2}\right)$ (c) $\left(\frac{\pi}{4}, \frac{\pi}{2}\right)$ (d) $\left(-\frac{\pi}{2}, \frac{\pi}{4}\right)$

22 Define $f(x) = \frac{1}{2}[|\sin x| + \sin x]$, $0 < x \le 2\pi$. Then, f is

(a) increasing in $\left(\frac{\pi}{2}, \frac{3\pi}{2}\right)$

(b) decreasing in $\left(0, \frac{\pi}{2}\right)$ and increasing in $\left(\frac{\pi}{2}, \pi\right)$

(c) increasing in $\left(0, \frac{\pi}{2}\right)$ and decreasing in $\left(\frac{\pi}{2}, \pi\right)$

(d) increasing in $\left(0, \frac{\pi}{4}\right)$ and decreasing in $\left(\frac{\pi}{4}, \pi\right)$

23 $f(x) = |x \log x|, x > 0$ is monotonically decreasing in

(a) (e, ∞) (b) $\left(0, \frac{1}{e}\right)$ (c) $\left[\frac{1}{e}, 2\right]$ (d) $\left[\frac{1}{e}, 1\right]$

24 Let f be a real valued function, defined on $R - \{-1, 1\}$ and given by $f(x) = 3 \log \left|\frac{x-1}{x+1}\right| - \frac{2}{x-1}$, then in which interval $f(x)$ is increasing **[JEE Main 2021]**

(a) $(-\infty, -1) \cup \left(\left[\frac{1}{2}, \infty\right) - \{1\}\right)$ (b) $(-\infty, -1) - \{-1, 1\}$

(c) $\left(-1, \frac{1}{2}\right)$ (d) $\left(-\infty, \frac{1}{2}\right] - \{-1\}$

25 Let $f(x) = \left(1 + \frac{1}{x}\right)^{x+1}$ is defined for $x > 0$, then

(a) $f(x)$ is increasing (b) $f(x)$ is decreasing
(c) $f(x)$ is non-monotonic (d) None of these

26 If $f(x)$ is differentiable function and $f(x) - x$ is strictly increasing, then $f(x) - x + x^3$ is

(a) increasing (b) decreasing
(c) not-monotonic (d) not depends on nature of $f(x)$

27 If $f(x) = x^3 + bx^2 + cx + d$ and $0 < b^2 < c$, then in $(-\infty, \infty)$

(a) $f(x)$ is strictly increasing function (b) $f(x)$ has local maxima
(c) $f(x)$ is strictly decreasing function (d) None of these

28 The interval of increase of the function $y = x - e^x + \tan \frac{\pi}{5}$ is

(a) $(-\infty, -1)$ (b) $(0, \infty)$ (c) $(-\infty, 0)$ (d) $(1, \infty)$

29 The number of points of non-monotonicity of $f(x) = \left|\sin x + \frac{1}{2}\right|$ for $x \in [-2\pi, 2\pi]$ is

(a) 7 (b) 8 (c) 9 (d) None of these

30 The function $f(x) = \sin^4 x + \cos^4 x$ increasing, if

(a) $0 < x < \frac{\pi}{8}$ (b) $\frac{\pi}{4} < x < \frac{3\pi}{8}$ (c) $\frac{3\pi}{8} < x < \frac{5\pi}{8}$ (d) None of these

31 The function $\frac{a \sin x + b \cos x}{c \sin x + d \cos x}$ is decreasing, if

(a) $ad - bc > 0$ (b) $ad - bc < 0$
(c) $ab - cd < 0$ (d) $ab - cd > 0$

32 Let $f : (-1, \infty) \to R$ be defined by $f(0) = 1$ and $f(x) = \frac{1}{x} \log_e (1 + x)$, $x \neq 0$. Then, the function f **[JEE Main 2020]**

(a) decreases in $(-1, 0)$ and increases in $(0, \infty)$
(b) increases in $(-1, \infty)$
(c) increases in $(-1, 0)$ and decreases in $(0, \infty)$
(d) decreases in $(-1, \infty)$

33 $f(x) = \frac{x}{\log x} - \frac{\log x}{x}$ is increasing in

(a) (e, ∞) (b) $(0, 1) \cup (1, e)$ (c) $(0, 1)$ (d) $(1, e)$

34 Let $f''(x) > 0$ for all x in R such that $f'(3) = 0$. If $g(x) = f(\tan^2 x - 2\tan x + 4)$ for all x in $\left(0, \frac{\pi}{2}\right)$, the interval, where $g(x)$ is increasing, is

(a) $\left(0, \frac{\pi}{4}\right)$ (b) $\left(\frac{\pi}{4}, \frac{\pi}{2}\right)$ (c) $\left(0, \frac{\pi}{2}\right)$ (d) None of these

35 If the function $f(x) = 2\tan x + (2a+1)\ln \sec x + (a-2)x + 2$ is non-decreasing on R, then the value of a is

(a) $-\frac{1}{2}$ (b) 1 (c) 2 (d) $\frac{1}{2}$

36 If $f : [1, 10] \to [1, 10]$ is non-decreasing function and $g : [1, 10] \to [1, 10]$ is non-increasing function. Let $h(x) = f\{g(x)\}$ with $h(1) = 1$, then $h(2)$ is

(a) less than 1 (b) more than 2 (c) equal to 2 (d) not defined

37 The function $f(x) = x^3 + \lambda x^2 + 5x + \sin 2x$ will be invertible function, if λ belongs to

(a) $(-\infty, -3)$ (b) $(-3, 3)$ (c) $(3, \infty)$ (d) None of these

38 Let $f(x) = x^3 + ax^2 + bx + 5\sin^2 x$ be an increasing function in the set of real numbers R. Then, a and b satisfy the condition

(a) $a^2 - 3b - 15 > 0$ (b) $a^2 - 3b + 15 < 0$

(c) $a^2 - 3b - 15 < 0$ (d) $a > 0$ and $b > 0$

39 If $f(x) = \frac{a^2 - 1}{a^2 + 1}x^3 - 3x + 2\log 5$ is a decreasing function of x for all $x \in R$, then the set of possible values of a is

(a) $[-1, 1]$ (b) $[1, \infty)$ (c) $(-\infty, -1)$ (d) None of these

40 $f(x)$ be a differentiable function such that $f'(x) = \frac{1}{\log_3 \{\log_{1/4}(\cos x + b)\}}$. If $f(x)$ be an increasing function, then the value of b is

(a) $(-1, 0)$ (b) $(0, 1)$ (c) 1 (d) None of these

41 The set of all x for which $\log(1 + x) \le x$ is

(a) $(0, \infty)$ (b) $(-1, \infty)$ (c) $(-1, 0)$ (d) None of these

42 If $0 \le x < \frac{\pi}{2}$, then

(a) $2\sin x + \tan x > 3x$ (b) $2\sin x + \tan x < 3x$

(c) $2\sin x + \tan x \ge 3x$ (d) $2\sin x + \tan x \le 3x$

43 The number of roots of the equation $\log(1 + x) - \frac{\tan^{-1} x}{1 + x} = 0$ is

(a) 0 (b) 1 (c) 2 (d) None of these

44 Let $f : [0, 2] \to R$ be a twice differentiable function such that $f''(x) > 0$, for all $x \in (0, 2)$. If $\phi(x) = f(x) + f(2 - x)$, then ϕ is **[JEE Main 2019]**

(a) increasing in (0, 1) and decreasing in (1, 2)

(b) decreasing in (0, 2)

(c) decreasing in (0, 1) and increasing in (1, 2)

(d) increasing in (0, 2)

45 If the function $f(x) = \sin(\ln x) - 2\cos(\ln x)$ is increasing in (e^p, e^q), then $\cos(p - q)$ is equal to

(a) 0 (b) 1 (c) -1 (d) None of these

46 Let $g(x) = 2f\left(\frac{x}{2}\right) + f(2-x)$ and $f''(x) < 0$, $\forall x \in (0, 2)$, then the intervals of increase of $g(x)$ is

(a) $(0, 2)$ (b) $\left(0, \frac{4}{3}\right)$ (c) $\left(\frac{4}{3}, 2\right)$ (d) None of these

47 Let $f(x) = e^x - x$ and $g(x) = x^2 - x$, $\forall x \in R$. Then, the set of all $x \in R$, where the function $h(x) = (fog)(x)$ is increasing, is **[JEE Main 2019]**

(a) $\left[0, \frac{1}{2}\right] \cup [1, \infty)$ (b) $\left[-1, \frac{-1}{2}\right] \cup \left[\frac{1}{2}, \infty\right)$ (c) $[0, \infty)$ (d) $\left[\frac{-1}{2}, 0\right] \cup [1, \infty)$

48 If $h(x) = f(x) - f^2(x) + f^3(x)$, then

(a) $h(x)$ is increasing, whenever $f(x)$ is increasing
(b) $h(x)$ is increasing, whenever $f(x)$ is decreasing
(c) $h(x)$ is decreasing, whenever $f(x)$ is increasing
(d) Nothing can be said in general

49 $f(x) = \int \left(2 - \frac{1}{1+x^2} - \frac{1}{\sqrt{1+x^2}}\right) dx$, then f is

(a) increasing in $(0, \infty)$ and decreasing in $(-\infty, 0)$ (b) increasing in $(-\infty, 0)$ and decrasing in $(0, \infty)$
(c) increasing in $(-\infty, \infty)$ (d) decreasing in $(-\infty, \infty)$

50 The interval, in which $f(x) = \int_0^x e^t (t-1)(t-2)\, dt$ is decreasing, is

(a) $(1, 3)$ (b) $(1, 4)$ (c) $\left(\frac{3}{2}, 4\right)$ (d) $(1, 2)$

Numerical Value Type Questions

51 The percentage error in 11th root of a number is approximately $\frac{1}{k}$ times the percentage error in number, then the value of k is ....

52 If the height of a cylinder is equal to the radius. If an error of 2% is made in the height, then the percentage error in volume is ....

53 A polynomial $y = f(x)$ of degree 4 increases in interval $(-\infty, 1) \cup (2, 3)$ and decreases in the interval $(1, 2) \cup (3, \infty)$ and satisfies $f(0) = 1$ and $f'(0) = 6$, then the value of $f(2)$ is ....

54 If m is the number of solution of the equation $3\tan x + x^3 = 2$, in $\left(0, \frac{\pi}{4}\right)$ and n is the number of inflection points of $f(x) = 3x^4 - 4x^3$, then the value of $(m + n + 2)$ is ....

55 If $f(x) = e^{2x} - (a+1)e^x + 2x$ is increasing for all real values of x in R, then the maximum value of a is ....

56 If $f(x) = 2x^3 + 9x^2 + \lambda x + 20$ is decreasing function of x in the largest possible interval $(-2, -1)$, then λ is equal to ....

57 Let a be an interger such that all the real roots of the polynomial $2x^5 + 5x^4 + 10x^3 + 10x^2 + 10x + 10$ lie in the interval $(a, a+1)$, then $|a|$ is equal to .... **[JEE Main 2021]**

58 Let $f(x) = \begin{cases} xe^{ax} & , x \le 0 \\ x + ax^2 - x^3 & , x > 0 \end{cases}$, where a is a positive constant, then the interval, in which $f'(x)$ is increasing, is $[p, q]$. Then, $-6pq$ is equal to ....

LEVEL 03 \ JEE Advanced Level

Single Choice Correct Questions

1 The period of oscillation, T of a pendulum of length, l at a place of acceleration due to gravity, g is given by $T = 2\pi\sqrt{\frac{l}{g}}$. If the calculated length is 0.992 times the actual length and if the value assume for g is 1.002 times the actual value. The relative error in the computed value of T is
(a) 0.005 (b) – 0.005 (c) 0.003 (d) – 0.003

2 If $f(x) = \ln(1+x)$ for all $x \in (-1, \infty)$ and $a = \sqrt{c+1} - \sqrt{c}$ and $b = \sqrt{c} - \sqrt{c-1}$ for all $c > 1$, then
(a) $a > b \Rightarrow f(a) > f(b)$
(b) $a > b \Rightarrow f(b) > f(a)$
(c) $a < b \Rightarrow f(a) > f(b)$
(d) $a < b \Rightarrow f(b) > f(a)$

3 The funtion $f(x) = \dfrac{ae^x + be^{-x}}{ce^x + de^{-x}}$ is increasing function of x, if
(a) $ab > cd$
(b) $ad \geq bc$
(c) $ac > bd$
(d) None of these

4 Let $f(x) = \{-c^2 + (b-1)c - 2\}x + \int(\sin^2 x + \cos^4 x)\, dx$, if $f(x)$ is increasing function of x, $\forall\, x \in R$, then the possible values of b (if $c \in R$) are
(a) $(1-\sqrt{5}, 1+\sqrt{5})$
(b) $(0, 1+\sqrt{5})$
(c) $(1-\sqrt{5}, 0)$
(d) $(-\infty, 1-\sqrt{5}) \cup (1+\sqrt{5}, \infty)$

5 For all $x \in (0, 1)$
(a) $e^x < 1 + x$
(b) $\log(1+x) < x$
(c) $\sin x > x$
(d) $\log x > x$

6 Let $f(x)$ be a function defined by $f(x) = (ab - a^2 - 2)x - \int_0^x (\cos^4 t + \sin^2 t - 2)\, dt$. If $f(x)$ is a decreasing function for all $x \in R$ and $a \in R$, where a is independent of x, then
(a) $b \in (1, \infty)$ (b) $b \in (-1, 1)$ (c) $b \in (-\infty, 1)$ (d) $b \in (-\sqrt{3}, \sqrt{3})$

7 Let $f(x) = (1-x)^2 \sin^2 x + x^2$ for all $x \in R$ and $g(x) = \int_1^x \left\{\dfrac{2(t-1)}{t+1} - \log t\right\} f(t)\, dt$ for all $x \in (1, \infty)$.
Then, which of the following is true?
(a) g is increasing in $(1, \infty)$
(b) g is decreasing in $(1, \infty)$
(c) g is increasing in (1, 2) and decreasing in $(2, \infty)$
(d) g is decreasing in (1, 2) and increasing in $(2, \infty)$

8 If $f(x) = a^{\{a^{|x|}\,\text{sgn}\,x\}}$: $g(x) = a^{[a^{|x|}\,\text{sgn}\,x]}$ for $a > 1$ and $x \in R$, where {} and [] denote the fractional part and integral part function respectively, then which of the following statement can hold good for the function $h(x)$, where $(\ln a)\, h(x) = \ln(f(x)) + \ln(g(x))$?
(a) h is odd (b) h is even (c) h is decreasing (d) h is increasing

9 Let $g(x) = \frac{1}{4} f(2x^2 - 1) + \frac{1}{2} f(1 - x^2)$, $\forall\, x \in R$, where $f''(x) > 0$, $\forall\, x \in R$, $g(x)$ is necessarily increasing in the interval
(a) $\left(-\sqrt{\frac{2}{3}}, \sqrt{\frac{2}{3}}\right)$
(b) $\left(-\sqrt{\frac{2}{3}}, 0\right) \cup \left(\sqrt{\frac{2}{3}}, \infty\right)$
(c) (– 1, 1)
(d) None of these

More than One Correct

10 Let $f(x) = x - [x]$. In which of the following interval $f(x)$ is increasing?

(a) [0, 1) (b) (1, 2) (c) [2, 3] (d) (2, 3]

11 The function $f(x) = x - \cot^{-1} x + \log(\sqrt{x^2 + 1} - x)$ is increasing in

(a) $(-\infty, 0)$ (b) $(0, \infty)$ (c) $(-\infty, \infty)$ (d) $[0, \infty)$

12 The point of non-monotonicity of $f(x) = \text{Max}\{1 + \sin x, 1 - \cos x, 1\}$ for $x \in [0, 2\pi]$ is

(a) $\frac{\pi}{2}$ (b) $\frac{3\pi}{4}$ (c) π (d) $\frac{3\pi}{2}$

13 A function $y = f(x)$ is given by $x = \frac{1}{1+t^2}$ and $y = \frac{1}{t(1+t^2)}$ for all $t > 0$, then f is

(a) increasing in $\left(0, \frac{3}{2}\right)$ and decreasing in $\left(\frac{3}{2}, \infty\right)$ (b) increasing in (0, 1)

(c) increasing in $(0, \infty)$ (d) decreasing in (0, 1)

14 The function $\sin^{-1}(2x\sqrt{1-x^2})$ is decreasing in

(a) $\left(-1, -\frac{1}{\sqrt{2}}\right)$ (b) $\left(-\frac{1}{\sqrt{2}}, \frac{1}{\sqrt{2}}\right)$ (c) $\left(\frac{1}{\sqrt{2}}, 1\right)$ (d) None of these

15 Which of the following is (are) correct?

(a) $3^\pi > \pi^3$ (b) $101^{202} > 202^{101}$

(c) $\left(\frac{4}{3}\right)^{\frac{9}{4}} > \left(\frac{9}{4}\right)^{\frac{3}{4}}$ (d) $\left(1 + \sin\frac{\pi}{3}\right)^{1+\cos\frac{\pi}{3}} > \left(1 + \cos\frac{\pi}{3}\right)^{1+\sin\frac{\pi}{3}}$

16 The function $f(x) = 2\log(x-2) - x^2 + 4x + 1$ increases in the interval

(a) (1, 2) (b) (2, 3) (c) $\left(\frac{5}{2}, 3\right)$ (d) (2, 4)

17 Let $f(x) = |x^2 - 3x - 4|, -1 \le x \le 4$, then

(a) increasing in $\left[-1, \frac{3}{2}\right]$ (b) decreasing in $\left[\frac{3}{2}, 4\right]$

(c) minimum value of $f(x)$ is 1 (d) None of these

18 If $f(x)$ and $g(x)$ are two positive and increasing functions, then which of the following is not always true?

(a) $\{f(x)\}^{g(x)}$ is always increasing (b) If $\{f(x)\}^{g(x)}$ is decreasing, then $f(x) \le 1$

(c) If $\{f(x)\}^{g(x)}$ is decreasing, then $f(x) > 1$ (d) If $f(x) > 1$, then $\{f(x)\}^{g(x)}$ is increasing

19 Let $g'(x) > 0$ and $f'(x) < 0, \forall x \in R$, then

(a) $g\{f(x+1)\} > g\{f(x-1)\}$ (b) $f\{g(x-1)\} > f\{g(x+1)\}$

(c) $g\{f(x+1)\} < g\{f(x-1)\}$ (d) $g\{g(x+1)\} < g\{g(x-1)\}$

20 If $f : R$ is a differentiable function such that $f'(x) > 2f(x)$ for all $x \in R$ and $f(0) = 1$, then

(a) $f(x) > e^{2x}$ in $(0, \infty)$ (b) $f(x)$ is decreasing in $(0, \infty)$

(c) $f(x)$ is increasing in $(0, \infty)$ (d) $f'(x) < e^{2x}$ in $(0, \infty)$

21 Let $f : (0, \infty) \to R$ be given by $f(x) = \int_{1/x}^{x} \frac{e^{-\left(t+\frac{1}{t}\right)}}{t}\,dt$, then

(a) $f(x)$ is monotonic increasing on $[1, \infty)$ (b) $f(x)$ is monotonic decreasing in (0, 1)

(c) $f(x) + f\left(\frac{1}{x}\right) = 0$ for all $x \in (0, \infty)$ (d) $f(2^x)$ is an odd function

22 If $f(x)=\dfrac{x}{\sin x}$ and $g(x)=\dfrac{x}{\tan x}$, where $0<x\le 1$, then

(a) both $f(x)$ and $g(x)$ are increasing function (b) both $f(x)$ and $g(x)$ are decreasing function
(c) $f(x)$ is an increasing function (d) $g(x)$ is an increasing function

Linked Comprehension Type

Paragraph I Directions (Q.Nos. 23-24)

If $Q(x)=2f\left(\dfrac{x^2}{2}\right)+f(6-x^2)$, $\forall x\in R$ and given that $f''(x)>0$, $\forall x\in R$.

23 $Q(x)$ is increasing in
(a) $(-2, 0)\cup(2,\infty)$ (b) $(-\infty,-2)\cup(0,2)$ (c) $(-\infty,\infty)$ (d) $(-2, 2)$

24 $Q(x)$ is decreasing in
(a) $(-2, 0)\cup(2,\infty)$ (b) $(-\infty,-2)\cup(0,2)$ (c) $(-2, 2)$ (d) $(-\infty,+2)$

Paragraph II Directions (Q.Nos. 25-27)

Let $p(x)=f(x)-a\{f(x)\}^2+a\{f(x)\}^3$ for every real number x

25 $p(x)$ increases as $f(x)$ increases for all real values of x, if
(a) $a\in[0,3]$ (b) $a\in(-2,2)$ (c) $a\in[3,\infty)$ (d) None of these

26 $p(x)$ increases as $f(x)$ decreases for all real values of x, if
(a) $a\in(0,3)$ (b) $a\in(-2,2)$ (c) $(3,\infty)$ (d) None of these

27 If $f(x)$ is strictly increasing function, then $p(x)$ is non-monotonic function given
(a) $a\in(0,3)$ (b) $a\in(-2,2)$ (c) $a\in(3,\infty)$ (d) $a\in(-\infty,0)\cup(3,\infty)$

Paragraph III Directions (Q.Nos. 28-30)

Let $f'(\sin x)<0$ and $f''(\sin x)>0$, $\forall x\in\left(0,\dfrac{\pi}{2}\right)$.

Now, consider a function $g(x)=f(\sin x)+f(\cos x)$.

28 $g(x)$ decreases, if x belongs to
(a) $\left(0,\dfrac{\pi}{4}\right)$ (b) $\left(\dfrac{\pi}{4},\dfrac{\pi}{2}\right)$ (c) $\left(\dfrac{\pi}{6},\dfrac{\pi}{3}\right)$ (d) None of these

29 $g(x)$ increases, if x belongs to
(a) $\left(0,\dfrac{\pi}{4}\right)$ (b) $\left(\dfrac{\pi}{4},\dfrac{\pi}{2}\right)$ (c) $\left(\dfrac{\pi}{8},\dfrac{\pi}{3}\right)$ (d) $\left(\dfrac{\pi}{6},\dfrac{\pi}{3}\right)$

30 The set of critical points of $g(x)$ is
(a) $\left\{\dfrac{\pi}{8},\dfrac{\pi}{3}\right\}$ (b) $\left\{\dfrac{\pi}{8},\dfrac{\pi}{6},\dfrac{\pi}{3}\right\}$ (c) $\left\{\dfrac{\pi}{8},\dfrac{\pi}{6},\dfrac{\pi}{4}\right\}$ (d) None of these

Matrix Match Type Questions

31 Match the following

	Column I		Column II
(i)	$f(x)=\log(x+\sqrt{x^2+1})$ is increasing in	(A)	(e,∞)
(ii)	$f(x)=\dfrac{\log x}{x}$, $x>0$ is increasing in	(B)	$\left(0,\dfrac{1}{e}\right)$
(iii)	$f(x)=\dfrac{x}{\log x}$ is increasing in	(C)	$(0,e)$
(iv)	$f(x)=x\log x$ is decreasing in	(D)	$(-\infty,\infty)$

32 Let $f(x) = \sin^{-1}\left(\frac{2x}{1+x^2}\right)$ and $g(x) = \cos^{-1}\left(\frac{1-x^2}{1+x^2}\right)$, then

	Column I		Column II
(i)	f is increasing in	(A)	$(0, \infty)$
(ii)	g is increasing in	(B)	$(-\infty, 0)$
(iii)	f is decreasing in	(C)	$(-1, 1)$
(iv)	g is decreasing in	(D)	$(1, \infty)$

33 Let $f(x) = |x-1| + |x-2| + |x-3| + |x-4|$ and $a, b, c, d \in Z^+$, then

	Column I		Column II
(i)	f is increasing in $[a, b]$, then $(a+b)$ is	(A)	3
(ii)	If f is decreasing in $[c, d]$, then $(c+d)$ is	(B)	0
(iii)	the number of points of discontinuity is	(C)	7
(iv)	the number of points of non-differentiability is	(D)	4

34 Answer the given questions by approximately matching the information given in three column of the following table. **[JEE Advanced 2017]**

Let $f(x) = x + \log x - x \log x$, $x \in (0, \infty)$

* Column I contains information about zero of $f(x)$, $f'(x)$ and $f''(x)$.

* Column II contains information about the limiting behaviour of $f(x)$, $f'(x)$ and $f''(x)$ at infinity.

* Column III contains information about increasing and decreasing nature of $f(x)$ and $f'(x)$.

Column I	Column II	Column III
(A) $f(x) = 0$ for some $x \in (1, e^2)$	(i) $\lim_{x \to 0} f(x) = 0$	(P) f is increasing in (0, 1)
(B) $f(x) = 0$ for some $x \in (1, e)$	(ii) $\lim_{x \to 0} f(x) = -\infty$	(Q) f is decreasing in (e, e^2)
(C) $f'(x) = 0$ for some $x \in (0, 1)$	(iii) $\lim_{x \to 0} f'(x) = 0$	(R) f' is increasing in (0, 1)
(D) $f''(x) = 0$ for some $x \in (1, e)$	(iv) $\lim_{x \to 0} f''(x) = 0$	(S) f' is decreasing in (e, e^2)

(i) Which of the following is the only correct combination?

(a) (A) (i) (S) (b) (A) (ii) (R) (c) (C) (iv) (P) (d) (B) (iii) (S)

(ii) Which of the following options is the only correct combination?

(a) (C) (iii) (R) (b) (A) (i) (P) (c) (D) (iv) (S) (d) (B) (ii) (Q)

(iii) Which of the following is the only incorrect combination?

(a) (B) (iii) (P) (b) (B) (iv) (Q) (c) (A) (iii) (P) (d) (C) (i) (R)

Numerical Value Type Questions

Question Stem Directions (Q.Nos. 35-37) The answer to each question is a **NUMERICAL VALUE.**

Consider the function $f(x) = x^3 - 3x + a$.

35 Values of a for which $f(x) = 0$ has one real root, are $(-\infty, -k) \cup (k, \infty)$, then k is equal to

36 Values of a for which $f(x) = 0$ has two real roots, are $\pm\alpha$, then α is equal to

37 Values of a for which $f(x) = 0$ has three real roots, are $(-k, k)$, then k is equal to

Directions (Q.Nos. 38-39)

If a, b and c are real numbers and $f(x) = \begin{vmatrix} x+a^2 & ab & ac \\ ab & x+b^2 & bc \\ ac & bc & x+c^2 \end{vmatrix}$

and $g(x) = x^3 - Ax^2 + 11x - 6$; a, b and c are in AP and roots of $g(x) = 0$ are a, b and c. Then,

38 the set of values of x such that $f'(x)$ is increasing, is $\left(\frac{-k}{3}, \infty\right)$, then k is equal to

39 the set of the values of x such that $f(x)$ is decreasing, is $\left(\frac{-\alpha}{3}, 0\right)$, then α is equal to

40 If $f(x) = \sin^2 x - 3\cos^2 x + 2ax - 4$ is increasing for all $x \geq 0$, then the least value of a is ....

41 The graph of the function $f(x) = 3x^4 + 2x^3 + ax^2 - x + 2$ is concave upward for all real x, then the least integral value of a is ....

42 The least integral value of x, where $f(x) = \log_{1/2}(x^2 - 2x - 3)$ is monotonically decreasing, is ....

43 The greatest value of m does the function $f(x) = (m+2)x^3 - 3mx^2 + 9mx - 1$ strictly decreases for all x, is ....

44 The real values of m for which the function $f(x) = \frac{x^3}{3} - (m-3)\frac{x^2}{2} + mx + 20$ is increasing in $x \in [0, \infty)$ is $[0, k]$, then the value of k is ....

45 Let $f(x) = x^3 + (2a+3)x^2 + 3(2a+1)x + 5$ is monotonic in R. The number of integral values in the set of values of a for which $f(x)$ is invertible, is ....

46 If $f(x) = \sin^{-1}\left(\frac{x}{\sqrt{x^2+1}}\right) - \ln x$ in $\left[\frac{1}{\sqrt{3}}, \sqrt{3}\right]$ and $f(x)$ is monotonically decreasing in the interval of $[a, b]$. Then, the value of $2ab$ is ....

47 Let $f(x)$ be an increasing function defined on $(0, \infty)$, if $f(2a^2 + a + 1) > f(3a^2 - 4a + 1)$, then the number of integers in the range of a is ....

48 Let $f(x) = 30 - 2x - x^3$, then the number of positive integral values of x, which satisfies $f(f(f(x))) > f(f(-x))$, is ....

49 The interval to which λ belongs so that the function $f(x) = \left(1 - \frac{\sqrt{21 - 4\lambda - \lambda^2}}{\lambda + 1}\right)x^3 + 5x + 7$ is increasing at every point of its domain is $[a, b] \cup [c, d]$, then $|a + b + c + d|$ is ....

ANSWER KEY

Level 1

1. (i) 7.0014 (ii) 4.0416 (iii) 1.3963 (iv) 0.48489
2. (i) 1.08 (ii) 0.02 (iii) 0.88 cm^2 (iv) 2%
3. 28.21
4. 0.03 $x^3 m^3$
5. (i) Strictly increasing (ii) increasing (iii) decreasing (iv) decreasing (v) decreasing (vi) increasing
6. (i) increasing (ii) increasing (iii) decreasing (iv) neither increasing nor decreasing
7. (i) Increasing in $(1, 2) \cup (3, \infty)$ and decreasing in $(-\infty, 1) \cup (2, 3)$
   (ii) Increasing in $(-\infty, -2) \cup (2, \infty)$ and decreasing in $(-2, 0) \cup (0, 2)$
   (iii) Increasing in $(-\infty, \log_2(4/3)) \cup (1, \infty)$ and decreasing in $(\log_2(4/3), 1)$
   (iv) Increasing in $\left(0, \frac{\pi}{4}\right) \cup \left(\frac{5\pi}{4}, 2\pi\right)$ and decreasing in $\left(\frac{\pi}{4}, \frac{5\pi}{4}\right)$
   (v) Increasing in $\left(\frac{1}{e}, \infty\right)$ and decreasing in $\left(0, \frac{1}{e}\right)$

Level 2

1. (a)	**2.** (c)	**3.** (b)	**4.** (c)	**5.** (b)	**6.** (b)	**7.** (d)
8. (a)	**9.** (c)	**10.** (b)	**11.** (c)	**12.** (d)	**13.** (d)	**14.** (c)
15. (c)	**16.** (c)	**17.** (a)	**18.** (c)	**19.** (a)	**20.** (a)	**21.** (d)
22. (c)	**23.** (d)	**24.** (a)	**25.** (b)	**26.** (a)	**27.** (a)	**28.** (c)
29. (b)	**30.** (b)	**31.** (b)	**32.** (d)	**33.** (a)	**34.** (b)	**35.** (d)
36. (a)	**37.** (b)	**38.** (b)	**39.** (a)	**40.** (d)	**41.** (b)	**42.** (c)
43. (b)	**44.** (c)	**45.** (c)	**46.** (b)	**47.** (a)	**48.** (a)	**49.** (c)
50. (d)	**51.** (11)	**52.** (6)	**53.** (3)	**54.** (5)	**55.** (3)	**56.** (12)
57. (2)	**58.** (4)					

Level 3

1. (b)	**2.** (d)	**3.** (b)	**4.** (a)	**5.** (b)	**6.** (d)	**7.** (b)
8. (d)	**9.** (b)	**10.** (a, b, c, d)	**11.** (a, b, c, d)	**12.** (a, b, c)	**13.** (b, c)	**14.** (a, c)
15. (a, b, d)	**16.** (b, c)	**17.** (a, b)	**18.** (a, b, c)	**19.** (b, c)	**20.** (a, c)	**21.** (a, c, d)
22. (c, d)	**23.** (a)	**24.** (b)	**25.** (a)	**26.** (d)	**27.** (d)	**28.** (a)
29. (b)	**30.** (d)					

31. (i)–D, (ii)–C, (iii)–A, (iv)–B
32. (i)–C, (ii)–A, (iii)–D, (iv)–B
33. (i)–C, (ii)–A, (iii)–B, (iv)–D
34. (i)–D, (ii)–D, (iii)–D

35. (2)	**36.** (2)	**37.** (2)	**38.** (14)	**39.** (28)	**40.** (2)	**41.** (1)
42. (4)	**43.** (–3)	**44.** (9)	**45.** (1)	**46.** (2)	**47.** (3)	**48.** (2)
49. (3)						

SOLUTIONS

Level 1

1. (i) Consider the function

$$y = f(x) = \sqrt{x}$$

Let $x = 49$ and $x + \Delta x = 49.02$

Then, $\Delta x = 0.02$

For $x = 49$, we obtain, $y = \sqrt{49} = 7$

Let $dx = \Delta x = 0.02$

Now, $y = \sqrt{x}$

$$\Rightarrow \quad \frac{dy}{dx} = \frac{1}{2\sqrt{x}}$$

$$\Rightarrow \quad \left(\frac{dy}{dx}\right)_{x=49} = \frac{1}{2\sqrt{49}} = \frac{1}{14}$$

$$\therefore \quad dy = \frac{dy}{dx} \cdot dx = \frac{1}{14}(0.02)$$

$$\Delta y = 0.0014 \qquad [\because \Delta y \approx dy]$$

Hence, $\sqrt{49.02} = 7 + 0.0014 = 7.0014$

$$[\because \sqrt{49.02} = y + \Delta y]$$

(ii) We consider a function

$$y = f(x) = x^{1/3}$$

Let $x = 64$ and $x + \Delta x = 66$

So, $\Delta x = 2$

For $x = 64$, we obtain, $y = (64)^{\frac{1}{3}} = 4$

Let $dx = \Delta x = 2$

Now, $y = x^{1/3}$

$$\Rightarrow \quad \frac{dy}{dx} = \frac{1}{3x^{\frac{2}{3}}}$$

$$\Rightarrow \quad \left(\frac{dy}{dx}\right)_{x=64} = \frac{1}{3(64)^{\frac{2}{3}}} = \frac{1}{3 \times 16} = 0.0208$$

Now, $dy = \frac{dy}{dx}\Delta x$

$$\Rightarrow \quad \Delta y = (0.0208)(2)$$

$$\Rightarrow \quad \Delta y = 0.0416$$

Hence, $(66)^{1/3} = 4 + 0.0416 = 4.0416$

(iii) Let $y = f(x) = \log_e x, x = 4$ and $\Delta x = 0.04$

Given, $y = \log_e 4 = 1.3863$

Now, $y = \log_e x$

$$\Rightarrow \quad \frac{dy}{dx} = \frac{1}{x}$$

$$\Rightarrow \quad \left(\frac{dy}{dx}\right)_{x=4} = \frac{1}{4}$$

$$\therefore \quad dy = \frac{dy}{dx} \cdot \Delta x$$

$$\Rightarrow \quad \Delta y = \frac{1}{4}(0.04)$$

$$\Rightarrow \quad \Delta y = 0.01$$

Hence, $\log 4.04 = 1.3863 + 0.01 = 1.3963$

(iv) Let $y = f(x) = \cos x, x = 60° = \left(\frac{\pi}{3}\right)^c$

and $\Delta x = 1° = 0.01745$

For $x = \frac{\pi}{3}$ rad,

$$y = \cos\frac{\pi}{3} = \frac{1}{2} = 0.5$$

Now, $y = \cos x$

$$\Rightarrow \quad \frac{dy}{dx} = -\sin x$$

$$\Rightarrow \quad \left(\frac{dy}{dx}\right)_{x=\frac{\pi}{3}} = -0.86603$$

[given, $\sin 60° = 0.86603$]

$$\therefore \quad dy = \frac{dy}{dx} \cdot dx$$

$$= (-0.86603)(0.01745)$$

$$\Delta y = -0.01511$$

Now, $\cos 61° = 0.5 + (-0.01511) = 0.48489$

2. (i) We have, $y = x^4 - 10, x = 3$ and $\Delta x = 0.01$

$$\frac{dy}{dx} = 4x^3$$

$$\Rightarrow \left(\frac{dy}{dx}\right)_{x=3} = 4(3)^3 = 108$$

Δy be the error in y, due to error in x.

$$\Delta y = \frac{dy}{dx} \cdot \Delta x = 108(0.01)$$

$$\Rightarrow \quad \Delta y = 1.08$$

(ii) We have, $y = \frac{1}{x^2}, x = 1$ and $\Delta x = -0.01$

$$\frac{dy}{dx} = \frac{-2}{x^3}$$

$$\Rightarrow \quad \left(\frac{dy}{dx}\right)_{x=1} = \frac{-2}{1} = -2$$

Now, $\Delta y = \frac{dy}{dx} \times (\Delta x)$

$$\Rightarrow \quad \Delta y = (-2)(-0.01)$$

$$\Rightarrow \quad \Delta y = 0.02$$

(iii) Let $r = 7$ cm and $\Delta r = 0.02$ cm

Now, area of circle $= \pi r^2$

$$\Rightarrow \quad A = \pi r^2$$

$$\Rightarrow \quad \frac{dA}{dr} = 2\pi r$$

$$\Rightarrow \quad \left(\frac{dA}{dr}\right)_{r=7} = 2 \times \frac{22}{7} \times 7 = 44$$

Now, $\Delta A = \frac{dA}{dr} \cdot \Delta r$

$$= (44)(0.02) = 0.88 \text{ cm}^2$$

(iv) Let x be the length of an edge of cube and S is its surface area.

So, $\quad \frac{\Delta x}{x} \times 100 = 1 \qquad$ (given)

Now, $\quad S = 6x^2$

$$\Rightarrow \quad \frac{dS}{dx} = 12x$$

$$\Rightarrow \quad \Delta S = \frac{dS}{dx} \times \Delta x$$

$$\Rightarrow \quad \frac{\Delta S}{S} = \frac{12x}{S} \times \Delta x$$

$$\Rightarrow \quad \frac{\Delta S}{S} = \frac{12x}{6x^2} \Delta x$$

$$\Rightarrow \quad \frac{\Delta S}{S} \times 100 = 2\frac{\Delta x}{x} \times 100 = 2(1)$$

$$\Rightarrow \quad \frac{\Delta S}{S} \times 100 = 2$$

So, there is 2% error in calculating the surface area of cube.

3. Let $y = f(x) = 4x^2 + 5x + 2$,

$x = 2$ and $x + \Delta x = 2.01$, then $\Delta x = 0.01$

We have, $\quad y = 4x^2 + 5x + 2$

When $x = 2$, $\quad y = 4(2)^2 + 5(2) + 2$

$\Rightarrow \quad y = 28$

Now, $\quad y = 4x^2 + 5x + 2$

$$\Rightarrow \quad \frac{dy}{dx} = 8x + 5$$

$$\Rightarrow \quad \left(\frac{dy}{dx}\right)_{x=2} = 8(2) + 5 = 21$$

$$\therefore \quad dy = \frac{dy}{dx} \cdot dx = 21(0.01)$$

$$\Rightarrow \quad \Delta y = dy = 0.21$$

Now, $\quad f(2.01) = y + \Delta y$

$$= 28 + 0.21 = 28.21$$

4. Let Δx be change in x and ΔV be the corresponding change in V.

It is given that

$$\frac{\Delta x}{x} \times 100 = 1 \Rightarrow \Delta x = \frac{x}{100}$$

We have, $\quad V = x^3 \Rightarrow \frac{dV}{dx} = 3x^2$

$$\therefore \quad \Delta V = \frac{dV}{dx} \times dx = 3x^2 \cdot \Delta x$$

$$\Rightarrow \quad \Delta V = 3x^2 \cdot \frac{x}{100} = x^3 \left(\frac{3}{100}\right)$$

$$\Rightarrow \quad \Delta V = 0.03x^3$$

Hence, the approximately changes in volume is $0.03x^3 \text{m}^3$.

5. (i) We have, $f(x) = e^x + \sin x$

$$\Rightarrow f'(x) = e^x + \cos x \begin{cases} \text{as } e^x > 1 \text{ for } R^+ \\ \text{and } -1 \le \cos x \le 1, x \in R^+ \end{cases}$$

Clearly, $f'(x) > 0, \forall\, x \in R^+$

Hence, $f(x)$ is strictly increasing function.

(ii) We have, $f(x) = x^3 - 3x^2 + 3x - 100, x \in R$

$$\therefore \quad f'(x) = 3x^2 - 6x + 3$$

$$= 3(x^2 - 2x + 1) = 3(x-1)^2 \ge 0$$

$$f'(x) \ge 0$$

Hence, $f(x)$ is increasing on R.

(iii) We have, $f(x) = \frac{1}{x} + 2$

$$\therefore \quad f'(x) = -\frac{1}{x^2}$$

Now, $x \in R - \{0\} \Rightarrow \frac{1}{x^2} > 0 \Rightarrow -\frac{1}{x^2} < 0$

$\Rightarrow f'(x) < 0$

Hence, $f(x)$ is decreasing for $x \in R - \{0\}$.

(iv) We have, $f(x) = \tan^{-1} x - x$

$$f'(x) = \frac{1}{1+x^2} - 1$$

$$= \frac{1 - (1+x^2)}{1+x^2}$$

$$= \frac{1 - 1 - x^2}{1+x^2}$$

$$= \frac{-x^2}{1+x^2} < 0$$

$$\therefore \quad f'(x) < 0$$

Hence, $f(x)$ is decreasing on R.

(v) We have, $f(x) = \cos^{-1}(x) + \log\left(x + \frac{1}{x}\right)$

$\cos^{-1} x$ is defined, if $x \in [-1, 1]$

and $\log\left(x + \frac{1}{x}\right)$ is defined, if $x > 0$.

So, domain of $f(x)$ is $(0, 1]$.

$$\text{Now,} \quad f'(x) = -\frac{1}{\sqrt{1-x^2}} + \frac{1 - \frac{1}{x^2}}{x + \frac{1}{x}}$$

For $x \in (0,1]$, $1-\frac{1}{x^2}<0$

So, $f'(x)<0$

Hence, $f(x)$ is decreasing function.

(vi) We have, $f(x)=\frac{4\sin x}{2+\cos x}-x$

$$f'(x)=\frac{(2+\cos x)(4\cos x)+4\sin^2 x}{(2+\cos x)^2}-1$$

$$=\frac{8\cos x+4}{(2+\cos x)^2}-1=\frac{4\cos x-\cos^2 x}{(2+\cos x)^2}$$

$$=\frac{\cos x(4-\cos x)}{(2+\cos x)^2}$$

$f'(x)>0, \forall x \in \left(0,\frac{\pi}{2}\right)$

Hence, $f(x)$ is increasing on $\left(0,\frac{\pi}{2}\right)$.

6. (i) $f(x)$ is increasing, as

$$f(a-h)<f(a)<f(a+h).$$

(ii) At $x=0$,

$$f(0-h)<f(0)<f(0+h)$$

Hence, $f(x)$ is increasing at $x=0$.

(iii) At $x=a$,

$$f(a-h)>f(a)>f(a+h)$$

So, $f(x)$ is decreasing.

(iv) Neither increasing nor decreasing as

$f(a-h)<f(a)$ and $f(a+h)<f(a)$.

7. (i) We have, $f(x)=x^4-8x^3+22x^2-24x+21$

$\therefore \quad f'(x)=4x^3-24x^2+44x-24$

$\Rightarrow \quad f'(x)=4(x^3-6x^2+11x-6)$

$\Rightarrow \quad f'(x)=4(x-1)(x-2)(x-3)$

Now, $f'(x)=0$, when $x=1, 2$ and 3.

Hence, critical points are 1, 2 and 3.

$\therefore$ The sign scheme for the derivative is

− + − +
1 2 3

Hence, function is increasing in $(1,2)\cup(3,\infty)$ and decreasing in $(-\infty,1)\cup(2,3)$.

(ii) We have, $\quad f(x)=\frac{x}{2}+\frac{2}{x}$

$$f'(x)=\frac{1}{2}-\frac{2}{x^2} \Rightarrow f'(x)=\frac{x^2-4}{2x^2}$$

Now, $f'(x)=0$, when $x=\pm 2$

Hence, critical points are 2 and −2.

The sign scheme for the derivative is

+ − +
−2 0 2

Hence, $f(x)$ is increasing in $(-\infty,-2)\cup(2,\infty)$ and decreasing in $(-2,0)\cup(0,2)$.

(iii) We have, $f(x)=(2^x-1)(2^x-2)^2$

$$f'(x)=2^x\log 2\,(2^x-2)^2+2(2^x-2)\,2^x\log 2\,(2^x-1)$$

$$=2^x\log 2\,(2^x-2)[(2^x-2)+2(2^x-1)]$$

$$=2^x\log 2\,(2^x-2)[3\times 2^x-4]$$

So, $2^x-2=0 \Rightarrow x=1$

and $3\cdot 2^x-4=0 \Rightarrow x=\log_2\left(\frac{4}{3}\right)$

The sign scheme for the derivative is

+ +
$\log_2\left(\frac{4}{3}\right)$ 1

Hence, $f(x)$ is increasing in $\left(-\infty,\log_2\frac{4}{3}\right)\cup(1,\infty)$

and decreasing in $\left(\log_2\frac{4}{3},1\right)$.

(iv) We have, $f(x)=\sin x+\cos x, x\in[0,2\pi]$

$$f'(x)=\cos x-\sin x$$

For critical points,

$$f'(x)=0$$

$\Rightarrow \quad \cos x-\sin x=0 \Rightarrow \tan x=1$

So, $x=\frac{\pi}{4}$ or $x=\pi+\frac{\pi}{4}$

$x=\frac{\pi}{4}$ or $x=\frac{5\pi}{4}$

The sign scheme for the derivative is

+ − +
0 π/4 5π/4 2π

Hence, $f(x)$ is increasing in $\left(0,\frac{\pi}{4}\right)\cup\left(\frac{5\pi}{4},2\pi\right)$ and

decreasing in $\left(\frac{\pi}{4},\frac{5\pi}{4}\right)$.

(v) We have, $f(x)=x^x, x>0$

$$f'(x)=x^x(1+\log x)$$

For critical points,

$$f'(x)=0$$

$\Rightarrow \quad (1+\log x)=0$

$\Rightarrow \quad \log x=-1$

$\Rightarrow \quad x=e^{-1} \Rightarrow x=\frac{1}{e}$

The sign scheme for $f'(x)$ is

− +
0 1/e

Hence, $f'(x)$ is increasing in $\left(\frac{1}{e},\infty\right)$ and decreasing in $\left(0,\frac{1}{e}\right)$.

Level II

1. (a) Given, $r = 9$ cm and $\Delta r = 0.03$ cm

$\therefore$ Volume of sphere $V = \frac{4}{3}\pi r^3$

$$\therefore \quad dV = 4\pi r^2 dr$$

$$\therefore \quad dV = 4\pi \times 81 \times 0.03$$

$$\Rightarrow \quad \Delta V = 9.72\pi \text{ cm}^3$$

2. (c) $\because$ Area of sector, $A = \frac{\theta}{360}\pi r^2$

$$\therefore \quad \frac{dA}{d\theta} = \frac{1}{360}\pi r^2$$

When $r = 30$ cm, $\frac{dA}{d\theta} = \frac{1}{360} \times \pi \times (900)$

Now, approximate error in A is

$$dA = \left(\frac{dA}{d\theta}\right) \cdot d\theta$$

$$\Rightarrow \quad dA = \frac{900}{360}\pi(1) \Rightarrow dA = 2.5\pi \text{ cm}^2$$

3. (b) $\because$ Area of equilateral triangle, $A = \frac{\sqrt{3}}{4}x^2$

$$\therefore \quad \frac{dA}{dx} = \frac{\sqrt{3}}{2}x$$

Given, percentage error in measuring side $= k\%$

$$\frac{dx}{x} \times 100 = k$$

So, $\quad dA = \frac{\sqrt{3}}{2}x \cdot dx$

$$\Rightarrow \quad \frac{dA}{A} \times 100 = \frac{\frac{\sqrt{3}}{2}x\,dx}{\frac{\sqrt{3}}{4}x^2} \times 100 = 2 \cdot \frac{dx}{x} \times 100$$

$$\Rightarrow \quad \frac{dA}{A} \times 100 = 2k$$

Hence, error in measuring $A = 2k\%$.

4. (c) Given that $\frac{r}{h} = \frac{1}{2} \Rightarrow 2r = h$

Volume of cone, $V = \frac{1}{3}\pi r^2 h = \frac{1}{3}\pi r^2(2r)$

$$V = \frac{2}{3}\pi r^3$$

On differentiating w.r.t. r, we get

$$\frac{dV}{dr} = 2\pi r^2 \Rightarrow dV = 2\pi r^2 dr$$

$$\Rightarrow \quad \frac{dV}{V} \times 100 = \frac{2\pi r^2 dr}{\frac{2}{3}\pi r^3} \times 100$$

$$= 3\frac{dr}{r} \times 100 = 3\%$$

5. (b) Given that $\frac{dh}{h} \times 100 = k$,

semi-vertical angle doesn't change.

$$\Rightarrow \quad \frac{h}{r} = t \quad \text{[constant]}$$

$$\Rightarrow \quad h = rt$$

$$\text{TSA }(A) = \pi r(l + r)$$
$$= \pi r(r\sqrt{1+t^2} + r)$$
$$= \pi r^2(1 + \sqrt{1+t^2}) \quad \text{...(i)}$$

$$\left[\because l = \sqrt{r^2+h^2} \Rightarrow l = \sqrt{r^2 + r^2t^2} \Rightarrow l = r\sqrt{1+t^2}\right.$$

$$\therefore \quad A = \pi\frac{h^2}{t^2}(1 + \sqrt{1+t^2})$$

$$\Rightarrow \quad \frac{dA}{dh} = \frac{\pi(2h)}{t^2}(1 + \sqrt{1+t^2})$$

$$\Rightarrow \quad \frac{dA}{A} = \frac{2}{h}dh \quad \text{[from Eq. (i)]}$$

$$\Rightarrow \quad \frac{dA}{A} \times 100 = 2\frac{dh}{h} \times 100$$

So, percentage error in A is $2k\%$.

6. (b) Given that $\frac{dl}{l} \times 100 = 1$

Now, $\quad T = 2\pi\sqrt{\frac{l}{g}}$

$$\Rightarrow \quad \ln T = \ln 2\pi + \frac{1}{2}\ln l - \frac{1}{2}\ln g$$

$$\Rightarrow \quad \frac{1}{T}\frac{dT}{dl} = \frac{1}{2}\frac{1}{l}$$

$$\Rightarrow \quad \frac{dT}{dl} = \frac{T}{2l} \Rightarrow \frac{dT}{T} = \frac{dl}{2l}$$

$$\Rightarrow \quad \frac{dT}{T} \times 100 = \frac{1}{2}\frac{dl}{l} \times 100$$

$$\Rightarrow \quad \frac{dT}{T} \times 100 = \frac{1}{2}(1) \quad \left[\text{given, } \frac{dl}{l} \times 100 = 1\right]$$

$$= \frac{1}{2} = 0.5$$

So, there is 0.5% error in calculating the time period T.

7. (d) Given, $\frac{dr}{r} = 0.05 \Rightarrow dr = (0.05)r$

Area of circle $= \pi r^2$

$$\Rightarrow \quad A = \pi r^2$$

$$\Rightarrow \quad \frac{dA}{dr} = 2\pi r$$

$$\Rightarrow \quad dA = 2\pi r\,dr$$

$$\Rightarrow \quad \frac{dA}{A} = \frac{2\pi r\,dr}{\pi r^2}$$

$$\therefore \quad \frac{dA}{A} = \frac{2dr}{r}$$

$$\therefore \quad \frac{dA}{A} = 2(0.05) \Rightarrow \frac{dA}{A} = 0.1$$

$\therefore$ Corresponding error in area $= 0.1\%$.

8. (a) The given relation, $PV^{1.4} = C \Rightarrow P = \dfrac{C}{V^{1.4}}$

On differentiating w.r.t. V, we get

$$\frac{dP}{dV} = \frac{-1.4C}{V^{2.4}}$$

Rearranging above and dividing both sides by P,

$$\frac{dP}{P} = \frac{-C(1.4)}{V^{2.4}} \cdot \frac{dV}{P}$$

Substituting value of P in above from given equation,

$$\frac{dP}{P} = \frac{-1.4C}{V^{2.4}} \cdot \frac{dV \cdot V^{1.4}}{C}$$

$$\Rightarrow \frac{dP}{P} = (-1.4)\frac{dV}{V}$$

$$\Rightarrow \frac{dP}{P} \times 100 = -1.4 \times \left(-\frac{1}{200} \times 100\right)$$

$$= \frac{7}{10} = 0.7\%$$

9. (c) Let r be the radius of the sphere and Δr be the error in measuring the radius, then $r = 7$ cm and $\Delta r = 0.02$ cm.

Let V be the volume of sphere.

$$\therefore \quad V = \frac{4}{3}\pi r^3$$

$$\Rightarrow \frac{dV}{dr} = 4\pi r^2$$

$$\Rightarrow \left(\frac{dV}{dr}\right)_{r=7} = 4 \cdot \frac{22}{7} \times 7 \times 7 = 616$$

So, $\Delta V = \dfrac{dV}{dr} \cdot \Delta r = 616 \times (0.02)$

$= 12.32 \text{ cm}^3$

Hence, the approximate error calculating the volume is 12.32 cm^3.

10. (b) Given, $\dfrac{1}{v} - \dfrac{1}{u} = \dfrac{2}{f}$

$$\Rightarrow \frac{-\Delta v}{v^2} + \frac{\Delta u}{u^2} = \frac{-2\Delta f}{f^2}$$

Since, given equal errors in measuring u and v

i.e. $\Delta u = \Delta v = \alpha \Rightarrow \alpha\left(\dfrac{1}{u} - \dfrac{1}{v}\right)\left(\dfrac{1}{u} + \dfrac{1}{v}\right) = \dfrac{-2\Delta f}{f^2}$

But $\dfrac{1}{v} - \dfrac{1}{u} = \dfrac{2}{f}$

So, $\dfrac{\Delta f}{f} = \alpha\left(\dfrac{1}{u} + \dfrac{1}{v}\right)$

11. (c) We have, $f\{f(x)\} = k(x^5 + x)$

$$\Rightarrow f'\{f(x)\} \cdot f'(x) = k(5x^4 + 1)$$

Thus, $f(x)$ is always increasing or decreasing, as $k(5x^4 + 1)$ is either always positive or negative.

12. (d) $f(x) = 3\sin^4 x + 10\sin^3 x + 6\sin^2 x - 3$

$\Rightarrow f'(x) = \cos x(12\sin^3 x + 30\sin^2 x + 12\sin x)$

$\Rightarrow f'(x) = 6\sin x \cos x(2\sin^2 x + 5\sin x + 2)$

$\Rightarrow f'(x) = 3\sin 2x(2\sin x + 1)(\sin x + 2)$

$f'(x) = 0$

$\sin 2x = 0$ or $2\sin x + 1 = 0$ $\quad [\because \sin x \neq -2]$

$\Rightarrow \quad x = 0$ or $x = n\pi + (-1)^n\left(-\dfrac{\pi}{6}\right)$

As, $x \in \left[\dfrac{-\pi}{6}, \dfrac{\pi}{2}\right] \Rightarrow x = 0, \dfrac{-\pi}{6}$

+	−	+	
$-\frac{\pi}{6}$	0	$\pi/2$	

So, $f(x)$ is increasing in the interval

$$x \in \left(0, \frac{\pi}{2}\right)$$

and $f(x)$ is decreasing in the interval

$$x \in \left(-\frac{\pi}{6}, 0\right)$$

13. (d) If $f(x)$ is increasing, then $f'(x)$ also increases.

If $f(x)$ is increasing, then $f'(x) > 0$.

$$\therefore \frac{d}{dx}\left(\frac{1}{f(x)}\right) = -\frac{f'(x)}{f^2(x)} < 0$$

$$\Rightarrow \frac{1}{f(x)} \text{ decreases.}$$

Now, $\dfrac{d}{dx}\left(\dfrac{f}{g}\right) = \dfrac{f'g - g'f}{g^2}$

If f and g are positive functions and $f' < 0$ and $g' > 0$, then $\dfrac{d}{dx}\left(\dfrac{f}{g}\right) < 0$.

14. (c) $f(x) = \begin{cases} \dfrac{-4}{3}x^3 + 2x^2 + 3x & , x > 0 \\ 3xe^x & , x \le 0 \end{cases}$

$$f'(x) = \begin{cases} -4x^2 + 4x + 3 & , x > 0 \\ 3e^x(x+1) & , x \le 0 \end{cases}$$

$$f'(x) = \begin{cases} -(2x+1)(2x-3) & , x > 0 \\ 3e^x(x+1) & , x \le 0 \end{cases}$$

$f'(x)<0$	$f'(x)>0$	$f'(x)>0$	$f'(x)<0$
−1	−1/2	0	3/2

$$f'(x) > 0 \Rightarrow x \in \left(-1, \frac{3}{2}\right)$$

15. (c) We have, $f(x) = kx^3 - 9x^2 + 9x + 3$

On differentiating w.r.t. 'x', we get

$$f'(x) = 3kx^2 - 18x + 9$$

For the function to be monotonically increasing,

$$f'(x) > 0$$

$$\Rightarrow \quad 3kx^2 - 18x + 9 > 0$$

$$\Rightarrow \quad kx^2 - 6x + 3 > 0$$

$$\Rightarrow \quad D = b^2 - 4ac < 0 \Rightarrow 36 - 12k < 0$$

$$\therefore \quad k > 3.$$

16. (c) Given, $f(x) = \sin x - \cos x - kx + 5$

$f(x)$ decreases for all x, if $f'(x) < 0$

$$f'(x) = \cos x + \sin x - k < 0$$

$$\therefore \quad k > \cos x + \sin x$$

We know that

$$\sin x + \cos x = \sqrt{2}\left(\frac{1}{\sqrt{2}}\cos x + \frac{1}{\sqrt{2}}\sin x\right)$$

$$= \sqrt{2}\sin\left(\frac{\pi}{4} + x\right)$$

and $-1 \le \sin\left(\frac{\pi}{4} + x\right) \le 1$

$$\Rightarrow \quad -\sqrt{2} \le \sqrt{2}\sin\left(\frac{\pi}{4} + x\right) \le \sqrt{2}$$

$\therefore$ Maximum value of $(\sin x + \cos x) = \sqrt{2}$

$$\therefore \quad k > \sqrt{2}.$$

17. (a) We have,

$$f(x) = \frac{4x^3 - 3x^2}{6} - 2\sin x + (2x - 1)\cos x$$

$$f'(x) = \frac{1}{6}(12x^2 - 6x) - 2\cos x + 2\cos x - (2x - 1)\sin x$$

$$= (2x^2 - x) - (2x - 1)\sin x$$

$$= x(2x - 1) - (2x - 1)\sin x$$

$$= (2x - 1)(x - \sin x)$$

$$f'(x) \ge 0 \text{ in } x \in \left[\frac{1}{2}, \infty\right)$$

18. (c) We have, $f(x) = \dfrac{x^2}{2 - 2\cos x}$

$$f'(x) = \frac{(2 - 2\cos x)(2x) - x^2(2\sin x)}{(2 - 2\cos x)^2}$$

$$= \frac{4x - 4x\cos x - 2x^2\sin x}{(2 - 2\cos x)^2}$$

For $0 < x < 1$,

$f'(x) > 0$

So, $f(x)$ is an increasing function

Now, $g(x) = \dfrac{x^2}{6x - 6\sin x}$

$$\Rightarrow \quad g'(x) = \frac{(6x - 6\sin x)2x - x^2(6 - 6\cos x)}{(6x - 6\sin x)^2}$$

$$g'(x) = \frac{6x^2 - 6x(2\sin x - x\cos x)}{(6x - 6\sin x)^2}$$

So, in $0 < x < 1$, $6x^2 < 6x(-2\sin x - x\cos x)$

$\therefore$ $g'(x) < 0$, so g is a decreasing function.

19. (a) Given, $P'(x) - P(x) > 0, \forall\, x \ge 1$

or $\quad e^{-x}\{P'(x) - P(x)\} > 0, \forall\, x \ge 1$

or $\quad \dfrac{d}{dx}e^{-x}P(x) > 0, \forall\, x \ge 1$

Therefore, $e^{-x}P(x)$ is an increasing function, $\forall\, x \ge 1$.

Since, $P(x)$ is a polynomial.

$$e^{-x}P(x) > e^{-1}P(1)$$

$$\Rightarrow \quad e^{-x}P(x) > 0 \qquad \{\because P(1) = 0\}$$

$$\Rightarrow \quad P(x) > 0$$

Hence, $P(x) \ge 0\, \forall\, x \ge 1$.

20. (a) We have, $f'(x) = |x| - \{x\}$

$$= |x| - (x - [x])$$

$$= |x| - x + [x]$$

For $x \in \left(-\frac{1}{2}, 0\right)$, $f'(x) = -x - x - 1 = -2x - 1$

Also, for $-\frac{1}{2} < x < 0$,

$$\Rightarrow \quad 0 < -2x < 1$$

$$\Rightarrow \quad -1 < -2x - 1 < 0$$

or $f'(x) < 0$, $f(x)$ is decreasing in $\left(-\frac{1}{2}, 0\right)$.

Similarly, we can check for other given options say for $\left(-\frac{1}{2}, 2\right)$.

$$f'(x) = \begin{cases} -2x - 1, & \frac{-1}{2} < x < 0 \\ x - x + 0, & 0 \le x < 1 \\ x - x + 1, & 1 \le x < 2 \end{cases}$$

Here, $f(x)$ decreases only in $\left(-\frac{1}{2}, 0\right)$ otherwise $f(x)$ in other intervals is constant.

21. (d) We have,

$$f(x) = \tan^{-1}(\sin x + \cos x)$$

$$\Rightarrow \quad f'(x) = \frac{1}{1 + (\sin x + \cos x)^2}(\cos x - \sin x)$$

$$\Rightarrow \quad f'(x) = \frac{\cos x}{1 + (\sin x + \cos x)^2}(1 - \tan x)$$

Clearly, $f'(x) > 0, \forall\, x \in \left(-\frac{\pi}{2}, \frac{\pi}{4}\right)$.

Hence, $f(x)$ is increasing in $\left(-\frac{\pi}{2}, \frac{\pi}{4}\right)$.

22. (c) Given, $f(x)=\frac{1}{2}[|\sin x|+\sin x],\ 0<x\le 2\pi$

As we know that

$$|\sin x|=\begin{cases}\sin x, & x\in[0,\pi)\\ -\sin x, & x\in(\pi,2\pi]\end{cases}\quad \text{...(i)}$$

Consider, $f(x)=\frac{1}{2}[|\sin x|+\sin x]$

Case I When $0<x\le\pi$

$$f(x)=\frac{1}{2}[\sin x+\sin x]\qquad\text{[from Eq. (i)]}$$

$$\Rightarrow\quad f(x)=\frac{1}{2}[2\sin x]\Rightarrow f(x)=\sin x$$

$$\therefore\quad f'(x)=\cos x$$

$$\cos x>0 \text{ for } 0<x<\frac{\pi}{2}\ \text{(increasing)}\quad\text{...(ii)}$$

$$\cos x<0 \text{ for } \frac{\pi}{2}<x<\pi\ \text{(decreasing)}\quad\text{...(iii)}$$

Case II When $\pi<x\le 2\pi$

$$f(x)=\frac{1}{2}[-\sin x+\sin x]\qquad\text{[from Eq. (i)]}$$

$$\Rightarrow\quad f(x)=0$$

Therefore, from Eqs. (ii) and (iii), we get

f is increasing in $\left(0,\frac{\pi}{2}\right)$ and decreasing in $\left(\frac{\pi}{2},\pi\right)$.

23. (d) $f(x)=|x\log x|=\begin{cases}-x\log x, & 0<x<1\\ x\log x, & x>1\end{cases}$

$$\Rightarrow f'(x)=\begin{cases}-(\log x+1), & 0<x<1\\ (\log x+1), & x>1\end{cases}$$

For $f(x)$ to be decreasing, we must have $f'(x)<0$

$$=\begin{cases}-(\log x+1)<0, & 0<x<1\\ (\log x+1)<0, & x>1\end{cases}$$

$$=\begin{cases}\log x+1>0, & 0<x<1\\ \log x+1<0, & x>1\end{cases}$$

$$=\begin{cases}x>\frac{1}{e}, & 0<x<1\\ x<\frac{1}{e}, & x>1\end{cases}$$

$$=x>\frac{1}{e},\ 0<x<1$$

$$=x\in\left(\frac{1}{e},1\right)$$

Hence, $f(x)$ is decreasing in $\left[\frac{1}{e},1\right]$.

24. (a) We have, $f(x)=3\log(x-1)-3\log(x+1)-\frac{2}{x-1}$

$$f'(x)=\frac{3}{x-1}-\frac{3}{x+1}+\frac{2}{(x-1)^2}$$

$$\Rightarrow\quad f'(x)=\frac{4(2x-1)}{(x-1)^2(x+1)}$$

Sign scheme of $f'(x)$ is

$+$ | -1 | $-$ | $1/2$ | $+$ | 1 | $+$

So, $f(x)$ is increasing, when $f'(x)\ge 0$.

$$x\in(-\infty,-1)\cup\left(\frac{1}{2},1\right)\cup(1,\infty).$$

25. (b) We have, $f(x)=\left(1+\frac{1}{x}\right)^{x+1},\ x>0$

$$=e^{(x+1)\ln\left(\frac{x+1}{x}\right)}$$

$$\Rightarrow f'(x)=\left(1+\frac{1}{x}\right)^{x+1}\left[(x+1)\left(\frac{1}{x+1}-\frac{1}{x}\right)+\ln\left(\frac{x+1}{x}\right)\right]$$

$$=\left(1+\frac{1}{x}\right)^{x+1}\left[\ln\left(\frac{x+1}{x}\right)+1-\frac{x+1}{x}\right]$$

Let $\frac{x+1}{x}=u$ for $x>0$,

$$u\in(1,\infty)$$

Consider, $h(u)=\ln u+1-u$

$$h'(u)=\frac{1}{u}-1<0 \text{ in } (1,\infty)$$

Hence, $h(u)$ is a decreasing function in $(1,\infty)$.

$$h(u)<h(1)$$

$$h(u)<0\qquad[\because h(1)=0]$$

$$\therefore\quad f'(x)<0$$

$\Rightarrow$ $f(x)$ is decreasing function.

26. (a) Let $H(x)=f(x)-x$

$$H'(x)=f'(x)-1>0\Rightarrow f'(x)>1\quad\text{...(i)}$$

Now, Let $F(x)=f(x)-x+x^3$

$$\Rightarrow\quad F'(x)=f'(x)-1+3x^2$$

From Eq. (i), $f'(x)>1$,

then $F'(x)>0$

Hence, $f(x)-x+x^3$ is increasing function.

27. (a) We have, $f(x)=x^3+bx^2+cx+d$

$$\Rightarrow\quad f'(x)=3x^2+2bx+c$$

Now, $D=(2b)^2-4(3)(c)=4(b^2-3c)$

$$=4(b^2-c)-8c<0\qquad[\because b^2-c<0 \text{ and } c>0]$$

$\Rightarrow$ $f'(x)>0\ \forall\ x\in R$

$\Rightarrow$ $f(x)$ is strictly increasing function.

28. (c) We have, $y=x-e^x+\tan\frac{\pi}{5}$

$$\Rightarrow\quad \frac{dy}{dx}=1-e^x$$

For increasing, $\frac{dy}{dx}>0$

$$\Rightarrow\quad 1-e^x>0\Rightarrow e^x<1$$

$$\Rightarrow\quad x<0\Rightarrow x\in(-\infty,0)$$

29. (b) We have, $f(x) = \left|\sin x + \frac{1}{2}\right|$ for $x \in [-2\pi, 2\pi]$

The graph of $y = f(x)$ is as shown below

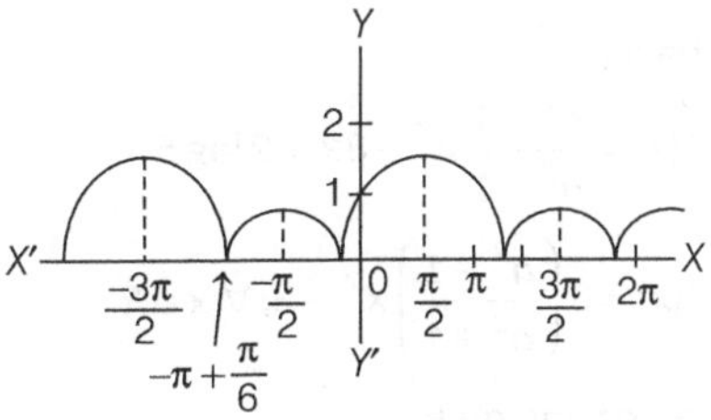

Clearly $f(x)$ is non-monotonic at

$$x = \frac{-3\pi}{2}, -\pi + \frac{\pi}{6}, -\frac{\pi}{2}, -\frac{\pi}{6}, \frac{\pi}{2}, \pi + \frac{\pi}{6}, \frac{3\pi}{2}, 2\pi - \frac{\pi}{6}$$

i.e. at $x = \frac{-3\pi}{2}, \frac{-5\pi}{6}, \frac{-\pi}{2}, \frac{-\pi}{6}, \frac{\pi}{2}, \frac{7\pi}{6}, \frac{3\pi}{2}, \frac{11\pi}{6}$

Hence, number of points of non-monotonocity is 8.

30. (b) Here, $f(x) = \sin^4 x + \cos^4 x$

$$\Rightarrow \quad f'(x) = 4\sin^3 x(\cos x) + 4\cos^3 x(-\sin x)$$

$$\Rightarrow \quad f'(x) = 4\sin x \cos x(\sin^2 x - \cos^2 x)$$

$$= 2\sin 2x(-\cos 2x)$$

$$\Rightarrow \quad f'(x) = -\sin 4x$$

Now, $f'(x) > 0$, if

$$\sin 4x < 0$$

$$\Rightarrow \quad \pi < 4x < 2\pi$$

$$\Rightarrow \quad \frac{\pi}{4} < x < \frac{\pi}{2}$$

Here, (b) is only subset of $\left(\frac{\pi}{4}, \frac{\pi}{2}\right)$

31. (b) Let $y = \frac{a\sin x + b\cos x}{c\sin x + d\cos x}$

The function is decreasing, then

$$\frac{dy}{dx} < 0$$

$$\Rightarrow \quad \frac{(c\sin x + d\cos x)(a\cos x - b\sin x) - (a\sin x + b\cos x)(c\cos x - d\sin x)}{(c\sin x + d\cos x)^2} < 0$$

$$\Rightarrow \quad \frac{ad(\sin^2 x + \cos^2 x) - bc(\sin^2 x + \cos^2 x)}{(c\sin x + d\cos x)^2} < 0$$

$$\Rightarrow \quad \frac{ad - bc}{(c\sin x + d\cos x)^2} < 0$$

So, $\quad ad - bc < 0$

32. (d) Given function

$$f(x) = \begin{cases} \frac{1}{x}\log_e(1+x), & x \neq 0 \\ 1, & x = 0 \end{cases},$$

for $x \in (-1, \infty)$

Now, $f'(x) = \frac{1}{x(1+x)} - \frac{\log_e(1+x)}{x^2}$,

for $x \in (-1, \infty) - \{0\}$

$$= \frac{x - (1+x)\log_e(1+x)}{x^2(1+x)}$$

Let another function

$$g(x) = x - (1+x)\log_e(1+x)$$

$$\because \quad g'(x) = 1 - 1 - \log_e(1+x) = -\log_e(1+x)$$

Since, for $x \in (-1, 0)$, $g'(x) > 0$.

So, $g(x)$ is increasing function for $x \in (-1, 0)$ but as $g(x) < g(0), \forall x \in (-1, 0)$

$\therefore \quad g(x) < 0, \ \forall x \in (-1, 0)$

$\therefore \quad f'(x) = \frac{g(x)}{x^2} < 0$

$\Rightarrow f(x)$ is decreasing function for $x \in (-1, 0)$.

Similarly, for $x \in (0, \infty)$, $g'(x) < 0$, so $g(x)$ is decreasing function for $x \in (0, \infty)$.

So, $\quad g(x) < g(0) \Rightarrow g(x) < 0, \forall x \in (0, \infty)$

$\therefore \ f'(x) = \frac{g(x)}{x^2} < 0 \Rightarrow f(x)$ is decreasing function for $x \in (0, \infty)$.

$\therefore$ The given function $f(x)$ is decreasing function for $(-1, \infty)$.

33. (a) We have, $f(x) = \frac{x}{\log x} - \frac{\log x}{x}$

$$\Rightarrow \quad f'(x) = \frac{\log x - \frac{1}{x} \cdot x}{(\log x)^2} - \frac{\frac{1}{x} \cdot x - \log x(1)}{x^2}$$

$$= \frac{\log x - 1}{(\log x)^2} - \frac{1 - \log x}{x^2}$$

$$\because \quad f'(x) > 0$$

$$\Rightarrow \quad \log x - 1 > 1 - \log x$$

$$\Rightarrow \quad 2\log x > 2$$

$$\Rightarrow \quad \log x > 1$$

$$\Rightarrow \quad x > e$$

$\therefore \quad f(x)$ is increasing in (e, ∞).

34. (b) We have,

$$g(x) = f(\tan^2 x - 2\tan x + 4)$$

$$\Rightarrow \quad g'(x) = f'(\tan^2 x - 2\tan x + 4)(2\tan x - 2)\sec^2 x$$

$$\Rightarrow \quad g'(x) = 2f'(\tan^2 x - 2\tan x + 4)(\tan x - 1)\sec^2 x$$

Since, $f''(x) > 0$, so $f'(x)$ is increasing

Now, $g'(x) = 2f'(\tan^2 x - 2\tan x + 4)(\tan x - 1)\sec^2 x$

$$= 2f'\{(\tan x - 1)^2 + 3\}(\tan x - 1)\sec^2 x > 0,$$

$$\forall x \text{ in } \left(\frac{\pi}{4}, \frac{\pi}{2}\right)$$

Thus, $g(x)$ is increasing in $\left(\frac{\pi}{4}, \frac{\pi}{2}\right)$.

35. (d) We have,

$$f(x)=2\tan x+(2a+1)\ln\sec x+(a-2)x+2$$

$$f'(x)=2\sec^2x+(2a+1)\frac{\tan x\sec x}{\sec x}+(a-2)$$

$$f'(x)=2\sec^2x+(2a+1)\tan x+(a-2)$$

Since, $f(x)$ is non-decreasing function on R,
then $f'(x)\geq 0, \forall\, x\in R$

$\Rightarrow\ 2\sec^2x+(2a+1)\tan x+(a-2)\geq 0, \forall\, x\in R$

$\Rightarrow\ 2\tan^2x+(2a+1)\tan x+a\geq 0, \forall\, x\in R$

$\therefore\quad (2a+1)^2-4\cdot 2\cdot a\leq 0$

$\Rightarrow\quad (2a-1)^2\leq 0$

$\Rightarrow\quad (2a-1)=0$

$\Rightarrow\quad a=\frac{1}{2}$

36. (a) We have,

$f:[1,10]\to[1,10]$ is non-decreasing function and
$g:[1,10]\to[1,10]$ is non-increasing function
So, $f(x)$ is increasing or constant and $g(x)$ is decreasing or constant.

Now, $h(x)=f\{g(x)\}\ \Rightarrow h(1)=f\{g(1)\}$

$1=f\{g(1)\}$ and $h(2)=f\{g(2)\}$

Since, $2\geq 1$

$\Rightarrow\ g(2)\leq g(1)$ as g is non-increasing function.

$\Rightarrow\ f\{g(2)\}\leq f\{g(1)\}$ as f is non-decreasing function.

So, $h(2)=f\{g(2)\}\leq f\{g(1)\}$

$h(2)\leq f\{g(1)\}$

$\Rightarrow\quad h(2)\leq 1$

37. (b) We have,

$$f(x)=x^3+\lambda x^2+5x+\sin 2x$$

$$f'(x)=3x^2+2\lambda x+5+2\cos 2x$$

For $f(x)$ to be invertible,

$$f'(x)>0, \forall\, x\in R$$

$\Rightarrow\ 3x^2+2\lambda x+5+2\cos 2x>0$

$\Rightarrow\quad 3x^2+2\lambda x+3>0$

For this discriminant of above quadratic should be negative.

So, $\quad (2\lambda)^2-4(3)(3)<0$

$\Rightarrow\quad 4\lambda^2-36<0$

$\Rightarrow\quad \lambda^2-9<0\Rightarrow\lambda\in(-3,3)$

38. (b) We have, $f(x)=x^3+ax^2+bx+5\sin^2x$

$$f'(x)=3x^2+2ax+b+5\sin 2x$$

$\because$ $f(x)$ is an increasing function.

$\therefore$ $f'(x)>0, \forall\, x\in R$

$\Rightarrow\ 3x^2+2ax+b+5\sin 2x>0$

$\Rightarrow\ 3x^2+2ax+(b-5)>0, \forall\, x\in R$

So, $(2a)^2-4(3)(b-5)<0$

$\Rightarrow\quad 4a^2-(12)(b-5)<0$

$\Rightarrow\quad a^2-3b+15<0$

39. (a) We have,

$$f(x)=\frac{a^2-1}{a^2+1}x^3-3x+2\log 5$$

$$f'(x)=3\left(\frac{a^2-1}{a^2+1}\right)x^2-3, \forall\, x\in R$$

$f(x)$ is decreasing on R.
So, $f'(x)<0, \forall\, x\in R$

$\Rightarrow\quad 3\left(\frac{a^2-1}{a^2+1}\right)x^2-3<0$

$\Rightarrow\quad \frac{a^2-1}{a^2+1}x^2-1<0, \forall\, x\in R$

$\Rightarrow\quad \frac{a^2-1}{a^2+1}\leq 0$

$\Rightarrow\quad a^2-1\leq 0\Rightarrow a\in[-1,1]$

40. (d) Since, $f(x)$ be an increasing funciton, we have

$$\log_3\{\log_{1/4}(\cos x+b)\}>0$$

$\Rightarrow\quad \{\log_{1/4}(\cos x+b)\}>1$

$\Rightarrow\quad 0<(\cos x+b)<\frac{1}{4}, \forall\, x$ in R.

$\Rightarrow\quad -\cos x<b<\frac{1}{4}-\cos x$

$\Rightarrow\quad 1<b<\frac{1}{4}-1\Rightarrow 1<b<-\frac{3}{4},$

which is not possible.

So, $\quad b=\phi$

41. (b) Let $f(x)=\log(1+x)-x$
Clearly, $f(x)$ is defined for $x>-1$

Now, $f'(x)=\frac{1}{1+x}-1=\frac{-x}{1+x}$

$\Rightarrow\ f'(x)<0$ for $x>0$ and $f'(x)>0$ for $-1<x<0$

$\Rightarrow\ f(x)$ is decreasing on $[0,\infty)$ and increasing on $(-1,0]$

$\Rightarrow\ f(x)\leq f(0), 0\leq x<\infty$ and $f(x)\leq f(0)$ for $-1<x\leq 0$.

$\Rightarrow\ f(x)\leq f(0)$ for $-1<x<\infty$.

$\Rightarrow\ \log(1+x)-x\leq 0$ for $(-1)<x<\infty$.

$\Rightarrow\ \log(1+x)\leq x$ for $x\in(-1,\infty)$

42. (c) Consider, the function

$$f(x)=2\sin x+\tan x-3x, \forall\, x\in\left[0,\frac{\pi}{2}\right)$$

$$f'(x)=2\cos x+\sec^2x-3$$

$\Rightarrow\ f''(x)=-2\sin x+2\sec^2x\tan x$

$\Rightarrow\ f''(x)=-2\sin x+\frac{2\sin x}{\cos^3x}$

$$\Rightarrow \quad f''(x) = \frac{2\sin x\,(1-\cos^3 x)}{\cos^3 x} > 0, \forall \left(0, \frac{\pi}{2}\right)$$

$$\Rightarrow \quad f'(x) \text{ is increasing on } \left(0, \frac{\pi}{2}\right).$$

$$\Rightarrow \quad f'(x) > f'(0), \forall\, x \in \left(0, \frac{\pi}{2}\right).$$

$$\Rightarrow \quad f'(x) > 0, \forall\, x \in \left(0, \frac{\pi}{2}\right).$$

$$\Rightarrow \quad f(x) \text{ is inceasing on } \left(0, \frac{\pi}{2}\right).$$

$$\Rightarrow \quad f(x) > f(0), \forall\, x \in \left(0, \frac{\pi}{2}\right).$$

$$\Rightarrow \quad 2\sin x + \tan x - 3x > 0, \forall\, x \in \left(0, \frac{\pi}{2}\right).$$

$$\Rightarrow \quad 2\sin x + \tan x > 3x, \forall\, x \in \left(0, \frac{\pi}{2}\right).$$

At $x = 0$, we have

$$f(x) = 0$$

$$\therefore \quad 2\sin x + \tan x \geq 3x, \forall\, x \in \left[0, \frac{\pi}{2}\right).$$

43. (b) Consider, $f(x) = (1+x)\log(1+x) - \tan^{-1} x, x > -1$

$$\therefore \quad f'(x) = \log(1+x) + 1 - \frac{1}{1+x^2}$$

$$\Rightarrow \quad f''(x) = \frac{1}{1+x} + \frac{2x}{(1+x^2)^2}$$

$$= \frac{(1+x)^2 + 3x^2 + x^4}{(1+x)(1+x^2)^2} > 0$$

So $f'(x)$ is increasing.

$\therefore$ For $x < 0$, $f'(x) < f'(0)$

$$\Rightarrow \quad f'(x) < 0 \qquad [\text{as } f'(0) = 0]$$

$\therefore$ $f(x)$ is decreasing function

For $x < 0$, $f(x) > f(0)$

$f(x) > 0$

$\Rightarrow$ $(1+x)\log(1+x) - \tan^{-1} x > 0$

$$\Rightarrow \quad \log(1+x) > \frac{\tan^{-1} x}{1+x}$$

For $x > 0$, $f'(x) > f'(0)$

$\Rightarrow$ $f'(x) > 0$

$\Rightarrow$ $f(x)$ is increasing function

$\Rightarrow$ $f(x) > f(0)$

$\Rightarrow$ $f(x) > 0$

Thus, $\log(1+x) > \dfrac{\tan^{-1} x}{1+x}$

Since, $f(x)$ is continuous and increasing it crosses X-axis only once at $x = 0$.

Therefore, equation has only one real root.

44. (c) Given, $\phi(x) = f(x) + f(2-x), \forall\, x \in (0,2)$

$$\Rightarrow \quad \phi'(x) = f'(x) - f'(2-x) \qquad \text{...(i)}$$

Also, we have $f''(x) > 0, \forall\, x \in (0,2)$

$\Rightarrow f'(x)$ is a strictly increasing function, $\forall\, x \in (0,2)$.

Now, for $\phi(x)$ to be increasing,

$\phi'(x) \geq 0$

$\Rightarrow$ $f'(x) - f'(2-x) \geq 0$ [using Eq. (i)]

$\Rightarrow$ $f'(x) \geq f'(2-x)$

$\Rightarrow$ $x > 2 - x$

[$\because f'$ is a strictly increasing function]

$\Rightarrow$ $2x > 2 \Rightarrow x > 1$

Thus, $\phi(x)$ is increasing on (1, 2).

Similarly, for $\phi(x)$ to be decreasing,

$\phi'(x) \leq 0$

$\Rightarrow$ $f'(x) - f'(2-x) \leq 0$ [using Eq. (i)]

$\Rightarrow$ $f'(x) \leq f'(2-x)$

$\Rightarrow$ $x < 2 - x$ [$\because f'$ is a strictly increasing function]

$\Rightarrow$ $2x < 2 \Rightarrow x < 1$

Thus, $\phi(x)$ is decreasing on (0, 1).

45. (c) We have, $f(x) = \sin(\ln x) - 2\cos(\ln x)$

$$f'(x) = \frac{\cos(\ln x) + 2\sin(\ln x)}{x}$$

$$= \frac{\sqrt{5}}{x}\left(\cos(\ln x)\frac{1}{\sqrt{5}} + \frac{2}{\sqrt{5}}\sin \ln x\right)$$

$$= \frac{\sqrt{5}}{x}\sin(\ln x + \alpha), \text{ where } \tan\alpha = \frac{1}{2}$$

For increasing, $f'(x) > 0$

$\Rightarrow$ $0 < \ln x + \alpha < \pi$

$\Rightarrow$ $-\alpha < \ln x < \pi - \alpha$

$\Rightarrow$ $n\pi - \alpha < \ln x < n\pi + \pi - \alpha$

$\Rightarrow$ $e^{(n\pi - \alpha)} < x < e^{(n\pi + \pi - \alpha)}$

Hence, $p = n\pi - \alpha$

$q = n\pi + \pi - \alpha$

So, $\cos\{n\pi - \alpha - (n\pi + \pi - \alpha)\}$

$= \cos(n\pi - \alpha - n\pi - \pi + \alpha)$

$= \cos\pi = -1$

46. (b) We have,

$$g(x) = 2f\left(\frac{x}{2}\right) + f(2-x)$$

$$\Rightarrow \quad g'(x) = 2f'\left(\frac{x}{2}\right)\cdot\frac{1}{2} + f'(2-x)(-1)$$

$$\Rightarrow \quad g'(x) = f'\left(\frac{x}{2}\right) - f'(2-x)$$

Given that, $f''(x) < 0, \forall\, x \in (0,2)$.

So, $f'(x)$ is decreasing in (0, 2), which gives arises two cases.

Case I $\frac{x}{2} > (2-x)$ and $f'(x)$ is decreasing.

$\Rightarrow \quad f'\left(\frac{x}{2}\right) < f'(2-x), \forall x > \frac{4}{3}$

or $\quad g'(x) = f'\left(\frac{x}{2}\right) - f'(2-x) < 0, \forall x \in \left(\frac{4}{3}, 2\right)$

Case II $\frac{x}{2} < (2-x)$ and $f'(x)$ is decreasing.

$f'\left(\frac{x}{2}\right) > f'(2-x), \forall x < \frac{4}{3}$

or $g'(x) = f'\left(\frac{x}{2}\right) - f'(2-x) > 0, \forall x \in \left(0, \frac{4}{3}\right)$

$\therefore$ $g(x)$ is increasing in $\left(0, \frac{4}{3}\right)$.

47. (a) The given functions are

$$f(x) = e^x - x,$$

and $\quad g(x) = x^2 - x, \forall x \in R$

Then, $h(x) = (fog)(x) = f\{g(x)\}$

Now, $h'(x) = f'\{g(x)\} \cdot g'(x)$

$$= (e^{g(x)} - 1) \cdot (2x - 1) = (e^{(x^2 - x)} - 1)(2x - 1)$$

$$= (e^{x(x-1)} - 1)(2x - 1)$$

$\because$ It is given that $h(x)$ is an increasing function, so $h'(x) \geq 0$

$\Rightarrow \quad (e^{x(x-1)} - 1)(2x - 1) \geq 0$

Case I $(2x - 1) \geq 0$ and $(e^{x(x-1)} - 1) \geq 0$

$\Rightarrow \quad x \geq \frac{1}{2}$ and $x(x-1) \geq 0$

$\Rightarrow x \in [1/2, \infty)$ and $x \in (-\infty, 0] \cup [1, \infty)$, so $x \in [1, \infty)$

Case II $(2x - 1) \leq 0$ and $[e^{x(x-1)} - 1] \leq 0$

$\Rightarrow \quad x \leq \frac{1}{2}$ and $x(x-1) \leq 0$

$\Rightarrow \quad x \in \left(-\infty, \frac{1}{2}\right]$ and $x \in [0, 1]$

So, $\quad x \in \left[0, \frac{1}{2}\right]$

From, the above cases, $x \in \left[0, \frac{1}{2}\right] \cup [1, \infty)$.

48. (a) We have, $h(x) = f(x) - f^2(x) + f^3(x)$

$$h'(x) = f'(x) - 2f(x) \cdot f'(x) + 3f^2(x) f'(x)$$

$$= f'(x)[1 - 2f(x) + 3f^2(x)]$$

$$h'(x) = f'(x)[3f^2(x) - 2f(x) + 1]$$

Here, $3f^2(x) - 2f(x) + 1$ is quadratic in $f(x)$ with $a = 3$ and $D < 0$.

So, $3f^2(x) - 2f(x) + 1 > 0$.

Hence, $h'(x)$ is positive or negative according as $f'(x)$ is positive or negative.

If $f'(x)$ is positive, then $h'(x)$ is also positive.

So, if $f(x)$ is increasing, then $h'(x)$ is also increasing function.

49. (c) $f'(x) = 2 - \frac{1}{1+x^2} - \frac{1}{\sqrt{1+x^2}}$

$$= \frac{2(1+x^2) - 1 - \sqrt{1+x^2}}{(1+x^2)}$$

$$= \frac{1 + 2x^2 - \sqrt{1+x^2}}{(1+x^2)} > 0, \forall x \in R$$

So, $f(x)$ is increasing in $(-\infty, \infty)$.

50. (d) We have, $f(x) = \int_0^x e^t (t-1)(t-2)\, dt$

$\Rightarrow \quad f'(x) = e^x (x-1)(x-2)$

For decreasing function,

$$f'(x) < 0$$

$\Rightarrow \quad e^x (x-1)(x-2) < 0 \Rightarrow (x-1)(x-2) < 0$

$\therefore \quad x \in (1, 2)$

51. (11) Let x be the number.

$$y = f(x) = x^{1/n}$$

$\Rightarrow \quad \log y = \frac{1}{n} \log x \quad$ [taking log on both sides]

$\Rightarrow \quad \frac{1}{y} dy = \frac{1}{n} \cdot \frac{1}{x} dx$

$\Rightarrow \quad \frac{\Delta y}{y} = \frac{1}{n} \cdot \frac{1}{x} dx$

$\Rightarrow \quad \frac{\Delta y}{y} \times 100 = \frac{1}{n}\left(\frac{dx}{x} \times 100\right)$

$= \frac{1}{n}$ times the percentage error in the number

Hence, 11th root of a number is approximately $\frac{1}{11}$ times the percentage error in the number.

Hence, $k = 11$.

52. (6) Volume of cylinder $= \pi r^2 h$

Since, $\quad h = r$

$\therefore \quad V = \pi h^3 \Rightarrow \frac{dV}{dh} = 3\pi h^2$

Given, % error in measuring height $= 2\%$

$\Rightarrow \quad \frac{\Delta h}{h} = \frac{2}{100}$

$\Rightarrow \quad \Delta h = \frac{2h}{100}$

Now, approximate error in V,

$$\Delta V = \left(\frac{dV}{dh}\right)\Delta h = 3\pi h^2 \cdot \left(\frac{2h}{100}\right)$$

$$\Rightarrow \quad \frac{\Delta V}{V} \times 100 = \frac{3\pi h^2 \left(\frac{2h}{100}\right)}{\pi h^3} \times 100 = 6\%.$$

53. (3) Given, $y = f(x)$ is a polynomial of degree 4.

Then, $\frac{dy}{dx} = f'(x)$ is a polynomial of degree 3.

and given, $f'(x) > 0$ for $x \in (-\infty, 1) \cup (2, 3)$

and $f'(x) < 0$ for $x \in (1, 2) \cup (3, \infty)$.

It is clear that

$$f'(x) = -(x-1)(x-2)(x-3)$$
$$= -x^3 + 6x^2 - 11x + 6$$

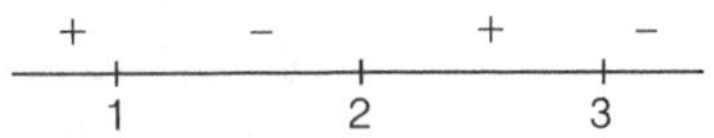

$$\therefore \quad f(x) = -\frac{x^4}{4} + 2x^3 - \frac{11x^2}{2} + 6x + \lambda \quad \text{...(i)}$$

Given, $\quad f(0) = 1$

$\Rightarrow \quad \lambda = 1$

From Eq. (i),

$$f(x) = -\frac{x^4}{4} + 2x^3 - \frac{11x^2}{2} + 6x + 1$$

$\therefore \quad f(2) = -4 + 16 - 22 + 12 + 1 = 3$

54. (5) Clearly, the number of solutions of

$3\tan x + x^3 = 2$, in $\left(0, \frac{\pi}{4}\right)$ is 1.

Thus, $m = 1$

Also, $f(x) = 3x^4 - 4x^3$

$\Rightarrow \quad f'(x) = 12x^3 - 12x^2$

$\Rightarrow \quad f''(x) = 36x^2 - 24x = 12x(3x-2)$

Now, $f''(x) = 0$, gives $x = 0, \frac{2}{3}$

So, the number of points of inflection is 2.

Thus, $n = 2$

Hence, the value of $(m + n + 2)$ is 5.

55. (3) Given, $f(x) = e^{2x} - (a+1)e^x + 2x$

$\Rightarrow \quad f'(x) = 2e^{2x} - (a+1)e^x + 2$

Since, $f(x)$ is increasing, so

$$2e^{2x} - (a+1)e^x + 2 \geq 0$$

Thus, its discriminant $D < 0$

$\Rightarrow \quad (a+1)^2 - 16 < 0$

$\Rightarrow \quad (a+1)^2 - 4^2 < 0$

$\Rightarrow (a+1+4)(a+1-4) < 0$

$\Rightarrow \quad (a+5)(a-3) < 0 \Rightarrow -5 < a < 3$

Hence, the maximum value of a is 3.

56. (12) We have, $f(x) = 2x^3 + 9x^2 + \lambda x + 20$

Since, $f(x)$ is decreasing in interval $(-2, -1)$.

Therefore, $\quad f'(x) < 0 \Rightarrow \quad 6x^2 + 18x + \lambda < 0$

The value of λ should be such that the equation $6x^2 + 18x + \lambda = 0$ has roots -2 and -1.

Therefore, $(-2)(-1) = \frac{\lambda}{6} \Rightarrow \lambda = 12$

57. (2) Let $f(x) = 2x^5 + 5x^4 + 10x^3 + 10x^2 + 10x + 10$

$$\Rightarrow \quad f'(x) = 10(x^4 + 2x^3 + 3x^2 + 2x + 1)$$
$$= 10x^2\left(x^2 + \frac{1}{x^2} + 2\left(x + \frac{1}{x}\right) + 3\right)$$
$$= 10x^2\left(\left(x + \frac{1}{x}\right)^2 + 2\left(x + \frac{1}{x}\right) + 1\right)$$
$$= 10x^2\left(x + \frac{1}{x} + 1\right)^2 > 0, \forall\, x \in R$$

$\therefore \quad f(x)$ is strictly increasing function. Since, it is an odd degree polynomial, it will have exactly one real root.

Now, by observation $f(-1) = 3 > 0$

$$f(-2) = -64 + 80 - 80 + 40 - 20 + 10 = -34 < 0$$

$\Rightarrow \quad f(x)$ has atleast one root in $(-2, -1) = (a, a+1)$

$$a = -2$$

$\Rightarrow \quad |a| = 2$

58. (4) Given, $f(x) = \begin{cases} xe^{ax} & , x \leq 0 \\ x + ax^2 - x^3 & , x > 0 \end{cases}$

Clearly, f is continuous at $x = 0$

$$\Rightarrow \quad f'(x) = \begin{cases} (ax+1)e^{ax} & , x < 0 \\ 1 & , x = 0 \\ 1 + 2ax - 3x^2 & , x > 0 \end{cases}$$

$$\Rightarrow \quad f''(x) = \begin{cases} a(ax+2)e^{ax} & , x < 0 \\ 2a & , x = 0 \\ 2a - 6x & , x > 0 \end{cases}$$

Now, for $x < 0, f''(x) > 0$ [$\because$ given a is a positive constant]

$\Rightarrow \quad ax + 2 > 0 \Rightarrow x > -\frac{2}{a}$

$\Rightarrow \quad f'(x)$ increases on $\left[-\frac{2}{a}, 0\right]$

Also, $x > 0, f''(x) > 0$

$\Rightarrow \quad 2a - 6x > 0 \Rightarrow x < \frac{a}{3}$

Thus, $f'(x)$ increases on $\left[0, \frac{a}{3}\right]$.

Hence, $f'(x)$ increases on $\left[-\frac{2}{a}, \frac{a}{3}\right]$.

$$\therefore \quad p = \frac{-2}{a}, q = \frac{a}{3}$$

$$-6pq = -6\left(\frac{-2}{a}\right)\left(\frac{a}{3}\right) = 4$$

Level III

1. (b) We have,

$$l + \Delta l = 0.992 l \text{ and } g + \Delta g = 1.002 g$$

$$\Rightarrow \quad \Delta l = -0.008 l \text{ and } \Delta g = 0.002 g$$

$$\Rightarrow \quad \frac{\Delta l}{l} = -0.008 \text{ and } \frac{\Delta g}{g} = 0.002$$

Now, $T = 2\pi\sqrt{\dfrac{l}{g}}$

$$\Rightarrow \log T = \log 2\pi + \frac{1}{2}(\log l - \log g)$$

$$\Rightarrow \quad \frac{dT}{T} = \frac{1}{2}\left(\frac{dl}{l} - \frac{dg}{g}\right)$$

$$\Rightarrow \quad \frac{\Delta T}{T} = \frac{1}{2}\left(\frac{\Delta l}{l} - \frac{\Delta g}{g}\right) \quad [\because dT \approx \Delta t, dl \approx \Delta l \text{ and } dg \approx \Delta g]$$

$$\therefore \quad \frac{\Delta T}{T} = \frac{1}{2}(-0.008 - 0.002) = -0.005$$

2. (d) Given, $a = \sqrt{c+1} - \sqrt{c} = \dfrac{1}{\sqrt{c+1} + \sqrt{c}}$

and $\quad b = \sqrt{c} - \sqrt{c-1} = \dfrac{1}{\sqrt{c} + \sqrt{c-1}}$

Clearly, $a < b$

Also, $f'(x) = \dfrac{1}{1+x} > 0, \forall x \in (-1, \infty)$

$\Rightarrow f$ is increasing function.

Thus, $a < b \Rightarrow f(a) < f(b)$.

3. (b) We have, $f(x) = \dfrac{ae^x + be^{-x}}{ce^x + de^{-x}}$

$$\Rightarrow \quad f'(x) = \frac{(ce^x + de^{-x})(ae^x - be^{-x}) - (ae^x + be^{-x})(ce^x - de^{-x})}{(ce^x + de^{-x})^2}$$

$$\Rightarrow \quad f'(x) = \frac{(ace^{2x} - bde^{-2x} - bc + ad) - (ace^{2x} - bde^{-2x} + bc - ad)}{(ce^x + de^{-x})^2}$$

$$\Rightarrow \quad f'(x) = \frac{2(ad - bc)}{(ce^x + de^{-x})^2}$$

$\Rightarrow f(x)$ is increasing, if $f'(x) \geq 0$

$$\Rightarrow \quad ad - bc \geq 0$$

$$\Rightarrow \quad ad \geq bc$$

4. (a) Given,

$$f(x) = \{-c^2 + (b-1)c - 2\}x + \int (\sin^2 x + \cos^4 x)\, dx$$

$$\Rightarrow \quad f'(x) = -c^2 + (b-1)c - 2 + \sin^2 x + \cos^4 x$$

$$\Rightarrow \quad f'(x) = (1 - \sin^2 x)^2 + \sin^2 x - c^2 + (b-1)c - 2$$

$$\Rightarrow \quad f'(x) = \sin^4 x - \sin^2 x - c^2 + (b-1)c - 1$$

$$= \left(\sin^2 x - \frac{1}{2}\right)^2 - c^2 + (b-1)c - \frac{5}{4}$$

Given, $f'(x) > 0, \forall x \in R$

$$\therefore \quad \left(\sin^2 x - \frac{1}{2}\right)^2 - c^2 + (b-1)c - \frac{5}{4} > 0$$

$$0 - c^2 + (b-1)c - \frac{5}{4} > 0$$

$$\left[\because \text{ least value of } \left(\sin^2 x - \frac{1}{2}\right) = 0\right]$$

$$\Rightarrow \quad c^2 - (b-1)c + \frac{5}{4} < 0 \quad [\because D < 0]$$

$$\Rightarrow \quad (b-1)^2 - 4 \cdot 1 \cdot \frac{5}{4} < 0$$

$$\Rightarrow \quad (b-1)^2 < 5$$

$$\Rightarrow \quad -\sqrt{5} < b - 1 < \sqrt{5}$$

$$\Rightarrow \quad 1 - \sqrt{5} < b < 1 + \sqrt{5}$$

Hence, $b \in (1 - \sqrt{5}, 1 + \sqrt{5})$

5. (b) (a) Let $f_1(x) = e^x - 1 - x$, then

$$f_1'(x) = e^x - 1 > 0 \text{ for } x > 0$$

So, $f_1(x)$ increases in (0, 1), therefore

$$0 < x$$

$$\Rightarrow \quad f_1(0) < f_1(x)$$

$$\Rightarrow \quad 0 < f_1(x)$$

$$\Rightarrow \quad 0 < e^x - 1 - x$$

$$\Rightarrow \quad e^x > 1 + x$$

(b) Let $f_2(x) = \log(1+x) - x$

Then, $f_2'(x) = \dfrac{1}{1+x} - 1 = \dfrac{-x}{1+x} < 0$ for $x > 0$

So, $f_2(x)$ is decreasing for $x > 0$, therefore

$$0 < x$$

$$\Rightarrow \quad f_2(0) > f_2(x)$$

$$\Rightarrow \quad 0 > f_2(x)$$

$$\Rightarrow \quad 0 > \log(1+x) - x$$

$$\Rightarrow \quad \log(1+x) < x$$

$$\Rightarrow \quad x > 0$$

(c) Let $f_3(x) = \sin x - x$

Then, $f_3'(x) = \cos x - 1 < 0$

So, $f_3(x)$ is decreasing for $x > 0$, therefore

$$x > 0$$

$$\Rightarrow \quad f_3(x) < f_3(0)$$

$$\Rightarrow \quad \sin x - x < 0$$

$$\Rightarrow \quad \sin x < x$$

(d) Let $f_4(x) = \log x - x$

$$f_4'(x) = \frac{1}{x} - 1 = \frac{1-x}{x} > 0 \quad [\because 0 < x < 1]$$

Therefore, $f_4(x)$ is increasing in (0, 1), so

$$0<x<1$$

$$\Rightarrow \quad \lim_{x\to 0^+} \log x = -\infty$$

So, $$\lim_{x\to 0^+} f(x) = -\infty$$

Then, $f_4(x)<0$ in (0, 1)

$$\Rightarrow \quad \log x - x < 0$$

$$\Rightarrow \quad \log x < x$$

6. (d) We have,

$$f(x) = (ab-a^2-2)\,x - \int_0^x (\cos^4 t + \sin^2 t - 2)\,dt$$

$$f'(x) = (ab-a^2-2) - (\cos^4 x + \sin^2 x - 2)$$

For $f(x)$ to be dcreasing,

$$f'(x)<0, \forall\, x\in R$$

$$\Rightarrow (ab-a^2-2)-(\cos^4 x+\sin^2 x-2)<0, \forall\, x\in R \text{ and } a\in R$$

$$\Rightarrow (ab-a^2-2)-(\cos^4 x-\cos^2 x-1)<0, \forall\, x\in R \text{ and } a\in R$$

$$\Rightarrow (ab-a^2-2)<\left(\cos^2 x-\frac{1}{2}\right)^2-\frac{5}{4}, \forall\, x\in R \text{ and } a\in R$$

$$\Rightarrow \left(ab-a^2-\frac{3}{4}\right)<\left(\cos^2 x-\frac{1}{2}\right)^2, \forall\, x\in R \text{ and } a\in R$$

$$\Rightarrow \quad ab-a^2-\frac{3}{4}<0, \forall\, a\in R$$

$$\Rightarrow 4a^2-4ab+3>0$$

$$\Rightarrow \quad 16b^2-48<0$$

$$\Rightarrow \quad b^2-3<0$$

$$\Rightarrow \quad b\in(-\sqrt{3},\sqrt{3})$$

7. (b) We have, $f(x)=(1-x)^2\sin^2 x+x^2$

and $$g(x)=\int_1^x \left\{\frac{2(t-1)}{t+1}-\log t\right\} f(t)\,dt$$

$$g'(x)=\left\{\frac{2(x-1)}{x+1}-\log x\right\} f(x)$$

Let $h(x)=\left\{\dfrac{2(x-1)}{x+1}-\log x\right\}=\left\{2-\dfrac{4}{x+1}-\log x\right\}$

Then, $h'(x)=\dfrac{4}{(x+1)^2}-\dfrac{1}{x}=\dfrac{(x-1)^2}{(x+1)^2 x}<0, \forall\, x>1$

So, $h(x)$ is decreasing, $\forall\, x>1$.

$$h(x)<h(1), \forall\, x>1$$

$$\Rightarrow \quad h(x)<0, \forall\, x>1$$

$$\Rightarrow \quad h(x)\,f(x)<0, \forall\, x>1$$

$$[\because f(x)=(1-x^2)\sin^2 x+x^2>0, \forall\, x\in(1,\infty)]$$

$$\Rightarrow \quad g'(x)<0, \forall\, x>1$$

$\Rightarrow g$ is decreasing in $(1,\infty)$.

8. (d) We have, $f(x)=a^{\{a^{|x|}\operatorname{sgn} x\}}$

and $g(x)=a^{[a^{|x|}\operatorname{sgn} x]}$, $a>1, x\in R$

$$(\ln a)\,h(x)=\ln(f(x))+\ln(g(x))$$

$$\Rightarrow \quad h(x)=\frac{1}{\ln a}[\ln(f(x)\,g(x))]$$

$$=\frac{1}{\ln a}\ln\left(a^{\{a^{|x|}\operatorname{sgn} x\}+[a^{|x|}\operatorname{sgn} x]}\right)$$

$$=\frac{1}{\ln a}\ln\left(a^{a^{|x|}\operatorname{sgn} x}\right)=\frac{1}{\ln a}a^{|x|}\operatorname{sgn} x\ln a$$

$$=a^{|x|}\operatorname{sgn} x$$

$$\Rightarrow \quad h(x)=\begin{cases}-a^{(-x)}, & x<0\\ 0, & x=0\\ a^x, & x>0\end{cases}$$

$\therefore \quad h(-x)=-a^{-(-x)}$ for $x>0=-a^x=-h(x)$

and $\quad h(-x)=a^{-x}$ for $x<0=-(-a^{-x})=-h(x)$

Thus, $h(x)$ is an odd function.

Also, $h'(x)=\begin{cases}a^{-x}\ln a & \text{for } x<0\\ a^x\ln a & \text{for } x>0\end{cases}$

$\Rightarrow h'(x)\geq 0, \forall\, x\in R \Rightarrow h(x)$ is increasing.

9. (b) We have, $f''(x)>0$

$\Rightarrow f'(x)$ is an increasing function.

Now, $g(x)=\dfrac{1}{4}f(2x^2-1)+\dfrac{1}{2}f(1-x^2)$

$$g'(x)=\frac{1}{4}f'(2x^2-1)(4x)+\frac{1}{2}f'(1-x^2)(-2x)$$

$g(x)$ is increasing.

So, $g'(x)>0 \Rightarrow x\,\{f'(2x^2-1)-f'(1-x^2)\}>0$

Case I If $x>0$

$$f'(2x^2-1)-f'(1-x^2)>0$$

$$\Rightarrow \quad f'(2x^2-1)>f'(1-x^2)\Rightarrow 2x^2-1>1-x^2$$

$$\Rightarrow \quad x\in\left(-\infty,-\sqrt{\frac{2}{3}}\right)\cup\left(\sqrt{\frac{2}{3}},\infty\right)$$

So, $x\in\left(\sqrt{\dfrac{2}{3}},\infty\right)$ $\quad [\because x>0]$

Case II If $x<0$

$$\Rightarrow f'(2x^2-1)<f'(1-x^2)\Rightarrow 2x^2-1<1-x^2$$

$$\Rightarrow \quad x\in\left(-\sqrt{\frac{2}{3}},\sqrt{\frac{2}{3}}\right)$$

So, $$x\in\left(-\sqrt{\frac{2}{3}},0\right)$$

$\therefore$ g is increasing in $\left(-\sqrt{\dfrac{2}{3}},0\right)\cup\left(\sqrt{\dfrac{2}{3}},\infty\right)$.

10. (a, b, c, d) We have, $f(x)=x-[x]=\{x\}$

$\Rightarrow f(x)$ is increasing in $(k,k+1)$ for each $k\in Z$ also in $[k,k+1)$.

11. (a, b, c, d) We have,

$$f(x) = x - \cot^{-1} x + \log(\sqrt{x^2+1} - x)$$

$$f'(x) = 1 + \frac{1}{\sqrt{1+x^2}} + \frac{1}{(\sqrt{x^2+1} - x)} \times \left[\frac{x}{\sqrt{x^2+1}} - 1\right]$$

$$= 1 + \frac{1}{\sqrt{1+x^2}} + \left(-\frac{1}{\sqrt{1+x^2}}\right) = 1 > 0$$

$\Rightarrow f(x)$ is increasing function, $\forall x \in (-\infty, \infty)$

12. (a, b, c) We have, $f(x) = \text{Max}\{1 + \sin x, 1 - \cos x, 1\}$

The graph of $f(x)$ is as shown below

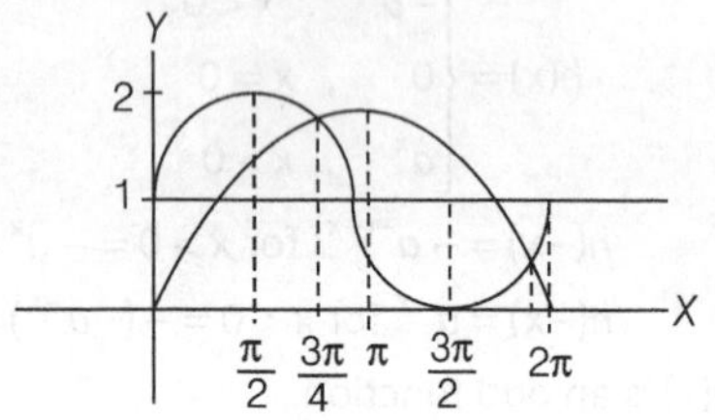

Clearly, $f(x)$ is non-monotonic at $x = \frac{\pi}{2}, \frac{3\pi}{4}$ and π.

13. (b, c) Given, $x = \frac{1}{1+t^2}$ and $y = \frac{1}{t(1+t^2)}, \forall t > 0$

$$x = (1+t^2)^{-1};\ y = (t+t^3)^{-1}$$

$$\frac{dx}{dt} = \frac{-1}{(1+t^2)^2}(2t);\ \frac{dy}{dt} = \frac{-1}{(t+t^3)^2}(1+3t^2)$$

$$\Rightarrow \frac{dy}{dx} = \frac{(1+3t^2)}{t^2(2t)} = \frac{1+3t^2}{2t^3} > 0 \text{ as } t > 0$$

$\Rightarrow f$ is increasing in $(0, \infty)$

14. (a, c) We have, $f(x) = \sin^{-1}(2x\sqrt{1-x^2})$

$$\Rightarrow 2\sin^{-1} x = \begin{cases} \sin^{-1}(2x\sqrt{1-x^2}), & -\frac{1}{\sqrt{2}} \le x \le \frac{1}{\sqrt{2}} \\ -\pi - \sin^{-1}(2x\sqrt{1-x^2}), & -1 \le x \le \frac{-1}{\sqrt{2}} \\ \pi - \sin^{-1}(2x\sqrt{1-x^2}), & \frac{1}{\sqrt{2}} \le x \le 1 \end{cases}$$

$$\Rightarrow f(x) = \begin{cases} 2\sin^{-1} x, & \frac{-1}{\sqrt{2}} \le x \le \frac{1}{\sqrt{2}} \\ -\pi - 2\sin^{-1} x, & -1 \le x \le \frac{-1}{\sqrt{2}} \\ \pi - 2\sin^{-1} x, & \frac{1}{\sqrt{2}} \le x \le 1 \end{cases}$$

$$\Rightarrow f'(x) = \begin{cases} \frac{2}{\sqrt{1-x^2}}, & \frac{-1}{\sqrt{2}} < x < \frac{1}{\sqrt{2}} \\ \frac{-2}{\sqrt{1-x^2}}, & x \in \left(-1, -\frac{1}{\sqrt{2}}\right) \cup \left(\frac{1}{\sqrt{2}}, 1\right) \end{cases}$$

Hence, $f(x)$ is decreasing in $\left(-1, \frac{-1}{\sqrt{2}}\right) \cup \left(\frac{1}{\sqrt{2}}, 1\right)$.

15. (a, b, d) Let $f(x) = x^{1/x}, \forall x > 0$.

$$\therefore \quad f'(x) = \frac{x^{1/x}(1 - \log x)}{x^2}$$

$\Rightarrow f'(x) > 0$, if $1 - \ln x > 0 \Rightarrow \ln x < 1 \Rightarrow x < e$

or $f(x)$ is increasing in $(0, e)$ and $f'(x) < 0$, if $1 - \ln x < 0$

$\Rightarrow \ln x > 1 \Rightarrow x > e$

or $f(x)$ is decreasing in (e, ∞).

(a) $\pi > 3 > e \Rightarrow f(\pi) < f(3) \Rightarrow \pi^{\frac{1}{\pi}} < 3^{\frac{1}{3}}$ [decreasing function]

$\Rightarrow \pi^3 < 3^\pi$ or $3^\pi > \pi^3$

(b) $e < 101 < 202 \Rightarrow f(101) > f(202)$ [decreasing function]

$\Rightarrow (101)^{\frac{1}{101}} > (202)^{\frac{1}{202}}$

$\Rightarrow (101)^{202} > (202)^{101}$

(c) $\because \frac{4}{3} < \frac{9}{4} < e$

$\Rightarrow f\left(\frac{4}{3}\right) < f\left(\frac{9}{4}\right)$ [increasing function]

$\Rightarrow \left(\frac{4}{3}\right)^{3/4} < \left(\frac{9}{4}\right)^{4/9}$

$\Rightarrow \left(\frac{4}{3}\right)^{9/4} < \left(\frac{9}{4}\right)^{4/3}$

(d) $\because e > 1 + \sin\frac{\pi}{3} > 1 + \cos\frac{\pi}{3}$ [increasing function]

$\Rightarrow f\left(1 + \sin\frac{\pi}{3}\right) > f\left(1 + \cos\frac{\pi}{3}\right)$

$\Rightarrow \left(1 + \sin\frac{\pi}{3}\right)^{\frac{1}{1+\sin\frac{\pi}{3}}} > \left(1 + \cos\frac{\pi}{3}\right)^{\frac{1}{1+\cos\frac{\pi}{3}}}$

$\Rightarrow \left(1 + \sin\frac{\pi}{3}\right)^{1+\cos\frac{\pi}{3}} > \left(1 + \cos\frac{\pi}{3}\right)^{1+\sin\frac{\pi}{3}}$

16. (b, c) We have, $f(x) = 2\log(x-2) - x^2 + 4x + 1$

$$\Rightarrow f'(x) = \frac{2}{x-2} - 2x + 4$$

$$= \frac{-2x^2 + 8x - 6}{x-2}$$

$$= \frac{-2(x^2 - 4x + 3)}{x-2}$$

$$= \frac{-2(x-1)(x-3)}{x-2}$$

$\because f(x)$ is increasing.

$\Rightarrow f'(x) > 0 \Rightarrow \frac{(x-1)(x-3)}{x-2} < 0$

$\Rightarrow x \in (-\infty, 1) \cup (2, 3)$ but D_f (Domain of f) $= (2, \infty)$

$\therefore f(x)$ increases in $(2, 3)$.

17. (a, b) The graph of $f(x)=|x^2-3x-4|$ is

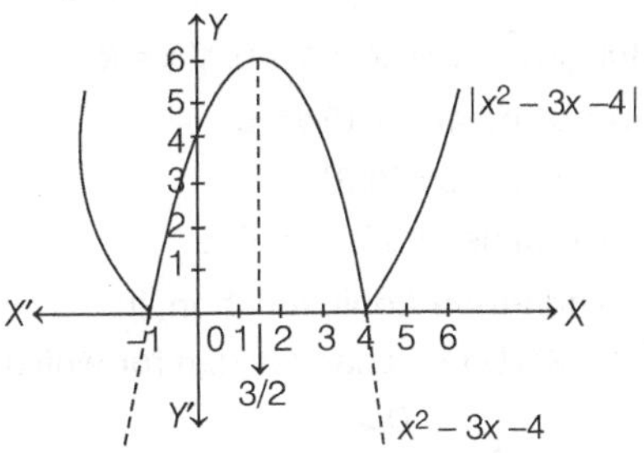

From graph,

$f(x)$ is increasing in $\left[-1,\dfrac{3}{2}\right]$ and decreasing in $\left[\dfrac{3}{2},4\right]$.

Minimum value of $f(x)$ is 0.

18. (a, b, c) If $f(x)$ and $g(x)$ are two positive and increasing function.

Now, let $y=[f(x)]^{g(x)}$

$$\therefore\ \frac{dy}{dx}=[f(x)]^{g(x)}\left[g(x)\frac{f'(x)}{f(x)}+g'(x)\log f(x)\right]$$

$[f(x)]^{g(x)}, g(x), f(x), f'(x)$ and $g'(x)$ are positive but $\log f(x)$ can be negative, which can cause $\dfrac{dy}{dx}<0$.

Hence, statement (a) is false.

If $f(x)<1$, then $\log f(x)<0$, which does not necessarily make $\dfrac{dy}{dx}<0$.

Hence, statement (b) is false.

$f(x)<0$ can also cause $\dfrac{dy}{dx}>0$

Hence, statement (c) is false.

But converse of (c) is true.

19. (b, c) We have, $g'(x)>0$ and $f'(x)<0$

So, $g(x)$ is increasing and $f(x)$ is decreasing function.

$$g(x+1)>g(x-1) \text{ and } f(x+1)<f(x-1)$$

or $\quad f\{g(x+1)\}<f\{g(x-1)\}$

and $g\{f(x+1)\}<g\{f(x-1)\}$

20. (a, c) Given that $f'x)>2f(x), \forall\, x\in R$

$\Rightarrow \quad f'(x)-2f(x)>0,\ \forall\, x\in R$

$\therefore\ e^{-2x}[f'(x)-2f(x)]>0, \forall\, x\in R$

$\Rightarrow \quad \dfrac{d}{dx}[e^{-2x} f(x)]>0, \forall\, x\in R$

Let $g(x)=e^{-2x} f(x)$

Now, $g'(x)>0, \forall\, x\in R$

i.e. $g(x)$ is strictly increasing, $\forall\, x\in R$.

Also, $g(0)=1 \Rightarrow g(x)>g(0)=1$

$\therefore\ e^{-2x} f(x)>1, \forall\, x\in(0,\infty) \Rightarrow f(x)>e^{2x}, \forall\, x\in(0,\infty)$

So, option (a) is correct.

As, $f'(x)>2f(x)>2e^{2x}>2, \forall\, x\in(0,\infty)$

i.e. $f(x)$ is strictly increasing in $x\in(0,\infty)$.

21. (a, c, d) Given, $f(x)=\displaystyle\int_{1/x}^{x}\frac{e^{-\left(t+\frac{1}{t}\right)}}{t}\,dt$

$$f'(x)=e^{-\left(x+\frac{1}{x}\right)}\cdot\frac{1}{x}-e^{-\left(\frac{1}{x}+x\right)}\left(-\frac{1}{x^2}\right)\cdot x$$

$$=\frac{2e^{-\left(x+\frac{1}{x}\right)}}{x}>0, \forall\, x>1$$

Thus, f is monotonic increasing on $[1,\infty)$.

$$\text{Also, } f(x)+f\left(\frac{1}{x}\right)=\int_{1/x}^{1} e^{-\left(t+\frac{1}{t}\right)}\frac{dt}{t}+\int_{x}^{1/x} e^{-\left(t+\frac{1}{t}\right)}\frac{dt}{t}$$

$$=\int_{1/x}^{x} e^{-\left(t+\frac{1}{t}\right)}\frac{dt}{t}-\int_{1/x}^{x} e^{-\left(t+\frac{1}{t}\right)}\frac{dt}{t}$$

$$=0$$

$$\text{Let } g(x)=f(2^x)=\int_{2^{-x}}^{2x}\left(\frac{e^{\left(-t+\frac{1}{t}\right)}}{t}\,dt\right)$$

$$g(-x)=f(2^{-x})=\int_{2^x}^{2^{-x}}\left(\frac{e^{-t+\frac{1}{t}}}{t}\,dt\right)=-g(x)$$

Thus, $g(x)$ is an odd function.

So, $f(2^x)$ is an odd function.

22. (c, d) Let $f(x)=\begin{cases}\dfrac{x}{\sin x}, & 0<x\le 1\\ 0, & x=0\end{cases}$

Clearly, f is continuous on $[0,1]$ and differentiable in $(0,1)$.

Now, $\quad f'(x)=\dfrac{\sin x-x\cos x}{\sin^2 x}=\dfrac{\cos x\,(\tan x-x)}{\sin^2 x}$

$\Rightarrow f'(x)>0$, for $0<x<1$

Thus, $f(x)$ increases on $[0,1]$.

Hence, $f(x)=\dfrac{\sin x}{x}$ increases on $0<x\le 1$.

Again, let $g(x)=\begin{cases}\dfrac{x}{\tan x}, & 0<x\le 1\\ 0, & x=0\end{cases}$

Clearly, g is continuous on $[0,1]$ and differentiable in $(0,1)$.

Now, $g'(x)=\dfrac{\tan x-x\sec^2 x}{\tan^2 x}$

Again, let $h(x)=\tan x-x\sec^2 x$

So, h is continuous on $[0,1]$ and differentiable in $(0,1)$.

$$h'(x)=\sec^2 x-\sec^2 x-x\sec^2 x\tan x$$

$$h'(x)=-x\sec^2 x\tan x<0 \text{ for } 0<x<1$$

h is a decreasing function on $[0,1]$.

Thus, $h(x) < h(0)$

$\Rightarrow \quad h(x) < 0 \Rightarrow g'(x) < 0$

Thus, $g(x)$ is a decreasing function on $[0, 1]$.

Therefore, $g(x) = \dfrac{x}{\tan x}$ is a decreasing function on $0 < x \le 1$.

Sol. (Q.Nos. 23-24)

23. (a) **24.** (b)

Given, $Q(x) = 2f\left(\dfrac{x^2}{2}\right) + f(6 - x^2)$

$\therefore \quad Q'(x) = 2xf'\left(\dfrac{x^2}{2}\right) - 2xf'(6 - x^2)$

$$= 2x\left\{f'\left(\frac{x^2}{2}\right) - f'(6 - x^2)\right\}$$

But given that $f''(x) > 0$

$\Rightarrow f'(x)$ is increasing for all $x \in R$.

Case I Let $\dfrac{x^2}{2} > 6 - x^2 \Rightarrow x^2 > 4$

$\therefore \quad x \in (-\infty, -2) \cup (2, \infty)$

$\Rightarrow \quad f\left(\dfrac{x^2}{2}\right) > f(6 - x^2) \Rightarrow f\left(\dfrac{x^2}{2}\right) - f(6 - x^2) > 0$

If $x > 0$, then $Q'(x) > 0 \Rightarrow x \in (2, \infty)$

and if $x < 0$, then $Q'(x) < 0 \Rightarrow x \in (-\infty, -2)$

Case II Let $\dfrac{x^2}{2} < (6 - x^2) \Rightarrow x^2 < 4$

$\therefore \quad x \in (-2, 2)$

$\Rightarrow \quad f'\left(\dfrac{x^2}{2}\right) < f'(6 - x^2)$

$\Rightarrow \quad f'\left(\dfrac{x^2}{2}\right) - f'(6 - x^2) < 0$

If $x > 0$, then $Q'(x) > 0 \Rightarrow x \in (0, 2)$

and $x < 0$, then $Q'(x) > 0 \Rightarrow x \in (-2, 0)$

Combining both cases $Q(x)$ is increasing in $x \in (-2, 0) \cup (2, \infty)$ and $Q(x)$ is decreasing in $x \in (-\infty, -2) \cup (0, 2)$.

Sol. (Q.Nos. 25-27)

25. (a) We have,

$$p(x) = f(x) - a[f(x)]^2 + a[f(x)]^3$$

$\Rightarrow \quad p'(x) = f'(x) - 2af(x)f'(x) + 3a\{f(x)\}^2 \cdot f'(x)$

$$= f'(x)[3a\{f(x)\}^2 - 2af(x) + 1]$$

Now, $p(x)$ increases, if $f(x)$ increases and

$3a[f(x)]^2 - 2af(x) + 1 > 0, \forall x \in R$

or $\quad 3a > 0$ and $4a^2 - 12a \le 0$

or $\quad a > 0$ and $a \in [0, 3]$

26. (d) $p(x)$ increases as $f(x)$ decreases for all real values of x, if

$3a[f(x)]^2 - 2af(x) + 1 \le 0, \forall x \in R$

or $\quad 3a < 0$ and $4a^2 - 12a \le 0$

or $\quad a < 0$ and $a \in [0, 3]$

So, no such value of a is possible.

27. (d) $p(x)$ is non-monotonic function, if $3a[f(x)]^2 - 2af(x) + 1$ changes sign for which

$$D > 0$$

$\Rightarrow \quad 4a^2 - 12a > 0 \Rightarrow 4a(a - 3) > 0$

$\Rightarrow \quad a \in (-\infty, 0) \cup (3, \infty)$.

Sol. (Q.Nos. 28-30)

We have, $f'(\sin x) < 0, \forall x \in \left(0, \dfrac{\pi}{2}\right)$ ...(i)

$\Rightarrow f'\left(\sin\left(\dfrac{\pi}{2} - x\right)\right) < 0, \forall x \in \left(0, \dfrac{\pi}{2}\right)$

$\Rightarrow \quad f'(\cos x) < 0, \forall x \in \left(0, \dfrac{\pi}{2}\right)$ ...(ii)

Similarly, $f''(\sin x) > 0, \forall x \in \left(0, \dfrac{\pi}{2}\right)$ ...(iii)

$\Rightarrow \quad f''(\cos x) > 0, \forall x \in \left(0, \dfrac{\pi}{2}\right)$ ...(iv)

Now, $g'(x) = f'(\sin x) \cdot \cos x - f'(\cos x) \cdot \sin x$

$$g''(x) = f'(\sin x)\cos^2 x - f'(\sin x)\sin x$$
$$- f''(\cos x)(-\sin^2 x) - f'(\cos x)\cos x$$

$\Rightarrow \quad g''(x) > 0, \forall x \in \left(0, \dfrac{\pi}{2}\right)$

[from Eqs. (i), (ii), (iii) and (iv)]

$\Rightarrow g'(x)$ is increasing in $x \in \left(0, \dfrac{\pi}{2}\right)$ with $g'(x) = 0 \Rightarrow x = \dfrac{\pi}{4}$

$\Rightarrow g'(x) < 0, \forall x \in \left(0, \dfrac{\pi}{4}\right)$

$\Rightarrow g(x)$ is decreasing in $\left(0, \dfrac{\pi}{4}\right)$

and $g'(x) > 0$ for $x \in \left(\dfrac{\pi}{4}, \dfrac{\pi}{2}\right)$

$\Rightarrow g(x)$ is increasing function for $x \in \left(\dfrac{\pi}{4}, \dfrac{\pi}{2}\right)$.

Hence,

28. (a) $g(x)$ is decreasing, when $x \in \left(0, \dfrac{\pi}{4}\right)$

29. (b) $g(x)$ increases, if $x \in \left(\dfrac{\pi}{4}, \dfrac{\pi}{2}\right)$

30. (d) Set of critical point of $g(x)$ is $\dfrac{\pi}{2}$.

31. (i) → (D), (ii) → (C), (iii) → (A), (iv) → (B)

(i) $f(x) = \log(x + \sqrt{x^2+1})$

$$\Rightarrow f'(x) = \frac{1}{x+\sqrt{x^2+1}}\left(1 + \frac{1}{2\sqrt{x^2+1}} \times 2x\right)$$

$$\Rightarrow f'(x) = \frac{1}{x+\sqrt{x^2+1}}\left(\frac{x+\sqrt{x^2+1}}{\sqrt{x^2+1}}\right)$$

$$\Rightarrow f'(x) = \frac{1}{\sqrt{x^2+1}} > 0, \forall x \in R$$

Hence, $f(x)$ is increasing on $(-\infty, \infty)$.

(ii) $f(x) = \frac{\log x}{x}$

$$\Rightarrow f'(x) = \frac{1-\log x}{x^2}$$

Now, $f'(x) = 0$

$$\Rightarrow \frac{1-\log x}{x^2} = 0 \Rightarrow x = e$$

Sign scheme for $f'(x)$ for $x > 0$ is

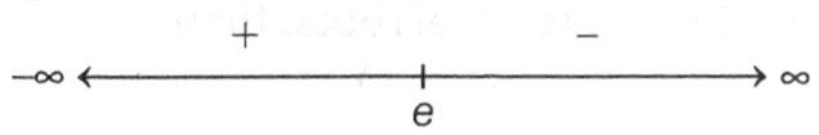

So, $f(x)$ is increasing on $(0, e)$.

(iii) $f(x) = \frac{x}{\log x}$

$$\Rightarrow f'(x) = \frac{\log x\,(1) - x\cdot\frac{1}{x}}{(\log x)^2} \Rightarrow f'(x) = \frac{\log x - 1}{(\log x)^2}$$

Now, $f'(x) = 0 \Rightarrow x = e$

Now, sign scheme is

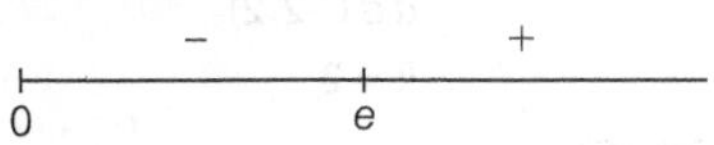

$f(x)$ is increasing on (e, ∞).

(iv) $f(x) = x\log x$

$$f'(x) = x\cdot\frac{1}{x} + \log x\,(1)$$

$$f'(x) = 1 + \log x$$

Now, $f'(x) = 0 \Rightarrow 1 + \log x = 0 \Rightarrow x = \frac{1}{e}$

Sign scheme for $f'(x)$ is

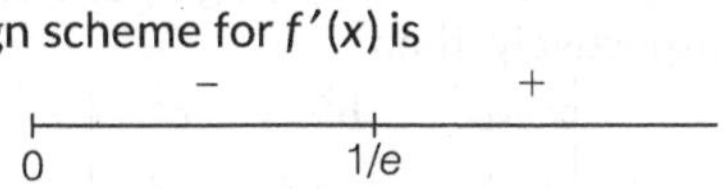

$f(x)$ is decreasing on $\left(0, \frac{1}{e}\right)$.

32. (i) → (C), (ii) → (A), (iii) → (D), (iv) → (B)

We have,

$$f(x) = \sin^{-1}\left(\frac{2x}{1+x^2}\right)$$

The graph of $f(x)$ is

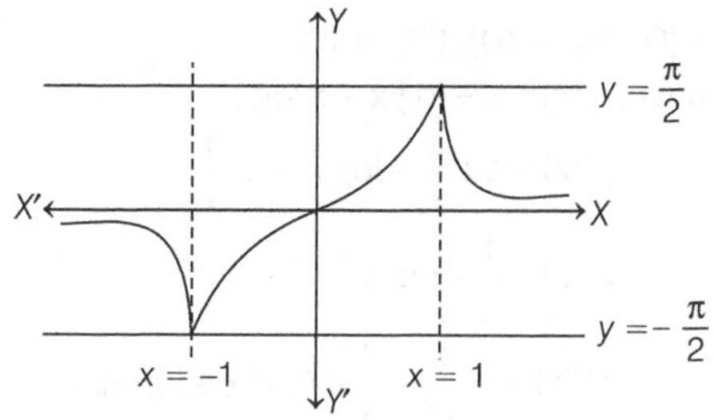

Clearly, f is increasing in $(-1, 1)$ and f is decreasing in $(-\infty, -1) \cup (1, \infty)$.

Now, graph of $g(x) = \cos^{-1}\left(\frac{1-x^2}{1+x^2}\right)$

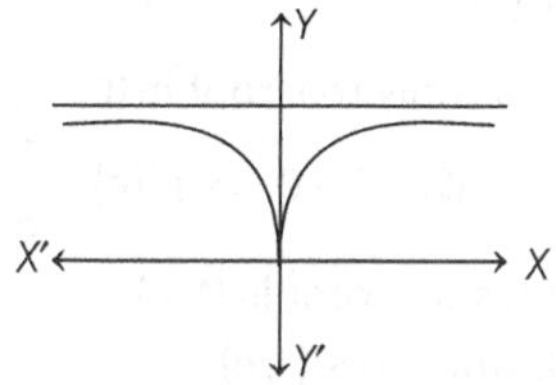

Clearly, $g(x)$ is increasing in $(0, \infty)$ and decreasing in $(-\infty, 0)$.

33. (i) → (C), (ii) → (A), (iii) → (B), (iv) → (D)

We have,

$$f(x) = |x-1| + |x-2| + |x-3| + |x-4|$$

$$f(x) = \begin{cases} -4x+10, & x<1 \\ -2x+8, & 1 \le x < 2 \\ 4, & 2 \le x < 3 \\ 2x-2, & 3 \le x < 4 \\ 4x-10, & x \ge 4 \end{cases}$$

The graph of $f(x)$ is

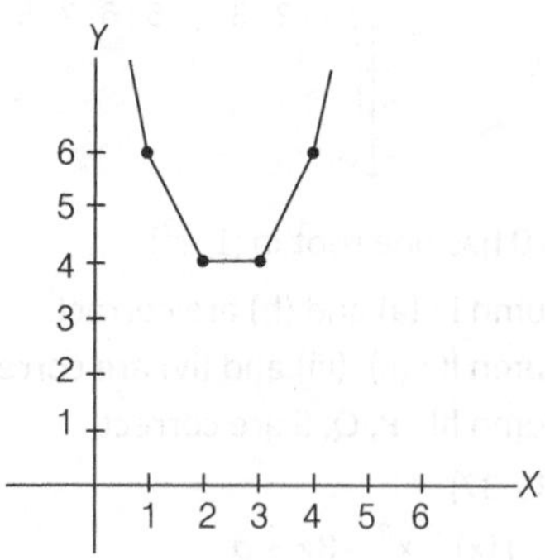

Now,

(i) $f(x)$ is increasing on $[3, 4]$

So, $a + b = 3 + 4 = 7$

(ii) f is decreasing on $[1, 2]$

So, $c + d = 1 + 2 = 3$

(iii) The f is continuous, $\forall x \in R$.

So, the number of points of discontinuity = 0

(iv) The nubmer of points of non-differentiability = 4

34. (i) → (D), (ii) → (D), (iii) → (D)

We have, $f(x) = x + \log x - x \log x$

$$f'(x) = 1 + \frac{1}{x} - \log x - x \cdot \frac{1}{x}$$

$$f'(x) = \frac{1}{x} - \log x$$

$$f''(x) = -\frac{1}{x^2} - \frac{1}{x} < 0, \forall\, x \in (0, \infty)$$

Thus, $f'(x)$ is strictly decreasing function for $x \in (0, \infty)$.

$$\lim_{x \to \infty} \left(\frac{1}{x} - \log x\right) = \lim_{x \to 0} f'(x) = -\infty$$

and $$\lim_{x \to 0^+} \left(\frac{1}{x} - \log x\right) = \lim_{x \to 0^+} f'(x) = \infty$$

i.e. $f'(x) = 0$ has only real root in $(0, \infty)$.

$$f'(1) = 1 > 0 \Rightarrow f'(e) = \frac{1}{e} - 1 < 0$$

So, $f'(x) = 0$ has one root in $(1, e)$.

Let $f'(\alpha) = 0$, where $\alpha \in (1, e)$.

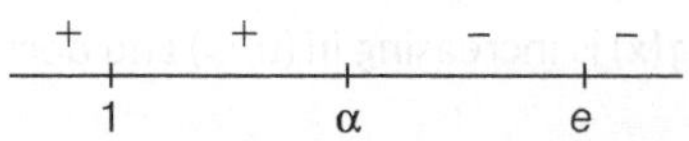

Therefore, $f(x)$ is increasing in $(0, \infty)$ and decreasing in (α, ∞).

$f(1) = 1$ and $f(e^2) = e^2 + 2 - 2e^2 = 2 - e^2 < 0$

The graph of the function is as shown in the given figure.

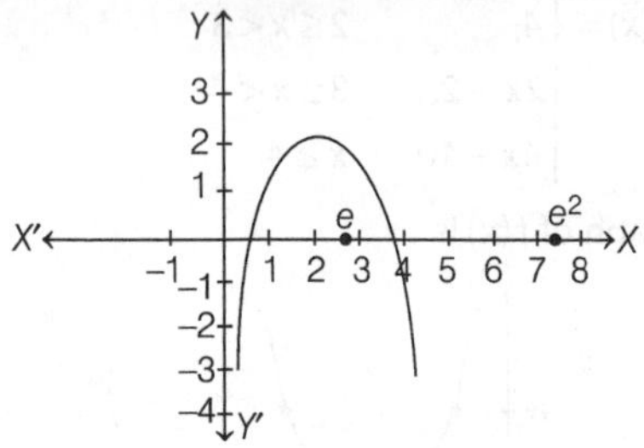

So, $f(x) = 0$ has one root in $(1, e^2)$.

From column I : (a) and (b) are correct.

From column II : (ii), (iii) and (iv) are correct.

From column III : P, Q, S are correct.

Sol. (Q.Nos. 35-37)

We have, $f(x) = x^3 - 3x + a$

$$f'(x) = 3x^2 - 3$$

$$= 3(x^2 - 1) = 3(x + 1)(x - 1)$$

35. (2) $f(x) = 0$ has one real root, then

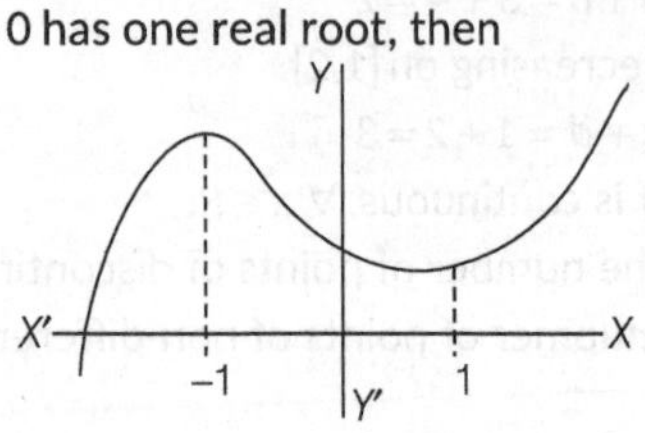

$\therefore \quad f(-1) \cdot f(1) > 0$

$\Rightarrow \quad (a + 2)(a - 2) > 0$

$\Rightarrow \quad a < -2$ or $a > 2$

$\Rightarrow \quad a \in (-\infty, -2) \cup (2, \infty)$

$\Rightarrow \quad k = 2$

36. (2) $f(x) = 0$ has two real roots, then

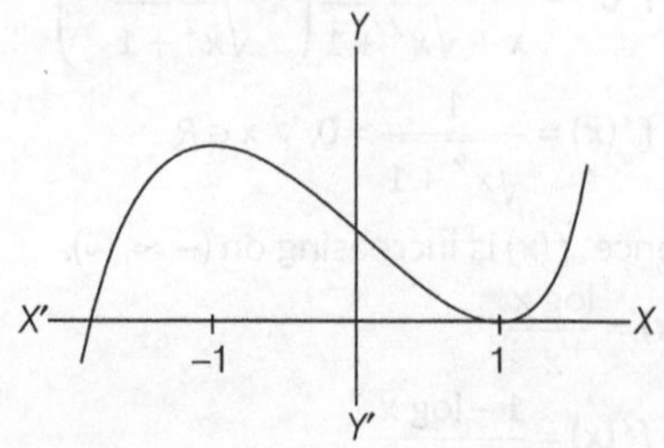

$\therefore \quad f(-1) \cdot f(1) = 0$

$\Rightarrow \quad (a - 2)(a + 2) = 0$

$\Rightarrow \quad a = \pm 2$

$\Rightarrow \quad \alpha = 2$

37. (2) $f(x) = 0$ has three real roots, then

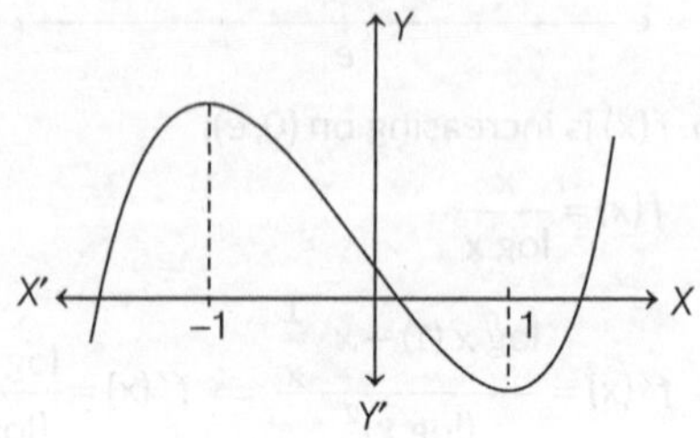

$\therefore \quad f(-1) f(1) < 0$

$\Rightarrow \quad (a - 2)(a + 2) < 0$

$\Rightarrow \quad a \in (-2, 2)$

$\Rightarrow \quad k = 2$

Sol. (Q.Nos. 38-39)

38. (14) **39.** (28)

We have, $f(x) = \begin{vmatrix} x + a^2 & ab & ac \\ ab & x + b^2 & bc \\ ac & bc & x + c^2 \end{vmatrix}$

Taking a, b and c common from R_1, R_2 and R_3, respectively and then multiplying by a, b and c in C_1, C_2 and C_3 respectively, then

$$f(x) = \begin{vmatrix} x + a^2 & b^2 & c^2 \\ a^2 & x + b^2 & c^2 \\ a^2 & b^2 & x + c^2 \end{vmatrix}$$

Applying $C_1 \to C_1 + C_2 + C_3$, then

$$f(x) = \begin{vmatrix} x + a^2 + b^2 + c^2 & b^2 & c^2 \\ x + a^2 + b^2 + c^2 & x + b^2 & c^2 \\ x + a^2 + b^2 + c^2 & b^2 & x + c^2 \end{vmatrix}$$

Applying $R_2 \to R_2 - R_1$ and $R_3 \to R_3 - R_1$, then

$$f(x) = (x + a^2 + b^2 + c^2)\begin{vmatrix} 1 & b^2 & c^2 \\ 0 & x & 0 \\ 0 & 0 & x \end{vmatrix}$$

$$f(x) = x^2 (x + a^2 + b^2 + c^2)$$
$$= x^3 + (a^2 + b^2 + c^2) x^2$$

Then, $f'(x) = 3x^2 + 2(a^2 + b^2 + c^2) x$ ...(i)

$f''(x) = 6x + 2(a^2 + b^2 + c^2)$ ...(ii)

Now, a, b and c are the roots of

$$x^3 - Ax^2 + 11x - 6 = 0 \quad ...(iii)$$

$$a + b + c = A$$

$\Rightarrow \quad 2b + b = A \quad \left[\because a, b \text{ and } c \text{ are in AP} \; 2b = a + c\right]$

$\Rightarrow \quad b = \dfrac{A}{3}$ ...(iv)

From Eqs. (iii) and (iv),

$$\left(\frac{A}{3}\right)^3 - A\left(\frac{A}{3}\right)^2 + 11\left(\frac{A}{3}\right) - 6 = 0$$

$\Rightarrow \quad 2A^3 - 99A + 162 = 0$

$\Rightarrow \quad (A - 6)(2A^2 + 12A - 27) = 0$

$\therefore \quad A = 6$ and $2A^2 + 122A - 27 \neq 0$ $\quad [\because A > 3\sqrt{6}]$

From Eq. (iii), $x^3 - 6x^2 + 11x - 6 = 0$

$\Rightarrow \quad (x - 1)(x - 2)(x - 3)$

$\therefore \quad x = 1, 2$ and 3

i.e. $a = 1, b = 2$ and $c = 3$

From Eqs. (i) and (ii), we get

$$f'(x) = 3x^2 + 28x$$

and $f''(x) = 6x + 28$

Now, $f'(x)$ increases if $f''(x) > 0$

$$6x + 28 > 0 \Rightarrow x > -\frac{14}{3}$$

$\therefore \quad x \in \left(-\dfrac{14}{3}, \infty\right)$

$\Rightarrow k = 14$

and $f(x)$ decreases, if $f'(x) < 0$

$\Rightarrow \quad 3x^2 + 28x < 0 \Rightarrow x(3x + 28) < 0$

$\therefore \quad x \in \left(-\dfrac{28}{3}, 0\right) \Rightarrow \alpha = 28$

40. (2) We have,

$$f(x) = \sin^2 x - 3\cos^2 x + 2ax - 4$$
$$f'(x) = \sin 2x + 3\sin 2x + 2a$$
$$f'(x) = 4\sin 2x + 2a$$

Since, $f(x)$ is increasing for all $x \geq 0$.

So, $\quad 4\sin 2x + 2a \geq 0$

$\Rightarrow \quad 2a \geq -4\sin 2x \Rightarrow a \geq -2\sin 2x$

$\Rightarrow \quad a \geq 2$

Hence, least value of a is 2.

41. (1) We have, $f(x) = 3x^4 + 2x^3 + ax^2 - x + 2$

$$f'(x) = 12x^3 + 6x^2 + 2ax - 1$$

and $\quad f''(x) = 36x^2 + 12x + 2a$

Now, $f(x)$ is concave upward for all $x \in R$.

$\therefore \quad f''(x) > 0$

$\Rightarrow \quad 36x^2 + 12x + 2a > 0$

$\therefore \quad D < 0$

$\Rightarrow \quad (12)^2 - 4(36)(2a) < 0$

$\Rightarrow \quad 1 - 2a < 0$

$\Rightarrow \quad a > \dfrac{1}{2} \Rightarrow a \in \left(\dfrac{1}{2}, \infty\right)$

Least integral value of a is 1.

42. (4) We have, $f(x) = \log_{1/2}(x^2 - 2x - 3)$

$\Rightarrow \quad x^2 - 2x - 3 > 0$

$\Rightarrow \quad (x - 3)(x + 1) > 0$

$\Rightarrow \quad x \in (-\infty, -1) \cup (3, \infty)$ ...(i)

Now, $f(x) = \log_{1/2}(x^2 - 2x - 3)$

$$= \frac{\log_e (x^2 - 2x - 3)}{\log_e \frac{1}{2}}$$

$\Rightarrow \quad f'(x) = \dfrac{1}{\log_e \frac{1}{2}} \cdot \dfrac{2x - 2}{(x^2 - 2x - 3)}$

$\because f(x)$ is monotonically decreasing.

$\therefore \quad f'(x) < 0$

$\Rightarrow \quad \dfrac{1}{\log_e \frac{1}{2}} \cdot \dfrac{2x - 2}{(x^2 - 2x - 3)} < 0$

$\Rightarrow \quad \dfrac{2(x - 1)}{(x - 3)(x + 1)} > 0 \quad \left[\because \log_e \dfrac{1}{2} < 0\right]$

– + – +
—|—|—|—
–1 1 3

$x \in (-1, 1) \cup (3, \infty)$ ...(ii)

From Eqs. (i) and (ii), $\quad x \in (3, \infty)$

Hence, least integral value is 4.

43. (–3) We have, $f(x) = (m + 2)x^3 - 3mx^2 + 9mx - 1$

$\Rightarrow \quad f'(x) = 3(m + 2)x^2 - 6mx + 9m$

$\therefore \quad f(x)$ decreases, $\forall x \in R$

$\Rightarrow (m + 2) < 0, 36m^2 - 108m(m + 2) \leq 0$

$\Rightarrow \quad m < -2, m^2 - 3m(m + 2) \leq 0$

$\Rightarrow \quad m < -2, -2m^2 - 6m \leq 0$

$\Rightarrow \quad m < -2, m^2 + 3m \geq 0$

$\Rightarrow \quad m < -2, m \in (-\infty, -3] \cup [0, \infty)$

$\Rightarrow \quad m \in (-\infty, -3]$

$\therefore$ Greatest value of m is -3.

44. (9) We have,

$$f(x)=\frac{x^3}{3}-(m-3)\frac{x^2}{2}+mx+20$$

$\therefore \quad f'(x)=x^2-(m-3)x+m\geq 0, \forall x\in[0,\infty)$

Case I

When $D\leq 0$

$\Rightarrow \quad (m-3)^2-4m\leq 0$

$\Rightarrow \quad m^2-6m+9-4m\leq 0$

$\Rightarrow \quad m^2-10m+9\leq 0$

$\Rightarrow \quad (m-1)(m-9)\leq 0$

$\Rightarrow \quad m\in[1,9]$ ...(i)

Case II

When $D\geq 0$

$m\in(-\infty,1]\cup[9,\infty)$ ...(ii)

and $\quad \frac{-b}{2a}\leq 0$

$\Rightarrow \quad (m-3)\leq 0$ ...(iii)

$\Rightarrow \quad m\leq 3$

and $\quad f'(0)\geq 0$

$\Rightarrow \quad m\geq 0$ ...(iv)

From Eqs. (ii), (iii) and (iv),

$m\in[0,1]$ ...(v)

From Eqs. (i) and (v),

$m\in[0,9]\equiv[0,k]$

$\therefore \quad k=9$

45. (1) We have,

$$f(x)=x^3+(2a+3)x^2+3(2a+1)x+5$$

$\Rightarrow f'(x)=3x^2+2(2a+3)x+3(2a+1)$

For $f(x)$ to be monotonic on R,

$$f'(x)\geq 0, \forall x\in R$$

$$4(2a+3)^2-36(2a+1)\leq 0$$

$\Rightarrow \quad 4a^2-6a\leq 0$

$\Rightarrow \quad 2a(2a-3)\leq 0$

$\Rightarrow \quad a\in\left[0,\frac{3}{2}\right]$

Also, $f'(x)\geq 0, \forall x\in R$ and $a\in\left[0,\frac{3}{2}\right]$

$\Rightarrow f(x)$ is invertible for $a\in\left[0,\frac{3}{2}\right]$

Hence, $a\in\left[0,\frac{3}{2}\right]$

The number of integral values on the set of a is 1.

46. (2) We have,

$$f(x)=\sin^{-1}\left(\frac{x}{\sqrt{x^2+1}}\right)-\ln x$$

$$f'(x)=\frac{1}{\sqrt{1-\left(\frac{x}{\sqrt{x^2+1}}\right)^2}}\times\frac{d}{dx}\left(\frac{x}{\sqrt{x^2+1}}\right)-\frac{1}{x}$$

$$=\sqrt{x^2+1}\left(\frac{\sqrt{x^2+1}-\frac{x}{\sqrt{x^2+1}}\cdot x}{(x^2+1)}\right)-\frac{1}{x}$$

$$=\frac{1}{x^2+1}-\frac{1}{x}$$

$$=\frac{x-(x^2+1)}{x(x^2+1)}=-\frac{x^2-x+1}{x(x^2+1)}$$

$$=-\left(\frac{\left(x-\frac{1}{2}\right)^2+\frac{3}{4}}{x(x^2+1)}\right)<0 \quad \left[\because x\in\left[\frac{1}{\sqrt{3}},\sqrt{3}\right]\right]$$

$\Rightarrow f(x)$ is monotonically decreasing function in the interval $\left[\frac{1}{\sqrt{3}},\sqrt{3}\right]$.

So, $2ab=2\cdot\frac{1}{\sqrt{3}}\cdot\sqrt{3}=2$

47. (3) Since, f is defined on $(0,\infty)$, $2a^2+a+1>0$, which true as $D<0$.

Also, $3a^2-4a+1>0$

$(3a-1)(a-1)>0$

i.e. $\quad a>\frac{1}{3}$ or $a>1$ ...(i)

As, f is increasing

$$f(2a^2+a+1)>f(3a^2-4a+1)$$

$\Rightarrow \quad 2a^2+a+1>3a^2-4a+1$

or $\quad 0>a^2-5a$

$\Rightarrow \quad a(a-5)<0$

$\Rightarrow \quad a\in(0,5)$

From Eqs. (i) and (ii),

$$a\in\left(0,\frac{1}{3}\right)\cup(1,5)$$

Therefore, the possible integers are {2, 3, 4}

Then, the number of integers in range of a is 3.

48. (2) We have,

$$f(x)=30-2x-x^3$$

$$f'(x)=-2-3x^2<0$$

$\Rightarrow f(x)$ is decreasing function.

Now, $\quad f(f(f(x)))>f(f(-x))$

$\Rightarrow \quad f(f(x))<f(-x)$

$\Rightarrow \quad f(x)>-x$

$\Rightarrow \quad 30-2x-x^3>-x$

$\Rightarrow \quad x^3+x-30<0$

$\Rightarrow \quad (x-3)(x^2+3x+10)<0$

$\Rightarrow \quad x<3$

Hence, number of positive integer values of x is 2.

49. (3) We have,

$$f(x)=\left(1-\frac{\sqrt{21-4\lambda-\lambda^2}}{\lambda+1}\right)x^3+5x+7$$

$$f'(x)=3\left(1-\frac{\sqrt{21-4\lambda-\lambda^2}}{\lambda+1}\right)x^2+5\geq 0$$

$$\Rightarrow \quad 0-4(3)(5)\left(1-\frac{\sqrt{21-4\lambda-\lambda^2}}{\lambda+1}\right)\leq 0 \quad [\because D\leq 0]$$

$$\Rightarrow \quad \left(1-\frac{\sqrt{21-4\lambda-\lambda^2}}{\lambda+1}\right)\geq 0$$

It is trivially true for $\lambda+1<0$

i.e. $\lambda\in(-\infty,-1)$ and $\lambda^2+4\lambda-21\leq 0$

$\Rightarrow \lambda\in(-\infty,-1)$ and $\lambda\in[-7,3]$

$\Rightarrow \lambda\in[-7,-1]$

For $\lambda+1>0$ and $\lambda\in[-7,3]$

i.e. $\lambda\in[-1,3]$

$$1\geq\frac{\sqrt{21-4\lambda-\lambda^2}}{\lambda+1}$$

$\Rightarrow \quad \lambda+1\geq\sqrt{21-4\lambda-\lambda^2}$

$\Rightarrow \quad \lambda^2+2\lambda+1\geq 21-4\lambda-\lambda^2$

$\Rightarrow \quad \lambda^2+3\lambda-10\geq 0$

$\Rightarrow \quad (\lambda+5)(\lambda-2)\geq 0$

$\Rightarrow \quad \lambda\in(-\infty,-5]\cup[2,\infty)$

But $\quad \lambda\in(-1,3]$

$\Rightarrow \quad \lambda\in[2,3]$

$\therefore \quad f(x)$ increases for

$$\lambda\in[-7,-1]\cup[2,3]=[a,b]\cup[c,d]$$

$\Rightarrow \quad a=-7, b=-1, c=2$ and $d=3$

$\therefore \quad |a+b+c+d|=|-7-1+2+3|=3$

CHAPTER

07

TANGENT AND NORMAL

LEVEL 01 \ Starter Level

7.1 Slope of Tangent and Normal

1 (i) Find the slope of the tangent to the curve $y = \frac{x-1}{x-2}, x \neq 2$ at $x = 10$.

(ii) Find the slope of tangent and normal to the curve $x = a\cos^3\theta$ and $y = a\sin^3\theta$ at $\theta = \frac{\pi}{4}$.

(iii) Find the slope of tangent and normal to the curve $y = |x^2 - |x||$ at $x = -3$.

(iv) Find the point on the curve $y = 2x^2 - 6x - 4$ at which the tangent is parallel to the X-axis.

(v) Find the point on the curve $2y^2 = x$, where tangent makes 45° angle with the positive direction of X-axis.

(vi) Find the point on curve $y = (x-3)^2$, where the tangent is parallel to the joining (4, 1) and (3, 0).

7.2 Equations of Tangent and Normal

2 (i) Find the equations of the tangent and normal to the curve $y = 2x^2 + 3\sin x$ at $x = 0$.

(ii) Find the equation of the tangent line to the curve $x = 1 - \cos\theta$ and $y = \theta - \sin\theta$ at $\theta = \frac{\pi}{4}$.

(iii) Find the equation of tangent line to $y = 2x^2 + 7$, which is parallel to the line $4x - y + 3 = 0$.

(iv) Find the equation of the normal to the curve $y = \cos\frac{\pi x}{2}$ at (1, 1).

(v) Find the equation of the normal to the curve $x = t^2$ and $y = 2at$ at $t = 1$.

(vi) Find the equation of normal at (a, a) to the curve $x^2y^3 = a^5$.

(vii) Find the equation of normal to the curve $x + y = x^y$, where it cuts X-axis.

(viii) Find the equation of all possible normals to the parabola $x^2 = 4y$ drawn from (1, 2).

7.3 Angle of Intersection of Two Curves

3 (i) Find the angle of intersection between the curves $y = x$ and $y^2 = 4x$ at (4, 4).

(ii) Find the angle of intersection between the curves $xy = 2$ and $y^2 = 4x$.

(iii) Find the angle of intersection of the curves $y^2 = x$ and $x^2 = y$ at (0, 0).

(iv) Find the angle of intersection of the curve $x^2 + 4y^2 = 32$ and $x^2 - y^2 = 12$ at any point of their intersection.

7.4 Length of Tangent, Normal, Subtangent and Subnormal

4 (i) Find the length of subtangent at any point (x_1, y_1) on the curve $y = 5^x$.

(ii) Find the length of subtangent to the curve $y = e^{x/a}$.

(iii) Find the length of tangent, subtangent, normal and subnormal for the curve $y = x^2 + x - 1$ at (1, 1).

(iv) Find the length of the normal to the curve $x = a(\theta + \sin\theta)$ and $y = a(1 - \cos\theta)$ at $\theta = \frac{\pi}{2}$.

LEVEL 02 \ JEE Main Level

Single Choice Correct Questions

7.1 Slope of Tangent and Normal

1 The slope of the tangent at $\left(\frac{\pi}{4}, 0\right)$ to the curve $1 + 16x^2y = \tan(x - 2y)$ is

(a) $\frac{2}{\pi + 4}$ (b) $\frac{1}{\pi^2 + 4}$ (c) $\frac{2}{\pi^2 + 4}$ (d) $\frac{1}{\pi + 4}$

2 If the tangent to the curve $y = x + \sin y$ at a point (a, b) is parallel to the line joining $\left(0, \frac{3}{2}\right)$ and $\left(\frac{1}{2}, 2\right)$, then [JEE Main 2020]

(a) $b = a$ (b) $|b - a| = 1$ (c) $b = \frac{\pi}{2} + a$ (d) $|a + b| = 1$

3 The curve $y - e^{xy} + x = 0$ has a vertical tangent at the point

(a) (1, 1) (b) at no point (c) (0, 1) (d) (1, 0)

4 If the curve $y = ax^2 + bx + c, x \in R$ passes through the point (1, 2) and the tangent line to this curve at origin is $y = x$, then the possible values of a, b and c are

(a) $a = -1, b = 1$ and $c = 1$ (b) $a = 1, b = 1$ and $c = 0$

(c) $a = \frac{1}{2}, b = \frac{1}{2}$ and $c = 1$ (d) $a = 1, b = 0$ and $c = 1$

5 If the tangent to the curve $y = f(x) = x \log x, x > 0$ at a point $\{c, f(c)\}$ is parallel to the line-segment joining the points (1, 0) and (e, e), then c is equal to **[JEE Main 2020]**

(a) $\frac{e-1}{e}$ (b) $e^{\left(\frac{1}{1-e}\right)}$ (c) $e^{\left(\frac{1}{e-1}\right)}$ (d) $\frac{1}{e-1}$

6 Tangent to the curve $x = a\sqrt{\cos 2\theta}\cos\theta$ and $y = a\sqrt{\cos 2\theta}\sin\theta$ at a point corresponding to $\theta = \frac{\pi}{6}$ is

(a) parallel to X-axis (b) parallel to Y-axis

(c) parallel to the line $y = x$ (d) None of these

7 The point on the curve $y^2 = x^2 + ax + 25$ touches the X-axis, then the value of a is

(a) ± 5 (b) ± 10 (c) ± 15 (d) None of these

7.2 Equations of Tangent and Normal

8 The equation of the tangent to the curve $y = \sin x$ at the point $(\pi, 0)$ is

(a) $x + y = 0$ (b) $x + y = \pi$ (c) $x - y = \pi$ (d) $x - y = 0$

9 The curve $y = be^{-x/a}$ crosses the Y-axis at P. The equation of the tangent at P is

(a) $\frac{x}{a} + \frac{y}{b} = 2$ (b) $\frac{x}{a} - \frac{y}{b} = 2$ (c) $\frac{x}{a} + \frac{y}{b} = 1$ (d) $\frac{x}{a} - \frac{y}{b} = 1$

10 Which of the following points lies on the tangent to the curve $x^4 e^y + 2\sqrt{y+1} = 3$ at the point (1, 0)? **[JEE Main 2020]**

(a) (2, 2) (b) (– 2, 4) (c) (2, 6) (d) (– 2, 6)

11 The tangent to the curve $y = e^{kx}$ at a point (0, 1) meets the X-axis, where $x \in [-2, -1]$, then $k \in$

(a) $\left[-\frac{1}{2}, 0\right]$ (b) $\left[-1, -\frac{1}{2}\right]$ (c) [0, 1] (d) $\left[\frac{1}{2}, 1\right]$

12 A curve in XY-plane is parametrically represented by the equations $x = t^2 + t + 1$ and $y = t^2 - t + 1$, where $t \geq 0$. The number of straight lines passing through the point (1, 1) which are tangent to the curve, is

(a) 0 (b) 1 (c) 2 (d) 3

13 If $y = 4x - 5$ is a tangent to the curve $y^2 = px^3 + q$ at (2, 3), then

(a) $p = 2$ and $q = 7$ (b) $p = -2$ and $q = 7$

(c) $p = -2$ and $q = -7$ (d) None of these

14 If the tangent to the curve $y = \frac{x}{x^2 - 3}$, $x \in R$, $(x \neq \pm\sqrt{3})$, at a point $(\alpha, \beta) \neq (0, 0)$ on it is parallel to the line $2x + 6y - 11 = 0$, then **[JEE Main 2019]**

(a) $|6\alpha + 2\beta| = 19$ (b) $|6\alpha + 2\beta| = 9$

(c) $|2\alpha + 6\beta| = 19$ (d) $|2\alpha + 6\beta| = 11$

15 Equation of a normal to the curve $y = x^2 - 6x + 6$ which is perpendicular to the straight line, joining the origin to the vertex of the parabola, is

(a) $4x - 4y - 1 = 0$ (b) $4x - 4y + 1 = 0$

(c) $4x - 4y - 21 = 0$ (d) $4x - 4y + 21 = 0$

16 Consider, $f(x)=\tan^{-1}\left(\sqrt{\frac{1+\sin x}{1-\sin x}}\right), x\in\left(0,\frac{\pi}{2}\right)$. A normal to $y=f(x)$ at $x=\frac{\pi}{6}$ also passes through point

(a) (0, 0) (b) $\left(0,\frac{2\pi}{3}\right)$ (c) $\left(\frac{\pi}{6},0\right)$ (d) $\left(\frac{\pi}{4},0\right)$

17 The normal to the curve $y(x-2)(x-3)=x+6$ at the point, where the curve intersects the Y-axis, passes through the point **[JEE Main 2017]**

(a) $\left(-\frac{1}{2},-\frac{1}{2}\right)$ (b) $\left(\frac{1}{2},\frac{1}{2}\right)$ (c) $\left(\frac{1}{2},-\frac{1}{3}\right)$ (d) $\left(\frac{1}{2},\frac{1}{3}\right)$

7.3 Angle of Intersection of Two Curves

18 The angle of intersection of the curves $y=x^2$ and $6y=7-x^2$ at (1, 1) is

(a) $\frac{\pi}{4}$ (b) $\frac{\pi}{2}$ (c) $\frac{\pi}{6}$ (d) $\tan^{-1}(7)$

19 The angle of intersection of curves $y=2\sin^2 x$ and $y=\cos 2x$ is

(a) $\frac{\pi}{4}$ (b) $\frac{\pi}{3}$ (c) $\frac{\pi}{6}$ (d) $\frac{\pi}{2}$

20 The curves $y=a^x$ and $y=b^x$, where $a, b>0$, intersect each other at θ, then the value of $\tan\theta$ is

(a) $\left(\frac{\log(a/b)}{\log a+\log b}\right)$ (b) $\frac{\log(a/b)}{1-\log a\log b}$

(c) $\frac{\log(a/b)}{1+\log a\log b}$ (d) $\frac{\log(ab)}{1+\log a\log b}$

21 If θ denotes the acute angle between the curves, $y=10-x^2$ and $y=2+x^2$ at a point of their intersection, then $|\tan\theta|$ is equal to **[JEE Main 2019]**

(a) $\frac{7}{17}$ (b) $\frac{8}{15}$ (c) $\frac{4}{9}$ (d) $\frac{8}{17}$

22 The curves $\frac{x^2}{a}+\frac{y^2}{b}=1$ and $\frac{x^2}{c}+\frac{y^2}{d}=1$ intersect each other at an angle of 90°. Then, which of the following is true? **[JEE Main 2020]**

(a) $a-c=b-d$ (b) $a+b=c+d$

(c) $ab=\frac{c+d}{a+b}$ (d) $a-b=c-d$

7.4 Length of Tangent, Normal, Subtangent and Subnormal

23 For the parabola $y^2=4ax$, the ratio of the subtangent to the abscissa is

(a) 1 : 1 (b) 2 : 1 (c) $x:y$ (d) $x^2:y^2$

24 If at any point on the curve the subtangent and subnormal are equal, then the tangent is equal to

(a) ordinate (b) $\sqrt{2}$ ordinate (c) $\sqrt{2\text{ (ordinate)}}$ (d) None of these

25 For the curve $xy=c^2$, the subnormal at any point varies as

(a) x^2 (b) y^2 (c) x^3 (d) y^3

26 The subnormal at any point of the curve $x^2y^2 = a^2(x^2 - a^2)$ varies inversely as

(a) cube of the abscissa of the point (b) square of the abscissa

(c) $\frac{3}{2}$ th power of abscissa (d) $\frac{2}{3}$ rd power of abscissa

27 The subtangent, ordinate and subnormal to the parabola $y^2 = 4ax$ at a point (different from origin) are in

(a) GP (b) HP (c) AP (d) None of these

28 The length of the normal to the curve $y = a\left(\frac{e^{-x/a} + e^{x/a}}{2}\right)$ at any point varies as the

(a) abscissa of the point (b) ordinate of the point

(c) square of the abscissa of the point (d) square of the ordinate of the point

29 If the length of subnormal at any point on the curve $xy^n = a^{n+1}$ is constant, then n equal to

(a) $\frac{1}{2}$ (b) 1 (c) 2 (d) -2

30 The length of subtangent at (a, a) on the curve $y^2 = \frac{x^2}{2a + x}$ is equal to

(a) $\frac{18}{5}$ (b) $\frac{18a}{5}$ (c) $-\frac{18a^2}{5}$ (d) $\frac{18a^2}{5}$

31 The length of subtangent at any point of the curve $x^m y^n = a^{m+n}$ is proportional to

(a) ordinate (b) abscissa (c) $(\text{abscissa})^2$ (d) $(\text{ordinate})^2$

Numerical Value Type Questions

32 The slope of the tangent to the curve $(y - x^5)^2 = x(1 + x^2)^2$ at the point (1, 3) is ....

33 The curve $x + y - \log(x + y) = 2x + 5$ has a vertical tangent at the point (k, l), then $k + l$ is ....

34 If at each point of the curve $y = x^3 - ax^2 + x + 1$, the tangent is inclined at a positive acute angle with positive direction of X-axis, then number of possible integral values of a is ....

35 If length of perpendicular from origin upon the tangent drawn to curve $x^2 - xy + y^2 + \alpha(x - 2) = 4$ at (2, 2), is equal to 2, then $|\alpha|$ equals ....

36 If m is the slope of the common tangents $y = x^2 - x + 1$ and $y = x^2 - 3x + 1$, then the value of $(|m| + 3)$ is ....

37 Let the normal at a point P on the curve $y^2 - 3x^2 + y + 10 = 0$ intersect the Y-axis at $\left(0, \frac{3}{2}\right)$. If m is the slope of the tangent at P to the curve, then $|m|$ is equal to .... **[JEE Main 2020]**

38 The sum of the intercepts made on the axes of coordinates by any tangent to the curve $\sqrt{x} + \sqrt{y} = 2$, is ....

39 Let the parabola $y = x^2 + ax + b$ and $y = x(c - x)$ touch each other at the point (1, 0), then the value of $(a + b + c + 4)$ ....

40 If the lines $x + y = a$ and $x - y = b$ touch the curve $y = x^2 - 3x + 2$ at the point, where the curve intersects the X-axis, then $\frac{b}{a}$ is equal to ....

41 If the curve $x = y^4$ and $xy = k$ and at right angle, then $(4k)^6$ is equal to .... **[JEE Main 2021]**

LEVEL 03 \ JEE Advanced Level

Single Choice Correct Questions

1 If the tangent to the curve $y = x^3$ at the point $P(t, t^3)$ meets the curve again at Q, then the ordinate of the point which divides PQ internally in the ratio of $1 : 2$, is **[JEE Main 2021]**

(a) 0 (b) $-2t^3$ (c) $2t^3$ (d) $-t^3$

2 The possible values of P such that the equation $Px^2 = \ln x$ has exactly one solution is

(a) $(-\infty, 0)$ (b) $(-\infty, 0) \cup \left\{\frac{1}{2e}\right\}$

(c) $(-\infty, 0] \cup \left\{\frac{1}{2e}\right\}$ (d) $(-\infty, 0] \cup \left\{\frac{1}{e}\right\}$

3 The area of triangle formed by a tangent to the curve $2xy = a^2$ and the coordinate axis, is

(a) $2a^2$ (b) a^2

(c) $3a^2$ (d) None of these

4 A curve is given by the equations $x = \sec^2\theta$ and $y = \cot\theta$. If the tangent at P, where $\theta = \frac{\pi}{4}$, meets the curve again at Q, then length PQ equals

(a) $\sqrt{15}$ units (b) $\frac{3}{2}\sqrt{5}$ units

(c) $\frac{1}{2}\sqrt{5}$ units (d) None of these

5 If the length of subnormal is equal to length of subtangent at any point (3, 4) on the curve $y = f(x)$ and the tangent at (3, 4) to $y = f(x)$ meets the ordinate axes at A and B, then maximum area of ΔOAB, where O is origin, is

(a) $\frac{45}{2}$ (b) $\frac{49}{2}$ (c) $\frac{25}{2}$ (d) $\frac{81}{2}$

6 The equation of the straight line which is tangent at one point and normal at another point of the curve $x = 3t^2$ and $y = 2t^3$, is

(a) $y = \pm\sqrt{2}(x-2)$ (b) $y = \pm 2(x-2)$

(c) $y = \pm 2\sqrt{2}(x-2)$ (d) $\sqrt{2}y = \pm(x-2)$

7 The acute angle between the curves $y = |x^2 - 1|$ and $y = |x^2 - 3|$ at their points of intersection, is

(a) $\tan^{-1}\left(\frac{\sqrt{2}}{7}\right)$ (b) $\tan^{-1}\left(\frac{2\sqrt{2}}{7}\right)$

(c) $\tan^{-1}\left(\frac{4\sqrt{2}}{7}\right)$ (d) None of these

8 The value of α such that the subtangent and subnormal are equal for the curve $y = e^{\alpha x} + \alpha x$ at the point (0, 1) is

(a) ± 1 (b) ± 2

(c) $\pm\frac{1}{2}$ (d) None of these

More than One Correct

9 The points on the curve $y^3 + 3x^2 = 12y$, where the tangent is vertical, is (are)

(a) $\left(\pm\frac{4}{\sqrt{3}}, -2\right)$ (b) $\left(\pm\sqrt{\frac{11}{3}}, 1\right)$ (c) (0, 0) (d) $\left(\pm\frac{4}{\sqrt{3}}, 2\right)$

10 For the curve represented parametrically by the equations $x = 2\ln \cot t + 1$ and $y = \tan t + \cot t$, then

(a) tangent at $t = \frac{\pi}{4}$ is parallel to X-axis

(b) normal at $t = \frac{\pi}{4}$ is parallel to Y-axis

(c) tangent at $t = \frac{\pi}{4}$ is parallel to $y = x$

(d) tangent and normal intersect at the point (2, 1)

11 For function $f(x) = \frac{\ln x}{x}$, which of the following is/are true?

(a) $f(x)$ has horizontal tangent at $x = e$
(b) $f(x)$ cuts the X-axis only at one point
(c) $f(x)$ is many one function
(d) $f(x)$ has one vertical tangent

12 Tangent to the curve $x = t^3 - 6t^2 + 9t$ and $y = t^2$ perpendicular to X-axis at

(a) $t = 1$ (b) $t = 0$ (c) $t = 3$ (d) $t = 4$

13 If the line $ax + by + c = 0$ is a normal to the curve $xy = 1$, then

(a) $a > 0$ and $b > 0$ (b) $a > 0$ and $b < 0$ (c) $a < 0$ and $b > 0$ (d) $a < 0$ and $b < 0$

14 The tangent to the curve $y = x^2 + 3x$ will pass through the point (0, – 9), if it is drawn as the point

(a) (3, 18) (b) (1, 4) (c) (– 4, 4) (d) (–3, 0)

15 If the tangent to the curve $2y^3 = ax^2 + x^3$ at the point (a, a) cuts-off intercepts α and β on coordinate axes, where $\alpha^2 + \beta^2 = 61$, then the value of a is

(a) 20 (b) 25 (c) 30 (d) – 30

16 Tangents are drawn to the hyperbola $\frac{x^2}{9} - \frac{y^2}{4} = 1$ parallel to the straight line $2x - y = 1$. The points of contact to the tangent and the hyperbola are

(a) $\left(\frac{9}{2\sqrt{2}}, \frac{1}{\sqrt{2}}\right)$ (b) $\left(-\frac{9}{2\sqrt{2}}, -\frac{1}{\sqrt{2}}\right)$

(c) $(3\sqrt{3}, -2\sqrt{2})$ (d) $(-3\sqrt{3}, 2\sqrt{2})$

17 Two tangents to the graph of function $f(x) = \sqrt{17(1 + x^2)}$ intersect at right angles at a certain point on the Y-axis. Then, equations of tangents are

(a) $y = -x + 4$ (b) $y = -x - 4$
(c) $y = x - 4$ (d) $y = 2x + 4$

18 Which of the following pair(s) of curves is/are orthogonal?

(a) $y^2 = 4ax$ and $y = e^{-x/2a}$ (b) $y^2 = 4ax$ and $x^2 = 4ay$ at (0, 0)
(c) $xy = a^2$ and $x^2 - y^2 = b^2$ (d) $y = ax$ and $x^2 + y^2 = c^2$

Linked Comprehension Type

Paragraph I Directions (Q.Nos. 19-21)

Consider the function $f(x) = x^2 f(1) - xf'(2) + f''(3)$ such that $f(0) = 2$.

19 The value of $f'(1)$ is

(a) 0 (b) 1 (c) -1 (d) 2

20 Equation of the tangent to $y = f(x)$ at $x = 3$ is

(a) $y = x - 7$ (b) $y = \frac{x}{4} - 7$

(c) $y = 4x - 7$ (d) None of these

21 The angle of intersection of $y = f(x)$ and $y = e^{2(x-1)}$ is

(a) $\tan^{-1}\left(\frac{3}{4}\right)$ (b) $\tan^{-1}(2)$

(c) 0 (d) None of these

Matrix Match Type Questions

22 Match the following.

	Column I		Column II
(i)	The angle of intersection of $y^2 = 4x$ and $x^2 = 4y$ is $\frac{\pi}{2}$ and $\tan^{-1}\left(\frac{m}{n}\right)$, then $\lvert m + n \rvert$ is equal to (m and n are coprime)	(A)	0
(ii)	The area of triangle formed by normal at the point (1, 0) in the curve $x = e^{\sin y}$ with axis is	(B)	$\frac{1}{2}$
(iii)	The angle between curves $x^2 y = 1$ and $y = e^{2(1-x)}$	(C)	7
(iv)	The length of subtangent at any point on the curve $y = be^{x/3}$ is equal to	(D)	3

23 Let $x = a\cos t + at\sin t$ and $y = a\sin t - at\cos t$

Match the columns for $t = \frac{\pi}{4}$.

	Column I		Column II
(i)	Length of subtangent	(A)	$\frac{a\pi}{2}$
(ii)	Length of normal	(B)	$\frac{a}{\sqrt{2}}\left(1 - \frac{\pi}{4}\right)$
(iii)	Length of intercepts by tangent between coordinate axes	(C)	$\frac{a\pi}{4}$
(iv)	Perpendicular distance of tangent from origin	(D)	$a\left(1 - \frac{\pi}{4}\right)$

Numerical Value Type Questions

Question Stem Directions (Q.Nos. 24-25) The answer to each question is a **NUMERICAL VALUE.**

The curve $y = ax^3 + bx^2 + cx + 5$ touches the X-axis at $P(-2, 0)$ and cuts the Y-axis at a point Q, where its gradient is 3, then

24 The value of $\left|\frac{c}{a}\right|$ is ....

25 The value of $\left|\frac{c}{b}\right|$ is ....

Question Stem Directions (Q.Nos. 26-28) The answer to each question is a **NUMERICAL VALUE.**

If $y = f(x)$ is a curve and if there exists two points $A(x_1, f(x_1))$ and $B(x_2, f(x_2))$ on it such that $f'(x_1) = \frac{1}{f'(x_2)} = \frac{f(x_2) - f(x_1)}{x_2 - x_1}$, then the tangent at x_1 is normal at x_2 for that curve.

26 Number of such lines on the curve $y = \sin x$ is ....

27 Numer of such lines on the curve $y = |\ln x|$ is ....

28 Number of such lines on the curve $y^2 = x^3$ is ....

29 If slope of line through the origin which is tangent to the curve $y = x^3 + x + 16$, is m, then the value of m is ....

30 Let $y = f(x)$ be drawn with $f(0) = 2$ and for each real number. The tangent to $y = f(x)$ at $\{a, f(a)\}$ has x-intercept $(a - 2)$. If $f(x)$ is of the form of ke^{px}, then $\left(\frac{k}{p}\right)$ equals ....

31 Let C be the curve defined by $y = e^{a + bx^2}$. The curve C passes through the point $P(1, 1)$ and the slope of the tangent at P is -2, then the value of $(2a - 3b)$ is ....

32 A curve is defined parametrically by the equation $x = t^2$ and $y = t^3$. A variable pair of perpendicular lines through the origin O meet the curve P and Q. If the locus of the point of intersection of the tangent at P and Q is $ay^2 = bx - 1$, then the value of $(a + b)$ is ....

33 At the point $P(a, a^n)$ on the graph of $y = x^n, (n \in N)$, in the first quadrant, a normal is drawn. The normal intersects the Y-axis at the point $(0, b)$. If $\lim_{a \to 0} b = \frac{1}{2}$, then n equals ....

34 The abscissa of the point on the curve $4y^2 = x^3$, the normal at which cuts intercepts (x-intercept and y-intercept) in the ratio of 3 : 4, is ....

35 If the tangent at any point $P(4m^2, 8m^3)$ of $x^3 - y^2 = 0$ is a normal also the curve $x^3 - y^2 = 0$, then the value of $(9m^2 + 2)$ is ....

36 If the tangent at P of the curve $y^2 = x^3$ intersects the curve again at Q and the straight lines OP and OQ, makes angle α and β with the X-axis, where 'O' is the origin, then $\left|\frac{\tan \alpha}{\tan \beta}\right|$ is equal to ....

37 If the curves $y^2 = 4ax$ and $xy = c^2$ cut orthogonally, then $\frac{c^4}{a^4}$ equals ....

ANSWER KEY

Level 1

1. (i) $-\frac{1}{64}$ (ii) Slope of tangent $= -1$, Slope of normal $= 1$

(iii) Slope of tangent $= -3$, Slope of normal $= \frac{1}{5}$

(iv) $\left(\frac{3}{2}, \frac{-17}{2}\right)$ (v) $\left(\frac{1}{8}, \frac{1}{4}\right)$ (vi) $\left(\frac{7}{2}, \frac{1}{4}\right)$

2. (i) Equation of tangent $y = 3x$

and equation of normal : $x + 3y = 0$

(ii) $(\sqrt{2} - 1)\, x - y = 2(\sqrt{2} - 1) - \frac{\pi}{4}$

(iii) $4x - y + 5 = 0$

(iv) $y - 1 = \frac{2}{\pi}(x - 1)$

(v) $x + ay - 2a^2 - 1 = 0$

(vi) $3x - 2y - a = 0$

(vii) $y = x + 1$

(viii) $x + y = 3$

3. (i) $\tan^{-1}\left(\frac{1}{3}\right)$ (ii) $\tan^{-1}(3)$ (iii) $\tan^{-1}\left(\frac{3}{4}\right)$ (iv) $\frac{\pi}{2}$

4. (i) $\frac{1}{\log 5}$ (ii) a

(iii) Length of tangent $= (\sqrt{10}/3)$

Length of subtangent $= (1/3)$

Length of normal $= \sqrt{10}$

Length of subnormal $= 3$

(iv) $a\sqrt{2}$

Level 2

1. (c)	**2.** (b)	**3.** (d)	**4.** (b)	**5.** (c)	**6.** (a)	**7.** (b)
8. (b)	**9.** (c)	**10.** (d)	**11.** (d)	**12.** (b)	**13.** (d)	**14.** (a)
15. (c)	**16.** (b)	**17.** (b)	**18.** (d)	**19.** (b)	**20.** (c)	**21.** (b)
22. (d)	**23.** (b)	**24.** (b)	**25.** (d)	**26.** (a)	**27.** (a)	**28.** (d)
29. (d)	**30.** (d)	**31.** (b)	**32.** (8)	**33.** (1)	**34.** (3)	**35.** (2)
36. (5)	**37.** (4)	**38.** (4)	**39.** (4)	**40.** (2)	**41.** (4)	

Level 3

1. (b)	**2.** (c)	**3.** (b)	**4.** (b)	**5.** (b)	**6.** (a)	**7.** (c)
8. (c)	**9.** (d)	**10.** (a, b)	**11.** (a, b, c, d)		**12.** (a, c)	**13.** (b, c)
14. (a, d)	**15.** (c, d)	**16.** (a, b)	**17.** (b, c)	**18.** (a, b, c, d)	**19.** (a)	**20.** (c)
21. (b)						

22. (i)–C, (ii)–B, (iii)–A, (iv)–D

23. (i)–B, (ii)–D, (iii)–A, (iv)–C

24. (6)	**25.** (4)	**26.** (0)	**27.** (0)	**28.** (2)	**29.** (13)	**30.** (4)
31. (5)	**32.** (7)	**33.** (2)	**34.** (1)	**35.** (4)	**36.** (2)	**37.** (32)

SOLUTIONS

Level 1

1. (i) Given, equation of the curve

$$y = \frac{x-1}{x-2} \quad \text{...(i)}$$

$$\therefore \quad \frac{dy}{dx} = \frac{(x-2)(1)-(x-1)(1)}{(x-2)^2}$$

$$\Rightarrow \quad \frac{dy}{dx} = \frac{-1}{(x-2)^2}$$

$\therefore$ Slope of the tangent at point $x = 10$ to the curve

$$= \frac{-1}{(10-2)^2} = -\frac{1}{64}$$

(ii) Given, $y = a\sin^3\theta$ and $x = a\cos^3\theta$

$$\therefore \quad \frac{dy}{d\theta} = 3a\sin^2\theta\cos\theta$$

and $\dfrac{dx}{d\theta} = -3a\cos^2\theta\sin\theta$

$$\therefore \quad \frac{dy}{dx} = \frac{dy/d\theta}{dx/d\theta} = \frac{3a\sin^2\theta\cos\theta}{-3a\cos^2\theta\sin\theta} = -\tan\theta$$

At $\theta = \dfrac{\pi}{4}$,

Slope of tangent $(m) = -\tan\dfrac{\pi}{4} = -1$

Now, slope of normal $= -\dfrac{1}{m} = -\left(\dfrac{1}{-1}\right) = 1$

(iii) We have, $y = |x^2 - |x||$

In the neighbourhood of $x = -3$,

$$y = |x^2 + x|$$

$$\Rightarrow \quad y = x^2 + x$$

$$\Rightarrow \quad \frac{dy}{dx} = 2x + 1$$

Slope of the tangent at $x = -3$,

$$m = \left.\frac{dy}{dx}\right|_{x=-3}$$

$$= 2(-3) + 1 = -5$$

So, slope of normal at $x = -3$ is $\dfrac{1}{5}$.

(iv) Let the required point be $P(x_1, y_1)$.

The given curve is

$$y = 2x^2 - 6x - 4$$

$$\Rightarrow \quad \frac{dy}{dx} = 4x - 6$$

$$\left.\frac{dy}{dx}\right|_{(x_1, y_1)} = 4x_1 - 6$$

Since, tangent is parallel to X-axis.

So, $\quad 4x_1 - 6 = 0$

$$\Rightarrow \quad x_1 = \frac{3}{2}$$

Since, (x_1, y_1) lies on the curve, therefore

$$y_1 = 2x_1^2 - 6x_1 - 4$$

$$\Rightarrow \quad y_1 = 2\left(\frac{3}{2}\right)^2 - 6\left(\frac{3}{2}\right) - 4$$

$$\Rightarrow \quad y_1 = -\frac{17}{2}$$

So, the required point is $\left(\dfrac{3}{2}, -\dfrac{17}{2}\right)$.

(v) Let (x_1, y_1) be the required point on curve $2y^2 = x$.

$$4y\frac{dy}{dx} = 1$$

$$\Rightarrow \quad \frac{dy}{dx} = \frac{1}{4y}$$

$$\left.\frac{dy}{dx}\right|_{(x_1, y_1)} = \frac{1}{4y_1}$$

Slope of the tangent $= \tan 45° = 1$

So, $\quad \dfrac{1}{4y_1} = 1$

$$\Rightarrow \quad y_1 = \frac{1}{4}$$

Since, (x_1, y_1) lies on $2y^2 = x$

So, $\quad 2y_1^2 = x_1$

$$\Rightarrow \quad 2\left(\frac{1}{4}\right)^2 = x_1$$

$$\Rightarrow \quad x_1 = \frac{1}{8}$$

Hence, required point is $\left(\dfrac{1}{8}, \dfrac{1}{4}\right)$.

(vi) Let the required point be $P(x_1, y_1)$.

Given curve is

$$y = (x-3)^2$$

$$\Rightarrow \quad \frac{dy}{dx} = 2(x-3)$$

$$\Rightarrow \quad \left.\frac{dy}{dx}\right|_{(x_1, y_1)} = 2(x_1 - 3)$$

Since, tangent at P is parallel to the line joining (4, 1) and (3, 0).

Therefore, slope of the tangent at P = Slope of the line joining (4, 1) and (3, 0)

$$\left.\frac{dy}{dx}\right|_{(x_1, y_1)} = \frac{0-1}{3-4}$$

$$\Rightarrow \quad 2(x_1 - 3) = 1$$

$$\Rightarrow \quad x_1 = \frac{7}{2}$$

Point (x_1, y_1) lies on curve $y = (x-3)^2$

$$\Rightarrow \quad y_1 = (x_1 - 3)^2$$

$$\Rightarrow \quad y_1 = \left(\frac{7}{2} - 3\right)^2$$

$$\Rightarrow \quad y_1 = \frac{1}{4}$$

Hence, the required point is $\left(\frac{7}{2}, \frac{1}{4}\right)$.

2. (i) The equation of curve is

$$y = 2x^2 + 3\sin x$$

On putting $x = 0$ and $y = 0$, the point of contact is (0, 0).

Now, $\quad y = 2x^2 + 3\sin x$

$$\frac{dy}{dx} = 4x + 3\cos x$$

$$\left.\frac{dy}{dx}\right|_{(0,0)} = 3$$

So, slope of tangent $= 3$ and slope of normal $= -\frac{1}{3}$

Equation of tangent at (0, 0) is

$$y - 0 = 3(x - 0)$$

$$\Rightarrow \quad y = 3x$$

and equation of the normal

$$y - 0 = -\frac{1}{3}(x - 0)$$

$$\Rightarrow \quad x + 3y = 0$$

(ii) We have,

$$x = 1 - \cos\theta \text{ and } y = \theta - \sin\theta$$

On putting $\theta = \frac{\pi}{4}$,

$$x = 1 - \frac{1}{\sqrt{2}} \text{ and } y = \frac{\pi}{4} - \frac{1}{\sqrt{2}}$$

So, the coordinate of point of contact is $\left(1 - \frac{1}{\sqrt{2}}, \frac{\pi}{4} - \frac{1}{\sqrt{2}}\right)$.

Now, $x = 1 - \cos\theta$

$$\frac{dx}{d\theta} = \sin\theta$$

and $\quad y = \theta - \sin\theta$

$$\Rightarrow \quad \frac{dy}{d\theta} = 1 - \cos\theta$$

$$\therefore \quad \frac{dy}{dx} = \frac{dy/d\theta}{dx/d\theta} = \frac{1-\cos\theta}{\sin\theta}$$

$$\Rightarrow \quad \left.\frac{dy}{dx}\right|_{\theta = \frac{\pi}{4}} = \frac{1 - \frac{1}{\sqrt{2}}}{\frac{1}{\sqrt{2}}} = \sqrt{2} - 1$$

So, the equation of the tangent at $\theta = \frac{\pi}{4}$,

$$y - \left(\frac{\pi}{4} - \frac{1}{\sqrt{2}}\right) = (\sqrt{2} - 1)\left\{x - \left(1 - \frac{1}{\sqrt{2}}\right)\right\}$$

$$\Rightarrow \quad (\sqrt{2} - 1)x - y = 2(\sqrt{2} - 1) - \frac{\pi}{4}$$

(iii) Let the point of contact to the required tangent be (x_1, y_1).

The equation of given curve is

$$y = 2x^2 + 7$$

$$\Rightarrow \quad \frac{dy}{dx} = 4x$$

$$\Rightarrow \quad \left.\frac{dy}{dx}\right|_{(x_1, y_1)} = 4x_1$$

Since, the line $4x - y + 3 = 0$ is parallel to the tangent.

So, slope of tangent = Slope of line $4x - y + 3 = 0$

$$4x_1 = -\frac{4}{(-1)}$$

$$\Rightarrow \quad x_1 = 1$$

Now, point (x_1, y_1) lies on the curve

$$y_1 = 2(1)^2 + 7$$

$$\Rightarrow \quad y_1 = 9$$

So, the coordinate of required point is (1, 9).

Then, the equation of tangent line at point (1, 9) with slope 4 is

$$(y - 9) = 4(x - 1)$$

$$= y - 4x - 5 = 0$$

$$\Rightarrow \quad 4x - y + 5 = 0$$

(iv) We have, $y = \cos\frac{\pi x}{2}$

On differentiating w.r.t. x, we get

$$\frac{dy}{dx} = -\frac{\pi}{2}\sin\frac{\pi}{2}x$$

$$\left.\frac{dy}{dx}\right|_{(1,1)} = -\frac{\pi}{2} = m_1 \text{ (let)}$$

Let slope of normal $= m_2$, then $m_1 m_2 = -1$

$$m_2 = \frac{1}{\pi/2} = \frac{2}{\pi}$$

Then, equation of normal at (1, 1)

$$y - 1 = \frac{2}{\pi}(x - 1)$$

(v) Given, $x = t^2$ and $y = 2at$

At $t = 1$,

$$x = 1 \text{ and } y = 2a$$

Now, $\frac{dx}{dt} = 2t$ and $\frac{dy}{dt} = 2a$

$$\frac{dy}{dx} = \frac{dy/dt}{dx/dt} = \frac{2a}{2t} = \frac{a}{t}$$

$$\left.\frac{dy}{dx}\right|_{t=1} = a = m_1 \text{ (let)}$$

Let slope of normal $= m_2$

$\therefore \quad m_1 m_2 = -1$

$$m_2 = -\frac{1}{a}$$

Then, equation of normal at $t = 1$ is

$$(y - 2a) = -\frac{1}{a}(x - 1)$$

$\Rightarrow \quad ay - 2a^2 = -x + 1$

$\Rightarrow \quad x + ay - 2a^2 - 1 = 0$

(vi) The given equation of the curve

$$x^2 y^3 = a^5$$

On differentiating w.r.t. x, we get

$$2xy^3 + 3x^2y^2 \frac{dy}{dx} = 0$$

$\Rightarrow \quad \frac{dy}{dx} = -\frac{2y}{3x}$

$\Rightarrow \quad \left(\frac{dy}{dx}\right)_{(a,\, a)} = -\frac{2a}{3a} = -\frac{2}{3}$

$\therefore$ Slope of tangent at $(a, a) = -\frac{2}{3}$

Slope of normal at $(a, a) = \frac{3}{2}$

Equation of normal at (a, a) is

$$y - a = \frac{3}{2}(x - a)$$

$\Rightarrow \quad 3x - 2y - a = 0$

(vii) Given curve is

$$x + y = x^y \quad \text{...(i)}$$

At X-axis, $y = 0$

$\therefore \quad x + 0 = x^0 \Rightarrow x = 1$

Now, taking log on both sides in Eq. (i), we get

$$\log(x + y) = y \log x$$

$\Rightarrow \quad \frac{1}{x+y}\left(1 + \frac{dy}{dx}\right) = \frac{y}{x} + \log x\left(\frac{dy}{dx}\right)$

On putting $x = 1$ and $y = 0$,

$$\left(1 + \frac{dy}{dx}\right) = 0 \Rightarrow \frac{dy}{dx} = -1$$

$\therefore$ Slope of tangent $= -1$

Slope of normal $= 1$

Hence, equation of normal at (1, 0),

$$y - 1 = 1(x - 0)$$

$$y - 1 = x$$

$\Rightarrow \quad y = x + 1$

(viii) Let the normal through $P(1, 2)$ to the curve $4y = x^2$ meet the curve at $Q(x_1, y_1)$.

Point Q lies on the curve.

So, $\quad x_1^2 = 4y_1 \quad \text{...(i)}$

Now, $\quad x^2 = 4y$

$\Rightarrow \quad 2x = 4\frac{dy}{dx} \Rightarrow \frac{dy}{dx} = \frac{x}{2}$

$\therefore \quad \left.\frac{dy}{dx}\right|_{(x_1,\, y_1)} = \frac{x_1}{2}$

$\therefore$ Slope of normal $= -\frac{2}{x_1}$

So, slope of $PQ =$ Slope of normal

$$\frac{y_1 - 2}{x_1 - 1} = -\frac{2}{x_1} \quad \text{...(ii)}$$

From Eqs. (i) and (ii), we get

$$\frac{\frac{x_1^2}{4} - 2}{x_1 - 1} = -\frac{2}{x_1}$$

$\Rightarrow \quad x_1^3 = 8$

$\therefore \quad x_1 = 2$ and $y_1 = 1$

So, the coordinate of $Q\,(2, 1)$.

Slope of $PQ = \frac{2-1}{1-2} = -1$

Hence, equation of normal

$$y - 1 = -1(x - 2)$$

$\Rightarrow \quad x + y = 3$

3. (i) We have,

$$y = x \text{ and } y^2 = 4x$$

On differentiating given equations w.r.t. x, we get

$$\left(\frac{dy}{dx}\right)_1 = 1 \text{ and } 2y\left(\frac{dy}{dx}\right)_2 = 4$$

$\Rightarrow \quad \left(\frac{dy}{dx}\right)_2 = \frac{2}{y}$

$$\left.\left(\frac{dy}{dx}\right)_1\right|_{(4,\,4)} = 1 \text{ and } \left.\left(\frac{dy}{dx}\right)_2\right|_{(4,\,4)} = \frac{2}{4} = \frac{1}{2}$$

Let the angle of intersection is θ.

So, $\quad \tan\theta = \left|\frac{1 - \frac{1}{2}}{1 + 1\left(\frac{1}{2}\right)}\right|$

$\Rightarrow \quad \tan\theta = \frac{1}{3} \Rightarrow \theta = \tan^{-1}\left(\frac{1}{3}\right)$

(ii) We have, $xy = 2$ and $y^2 = 4x$ ...(i)

For intersection point of $xy = 2$ and $y^2 = 4x$

$$y^2 = 4\cdot\frac{2}{y}$$

$$\Rightarrow \quad y^3 = 8 \Rightarrow y = 2$$

Then, $x = 1$

Hence, point of intersection is (1, 2).

On differentiating both equations w.r.t. x, we get

$$x\left(\frac{dy}{dx}\right)_1 + y = 0$$

and $$2y\left(\frac{dy}{dx}\right)_2 = 4$$

$$\Rightarrow \quad \left(\frac{dy}{dx}\right)_1 = -\frac{y}{x}$$

and $$\left(\frac{dy}{dx}\right)_2 = \frac{2}{y}$$

$$\left(\frac{dy}{dx}\right)_1\Bigg|_{(1,\,2)} = -2 \text{ and } \left(\frac{dy}{dx}\right)_2\Bigg|_{(1,\,2)} = \frac{2}{2} = 1$$

Let angle of intersection is θ

$$\tan\theta = \left|\frac{1+2}{1-2}\right| = |-3|$$

$$\theta = \tan^{-1}(3)$$

(iii) Given, curves $y^2 = x$ and $x^2 = y$

Intersection points of curves are (0, 0) and (1, 1).

On differentiating both equations w.r.t. x, we get

$$2y\left(\frac{dy}{dx}\right)_1 = 1 \text{ and } 2x = \left(\frac{dy}{dx}\right)_2$$

$$\Rightarrow \quad \left(\frac{dy}{dx}\right)_1 = \frac{1}{2y} \text{ and } \left(\frac{dy}{dx}\right)_2 = 2x$$

$$\left(\frac{dy}{dx}\right)_1\Bigg|_{(0,\,0)} = \infty \text{ and } \left(\frac{dy}{dx}\right)_2\Bigg|_{(0,\,0)} = 0$$

Tangent to the curve $y^2 = x$ is parallel to the Y-axis and tangent to the curve $x^2 = y$ is parallel to X-axis

$\therefore$ Angle of intersection $= \frac{\pi}{2}$

Also, $$\left(\frac{dy}{dx}\right)_1\Big|_{(1,\,1)} = \frac{1}{2}$$

and $$\left(\frac{dy}{dx}\right)_2\Big|_{(11)} = 2$$

$$\because \tan\theta = \left|\frac{2-\frac{1}{2}}{1+\frac{1}{2}\cdot 2}\right| = \left|\frac{3}{4}\right|$$

$$\Rightarrow \quad \theta = \tan^{-1}\frac{3}{4}$$

(iv) The given curves are

$$x^2 + 4y^2 = 32 \quad \text{...(i)}$$

and $$x^2 - y^2 = 12 \quad \text{...(ii)}$$

Solving Eqs. (i) and (ii), we get

$$x = \pm 4 \text{ and } y = \pm 2$$

The points of intersection are $(\pm 4, \pm 2)$ (four points)

Since, both curves are symmetric about both axes. The angle of intersection is same at any of these points.

Now, from Eq. (i),

$$\left(\frac{dy}{dx}\right) = -\frac{x}{4y}$$

$$\Rightarrow \quad \left(\frac{dy}{dx}\right)_{(4,\,2)} = -\frac{1}{2}$$

From Eq. (ii),

$$\frac{dy}{dx} = \frac{x}{y}$$

$$\Rightarrow \quad \left(\frac{dy}{dx}\right)_{(4,\,2)} = \frac{4}{2} = 2$$

Therefore, product of the slopes of the tangents to the curves at (4, 2) is -1.

Hence, the angle of intersection of the curve is $\frac{\pi}{2}$.

4. (i) Given, $y = 5^x$

$$\Rightarrow \quad \frac{dy}{dx} = 5^x \log 5$$

$$\frac{dy}{dx}\Bigg|_{(x_1,\,y_1)} = 5^{x_1}\log 5$$

Hence, length of subtangent $= \dfrac{y_1}{(dy/dx)_{(x_1,\,y_1)}}$

$$= \frac{5^{x_1}}{5^{x_1}\log 5}$$

$$= \frac{1}{\log 5}$$

(ii) We have, $y = e^{x/a}$

Let $P(x_1, y_1)$ be any point on the curve.

$$\frac{dy}{dx} = \frac{1}{a}e^{x/a}$$

$$\frac{dy}{dx}\Bigg|_{(x_1,\,y_1)} = \frac{1}{a}e^{\frac{x_1}{a}} = \frac{y_1}{a}$$

So, length of subtangent $= \dfrac{y_1}{\left(\frac{dy}{dx}\right)_{(x_1,\,y_1)}} = \dfrac{y_1}{(y_1/a)}$

$= a$

(iii) The equation of the curve is

$$y = x^2 + x - 1$$

$$\therefore \quad \frac{dy}{dx} = 2x + 1$$

$$\Rightarrow \quad \left(\frac{dy}{dx}\right)_{(1,1)} = 3$$

Now, length of tangent $= (y)_{(1,1)} = \dfrac{\sqrt{1+\left(\frac{dy}{dx}\right)^2_{(1,1)}}}{\left(\frac{dy}{dx}\right)_{(1,1)}}$

$$= \frac{\sqrt{1+3^2}}{3} = \frac{\sqrt{10}}{3}$$

Length of subtangent $= \dfrac{y}{\left(\frac{dy}{dx}\right)_{(1,1)}} = \dfrac{1}{3}$

Length of normal $= (y)_{(1,1)} = \sqrt{1+\left(\frac{dy}{dx}\right)^2_{(1,1)}}$

$$= \sqrt{1+(3)^2} = \sqrt{10}$$

Length of subnormal at (1, 1)

$$= y\left(\frac{dy}{dx}\right)_{(1,1)} = 1 \times 3 = 3$$

(iv) We have, $x = a(\theta + \sin\theta)$ and $y = a(1 - \cos\theta)$

On differentiating given equations w.r.t. θ, we get

$$\frac{dx}{d\theta} = a(1 + \cos\theta)$$

and $\quad \dfrac{dy}{d\theta} = a\sin\theta$

Therefore, $\dfrac{dy}{dx} = \dfrac{dy/d\theta}{dx/d\theta} = \dfrac{\sin\theta}{1+\cos\theta}$

So, $\left(\dfrac{dy}{dx}\right)_{\theta=\frac{\pi}{2}} = 1$

Therefore, length of normal

$$= (y)_{\theta=\frac{\pi}{2}}\sqrt{1+\left(\frac{dy}{dx}\right)^2_{\theta=\frac{\pi}{2}}} = a\sqrt{1+1}$$

$$= \sqrt{2}a$$

Level 2

1. (c) On differentiating the curve equation w.r.t. x, we get

$$32xy + 16x^2\frac{dy}{dx} = \sec^2(x-2y)\left(1 - 2\frac{dy}{dx}\right)$$

On putting $x = \dfrac{\pi}{4}$ and $y = 0$,

$$16\cdot\frac{\pi^2}{16}\left(\frac{dy}{dx}\right)_{\left(\frac{\pi}{4},0\right)} = \sec^2\left(\frac{\pi}{4} - 0\right)\left[1 - 2\left(\frac{dy}{dx}\right)_{\left(\frac{\pi}{4},0\right)}\right]$$

$$\Rightarrow \quad \pi^2\left(\frac{dy}{dx}\right)_{\left(\frac{\pi}{4},0\right)} = 2 - 4\left(\frac{dy}{dx}\right)_{\left(\frac{\pi}{4},0\right)}$$

$$\Rightarrow (\pi^2 + 4)\left(\frac{dy}{dx}\right)_{\left(\frac{\pi}{4},0\right)} = 2$$

$$\Rightarrow \quad \left(\frac{dy}{dx}\right)_{\left(\frac{\pi}{4},0\right)} = \frac{2}{\pi^2 + 4}$$

2. (b) Slope of tangent to the curve $y = x + \sin y$ is

$$\frac{dy}{dx} = 1 + \cos y\frac{dy}{dx}$$

$$\Rightarrow \quad (1 - \cos y)\frac{dy}{dx} = 1$$

$$\Rightarrow \quad \frac{dy}{dx} = \frac{1}{1 - \cos y}$$

$$\Rightarrow \quad \left(\frac{dy}{dx}\right)_{(a,b)} = \frac{1}{1 - \cos b}$$

Now, slope of line joining the points $\left(\dfrac{1}{2}, 2\right)$ and $\left(0, \dfrac{3}{2}\right)$

$$= \frac{2 - \frac{3}{2}}{\frac{1}{2} - 0} = 1$$

According to question,

$$\frac{1}{1 - \cos b} = 1$$

$$\Rightarrow \quad 1 - \cos b = 1$$

$$\Rightarrow \quad \cos b = 0$$

So, $\sin b = \pm 1$

Point (a, b) lies on curve $y = x + \sin y$

$$\therefore \quad b = a + \sin b$$

$$\Rightarrow b - a = \sin b$$

So, $|b - a| = 1 \qquad (\because \sin b = \pm 1)$

3. (d) Given, curve is $y - e^{xy} + x = 0$

$$\Rightarrow \quad \frac{dy}{dx} - e^{xy}\left(x\frac{dy}{dx} + y\cdot 1\right) + 1 = 0$$

$$\Rightarrow \quad \frac{dy}{dx}(1 - xe^{xy}) = (ye^{xy} - 1)$$

$$\Rightarrow \quad \frac{dy}{dx} = \left(\frac{ye^{xy} - 1}{1 - xe^{xy}}\right)$$

Since, it has a vertical tangent, so $\dfrac{dx}{dy} = 0$

$$\Rightarrow \quad \left(\frac{1-xe^{xy}}{ye^{xy}-1}\right)=0$$

$$\Rightarrow \quad 1-xe^{xy}=0$$

When $y=0$, then $x=1$

So, the point is (1, 0).

4. (b) Given curve, $y=ax^2+bx+c, x\in R$

This curve passes through the point (1, 2)

$$\therefore \quad 2=a+b+c \quad \text{...(i)}$$

Given, slope of tangent at origin is 1.

$$\therefore \quad \frac{dy}{dx}=2ax+b$$

$$\left.\frac{dy}{dx}\right|_{(0,\,0)}=b=1$$

Put in Eq. (i),

$$a+c=1$$

Since, curve passes through origin.

So, $c=0$, then $a=1$

5. (c) We have, $y=f(x)=x\log x$

$$\Rightarrow \quad \frac{dy}{dx}=1+\log x$$

$$\Rightarrow \quad \left.\frac{dy}{dx}\right|_{\{c,\,f(c)\}}=1+\log c$$

This tangent is parallel to the line segment joining the points (1, 0) and (e, e).

So, $$1+\log c=\frac{e-0}{e-1}$$

$$\Rightarrow \quad \log c=\frac{e}{e-1}-1$$

$$=\frac{e-e+1}{e-1}$$

$$\Rightarrow \quad \log c=\frac{1}{e-1}$$

$$\Rightarrow \quad c=e^{\left(\frac{1}{e-1}\right)}$$

6. (a) We have,

$$x=a\sqrt{\cos 2\theta}\cos\theta$$

and $$y=a\sqrt{\cos 2\theta}\sin\theta$$

Now,

$$\frac{dx}{d\theta}=a\sqrt{\cos 2\theta}(-\sin\theta)+a\cos\theta\cdot\frac{1}{2\sqrt{\cos 2\theta}}(-2\sin 2\theta)$$

$$=\frac{-a\sin\theta\cos 2\theta-a\cos\theta\sin 2\theta}{\sqrt{\cos 2\theta}}$$

$$=\frac{-a\sin(\theta+2\theta)}{\sqrt{\cos 2\theta}}=\frac{-a\sin 3\theta}{\sqrt{\cos 2\theta}}$$

and $$\frac{dy}{d\theta}=a\sqrt{\cos 2\theta}(\cos\theta)+a\sin\theta\left(\frac{1}{2\sqrt{\cos 2\theta}}(-2\sin 2\theta)\right)$$

$$=\frac{a\cos 3\theta}{\sqrt{\cos 2\theta}}$$

$$\Rightarrow \quad \frac{dy}{dx}=\frac{dy/d\theta}{dx/d\theta}=-\cot 3\theta$$

$$\left(\frac{dy}{dx}\right)_{\theta=\frac{\pi}{6}}=-\cot\frac{\pi}{2}=0$$

i.e. parallel to X-axis.

7. (b) We have,

$$y^2=x^2+ax+25$$

$$2y\frac{dy}{dx}=2x+a$$

Given, tangent is X-axis, so $\frac{dy}{dx}=0$

$$2x+a=0$$

$$\Rightarrow \quad x=-\frac{a}{2}$$

Point lies on the curve, so point is $\left(-\frac{a}{2},0\right)$

$$y=x^2+ax+25$$

$$\Rightarrow \quad 0=\frac{a^2}{4}-\frac{a^2}{2}+25$$

$$\Rightarrow \quad 0=\frac{100-a^2}{4}$$

$$\Rightarrow \quad a^2=100$$

$$\Rightarrow \quad a=\pm 10$$

8. (b) We have,

$$y=\sin x$$

$$\Rightarrow \quad \frac{dy}{dx}=\cos x$$

$$\Rightarrow \quad \left.\frac{dy}{dx}\right|_{(\pi,\,0)}=\cos\pi=-1$$

Therefore, the equation of tangent at $(\pi, 0)$ is given by

$$y-0=-1(x-\pi)$$

$$\Rightarrow \quad x+y=\pi$$

9. (c) $x=0 \Rightarrow y=b$.

Therefore, $P=(0,b)$

Now, $$\frac{dy}{dx}=\frac{-b}{a}e^{-x/a}$$

$$\Rightarrow \quad \left(\frac{dy}{dx}\right)_{(0,\,b)}=\frac{-b}{a}$$

Equation of the tangent at $P(0,b)$ is

$$y-b=\frac{-b}{a}(x-0) \Rightarrow \frac{x}{a}+\frac{y}{b}=1$$

10. (d) Given curve,

$$x^4e^y+2\sqrt{y+1}=3$$

On differentiating both sides w.r.t. x, we get

$$x^4e^y\frac{dy}{dx}+e^y(4x^3)+\frac{2\frac{dy}{dx}}{2\sqrt{y+1}}=0$$

At $P(1,0)$, let $\frac{dy}{dx}=m$

$$\Rightarrow \quad m+4+m=0$$

$$\Rightarrow \quad 2m=-4$$

$$\Rightarrow \quad m=-2$$

Tangent at $P(1,0)$ is

$$y-0=-2(x-1)$$

$$2x+y=2$$

Only $(-2,6)$ satisfies the equation of tangent, so $(-2,6)$ lies on the tangent.

11. (d) We have, $\quad y=e^{kx}$

$$\frac{dy}{dx}=ke^{kx}$$

$$\left(\frac{dy}{dx}\right)_{(0,\,1)}=k$$

Equation of tangent at (0, 1) is

$$y-1=k(x-0)$$

$$y-1=kx$$

At point of intersection with X-axis, $x=-\frac{1}{k}$,

where $\quad -2\le-\frac{1}{k}\le-1$

$$\Rightarrow \quad \frac{1}{2}\le k\le1$$

Hence, $k\in\left[\frac{1}{2},1\right]$

12. (b) $t=0\Rightarrow x=1$ and $y=1$.

Therefore, (1, 1) is a point on the given curve.

Also, $\quad \frac{dy}{dx}=\frac{dy}{dt}\div\frac{dx}{dt}=\frac{2t-1}{2t+1}$

Now, $\quad \left(\frac{dy}{dx}\right)_{(1,\,1)}=\left(\frac{dy}{dx}\right)_{t=0}=\frac{-1}{1}=-1$

Equation of the tangent at (1, 1) is

$$y-1=-1(x-1)$$

$$\Rightarrow \quad x+y=2$$

Thus, one straight line passing through the point (1,1) which are tangent to the curve.

13. (d) Given, curve is $y^2=px^3+q$

$$\Rightarrow \quad 2y\frac{dy}{dx}=3px^2 \Rightarrow \frac{dy}{dx}=\frac{3px^2}{2y}$$

Slope of the tangent $=m=\left(\frac{dy}{dx}\right)_{(2,\,3)}=\frac{3p\cdot4}{2\cdot3}=2p$

Equation of the tangent at (2, 3) is

$$y-3=2p(x-2)$$

$$\Rightarrow \quad y-3=2px-4p$$

$$\Rightarrow \quad y=2px+(3-4p)$$

Given tangent is $y=4x-5$

On comparing the coefficients of x and the constant term, we get

$$p=2$$

Also, the point (2, 3) lies on the curve.

So, $\quad 9=16+q \Rightarrow q=-7$

Hence, the values of p and q are $p=2$ and $q=-7$

14. (a) Equation of given curve is

$$y=\frac{x}{x^2-3},\ x\in R,\ (x\ne\pm\sqrt{3}) \quad \text{...(i)}$$

On differentiating Eq. (i) w.r.t. x, we get

$$\frac{dy}{dx}=\frac{(x^2-3)-x(2x)}{(x^2-3)^2}=\frac{(-x^2-3)}{(x^2-3)^2}$$

It is given that tangent at a point $(\alpha,\beta)\ne(0,0)$ on it is parallel to the line

$$2x+6y-11=0$$

$\therefore$ Slope of this line $=-\frac{2}{6}=\left.\frac{dy}{dx}\right|_{(\alpha,\,\beta)}$

$$\Rightarrow \quad -\frac{(\alpha^2+3)}{(\alpha^2-3)^2}=-\frac{1}{3}$$

$$\Rightarrow \quad 3\alpha^2+9=\alpha^4-6\alpha^2+9$$

$$\Rightarrow \quad \alpha^4-9\alpha^2=0 \Rightarrow \alpha=0,-3,3$$

$$\Rightarrow \quad \alpha=3 \text{ or } -3, \qquad [\because \alpha\ne0]$$

Now, from Eq. (i),

$$\beta=\frac{\alpha}{\alpha^2-3} \Rightarrow \beta=\frac{3}{9-3} \text{ or } \frac{-3}{9-3}=\frac{1}{2} \text{ or } -\frac{1}{2}$$

According to the options, $|6\alpha+2\beta|=19$

at $(\alpha,\beta)=\left(\pm3,\pm\frac{1}{2}\right)$

15. (c) Given parabola,

$$y=x^2-6x+6$$

$$\Rightarrow \quad y=(x-3)^2-3$$

$$\Rightarrow \quad y+3=(x-3)^2$$

$\therefore$ Vertex of parabola $=(3,-3)$

$\therefore$ Slope of straight line join vertex and origin

$$= m = \frac{3}{-3} = -1$$

$\Rightarrow \quad m_N = 1$ and $\frac{dy}{dx} = 2x - 6$

So, $\frac{-1}{2x-6} = 1$

$\Rightarrow \quad x = \frac{5}{2}$ and $y = -\frac{11}{4}$

$\therefore$ Equation of normal

$$y + \frac{11}{4} = 1\left(x - \frac{5}{2}\right)$$

$\Rightarrow \quad 4x - 4y - 21 = 0$

16. (b) We have,

$$y = \tan^{-1}\left(\sqrt{\frac{1+\sin x}{1-\sin x}}\right)$$

$$\Rightarrow \quad y = \tan^{-1}\left(\frac{\cos\frac{x}{2} + \sin\frac{x}{2}}{\cos\frac{x}{2} - \sin\frac{x}{2}}\right)$$

$$\Rightarrow \quad y = \tan^{-1}\left\{\tan\left(\frac{\pi}{4} + \frac{x}{2}\right)\right\}$$

$$\Rightarrow \quad y = \frac{\pi}{4} + \frac{x}{2}$$

$$\Rightarrow \quad \frac{dy}{dx} = \frac{1}{2}$$

So, slope of normal $(m) = -\frac{1}{dy/dx} = -2$

When $x = \frac{\pi}{6}$, then $y = \frac{\pi}{4} + \frac{\pi}{12} = \frac{\pi}{3}$

The equation of normal at $\left(\frac{\pi}{6}, \frac{\pi}{3}\right)$ is

$$\left(y - \frac{\pi}{3}\right) = -2\left(x - \frac{\pi}{6}\right)$$

$$\Rightarrow \quad 2x + y = \frac{2\pi}{3}$$

Clearly, it passes through $\left(0, \frac{2\pi}{3}\right)$.

17. (b) The equation of the curve

$$y(x-2)(x-3) = x + 6 \quad \text{...(i)}$$

It intersects the Y-axis at $x = 0$.

On putting $x = 0$ in Eq. (i), we obtain

$$y = 1.$$

Hence, curve intersects Y-axis at (0, 1).

Now, $y(x-2)(x-3) = x + 6$

$$y = \frac{x+6}{x^2 - 5x + 6}$$

On differentiating both sides w.r.t. x, we obtain

$$\frac{dy}{dx} = \frac{(x^2 - 5x + 6) - (x+6)(2x-5)}{(x^2 - 5x + 6)^2}$$

$$\therefore \quad \left(\frac{dy}{dx}\right)_{(0,\,1)} = 1$$

The equation of the normal at (0, 1) is

$$y - 1 = (-1)(x - 0) \Rightarrow x + y = 1$$

Clearly, it passes through $\left(\frac{1}{2}, \frac{1}{2}\right)$.

18. (d) On differentiating $y = x^2$ w.r.t. x, we get

$$\frac{dy}{dx} = 2x$$

$$\Rightarrow \quad \left(\frac{dy}{dx}\right)_{(1,\,1)} = 2 = m_1 \quad \text{(let)}$$

On differentiating $6y = 7 - x^2$ w.r.t. x, we get

$$\frac{dy}{dx} = \frac{-x}{3}$$

$$\Rightarrow \quad \left(\frac{dy}{dx}\right)_{(1,\,1)} = -\frac{1}{3} = m_2 \quad \text{(let)}$$

Suppose the acute angle of intersection is θ.

Therefore, $\tan\theta = \left|\frac{m_1 - m_2}{1 + m_1 m_2}\right| = \left|\frac{2 + \frac{1}{3}}{1 - \frac{2}{3}}\right| = 7$

$$\Rightarrow \quad \theta = \tan^{-1}(7)$$

19. (b) Given, curves are

$$y = 2\sin^2 x \quad \text{...(i)}$$

and $\quad y = \cos 2x \quad$ ...(ii)

Solving Eqs. (i) and (ii), we get

$$2\sin^2 x = \cos 2x$$

$$\Rightarrow \quad 1 - \cos 2x = \cos 2x$$

$$\Rightarrow \quad \cos 2x = \frac{1}{2}$$

$$\Rightarrow \quad \cos 2x = \cos\frac{\pi}{3} \Rightarrow 2x = \pm\frac{\pi}{3}$$

$\therefore x = \pm\frac{\pi}{6}$ are the points of intersection.

From Eq. (i),

$$\frac{dy}{dx} = 4\sin x\cos x = 2\sin 2x = m_1 \quad \text{(let)}$$

From Eq. (ii),

$$\frac{dy}{dx} = -2\sin 2x = m_2 \quad \text{(let)}$$

If the angle of intersection is θ, then

$$\tan\theta = \left|\frac{m_1 - m_2}{1 + m_1 \cdot m_2}\right| = \left|\frac{4\sin 2x}{1 - 4\sin^2 2x}\right|$$

$$\Rightarrow \quad (\tan\theta)_{x=\pm\pi/6} = \left|\frac{4\times\left(\pm\frac{\sqrt{3}}{2}\right)}{1-4\times\frac{3}{4}}\right| = \left|\frac{\pm 2\sqrt{3}}{-2}\right| = \sqrt{3}$$

$$\therefore \quad \theta = \frac{\pi}{3}$$

20. (c) For finding point of intersection P,

$$a^x = b^x$$

$$\left(\frac{a}{b}\right)^x = 1$$

$$\Rightarrow \quad x = 0$$

Then, $y = 1$

Hence, point of intersection is (0, 1)

Now, $y = a^x$

$$\Rightarrow \quad \frac{dy}{dx} = a^x \log a$$

$$\Rightarrow \quad \left(\frac{dy}{dx}\right)_{(0,1)} = m_1 = \log a$$

and $y = b^x$

$$\Rightarrow \quad \frac{dy}{dx} = b^x \log b$$

$$\Rightarrow \quad \left(\frac{dy}{dx}\right)_{(0,1)} = m_2 = \log b$$

$$\tan\theta = \left|\frac{m_1 - m_2}{1 + m_1 m_2}\right| = \frac{\log a - \log b}{1 + \log a \log b}$$

$$= \frac{\log(a/b)}{1 + \log a \log b}$$

21. (b) Given, equation of curves are

$$y = 10 - x^2 \quad \text{...(i)}$$

and $$y = 2 + x^2 \quad \text{...(ii)}$$

For point of intersection, consider

$$10 - x^2 = 2 + x^2$$

$$\Rightarrow \quad 2x^2 = 8$$

$$\Rightarrow \quad x^2 = 4$$

$$\Rightarrow \quad x = \pm 2$$

Clearly, when $x = 2$, then $y = 6$ (using Eq. (i)) and when $x = -2$, then $y = 6$

Thus, the point of intersection are (2, 6) and (−2,6).

Let m_1 be the slope of tangent to the curve (i) and m_2 be the slope of tangent to the curve (ii)

For curve (i) $\frac{dy}{dx} = -2x$ and for curve (ii) $\frac{dy}{dx} = 2x$

$\therefore$ At (2, 6), slopes $m_1 = -4$ and $m_2 = 4$, and in that case

$$|\tan\theta| = \left|\frac{m_2 - m_1}{1 + m_1 m_2}\right| = \left|\frac{4+4}{1-16}\right| = \frac{8}{15}$$

At (−2,6), slopes $m_1 = 4$ and $m_2 = -4$ and in that case

$$|\tan\theta| = \left|\frac{m_2 - m_1}{1 + m_1 m_2}\right| = \left|\frac{-4-4}{1-16}\right| = \frac{8}{15}$$

22. (d) We have, $\frac{x^2}{a} + \frac{y^2}{b} = 1$ ...(i)

On differentiating both sides w.r.t. x, we get

$$\frac{2x}{a} + \frac{2y}{b}\frac{dy}{dx} = 0$$

$$\Rightarrow \quad \frac{y}{b}\frac{dy}{dx} = -\frac{x}{a}$$

$$\Rightarrow \quad \frac{dy}{dx} = -\frac{bx}{ay} \quad \text{...(ii)}$$

Similarly, for curve

$$\frac{x^2}{c} + \frac{y^2}{d} = 1 \quad \text{...(iii)}$$

$$\frac{dy}{dx} = \frac{-xd}{cy} \quad \text{...(iv)}$$

Both curves intersect at right angle

$$\left(\frac{-bx}{ay}\right)\left(\frac{-xd}{cy}\right) = -1 \quad [\because m_1 m_2 = -1]$$

$$\Rightarrow \quad bdx^2 = -acy^2 \quad \text{...(v)}$$

Subtracting Eq. (iii) from Eq. (i), we get

$$\left(\frac{1}{a} - \frac{1}{c}\right)x^2 + \left(\frac{1}{b} - \frac{1}{d}\right)y^2 = 0$$

$$\Rightarrow \quad \frac{c-a}{ac}x^2 + \frac{d-b}{bd}\left(\frac{-bd}{ac}\right)x^2 = 0 \quad \text{[from Eq. (v)]}$$

$$\Rightarrow \quad (c-a) - (d-b) = 0$$

$$\Rightarrow \quad c - a = d - b$$

$$\Rightarrow \quad c - d = a - b$$

23. (b) We have,

$$\Rightarrow \quad y^2 = 4ax$$

$$\Rightarrow \quad 2y\frac{dy}{dx} = 4a$$

$$\Rightarrow \quad \frac{dy}{dx} = \frac{2a}{y}$$

$$\text{Subtangent} = \frac{y}{dy/dx} = \frac{y}{2a/y} = \frac{y^2}{2a} = \frac{4ax}{2a} = 2x$$

$$\text{Ratio of subtangent to the abscissa} = \frac{2x}{x} = \frac{2}{1}$$

So, required ratio is 2 : 1.

24. (b) Given,

Subtangent = Subnormal

$$\Rightarrow \quad y\frac{dx}{dy} = y\frac{dy}{dx}$$

$$\Rightarrow \quad \left(\frac{dy}{dx}\right)^2 = 1 \Rightarrow \frac{dy}{dx} = \pm 1$$

$$\therefore \text{ Length of tangent} = \frac{y\sqrt{1+\left(\frac{dx}{dy}\right)^2}}{\frac{dy}{dx}}$$

$$= y\sqrt{2} = \sqrt{2} \text{ ordinate}$$

25. (d) We have, $xy = c^2$

$$\Rightarrow \quad x\frac{dy}{dx} + y = 0 \Rightarrow \frac{dy}{dx} = -\frac{y}{x}$$

Now, length of subnormal $= \left|y\frac{dy}{dx}\right|$

$$= y\left|-\left(\frac{y}{x}\right)\right| = \left|\frac{-y^2}{x}\right|$$

$$= \left|-\frac{y^2}{c^2/y}\right| = \frac{y^3}{c^2}$$

Hence, subnormal $\propto y^3$.

26. (a) On differentiating both sides w.r.t. x, we get

$$2xy^2 + 2x^2y\frac{dy}{dx} = 2a^2x$$

$$\Rightarrow \quad \frac{dy}{dx} = \frac{a^2 - y^2}{xy}$$

Now, the subnormal is given by

$$\left|y \cdot \frac{dy}{dx}\right| = \left|\frac{a^2 - y^2}{x}\right| = \left|\frac{a^2 - \frac{a^2(x^2 - a^2)}{x^2}}{x}\right| = \frac{a^4}{|x^3|}$$

27. (a) Given, parabola is $y^2 = 4ax$

$$\Rightarrow \quad 2y\frac{dy}{dx} = 4a \Rightarrow \frac{dy}{dx} = \frac{2a}{y}$$

$$\text{Subtangent} = y \cdot \frac{dx}{dy} = y \cdot \left(\frac{y}{2a}\right) = \frac{y^2}{2a}$$

Ordinate $= y$

$$\text{Subnormal} = y\left(\frac{dy}{dx}\right) = 2a$$

$$\therefore \text{ Subtangent} \times \text{Subnormal} = \frac{y^2}{2a} \times 2a$$

$\Rightarrow$ (Ordinate)^2 = Subtangent $\times$ Subnormal

Hence, they are in GP.

28. (d) We have, $y = a\left(\frac{e^{-x/a} + e^{x/a}}{2}\right)$

$$\frac{dy}{dx} = \frac{e^{x/a} - e^{-x/a}}{2}$$

$\therefore$ Length of normal at any point

$$= y\sqrt{1+\left(\frac{dy}{dx}\right)^2}$$

$$= y\sqrt{1+\left(\frac{e^{x/a} - e^{-x/a}}{2}\right)^2}$$

$$= y\sqrt{\left(\frac{e^{\frac{x}{a}} + e^{-\frac{x}{a}}}{2}\right)^2}$$

$$= y\left(\frac{e^{x/a} + e^{-x/a}}{2}\right)$$

$$= \frac{y^2}{a} \qquad \left[\because \frac{e^{x/a} + e^{-x/a}}{2} = \frac{y}{a}\right]$$

Hence, the length of the normal at any point varies as the square of the ordinate of the point.

29. (d) Given, $xy^n = a^{n+1}$ ...(i)

On differentiating both sides w.r.t. x, we get

$$y^n + nxy^{n-1}\frac{dy}{dx} = 0$$

$$\Rightarrow \quad \frac{dy}{dx} = -\frac{y}{nx}$$

Now, length of subnormal

$$= y\frac{dy}{dx} = y\left(\frac{-y}{nx}\right)$$

$$= \frac{-y^2}{nx} = \frac{-y^2}{n} \cdot \frac{y^n}{a^{n+1}} = -\frac{y^{n+2}}{n(a^{n+1})}$$

Length of subnormal is constant, if $n + 2 = 0$

$n = -2$

30. (d) Given, $y^2 = \frac{x^2}{2a+x}$

$$\Rightarrow \quad 2y\frac{dy}{dx} = \frac{(2a+x)2x - x^2}{(2a+x)^2}$$

$$\Rightarrow \quad \frac{dy}{dx} = \frac{4ax + x^2}{2y(2a+x)^2}$$

$$\left(\frac{dy}{dx}\right)_{(a,\,a)} = \frac{5a^2}{18a^3} = \frac{5}{18a}$$

$$\text{Length of subtangent} = \frac{y}{(dy/dx)} = \frac{a}{(5/18a)} = \frac{18a^2}{5}.$$

31. (b) We have, $x^m y^n = a^{m+n}$

On differentiating w.r.t. x, we get

$$x^m \cdot ny^{n-1}\frac{dy}{dx} + y^n mx^{m-1} = 0$$

$$\Rightarrow \quad \frac{dy}{dx} = -\frac{my}{nx}$$

Then, length of subtangent

$$= \left|\frac{y}{dy/dx}\right| = \left|\frac{y}{-my/nx}\right|$$

$$= \frac{nx}{m}$$

which is proportional to abscissa.

32. (8) The equation of the curve is

$$(y - x^5)^2 = x(1+x^2)^2$$

On differentiating both sides, w.r.t. x, we get

$$2(y - x^5)\left(\frac{dy}{dx} - 5x^4\right) = (1+x^2)^2 + 4x^2(1+x^2)$$

On putting $x = 1$ and $y = 3$, we get

$$4\left(\frac{dy}{dx} - 5\right) = 4 + 8$$

$$\Rightarrow \quad \frac{dy}{dx} = 8$$

Hence, required slope of tangent is 8.

33. (1) We have, $x + y - \log(x+y) = 2x + 5$

On differentiating both sides w.r.t. x, we get

$$1 + y' - \frac{1}{x+y}(1+y') = 2$$

$$\Rightarrow \quad (1+y')(x+y) - (1+y') = 2x + 2y$$

$$\Rightarrow \quad (x+y-1) + (x+y-1)y' = 2x + 2y$$

$$\Rightarrow \quad y' = \frac{x+y+1}{x+y-1}$$

For vertical tangent at (k, l),

$$k + l - 1 = 0$$

$$\therefore \quad k + l = 1$$

34. (3) We have, $y = x^3 - ax^2 + x + 1$

On differentiating both sides w.r.t. x, we get

$$\frac{dy}{dx} = 3x^2 - 2ax + 1 > 0 \,\forall\, x$$

$$\Rightarrow \quad (-2a)^2 - 4(3)(1) < 0$$

$$\Rightarrow \quad a^2 < 3$$

$$\Rightarrow \quad -\sqrt{3} < a < \sqrt{3}$$

$\therefore$ Integral values of a are $-1, 0$ and 1.

So, number of integral values of 3.

35. (2) We have,

$$x^2 - xy + y^2 + \alpha(x-2) = 4$$

On differentiating both sides w.r.t. x, we get

$$2x - xy' - y + 2yy' + \alpha = 0$$

$$\therefore \quad y' = \frac{2x - y - \alpha}{x - 2y}$$

$$(y')_{(2,\,2)} = \frac{2+\alpha}{-2}$$

Therefore, the equation of tangent at (2, 2) is

$$y - 2 = \frac{2+\alpha}{-2}(x-2)$$

$$\text{or} \quad \left(1 + \frac{\alpha}{2}\right)x + y - \alpha - 4 = 0$$

It is given that the distance of (0, 0) from the line is 2.

$$\frac{|\alpha + 4|}{\sqrt{(1+\alpha)^2 + 1}} = 2$$

$$\Rightarrow \quad (\alpha+4)^2 = 4\left[1 + \alpha + \frac{\alpha^2}{4} + 1\right]$$

$$\Rightarrow \quad \alpha^2 + 8\alpha + 16 = 8 + \alpha^2 + 4\alpha$$

$$\Rightarrow \quad 4\alpha = -8 \Rightarrow \alpha = -2$$

Hence, $|\alpha| = 2$

36. (5) Let the common tangent be $y = ax + b$

On solving with both the curves, we get

$$x^2 - x + 1 = ax + b$$

$$\text{and} \quad x^2 - 3x + 1 = ax + b$$

$$\Rightarrow \quad x^2 - (1+a)x + (1-b) = 0$$

$$\text{and} \quad x^2 - (a+3)x + (1-b) = 0$$

On putting $D = 0$, we get

$$(a+1)^2 - 4(1-b) = 0$$

$$\text{and} \quad (a+3)^2 - 4(1-b) = 0$$

$$\Rightarrow \quad a^2 + 2a + 4b - 3 = 0$$

$$\text{and} \quad a^2 + 6a + 4b + 5 = 0$$

On solving these equations, we get,

$$a = -2$$

Thus, slope of the common tangent $= m = -2$

Hence, the value of $(|m| + 3)$ is 5.

37. (4) Let a point $P(x_1, y_1)$ on the curve $y^2 - 3x^2 + y + 10 = 0$, then slope of normal to the curve at point P is $m_N = -\dfrac{1}{\left(\dfrac{dy}{dx}\right)_{(x_1,\,y_1)}}$

On differentiating the curve w.r.t. x, we get

$$2y\frac{dy}{dx} - 6x + \frac{dy}{dx} = 0 \Rightarrow \frac{dy}{dx} = \frac{6x}{2y+1}$$

$$\therefore \quad m_N = -\frac{2y_1 + 1}{6x_1}$$

Now, the normal to the curve at point P intersect the Y-axis $\left(0, \frac{3}{2}\right)$, so slope of normal is $m_N = \frac{\frac{3}{2} - y_1}{0 - x_1}$

On equating the value of m_N, we get

$$\frac{3 - 2y_1}{x_1} = \frac{2y_1 + 1}{3x_1}$$

$$\Rightarrow \quad 9 - 6y_1 = 2y_1 + 1 \quad (\because x_1 \neq 0)$$

$$\Rightarrow \quad 8y_1 = 8 \Rightarrow y_1 = 1$$

$\because$ Point $P(x_1, y_1)$ on the curve, so

$$(1)^2 - 3x_1^2 + 1 + 10 = 0$$

$$\Rightarrow \quad 3x_1^2 = 12$$

$$\Rightarrow \quad x_1^2 = 4 \Rightarrow x_1 = \pm 2$$

Now, slope of tangent to the curve at $P(x_1, y_1)$ is

$$m = -\frac{1}{m_N} = \left.\frac{dy}{dx}\right|_{(x_1, y_1)} = \frac{6x_1}{2y_1 + 1}$$

$$\Rightarrow |m| = \left|\frac{6x_1}{2y_1 + 1}\right| = \left|\frac{6(\pm 2)}{2(1) + 1}\right| = \frac{12}{3} = 4$$

38. (4) Given, curve is $\sqrt{x} + \sqrt{y} = 2$

$$\Rightarrow \quad \frac{dy}{dx} = -\sqrt{\frac{y}{x}}$$

Let $P(\alpha, \beta)$ be a variable point on the given curve

$$\therefore \quad \left(\frac{dy}{dx}\right)_{(\alpha, \beta)} = -\sqrt{\frac{\beta}{\alpha}}$$

$\therefore$ Equation of tangent is

$$y - \beta = -\frac{\sqrt{\beta}}{\sqrt{\alpha}}(x - \alpha)$$

$$\Rightarrow \quad \frac{y - \beta}{\sqrt{\beta}} = -\frac{(x - \alpha)}{\sqrt{\alpha}}$$

$$\Rightarrow \quad \frac{y}{\sqrt{\beta}} - \sqrt{\beta} = -\frac{x}{\sqrt{\alpha}} + \sqrt{\alpha}$$

$$\Rightarrow \quad \frac{x}{\sqrt{\alpha}} + \frac{y}{\sqrt{\beta}} = \sqrt{\alpha} + \sqrt{\beta}$$

$$\Rightarrow \quad \frac{x}{\sqrt{\alpha}} + \frac{y}{\sqrt{\beta}} = 2 \Rightarrow \frac{x}{2\sqrt{\alpha}} + \frac{y}{2\sqrt{\beta}} = 1$$

Thus, the sum of the intercepts $= 2(\sqrt{\alpha} + \sqrt{\beta}) = 2 \times 2 = 4$

39. (4) Given, curves are $y = x^2 + ax + b$

and $\quad y = x(c - x)$

Since, the point (1, 0) lie on the both curves.

So, $\quad a + b + 1 = 0$ and $c - 1 = 0$

$$a + b = -1 \text{ and } c = 1 \quad \text{...(i)}$$

Again, both the curves touch each other at (1, 0)

So, $(2x + a)_{(1, 0)} = (c - 2x)_{(1, 0)}$

$$2 + a = c - 2$$

$$a = c - 4 \quad \text{...(ii)}$$

From Eqs. (i) and (ii), we get

$$a = -3, b = 2 \text{ and } c = 1$$

Hence, the value of $(a + b + c + 4)$ is 4.

40. (2) We have, $\quad y = x^2 - 3x + 2 \quad$ ...(i)

$$\Rightarrow \quad y = (x - 1)(x - 2)$$

Curve intersects the X-axis, then $y = 0$

$$0 = (x - 1)(x - 2) \Rightarrow x = 1, 2$$

So, the curve intersects the X-axis at $A(1, 0)$ and $B(2, 0)$

Now, on differentiating Eq. (i) w.r.t. x, we get

$$\frac{dy}{dx} = 2x - 3$$

$$\therefore \quad \left(\frac{dy}{dx}\right)_{x=1} = -1 \text{ and } \left(\frac{dy}{dx}\right)_{x=2} = 1$$

Equation of tangent at (1, 0) is

$$y = -1(x - 1)$$

$$y = -x + 1$$

$$\Rightarrow \quad x + y = 1$$

and equation of tangent at (2, 0) is

$$y = 1(x - 2)$$

$$\Rightarrow \quad x - y = 2$$

$$\therefore \quad \frac{b}{a} = 2$$

41. (4) We have, $x = y^4$ and $xy = k$

For intersection, $\quad y^5 = k \quad$ ...(i)

Also, $x = y^4$

$$\Rightarrow \quad 1 = 4y^3 \frac{dy}{dx}$$

$$\Rightarrow \quad \frac{dy}{dx} = \frac{1}{4y^3}$$

For $\; xy = k \Rightarrow x = \frac{k}{y}$

On differentiating both sides w.r.t. x, we get

$$\Rightarrow \quad 1 = \frac{-k}{y^2}\frac{dy}{dx}$$

$$\Rightarrow \quad \frac{dy}{dx} = \frac{-y^2}{k}$$

$\because$ Curves cut at right angles

So, $\quad \frac{1}{4y^3} \times \left(-\frac{y^2}{k}\right) = -1 \Rightarrow y = \frac{1}{4k}$

$\therefore$ From Eq. (i), $\quad y^5 = k$

$$\Rightarrow \quad k = \frac{1}{(4k)^5}$$

$$\Rightarrow \quad (4k)^6 = 4$$

Level 3

1. (b) Given, $P(t, t^3)$ and let $Q(t_1, t_1^3)$

Slope of tangent at point $P(t, t^3)$ is

$$\frac{dy}{dx} = 3x^2$$

$$\left.\frac{dy}{dx}\right|_{(t,\, t^3)} = 3t^2$$

Now, this slope is equal to the slope of PQ.

$$\text{Slope of PQ} = \frac{t_1^3 - t^3}{t_1 - t}$$

$$3t^2 = \frac{(t_1 - t)(t_1^2 + tt_1 + t^2)}{(t_1 - t)}$$

$$\Rightarrow \quad t_1^2 + tt_1 + t^2 = 3t^2$$

$$\Rightarrow \quad t_1^2 + tt_1 - 2t^2 = 0$$

$$\Rightarrow \quad t_1^2 + 2tt_1 - tt_1 - 2t^2 = 0$$

$$\Rightarrow \quad t_1(t_1 + 2t) - t(t_1 + 2t) = 0$$

$$\Rightarrow \quad (t_1 + 2t)(t_1 - t) = 0$$

$$\Rightarrow \quad t_1 + 2t = 0$$

$$\Rightarrow \quad t_1 = -2t$$

So, $\quad Q = (-2t, -8t^3)$

Now, let M divides PQ in the ratio of 1 : 2.

So,

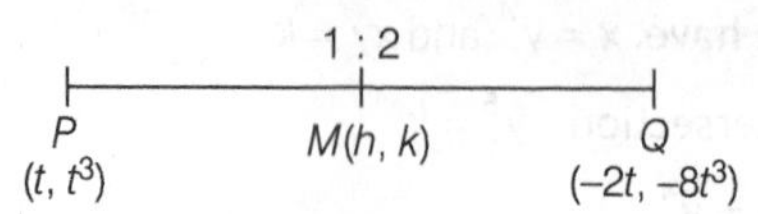

$$\therefore \quad h = \frac{2t - 2t}{3} = 0$$

$$k = \frac{2t^3 - 8t^3}{3} = -2t^3.$$

$\therefore$ Point $M = (0, -2t^3)$

So ordinate of $M = -2t^3$

2. (c) Two curves must intersect at only one point. Hence,

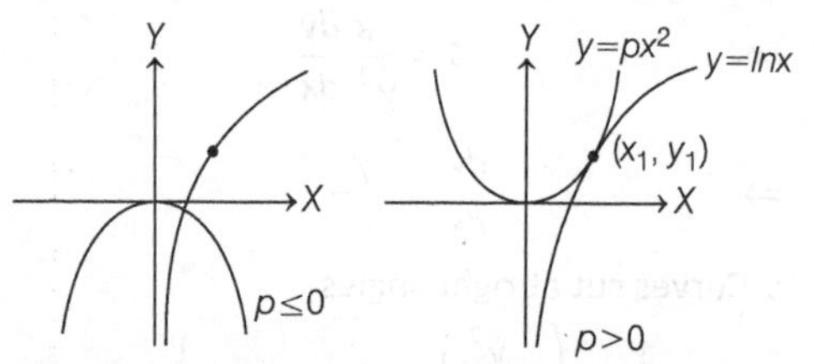

I. If $p \le 0$, then only one solution (see graph)

II. If $p > 0$, then the two curves must only touch each other i.e. tangent at $y = px^2$ and $y = \ln x$ have same slope at point (x_1, y_1)

$$\Rightarrow \quad 2px_1 = \frac{1}{x_1}$$

$$\Rightarrow \quad x_1^2 = \frac{1}{2p} \qquad \text{...(i)}$$

Also, $\quad y_1 = px_1^2$

$$\Rightarrow \quad y_1 = p\left(\frac{1}{2p}\right)$$

$$\Rightarrow \quad y_1 = \frac{1}{2} \qquad \text{...(ii)}$$

and $\quad y_1 = \ln x_1$

$$\Rightarrow \quad \frac{1}{2} = \ln x_1$$

$$\Rightarrow \quad x_1 = e^{1/2} \qquad \text{...(iii)}$$

Hence, $x_1^2 = \dfrac{1}{2p}$

$$\Rightarrow \quad e = \frac{1}{2p}$$

$$\Rightarrow \quad p = \frac{1}{2e}$$

Hence, possible values of p are $(-\infty, 0] \cup \left\{\dfrac{1}{2e}\right\}$

3. (b) Let $P(x_1, y_1)$ be a point on the curve

$$2xy = a^2$$

Then, $\quad 2x_1y_1 = a^2 \qquad \text{...(i)}$

Now, $\quad 2xy = a^2$

On differentiating both sides w.r.t. x, we get

$$2\left(x\frac{dy}{dx} + y\right) = 0$$

$$\Rightarrow \quad \frac{dy}{dx} = -\frac{y}{x}$$

$$\left(\frac{dy}{dx}\right)_{(x_1,\, y_1)} = -\frac{y_1}{x_1}$$

The equation of tangent to $P(x_1, y_1)$ is

$$y - y_1 = -\frac{y_1}{x_1}(x - x_1)$$

$$\Rightarrow \quad x_1y - x_1y_1 = -xy_1 + x_1y_1$$

$$\Rightarrow \quad xy_1 + yx_1 = 2x_1y_1$$

$$\Rightarrow \quad xy_1 + yx_1 = a^2 \qquad \text{[using Eq. (i)]}$$

This curve meets the coordinate axis at $A\left(\dfrac{a^2}{y_1}, 0\right)$ and $B\left(0, \dfrac{a^2}{x_1}\right)$

$$\therefore \text{ Area of } \Delta OAB = \frac{1}{2} \times OA \times OB$$

$$= \frac{1}{2}\frac{a^2}{y_1} \cdot \frac{a^2}{x_1} = \frac{a^4}{2x_1y_1} = \frac{a^4}{a^2} = a^2$$

4. (b) Given, $x = \sec^2\theta$ and $y = \cot\theta$

$$\frac{dy}{dx} = \frac{dy/d\theta}{dx/d\theta} = -\frac{1}{2}\cot^3\theta$$

$$\left(\frac{dy}{dx}\right)_{\theta=\frac{\pi}{4}} = -\frac{1}{2}\cot^3\left(\frac{\pi}{4}\right) = -\frac{1}{2}$$

Now, when $\theta = \frac{\pi}{4}$,

$$x = \sec^2\left(\frac{\pi}{4}\right) = 2 \text{ and } y = \cot\left(\frac{\pi}{4}\right) = 1$$

So, coordinate of $P(2, 1)$.

Equation of tangent at (2, 1) is

$$y - 1 = -\frac{1}{2}(x - 2)$$

$$\Rightarrow \quad x + 2y - 4 = 0 \quad \text{...(i)}$$

Now, the cartesian equation of parametric function can be obtained by eliminating θ.

Using $\sec^2\theta - \tan^2\theta = 1$, we get

$$y^2 = \frac{1}{x-1} \quad \text{...(ii)}$$

From Eq. (i), $y = \frac{4-x}{2}$

Substituting in Eq. (ii),

$$\left(\frac{4-x}{2}\right)^2 = \frac{1}{x-1}$$

On solving, we get

$$x = 2, 5$$

If $x = 2$, then $y = 1$

and $x = 5$, then $y = -\frac{1}{2}$.

Hence, $P(2, 1)$ and $Q\left(5, -\frac{1}{2}\right)$

$$\therefore PQ = \sqrt{(5-2)^2 + \left(-\frac{1}{2} - 1\right)^2} = \sqrt{9 + \frac{9}{4}} = \sqrt{\frac{45}{4}} = \frac{3\sqrt{5}}{2}$$

5. (b) Given that at (3, 4),

Length of subtangent = Length of subnormal

$$\frac{y}{\left(\frac{dy}{dx}\right)} = y \cdot \left(\frac{dy}{dx}\right)$$

$$\Rightarrow \quad \left(\frac{dy}{dx}\right)^2_{x=3} = 1 \Rightarrow f'(3) = \pm 1$$

If $f'(3) = 1$, then the equations of tangent as

$$y - 4 = 1(x - 3)$$

$$\Rightarrow \quad y = x + 1$$

Hence, $A = (-1, 0)$ and $B = (0, 1)$

Thus, area of $\Delta OAB = \frac{1}{2} \times 1 \times 1 = \frac{1}{2}$ sq unit

Considering $f'(3) = -1$, we get

$$y - 4 = -1(x - 3)$$

$$\Rightarrow \quad x + y = 7$$

Hence, $A = (7, 0)$ and $B(0, 7)$.

Thus, area of $\Delta OAB = \frac{1}{2} \times 7 \times 7 = \frac{49}{2}$ sq units

Hence, maximum area is $\frac{49}{2}$ sq units.

6. (a) Let tangent be drawn at $A(3t^2, 2t^3)$ and it is normal at $B(3t_1^2, 2t_1^3)$.

Slope of tangent at $A = \frac{dy}{dt} \div \frac{dx}{dt} = t$

Its equation is $(y - 2t^3) = t(x - 3t^2)$

$$\text{or} \quad y = tx - t^3 \quad \text{...(i)}$$

Slope of normal at $B = -\frac{1}{t_1}$

But they represent the same line.

$$\therefore \quad t = -\frac{1}{t_1}$$

$$\text{or} \quad tt_1 = -1 \quad \text{...(ii)}$$

Now, tangent at A is $tx - y - t^3 = 0$

It passes through $B(3t_1^2, 2t_1^3)$

$$\therefore \quad t \cdot 3t_1^2 - 2t_1^3 - t^3 = 0$$

$$\text{or} \quad 3t_1^2(t - t_1) - (t^3 - t_1^3) = 0$$

$$\text{or} \quad 3t_1^2 - (t^2 + tt_1 + t_1^2) = 0$$

$$\text{or} \quad 2t_1^2 - tt_1 - t^2 = 0$$

$$\text{or} \quad (t_1 - t)(2t_1 + t) = 0$$

$$\therefore \quad t = -2t_1 = \frac{2}{t} \quad \text{[from Eq. (ii)]}$$

$$\therefore \quad t^2 = 2 \text{ or } t = \pm\sqrt{2}$$

Putting the value of t in Eq. (i), equation of tangent is

$$y = \pm\sqrt{2}(x - 2)$$

7. (c) The given curves are $y = |x^2 - 1|$

and $y = |x^2 - 3|$

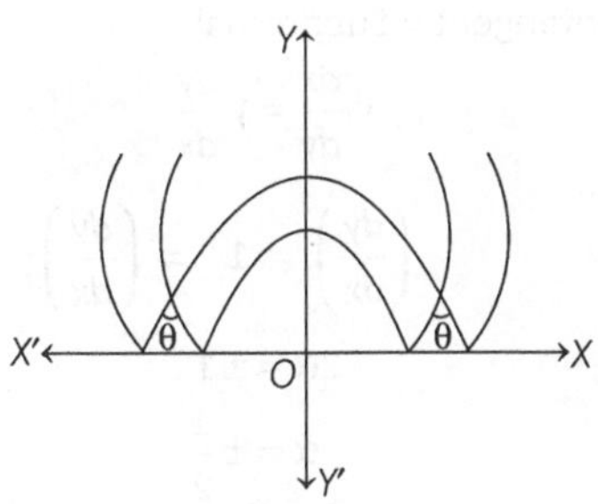

On solving these curves, we get

$$x^2 - 1 = -x^2 + 3$$
$$\Rightarrow \quad 2x^2 = 4$$
$$\Rightarrow \quad x^2 = 2$$
$$\Rightarrow \quad x = \pm\sqrt{2}$$

When $x = \pm\sqrt{2}$, then $y = 2 - 1 = 1$

So, the point of intersection is $(\pm\sqrt{2}, 1)$.

Now, consider the point of intersection is $P(\sqrt{2}, 1)$

Now, $\quad y = |x^2 - 1|$

$$\Rightarrow \quad y = x^2 - 1$$

On differentiating both sides w.r.t. x, we get

$$\Rightarrow \quad \frac{dy}{dx} = 2x$$
$$\Rightarrow \quad m_1 = \left(\frac{dy}{dx}\right)_P = 2\sqrt{2}$$

Also, $y = |x^2 - 3|$

$$\Rightarrow \quad y = -x^2 + 3$$
$$\therefore \quad \frac{dy}{dx} = -2x$$
$$\Rightarrow \quad m_2 = \left(\frac{dy}{dx}\right)_P = -2\sqrt{2}$$

Let θ be the angle between them.

Then, $\tan\theta = \left|\frac{m_2 - m_1}{1 + m_1 m_2}\right| = \left|\frac{-2\sqrt{2} - 2\sqrt{2}}{1 + 2\sqrt{2}\cdot(-2\sqrt{2})}\right|$

$$\Rightarrow \quad \tan\theta = \frac{4\sqrt{2}}{7}$$
$$\Rightarrow \quad \theta = \tan^{-1}\left(\frac{4\sqrt{2}}{7}\right)$$

8. (c) We have, $\quad y = e^{\alpha x} + \alpha x$

On differentiating both sides w.r.t. x, we get

$$\frac{dy}{dx} = \alpha e^{\alpha x} + \alpha$$
$$\Rightarrow \quad \left(\frac{dy}{dx}\right)_{(0,\,1)} = \alpha e^0 + \alpha = 2\alpha$$

Now, subtangent = Subnormal

$$y\frac{dx}{dy} = y\frac{dy}{dx}$$
$$\Rightarrow \quad \left(\frac{dy}{dx}\right)^2 = 1 \Rightarrow \left(\frac{dy}{dx}\right) = \pm 1$$
$$\Rightarrow \quad 2\alpha = \pm 1$$
$$\Rightarrow \quad \alpha = \pm\frac{1}{2}$$

9. (d) Given, curve is $y^3 + 3x^2 = 12y$

On differentiating both sides w.r.t. x, we get

$$\Rightarrow \quad 3y^2\frac{dy}{dx} + 6x = 12\frac{dy}{dx}$$
$$\Rightarrow \quad y^2\frac{dy}{dx} + 2x = 4\frac{dy}{dx}$$
$$\Rightarrow \quad \frac{dy}{dx} = \left(\frac{-2x}{y^2 - 4}\right)$$

Since, the tangent is vertical.

So, $\quad \frac{dx}{dy} = 0$

$$\Rightarrow \quad \left(\frac{y^2 - 4}{-2x}\right) = 0 \Rightarrow y = \pm 2$$

When $y = 2$, then

$$3x^2 = 12y - y^3 = (24 - 8) = 16$$
$$\Rightarrow \quad x^2 = \frac{16}{3} \Rightarrow x = \pm\frac{4}{\sqrt{3}}$$

If $y = -2$, then $3x^2 = -16$,

we get no real solution

Thus, the points are $\left(\frac{4}{\sqrt{3}}, 2\right)$ and $\left(-\frac{4}{\sqrt{3}}, 2\right)$.

10. (a, b) We have, $x = 2\ln\cot t + 1$

On differentiating both sides w.r.t. x, we get

$$\Rightarrow \quad \frac{dx}{dt} = \frac{2}{\cot t}(-\operatorname{cosec}^2 t)$$
$$= \frac{-2}{\sin^2 t\cot t}$$
$$\left(\frac{dx}{dt}\right)_{t=\frac{\pi}{4}} = -4$$

and $\quad y = \tan t + \cot t$

$$\frac{dy}{dt} = \sec^2 t - \operatorname{cosec}^2 t$$
$$= \frac{\sin^2 t - \cos^2 t}{\sin^2 t\cos^2 t}$$
$$\frac{dy}{dt} = 0$$

Hence, $\quad \left(\frac{dy}{dx}\right)_{t=\frac{\pi}{4}} = \frac{dy/dt}{dx/dt} = 0$

So, slope of tangent = 0

Tangent is parallel to X-axis.

Slope of normal $= -\frac{1}{\left(\frac{dy}{dx}\right)} = -\infty$

So, normal is parallel to the Y-axis.

11. (a, b, c, d) We have,

$$y = f(x) = \frac{\ln x}{x}$$

$$\therefore \quad \frac{dy}{dx} = \frac{x\left(\frac{1}{x}\right) - \ln x}{x^2} = \frac{1 - \ln x}{x^2}$$

For horizontal tangent,

$$\frac{dy}{dx} = 0$$

$$\Rightarrow \quad \ln x = 1$$

$$\Rightarrow \quad x = e$$

$f(x)$ intersects X-axis, where $x = 1$ i.e. at (1, 0)

$\therefore$ $f(x)$ is decreases for $x > e$ and increases for $x \in (0, e)$ and $f(x)$ is continuous on $(0, \infty)$

$\Rightarrow f(x)$ is many one function.

Also, $\quad \frac{dy}{dx} = \frac{1 - \ln x}{x^2} \to \infty$ as $x \to 0$

$\Rightarrow f(x)$ has one vertical tangent (asymptote)

12. (a, c) We have,

$$x = t^3 - 6t^2 + 9t$$

$$\therefore \quad \frac{dx}{dt} = 3t^2 - 12t + 9$$

and $\quad y = t^2$

$$\therefore \quad \frac{dy}{dt} = 2t$$

Now, $\quad \frac{dy}{dx} = \frac{dy/dt}{dx/dt} = \frac{2t}{3t^2 - 12t + 9}$

When tangent is perpendicular to X-axis, then

$$\frac{dx}{dy} = 0$$

$$\Rightarrow \quad \frac{3t^2 - 12t + 9}{2t} = 0$$

$$\Rightarrow \quad t^2 - 4t + 3 = 0$$

$$\Rightarrow \quad (t-1)(t-3) = 0$$

$$\Rightarrow \quad t = 1, 3$$

13. (b, c) We have, $\quad ax + by + c = 0$

Slope, $m = -\frac{a}{b}$

Hence, slope of normal $= -\frac{a}{b}$

Now, given curve is $\quad xy = 1$

$$\therefore \quad x\frac{dy}{dx} + y = 0$$

$$\Rightarrow \quad \frac{dy}{dx} = -\frac{y}{x}$$

$$\Rightarrow \quad \frac{dy}{dx} = -\frac{1}{x^2} \quad [\because xy = 1]$$

So, slope of normal $= -\frac{1}{dy/dx} = \frac{-1}{-1/x^2} = x^2$

Hence, $x^2 = -\frac{a}{b}$

Now, $x^2 > 0$

$\Rightarrow -\frac{a}{b}$ must be a positive number, it is only possible, when $a < 0$ and $b > 0$

or $\quad a > 0$ and $b < 0$

14. (a, d) Let (x_1, y_1) be the point of contact from $(0, -9)$ to the curve $y = x^2 + 3x$.

Then, the slope of tangent $= \frac{y_1 + 9}{x_1}$

Also, $\quad \frac{dy}{dx} = 2x + 3$

$$\left.\frac{dy}{dx}\right|_{(x_1, y_1)} = 2x_1 + 3$$

So, $\quad \frac{y_1 + 9}{x_1} = 2x_1 + 3$

$$\Rightarrow \quad y_1 = 2x_1^2 + 3x_1 - 9$$

Also, since (x_1, y_1) satisfies the equation of the curve

$$\therefore \quad y_1 = x_1^2 + 3x_1$$

On comparing Eqs. (i) and (ii), we get

$$x_1^2 + 3x_1 = 2x_1^2 + 3x_1 - 9$$

$$\Rightarrow \quad x_1^2 = 9$$

$$\Rightarrow \quad x_1 = \pm 3$$

So, points of contact are (3, 18) and $(-3, 0)$.

15. (c, d) We have,

$$2y^3 = ax^2 + x^3$$

$$\therefore \quad 6y^2\frac{dy}{dx} = 2ax + 3x^2$$

$$\Rightarrow \quad \left.\frac{dy}{dx}\right|_{(a, a)} = \frac{2a^2 + 3a^2}{6a^2} = \frac{5}{6}$$

Let the tangent be $\frac{x}{\alpha} + \frac{y}{\beta} = 1$

$$\Rightarrow \quad \beta x + \alpha y = \alpha\beta$$

So, slope of tangent $= -\frac{\beta}{\alpha}$

$$\Rightarrow \quad \frac{-\beta}{\alpha} = \frac{5}{6} \Rightarrow \beta = -\frac{5}{6}\alpha$$

Now, $\alpha^2 + \beta^2 = 61$

$$\Rightarrow \quad \alpha^2 + \frac{25}{36}\alpha^2 = 61$$

$$\Rightarrow \quad \alpha^2 = 36$$

So, $\alpha = \pm 6$, then $\beta = \mp 6$

Also, (a, a) lies on $\frac{x}{\alpha} + \frac{y}{\beta} = 1$

So, $\frac{a}{\alpha} + \frac{a}{\beta} = 1$

Case I. If $\alpha = 6$ and $\beta = -6$

$\Rightarrow \quad \frac{a}{6} - \frac{a}{5} = 1$

$\Rightarrow \quad 5a - 6a = 30$

$\Rightarrow \quad a = -30$

Case II. If $\alpha = -6$ and $\beta = 5$

$\Rightarrow \quad -\frac{a}{6} + \frac{a}{5} = 1$

$\Rightarrow \quad a = 30$

16. (a, b) Given, curve is $\frac{x^2}{9} - \frac{y^2}{4} = 1$

$\Rightarrow \quad \frac{2x}{9} = \frac{2y}{4}\frac{dy}{dx}$

$\Rightarrow \quad \frac{dy}{dx} = \frac{4x}{9y}$

Since, the tangent is parallel to the straight line $2x - y = 1$.

So, $\quad \frac{4x}{9y} = 2$

$\Rightarrow \quad x = \frac{9y}{2}$

When $x = \frac{9y}{2}$, then $\frac{\left(\frac{9y}{2}\right)^2}{9} - \frac{y^2}{4} = 1$

$\Rightarrow \quad \frac{9y^2}{4} - \frac{y^2}{4} = 1$

$\Rightarrow \quad \frac{8y^2}{4} = 1$

$\Rightarrow \quad y^2 = \frac{1}{2}$

$\Rightarrow \quad y = \pm \frac{1}{\sqrt{2}}$

When $y = \pm \frac{1}{\sqrt{2}}$, then $x = \pm \frac{9}{2\sqrt{2}}$

Hence, the points of contact are $\left(\frac{9}{2\sqrt{2}}, \frac{1}{\sqrt{2}}\right)$ and $\left(-\frac{9}{2\sqrt{2}}, -\frac{1}{\sqrt{2}}\right)$

17. (b, c) Let the point of intersection of tangents at Y-axis is $P(0, h)$

Equation of tangent through the point $(0, h)$ is

$$(y - h) = m(x - 0)$$

$\Rightarrow \quad y = mx + h$

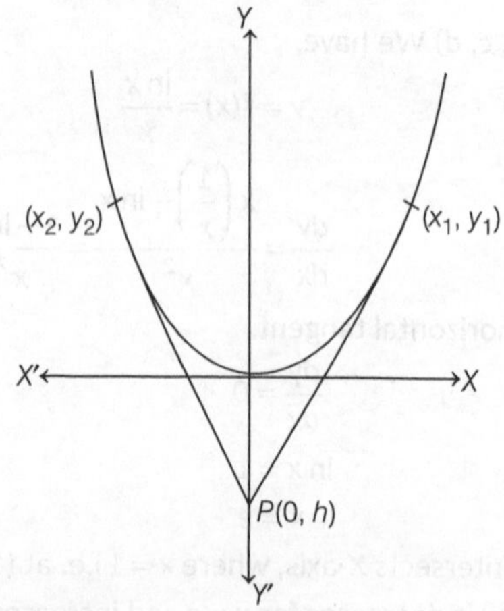

It intersects curve

$$y = \sqrt{17(1 + x^2)}$$

$\Rightarrow \quad mx + h = \sqrt{17(1 + x^2)}$

$\Rightarrow \quad m^2x^2 + 2mhx + h^2 = 17 + 17x^2$

$\Rightarrow \quad (m^2 - 17)x^2 + 2mhx + (h^2 - 17) = 0$

Its discriminant must be zero.

$\Rightarrow \quad 4m^2h^2 - 4(m^2 - 17)(h^2 - 17) = 0$

$\Rightarrow \quad m^2h^2 - (m^2h^2 - 17h^2 - 17m^2 + 289) = 0$

$\Rightarrow \quad 17h^2 + 17m^2 - 289 = 0$

$\Rightarrow \quad h^2 + m^2 = 17$

$\Rightarrow \quad m^2 + (h^2 - 17) = 0$

For perpendicular tangents,

$$h^2 - 17 = -1$$

$\Rightarrow \quad h^2 = 16$

$\Rightarrow \quad h = \pm 4$

But $h < 0$

So, $\quad h = -4$

$\Rightarrow \quad m^2 = 1 \Rightarrow m = \pm 1$

Hence, equation of tangents are $y = x - 4$ and $y = -x - 4$.

18. (a, b, c, d)

(a) $\quad y^2 = 4ax$

$2y\frac{dy}{dx} = 4a$

$\Rightarrow \quad \frac{dy}{dx} = \frac{2a}{y} = m_1$ (let)

and $\quad y = e^{-x/2a}$

$\therefore \quad \frac{dy}{dx} = e^{-x/2a}\left(-\frac{1}{2a}\right)$

$\frac{dy}{dx} = -\frac{1}{2a}y = m_2 \quad$ (let)

So, $m_1 m_2 = \left(\frac{2a}{y}\right)\left(-\frac{y}{2a}\right) = -1$

So, given curves are orthogonal.

(b) $y^2 = 4ax$

$\therefore \frac{dy}{dx} = \frac{2a}{y}$

$\left(\frac{dy}{dx}\right)_{(0,\,0)} = \infty$

So, tangent is parallel to Y-axis

and $x^2 = 4ay$

$2x = 4a\frac{dy}{dx}$

$\frac{dy}{dx} = \frac{x}{2a}$

$\left(\frac{dy}{dx}\right)_{(0,\,0)} = 0$

So, tangent is parallel to X-axis.

Hence, both curves intersect at right angle.

So, given curves are orthogonal at $(0, 0)$.

(c) $xy = a^2$

$x\frac{dy}{dx} + y(1) = 0$

$\therefore \frac{dy}{dx} = -\frac{y}{x} = m_1$ (let)

and $x^2 - y^2 = b^2$

$\therefore \frac{dy}{dx} = \frac{x}{y} = m_2$ (let)

So, $m_1 m_2 = \left(-\frac{y}{x}\right)\left(\frac{x}{y}\right) = -1$

So, given curves are orthogonal.

(d) $y = ax$

$\therefore \frac{dy}{dx} = a = m_1$ (let)

and $x^2 + y^2 = c^2$

$\therefore 2x + 2y\frac{dy}{dx} = 0$

$\Rightarrow \frac{dy}{dx} = -\frac{x}{y} = m_2$ (let)

Now, $m_1 m_2 = -\frac{ax}{y} = -\frac{y}{y} = -1$

Hence, given curves are orthogonal.

Sol. (Q. Nos. 19-21)

We have, $f(x) = x^2 f(1) - x f'(2) + f''(3)$

$\Rightarrow f'(x) = 2x f(1) - f'(2)$ ...(i)

$\Rightarrow f''(x) = 2f(1)$ ...(ii)

$\Rightarrow f''(3) = 2f(1)$ ...(iii)

Also, $f(0) = 2$

and $f''(3) = 2$ ...(iv)

$\Rightarrow f(1) = 1$ [using Eqs. (i), (iii) and (iv)]

$\Rightarrow f'(2) = 2f(1) = 2$ [$\because f(1) = 1$]

$\Rightarrow f(x) = x^2 - 2x + 2$

19. (a) $f'(x) = 2x - 2$

So, $f'(1) = 2 - 2 = 0$

20. (c) Slope of tangent at $(3, 5)$ is $f'(3) = 4$

So, equation of tangent

$y - 5 = 4(x - 3)$

$\Rightarrow y - 5 = 4x - 12 \Rightarrow y = 4x - 7$

21. (b) Let $f(x) = y = x^2 - 2x + 2$ and $g(x) = y = e^{2(x-1)}$

Clearly, $x^2 - 2x + 2 = e^{2(x-1)}$ holds at $x = 1$

$\therefore$ (1, 1) is the point of intersection of two curves.

Now, $f'(x) = 2x - 2$

$f'(1) = 0 = m_1$ (let)

and $g'(x) = 2e^{2(x-1)}$

$g'(1) = 2 = m_2$ (let)

$\therefore \tan\theta = \left|\frac{2-0}{1+2(0)}\right|$

$\Rightarrow \tan\theta = 2$

$\Rightarrow \theta = \tan^{-1}(2)$

22. (i) $\rightarrow$ (C), (ii) $\rightarrow$ (B), (iii) $\rightarrow$ (A), (iv) $\rightarrow$ (D)

(i) Let the point of intersection be (x_0, y_0)

Now, $x^2 = 4y$

$\Rightarrow \frac{dy}{dx} = \frac{x}{2}$

$\therefore m_1 = \left(\frac{dy}{dx}\right)_{(x_0,\,y_0)} = \frac{x_0}{2}$

and for $y^2 = 4x$,

$\frac{dy}{dx} = \frac{2}{y}$

$\therefore m_2 = \left(\frac{dy}{dx}\right)_{(x_0,\,y_0)} = \frac{2}{y_0}$

$\therefore \theta = \tan^{-1}\left|\frac{\frac{x_0}{2} - \frac{2}{y_0}}{1 + \frac{x_0}{2}\cdot\frac{2}{y_0}}\right| = \tan^{-1}\left|\frac{x_0 y_0 - 4}{2(x_0 + y_0)}\right|$

Clearly, $\theta = 90^\circ$ at $x_0 = y_0 = 0$

At the point of intersection,

$y_0 = \frac{x_0^2}{4}$ and $y_0^2 = 4x_0$

$\Rightarrow \frac{x_0^4}{16} = 4x_0$

$$\Rightarrow \quad x_0 = 4 \text{ and } y_0 = 4$$

$$\Rightarrow \quad \theta = \tan^{-1}\left|\frac{3}{4}\right|$$

$$\Rightarrow \quad |m+n| = 7$$

(ii) Given curve is, $x = e^{\sin y}$

$$\Rightarrow \quad \ln x = \sin y$$

$$\Rightarrow \quad \frac{1}{x} = (\cos y)\frac{dy}{dx}$$

$$\Rightarrow \quad \frac{dy}{dx} = \frac{1}{x}\sec y \Rightarrow \left(\frac{dy}{dx}\right)_{(1,\,0)} = 1$$

$\therefore$ Equation of normal at (1, 0) is

$$(y-0) = -1(x-1)$$

$$\Rightarrow \quad x + y = 1$$

$\therefore$ Area of triangle $= \frac{1}{2}(1)(1) = \frac{1}{2}$ sq units

(iii) Given curve is

$$x^2y = 1 \Rightarrow x^2y' + y(2x) = 0$$

$$\Rightarrow \quad y' = \frac{-2y}{x}$$

$$\Rightarrow \quad m_1 = \frac{-2y_0}{x_0} \quad \text{...(i)}$$

Also, second curve is

$$y = e^{2(1-x)}$$

$$\Rightarrow \quad y' = -2e^{2(1-x)}$$

$$\Rightarrow \quad m_2 = -2e^{2(1-x_0)} \quad \text{...(ii)}$$

$$\therefore \quad \theta = \tan^{-1}\left|\frac{\frac{-2y_0}{x_0} + 2e^{2(1-x_0)}}{1 + \frac{4y_0e^{2(1-x_0)}}{x_0}}\right|$$

$$= \tan^{-1}\left|\frac{2x_0e^{2(1-x_0)} - 2y_0}{x_0 + 4y_0e^{2(1-x_0)}}\right|$$

At the point of intersection,

$$x_0^2y_0 = 1 \text{ and } y_0 = e^{2(1-x_0)}$$

$$\Rightarrow \quad x_0 = y_0 = 1$$

$$\therefore \quad \theta = \tan^{-1}\left|\frac{2x_0y_0 - 2y_0}{x_0 + 4y_0^2}\right| = 0$$

(iv) Given curve is,

$$y = be^{x/3}$$

$$\Rightarrow \quad y' = \frac{b}{3}e^{x/3}$$

Length of sub-tangent $\frac{y}{y'} = \frac{y}{\frac{b}{3}e^{x/3}}$

$$= \frac{3be^{x/3}}{be^{x/3}} = 3$$

23. (i) → (B), (ii) → (D), (iii) → (A), (iv) → (C)

We have, $\quad x = a\cos t + at\sin t$

and $\quad y = a\sin t - at\cos t$

$$\therefore \quad \frac{dy}{dt} = a\cos t + at\sin t - a\cos t$$

and $\quad \frac{dx}{dt} = -a\sin t + at\cos t + a\sin t$

So, $\quad \frac{dy}{dx} = \frac{a\cos t + at\sin t - a\cos t}{-a\sin t + at\cos t + a\sin t}$

$$\Rightarrow \quad \frac{dy}{dx} = \tan t$$

(i) $\therefore$ Length of subtangent $= \frac{y}{dy/dx}$

$$= \frac{(a\sin t - at\cos t)}{\tan t}$$

$$= a\cos t - at\cos t\cot t$$

$\therefore$ At $t = \frac{\pi}{4}$,

$$= \frac{a}{\sqrt{2}} - a\left(\frac{\pi}{4}\right)\left(\frac{1}{\sqrt{2}}\right)(1) = \frac{a}{\sqrt{2}}\left(1 - \frac{\pi}{4}\right) \text{ units}$$

(ii) $\therefore$ Length of normal $= y\sqrt{1 + \left(\frac{dy}{dx}\right)^2}$

at $x = \frac{\pi}{4}$, $\quad = \left(\frac{a}{\sqrt{2}} - \frac{a\pi}{4\sqrt{2}}\right)\sqrt{1+(1)^2}$

$$= \left(\frac{a}{\sqrt{2}} - \frac{a\pi}{4\sqrt{2}}\right)(\sqrt{2})$$

$$= a\left(1 - \frac{\pi}{4}\right) \text{ units}$$

(iii) Equation of tangent at $t = \frac{\pi}{4}$ will be

$$y - \left(\frac{a}{\sqrt{2}} - \frac{a\pi}{4\sqrt{2}}\right) = 1\left\{x - \left(\frac{a}{\sqrt{2}} + \frac{a\pi}{4\sqrt{2}}\right)\right\}$$

On Y-axis, $x = 0$

So, $\quad y = \frac{-a\pi}{2\sqrt{2}}$

and on X-axis, $y = 0$

So, $\quad x = \frac{a\pi}{2\sqrt{2}}$

Length of intersect of tangent between axes $= \frac{a\pi}{2}$

(iv) $\therefore$ Perpendicular distance of tangent from origin

$$= \left|\frac{0 - 0 - \frac{a\pi}{2\sqrt{2}}}{\sqrt{2}}\right|$$

$$= \frac{a\pi}{4} \text{ units}$$

Sol. (Q. Nos. 24-25)

Since, the curve $y = ax^3 + bx^2 + cx + 5$ touches X-axis at $P(-2, 0)$, then X-axis is the tangent at $(-2, 0)$.

The curve meets the Y-axis at $Q(0, 5)$, we have

$$\frac{dy}{dx} = 3ax^2 + 2bx + c$$

$$\left.\frac{dy}{dx}\right|_{(0,5)} = 0 + 0 + c = 3 \qquad \text{[given]}$$

$\Rightarrow \quad c = 3$

and $\quad \left.\frac{dy}{dx}\right|_{(-2,0)} = 0$

$\Rightarrow \quad 12a - 4b + c = 0$

$\Rightarrow \quad 12a - 4b + 3 = 0 \qquad$ ...(i) [$\because c = 3$]

Point $(-2, 0)$ lies on the curve,

then $\quad 0 = -8a + 4b - 2c + 5$

$\Rightarrow \quad -8a + 4b - 6 + 5 = 0$

$\Rightarrow \quad 8a - 4b + 1 = 0 \qquad$ ...(ii)

From Eqs. (i) and (ii), we get

$$a = -\frac{1}{2} \text{ and } b = -\frac{3}{4}$$

So,

24. (6) $\left|\frac{c}{a}\right| = \left|\frac{3}{-1/2}\right| = 6$

25. (4) $\left|\frac{c}{b}\right| = \left|\frac{3}{-3/4}\right| = 4$

26. (0) $f(x) = y = \sin x$

$$f'(x) = \frac{dy}{dx} = \cos x$$

$\therefore \cos x_1 = \frac{1}{\cos x_2} = \frac{\sin x_2 - \sin x_1}{x_2 - x_1}$

i.e. $\cos x_1 \cos x_2 = -1$

$\therefore \sin x_1 = \sin x_2 = 0$

$\therefore$ There is no solution.

27. (0) $f(x) = y = |\ln x|$

$\therefore \quad f'(x_1) = -\frac{1}{f'(x_2)} = \frac{f(x_2) - f(x_1)}{x_2 - x_1}$

$\Rightarrow \frac{\ln x_1}{x_1 |\ln x_1|} = -\frac{x_2 |\ln x_2|}{\ln x_2} = \frac{|\ln x_2| - |\ln x_1|}{x_2 - x_1} \qquad$ ...(i)

$\Rightarrow \ln x_1 \cdot \ln x_2 < 0$

Let $0 < x_1 < 1$, then $1 < x_2$ and $x_1 \cdot x_2 = 1$

From Eq. (i), we get

$$-\frac{1}{x^1} = -x_2 = \frac{\ln x_2 + \ln x_1}{x_2 - x_1} = \frac{\ln x_1 x_2}{x_2 - x_1} = 0$$

which is not possible

$\therefore$ There is no solution.

28. (2) $\quad y^2 = x^3$

$\therefore \quad y = \sqrt{x^3}$ or $-\sqrt{x^3}$

$$2y\frac{dy}{dx} = 3x^2$$

$$\frac{3x_1^2}{2y_1} = -\frac{2y_2}{3x_2^2}$$

$\Rightarrow \quad \frac{9}{4}x_1^2 x_2^2 = -y_1 y_2$

$\therefore \quad y_1 y_2 < 0$

Let $\quad y_1 = \sqrt{x_1^3}$ and $y_2 = -\sqrt{x_2^3}$

Thus, $\quad \frac{3x_1^2}{2\sqrt{x_1^3}} = -\frac{2\sqrt{x_2^3}}{3x_2^2} = -\frac{-\sqrt{x^3} - \sqrt{x_1^3}}{x} \quad \frac{}{x_1}$

$\Rightarrow \quad \frac{3\sqrt{x_1}}{2} = \frac{2}{3\sqrt{x_2}} = \frac{\sqrt{x_2^3} + \sqrt{x_1^3}}{x_1 - x_2}$

$\Rightarrow \quad \sqrt{x_1 x_2} = \frac{4}{9}$

$\Rightarrow \quad 3x_1\sqrt{x_1} - 3\sqrt{x_1}x_2 = 2\sqrt{x_2^3} + 2\sqrt{x_1^3}$

$\Rightarrow \quad 3(\sqrt{x_1})^3 - \frac{3 \times 16}{31\sqrt{x_1}} = 2 \cdot \frac{64}{729\sqrt{x_1^3}} + 2\sqrt{x_1^3}$

$\Rightarrow \quad 3x_1^3 - \frac{16}{27}x_1 = \frac{128}{729} + 2x_1^2$

$\Rightarrow \quad x_1^3 - \frac{16}{27}x_1 = \frac{128}{729}$

$\Rightarrow \quad 739x_1^3 - 432x_1 - 12 = 0$

Consider, $h(t) = 729x^3 - 432t - 128$

$$h'(t) = 3 \times 729t^2 - 432 = 0$$

Gives $\quad t = \pm\frac{4}{9}$

$$h\left(-\frac{4}{9}\right) = 0$$

$\therefore$ There are two distinct solutions of

$$729x_1^3 - 432x_1 - 128 = 0.$$

29. (13) Let the coordinate of point of contact is $P(x_1, y_1)$.

Given curve, $\quad y = x^3 + x + 16$

$\Rightarrow \quad \frac{dy}{dx} = 3x^2 + 1$

$$\left(\frac{dy}{dx}\right)_{(x_1, y_1)} = 3x_1^2 + 1 \qquad \text{...(i)}$$

Slope of tangent line joining origin and (x_1, y_1) is

$$= \frac{y_1 - 0}{x_1 - 0} = \frac{y_1}{x_1} \qquad \text{...(i)}$$

So, $3x_1^2 + 1 = \frac{y_1}{x_1}$

$$\Rightarrow \quad 3x_1^2 + 1 = \frac{x_1^3 + x_1 + 16}{x_1}$$

$$\Rightarrow \quad 3x_1^3 + x_1 = x_1^3 + x_1 + 16$$

$$\Rightarrow \quad 2x_1^3 = 16$$

$$\Rightarrow \quad x_1^3 = 8 \Rightarrow x_1 = 2$$

Then, $\quad y_1 = 26$

$\therefore$ Slope of tangent $= \dfrac{y_1}{x_1} = \dfrac{26}{2} = 13$

Hence, value of m is 13.

30. (4) We have, $f(0) = 2$

Now, $y - f(a) = f'(a)[x - a]$

For x-intercept, $y = 0$

So, $\quad x = a - \dfrac{f(a)}{f'(a)} = a - 2$

$$\Rightarrow \quad \frac{f(a)}{f'(a)} = 2$$

$$\Rightarrow \quad \frac{f'(a)}{f(a)} = \frac{1}{2}$$

On integrating both sides w.r.t. a,

$$\ln\{f(a)\} = \frac{a}{2} + \log C$$

$$\Rightarrow \quad f(a) = Ce^{a/2}$$

So, $\quad f(x) = Ce^{x/2}$

$$\Rightarrow \quad f(0) = C = 2$$

So, $\quad f(x) = 2e^{x/2}$

Hence, $k = 2$ and $p = \dfrac{1}{2}$

$$\frac{k}{p} = 4$$

31. (5) We have, $y = e^{a + bx^2}$

passes through (1, 1)

$$\therefore \quad 1 = e^{a+b}$$

$$\Rightarrow \quad a + b = 0$$

Also, $\quad \left(\dfrac{dy}{dx}\right)_{(1,1)} = -2$

$$\therefore \quad e^{a+bx^2} \cdot 2bx = -2$$

or $\quad e^{a+b}\,2b(1) = -2$

$$\Rightarrow \quad b = -1 \text{ and } a = 1$$

$$\Rightarrow \quad 2a - 3b = 5$$

32. (7) We have,

$$x = t^2 \text{ and } y = t^3$$

$$\frac{dx}{dt} = 2t \text{ and } \frac{dy}{dt} = 3t^2$$

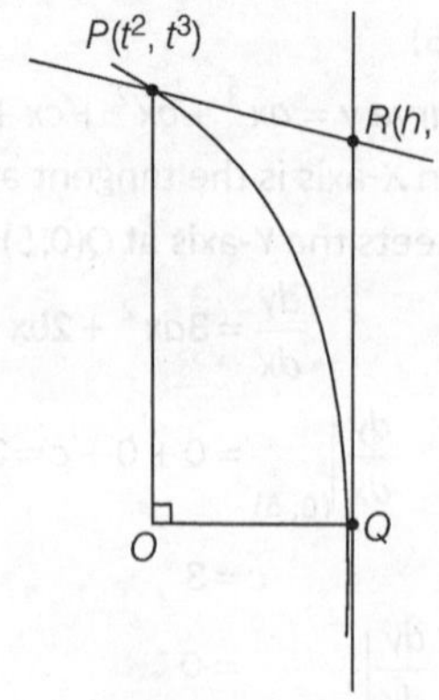

Now, $\quad \dfrac{dy}{dx} = \dfrac{3t}{2}$

Equation of tangent at P is

$$y - t^3 = \frac{3t}{2}(x - t^2)$$

$$\Rightarrow \quad 2y - 2t^3 = 3t(x - t^2)$$

Passing through (h, k)

$$\therefore \quad 2k - 2t^3 = 3th - 3t^3$$

$$\Rightarrow \quad t^3 - 3th + 2k = 0 \qquad \text{...(i)}$$

Products of roots $t_1t_2t_3 = -2k$

On putting $t_1t_2 = -1$,

$$t_3 = 2k$$

Now, t_3 must satisfy Eq. (i), therefore

$$(2k)^3 - 3(2k)h + 2k = 0$$

i.e. $\quad 4y^2 - 3x + 1 = 0$

$$\Rightarrow \quad 4y^2 = 3x - 1$$

or $\quad a + b = 7$

33. (2) We have, $y = x^n$

$$\therefore \quad \frac{dy}{dx} = nx^{n-1} = na^{n-1}$$

Slope of normal $= -\dfrac{1}{na^{n-1}}$

Equation of normal is

$$y - a^n = -\frac{1}{na^{n-1}}(x - a)$$

On putting $x = 0$ to get y-intercept.

Then, $\quad y = a^n + \dfrac{1}{na^{n-2}}$

Hence, $b = a^n + \dfrac{1}{na^{n-2}}$

or $$\lim_{a \to 0} b = \begin{cases} 0, & \text{if } n < 2 \\ \dfrac{1}{2}, & \text{if } n = 2 \\ \infty, & \text{if } n > 2 \end{cases}$$

34. (1) Given, curve is $4y^2 = x^3$ ...(i)

Normal cuts intercept in the ratio of 3 : 4, then

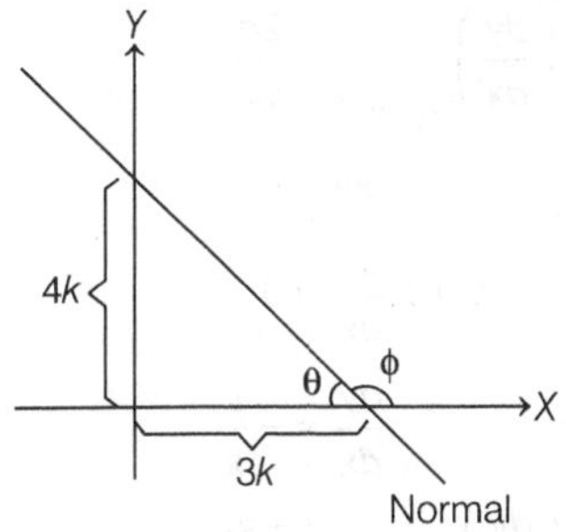

$$\tan\phi = \tan(180^\circ - \theta) = -\tan\theta$$

$$\tan\phi = -\frac{4}{3}$$

So, slope of normal $= -\frac{4}{3}$

Now, given curve is

$$4y^2 = x^3$$

$$\therefore \quad 8y\frac{dy}{dx} = 3x^2$$

$$\Rightarrow \quad \frac{dy}{dx} = \frac{3x^2}{8y}$$

$\therefore$ Slope of normal $= -\frac{8y}{3x^2}$

$$\text{So,} \quad -\frac{8y}{3x^2} = -\frac{4}{3}$$

$$\Rightarrow \quad \frac{64y^2}{9x^2} = \frac{16}{9}$$

$$\Rightarrow \quad 4y^2 = x^2$$

From Eq. (i),

$$x^3 = x^2$$

$$x = 1$$

Hence, abscissa of required point is 1.

35. (4) Given, curve is

$$x^3 - y^2 = 0 \quad \text{...(i)}$$

$$\therefore \quad 3x^2 - 2y\frac{dy}{dx} = 0$$

$$\Rightarrow \quad \frac{dy}{dx} = \frac{3x^2}{2y}$$

$\therefore$ Slope of the tangent

$$\left(\frac{dy}{dx}\right)_{(4m^r,\, 8m^3)}$$

$$= \frac{3 \times 16m^4}{2 \times 8m^2} = 3m$$

Hence, the equation of the tangent is

$$y - 8m^3 = 3m(x - 4m^2)$$

$$\Rightarrow \quad y = 3mx - 4m^3 \quad \text{...(ii)}$$

On solving Eqs. (i) and (ii), we get

$$x = 4m^2, m^2$$

Thus, the new point is $Q(m^2, -m^3)$

Slope of the tangent at $Q(m^2, -m^3) = -\frac{3}{2}m$

Slope of the normal at $Q(m^2, -m^3) = \frac{2}{3m}$

Since, tangent at $P(4m^2, 8m^3)$ and normal at $Q(m^2, -m^3)$ coincide, we get

$$\frac{2}{3m} = 3m$$

$$\Rightarrow \quad 9m^2 = 2$$

$$\Rightarrow \quad 9m^2 + 2 = 2 + 2 = 4$$

36. (2) Given, curve is $y^2 = x^3$

$$\Rightarrow \quad y = \pm\sqrt{3}$$

$$\Rightarrow \quad x^3 \geq 0$$

$$\Rightarrow \quad x \geq 0$$

Let $P(t, t^{3/2})$ be any point on the curve $y^2 = x^3$ ...(i)

$\therefore$ Slope of tangent at $P = \frac{3}{2}t^{1/2}$

$\therefore$ Equation of tangent to curve at P is

$$(y - t^{3/2}) = \frac{3}{2}t^{1/2}(x - t)$$

$$\Rightarrow \quad y = \frac{3}{2}t^{1/2}(x - t) + t^{3/2}$$

$$\Rightarrow \quad y = \frac{3}{2}t^{1/2}x - \frac{3}{2}t^{3/2} + t^{3/2}$$

$$\Rightarrow \quad y = \frac{3}{2}t^{1/2}x - \frac{1}{2}t^{3/2}$$

$$\Rightarrow \quad y = \frac{1}{2}\sqrt{t}(3x - t) \quad \text{...(ii)}$$

$\therefore$ At the point of intersection of Eqs. (i) and (ii), we get

$$\frac{1}{4}t(3x - t)^2 = x^3$$

$$\Rightarrow \quad t(9x^2 + t^2 - 6xt) = 4x^3$$

$$\Rightarrow \quad 4x^3 - 9tx^2 + 6t^2x - t^3 = 0$$

Clearly, $(x - t)^2$ is its factor as line (ii) is tangent to curve at $(t, t^{3/2})$

$$\Rightarrow \quad (x - t)^2(4x - t) = 0$$

The tangent will again intersect the curve at

$$Q\left(\frac{1}{t},\left(\frac{1}{4}\right)^{3/2}\right) \text{ or } \left(\frac{t}{4},-\frac{t^{3/2}}{8}\right)$$

From Eq. (ii), put $x=\frac{1}{4}$

$$y=\frac{1}{2}\sqrt{t}\left(\frac{3t}{4}-t\right)$$

$$\Rightarrow \quad y=\frac{1}{2}\sqrt{t}\left(\frac{-t}{4}\right)=-\frac{1}{8}(t)^{3/2}$$

$$\Rightarrow \quad Q\left(\frac{t}{4},-\frac{1}{8}(t)^{3/2}\right)$$

$$\Rightarrow \quad \tan\alpha=\frac{t^{3/2}}{t}=\sqrt{t}$$

and

$$\tan\beta=\frac{-\frac{1}{8}t^{3/2}}{\left(\frac{t}{4}\right)}=\frac{-t^{3/2}}{8}\times\frac{4}{t}=\frac{-\sqrt{t}}{2}$$

$$\Rightarrow \quad \left|\frac{\tan\alpha}{\tan\beta}\right|=\left|\frac{\sqrt{t}}{\left(\frac{-1}{2}\sqrt{t}\right)}\right|=|-2|=2$$

37. (32) We have,

$$y^2=4ax \text{ and } xy=c^2$$

Let they intersect at (x_1, y_1).

On differentiating both curves w.r.t. x, we get

$$y^2=4ax$$

$$\Rightarrow \quad 2y\frac{dy}{dx}=4a$$

$$\Rightarrow \quad \frac{dy}{dx}=\frac{2a}{y}$$

$$\Rightarrow \quad \left(\frac{dy}{dx}\right)_{(x_1,\, y_1)}=\frac{2a}{y_1}$$

and

$$xy=c^2$$

$$\Rightarrow \quad y+x\frac{dy}{dx}=0$$

$$\Rightarrow \quad \frac{dy}{dx}=-\frac{y}{x}$$

$$\Rightarrow \quad \left(\frac{dy}{dx}\right)_{(x_1,\, y_1)}=\frac{-y_1}{x_1}$$

When curve intersects orthogonally, then

$$\frac{2a}{y_1}\times\left(\frac{-y_1}{x_1}\right)=-1$$

$$x_1=2a$$

From $y^2=4ax$,

$$y_1=\sqrt{8a^2}$$

$$y_1=2\sqrt{2}a$$

and

$$xy=c^2$$

$$\Rightarrow \quad x_1y_1=c^2$$

$$\Rightarrow \quad 2a(2\sqrt{2}a)=c^2$$

$$\Rightarrow \quad \frac{c^2}{a^2}=4\sqrt{2}$$

$$\Rightarrow \quad \frac{c^4}{a^4}=32$$

CHAPTER

08

MAXIMA AND MINIMA

LEVEL 01 \ Starter Level

8.1 Local Maxima and Minima

1 Find the maximum and minimum value, if any

(i) $f(x) = (x-2)^2 + 4$.

(ii) $f(x) = |x| + 2$.

(iii) $f(x) = -|x-2| + 4$.

(iv) $f(x) = |\sin 3x + 2|$.

2 Consider the following graphs of the functions, check each for the extrema at $x = a$.

(i)

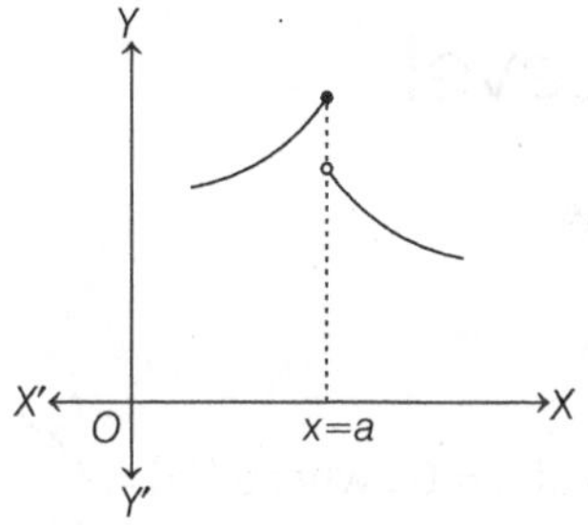

(ii)

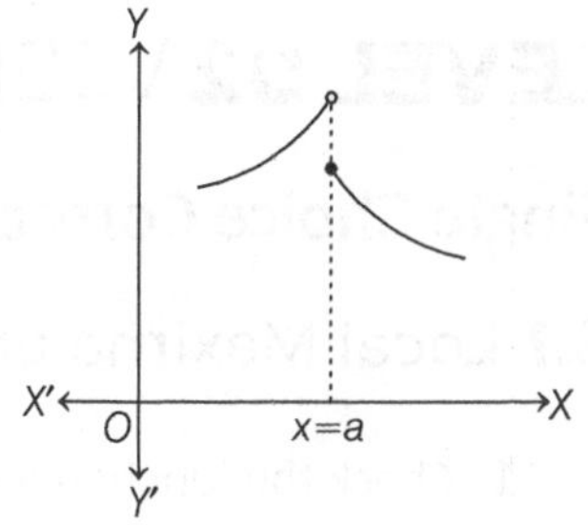

(iii)

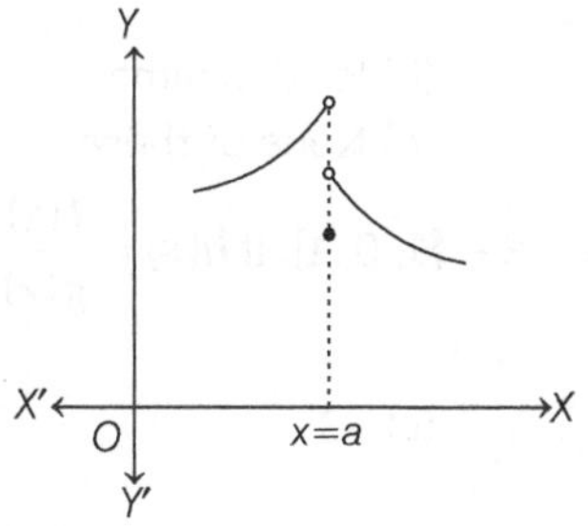

3 Find the points of local maxima or local minima and corresponding local maximum and local minimum values of each of the following functions.

(i) $f(x) = x^2 - 6x + 5$

(ii) $f(x) = x^4 - 62x^2 + 120x + 9$

(iii) $f(x) = x + \frac{a^2}{x}, a > 0, x \neq 0$

(iv) $f(x) = \frac{\log x}{x}$

(v) $f(x) = \sin 2x$, when $0 < x < \pi$

4 Find the points of inflection of the following functions.

(i) $f(x) = 3x^4 - 4x^3$

(ii) $f(x) = x^2 - \frac{1}{6x^3}$

(iii) $f(x) = \sin x$

(iv) $f(x) = e^{-x^2}$

(v) $y = (x - 2)^{2/3} + 10$

8.2 Global/Absolute Maximum and Minimum Value

5 Find the absolute maximum and absolute minimum values of the following function in the given intervals.

(i) $f(x) = 2x^3 - 24x + 100$ in $[1, 3]$

(ii) $f(x) = x + \sin 2x$ in $[0, \pi]$

(iii) $f(x) = \sin x + \cos x$ in $[0, 2\pi]$

(iv) $f(x) = 12x^{4/3} - 6x^{1/3}$ in $[-1, 1]$

8.3 Application of Maxima and Minima

6 A wire of length 25 m is to be cut into two pieces. One of the wires is to be made into a square and the other into a circle. What should be the lengths of the two pieces, so that the combined area of the square and the circle is minimum?

7 A tank with rectangular base and rectangular sides open at the top is to be constructed, so that its depth is 3 m and volume is 75 m^3. If building of tank costs ₹ $100/\text{m}^2$ for the base and ₹ $50/\text{m}^2$ for the sides, then find the cost of least expensive tank.

8 Find the point p on the curve $y^2 = 4ax$, which is nearest to the point $(11a, 0)$.

9 Find the area of greatest rectangle that can be inscribed in an ellipse $\frac{x^2}{a^2} + \frac{y^2}{b^2} = 1$.

LEVEL 02 \ JEE Main Level

Single Choice Correct Questions

8.1 Local Maxima and Minima

1 Check the function for local extrema at $x = 0$, where $f(x) = \begin{cases} x^3 + x^2 + 5x, & x < 0 \\ 1 - xe^x, & x \geq 0 \end{cases}$

(a) local maxima

(b) local minima

(c) point of inflection

(d) None of these

2 Let $f(x) = x^2 + \frac{1}{x^2}$ and $g(x) = x - \frac{1}{x}$, $x \in R - \{1, 0, 1\}$. If $h(x) = \frac{f(x)}{g(x)}$, then local minimum value of $h(x)$ is

(a) $2\sqrt{2}$

(b) 3

(c) -3

(d) $-2\sqrt{2}$

3 Let $f(x) = \begin{cases} |x^2 - 3x| + a, & 0 \le x < \frac{3}{2} \\ -2x + 3, & x \ge \frac{3}{2} \end{cases}$. If $f(x)$ has a local maxima at $x = \frac{3}{2}$, then the greatest value of a is

(a) $-\frac{9}{2}$ (b) $-\frac{9}{4}$ (c) $\frac{3}{4}$ (d) $-\frac{3}{4}$

4 A function f is defined by $f(x) = |x|^m |x-1|^n \ \forall x \in R \ (m, n \in N)$, local maximum value of the function is

(a) 1 (b) $m^n n^m$ (c) $\frac{m^m n^n}{(m+n)^{m+n}}$ (d) $\frac{(mn)^{mn}}{(m+n)^{m+n}}$

5 The function $f(x) = 2x^3 - 9ax^2 + 12a^2x + 1$ has a local maximum value at $x = \alpha$ and a local minimum value at $x = \beta$, such that $\beta = \alpha^2$, then α is equal to

(a) 0 (b) $\frac{1}{4}$ (c) 2 (d) either 0 or 2

6 The local maximum value of the function $f(x) = \left(\frac{2}{x}\right)^{x^2}$, $x > 0$.

[JEE Main 2021]

(a) $(2\sqrt{e})^{\frac{1}{e}}$ (b) $\left(-\frac{4}{\sqrt{e}}\right)^{\frac{e}{4}}$ (c) $(e)^{\frac{2}{e}}$ (d) 1

7 If $a, b > 0$ and $\Delta = \begin{vmatrix} x & a & a \\ b & x & a \\ b & b & x \end{vmatrix}$, then Δ has a local minimum value at x equal to

(a) $\sqrt{ab}$ (b) $-\sqrt{ab}$ (c) $\sqrt{\frac{a}{b}}$ (d) $-\sqrt{\frac{a}{b}}$

8 The sum of all the local minimum values of the twice diffrentiable function $f: R \to R$ defined by $f(x) = x^3 - 3x^2 - \frac{3f''(2)}{2}x + f''(1)$ is

[JEE Main 2021]

(a) -22 (b) 5 (c) -27 (d) 0

9 The point of inflection for the curve $y = x^3 - 6x^2 + 12x + 5$ is

(a) $x = 0$ (b) $x = 1$ (c) $x = -2$ (d) $x = 2$

10 The condition for $y = ax^4 + bx^3 + cx^2 + dx + e$ to have points of inflection is

(a) $b^2 - 4ac > 0$ (b) $3b^2 - 8ac = 0$ (c) $3b^2 - 8ac > 0$ (d) $3b^2 - 8ab < 0$

11 The set of values of a, for which the function $f(x) = \frac{ax^3}{3} + (a+2)x^2 + (a-1)x + 2$, possess a negative point of inflection.

(a) $(-\infty, -2) \cup (0, \infty)$ (b) $\left\{-\frac{4}{5}\right\}$ (c) $(-2, 0)$ (d) Empty set

12 Let $f(x) = x^2 + \frac{1}{x^2}$ and $g(x) = x - \frac{1}{x}$, $x \in R - \{-1, 0, 1\}$. If $h(x) = \frac{f(x)}{g(x)}$, then the local minimum value of $h(x)$ is

[JEE Main 2018]

(a) 3 (b) -3 (c) $-2\sqrt{2}$ (d) $2\sqrt{2}$

8.2 Maximum and Minimum Values

13 The function $\frac{\sin(x+\alpha)}{\sin(x+\beta)}$ has no maximum or minimum value, if (k is an integer)

(a) $\beta-\alpha=k\pi$ (b) $\beta-\alpha\neq k\pi$ (c) $\beta-\alpha=(k+1)\frac{\pi}{2}$ (d) None of these

14 The minimum value of $e^{(2x^2-2x+1)\sin^2 x}$ is

(a) e (b) $\frac{1}{e}$ (c) 1 (d) 0

15 For $x\geq 0$ the least value of expression $\frac{1+x^2}{1+x}$ is

(a) $\sqrt{2}$ (b) 1 (c) $2(\sqrt{2}+1)$ (d) $2(\sqrt{2}-1)$

16 On the interval $[0, 1]$ the function $x^{25}(1-x)^{75}$ takes its maximum value at the point

(a) 0 (b) $\frac{1}{4}$ (c) $\frac{1}{2}$ (d) $\frac{1}{3}$

17 The range of $a\in R$ for which the function $f(x)=(4a-3)(x+\log_e 5)+2(a-7)\cot\left(\frac{x}{2}\right)\sin^2\left(\frac{x}{2}\right)$, $x\neq 2n\pi, n\in N$, has critical points, is **[JEE Main 2021]**

(a) $(-3, 1)$ (b) $\left[-\frac{4}{3}, 2\right]$ (c) $[1, \infty)$ (d) $(-\infty, -1]$

18 Let the function $f(x)$ be defined as $f(x)=\begin{cases}\tan^{-1}\alpha-3x^2, & 0<x<1\\ -6x, & x\geq 1\end{cases}$ $f(x)$ can have a maximum value at $x=1$, if the value of α is

(a) 0 (b) 2 (c) 1 (d) None of these

19 The global maximum value of $f(x)=\log_{10}(4x^3-12x^2+11x-3)$, $x\in[2, 3]$ is

(a) $-\frac{3}{2}\log_{10}3$ (b) $1+\log_{10}3$ (c) $\log_{10}3$ (d) $\frac{3}{2}\log_{10}3$

20 The minimum value of the function $y=\ln(3x^4-2x^3-6x^2+6x+1)$ in $(0, 2)$ is

(a) $\ln\left(\frac{39}{16}\right)$ (b) $\ln 2$ (c) $\ln 3$ (d) $\ln\frac{1}{2}$

21 If $f(x)$ is a non-zero polynomial of degree four, having local extreme points at $x=-1, 0, 1$, then the set $S=\{x\in R: f(x)=f(0)\}$ contains exactly **[JEE Main 2019]**

(a) four rational numbers
(b) two irrational and two rational numbers
(c) four irrational numbers
(d) two irrational and one rational number

22 Suppose, $f(x)=\begin{cases}-x^3+k^2-3k+2, & 0\leq x<1\\ 2x-3 & 1\leq x\leq 3\end{cases}$, if $f(x)$ is smallest at $x=1$, then the interval of k is

(a) $k\in(1, 2)$ (b) $k\in[1, \infty)$
(c) $k\in(-\infty, 2)$ (d) $k\in(-\infty, 1]\cup[2, \infty)$

23 The minimum value of the function $f(x)=2|x-1|+5|x-2|$ for all $x\in R$ is

(a) 2 (b) 5 (c) 1 (d) 7

24 If minimum value of $f(x) = x^2 + 2bx + 2c^2$ is greater than maximum value of $g(x) = -x^2 - 2cx + b^2$, then for x is real

(a) $|c| > |b|\sqrt{2}$ (b) $|c|\sqrt{2} > b$ (c) $0 < c < \sqrt{2}b$ (d) None of these

25 Let $f(x) = \begin{cases} x^3 - x^2 + 10x - 7, & x \le 1 \\ -2x + \log_2(b^2 - 1), & x > 1 \end{cases}$. The set of values of b, for which $f(x)$ has greatest value at $x = 1$, is

(a) $(-6, -2)$ (b) $(2, 6)$ (c) $(-6, -2) \cup (2, 6)$ (d) $(-6, 6)$

26 If $x = -1$ and $x = 2$ are extreme points of $f(x) = \alpha \log|x| + \beta x^2 + x$, then

(a) $\alpha = -6$ and $\beta = \frac{1}{3}$ (b) $\alpha = -6$ and $\beta = -\frac{1}{2}$

(c) $\alpha = 2$ and $\beta = -\frac{1}{2}$ (d) $\alpha = 2$ and $\beta = \frac{1}{2}$

27 The range of the function $f(x) = x(\log x)^2$ is

(a) $[1, e^2]$ (b) $[0, 4e^{-2}]$ (c) $[0, e^2]$ (d) $[0, e]$

28 The maximum slope of the curve $y = \frac{x^4}{2} - 5x^3 + 18x^2 - 19x$ occurs at the point **[JEE Main 2021]**

(a) $(2, 9)$ (b) $(2, 2)$ (c) $\left(3, \frac{21}{2}\right)$ (d) $(0, 0)$

29 Suppose, the cubic polynomial $x^3 - ax + b$ has three real roots, where $a > 0$ and $b > 0$, then which one of the following holds?

(a) Minimum value at both $\sqrt{\frac{a}{3}}$ and $-\sqrt{\frac{a}{3}}$

(b) Maximum value at both $\sqrt{\frac{a}{3}}$ and $-\sqrt{\frac{a}{3}}$

(c) Minimum value at $\sqrt{\frac{a}{3}}$ and maximum value at $-\sqrt{\frac{a}{3}}$

(d) Minimum value at $-\sqrt{\frac{a}{3}}$ and maximum value at $\sqrt{\frac{a}{3}}$

30 Given, $p(x) = x^4 + ax^3 + bx^2 + cx + d$, such that $x = 0$ is the only real root of $p'(x) = 0$, if $p(-1) < p(1)$, then in the interval $[-1, 1]$

(a) $p(-1)$ is the minimum value and $p(1)$ is the maximum value of p

(b) $p(-1)$ is not minimum value but $p(1)$ is the maximum value of p

(c) $p(-1)$ is the minimum value but $p(1)$ is not maximum value of p

(d) None of the above

31 The number of solution of the equation $a^{f(x)} + g(x) = 0$ where $a > 0$, $g(x) \ne 0$ and $g(x)$ has minimum value $\frac{1}{2}$, is

(a) 1 (b) 2 (c) 0 (d) Infinitely many

32 The sum of an infinitely decreasing geometric progression is equal to the least value of the function $f(x) = 3x^2 - x + \frac{25}{12}$ and the first term of the progression is equal to the square of its common ratio. Then, the common ratio is

(a) $\sqrt{2} - 1$ (b) $\sqrt{3} - 1$ (c) $2 - \sqrt{3}$ (d) $\frac{2\sqrt{2}}{3}$

8.3 Application of Maxima and Minima

33 The coordinate of a point on the parabola $y = x^2 + 7x + 2$, which is closest to the straight line $y = 3x - 3$ is

(a) (0, 2) (b) (2, 20)
(c) (−2, −8) (d) None of these

34 Let a tangent be drawn to the ellipse $\frac{x^2}{27} + y^2 = 1$ at $(3\sqrt{3}\cos\theta, \sin\theta)$, where $\theta \in \left(0, \frac{\pi}{2}\right)$, then the value of θ such that the sum of intercepts on the axis made by this tangent is minimum, is equal to

(a) $\frac{\pi}{8}$ (b) $\frac{\pi}{4}$ (c) $\frac{\pi}{6}$ (d) $\frac{\pi}{3}$

35 The sum of length of the hypotenuse and another side of a right angled triangle is given. The area of triangle will be maximum, if the angle between them is

(a) $\frac{\pi}{6}$ (b) $\frac{\pi}{4}$ (c) $\frac{\pi}{3}$ (d) $\frac{5\pi}{12}$

36 A box open from top is made from a rectangular sheet of dimension $a \times b$ by cutting squares each of side x from each of the four corners and folding up the flaps. If the volume of the box is maximum, then x is equal to **[JEE Main 2021]**

(a) $\frac{a+b-\sqrt{a^2+b^2-ab}}{12}$ (b) $\frac{a+b-\sqrt{a^2+b^2+ab}}{6}$

(c) $\frac{a+b-\sqrt{a^2+b^2-ab}}{6}$ (d) $\frac{a+b+\sqrt{a^2+b^2-ab}}{6}$

37 A wire of length 2 units is cut into two parts, which are bent respectively to form a square of side x units and a circle of radius r units, if the sum of the areas of the square and the circle, so formed is minimum, then

(a) $(4-\pi)x = \pi r$ (b) $x = 2r$
(c) $2x = r$ (d) $2x = (\pi + 4)r$

38 The height of a right circular cylinder of maximum volume inscribed in a sphere of radius 3 units is **[JEE Main 2019]**

(a) $\sqrt{6}$ units (b) $2\sqrt{3}$ units (c) $\sqrt{3}$ units (d) $\frac{2}{3}\sqrt{3}$ units

39 A rectangle is inscribed in an equilateral triangle of side $2a$ units. The maximum area of rectangle can be

(a) $\sqrt{3}\,a^2$ sq units (b) $\frac{\sqrt{3}a^2}{4}$ sq units (c) a^2 sq units (d) $\frac{\sqrt{3}a^2}{2}$ sq units

40 The triangle of maximum area that can be inscribed in a given circle of radius r, is **[JEE Main 2021]**

(a) a right angled triangle having two of its sides of length $2r$ and r units
(b) an equilateral triangle of height $\frac{2r}{3}$ units
(c) an isosceles triangle with base equal to $2r$
(d) an equilateral triangle having each of its side of length $\sqrt{3}r$ units

41 A closed circular cylinder of volume V cubic units is to be formed with minimum amount of material. The ratio of its height to the radius of the base is

(a) 4 : 1 (b) 3 : 1 (c) 2 : 1 (d) 3 : 2

42 From a rectangular cardboard of size 3×8 equal square pieces are removed from the four corners and an open rectangular box is formed from the remaining, the maximum volume of the box is

(a) $\frac{250}{6}$ cu units (b) $\frac{250}{3}$ cu units (c) 125 cu units (d) $\frac{200}{27}$ cu units

43 A right triangle is drawn in a semi-circle of radius $\frac{1}{2}$ unit with one of its legs along the diameter. The maximum area of the triangle is

(a) $\frac{1}{4}$ sq unit (b) $\frac{3\sqrt{3}}{32}$ sq unit (c) $\frac{3\sqrt{3}}{16}$ sq unit (d) $\frac{1}{8}$ sq unit

44 The maximum volume (in cu.m) of the right circular cone having slant height 3m is **[JEE Main 2019]**

(a) $\frac{4}{3}\pi$ (b) $2\sqrt{3}\pi$ (c) $3\sqrt{3}\pi$ (d) 6π

Numerical Value Type Questions

45 Let $p(x)$ be a real polynomial of least degree, which has a local maxima at $x = 1$ and local minima at $x = 3$, if $p(1) = 6$ and $p(3) = 2$, then $p'(0)$ is ....

46 Let $f(x) = -\sin^3 x + 3\sin^2 x + 5$ in $\left[0, \frac{\pi}{2}\right]$, then sum of local maximum and local minimum value is ....

47 Let $f : R \to R$ be defined as $f(x) = |x| + |x^2 - 1|$. The total number of points at which f attains either a local maxima or a local minima is ....

48 If the point (1, 3) serves as the point of inflection of the curve $y = ax^3 + bx^2$, then $-3a + b$ is ....

49 Let $f(x)$ be a polynomial of degree 4 having extreme values at $x = 1$ and $x = 2$. If $\lim_{x \to 0}\left[1 + \frac{f(x)}{x^2}\right] = 3$, then $f(2)$ is equal to ....

50 Let $f(x)$ be a polynomial of degree 3 such that $f(-1) = 10$, $f(1) = -6$, $f(x)$ has a critical point at $x = -1$ and $f'(x)$ has a critical point at $x = 1$. Then, $f(x)$ has a local minima at x is equal to .... **[JEE Main 2020]**

51 Let $f(\alpha, \beta, \gamma, \delta) = \sum \alpha^2 - 2\sum \alpha + 10$ and $\alpha, \gamma \in [-1, 2]$ and $\beta, \delta, \in [1, 3]$, then the maximum value of f is ....

52 If the function $y = f(x)$ is represented parametrically by $x = t^5 - 5t^3 - 20t + 7$ and $y = 4t^3 - 3t^2 - 18t + 3\,(-2 < t < 2)$, then the maximum value of $y = f(x)$ is ....

53 Let $f(x)$ be a cubic polynomial with $f(1) = -10$, $f(-1) = 6$, and has a local minima at $x = 1$, and $f'(x)$ has a local minima at $x = -1$. Then, $f(3)$ .... **[JEE Main 2021]**

54 The minimum value of $f(x) = |3 - x| + |2 + x| + |5 - x|$ is ....

55 The difference between the greatest and least value of the function $f(x) = \int_0^x (\cos^2 t + \cos t + 2)\, dt$ in the interval $[0, 2\pi]$ is $k\pi$, then k is equal to ....

56 Let $f(x)$ be a polynomial of degree 6 in x, in which the coefficient of x^6 is unity and it has extrema at $x = -1$ and $x = 1$. If $\lim_{x \to 0} \frac{f(x)}{x^3} = 1$, then $5f(2)$ is equal to ....

57 From a given solid cone of height H another inverted cone is curved, whose height is h such that its volume is maximum. Then, the ratio of $\frac{H}{h}$ is ....

58 The three sides of trapezium are equal, each being 6 m long, then the area of the trapezium, when its area is maximum is $k\sqrt{3}\ m^2$, the value of k is ....

59 A straight line L with negative slope passes through the point (8, 2) and cuts the positive coordinate axes at P and Q. Then, the absolute minimum value of $OP + OQ$ as L varies, where O is the origin, is ....

60 The altitude of right circular cone of minimum volume circumscribed about a sphere of radius r is λr, then the value of λ is ....

61 A wire of length 36 m is cut into two pieces, one of the pieces is bent to form a square and the other is bent to form a circle. If the sum of the areas of the two figures is minimum and the circumference of the circle is km, then $\left(\frac{4}{\pi}+1\right)k$ ....

[JEE Main 2021]

LEVEL 03 \ JEE Advanced Level

Single Choice Correct Questions

1 Let $f(x) = \cos 2\pi x + x - [x]$, where [] denotes the greatest integer function. Then, the number of points in [0, 10] at which $f(x)$ assumes its local maximum value is

(a) 10 (b) 9 (c) 1 (d) infinite

2 Let $f(x) = (1 + b^2)x^2 + 2bx + 1$ and let $m(b)$ be the minimum value of $f(x)$. As, b varies the range of $m(b)$ is

(a) [0, 1] (b) $\left(0, \frac{1}{2}\right]$ (c) $\left[\frac{1}{2}, 1\right]$ (d) (0, 1]

3 If $a > b > 0$ and $f(\theta) = \frac{(a^2 - b^2)\cos\theta}{a - b\sin\theta}$, then the maximum value of $f(\theta)$ is

(a) $\sqrt{ab}$ (b) $\sqrt{a^2 + b^2}$ (c) $\sqrt{a^2 - b^2}$ (d) None of these

4 For any real number b, let $f(b)$ denotes the maximum of $\left|\sin x + \frac{2}{3 + \sin x} + b\right| \forall x \in R$. Then, the minimum value of $f(b)\ \forall\ b \in R$ is

(a) $\frac{1}{2}$ (b) $\frac{3}{4}$ (c) $\frac{1}{4}$ (d) 1

5 If $a^2x^4 + b^2y^4 = c^6$, then maximum value of xy is

(a) $\frac{c^2}{\sqrt{ab}}$ (b) $\frac{c^3}{ab}$ (c) $\frac{c^3}{\sqrt{2ab}}$ (d) $\frac{c^3}{2ab}$

6 A circular sector of perimeter 60 m with maximum area is to be constructed. The radius of circular arc in metre must be

(a) 10 m (b) 15 m (c) 20 m (d) None of these

7 If θ is the semivertical angle of a cone of maximum volume and given slant height, then $\tan\theta$ is given by

(a) 2 (b) 1 (c) $\sqrt{2}$ (d) $\sqrt{3}$

8 The least perimeter of an isosceles triangle, in which a circle of radius r can be inscribed, is
(a) $4\sqrt{3}\,r$ (b) $2\sqrt{3}\,r$ (c) $6\sqrt{3}\,r$ (d) $8\sqrt{3}\,r$

More than One Correct

9 If $f(x)=\begin{vmatrix} \cos 2x & \cos 2x & \sin 2x \\ -\cos x & \cos x & -\sin x \\ \sin x & \sin x & \cos x \end{vmatrix}$, then **[JEE Advanced 2017]**

(a) $f'(x)=0$ at exactly three points in $(-\pi, \pi)$ (b) $f(x)$ attains its maximum value at $x=0$
(c) $f(x)$ attains its minimum value at $x=0$ (d) $f'(x)=0$ at more than three points in $(-\pi, \pi)$

10 If $f(x)=\begin{cases} e^x, & 0\le x\le 1 \\ 2-e^{x-1} & 1<x\le 2 \\ x-e, & 2<x\le 3 \end{cases}$ and $g(x)=\int_0^x f(t)\,dt,\ x\in[1, 3]$, then $g(x)$ has

(a) local maxima at $x=1+\log 2$ (b) local maxima at $x=1$ and local minima at $x=2$
(c) local maxima at $x=e$ (d) local minima at $x=e$

11 The function $f(x)=\int_{-1}^{x} t(e^t-1)(t-1)(t-2)^3(t-3)^5\,dt$ has a local minimum value at x is equal to
(a) 0 (b) 1 (c) 2 (d) 3

12 The function $\sin^{-1}(2x\sqrt{1-x^2})$

(a) has $x=-\dfrac{1}{\sqrt{2}}$ as a point of minima (b) has $x=\dfrac{1}{\sqrt{2}}$ as a point of maxima
(c) attains its least value at $x=0$ (d) attains its least value at $x=1$

13 The point (0, 3) is closest to the curve $x^2=2y$ at
(a) $(2\sqrt{2}, 0)$ (b) (0, 0) (c) (2, 2) (d) (−2, 2)

14 Let $f(x)=(x-1)^p(x-2)^q$, where $p>1$ and $q>1$. Each critical point of $f(x)$ is a point of extremum, when
(a) $p=3, q=4$ (b) $p=4, q=2$ (c) $p=2, q=3$ (d) $p=2, q=4$

15 Suppose, $f(x)=\int_0^{\sqrt{9-x^2}} e^{\frac{t^2}{2}}\,dt$, then

(a) $f(x)$ is increasing for $x\in(-3, 0)$ (b) $f(0)$ is maximum value of $f(x)$
(c) $f'(x)$ is an odd function (d) $f'(x)$ is an even function

16 The parabola $y=x^2+px+q$ cuts the straight line $y=2x-3$ at a point with abscissa 1. If the distance between the vertex of parabola and the X-axis is least, then
(a) $p=0$ and $q=-2$
(b) $p=-2$ and $q=0$
(c) least distance between parabola and X-axis is 2
(d) least distance between parabola and X-axis is 1

17 $f(x)$ is cubic polynomial, which has local maxima at $x=-1$, If $f(2)=18$, $f(1)=-1$ and $f'(x)$ has local minima at $x=0$, then
(a) the distance between point of maxima and minima is $2\sqrt{5}$
(b) $f(x)$ is increasing for $x\in[1, 2\sqrt{5}]$
(c) $f(x)$ has local minima at $x=1$
(d) the value of $f(0)=5$

Linked Comprehension Type

Paragraph I Directions (Q.Nos. 18-20)

Let $f(x)$ be a polynomial of degree 4 vanishes $x = 0$. Given, $f(-2) = 216$ and $f(x)$ has relative maxima/relative minima at $x = 1$, 2 and 3.

18 Area of triangle formed by extremum point of $f(x)$ is

(a) $\frac{1}{8}$ sq unit (b) $\frac{1}{4}$ sq unit

(c) $\frac{1}{2}$ sq unit (d) 1 sq unit

19 The value of $f'(0)$ is

(a) 24 (b) – 30 (c) – 18 (d) – 24

20 Which of the following statement is correct?

(a) $f(x)$ has two relative maximum points and one relative minimum point
(b) $f(x)$ has exactly one inflection point
(c) Range of $f(x)$ contains nine negative integers
(d) Sum of real roots of $f(x) = 0$ is 6

Paragraph II Directions (Q.Nos. 21-23)

Let $f(x) = x^3 - 3(7-a)x^2 - 3(9-a^2)x + 2$

21 The value of parameter a, if $f(x)$ has a positive point of local minima are

(a) ϕ (b) $(3, 3)$

(c) $\left(-\infty, \frac{58}{14}\right)$ (d) None of these

22 The value of parameter a, if $f(x)$ has a positive point of local maxima are

(a) ϕ (b) $(-\infty, -3) \cup \left(3, \frac{29}{7}\right)$

(c) $\left(-\infty, \frac{58}{14}\right)$ (d) None of these

23 The value of parameter a, if $f(x)$ has point of extrema, which are opposite in sign are

(a) ϕ (b) $(-3, 3)$

(c) $\left(-\infty, \frac{58}{14}\right)$ (d) None of these

Matrix Match Type Questions

24 Match the following.

	Column I		Column II
(i)	$f(x) = \|2x-1\| + \|2x-3\|$	(A)	has infinite points of extrema
(ii)	$f(x) = 2\sin x - x$	(B)	has one point of maxima
(iii)	$f(x) = \|x-1\| + \|2x-3\|$	(C)	has one point of minima
(iv)	$f(x) = \|x\| - \|2x-3\|$	(D)	has infinite points of minima

25 Match the following.

	Column I		Column II
(i)	If the slope of $y = 3x^2 + ax^3$ is maximum at $x = \frac{1}{2}$, then value of a is	(A)	3
(ii)	The maximum slope of the curve $y = -x^3 + 3x^2 + 2x - 27$ is	(B)	-2
(iii)	The number of distinct real roots of the equation $(x+1)^5 = 2(x^5 + 1)$ is	(C)	5
(iv)	Let $f(x) = ab\sin x + \sqrt{1-a^2}\cos x + C$, where $\|a\|, \|b\| < 1$, then difference of maximum and minimum value of $f(x)$ is	(D)	2

26 Match the following.

	Column I		Column II
(i)	The point on the curve $y = \frac{x}{1+x^2}$, where the tangent to the curve has the greatest slope.	(A)	(1, 2)
(ii)	The point on the curve $y^2 = 4x$, which is nearest to the point (2, 1) is	(B)	$\left(\frac{17}{2}, \frac{-9}{2}\right)$
(iii)	Coordinate of the point on the straight line $x + y = 4$, which is nearest to the parabola $y^2 = 4(x-10)$ is	(C)	$\left(\sqrt{\frac{7}{2}}, \frac{-1}{2}\right)$
(iv)	The point on the curve $y = x^2 - 4$, which is nearest to the origin.	(D)	(0, 0)

27 For the function $f(x) = x^4(12\ln x - 7)$. Match the following.

	Column I		Column II
(i)	If (a, b) is the point of inflection, then $(a - b)$ is equal to	(A)	3
(ii)	If e^t is a point of minima, then $12t$ is equal to	(B)	1
(iii)	If graph is concave downward in (d, e), then $d + 3e$ is equal to	(C)	4
(iv)	The graph is concave upward in (p, ∞), then the least value of p is equal to	(D)	8

Numerical Value Type Questions

Question Stem Directions (Q.Nos. 28-30) The answer to each question is a **NUMERICAL VALUE.**

Let $f(x) = \frac{x^4 - x^2}{x^6 + 2x^3 - 1}$, $x > 1$. If the maximum value of y is $\frac{a}{b}$, where a and b are in their lowest form and extremum occurs at $x = \frac{m + \sqrt{n}}{p}$, where m, n and p are pairwise relatively prime number, then

28 The maximum value of y is $\frac{1}{k}$, then k is equal to ....

29 The value of $(a + b + 3)$ is ....

30 The value of $m + n + p$ is ....

Question Stem Directions (Q.Nos. 31-33) The answer to each question is a **NUMERICAL VALUE.**

Consider a function $f(x) = \left(\alpha - \frac{1}{\alpha} - x\right)(4 - 3x^2)$, where α is a positive parameter.

31 Number of points of extrema of $f(x)$ for a given value of α is ....

32 Absolute diference between local maximum and local minimum values of $f(x)$ in terms of α is $\frac{a}{b}\left(\alpha + \frac{1}{\alpha}\right)^3$, then $(b - a)$ is equal to ....

33 Least possible value of absolute difference between local maximum and local minimum values of $f(x)$ is $\frac{k_1}{k_2}$, then $k_1 - k_2$ is equal to ....

34 Let $f(x) = x^3 - 3x$. If m is minimum value and n is maximum value of $f(x)$, the value of $(m + n + 4)$ is ....

35 The minimum value of $f(x, y) = x^2 + y^2$, where $x^2(xy - 1) = y^2(1 + xy)$, is ....

36 The least value of a, for which the equation $\frac{4}{\sin x} + \frac{1}{1 - \sin x} = a$ has atleast one solution in the interval $\left(0, \frac{\pi}{2}\right)$, is ....

37 A cubic function $f(x)$ vanishes at $x = -2$ and has relative minimum/maximum value at $x = -1$ and $x = \frac{1}{3}$, if $\int_{-1}^{1} f(x)\,dx = \frac{14}{3}$, then $f(2)$ is ....

38 A window in the form of a rectangle is surmounted by a semicircular opening. The total perimeter of the window is 20 m. The length of the rectangular part of the window to admit maximum light through the whole opening is $\frac{k}{\pi + 4}$, then k is ....

39 AB is the chord of contact of tangents drawn from a point $P(6, 8)$ to the circle $x^2 + y^2 = r^2$. If the area of ΔPAB is maximum, then radius r is ....

40 Let s be a square of unit area. Consider any quadrilateral, which has one vertex on each side of s. If a, b, c and d denotes the length of the sides of the quadrilateral and $a^2 + b^2 + c^2 + d^2 \in [m, n]$, then $m + n$ is equal to ....

41 Two towns are to get their water supply from a river. Both towns are on the same side of the river at a distance of 6 km and 18 km from the river bank. If the distance between the points on the river bank nearest to the two towns be 10 km. A single pumping station may be located requiring the least length of pipe, then the length of pipe is ....

42 The greatest area of the rectangular plot which can be made out within a triangle of base 36 ft and altitude 12 ft, is ....

43 When travelling at x km/h a bus, burns diesel at the rate of $\frac{1}{300}\left(\frac{900}{x} + x\right)$ L/km. If the diesel oil costs 40 paise per litre and driver is paid ₹ 1.5 per hour. The steady speed that will minimise the total cost of the trip of 500 km, is ....

ANSWER KEY

Level 1

1. (i) Minimum value = 4
   (ii) Minimum value = 2
   Maximum value does not exist
   (iii) Minimum value does not exist
   Maximum value = 4
   (iv) Minimum value = 1
   Maximum value = 3
2. (i) $x = a$ is the point of maxima
   (ii) $x = a$ is neither point of maxima nor point of minima
   (iii) $x = a$ is the point of minima
3. (i) Point of local minima = 3
   Local minimum value = – 4
   (ii) Point of Local minima = 5
   Local minimum value= – 316
   Point of Local maxima = 1, – 6
   Local maximum value = 68 , –1647
   (iii) Point of Local minima = a
   Local minimum value = $2a$
   Point of Local maxima = $-a$
   Local maximum value = $-2a$
   (iv) Point of local maxima = e
   Local maximum value = $\frac{1}{e}$
   (v) Point of local minima = $\frac{3\pi}{4}$
   Local minimum value = – 1
   Point of local maxima = $\frac{\pi}{4}$
   Local maximum value = 1
4. (i) $0, \frac{2}{3}$ (ii) $\left(1, \frac{5}{6}\right)$ (iii) $x = n\pi, n \in Z$ (iv) $\left(\pm \frac{1}{\sqrt{2}}, \frac{1}{\sqrt{e}}\right)$ (v) (2, 10)
5. (i) Absolute maximum value = 82
   and absolute minimum value = 68
   (ii) Absolute maximum value = π
   and absolute minimum value = $\frac{2\pi}{3} - \frac{\sqrt{3}}{2}$
   (iii) Absolute maximum value = $\sqrt{2}$
   and absolute minimum value = $-\sqrt{2}$
   (iv) Absolute maximum value = 18
   and absolute minimum value = $-\frac{9}{4}$
6. $\left(\frac{100}{\pi + 4}\right)$m and $\left(\frac{25\pi}{\pi + 4}\right)$m 7. ₹ 5500
8. (9a, 6a) 9. 2ab sq units

Level 2

1. (a)	**2.** (a)	**3.** (b)	**4.** (c)	**5.** (c)	**6.** (c)	**7.** (a)
8. (c)	**9.** (d)	**10.** (c)	**11.** (a)	**12.** (d)	**13.** (b)	**14.** (c)
15. (d)	**16.** (b)	**17.** (b)	**18.** (d)	**19.** (b)	**20.** (b)	**21.** (d)
22. (d)	**23.** (a)	**24.** (a)	**25.** (c)	**26.** (c)	**27.** (b)	**28.** (b)
29. (c)	**30.** (b)	**31.** (c)	**32.** (b)	**33.** (c)	**34.** (c)	**35.** (c)
36. (c)	**37.** (b)	**38.** (b)	**39.** (d)	**40.** (d)	**41.** (c)	**42.** (d)
43. (b)	**44.** (b)	**45.** (9)	**46.** (12)	**47.** (5)	**48.** (9)	**49.** (0)
50. (3)	**51.** (22)	**52.** (14)	**53.** (22)	**54.** (7)	**55.** (5)	**56.** (144)
57. (3)	**58.** (27)	**59.** (18)	**60.** (4)			
61. (36)						

Level 3

1. (c)	**2.** (d)	**3.** (c)	**4.** (b)	**5.** (c)	**6.** (b)	**7.** (c)
8. (c)	**9.** (b, d)	**10.** (a, d)	**11.** (b, d)	**12.** (a, d)	**13.** (c, d)	**14.** (b, d)
15. (a, b, c)	**16.** (b, d)	**17.** (b, c)	**18.** (d)	**19.** (d)	**20.** (c)	**21.** (a)
22. (d)	**23.** (b)					

24. (i)–D, (ii)–A, (iii)–C, (iv)–B

25. (i)–B, (ii)–C, (iii)–A, (iv)–D

26. (i)–D, (ii)–A, (iii)–B, (iv)–C

27. (i)–D, (ii)–C, (iii)–A, (iv)–B

28. (6)	**29.** (10)	**30.** (6)	**31.** (2)	**32.** (5)	**33.** (23)	**34.** (4)
35. (4)	**36.** (9)	**37.** (12)	**38.** (40)	**39.** (5)	**40.** (6)	**41.** (26)
42. (108)	**43.** (45)					

SOLUTIONS

Level 1

1. (i) We know that

$$(x-2)^2 \geq 0$$

$$\Rightarrow \quad (x-2)^2 + 4 \geq 4$$

Hence, the minimum value of given function is 4, when $x - 2 = 0$

$$\Rightarrow \quad x = 2$$

(ii) We know that

$$|x| \geq 0$$

$$\Rightarrow \quad |x| + 2 \geq 2$$

Hence, minimum value of $f(x)$ is 2, when $x = 0$ and maximum value of $f(x)$ does not exist.

(iii) We know that

$$|x-2| \geq 0$$

$$\Rightarrow \quad -|x-2| \leq 0$$

$$\Rightarrow \quad -|x-2| + 4 \leq 4$$

$$\Rightarrow \quad f(x) \leq 4$$

Hence, minimum value of $f(x)$ does not exist and maximum value is 4, when $x = 2$.

(iv) We know that for all $x \in R$

$$-1 \leq \sin 3x \leq 1$$

$$\Rightarrow \quad -1+2 \leq \sin 3x + 2 \leq 1+2$$

$$\Rightarrow \quad 1 \leq \sin 3x + 2 \leq 3$$

$$\Rightarrow \quad 1 \leq |\sin 3x + 2| \leq 3$$

Hence, minimum value of $f(x)$ is 1 and maximum value of $f(x)$ is 3.

2. (i) Here, $f(a) > f(a-h)$ and $f(a) > f(a+h)$.

So, $x = a$ is the point of maxima.

(ii) Here, $f(a) < f(a-h)$ and $f(a) > f(a+h)$.

So, $x = a$ is neither point of maxima nor point of minima.

(iii) Here, $f(a) < f(a-h)$ and $f(a) < f(a+h)$

So, $x = a$ is the point of minima.

3. (i) We have, $f(x) = x^2 - 6x + 5$

$$\Rightarrow \quad f'(x) = 2x - 6$$

The critical points of $f(x)$ are given by

$$f'(x) = 0$$

$$\Rightarrow \quad 2x - 6 = 0$$

$$\Rightarrow \quad 2x = 6 \Rightarrow x = 3$$

Now, $f''(x) = 2 > 0$

So, $x = 3$ is the point of local minima.

$\therefore$ Local minimum value at $x = 3$ is

$$f(3) = (3)^2 - 6(3) + 5$$

$$= 9 - 18 + 5$$

$$\Rightarrow \quad f(3) = -4$$

(ii) We have, $f(x) = x^4 - 62x^2 + 120x + 9$

$$\Rightarrow \quad f'(x) = 4x^3 - 124x + 120$$

$$\Rightarrow \quad f'(x) = 4(x^3 - 31x + 30)$$

Now, $f'(x) = 0$

$$\Rightarrow \quad 4(x^3 - 31x + 30) = 0$$

$$\Rightarrow \quad (x-1)(x-5)(x+6) = 0$$

$$\Rightarrow \quad x = 1, 5, -6$$

Now, $f''(x) = 4(3x^2 - 31)$

At $x = 1$,

$f''(1) = 4(3 - 31) = -112 < 0$

Hence, $x = 1$ is the point of local maxima.

Now, local maximum value at $x = 1$ is

$$f(1) = (1)^4 - 62(1)^2 + 120(1) + 9 = 68$$

At $x = 5$,

$$f''(5) = 4\{3(5)^2 - 31\}$$

$$= 4(75 - 31)$$

$$= 4(44) = 176 > 0$$

$\therefore$ $x = 5$ is the point of local minima

So, local minimum value is

$$f(5) = (5)^4 - 62(5)^2 + 120(5) + 9$$

$$= -316$$

Now, at $x = -6$

$$f''(-6) = 4\{3(-6)^2 - 31\} = 308 > 0$$

$x = -6$ is the point of local maxima

So, local maximum value at $x = -6$ is

$$f(-6) = (-6)^4 - 62(-6)^2 + 120(-6) + 9$$

$$= -1647$$

(iii) We have, $f(x) = x + \dfrac{a^2}{x}$

$$\Rightarrow \quad f'(x) = 1 - \frac{a^2}{x^2}$$

Now, $f'(x) = 0$

$$\Rightarrow \quad 1 - \frac{a^2}{x^2} = 0$$

$$\Rightarrow \quad \frac{a^2}{x^2} = 1$$

$$\Rightarrow \quad x = \pm a$$

Now, $f''(x) = \dfrac{2a^2}{x^3}$

At $x = a$,

$$f''(a) = \frac{2a^2}{a^3} = \frac{2}{a} > 0 \qquad [\because a > 0]$$

$\therefore$ $x = a$ is the point of local minima.

Local minimum value at $x = a$ is

$$f(a) = a + \frac{a^2}{a} = 2a$$

At $x = -a$,

$$f''(-a) = \frac{2a^2}{(-a)^3} = -\frac{2}{a} < 0 \qquad [\because a > 0]$$

$\therefore$ $x = -a$ is the point of local maxima.

Local maximum value at $x = -a$ is

$$f(-a) = -a + \frac{a^2}{-a} = -a - a = -2a$$

(iv) We have, $f(x) = \frac{\log x}{x}$

$\Rightarrow$ $f'(x) = \frac{1 - \log x}{x^2}$

Now, $f'(x) = 0$

$\Rightarrow$ $\frac{1 - \log x}{x^2} = 0$

$\Rightarrow$ $\log x = 1 \Rightarrow x = e$

Now, $f'(x) = \frac{1 - \log x}{x^2}$

$\Rightarrow$ $f'(x) = x^{-2}(1 - \log x)$

$\Rightarrow$ $f''(x) = -2x^{-3}(1 - \log x) - x^{-3}$

$= -x^{-3}(3 - 2\log x)$

At $x = e$,

$$f''(e) = -e^{-3}(3 - 2\log e)$$

$$= -\frac{1}{e^3} < 0$$

Hence, $x = e$ is the point of local maxima.

Local maximum value is

$$f(e) = \frac{\log e}{e} = \frac{1}{e}$$

(v) We have, $f(x) = \sin 2x$

$\Rightarrow$ $f'(x) = 2\cos 2x$

At critical point,

$f'(x) = 0$

$\Rightarrow$ $2\cos 2x = 0$

$\Rightarrow$ $2x = \frac{\pi}{2}, \frac{3\pi}{2}$ $\left[\begin{array}{l}\because 0 < x < \pi \\ \Rightarrow 0 < 2x < 2\pi\end{array}\right]$

$\Rightarrow$ $x = \frac{\pi}{4}, \frac{3\pi}{4}$

Now, $f''(x) = -4\sin 2x$

At $x = \frac{\pi}{4}$,

$$f''\left(\frac{\pi}{4}\right) = -4\sin\frac{\pi}{2} = -4 < 0$$

$\therefore$ $x = \frac{\pi}{4}$ is the point of local maxima.

Local maximum value is

$$f\left(\frac{\pi}{4}\right) = \sin 2\left(\frac{\pi}{4}\right) = 1$$

At $x = \frac{3\pi}{4}$,

$$f''\left(\frac{3\pi}{4}\right) = -4\sin 2\left(\frac{3\pi}{4}\right)$$

$$= -4\sin\frac{3\pi}{2} = 4 > 0$$

$\therefore$ $x = \frac{3\pi}{4}$ is the point of local minima.

Local minimum value at $x = \frac{3\pi}{4}$ is

$$f\left(\frac{3\pi}{4}\right) = \sin 2\left(\frac{3\pi}{4}\right)$$

$$= \sin\frac{3\pi}{2} = -1$$

4. (i) We have, $f(x) = 3x^4 - 4x^3$

$\therefore$ $f'(x) = 12x^3 - 12x^2$

and $f''(x) = 36x^2 - 24x$

Now, $f''(x) = 0$

$x = 0, \frac{2}{3}$, which are the points of inflection.

(ii) Given, $f(x) = x^2 - \frac{1}{6x^3}$

$\Rightarrow$ $f'(x) = 2x + \frac{1}{2x^4}$

$\Rightarrow$ $f''(x) = 2 - \frac{2}{x^5}$

Now, $f''(x) = 0$, gives $2 - \frac{2}{x^5} = 0$

Thus, $x = 1$

When $x = 1$, then $y = \frac{5}{6}$

Thus, point of inflection is $\left(1, \frac{5}{6}\right)$.

(iii) Given, $f(x) = \sin x$

$\Rightarrow$ $f'(x) = \cos x \Rightarrow f''(x) = -\sin x$

Now, $f''(x) = 0$

$\therefore$ $x = n\pi, n \in Z$

So, points of inflection are $x = n\pi$, $n \in Z$.

(iv) We have, $f(x) = e^{-x^2}$

$\Rightarrow$ $f'(x) = -2xe^{-x^2}$

$\Rightarrow$ $f''(x) = -2\{e^{-x^2}(1) + x \cdot e^{-x^2}(-2x)\}$

$= -2(e^{-x^2} - 2x^2 e^{-x^2})$

If $f''(x) = 0$

$\Rightarrow$ $(2x^2 - 1)e^{-x^2} = 0$

$\Rightarrow \quad x = \pm \frac{1}{\sqrt{2}}$

When $x = \pm \frac{1}{\sqrt{2}}$, then $y = e^{-\frac{1}{2}} = \frac{1}{\sqrt{e}}$

Thus, the points of inflection are $\left(\pm \frac{1}{\sqrt{2}}, \frac{1}{\sqrt{e}}\right)$.

(v) We have, $y = f(x) = (x-2)^{2/3} + 10$

$\Rightarrow \quad f'(x) = \frac{2}{3}(x-2)^{-1/3}$

$= \frac{2}{3(x-2)^{1/3}}$

$\Rightarrow \quad f''(x) = \frac{2}{9(x-2)^{4/3}}$

Thus, $f''(x)$ does not exist at $x = 2$.

When $x = 2$, then $y = 0 + 10 = 10$

Thus, the point of inflection is (2, 10).

5. (i) We have, $f(x) = 2x^3 - 24x + 100$

$\Rightarrow \quad f'(x) = 6x^2 - 24$

Now, $\quad f'(x) = 0$

$\Rightarrow \quad 6x^2 - 24 = 0$

$\Rightarrow \quad x^2 - 4 = 0$

$\Rightarrow \quad x = \pm 2$

But $x = -2 \notin [1, 3]$, so $x = 2$ is the only stationary point.

Now, $f(1) = 2(1)^3 - 24(1) + 100 = 78$

$f(2) = 2(2)^3 - 24(2) + 100 = 68$

$f(3) = 2(3)^3 - 24(3) + 100 = 82$

The absolute maximum value of $f(x)$ is 82, which attains at $x = 3$ and the minimum value is 68, which attains at $x = 2$.

(ii) We have, $f(x) = x + \sin 2x$

$\Rightarrow \quad f'(x) = 1 + 2\cos 2x$

Now, $\quad f'(x) = 0$

$\Rightarrow 1 + 2\cos 2x = 0$

$\Rightarrow \quad \cos 2x = \frac{-1}{2}$

$\Rightarrow \quad 2x = \frac{2\pi}{3}$

or $\quad 2x = \frac{4\pi}{3} \qquad \left[\begin{array}{l} \because x \in [0, \pi] \\ \Rightarrow 2x \in [0, 2\pi] \end{array}\right]$

$\Rightarrow \quad x = \frac{\pi}{3}$

or $\quad x = \frac{2\pi}{3}$

Now, $f(0) = 0 + \sin 0 = 0$

$f\left(\frac{\pi}{3}\right) = \frac{\pi}{3} + \sin 2\left(\frac{\pi}{3}\right) = \frac{\pi}{3} + \frac{\sqrt{3}}{2}$

$\Rightarrow \quad f\left(\frac{2\pi}{3}\right) = \frac{2\pi}{3} + \sin 2\left(\frac{2\pi}{3}\right)$

$= \frac{2\pi}{3} - \frac{\sqrt{3}}{2}$

$\Rightarrow \quad f(\pi) = \pi + \sin 2(\pi) = \pi$

Hence, the absolute maximum value is π at $x = \pi$ and absolute minimum value is $\frac{2\pi}{3} - \frac{\sqrt{3}}{2}$ at $x = \frac{2\pi}{3}$.

(iii) We have, $\quad f(x) = \sin x + \cos x$

$\Rightarrow \quad f'(x) = \cos x - \sin x$

Now, $\quad f'(x) = 0$

$\Rightarrow \cos x - \sin x = 0 \Rightarrow \tan x = 1$

$\Rightarrow \quad x = \frac{\pi}{4}$

or $\quad x = \pi + \frac{\pi}{4} = \frac{5\pi}{4}$

Now, $f(0) = \sin 0 + \cos 0 = 1$

$f\left(\frac{\pi}{4}\right) = \sin\frac{\pi}{4} + \cos\frac{\pi}{4}$

$= \frac{1}{\sqrt{2}} + \frac{1}{\sqrt{2}} = \frac{2}{\sqrt{2}} = \sqrt{2}$

$\Rightarrow \quad f\left(\frac{5\pi}{4}\right) = \sin\frac{5\pi}{4} + \cos\frac{5\pi}{4}$

$= -\frac{1}{\sqrt{2}} - \frac{1}{\sqrt{2}} = -\sqrt{2}$

$\Rightarrow \quad f(2\pi) = \sin 2\pi + \cos 2\pi = 1$

Hence, absolute maximum value is $\sqrt{2}$ at $x = \frac{\pi}{4}$ and absolute minimum value is $-\sqrt{2}$ at $x = \frac{5\pi}{4}$.

(iv) We have, $\quad f(x) = 12x^{4/3} - 6x^{1/3}$

$\Rightarrow \quad f'(x) = 16x^{1/3} - 2x^{-2/3} = \frac{2(8x-1)}{x^{2/3}}$

At critical point $f'(x) = 0$

$\Rightarrow \quad \frac{2(8x-1)}{x^{2/3}} = 0$

$\Rightarrow \quad x = \frac{1}{8}$

Now, $f(-1) = 12(-1)^{4/3} - 6(-1)^{1/3}$

$= 12 + 6 = 18$

$\Rightarrow \quad f\left(\frac{1}{8}\right) = 12\left(\frac{1}{8}\right)^{4/3} - 6\left(\frac{1}{8}\right)^{1/3} = -\frac{9}{4}$

$\Rightarrow \quad f(1) = 12(1)^{4/3} - 6(1)^{1/3} = 6$

Hence, absolute maximum value is 18 at $x = -1$ and absolute minimum value is $\frac{-9}{4}$ at $x = \frac{1}{8}$.

6. Let x m and $(25-x)$ m be the required lengths.
Again, let a be the side of the square formed and r be the radius of the circle formed. Then,

$$4a = x \text{ and } 2\pi r = 25 - x \Rightarrow a = \frac{x}{4} \text{ and } r = \frac{25-x}{2\pi}$$

$\therefore$ Area of the square $= a^2 = \left(\frac{x^2}{16}\right) m^2$

and area of the circle $= \pi r^2 = \pi\left(\frac{25-x}{2\pi}\right)^2 = \frac{(25-x)^2}{4\pi}$

Now, combined area, $A = \frac{x^2}{16} + \frac{(25-x)^2}{4\pi}$

On differentiating both sides w.r.t. x, we get

$$\frac{dA}{dx} = \frac{x}{8} - \frac{(25-x)}{2\pi} = \frac{(\pi+4)x - 100}{8\pi}$$

and $\frac{d^2A}{dx^2} = \frac{(\pi+4)}{8\pi}$

For maxima or minima, put $\frac{dA}{dx} = 0$

$$\Rightarrow \quad \frac{(\pi+4)x - 100}{8\pi} = 0 \Rightarrow x = \frac{100}{(\pi+4)}$$

and $\frac{d^2A}{dx^2} = \frac{(\pi+4)}{8\pi} > 0$ for all values of x.

$\therefore \quad x = \frac{100}{(\pi+4)}$ is a point of local minima.

Hence, the lengths of pieces are $\left(\frac{100}{\pi+4}\right)$ m and $\left(\frac{25\pi}{\pi+4}\right)$ m, respectively.

7. Let the length and breadth of the tank be x and y m, respectively.
Then, volume $= 75 \text{m}^3$

$\Rightarrow \quad 3xy = 75 \quad$ [$\because$ depth of tank $= 3$ m]

$\Rightarrow \quad y = \frac{25}{x}$

Let C be the cost of the tank.

Then, $C = 100xy + 50(3 \times 2x + 3 \times 2y)$

$= 100xy + 300x + 300y$

$= 100x \times \frac{25}{x} + 300x + 300 \times \frac{25}{x}$

$\Rightarrow \quad C = 2500 + 300x + \frac{7500}{x}$

On differentiating twice w.r.t. x, we get

$$\frac{dC}{dx} = 300 - \frac{7500}{x^2}$$

and $\frac{d^2C}{dx^2} = \frac{15000}{x^3} > 0$

For minimum value, put $\frac{dC}{dx} = 0$

$\Rightarrow \quad 300 - \frac{7500}{x^2} = 0 \Rightarrow x^2 = 25$

$\Rightarrow \quad x = 5$ [$\because$ length cannot be negative]

At $x = 5$, $\frac{d^2C}{dx^2} = \frac{15000}{5^3} = 120 > 0$

So, C is minimum.

When $x = 5$, then $C = 2500 + 1500 + 1500 = 5500$

Hence, the cost of least expensive tank is ₹ 5500.

8. Let the point on $y^2 = 4ax$ be (x_1, y_1).

Then, $\quad y_1^2 = 4ax_1 \quad$ ...(i)

Distance between (x_1, y_1) and $(11a, 0)$ is given by

$$D = \sqrt{(x_1 - 11a)^2 + (y_1 - 0)^2} = \sqrt{(x_1 - 11a)^2 + y_1^2}$$

$$= \sqrt{(x_1 - 11a)^2 + 4ax_1} \quad \text{[from Eq. (i)]}$$

On differentiating both sides w.r.t. x_1, we get

$$\frac{dD}{dx_1} = \frac{1}{2\sqrt{(x_1 - 11a)^2 + 4ax_1}}[2(x_1 - 11a) + 4a]$$

$$= \frac{2x_1 - 22a + 4a}{2\sqrt{(x_1 - 11a)^2 + 4ax_1}} = \frac{x_1 - 9a}{\sqrt{(x_1 - 11a)^2 + 4ax_1}}$$

For critical points, put $\frac{dD}{dx_1} = 0 \Rightarrow x_1 - 9a = 0 \Rightarrow x_1 = 9a$

If $x_1 = 9a$, then $y_1^2 = 36a^2 \Rightarrow y_1 = \pm 6a$

Hence, required points are $(9a, 6a)$ and $(9a, -6a)$.

Now, $\frac{d^2D}{dx_1^2} = \frac{d}{dx_1}\left(\frac{dD}{dx_1}\right) = \frac{d}{dx_1}\left(\frac{x_1 - 9a}{\sqrt{(x_1 - 11a)^2 + 4ax_1}}\right)$

$$= \frac{\sqrt{(x_1 - 11a)^2 + 4ax_1} - (x_1 - 9a)\frac{1}{2} \cdot \frac{[2(x_1 - 11a) + 4a]}{\sqrt{(x_1 - 11a)^2 + 4ax_1}}}{(x_1 - 11a)^2 + 4ax_1}$$

At $(9a, 6a)$, $\frac{d^2D}{dx_1^2} > 0$

So, at $(9a, 6a)$, D is minimum.

Hence, the required point is $(9a, 6a)$.

9. Let $PQRS$ be a triangle which inscribed in the ellipse $\frac{x^2}{a^2} + \frac{y^2}{b^2} = 1$.

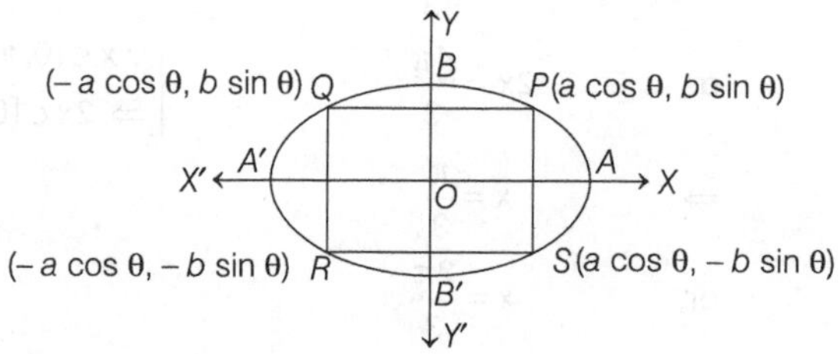

Length of rectangle, $l = 2a\cos\theta$

and breadth of rectangle, $b = 2b\sin\theta$

$\because$ Area of rectangle, $A = l \cdot b$

$\therefore A = 2ab \cdot 2\sin\theta\cos\theta = 2ab\sin 2\theta$

On differentiating both sides w.r.t. θ, we get

$$\frac{dA}{d\theta} = 2ab \cdot \cos 2\theta \cdot 2$$

For critical points, put $\frac{dA}{d\theta} = 0 \Rightarrow \cos 2\theta = 0$

$$\Rightarrow \quad 2\theta = \frac{\pi}{2} \Rightarrow \theta = \frac{\pi}{4}$$

Now, $\frac{d^2A}{d\theta^2} = -8ab\sin 2\theta$ which is negative for $\theta = \frac{\pi}{4}$.

Hence, area is maximum.

Now, maximum area $= 2ab\sin\frac{\pi}{2} = 2ab$ sq units

Level 2

1. (a) We have, $f(x) = \begin{cases} x^3 + x^2 + 5x, & x < 0 \\ 1 - xe^x, & x \geq 0 \end{cases}$

$$f(0^-) = 0; f(0) = f(0^+) = 1$$

$\Rightarrow f(x)$ is discontinuous at $x = 0$

$$\Rightarrow \quad f'(x) = \begin{cases} 3x^2 + 2x + 5, & x < 0 \\ -xe^x - e^x, & x > 0 \end{cases}$$

$\Rightarrow f'(x) > 0$ for $x < 0$ and $f'(x) < 0$ for $x > 0$ and

$$f(0^-) = 0 < f(0) = f(0^+) - 1 > f(x), \forall\, x > 0$$

$\Rightarrow \quad x = 0$ is the point of local maxima.

2. (a) We have, $f(x) = x^2 + \frac{1}{x^2}$ and $g(x) = x - \frac{1}{x}$.

Now, $$h(x) = \frac{f(x)}{g(x)} = \frac{x^2 + \frac{1}{x^2}}{x - \frac{1}{x}}$$

$$= \frac{\left(x - \frac{1}{x}\right)^2 + 2}{\left(x - \frac{1}{x}\right)} = \left(x - \frac{1}{x}\right) + \frac{2}{\left(x - \frac{1}{x}\right)}$$

Let $x - \frac{1}{x} = t$

When $t \geq 0$

$$h(x) = t + \frac{2}{t} = \left(\sqrt{t} - \sqrt{\frac{2}{t}}\right)^2 + 2\sqrt{2} \geq 2\sqrt{2}$$

So, local minimum value of $h(x)$ is $2\sqrt{2}$.

Now, when $t < 0$

$$h(x) = -\left[-t - \frac{2}{t}\right] = -\left[\sqrt{-t} + \sqrt{\frac{2}{-t}}\right]^2 - 2\sqrt{2} \leq -2\sqrt{2}$$

So, $-2\sqrt{2}$ is local maximum value.

3. (b) We have, $f(x) = \begin{cases} |x^2 - 3x| + a, & 0 \leq x < \frac{3}{2} \\ -2x + 3, & x \geq \frac{3}{2} \end{cases}$

$x = \frac{3}{2}$ is the point of local maxima.

So, $f\left(\frac{3}{2}\right) = 0$

So, $\lim_{x \to \frac{3}{2}} |x^2 - 3x| + a \leq 0$

$$\Rightarrow \quad \frac{9}{4} + a \leq 0$$

$$\Rightarrow \quad a \leq -\frac{9}{4}$$

Then, the greatest value of a is $-\frac{9}{4}$.

4. (c) We have, $f(x) = |x|^m |x - 1|^n$

$$\Rightarrow \quad f(x) = \begin{cases} (-1)^{m+n} x^m (x-1)^n, & \text{if } x < 0 \\ (-1)^n x^m (x-1)^n, & \text{if } 0 \leq x < 1 \\ x^m (x-1)^n, & \text{if } x \geq 1 \end{cases}$$

Let $g(x) = x^m (x-1)^n$

$\Rightarrow g'(x) = mx^{m-1}(x-1)^n + nx^m(x-1)^{n-1}$

$= x^{m-1}(x-1)^{n-1}\{mx - m + nx\}$

Now, $f'(x) = 0$ or $g'(x) = 0$ i.e. $x = 0, 1$ or $\frac{m}{m+n}$

$f(0) = 0, f(1) = 0$ and $f\left(\frac{m}{m+n}\right) = (-1)^n \frac{m^m n^n (-1)^n}{(m+n)^{m+n}}$

$$\left[\because 0 < \frac{m}{m+n} < 1\right]$$

$$= \frac{m^m n^n}{(m+n)^{m+n}} > 0$$

$\therefore$ Maximum value is $\frac{m^m n^n}{(m+n)^{m+n}}$.

5. (c) We have, $f(x) = 2x^3 - 9ax^2 + 12a^2x + 1$

$$\Rightarrow \quad f'(x) = 6x^2 - 18ax + 12a^2$$

$$= 6(x^2 - 3ax + 2a^2)$$

Now, $f'(x) = 0$

$\Rightarrow \quad 6(x^2 - 3ax + 2a^2) = 0$

or $\quad 6(x - a)(x - 2a) = 0$

$\therefore \quad x = a, 2a$

$f''(x) = 6(2x - 3a) > 0$ for $x = 2a$ and hence local minimum at $x = 2a = \beta$, say.

$f''(x) = 6(2x - 3a) < 0$ for $x = a$, hence local maximum at $x = a = \alpha$, say

But $\beta = \alpha^2 \Rightarrow 2a = a^2 \Rightarrow a(a - 2) = 0$

$\therefore \quad a = 2$

$a = 0$ is ruled out as $f(x) = 2x^3 + 1$ and $f'(x) = 6x^2$, which is always '+ve'.

Hence, $f(x)$ is an increasing function in this case.

6. (c) $f(x) = \left(\frac{2}{x}\right)^{x^2}, x > 0$

$$\therefore \quad \log f(x) = x^2(\log 2 - \log x)$$

$$f'(x) = f(x)[-x + (\log 2 - \log x)2x]$$

$$f'(x) = f(x) \cdot x\,(2\log 2 - 2\log x - 1)$$

For maxima or minima put $f'(x) = 0$, we get

$$2\log 2 - 2\log x - 1 = \log\left(\frac{4}{x^2}\right) - 1 = 0$$

$$\Rightarrow \quad x = \frac{2}{\sqrt{e}} \quad [\because x > 0]$$

Sign of $f'(x)$: $+$ before $\frac{2}{\sqrt{e}}$, $-$ after $\frac{2}{\sqrt{e}}$

$\therefore$ At $x = \frac{2}{\sqrt{e}}$, $f(x)$ has maximum value.

$$\text{Maximum} = \left(\frac{2\sqrt{e}}{2}\right)^{\frac{4}{e}} = e^{\frac{2}{e}}$$

7. (a) We have, $\Delta = \begin{vmatrix} x & a & a \\ b & x & a \\ b & b & x \end{vmatrix}$

So, $\quad \Delta = x(x^2 - ab) - a(bx - ab) + a(b^2 - bx)$

$$\Rightarrow \quad \Delta = x^3 - 3abx + a^2b + ab^2$$

Δ is cubic in x.

So, for local minimum of Δ,

$$\frac{d\Delta}{dx} = 0$$

$$\Rightarrow \quad 3x^2 - 3ab = 0$$

$$\Rightarrow \quad x = \sqrt{ab}$$

8. (c) Given,

$$f(x) = x^3 - 3x^2 - \frac{3}{2}f''(2) \cdot x + f''(1)$$

Then, $f'(x) = 3x^2 - 6x - \frac{3}{2}f''(2)$

$$f''(x) = 6x - 6$$

Put $\quad x = 1$ and $x = 2$,

$$f''(2) = 12 - 6 = 6 \text{ and } f''(1) = 0$$

Therefore, $f(x) = x^3 - 3x^2 - \frac{3}{2} \times 6x + 0$

$$\Rightarrow \quad f(x) = x^3 - 3x^2 - 9x \quad \text{...(i)}$$

and $\quad f'(x) = 3x^2 - 6x - 9$ and $f''(x) = 6x - 6$

Equate $\quad f'(x) = 0$ gives,

$$\Rightarrow \quad 3x^2 - 6x - 9 = 0$$

$$\Rightarrow \quad 3(x^2 - 2x - 3) = 0$$

$$3(x + 1)(x - 3) = 0$$

$$\Rightarrow \quad x = -1 \text{ or } 3$$

Now, $f''(-1) = 6(-1) - 6 = -12 < 0$

$$f''(3) = 6(3) - 6 = 12 > 0$$

$\therefore x = -1$ is point of maxima and $x = 3$ is point of minima.

Local minimum value $= f(3)$

$$f(3) = (3)^3 - 3(3)^2 - 9(3) \quad \text{[using Eq. (i)]}$$

$$= 27 - 27 - 27$$

$$f(3) = -27$$

9. (d) We have, $y = x^3 - 6x^2 + 12x + 5$

$$\Rightarrow \quad \frac{dy}{dx} = 3x^2 - 12x + 12$$

$$\Rightarrow \quad \frac{d^2y}{dx^2} = 6x - 12$$

For point of inflection,

$$\frac{d^2y}{dx^2} = 0$$

$$\Rightarrow \quad 6x - 12 = 0$$

$$\Rightarrow \quad x = 2$$

10. (c) We have, $f(x) = y = ax^4 + bx^3 + cx^2 + dx + e$

$$\Rightarrow \quad f'(x) = 4ax^3 + 3bx^2 + 2cx + d$$

$$\Rightarrow \quad f''(x) = 12ax^2 + 6bx + 2c$$

Now, for point of inflection $f''(x) = 0$

$$\Rightarrow \quad 12ax^2 + 6bx + 2c = 0$$

Roots of this equation will be

$$x = \frac{-3b \pm \sqrt{9b^2 - 24ac}}{12a}$$

So, $9b^2 - 24ac > 0$

$$\Rightarrow \quad 3b^2 - 8ac > 0$$

11. (a) We have, $f(x) = \frac{ax^3}{3} + (a + 2)x^2 + (a - 1)x + 2$

$$\Rightarrow \quad f'(x) = ax^2 + 2(a + 2)x + (a - 1)$$

$$\Rightarrow \quad f''(x) = 2ax + 2(a + 2)$$

For point of inflection,

$$f''(x) = 0$$

$$\Rightarrow \quad x = -\left(\frac{a + 2}{a}\right)$$

For negative point of inflection,

$$\frac{a + 2}{a} > 0$$

$$\Rightarrow \quad a \in (-\infty, -2) \cup (0, \infty).$$

12. (d) We have,

$$f(x)=x^2+\frac{1}{x^2} \text{ and } g(x)=x-\frac{1}{x}$$

$$\Rightarrow \quad h(x)=\frac{f(x)}{g(x)}$$

$$\therefore \quad h(x)=\frac{x^2+\frac{1}{x^2}}{x-\frac{1}{x}}=\frac{\left(x-\frac{1}{x}\right)^2+2}{x-\frac{1}{x}}$$

$$\Rightarrow \quad h(x)=\left(x-\frac{1}{x}\right)+\frac{2}{x-\frac{1}{x}}$$

$$x-\frac{1}{x}>0, \left(x-\frac{1}{x}\right)+\frac{2}{x-\frac{1}{x}}\in[2\sqrt{2},\infty]$$

$$x-\frac{1}{x}<0, \left(x-\frac{1}{x}\right)+\frac{2}{x-\frac{1}{x}}\in[-\infty,2\sqrt{2}]$$

$\therefore$ Local minimum value is $2\sqrt{2}$.

13. (b) Let $f(x)=\dfrac{\sin(x+\alpha)}{\sin(x+\beta)}$

$$\Rightarrow \quad f'(x)=\frac{\sin(x+\beta)\cos(x+\alpha)-\cos(x+\beta)\sin(x+\alpha)}{\sin^2(x+\beta)}$$

$$=\frac{\sin(\beta-\alpha)}{\sin^2(x+\beta)}$$

Thus, $f(x)$ has no maximum and minimum, if $f'(x)\neq 0$.

$\Rightarrow \sin(\beta-\alpha)\neq 0$

$\Rightarrow \quad \beta-\alpha\neq k\pi, k\in Z.$

14. (c) We have, $y=e^{(2x^2-2x+1)\sin^2 x}$

$$\Rightarrow \quad y=e^{2(x^2-x+1/2)\sin^2 x}$$

$$=e^{2\left(\left(x-\frac{1}{2}\right)^2+\frac{1}{4}\right)\sin^2 x}$$

Clearly, the minimum value occurs, when

$$\sin^2 x=0 \text{ as } \left[\left(x-\frac{1}{2}\right)^2+\frac{1}{4}\right]\geq 4$$

So, minimum value of y is

$$y=e^0 \Rightarrow y=1$$

15. (d) Let $f(x)=\dfrac{1+x^2}{1+x}, x\geq 0$

Differentiating, we get

$$f'(x)=\frac{2x(1+x)-1(1+x^2)}{(1+x)^2}$$

$$=\frac{x^2+2x-1}{(1+x)^2}$$

$$=1-\frac{2}{(1+x)^2}$$

Now, $f'(x)=0$

$\Rightarrow \quad (x+1)^2=2 \Rightarrow x=\pm\sqrt{2}-1$

Now, $x\geq 0 \Rightarrow x=\sqrt{2}-1$ is the only critical point.

Also, $\quad f''(x)=\dfrac{4}{(1+x)^3}$

$$\Rightarrow f''(\sqrt{2}-1)=\frac{4}{(1+\sqrt{2}-1)^3}=\frac{4}{2\sqrt{2}}>0$$

f is minimum at $x=\sqrt{2}-1$ and least value is

$$f(\sqrt{2}-1)=\frac{1+(\sqrt{2}-1)^2}{1+\sqrt{2}-1}$$

$$=\frac{4-2\sqrt{2}}{\sqrt{2}}=2\sqrt{2}-2=2(\sqrt{2}-1)$$

16. (b) Let $f(x)=x^{25}(1-x)^{75}$

On differentiating w.r.t. x, we get

$$f'(x)=25x^{24}(1-x)^{75}-75x^{25}(1-x)^{74}$$

$$=x^{24}(1-x)^{74}[25(1-x)-75x]$$

$$=x^{24}(1-x)^{74}[25-100x]$$

Also, $0<x<1$ and $f'(x)=0\Rightarrow x=\dfrac{1}{4}$

Now, $x<\dfrac{1}{4} \Rightarrow f'(x)>0$

and $\quad x>\dfrac{1}{4} \Rightarrow f'(x)<0$

i.e. $f'(x)$ changes sign from positive to negative.

Therefore, $f(x)$ has maximum at $x=\dfrac{1}{4}$.

17. (b) Given, $f(x)=(4a-3)(x+\log_e 5)+2(a-7)\cot\dfrac{x}{2}\cdot\sin^2\dfrac{x}{2}$

$$\Rightarrow \quad f(x)=(4a-3)(x+\log_e 5)+2(a-7)\cos\frac{x}{2}\sin\frac{x}{2}$$

$$\Rightarrow \quad f(x)=(4a-3)(x+\log_e 5)+(a-7)\sin x$$

$$\Rightarrow \quad f'(x)=(4a-3)(1+0)+(a-7)\cos x$$

$$\Rightarrow \quad f'(x)=(4a-3)+(a-7)\cos x$$

When $f'(x)=0$,

$$(4a-3)+(a-7)\cos x=0$$

$$\Rightarrow \quad \cos x=\frac{4a-3}{7-a}$$

As, $-1\leq\cos x\leq 1$

So, $-1\leq\dfrac{4a-3}{7-a}\leq 1$

$$\Rightarrow \quad \frac{4a-3}{7-a}+1\geq 0 \Rightarrow \frac{4a-3+7-a}{7-a}\geq 0$$

$$\Rightarrow \quad \frac{3a+4}{7-a}\geq 0$$

$-\quad | -4/3 \quad + \quad | 7 \quad -$

$$\Rightarrow \quad a\in[-4/3,7) \quad \text{...(i)}$$

Now, $\frac{4a-3}{7-a}-1\leq 0$

$\Rightarrow \frac{4a-3-7+a}{7-a}\leq 0 \Rightarrow \frac{5a-10}{7-a}\leq 0$

$-\quad 2 \quad + \quad 7 \quad -$

$\Rightarrow \qquad a\in(-\infty,2]\cup[7,\infty)$ ... (ii)

From Eqs. (i) and (ii),

$$a\in\left[-\frac{4}{3},2\right]$$

18. (d) We have, $f(x)=\begin{cases}\tan^{-1}\alpha-3x^2, & 0<x<1\\ -6x, & x\geq 1\end{cases}$

$\Rightarrow f(1^-)=\tan^{-1}\alpha-3 \Rightarrow f(1^+)=f(1)=-6$

$\therefore$ $f(x)$ is discontinuous at $x=1$

and $f'(x)=\begin{cases}-6x, & 0<x<1\\ -6 & x>1\end{cases}$

$\Rightarrow$ $f(x)$ is a decreasing function.

Thus, for $f(x)$ to have maximum at $x=1$

$$f(1^-)<f(1)$$

$\Rightarrow \qquad \tan^{-1}\alpha-3<-6$

$\Rightarrow \qquad \tan^{-1}\alpha<-3$

which is never possible as $\tan^{-1}\alpha>-\frac{\pi}{2}$ for every real value of α.

19. (b) Let $g(x)=4x^3-12x^2+11x-3$

$\therefore \qquad g'(x)=12x^2-24x+11$

$=12(x-1)^2-1>0$ for $x\in[2,3]$

Thus, $g(x)$ is increasing in $[2,3]$.

$f(x)_{max}=f(3)=\log_{10}\{4(27)-12(9)+11(3)-3\}$

$=\log_{10}30$

$=\log_{10}3+\log_{10}10$

$=1+\log_{10}3$

20. (b) Given, $y=\ln(3x^4-2x^3-6x^2+6x+1)$

Let $Z=3x^4-2x^3-6x^2+6x+1$

$\therefore \qquad \frac{dZ}{dx}=12x^3-6x^2-12x+6$

$=12\left(x-\frac{1}{2}\right)(x+1)(x-1)$

Now, for maximum or minimum of Z, $\frac{dZ}{dx}=0$

Then, $x=\frac{1}{2}$, -1 and 1 but $x\in(0,2)$

$\therefore \qquad x=\frac{1}{2},1$

Now, $\frac{d^2Z}{dx^2}=36x^2-12x-12$

Thus, $\left.\frac{d^2Z}{dx^2}\right|_{x=\frac{1}{2}}=-9<0$

$x=\frac{1}{2}$ is the point of maxima.

Then, at $x=1$

$$\left.\frac{d^2Z}{dx^2}\right|_{x=1}=12>0$$

Hence, $x=1$ is the point of minimum value of Z.

Hence, y is also minimum, when $x=1$

At $y=1$,

$$y=\ln(3-2-6+6+1)$$

$$y=\ln 2$$

Hence, minimum value of y is $\ln 2$.

21. (d) The non-zero four degree polynomial $f(x)$ has extremum points at $x=-1,0,1$, so we can assume $f'(x)=a(x+1)(x-0)(x-1)=ax(x^2-1)$

where, a is non-zero constant.

$$f'(x)=ax^3-ax$$

$\Rightarrow \qquad f(x)=\frac{a}{4}x^4-\frac{a}{2}x^2+C$

[integrating both sides]

where, C is constant of integration.

Now, since $f(x)=f(0)$

$\Rightarrow \frac{a}{4}x^4-\frac{a}{2}x^2+C=C \Rightarrow \frac{x^4}{4}=\frac{x^2}{2}$

$\Rightarrow \qquad x^2(x^2-2)=0 \Rightarrow x=-\sqrt{2},0,\sqrt{2}$

Thus, $f(x)=f(0)$ has one rational and two irrational roots.

22. (d) We have, $f(x)=\begin{cases}-x^3+k^2-3k+2, & 0\leq x<1\\ 2x-3, & 1\leq x\leq 3\end{cases}$

$f(1^-)=-1+k^2-3k+2$

$=k^2-3k+1$

and $f(1^+)=f(1)=-1$

$\Rightarrow \quad f'(x)=\begin{cases}-3x^2, & 0<x<1\\ 2, & 1<x<3\end{cases}$

$\therefore$ For $f(x)$ to have smallest value at 1,

$f(x)\geq -1$ for $0\leq x\leq 1$

$\Rightarrow \qquad -x^3+k^2-3k+2\geq -1$

$\Rightarrow \qquad -x^3+k^2-3k+3\geq 0 \,\forall\, 0\leq x\leq 1$

$\Rightarrow k^2-3k+3\geq x^3 \,\forall\, x\in[0,1]$

$\Rightarrow k^2-3k+3\geq 1$

$\Rightarrow k^2-3k+2\geq 0$

$\Rightarrow (k-2)(k-1)\geq 0$

$\Rightarrow k\leq 1$ or $k\geq 2$

23. (a) We have, $f(x)=2|x-1|+5|x-2|$

$$\Rightarrow \quad f(x)=\begin{cases}-2(x-1)-5(x-2), & x<1\\ 2(x-1)-5(x-2), & 1\le x<2\\ 2(x-1)+5(x-2), & x\ge 2\end{cases}$$

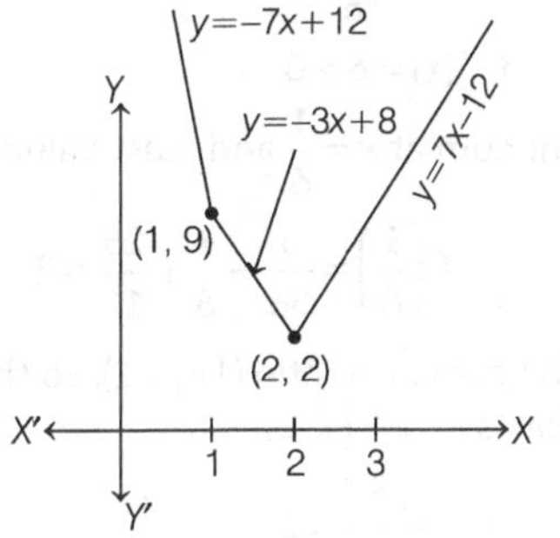

$$\Rightarrow \quad f(x)=\begin{cases}-7x+12, & x<1\\ -3x+8, & 1\le x<2\\ 7x-12, & x\ge 2\end{cases}$$

$$\Rightarrow \quad f'(x)=\begin{cases}-7, & x<1\\ -3, & 1<x<2\\ 7, & x>2\end{cases}$$

Clearly, $f(x)$ is increasing for all $x>2$ and decreasing for all $x<1$ and $x\in(1,2)$.

$\therefore$ Minimum value of $f(x)=2$, when $x=2$.

24. (a) $f(x)=(x+b)^2+(2c^2-b^2)$

$\Rightarrow g(x)=(b^2+c^2)-(x+c)^2$

Minimum of $f(x)=2c^2-b^2$, when $x=-b$

and maximum of $g(x)=b^2+c^2$, when $x=-c$

By the given condition,

$$2c^2-b^2>b^2+c^2$$

$$\Rightarrow \quad c^2>2b^2$$

$$\Rightarrow \quad |c|>\sqrt{2}|b|$$

25. (c) We have,

$$f'(x)=\begin{cases}3x^2-2x+10, & x\le 1\\ -2, & x>1\end{cases}$$

Clearly, $f'(x)>0$ for $x<1$ and $f'(x)<0$ for $x>1$.

Thus, $f(x)$ is increasing for all $x<1$ and decreasing for $x>1$.

Therefore, $f(x)$ will have greatest value at $x=1$, if

$$\lim_{x\to 1^+} f(x)<f(1)$$

$$\Rightarrow \lim_{x\to 1} -2x+\log_2(b^2-4)<1-1+10-7$$

$$\Rightarrow \quad -2+\log_2(b^2-4)<3$$

$$\Rightarrow \quad \log_2(b^2-4)<5$$

$$\Rightarrow \quad b^2-4<2^5 \text{ and } b^2-4>0$$

$$\Rightarrow \quad b^2-36<0 \text{ and } b^2-4>0$$

$$\Rightarrow \quad -6<b<6 \text{ and } b\in(-\infty,-2)\cup(2,\infty)$$

$$\Rightarrow \quad b\in(-6,-2)\cup(2,6)$$

26. (c) We have, $f(x)=\alpha\log|x|+\beta x^2+x$

$$\Rightarrow \quad f'(x)=\frac{\alpha}{x}+2\beta x+1$$

Now, $x=-1$ and $x=2$ are extreme points.

So, $\quad f'(-1)=0$

$$\Rightarrow \quad -\alpha-2\beta+1=0 \qquad \text{...(i)}$$

and $\quad f'(2)=0$

$$\Rightarrow \quad \frac{\alpha}{2}+4\beta+1=0$$

$$\Rightarrow \quad \alpha+8\beta+2=0 \qquad \text{...(ii)}$$

From Eqs. (i) and (ii),

$$\alpha=2 \text{ and } \beta=\frac{-1}{2}.$$

27. (b) We have, $f(x)=x(\log x)^2$ is defined for all $x>0$.

Now, $f'(x)=(\log x)^2+\dfrac{2(\log x)\,x}{x}$

$$=\log x(\log x+2)$$

Therefore, $f'(x)=0$

$$\Rightarrow \quad x=1 \text{ or } x=e^{-2}$$

Now, $f''(x)=\dfrac{2\log x}{x}+\dfrac{2}{x}$

$$\Rightarrow \quad f''(1)=2>0$$

and $f''(e^{-2})=-4e^2+2e^2<0$

Hence, f is minimum at $x=1$ and maximum at $x=e^{-2}$

Now, $f(1)=0$

and $f(e^{-2})=e^{-2}(\log e^{-2})^2=4e^{-2}$

Therefore, range of $f=[0,4e^{-2}]$.

28. (b) We have, $y=\dfrac{x^4}{2}-5x^3+18x^2-19x$

$$\Rightarrow \quad \frac{dy}{dx}=2x^3-15x^2+36x-19$$

Let $f(x)=2x^3-15x^2+36x-19$

$\Rightarrow f'(x)=6x^2-30x+36$

Now, $\quad f'(x)=0$

$\Rightarrow 6x^2-30x+36=0$

$\Rightarrow \quad 6(x^2-5x+6)=0$

$$\Rightarrow \quad x=2,3$$

Now, $\quad f''(x)=12x-30$

$\quad f''(x)<0$ for $x=2$

At $x=2$,

$$y=8-40+72-38=72-70=2$$

Hence, the required point is (2, 2).

29. (c) Let $f(x)=x^3-ax+b$

$\Rightarrow \quad f'(x)=3x^2-a=3\left(x-\sqrt{\frac{a}{3}}\right)\left(x+\sqrt{\frac{a}{3}}\right)$

The sign scheme for $f'(x)$ is

$+ \quad - \quad +$

$-\sqrt{a/3} \qquad \sqrt{a/3}$

As, $f'(x)$ changes its sign from positive to negative in neighbourhood of $-\sqrt{\frac{a}{3}}$.

So, $-\sqrt{\frac{a}{3}}$ is a point of local maxima.

Similarly, $x=\sqrt{\frac{a}{3}}$ is point of local minima.

30. (b) We have, $p(x)=x^4+ax^3+bx^2+cx+d$

$\Rightarrow \quad p'(x)=4x^3+3ax^2+2bx+c$

It is given that $x=0$ is a root of $p'(x)=0$

$p'(0)=0$

$\Rightarrow \quad c=0$

Putting $c=0$, we get

$p'(x)=x(4x^2+3ax+2b)$

Since, $x=0$ is the only real root of $p'(x)=0$

Therefore, $4x^2+3ax+2b=0$ has no real roots

$\Rightarrow \quad (3a)^2-4(4)(2b)<0$

$\Rightarrow \quad 9a^2-32b<0$

It is given that

$p(-1)<p(1)$

$\Rightarrow 1-a+b-c+d<1+a+b+c+d$

$\Rightarrow \quad a>0 \qquad [\because c=0]$

But $9a^2-32b<0$, therefore $b>0$

$\therefore\ p'(x)=x(4x^2+3ax+2b)>0\ \forall x\in(0,1]$

$\Rightarrow p(x)$ is increasing in $(0,1]$

$\therefore\ p(1)$ is the maximum value of $p(x)$.

Also, $p'(x)=x\,(4x^2+3ax+2b)<0\ \forall\ x\in[-1,0)$

$\Rightarrow p(x)$ is decreasing in $[-1,0)$

$\therefore\ p(-1)$ is not minimum value.

31. (c) We have, $a^{f(x)}+g(x)=0$

$\Rightarrow \quad a^{f(x)}=-g(x)$

$\Rightarrow \quad a^{f(x)}<-\frac{1}{2}\forall x \qquad \left[\because g(x)\geq\frac{1}{2}\right]$

But $a^{f(x)}>0$ for all x.

Thus, the number of solution is zero.

32. (b) We have, $f(x)=3x^2-x+\frac{25}{12}$

$\Rightarrow \quad f'(x)=6x-1=0$

$\Rightarrow \quad x=\frac{1}{6}$

Now, $\quad f''(x)=6>0$

$\Rightarrow f$ is minimum at $x=\frac{1}{6}$ and least value of f is

$$f\left(\frac{1}{6}\right)=\frac{3}{36}-\frac{1}{6}+\frac{25}{12}=2$$

Let r be the common ratio $(|r|<1)$, so that the progression is r^2, r^3, r^4......

$$\frac{r^2}{1-r}=2$$

$\Rightarrow \quad r^2+2r-2=0$

$\Rightarrow \quad r=\frac{-2\pm\sqrt{4+8}}{2}=-1\pm\sqrt{3}$

Now, $|r|<1 \Rightarrow r=-1+\sqrt{3}$

33. (c) Let the point $P(x,y)$ be on the parabola $y=x^2+7x+2$.

Its distance from the line,

$y=3x-3$ or $3x-y-3=0$ is

$$D=\left|\frac{3x-y-3}{\sqrt{10}}\right|$$

$$=\left|\frac{3x-(x^2+7x+2)-3}{\sqrt{10}}\right|=\left|\frac{-x^2-4x-5}{\sqrt{10}}\right|$$

$\Rightarrow \quad D=\left|\frac{x^2+4x+5}{\sqrt{10}}\right|=\left|\frac{(x+2)^2+1}{\sqrt{10}}\right|$

Hence, $D=\frac{(x+2)^2+1}{\sqrt{10}}$

Now, $\quad \frac{dD}{dx}=\frac{2(x+2)}{\sqrt{10}}$

$\therefore \quad \frac{dD}{dx}=0$

$\Rightarrow \quad \frac{2(x+2)}{\sqrt{10}}=0$

$\therefore \quad x=-2$

Hence, $\quad y=-8$

The point is $(-2,-8)$.

Now, $\frac{d^2D}{dx^2}=\frac{2}{\sqrt{10}}>0$

Hence, $(-2,-8)$ is the point of minima.

34. (c) Given, ellipse is

$$\frac{x^2}{27}+y^2=1$$

Equation of tangent at $(3\sqrt{3}\cos\theta, \sin\theta)$ is

$$\frac{x\cos\theta}{3\sqrt{3}} + \frac{y\sin\theta}{1} = 1$$

$y =$ sum of intercepts $= 3\sqrt{3}\sec\theta + \operatorname{cosec}\theta$

For minimum value, $\frac{dy}{d\theta} = 0$

$\therefore \quad 3\sqrt{3}\sec\theta\tan\theta - \operatorname{cosec}\theta\cot\theta = 0$

$\therefore \quad \tan^3\theta = \frac{1}{3\sqrt{3}}$

$\Rightarrow \quad \tan\theta = \frac{1}{\sqrt{3}}$

$\Rightarrow \quad \theta = \frac{\pi}{6}$

35. (c) Let $AB + AC = k$

If $AB = x$, $AC = k - x$

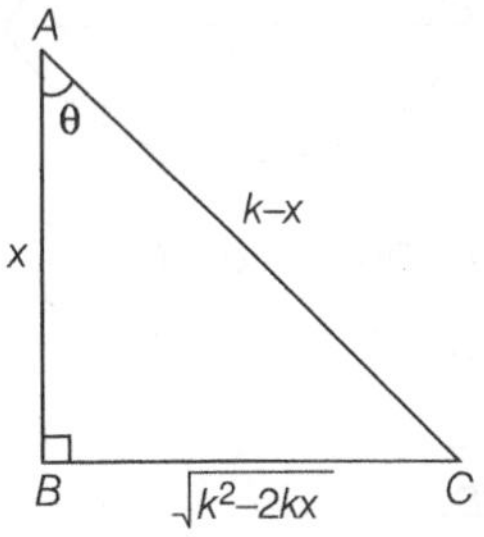

So, $\quad BC = \sqrt{k^2 - 2kx}$

So, Area $= \frac{x\sqrt{k^2-2kx}}{2}$

$\Rightarrow \quad 2A = x\sqrt{k^2-2kx}$

$\Rightarrow \quad f(x) = 4A^2 = x^2(k^2 - 2kx)$

$\Rightarrow \quad f'(x) = x^2(-2k) + 2x(k^2 - 2kx)$

$\Rightarrow \quad f'(x) = 2k^2x - 6kx^2$

$\because \quad f'(x) = 0$

$\therefore \quad 2kx(k - 3x) = 0$

$\Rightarrow \quad x = 0,$

$\Rightarrow \quad k - 3x = 0$

$\Rightarrow \quad x = \frac{k}{3}$

and $\quad f''(x) = 2k^2 - 12kx$

$\Rightarrow \quad f''(0) = 2k^2 > 0$

$\Rightarrow \quad f''\left(\frac{k}{3}\right) = -2k^2 < 0$

$\therefore$ $f(x)$ is maximum at $x = \frac{k}{3}$

$\Rightarrow \quad \cos\theta = \frac{1}{2}$

$\Rightarrow \quad \theta = \frac{\pi}{3}$

36. (c) Let length of box $= a - 2x$,
breadth of box $= b - 2x$
and height of box $= x$

$\therefore$ Volume of box, $(V) = (a - 2x)(b - 2x)x$

$\Rightarrow \quad V = 4x^3 - 2ax^2 - 2bx^2 + abx$

On differentiating V w.r.t. x, we get

$$V'(x) = 12x^2 - 4(a+b)x + ab$$

Critical Point, $V'(x) = 0$

$\Rightarrow \quad 12x^2 - 4(a+b)x + ab = 0$

$\Rightarrow \quad x = \frac{4(a+b) \pm \sqrt{16(a+b)^2 - 4\cdot 12\cdot ab}}{2(12)}$

$$x = \frac{(a+b) \pm \sqrt{a^2+b^2-ab}}{6}$$

Again, differentiating V w.r.t. x, we get

$$V''(x) = 24x - 4(a+b)$$

For $\quad x = \frac{(a+b) - \sqrt{a^2+b^2-ab}}{6}, V''(x) < 0$

Hence, for maximum volume

$$x = \frac{(a+b) - \sqrt{a^2+b^2-ab}}{6}$$

37. (b) Given that side of square is x units and radius of circle is r units.

So, $4x + 2\pi r = 2$ ...(i)

Now, sum of areas of square and circle is

$$A = x^2 + \pi r^2$$

$\Rightarrow \quad A = \left(\frac{1-\pi r}{2}\right)^2 + \pi r^2$

$\Rightarrow \quad \frac{dA}{dr} = 2\left(\frac{\pi r - 1}{2}\right)\left(\frac{\pi}{2}\right) + 2\pi r$

Now, $\quad \frac{dA}{dr} = 0$

$\Rightarrow \quad r = \frac{1}{\pi + 4}$

From Eq. (i), $\quad 2x = \frac{4}{\pi + 4}$

$\Rightarrow \quad x = 2r$

38. (b)

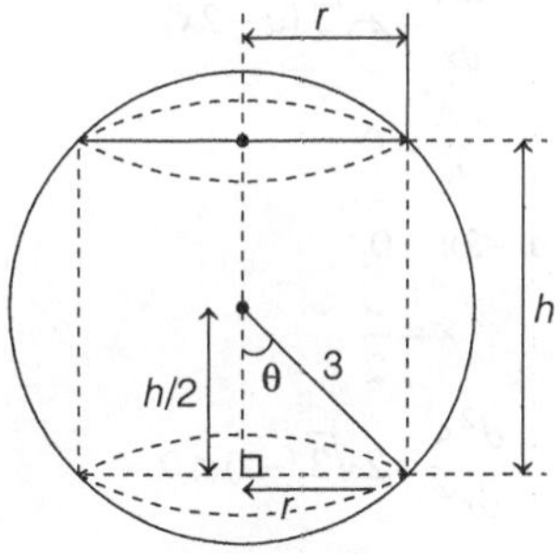

Let a sphere of radius 3 units, which inscribed a right circular cylinder having radius r and height is h, so

From the figure, $\frac{h}{2}=3\cos\theta$

$\Rightarrow \quad h=6\cos\theta$

and $\quad r=3\sin\theta$ ...(i)

$\because$ Volume of cylinder $(V)=\pi r^2 h$

$=\pi(3\sin\theta)^2(6\cos\theta)=54\pi\sin^2\theta\cos\theta$.

For maxima or minima, $\frac{dV}{d\theta}=0$

$\Rightarrow \quad 54\pi[2\sin\theta\cos^2\theta-\sin^3\theta]=0$

$\Rightarrow \quad \sin\theta[2\cos^2\theta-\sin^2\theta]=0$

$\Rightarrow \quad \tan^2\theta=2 \quad \left[\because \theta\in\left(0,\frac{\pi}{2}\right)\right]$

$\Rightarrow \quad \tan\theta=\sqrt{2}$

$\Rightarrow \quad \sin\theta=\sqrt{\frac{2}{3}}$ and $\cos\theta=\frac{1}{\sqrt{3}}$ ...(ii)

From Eqs. (i) and (ii), we get

$$h=6\frac{1}{\sqrt{3}}=2\sqrt{3}$$

39. (d) Let $BD_1=x$ or $BC_1=(a-x)$

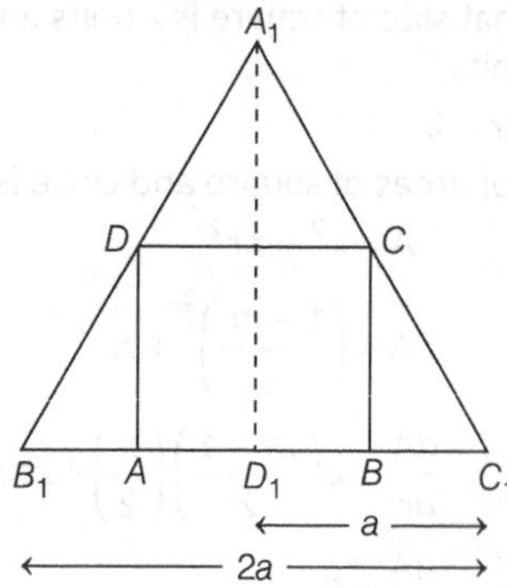

$\therefore \quad BC=(a-x)\tan\frac{\pi}{3}$

$\Rightarrow BC=\sqrt{3}(a-x)$

Now, area of rectangle, $ABCD=(AB)(BC)$

$A=2\sqrt{3}x(a-x)$ ...(i)

$\Rightarrow \quad A=2\sqrt{3}(ax-x^2)$

$\Rightarrow \quad \frac{dA}{dx}=2\sqrt{3}\,(a-2x)$

Now, $\quad \frac{dA}{dx}=0$

$\Rightarrow \quad a-2x=0$

$\Rightarrow \quad x=\frac{a}{2}$

$\Rightarrow \quad \frac{d^2A}{dx^2}=2\sqrt{3}\,(-2)<0$

So, $x=\frac{a}{2}$ is the point of maxima.

From Eq. (i),

Maximum area, $A=2\sqrt{3}\left(\frac{a}{2}\right)\left(a-\frac{a}{2}\right)$

$$=2\sqrt{3}\,\frac{a^2}{4}=\frac{\sqrt{3}a^2}{2}$$

40. (d) Triangle of maximum area that can be inscribed in a circle is an equilateral triangle. Let ΔABC be inscribed in circle.

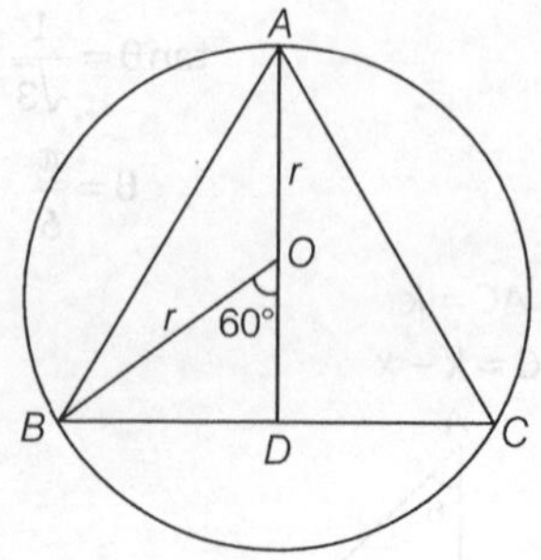

Now, in ΔOBD, $\quad OD=r\cos 60°=\frac{r}{2}$

Height, $AD=\frac{3r}{2}$

Again, in ΔABD,

$$\sin 60°=\frac{\frac{3r}{2}}{AB}$$

$\Rightarrow \quad AB=\sqrt{3}r$

41. (c) Let h be height and r the radius of its base

$\pi r^2 h=k$ (constant)

Let S be the total surface area of the cylinder.

Therefore, $S=2\pi r^2+2\pi rh$

$$=2\pi r^2+2\pi r\left(\frac{k}{\pi r^2}\right)=2\pi r^2+\frac{2k}{r},$$

where, $2\pi r^2$ = sum of the areas of the base and the top and $2\pi rh$ is the lateral surface area.

Therefore, differentiating, we get

$$\frac{dS}{dr}=4\pi r-\frac{2k}{r^2}$$

Now, $\quad \frac{dS}{dr}=0$

$\Rightarrow \quad r^3=\frac{k}{2\pi}$

$\Rightarrow \quad r=\left(\frac{k}{2\pi}\right)^{1/3}$

Since, $r>0$ and $k>0$, we have

$$\frac{d^2S}{dr^2}=4\pi+\frac{4k}{r^3}>0$$

So, S is minimum, when $r=\left(\frac{k}{2\pi}\right)^{1/3}$.

Hence, $\frac{h}{r}=\frac{\frac{k}{\pi r^2}}{r}=\frac{k}{\pi}\cdot\frac{1}{r^3}$

$=\frac{k}{\pi}\cdot\frac{2\pi}{k}=2$

42. (d) Let x be the side of the square piece removed. Therefore, the dimensions of the box are $3-2x, 8-2x$ and x.

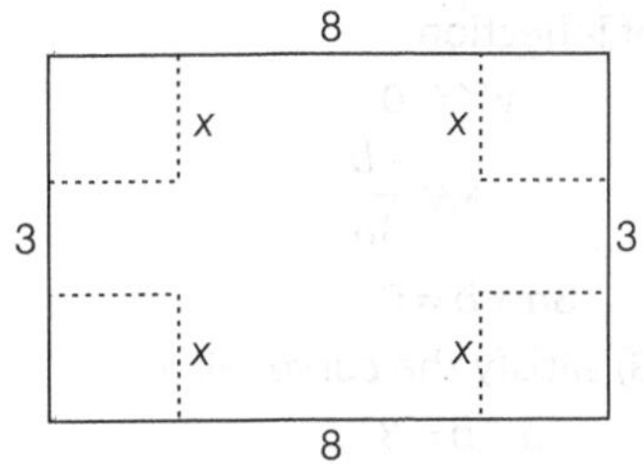

Let V be the volume of the box, so that

$$V=x(3-2x)\cdot(8-2x)$$

Now, $\frac{dV}{dx}=(3-2x)(8-2x)+x(8-2x)(-2)+x(3-2x)(-2)$

$$=(24-22x+4x^2)+(4x^2-16x)+(4x^2-6x)$$
$$=12x^2-44x+24=4(3x^2-11x+6)$$
$$=4(3x-2)(x-3)$$

Therefore, the critical points are $x=\frac{2}{3}$ and $x=3$.

Now, $x=3$ is not possible. Hence, $x=\frac{2}{3}$ is the only critical point.

So, $\frac{d^2V}{dx^2}=24x-44$

$\Rightarrow \left(\frac{d^2V}{dx^2}\right)_{x=\frac{2}{3}}=24\left(\frac{2}{3}\right)-44=-28<0$

Hence, V is maximum at $x=\frac{2}{3}$ and the maximum volume is

$$V=\frac{2}{3}\left(3-\frac{4}{3}\right)\left(8-\frac{4}{3}\right)=\frac{2}{3}\left(\frac{5}{3}\right)\left(\frac{20}{3}\right)=\frac{200}{27} \text{ cu units}$$

43. (b) Construct an altitude from B to AC meeting the diameter AC at D and produced BD intersects circle at E.

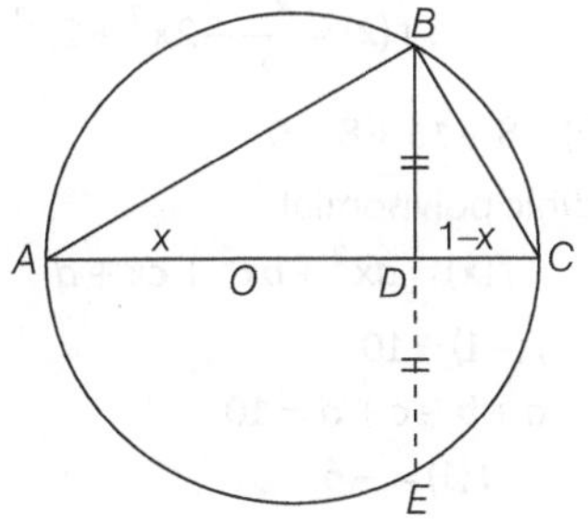

$(BD)(DE)=x(1-x)$ [by property of circle]

But $BD=DE$

$\therefore BD=\sqrt{x(1-x)}$

$\therefore$ Area of a right triangle, $A=\frac{x\sqrt{x-x^2}}{2}$ ...(i)

$\Rightarrow A=\frac{\sqrt{x^3-x^4}}{2}$

$\Rightarrow A^2=\frac{x^3-x^4}{4}$

Let $f(x)=4A^2=x^3-x^4$

$\Rightarrow f'(x)=3x^2-4x^3$

$\because f'(x)=0$

So, $3x^2-4x^3=0$

$\Rightarrow x^2(3-4x)=0$

$\Rightarrow x=0 \text{ or } x=\frac{3}{4}$

and $f''(x)=6x-12x^2$

At $x=\frac{3}{4}$,

$$f''\left(\frac{3}{4}\right)=6\left(\frac{3}{4}\right)-12\left(\frac{3}{4}\right)^2$$
$$=\frac{9}{2}-\frac{27}{4}$$
$$=\frac{18-27}{4}=-\frac{9}{4}<0$$

$\therefore$ Maximum area of triangle is achieved, when $x=\frac{3}{4}$

$A_{max}=\frac{\frac{3}{4}\sqrt{\frac{3}{4}-\frac{9}{16}}}{2}=\frac{3\sqrt{3}}{32}$ [from Eq. (i)]

44. (b) Let $h=$ height of the cone,

$r=$ radius of circular base

$=\sqrt{(3)^2-h^2}$

$=\sqrt{9-h^2}$ ...(i)

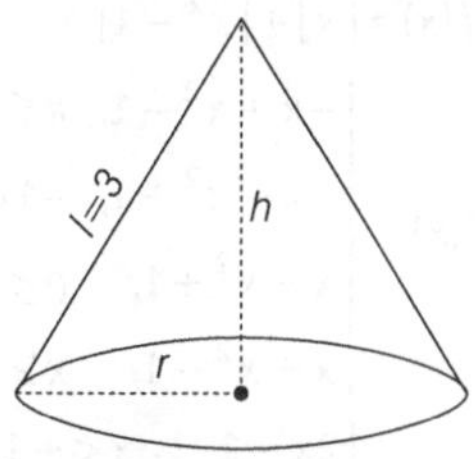

Now, volume (V) of cone $=\frac{1}{3}\pi r^2h$

$\Rightarrow V(h)=\frac{1}{3}\pi(9-h^2)h$ [from Eq. (i)]

$$= \frac{1}{3}\pi[9h - h^3] \quad \text{...(ii)}$$

For maximum volume $V'(h) = 0$ and $V''(h) < 0$.

Here, $\quad V'(h) = 0$

$\Rightarrow \quad (9 - 3h^2) = 0$

$\Rightarrow \quad h = \sqrt{3} \quad [\because h \nless 0]$

and $V''(h) = \frac{1}{3}\pi(-6h) < 0$ for $h = \sqrt{3}$

Thus, volume is maximum when $h = \sqrt{3}$

Now, maximum volume

$$V(\sqrt{3}) = \frac{1}{3}\pi(9\sqrt{3} - 3\sqrt{3}) \quad \text{[from Eq. (ii)]}$$

$$= 2\sqrt{3}\pi \text{ cu. m}$$

45. (9) Let $p'(x) = k(x-1)(x-3)$

$\therefore \quad p(x) = k\left(\frac{x^3}{3} - 2x^2 + 3x\right) + C$

Now, $\quad p(1) = 6 \Rightarrow \frac{4}{3}k + C = 6$

Also, $\quad p(3) = 2 \Rightarrow C = 2$

So, $\quad k = 3$

Thus, $p'(0) = 3k = 9$

46. (12) Given, $f(x) = -\sin^3 x + 3\sin^2 x + 5$

$\therefore \quad f'(x) = -3\sin^2 x \cos x + 6 \sin x \cos x$

$$= 3 \sin x \cos x (2 - \sin x) \geq 0$$

$\Rightarrow f(x)$ is increasing in $\left[0, \frac{\pi}{2}\right]$.

Hence, local minimum value at $x = 0$ is $f(0) = 5$

and local maximum value is at $x = \frac{\pi}{2}$.

$\therefore \quad f\left(\frac{\pi}{2}\right) = -\sin^3\left(\frac{\pi}{2}\right) + 3\sin^2\left(\frac{\pi}{2}\right) + 5 = 7$

So, sum of local maximum and minimum value is $5 + 7 = 12$.

47. (5) We have,

$$f(x) = |x| + |x^2 - 1|$$

$$\Rightarrow \quad f(x) = \begin{cases} -x + x^2 - 1, & x \leq -1 \\ -x - x^2 + 1, & -1 < x < 0 \\ x - x^2 + 1, & 0 \leq x < 1 \\ x + x^2 - 1, & x \geq 1 \end{cases}$$

$$\Rightarrow \quad f'(x) = \begin{cases} 2x - 1, & x < -1 \\ -2x - 1, & -1 < x < 0 \\ -2x + 1, & 0 < x < 1 \\ 2x + 1, & x > 1 \end{cases}$$

The sign scheme for $f'(x)$ is

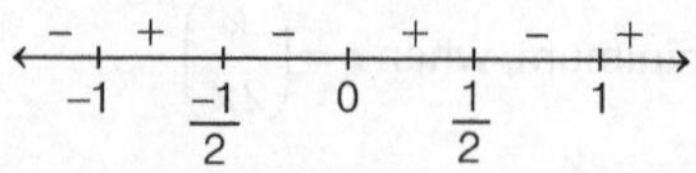

$f(x)$ changes its sign at 5 points.

Hence, total number of points of local maximum and minimum of $f(x)$ is 5.

48. (9) We have, $y = ax^3 + bx^2$

$\Rightarrow \quad y' = 3ax^2 + 2bx$

$\Rightarrow \quad y'' = 6ax + 2b$

At point of inflection

$$y'' = 0$$

$$x = \frac{-b}{3a}$$

$$3a + b = 0 \quad \text{[as } x = 1\text{] ...(i)}$$

Point (1, 3) satisfy the curve, also

So, $\quad a + b = 3 \quad$ ...(ii)

From Eqs. (i) and (ii),

$$a = \frac{-3}{2} \text{ and } b = \frac{9}{2}$$

So, $-3a + b = -3\left(-\frac{3}{2}\right) + \frac{9}{2}$

$$= \frac{9}{2} + \frac{9}{2} = 9$$

49. (0) Given, extreme values of $f(x)$ are 1 and 2.

$\Rightarrow f'(x) = 0$ at $x = 1$ and $x = 2$

or $f'(1) = 0$ and $f'(2) = 0$

Also, $\lim\limits_{x \to 0} \frac{f(x)}{x^2} = 2$

$\therefore \quad f(x) = ax^4 + bx^3 + 2x^2$

$\Rightarrow \quad f'(x) = 4ax^3 + 3bx^2 + 4x$

$\therefore \quad f'(1) = 4a + 3b + 4$

and $\quad f'(2) = 32a + 12b + 8$

Thus, we have

$$8a + 3b + 2 = 0 \quad \text{...(i)}$$

$$\Rightarrow \quad 4a + 3b + 4 = 0 \quad \text{...(ii)}$$

On solving Eqs. (i) and (ii), we get

$$4a = 2$$

$$\Rightarrow \quad a = \frac{1}{2} \text{ and } b = -2$$

$$\therefore \quad f(x) = \frac{x^4}{2} - 2x^3 + 2x^2$$

Hence, $f(2) = 8 - 16 + 8 = 0$

50. (3) Let a cubic polynomial

$$f(x) = ax^3 + bx^2 + cx + d$$

$\because \quad f(-1) = 10$

$\Rightarrow \quad -a + b - c + d = 10 \quad$ ...(i)

$\because \quad f(1) = -6$

$\Rightarrow \quad a+b+c+d=-6 \qquad \ldots(ii)$

$\because \quad f'(-1)=0$

$\Rightarrow \quad 3a-2b+c=0 \qquad \ldots(iii)$

$\because \quad f''(1)=0$

$\Rightarrow \quad 6a+2b=0$

$\Rightarrow \quad 3a+b=0 \qquad \ldots(iv)$

From Eqs. (i) and (ii), we get

$-2a-2c=16$

$\Rightarrow \quad a+c=-8 \qquad \ldots(v)$

From Eqs. (iii), (iv) and (v), we get

$3a-2(-3a)+(-a-8)=0$

$\Rightarrow \quad 8a-8=0 \Rightarrow a=1$

So, $\quad b=-3, c=-9$ and $d=5$

$\therefore \quad f(x)=x^3-3x^2-9x+5$

$\because \quad f'(x)=3x^2-6x-9=0$

(for local maxima and minima)

$\Rightarrow \quad x^2-2x-3=0 \Rightarrow x^2-3x+x-3=0$

$\Rightarrow (x+1)(x-3)=0 \Rightarrow f'(x)=0$

$\Rightarrow \quad x=-1,3$

$\because \quad f''(x)=6x-6$

$\because f''(-1)=-12$ and $f''(3)=12$

$\therefore x=3$ is point of local minima.

51. (22) We have,

$$f(\alpha,\beta,\gamma,\delta)=\Sigma\alpha^2-2\,\Sigma\alpha+10$$
$$=\alpha^2+\beta^2+\gamma^2+\delta^2-2(\alpha+\beta+\gamma+\delta)+10$$
$$=(\alpha-1)^2+(\beta-1)^2+(\gamma-1)^2+(\delta-1)^2+6$$

Clearly, f is maximum, when

$\alpha,\gamma=-1$ and $\beta,\delta=3$

$$\therefore f_{max}=(-1-1)^2+(3-1)^2+(-1-1)^2+(3-1)^2+6$$
$$=4+4+4+4+6=22$$

52. (14) Given, $x=\phi(t)=t^5-5t^3-20t+7$

$$\Rightarrow \quad \frac{dx}{dt}=\phi'(t)=5t^4-15t^2-20$$
$$=5(t^4-3t^2-4)$$
$$=5(t^2-4)(t^2+1)\neq 0, \text{ if } -2<t<2$$

and $\quad y=\psi(t)=4t^3-3t^2-18t+3$

$$\frac{dy}{dt}=\psi'(t)=12t^2-6t-18$$

For maxima and minima $\frac{dy}{dt}=0$

$\Rightarrow \quad 12t^2-6t-18=0$

$\Rightarrow \quad 6(2t^2-t-3)=0$

$\Rightarrow \quad t=-1,\frac{3}{2}$

Both values of $t\in(-2,2)$.

Now, $\frac{d^2y}{dt^2}=\psi''(t)=24t-6$

$\therefore\ \psi''(-1)=-24-6=-30<0, t=1$ is the point of maxima.

and $\psi''\left(\frac{3}{2}\right)=36-6=30>0$

Hence, $t=\frac{3}{2}$ is the point of minima.

$\therefore$ Maximum value of $y=14$ at $t=-1$.

53. (22) Let $f(x)=ax^3+bx^2+cx+d$

$f'(x)=3ax^2+2bx+c$

$f''(x)=6ax+2b$

$f'(x)$ has local minima at $x=-1$, so

$\because \quad f''(-1)=0$

$\Rightarrow \quad -6a+2b=0$

$\Rightarrow \quad b=3a \qquad \ldots(i)$

$f(x)$ has local minima at $x=1$

$f'(1)=0$

$\Rightarrow \quad 3a+6a+c=0$

$\Rightarrow \quad c=-9a \qquad \ldots(ii)$

$f(1)=-10$

$\Rightarrow \quad -5a+d=-10 \qquad \ldots(iii)$

$f(-1)=6$

$\Rightarrow \quad 11a+d=6 \qquad \ldots(iv)$

Solving Eqs. (iii) and (iv)

$a=1$ and $d=-5$

From Eqs. (i) and (ii),

$b=3$ and $c=-9$

$\therefore \quad f(x)=x^3+3x^2-9x-5$

So, $\quad f(3)=27+27-27-5=22$

54. (7) We have, $f(x)=|3-x|+|2+x|+|5-x|$

$$\text{So,} \quad f(x)=\begin{cases}6-3x, & x<-2\\ 10-x, & -2\le x<3\\ x+4, & 3\le x<5\\ 3x-4, & x\ge 5\end{cases}$$

$\frac{dy}{dx}$ is '–ve' for $x\in(-\infty,3]$.

So, $f(x)$ is decreasing function in $(-\infty,3)$

and $\frac{dy}{dx}$ is '+ve' for $(3,\infty)$.

So, $f(x)$ is increasing function.

Hence, $x=3$ is the point of minimum value of $f(x)$ as the function changes from –ve to '+ve'.

$\therefore \quad f(3)=0+5+2=7.$

Minimum value of $f(x)$ is 7.

55. (5) We have, $f(x)=\int_0^x(\cos^2 t+\cos t+2)\,dt$

$$\Rightarrow \quad f'(x)=\cos^2 x+\cos x+2>0$$

$\Rightarrow f'(x)$ is an increasing function.

So, $f(x)$ is minimum at $x=0$ and maximum at $x=2\pi$.

$$f(x)_{\min}=f(0)=0$$
$$f(x)_{\max}=f(2\pi)=5\pi$$

Hence, difference between greatest and least value is 5π.

So, $\quad k=5$

56. (144) $f(x)=x^6+ax^5+bx^4+cx^3+dx^2+ex+f$

As, $\lim\limits_{x\to 0}\dfrac{f(x)}{x^3}=1$ (non-zero finite)

So, $d=e=f=0$

and $f(x)=x^3(x^3+ax^2+bx+c)$

Hence, $\lim\limits_{x\to 0}\dfrac{f(x)}{x^3}=c=1 \qquad$ [given]

Now, as $f(x)=x^6+ax^5+bx^4+x^3$

and $\quad f'(x)=0$ at $x=1$ and $x=-1$

i.e. $\quad f'(x)=6x^5+5ax^4+4bx^3+3x^2$

Now, $f'(1)=0$

$\Rightarrow \quad 6+5a+4b+3=0$

$\Rightarrow \quad 5a+4b=-9 \qquad$ ... (i)

and $\quad f'(-1)=0$

$\Rightarrow \quad -6+5a-4b+3=0$

$\Rightarrow \quad 5a-4b=3 \qquad$ ... (ii)

From Eqs. (i) and (ii),

$$a=-3/5 \text{ and } b=-3/2$$

$$\therefore \quad f(x)=x^6-\frac{3}{5}x^5-\frac{3}{2}x^4+x^3$$

$$\therefore \quad 5f(2)=5\left[2^6-\frac{3}{5}(2)^5-\frac{3}{2}(2)^4+(2)^3\right]$$

$$=5\left[64-\frac{3\times 32}{5}-\frac{3\times 16}{2}+8\right]$$

$$=320-96-120+40=144$$

57. (3) From ΔADE and ΔABC,

$$\frac{r}{R}=\frac{H-h}{H}\Rightarrow r=\frac{R(H-h)}{H}$$

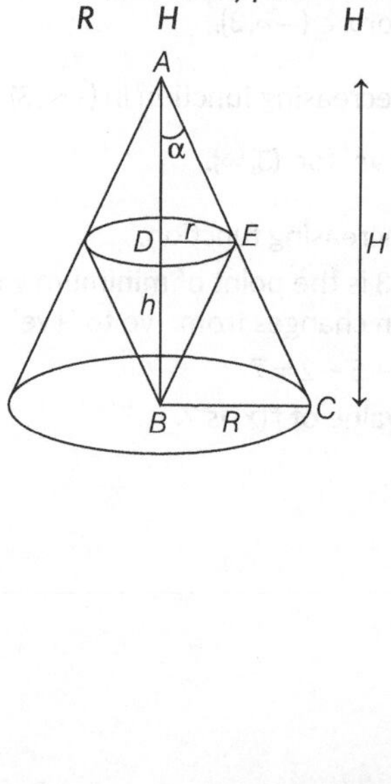

Volume, $V=\dfrac{1}{3}\pi\dfrac{R^2(H-h)^2}{H^2}h$

$$=\frac{\pi R^2}{3H^2}(H-h)^2 h$$

$$\therefore \quad \frac{dV}{dh}=\frac{\pi R^2}{3H^2}[(H-h)^2-2h(H-h)]$$

Therefore, $\dfrac{dV}{dh}=0$, if $h=\dfrac{H}{3}$

Also, $h=\dfrac{H}{3}$ is a point of maxima

Thus, $\dfrac{H}{h}=3$

58. (27) $AP=BQ=6\sin\alpha$

$$DP=QC=6\cos\alpha$$

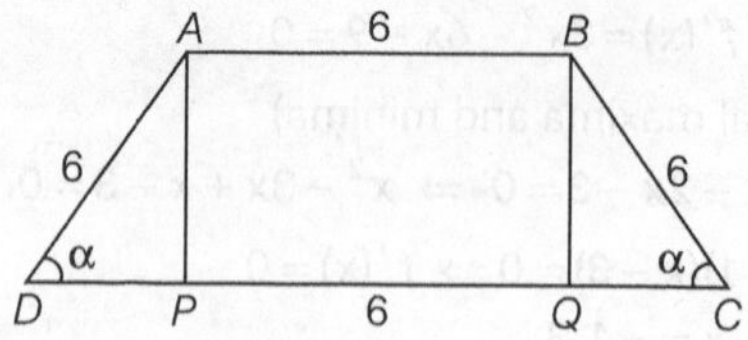

$\therefore \; A=$ Area of trapezium

$$=\frac{1}{2}(AB+DC)\cdot BQ=\frac{1}{2}(6+6+12\cos\alpha)\cdot 6\sin\alpha$$

$$A=36\sin\alpha(1+\cos\alpha)$$

Now, $\dfrac{dA}{d\alpha}=36(\cos\alpha+\cos 2\alpha)\Rightarrow\dfrac{dA}{d\alpha}=0$

$\Rightarrow \quad 36(\cos\alpha+\cos 2\alpha)=0$

$\Rightarrow \quad 2\cos^2\alpha+\cos\alpha-1=0$

$\Rightarrow \quad (2\cos\alpha-1)(\cos\alpha+1)=0$

$\Rightarrow \quad \cos\alpha=\dfrac{1}{2}$ or $\cos\alpha=-1$ [not possible]

$\Rightarrow \quad \alpha=\dfrac{\pi}{3}$

Again, $\dfrac{d^2A}{d\alpha^2}=-36(\sin\alpha+2\sin 2\alpha)<0$

Then, area is maximum, when $\alpha=\dfrac{\pi}{3}$.

$$A=36\left(\sin 60^\circ+\frac{1}{2}\sin 120^\circ\right)$$

$$=36\left(\frac{\sqrt{3}}{2}+\frac{1}{2}\frac{\sqrt{3}}{2}\right)$$

$$=27\sqrt{3}$$

$\therefore \quad k=27$

59. (18) Let equation of straight line, which passes through $(8,2)$ with negative slope, say $-m, m>0$ is

$$y-2=-m(x-8)$$

If it cuts the axes in point P and Q, then

$$OP = x = 8 + \frac{2}{m},$$
$$OQ = y = 2 + 8m$$

$\therefore$ $$OP + OQ = 10 + \frac{2}{m} + 8m = Z \text{ (let)}$$

For maximum or minimum,

$$\frac{dZ}{dm} = 0$$

$\Rightarrow$ $$-\frac{2}{m^2} + 8 = 0$$

$\Rightarrow$ $$m^2 = \frac{1}{4}$$

$\Rightarrow$ $$m = \frac{1}{2} \qquad [\because m > 0]$$

Now, $$\frac{d^2Z}{dm^2} = \frac{4}{m^3}$$

At $m = \frac{1}{2}$, $$\frac{d^2Z}{dm^2} = 32 > 0$$

Hence, Z is minimum, when $m = \frac{1}{2}$

Then, $Z = 10 + \dfrac{2}{\frac{1}{2}} + 8 \times \dfrac{1}{2} = 10 + 4 + 4 = 18$

60. (4) Let R be the radius of cone, l be its slant height and h be the height.

$\therefore$ $$V = \frac{1}{3}\pi R^2 h$$

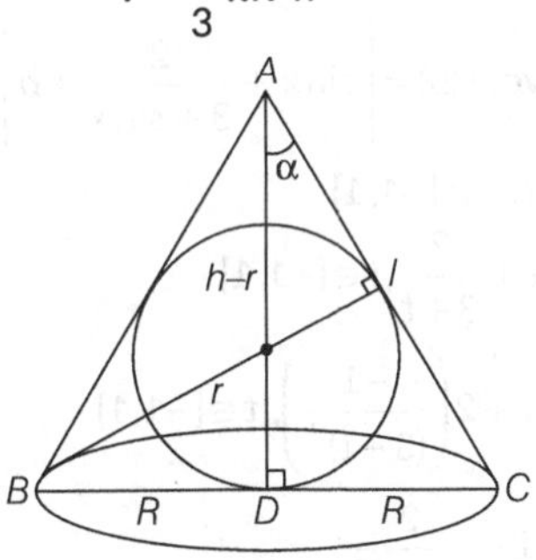

We have to make V a function of single variable

$$\frac{r}{h-r} = \frac{R}{l} = \sin\alpha$$

$\Rightarrow$ $$\frac{r}{h-r} = \frac{R}{\sqrt{h^2 + R^2}}$$

$\therefore$ $$r^2(R^2 + h^2) = R^2(h^2 - 2hr + r^2)$$

$\therefore$ $$R^2 h = \frac{r^2 h^2}{h - 2r}$$

$\Rightarrow$ $$V = \frac{1}{3}\frac{\pi r^2 h^2}{(h-2r)}$$

$\Rightarrow$ $$V = \frac{\pi}{3}\frac{r^2}{\frac{1}{h} - \frac{2r}{h^2}}$$

Now, V is minimum, if $Z = \frac{1}{h} - \frac{2r}{h^2}$ is maximum.

$$\frac{dZ}{dh} = -\frac{1}{h^2} + \frac{4r}{h^3} = 0$$

$\Rightarrow$ $$h = 4r$$

$\Rightarrow$ $$\frac{d^2Z}{dh^2} = \frac{2}{h^3} - \frac{12r}{h^4}$$
$$= \frac{2}{h^3}\left(1 - \frac{6r}{h}\right)$$
$$= \frac{2}{h^3}\left(1 - \frac{6}{4}\right)$$
$$= \frac{2}{h^3}\left(\frac{-2}{4}\right)$$
$$= -\frac{1}{h^3} < 0$$

Z is maximum and hence V is minimum, when $h = 4r$

Hence, $\lambda = 4$.

61. (36) Let $x + y = 36$ ...(i)

where, x is perimeter of square and y is perimeter of circle.

Then, side of square $= \frac{x}{4}$ and radius of circle $= \frac{y}{2\pi}$

Now, sum of areas of square and circle,

$$A = \frac{x^2}{16} + \frac{y^2}{4\pi}$$

$\Rightarrow$ $$A = \frac{x^2}{16} + \frac{(36 - x)^2}{4\pi} \qquad \text{[from Eq. (i)]}$$

For minimum area,

$$\frac{dA}{dx} = 0$$

Now, $$\frac{dA}{dx} = \frac{2x}{16} + \frac{-2(36 - x)}{4\pi} = 0$$

$\Rightarrow$ $$x = \frac{144}{\pi + 4}$$

Circumference of circle $= y$

$$= (36 - x)$$
$$= 36 - \frac{144}{\pi + 4} = \frac{36\pi}{\pi + 4}$$

According to the question,

$$k = \frac{36\pi}{\pi + 4}$$

$\Rightarrow$ $$\left(\frac{4}{\pi} + 1\right)k = \left(\frac{4}{\pi} + 1\right)\frac{36\pi}{\pi + 4} = 36$$

Level 3

1. (c) We have, $f(x) = \cos 2\pi x + x - [x]$

$\Rightarrow \quad f(x) = \cos 2\pi x + \{x\}$

$f(x)$ is a periodic function with period 1 and

$$f(x) = \begin{cases} 1, & \text{at } x = 0 \\ x + \cos 2\pi x, & \text{for } 0 < x < 1 \\ 1, & \text{at } x = 1 \end{cases}$$

Graphically $f(x)$ is shown as below

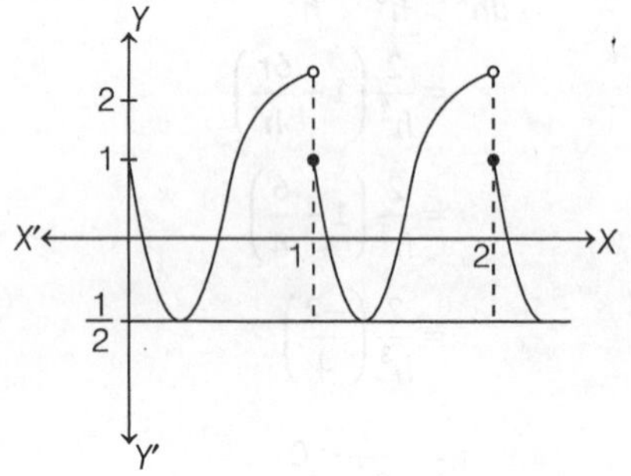

$\therefore$ In [0, 10] $f(x)$ attains its local maximum value only at $x = 0$ i.e. only 1.

2. (d) Given, $f(x) = (1 + b^2)x^2 + 2bx + 1$

$\Rightarrow \quad f'(x) = 2(1 + b^2)x + 2b$

$\Rightarrow \quad f''(x) = 2(1 + b^2) > 0$

So, $f(x)$ will provide us the minimum value.

Now, $f'(x) = 0$ gives $2(1 + b^2)x + 2b = 0$

$\Rightarrow \quad x = -\frac{2b}{2(1 + b^2)} = -\frac{b}{(1 + b^2)}$

Thus, the minimum value of $f(x) = m(b)$

$$= (1 + b^2)\left(-\frac{b}{(1 + b^2)}\right)^2 + 2b\left(-\frac{b}{(1 + b^2)}\right) + 1$$

$$= \left(\frac{b^2}{(1 + b^2)}\right) - \left(\frac{2b^2}{(1 + b^2)}\right) + 1$$

$$= \left(1 - \frac{b^2}{1 + b^2}\right) = \frac{1}{1 + b^2}$$

Thus, the range of $m(b) = (0, 1]$.

3. (c) We have, $f(\theta) = \frac{(a^2 - b^2)\cos\theta}{a - b\sin\theta}$

$$= \frac{(a^2 - b^2)}{a\sec\theta - b\tan\theta}$$

$$= \frac{(a^2 - b^2)}{g(\theta)}, \text{ where } g(\theta) = a\sec\theta - b\tan\theta$$

$f(\theta)$ is maximum or minimum value according as $g(\theta)$ is minimum and maximum.

Now, $g(\theta) = a\sec\theta - b\tan\theta$

$\Rightarrow \quad g'(\theta) = a\sec\theta\tan\theta - b\sec^2\theta$

$\Rightarrow \quad g''(\theta) = a\sec^3\theta + a\sec\theta\tan^2\theta - 2b\sec^2\theta\tan\theta$

$$= a\sec^3\theta + a\sec\theta(\sec^2\theta - 1) - 2b\sec^2\theta\tan\theta$$

$$= 2a\sec^3\theta - a\sec\theta - 2b\sec^2\theta\tan\theta$$

$$= \frac{a + a\sin^2\theta - 2b\sin\theta}{\cos^3\theta}$$

For maxima or minima $g'(\theta)$ gives,

$a\sec\theta\tan\theta - b\sec^2\theta = 0$

$\Rightarrow \quad a\sin\theta - b = 0 \Rightarrow \sin\theta = \frac{b}{a}$

Now, $\{g''(\theta)\}_{\sin\theta = \frac{b}{a}} = \frac{a^2 - b^2}{a\cos^3\theta} > 0$

$\Rightarrow$ $g(\theta)$ will provide us the minimum value.

$\Rightarrow$ $f(\theta)$ will provide us the maximum value.

Now, the maximum value of $f(\theta)$

$$= \frac{(a^2 - b^2)\frac{\sqrt{a^2 - b^2}}{a}}{a - b\left(\frac{b}{a}\right)}$$

$$= \frac{(a^2 - b^2)\sqrt{a^2 - b^2}}{(a^2 - b^2)}$$

$$= \sqrt{a^2 - b^2}$$

4. (b) We have, $f(b) = \left|\sin x + \frac{2}{3 + \sin x} + b\right|$

Let $\sin x = t, \ t \in [-1, 1]$

and $g(t) = t + \frac{2}{3 + t}, t \in [-1, 1]$

$$g'(t) = 1 + 2\left(\frac{-1}{(3 + t)^2}\right), \ t \in [-1, 1]$$

$$= 1 - \frac{2}{(3 + t)^2} \qquad \left[\begin{array}{l} \because t \in [-1, 1] \\ 3 + t \in [2, 4] \end{array}\right]$$

$g'(t) > 0$

$\Rightarrow \quad 1 - \frac{2}{(3 + t)^2} > 0$

$\Rightarrow \quad (3 + t)^2 - (\sqrt{2})^2 > 0$

$\Rightarrow \quad (t + 3 - \sqrt{2})(t + 3 + \sqrt{2}) > 0$

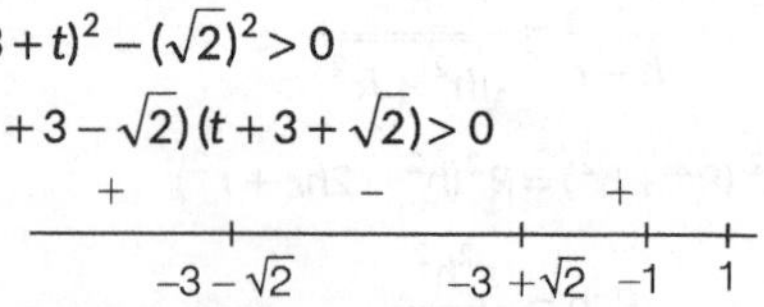

Hence, $g(t)$ is increasing $\forall \ t \in [-1, 1]$.

$$g(t)_{max} = g(1) = 1 + \frac{2}{3 + 1} = \frac{3}{2}$$

$$g(t)_{min} = g(-1) = -1 + \frac{2}{3 - 1} = 0$$

$$\text{So, } f(b) = \begin{cases} \frac{3}{2} + b, & \text{if } b \geq \frac{-3}{4} \\ -b, & \text{if } b < -\frac{3}{4} \end{cases}$$

$\therefore$ Minimum value of $f(b) = -\left(-\frac{3}{4}\right) = \frac{3}{4}$

5. (c) We have, $a^2x^4 + b^2y^4 = c^6$

$$P = xy$$

P will be maximum, if P^4 is maximum.

$$\therefore \quad P^4 = x^4y^4 = \frac{1}{b^2}x^4(b^2y^4)$$

$$\Rightarrow \quad P^4 = \frac{x^4}{b^2}(c^6 - a^2x^4)$$

$$\Rightarrow \quad P^4 = \frac{c^6}{b^2}x^4 - \frac{a^2}{b^2}x^8$$

$$\Rightarrow \quad \frac{d(P^4)}{dx} = \frac{c^6}{b^2}(4x^3) - \frac{a^2}{b^2}(8x^7)$$

$$\therefore \quad \frac{dP^4}{dx} = 0$$

$$\Rightarrow \quad 4x^3\left(\frac{c^6}{b^2} - 2\frac{a^2}{b^2}x^4\right) = 0$$

$$\Rightarrow \quad x = 0 \text{ or } x^4 = \frac{c^6}{2a^2}$$

$$\Rightarrow \quad x = \left(\frac{c^6}{2a^2}\right)^{1/4} = \frac{c^{3/2}}{2^{1/4}a^{1/2}}$$

and $b^2y^4 = c^6 - \frac{c^6}{2} = \frac{c^6}{2}$.

$$\Rightarrow \quad y^4 = \frac{c^6}{2b^2}$$

$$\therefore \quad x^4y^4 = \frac{c^{12}}{4a^2b^2}$$

$$\Rightarrow \quad xy = \frac{c^3}{(4a^2b^2)^{1/4}} = \frac{c^3}{\sqrt{2ab}} = P_{max}.$$

6. (b) Given that, perimeter, $P = 60$

$$\Rightarrow \quad l + 2r = 60 \quad \text{...(i)}$$

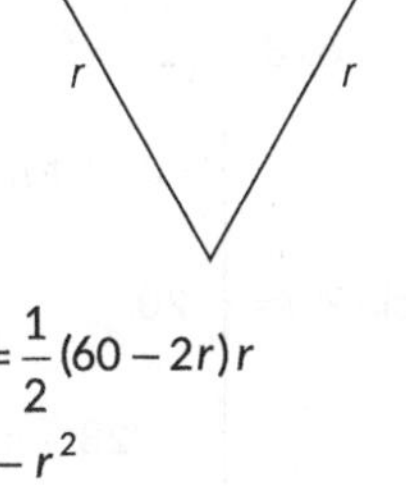

Also, $A = \frac{1}{2}lr = \frac{1}{2}(60 - 2r)r$

$$\Rightarrow \quad A = 30r - r^2$$

$$\Rightarrow \quad \frac{dA}{dr} = 30 - 2r$$

Now, $\frac{dA}{dr} = 0 \Rightarrow 30 - 2r = 0$

$$\Rightarrow \quad r = 15$$

Now, $\frac{d^2A}{dr^2} = -2 < 0$

So, maximum area, when $r = 15$.

7. (c) Volume of cone $= \frac{1}{3}\pi r^2 h$

$$\Rightarrow \quad V = \frac{1}{3}\pi r^2(\sqrt{l^2 - r^2})$$

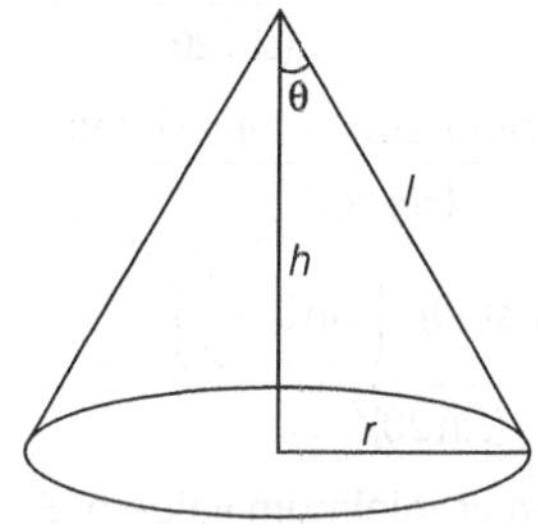

$$\frac{dV}{dr} = \frac{\pi}{3}\left[r^2\left(\frac{-2r}{2\sqrt{l^2 - r^2}}\right) + 2r\sqrt{l^2 - r^2}\right]$$

$$= \frac{\pi}{3}\left[\frac{-r^3 + 2r(l^2 - r^2)}{\sqrt{l^2 - r^2}}\right] = \frac{\pi}{3}\left[\frac{2rl^2 - 3r^3}{\sqrt{l^2 - r^2}}\right]$$

For maximum volume, $\frac{dV}{dr} = 0$

$$\Rightarrow \quad r(2l^2 - 3r^2) = 0$$

$$\Rightarrow \quad \frac{l^2}{r^2} = \frac{3}{2}$$

$$\Rightarrow \quad \frac{l}{r} = \frac{\sqrt{3}}{\sqrt{2}} \quad \left[\because \frac{l}{r} > 0\right]$$

$$\Rightarrow \quad \text{cosec}\,\theta = \sqrt{\frac{3}{2}}$$

$$\Rightarrow \quad \cot\theta = \frac{1}{\sqrt{2}} \Rightarrow \tan\theta = \sqrt{2}$$

8. (c) Let ABC be an isosceles triangle, in which a circle of radius r is inscribed. $\angle BAD = \theta$

In ΔOAE, $OA = OE$ cosec$\theta = r$ cosecθ and $AE = r\cot\theta$

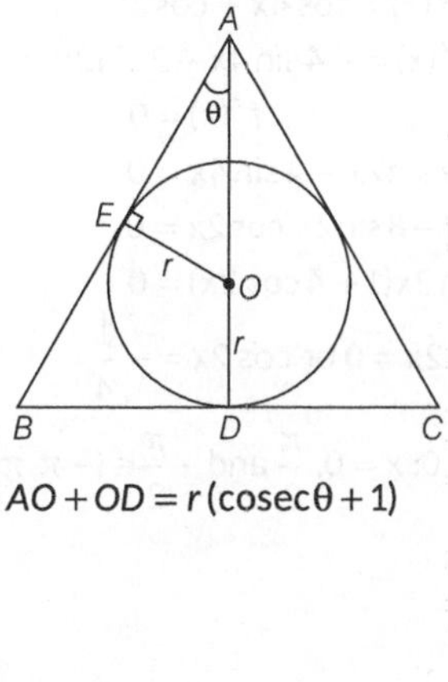

Thus, $AD = AO + OD = r(\text{cosec}\,\theta + 1)$

In ΔABD, $BD = AD\tan\theta = r(\text{cosec}\theta + 1)\tan\theta$

$\Rightarrow \quad AB = AD\sec\theta = r(\text{cosec}\theta + 1)\sec\theta$

Now, perimeter of ΔABC,

$$S = AB + AC + BC = 2AB + 2BD \quad [\because AC = AB]$$

$$= 2r(\text{cosec}\theta + 1)(\sec\theta + \tan\theta)$$

$$= \frac{4r(1+\sin\theta)}{\sin 2\theta}$$

$$\frac{dS}{d\theta} = \frac{4r[2(1+\sin\theta)(\cos\theta\sin 2\theta) - (1-\sin\theta)^2 2\cos 2\theta]}{(\sin 2\theta)^2}$$

$$= \frac{8r(1+\sin\theta)[\sin 2\theta\cos\theta - \cos 2\theta\sin\theta - \cos 2\theta]}{(\sin 2\theta)^2}$$

$$= \frac{8r(1+\sin\theta)(\sin\theta - 1 + 2\sin^2\theta)}{(\sin 2\theta)^2}$$

$$= \frac{16r(1+\sin\theta)^2\left(\sin\theta - \frac{1}{2}\right)}{(\sin 2\theta)^2}$$

For maximum or minimum value of S,

$$\frac{dS}{d\theta} = 0$$

$\Rightarrow \quad \sin\theta = \frac{1}{2} \quad [\because \sin\theta \neq -1 \text{ as } \theta \text{ is an acute angle}]$

$\therefore \quad \theta = \frac{\pi}{6}$

Now, if θ is little less or greater than $\frac{\pi}{6}$, the sign of $\frac{dS}{d\theta}$ changes from '–ve' to '+ve'. Hence, S is minimum when, $\theta = \frac{\pi}{6}$.

Hence, least perimeter of triangle

$$= \frac{4r\left(1+\sin\frac{\pi}{6}\right)^2}{\sin\left(\frac{\pi}{3}\right)}$$

$$= 6\sqrt{3}\, r$$

9. (b, d) We have, $f(x) = \begin{vmatrix} \cos 2x & \cos 2x & \sin 2x \\ -\cos x & \cos x & -\sin x \\ \sin x & \sin x & \cos x \end{vmatrix}$

On solving, $f(x) = \cos 4x + \cos 2x$

Now, $\quad f'(x) = -4\sin 4x - 2\sin 2x$

$\Rightarrow \quad f'(x) = 0$

$\Rightarrow \quad -2\sin 2x - 4\sin 4x = 0$

$\Rightarrow -2\sin 2x - 8\sin 2x\cos 2x = 0$

$\Rightarrow \quad 2\sin 2x(1 + 4\cos 2x) = 0$

$\Rightarrow \quad \sin 2x = 0 \text{ or } \cos 2x = -\frac{1}{4}$

For $\sin 2x = 0$; $x = 0, \frac{\pi}{2}$ and $-\frac{\pi}{2} \in (-\pi, \pi)$

For $\cos 2x = -\frac{1}{4}$, there are three solutions.

So, $f'(x) = 0$ has more than three solutions.

Now, $f''(x) = -16\cos 4x - 4\cos 2x$

At $x = 0$, $f''(0) < 0$

Hence, $f(x)$ attains maximum at $x = 0$.

10. (a, d) Given, $g(x) = \int_0^x f(t)\,dt, x \in [1, 3]$

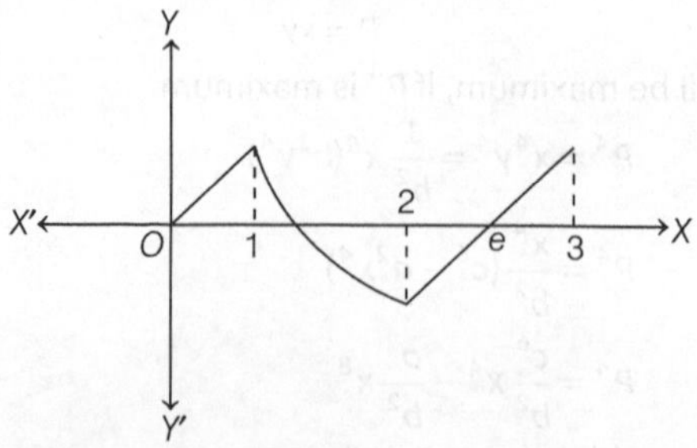

$\Rightarrow \quad g'(x) = f(x)$

$\Rightarrow \quad g'(x) = 0 \text{ gives } f(x) = 0$

$\Rightarrow \quad 2 = e^{x-1} \text{ or } x = e$

$\Rightarrow \quad x - 1 = \log 2 \text{ or } x = e$

$\Rightarrow \quad x = 1 + \log 2 \text{ or } x = e$

$$\text{As, } g'(x) = \begin{cases} > 0 : 0 < x < 1 + \log 2 \\ < 0 : 1 + \log 2 < x < e \\ > 0 : e < x < 3 \end{cases}$$

Thus, $g(x)$ has a local maxima at $x = 1 + \log 2$ and a local minima at $x = e$.

11. (b, d) Given, function is

$$f(x) = \int_{-1}^{x} t(e^t - 1)(t-1)(t-2)^3(t-3)^5\,dt$$

$\Rightarrow \quad f'(x) = x(e^x - 1)(x-1)(x-2)^3(x-3)^5$

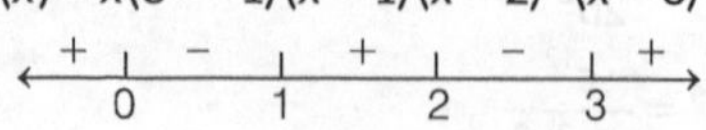

From the sign scheme, it is clear that, $f(x)$ has a local maxima at $x = 0$ and $x = 2$ and local minima at $x = 1$ and $x = 3$.

12. (a, d) $y = \sin^{-1}(2x\sqrt{1-x^2})$

Let $\quad \theta = \sin^{-1} x$

$\Rightarrow \quad \theta \in \left[-\frac{\pi}{2}, \frac{\pi}{2}\right], x \in [-1, 1]$

$\Rightarrow \quad x = \sin\theta \Rightarrow \cos\theta = \sqrt{1-x^2}$

$$\Rightarrow \quad y = \sin^{-1}(\sin 2\theta) = \begin{cases} (-2\theta + \pi), & 2\theta \in \left[-\pi, -\frac{\pi}{2}\right] \\ 2\theta, & 2\theta \in \left[-\frac{\pi}{2}, \frac{\pi}{2}\right] \\ -(2\theta - \pi), & 2\theta \in \left[\frac{\pi}{2}, \pi\right] \end{cases}$$

$$= \begin{cases} -2\sin^{-1}x - \pi, \ \theta \in \left[-\frac{\pi}{2}, -\frac{\pi}{4}\right], x \in \left[-1, -\frac{1}{\sqrt{2}}\right] \\ 2\sin^{-1}x, \ \theta \in \left[-\frac{\pi}{4}, \frac{\pi}{4}\right], \ x \in \left[-\frac{1}{\sqrt{2}}, \frac{1}{\sqrt{2}}\right] \\ -2\sin^{-1}x + \pi, \ \theta \in \left[\frac{\pi}{4}, \frac{\pi}{2}\right], x \in \left[\frac{1}{\sqrt{2}}, 1\right] \end{cases}$$

$$\Rightarrow \ f'(x) = \begin{cases} \frac{-2}{\sqrt{1-x^2}}, \ x \in \left[-1, -\frac{1}{\sqrt{2}}\right] \\ \frac{2}{\sqrt{1-x^2}}, \ x \in \left[-\frac{1}{\sqrt{2}}, \frac{1}{\sqrt{2}}\right] \\ \frac{-2}{\sqrt{1-x^2}}, \ x \in \left[\frac{1}{\sqrt{2}}, 1\right] \end{cases}$$

$f(x)$ is decreasing on $x \in \left[-1, \frac{-1}{\sqrt{2}}\right]$ and increasing on $x \in \left[-\frac{1}{\sqrt{2}}, \frac{1}{\sqrt{2}}\right]$ and decreasing on $\left[\frac{1}{\sqrt{2}}, 1\right]$.

$\Rightarrow$ $f(x)$ has local maxima at $x = -1, \frac{1}{\sqrt{2}}$ and local minima at $x = \frac{-1}{\sqrt{2}}, 1$.

Clearly, $f(x)$ has its least value at $x = \frac{-1}{\sqrt{2}}$ or $x = 1$

$$\Rightarrow \ f\left(\frac{-1}{\sqrt{2}}\right) = \sin^{-1}\left(-\sqrt{2} \times \frac{1}{\sqrt{2}}\right) = \frac{-\pi}{2}$$

and $\quad f(1) = 0$

$\Rightarrow$ The least value of $f(x)$ is $-\frac{\pi}{2}$ at $x = \frac{-1}{\sqrt{2}}$.

13. (c, d) Let (a, b) is the point on the parabola from which its distance from (0, 3) is minimum.

$$\Rightarrow \quad a^2 = 2b$$

So, distance, $d^2 = a^2 + (b-3)^2$

$$\Rightarrow \quad d^2 = 2b + (b-3)^2$$

For minimum value of d, $\frac{d(d^2)}{db^2} = 0$

$$\Rightarrow \quad 2 + 2(b-3) = 0$$

$$\Rightarrow \quad b = 2$$

and $\quad a^2 = 2b = 4$

$$\Rightarrow \quad a = \pm 2$$

Hence, required points on parabola are (2, −2) and (2, 2).

14. (b, d) We have,

$$f(x) = (x-1)^p (x-2)^q$$

$$\Rightarrow \ f'(x) = p(x-1)^{p-1}(x-2)^q + q(x-1)^p (x-2)^{q-1}$$

$$\Rightarrow \ f'(x) = (x-1)^{p-1}(x-2)^{q-1}\{p(x-2) + q(x-1)\}$$

Now, $f'(x) = 0$

$$\Rightarrow \ f'(x) = (x-1)^{p-1}(x-2)^{q-1}\{x(p+q) - (2p+q)\} = 0$$

$$\Rightarrow \quad x = 1, 2 \text{ and } \frac{2p+q}{p+q}$$

Each critical point of $f(x)$ is a point of extremum.

So, $p-1 \in$ odd and $q-1 \in$ odd.

$p \in$ even number and $q \in$ even number.

So, option (b) and (d) are correct.

15. (a, b, c) We have,

$$f(x) = \int_0^{\sqrt{9-x^2}} e^{t^2/2}\, dt$$

$$\Rightarrow \quad f'(x) = e^{(9-x^2)/2} \frac{1}{2\sqrt{9-x^2}} \times (-2x)$$

$$\Rightarrow \quad f'(x) = \frac{-x}{\sqrt{9-x^2}} e^{(9-x^2)/2}$$

$\Rightarrow$ $f'(x) > 0$ for $x \in (-3, 0)$ and $f'(x) < 0$ for $x \in (0, 3)$

$\Rightarrow$ $f'(x) = 0$ at $x = 0$

and $f(0) = \int_0^3 e^{t^2/2}\, dt$ is maximum value of $f(x)$.

$\Rightarrow$ $f'(-x) = -f(x)$

$\Rightarrow$ $f'(x)$ is an odd function.

16. (b, d) We have, $y = x^2 + px + q$ ...(i)

Cuts the straight line $y = (2x - 3)$, where $x = 1$ i.e. $y = -1$

$\Rightarrow$ (1, −1) lies on $y = x^2 + px + q$

$$\Rightarrow \quad -1 = 1 + p + q$$

$\Rightarrow$ $p + q = -2$ ...(ii)

Vertex of parabola is

$$\left(-\frac{p}{2}, \frac{-(p^2 - 4q)}{4}\right) = \left(\frac{-p}{2}, \frac{-p^2 + 4q}{4}\right)$$

Distance between vertex and X-axis

$$\frac{|p^2 - 4q|}{4} = \frac{1}{4}|p^2 - 4(-2-p)| = \frac{1}{4}|p^2 + 4p + 8|$$

$$= \frac{p^2 + 4p + 8}{4} \text{ as } p^2 + 4p + 8 > 0$$

Let $L = \frac{1}{4}(p^2 + 4p + 8)$

$$\frac{dL}{dp} = \frac{1}{4}(2p + 4) = \frac{p}{2} + 1$$

For least value of L, $\frac{dL}{dp} = 0$

$$\Rightarrow \quad \frac{p}{2} + 1 = 0 \Rightarrow p = -2$$

Also, $p + q = -2$

$$\Rightarrow \quad q = -2 - p$$

$$\Rightarrow \quad q = -2 - (-2) = 0$$

$$\therefore \quad p = -2 \text{ and } q = 0$$

and least distance $= \frac{1}{4}(4 - 8 + 8) = 1$

17. (b, c) Since, $f(x)$ has local maxima at $x=-1$ and $f'(x)$ has local minima at $x=0$

$$\therefore \quad f''(x)=\lambda x$$

$$\Rightarrow \quad f'(x)=\lambda\frac{x^2}{2}+C$$

$$\Rightarrow \quad \frac{\lambda}{2}+C=0 \qquad [f'(-1)=0]$$

$$\Rightarrow \quad \lambda=-2C \qquad \text{...(i)}$$

Again, integrating both sides, we get

$$f(x)=\lambda\frac{x^3}{6}+Cx+D \qquad \text{...(ii)}$$

$$\Rightarrow \quad f(2)=\lambda\left(\frac{8}{6}\right)+2C+D=18$$

and
$$f(1)=\frac{\lambda}{6}+C+D=-1 \qquad \text{...(iii)}$$

$\therefore$ Using Eqs. (i), (ii) and (iii), we get

$$f(x)=\frac{1}{4}(19x^3-57x+34)$$

$$\therefore \quad f'(x)=\frac{1}{4}(57x^2-57)$$

$$=\frac{57}{4}(x-1)(x+1) \qquad [\text{using number line rule}]$$

$\therefore$ $f(x)$ is increasing for $[1, 2\sqrt{5}]$

and $f(x)$ has local maximum value at $x=-1$ and $f(x)$ has local minimum value at $x=1$.

Also, $f(0)=\frac{34}{4}$

Sol. (Q.Nos. 18-20)

18. (d) **19.** (d) **20.** (c)

Since, $f(x)$ is a polynomial of degree 4, $f'(x)$ is a cubic polynomial. $f(x)$ has relative maximum/relative minimum at $x=1$, 2 and 3.

So, $f'(1)=f'(2)=f'(3)=0$

Then, $f'(x)=a(x-1)(x-2)(x-3)$

$$=a(x^3-6x^2+11x-6)$$

Integrating w.r.t. x, we get

$$f(x)=a\left(\frac{x^4}{4}-2x^3+\frac{11x^2}{2}-6x\right)+c \qquad \text{...(i)}$$

Now, $f(0)=0$

$\therefore \quad c=0$

From Eq. (i),

$$f(x)=a\left(\frac{x^4}{4}-2x^3+\frac{11x^2}{2}-6x\right) \qquad \text{...(ii)}$$

Also, $f(-2)=216 \Rightarrow a=4$

From Eq. (ii), $f(x)=x^4-8x^3+22x^2-24x$ ...(iii)

and $\quad f'(x)=4(x-1)(x-2)(x-3)$ ...(iv)

18. Extremum points of $f(x)$ are $\{1, f(1)\}, \{2, f(2)\}$ and $\{3, f(3)\}$ i.e. $(1, -9)$, $(2, -8)$ and $(3, -9)$.

$\therefore$ Area of triangle formed by extremum points of

$$f(x)=\frac{1}{2}\begin{vmatrix}1 & -9 & 1\\ 2 & -8 & 1\\ 3 & -9 & 1\end{vmatrix}$$

$$=\frac{1}{2}|-2|$$

$$=1 \text{ sq units}$$

19. From Eq. (iv),

$$f'(x)=4(x-1)(x-2)(x-3)$$

$$f'(0)=4(-1)(-2)(-3)=-24$$

20. $f'(x)=4(x-1)(x-2)(x-3)$

$f(x)$ has two relative minima at $x=1$, 3 and one relative maxima at $x=2$.

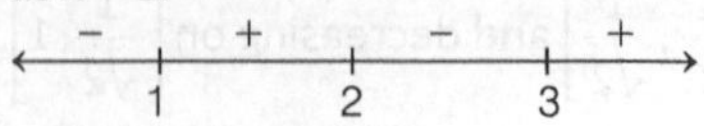

Hence, option (a) is incorrect.

Now, $f''(x)=4(3x^2-12x+11)=0$

$$\Rightarrow \quad x=2\pm\frac{1}{\sqrt{3}}$$

$$\Rightarrow \quad f'''(x)=24(x-2)=\pm\frac{24}{\sqrt{3}} \qquad \left(\text{at } x=2\pm\frac{1}{\sqrt{3}}\right)$$

i.e. two points of inflection.

$\because \quad f(0)=0, f(1)=-9, f(2)=-8$

and $f(3)=-9$

$\therefore$ Range of $f(x)=[-9, \infty)$

Range of $f(x)$ contains nine negative integers

$$f(x)=0$$

$$x^4-8x^3+22x^2-24x=0$$

Sum of roots $=8$

Sol. (Q.Nos. 21-23)

21. (a) **22.** (d) **23.** (b)

We have, $f(x)=x^3-3(7-a)x^2-3(9-a^2)x+2$

$$\Rightarrow \quad f'(x)=3x^2-6(7-a)x-3(9-a^2)$$

For real roots,

$$D\geq 0$$

$$\Rightarrow \quad 49+a^2-14a+9-a^2\geq 0$$

$$\Rightarrow \quad a\leq\frac{58}{14} \qquad \text{...(i)}$$

When point of minima is negative, point of maxima is also negative.

Hence, equation $f'(x)=3x^2-6(7-a)x-3(9-a^2)=0$ has both roots negative.

Sum of roots $=2(7-a)<0$

$\Rightarrow \quad a > 7$, which is not possible as $a \le \dfrac{58}{14}$.

When the point of maxima is positive, point of minima is also positive.

Hence, equation $f'(x) = 3x^2 - 6(7 - a)x - 3(9 - a^2) = 0$ has both roots positive.

Sum of roots $= 2(7 - a) > 0$ or $a < 7$ ...(ii)

Also, product of roots is positive.

$$-(9 - a^2) > 0 \Rightarrow a^2 > 9$$

or $\qquad a \in (-\infty, -3) \cup (3, \infty)$ ...(iii)

From Eqs. (i), (ii) and (iii),

$$a \in (-\infty, -3) \cup \left(3, \frac{58}{14}\right)$$

For points of extrema of opposite sign, Eq. (i) has roots of opposite sign.

Thus, $a \in (-3, 3)$

24. (i) → (D), (ii) → (A), (iii) → (C), (iv) → (B)

(i) Graph of $f(x) = |2x - 1| + |2x - 3|$

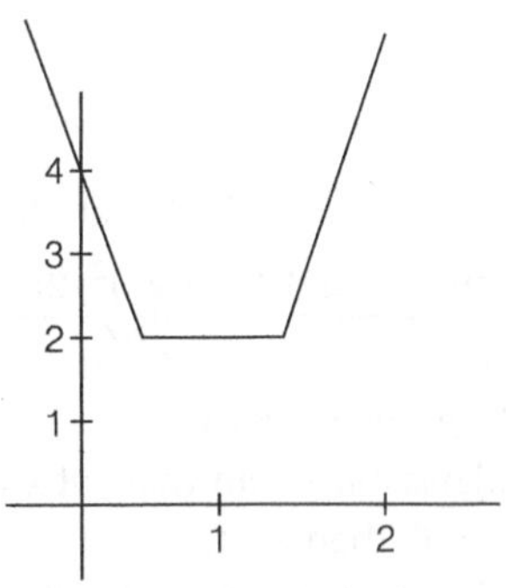

From the graph, $f(x)$ has infinite points of minima.

(ii) $f(x) = 2\sin x - x$

$\Rightarrow \quad f'(x) = 2\cos x - 1 = 0$

$\Rightarrow \quad \cos x = \dfrac{1}{2}$

which has infinite points of extrema.

(iii) Graph of $f(x) = |x - 1| + |2x - 3|$

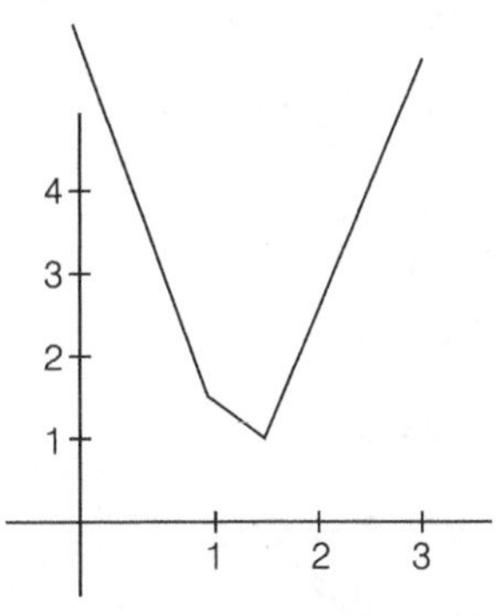

From the graph, $f(x)$ has one point of minima.

(iv) Graph of $f(x) = |x| - |2x - 3|$

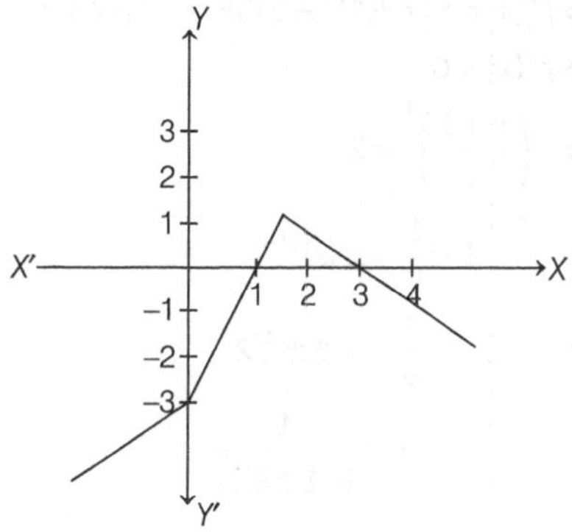

From the graph, $f(x)$ has one point of maxima.

25. (i) → (B), (ii) → (C), (iii) → (A), (iv) → (D)

(i) Given, curve is

$$y = 3x^2 + ax^3$$

$\Rightarrow \qquad \dfrac{dy}{dx} = 6x + 3ax^2 = m$ (say) ...(i)

$\Rightarrow \qquad \dfrac{dm}{dx} = 6 + 6ax$

and $\qquad \dfrac{d^2m}{dx^2} = 6a$

For maxima or minima,

$$\frac{dm}{dx} = 0$$

$\Rightarrow \qquad 6 + 6ax = 0$

$\Rightarrow \qquad x = \dfrac{-1}{a}$

But given m is maximum value at $x = \dfrac{1}{2}$

Therefore, $\dfrac{-1}{a} = \dfrac{1}{2}$

$$a = -2$$

(ii) Given, $y = -x^3 + 3x^2 + 2x - 27$

$\Rightarrow \qquad m = \dfrac{dy}{dx} = -3x^2 + 6x + 2$ (say)

$\Rightarrow \qquad \dfrac{dm}{dx} = -6x + 6$

and $\qquad \dfrac{d^2m}{dx^2} = -6$

Now, $\qquad \dfrac{dm}{dx} = 0$

$\Rightarrow \qquad -6x + 6 = 0$

$\Rightarrow \qquad x = 1$

Clearly, $\dfrac{d^2m}{dx^2} = -6 < 0 \; \forall \; x$

Put $x = 1$ in curve, we get $y = -23$

Thus, maximum slope at $(+1, -23)$.

So, maximum slope is $m = 5$.

(iii) Let $f(x)=(x+1)^5-2(x^5+1)$

$\Rightarrow f'(x)=5(x+1)^4-2(5x^4)=5[(x+1)^4-2x^4]$

$\Rightarrow f'(x)=0$

$$\Rightarrow \left(\frac{x+1}{x}\right)^4=2$$

$$\Rightarrow \quad 1+\frac{1}{x}=\pm(2)^{1/4}$$

$$\Rightarrow \quad \frac{1}{x}=-1\pm\sqrt[4]{2}$$

$$\Rightarrow \quad x=\frac{1}{(-1\pm\sqrt[4]{2})}$$

i.e. $\dfrac{1}{(-1-(2)^{1/4})}$ and $\dfrac{1}{-1+(2)^{1/4}}=\alpha,\beta$ (say)

And $f(-\infty)=\lim\limits_{x\to-\infty}(x+1)^5-2(x^5+1)$

$$=\lim_{x\to-\infty}x^5\left[\left(1+\frac{1}{x}\right)^5-2\left(1+\frac{1}{x^5}\right)\right]$$

$$=\lim_{x\to-\infty}x^5[1-2]=\infty$$

$\Rightarrow \quad f(\infty)=-\infty, f(0)=-1$ and $f(-1)=0$

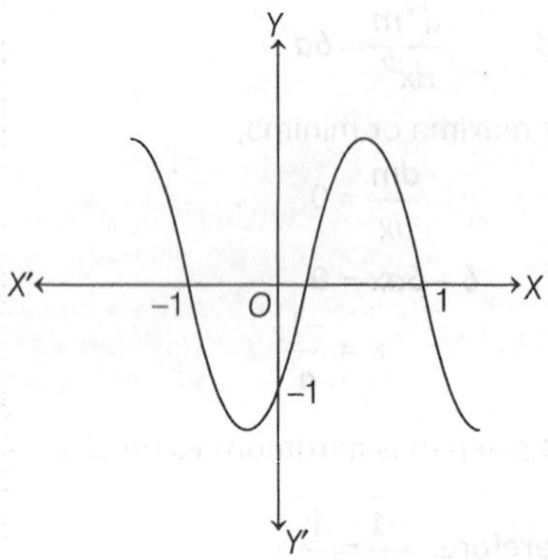

$\because f'(x)=0$ has two real roots and $f(x)$ is a polynomial of 5th degree and $f'(\alpha)=f'(\beta)=0$ but $f(\alpha)\neq 0$ and $f(\beta)\neq 0$

$\Rightarrow \quad f(x)=0$ has exactly 3 distinct real roots.

(iv) We have, $f(x)=ab\sin x+\sqrt{1-a^2}\cos x+C; |a|<1,$

$\Rightarrow f(x)\in[c-\sqrt{a^2b^2+(1-a^2)}, c+\sqrt{a^2b^2+(1-a^2)}]$

$\Rightarrow (f_{max}-f_{min})=2\sqrt{a^2b^2+(1-a^2)}$

$$=2\sqrt{a^2(b^2-1)+1}$$

Now, $a^2(b^2-1)+1\leq-\dfrac{[(0)^2-4(b^2-1)(1)]}{4(b^2-1)}$

$$\left(\because ax^2+bx+c\leq\frac{-D}{4a}\text{ for } a<0\right)$$

$\Rightarrow \quad a^2(b^2-1)+1\leq 1$ at $a=0$

$\Rightarrow \quad (f_{max}-f_{min})\leq 2$

26. (i) $\rightarrow$ (D), (ii) $\rightarrow$ (A), (iii) $\rightarrow$ (B), (iv) $\rightarrow$ (C)

(i) We have, $y=\dfrac{x}{1+x^2}$

$$\Rightarrow \frac{dy}{dx}=\frac{(1+x^2)(1)-x(2x)}{(1+x^2)^2}$$

$$=\frac{1+x^2-2x^2}{(1+x^2)^2}=\frac{1-x^2}{(1+x^2)^2}$$

Let $m=\dfrac{dy}{dx}=\dfrac{1-x^2}{(1+x)^2}$

$$\Rightarrow \frac{dm}{dx}=\frac{(1+x^2)^2(-2x)-(1-x^2)\{2(1+x^2)(2x)\}}{(1+x^2)^4}$$

$$=\frac{-2x(1+x^2)\{(1+x^2)+2(1-x^2)\}}{(1+x^2)^4}$$

$$\Rightarrow \frac{dm}{dx}=\frac{-2x(1+x^2)(3-x^2)}{(1+x^2)^4}$$

$$\Rightarrow \frac{dm}{dx}=\frac{-2x(3-x^2)}{(1+x^2)^3}$$

Now, $\dfrac{dm}{dx}=0$

$\Rightarrow \quad x=0,\pm\sqrt{3}$

$$\text{Now, } \frac{d^2m}{dx^2}=\frac{(1+x^2)^3(6x^2-6)-(2x^3-6x)\cdot 3(1+x^2)^2\cdot 2x}{(1+x^2)^6}$$

At $x=0$, $g''(x)=-6<0$

$\therefore \quad g'(x)$ has maximum value at $x=0$.

When, $x=0$, then $y=0$

Hence, (0, 0) is the required point.

(ii) Let $P(x,y)$ be a point on the parabola $y^2=4x$.

Distance between $P(x,y)$ and (2, 1) is S.

So, $S^2=(x-2)^2+(y-1)^2$

$$\Rightarrow S^2=\left(\frac{y^2}{4}-2\right)^2+(y-1)^2$$

$$\Rightarrow S^2=\frac{y^4}{16}+4-y^2+y^2+1-2y$$

Let $f(y)=S^2=\dfrac{y^4}{16}-2y+5$

$$\Rightarrow f'(y)=\frac{y^3}{4}-2$$

Now, $f'(y)=0$

$$\Rightarrow \quad \frac{y^3}{4}-2=0$$

$\Rightarrow \quad y^3=8 \Rightarrow y=2$

$$\Rightarrow \quad f''(y)=\frac{3y^2}{4}$$

At $y = 2$,

$$f''(2) = 3 > 0$$

So, $y = 2$ is the point of minima.

Now, $(2)^2 = 4(x)$

$\Rightarrow \quad x = 1$

Hence, required point is (1, 2).

(iii) We have parabola,

$$y^2 = 4(x - 10)$$

$\Rightarrow \quad 2y\dfrac{dy}{dx} = 4$

$\Rightarrow \quad \dfrac{dy}{dx} = \dfrac{2}{y}$

Slope of line $x + y = 4$ is -1.

Hence, slope of tangent of parabola

= Slope of line $x + y = 4$

[as shortest distance would be perpendicular]

$\Rightarrow \quad \dfrac{dy}{dx} = \dfrac{2}{y} = -1$

$\Rightarrow \quad y = -2$ and $(-2)^2 = 4(x - 10)$

$\Rightarrow \quad 1 = x - 10$

$\Rightarrow \quad x = 11$

(11, 2) is the point on parabola. Slope of perpendicular line = 1, then equation of perpendicular line is

$$y + 2 = (x - 11)$$

$\Rightarrow \quad x - y = 13$

Now, point on line is the intersection of $x - y = 13$ and $x + y = 4$ is

$$x = \frac{17}{2} \text{ and } y = -\frac{9}{2}$$

Hence, coordinate is $\left(\dfrac{17}{2}, \dfrac{-9}{2}\right)$

(iv) Let (x, y) be a point on the curve

$$y = x^2 - 4$$

Distance from (0, 0) is

$$D^2 = x^2 + y^2$$

So, $f(y) = D^2 = y + 4 + y^2$

$\Rightarrow \quad f'(y) = 2y + 1$

$\Rightarrow \quad f'(y) = 0$

$\Rightarrow 2y + 1 = 0$

$\Rightarrow \quad y = \dfrac{-1}{2}$

From Eq. (i), $-\dfrac{1}{2} + 4 = x^2$

$\Rightarrow \quad x = \pm\sqrt{\dfrac{7}{2}}$

So, point is $\left(\pm\sqrt{\dfrac{7}{2}}, -\dfrac{1}{2}\right)$.

27. (i) → (D), (ii) → (C), (iii) → (A), (iv) → (B)

We have, $f(x) = x^4(12\ln x - 7)$

$\Rightarrow \quad f'(x) = 4x^3(12\ln x - 7) + x^4\left(\dfrac{12}{x}\right)$

$\Rightarrow \quad f'(x) = 4x^3(12\ln x) - 28x^3 + 12x^3$

$\Rightarrow \quad f'(x) = 48x^3\ln x - 16x^3$

$\Rightarrow \quad f'(x) = 16x^3(3\ln x - 1)$

Now, $f'(x) = 0$

$\Rightarrow \quad 16x^3(3\ln x - 1) = 0$

$\Rightarrow \quad x = 0$ or $\ln x = \dfrac{1}{3}$

$\Rightarrow \quad x = e^{\frac{1}{3}}$

$\Rightarrow f''(x) = 48x^2(3\ln x - 1) + 16x^3\left(\dfrac{3}{x}\right)$

$\Rightarrow f''(x) = 48x^2(3\ln x - 1) + 48x^2$

$\Rightarrow f''(x) = 144x^2\ln x$

$\Rightarrow f''(x) = 0$, when $x = 1$.

When $x = 1$, then $f(x) = -7$

Hence, $(1, -7)$ is the point of inflection.

Now, $a - b = 1 - (-7) = 8$

$e^{\frac{1}{3}}$ is the point of minima.

So, $e^t = e^{\frac{1}{3}}$

Hence, $t = \dfrac{1}{3}$

So, $12(t) = 12 \times \dfrac{1}{3} = 4$

The graph of $f(x)$ is concave downwards in (0, 1) and concave upward in $(1, \infty)$.

So, $(d, e) = (0, 1)$

$\Rightarrow \quad d + 3e = 0 + 3(1) = 3$

and $(p, \infty) = (1, \infty)$

$\therefore \quad p = 1$

Sol. (Q.Nos. 28-30)

28. (6) **29.** (10) **30.** (6)

We have, $y = f(x) = \dfrac{x^4 - x^2}{x^6 + 2x^3 - 1}, x > 1$

$$= \frac{x^3\left(x - \frac{1}{x}\right)}{x^3\left(x^3 - \frac{1}{x^3} + 2\right)} = \frac{x - \frac{1}{x}}{\left(x^3 - \frac{1}{x^3} + 2\right)}$$

$$= \frac{x - \frac{1}{x}}{\left(x - \frac{1}{x}\right)^3 + 3\left(x - \frac{1}{x}\right) - 2}$$

Let $x - \frac{1}{x} = t$

$\Rightarrow \quad y = \frac{t}{t^3 + 3t - 2}$

$\Rightarrow \quad \frac{dy}{dt} = \frac{(t^3 + 3t - 2) - t(3t^2 + 3)}{(t^3 + 3t - 2)^2} \Rightarrow \frac{dy}{dt} = \frac{-2(t^3 + 1)}{(t^3 + 3t - 2)^2}$

For maxima or minima,

$$\frac{dy}{dt} = 0$$

$\Rightarrow \quad \frac{-2(t^3 + 1)}{(t^3 + 3t - 2)^2} = 0$

$\Rightarrow \quad t^3 + 1 = 0 \Rightarrow t = -1$

Hence, the maximum value is $\frac{1}{6}$

$\Rightarrow \quad k = 6$

At $x - \frac{1}{x} = -1$,

$\Rightarrow \quad x^2 + x - 1 = 0$

$\Rightarrow \quad x = \frac{-1 \pm \sqrt{1+4}}{2}$

$\Rightarrow \quad x = \frac{-1 + \sqrt{5}}{2}$, since $x > 1$

Therefore, $a + b + 3 = 1 + 6 + 3 = 10$

and $\quad m + n + p = (-1) + (5) + 2 = 6$

Sol. (Q.Nos. 31-33)

31. (2) We have, $f(x) = \left(\alpha - \frac{1}{\alpha} - x\right)(4 - 3x^2)$

$\Rightarrow \quad f'(x) = -(4 - 3x^2) + \left(\alpha - \frac{1}{\alpha} - x\right)(-6x)$

$$= 9x^2 - 6\left(\alpha - \frac{1}{\alpha}\right)x - 4$$

$\Rightarrow \quad D = 36\left(\alpha - \frac{1}{\alpha}\right)^2 + 144 > 0$

$\therefore$ There are two critical points, which are given by

$$x = \frac{6\left(\alpha - \frac{1}{\alpha}\right) \pm 6\sqrt{\left(\alpha - \frac{1}{\alpha}\right)^2 + 4}}{18}$$

$$= \frac{6\left(\alpha - \frac{1}{\alpha}\right) \pm 6\left(\alpha + \frac{1}{\alpha}\right)}{18}$$

$$= \frac{2\alpha}{3}, \frac{-2}{3\alpha}$$

$x = \frac{-2}{3\alpha}$ is point of local maxima and $x = \frac{2\alpha}{3}$ is point of local minima.

32. (5) $f\left(\frac{2\alpha}{3}\right) = \left(\alpha - \frac{1}{\alpha} - \frac{2\alpha}{3}\right)\left(4 - 3\frac{4\alpha^2}{9}\right)$

$$= \left(\frac{\alpha}{3} - \frac{1}{\alpha}\right)\left(4 - \frac{4\alpha^2}{3}\right)$$

$$= 4\left(\frac{\alpha^2 - 3}{3\alpha}\right)\left(\frac{3 - \alpha^2}{3}\right)$$

$$= \frac{4}{9\alpha}(\alpha^2 - 3)(3 - \alpha^2)$$

$$= \frac{-4(\alpha^2 - 3)^2}{9\alpha}$$

$\Rightarrow f\left(\frac{-2}{3\alpha}\right) = \left(\alpha - \frac{1}{\alpha} + \frac{2}{3\alpha}\right)\left(4 - \frac{4}{3\alpha^2}\right)$

$$= 4\left(\alpha - \frac{1}{3\alpha}\right)\left(\frac{3\alpha^2 - 1}{3\alpha^2}\right)$$

$$= \frac{4(3\alpha^2 - 1)^2}{9\alpha^3}$$

$\therefore f\left(-\frac{2}{3\alpha}\right) - f\left(\frac{2\alpha}{3}\right) = \frac{4(3\alpha^2 - 1)^2}{9\alpha^3} + \frac{4(\alpha^2 - 3)^2}{9\alpha}$

$$= \frac{4(3\alpha^2 - 1)^2 + 4\alpha^2(\alpha^2 - 3)^2}{9\alpha^3}$$

$$= \frac{4}{9\alpha^3}[9\alpha^4 - 6\alpha^2 + 1 + \alpha^6 - 6\alpha^4 + 9\alpha^2]$$

$$= \frac{4}{9\alpha^3}(\alpha^6 + 3\alpha^4 + 3\alpha^2 + 1)$$

$$= \frac{4(\alpha^2 + 1)^3}{9\alpha^3} = \frac{4}{9}\left(\alpha + \frac{1}{\alpha}\right)^3$$

$\Rightarrow b - a = 9 - 4 = 5$

33. (23) Least possible value of the absolute difference between local maximum and local minimum is $\frac{32}{9}$.

$\Rightarrow k_1 - k_2 = 32 - 9 = 23$

34. (4) We have, $f(x) = x^3 - 3x$

$\Rightarrow \quad f'(x) = 3x^2 - 3 = 3(x^2 - 1)$

For maxima or minima, $f'(x) = 0$

$\Rightarrow \quad 3(x^2 - 1) = 0$

$\Rightarrow \quad (x^2 - 1) = 0 \Rightarrow x = \pm 1$

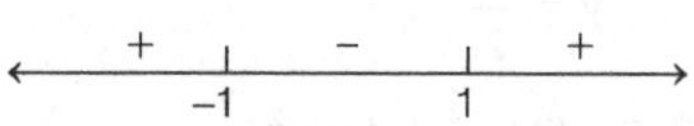

Clearly, the function $f(x)$ have the point of maxima at $x = -1$ and the point of minima at $x = 1$.

Thus, $m =$ minimum value of $f(x) = 1 - 3 = -2$

and $n =$ maximum value of $f(x) = -1 + 3 = 2$

Hence, the value of $(m + n + 4) = -2 + 2 + 4 = 4$

35. (4) Put $x = r\cos\theta$ and $y = r\sin\theta$

We have, $x^2(xy - 1) = y^2(1 + xy)$

$$\Rightarrow \quad xy(x^2 - y^2) = x^2 + y^2$$

$$\Rightarrow \quad r^4 \sin\theta \cdot \cos\theta \cdot \cos 2\theta = r^2$$

$$\Rightarrow \quad r^2(2\sin\theta \cdot \cos\theta) \cdot \cos 2\theta = 2$$

$$\Rightarrow \quad r^2(2\sin 2\theta \cos 2\theta) = 4$$

$$\Rightarrow \quad r^2(\sin 4\theta) = 4$$

$$\Rightarrow \quad r^2 = \frac{4}{\sin 4\theta} = 4\,\text{cosec}\, 4\theta$$

$$\Rightarrow \quad f(x, y) = x^2 + y^2$$

Thus, $f(x, y) = x^2 + y^2 = r^2 = 4\,\text{cosec}\, 4\theta$

Hence, minimum value is 4.

36. (9) Given, $a = \dfrac{4}{\sin x} + \dfrac{1}{1 - \sin x}$

$$\therefore \quad \frac{da}{dx} = \cos x\left(-\frac{4}{\sin^2 x} + \frac{1}{(1 - \sin x)^2}\right)$$

For maximum and minimum value of a, $\dfrac{da}{dx} = 0$, we get

$$\cos x = 0 \text{ or } -\frac{4}{\sin^2 x} + \frac{1}{(1 - \sin x)^2} = 0$$

$$\Rightarrow \quad x = \frac{\pi}{2} \text{ or } 4(1 - \sin x)^2 - \sin^2 x = 0$$

$$\because \quad x \in \left(0, \frac{\pi}{2}\right)$$

$$\therefore \quad x \neq \frac{\pi}{2}$$

Then, $4(1 - \sin x)^2 = \sin^2 x$

$$\Rightarrow \quad 2(1 - \sin x) = \pm \sin x, \text{ we get}$$

$$\sin x = 2, \frac{2}{3} \qquad [\because \sin x \neq 2]$$

$$\therefore \quad \sin x = \frac{2}{3} \Rightarrow x = \sin^{-1}\frac{2}{3}$$

Again, $$\frac{d^2a}{dx^2} = -\sin x\left(-\frac{4}{\sin^2 x} + \frac{1}{(1 - \sin x)^2}\right) + \cos x\left(\frac{8\cos x}{\sin^3 x} + \frac{2\cos x}{(1 - \sin x)^3}\right)$$

$$\therefore \quad \left.\frac{d^2a}{dx^2}\right|_{x = \sin^{-1}\left(\frac{2}{3}\right)} = -\frac{2}{3}(-9 + 9) + \left(1 - \frac{4}{9}\right)(27 + 54)$$

$$= 45 > 0$$

Thus, a is minimum at $x = \sin^{-1}\left(\dfrac{2}{3}\right)$ in $\left(0, \dfrac{\pi}{2}\right)$.

Hence, least value of a is $\dfrac{4}{\frac{2}{3}} + \dfrac{1}{1 - \frac{2}{3}} = 9$

$$\therefore \quad a = 9$$

37. (12) Since, $f(x)$ being a cubic function, $f'(x)$ is a quadratic function, $f(x)$ has relative maximum and minimum value at $x = -1$ and $x = \dfrac{1}{3}$.

So, $f'(-1) = f\left(\dfrac{1}{3}\right) = 0$

Then, $$f'(x) = a(x + 1)\left(x - \frac{1}{3}\right) = a\left(x^2 + \frac{2x}{3} - \frac{1}{3}\right),$$

where a is any constant.

Integrating w.r.t. x, we get

$$f(x) = a\left(\frac{x^3}{3} + \frac{x^2}{3} - \frac{x}{3}\right) + b, \qquad \text{...(i)}$$

where b is the constant of integration and $f(-2) = 0$.

Then, $a\left(\dfrac{-8}{3} + \dfrac{4}{3} + \dfrac{2}{3}\right) + b = 0$

$$\therefore \quad b = \frac{2a}{3}$$

From Eq. (i),

$$f(x) = \frac{a}{3}(x^3 + x^2 - x + 2) \qquad \text{...(ii)}$$

Now, $$\int_{-1}^{1} f(x)\, dx = \frac{14}{3}$$

$$\Rightarrow \quad \frac{a}{3}\int_{-1}^{1}(x^3 + x^2 - x + 2)\, dx = \frac{14}{3}$$

$$\Rightarrow \quad \frac{a}{3}\int_{-1}^{1}(x^3 - x)\, dx + \frac{a}{3}\int_{-1}^{1}(x^2 + 2)\, dx = \frac{14}{3}$$

$$\Rightarrow \quad 0 + \frac{2a}{3}\int_{0}^{1}(x^2 + 2)\, dx = \frac{14}{3}$$

[$\because$ 1st integral is odd and 2nd is even]

$$\Rightarrow \quad \frac{2a}{3}\left[\frac{x^3}{3} + 2x\right]_0^1 = \frac{14}{3}$$

$$\Rightarrow \quad \frac{2a}{3}\left(\frac{7}{3}\right) = \frac{14}{3}$$

$$\Rightarrow \quad a = 3$$

From Eq. (ii),

$$f(x) = x^3 + x^2 - x + 2$$

Hence, $$f(2) = 2^3 + 2^2 - 2 + 2 = 8 + 4 = 12$$

38. (40) Given, perimeter of window = 20 m

$$\Rightarrow \quad 2x + 2r + \frac{1}{2}(2\pi r) = 20$$

$$\Rightarrow \quad 2x = 20 - \pi r - 2r \qquad \text{...(i)}$$

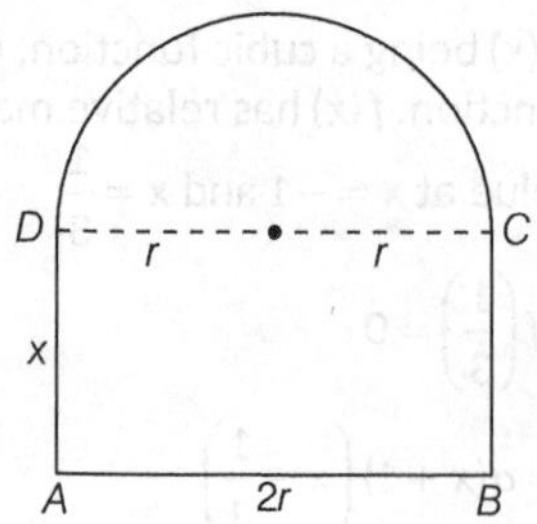

$\therefore \quad A = \dfrac{\pi r^2}{2} + 2rx$

$= \dfrac{1}{2}\pi r^2 + r(20 - \pi r - 2r)$ [from Eq. (i)]

$\Rightarrow \quad A = 20r - \dfrac{1}{2}\pi r^2 - 2r^2$

For maximum value of A,

$\dfrac{dA}{dr} = 20 - \pi r - 4r = 0$

$\therefore \quad r = \dfrac{20}{\pi + 4}$

Now, $\dfrac{d^2A}{dr^2} = -\pi - 4 < 0$

$\therefore$ A is maximum.

Then, length of rectangular part $= 2r = 2 \times \dfrac{20}{\pi + 4} = \dfrac{40}{\pi + 4}$

On comparing, we get

$k = 40$

39. (5) Let $\angle APB = 2\theta$ and $CP = \sqrt{36 + 64} = 10$

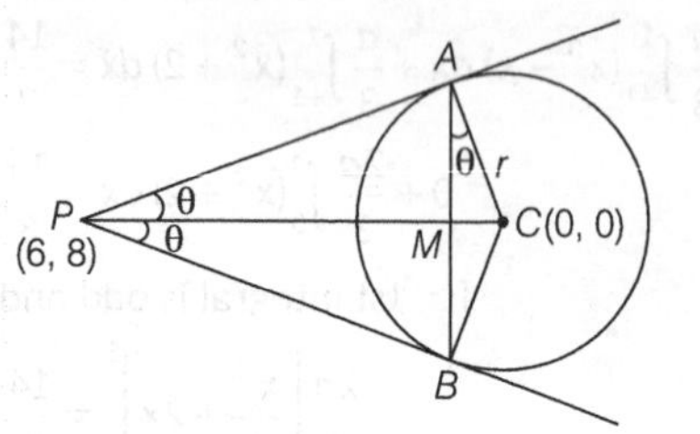

$\therefore \; r = 10 \sin\theta$ and $AP = 10 \cos\theta$

$PM = CP - CM = 10 - r \sin\theta$

$A =$ Area of $\Delta APB = 2\Delta APM$

$A = 2 \cdot \dfrac{1}{2} AM \cdot PM = r \cos\theta (10 - r \sin\theta)$

Now, $r = 10 \sin\theta$

$A = 100 \{\sin\theta \cos\theta (1 - \sin^2\theta)\}$

$= 100 \sin\theta \cdot \cos^3\theta$

Now, A will be maximum, when $z = \sin\theta \cos^3\theta$ is maximum.

$\dfrac{dz}{d\theta} = \cos^4\theta - 3\sin^2\theta \cos^2\theta = 0$

$\cos^2\theta (\cos^2\theta - 3\sin^2\theta) = 0$

$\therefore \quad \cos\theta = 0$ or $\tan^2\theta = \dfrac{1}{3}$

$\therefore \quad \theta = 90°$ or $\theta = 30°$

Clearly, $\theta = 30°$

$\dfrac{d^2z}{d\theta^2} = -2\cos\theta \sin\theta (\cos^2\theta - 3\sin^2\theta)$

$+ \cos^2\theta(-2\cos\theta \sin\theta - 6\sin\theta \cos\theta)$

$= 0 - 8\sin\theta \cos^3\theta < 0$ for $\theta = 30°$

$\therefore$ z is maximum and hence area A is maximum.

Hence, $r = 10 \sin\theta$

$= 10 \sin 30°$

$= 10 \times \dfrac{1}{2}$

$\Rightarrow \quad r = 5$

40. (6) Let quadrilateral is *PQRS*. Let coordinates of *P*, *Q*, *R* and *S* are $(p, 0)$, $(1, q)$, $(r, 1)$ and $(0, s)$ respectively.

$\therefore \quad a^2 = p^2 + s^2$

$\Rightarrow \quad b^2 = (1 - p)^2 + q^2$

$\Rightarrow \quad c^2 = (1 - r)^2 + (1 - q)^2$

$\Rightarrow \quad d^2 = r^2 + (1 - s)^2$

So, $a^2 + b^2 + c^2 + d^2$

$= p^2 + (1 - p)^2 + q^2 + (1 - q)^2 + r^2 + (1 - r)^2 + s^2 + (1 - s)^2$ ...(i)

Now, we know that

$0 \le p, q, r$ and $s \le 1$

Consider, the function

$f(x) = x^2 + (1 - x)^2, \quad 0 \le x \le 1$

$\Rightarrow \quad f'(x) = 2x - 2(1 - x)$

Now, $f'(x) = 0$

$\Rightarrow \quad x = \dfrac{1}{2}$

$\Rightarrow \quad f''(x) = 4 > 0$

$\therefore \quad f(x)$ is minimum, when $x = \dfrac{1}{2}$.

Minimum value of $f(x) = \dfrac{1}{2}$ and maximum value of $f(x) = 1$, at end points.

$\Rightarrow \quad \dfrac{1}{2} \le p^2 + (1 - p)^2 \le 1$

$\Rightarrow \quad \dfrac{1}{2} \le q^2 + (1 - q)^2 \le 1$

$\Rightarrow \quad \dfrac{1}{2} \le r^2 + (1 - r)^2 \le 1$

$\Rightarrow \quad \dfrac{1}{2} \le s^2 + (1 - s)^2 \le 1$

Adding all, we get

$2 \le a^2 + b^2 + c^2 + d^2 \le 4$ [from Eq. (i)]

$\therefore \quad a^2 + b^2 + c^2 + d^2 \in [2, 4]$

Here, $m = 2$ and $n = 4$

Hence, $m + n = 2 + 4 = 6$

41. (26) Let P be the pumping station at a distance x from A, the point nearest to the town T_1.

Also, $AB = 10$

$\therefore \quad PB = 10 - x$

Let L be the length of pipe required.

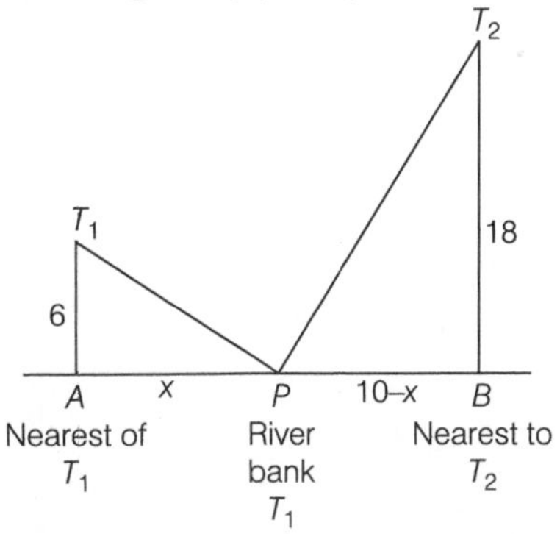

$$L = T_1P + T_2P = \sqrt{36 + x^2} + \sqrt{18^2 + (10 - x)^2}$$

$$\therefore \quad \frac{dL}{dx} = \frac{2x}{2\sqrt{36 + x^2}} - \frac{2(10 - x)}{2\sqrt{18^2 + (10 - x)^2}} = 0$$

$$\therefore \quad \frac{x}{\sqrt{36 + x^2}} = \frac{10 - x}{\sqrt{18^2 + (10 - x)^2}}$$

$$\Rightarrow \quad x^2[18^2 + (10 - x)^2] = (10 - x)^2 (36 + x^2)$$

$$\Rightarrow \quad 18^2 x^2 = (10 - x)^2 (36 + x^2 - x^2)$$

$$= 36(10 - x)^2$$

$$\Rightarrow \quad 18x = \pm 6(10 - x)$$

$$\Rightarrow \quad 18x = 60 - 6x \text{ or } 18x = -60 + 6x$$

$$\Rightarrow \quad 24x = 60 \Rightarrow x = \frac{5}{2}$$

$$\text{or} \quad 12x = -60 \Rightarrow x = -5$$

The value of x cannot be negative.

Hence, we shall consider $x = \frac{5}{2}$.

Now, $\frac{d^2L}{dx^2} = \frac{36}{(36 + x^2)^{3/2}} + \frac{18}{\{18^2 + (10 - x)^2\}^{3/2}}$,

which is '+ve' for $x = \frac{5}{2}$.

Hence, P is minimum

Putting $x = \frac{5}{2}$ in Eq. (i)

$$L = \sqrt{\left(36 + \frac{25}{4}\right)} + \sqrt{\left(324 + \frac{225}{4}\right)}$$

$$\Rightarrow \quad L = \frac{13}{2} + \frac{39}{2} = \frac{52}{2} = 26$$

42. (108) Let the sides of a rectangle be x and y, respectively.

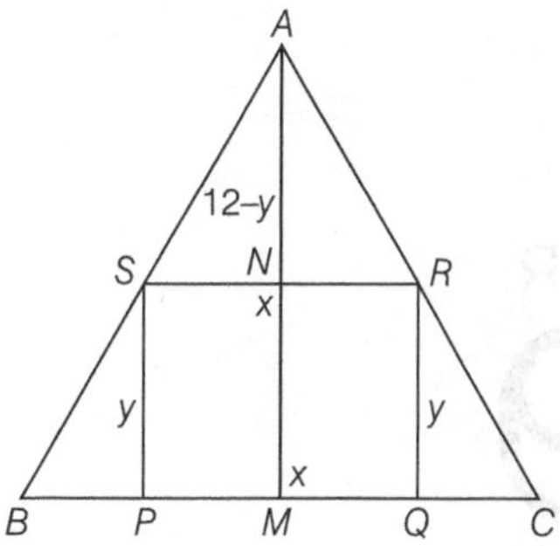

Here, $AM = 12$, $BC = 36$, $AN = 12 - y$

$PQ = x$ and $QR = y$

Also, $\frac{36}{x} = \frac{12}{12 - y}$

$$\Rightarrow \quad x = 36 - 3y$$

$$\Rightarrow \quad y = \frac{(36 - x)}{3}$$

Now, $A = xy = x\left(\frac{36 - x}{3}\right)$

$$\Rightarrow \quad A = \frac{1}{3}(36x - x^2)$$

$$\Rightarrow \quad \frac{dA}{dx} = \frac{1}{3}(36 - 2x) \Rightarrow \frac{d^2A}{dx^2} = -\frac{2}{3} < 0$$

So, area is maximum.

For maxima or minima, $\frac{dA}{dx} = 0$

$$\Rightarrow \quad \frac{1}{3}(36 - 2x) = 0 \Rightarrow x = 18$$

when, $x = 18$, then $y = 6$.

Maximum area $= A = 18.6 = 108$ sq ft.

43. (45) Speed of bus is x km/h.

$\therefore$ Time taken to cover 500 km, $t = \left(\frac{500}{x}\right)$ h

Total cost of trip = cost of diesel + payment to the driver

$$\Rightarrow \quad C = \frac{1}{300}\left(\frac{900}{x} + x\right) \times 500 \times \frac{40}{100} + \frac{500}{x} \times 1.5$$

$$\Rightarrow \quad C = \frac{1350}{x} + \frac{2x}{3} \Rightarrow \frac{dC}{dx} = -\frac{1350}{x^2} + \frac{2}{3}$$

For maxima or minima, $\frac{dC}{dx} = 0$

$$\therefore \quad x = 45$$

Now, $\frac{d^2C}{dx^2} = \frac{2700}{x^3}$

$$\Rightarrow \quad \left.\frac{d^2C}{dx^2}\right|_{x = 45} = \frac{2700}{(45)^3} > 0$$

Total cost C is minimum, when $x = 45$ km/h.

Hence, the most economical speed is 45 km/h.

CHAPTER

09

INDEFINITE INTEGRALS

LEVEL 01 \ Starter Level

9.1 Problems Based on Fundamentals

1 Evaluate following integrals.

(i) $\int\left(\sqrt{x}+\frac{1}{\sqrt{x}}\right)^2 dx$

(ii) $\int\frac{x^4+x^2+1}{x^2-x+1}dx$

(iii) $\int\frac{\cos x-\cos 2x}{1-\cos x}dx$

(iv) $\int\frac{1+3\sin x}{\cos^2 x}dx$

(v) $\int(\tan x-\cot x)^2\, dx$

(vi) $\int\tan^{-1}(\sec x+\tan x)\,dx, -\frac{\pi}{2}<x<\frac{\pi}{2}$

(vii) $\int\frac{(2^x+3^x)^2}{2^x\cdot 3^x}dx$

(viii) $\int\frac{\sin^8 x-\cos^8 x}{1-2\sin^2 x\cos^2 x}dx$

(ix) $\int\frac{1}{\sqrt{1-2x}+\sqrt{3-2x}}dx$

(x) $\int\frac{x^2+1}{(x+1)^2}dx$

(xi) $\int\frac{x}{\sqrt{x+2}}dx$

(xii) $\int\cos^4 x\,dx$

(xiii) $\int\frac{\sin x}{\sin(x-a)}dx$

(xiv) $\int\frac{1}{\cos(x-a)\cos(x-b)}dx$

9.2 Integration by Substitutions

2 Evaluate following integrals.

(i) $\int \frac{e^{2x}-1}{e^{2x}+1}\,dx$ (ii) $\int \frac{(1+\log x)^2}{x}\,dx$

(iii) $\int \frac{x^2 \tan^{-1} x^3}{1+x^6}\,dx$ (iv) $\int \frac{1}{\sqrt{\sin^3 x \sin(x+\alpha)}}\,dx,\ \alpha \neq n\pi, n \in Z$

(v) $\int \frac{\sin 2x}{a^2 \sin^2 x + b^2 \cos^2 x}\,dx$ (vi) $\int 2^{2^{2^x}} \cdot 2^{2^x} \cdot 2^x\,dx$

(vii) $\int \frac{1}{\sin x \cos^3 x}\,dx$

9.3 Problems Based on Standard Formulae and Partial Fractions

3 Evaluate following integrals.

(i) $\int \frac{1}{25+16x^2}\,dx$ (ii) $\int \frac{1}{16-9x^2}\,dx$

(iii) $\int \frac{1}{\sqrt{16x^2+25}}\,dx$ (iv) $\int \frac{x^4}{x^2+1}\,dx$

(v) $\int \frac{1}{3+2x-x^2}\,dx$ (vi) $\int \frac{1}{x(x^n+1)}\,dx$

(vii) $\int \frac{e^x}{\sqrt{5-4e^x-e^{2x}}}\,dx$ (viii) $\int \frac{2\sin 2\theta - \cos\theta}{6-\cos^2\theta - 4\sin\theta}\,d\theta$

4 Evaluate following integrals.

(i) $\int \frac{1}{a^2 \sin^2 x + b^2 \cos^2 x}\,dx$ (ii) $\int \frac{1}{1+3\sin^2 x}\,dx$

(iii) $\int \frac{1}{\cos 2x + 3\sin^2 x}\,dx$ (iv) $\int \frac{1}{1-2\sin x}\,dx$

(v) $\int \frac{1+\sin x}{\sin x(1+\cos x)}\,dx$ (vi) $\int \frac{1}{\sqrt{3}\sin x + \cos x}\,dx$

(vii) $\int \frac{3\sin x + 2\cos x}{3\cos x + 2\sin x}\,dx$ (viii) $\int \frac{1}{1+\cot x}\,dx$

5 Evaluate following integrals.

(i) $\int \frac{x-1}{(x+1)(x-2)}\,dx$

(ii) $\int \frac{2x}{(x^2+1)(x^2+2)}\,dx$

(iii) $\int \frac{x^2+1}{(x-1)^2(x+3)}\,dx$

(iv) $\int \frac{x}{(x-1)(x^2+4)}\,dx$

(v) $\int \frac{x^2}{(x^2+1)(x^2+4)}\,dx$

9.4 Integration by Parts

6 Evaluate following integrals.

(i) $\int x \log x \, dx$ (ii) $\int \log(1+x^2) dx$

(iii) $\int e^{-x} \cos x \, dx$ (iv) $\int \tan^{-1} x \, dx$

(v) $\int x \sin^{-1} x \, dx$

7 Evaluate following integrals.

(i) $\int e^x \left(\frac{1}{x} - \frac{1}{x^2}\right) dx$ (ii) $\int e^x (\sin x + \cos x) \, dx$

(iii) $\int e^x \left(\frac{x-3}{(x-1)^3}\right) dx$

8 Evaluate following integrals.

(i) $\int \sqrt{4x^2 + 9} \, dx$ (ii) $\int \sqrt{x^2 + 2x + 5} \, dx$

(iii) $\int (x+1) \sqrt{1 - x - x^2} \, dx$

LEVEL 02 \ JEE Main Level

Single Choice Correct Questions

9.1 Problems Based on Basic Fundamentals

1 $\int \frac{dx}{\sqrt{ax+b} - \sqrt{ax+c}} = \frac{2}{3k}[(ax+b)^{3/2} + (ax+c)^{3/2}] + C$, where k is equal to

(a) $a(b-c)$ (b) $a(c-b)$ (c) $b(c-a)$ (d) $b(a-c)$

2 $\int \frac{\sqrt{x-1}}{x\sqrt{x+1}} dx$ is equal to

(a) $\ln|x - \sqrt{x^2-1}| - \tan^{-1} x + C$

(b) $\ln|x + \sqrt{x^2-1}| - \tan^{-1} x + C$

(c) $\ln|x - \sqrt{x^2-1}| - \sec^{-1} x + C$

(d) $\ln|x + \sqrt{x^2-1}| - \sec^{-1} x + C$

3 $\int \frac{\sin x}{\sin 3x \sin 2x} dx$ is equal to

(a) $\log|\sin 2x| - \log|\sin 3x| + C$

(b) $\frac{1}{2}\log|\sin 2x| + \frac{1}{3}\log|\sin 3x| + C$

(c) $\frac{1}{2}\log|\sin 2x| - \frac{1}{3}\log|\sin 3x| + C$

(d) $2\log|\sin 2x| - 3\log|\sin 3x| + C$

4 If $\int (\sin 2x - \cos 2x) \, dx = \frac{1}{\sqrt{2}} \sin(2x - a) + C$, then a is equal to

(a) $\frac{7\pi}{4}$ (b) $-\frac{5\pi}{4}$ (c) $\frac{5\pi}{4}$ (d) $-\frac{3\pi}{4}$

5 $\int \frac{\sin x}{\sin\left(x-\frac{\pi}{4}\right)} dx$ is equal to

(a) $\frac{1}{\sqrt{2}}\left\{x+\log\left|\cos\left(x-\frac{\pi}{4}\right)\right|\right\}+C$

(b) $\frac{1}{\sqrt{2}}\left\{x-\log\left|\sin\left(x-\frac{\pi}{4}\right)\right|\right\}+C$

(c) $\sqrt{2}\left\{x+\log\left|\sin\left(x-\frac{\pi}{4}\right)\right|\right\}+C$

(d) $\frac{1}{\sqrt{2}}\left\{x+\log\left|\sin\left(x-\frac{\pi}{4}\right)\right|\right\}+C$

6 $\int \frac{\sin\frac{5x}{2}}{\sin\frac{x}{2}} dx$ is equal to (where, C is a constant of integration) [JEE Main 2019]

(a) $2x+\sin x+2\sin 2x+C$

(b) $x+2\sin x+2\sin 2x+C$

(c) $x+2\sin x+\sin 2x+C$

(d) $2x+\sin x+\sin 2x+C$

7 $\int\left\{\left(\frac{\cot^2 2x-1}{2\cot 2x}\right)-\cos 8x\cot 4x\right\} dx$ is equal to

(a) $\frac{\cos 8x}{8}+C$

(b) $\frac{\sin 8x}{8}+C$

(c) $-\frac{\cos 8x}{8}+C$

(d) $-\frac{\sin 8x}{8}+C$

9.2 Integration by Substitution

8 $\int 5^{5^{5^x}}\cdot 5^{5^x}\cdot 5^x\, dx$ is equal to

(a) $\frac{5^{5^{5^x}}}{(\log 5)^2}+C$

(b) $\frac{5^{5^{5^x}}}{(\log 5)^3}+C$

(c) $5^{5^5}(\log 5)^3+C$

(d) None of these

9 The integral $\int \frac{e^{3\log_e 2x}+5e^{2\log_e 2x}}{e^{4\log_e x}+5e^{3\log_e x}-7e^{2\log_e x}} dx$, $x>0$ is equal to

(a) $\log|x^2+5x-7|+C$

(b) $\frac{1}{4}\log|x^2+5x-7|+C$

(c) $4\log|x^2+5x-7|+C$

(d) $\log\sqrt{x^2+5x-7}+C$

10 The integral $\int \frac{2x^3-1}{x^4+x} dx$ is equal to (here, C is a constant of integration) [JEE Main 2019]

(a) $\frac{1}{2}\log_e\frac{|x^3+1|}{x^2}+C$

(b) $\frac{1}{2}\log_e\frac{(x^3+1)^2}{|x^3|}+C$

(c) $\log_e\left|\frac{x^3+1}{x}\right|+C$

(d) $\log_e\frac{|x^3+1|}{x^2}+C$

11 $\int (x^4+1)^{1/3}\cdot x^7\, dx = k(x^4+1)^{4/3}(4x^4-3)+C$, where k is equal to

(a) $\frac{1}{112}$

(b) $\frac{1}{28}$

(c) $\frac{3}{112}$

(d) $\frac{1}{56}$

12 The value of $\int \frac{1}{x^n(1+x^n)^{1/n}}\,dx\ (n \in N)$ is

(a) $\frac{1}{1-n}\left\{1+\frac{1}{x^n}\right\}^{1-\frac{1}{n}}+C$ (b) $\frac{1}{1+n}\left\{1-\frac{1}{x^n}\right\}^{1-\frac{1}{n}}+C$

(c) $-\frac{1}{1-n}\left\{1-\frac{1}{x^n}\right\}^{1-\frac{1}{n}}+C$ (d) $-\frac{1}{1+n}\left\{1+\frac{1}{x^n}\right\}^{1-\frac{1}{n}}+C$

13 $\int \sqrt{e^x-1}\,dx$ is equal to

(a) $2[\sqrt{e^x-1}-\tan^{-1}\sqrt{e^x-1}]+C$ (b) $\sqrt{e^x-1}-\tan^{-1}\sqrt{e^x-1}+C$

(c) $\sqrt{e^x-1}+\tan^{-1}\sqrt{e^x-1}+C$ (d) $2[\sqrt{e^x-1}+\tan^{-1}\sqrt{e^x-1}]+C$

14 The value of $\int \frac{1}{x^2(x^4+1)^{3/4}}\,dx$ is

(a) $\left(1+\frac{1}{x^4}\right)^{1/4}+C$ (b) $-\left(1+\frac{1}{x^4}\right)^{1/4}+C$ (c) $-\frac{1}{4}\left(1+\frac{1}{x^4}\right)^{1/4}+C$ (d) None of these

15 $\int \frac{x^4-1}{x^2\sqrt{x^4+x^2+1}}\,dx$ is equal to

(a) $\sqrt{x+\frac{1}{x}+1}+C$ (b) $\frac{\sqrt{x^4+x^2+1}}{x^2}+C$

(c) $\frac{\sqrt{x^4+x^2+1}}{x}+C$ (d) None of these

16 $\int x^{-6}(1+2x^3)^{2/3}\,dx$ is equal to

(a) $-\frac{1}{5}(2+x^{-3})^{5/3}+C$ (b) $\frac{1}{5}(x^{-3}+2)^{5/3}+C$

(c) $\frac{3}{5}(2+x^{-3})^{5/3}+C$ (d) $\frac{-3}{5}(x^{-3}+2)^{-5/3}+C$

17 $\int \frac{x}{\sqrt{1+x^2+\sqrt{(1+x^2)^3}}}\,dx$ is equal to

(a) $\frac{1}{2}\ln(1+\sqrt{1+x^2})+C$ (b) $\frac{-2}{3(1+\sqrt{1+x^2})^{3/2}}+C$

(c) $2(1+\sqrt{1+x^2})+C$ (d) $2\sqrt{1+\sqrt{1+x^2}}+C$

18 $\int (x^{3m}+x^{2m}+x^m)(2x^{2m}+3x^m+6)^{1/m}\,dx,\ x>0$ is equal to

(a) $\frac{1}{6(m+1)}(2x^{3m}+3x^{2m}+6x^m)^{\frac{1+m}{m}}+C$ (b) $\frac{1}{m+1}(2x^{2m}+3x^m+6)^{\frac{1+m}{m}}+C$

(c) $\frac{1}{m+1}(x^{3m}+x^{2m}+x^m)^{\frac{1+m}{m}}+C$ (d) $\frac{1}{6m+1}(x^{3m}+x^{2m}+m^m)^{\frac{1+m}{m}}+C$

19 If $\int \frac{\cos x\,dx}{\sin^3 x(1+\sin^6 x)^{2/3}}=f(x)(1+\sin^6 x)^{1/\lambda}+C$ (where, C is a constant of integration), then $\lambda f\left(\frac{\pi}{3}\right)$ is equal to **[JEE Main 2020]**

(a) 2 (b) $-\frac{9}{8}$ (c) -2 (d) $\frac{9}{8}$

20 The value of $\int \frac{1+\log x}{\sqrt{(x^x)^2-1}}dx$ is

(a) $\sec^{-1}(x^x)+C$ (b) $\tan^{-1}(x^x)+C$

(c) $\log(x^x+\sqrt{x^{2x}-1})+C$ (d) $\cot^{-1}(x^x)+C$

21 $\int \frac{\ln(\tan x)}{\sin x \cos x}dx$ is equal to

(a) $\frac{1}{2}\ln|\tan x|+C$ (b) $\frac{1}{2}\ln(\tan^2 x)+C$ (c) $\frac{1}{2}\{\ln(\tan x)\}^2+C$ (d) None of these

22 Let $I_n=\int \tan^n x\,dx (n>1)$. $I_4+I_6=a\tan^5 x+bx^5+C$, where C is a constant of integration, then the ordered pair (a, b) is equal to **[JEE Main 2017]**

(a) $\left(-\frac{1}{5},0\right)$ (b) $\left(-\frac{1}{5},1\right)$ (c) $\left(\frac{1}{5},0\right)$ (d) $\left(\frac{1}{5},-1\right)$

23 If $\int \cos^7 x\,dx = A\sin^7 x+B\sin^5 x+C\sin^3 x+\sin x+K$, then $7A+20B+2C$ is equal to

(a) 9 (b) 10 (c) 11 (d) 12

24 $\int \frac{\sin x}{\sin 3x}dx$ is equal to

(a) $\frac{1}{2\sqrt{3}}\log\left|\frac{\sqrt{3}+\tan x}{\sqrt{3}-\tan x}\right|+C$ (b) $\frac{1}{\sqrt{3}}\log\left|\frac{\sqrt{3}+\tan x}{\sqrt{3}-\tan x}\right|+C$

(c) $\frac{1}{2\sqrt{3}}\log\left|\frac{\sqrt{3}-\tan x}{\sqrt{3}+\tan x}\right|+C$ (d) None of these

25 The integral $\int \frac{\sin^2 x\cos^2 x}{(\sin^5 x+\cos^3 x\sin^2 x+\sin^3 x\cos^2 x+\cos^5 x)^2}dx$ is equal to **[JEE Main 2018]**

(a) $\frac{-1}{1+\cot^3 x}+C$ (b) $\frac{1}{3(1+\tan^3 x)}+C$

(c) $\frac{-1}{3(1+\tan^3 x)}+C$ (d) $\frac{1}{1+\cot^3 x}+C$

26 $\int \frac{\sec x}{\sqrt{\sin(2x+\alpha)+\sin\alpha}}dx$ is equal to

(a) $\sqrt{2\sec\alpha\,(\tan x+\tan\alpha)}+C$ (b) $\sqrt{2\sec\alpha\,(\tan x-\tan\alpha)}+C$

(c) $\sqrt{2\sec\alpha\,(\tan\alpha-\tan x)}+C$ (d) None of these

27 $\int \frac{\cos^{n-1}x}{\sin^{n+1}x}dx$, $n\neq 0$ is equal to

(a) $\frac{\cot^n x}{n}+C$ (b) $\frac{-\cot^{n-1}x}{n-1}+C$ (c) $\frac{-\cot^n x}{n}+C$ (d) $\frac{\cot^{n-1}x}{n-1}+C$

28 If $\int \frac{d\theta}{\cos^2\theta(\tan 2\theta+\sec 2\theta)}=\lambda\tan\theta+2\log_e|f(\theta)|+C$, where C is a constant of integration, then the ordered pair $\{\lambda, f(\theta)\}$ is equal to **[JEE Main 2020]**

(a) $(1, 1+\tan\theta)$ (b) $(1, 1-\tan\theta)$ (c) $(-1, 1+\tan\theta)$ (d) $(-1, 1-\tan\theta)$

29 If $\int \frac{\sin^4 x}{\cos^8 x}dx = a\tan^5 x+b\tan^7 x+C$, then

(a) $7a=5b$ (b) $5a=7b$ (c) $7a+5b=0$ (d) $5a+7b=0$

30 If $\int \frac{dx}{(x^2-2x+10)^2} = A\left[\tan^{-1}\left(\frac{x-1}{3}\right) + \frac{f(x)}{x^2-2x+10}\right] + C$, where C is a constant of integration, then

(a) $A = \frac{1}{27}$ and $f(x) = 9(x-1)$ (b) $A = \frac{1}{81}$ and $f(x) = 3(x-1)$

(c) $A = \frac{1}{54}$ and $f(x) = 3(x-1)$ (d) $A = \frac{1}{54}$ and $f(x) = 9(x-1)^2$

31 $\int \frac{3-2\cot^{n-2} x}{3\tan x + 2\cot x \cot^{n-2} x} dx$ is equal to

(a) $n \ln|3\sin^n x - 2\cos^n x| + C$ (b) $\frac{1}{n}\ln|3\cos^n x - 2\sin^n x| + C$

(c) $\frac{1}{n}\ln|3\sin^n x + 2\cos^n x| + C$ (d) None of these

32 $I = \int \tan\left[2\tan^{-1}\left(\sqrt{\frac{\sqrt{1+\sqrt{x}}-1}{\sqrt{1+\sqrt{x}}+1}}\right)\right] dx$ and $x \in (0, 1)$ is equal to

(a) $\frac{4}{5}x^{5/4} + C$ (b) $\frac{5}{4}x^{4/5} + C$ (c) $\frac{1}{2}\sqrt{x} + C$ (d) None of these

33 If $\int \frac{\cos x - \sin x}{\sqrt{8-\sin 2x}} dx = a\sin^{-1}\left(\frac{\sin x + \cos x}{b}\right) + C$, where C is a constant of integration, then the ordered pair (a, b) is

(a) (1, – 3) (b) (1, 3) (c) (– 1, 3) (d) (3, 1)

34 The integral of $\int \cos\left(2\cot^{-1}\sqrt{\frac{a-x}{a+x}}\right) dx$ is

(a) $\frac{2a^2-x^2}{4a} + C$ (b) $\frac{2a^2-x^2}{2a} + C$ (c) $\frac{a^2-2x^2}{4a} + C$ (d) $\frac{a^2-x^2}{4a} + C$

9.3 Problems Based on Standard Formulae and Partial Fraction

35 $\int \frac{1}{(x+4)^{8/7}(x-3)^{6/7}} dx$ is equal to **[JEE Main 2020]**

(a) $\left(\frac{x+4}{x-3}\right)^{\frac{1}{7}} + C$ (b) $\left(\frac{x-3}{x+4}\right)^{\frac{1}{7}} + C$ (c) $\frac{1}{7}\left(\frac{x+4}{x-3}\right)^{\frac{1}{7}} + C$ (d) $\frac{1}{7}\left(\frac{x-3}{x+4}\right)^{\frac{1}{7}} + C$

36 $\int \frac{dx}{\sqrt{(x-p)^3(x-q)}}$ is equal to

(a) $\frac{2}{(p-q)}\sqrt{\left(\frac{x-p}{x-q}\right)} + C$ (b) $-\frac{2}{(p-q)}\sqrt{\left(\frac{x-q}{x-p}\right)} + C$

(c) $\frac{2}{\sqrt{(x-p)(x-q)}} + C$ (d) None of these

37 If $\frac{d}{dx}f(x) = f'(x)$, then $\int \frac{xf'(x) - 2f(x)}{\sqrt{x^4 \cdot f(x)}} dx$ is equal to

(a) $\frac{x^2}{f(x)} + C$ (b) $|x| f'(x) + C$ (c) $\frac{2\sqrt{f(x)}}{|x|} + C$ (d) $|x|\sqrt{f(x)} + C$

38 $4\int \frac{\sqrt{a^6+x^8}}{x}\,dx$ is equal to

(a) $\sqrt{a^6+x^8}+\frac{a^3}{2}\ln\left|\frac{\sqrt{a^6+x^8}+a^3}{\sqrt{a^6+x^8}-a^3}\right|+C$

(b) $a^6\ln\left|\frac{\sqrt{a^6+x^8}-a^3}{\sqrt{a^6+x^8}+a^3}\right|+C$

(c) $\sqrt{a^6+x^8}+\frac{a^3}{2}\ln\left|\frac{\sqrt{a^6+x^8}-a^3}{\sqrt{a^6+x^8}+a^3}\right|+C$

(d) $a^6\ln\left|\frac{\sqrt{a^6+x^8}+a^3}{\sqrt{a^6+x^8}-a^3}\right|+C$

39 $\int \frac{x^2+x+1}{\sqrt{x^2+2x+3}}\,dx$ is equal to

(a) $(x-1)\sqrt{x^2+2x+3}+C$

(b) $\frac{1}{2}(x+1)\sqrt{x^2+2x+3}+C$

(c) $\frac{1}{2}(x-1)\sqrt{x^2+2x+3}+C$

(d) $2(x+1)\sqrt{x^2+2x+3}+C$

40 $\int \frac{x^4+1}{x^6+1}\,dx$ is equal to

(a) $\tan^{-1}x-\frac{1}{3}\tan^{-1}(x^3)+C$

(b) $\tan^{-1}x+\frac{1}{3}\tan^{-1}(x^3)+C$

(c) $\tan^{-1}x+\frac{1}{2}\tan^{-1}(x^2)+C$

(d) $\tan^{-1}x-\frac{1}{2}\tan^{-1}(x^2)+C$

41 The value of $\int \frac{1}{x+\sqrt{x-1}}\,dx$ is

(a) $\log(x+\sqrt{x-1})+\sin^{-1}\sqrt{\frac{x-1}{x}}+C$

(b) $\log(x+\sqrt{x-1})+C$

(c) $\log(x+\sqrt{x-1})-\frac{2}{3}\tan^{-1}\left(\frac{2\sqrt{x-1}+1}{\sqrt{3}}\right)+C$

(d) None of the above

42 The value of $\int \frac{1-\cos x}{\cos x(1+\cos x)}\,dx$ is

(a) $\ln|\sin x|-2\tan\frac{x}{2}+C$

(b) $\ln|\sec x+\tan x|-\tan\frac{x}{2}+C$

(c) $\ln|\sec x+\tan x|-2\tan\frac{x}{2}+C$

(d) $\ln|\sec x+\tan x|+C$

43 If $\int \frac{\cos\theta}{5+7\sin\theta-2\cos^2\theta}\,d\theta = A\log_e|B(\theta)|+C$, where C is a constant of integration, then $\frac{B(\theta)}{A}$ can be **[JEE Main 2020]**

(a) $\frac{2\sin\theta+1}{\sin\theta+3}$

(b) $\frac{2\sin\theta+1}{5(\sin\theta+3)}$

(c) $\frac{5(\sin\theta+3)}{2\sin\theta+1}$

(d) $\frac{5(2\sin\theta+1)}{\sin\theta+3}$

44 The value of $\int \frac{dx}{x^{13}(x^6-1)}$ is

(a) $\log\left|\frac{x^6-1}{x^6}\right|+x^{-6}+x^{-12}+C$

(b) $\frac{1}{6}\log\left|\frac{x^6-1}{x^6}\right|+\frac{1}{6}x^{-6}+\frac{1}{4}x^{-12}+C$

(c) $\frac{1}{6}\log\left|\frac{x^6-1}{x^6}\right|+x^{-6}+\frac{1}{2}x^{-12}+C$

(d) None of these

45 $\int \frac{d\theta}{(\sin\theta - 2\cos\theta)(2\sin\theta + \cos\theta)}$ is equal to

(a) $\frac{1}{5}\log\left|\frac{\tan\theta - 2}{2\tan\theta + 1}\right| + C$
(b) $\frac{1}{5}\log\left|\frac{2\tan\theta + 1}{\tan\theta - 2}\right| + C$
(c) $5\log\left|\frac{\tan\theta - 2}{2\tan\theta + 1}\right| + C$
(d) $\log\left|\frac{\tan\theta - 2}{2\tan\theta + 1}\right| + C$

46 $\int \frac{dx}{\sin x + \sin 2x} = A\log(1-\cos x) + B\log(1+\cos x) + C\log(1+2\cos x) + K$, then $6(A+B+C)$ is

(a) 0 (b) 1 (c) 5 (d) 2

47 If $\int \frac{1}{(e^x - 1)^2}\, dx = f(x) - \log|g(x)| + C$, then

(a) $f(x) = (1 - e^x)$ and $g(x) = 1 - e^{-x}$
(b) $f(x) = (1 - e^x)^{-1}$ and $g(x) = 1 - e^{-x}$
(c) $f(x) = g(x) = (1 - e^x)^{-1}$
(d) $f(x) = (1 - e^x)^{-1}$ and $g(x) = (1 - e^{-x})^{-1}$

48 $f(x) = \frac{x+2}{2x+3}$, then $\int \left(\frac{f(x)}{x^2}\right)^{1/2} dx$ is equal to $\frac{1}{\sqrt{2}}\, g\left(\frac{1+\sqrt{2f(x)}}{1-\sqrt{2f(x)}}\right) - \sqrt{\frac{2}{3}}\, h\left(\frac{\sqrt{3f(x)}+\sqrt{2}}{\sqrt{3f(x)}-\sqrt{2}}\right) + C$, where

(a) $g(x) = \tan^{-1} x$ and $h(x) = \log|x|$
(b) $g(x) = \log|x|$ and $h(x) = \tan^{-1} x$
(c) $g(x) = h(x) = \tan^{-1} x$
(d) $g(x) = h(x) = \log|x|$

9.4 Integration by Parts

49 $\int \left(\frac{x - \sin x}{1 - \cos x}\right) dx$ is equal to

(a) $x\cot\frac{x}{2} + C$ (b) $-x\cot\frac{x}{2} + C$ (c) $\frac{x}{2}\cot x + C$ (d) None of these

50 If $\int x \cdot \frac{\ln(x + \sqrt{1+x^2})}{\sqrt{1+x^2}}\, dx = a\sqrt{1+x^2}\ln(x + \sqrt{1+x^2}) + bx + C$, then the value of a is

(a) 0 (b) 1 (c) 2 (d) 3

51 $\int \frac{\sin^{-1}\sqrt{x} - \cos^{-1}\sqrt{x}}{\sin^{-1}\sqrt{x} + \cos^{-1}\sqrt{x}}\, dx$ is equal to

(a) $\frac{4}{\pi}\left[\frac{\sqrt{x-x^2}}{2} + x\sin^{-1}\sqrt{x} + \frac{1}{2}\sin^{-1}\sqrt{1-x}\right] - x + C$
(b) $\frac{2}{\pi}[\sqrt{x-x^2} + x\sin^{-1} x - \sin^{-1}\sqrt{1-x}] - x + C$
(c) $\frac{2}{\pi}\left[\sqrt{x-x^2} + x\sin^{-1} x - \frac{1}{2}\sin^{-1}\sqrt{1-x}\right] - x + C$
(d) $\frac{2}{\pi}\left[\sqrt{x-x^2} + x\sin^{-1} x + \frac{1}{2}\sin^{-1}\sqrt{1-x}\right] + C$

52 If $\int \sin^{-1}\left(\sqrt{\frac{x}{1+x}}\right) dx = A(x)\tan^{-1}(\sqrt{x}) + B(x) + C$, where C is constant of integration, then the ordered pair $\{A(x), B(x)\}$ can be **[JEE Main 2020]**

(a) $(x-1, \sqrt{x})$
(b) $(x-1, -\sqrt{x})$
(c) $(x+1, \sqrt{x})$
(d) $(x+1, -\sqrt{x})$

53 If $\int\left\{\log(\log x)+\frac{1}{(\log x)^2}\right\}dx = x\{f(x)-g(x)\}+C$, then

(a) $f(x)=\log(\log x)$ and $g(x)=\frac{1}{\log x}$ (b) $f(x)=\log x$ and $g(x)=\frac{1}{\log(x)}$

(c) $f(x)=\frac{1}{\log x}$ and $g(x)=\log(\log x)$ (d) $f(x)=\frac{1}{x\log x}$ and $g(x)=\frac{1}{\log x}$

54 $\int \sin(\log x)dx$ is equal to

(a) $\frac{x}{2}[\sin(\log x)+\cos(\log x)]+C$ (b) $\frac{x}{2}[\cos(\ln x)-\sin(\ln x)]+C$

(c) $\frac{x}{2}[\sin(\ln x)-\cos(\ln x)]+C$ (d) $x[\sin(\ln x)-\cos(\ln x)]+C$

55 $\int \sqrt{x}\, e^{\sqrt{x}}\, dx$ is equal to

(a) $2e^{\sqrt{x}}(x-2\sqrt{x}+2)+C$ (b) $2e^{\sqrt{x}}(x+2\sqrt{x}+2)+C$

(c) $2e^{\sqrt{x}}(x+2\sqrt{x}-2)+C$ (d) $2e^{\sqrt{x}}(x-2\sqrt{x}-2)+C$

56 $\int 4^x\{g'(x)+g(x)\log 4\}\,dx$ is equal to

(a) $\frac{4^x}{\log 4}g(x)$ (b) 4^x (c) $4x\log 4\cdot g(x)$ (d) $4^x g(x)$

57 If $\int (e^{2x}+2e^x-e^{-x}-1)e^{(e^x+e^{-x})}dx = g(x)e^{(e^x+e^{-x})}+C$, where C is a constant of integration, then $g(0)$ is equal to **[JEE Main 2020]**

(a) 2 (b) e^2 (c) e (d) 1

58 $\int\left(1+2x^2+\frac{1}{x}\right)e^{\left(x^2-\frac{1}{x}\right)}dx$ is equal to

(a) $-xe^{x^2-\frac{1}{x}}+C$ (b) $xe^{x^2-\frac{1}{x}}+C$ (c) $(2x-1)e^{x^2-\frac{1}{x}}+C$ (d) $(2x+1)e^{x^2-\frac{1}{x}}+C$

59 If $\int e^{\sec x}[(\sec x\tan x)f(x)+(\sec x\tan x+\sec^2 x)]dx = e^{\sec x}f(x)+C$, then a possible choice of $f(x)$ is **[JEE Main 2019]**

(a) $x\sec x+\tan x+\frac{1}{2}$ (b) $\sec x+\tan x+\frac{1}{2}$ (c) $\sec x+x\tan x-\frac{1}{2}$ (d) $\sec x-\tan x-\frac{1}{2}$

Numerical Value Type Questions

60 If $\int \frac{5\tan x}{\tan x-2}dx = x+a\ln(\sin x-2\cos x)+C$, then a is equal to ....

61 If $\int \frac{\cos x+\cos 2x}{2\cos x-1}dx = A\sin x+Bx+C$, then the value of $A+B+3$ is ....

62 $\int \frac{2\cos x-\sin x+\lambda}{\cos x+\sin x-2}dx = A\ln|\cos x+\sin x-2|+Bx+C$, then the value of $A+B+|\lambda|$ is ....

63 If $\int \frac{dx}{\sin^6 x+\cos^6 x} = \tan^{-1}(L\tan x+M\cot x)+C$, then the value of $L+M+4$ is ....

64 If $\int \frac{1-(\cot x)^{2010}}{\tan x+(\cot x)^{2011}}dx = \frac{1}{k}\log|(\sin x)^k+(\cos x)^k|+C$, then the value of k is ....

65 If $\int\left(\frac{3x^2+2x}{x^6+2x^5+x^4+2x^3+2x^2+2}\right)dx = \frac{1}{L}\tan^{-1}\left(\frac{x^M+x^N+P}{Q}\right)+k$, then the value of $L+M+N+P+Q$ is ....

66 If $\int \frac{x+\sqrt[3]{x^2}+\sqrt[6]{x}}{x(1+\sqrt[3]{x})}dx = ax^{2/3}+b\tan^{-1}(\sqrt[6]{x})+C$, then ab is equal to ....

67 If $\int \frac{2x+5}{\sqrt{9x^2+6x+2}}dx = a\sqrt{9x^2+6x+2}+b\log(3x+1+\sqrt{9x^2+6x+2})+C$, then $9(a+b)$ is equal to ....

68 If $f(x)=\int \frac{5x^8+7x^6}{(x^2+1+2x^7)^2}dx, x \ge 0$, if $f(0)=0$ and $f(1)=\frac{1}{k}$, then the value of k is ....

69 $\int e^{x^3+x^2-1}(3x^4+2x^3+2x)\,dx = f(x)+C$, then the value of $f(1)\times f(-1)$ is ....

70 If $\int \frac{e^x(2-x^2)}{(1-x)\sqrt{1-x^2}}dx = Ae^x\left(\frac{1+x}{1-x}\right)^B + C$, then value of $2(A+B)$ is ....

LEVEL 03 \ JEE Advanced Level

Single Choice Correct Questions

1 The value of $\int \{1+\tan x\cdot\tan(x+A)\}dx$ is

(a) $\cot A\cdot\log\left|\frac{\sec x}{\sec(x+A)}\right|+C$
(b) $\tan A\cdot\log|\sec(x+A)|+C$
(c) $\cot A\cdot\log\left|\frac{\sec(x+A)}{\sec x}\right|+C$
(d) None of these

2 The integral $\int \frac{\sec^2 x}{(\sec x+\tan x)^{9/2}}dx$ is equal to

(a) $-\frac{1}{(\sec x+\tan x)^{11/2}}\left\{\frac{1}{11}-\frac{1}{7}(\sec x+\tan x)^2\right\}+C$

(b) $\frac{1}{(\sec x+\tan x)^{11/2}}\left\{\frac{1}{11}-\frac{1}{7}(\sec x+\tan x)^2\right\}+C$

(c) $-\frac{1}{(\sec x+\tan x)^{11/2}}\left\{\frac{1}{11}+\frac{1}{7}(\sec x+\tan x)^2\right\}+C$

(d) $\frac{1}{(\sec x+\tan x)^{11/2}}\left\{\frac{1}{11}+\frac{1}{7}(\sec x+\tan x)^2\right\}+C$

3 If $\int\left(\frac{1-\sqrt{x}}{1+\sqrt{x}}\right)^{1/2}\frac{dx}{x}=2\cos^{-1}\sqrt{x}-f(x)+C$, then $f(x)$ is equal to

(a) $\log\left(\frac{1+\sqrt{1-x}}{\sqrt{x}}\right)$
(b) $\frac{1}{2}\log\left(\frac{1+\sqrt{1-x}}{\sqrt{x}}\right)$
(c) $2\log\left(\frac{1-\sqrt{1-x}}{\sqrt{x}}\right)$
(d) $2\log\left(\frac{1+\sqrt{1-x}}{\sqrt{x}}\right)$

4 $\int \frac{\sin x+\cos x}{9+16\sin 2x}dx$ is equal to

(a) $\frac{1}{40}\ln\left|\frac{5+4(\sin x-\cos x)}{5-4(\sin x-\cos x)}\right|+C$
(b) $\frac{1}{40}\ln\left|\frac{5-4(\sin x-\cos x)}{5+4(\sin x-\cos x)}\right|+C$
(c) $\frac{1}{40}\ln\left|\frac{5+4(\sin x+\cos x)}{5-4(\sin x+\cos x)}\right|+C$
(d) $\frac{1}{40}\ln\left|\frac{5-4(\sin x+\cos x)}{5+4(\sin x-\cos x)}\right|+C$

5 $\int \frac{1+x\cos x}{x(1-x^2e^{2\sin x})}dx$ is equal to

(a) $\log|xe^{\sin x}|+\frac{1}{2}\log|1+x^2e^{2\sin x}|+C$

(b) $\log|xe^{\sin x}|-\frac{1}{2}\log|1-x^2e^{2\sin x}|+C$

(c) $\frac{1}{2}\{\log|xe^{\sin x}|-\log|1-x^2e^{2\sin x}|\}+C$

(d) None of these

6 $\int \frac{x}{(x^2-3x+2)\sqrt{x^2-4x+3}}dx$ is equal to

(a) $-2\sin^{-1}\frac{1}{x-2}+\frac{\sqrt{x^2-4x+3}}{x-1}+C$

(b) $2\sin^{-1}\frac{1}{x-2}+\frac{\sqrt{x^2-4x+3}}{x-1}+C$

(c) $-2\sin^{-1}\frac{1}{x-2}-\frac{\sqrt{x^2-4x+3}}{x-1}+C$

(d) $2\sin^{-1}\frac{1}{x-2}-\frac{\sqrt{x^2-4x+3}}{x-1}+C$

7 Let $I=\int \frac{e^x}{e^{4x}+e^{2x}+1}dx$ and $J=\int \frac{e^{3x}}{e^{4x}+e^{2x}+1}dx$. Then, for an arbitrary constant C, the value of J-I is equal to

(a) $\frac{1}{2}\log\left(\frac{e^{4x}-e^{2x}+1}{e^{4x}+e^{2x}+1}\right)+C$

(b) $\frac{1}{2}\log\left(\frac{e^{2x}+e^{x}+1}{e^{2x}-e^{x}+1}\right)+C$

(c) $\frac{1}{2}\log\left(\frac{e^{2x}-e^{x}+1}{e^{2x}+e^{x}+1}\right)+C$

(d) $\frac{1}{2}\log\left(\frac{e^{4x}+e^{2x}+1}{e^{4x}-e^{2x}+1}\right)+C$

8 The value of $\int \sqrt{\frac{3-x}{3+x}}\sin^{-1}\left(\frac{1}{\sqrt{6}}\sqrt{3-x}\right)dx$ is equal to

(a) $\frac{1}{4}\left\{-3\left(\cos^{-1}\left(\frac{x}{3}\right)\right)^2+2\sqrt{9-x^2}\cdot\cos^{-1}\left(\frac{x}{3}\right)+2x\right\}+C$

(b) $\frac{1}{4}\left\{-3\left(\cos^{-1}\left(\frac{x}{3}\right)\right)^2+2\sqrt{9-x^2}\cdot\sin^{-1}\left(\frac{x}{3}\right)+2x\right\}+C$

(c) $\frac{1}{4}\left\{-3\left(\sin^{-1}\left(\frac{x}{3}\right)\right)^2+2\sqrt{9-x^2}\cdot\sin^{-1}\left(\frac{x}{3}\right)+2x\right\}+C$

(d) None of the above

9 The integral $\int \left(\frac{x}{x\sin x+\cos x}\right)^2 dx$ is equal to (where, C is a constant of integration)

(a) $\tan x+\frac{x\sec x}{x\sin x+\cos x}+C$

(b) $\tan x-\frac{x\sec x}{x\sin x+\cos x}+C$

(c) $\sec x+\frac{x\tan x}{x\sin x+\cos x}+C$

(d) $\sec x-\frac{x\tan x}{x\sin x+\cos x}+C$

10 $\int \frac{e^{(x^2+4\ln x)}-x^3e^{x^2}}{x-1}dx$ is equal to

(a) $\left(\frac{x^3-x}{2}\right)e^{x^2}+C$

(b) $\frac{(x-1)xe^{x^2}}{2}+C$

(c) $\left(\frac{x^2-1}{2x}\right)e^{x^2}+C$

(d) $\frac{(x^2-1)e^{x^2}}{2}+C$

More than One Correct

11 If $\int \sin x(\sec x \tan x)\, dx = f(x) + g(x) + C$, then

(a) $f(x) = \tan x$ (b) $f(x) = \sec x$
(c) $g(x) = -2x$ (d) $g(x) = -x$

12 $\int \sqrt{\frac{a+x}{a-x}}\, dx$ is equal to

(a) $a\sin^{-1}\left(\frac{x}{a}\right) + \frac{\sqrt{a^2-x^2}}{2} + C$ (b) $a\sin^{-1}\left(\frac{x}{a}\right) - \sqrt{a^2-x^2} + C$
(c) $a\sin^{-1}\left(\frac{x}{a}\right) + \sqrt{a^2-x^2} + C$ (d) $-a\cos^{-1}\frac{x}{a} - \sqrt{a^2-x^2} + C$

13 $\int \frac{4e^x + 6e^{-x}}{9e^x - 4e^{-x}}\, dx$ is equal to

(a) $-\frac{19}{36}x + \frac{35}{36}\log(9e^x - 4e^{-x}) + C$ (b) $-\frac{3}{2}x + \frac{35}{36}\log(9e^{2x} - 4) + C$
(c) $\frac{4}{9}x + \frac{70}{9}\int \frac{dx}{9e^{2x} - 4}$ (d) $\frac{4}{9}x - \frac{70}{9}\int \frac{dx}{9e^{2x} - 4}$

14 Let $\int \frac{f'(x)g(x) - g'(x)f(x)}{\{f(x)+g(x)\}\cdot\sqrt{f(x)\cdot g(x) - g^2(x)}}\, dx = \sqrt{m}\tan^{-1}\left(\sqrt{\frac{f(x)-g(x)}{ng(x)}}\right) + C$, where m and n are natural number and C is constant of integration, then

(a) $m = 1$ (b) $m = 2$ (c) $n = 2$ (d) $n = 1$

15 For $a > 0$, if $I = \int \sqrt{\frac{x}{a^3 - x^3}}\, dx = A\sin^{-1}\left(\frac{x^{3/2}}{B}\right) + C$, where C is any arbitrary constant, then

(a) $A = \frac{2}{3}$ (b) $A = \frac{1}{3}$ (c) $B = a^{3/2}$ (d) $B = a^{1/3}$

16 If $\int \frac{dx}{1 - \sin^4 x} = a\tan x + b\tan^{-1}(c\tan x) + D$, then

(a) $a = \frac{1}{2}$ (b) $b = \sqrt{2}$ (c) $c = \sqrt{2}$ (d) $b = \frac{1}{2\sqrt{2}}$

17 If $I = \int \frac{\sin x + \sin^3 x}{\cos 2x}\, dx = P\cos x + Q\log|f(x)| + C$, then

(a) $P = \frac{1}{2}$ and $Q = -\frac{3}{4\sqrt{2}}$ (b) $P = \frac{1}{4}$ and $Q = -\frac{1}{\sqrt{2}}$
(c) $f(x) = \frac{\sqrt{2}\cos x + 1}{\sqrt{2}\cos x - 1}$ (d) $f(x) = \frac{\sqrt{2}\cos x - 1}{\sqrt{2}\cos x + 1}$

18 $\int(\sqrt{\tan x} + \sqrt{\cot x})\, dx$ is equal to

(a) $\sqrt{2}\tan^{-1}\left(\frac{\sqrt{\tan x} - \sqrt{\cot x}}{\sqrt{2}}\right) + C$
(b) $\sqrt{2}\tan^{-1}\left(\frac{\sqrt{\tan x} + \sqrt{\cot x}}{\sqrt{2}}\right) + C$
(c) $\sqrt{2}\sin^{-1}(\sin x - \cos x) + C$
(d) $\sqrt{2}\tan^{-1}(\sin x - \cos x) + C$

19 If $\int x \log\left(1+\frac{1}{x}\right) dx = f(x) \cdot \log(x+1) + g(x) \log x + Lx + C$, then

(a) $f(x) = \frac{x^2 - 1}{2}$

(b) $g(x) = -\frac{x^2}{2}$

(c) $L = 1$

(d) $L = \frac{1}{2}$

20 If $\int \frac{e^{x-1}}{(x^2 - 5x + 4)} \cdot 2x \, dx = Af(x-1) + Bf(x-4) + C$ and $f(x) = \int \frac{e^x}{x} dx$, then

(a) $A = -\frac{2}{3}$

(b) $B = \left(\frac{4}{3}\right) e^3$

(c) $A = \frac{2}{3}$

(d) $B = \left(\frac{8}{3}\right) e^3$

Linked Comprehension Type

Paragraph I Directions (Q.Nos. 21-23)

Consider the function $f''(x) = \sec^4 x + 4$ with $f(0) = 0$ and $f'(0) = 0$.

21 The value of $f'(x)$ is

(a) $\tan x - \frac{\tan^3 x}{3} + C$

(b) $\tan x + \frac{\tan^3 x}{3} + 4x$

(c) $\tan x + \frac{\sec^3 x}{3} + C$

(d) $-\tan x - \frac{\tan^3 x}{3} + C$

22 $f(x)$ is equal to

(a) $\frac{2 \ln \sec x}{3} + \frac{\tan^2 x}{6} + 2x^2$

(b) $\frac{3 \ln \sec x}{2} + \frac{\cot^2 x}{6} + 2x^2$

(c) $\frac{4 \ln \sec x}{3} + \frac{\sec^2 x}{6} + 2x^2$

(d) $\ln \sec x + \frac{\tan^4 x}{12} + 2x^2$

23 The value of $\int \tan x \, f''(x) \, dx = A \tan^2 x + B \tan^4 x + 4 \log \sec x + C$, then $A + B$ is equal to

(a) $\frac{1}{2}$ (b) $\frac{1}{4}$ (c) $\frac{3}{4}$ (d) $\frac{3}{8}$

Paragraph II Directions (Q.Nos. 24-26)

If $\int \frac{x(x-1)}{(x^2+1)(x+1)\sqrt{x^3+x^2+x}} dx = \frac{1}{2} \log \left| \frac{\sqrt{f(x)} - 1}{\sqrt{f(x)} + 1} \right| - \tan^{-1} \sqrt{f(x)} + C$, then

24 the value of $f(1)$ is

(a) 1 (b) 2 (c) 3 (d) 4

25 the value of $\lim_{x \to \infty} \tan^{-1} \sqrt{f(x)}$ is

(a) $\frac{\pi}{2}$ (b) $\frac{\pi}{4}$ (c) π (d) 2π

26 $f(x)$ is increasing in the interval

(a) $(-\infty, -1) \cup (1, \infty)$

(b) $(-1, 1)$

(c) $(-\infty, 0] \cup [1, \infty)$

(d) $(-\infty, -1) \cup (0, \infty)$

Paragraph III Directions (Q.Nos. 27-29)

Consider two differential functions $f(x)$, $g(x)$ satisfying $6\int f(x)\, g(x)\, dx = x^6 + 3x^4 + 3x^2 + C$ and $2\int \frac{g(x)\, dx}{f(x)} = x^2 + C$, where $f(x) > 0$, $\forall\, x \in R$.

27 $\int \{g(x) - f(x)\}\, dx$ is equal to

(a) $\frac{x^4}{4} - \frac{x^2}{2} + x + C$ (b) $\frac{x^4}{4} + \frac{x^2}{2} - \frac{x^3}{3} - x + C$

(c) $\frac{x^4}{4} - \frac{x^3}{3} + \frac{x^2}{2} - x + C$ (d) $\frac{x^4}{4} + \frac{x^3}{3} + C$

28 $\lim_{x \to 0} \frac{\log\{f(x)\}}{g(x)}$ is equal to

(a) e (b) 2 (c) 3 (d) 0

29 The number of real roots of $g(x) - f(x) = 0$ is

(a) 1 (b) 2

(c) 3 (d) None of these

Matrix Match Type Questions

30 Match the following.

	Column I		Column II
(i)	If $\int \frac{\sqrt{x^2+1}}{x^4}\, dx = -\frac{1}{3} \frac{(1+x^2)^{3/2}}{x^k}$, then k is equal to	(A)	3
(ii)	If $\int \frac{2+\sqrt{x}}{(x+\sqrt{x}+1)^2}\, dx = \frac{\lambda x}{x+\sqrt{x}+1} + C$, then λ is equal to	(B)	2
(iii)	If $\int \frac{\sqrt{x^{10}+2}}{x^{16}}\, dx = -\frac{1}{30} \frac{(x^{10}+2)^{3/2}}{x^m} + C$, then m is equal to	(C)	15
(iv)	If $\int \frac{dx}{x+\sqrt{x}} = n \log(1+\sqrt{x}) + C$, then n is equal to	(D)	5

31 Match the following.

	Column I		Column II
(i)	$\int \frac{1}{x(x^n+1)}\, dx$	(A)	$\frac{1}{3} \log \frac{x^3}{x^3+1} + C$
(ii)	$\int \frac{1}{x(x^3+1)}\, dx$	(B)	$\frac{1}{3} \log \left\{ \frac{\sqrt{1+x^3}-1}{\sqrt{1+x^3}+1} \right\} + C$
(iii)	$\int \frac{1}{x\sqrt{(x^n+1)}}\, dx$	(C)	$\frac{1}{n} \log \left(\frac{x^n}{x^n+1} \right) + C$
(iv)	$\int \frac{1}{x\sqrt{x^3+1}}\, dx$	(D)	$\frac{1}{n} \log \left\{ \frac{\sqrt{1+x^n}-1}{\sqrt{1+x^n}+1} \right\} + C$

32 Consider, $\int \frac{x^3+3x^2+2x+1}{\sqrt{x^2+x+1}} = (ax^2+bx+c)\sqrt{x^2+x+1}+\lambda\int\frac{dx}{\sqrt{x^2+x+1}}$, then match the following column.

Column I		Column II	
(i)	The value of a is	(A)	$-\frac{7}{24}$
(ii)	The value of b is	(B)	$\frac{1}{3}$
(iii)	The value of c is	(C)	$\frac{1}{16}$
(iv)	The value of λ is	(D)	$\frac{13}{12}$

33 Match the following.

Column I		Column II	
(i)	$\int e^x\left(\frac{1-x}{1+x^2}\right)^2 dx$	(A)	$e^x\left(\frac{x+1}{x+2}\right)+C$
(ii)	$\int e^x\frac{(x^3+x+1)}{(1+x^2)^{3/2}}dx$	(B)	$\frac{e^x}{1+x^2}+C$
(iii)	$\int e^x\frac{(x^2+3x+3)}{(x+2)^2}dx$	(C)	$\frac{e^x x}{\sqrt{1+x^2}}+C$
(iv)	$\int e^x\frac{(x^3-x+2)}{(x^2+1)^2}dx$	(D)	$\frac{e^x(x+1)}{x^2+1}+C$

Numerical Value Type Questions

Question Stem Directions (Q.Nos. 34-36) The answer to each question is a **NUMERICAL VALUE.**

$$\int f\{g(x)\}\cdot g'(x)dx = \int f(t)\,dt, \text{ where } t = g(x)$$

This rule is called substitution rule.

34 $\int \sec x \tan x\sqrt{\sec^2 x+1}\,dx = \frac{1}{2}\sin h^{-1}\{\sec(x)\}+\frac{1}{2}f(x)+C$, then $f(0)$ is equal to ....

35 $\int \frac{x^2-1}{x^2+1}\frac{dx}{\sqrt{1+x^4}}$ is equal to $\frac{k}{\sqrt{2}}\cos^{-1}\left(\frac{x\sqrt{2}}{1+x^2}\right)+C$, then k is equal to ....

36 If $\int\frac{dx}{(x+1)^3\sqrt{x^2+2x-3}} = \frac{\sqrt{x^2+2x-3}}{8(x+1)^2}-\frac{k}{32}\sin^{-1}\left(\frac{2}{x+1}\right)+C$, k is equal to ....

Question Stem Directions (Q.Nos. 37-39) The answer to each question is a **NUMERICAL VALUE.**

To evaluate $\int \frac{a\sin x + b\cos x}{c\sin x + d\cos x}\,dx$, where $c^2 + d^2 \neq 0$ express the numerator as

$a\sin x + b\cos x = \lambda$ (denominator) $+ \mu$ (derivative of the denominator)

37 $\int \frac{1}{1+\tan x}\,dx$ is equal to $kx + k\log|\sin x + \cos x| + C$, then k is equal to ....

38 $\int \frac{2\sin x + 3\cos x}{3\sin x + 4\cos x}\,dx$ is equal to $kx + 18k\log_e|2\sin x + 3\cos x| + C$ then k is equal to ....

39 If $\int \frac{\sin x + 2\cos x}{2\sin x + \cos x}\,dx = px + q\log|2\sin x + \cos x| + C$, then $p + q$ is equal to ....

40 If $\int \frac{dx}{\sqrt{x+7} - \sqrt[4]{x+7}} = P\sqrt{x+7} + Q(x+7)^{1/4} + R\ln|(x+7)^{1/4} - 1| + C$, then the value of $P + Q + R$ is ....

41 If $\int \frac{dx}{\sqrt{\sin^3 x \cos^5 x}} = a\sqrt{\cot x} + b\sqrt{\tan^3 x} + C$, then $a + 3b$ is equal to ....

42 If $\int \left(\frac{\cos^4 x}{\sin^3 x(\sin^5 x + \cos^5 x)^{3/5}}\right)dx = -\frac{1}{L}\left(\frac{1+\tan^5 x}{\tan^5 x}\right)^{\frac{M}{N}} + C$, then the value of $L + M + N$ is ....

43 If $\int \frac{1 - 5\sin^2 x}{\cos^5 x \sin^2 x}\,dx = \frac{f(x)}{\cos^5 x} + C$, then $f\left(\frac{\pi}{4}\right)$ is equal to ....

44 If $\int \frac{4x+1}{x^2 + 3x + 2}\,dx = a\log|x+1| + b\log|x+2| + C$, then $a + b$ is equal to ....

45 If $\int \frac{x^2 - 1}{(x^4 + 3x^2 + 1)\tan^{-1}\left(\frac{x^2+1}{x}\right)}\,dx = k\log\left|\tan^{-1}\frac{x^2+1}{x}\right| + C$, then k is equal to ....

46 Let $f(x)$ be a quadratic function such that $f(0) = 1$ and $\int \frac{f(x)}{x^2(x+1)^3}\,dx$ is a rational function, then the value of $f'(0)$ is ....

47 If $\int x^{-11}(1+x^4)^{-1/2}\,dx = -\frac{1}{10}x^{-a}(1+x^4)^{5/2} + \frac{1}{3}x^{-b}(1+x^4)^{3/2} - \frac{1}{2}x^{-c}(1+x^4)^{1/2} + K$, then $a + b + c$ is equal to ....

48 If $f'(x) = 3x^2\sin\frac{1}{x} - x\cos\frac{1}{x}$, $x \neq 0$, $f(0) = 0$, then the value of $f\left(\frac{1}{\pi}\right)$ is ....

49 If $\int (\sin 3x + \sin x)e^{\sin x}\cdot\cos x\,dx = (A\sin^3 x + B\cos^2 x + C\sin x + D\cos x + E)e^{\sin x} + F$, then $|A + B|$ is equal to ....

ANSWER KEY

Level 1

1. (i) $\frac{x^2}{2} + \log x + 2x + C$ (ii) $\frac{x^3}{3} + \frac{x^2}{2} + x + C$ (iii) $2\sin x + x + C$

(iv) $\tan x + 3\sec x + C$ (v) $\tan x - \cot x - 4x + C$ (vi) $\frac{\pi x}{4} + \frac{x^2}{4} + C$

(vii) $\frac{(2/3)^x}{\log(2/3)} + \frac{(3/2)^x}{\log(3/2)} + 2x + C$ (viii) $-\frac{1}{2}\sin 2x + C$

(ix) $\frac{1}{6}(1-2x)^{3/2} - \frac{1}{6}(3-2x)^{3/2} + C$ (x) $x - 2\log|x+1| - \frac{2}{x+1} + C$

(xi) $\frac{2}{3}(x+2)^{3/2} - 4(x+2)^{1/2} + C$ (xii) $\frac{1}{8}\left(3x + 2\sin 2x + \frac{\sin 4x}{4}\right) + C$

(xiii) $x\cos a + \sin a \log|\sin(x-a)| + C$ (xiv) $\frac{1}{\sin(a-b)} \log\left|\frac{\sec(x-b)}{\sec(x-a)}\right| + C$

2. (i) $\log|e^x + e^{-x}| + C$ (ii) $\frac{(1+\log x)^3}{3} + C$ (iii) $\frac{1}{6}(\tan^{-1} x^3)^2 + C$ (iv) $-2\operatorname{cosec}\alpha(\cos\alpha + \cot x \sin\alpha)^{1/2} + C$

(v) $\frac{1}{a^2 - b^2}\log|a^2\sin^2 x + b^2\cos^2 x| + C$ (vi) $\frac{1}{(\log 2)^3}(2^{2^{2^x}}) + C$ (vii) $\log|\tan x| + \frac{\tan^2 x}{2} + C$

3. (i) $\frac{1}{20}\tan^{-1}\left(\frac{4x}{5}\right) + C$ (ii) $\frac{1}{24}\log\left|\frac{4+3x}{4-3x}\right| + C$

(iii) $\frac{1}{4}\log\left|\frac{4x + \sqrt{16x^2+25}}{4}\right| + C$ (iv) $\frac{x^3}{3} - x + \tan^{-1} + C$

(v) $\frac{1}{4}\log\left|\frac{x+1}{3-x}\right| + C$ (vi) $\frac{1}{n}\log\left|\frac{x^n}{x^n+1}\right| + C$

(vii) $\sin^{-1}\left(\frac{e^x+2}{3}\right) + C$

(viii) $2\log|\sin^2\theta - 4\sin\theta + 5| + 7\tan^{-1}(\sin\theta - 2) + C$

4. (i) $\frac{1}{ab}\tan^{-1}\left(\frac{a\tan x}{b}\right) + C$ (ii) $\frac{1}{2}\tan^{-1}(2\tan x) + C$

(iii) $\frac{1}{\sqrt{2}}\tan^{-1}(\sqrt{2}\tan x) + C$ (iv) $\frac{1}{\sqrt{3}}\log\left|\frac{\log(x/2) - 2 - \sqrt{3}}{\tan(x/2) - 2 + \sqrt{3}}\right| + C$

(v) $\frac{1}{2}\left\{\log\left|\tan\frac{x}{2}\right| + \frac{\tan^2 x/2}{2} + 2\tan x/2\right\} + C$

(vi) $\frac{1}{2}\log\left|\tan\left(\frac{x}{2} + \frac{\pi}{12}\right)\right| + C$ (vii) $\frac{12}{13}x - \frac{5}{13}\log|3\cos x + 2\sin x| + C$

(viii) $-\frac{1}{2}\log|\cos x + \sin x| + \frac{x}{2} + C$

5. (i) $\frac{2}{3}\log|x+1| + \frac{1}{3}\log|x-2| + C$

(ii) $\log\left(\frac{x^2+1}{x^2+2}\right) + C$

(iii) $\frac{3}{8}\log|x-1| - \frac{1}{2(x-1)} + \frac{5}{8}\log|x+3| + C$

(iv) $\frac{1}{5}\log|x-1| - \frac{1}{10}\log(x^2+4) + \frac{2}{5}\tan^{-1}\frac{x}{2} + C$

(v) $-\frac{1}{3}\tan^{-1} x + \frac{2}{3}\tan^{-1}\frac{x}{2} + C$

6. (i) $\frac{x^2}{2}\log x - \frac{x^2}{4} + C$

(ii) $x\log(x^2+1) - 2x + 2\tan^{-1}x + C$

(iii) $\frac{e^{-x}}{2}(\sin x - \cos x) + C$

(iv) $x\tan^{-1}x - \frac{1}{2}\log|1+x^2| + C$

(v) $\frac{x^2}{2}\sin^{-1}x + \frac{x}{4}\sqrt{1-x^2} - \frac{1}{4}\sin^{-1}x + C$

7. (i) $\frac{e^x}{x} + C$

(ii) $e^x \sin x + C$

(iii) $\frac{e^x}{(x-1)^2} + C$

8. (i) $\frac{x}{2}\sqrt{4x^2+9} + \frac{9}{4}\log\left|\frac{2x+\sqrt{4x^2+9}}{2}\right| + C$

(ii) $\frac{x+1}{2}\sqrt{x^2+2x+5} + 2\log|(x+1)+\sqrt{x^2+2x+5}| + C$

(iii) $-\frac{1}{3}(1-x-x^2)^{3/2} + \frac{1}{8}(2x+1)\sqrt{1-x-x^2} + \frac{5}{16}\sin^{-1}\left(\frac{2x+1}{\sqrt{5}}\right) + C$

Level 2

1. (a) **2.** (d) **3.** (c) **4.** (b) **5.** (d) **6.** (c) **7.** (c)
8. (b) **9.** (c) **10.** (c) **11.** (c) **12.** (a) **13.** (a) **14.** (b)
15. (c) **16.** (a) **17.** (d) **18.** (a) **19.** (c) **20.** (a) **21.** (c)
22. (c) **23.** (a) **24.** (a) **25.** (c) **26.** (a) **27.** (c) **28.** (c)
29. (b) **30.** (c) **31.** (c) **32.** (a) **33.** (b) **34.** (c) **35.** (b)
36. (b) **37.** (c) **38.** (c) **39.** (c) **40.** (b) **41.** (c) **42.** (c)
43. (d) **44.** (d) **45.** (a) **46.** (a) **47.** (b) **48.** (d) **49.** (b)
50. (b) **51.** (a) **52.** (d) **53.** (a) **54.** (c) **55.** (a) **56.** (d)
57. (a) **58.** (b) **59.** (b) **60.** (2) **61.** (5) **62.** (3) **63.** (4)
64. (2012) **65.** (8) **66.** (9) **67.** (41) **68.** (4) **69.** (1) **70.** (3)

Level 3

1. (c) **2.** (c) **3.** (d) **4.** (a) **5.** (b) **6.** (c) **7.** (c)
8. (a) **9.** (b) **10.** (d) **11.** (a, d) **12.** (b, d) **13.** (a, c) **14.** (b, c)
15. (a, c) **16.** (a, c, d) **17.** (a, d) **18.** (a, c) **19.** (a, b, d) **20.** (a, d) **21.** (b)
22. (a) **23.** (c) **24.** (c) **25.** (a) **26.** (a) **27.** (b) **28.** (d)
29. (a)
30. (i)–A, (ii)–B, (iii)–C, (iv)–B
31. (i)–C, (ii)–A, (iii)–D, (iv)–B
32. (i)–B, (ii)–D, (iii)–A, (iv)–C
33. (i)–B, (ii)–C, (iii)–A, (iv)–D
34. (1.41) **35.** (1) **36.** (2) **37.** (0.5) **38.** (0.04) **39.** (1.4) **40.** (10)
41. (0) **42.** (9) **43.** (1) **44.** (4) **45.** (1) **46.** (3) **47.** (18)
48. (0) **49.** (16)

SOLUTIONS

Level 1

1. (i) $\int\left(\sqrt{x}+\frac{1}{\sqrt{x}}\right)^2 dx$

$$=\int\left(x+\frac{1}{x}+2\right)dx$$

$$=\frac{x^2}{2}+\log x+2x+C$$

(ii) $\int\frac{x^4+x^2+1}{x^2-x+1}dx$

$$=\int\frac{(x^2+1)^2-x^2}{x^2-x+1}dx$$

$$=\int\frac{(x^2+1-x)(x^2+1+x)}{x^2-x+1}dx$$

$$=\int(x^2+x+1)\,dx$$

$$=\frac{x^3}{3}+\frac{x^2}{2}+x+C$$

(iii) $\int\frac{\cos x-\cos 2x}{1-\cos x}dx$

$$=\int\frac{\cos x-(2\cos^2 x-1)}{1-\cos x}dx$$

$$=\int\frac{\cos x-2\cos^2 x+1}{1-\cos x}dx$$

$$=\int\frac{2\cos^2 x-\cos x-1}{(\cos x-1)}dx$$

$$=\int\frac{(2\cos x+1)(\cos x-1)}{(\cos x-1)}dx$$

$$=\int(2\cos x+1)\,dx$$

$$=2\sin x+x+C$$

(iv) $\int\frac{1+3\sin x}{\cos^2 x}dx$

$$=\int(\sec^2 x+3\sec x\tan x)\,dx$$

$$=\tan x+3\sec x+C$$

(v) $\int(\tan x-\cot x)^2 dx$

$$=\int(\tan^2 x+\cot^2 x-2\tan x\cot x)\,dx$$

$$=\int[(\sec^2 x-1)+(\text{cosec}^2 x-1)-2]\,dx$$

$$=\int(\sec^2 x+\text{cosec}^2 x-4)dx$$

$$=\tan x-\cot x-4x+C$$

(vi) $\int\tan^{-1}(\sec x+\tan x)\,dx$

$$=\int\tan^{-1}\left(\frac{1}{\cos x}+\frac{\sin x}{\cos x}\right)dx$$

$$=\int\tan^{-1}\left(\frac{1+\sin x}{\cos x}\right)dx$$

$$=\int\tan^{-1}\left\{\frac{1-\cos\left(\frac{\pi}{2}+x\right)}{\sin\left(\frac{\pi}{2}+x\right)}\right\}dx$$

$$=\int\tan^{-1}\left\{\frac{2\sin^2\left(\frac{\pi}{4}+\frac{x}{2}\right)}{2\sin\left(\frac{\pi}{4}+\frac{x}{2}\right)\cos\left(\frac{\pi}{4}+\frac{x}{2}\right)}\right\}dx$$

$$=\int\tan^{-1}\left\{\tan\left(\frac{\pi}{4}+\frac{x}{2}\right)\right\}dx$$

$$=\int\left(\frac{\pi}{4}+\frac{x}{2}\right)dx$$

$$=\frac{\pi}{4}x+\frac{x^2}{4}+C$$

(vii) $\int\frac{(2^x+3^x)^2}{2^x\cdot 3^x}dx$

$$=\int\left[\frac{(2^x)^2+(3^x)^2+2\cdot 2^x\cdot 3^x}{2^x\cdot 3^x}\right]dx$$

$$=\int\left[\left(\frac{2}{3}\right)^x+\left(\frac{3}{2}\right)^x+2\right]dx$$

$$=\frac{\left(\frac{2}{3}\right)^x}{\log(2/3)}+\frac{\left(\frac{3}{2}\right)^x}{\log(3/2)}+2x+C$$

(viii) $\int\frac{\sin^8 x-\cos^8 x}{1-2\sin^2 x\cos^2 x}dx$

$$=\int\frac{(\sin^4 x+\cos^4 x)(\sin^4 x-\cos^4 x)}{(\sin^2 x+\cos^2 x)^2-2\sin^2 x\cos^2 x}dx$$

$$=\int\frac{(\sin^4 x+\cos^4 x)(\sin^2 x+\cos^2 x)(\sin^2 x-\cos^2 x)}{(\sin^4 x+\cos^4 x)}dx$$

$$=-\int(\cos^2 x-\sin^2 x)dx$$

$$=-\int\cos 2x\,dx$$

$$=-\frac{1}{2}\sin 2x+C$$

(ix) $\int\frac{1}{\sqrt{1-2x}+\sqrt{3-2x}}dx$

$$=\int\frac{\sqrt{1-2x}-\sqrt{3-2x}}{(1-2x)-(3-2x)}dx$$

$$=\int\frac{\sqrt{1-2x}-\sqrt{3-2x}}{-2}dx$$

$$= -\frac{1}{2}\left[\left\{\frac{(1-2x)^{3/2}}{-2\times\frac{3}{2}}\right\} - \left\{\frac{(3-2x)^{3/2}}{-2\times\frac{3}{2}}\right\}\right] + C$$

$$= \frac{1}{6}(1-2x)^{3/2} - \frac{1}{6}(3-2x)^{3/2} + C$$

(x) $\int \frac{x^2+1}{(x+1)^2}\,dx$

$$= \int \frac{x^2+1+2x-2x}{(x+1)^2}\,dx$$

$$= \int \frac{(x+1)^2}{(x+1)^2}\,dx - 2\int \frac{x}{(x+1)^2}\,dx$$

$$= \int 1dx - 2\int \left[\frac{x+1-1}{(x+1)^2}\right] dx$$

$$= \int 1dx - 2\left[\int \frac{1}{x+1}\,dx - \int \frac{1}{(x+1)^2}\,dx\right]$$

$$= x - 2\log|x+1| - \frac{2}{x+1} + C$$

(xi) $\int \frac{x}{\sqrt{x+2}}\,dx$

$$= \int \left(\frac{x+2-2}{\sqrt{x+2}}\right) dx$$

$$= \int \left(\sqrt{x+2} - \frac{2}{\sqrt{x+2}}\right) dx$$

$$= \frac{(x+2)^{3/2}}{3/2} - 2\left[\frac{(x+2)^{1/2}}{1/2}\right] + C$$

$$= \frac{2}{3}(x+2)^{3/2} - 4(x+2)^{1/2} + C$$

(xii) $\int \cos^4 x\,dx$

$$= \int \left(\frac{1+\cos 2x}{2}\right)^2 dx$$

$$= \frac{1}{4}\int (1 + 2\cos 2x + \cos^2 2x)\,dx$$

$$= \frac{1}{4}\int \left(1 + 2\cos 2x + \frac{1+\cos 4x}{2}\right) dx$$

$$= \frac{1}{8}\int (3 + 4\cos 2x + \cos 4x)\,dx$$

$$= \frac{1}{8}\left(3x + 2\sin 2x + \frac{\sin 4x}{4}\right) + C$$

(xiii) $\int \frac{\sin x}{\sin(x-a)}\,dx$

$$= \int \frac{\sin(x-a+a)}{\sin(x-a)}\,dx$$

$$= \int \left(\frac{\sin(x-a)\cos a + \cos(x-a)\sin a}{\sin(x-a)}\right) dx$$

$$= \int \cos a\,dx + \int \sin a \cdot \cot(x-a)\,dx$$

$$= x\cos a + \sin a \log|\sin(x-a)| + C$$

(xiv) $\int \frac{1}{\cos(x-a)\cos(x-b)}\,dx$

$$= \frac{1}{\sin(a-b)}\int \frac{\sin(a-b)}{\cos(x-a)\cos(x-b)}\,dx$$

$$= \frac{1}{\sin(a-b)}\int \frac{\sin\{(x-b)-(x-a)\}}{\cos(x-a)\cos(x-b)}\,dx$$

$$= \frac{1}{\sin(a-b)}$$

$$\int \left[\frac{\sin(x-b)\cos(x-a) - \cos(x-b)\sin(x-a)}{\cos(x-a)\cos(x-b)}\right] dx$$

$$= \frac{1}{\sin(a-b)}\int \{\tan(x-b) - \tan(x-a)\}\,dx$$

$$= \frac{1}{\sin(a-b)}[\log|\sec(x-b)| - \log|\sec(x-a)|] + C$$

$$= \frac{1}{\sin(a-b)}\log\left|\frac{\sec(x-b)}{\sec(x-a)}\right| + C$$

2. (i) $\int \frac{e^{2x}-1}{e^{2x}+1}\,dx$

$$= \int \frac{e^x - e^{-x}}{e^x + e^{-x}}\,dx$$

Let $e^x + e^{-x} = t \Rightarrow (e^x - e^{-x})dx = dt$

$$= \int \frac{1}{t}\,dt$$

$$= \log|t| + C$$

$$= \log|e^x + e^{-x}| + C$$

(ii) $\int \frac{(1+\log x)^2}{x}\,dx$

Let $1 + \log x = t \Rightarrow \frac{1}{x}\,dx = dt$

$$= \int t^2 dt$$

$$= \frac{t^3}{3} + C$$

$$= \frac{(1+\log x)^3}{3} + C$$

(iii) $\int \frac{x^2 \tan^{-1} x^3}{1+x^6}\,dx$

Let $\tan^{-1} x^3 = t \Rightarrow \frac{1}{1+x^6}\cdot 3x^2\,dx = dt$

$$\Rightarrow \frac{x^2}{1+x^6}\,dx = \frac{1}{3}\,dt$$

$$= \frac{1}{3}\int t\,dt = \frac{1}{3}\left(\frac{t^2}{2}\right) + C$$

$$= \frac{1}{6}(\tan^{-1} x^3)^2 + C$$

(iv) Let $I=\int \frac{1}{\sqrt{\sin^3 x \sin(x+\alpha)}}dx$

$$=\int \frac{1}{\sqrt{\sin^3 x(\sin x\cos\alpha+\cos x\sin\alpha)}}dx$$

$$=\int \frac{1}{\sin^2 x\sqrt{\cos\alpha+\cot x\sin\alpha}}dx$$

$$\Rightarrow \quad I=\int \frac{\text{cosec}^2 x}{\sqrt{\cos\alpha+\cot x\sin\alpha}}dx$$

On putting $\cos\alpha+\cot x\sin\alpha=t$

$$\Rightarrow -\text{cosec}^2 x\, dx=\frac{dt}{\sin\alpha}$$

$$\therefore \quad I=\int -\frac{1}{\sin\alpha\sqrt{t}}dt$$

$$=-\frac{1}{\sin\alpha}\left(\frac{t^{1/2}}{1/2}\right)+C$$

$$\Rightarrow \quad I=-2\,\text{cosec}\,\alpha(\cos\alpha+\cot x\sin\alpha)^{1/2}+C$$

(v) Let $I=\int \frac{\sin 2x}{a^2\sin^2 x+b^2\cos^2 x}dx$

Let $a^2\sin^2 x+b^2\cos^2 x=t$

$$\Rightarrow [a^2(2\sin x\cos x)+b^2\{2\cos x(-\sin x)\}]\,dx=dt$$

$$\Rightarrow \quad (a^2-b^2)\sin 2x\, dx=dt$$

$$\Rightarrow \quad \sin 2x\, dx=\frac{1}{(a^2-b^2)}dt$$

$$\therefore \quad I=\frac{1}{a^2-b^2}\int\frac{dt}{t}$$

$$=\frac{1}{a^2-b^2}\log|t|+C$$

$$=\frac{1}{a^2-b^2}\log|a^2\sin^2 x+b^2\cos^2 x|+C$$

(vi) $\int 2^{2^{2^x}}\cdot 2^{2^x}\cdot 2^x\, dx$

Let $2^{2^{2^x}}=t \Rightarrow 2^{2^{2^x}}\cdot 2^{2^x}\cdot 2^x\cdot(\log 2)^3\, dx=dt$

$$=\int\frac{1}{(\log 2)^3}dt$$

$$=\frac{1}{(\log 2)^3}t+C$$

$$=\frac{1}{(\log 2)^3}(2^{2^{2^x}})+C$$

(vii) $\int\frac{1}{\sin x\cos^3 x}dx$

$$=\int\frac{\sec^4 x}{\tan x}dx$$

$$=\int\frac{(1+\tan^2 x)\sec^2 x}{\tan x}dx$$

On putting $\tan x=t \Rightarrow \sec^2 x\, dx=dt$

$$=\int\left(\frac{1+t^2}{t}\right)dt$$

$$=\int\left(\frac{1}{t}+t\right)dt$$

$$=\log|t|+\frac{t^2}{2}+C$$

$$=\log|\tan x|+\frac{\tan^2 x}{2}+C$$

3. (i) $\int\frac{1}{25+16x^2}dx=\frac{1}{16}\int\frac{1}{\frac{25}{16}+x^2}dx$

$$=\frac{1}{16}\int\frac{1}{\left(\frac{5}{4}\right)^2+x^2}dx$$

$$=\frac{1}{16}\cdot\frac{1}{5/4}\tan^{-1}\left(\frac{x}{5/4}\right)+C$$

$$=\frac{1}{20}\tan^{-1}\left(\frac{4x}{5}\right)+C$$

(ii) $\int\frac{1}{16-9x^2}dx$

$$=\frac{1}{9}\int\frac{1}{\frac{16}{9}-x^2}dx$$

$$=\frac{1}{9}\int\frac{1}{\left(\frac{4}{3}\right)^2-x^2}dx$$

$$=\frac{1}{9}\cdot\frac{1}{2\cdot\frac{4}{3}}\log\left|\frac{\frac{4}{3}+x}{\frac{4}{3}-x}\right|+C$$

$$=\frac{1}{24}\log\left|\frac{4+3x}{4-3x}\right|+C$$

(iii) $\int\frac{1}{\sqrt{16x^2+25}}dx$

$$=\frac{1}{4}\int\frac{1}{\sqrt{x^2+\left(\frac{5}{4}\right)^2}}dx$$

$$=\frac{1}{4}\log\left|x+\sqrt{x^2+\left(\frac{5}{4}\right)^2}\right|+C$$

$$\left[\because \int\frac{1}{\sqrt{x^2+a^2}}dx=\log|x+\sqrt{x^2+a^2}|+C\right]$$

$$=\frac{1}{4}\log\left|\frac{4x+\sqrt{16x^2+25}}{4}\right|+C$$

(iv) $\int\frac{x^4}{x^2+1}dx$

$$=\int\frac{x^4-1+1}{x^2+1}dx$$

$$= \int \frac{x^4 - 1}{x^2 + 1} dx + \int \frac{1}{x^2 + 1} dx$$

$$= \int (x^2 - 1)\, dx + \int \frac{1}{x^2 + 1} dx$$

$$= \frac{x^3}{3} - x + \tan^{-1} x + C$$

(v) $\int \frac{1}{3 + 2x - x^2} dx$

$$= \int \frac{1}{-(x^2 - 2x - 3)} dx$$

$$= \int \frac{1}{-(x^2 - 2x + 1 - 1 - 3)} dx$$

$$= \int \frac{1}{-\{(x - 1)^2 - 4\}} dx$$

$$= \int \frac{1}{(2)^2 - (x - 1)^2} dx$$

$$= \frac{1}{2(2)} \log \left| \frac{2 + (x - 1)}{2 - (x - 1)} \right| + C$$

$$= \frac{1}{4} \log \left| \frac{x + 1}{3 - x} \right| + C$$

(vi) $\int \frac{1}{x(x^n + 1)} dx$

$$= \int \frac{x^{n-1}}{x^n (x^n + 1)} dx$$

On putting $x^n = t \Rightarrow nx^{n-1}\, dx = dt$

$$= \frac{1}{n} \int \frac{dt}{t(t + 1)}$$

$$= \frac{1}{n} \int \frac{dt}{t^2 + t}$$

$$= \frac{1}{n} \int \frac{dt}{t^2 + t + \left(\frac{1}{2}\right)^2 - \left(\frac{1}{2}\right)^2}$$

$$= \frac{1}{n} \int \frac{dt}{\left(t + \frac{1}{2}\right)^2 - \left(\frac{1}{2}\right)^2}$$

$$= \frac{1}{n} \cdot \frac{1}{2\left(\frac{1}{2}\right)} \log \left| \frac{t + \frac{1}{2} - \frac{1}{2}}{t + \frac{1}{2} + \frac{1}{2}} \right| + C$$

$$= \frac{1}{n} \log \left| \frac{t}{t + 1} \right| + C$$

$$= \frac{1}{n} \log \left| \frac{x^n}{x^n + 1} \right| + C$$

(vii) $\int \frac{e^x}{\sqrt{5 - 4e^x - e^{2x}}} dx$

Let $e^x = t \Rightarrow e^x dx = dt$

$$= \int \frac{dt}{\sqrt{5 - 4t - t^2}}$$

$$= \int \frac{dt}{\sqrt{-(t^2 + 4t - 5)}}$$

$$= \int \frac{dt}{\sqrt{-\{t^2 + 4t + (2)^2 - (2)^2 - 5\}}}$$

$$= \int \frac{dt}{\sqrt{-\{(t + 2)^2 - (3)^2\}}}$$

$$= \int \frac{dt}{\sqrt{(3)^2 - (t + 2)^2}}$$

$$= \sin^{-1} \left(\frac{t + 2}{3} \right) + C$$

$$= \sin^{-1} \left(\frac{e^x + 2}{3} \right) + C$$

(viii) Let $I = \int \frac{2 \sin 2\theta - \cos\theta}{6 - \cos^2\theta - 4 \sin\theta} d\theta$

$$= \int \frac{\cos\theta\, (4 \sin\theta - 1)}{6 - (1 - \sin^2\theta) - 4 \sin\theta} d\theta$$

$$= \int \frac{(4 \sin\theta - 1) \cos\theta}{\sin^2\theta - 4 \sin\theta + 5} d\theta$$

Let $\sin\theta = t \Rightarrow \cos\theta\, d\theta = dt$

$\therefore \quad I = \int \frac{(4t - 1)}{t^2 - 4t + 5} dt$

Let $4t - 1 = A \frac{d}{dt} (t^2 - 4t + 5) + B$

$\Rightarrow 4t - 1 = A(2t - 4) + B$

On comparing coefficients of like powers of t, we get

$$A = 2 \text{ and } B = 7$$

So, $(4t - 1) = 2\,(2t - 4) + 7$

$$I = \int \frac{2(2t - 4) + 7}{t^2 - 4t + 5} dt$$

$$= 2 \int \frac{2t - 4}{t^2 - 4t + 5} dt + 7 \int \frac{1}{t^2 - 4t + 5} dt$$

$$= 2 \log |t^2 - 4t + 5| + 7 \int \frac{1}{(t - 2)^2 + (1)^2} dt$$

$$= 2 \log |t^2 - 4t + 5| + 7 \tan^{-1}(t - 2) + C$$

$$\Rightarrow \quad I = 2 \log |\sin^2\theta - 4 \sin\theta + 5| + 7 \tan^{-1} (\sin\theta - 2) + C$$

4. (i) $\int \frac{1}{a^2 \sin^2 x + b^2 \cos^2 x} dx$

$$= \int \frac{\sec^2 x\, dx}{a^2 \tan^2 x + b^2} dx$$

On putting $\tan x = t \Rightarrow \sec^2 x\, dx = dt$

$$= \int \frac{dt}{a^2 t^2 + b^2}$$

$$= \frac{1}{a^2}\int \frac{dt}{t^2 + \left(\frac{b}{a}\right)^2}$$

$$= \frac{1}{a^2} \times \frac{1}{b/a} \tan^{-1}\left(\frac{t}{b/a}\right) + C$$

$$= \frac{1}{ab} \tan^{-1}\left(\frac{at}{b}\right) + C$$

$$= \frac{1}{ab} \tan^{-1}\left(\frac{a \tan x}{b}\right) + C$$

(ii) $\int \frac{1}{1+3\sin^2 x} dx$

On dividing by $\cos^2 x$ in numerator and denominator, we get

$$= \int \frac{\sec^2 x}{\sec^2 x + 3\tan^2 x} dx$$

$$= \int \frac{\sec^2 x}{1 + \tan^2 x + 3\tan^2 x} dx$$

$$= \int \frac{\sec^2 x}{1 + 4\tan^2 x} dx$$

On putting $\tan x = t \Rightarrow \sec^2 x\, dx = dt$

$$= \int \frac{1}{1+4t^2} dt = \frac{1}{4}\int \frac{1}{\left(\frac{1}{2}\right)^2 + t^2} dt$$

$$= \frac{1}{4} \cdot \frac{1}{1/2} \tan^{-1}\left(\frac{t}{1/2}\right) + C$$

$$= \frac{1}{2}\tan^{-1}(2t) + C$$

$$= \frac{1}{2}\tan^{-1}(2\tan x) + C$$

(iii) $\int \frac{1}{\cos 2x + 3\sin^2 x} dx$

$$= \int \frac{1}{\cos^2 x - \sin^2 x + 3\sin^2 x} dx$$

$$= \int \frac{1}{\cos^2 x + 2\sin^2 x} dx$$

$$= \int \frac{\sec^2 x}{1 + 2\tan^2 x} dx$$

On putting $\tan x = t \Rightarrow \sec^2 x\, dx = dt$

$$= \int \frac{dt}{1+2t^2} = \frac{1}{2}\int \frac{dt}{\left(\frac{1}{\sqrt{2}}\right)^2 + t^2}$$

$$= \frac{1}{2} \cdot \frac{1}{1/\sqrt{2}} \tan^{-1}\left(\frac{t}{1/\sqrt{2}}\right) + C$$

$$= \frac{1}{\sqrt{2}}\{\tan^{-1}(\sqrt{2}t)\} + C$$

$$= \frac{1}{\sqrt{2}}\tan^{-1}(\sqrt{2}\tan x) + C$$

(iv) $\int \frac{1}{1-2\sin x} dx$

On putting $\sin x = \frac{2\tan x/2}{1+\tan^2 x/2}$, we get

$$= \int \frac{1}{1 - 2\left(\frac{2\tan\frac{x}{2}}{1+\tan^2\frac{x}{2}}\right)} dx$$

$$= \int \frac{\sec^2\frac{x}{2}}{1+\tan^2\frac{x}{2} - 4\tan\frac{x}{2}} dx$$

On putting $\tan\frac{x}{2} = t \Rightarrow \frac{1}{2}\sec^2\frac{x}{2} dx = dt$

$$= \int \frac{2}{1+t^2-4t} dt$$

$$= 2\int \frac{1}{(t-2)^2 - (\sqrt{3})^2} dt$$

$$= 2 \cdot \frac{1}{2\sqrt{3}} \log\left|\frac{t-2-\sqrt{3}}{t-2+\sqrt{3}}\right|$$

$$= \frac{1}{\sqrt{3}} \log\left|\frac{\tan\frac{x}{2} - 2 - \sqrt{3}}{\tan\frac{x}{2} - 2 + \sqrt{3}}\right| + C$$

(v) Let $I = \int \frac{1+\sin x}{\sin x(1+\cos x)} dx$

On putting $\sin x = \frac{2\tan\frac{x}{2}}{1+\tan^2\frac{x}{2}}$ and $\cos x = \frac{1-\tan^2\frac{x}{2}}{1+\tan^2\frac{x}{2}}$,

$$\therefore \quad I = \int \frac{1 + \frac{2\tan\frac{x}{2}}{1+\tan^2\frac{x}{2}}}{\frac{2\tan\frac{x}{2}}{1+\tan^2\frac{x}{2}} \cdot \left(1 + \frac{1-\tan^2\frac{x}{2}}{1+\tan^2\frac{x}{2}}\right)} dx$$

$$= \int \frac{\left(1+\tan\frac{x}{2}\right)^2 \sec^2\frac{x}{2}}{4\tan\frac{x}{2}} dx$$

On putting $\tan\frac{x}{2} = t \Rightarrow \frac{1}{2}\sec^2\frac{x}{2} dx = dt$

$$\Rightarrow \quad \sec^2\frac{x}{2} dx = 2dt$$

$$\therefore \quad I = \int \frac{(1+t)^2}{4t} 2dt$$

$$= \int \frac{1+t^2+2t}{2t}\,dt$$

$$= \frac{1}{2}\int\left(\frac{1}{t}+t+2\right)dt$$

$$= \frac{1}{2}\left\{\log|t| + \frac{t^2}{2} + 2t\right\} + C$$

$$= \frac{1}{2}\left\{\log\left|\tan\frac{x}{2}\right| + \frac{\tan^2\frac{x}{2}}{2} + 2\tan\frac{x}{2}\right\} + C$$

(vi) Let $I = \int \frac{1}{\sqrt{3}\sin x + \cos x}\,dx$

Let $\sqrt{3} = r\sin\theta$ and $1 = r\cos\theta$

$\Rightarrow \quad r = \sqrt{(\sqrt{3})^2 + (1)^2} = 2$ and $\theta = \frac{\pi}{3}$

$\therefore \quad I = \int \frac{1}{r\sin\theta\sin x + r\cos\theta\cos x}\,dx$

$$= \frac{1}{r}\int \frac{1}{\cos(x-\theta)}\,dx$$

$$= \frac{1}{r}\int \sec(x-\theta)\,dx$$

$$= \frac{1}{r}\log\left|\tan\left(\frac{\pi}{4}+\frac{x}{2}-\frac{\theta}{2}\right)\right| + C$$

$$= \frac{1}{2}\log\left|\tan\left(\frac{\pi}{4}+\frac{x}{2}-\frac{\pi}{6}\right)\right| + C$$

$$= \frac{1}{2}\log\left|\tan\left(\frac{x}{2}+\frac{\pi}{12}\right)\right| + C$$

(vii) Let $I = \int \frac{3\sin x + 2\cos x}{3\cos x + 2\sin x}\,dx$

Let $3\sin x + 2\cos x = \lambda\frac{d}{dx}(3\cos x + 2\sin x) + \mu(3\cos x + 2\sin x)$

$(3\sin x + 2\cos x)$

$$= \lambda(-3\sin x + 2\cos x) + \mu(3\cos x + 2\sin x)$$

By comparing coefficients of $\cos x$ and $\sin x$,

$$\lambda = -\frac{5}{13} \text{ and } \mu = \frac{12}{13}$$

$\therefore \quad I = \int \frac{\lambda(-3\sin x + 2\cos x) + \mu(3\cos x + 2\sin x)}{3\cos x + 2\sin x}\,dx$

$\Rightarrow I = \mu\int 1\,dx + \lambda\int \frac{-3\sin x + 2\cos x}{3\cos x + 2\sin x}\,dx$

$\Rightarrow I = \mu x + \lambda\int \frac{dt}{t}$,

where $t = 3\cos x + 2\sin x$

$\Rightarrow I = \mu x + \lambda\log|t| + C$

$$= \frac{12}{13}x - \frac{5}{13}\log|3\cos x + 2\sin x| + C$$

(viii) Let $I = \int \frac{1}{1+\cot x}\,dx = \int \frac{\sin x}{\sin x + \cos x}\,dx$

Let $\sin x = \lambda\frac{d}{dx}(\sin x + \cos x) + \mu(\sin x + \cos x)$

$$= \lambda(\cos x - \sin x) + \mu(\sin x + \cos x)$$

On solving, $\lambda = -\frac{1}{2}$ and $\mu = \frac{1}{2}$

$\therefore \; I = \lambda\int \frac{\cos x - \sin x}{\cos x + \sin x}\,dx + \mu\int \frac{\sin x + \cos x}{\sin x + \cos x}\,dx$

$= \lambda\int \frac{dt}{t} + \mu\int 1\,dx$, where $t = \sin x + \cos x$

$$= \lambda\log|t| + \mu x + C$$

$$= -\frac{1}{2}\log|\cos x + \sin x| + \frac{1}{2}x + C$$

5. (i) $\int \frac{x-1}{(x+1)(x-2)}\,dx$

Let $\frac{x-1}{(x+1)(x-2)} = \frac{A}{x+1} + \frac{B}{x-2}$

$\Rightarrow \quad x - 1 = A(x-2) + B(x+1)$

When $x = 2$, then $B = \frac{1}{3}$

When $x = -1$, then $A = \frac{2}{3}$

So, $\frac{x-1}{(x+1)(x-2)} = \frac{2}{3}\cdot\frac{1}{x+1} + \frac{1}{3}\cdot\frac{1}{x-2}$

$\therefore I = \int \frac{x-1}{(x+1)(x-2)}\,dx$

$$= \frac{2}{3}\int \frac{1}{x+1}\,dx + \frac{1}{3}\int \frac{1}{x-2}\,dx$$

$$= \frac{2}{3}\log|x+1| + \frac{1}{3}\log|x-2| + C$$

(ii) Let $I = \int \frac{2x}{(x^2+1)(x^2+2)}\,dx$

On putting $x^2 = t \Rightarrow 2x\,dx = dt$

$$= \int \frac{dt}{(t+1)(t+2)}$$

Let $\frac{1}{(t+1)(t+2)} = \frac{A}{t+1} + \frac{B}{t+2}$

$\Rightarrow 1 = A(t+2) + B(t+1)$

So, $A = 1$ and $B = -1$

Then, $\frac{1}{(t+1)(t+2)} = \frac{1}{t+1} - \frac{1}{t+2}$

$\therefore \; I = \int \frac{dt}{(t+1)(t+2)} = \int \frac{1}{t+1}\,dt - \int \frac{1}{t+2}\,dt$

$$= \log(t+1) - \log(t+2) + C$$

$$= \log\left(\frac{t+1}{t+2}\right) + C = \log\left(\frac{x^2+1}{x^2+2}\right) + C$$

(iii) $\int \frac{x^2+1}{(x-1)^2(x+3)}dx$

Let $\frac{x^2+1}{(x-1)^2(x+3)} = \frac{A}{x-1}+\frac{B}{(x-1)^2}+\frac{C}{x+3}$

$\Rightarrow x^2+1 = A(x-1)(x+3)+B(x+3)+C(x-1)^2$

On putting $x=1$, we get $B=\frac{1}{2}$

On putting $x=-3$, we get $C=\frac{5}{8}$

On equating the coefficients of x^2 on both sides, we get

$$A+C=1$$

$$\Rightarrow \quad A = 1-\frac{5}{8}=\frac{3}{8}$$

So, $\frac{x^2+1}{(x-1)^2(x+3)}$

$$= \frac{3}{8}\left(\frac{1}{x-1}\right)+\frac{1}{2}\cdot\frac{1}{(x-1)^2}+\frac{5}{8}\frac{1}{(x+3)}$$

$$\Rightarrow \int \frac{x^2+1}{(x-1)^2(x+3)}dx$$

$$= \frac{3}{8}\int\frac{1}{x-1}dx+\frac{1}{2}\int\frac{1}{(x-1)^2}dx+\frac{5}{8}\int\frac{1}{x+3}dx$$

$$= \frac{3}{8}\log|x-1|+\frac{1}{2}\left(\frac{-1}{x-1}\right)+\frac{5}{8}\log|x+3|+C$$

$$= \frac{3}{8}\log|x-1|-\frac{1}{2(x-1)}+\frac{5}{8}\log|x+3|+C$$

(iv) $\int \frac{x}{(x-1)(x^2+4)}dx$

Let $\frac{x}{(x-1)(x^2+4)} = \frac{A}{x-1}+\frac{Bx+C}{x^2+4}$

$\Rightarrow x = A(x^2+4)+(Bx+C)(x-1)$

On putting $x=1$, we get $A=\frac{1}{5}$

On putting $x=0$, we get $4A-C=0 \Rightarrow C=\frac{4}{5}$

On putting $x=-1$, we get

$$5A+2B-2C=-1$$

$$\Rightarrow \quad B=-\frac{1}{5}$$

So, $\frac{x}{(x-1)(x^2+4)}$

$$= \frac{1}{5(x-1)}+\frac{-\frac{1}{5}x+\frac{4}{5}}{x^2+4}$$

$$\int \frac{x}{(x-1)(x^2+4)}dx$$

$$= \frac{1}{5}\int\frac{1}{(x-1)}dx-\frac{1}{5}\int\frac{x-4}{x^2+4}dx$$

$$= \frac{1}{5}\int\frac{1}{x-1}dx-\frac{1}{10}\int\frac{2x}{x^2+4}dx+\frac{4}{5}\int\frac{1}{x^2+4}dx$$

$$= \frac{1}{5}\log|x-1|-\frac{1}{10}\log(x^2+4)+\frac{4}{5}\cdot\frac{1}{2}\tan^{-1}\frac{x}{2}+C$$

$$= \frac{1}{5}\log|x-1|-\frac{1}{10}\log(x^2+4)+\frac{2}{5}\tan^{-1}\frac{x}{2}+C$$

(v) $\int \frac{x^2}{(x^2+1)(x^2+4)}dx$

Let $x^2=y$, then $\frac{x^2}{(x^2+1)(x^2+4)} = \frac{y}{(y+1)(y+4)}$

Let $\frac{y}{(y+1)(y+4)} = \frac{A}{y+1}+\frac{B}{y+4}$

$\Rightarrow \quad y = A(y+4)+B(y+1)$

On putting $y=-1$ and $y=-4$, we get

$$A=-\frac{1}{3} \text{ and } B=\frac{4}{3}$$

On substituting the values of A and B, we get

$$\frac{y}{(y+1)(y+4)} = -\frac{1}{3(y+1)}+\frac{4}{3(y+4)}$$

$$\Rightarrow \quad \frac{x^2}{(x^2+1)(x^2+4)} = -\frac{1}{3(x^2+1)}+\frac{4}{3(x^2+4)}$$

$$\Rightarrow \int\frac{x^2}{(x^2+1)(x^2+4)}dx$$

$$= -\frac{1}{3}\int\frac{1}{x^2+1}dx+\frac{4}{3}\int\frac{1}{x^2+4}dx$$

$$= -\frac{1}{3}\tan^{-1}x+\frac{4}{3}\cdot\frac{1}{2}\tan^{-1}\left(\frac{x}{2}\right)+C$$

$$= -\frac{1}{3}\tan^{-1}x+\frac{2}{3}\tan^{-1}\frac{x}{2}+C$$

6. (i) $\int \underset{\text{II}}{x}\,\underset{\text{I}}{\log x}\,dx$

$$= \log x\int x\,dx-\int\left(\frac{d}{dx}(\log x)\int x\,dx\right)dx$$

$$= \frac{x^2}{2}\log x-\int\frac{1}{x}\cdot\frac{x^2}{2}dx$$

$$= \frac{x^2}{2}\log x-\frac{1}{2}\int x\,dx = \frac{x^2}{2}\log x-\frac{x^2}{4}+C$$

(ii) $\int(\underset{\text{I}}{\log(1+x^2)}\cdot\underset{\text{II}}{1}\,dx$

$$= x\log(1+x^2)-\int\left(\frac{1}{1+x^2}(2x)\right)x\,dx$$

$$= x\log(1+x^2)-2\int\frac{x^2}{1+x^2}dx$$

$$= x\log(1+x^2)-2\int\frac{x^2+1-1}{x^2+1}dx$$

$$= x\log(x^2+1)-2\int\left(1-\frac{1}{x^2+1}\right)dx$$

$$= x\log(x^2+1)-2x+2\tan^{-1}x+C$$

(iii) Let $I=\int e^{-x}\cos x\,dx$

$\Rightarrow\quad I=e^{-x}\sin x-\int -e^{-x}\sin x\,dx$

$\Rightarrow\quad I=e^{-x}\sin x+e^{-x}(-\cos x)-\int(-e^{-x})(-\cos x)\,dx$

$\Rightarrow\quad I=e^{-x}\sin x-e^{-x}\cos x-\int e^{-x}\cos x\,dx$

$\Rightarrow\quad I=e^{-x}\sin x-e^{-x}\cos x-I$

$\Rightarrow\quad 2I=e^{-x}(\sin x-\cos x)+C$

$\Rightarrow\quad I=\dfrac{e^{-x}}{2}(\sin x-\cos x)+C$

(iv) $\int\tan^{-1}x\,dx$

$$=\int\tan^{-1}x\cdot 1\,dx$$

$$=\tan^{-1}x\int 1\,dx-\int\left\{\left(\frac{d}{dx}\tan^{-1}x\right)\int 1\,dx\right\}dx$$

$$=x\tan^{-1}x-\int\frac{1}{1+x^2}\cdot x\,dx$$

$$=x\tan^{-1}x-\frac{1}{2}\int\frac{dt}{t}$$

[putting $1+x^2=t\Rightarrow 2x\,dx=dt$]

$$=x\tan^{-1}x-\frac{1}{2}\log|t|+C$$

$$=x\tan^{-1}x-\frac{1}{2}\log|1+x^2|+C$$

(v) Let $I=\int x\sin^{-1}x\,dx$

$$\Rightarrow\quad I=\sin^{-1}x\left(\frac{x^2}{2}\right)-\int\frac{1}{\sqrt{1-x^2}}\cdot\frac{x^2}{2}dx$$

$$=\frac{x^2}{2}\sin^{-1}x+\frac{1}{2}\int\frac{-x^2}{\sqrt{1-x^2}}dx$$

$$=\frac{x^2}{2}\sin^{-1}x+\frac{1}{2}\int\frac{1-x^2-1}{\sqrt{1-x^2}}dx$$

$$=\frac{x^2}{2}\sin^{-1}x+\frac{1}{2}\left[\int\sqrt{1-x^2}\,dx-\int\frac{1}{\sqrt{1-x^2}}dx\right]$$

$$=\frac{x^2}{2}\sin^{-1}x+\frac{1}{2}\left[\frac{x}{2}\sqrt{1-x^2}+\frac{1}{2}\sin^{-1}x\right]-\frac{1}{2}\sin^{-1}x+C$$

$$=\frac{x^2}{2}\sin^{-1}x+\frac{x}{4}\sqrt{1-x^2}-\frac{1}{4}\sin^{-1}x+C$$

7. (i) $\int e^x\left(\dfrac{1}{x}-\dfrac{1}{x^2}\right)dx$

Let $f(x)=\dfrac{1}{x}\Rightarrow f'(x)=-\dfrac{1}{x^2}$

So, given integral is in the form of

$\int e^x\{f(x)+f'(x)\}\,dx=e^x f(x)+C$

Hence, $\int e^x\left(\dfrac{1}{x}-\dfrac{1}{x^2}\right)dx=e^x\dfrac{1}{x}+C=\dfrac{e^x}{x}+C$

(ii) $\int e^x(\sin x+\cos x)\,dx$

Let $f(x)=\sin x\Rightarrow f'(x)=\cos x$

So, $\int e^x(\sin x+\cos x)\,dx=e^x\sin x+C$

(iii) $\int e^x\left(\dfrac{x-3}{(x-1)^3}\right)dx$

$$=\int e^x\left(\frac{x-1-2}{(x-1)^3}\right)dx$$

$$=\int e^x\left[\frac{1}{(x-1)^2}-\frac{2}{(x-1)^3}\right]dx$$

Let $f(x)=\dfrac{1}{(x-1)^2}\Rightarrow f'(x)=\dfrac{-2}{(x-1)^3}$

So, $\int e^x\left(\dfrac{x-3}{x-1^3}\right)dx=e^x\left[\dfrac{1}{(x-1)^2}\right]+C$

$$=\frac{e^x}{(x-1)^2}+C$$

8. (i) $\int\sqrt{4x^2+9}\,dx$

$$=2\int\sqrt{\left(x^2+\frac{9}{4}\right)}\,dx$$

$$=2\int\sqrt{x^2+\left(\frac{3}{2}\right)^2}\,dx$$

$$=2\left\{\frac{x}{2}\sqrt{x^2+\left(\frac{3}{2}\right)^2}+\frac{\left(\frac{3}{2}\right)^2}{2}\log\left|x+\sqrt{x^2+\left(\frac{3}{2}\right)^2}\right|\right\}+C$$

$$=\frac{x}{2}\sqrt{4x^2+9}+\frac{9}{4}\log\left|\frac{2x+\sqrt{4x^2+9}}{2}\right|+C$$

(ii) Let $I=\int\sqrt{x^2+2x+5}\,dx$

$$=\int\sqrt{x^2+2x+1+4}\,dx$$

$$=\int\sqrt{(x+1)^2+(2)^2}\,dx$$

$$=\frac{x+1}{2}\sqrt{(x+1)^2+(2)^2}+\frac{(2)^2}{2}\log|(x+1)+\sqrt{(x+1)^2+(2)^2}|+C$$

$$=\frac{x+1}{2}\sqrt{x^2+2x+5}+2\log|(x+1)+\sqrt{x^2+2x+5}|+C$$

(iii) Let $I=\int(x+1)\sqrt{1-x-x^2}\,dx$

Let $x+1=A\dfrac{d}{dx}(1-x-x^2)+B$

$\Rightarrow\ x+1=A(-1-2x)+B$

By comparing coefficients,

$$A=-\frac{1}{2}\text{ and }B=\frac{1}{2}$$

$$I=\int\left\{-\frac{1}{2}(-1-2x)+\frac{1}{2}\right\}\sqrt{1-x-x^2}\,dx$$

$$=-\frac{1}{2}\int(-1-2x)\sqrt{1-x-x^2}\,dx+\frac{1}{2}\int\sqrt{1-x-x^2}\,dx$$

Let $t=1-x-x^2 \Rightarrow dt=(-1-2x)\,dx$

$$=-\frac{1}{2}\int\sqrt{t}\,dt+\frac{1}{2}\int\sqrt{\left(\frac{\sqrt{5}}{2}\right)^2-\left(x+\frac{1}{2}\right)^2}\,dx$$

$$=-\frac{1}{2}\left(\frac{t^{3/2}}{3/2}\right)+\frac{1}{2}\left\{\frac{1}{2}\left(x+\frac{1}{2}\right)\sqrt{1-x-x^2}+\frac{1}{2}\cdot\frac{5}{4}\sin^{-1}\left(\frac{x+\frac{1}{2}}{\frac{\sqrt{5}}{2}}\right)\right\}+C$$

$$=-\frac{1}{3}(1-x-x^2)^{3/2}+\frac{1}{8}(2x+1)\sqrt{1-x-x^2}+\frac{5}{16}\sin^{-1}\left(\frac{2x+1}{\sqrt{5}}\right)+C$$

Level 2

1. (a) We have, $\int\frac{dx}{\sqrt{ax+b}-\sqrt{ax+c}}$

$$=\int\frac{\sqrt{ax+b}+\sqrt{ax+c}}{(ax+b)-(ax+c)}\,dx$$

$$=\frac{1}{b-c}\int\{(ax+b)^{1/2}+(ax+c)^{1/2}\}\,dx$$

$$=\frac{1}{b-c}\left[\frac{(ax+b)^{3/2}}{\frac{3}{2}a}+\frac{(ax+c)^{3/2}}{\frac{3}{2}a}\right]+C$$

$$=\frac{2}{3a(b-c)}[(ax+b)^{3/2}+(ax+c)^{3/2}]+C$$

Therefore, $k=a(b-c)$

2. (d) Let $I=\int\frac{\sqrt{x-1}}{x\sqrt{x+1}}\,dx$

$$=\int\frac{(x-1)}{x\sqrt{x^2-1}}\,dx$$

$$=\int\frac{dx}{\sqrt{x^2-1}}-\int\frac{dx}{x\sqrt{x^2-1}}$$

$$=\ln|x+\sqrt{x^2-1}|-\sec^{-1}x+C$$

3. (c) Let $I=\int\frac{\sin x}{\sin 3x\sin 2x}\,dx$

$$=\int\frac{\sin(3x-2x)}{\sin 3x\sin 2x}\,dx$$

$$=\int\frac{\sin 3x\cos 2x-\cos 3x\sin 2x}{\sin 3x\sin 2x}\,dx$$

$$=\int(\cot 2x-\cot 3x)\,dx$$

$$=\frac{1}{2}\log|\sin 2x|-\frac{1}{3}\log|\sin 3x|+C$$

4. (b) Let $I=\int(\sin 2x-\cos 2x)\,dx$

$$=-\sqrt{2}\int\left(-\frac{1}{\sqrt{2}}\sin 2x+\frac{1}{\sqrt{2}}\cos 2x\right)dx$$

$$=-\sqrt{2}\int\left(\sin 2x\sin\frac{5\pi}{4}-\cos\frac{5\pi}{4}\cos 2x\right)dx$$

$$=\sqrt{2}\int\cos\left(2x+\frac{5\pi}{4}\right)dx$$

$$=\frac{\sqrt{2}}{2}\sin\left(2x+\frac{5\pi}{4}\right)+C$$

$$=\frac{1}{\sqrt{2}}\sin\left(2x+\frac{5\pi}{4}\right)+C$$

$\therefore\quad a=-\frac{5\pi}{4}$

5. (d) Let $I=\int\frac{\sin x}{\sin\left(x-\frac{\pi}{4}\right)}\,dx$

$$=\int\frac{\sin\left\{\left(x-\frac{\pi}{4}\right)+\frac{\pi}{4}\right\}}{\sin\left(x-\frac{\pi}{4}\right)}\,dx$$

$$=\frac{1}{\sqrt{2}}\int\frac{\sin\left(x-\frac{\pi}{4}\right)+\cos\left(x-\frac{\pi}{4}\right)}{\sin\left(x-\frac{\pi}{4}\right)}\,dx$$

$$=\frac{1}{\sqrt{2}}\int\left[1+\cot\left(x-\frac{\pi}{4}\right)\right]dx$$

$$=\frac{1}{\sqrt{2}}\left[x+\log\left|\sin\left(x-\frac{\pi}{4}\right)\right|\right]+C$$

6. (c) Let $I=\int\frac{\sin\frac{5x}{2}}{\sin\frac{x}{2}}\,dx$

$$=\int\frac{2\sin\frac{5x}{2}\cos\frac{x}{2}}{2\sin\frac{x}{2}\cos\frac{x}{2}}\,dx$$

[multiplying by $2\cos\frac{x}{2}$ in numerator and denominator]

$$=\int\frac{\sin 3x+\sin 2x}{\sin x}\,dx$$

[$\because 2\sin A\cos B=\sin(A+B)+\sin(A-B)$]

$$=\int\frac{(3\sin x-4\sin^3 x)+2\sin x\cos x}{\sin x}\,dx$$

$$=\int(3-4\sin^2 x+2\cos x)\,dx$$

$$=\int[3-2(1-\cos 2x)+2\cos x]\,dx$$

[$\because 2\sin^2 x=1-\cos 2x$]

$$= \int [3 - 2 + 2\cos 2x + 2\cos x]dx$$
$$= \int [1 + 2\cos 2x + 2\cos x]dx$$
$$= x + 2\sin x + \sin 2x + C$$

7. (c) Let $I = \int \left\{ \left(\frac{\cot^2 2x - 1}{2\cot 2x} \right) - \cos 8x \cot 4x \right\} dx$
$$= \int (\cot 4x - \cos 8x \cot 4x)\, dx$$
$$= \int \cot 4x (1 - \cos 8x)\, dx$$
$$= \int \cot 4x \cdot 2\sin^2 4x\, dx$$
$$= \int \frac{\cos 4x}{\sin 4x} \cdot 2\sin^2 4x\, dx$$
$$= \int 2\sin 4x \cos 4x\, dx$$
$$= \int \sin 8x\, dx = -\frac{\cos 8x}{8} + C$$

8. (b) Let $I = \int 5^{5^{5^x}} \cdot 5^{5^x} \cdot 5^x\, dx$

Let $5^{5^{5^x}} = t$
$$\Rightarrow 5^{5^{5^x}} \cdot 5^{5^x} \cdot 5^x \cdot (\log 5)^3\, dx = dt$$
$$= \frac{1}{(\log 5)^3} \int dt = \frac{t}{(\log 5)^3} + C = \frac{5^{5^{5^x}}}{(\log 5)^3} + C$$

9. (c) We have, $I = \int \frac{e^{3\log_e 2x} + 5e^{2\log_e 2x}}{e^{4\log_e x} + 5e^{3\log_e x} - 7e^{2\log_e x}} dx$
$$= \int \frac{8x^3 + 5(4x^2)}{x^4 + 5\cdot x^3 - 7x^2} dx$$
$$= \int \frac{8x^3 + 20x^2}{x^4 + 5x^3 - 7x^2} dx$$
$$= \int \frac{8x + 20}{x^2 + 5x - 7} dx$$
$$= 4\int \frac{2x + 5}{x^2 + 5x - 7} dx$$

Let $x^2 + 5x - 7 = t \Rightarrow (2x + 5)\, dx = dt$
$$= 4\int \frac{1}{t} dt$$
$$= 4\log|t| + C = 4\log|x^2 + 5x - 7| + C$$

10. (c) Let $I = \int \frac{2x^3 - 1}{x^4 + x} dx = \int \frac{2x - (1/x^2)}{x^2 + \frac{1}{x}} dx$

[dividing each term of numerator and denominator by x^2]

Put $x^2 + \frac{1}{x} = t \Rightarrow \left\{ 2x + \left(-\frac{1}{x^2} \right) \right\} dx = dt$
$$\therefore \quad I = \int \frac{dt}{t} = \log_e |t| + C$$
$$= \log_e \left| \left(x^2 + \frac{1}{x} \right) \right| + C$$
$$= \log_e \left| \frac{x^3 + 1}{x} \right| + C$$

11. (c) Let $I = \int (x^4 + 1)^{1/3} x^7\, dx$

On putting $x^4 + 1 = t^3$

So, $4x^3 dx = 3t^2 dt$
$$\therefore \quad I = \int t \cdot (t^3 - 1) \frac{3}{4} t^2 dt$$
$$= \frac{3}{4} \int (t^6 - t^3)\, dt$$
$$= \frac{3}{4} \left[\frac{t^7}{7} - \frac{t^4}{4} \right] + C$$
$$= \frac{3}{112} t^4 (4t^3 - 7) + C$$
$$= \frac{3}{112} (x^4 + 1)^{4/3} [4(x^4 + 1) - 7] + C$$
$$= \frac{3}{112} (x^4 + 1)^{4/3} (4x^4 - 3) + C$$

Hence, $k = \frac{3}{112}$

12. (a) Let $I = \int \frac{1}{x^n (1 + x^n)^{1/n}} dx$
$$\Rightarrow \quad I = \int \frac{1}{x^{n+1} \left(1 + \frac{1}{x^n} \right)^{1/n}} dx$$
$$\Rightarrow \quad I = \int (1 + x^{-n})^{-1/n} \cdot x^{-n-1}\, dx$$

Let $(1 + x^{-n}) = t \Rightarrow -nx^{-n-1}\, dx = dt$
$$\therefore \quad I = -\frac{1}{n} \int t^{-1/n} dt$$
$$= -\frac{1}{n} \left(\frac{t^{\frac{-1}{n}+1}}{-\frac{1}{n}+1} \right)$$
$$= \frac{1}{1-n} \left(1 + \frac{1}{x^n} \right)^{1-\frac{1}{n}} + C$$

13. (a) Let $I = \int \sqrt{e^x - 1}\, dx$

Let $e^x - 1 = t^2 \Rightarrow e^x dx = 2t\, dt \Rightarrow dx = \frac{2t}{t^2 + 1} dt$
$$\therefore \quad I = \int t \cdot \frac{2t}{t^2 + 1} dt$$
$$= \int \frac{2t^2}{t^2 + 1} dt$$
$$= \int \frac{2(t^2 + 1) - 2}{t^2 + 1} dt$$
$$= 2\int dt - 2\int \frac{1}{t^2 + 1} dt$$
$$= 2t - 2\tan^{-1} t + C$$
$$= 2\sqrt{e^x - 1} - 2\tan^{-1} \sqrt{e^x - 1} + C$$

14. (b) Let $I=\int \frac{1}{x^2(x^4+1)^{3/4}}dx$

$\Rightarrow \quad I=\int \frac{1}{x^5\left(1+\frac{1}{x^4}\right)^{3/4}}dx$

$\Rightarrow \quad I=-\frac{1}{4}\int\left(1+\frac{1}{x^4}\right)^{-3/4}\left(-\frac{4}{x^5}\right)dx$

Let $1+\frac{1}{x^4}=t \Rightarrow -\frac{4}{x^5}dx=dt$

$\therefore \quad I=-\frac{1}{4}\int t^{-3/4}\,dt$

$=-\frac{1}{4}\left(\frac{t^{1/4}}{\frac{1}{4}}\right)+C$

$=-\left(1+\frac{1}{x^4}\right)^{1/4}+C$

15. (c) Let $I=\int \frac{x^4-1}{x^2\sqrt{x^4+x^2+1}}dx$

$=\int \frac{x^4-1}{x^2\cdot x\sqrt{x^2+\frac{1}{x^2}+1}}dx$

$=\int \frac{\left(x-\frac{1}{x^3}\right)}{\sqrt{x^2+\frac{1}{x^2}+1}}dx$

On putting $x^2+\frac{1}{x^2}+1=t$ or $2\left(x-\frac{1}{x^3}\right)dx=dt$

$\therefore \quad I=\frac{1}{2}\int\frac{dt}{\sqrt{t}}$

$=\frac{1}{2}\cdot 2\sqrt{t}+C$

$=\sqrt{t}+C$

$=\sqrt{x^2+\frac{1}{x^2}+1}+C$

$=\frac{\sqrt{x^4+x^2+1}}{x}+C$

16. (a) Let I be the given integral, which is binomial differential and $m=-6$, $n=3$ and $p=\frac{2}{3}$, so that

$\frac{m+1}{n}+p=-\frac{5}{3}+\frac{2}{3}=-1$ (integer)

On putting $1+2x^3=x^3t^3$

$\Rightarrow \quad x^3(t^3-2)=1$

$\Rightarrow \quad x=(t^3-2)^{-1/3}$

$\Rightarrow \quad dx=-\frac{1}{3}(t^3-2)^{-4/3}(3t^2)dt$

$\Rightarrow \quad dx=-t^2(t^3-2)^{-4/3}dt$

Hence, $I=\int[(t^3-2)^2\{(t^3-2)^{-1}t^3\}^{2/3}(-t^2)(t^3-2)^{-4/3}]\,dt$

$=-\int t^4(t^3-2)^{2-\frac{2}{3}-\frac{4}{3}}dt$

$=-\int t^4dt$

$=-\frac{t^5}{5}+C$

$=-\frac{1}{5}\left(\frac{1+2x^3}{x^3}\right)^{5/3}+C$

$=-\frac{1}{5}(2+x^{-3})^{5/3}+C$

17. (d) Let $I=\int \frac{x}{\sqrt{1+x^2+\sqrt{(1+x^2)^3}}}dx$

Let $(1+x^2)=t^2 \Rightarrow xdx=t\,dt$

$\therefore \quad I=\int\frac{t}{\sqrt{t^2+t^3}}dt=\int\frac{1}{\sqrt{1+t}}dt$

$=2\sqrt{1+t}+C$

$\Rightarrow \quad I=2\sqrt{1+\sqrt{1+x^2}}+C$

18. (a) Let $I=\int(x^{3m}+x^{2m}+x^m)(2x^{2m}+3x^m+6)^{\frac{1}{m}}dx$

$=\int(x^{2m}+x^m+1)(2x^{3m}+3x^{2m}+6x^m)^{\frac{1}{m}}\cdot x^{m-1}dx$

On putting $2x^{3m}+3x^{2m}+6x^m=t$

$\Rightarrow 6m(x^{3m-1}+x^{2m-1}+x^{m-1})\,dx=dt$

$\Rightarrow \quad 6m(x^{2m}+x^m+1)x^{m-1}dx=dt$

$\therefore \quad I=\int t^{1/m}\left(\frac{1}{6m}\right)dt$

$=\frac{1}{6m}\left(\frac{t^{(1/m)+1}}{\frac{1}{m}+1}\right)+C$

$=\frac{1}{6(m+1)}(2x^{3m}+3x^{2m}+6x^m)^{\frac{1+m}{m}}+C$

19. (c) Given that

$\int\frac{\cos x}{\sin^3 x(1+\sin^6 x)^{2/3}}dx$

$=f(x)(1+\sin^6 x)^{1/\lambda}+C \quad$ ...(i)

Let $\quad I=\int\frac{\cos x}{\sin^3 x(1+\sin^6 x)^{2/3}}dx$

$I=\int\frac{\cos x}{\sin^3 x\sin^4 x\left(\frac{1}{\sin^6 x}+1\right)^{2/3}}dx$

Let $\frac{1}{\sin^6 x}+1=t^3 \Rightarrow \frac{-6\cos x}{\sin^7 x}dx=3t^2dt$

So, $I=-\frac{1}{2}\int\frac{t^2dt}{t^2}=-\frac{1}{2}\int 1\,dt$

$$=-\frac{1}{2}t+C=-\frac{1}{2}\left(\frac{1}{\sin^6 x}+1\right)^{1/3}+C$$

$$I=-\frac{1}{2}\operatorname{cosec}^2x(1+\sin^6 x)^{1/3}+C \qquad \text{...(ii)}$$

On comparing Eqs. (i) and (ii), we get

$$f(x)=-\frac{1}{2}\operatorname{cosec}^2x \text{ and } \lambda=3$$

$$\therefore \lambda f\left(\frac{\pi}{3}\right)=3\left\{-\frac{1}{2}\operatorname{cosec}^2\left(\frac{\pi}{3}\right)\right\}=-\frac{3}{2}\times\frac{4}{3}=-2$$

20. (a) Let $I=\int\frac{1+\log x}{\sqrt{(x^x)^2-1}}dx$

$$=\int\frac{x^x(1+\log x)}{x^x\sqrt{(x^x)^2-1}}dx$$

Let $x^x=t$

$\therefore x\log x=\log t \Rightarrow (\log x+1)dx=\frac{1}{t}dt$

$\Rightarrow x^x(1+\log x)dx=dt$

$\therefore \qquad I=\int\frac{1}{t\sqrt{t^2-1}}dt$

$$=\sec^{-1}(t)+C=\sec^{-1}(x^x)+C$$

21. (c) $\int\frac{\ln(\tan x)}{\sin x\cos x}dx$

Let $t=\ln(\tan x) \Rightarrow dt=\frac{\sec^2 x}{\tan x}dx \Rightarrow dt=\frac{1}{\sin x\cos x}dx$

$\therefore \int\frac{\ln(\tan x)}{\sin x\cos x}dx=\int t\,dt=\frac{t^2}{2}+C$

$$=\frac{1}{2}(\ln\tan x)^2+C$$

22. (c) We have, $I_n=\int\tan^n x\,dx$

$\Rightarrow \qquad I_4+I_6=\int(\tan^4 x+\tan^6 x)\,dx$

$$=\int\tan^4 x(1+\tan^2 x)\,dx$$

$$=\int\tan^4 x\sec^2 x\,dx$$

Let $\tan x=t \Rightarrow \sec^2 x\,dx=dt$

$$=\int t^4dt$$

$$=\frac{t^5}{5}+C=\frac{\tan^5 x}{5}+C$$

$\therefore \qquad a=\frac{1}{5}$ and $b=0$

$\Rightarrow \qquad (a,b)=\left(\frac{1}{5},0\right)$

23. (a) Let $I=\int\cos^7 x\,dx$

$$=\int(1-\sin^2 x)^3\cos x\,dx$$

On putting $\sin x=t \Rightarrow \cos x\,dx=dt$

$$I=\int(1-t^2)^3\,dt$$

$$=\int\{1-(t^2)^3-3(t^2)+3(t^2)^2\}\,dt$$

$$=\int(1-t^6-3t^2+3t^4)\,dt$$

$$=\left[t-\frac{t^7}{7}-t^3+\frac{3t^5}{5}\right]+K$$

$$=-\frac{1}{7}\sin^7 x+\frac{3}{5}\sin^5 x-\sin^3 x+\sin x+K$$

Then, $A=-\frac{1}{7}, B=\frac{3}{5}$ and $C=-1$

$$\therefore 7A+20B+2C=7\left(-\frac{1}{7}\right)+20\left(\frac{3}{5}\right)+2(-1)=-1+12-2=9$$

24. (a) Let $I=\int\frac{\sin x}{\sin 3x}dx$

$$=\int\frac{\sin x}{3\sin x-4\sin^3 x}dx$$

$$=\int\frac{1}{3-4\sin^2 x}dx$$

$$=\int\frac{\sec^2 x}{3\sec^2 x-4\tan^2 x}dx$$

[$\because$ multiplying in $\sec^2 x$ in numerator and denominator]

$$=\int\frac{\sec^2 x}{3+3\tan^2 x-4\tan^2 x}dx$$

$$=\int\frac{\sec^2 x}{3-\tan^2 x}dx$$

On putting $\tan x=t \Rightarrow \sec^2 x\,dx=dt$

$\therefore \qquad I=\int\frac{dt}{3-t^2}=\frac{1}{2\sqrt{3}}\log\left|\frac{\sqrt{3}+t}{\sqrt{3}-t}\right|$

$$=\frac{1}{2\sqrt{3}}\log\left|\frac{\sqrt{3}+\tan x}{\sqrt{3}-\tan x}\right|+C$$

25. (c) Let $I=\int\frac{\sin^2 x\cos^2 x}{(\sin^5 x+\cos^3 x\sin^2 x+\sin^3 x\cos^2 x+\cos^5 x)^2}dx$

On dividing numerator and denominator by $\cos^{10} x$, we get

$$I=\int\frac{\tan^2 x\sec^6 x}{(\tan^5 x+\tan^2 x+\tan^3 x+1)^2}dx$$

$$=\int\frac{\tan^2 x\sec^6 x}{(1+\tan^2 x)^2(1+\tan^3 x)^2}dx$$

$$=\int\frac{\tan^2 x\sec^6 x}{(\sec^2 x)^2(1+\tan^3 x)^2}dx$$

$$=\int\frac{\tan^2 x\sec^2 x}{(1+\tan^3 x)^2}dx$$

$$=\frac{1}{3}\int\frac{3\tan^2 x\sec^2 x}{(1+\tan^3 x)^2}dx$$

$\therefore \qquad I=\frac{-1}{3(1+\tan^3 x)}+C$

26. (a) Let $I=\int \frac{\sec x}{\sqrt{\sin(2x+\alpha)+\sin\alpha}}\,dx$

$$\Rightarrow \quad I=\int \frac{\sec x}{\sqrt{2\sin(x+\alpha)\cos x}}\,dx$$

$$=\frac{1}{\sqrt{2}}\int \frac{\sec^2 x}{\sqrt{\tan x\cos\alpha+\sin\alpha}}\,dx$$

Let $\tan x\cos\alpha+\sin\alpha=t \Rightarrow \sec^2 x\cos\alpha\,dx=dt$

$$\therefore \quad I=\frac{1}{\sqrt{2}\cos\alpha}\int \frac{1}{\sqrt{t}}\,dt=\frac{1}{\sqrt{2}\cos\alpha}\cdot 2\sqrt{t}+C$$

$$=\frac{1}{\sqrt{2}\cos\alpha}\cdot 2\sqrt{\tan x\cos\alpha+\sin\alpha}+C$$

$$\Rightarrow \quad I=\sqrt{2\sec^2\alpha(\tan x\cos\alpha+\sin\alpha)}+C$$

$$\Rightarrow \quad I=\sqrt{2\sec\alpha(\tan x+\tan\alpha)}+C$$

27. (c) Let $I=\frac{\cos^{n-1}x}{\sin^{n+1}x}\,dx$

$$=\int\left(\frac{\cos x}{\sin x}\right)^n\cdot\frac{1}{\sin x\cos x}\,dx$$

$$=\int \cot^n x\cdot\frac{\sin x}{\cos x}\cdot\frac{1}{\sin^2 x}\,dx$$

$$=\int \frac{\cot^n x}{\cot x}\cdot \text{cosec}^2 x\,dx$$

$$=\int \cot^{n-1}x\cdot\text{cosec}^2 x\,dx$$

Let $\cot x=t \Rightarrow -\text{cosec}^2 x\,dx=dt$

$$=-\int t^{n-1}\,dt$$

$$=-\frac{t^n}{n}+C$$

$$\Rightarrow \quad I=-\frac{\cot^n x}{n}+C$$

28. (c) Let $I=\int \frac{d\theta}{\cos^2\theta(\tan 2\theta+\sec 2\theta)}$

$$=\int \frac{\sec^2\theta\cos 2\theta}{(\sin 2\theta+1)}\,d\theta$$

$$=\int \frac{\sec^2\theta(1-\tan^2\theta)}{(2\tan\theta+1+\tan^2\theta)}\,d\theta$$

$$=\int \frac{\sec^2\theta(1-\tan\theta)}{1+\tan\theta}\,d\theta$$

Let $\tan\theta=t \Rightarrow \sec^2\theta d\theta=dt$

Then, $I=\int\frac{1-t}{1+t}\,dt$

$$=\int\left(-1+\frac{2}{1+t}\right)dt$$

$$=-t+2\log_e|1+t|+C$$

$$=-\tan\theta+2\log_e|1+\tan\theta|+C$$

$=\lambda\tan\theta+2\log_e|f(\theta)|+C$ [given]

$\therefore \{\lambda, f(\theta)\}=(-1, 1+\tan\theta)$ [on comparing]

29. (b) Let $I=\int \frac{\sin^4 x}{\cos^8 x}\,dx$

$$\Rightarrow \quad I=\int\frac{\tan^4 x}{\cos^4 x}\,dx=\int \tan^4 x\sec^4 x\,dx$$

$$\Rightarrow \quad I=\int \tan^4 x\cdot(1+\tan^2 x)\sec^2 x\,dx$$

Let $\tan x=t \Rightarrow \sec^2 x\,dx=dt$

$$\Rightarrow \quad I=\int t^4(1+t^2)\,dt$$

$$\Rightarrow \quad I=\int (t^4+t^6)\,dt$$

$$\Rightarrow \quad I=\frac{t^5}{5}+\frac{t^7}{7}+C$$

$$\Rightarrow \quad I=\frac{1}{5}\tan^5 x+\frac{1}{7}\tan^7 x+C$$

$\therefore a=\frac{1}{5}$ and $b=\frac{1}{7}$

$\Rightarrow 5a=7b$

30. (c) Let $I=\int\frac{dx}{\{x^2-2x+10\}^2}=\int\frac{dx}{((x-1)^2+3^2)^2}$

Now, put $x-1=3\tan\theta$

$$\Rightarrow \quad dx=3\sec^2\theta\,d\theta$$

So, $I=\int\frac{3\sec^2\theta\,d\theta}{(3^2\tan^2\theta+3^2)^2}=\int\frac{3\sec^2\theta\,d\theta}{3^4\sec^4\theta}$

$$=\frac{1}{27}\int\cos^2\theta\,d\theta=\frac{1}{27}\int\frac{1+\cos 2\theta}{2}\,d\theta$$

$$=\frac{1}{54}\int(1+\cos 2\theta)\,d\theta=\frac{1}{54}\left(\theta+\frac{\sin 2\theta}{2}\right)+C$$

$$=\frac{1}{54}\theta+\frac{1}{108}\left(\frac{2\tan\theta}{1+\tan^2\theta}\right)+C$$

$$=\frac{1}{54}\tan^{-1}\left(\frac{x-1}{3}\right)+\frac{1}{54}\,\frac{\left(\frac{x-1}{3}\right)}{1+\left(\frac{x-1}{3}\right)^2}+C$$

$$\left[\because \text{put } \theta=\tan^{-1}\left(\frac{x-1}{3}\right)\right]$$

$$=\frac{1}{54}\tan^{-1}\left(\frac{x-1}{3}\right)+\frac{1}{18}\left(\frac{x-1}{(x-1)^2+3^2}\right)+C$$

$$=\frac{1}{54}\tan^{-1}\left(\frac{x-1}{3}\right)+\frac{1}{18}\left(\frac{x-1}{x^2-2x+10}\right)+C$$

$$=\frac{1}{54}\left[\tan^{-1}\left(\frac{x-1}{3}\right)+\frac{3(x-1)}{x^2-2x+10}\right]+C$$

It is given that

$$I=A\left[\tan^{-1}\left(\frac{x-1}{3}\right)+\frac{f(x)}{x^2-2x+10}\right]+C$$

On comparing, we get

$$A=\frac{1}{54} \text{ and } f(x)=3(x-1)$$

31. (c) Let $I=\int \frac{3-2\cot^{n-2}x}{3\tan x+2\cot x\cot^{n-2}x}dx$

$$=\int \frac{3\tan x-2\tan x\cot^{n-2}x}{3\tan^2 x+2\cot^{n-2}x}dx$$

$$=\int \frac{\tan x(3-2\cot^{n-2}x)}{3\tan^2 x+2\cot^{n-2}x}dx$$

$$=\int \frac{\tan x\cot^2 x(3-2\cot^{n-2}x)}{3\tan^2 x\cot^2 x+2\cot^n x}dx$$

$$=\int \frac{3\cot x-2\cot^{n-1}x}{3+2\cot^n x}dx$$

$$=\int \frac{3\cos x(\sin^{n-1}x)-2(\cos^{n-1}x)(\sin x)}{3\sin^n x+2\cos^n x}dx$$

On putting $m=3\sin^n x+2\cos^n x$

$\Rightarrow$ $[3n\sin^{n-1}x(\cos x)+2n\cos^{n-1}x(-\sin x)]dx=dm$

$\therefore$ $I=\frac{1}{n}\int \frac{1}{m}dm$

$$=\frac{1}{n}\log|m|+C$$

$$=\frac{1}{n}\log|3\sin^n x+2\cos^n x|+C$$

32. (a) We have, $I=\int \tan\left[2\tan^{-1}\left(\sqrt{\frac{\sqrt{1+\sqrt{x}}-1}{\sqrt{1+\sqrt{x}}+1}}\right)\right]dx$

Let $\sqrt{x}=\tan^2\theta$,

$x\in(0,1),\theta\in\left(0,\frac{\pi}{4}\right)$

$x=\tan^4\theta \Rightarrow \tan\theta=x^{1/4}$

$\therefore$ $dx=4\tan^3\theta\sec^2\theta\, d\theta$

Also, $\sqrt{1+\sqrt{x}}=\sec\theta$

Now, $\tan\left(2\tan^{-1}\sqrt{\frac{\sqrt{1+\sqrt{x}}-1}{\sqrt{1+\sqrt{x}}+1}}\right)$

$$=\tan\left(2\tan^{-1}\sqrt{\frac{\sec\theta-1}{\sec\theta+1}}\right)$$

$$=\tan\left\{2\tan^{-1}\left(\tan\frac{\theta}{2}\right)\right\}$$

$$=\tan\theta$$

Now, $I=\int \tan\theta\cdot 4\tan^3\theta\cdot\sec^2\theta\, d\theta$

$$=\frac{4}{5}\tan^5\theta+C$$

$$=\frac{4}{5}(x)^{5/4}+C$$

33. (b) Let $I=\int \frac{\cos x-\sin x}{\sqrt{8-\sin 2x}}dx$

On putting $\sin x+\cos x=t$

$\Rightarrow$ $(\cos x-\sin x)dx=dt$

and $1+\sin 2x=t^2$

$\therefore$ $I=\int \frac{dt}{\sqrt{8-(t^2-1)}}$

$$=\int \frac{dt}{\sqrt{9-t^2}}$$

$$=\sin^{-1}\left(\frac{t}{3}\right)+C$$

$$=\sin^{-1}\left(\frac{\sin x+\cos x}{3}\right)+C$$

Hence, $a=1$ and $b=3$

Then, ordered pair (a,b) is (1, 3).

34. (c) Let $I=\int \cos\left(2\cot^{-1}\sqrt{\frac{a-x}{a+x}}\right)dx$

On putting $x=a\cos 2\theta \Rightarrow dx=2a(-\sin 2\theta)\, d\theta$

$\therefore$ $I=\int \cos\left(2\cot^{-1}\sqrt{\frac{1-\cos 2\theta}{1+\cos 2\theta}}\right)2a(-\sin 2\theta)\, d\theta$

$\Rightarrow$ $I=\int \cos\{2\cot^{-1}(\tan\theta)\}(-2a\sin 2\theta)\, d\theta$

$$=\int \cos 2\left(\frac{\pi}{2}-\theta\right)(-2a\sin 2\theta)\, d\theta$$

$$=\int(-\cos 2\theta)(-2a\sin 2\theta)\, d\theta$$

$$=a\int \sin 4\theta\, d\theta$$

$$=-\frac{a\cos 4\theta}{4}+C \quad \text{...(i)}$$

We have, $x=a\cos 2\theta \Rightarrow \cos 2\theta=\frac{x}{a}$

$\Rightarrow$ $\cos 4\theta=2\cos^2 2\theta-1$

$$=\frac{2x^2}{a^2}-1=\frac{2x^2-a^2}{a^2}$$

From Eq. (i), $I=-\frac{a}{4}\left(\frac{2x^2-a^2}{a^2}\right)+C$

$\Rightarrow$ $I=\frac{a^2-2x^2}{4a}+C$

35. (b) Let $I=\int \frac{1}{(x+4)^{8/7}(x-3)^{6/7}}dx$

$$=\int \frac{1}{(x+4)^{8/7}(x+4)^{6/7}\left(\frac{x-3}{x+4}\right)^{6/7}}dx$$

$$=\int \frac{1}{(x+4)^2\left(\frac{x-3}{x+4}\right)^{6/7}}dx$$

On putting $t=\frac{x-3}{x+4}$

$\Rightarrow$ $\frac{dt}{dx}=\frac{(x+4)(1)-(x-3)(1)}{(x+4)^2}$

$$\Rightarrow \quad dt = \frac{7}{(x+4)^2}\,dx$$

$$\therefore \quad I = \frac{1}{7}\int \frac{dt}{t^{6/7}}$$

$$= \frac{1}{7}\left(\frac{t^{-6/7+1}}{-\frac{6}{7}+1}\right) + C = t^{1/7} + C$$

$$= \left(\frac{x-3}{x+4}\right)^{1/7} + C$$

36. (b) Let $I = \int \frac{dx}{\sqrt{(x-p)^3(x-q)}}$

$$= \int \frac{dx}{(x-p)^{3/2}(x-q)^{1/2}}$$

$$= \int \frac{dx}{(x-p)^2\left[\frac{(x-q)}{(x-p)}\right]^{1/2}}$$

Let $\frac{(x-q)}{(x-p)} = t$

$$\Rightarrow \left[\frac{(x-p)\cdot 1 - (x-q)\cdot 1}{(x-p)^2}\right]dx = dt$$

$$\Rightarrow \quad \frac{dx}{(x-p)^2} = \frac{dt}{(q-p)}$$

$$\therefore \quad I = \frac{1}{(q-p)}\int \frac{dt}{t^{1/2}}$$

$$= \frac{1}{(q-p)} \times 2\sqrt{t} + C$$

$$= \frac{2}{(q-p)}\sqrt{\frac{x-q}{x-p}} + C$$

37. (c) Let $I = \int \frac{xf'(x) - 2f(x)}{\sqrt{x^4 \cdot f(x)}}\,dx$

$$= \int \frac{xf'(x) - 2f(x)}{x^2\sqrt{f(x)}}\,dx$$

$$= \int \frac{\frac{x^2 f'(x) - 2x\,f(x)}{x^4}}{\sqrt{\frac{f(x)}{x^2}}}\,dx$$

On putting $\frac{f(x)}{x^2} = t^2$

$$\Rightarrow \frac{x^2 f'(x) - 2x\,f(x)}{x^4}\,dx = 2t\,dt$$

$$\therefore \quad I = \int \frac{2t\,dt}{t}$$

$$\Rightarrow \quad I = 2t + C$$

$$\Rightarrow \quad I = 2\sqrt{\frac{f(x)}{x^2}} + C$$

$$\Rightarrow \quad I = \frac{2\sqrt{f(x)}}{|x|} + C$$

38. (c) Let $I = 4\int \frac{\sqrt{a^6 + x^8}}{x}\,dx$

On putting $a^6 + x^8 = t^2 \Rightarrow 8x^7 dx = 2t\,dt$

$$\therefore \quad I = \int \frac{t^2}{t^2 - a^6}\,dt$$

$$= \int \left(1 + \frac{a^6}{t^2 - a^6}\right)dt$$

$$= t + \frac{a^3}{2}\ln\left|\frac{t-a^3}{t+a^3}\right| + C$$

$$\Rightarrow \quad I = \sqrt{a^6 + x^8} + \frac{a^3}{2}\ln\left|\frac{\sqrt{a^6+x^8} - a^3}{\sqrt{a^6+x^8} + a^3}\right| + C$$

39. (c) Let $I = \int \frac{x^2 + x + 1}{\sqrt{x^2 + 2x + 3}}\,dx$

$$\Rightarrow \quad I = \int \frac{(x^2+2x+3) - (x+2)}{\sqrt{x^2+2x+3}}\,dx$$

$$= \int \frac{x^2+2x+3}{\sqrt{x^2+2x+3}}\,dx - \int \frac{x+2}{\sqrt{x^2+2x+3}}\,dx$$

$$= \int \sqrt{x^2+2x+3}\,dx - \int \frac{\frac{1}{2}(2x+2)+1}{\sqrt{(x+1)^2+2}}\,dx$$

$$= \int \sqrt{(x+1)^2+2}\,dx - \frac{1}{2}\int \frac{(2x+2)\,dx}{\sqrt{(x+1)^2+2}}$$

$$- 1\int \frac{dx}{\sqrt{(x+1)^2+2}}$$

$$= \frac{1}{2}(x+1)\sqrt{x^2+2x+3} - \sqrt{x^2+2x+3} + C$$

$$\Rightarrow \quad I = \frac{1}{2}(x-1)\sqrt{x^2+2x+3} + C$$

40. (b) Let $I = \int \frac{x^4+1}{x^6+1}\,dx$

$$= \int \frac{x^4+1}{(x^2+1)(x^4-x^2+1)}\,dx$$

$$= \int \frac{(x^4-x^2+1)+x^2}{(x^2+1)(x^4-x^2+1)}\,dx$$

$$= \int \frac{dx}{x^2+1} + \int \frac{x^2}{x^6+1}\,dx$$

$$= \tan^{-1}x + \frac{1}{3}\int \frac{3x^2}{(x^3)^2+1}\,dx$$

$$= \tan^{-1}x + \frac{1}{3}\int \frac{dt}{1+t^2}, \text{ where } t = x^3$$

$$= \tan^{-1}x + \frac{1}{3}\tan^{-1}t + C$$

$$= \tan^{-1}x + \frac{1}{3}\tan^{-1}(x^3) + C$$

41. (c) Let $I=\int\frac{1}{x+\sqrt{x-1}}dx$

On putting $x-1=t^2 \Rightarrow dx=2t\,dt$

$$\therefore \quad I=2\int\frac{t}{t^2+t+1}dt$$

$$=\int\frac{(2t+1)-1}{t^2+t+1}dt$$

$$=\int\frac{2t+1}{t^2+t+1}dt-\int\frac{1}{t^2+t+1}dt$$

$$=\log(t^2+t+1)-\int\frac{1}{\left(t+\frac{1}{2}\right)^2+\left(\frac{\sqrt{3}}{2}\right)^2}dt$$

$$=\log(t^2+t+1)-\frac{2}{\sqrt{3}}\tan^{-1}\left(\frac{2t+1}{\sqrt{3}}\right)+C$$

$$=\log(x+\sqrt{x-1})-\frac{2}{\sqrt{3}}\tan^{-1}\left(\frac{2\sqrt{x-1}+1}{\sqrt{3}}\right)+C$$

42. (c) Let $I=\int\frac{1-\cos x}{\cos x(1+\cos x)}dx$

Let $\cos x=y$

Then, $\frac{1-\cos x}{\cos x\,(1+\cos x)}=\frac{1-y}{y(1+y)}$

By partial fraction,

$$=\frac{1}{y}-\frac{2}{1+y}$$

$$=\frac{1}{\cos x}-\frac{2}{1+\cos x}$$

$$\therefore \quad I=\int\frac{1-\cos x}{\cos x(1+\cos x)}dx$$

$$=\int\frac{1}{\cos x}dx-\int\frac{2}{1+\cos x}dx$$

$$=\int\sec x\,dx-\int\frac{2}{2\cos^2\frac{x}{2}}dx$$

$$=\ln|\sec x+\tan x|-\int\sec^2\frac{x}{2}dx$$

$$\Rightarrow I=\ln|\sec x+\tan x|-2\tan\frac{x}{2}+C$$

43. (d) Let $I=\int\frac{\cos\theta}{5+7\sin\theta-2\cos^2\theta}d\theta$

$$=\int\frac{\cos\theta}{3+7\sin\theta+2\sin^2\theta}d\theta$$

Let $\sin\theta=t \Rightarrow \cos\theta\,d\theta=dt$

$$\therefore \quad I=\int\frac{dt}{2t^2+7t+3}$$

$$=\frac{1}{2}\int\frac{dt}{t^2+\frac{7}{2}t+\frac{3}{2}}$$

$$=\frac{1}{2}\int\frac{dt}{\left(t+\frac{7}{4}\right)^2+\left(\frac{3}{2}-\frac{49}{16}\right)}=\frac{1}{2}\int\frac{dt}{\left(t+\frac{7}{4}\right)^2-\left(\frac{5}{4}\right)^2}$$

$$=\frac{1}{2}\times\frac{1}{2\times\frac{5}{4}}\log_e\left|\frac{t+\frac{7}{4}-\frac{5}{4}}{t+\frac{7}{4}+\frac{5}{4}}\right|+C'$$

$$=\frac{1}{5}\log_e\left|\frac{2t+1}{2t+6}\right|+C'$$

$$=\frac{1}{5}\log_e\left|\frac{2\sin\theta+1}{\sin\theta+3}\right|+C, \text{ where } C=C'-\frac{1}{5}\log_e 2$$

$$=A\log_e|B(\theta)|+C \qquad \text{[given]}$$

$$\therefore \quad \frac{B(\theta)}{A}=\frac{5(2\sin\theta+1)}{(\sin\theta+3)}$$

44. (d) Let $I=\int\frac{1}{x^{13}(x^6-1)}dx$

$$=\int\frac{x^5}{x^{18}(x^6-1)}dx$$

On putting $x^6=t \Rightarrow 6x^5dx=dt$

$$\therefore \quad I=\frac{1}{6}\int\frac{dt}{t^3(t-1)}$$

$$=\frac{1}{6}\int\left(\frac{1}{t-1}-\frac{1}{t}-\frac{1}{t^2}-\frac{1}{t^3}\right)dt$$

$$=\frac{1}{6}\left[\log|t-1|-\log|t|+\frac{1}{t}+\frac{1}{2t^2}\right]+C$$

$$=\frac{1}{6}\left[\log\left|\frac{t-1}{t}\right|+\frac{1}{t}+\frac{1}{2t^2}\right]+C$$

$$=\frac{1}{6}\left[\log\left|\frac{x^6-1}{x^6}\right|+x^{-6}+\frac{1}{2}x^{-12}\right]+C$$

45. (a) Let $I=\int\frac{d\theta}{(\sin\theta-2\cos\theta)(2\sin\theta+\cos\theta)}$

$$\Rightarrow \quad I=\int\frac{\sec^2\theta\,d\theta}{(\tan\theta-2)(2\tan\theta+1)}$$

On putting $\tan\theta=t \Rightarrow \sec^2\theta\,d\theta=dt$

$$\therefore \quad I=\int\frac{dt}{(t-2)(2t+1)}$$

Split into partial fractions, we get

$$=\int\left[\frac{1}{5(t-2)}-\frac{2}{5(2t+1)}\right]dt$$

$$=\frac{1}{5}\log|t-2|-\frac{1}{5}\log|2t+1|+C$$

$$=\frac{1}{5}\log\left|\frac{t-2}{2t+1}\right|+C$$

$$=\frac{1}{5}\log\left|\frac{\tan\theta-2}{2\tan\theta+1}\right|+C$$

46. (a) Let $I = \int \frac{dx}{\sin x + \sin 2x}$

$$= \int \frac{\sin x}{\sin^2 x(1 + 2\cos x)} dx$$

$$= \int \frac{\sin x\, dx}{(1 - \cos^2 x)(1 + 2\cos x)}$$

$$= \int \frac{\sin x\, dx}{(1 - \cos x)(1 + \cos x)(1 + 2\cos x)}$$

Let $\cos x = t \Rightarrow -\sin x\, dx = dt$

$$\therefore I = -\int \frac{dt}{(1-t)(1+t)(1+2t)}$$

Split into partial fractions, we get

$$= -\int \left[\frac{1}{6(1-t)} - \frac{1}{2(1+t)} + \frac{4}{3(1+2t)}\right] dt$$

$$= \int \left[\frac{-1}{6(1-t)} + \frac{1}{2(1+t)} - \frac{4}{3}\frac{1}{(1+2t)}\right] dt$$

$$= \frac{1}{6}\log(1 - \cos x) + \frac{1}{2}\log(1 + \cos x) - \frac{2}{3}\log(1 + 2\cos x) + K$$

So, $A = \frac{1}{6}, B = \frac{1}{2}$ and $C = -\frac{2}{3}$

$$\therefore \quad 6(A + B + C) = 6\left(\frac{1}{6} + \frac{1}{2} - \frac{2}{3}\right) = 6\left(\frac{1+3-4}{6}\right) = 0$$

47. (b) We have, $f(x) = \int \frac{1}{(e^x - 1)^2} dx = \int \frac{e^x}{e^x(e^x - 1)^2} dx$

On putting $e^x = t \Rightarrow e^x dx = dt$

$$\therefore \quad f(x) = \int \frac{dt}{t(t-1)^2}$$

$$= \int \left[\frac{1}{(t-1)^2} - \frac{1}{(t-1)} + \frac{1}{t}\right] dt$$

$$= -\frac{1}{(t-1)} - \log|t - 1| + \log|t| + C$$

$$= -\frac{1}{(e^x - 1)} - [\log|t - 1| - \log|t|] + C$$

$$= \frac{1}{1 - e^x} - \log\left|\frac{t-1}{t}\right| + C$$

$$= (1 - e^x)^{-1} - \log\left|\frac{e^x - 1}{e^x}\right| + C$$

$$= (1 - e^x)^{-1} - \log(1 - e^{-x}) + C$$

So, $f(x) = (1 - e^x)^{-1}$ and $g(x) = 1 - e^{-x}$

48. (d) Let $y^2 = f(x) = \frac{x+2}{2x+3}$, then

$$x = \frac{3y^2 - 2}{1 - 2y^2}$$

$$\Rightarrow \quad dx = -\frac{2y}{(1 - 2y^2)^2} dy$$

So, $\int \{f(x)\}^{1/2} \frac{dx}{x} = -\int y.\frac{2y}{(1-2y^2)^2} \cdot \frac{1-2y^2}{3y^2 - 2} dy$

$$= 2\int \frac{y^2}{(2y^2 - 1)(3y^2 - 2)} dy$$

By partial fraction,

$$= -2\int \left[\frac{1}{2y^2 - 1} - \frac{2}{3y^2 - 2}\right] dy$$

$$= \frac{1}{\sqrt{2}}\log\left|\frac{1 + \sqrt{2}y}{1 - \sqrt{2}y}\right| - \sqrt{\frac{2}{3}}\log\left|\frac{\sqrt{3}y + \sqrt{2}}{\sqrt{3}y - \sqrt{2}}\right| + C$$

Thus, $g(x) = \log|x|$ and $h(x) = \log|x|$

49. (b) Let $I = \int \left(\frac{x - \sin x}{1 - \cos x}\right) dx$

$$= \int \frac{x - \sin x}{2\sin^2 \frac{x}{2}} dx$$

$$= \frac{1}{2}\int x \operatorname{cosec}^2 \frac{x}{2} dx - \int \frac{2\sin\frac{x}{2}\cos\frac{x}{2}}{2\sin^2\frac{x}{2}} dx$$

$$= \frac{1}{2}\left[x\int \operatorname{cosec}^2\frac{x}{2}dx - \int\left(\frac{d}{dx}(x)\int \operatorname{cosec}^2\frac{x}{2}dx\right)dx\right]$$

$$- \int \cot\frac{x}{2}dx$$

$$= \frac{1}{2}\left[x\left(-2\cot\frac{x}{2}\right) - \int\left(-2\cot\frac{x}{2}\right)dx\right] - \int \cot\frac{x}{2}dx$$

$$= -x\cot\frac{x}{2} + \int \cot\frac{x}{2}dx - \int \cot\frac{x}{2}dx$$

$$= -x\cot\frac{x}{2} + C$$

50. (b) Let $I = \int x \cdot \frac{\ln(x + \sqrt{1 + x^2})}{\sqrt{1 + x^2}} dx$

On putting $\sqrt{1 + x^2} = t \Rightarrow \frac{x}{\sqrt{1 + x^2}} dx = dt$

$$\therefore \quad I = \int \ln(t + \sqrt{t^2 - 1})\, dt$$

$$\Rightarrow \quad I = \ln(t + \sqrt{t^2 - 1}) \cdot t - \int \frac{1 + \frac{t}{\sqrt{t^2 - 1}}}{t + \sqrt{t^2 - 1}} t\, dt$$

$$= t\ln(t + \sqrt{t^2 - 1}) - \frac{1}{2}\int \frac{2t}{\sqrt{t^2 - 1}} dt$$

$$\Rightarrow \quad I = \sqrt{1 + x^2}\ln(x + \sqrt{1 + x^2}) - x + C$$

$$\Rightarrow \quad a = 1$$

51. (a) Let $I = \int \frac{\sin^{-1}\sqrt{x} - \cos^{-1}\sqrt{x}}{\sin^{-1}\sqrt{x} + \cos^{-1}\sqrt{x}} dx$

$$\Rightarrow \quad I = \frac{2}{\pi}\int\left(2\sin^{-1}\sqrt{x} - \frac{\pi}{2}\right)dx \left[\because \sin^{-1}A + \cos^{-1}A = \frac{\pi}{2}\right]$$

$$\Rightarrow \quad I = \frac{4}{\pi}\int \sin^{-1}\sqrt{x}\,dx - x \quad \text{...(i)}$$

Now, let $I_1 = \int \sin^{-1}\sqrt{x}\,dx$, using integration by parts

$$I_1 = x\sin^{-1}\sqrt{x} - \int \frac{x}{\sqrt{1-x}} \cdot \frac{1}{2\sqrt{x}}\,dx$$

$$= x\sin^{-1}\sqrt{x} - \frac{1}{2}\int x^{1/2}(1-x)^{-1/2}dx$$

Let $1 - x = t^2 \Rightarrow dx = -2t\,dt$

$$\Rightarrow \quad I_1 = x\sin^{-1}\sqrt{x} - \frac{1}{2}\int \frac{\sqrt{1-t^2}}{t}(-2t\,dt)$$

$$= x\sin^{-1}\sqrt{x} + \int \sqrt{1-t^2}\,dt$$

$$= x\sin^{-1}\sqrt{x} + \left[\frac{t}{2}\sqrt{1-t^2} + \frac{1}{2}\sin^{-1}t\right]$$

$$\Rightarrow \quad I_1 = x\sin^{-1}\sqrt{x} + \left[\frac{\sqrt{x-x^2}}{2} + \frac{1}{2}\sin^{-1}\sqrt{1-x}\right]$$

On substituting the value of I_1 in Eq. (i), we get

$$I = \frac{4}{\pi}\left[x\sin^{-1}\sqrt{x} + \frac{1}{2}\sqrt{x-x^2} + \frac{1}{2}\sin^{-1}\sqrt{1-x}\right] - x + C$$

52. (d) Let $I = \int \sin^{-1}\left(\sqrt{\frac{x}{1+x}}\right)dx$

Let $\sin^{-1}\left(\sqrt{\frac{x}{1+x}}\right) = \theta$

$$\Rightarrow \quad \sin\theta = \sqrt{\frac{x}{1+x}}$$

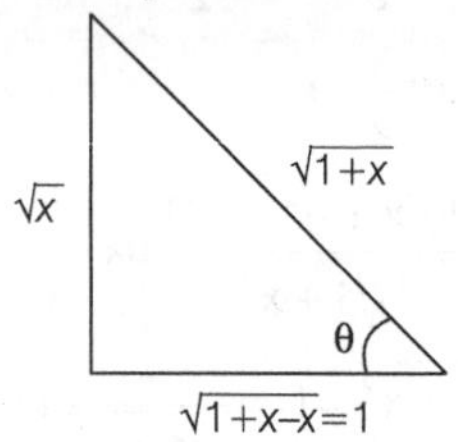

$\therefore \quad \tan\theta = \sqrt{x} \Rightarrow \theta = \tan^{-1}\sqrt{x}$

$\therefore \quad I = \int \tan^{-1}\sqrt{x}\,dx$

$$\Rightarrow \quad I = (\tan^{-1}\sqrt{x})x - \int \left(\frac{1}{1+x}\right)\left(\frac{x}{2\sqrt{x}}\right)dx \quad \text{...(i)}$$

Now, let $I_1 = \frac{1}{2}\int \frac{x}{(x+1)\sqrt{x}}\,dx$

Let $\sqrt{x} = t \Rightarrow \frac{1}{2\sqrt{x}}\,dx = dt$

$$\Rightarrow \quad I_1 = \int \frac{t^2}{t^2+1}\,dt$$

$$= \int \frac{t^2+1-1}{t^2+1}\,dt$$

$$= \int 1\,dt - \int \frac{1}{1+t^2}\,dt$$

$$\Rightarrow \quad I_1 = t - \tan^{-1}t$$

$$\Rightarrow \quad I_1 = \sqrt{x} - \tan^{-1}\sqrt{x}$$

From Eq. (i), we get

$$I = (\tan^{-1}\sqrt{x})\,x - (\sqrt{x} - \tan^{-1}\sqrt{x}) + C$$

$$\Rightarrow \quad I = \tan^{-1}\sqrt{x}\,(x+1) - \sqrt{x} + C$$

By comparing with $A(x)\tan^{-1}\sqrt{x} + B(x) + C$,

$$A(x) = x + 1 \text{ and } B(x) = -\sqrt{x}$$

Thus, the ordered pair is $(x+1, -\sqrt{x})$.

53. (a) Let $I = \int \left\{\log(\log x) + \frac{1}{(\log x)^2}\right\}dx$

Let $\log x = t \Rightarrow \frac{1}{x}dx = dt$

$$\Rightarrow \quad dx = x\,dt \Rightarrow dx = e^t dt$$

$$= \int e^t\left(\log t + \frac{1}{t^2}\right)dt$$

$$= \int e^t\left\{\left(\log t - \frac{1}{t}\right) + \left(\frac{1}{t} + \frac{1}{t^2}\right)\right\}dt$$

$$\Rightarrow \quad I = e^t\left(\log t - \frac{1}{t}\right) + C$$

$$= x\left[\log(\log x) - \frac{1}{\log x}\right] + C$$

$$\therefore \quad f(x) = \log(\log x) \text{ and } g(x) = \frac{1}{\log x}$$

54. (c) Let $I = \int \sin(\log x)\,dx$

Let $\log x = t$

$$\Rightarrow \quad x = e^t \Rightarrow dx = e^t dt$$

$$\therefore \quad I = \int e^t \sin t\,dt$$

$$= \sin t\,e^t - \int \cos t\,e^t\,dt$$

$$= \sin t \cdot e^t - [\cos t \cdot e^t + \int \sin t\,e^t dt]$$

$$\Rightarrow \quad I = e^t\sin t - e^t\cos t - I$$

$$\Rightarrow \quad 2I = e^t(\sin t - \cos t) + C$$

$$\Rightarrow \quad I = \frac{e^t}{2}(\sin t - \cos t) + C$$

$$\Rightarrow \quad I = \frac{x}{2}[\sin(\log x) - \cos(\log x)] + C$$

$$= \frac{x}{2}[\sin\{\ln(x)\} - \cos\{\ln(x)\}] + C$$

55. (a) Let $I = \int \sqrt{x}\,e^{\sqrt{x}}dx$

Let $\sqrt{x} = t \Rightarrow \frac{1}{2\sqrt{x}}dx = dt \Rightarrow dx = 2t\,dt$

$$\therefore \quad I = 2\int t^2 e^t dt$$

$$= 2[t^2 \cdot e^t - \int 2te^t dt]$$

$$= 2[t^2 e^t - 2(te^t - \int e^t dt)]$$

$$= 2[t^2e^t - 2te^t + 2e^t] + C$$
$$= 2[xe^{\sqrt{x}} - 2\sqrt{x}\,e^{\sqrt{x}} + 2e^{\sqrt{x}}] + C$$
$$= 2e^{\sqrt{x}}(x - 2\sqrt{x} + 2) + C$$

56. (d) Let $I = \int 4^x\{g'(x) + g(x)\log 4\}\,dx$

$$= \int 4^x g'(x)dx + \int 4^x \cdot g(x)\log 4\,dx$$
$$= 4^x \cdot g(x) - \int (4^x \log 4)\cdot g(x)\,dx + \int 4^x g(x)\log 4\,dx$$
$$\Rightarrow \quad I = 4^x g(x) + C$$

57. (a) Let $I = \int (e^{2x} + 2e^x - e^{-x} - 1)e^{(e^x + e^{-x})}dx$

Let $e^x = t \Rightarrow dx = dt/t$

So, $I = \int \left(t^2 + 2t - \frac{1}{t} - 1\right) e^{\left(t+\frac{1}{t}\right)} \frac{dt}{t}$

$$= \int \left(t + 2 - \frac{1}{t^2} - \frac{1}{t}\right) e^{\left(t+\frac{1}{t}\right)} dt$$
$$= \int \left[\left(t - \frac{1}{t}\right) + \left(1 - \frac{1}{t^2}\right) + 1\right] e^{\left(t+\frac{1}{t}\right)} dt$$
$$= \int \left[(1+t)\left(1 - \frac{1}{t^2}\right) + 1\right] e^{\left(t+\frac{1}{t}\right)} dt$$
$$= \int (1+t)\left(1 - \frac{1}{t^2}\right) e^{\left(t+\frac{1}{t}\right)} dt + \int e^{\left(t+\frac{1}{t}\right)} dt$$
$$= (1+t)e^{\left(t+\frac{1}{t}\right)} - \int 1\cdot e^{\left(t+\frac{1}{t}\right)} dt + \int e^{\left(t+\frac{1}{t}\right)} dt$$
$$= (1+t)e^{t+\frac{1}{t}} + C = (1+e^x)e^{(e^x + e^{-x})} + C$$
$$= g(x)e^{(e^x + e^{-x})} + C \qquad \text{[given]}$$

$\therefore\ g(x) = 1 + e^x \Rightarrow g(0) = 1 + 1 = 2$

58. (b) Let $I = \int \left(1 + 2x^2 + \frac{1}{x}\right) e^{\left(x^2 - \frac{1}{x}\right)} dx$

$$= \int \left[e^{x^2 - \frac{1}{x}} + x\left(2x + \frac{1}{x^2}\right) e^{x^2 - \frac{1}{x^2}}\right] dx$$
$$= xe^{x^2 - \frac{1}{x}} - \int e^{x^2 - \frac{1}{x}}\left(2x + \frac{1}{x^2}\right) x\,dx + \int x\left(2x + \frac{1}{x^2}\right) e^{x^2 - \frac{1}{x}} dx$$
$$= xe^{x^2 - \frac{1}{x}} + C$$

59. (b) Given,

$\int e^{\sec x}[(\sec x \tan x)f(x) + (\sec x \tan x + \sec^2 x)]dx$

$$= e^{\sec x}\cdot f(x) + C$$

On differentiating both sides w.r.t. x, we get

$e^{\sec x}[(\sec x \tan x)f(x) + (\sec x \tan x + \sec^2 x)]$

$$= e^{\sec x} f'(x) + e^{\sec x}(\sec x \tan x)f(x)$$

$\Rightarrow\ e^{\sec x}(\sec x \tan x + \sec^2 x) = e^{\sec x} f'(x)$,

where $f'(x) = \sec x \tan x + \sec^2 x$

So, $\quad f(x) = \int f'(x)dx = \int (\sec x \tan x + \sec^2 x)dx$

$$= \sec x + \tan x + C$$

So, possible value of $f(x)$ from options, is

$$f(x) = \sec x + \tan x + \frac{1}{2}$$

60. (2) $\int \frac{5\tan x}{\tan x - 2}dx = \int \frac{5\sin x}{\sin x - 2\cos x}dx$

$$= \int \frac{(\sin x - 2\cos x) + 2(\cos x + 2\sin x)}{(\sin x - 2\cos x)}dx$$
$$= \int 1\,dx + 2\int \frac{\cos x + 2\sin x}{\sin x - 2\cos x}dx$$
$$= x + 2\ln|\sin x - 2\cos x| + C$$

Hence, $a = 2$

61. (5) The given integral is

$\int \frac{\cos x + \cos 2x}{2\cos x - 1}dx$

$$= \int \frac{2\cos^2 x + \cos x - 1}{2\cos x - 1}dx \quad [\because \cos 2x = 2\cos^2 x - 1]$$
$$= \int \frac{(2\cos x - 1)(\cos x + 1)}{2\cos x - 1}dx$$
$$= \int (\cos x + 1)\,dx$$
$$= \sin x + x + C$$

Clearly, $A = 1$ and $B = 1$

Hence, the value of $A + B + 3 = 5$

62. (3) $\frac{d}{dx}(A\ln|\cos x + \sin x - 2| + Bx + C)$

$$= A\frac{\cos x - \sin x}{\cos x + \sin x - 2} + B$$
$$= \frac{A\cos x - A\sin x + B\cos x + B\sin x - 2B}{\cos x + \sin x - 2}$$

$\therefore\ A + B = 2,\ -1 = -A + B$ and $\lambda = -2B$

$\therefore\ A = \frac{3}{2}, B = \frac{1}{2}$ and $\lambda = -1$

$\therefore\ A + B + |\lambda| = \frac{3}{2} + \frac{1}{2} + |-1| = 3$

63. (4) The given integral is

$\int \frac{dx}{\sin^6 x + \cos^6 x} = \int \frac{dx}{1 - 3\sin^2 x \cos^2 x}$

$$= \int \frac{\sec^4 x\,dx}{\sec^4 x - 3\tan^2 x}$$
$$= \int \frac{(1 + \tan^2 x)\sec^2 x}{(1 + \tan^2 x)^2 - 3\tan^2 x}dx$$
$$= \int \frac{(1 + t^2)dt}{(1 + t^2)^2 - 3t^2}$$
$$= \int \left(\frac{t^2 + 1}{t^4 - t^2 + 1}\right)dt$$

$$=\int\left(\frac{1+\frac{1}{t^2}}{\left(t^2+\frac{1}{t^2}\right)-1}\right)dt$$

$$=\int\left(\frac{1+\frac{1}{t^2}}{\left(t-\frac{1}{t}\right)^2+1}\right)dt$$

$$=\tan^{-1}\left(t-\frac{1}{t}\right)+C$$

$$=\tan^{-1}(\tan x-\cot x)+C$$

Clearly, $L=1$ and $M=-1$

Hence, the value of $L+M+4=4$

64. (2012) Let $I=\int\frac{1-(\cot x)^{2010}}{\tan x+(\cot x)^{2011}}dx$

$$=\int\frac{(\sin x)^{2010}-(\cos x)^{2010}}{(\sin x)^{2010}}\cdot\frac{(\sin x)^{2011}\cdot\cos x}{(\sin x)^{2012}+(\cos x)^{2012}}dx$$

$$=\int\frac{[(\sin x)^{2010}-(\cos x)^{2010}]\sin x\cos x}{(\sin x)^{2012}+(\cos x)^{2012}}dx$$

On putting $t=(\sin x)^{2012}+(\cos x)^{2012}$

$\Rightarrow dt=[2012(\sin x)^{2011}\cos x+2012(\cos x)^{2011}(-\sin x)]dx$

$=2012[(\sin x)^{2010}-(\cos x)^{2010}]\sin x\cos x\,dx$

Therefore, $I=\int\frac{1}{t}\left(\frac{1}{2012}\right)dt$

$$=\frac{1}{2012}\log|t|+C$$

$$=\frac{1}{2012}\log|(\sin x)^{2012}+(\cos x)^{2012}|+C$$

Hence, $k=2012$

65. (8) Let $I=\int\left(\frac{3x^2+2x}{x^6+2x^5+x^4+2x^3+2x^2+2}\right)dx$

$$=\int\left(\frac{3x^2+2x}{1+\{x^6+x^4+1+2(x^5+x^3+x^2)\}}\right)dx$$

$$=\int\frac{3x^2+2x}{1+(x^3+x^2+1)^2}dx$$

$$=\tan^{-1}(x^3+x^2+1)+C$$

Thus, $L=1, M=3, N=2, P=1$ and $Q=1$

Hence, the value of $L+M+N+P+Q=8$

66. (9) Let $I=\int\frac{x+\sqrt[3]{x^2}+\sqrt[6]{x}}{x(1+\sqrt[3]{x})}dx$

Let $x^{1/6}=u\Rightarrow\frac{1}{6}\frac{1}{x^{5/6}}dx=du$

$\therefore\quad I=6\int\frac{u^5+u^3+1}{u^2+1}du=6\int\left(u^3+\frac{1}{u^2+1}\right)du$

$$=6\left[\frac{u^4}{4}+\tan^{-1}(u)\right]+C$$

$$=6\frac{(x^{1/6})^4}{4}+6\tan^{-1}(x^{1/6})+C$$

$$=\frac{3}{2}x^{2/3}+6\tan^{-1}(x^{1/6})+C$$

So, $a=\frac{3}{2}$ and $b=6$

Hence, $ab=\frac{3}{2}\times6=9$

67. (41) Let $2x+5=A\frac{d}{dx}(9x^2+6x+2)+B$

$\Rightarrow\quad 2x+5=A(18x+6)+B$

Therefore, $A=\frac{2}{18}=\frac{1}{9}$ and $B=\frac{13}{3}$

Hence, $\int\frac{2x+5}{\sqrt{9x^2+6x+2}}dx$

$$=\frac{1}{9}\int\frac{18x+6}{\sqrt{9x^2+6x+2}}dx+\frac{13}{3}\int\left(\frac{1}{\sqrt{(3x+1)^2+1}}\right)dx$$

$$=\frac{2}{9}\sqrt{9x^2+6x+2}+\frac{13}{3}\log(3x+1+\sqrt{9x^2+6x+2})+C$$

Therefore, $9(a+b)=9\left(\frac{2}{9}+\frac{13}{3}\right)=41$

68. (4) We have, $f(x)=\int\frac{5x^8+7x^6}{(x^2+1+2x^7)^2}dx$

$$=\int\frac{5x^8+7x^6}{x^{14}(x^{-5}+x^{-7}+2)^2}dx$$

$$=\int\frac{5x^{-6}+7x^{-8}}{(x^{-5}+x^{-7}+2)^2}dx$$

Let $x^{-5}+x^{-7}+2=t\Rightarrow(-5x^{-6}-7x^{-8})dx=dt$

$\therefore\quad f(x)=\int-\frac{1}{t^2}dt$

$\Rightarrow\quad f(x)=\frac{1}{t}+C$

$\Rightarrow\quad f(x)=\frac{1}{x^{-5}+x^{-7}+2}+C$

Now, $x=0$, $f(0)=0\Rightarrow C=0$

$\Rightarrow\quad f(x)=\frac{1}{x^{-5}+x^{-7}+2}$

$\Rightarrow\quad f(x)=\frac{x^7}{x^2+1+2x^7}$

Hence, $f(1)=\frac{1}{1+1+2}=\frac{1}{4}$,

so $\quad k=4$

69. (1) $\int e^{x^3+x^2-1}(3x^4+2x^3+2x)\,dx$

$$=\int \underbrace{x^2}_{\text{I}}\underbrace{e^{x^3+x^2-1}(3x^2+2x)\,dx}_{\text{II}}+\int e^{x^3+x^2-1}(2x)\,dx$$

$$=x^2e^{x^3+x^2-1}-\int 2x\,e^{x^3+x^2-1}dx+\int e^{x^3+x^2-1}(2x)\,dx$$

$$=x^2e^{x^3+x^2-1}+C$$

So, $\quad f(x)=x^2e^{x^3+x^2-1}$

$\Rightarrow \quad f(1)=1e^{1+1-1}=e$

$\Rightarrow \quad f(-1)=1\,e^{-1+1-1}=e^{-1}=\frac{1}{e}$

$\Rightarrow f(1)\times f(-1)=e\cdot\frac{1}{e}=1$

70. (3) Let $I=\int\frac{e^x(2-x^2)}{(1-x)\sqrt{1-x^2}}dx$

$$\Rightarrow \quad I=\int e^x\left(\frac{1+(1-x^2)}{(1-x)\sqrt{1-x^2}}\right)dx$$

$$=\int e^x\left(\frac{1}{(1-x)\sqrt{1-x^2}}+\sqrt{\frac{1+x}{1-x}}\right)dx$$

Here, $f(x)=\sqrt{\frac{1+x}{1-x}}$

Then, $f'(x)=\frac{1}{(1-x)\sqrt{1-x^2}}$

So, $I=e^x\sqrt{\frac{1+x}{1-x}}+C$

So, $A=1$ and $B=\frac{1}{2}$

Then, $2(A+B)=2\left(1+\frac{1}{2}\right)=2\times\frac{3}{2}=3$

Level 3

1. (c) Let $I=\int\{1+\tan x\cdot\tan(x+A)\}\,dx$

$$=\int\left\{1+\frac{\sin x\sin(x+A)}{\cos x\cos(x+A)}\right\}dx$$

$$=\int\frac{\cos x\cos(x+A)+\sin x\sin(x+A)}{\cos x\cos(x+A)}dx$$

$$=\int\frac{\cos(x+A-x)\,dx}{\cos x\cos(x+A)}$$

$$=\cos A\int\frac{dx}{\cos x\cos(x+A)}$$

On multiplying and dividing by $\sin A$, we get

$$=\cot A\int\frac{\sin A\,dx}{\cos x\cos(x+A)}$$

$$=\cot A\int\frac{\sin(x+A-x)\,dx}{\cos x\cos(x+A)}$$

$$=\cot A\int\left\{\frac{\sin(x+A)\cos x}{\cos x\cos(x+A)}-\frac{\cos(x+A)\sin x}{\cos x\cos(x+A)}\right\}dx$$

$$=\cot A\int\{\tan(x+A)-\tan x\}\,dx$$

$$=\cot A\,\{\log|\sec(x+A)|-\log|\sec x|\}+C$$

$$=\cot A\log\left|\frac{\sec(x+A)}{\sec x}\right|+C$$

2. (c) Let $I=\int\frac{\sec^2 x}{(\sec x+\tan x)^{9/2}}dx$

$$=\int\frac{\sec x\cdot\sec x}{(\sec x+\tan x)^{9/2}}dx$$

On putting $\sec x+\tan x=t$

$\Rightarrow \sec x(\sec x+\tan x)dx=dt$

$\Rightarrow \quad \sec x\,dx=\frac{dt}{t}$

Also, $\sec x-\tan x=\frac{1}{(\sec x+\tan x)}=\frac{1}{t}$

Thus, $\sec x=\frac{1}{2}\left(t+\frac{1}{t}\right)$

$$\therefore \quad I=\int\frac{\frac{1}{2}\left(t+\frac{1}{t}\right)\frac{dt}{t}}{t^{9/2}}=\frac{1}{2}\int\frac{(t^2+1)}{t^{13/2}}dt$$

$$=\frac{1}{2}\int\left(\frac{1}{t^{9/2}}+\frac{1}{t^{13/2}}\right)dt$$

$$=\frac{-1}{2}\left(\frac{2}{7t^{7/2}}+\frac{2}{11t^{11/2}}\right)+C$$

$$=-\left(\frac{1}{7t^{7/2}}+\frac{1}{11t^{11/2}}\right)+C$$

$$=-\left(\frac{1}{7(\sec x+\tan x)^{7/2}}+\frac{1}{11(\sec x+\tan x)^{11/2}}\right)+C$$

$$=-\frac{1}{(\sec x+\tan x)^{11/2}}\left\{\frac{1}{11}+\frac{1}{7}(\sec x+\tan x)^2\right\}+C$$

3. (d) Let $I=\int\left(\frac{1-\sqrt{x}}{1+\sqrt{x}}\right)^{\frac{1}{2}}\frac{dx}{x}$

On putting $x=\cos^2 2\theta\Rightarrow dx=-2\sin 4\theta\,d\theta$

$$\therefore \quad I=\int\left(\frac{1-\cos 2\theta}{1+\cos 2\theta}\right)^{1/2}\frac{1}{\cos^2 2\theta}(-2\sin 4\theta)\,d\theta$$

$$=\int\frac{\tan\theta}{\cos^2 2\theta}(-4\sin 2\theta\cos 2\theta)\,d\theta$$

$$=\int\frac{\sin\theta}{\cos\theta\cdot\cos 2\theta}(-8\sin\theta\cos\theta)\,d\theta$$

$$=-8\int\frac{\sin^2\theta}{\cos 2\theta}d\theta$$

$$=-8\int\frac{1-\cos^2\theta}{2\cos^2\theta-1}d\theta$$

$$= 4\int \frac{2\cos^2\theta - 1 - 1}{2\cos^2\theta - 1} d\theta$$

$$= 4\int \left(1 - \frac{1}{\cos 2\theta}\right) d\theta$$

$$= 4\int (1 - \sec 2\theta) d\theta$$

$$= 4\left[\theta - \frac{1}{2}\log|\sec 2\theta + \tan 2\theta| + C\right]$$

$$= 4\left(\frac{1}{2}\cos^{-1}\sqrt{x}\right) - 2\log\left|\frac{1}{\sqrt{x}} + \frac{\sqrt{1-x}}{\sqrt{x}}\right| + C$$

$$= 2\cos^{-1}\sqrt{x} - 2\log\left(\frac{1+\sqrt{1-x}}{\sqrt{x}}\right) + C$$

So, $f(x) = 2\log\left(\frac{1+\sqrt{1-x}}{\sqrt{x}}\right)$

4. (a) Let $I = \int \frac{\sin x + \cos x}{9 + 16\sin 2x} dx$

Let $\sin x - \cos x = t \Rightarrow (\cos x + \sin x) dx = dt$

and $t^2 = 1 - \sin 2x \Rightarrow \sin 2x = (1 - t^2)$

$$\therefore \quad I = \int \frac{dt}{9 + 16(1-t^2)} = \int \frac{dt}{25 - 16t^2}$$

$$= \frac{1}{16}\int \frac{dt}{\left(\frac{5}{4}\right)^2 - t^2}$$

$$= \frac{1}{16} \times \frac{1}{2\left(\frac{5}{4}\right)} \ln\left|\frac{\frac{5}{4} + t}{\frac{5}{4} - t}\right| + C$$

$$= \frac{1}{40}\ln\left|\frac{5+4t}{5-4t}\right| + C$$

$$= \frac{1}{40}\ln\left|\frac{5 + 4(\sin x - \cos x)}{5 - 4(\sin x - \cos x)}\right| + C$$

5. (b) Let $I = \int \frac{1 + x\cos x}{x(1 - x^2 e^{2\sin x})} dx$

On putting $x e^{\sin x} = t \Rightarrow (x e^{\sin x}\cos x + e^{\sin x})\, dx = dt$

$\Rightarrow \quad e^{\sin x}(x\cos x + 1)\, dx = dt$

$$\therefore \quad I = \int \frac{dt}{t(1-t^2)} = \int \frac{dt}{t(1-t)(1+t)}$$

Using partial fraction

$$= \int \left(\frac{1}{t} + \frac{1}{2(1-t)} - \frac{1}{2(1+t)}\right) dt$$

$$= \log|t| - \frac{1}{2}\log|1-t| - \frac{1}{2}\log|1+t| + C$$

$$= \log|t| - \frac{1}{2}\log|1-t^2| + C$$

$$= \log|x e^{\sin x}| - \frac{1}{2}\log|1 - x^2 e^{2\sin x}| + C$$

6. (c) Let $I = \int \frac{x}{(x^2 - 3x + 2)\sqrt{x^2 - 4x + 3}} dx$

$$= \int \frac{x}{(x-2)(x-1)\sqrt{x^2 - 4x + 3}} dx$$

$$= \int \left(\frac{2}{x-2} - \frac{1}{x-1}\right) \frac{dx}{\sqrt{x^2 - 4x + 3}}$$

Therefore,

$$I = 2\int \frac{dx}{(x-2)\sqrt{x^2-4x+3}} - \int \frac{dx}{(x-1)\sqrt{x^2-4x+3}} \quad \text{...(i)}$$

Let $I_1 = \int \frac{dx}{(x-2)\sqrt{x^2-4x+3}}$

$$= \int \frac{dx}{(x-2)\sqrt{(x-2)^2 - 1}}$$

$$= \int \frac{\frac{1}{(x-2)^2}}{\sqrt{1 - \left(\frac{1}{x-2}\right)^2}} dx = -\sin^{-1}\left(\frac{1}{x-2}\right) + C$$

Now, let $I_2 = \int \frac{dx}{(x-1)\sqrt{x^2-4x+3}}$

$$= \int \frac{dx}{(x-1)^{3/2}(x-3)^{1/2}} = \int \frac{dx}{\left(\frac{x-1}{x-3}\right)^{3/2}(x-3)^2}$$

On putting $\frac{x-1}{x-3} = t^2$

$$\Rightarrow \quad x = \frac{1-3t^2}{1-t^2} = 3 - \frac{2}{1-t^2}$$

$$\Rightarrow \quad x - 3 = \frac{-2}{1-t^2}$$

Therefore, $dx = \frac{-4t}{(1-t^2)^2} dt$

This gives, $I_2 = \int \frac{1}{t^3\left(\frac{4}{(1-t^2)^2}\right)} \cdot \frac{-4t}{(1-t^2)^2} dt$

$$I_2 = -\int \frac{1}{t^2} dt$$

$$\Rightarrow \quad I_2 = \frac{1}{t}$$

$$\Rightarrow \quad I_2 = \sqrt{\frac{x-3}{x-1}}$$

$$\Rightarrow \quad I_2 = \frac{\sqrt{x^2 - 4x + 3}}{x-1} + C$$

On substituting the values of I_1 and I_2 in Eq. (i), we get

$$I = -2\sin^{-1}\left(\frac{1}{x-2}\right) - \frac{\sqrt{x^2-4x+3}}{x-1} + C$$

7. (c) We have, $J - I = \int \left(\frac{e^{3x}}{e^{4x} + e^{2x} + 1} - \frac{e^x}{e^{4x} + e^{2x} + 1} \right) dx$

$$\Rightarrow \quad J - I = \int \frac{e^{3x} - e^x}{e^{4x} + e^{2x} + 1} dx$$

Let $e^x = t \Rightarrow e^x dx = dt$

$$\Rightarrow \quad J - I = \int \frac{t^2 - 1}{t^4 + t^2 + 1} dt$$

$$\Rightarrow \quad J - I = \int \frac{\left(1 - \frac{1}{t^2}\right)}{\left(t + \frac{1}{t}\right)^2 - 1} dt$$

Let $t + \frac{1}{t} = P \Rightarrow \left(1 - \frac{1}{t^2}\right) dt = dP$

$$\therefore \quad J - I = \int \frac{dP}{P^2 - 1}$$

$$\Rightarrow \quad J - I = \frac{1}{2} \log \left| \frac{P - 1}{P + 1} \right| + C$$

$$= \frac{1}{2} \log \left(\frac{t + \frac{1}{t} - 1}{t + \frac{1}{t} + 1} \right) + C$$

$$= \frac{1}{2} \log \left(\frac{t^2 - t + 1}{t^2 + t + 1} \right) + C$$

$$= \frac{1}{2} \log \left(\frac{e^{2x} - e^x + 1}{e^{2x} + e^x + 1} \right) + C$$

8. (a) Let $I = \int \sqrt{\frac{3 - x}{3 + x}} \sin^{-1} \left(\frac{1}{\sqrt{6}} \sqrt{3 - x} \right) dx$

On putting $x = 3 \cos 2\theta \Rightarrow dx = -6 \sin 2\theta \, d\theta$

$$I = \int \sqrt{\frac{3 - 3\cos 2\theta}{3 + 3\cos 2\theta}} \sin^{-1} \left(\frac{1}{\sqrt{6}} \sqrt{3 - 3\cos 2\theta} \right) (-6 \sin 2\theta) \, d\theta$$

$$= \int \frac{\sin\theta}{\cos\theta} \sin^{-1}(\sin\theta) (-6 \sin 2\theta) \, d\theta$$

$$= -6 \int \theta (2 \sin^2 \theta) \, d\theta = -6 \int \theta (1 - \cos 2\theta) \, d\theta$$

$$= -6 \left\{ \frac{\theta^2}{2} - \int \theta \cos 2\theta \, d\theta \right\}$$

$$= -6 \left\{ \frac{\theta^2}{2} - \left(\theta \cdot \frac{\sin 2\theta}{2} - \int 1 \cdot \left(\frac{\sin 2\theta}{2} \right) d\theta \right) \right\}$$

$$= -6 \left\{ \frac{\theta^2}{2} - \left(\frac{\theta \sin 2\theta}{2} + \frac{\cos 2\theta}{4} \right) \right\} + C$$

$$= -3\theta^2 + 6 \left\{ \frac{\theta \sin 2\theta}{2} + \frac{\cos 2\theta}{4} \right\} + C$$

$$= \frac{1}{4} \left\{ -3 \left(\cos^{-1} \left(\frac{x}{3} \right) \right)^2 + 2\sqrt{9 - x^2} \cos^{-1} \left(\frac{x}{3} \right) + 2x \right\} + C$$

9. (b) Let $I = \int \left(\frac{x}{x \sin x + \cos x} \right)^2 dx$

$$I = \int \frac{x}{\cos x} \cdot \frac{x \cos x}{(x \sin x + \cos x)^2} dx$$

$$= \frac{x}{\cos x} \int \frac{x \cos x}{(x \sin x + \cos x)^2} dx$$

$$- \int \left(\frac{\cos x + x \sin x}{\cos^2 x} \int \frac{x \cos x}{(x \cos x + \cos x)^2} dx \right) dx \quad \text{...(i)}$$

Let $I_1 = \int \frac{x \cos x}{(x \sin x + \cos x)^2} dx$

Let $x \sin x + \cos x = t$

$\Rightarrow (x \cos x + \sin x - \sin x) \, dx = dt$

$\Rightarrow \quad x \cos x \, dx = dt$

$$\therefore \quad I_1 = \int \frac{1}{t^2} dt = -\frac{1}{t} = -\frac{1}{x \sin x + \cos x}$$

From Eq. (i),

$$I = \frac{x}{\cos x} \left(-\frac{1}{x \sin x + \cos x} \right) - \int \frac{\cos x + x \sin x}{\cos^2 x} \times \left(-\frac{1}{x \sin x + \cos x} \right) dx$$

$$= \frac{x}{\cos x} \left(-\frac{1}{x \sin x + \cos x} \right) + \int \sec^2 x \, dx$$

$$= \frac{-x \sec x}{x \sin x + \cos x} + \tan x + C$$

$$= \tan x - \frac{x \sec x}{x \sin x + \cos x} + C$$

10. (d) Let $I = \int \frac{e^{(x^2 + 4 \ln x)} - x^3 e^{x^2}}{x - 1} dx$

$$= \int \frac{e^{x^2} \cdot x^4 - x^3 \cdot e^{x^2}}{x - 1} dx$$

$$= \int \frac{x^3 \cdot e^{x^2} (x - 1)}{x - 1} dx$$

$$= \int x^3 \cdot e^{x^2} dx$$

On putting $x^2 = t \Rightarrow 2x \, dx = dt$

$$\therefore \quad I = \frac{1}{2} \int t e^t dt$$

$$= \frac{1}{2} [t \cdot e^t - \int e^t dt]$$

$$= \frac{1}{2} [t e^t - e^t] + C$$

$$= \frac{1}{2} (t - 1) e^t + C$$

$$= \frac{1}{2} (x^2 - 1) e^{x^2} + C$$

11. (a, d) Let $I=\int \sin x \sec x \tan x\, dx$

$$=\int \sin x \cdot \frac{1}{\cos x}\cdot \frac{\sin x}{\cos x}\, dx$$
$$=\int \tan^2 x\, dx$$
$$=\int (\sec^2 x-1)\, dx$$
$$=\tan x - x + C$$

Hence, $f(x)=\tan x$ and $g(x)=-x$

12. (b, d) Let $I=\int \sqrt{\frac{a+x}{a-x}}\, dx$

$$=\int \sqrt{\frac{(a+x)(a+x)}{(a-x)(a+x)}}\, dx$$
$$=\int \frac{a+x}{\sqrt{a^2-x^2}}\, dx$$
$$=a\int \frac{dx}{\sqrt{a^2-x^2}}+\int \frac{x}{\sqrt{a^2-x^2}}\, dx \quad \text{...(i)}$$
$$=a\sin^{-1}\frac{x}{a}-\sqrt{a^2-x^2}+C$$

Also, again from Eq. (i),

$$=a\int \frac{dx}{\sqrt{a^2-x^2}}+\int \frac{x}{\sqrt{a^2-x^2}}\, dx$$
$$=-a\cos^{-1}\frac{x}{a}-\sqrt{a^2-x^2}+C$$

13. (a, c) Let $I=\int \frac{4e^x+6e^{-x}}{9e^x-4e^{-x}}\, dx$

Write $4e^x+6e^{-x}=\lambda(9e^x-4e^{-x})+\mu\frac{d}{dx}(9e^x-4e^{-x})$

$\Rightarrow \quad 4e^x+6e^{-x}=\lambda(9e^x-4e^{-x})+\mu(9e^x+4e^{-x})$

$\therefore 9\lambda+9\mu=4$ and $-4\lambda+4\mu=6$

On solving, we get

$$\lambda=-\frac{19}{36} \text{ and } \mu=\frac{35}{36}$$

Therefore, $I=\int \lambda dx+\mu\int \frac{9e^x+4e^{-x}}{9e^x-4e^{-x}}\, dx$

$$=\frac{-19}{36}x+\frac{35}{36}\log(9e^x-4e^{-x})+C$$

If we write $e^{-x}=\frac{1}{e^x}$, then

$$I=\int \frac{4e^{2x}+6}{9e^{2x}-4}\, dx$$

i.e. $\frac{4e^{2x}+6}{9e^{2x}-4}=\frac{4}{9}+\frac{70}{9(9e^{2x}-4)}$

Therefore, $\int \frac{4e^{2x}+6}{9e^{2x}-4}\, dx=\frac{4}{9}x+\int \frac{70}{9(9e^{2x}-4)}\, dx$

$$=\frac{4}{9}x+\frac{70}{9}\int \frac{dx}{9e^{2x}-4}$$

14. (b, c) Let $I=\int \frac{\{f'(x)\cdot g(x)-g'(x)\cdot f(x)\}\, dx}{\{f(x)+g(x)\}\sqrt{g(x)f(x)-g^2(x)}}$

$$=\int \frac{\left[\frac{f'(x)\cdot g(x)-g'(x)\cdot f(x)}{\{g(x)\}^2}\right]dx}{\left(\frac{f(x)}{g(x)}+1\right)\sqrt{\frac{f(x)}{g(x)}-1}}$$

On putting $\frac{f(x)}{g(x)}-1=t^2$

$$\Rightarrow \left(\frac{f'(x)\cdot g(x)-g'(x)\cdot f(x)}{\{g(x)\}^2}\right)dx=2t\,dt$$

$\therefore \quad I=\int \frac{2t\,dt}{(t^2+2)t}$

$$=2\cdot\frac{1}{\sqrt{2}}\tan^{-1}\left(\frac{t}{\sqrt{2}}\right)$$
$$=\sqrt{2}\tan^{-1}\left(\frac{1}{\sqrt{2}}\sqrt{\frac{f(x)}{g(x)}-1}\right)+C$$
$$=\sqrt{2}\tan^{-1}\left(\sqrt{\frac{f(x)-g(x)}{2g(x)}}\right)+C$$

Hence, $m=2$ and $n=2$

15. (a, c) We have, $I=\int \sqrt{\frac{x}{a^3-x^3}}\, dx$

$$=\int \frac{\sqrt{x}}{\sqrt{(a^{3/2})^2-(x^{3/2})^2}}\, dx$$

Let $x^{3/2}=t \Rightarrow \frac{3}{2}x^{1/2}\, dx=dt \Rightarrow x^{1/2}\, dx=\frac{2}{3}\, dt$

$\therefore \quad I=\frac{2}{3}\int \frac{dt}{\sqrt{(a^{3/2})^2-t^2}}$

$$=\frac{2}{3}\sin^{-1}\left(\frac{t}{a^{3/2}}\right)+C$$
$$=\frac{2}{3}\sin^{-1}\left(\frac{x^{3/2}}{a^{3/2}}\right)+C$$

Hence, $A=\frac{2}{3}$ and $B=a^{3/2}$

16. (a, c, d) Let $I=\int \frac{dx}{1-\sin^4 x}$

Numerator and denominator divided by $\cos^4 x$,

$$=\int \frac{\sec^4 x}{\sec^4 x-\tan^4 x}\, dx$$
$$=\int \frac{\sec^2 x\cdot\sec^2 x}{(1+\tan^2 x)^2-\tan^4 x}\, dx$$
$$=\int \frac{\sec^2 x\cdot\sec^2 x}{1+\tan^4 x+2\tan^2 x-\tan^4 x}\, dx$$
$$I=\int \frac{(1+\tan^2 x)\sec^2 x\, dx}{1+2\tan^2 x}$$

On putting $\tan x = t \Rightarrow \sec^2 x\, dx = dt$

$$\therefore \quad I = \int \frac{1+t^2}{1+2t^2}\, dt$$

$$= \int \frac{1}{1+2t^2}\, dt + \int \frac{t^2}{1+2t^2}\, dt$$

$$= \frac{1}{\sqrt{2}} \tan^{-1}(\sqrt{2}t) - \frac{1}{2\sqrt{2}} \tan^{-1}(\sqrt{2}\, t) + \frac{1}{2} t + D$$

$$= \frac{1}{2} t + \frac{1}{2\sqrt{2}} \tan^{-1}(\sqrt{2}t) + D$$

$$= \frac{1}{2} \tan x + \frac{1}{2\sqrt{2}} \tan^{-1}(\sqrt{2} \tan x) + D$$

Hence, $a = \frac{1}{2}, b = \frac{1}{2\sqrt{2}}$ and $c = \sqrt{2}$

17. (a, d) We have, $I = \int \frac{\sin x + \sin^3 x}{\cos 2x}\, dx$

Let $\cos x = t$,

so $\cos 2x = 2t^2 - 1$ and $dt = -\sin x\, dx$

$$\therefore \quad I = \int \frac{t^2 - 2}{2t^2 - 1}\, dt = \frac{1}{2} \int \frac{2t^2 - 4}{2t^2 - 1}\, dt$$

$$= \frac{1}{2} \int 1\, dt - \frac{3}{2} \int \frac{dt}{2t^2 - 1}$$

$$= \frac{t}{2} - \frac{3}{2\sqrt{2}} \cdot \frac{1}{2} \log \left| \frac{\sqrt{2}t - 1}{\sqrt{2}t + 1} \right| + C$$

$$= \frac{1}{2} \cos x - \frac{3}{4\sqrt{2}} \log \left| \frac{\sqrt{2} \cos x - 1}{\sqrt{2} \cos x + 1} \right| + C$$

So, $P = \frac{1}{2}, Q = -\frac{3}{4\sqrt{2}}$ and $f(x) = \frac{\sqrt{2} \cos x - 1}{\sqrt{2} \cos x + 1}$

18. (a, c) Let $I = \int (\sqrt{\tan x} + \sqrt{\cot x})\, dx$

On putting $\tan x = t^2 \Rightarrow \sec^2 x\, dx = 2t\, dt$

$$\Rightarrow \quad dx = \frac{2t}{1+t^4}\, dt$$

Now, $$I = \int \left(t + \frac{1}{t} \right) \frac{2t}{1+t^4}\, dt$$

$$= \int \frac{2(t^2+1)}{t^4+1}\, dt$$

$$= 2 \int \frac{1 + \left(\frac{1}{t^2} \right)}{t^2 + \left(\frac{1}{t^2} \right)}\, dt$$

Let $t - \frac{1}{t} = z \Rightarrow \left(1 + \frac{1}{t^2} \right) dt = dz$

$$\therefore \quad I = 2 \int \frac{dz}{z^2 + 2} = \frac{2}{\sqrt{2}} \tan^{-1} \left(\frac{z}{\sqrt{2}} \right) + C$$

$$= \sqrt{2} \tan^{-1} \left(\frac{\sqrt{\tan x} - \sqrt{\cot x}}{\sqrt{2}} \right) + C$$

Also, $I = \int \frac{\sin x + \cos x}{\sqrt{\sin x \cos x}}\, dx$

$$= \sqrt{2} \int \frac{\sin x + \cos x}{\sqrt{1 - (\sin x - \cos x)^2}}\, dx$$

Let $\sin x - \cos x = t \Rightarrow (\cos x + \sin x)\, dx = dt$

$$I = \sqrt{2} \int \frac{dt}{\sqrt{1-t^2}}$$

$$= \sqrt{2} \sin^{-1} t + C$$

$$= \sqrt{2} \sin^{-1}(\sin x - \cos x) + C$$

19. (a, b, d) Let $I = \int x \log \left(1 + \frac{1}{x} \right) dx$

$$= \int \underset{\text{II}}{x} \underset{\text{I}}{\log(1+x)}\, dx - \int \underset{\text{II}}{x} \underset{\text{I}}{\log x}\, dx$$

$$\Rightarrow \quad I = \frac{x^2}{2} \log(1+x) - \frac{1}{2} \int \frac{x^2}{x+1}\, dx - \left\{ \frac{x^2}{2} \log x - \int \frac{1}{x} \cdot \frac{x^2}{2}\, dx \right\}$$

$$\Rightarrow \quad I = \frac{x^2}{2} \log(1+x) - \frac{1}{2} \int \left(x - 1 + \frac{1}{x+1} \right) dx - \frac{x^2}{2} \log x + \frac{x^2}{4} + C$$

$$= \frac{x^2}{2} \log(1+x) - \frac{1}{2} \left(\frac{x^2}{2} - x \right) - \frac{1}{2} \log(x+1) - \frac{x^2}{2} \log x + \frac{x^2}{4} + C$$

$$\Rightarrow \quad I = \left(\frac{x^2 - 1}{2} \right) \log(1+x) - \frac{x^2}{2} \log x + \frac{x}{2} + C$$

Hence, $f(x) = \frac{x^2 - 1}{2}, g(x) = -\frac{x^2}{2}$ and $L = \frac{1}{2}$

20. (a, d) We have,

$$\frac{2x}{x^2 - 5x + 4} = \frac{2x}{(x-1)(x-4)} = \frac{P}{x-1} + \frac{Q}{x-4}$$

$$2x = P(x-4) + Q(x-1)$$

$$\therefore \quad P = -\frac{2}{3} \text{ and } Q = \frac{8}{3}$$

$$\therefore \int \frac{e^{x-1}}{(x-1)(x-4)} 2x\, dx = \int e^{x-1} \left(-\frac{2/3}{x-1} + \frac{8/3}{x-4} \right) dx$$

$$= -\frac{2}{3} f(x-1) + \frac{8}{3} e^3 f(x-4) + C$$

$$\therefore A = -\frac{2}{3} \text{ and } B = \frac{8}{3} e^3$$

Sol. (Q.Nos. 21-23)

21. (b) Clearly, $f'(x) = \int f''(x)dx + C_1$

$$= \int (\sec^4 x + 4)\, dx + C_1$$

$$= \int \sec^2 x \sec^2 x\, dx + \int 4\, dx + C_1$$

$$= \int (1 + \tan^2 x) \sec^2 x\, dx + 4x + C_1$$

$$= I_1 + 4x + C_1$$

On putting $\tan x = t$ in the integral I_1, then

$$\sec^2 x\, dx = dt$$

$$\therefore I_1 = \int (1 + t^2) dt = t + \frac{t^3}{3} + C_2$$

$$= \tan x + \frac{\tan^3 x}{3} + C_2$$

$$\therefore f'(x) = \tan x + \frac{\tan^3 x}{3} + 4x + C \text{ where, } C = C_1 + C_2$$

$\because f'(0) = 0$

$\Rightarrow \quad C = 0$

Thus, $f'(x) = \tan x + \dfrac{\tan^3 x}{3} + 4x$

22. (a) Clearly,

$$f(x) = \int f'(x)\, dx + C_3$$

$$= \int \left(\tan x + \frac{\tan^3 x}{3} + 4x \right) dx + C_3$$

$$= \int \tan x\, dx + \frac{1}{3} \int \tan^3 x\, dx + 4 \int x\, dx + C_3$$

$$= \int \tan x\, dx + \frac{1}{3} \int \tan x (\sec^2 x - 1) dx + 4 \cdot \frac{x^2}{2} + C_3$$

$$= \frac{2}{3} \int \tan x\, dx + \frac{1}{3} \int \tan x \cdot \sec^2 x\, dx + 2x^2 + C_3$$

$$= \frac{2}{3} \ln(\sec x) + \frac{1}{3} I_2 + 2x^2 + C_3$$

Consider, $I_2 = \int \tan x \sec^2 x\, dx$

On putting $\tan x = t$

$\Rightarrow \quad \sec^2 x\, dx = dt$

$$\Rightarrow I_2 = \int t\, dt = \frac{t^2}{2} + C_4 = \frac{\tan^2 x}{2} + C_4$$

$$\therefore f(x) = \frac{2}{3} \ln(\sec x) + \frac{1}{6} \tan^2 x + 2x^2 + C_5$$

where, $C_5 = C_3 + \left(\dfrac{C_4}{3} \right)$

$\because f(0) = 0$

$$\therefore \quad 0 = \frac{2}{3} \ln(1) + 0 + 0 + C_5$$

$\Rightarrow C_5 = 0$

$$\therefore f(x) = \frac{2}{3} \ln(\sec x) + \frac{1}{6} \tan^2 x + 2x^2$$

23. (c) Let $I = \int \tan x\, f''(x)\, dx$

$$I = \int \tan x\, (\sec^4 x + 4)\, dx$$

$$I = \int \tan x \sec^4 x\, dx + 4 \int \tan x\, dx$$

$$I = \int \tan x\, (1 + \tan^2 x) \sec^2 x\, dx + 4 \log \sec x + C$$

Let $\tan x = t \Rightarrow \sec^2 x\, dx = dt$

$$I = \int (t + t^3)\, dt + 4 \log \sec x + C$$

$$I = \frac{t^2}{2} + \frac{t^4}{4} + 4 \log \sec x + C$$

$$I = \frac{\tan^2 x}{2} + \frac{\tan^4 x}{4} + 4 \log \sec x + C$$

$$\therefore \quad A = \frac{1}{2} \text{ and } B = \frac{1}{4} \Rightarrow A + B = \frac{1}{2} + \frac{1}{4} = \frac{3}{4}$$

Sol. (Q.Nos. 24-26)

Let $\quad I = \displaystyle\int \frac{x(x-1)}{(x^2+1)(x+1)\sqrt{x^3+x^2+x}}\, dx$

$$= \int \frac{x^2 - x}{(x^2+1)(x+1)x\sqrt{x+1+\frac{1}{x}}}\, dx$$

$$= \int \frac{x^2 - 1}{(x^2+1)(x+1)^2 \sqrt{x+1+\frac{1}{x}}}\, dx$$

$$= \int \frac{1 - \frac{1}{x^2}}{\left(x + \frac{1}{x}\right)\left(x + 2 + \frac{1}{x}\right)\sqrt{x+1+\frac{1}{x}}}\, dx$$

On putting $x + 1 + \dfrac{1}{x} = t^2$

$$= \int \frac{2t\, dt}{(t^2-1)(t^2+1)t}$$

$$= 2 \int \frac{1}{(t^2-1)(t^2+1)}\, dt$$

$$= \int \left(\frac{1}{t^2-1} - \frac{1}{t^2+1} \right) dt$$

$$= \frac{1}{2} \log \left| \frac{t-1}{t+1} \right| - \tan^{-1} t + C$$

$$= \frac{1}{2} \log \left| \frac{\sqrt{x + \frac{1}{x} + 1} - 1}{\sqrt{x + \frac{1}{x} + 1} + 1} \right| - \tan^{-1} \sqrt{x + \frac{1}{x} + 1} + C$$

24. (c) So, $f(x) = x + \dfrac{1}{x} + 1$

$\therefore \quad f(1) = 3$

25. (a) $\displaystyle\lim_{x \to \infty} \tan^{-1} \sqrt{x + \frac{1}{x} + 1} = \tan^{-1} \infty$

$$= \frac{\pi}{2}$$

26. (a) $f(x) = x + \frac{1}{x} + 1$

$\Rightarrow f'(x) = 1 - \frac{1}{x^2}$

For increasing $f'(x) > 0$,

$$1 - \frac{1}{x^2} > 0$$

$$\Rightarrow \quad \frac{1}{x^2} < 1$$

$$\Rightarrow \quad x^2 > 1$$

$\because \quad x \in (-\infty, -1) \cup (1, \infty)$

$\therefore$ $f(x)$ is increasing in the interval

$x \in (-\infty, -1) \cup (1, \infty)$

Sol. (Q.Nos. 27-29)

We have, $6\int f(x) g(x)\, dx = x^6 + 3x^4 + 3x^2 + C$

On differentiating both sides, we get

$$6f(x)\, g(x) = 6x^5 + 12x^3 + 6x$$

$$\therefore \quad f(x)\, g(x) = x^5 + 2x^3 + x \quad \text{...(i)}$$

$$2\int \frac{g(x)\, dx}{f(x)} = x^2 + C$$

On differentiating both sides, we get

$$\frac{g(x)}{f(x)} = x \quad \text{...(ii)}$$

Multiplying Eqs. (i) and (ii), we get

$$\{g(x)\}^2 = x^6 + 2x^4 + x^2$$

$$\{g(x)\}^2 = (x^3 + x)^2$$

$$\therefore \quad g(x) = x^3 + x$$

$$\therefore \quad f(x) = x^2 + 1, \text{ as } f(x) > 0$$

27. (b) Now, $\int \{g(x) - f(x)\}\, dx$

$$= \int (x^3 + x - x^2 - 1)\, dx$$

$$= \frac{x^4}{4} + \frac{x^2}{2} - \frac{x^3}{3} - x + C$$

28. (d) $\lim_{x \to 0} \frac{\log\{f(x)\}}{g(x)} = \lim_{x \to 0} \frac{\log(1 + x^2)}{x + x^3}$

$$= \lim_{x \to 0} \frac{\log(1 + x^2)}{x^2} \cdot \frac{x^2}{x + x^3}$$

$$= \lim_{x \to 0} \frac{x}{1 + x^2} = 0$$

29. (a) $g(x) = f(x)$

$x^3 + x = x^2 + 1$

$\Rightarrow \quad x^3 + x - x^2 - 1 = 0$

$\Rightarrow \quad x^3 - x^2 + x - 1 = 0$

$\Rightarrow \quad x^2(x - 1) + 1(x - 1) = 0$

$\Rightarrow \quad (x^2 + 1)(x - 1) = 0$

$\Rightarrow \quad x - 1 = 0$

$\Rightarrow \quad x^2 + 1 \neq 0$

$\because$ Real root of $g(x) - f(x) = 0$ is 1.

30. (i) $\rightarrow$ (A), (ii) $\rightarrow$ (B), (iii) $\rightarrow$ (C), (iv) $\rightarrow$ (B)

(i) Let $I = \int \frac{\sqrt{x^2 + 1}}{x^4}\, dx$

On putting $x = \tan\theta$,

then $I = \int \frac{\sec\theta \cdot \sec^2\theta}{\tan^4\theta}\, d\theta$

$$= \int \frac{\cos\theta}{\sin^4\theta}\, d\theta$$

$$= \int t^{-4} dt, \text{ where } t = \sin\theta$$

$$= \frac{t^{-4+1}}{-4+1} + C$$

$$= -\frac{1}{3} t^{-3} + C$$

$$= -\frac{1}{3} \frac{1}{\sin^3\theta} + C$$

$$= -\frac{1}{3}\left(\frac{1}{(\sin^2\theta)^{3/2}}\right) + C$$

$$= -\frac{1}{3}(\text{cosec}^2\theta)^{3/2} + C$$

$$= -\frac{1}{3}(1 + \cot^2\theta)^{3/2} + C$$

$$= -\frac{1}{3}\left(1 + \frac{1}{x^2}\right)^{3/2} + C = -\frac{1}{3}\frac{(x^2 + 1)^{3/2}}{x^3} + C$$

Therefore, $k = 3$

(ii) We have, $I = \int \frac{2 + \sqrt{x}}{(x + \sqrt{x} + 1)^2}\, dx$

$$= \int \frac{2 + \sqrt{x}}{x^2\left(1 + \frac{1}{\sqrt{x}} + \frac{1}{x}\right)^2}\, dx$$

$$= \int \frac{\frac{2}{x^2} + \frac{1}{x\sqrt{x}}}{\left(1 + \frac{1}{\sqrt{x}} + \frac{1}{x}\right)^2}\, dx$$

On putting $1 + \frac{1}{\sqrt{x}} + \frac{1}{x} = t$

$$\Rightarrow \left(-\frac{1}{2x\sqrt{x}} - \frac{1}{x^2}\right) dx = dt$$

$$\Rightarrow \left(\frac{2}{x^2} + \frac{1}{x\sqrt{x}}\right) dx = -2dt$$

Therefore, $I = \int \frac{1}{t^2}(-2dt) = \frac{2}{t} + C$

$$= \frac{2}{\frac{1}{x} + \frac{1}{\sqrt{x}} + 1} + C$$

$$= \frac{2x}{1 + \sqrt{x} + x} + C$$

Therefore, $\lambda = 2$

(iii) We have, $I=\int \frac{\sqrt{x^{10}+2}}{x^{16}}dx$

$$=\int \frac{\sqrt{x^{10}+2}}{x^{11}\cdot x^5}dx$$

$$=\int \frac{\sqrt{1+\frac{2}{x^{10}}}}{x^{11}}dx$$

On putting $1+\frac{2}{x^{10}}=t^2$

$\Rightarrow \quad \frac{-20}{x^{11}}dx=2t\,dt \Rightarrow \frac{dx}{x^{11}}=-\frac{1}{10}t\,dt$

Therefore, $I=\int t\left(-\frac{1}{10}\right)t\,dt$

$$=-\frac{t^3}{30}+C$$

$$=-\frac{1}{30}\left(1+\frac{2}{x^{10}}\right)^{3/2}+C$$

$$=-\frac{1}{30}\frac{(x^{10}+2)^{3/2}}{x^{15}}+C$$

So, $m=15$

(iv) We have, $I=\int \frac{dx}{x+\sqrt{x}}$

$$=\int \frac{dx}{\sqrt{x}(\sqrt{x}+1)}$$

On putting $\sqrt{x}=t \Rightarrow \frac{1}{2\sqrt{x}}dx=dt$

Therefore, $I=\int \frac{2}{t+1}dt$

$$=2\log(t+1)+C$$

$$=2\log(\sqrt{x}+1)+C$$

So, $n=2$

31. (i) $\rightarrow$ (C), (ii) $\rightarrow$ (A), (iii) $\rightarrow$ (D), (iv) $\rightarrow$ (B)

(i) $\int \frac{1}{x(x^n+1)}dx=\int \frac{x^{n-1}}{x^n(x^n+1)}dx$

On putting $x^n=t \Rightarrow nx^{n-1}dx=dt$

$$=\frac{1}{n}\int \frac{1}{t(t+1)}dt$$

$$=\frac{1}{n}\int\left(\frac{1}{t}-\frac{1}{t+1}\right)dt$$

$$=\frac{1}{n}[\log t-\log(t+1)]$$

$$=\frac{1}{n}\log\left(\frac{t}{t+1}\right)+C$$

$$=\frac{1}{n}\log\left(\frac{x^n}{x^n+1}\right)+C$$

(ii) Similarly, $\int \frac{1}{x(x^3+1)}=\frac{1}{3}\log\left(\frac{x^3}{x^3+1}\right)+C$

[putting $n=3$ in part (i)]

(iii) $\int \frac{1}{x\sqrt{x^n+1}}dx=\int \frac{x^{n-1}}{x^n\sqrt{x^n+1}}dx$

On putting $x^n=t^2 \Rightarrow nx^{n-1}dx=2t\,dt$

$$=\frac{1}{n}\int \frac{2t\,dt}{(t^2-1)t}$$

$$=\frac{2}{n}\int \frac{dt}{t^2-1}$$

$$=\frac{2}{n}\left(\frac{1}{2}\log\frac{t-1}{t+1}\right)$$

$$=\frac{1}{n}\log\left(\frac{t-1}{t+1}\right)+C$$

$$=\frac{1}{n}\log\left(\frac{\sqrt{1+x^n}-1}{\sqrt{1+x^n}+1}\right)+C$$

(iv) Similarly, $\int \frac{1}{x\sqrt{x^3+1}}dx=\frac{1}{3}\log\left(\frac{\sqrt{1+x^3}-1}{\sqrt{1+x^3}+1}\right)+C$

[putting $n=3$ in part (iii)]

32. (i) $\rightarrow$ (B), (ii) $\rightarrow$ (D), (iii) $\rightarrow$ (A), (iv) $\rightarrow$ (C)

We have, $\int \frac{x^3+3x^2+2x+1}{\sqrt{x^2+x+1}}dx$

$$=(ax^2+bx+c)\sqrt{x^2+x+1}+\lambda\int \frac{dx}{\sqrt{x^2+x+1}}$$

On differentiating both sides, we get

$$\frac{x^3+3x^2+2x+1}{\sqrt{x^2+x+1}}=\frac{(ax^2+bx+c)(2x+1)}{2\sqrt{x^2+x+1}}+(2ax+b)\sqrt{x^2+x+1}+\lambda\left(\frac{1}{\sqrt{x^2+x+1}}\right)$$

$$\Rightarrow \frac{x^3+3x^2+2x+1}{\sqrt{x^2+x+1}}=\frac{6ax^3+(5a+4b)x^2+(4a+3b+2c)x+(c+2b+2\lambda)}{2\sqrt{x^2+x+1}}$$

On comparing coefficients of both sides, we get

$$1=3a \Rightarrow a=\frac{1}{3}$$

and $\frac{5a+4b}{2}=3 \Rightarrow 5a+4b=6$

$\Rightarrow \quad 4b=6-\frac{5}{3}$

$\Rightarrow \quad 4b=\frac{13}{3} \Rightarrow b=\frac{13}{12}$

and $\quad \frac{4a+3b+2c}{2}=2$

$$\Rightarrow \quad 4\times\frac{1}{3}+3\times\frac{13}{12}+2c=4$$

$$\Rightarrow \quad \frac{4}{3}+\frac{13}{4}+2c=4 \Rightarrow c=-\frac{7}{24}$$

and $\frac{c+2b+2\lambda}{2}=1$

$$\Rightarrow \quad \frac{-7}{24}+2\times\frac{13}{12}+2\lambda=2 \Rightarrow \lambda=\frac{1}{16}$$

33. (i) $\rightarrow$ (B), (ii) $\rightarrow$ (C), (iii) $\rightarrow$ (A), (iv) $\rightarrow$ (D)

(i) We have,

$$\int e^x\left(\frac{1-x}{1+x^2}\right)^2 dx = \int e^x\left(\frac{1}{1+x^2}-\frac{2x}{(1+x^2)^2}\right)dx$$

$$= e^x\left(\frac{1}{1+x^2}\right)+C$$

$$= \frac{e^x}{1+x^2}+C$$

(ii) $\int \frac{e^x(x^3+x+1)}{(1+x^2)^{3/2}}dx = \int \frac{e^x[x(x^2+1)+1]}{(1+x^2)^{3/2}}dx$

$$= \int e^x\left[\frac{x}{\sqrt{1+x^2}}+\frac{1}{(1+x^2)^{3/2}}\right]dx$$

$$= \frac{e^x\cdot x}{\sqrt{1+x^2}}+C$$

$$[\because \int e^x\{f(x)+f'(x)\}\,dx = e^x f(x)+C]$$

$$= \frac{xe^x}{\sqrt{1+x^2}}+C$$

(iii) $\int \frac{e^x(x^2+3x+3)}{(x+2)^2}dx$

$$= \int e^x\left(\frac{(x+1)(x+2)+1}{(x+2)^2}\right)dx$$

$$= \int e^x\left(\frac{x+1}{x+2}+\frac{1}{(x+2)^2}\right)dx$$

$$= \frac{e^x(x+1)}{x+2}+C$$

(iv) $\int e^x\frac{(x^3-x+2)}{(x^2+1)^2}dx$

$$= \int e^x\frac{[(x+1)(x^2+1)+1-2x-x^2]}{(x^2+1)^2}dx$$

$$= \int e^x\left[\frac{x+1}{x^2+1}+\frac{1-2x-x^2}{(x^2+1)^2}\right]dx$$

$$= \frac{e^x(x+1)}{(x^2+1)}+C$$

Sol. (Q.Nos. 34-36)

34. (1.41) Let $I=\int \sec x\tan x\sqrt{\sec^2 x+1}\,dx$

On putting $\sec x=t \Rightarrow \sec x\cdot\tan x\,dx=dt$

$$\therefore \quad I=\int\sqrt{t^2+1}\,dt$$

$$=\frac{t}{2}\sqrt{t^2+1}+\frac{1}{2}\sinh^{-1}t+C$$

$$=\frac{\sec x\sqrt{\sec^2 x+1}}{2}+\frac{1}{2}\sinh^{-1}(\sec x)+C$$

So, $f(x)=\sec x\sqrt{\sec^2 x+1}$

$$\Rightarrow f(\theta)=\sqrt{2}=1.41$$

35. (1) Let $I=\int\frac{x^2-1}{x^2+1}\frac{dx}{\sqrt{x^4+1}}$

Dividing numerator and denominator by x^2, we get

$$I=\int\frac{1-\left(\frac{1}{x}\right)^2}{\left(x+\frac{1}{x}\right)\sqrt{x^2+\frac{1}{x^2}}}dx$$

Let $\quad t=x+\frac{1}{x}\Rightarrow dt=\left(1-\frac{1}{x^2}\right)dx$

$$I=\int\frac{dt}{t\sqrt{t^2-2}}=\int\frac{t\,dt}{t^2\sqrt{t^2-2}}$$

On putting $t^2-2=z^2$

$$\Rightarrow \quad 2t\,dt=2z\,dz$$

$$\Rightarrow \quad t\,dt=z\,dz$$

$$\therefore \quad I=\int\frac{1}{(z^2+2)\cdot z}(z)\,dz$$

$$=\frac{1}{\sqrt{2}}\tan^{-1}\left(\frac{z}{\sqrt{2}}\right)+C$$

$$=\frac{1}{\sqrt{2}}\tan^{-1}\left(\frac{\sqrt{t^2-2}}{\sqrt{2}}\right)+C$$

$$=\frac{1}{\sqrt{2}}\tan^{-1}\left(\sqrt{\frac{x^2+\frac{1}{x^2}}{2}}\right)+C$$

If $\tan^{-1}\sqrt{\frac{x^2+\frac{1}{x^2}}{2}}=\theta$, then

$$\tan^2\theta=\frac{x^2+\left(\frac{1}{x^2}\right)}{2}\Rightarrow\sec^2\theta=\frac{\left(x+\frac{1}{x}\right)^2}{2}$$

$$\Rightarrow\cos\theta=\frac{\sqrt{2}}{x+\frac{1}{x}}\Rightarrow\cos\theta=\frac{\sqrt{2}\,x}{x^2+1}\Rightarrow\theta=\cos^{-1}\left(\frac{x\sqrt{2}}{x^2+1}\right)$$

$$\therefore \quad I=\frac{1}{\sqrt{2}}\cos^{-1}\frac{x\sqrt{2}}{x^2+1}+C$$

$$\Rightarrow \quad k=1$$

36. (2) Let $I = \int \frac{dx}{(x+1)^3\sqrt{x^2+2x-3}}$

On putting $x+1 = \frac{1}{t} \Rightarrow dx = -\frac{1}{t^2}dt$

$$I = \int \frac{t^3}{\sqrt{\frac{1}{t^2}-4}}\left(-\frac{1}{t^2}\right)dt$$

$$= -\int \frac{t^2}{\sqrt{1-4t^2}}dt$$

$$= \frac{1}{4}\int \frac{1-4t^2-1}{\sqrt{1-4t^2}}dt$$

$$= \frac{1}{4}\int \sqrt{1-4t^2}\,dt - \frac{1}{4}\int \frac{dt}{\sqrt{1-4t^2}}$$

$$= \frac{1}{4}\times 2\int \sqrt{\frac{1}{4}-t^2}\,dt - \frac{1}{8}\int \frac{dt}{\sqrt{\left(\frac{1}{2}\right)^2 - t^2}}$$

$$= \frac{1}{2}\left[\frac{t\sqrt{\frac{1}{4}-t^2}}{2} + \frac{1}{8}\sin^{-1}2t\right] - \frac{1}{8}\sin^{-1}(2t) + C$$

$$= \frac{1}{8}\left[\left(\frac{1}{x+1}\right)\sqrt{1-\frac{4}{(x+1)^2}}\right] + \frac{1}{16}\sin^{-1}(2t) - \frac{1}{8}\sin^{-1}(2t) + C$$

$$= \frac{\sqrt{x^2+2x-3}}{8(x+1)^2} - \frac{1}{16}\sin^{-1}\left(\frac{2}{x+1}\right) + C$$

Therefore, $k = 2$

Sol. (Q.Nos. 37-39)

37. (0.5) Let $I = \int \frac{dx}{1+\tan x}$

$$I = \int \frac{\cos x}{\sin x + \cos x}dx$$

Let $\cos x = \lambda(\sin x + \cos x) + \mu \frac{d}{dx}(\sin x + \cos x)$

$\Rightarrow \cos x = \mu \cos x - \mu \sin x + \lambda \sin x + \lambda \cos x$

Equating the coefficients of $\sin x$ and $\cos x$, we get

$\lambda + \mu = 1$ and $\lambda - \mu = 0$

Solving these equations, we get

$$\lambda = \mu = \frac{1}{2}$$

$$\therefore \quad I = \int \frac{1}{2}\frac{(\cos x - \sin x)}{\sin x + \cos x}dx + \frac{1}{2}\int \frac{\sin x + \cos x}{\sin x + \cos x}dx$$

$$I = \frac{1}{2}\log|\sin x + \cos x| + \frac{1}{2}x + C$$

$$I = \frac{x}{2} + \frac{1}{2}\log|\sin x + \cos x| + C$$

$\Rightarrow \quad k = 0.5$

38. (0.04) Let $I = \int \frac{2\sin x + 3\cos x}{3\sin x + 4\cos x}dx$

Let $2\sin x + 3\cos x = \lambda(3\sin x + 4\cos x) + \mu(3\cos x - 4\sin x)$

Now, equating the coefficients of $\sin x$ and $\cos x$ on both sides, we get

$$3\lambda - 4\mu = 2$$
$$4\lambda + 3\mu = 3$$

Solving these equations, we get

$$\lambda = 18/25 \text{ and } \mu = 1/25$$

Therefore, $I = \int \frac{18}{25}dx + \frac{1}{25}\int \frac{3\cos x - 4\sin x}{3\sin x + 4\cos x}dx$

$$= \frac{18}{25}x + \frac{1}{25}\log_e|3\sin x + 4\cos x| + C$$

$\Rightarrow \quad k = 0.04$

39. (1.4) Let $I = \int \frac{\sin x + 2\cos x}{2\sin x + \cos x}dx$

Let $\sin x + 2\cos x = \lambda(2\sin x + \cos x) + \mu(2\cos x - \sin x)$

Therefore, $\quad 2\lambda - \mu = 1$

$\quad \lambda + 2\mu = 2$

Solving these equations, we get

$$\lambda = 4/5 \text{ and } \mu = 3/5$$

$$\therefore \quad I = \frac{4}{5}\int 1\,dx + \frac{3}{5}\int \frac{2\cos x - \sin x}{2\sin x + \cos x}dx$$

$$\Rightarrow \quad I = \frac{4}{5}x + \frac{3}{5}\log|2\sin x + \cos x| + C$$

Hence, $p = \frac{4}{5}$ and $q = \frac{3}{5}$ and so, $p + q = \frac{7}{5} = 1.4$

40. (10) Let $I = \int \frac{dx}{\sqrt{x+7} - \sqrt[4]{x+7}}$

Let $x + 7 = t^4$

$\Rightarrow \quad dx = 4t^3\,dt$

$$\Rightarrow \quad I = \int \frac{4t^3}{t^2 - t}dt = 4\int \frac{t^2}{t-1}dt$$

$$= 4\int \frac{t^2 - 1 + 1}{t-1}dt$$

$$= 4\int \frac{(t+1)(t-1)}{t-1}dt + 4\int \frac{1}{t-1}dt$$

$$= 4\int (t+1)\,dt + 4\int \frac{1}{t-1}dt$$

$$= 4\left(\frac{t^2}{2} + t\right) + 4\ln|t-1| + C$$

$$= 2(x+7)^{1/2} + 4(x+7)^{1/4} + 4\ln|(x+7)^{1/4} - 1| + C$$

So, $P = 2, Q = 4$ and $R = 4$

Hence, $P + Q + R = 2 + 4 + 4 = 10$

41. (0) Let $I=\int \frac{dx}{\sqrt{\sin^3 x\cos^5 x}}$

$$=\int \frac{dx}{\sqrt{\frac{\sin^3 x}{\cos^3 x}\cos^8 x}}$$

$$=\int \frac{\sec^4 x}{\sqrt{\tan^3 x}}dx$$

$$=\int \frac{(1+\tan^2 x)\sec^2 x}{\sqrt{\tan^3 x}}dx$$

Let $t=\tan x \Rightarrow dt=\sec^2 x\,dx$

$\therefore \quad I=\int \frac{1+t^2}{t^{3/2}}dt$

$$=\int (t^{-3/2}+t^{1/2})\,dt$$

$$=-2t^{-1/2}+\frac{2}{3}t^{3/2}+C$$

$$=-2\sqrt{\cot x}+\frac{2}{3}\sqrt{\tan^3 x}+C$$

$\therefore\ a=-2$ and $b=\frac{2}{3}$

Hence, $a+3b=-2+3\left(\frac{2}{3}\right)=0$

42. (9) Let $I=\int \frac{\cos^4 x}{\sin^6 x(1+\cot^5 x)^{3/5}}dx$

$$=\int \frac{\cot^4 x\,\text{cosec}^2 x\,dx}{(1+\cot^5 x)^{3/5}}$$

Let $\cot x=t \Rightarrow -\text{cosec}^2 x\,dx=dt$

$\therefore \quad I=-\int \frac{t^4 dt}{(1+t^5)^{3/5}}=-\frac{1}{5}\int \frac{5t^4 dt}{(1+t^5)^{3/5}}$

On putting $(1+t^5)=y^5 \Rightarrow 5t^4 dt=5y^4 dy$

$$I=-\frac{1}{5}\int \frac{5y^4 dy}{y^3}=-\int y\,dy$$

$$=-\left(\frac{y^2}{2}\right)+C$$

$$=-\frac{1}{2}(1+t^5)^{2/5}+C$$

$$=-\frac{1}{2}(1+\cot^5 x)^{2/5}+C$$

$$=-\frac{1}{2}\left(\frac{\tan^5 x+1}{\tan^5 x}\right)^{2/5}+C$$

Thus, $L=2, M=2$ and $N=5$

Hence, the value of $L+M+N=9$

43. (1) Let $I=\int \frac{1-5\sin^2 x}{\cos^5 x\sin^2 x}dx$

$\Rightarrow \quad I=\int \frac{(\cos^2 x+\sin^2 x)-5\sin^2 x}{\cos^5 x\sin^2 x}dx$

$$=\int \frac{\cos^2 x-4\sin^2 x}{\cos^5 x\sin^2 x}dx$$

$$=\int \frac{\cos^5 x-4\sin^2 x\cos^3 x}{\cos^8 x\sin^2 x}dx$$

$\Rightarrow \quad I=\int \frac{d}{dx}\left(\frac{1}{\cos^4 x\sin x}\right)dx$

$$=\frac{1}{\cos^4 x\sin x}+C \Rightarrow I=\frac{\cot x}{\cos^5 x}+C$$

Hence, $f(x)=\cot x \Rightarrow f\left(\frac{\pi}{4}\right)=\cot\frac{\pi}{4}=1$

44. (4) Let $I=\int \frac{4x+1}{x^2+3x+2}dx$

Let $4x+1=A(2x+3)+B$

On solving, we get

$A=2$ and $B=-5$

$$I=\int \frac{2(2x+3)-5}{x^2+3x+2}dx$$

$\Rightarrow \quad I=2\int \frac{(2x+3)}{x^2+3x+2}dx-5\int \frac{1}{x^2+3x+2}dx$

$\Rightarrow \quad I=2\log|x^2+3x+2|-5\int \frac{1}{\left(x+\frac{3}{2}\right)^2-\left(\frac{1}{2}\right)^2}dx$

$\Rightarrow \quad I=2\log|x^2+3x+2|-5\log\left|\frac{x+1}{x+2}\right|+C$

$\Rightarrow \quad I=2\log|x+1|+2\log|x+2|-5\log|x+1|+5\log|x+2|+C$

$\Rightarrow \quad I=-3\log|x+1|+7\log|x+2|+C$

$\therefore \quad a=-3$ and $b=7$

Hence, $a+b=-3+7=4$

45. (1) We have, $I=\int \frac{x^2-1}{(x^4+3x^2+1)\tan^{-1}\left(\frac{x^2+1}{x}\right)}dx$

Numerator and denominator divided by x^2,

$$I=\int \frac{\left(1-\frac{1}{x^2}\right)}{\left(x^2+\frac{1}{x^2}+2+1\right)\tan^{-1}\left(x+\frac{1}{x}\right)}dx$$

$\Rightarrow \quad I=\int \frac{\left(1-\frac{1}{x^2}\right)}{\left[\left(x+\frac{1}{x}\right)^2+1\right]\tan^{-1}\left(x+\frac{1}{x}\right)}dx$

Let $x+\frac{1}{x}=t \Rightarrow \left(1-\frac{1}{x^2}\right)dx=dt$

$\Rightarrow \quad I=\int \frac{dt}{(t^2+1)\tan^{-1}t}$

Let $\tan^{-1}t=v$

$\Rightarrow \frac{1}{1+t^2}dt=dv$

$\Rightarrow \quad I = \int \frac{dv}{v}$

$= \ln|v| + C$

$= \ln|\tan^{-1} t| + C$

$\Rightarrow \quad I = \ln\left|\tan^{-1}\left(x + \frac{1}{x}\right)\right| + C$

Hence, $k = 1$

46. (3) Let $g(x) = \int \frac{f(x)}{x^2(x+1)^3} dx$

It is given that $f(x)$ is a quadratic function.

So, $\frac{f(x)}{x^2(x+1)^3} = \frac{A}{x} + \frac{B}{x^2} + \frac{C}{x+1} + \frac{D}{(x+1)^2} + \frac{E}{(x+1)^3}$

$\Rightarrow \quad g(x) = \int \left(\frac{A}{x} + \frac{B}{x^2} + \frac{C}{x+1} + \frac{D}{(x+1)^2} + \frac{E}{(x+1)^3} \right) dx \quad \text{...(i)}$

$\Rightarrow \quad g(x) = A \ln x - \frac{B}{x} + C \ln(x+1) - \frac{D}{x+1} - \frac{E}{2(x+1)^2} + K$

It is given that $g(x)$ is a rational function.

$\therefore \quad A = C = 0$

Putting values of A and C in Eq. (i), we get

$g(x) = \int \left\{ \frac{B}{x^2} + \frac{D}{(x+1)^2} + \frac{E}{(x+1)^3} \right\} dx$

$g(x) = \int \left\{ \frac{B(x+1)^3 + Dx^2(x+1) + Ex^2}{x^2(x+1)^3} \right\} dx$

$\therefore \quad f(x) = B(x+1)^3 + Dx^2(x+1) + Ex^2$

$\Rightarrow \quad f(x) = (B+D)x^3 + (3B+D+E)x^2 + 3Bx + B$

But $f(x)$ is a quadratic function such that $f(0) = 1$

Therefore, $B + D = 0$ and $B = 1 \Rightarrow B = 1, D = -1$

$\therefore \quad f(x) = (2+E)x^2 + 3x + 1$

$\Rightarrow \quad f'(x) = 2(2+E)x + 3$

$\Rightarrow \quad f'(0) = 3$

47. (18) Let $I = \int x^{-11}(1+x^4)^{-\frac{1}{2}} dx$

Given, integral is integration of binomial differential in which $m = -11$, $n = 4$ and $p = -\frac{1}{2}$ so that

$\frac{m+1}{n} + p = \frac{-11+1}{4} - \frac{1}{2} = -\frac{5}{2} - \frac{1}{2} = -3$ (integer)

Therefore, put $1 + x^4 = x^4 t^2$

$\Rightarrow \quad x^4 = \frac{1}{t^2 - 1}$

$\Rightarrow \quad x = (t^2-1)^{-1/4}$

Thus, $dx = -\frac{1}{4}(t^2-1)^{-5/4}(2t)\, dt$

$= -\frac{t}{2}(t^2-1)^{-5/4}\, dt$

Therefore, $x^{-11} = (t^2-1)^{11/4}$

and $\quad 1 + x^4 = x^4 t^2 = (t^2-1)^{-1} t^2$

Hence, $I = \int (t^2-1)^{11/4}(t^2-1)^{1/2} \frac{1}{t}\left[-\frac{t}{2}(t^2-1)^{-5/2} \right] dt$

$= -\frac{1}{2}\int (t^2-1)^2\, dt$

$= -\frac{1}{2}\left[\frac{t^5}{5} - \frac{2}{3}t^3 + t \right] + K$

$= -\frac{1}{10}\left(\frac{1+x^4}{x^4}\right)^{5/2} + \frac{1}{3}\left(\frac{1+x^4}{x^4}\right)^{3/2} - \frac{1}{2}\left(\frac{1+x^4}{x^4}\right)^{1/2} + K$

$= -\frac{1}{10}x^{-10}(1+x^4)^{5/2} + \frac{1}{3}x^{-6}(1+x^4)^{3/2} - \frac{1}{2}x^{-2}(1+x^4)^{1/2} + K$

Therefore, $a + b + c = 10 + 6 + 2 = 18$

48. (0) $f'(x) = 3x^2 \sin\frac{1}{x} - x\cos\frac{1}{x}$

$\Rightarrow \quad f(x) = \int \left(3x^2 \sin\frac{1}{x} \right) dx - \int x \cos\frac{1}{x} dx$

$\Rightarrow \quad f(x) = \sin\frac{1}{x}\cdot x^3 - \int \cos\frac{1}{x}\left(-\frac{1}{x^2}\cdot x^3 \right) dx - \int x\cos\frac{1}{x} dx$

$= x^3 \sin\frac{1}{x} + \int x\cos\frac{1}{x} dx - \int x\cos\frac{1}{x} dx$

$\Rightarrow \quad f(x) = x^3 \sin\frac{1}{x} + C$

Now, $f(0) = 0 \Rightarrow C = 0$

$\Rightarrow \quad f(x) = x^3 \sin\frac{1}{x}$

$\therefore \quad f\left(\frac{1}{\pi}\right) = \frac{1}{\pi^3}\cdot \sin(\pi) = 0$

49. (16) Let $I = \int (\sin 3x + \sin x) e^{\sin x} \cdot \cos x\, dx$

$= \int \{3\sin x - 4\sin^3 x) + \sin x\}\, e^{\sin x} \cos x\, dx$

$= \int (4\sin x - 4\sin^3 x) e^{\sin x} \cos x\, dx$

Let $\sin x = t \Rightarrow \cos x\, dx = dt$

$\therefore \quad I = \int (4t - 4t^3)\, e^t dt$

$= 4[\int te^t dt - \int t^3 e^t dt]$

$= 4[\int te^t dt - \{t^3 e^t - 3\int t^2 \cdot e^t dt\}]$

$= 4[\int te^t dt - t^3 e^t + 3\{t^2 e^t - 2\int te^t dt\}]$

$= 4[-5\int te^t dt - t^3 e^t + 3t^2 e^t]$

$= 4[-5[te^t - e^t] - t^3 e^t + 3t^2 e^t + C$

$= e^t[-20t + 20 - 4t^3 + 12t^2] + C$

$= e^{\sin x}(-20\sin x + 20 - 4\sin^3 x + 12\sin^2 x) + C$ [put $t = \sin x$]

$= e^{\sin x}(-20\sin x + 20 - 4\sin^3 x + 12 - 12\cos^2 x) + C$

$= e^{\sin x}(-4\sin^3 x - 12\cos^2 x - 20\sin x + 32) + C$

Hence, $A = -4$ and $B = -12$

$|A + B| = |-4 - 12| = |-16| = 16$

CHAPTER

10

DEFINITE INTEGRALS

LEVEL 01 \ Starter Level

10.1 Fundamental Theorem of Integral

1 Evaluate the following.

(i) $\int_0^4 \sqrt{x}(1-x)\,dx$

(ii) $\int_{-4}^{-1} \frac{1}{x}\,dx$

(iii) $\int_0^1 \frac{1}{\sqrt{1+x}+\sqrt{x}}\,dx$

(iv) $\int_0^{\pi/2} \sin^2 x\,dx$

(v) $\int_0^{\pi/4} \sqrt{1-\sin 2x}\,dx$

(vi) $\int_0^2 \frac{5x+1}{x^2+4}\,dx$

(vii) $\int_0^a \frac{1}{\sqrt{ax-x^2}}\,dx$

(viii) $\int_1^3 \frac{1}{x^2(x+1)}\,dx$

(ix) $\int_0^{\pi/2} x \sin x\,dx$

(x) $\int_0^1 \left\{xe^x + \sin \frac{\pi}{4}x\right\} dx$

(xi) $\int_0^{\sqrt{2}} \sqrt{2-x^2}\,dx$

10.2 Definite Integral by Substitution

2 Evaluate the following integral.

(i) $\int_0^1 \frac{2x}{5x^2+1}\,dx$

(ii) $\int_0^1 x\,e^{x^2}\,dx$

(iii) $\int_0^{\pi/2} \sqrt{\cos\theta}\cdot\sin^5\theta\,d\theta$

(iv) $\int_0^\infty \frac{1}{a^2+b^2x^2}\,dx$

(v) $\int_0^\pi \frac{1}{5+4\cos x}\,dx$

(vi) $\int_0^{\pi/2} \frac{1}{4\sin^2 x+5\cos^2 x}\,dx$

(vii) $\int_0^{1/\sqrt{2}} \frac{\sin^{-1} x}{(1-x^2)^{3/2}}\,dx$

10.3 Sum of Infinite Series using Integrations

3 Evaluate the following.

(i) $\lim_{n\to\infty}\left(\frac{1}{n+1}+\frac{1}{n+2}+\frac{1}{n+3}+\ldots+\frac{1}{2n}\right)$

(ii) $\lim_{n\to\infty}\left(\frac{n}{n^2+1^2}+\frac{n}{n^2+2^2}+\ldots+\frac{n}{2n^2}\right)$

(iii) $\lim_{n\to\infty}\left(\frac{1}{2n+1}+\frac{1}{2n+2}+\ldots+\frac{1}{6n}\right)$

(iv) $\lim_{n\to\infty}\sum_{r=0}^{n-1}\frac{1}{\sqrt{4n^2-r^2}}$

(v) $\lim_{n\to\infty}\frac{(1^p+2^p+\ldots+n^p)}{n^{p+1}}, p>0$

10.4 Properties of Definite Integral

4 Evaluate following integrals.

(i) $\int_{-1}^{1}e^{|x|}\,dx$ (ii) $\int_{-1}^{2}|x^3-x|\,dx$

(iii) $\int_0^4 [x]\,dx$ (iv) $\int_1^2 [3x]\,dx$, $[\cdot]$ is greatest integer function

(v) $\int_{-2}^{1}\left[\left(1+\cos\left(\frac{\pi x}{2}\right)\right)\right]dx$, $[\cdot]$ is greatest integer function

(vi) $\int_0^{100}[\tan^{-1}x]\,dx$

5 Evaluate the following integrals

(i) $\int_0^{\pi/2}\frac{\sin x}{\sin x+\cos x}\,dx$ (ii) $\int_0^{\pi/2}\log\tan x\,dx$

(iii) $\int_0^1 x(1-x)^n\,dx$ (iv) $\int_0^{\pi}\frac{x}{1+\sin x}\,dx$

(v) $\int_0^{\pi/2}\frac{dx}{\sqrt{\tan x}-\sqrt{\cot x}}$ (vi) $\int_a^b\frac{f(x)}{f(a+b-x)+f(x)}\,dx$

6 Evaluate the following integrals

(i) $\int_0^{2\pi}\cos^5 x\,dx$ (ii) $\int_0^{\pi}\frac{x}{1+\cos^2 x}\,dx$

(iii) $I=\int_0^{\pi}\frac{dx}{3\sin^2 x+4\cos^2 x}$ (iv) $\int_0^{\pi}x\log(\sin x)\,dx$

7 Evaluate the following integrals

(i) $\int_{-\pi/2}^{\pi/2}\sin^3 x\,dx$ (ii) $\int_{-\pi/2}^{\pi/2}\cos^2 x\,dx$

(iii) $I=\int_{-1}^{1}e^{|x|}\,dx$ (iv) $\int_{-\pi}^{\pi}x^{10}\sin^7 x\,dx$

(v) $I=\int_{-1}^{1}\log\left(\frac{2-x}{2+x}\right)dx$

10.5 Leibnitz's Rule (Differentiation Under Integral Sign)

8 Find the derivative of the following with respect to x.

(i) $\int_0^{x^2} \cos t\,dt$ (ii) $\int_{1/x}^{\sqrt{x}} \cos t^2\,dt$ (iii) $\int_{x^2}^{x^3} \frac{1}{\log t}\,dt$ (iv) $\int_0^{x^2} \frac{t^2 - 5t + 4}{2 + e^t}\,dt$

9 If $\int_{\pi/3}^{x} \sqrt{3 - \sin^2 t}\,dt + \int_0^{y} \cos t\,dt = 0$, then evaluate $\frac{dy}{dx}$.

10 If $F(x) = \int_1^x f(t)dt$, where $f(t) = \int_1^{t^2} \frac{\sqrt{1+u^4}}{4}du$, then find the value of $F''(2)$.

LEVEL 02 \ JEE Main Level

Single Choice Correct Questions

10.1 Basic Fundamentals of Definite Integral

1 The value of $\int_0^1 \frac{2^{2x+1} - 5^{2x-1}}{10^x}dx$ is

(a) $\frac{3}{5}\left[\frac{2}{\log\left(\frac{2}{5}\right)} + \frac{1}{2\log\left(\frac{5}{2}\right)}\right]$ (b) $-\frac{3}{5}\left[\frac{2}{\log\left(\frac{2}{5}\right)} + \frac{1}{2\log\left(\frac{5}{2}\right)}\right]$

(c) $\frac{3}{5}\left[\frac{2}{\log\left(\frac{2}{5}\right)} - \frac{1}{2\log\left(\frac{5}{2}\right)}\right]$ (d) $\frac{1}{2}[\log 2 - \log 3]$

2 If $2\int_0^1 \tan^{-1} x\,dx = \int_0^1 \cot^{-1}(1 - x + x^2)\,dx$, then $\int_0^1 \tan^{-1}(1 - x + x^2)\,dx$ is equal to

(a) log 2 (b) $\frac{\pi}{2} + \log 2$ (c) log 4 (d) $\frac{\pi}{2} - \log 4$

3 If $f(y) = e^y$, $g(y) = y$, $y > 0$ and $F(t) = \int_0^t f(t - y)\,g(y)\,dy$, then

(a) $F(t) = e^t - (1 + t)$ (b) $F(t) = te^t$ (c) $F(t) = te^{-t}$ (d) $F(t) = 1 - e^t(1 + t)$

4 If $I(m, n) = \int_0^1 t^m (1 - t)^n\,dt$, then the expression for $I(m, n)$ in terms of $I(m + 1, n - 1)$ is

(a) $\frac{2^n}{m+1} + \frac{n}{m+1} I(m + 1, n - 1)$ (b) $\frac{n}{m+1} I(m + 1, n - 1)$

(c) $\frac{2^n}{m+1} + \frac{n}{m+1} I(m + 1, n - 1)$ (d) $\frac{m}{n+1} I(m + 1, n - 1)$

5 The value of $\frac{\int_0^{\pi/2} (\sin x)^{\sqrt{3}+1}\,dx}{\int_0^{\pi/2} (\sin x)^{\sqrt{3}-1}\,dx}$ is

(a) $\frac{\sqrt{3}+1}{\sqrt{3}-1}$ (b) $\frac{\sqrt{3}-1}{\sqrt{3}+1}$ (c) $\frac{\sqrt{3}+1}{\sqrt{3}}$ (d) $\frac{\sqrt{3}}{\sqrt{3}+1}$

6 A value of α such that $\int_{\alpha}^{\alpha+1} \frac{dx}{(x+\alpha)(x+\alpha+1)} = \log_e\left(\frac{9}{8}\right)$ is **[JEE Main 2019]**

(a) -2 (b) $\frac{1}{2}$ (c) $-\frac{1}{2}$ (d) 2

10.2 Definite Integral by Substitution

7 $\int_1^{e^{37}} \frac{\pi \sin(\pi \log x)}{x} dx$ is equal to

(a) 2 (b) -2 (c) $\frac{2}{\pi}$ (d) 2π

8 $\int_0^{\log 5} \frac{e^x \sqrt{e^x - 1}}{e^x + 3} dx$ is equal to

(a) $3 + \pi$ (b) $3 - \pi$ (c) $4 + \pi$ (d) $4 - \pi$

9 For $x > 0$, let $F(x) = \int_1^x \frac{\ln t}{1+t} dt$, then $f(e) + f\left(\frac{1}{e}\right)$ is

(a) 2 (b) 1 (c) $\frac{1}{2}$ (d) None of these

10 If $f(x) = x^5 + 5x - 1$, then $\int_5^{41} \frac{dx}{[f^{-1}(x)]^5 + 5f^{-1}(x)}$ equals

(a) 0 (b) log 3 (c) log 4 (d) log 7

11 If the value of the integral $\int_0^{1/2} \frac{x^2}{(1-x^2)^{3/2}} dx$ is $\frac{k}{6}$, then k is equal to **[JEE Main 2020]**

(a) $3\sqrt{2} + \pi$ (b) $2\sqrt{3} - \pi$ (c) $2\sqrt{3} + \pi$ (d) $3\sqrt{2} - \pi$

12 If $\int_0^x e^{zx} e^{-z^2} dx = f(x) \int_0^x e^{-z^2/4} dz$, then $\int e^x \left[\log(f(x)) + \frac{x}{2}\right] dx$ equals

(a) $\frac{xe^x}{2} + C$ (b) $\frac{x^2 e^x}{4} + C$ (c) $\frac{x^2 e^x}{2} + C$ (d) $\frac{xe^x}{4} + C$

13 The value of $\int_0^{\pi} \sqrt{(|\cos x| - |\cos^3 x|} dx$ is

(a) $\frac{4}{3}$ (b) 1 (c) 3 (d) 4

14 If $\int_{-\pi/4}^{3\pi/4} \frac{e^{\pi/4} dx}{(e^x + e^{\pi/4})(\sin x + \cos x)} = K \int_{-\pi/2}^{\pi/2} \sec x \, dx$, then the value of K is

(a) $\frac{1}{2}$ (b) $\frac{1}{\sqrt{2}}$ (c) $\frac{1}{2\sqrt{2}}$ (d) $-\frac{1}{\sqrt{2}}$

15 If $\int_0^{\pi/3} \frac{\tan \theta}{\sqrt{2k \sec \theta}} d\theta = 1 - \frac{1}{\sqrt{2}}$, $(k > 0)$, then the value of k is **[JEE Main 2019]**

(a) 1 (b) $\frac{1}{2}$ (c) 2 (d) 4

16 The value of the definite integral $\int_0^{\pi/2} \sqrt{\tan x} \, dx$ is

(a) $\sqrt{2}\pi$ (b) $\frac{\pi}{\sqrt{2}}$ (c) $2\sqrt{2}\pi$ (d) $\frac{\pi}{2\sqrt{2}}$

17 If $\text{In} = \int_0^{\pi/4} \tan^n x \sec^2 x\, dx$, then I_1, I_2, I_3 are in

(a) AP (b) GP (c) HP (d) None of these

18 The integral $\int_{\pi/6}^{\pi/3} \tan^3 x \cdot \sin^2 3x (2\sec^2 x \cdot \sin^2 3x + 3\tan x \cdot \sin 6x) dx$ is equal to **[JEE Main 2020]**

(a) $-\frac{1}{9}$ (b) $\frac{7}{18}$ (c) $-\frac{1}{18}$ (d) $\frac{9}{2}$

19 The value of $\int_0^1 \cos\left(2 \cot^{-1} \sqrt{\frac{1-x}{1+x}}\right) dx$ is equal to

(a) $-\frac{1}{2}$ (b) $\frac{1}{2}$ (c) 0 (d) 1

20 The value of $\int_0^{\pi/4} \frac{\sin\theta + \cos\theta}{9 + 16\sin 2\theta} d\theta$ is equal to

(a) $\frac{1}{20}\log 2$ (b) $\frac{1}{20}\log 3$ (c) $\log 3$ (d) $\log 2$

21 $\int_0^{\pi/2} \frac{dx}{3 + 4\sin x}$ is equal to

(a) $\frac{1}{\sqrt{7}}\log\left(\frac{4+\sqrt{7}}{3}\right)$ (b) $\frac{1}{\sqrt{7}}\log\left(\frac{4-\sqrt{7}}{3}\right)$

(c) $\frac{2}{\sqrt{7}}\log\left(\frac{4-\sqrt{7}}{3}\right)$ (d) $\sqrt{7}\log\left(\frac{4+\sqrt{7}}{3}\right)$

22 The value of $\int_0^{\pi/2} \frac{dx}{4\cos^2 x + 9\sin^2 x}$ is equal to

(a) $\frac{\pi}{12}$ (b) $\frac{\pi}{4}$ (c) $\frac{\pi}{9}$ (d) $\frac{\pi}{6}$

23 Evaluate $\int_0^1 \frac{1-x}{1+x} \frac{dx}{\sqrt{x + x^2 + x^3}}$

(a) $\frac{\pi}{3}$ (b) $\frac{\pi}{6}$ (c) $\frac{\pi}{12}$ (d) $\frac{\pi}{4}$

24 If $I = \int_0^1 \frac{dx}{(1+x)(2+x)\sqrt{x(1-x)}}$, then I equals

(a) 2π (b) π (c) $\frac{\pi}{\sqrt{6}}(\sqrt{3}+1)$ (d) $\frac{\pi}{\sqrt{6}}(\sqrt{3}-1)$

25 If $\int_{\log 2}^{x} \frac{dy}{\sqrt{e^y - 1}} = \frac{\pi}{6}$, then x is equal to

(a) $\log_e 4$ (b) $\log_e 2$ (c) 4 (d) 2

26 The value of $\int_0^1 \frac{\sqrt{x}\, dx}{(1+x)(1+3x)(3+x)}$ is

(a) $\frac{\pi}{8}\left(1 - \frac{\sqrt{3}}{2}\right)$ (b) $\frac{\pi}{4}\left(1 - \frac{\sqrt{3}}{6}\right)$

(c) $\frac{\pi}{8}\left(1 - \frac{\sqrt{3}}{6}\right)$ (d) $\frac{\pi}{4}\left(1 - \frac{\sqrt{3}}{2}\right)$

10.3 Sum of Infinite Series using Integration

27 $\lim_{n\to\infty} \frac{1}{\sqrt{n}}\left(1+\frac{1}{\sqrt{2}}+\frac{1}{\sqrt{3}}+\ldots+\frac{1}{\sqrt{n}}\right)$ is equal to

(a) 2 (b) $\sqrt{2}$ (c) 3 (d) $\sqrt{3}$

28 $\lim_{n\to\infty} \frac{(n!)^{1/n}}{n}$ equals

(a) e (b) e^{-1} (c) 1 (d) None of these

29 $\lim_{n\to\infty} \frac{1}{n}[(n+1)(n+2)\ldots(n+n)]^{1/n}$ is equal to

(a) $\frac{2}{e}$ (b) $\frac{3}{e}$ (c) $\frac{4}{e}$ (d) None of these

30 If $S_n = \left[\frac{1}{n}+\frac{n^2}{(n+1)^3}+\frac{n^2}{(n+2)^3}+\ldots+\frac{1}{8^n}\right]$, then $\lim_{n\to\infty} S_n$ is

(a) $\frac{3}{8}$ (b) $\frac{1}{8}$ (c) $\frac{1}{4}$ (d) $\frac{5}{8}$

31 $\lim_{n\to\infty}\left(\frac{(n+1)^{1/3}}{n^{4/3}}+\frac{(n+2)^{1/3}}{n^{4/3}}+\ldots..+\frac{(2n)^{1/3}}{n^{4/3}}\right)$ is equal to **[JEE Main 2019]**

(a) $\frac{4}{3}(2)^{4/3}$ (b) $\frac{3}{4}(2)^{4/3}-\frac{4}{3}$ (c) $\frac{3}{4}(2)^{4/3}-\frac{3}{4}$ (d) $\frac{4}{3}(2)^{3/4}$

32 The value of $\lim_{n\to\infty}\sum_{r=1}^{n}\left[\frac{\sqrt{n}}{\sqrt{r}(\sqrt{r}+4\sqrt{n})^2}\right]$ is

(a) $\frac{1}{10}$ (b) $\frac{2}{10}$ (c) $\frac{1}{5}$ (d) $-\frac{1}{10}$

33 The value of $\lim_{n\to\infty}\frac{1}{n}\sum_{i=1}^{n}\frac{(2i-1)+8n}{(2i-1)+4n}$ is equal to

(a) $5+\log_e\left(\frac{3}{2}\right)$ (b) $2-\log\left(\frac{2}{3}\right)$ (c) $3+2\log\left(\frac{2}{3}\right)$ (d) $1+2\log\left(\frac{3}{2}\right)$

10.4 Properties of Definite Integral

34 If $\int_0^\infty \frac{\sin x}{x}\,dx=\frac{\pi}{2}$, then $\int_0^\infty \frac{\sin^3 x}{x}\,dx$ is equal to

(a) $\frac{\pi}{2}$ (b) $\frac{\pi}{4}$ (c) $\frac{\pi}{6}$ (d) $\frac{3\pi}{2}$

35 The value of $\int_0^2 [x^2]\,dx$ is

(a) $5-\sqrt{2}-\sqrt{3}$ (b) $2-\sqrt{2}+\sqrt{3}$ (c) $2+\sqrt{2}-\sqrt{3}$ (d) $\sqrt{2}+\sqrt{3}$

36 The value of $\int_{-2}^{1}\left[x\left[1+\cos\left(\frac{\pi x}{2}\right)\right]+1\right]dx$, where [ ] denotes the greatest integer function, is

(a) 1 (b) $\frac{1}{2}$ (c) 2 (d) None of these

37 If $f(x)$ is continuous for all real values of x, then $\sum_{r=1}^{n} \int_0^1 f(r-1+x)\,dx$ is equal to

(a) $\int_0^n f(x)\,dx$ (b) $\int_0^1 f(x)\,dx$ (c) $n\int_0^1 f(x)\,dx$ (d) $(n-1)\int_0^1 f(x)\,dx$

38 The value of α for which $4\alpha\int_{-1}^{2} e^{-\alpha|x|}\,dx = 5$, is **[JEE Main 2020]**

(a) $\log_e \sqrt{2}$ (b) $\log_e \left(\frac{3}{2}\right)$ (c) $\log_e 2$ (d) $\log_e \left(\frac{4}{3}\right)$

39 If $\{x\}$ denotes the fractional part of x, then $\int_0^x \left(\{x\} - \frac{1}{2}\right) dx$ is equal to

(a) $\frac{1}{2}\{x\}(\{x\}+1)$ (b) $\frac{1}{2}\{x\}(\{x\}-1)$ (c) $\{x\}(\{x\}-1)$ (d) $\{x\}(\{x\}+1)$

40 $\int_6^{16} \frac{\log x^2}{\log x^2 + \log(x^2 - 44x + 484)}\,dx$ is equal to

(a) 10 (b) 5 (c) 0 (d) None of these

41 If $f(x) = \frac{2 - x\cos x}{2 + x\cos x}$ and $g(x) = \log_e x$, $(x > 0)$, then the value of the integral $\int_{-\pi/4}^{\pi/4} g(f(x))dx$ is **[JEE Main 2019]**

(a) $\log_e 3$ (b) $\log_e e$ (c) $\log_e 2$ (d) $\log_e 1$

42 The value of $I = \int_{\sqrt{\ln 2}}^{\sqrt{\ln 3}} \left[\frac{x\sin(x^2)}{\sin(x^2) + \sin[\ln(6 - x^2)]}\right] dx$ is

(a) $\frac{1}{4}\ln\left(\frac{3}{2}\right)$ (b) $\frac{1}{2}\ln\left(\frac{3}{2}\right)$ (c) $\ln\left(\frac{3}{2}\right)$ (d) $\frac{1}{6}\ln\left(\frac{3}{2}\right)$

43 The value of $\int_0^{\pi} \frac{1}{e^{\cos x} + 1}\,dx$ is

(a) π (b) 0 (c) 2π (d) $\frac{\pi}{2}$

44 If $f(a + b + 1 - x) = f(x)$, for all x, where a and b are fixed positive real numbers, then $\frac{1}{a+b}\int_a^b x(f(x) + f(x+1))dx$ is equal to **[JEE Main 2020]**

(a) $\int_{a+1}^{b+1} f(x+1)dx$ (b) $\int_{a+1}^{b+1} f(x)dx$ (c) $\int_{a-1}^{b-1} f(x+1)dx$ (d) $\int_{a-1}^{b-1} f(x)dx$

45 If $I = \int_0^{\infty} \frac{x\,dx}{(1+x)(1+x^2)}$ is equal to

(a) $\frac{\pi}{4}$ (b) $\frac{\pi}{2}$ (c) π (d) None of these

46 The integral $\int_0^a \frac{g(x)}{f(x) + f(a-x)}\,dx$ vanishes, if

(a) $g(x)$ is odd (b) $f(x) = f(a-x)$
(c) $g(x) = -g(a-x)$ (d) $f(a-x) = g(x)$

47 If $I_1 = \int_0^1 (1 - x^{50})^{100}\,dx$ and $I_2 = \int_0^1 (1 - x^{50})^{101}\,dx$ such that $I_2 = \alpha I_1$, then α is equal to **[JEE Main 2020]**

(a) $\frac{5049}{5050}$ (b) $\frac{5050}{5049}$ (c) $\frac{5050}{5051}$ (d) $\frac{5051}{5050}$

48 The value of $\int_0^{2\pi} \frac{x \sin^8 x}{\sin^8 x + \cos^8 x} dx$ is equal to **[JEE Main 2020]**

(a) 2π (b) π^2 (c) $2\pi^2$ (d) 4π

49 The value of $\int_0^{\pi} \frac{x \sin x}{1+\cos^2 x} dx$ is

(a) $\frac{\pi^2}{2}$ (b) $\frac{\pi^2}{4}$ (c) $\frac{\pi^2}{8}$ (d) $\frac{\pi^2}{16}$

50 Let $I = \int_0^{\pi/2} \frac{dx}{1+\sin x}$, then $\int_0^{\pi} \frac{x^2 \cos x}{(1+\sin x)^2} dx$ is

(a) $-\pi^2$ (b) $\pi^2 - 2\pi I$ (c) $2\pi I$ (d) $2\pi I - \pi^2$

51 The value of $\int_{-\pi/2}^{\pi/2} \frac{dx}{[x]+[\sin x]+4}$, where $[t]$ denotes the greatest integer less than or equal to t, is **[JEE Main 2019]**

(a) $\frac{1}{12}(7\pi-5)$ (b) $\frac{1}{12}(7\pi+5)$ (c) $\frac{3}{10}(4\pi-3)$ (d) $\frac{3}{20}(4\pi-3)$

52 The value of $\int_{-\pi}^{\pi} \left[\frac{2x(1+\sin x)}{1+\cos^2 x}\right] dx$ is

(a) π^2 (b) $2\pi^2$ (c) $\frac{\pi^2}{2}$ (d) None of these

53 The value of $\int_{-1/\sqrt{2}}^{1/\sqrt{2}} \left[\left(\frac{x+1}{x-1}\right)^2 + \left(\frac{x-1}{x+1}\right)^2 - 2\right]^{1/2} dx$ is

(a) $\log 4$ (b) $\log 16$
(c) $2 \log 16$ (d) $4 \log (3+2\sqrt{2})$

54 $\int_{-\pi}^{\pi} |\pi - |x|| dx$ is equal to **[JEE Main 2020]**

(a) $\sqrt{2}\pi^2$ (b) $2\pi^2$ (c) π^2 (d) $\frac{\pi^2}{2}$

55 The value of $\int_{-\pi}^{\pi} \left(\frac{\cos^2 x}{1+a^x}\right) dt$ is

(a) π (b) πa (c) $\frac{\pi}{2}$ (d) 2π

56 The value of $\frac{\left\{\int_{-1/2}^{1/2} \cos 2x \log\left(\frac{1+x}{1-x}\right) dx\right\}}{\left\{\int_0^{1/2} \cos 2x \log\left(\frac{1+x}{1-x}\right) dx\right\}}$ is

(a) 0 (b) 4 (c) 2 (d) 8

57 The value of $\int_{-\pi/2}^{\pi/2} \frac{1}{1+e^{\sin x}} dx$ is **[JEE Main 2020]**

(a) $\frac{\pi}{2}$ (b) $\frac{\pi}{4}$ (c) π (d) $\frac{3\pi}{2}$

58 $\int_{-3\pi/2}^{-\pi/2} \{(x+\pi)^3 + \cos^2(x+3\pi)\}\, dx$ is equal to

(a) $\frac{\pi}{4}$ (b) $\frac{\pi}{2}$ (c) $\frac{\pi}{3}$ (d) $\frac{\pi}{4}$

59 The value of the integral $\int_{-2}^{2} \frac{\sin^2 x}{\left[\frac{x}{\pi}\right]+\frac{1}{2}} dx$

(where, $[x]$ denotes the greatest integer less than or equal to x) is [JEE Main 2019]

(a) $4-\sin 4$ (b) 4 (c) $\sin 4$ (d) 0

60 Let $f: R \to R$ be a continuous function such that $f(x)+f(x+1)=2$ for all $x \in R$.

If $I_1 = \int_0^8 f(x)\, dx$ and $I_2 = \int_{-1}^3 f(x)\, dx$, then the value of $I_1 + 2I_2$ is equal to

(a) 16 (b) 18 (c) 21 (d) -16

61 $\int_0^5 \frac{\tan^{-1}(x-[x])}{1+(x-[x])^2} dx$ is equal to

(a) $\frac{\pi^2}{32}$ (b) $\frac{\pi^2}{16}$ (c) $\frac{5\pi^2}{16}$ (d) $\frac{5\pi^2}{32}$

62 The value of $\int_{-\pi/2}^{\pi/2} \frac{x^2 \cos x}{1+e^x} dx$ is equal to

(a) $\frac{\pi^2}{4}-2$ (b) $\frac{\pi^2}{4}+2$ (c) $\pi^2 - e^{\pi/2}$ (d) $\pi^2 + e^{\pi/2}$

63 If $I = \int_0^1 \frac{dx}{1+x^{\pi/2}}$, then

(a) $\log_e 2 < I < \frac{\pi}{4}$ (b) $\log_e 2 > I$ (c) $I = \frac{\pi}{4}$ (d) $I = \log_e 2$

64 The minimum odd value of $a\ (a>0)$ for which $\int_{10}^{19} \frac{\sin x}{1+x^a} dx < \frac{1}{9}$, is equal to

(a) 1 (b) 3 (c) 5 (d) 9

65 If $I = \int_1^2 \frac{dx}{\sqrt{2x^3-9x^2+12x+4}}$, then [JEE Main 2020]

(a) $\frac{1}{6} < I^2 < \frac{1}{2}$ (b) $\frac{1}{8} < I^2 < \frac{1}{4}$ (c) $\frac{1}{9} < I^2 < \frac{1}{8}$ (d) $\frac{1}{16} < I^2 < \frac{1}{9}$

10.5 Leibnitz's Rule (Differentiation under Integral Sign)

66 If $f(x)$ is a continuous function for all real values of x and satisfies

$\int_0^x f(t)\, dt = \int_x^1 t^2 f(t)\, dt + \frac{x^{16}}{8} + \frac{x^6}{3} + a$, then the value of a is

(a) $-\frac{1}{24}$ (b) $\frac{17}{168}$ (c) $\frac{1}{7}$ (d) $-\frac{167}{840}$

67 If $f(x)$ is differentiable and $\int_0^{t^2} x\, f(x)\, dx = \frac{2}{5}t^5$, then $f\left(\frac{4}{25}\right)$ equals

(a) $\frac{2}{5}$ (b) $-\frac{5}{2}$ (c) 1 (d) $\frac{5}{2}$

68 The total number of distinct $x \in [0, 1]$ for which $\int_0^x \frac{t^2}{1+t^4} dt = 2x - 1$, is

(a) 1 (b) 2 (c) 3 (d) infinitely many

69 Let $f(x) = \int_0^x g(t)dt$, where g is a non-zero even function. If $f(x+5) = g(x)$, then $\int_0^x f(t)dt$ equals **[JEE Main 2019]**

(a) $5 \int_{x+5}^{5} g(t)dt$ (b) $\int_{5}^{x+5} g(t)dt$ (c) $2 \int_{5}^{x+5} g(t)dt$ (d) $\int_{x+5}^{5} g(t)dt$

70 If $f : R \to R$ is continuous and differentiable function such that
$\int_{-1}^{x} f(t)\, dt + f'''(3) \int_x^0 dt = \int_1^x t^3\, dt - f'(1) \int_0^x t^2\, dt + f''(2) \int_x^3 t\, dt$, then the value of $f'(4)$ is

(a) $48 - 8f'(1) + f''(2)$ (b) $48 - 8f'(1) - f''(2)$
(c) $48 + 8f'(1) - f''(2)$ (d) None of these

71 If $f(x) = e^{g(x)}$ and $g(x) = \int_2^x \frac{t\, dt}{1+t^4}$, then the value of $f'(2)$ is

(a) $\frac{1}{17}$ (b) $\frac{2}{17}$ (c) $\frac{3}{17}$ (d) None of these

72 Let $f(x)$ be a continuous function $\forall\, x \in R$ except at $x = 0$ such that $\int_0^a f(x)\, dx$, $a \in R^+$ exists. If $g(x) = \int_x^a \frac{f(t)}{t} dt$, then $\int_0^a f(x)\, dx$ is equal to

(a) $-\int_0^a g(x)\, dx$ (b) $-2 \int_0^a g(x)\, dx$ (c) $2 \int_0^a g(x)\, dx$ (d) $\int_0^a g(x)\, dx$

73 If $\int_0^x f(t)\, dt = x^2 + \int_x^1 t^2 f(t) dt$, then $f'\left(\frac{1}{2}\right)$ is **[JEE Main 2019]**

(a) $\frac{24}{25}$ (b) $\frac{18}{25}$ (c) $\frac{6}{25}$ (d) $\frac{4}{5}$

Numerical Value Type Questions

74 If $f(x) = \begin{vmatrix} \sin x + \sin 2x + \sin 3x & \sin 2x & \sin 3x \\ 3 + 4 \sin x & 3 & 4 \sin x \\ 1 + \sin x & \sin x & 1 \end{vmatrix}$, then the value of $\int_0^{\pi/2} f(x)\, dx$ is $\frac{1}{K}$, then the value of K is ……… .

75 If $I = \int_0^1 \frac{dx}{\sqrt{x} + \sqrt{2-x}}$, then the value of $[I + \log(\sqrt{2} - 1)]^2$ is ……… .

76 The value of $5050 \times \left[\frac{\int_0^1 (1-x^{50})^{100}\, dx}{\int_0^1 (1-x^{50})^{101}\, dx}\right]$ is ……… .

77 Let [ ] denotes the greatest integer function, then the value of $\int_0^{1.5} x\, [x^2]\, dx$ is $\frac{p}{q}$, then $p + q$ is ……… .

78 Let $f : (0, 2) \to R$ be defined as $f(x) = \log_2 \left[1 + \tan\left(\frac{\pi x}{4}\right)\right]$, then

$\lim_{n \to \infty} \frac{2}{n}\left[f\left(\frac{1}{n}\right) + f\left(\frac{2}{n}\right) + \ldots + f(1)\right]$ is equal to ……… .

79 Let $[t]$ denote the greatest integer less than or equal to t. Then, the value of $\int_1^2 |2x-[3x]|\,dx$ is .... **[JEE Main 2020]**

80 If $\int_0^1 \cot^{-1}(1+x^2-x)\,dx = K\left(\frac{\pi}{4}-\log\sqrt{2}\right)$, then the value of K is ....

81 For any real number x, let $[x]$ denote the largest integer less than or equal to x. Let f be a real function defined on $[-10, 10]$ by

$$f(x)=\begin{cases} x-[x] & : \text{if } [x] \text{ is odd} \\ 1+[x]-x & : \text{if } [x] \text{ is even} \end{cases}$$

Then, the value of $\frac{\pi^2}{10}\times\int_{-10}^{10} f(x)\cos(\pi x)\,dx$ is ....

82 The integral $\int_0^2 ||x-1|-x|\,dx$ is equal to .... **[JEE Main 2020]**

83 $\int_{-2}^{2} \frac{3x^5+4x^3+2x^2+x+20}{x^2+4}\,dx = A+B\pi$, then $A+B$ is equal to ....

84 Let $f(x)=\int_1^x \sqrt{2-t^2}\,dt$. Then, the number of real roots of the equation $x^2-f'(x)=0$ is ....

85 A function f is continuous for all x (and not everywhere zero) such that $f^2(x)=\int_0^x f(t)\frac{\cos t}{2+\sin t}\,dt$, then $f(x)$ is $\frac{1}{K}\ln\left(\frac{K+\sin x}{K}\right)$, then the value of K is ....

86 Let $\{x\}$ and $[x]$ denote the fractional part of x and the greatest integer $\le x$ respectively of a real number x. If $\int_0^n \{x\}\,dx, \int_0^n [x]\,dx$ and $10(n^2-n)$, $(n\in N, n>1)$ are three consecutive terms of a GP, then n is equal to .... **[JEE Main 2020]**

87 If $x\int_0^x \sin(f(t))\,dt=(x+2)\int_0^x t\sin(f(t))\,dt$, where $x>0$ then the value of $f'(x)\cot f(x)+\frac{3}{1+x}$ is ....

LEVEL 03 \ JEE Advanced Level

Single Choice Correct Questions

1 $\int_0^x \frac{2^t}{2^{[t]}}\,dt$ where [ ] denotes the greatest integer function, and $x\in R^+$ is equal to

(a) $\frac{1}{\ln 2}([x]+2^{\{x\}}-1)$
(b) $\frac{1}{\ln 2}([x]+2^{\{x\}})$
(c) $\frac{1}{\ln 2}([x]-2^{\{x\}})$
(d) $\frac{1}{\ln 2}([x]+2^{\{x\}}+1)$

2 The value of $\int_0^\pi \frac{|x|\sin^2 x}{1+2|\cos x|\sin x}\,dx$ is equal to

(a) $\frac{\pi}{4}$ (b) $\frac{\pi}{2}$ (c) π (d) 2π

3 If $\alpha, \beta\ (\beta > \alpha)$ are the roots of $f(x) = ax^2 + bx + c = 0$ and $f(x)$ is an even function, then

$$\int_{\alpha}^{\beta} \frac{e^{f\left(\frac{f(x)}{x-\alpha}\right)}}{e^{f\left(\frac{f(x)}{x-\alpha}\right)} + e^{f\left(\frac{f(x)}{x-\beta}\right)}} dx \text{ is equal to}$$

(a) $\left|\frac{b}{a}\right|$ (b) $\left|\frac{a}{b}\right|$

(c) $\frac{\sqrt{b^2 - 4ac}}{|2a|}$ (d) None of these

4 The function $f(x)$ that satisfies the condition $f(x) = x + \int_0^{\frac{\pi}{2}} \sin x \cos y\, f(y)\, dy$ is

(a) $x + \frac{2}{3}(\pi - 2) \sin x$ (b) $x + (\pi + 2) \sin x$

(c) $x + \frac{\pi}{2} \sin x$ (d) $x + (\pi - 2) \sin x$

5 The integral $\int_{\pi/4}^{\pi/2} (2 \operatorname{cosec} x)^{17}\, dx$ is equal to

(a) $\int_0^{\log(1+\sqrt{2})} 2\,(e^u + e^{-u})^{16}\, du$ (b) $\int_0^{\log(1+\sqrt{2})} (e^u + e^{-u})^{17}\, du$

(c) $\int_0^{\log(1+\sqrt{2})} (e^u - e^{-u})^{17}\, du$ (d) $\int_0^{\log(1+\sqrt{2})} 2\,(e^u - e^{-u})^{16}\, du$

6 If $[x]$ is the greatest integer function, then

$\pi^2 \int_0^2 \left(\sin \frac{\pi x}{2}\right)(x - [x])^{[x]}\, dx$ is equal to

(a) $2(\pi - 1)$ (b) $4(\pi - 1)$ (c) $4(\pi + 1)$ (d) $2(\pi + 1)$

7 If $f'(x) = f(x) + \int_0^1 f(x)\, dx$, given $f(0) = 1$ then the value of $f(\log 2)$ is

(a) $\frac{1}{3+e}$ (b) $\frac{5-e}{3-e}$ (c) $\frac{2+e}{e-2}$ (d) None of these

8 If the function $f : [0, 8] \to R$ is differentiable, then for $0 < \beta < 1$ and $0 < \alpha < 2$. $\int_0^8 f(t)\, dt$ is equal to

(a) $3\,[\alpha^3 f(\alpha^2) + \beta^2 f(\beta^2)]$ (b) $3\,[\alpha^3 f(\alpha) + \beta^3 f(\beta)]$

(c) $3\,[\alpha^2 f(\alpha^3) + \beta^2 f(\beta^3)]$ (d) $3\,[\alpha^2 f(\alpha^2) + \beta^2 f(\beta^2)]$

9 Let $f : R \to R$ be a differentiable function satisfying

$f(x) = x^2 + 3\int_0^{x^{\frac{1}{3}}} e^{-t^3} t^2 f(x - t^3) dt$, then $f(x)$ is

(a) $\frac{x^2}{2} + x$ (b) $\frac{x^3}{3} + x^2$ (c) $\frac{x^3}{3} - x^2$ (d) None of these

10 For $x \in R$, $x \neq 0$, if $y(x)$ is differentiable function such that $x\int_1^x y(t)\, dt = (x+1)\int_1^x t \cdot y(t)\, dt$, then $y(x)$ equals where C is a constant

(a) $Cx^3 e^{\frac{1}{x}}$ (b) $\frac{C}{x} e^{-\frac{1}{x}}$ (c) $\frac{C}{x^2} e^{-\frac{1}{x}}$ (d) $\frac{C}{x^3} e^{-\frac{1}{x}}$

11 $\lim_{n\to\infty}\left[\left(1+\frac{1}{n^2}\right)^{\frac{2}{n^2}}\left(1+\frac{2^2}{n^2}\right)^{\frac{4}{n^2}}\left(1+\frac{3^2}{n^2}\right)^{\frac{6}{n^2}}\ldots(2)^{\frac{2}{n}}\right]$ is equal to

(a) e^4 (b) $2e^{-1}$

(c) $4e^{-1}$ (d) $4e^{-2}$

12 If $np = 1$, and $L = \lim_{n\to\infty}\prod_{r=1}^{n}[1+(rp)^2]^{1/r}$, then the value of $\ln L$ is

(a) $\frac{\pi^2}{12}$ (b) $\frac{\pi^2}{24}$

(c) $\frac{\pi^2}{6}$ (d) $\frac{\pi^2}{4}$

More than One Correct

13 Let $u=\int_0^\infty \frac{dx}{x^4+7x^2+1}$ and $v=\int_0^\infty \frac{x^2\,dx}{x^4+7x^2+1}$ then

(a) $v > u$ (b) $6v = \pi$

(c) $3u + 2v = \frac{5\pi}{6}$ (d) $u + v = \frac{\pi}{3}$

14 If $\int_0^\alpha \frac{dx}{1-\cos\alpha\cos x} = \frac{A}{\sin\alpha} + B$ $(\alpha \neq 0)$ then possible values of A and B are

(a) $A=\frac{\pi}{2}, B=0$ (b) $A=\frac{\pi}{4}, B=\frac{\pi}{4\sin\alpha}$

(c) $A=\frac{\pi}{6}, B=\frac{\pi}{\sin\alpha}$ (d) $A=\pi, B=\frac{\pi}{\sin\alpha}$

15 The value of $\int_0^1 \frac{2x^2+3x+3}{(x+1)(x^2+2x+2)}\,dx$ is

(a) $\frac{\pi}{4}+2\log 2-\tan^{-1}2$ (b) $\frac{\pi}{4}+2\log 2-\tan^{-1}\frac{1}{3}$

(c) $2\log 2-\cot^{-1}3$ (d) $-\frac{\pi}{4}+\log 4+\cot^{-1}2$

16 If $I_n=\int_{-\pi}^{\pi}\left(\frac{\sin nx}{(1+\pi^x)\sin x}\right)dx, n=0,1,2,\ldots$ then

(a) $I_n = I_{n+2}$ (b) $\sum_{m=1}^{10}(I_{2m+1}) = 10\pi$

(c) $\sum_{m=1}^{10}(I_{2m}) = 0$ (d) $I_n = I_{n+1}$

17 If $g(x)=\int_{\sin x}^{\sin 2x}\sin^{-1}t\,dt$ then [JEE Adv. 2017]

(a) $g'\left(\frac{\pi}{2}\right) = -2\pi$ (b) $g'\left(-\frac{\pi}{2}\right) = 2\pi$

(c) $g'\left(\frac{\pi}{2}\right) = 0$ (d) $g'\left(-\frac{\pi}{2}\right) = 0$

18 Let $f : R \to (0, 1)$ be continuous function then which of the following function(s) has (have) the value zero at some point in the interval (0, 1). **[JEE Adv. 2017]**

(a) $e^x - \int_0^x f(t) \sin t \, dt$ (b) $x^9 - f(x)$

(c) $f(x) + \int_0^{\frac{\pi}{2}} f(t) \sin t \, dt$ (d) $x - \int_0^{\frac{\pi}{2} - x} f(t) \cos t \, dt$

19 Let $f : [a, b] \to [1, \infty)$ be a continuous function and let $g : R \to R$ be defined as

$$g(x) = \begin{cases} 0, & \text{if } x < a \\ \int_a^x f(t)\, dt, & \text{if } a \le x \le b \\ \int_a^b f(t)\, dt, & \text{if } x > b \end{cases}$$

(a) $g(x)$ is continuous but not differentiable at a

(b) $g(x)$ is differentiable on R

(c) $g(x)$ is continuous but not differentiable at b

(d) $g(x)$ is continuous and differentiable at either a or b but not both.

20 $f : [0, 1] \to R$ be a non-increasing function then for $\alpha \in (0, 1)$

(a) $\alpha \int_0^1 f(x)\, dx \le \int_0^\alpha f(x)\, dx$ (b) $\alpha \int_0^1 f(x)\, dx \ge \int_0^\alpha f(x)\, dx$

(c) $\alpha^2 \int_0^1 f(x)\, dx \le \int_0^\alpha f(x)\, dx$ (d) $\sqrt{\alpha} \int_0^1 f(x)\, dx \le \int_0^\alpha f(x)\, dx$

21 $\int_0^x \{\int_0^u f(t)\, dt\}\, du$ is equal to

(a) $\int_0^x (x - u) f(u)\, du$ (b) $\int_0^x u f(x - u)\, du$

(c) $x \int_0^x f(u)\, du$ (d) $x \int_0^x u f(u - x)\, du$

Linked Comprehension Type Questions

Paragraph I Directions (Q.Nos. 22-24)

Given that for each $Q \in (0, 1)$ $\lim_{h \to 0^+} \int_h^{1-h} t^{-a} (1 - t)^{a - 1}\, dt$ exists let this limit be $g(a)$. In addition it is given that the function $g(a)$ is differentiable on (0, 1)

22 The value of $g\left(\frac{1}{2}\right)$ is

(a) π (b) 2π (c) $\frac{\pi}{2}$ (d) $\frac{\pi}{4}$

23 The value of $g(a) - g(1 - a)$ is

(a) 0 (b) $g(1 + a)$

(c) 1 (d) None of these

24 The value of $g'\left(\frac{1}{2}\right)$ is

(a) $\frac{\pi}{2}$ (b) π (c) $-\frac{\pi}{2}$ (d) 0

Paragraph II Directions (Q.Nos. 25-27)

f is continuous for all $x \geq 0$ then answer the following questions

25 If $g : [0, a]$ is continuous and $f(x) = f(a - x)$, $g(x) + g(a - x) = 2$ then $\int_0^a f(x) . g(x)\, dx =$

(a) $\int_0^a g(x)\, dx$ (b) $\int_0^a f(x)\, dx$ (c) $2\int_0^a f(x)\, dx$ (d) 0

26 $\int_0^\pi e^{|\cos x|} \left[2 \sin\left(\frac{1}{2} \cos x\right) dx\right]$ is equal to

(a) π (b) $\frac{\pi}{2}$ (c) $\frac{\pi^2}{2}$ (d) 0

27 $\int_{\pi/4}^{3\pi/4} \frac{x}{1 + \sin x}\, dx$

(a) $\pi(\sqrt{2} - 1)$ (b) $\pi(\sqrt{2} + 1)$

(c) $2\pi(-\sqrt{2})$ (d) $\pi\sqrt{2}$

Matrix Match Type Questions

28 Match the columns

	Column I		Column II
(i)	If $f(x+1) = f(3+x)\ \forall\ x$ and the value of $\int_a^{a+b} f(x)\, dx$ is independent of a, then the value of b can be	(A)	2
(ii)	$25\int_0^{\pi/4} (\tan^6 (x - [x]) + \tan^4 (x - [x])\, dx$	(B)	1
(iii)	The value of $I = \int_1^4 \frac{\tan^{-1} [x^2]}{\tan^{-1}[x^2] + \tan^{-1} [25 + x^2 - 10x]} dx$ (where [] denotes the greatest integer function) then $[I]$ is	(C)	3
(iv)	If $I = \int_0^2 \sqrt{x + \sqrt{x + \sqrt{x + \ldots \infty}}}\, dx$(where $x > 0$), then $[I]$ is equal to (where [] denotes the greatest integer function)	(D)	5

29 Match the columns

	Column I		Column II
(i)	The value of $\int_{1/e}^{e^2} \left\| \frac{\log x}{x} \right\| dx$ is	(A)	5
(ii)	$\int_0^9 \{\sqrt{x}\}$ where $\{t\}$ denotes the fractional part of t is	(B)	2
(iii)	$\int_0^{\pi/2} \frac{f(x)}{f(x) + f\left(\frac{\pi}{2} - x\right)} dx = \frac{\pi}{k}$ where k is	(C)	$\frac{5}{2}$
(iv)	$\lim_{n \to \infty} \left[\frac{1}{1+n^2} + \frac{2}{2^2 + n^2} + \ldots \frac{n}{n^2 + n^2}\right]$ is $\frac{1}{k} \log k$, then k is	(D)	4

30 Match the columns

Column I		Column II
(i) The number of polynomials $f(x)$ with non-negative integer coefficients of degree ≤ 2 satisfying $f(0)=0$ and $\int_0^1 f(x)\,dx = 1$ is	(A)	8
(ii) $\int_{-2}^{2} \frac{3x^2}{1+e^x}\,dx$ equals	(B)	2
(iii) $\dfrac{\int_{-1/2}^{1/2} \cos 2x \log\left(\frac{1+x}{1-x}\right)dx}{\int_0^{1/2} \cos 2x \log\left(\frac{1+x}{1-x}\right)dx}$ equals	(C)	4
(iv) $\int_{-1}^{3} \text{sgn}(x-[x])\,dx$ is	(D)	0

31 Match the columns

Column I		Column II
(i) $\lim_{n\to\infty} \sum_{r=1}^{n} \frac{n}{(n+r)\sqrt{r(2n+r)}}$ is	(A)	0
(ii) $\lim_{n\to\infty} \sum_{r=1}^{n} \frac{1}{\sqrt{2rn-r^2}}$ is	(B)	$\frac{\pi}{4}$
(iii) $\int_{\log\frac{1}{2}}^{\log 2} \sin\left(\frac{e^x-1}{e^x+1}\right)dx$ is	(C)	$\frac{\pi}{2}$
(iv) If $\int_0^\infty \frac{\sin x}{x}\,dx = \frac{\pi}{2}$, then $\int_0^\infty \frac{\sin^3 x}{x}\,dx$ is equal to	(D)	$\frac{\pi}{3}$

Numerical Value Type Questions

Question Stem Directions (Q.Nos. 32-34) The answer to each question is a **NUMERICAL VALUE.**

Let $g_i : \left[\frac{\pi}{8}, \frac{3\pi}{3}\right] \to R, i = 1, 2$ and $f\left[\frac{\pi}{8}, \frac{3\pi}{8}\right] \to R$ be function such that $g_1(x) = 1$, $g_2(x) = |4x - \pi|$ and $f(x) = \sin^2 x$ for all $x \in \left[\frac{\pi}{8}, \frac{3\pi}{8}\right]$

Define $S_i = \int_{\frac{\pi}{8}}^{\frac{3\pi}{8}} f(x) \cdot g_i(x)\,dx, i = 1, 2$ then

32 The value of $\frac{16S_1}{\pi}$ is ....

33 The value of $\frac{48S_2}{\pi^2}$ is ....

34 The value of $\int_0^{\pi/2} g_2(x)dx$ is $\frac{\pi^2}{k}$, then k is equal to ....

Question Stem Directions (Q.Nos. 35-37) The answer to each question is a **NUMERICAL VALUE.**

Let $g(x) = x^c\ e^{cx}$ and $f(x) = \int_0^x t \cdot e^{2t}(1+3t^2)^{1/2}\,dt$.

If $L = \lim\limits_{x \to \infty} \dfrac{f'(x)}{g'(x)}$ is non-zero finite number, then

35 The value of c is ....

36 The value of L is ....

37 Number of real roots of the equation $g(x) = x$ is ....

38 The value of integral $\int_0^{1/2} \dfrac{1+\sqrt{3}}{((x+1)^2(1-x)^6)^{1/4}}\,dx$ is ....

39 Let $I = \int_0^{\pi/4} (\pi x - 4x^2) \ln(1+\tan x)\,dx$. If value of $I = \dfrac{\pi^3 \ln 2}{k}$ then k is ....

40 If $\alpha = \int_0^1 e^{9x+3\tan^{-1}x}\left(\dfrac{12+9x^2}{1+x^2}\right)dx$ where $\tan^{-1}x$ takes only principal values, then the value of $\left(\log|1+\alpha| - \dfrac{3\pi}{4}\right)$ is ....

41 If $f(x) = \dfrac{e^x}{1+e^x}$. $I_1 = \int_{f(-a)}^{f(a)} xg[x(1-x)]dx$ and $I_2 = \int_{f(-a)}^{f(a)} g[x(1-x)]dx$, where g is not an identity function, then the value of $\dfrac{I_2}{I_1}$ is ....

42 Let $f(x) = x^3 - \dfrac{3x^2}{2} + x + \dfrac{1}{4}$, then the value of $\int_{1/4}^{3/4} f(f(x))\,dx$ is P, then $8P$ is ....

43 The value of $\int_0^1 4x^3 \left\{\dfrac{d^2}{dx^2}(1-x^2)^5\right\} dx$ is ....

44 Let $f: R \to R$ be a differentiable function such that $f(0) = 0$, $f\left(\dfrac{\pi}{2}\right) = 3$ and $f'(0) = 1$, if $g(x) = \int_x^{\pi/2} [f'(t)\operatorname{cosec} t - \cot t \operatorname{cosec} t\, f(t)]\,dt$ for $x \in \left(0, \dfrac{\pi}{2}\right]$, then $\lim\limits_{x \to 0} g(x)$ ....

45 Let $y(x) = \int_{\pi^2/16}^{x^2} \dfrac{\cos x \cos\sqrt{\theta}}{1+\sin^2\sqrt{\theta}}\,d\theta$ then $\left(\dfrac{dy}{dx}\right)_{x=\pi} = k\pi$, then the value of k is ....

46 If $I_1 = \int_1^{\sin\theta} \dfrac{x}{1+x^2}\,dx$ and $I_2 = \int_1^{\operatorname{cosec}\theta} \dfrac{1}{x(x^2+1)}\,dx$, then the value of $\begin{vmatrix} I_1 & I_1^2 & I_2 \\ e^{I_1+I_2} & I_2^2 & -1 \\ 1 & I_1^2 + I_2^2 & -1 \end{vmatrix}$ is ....

ANSWER KEY

Level 1

1. (i) $\frac{-112}{15}$ (ii) $-\log 4$ (iii) $\frac{4}{3}(\sqrt{2}-1)$ (iv) $\frac{\pi}{4}$
(v) $\sqrt{2}-1$ (vi) $\frac{5}{2}\log 2+\frac{\pi}{8}$ (vii) π (viii) $\log\left(\frac{2}{3}\right)+\frac{2}{3}$
(ix) 1 (x) $1-\frac{2\sqrt{2}}{\pi}+\frac{4}{\pi}$ (xi) $\frac{\pi}{2}$

2. (i) $\frac{1}{5}\log 6$ (ii) $\frac{e-1}{2}$ (iii) $\frac{64}{221}$ (iv) $\frac{\pi}{2ab}$
(v) $\frac{\pi}{3}$ (vi) $\frac{\pi}{4\sqrt{5}}$ (vii) $\frac{\pi}{4}-\frac{1}{2}\log 2$

3. (i) $\log 2$ (ii) $\frac{\pi}{4}$ (iii) $\log 3$ (iv) $\frac{\pi}{6}$ (v) $\frac{1}{p+1}$

4. (i) $2e-2$ (ii) $\frac{11}{4}$ (iii) 6
(iv) 4 (v) 2 (vi) $100-\tan 1$

5. (i) $\frac{\pi}{4}$ (ii) 0 (iii) $\frac{1}{(n+1)(n+2)}$
(iv) 2π (v) 0 (vi) $\frac{b-a}{2}$

6. (i) 0 (ii) $\frac{\pi^2}{2\sqrt{2}}$ (iii) $\frac{\pi}{2\sqrt{3}}$ (iv) $\frac{\pi^2}{2}\log\left(\frac{1}{2}\right)$

7. (i) 0 (ii) $\frac{\pi}{2}$ (iii) $2(e-1)$ (iv) 0 (v) 0

8. (i) $2x\cos x^2$ (ii) $\frac{1}{2\sqrt{x}}\cos x+\frac{1}{x^2}\cos\left(\frac{1}{x^2}\right)$ (iii) $\frac{1}{\log x}(x^2-x)$
(iv) $\frac{2x(x-1)(x+1)(x-2)(x+2)}{2+e^{x^2}}$

9. $\frac{-\sqrt{3-\sin^2 x}}{\cos y}$

10. $\sqrt{257}$

Level 2

1. (b)	**2.** (a)	**3.** (a)	**4.** (b)	**5.** (d)	**6.** (a)	**7.** (a)
8. (d)	**9.** (c)	**10.** (d)	**11.** (b)	**12.** (b)	**13.** (a)	**14.** (c)
15. (c)	**16.** (a)	**17.** (c)	**18.** (c)	**19.** (a)	**20.** (b)	**21.** (a)
22. (a)	**23.** (a)	**24.** (d)	**25.** (a)	**26.** (a)	**27.** (a)	**28.** (b)
29. (c)	**30.** (a)	**31.** (c)	**32.** (a)	**33.** (d)	**34.** (b)	**35.** (a)
36. (c)	**37.** (a)	**38.** (c)	**39.** (b)	**40.** (b)	**41.** (d)	**42.** (a)
43. (d)	**44.** (c)	**45.** (a)	**46.** (c)	**47.** (c)	**48.** (b)	**49.** (b)
50. (d)	**51.** (d)	**52.** (a)	**53.** (b)	**54.** (c)	**55.** (c)	**56.** (a)
57. (a)	**58.** (b)	**59.** (d)	**60.** (a)	**61.** (d)	**62.** (a)	**63.** (a)
64. (b)	**65.** (c)	**66.** (d)	**67.** (a)	**68.** (a)	**69.** (d)	**70.** (b)
71. (b)	**72.** (d)	**73.** (a)	**74.** (3)	**75.** (2)	**76.** (5051)	**77.** (7)
78. (1)	**79.** (1)	**80.** (2)	**81.** (4)	**82.** (1.5)	**83.** (11)	**84.** (2)
85. (2)	**86.** (21)	**87.** (0)				

Level 3

1. (a)	**2.** (b)	**3.** (c)	**4.** (d)	**5.** (a)	**6.** (b)	**7.** (b)
8. (c)	**9.** (b)	**10.** (d)	**11.** (c)	**12.** (b)	**13.** (b, d)	**14.** (a. b)
15. (a, d)	**16.** (a, b, c)	**17.** (c, d)	**18.** (b, d)	**19.** (a, c)	**20.** (a, c)	**21.** (a, b)
22. (a)	**23.** (a)	**24.** (d)	**25.** (b)	**26.** (d)	**27.** (a)	

28. (i)–A, (ii)–D, (iii)–B, (iv)–C

29. (i)–C, (ii)–A, (iii)–D, (iv)–B

30. (i)–B, (ii)–A, (iii)–D, (iv)–C

31. (i)–D, (ii)–C, (iii)–A, (iv)–B

32. (2)	**33.** (1.5)	**34.** (4)	**35.** (2)	**36.** (0.86)	**37.** (2)	**38.** (2)
39. (192)	**40.** (9)	**41.** (2)	**42.** (2)	**43.** (2)	**44.** (2)	**45.** (2)
46. (0)						

SOLUTIONS

Level I

1. (i) We have, $\int_0^4 \sqrt{x}(1-x)\,dx$

$$= \int_0^4 (\sqrt{x} - x^{3/2})\,dx$$

$$= \left[\frac{2}{3}x^{3/2} - \frac{2}{5}x^{5/2}\right]_0^4$$

$$= \frac{2}{3}[4]^{3/2} - \frac{2}{5}[4]^{5/2}$$

$$= \frac{2}{3}(8) - \frac{2}{5}[32]$$

$$= \frac{16}{3} - \frac{64}{5} = \frac{-112}{15}$$

(ii) We have, $\int_{-4}^{-1} \frac{1}{x}\,dx = [\log|x|]_{-4}^{-1}$

$$= \log 1 - \log 4 = 0 - \log 4 = -\log 4$$

(iii) Given, $\int_0^1 \frac{1}{\sqrt{1+x}+\sqrt{x}}\,dx$

$$= \int_0^1 \left(\frac{1}{\sqrt{1+x}+\sqrt{x}} \times \frac{\sqrt{1+x}-\sqrt{x}}{\sqrt{1+x}-\sqrt{x}}\right) dx$$

$$= \int_0^1 \frac{\sqrt{1+x}-\sqrt{x}}{(1+x)-x}\,dx$$

$$= \int_0^1 (\sqrt{1+x}-\sqrt{x})\,dx$$

$$= \left[\frac{2}{3}(1+x)^{3/2} - \frac{2}{3}(x)^{3/2}\right]_0^1$$

$$= \left[\frac{2}{3}(2)^{3/2} - \frac{2}{3}(1)^{3/2}\right] - \left[\frac{2}{3}(1)^{3/2} - 0\right]$$

$$= \frac{2}{3}(2)^{3/2} - \frac{2}{3} - \frac{2}{3}$$

$$= \frac{2}{3}[2\sqrt{2}-2] = \frac{4}{3}[\sqrt{2}-1]$$

(iv) We have, $\int_0^{\pi/2} \sin^2 x\,dx$

$$= \int_0^{\pi/2} \frac{1-\cos 2x}{2}\,dx$$

$$= \frac{1}{2}\int_0^{\pi/2} (1-\cos 2x)\,dx$$

$$= \frac{1}{2}\left[x - \frac{\sin 2x}{2}\right]_0^{\pi/2}$$

$$= \frac{1}{2}\left[\left(\frac{\pi}{2} - \frac{\sin \pi}{2}\right) - \left(0 - \frac{\sin 0}{2}\right)\right]$$

$$= \frac{1}{2}\left(\frac{\pi}{2}\right) = \frac{\pi}{4}.$$

(v) Given, $\int_0^{\pi/4} \sqrt{1-\sin 2x}\,dx$

$$= \int_0^{\pi/4} \sqrt{\sin^2 x + \cos^2 x - 2\sin x \cos x}\,dx$$

$$= \int_0^{\pi/4} |\cos x - \sin x|\,dx$$

$$= \int_0^{\pi/4} (\cos x - \sin x)\,dx \quad \left\{\begin{array}{l} \because\ 0 < x < \frac{\pi}{4} \\ \therefore\ \cos x > \sin x \\ \cos x - \sin x > 0 \end{array}\right.$$

$$= [\sin x + \cos x]_0^{\pi/4}$$

$$= \left(\sin\frac{\pi}{4} + \cos\frac{\pi}{4}\right) - (\sin 0 + \cos 0)$$

$$= \frac{1}{\sqrt{2}} + \frac{1}{\sqrt{2}} - (0+1)$$

$$= \frac{2}{\sqrt{2}} - 1 = \sqrt{2} - 1$$

(vi) We have, $\int_0^2 \frac{5x}{x^2+4}\,dx + \int_0^2 \frac{1}{x^2+4}\,dx$

$$= \frac{5}{2}\int_0^2 \frac{2x}{x^2+4}\,dx + \int_0^2 \frac{1}{x^2+(2)^2}\,dx$$

$$= \frac{5}{2}[\ln(x^2+4)]_0^2 + \left[\frac{1}{2}\tan^{-1}\frac{x}{2}\right]_0^2$$

$$= \frac{5}{2}(\log 8 - \log 4) + \frac{1}{2}[\tan^{-1}(1) - \tan^{-1}(0)]$$

$$= \frac{5}{2}\log 2 + \frac{1}{2}\left(\frac{\pi}{4}\right) = \frac{5}{2}\log 2 + \frac{\pi}{8}$$

(vii) Given, $\int_0^a \frac{1}{\sqrt{ax-x^2}}\,dx$

$$= \int_0^a \frac{1}{\sqrt{\left(\frac{a}{2}\right)^2 - \left(x-\frac{a}{2}\right)^2}}\,dx$$

$$= \left[\sin^{-1}\left(\frac{x-\frac{a}{2}}{\frac{a}{2}}\right)\right]_0^a$$

$$= \left[\sin^{-1}\left(\frac{2x-a}{a}\right)\right]_0^a$$

$$= \sin^{-1}1 - \sin^{-1}(-1)$$

$$= 2\left(\frac{\pi}{2}\right) = \pi$$

(viii) Given, $\int_1^3 \frac{1}{x^2(x+1)}dx$

Let $\frac{1}{x^2(x+1)} = \frac{A}{x} + \frac{B}{x^2} + \frac{C}{x+1}$

$\Rightarrow \quad \frac{1}{x^2(x+1)} = \frac{Ax(x+1)+B(x+1)+Cx^2}{x^2(x+1)}$

$\Rightarrow \quad 1 = Ax(x+1) + B(x+1) + Cx^2$

So, $B = 1, A = -1, C = 1$

$$\int_1^3 \frac{1}{x^2(x+1)}dx = \int_1^3 \left(-\frac{1}{x} + \frac{1}{x^2} + \frac{1}{x+1}\right)dx$$

$$= \left[-\log|x| - \frac{1}{x} + \log|x+1|\right]_1^3$$

$$= \left(-\log 3 - \frac{1}{3} + \log 4\right) - (-\log 1 - 1 + \log 2)$$

$$= \log\left(\frac{4}{2\times 3}\right) - \frac{1}{3} + 1 = \log\left(\frac{2}{3}\right) + \frac{2}{3}$$

(ix) We have, $\int_0^{\pi/2} x \sin x\, dx$

$$= [-x\cos x]_0^{\pi/2} - \int_0^{\pi/2} 1(-\cos x)\, dx$$

$$= -[x\cos x]_0^{\pi/2} + [\sin x]_0^{\pi/2}$$

$$= -\left(\frac{\pi}{2}\cos\frac{\pi}{2} - 0\cos 0\right) + \left(\sin\frac{\pi}{2} - \sin 0\right)$$

$$= 0 + (1 - 0) = 1$$

(x) We have, $\int_0^1 \left\{xe^x + \sin\frac{\pi}{4}x\right\} dx$

$$= [xe^x]_0^1 - \int_0^1 1\cdot e^x\, dx - \frac{4}{\pi}\left[\cos\frac{\pi}{4}x\right]_0^1$$

$$= (1\cdot e - 0) - [e^x]_0^1 - \frac{4}{\pi}\left[\cos\frac{\pi}{4} - \cos 0\right]$$

$$= e - (e-1) - \frac{4}{\pi}\left(\frac{1}{\sqrt{2}} - 1\right)$$

$$= e - e + 1 - \frac{4}{\pi}\left(\frac{1-\sqrt{2}}{\sqrt{2}}\right)$$

$$= 1 - \frac{2\sqrt{2}}{\pi} + \frac{4}{\pi}$$

(xi) Given, $\int_0^{\sqrt{2}} \sqrt{2-x^2}\, dx$

$$= \int_0^{\sqrt{2}} \sqrt{(\sqrt{2})^2 - x^2}\, dx$$

$$= \left[\frac{x}{2}\sqrt{2-x^2} + \frac{(\sqrt{2})^2}{2}\sin^{-1}\frac{x}{\sqrt{2}}\right]_0^{\sqrt{2}}$$

$$= \{0 + \sin^{-1} 1\} - \{0 + \sin^{-1} 0\}$$

$$= \frac{\pi}{2}$$

2. (i) Let $I = \int_0^1 \frac{2x}{5x^2+1}dx$

Put $5x^2 + 1 = t$

$\Rightarrow \quad 10x\, dx = dt$

$\Rightarrow \quad 2x\, dx = \frac{1}{5}dt$

when $x = 0, t = 1$

when $x = 1, t = 6$

$$\int_1^6 \frac{1}{t}\cdot\frac{dt}{5}$$

$$= \frac{1}{5}[\log t]_1^6$$

$$= \frac{1}{5}(\log 6 - \log 1)$$

$$= \frac{1}{5}\log 6$$

(ii) Let $I = \int_0^1 x\, e^{x^2}\, dx$

Put $x^2 = t$

$\Rightarrow \quad 2x\, dx = dt$

$\Rightarrow \quad x\, dx = \frac{dt}{2}$

when $x = 0$, then $t = 0$

$x = 1$, then $t = 1$

$$= \frac{1}{2}\int_0^1 e^t\, dt = \frac{1}{2}[e^t]_0^1$$

$$= \frac{1}{2}(e^1 - e^0) = \frac{e-1}{2}$$

(iii) We have, $\int_0^{\pi/2} \sqrt{\cos\theta}\cdot\sin^5\theta\, d\theta$

$$= \int_0^{\pi} \sqrt{\cos\theta}\,(1-\cos^2\theta)^2 \sin\theta\, d\theta$$

Let $\cos\theta = t \Rightarrow -\sin\theta\, d\theta = dt$

when $\theta = 0, t = 1$

$\theta = \frac{\pi}{2}, t = 0$

$$= -\int_1^0 \sqrt{t}(1-t^2)^2\, dt$$

$$= -\int_1^0 \sqrt{t}(1+t^4-2t^2)\, dt$$

$$= -\int_1^0 (t^{1/2} + t^{9/2} - 2t^{5/2})\, dt$$

$$= -\left[\frac{2t^{3/2}}{3} + \frac{2}{11}t^{11/2} - 2\frac{2}{7}t^{7/2}\right]_1^0$$

$$= -\left[0 - \left(\frac{2}{3} + \frac{2}{11} - \frac{4}{7}\right)\right]$$

$$= \frac{64}{221}$$

(iv) Let $I = \int_0^\infty \frac{1}{a^2 + b^2 x^2} dx$

$$= \frac{1}{b^2} \int_0^\infty \frac{1}{\left(\frac{a}{b}\right)^2 + x^2} dx$$

$$= \frac{1}{b^2} \left[\frac{1}{a/b} \tan^{-1} \left(\frac{x}{a/b} \right) \right]_0^\infty$$

$$= \frac{1}{b^2} \frac{b}{a} \left[\tan^{-1} \frac{xb}{a} \right]_0^\infty$$

$$= \frac{1}{ab} \left[\frac{\pi}{2} - 0 \right] = \frac{\pi}{2ab}$$

(v) $I = \int_0^\pi \frac{1}{5 + 4\cos x} dx$

$$= \int_0^\pi \frac{1}{5 + 4\left(\frac{1 - \tan^2 \frac{x}{2}}{1 + \tan^2 \frac{x}{2}} \right)} dx$$

$$= \int_0^\pi \frac{\sec^2 \frac{x}{2}}{5\left(1 + \tan^2 \frac{x}{2}\right) + 4\left(1 - \tan^2 \frac{x}{2}\right)} dx$$

$$= \int_0^\pi \frac{\sec^2 \frac{x}{2}}{9 + \tan^2 \frac{x}{2}} dx$$

Put $\tan \frac{x}{2} = t$

$\Rightarrow \quad \sec^2 \frac{x}{2} \cdot \frac{1}{2} dx = dt$

$\Rightarrow \quad \sec^2 \frac{x}{2} dx = 2\, dt$

If $x = 0$, then $t = 0$

If $x = \pi$, then $t = \infty$

$$= 2 \int_0^\infty \frac{dt}{(3)^2 + t^2}$$

$$= \frac{2}{3} \left[\tan^{-1} \frac{t}{3} \right]_0^\infty$$

$$= \frac{2}{3} \cdot \frac{\pi}{2}$$

$\Rightarrow \quad I = \frac{\pi}{3}$

(vi) $I = \int_0^{\pi/2} \frac{1}{4\sin^2 x + 5\cos^2 x} dx$

$$= \int_0^{\pi/2} \frac{\sec^2 x}{4\tan^2 x + 5} dx$$

$$= \frac{1}{4} \int_0^{\pi/2} \frac{\sec^2 x}{\tan^2 x + \frac{5}{4}} dx$$

Put $\tan x = t$

$\Rightarrow \sec^2 x\, dx = dt$

when $x = 0$, then $t = 0$

$x = \frac{\pi}{2}$, then $t = \infty$

$$= \frac{1}{4} \int_0^\infty \frac{dt}{t^2 + \left(\frac{\sqrt{5}}{2} \right)^2}$$

$$= \frac{1}{4} \cdot \frac{1}{\left(\frac{\sqrt{5}}{2} \right)} \left[\tan^{-1} \left(\frac{2t}{\sqrt{5}} \right) \right]_0^\infty$$

$$= \frac{1}{2\sqrt{5}} \left[\frac{\pi}{2} - 0 \right] = \frac{\pi}{4\sqrt{5}}$$

(vii) $I = \int_0^{1/\sqrt{2}} \frac{\sin^{-1} x}{(1 - x^2)^{3/2}} dx$

Put $\sin^{-1} x = t \Rightarrow x = \sin\theta$

$$\frac{1}{\sqrt{1 - x^2}} dx = dt$$

when $x = 0$, then $t = 0$

$x = \frac{1}{\sqrt{2}}$, then $t = \frac{\pi}{4}$

$$= \int_0^{\pi/4} \frac{t}{\cos^2 t} dt$$

$$= \int_0^{\pi/4} \underset{\text{I}}{t} \cdot \underset{\text{II}}{\sec^2 t}\, dt$$

$$= [t \cdot \tan t]_0^{\pi/4} - \int_0^{\pi/4} 1 \cdot \tan t\, dt$$

$$= \frac{\pi}{4} - [\log \sec t]_0^{\pi/4} \quad \left[\because \int \tan t\, dt = \log|\sec t| \right]$$

$$= \frac{\pi}{4} - [\log \sqrt{2} - \log 1]$$

$$= \frac{\pi}{4} - \log \sqrt{2}$$

$$= \frac{\pi}{4} - \frac{1}{2} \log 2$$

3. (i) Let $S_n = \frac{1}{n+1} + \frac{1}{n+2} + \ldots \frac{1}{2n}$

$$S_n = \sum_{r=1}^{n} \frac{1}{n+r} = \sum_{r=1}^{n} \frac{1}{n\left(1 + \frac{r}{n}\right)}$$

Hence, $S = \lim_{n \to \infty} S_n = \lim_{n \to \infty} \frac{1}{n} \sum_{r=1}^{n} \frac{1}{1 + \frac{r}{n}}$

$$= \int_0^1 \frac{dx}{1+x}$$

$$= [\log(1+x)]_0^1 = \log 2$$

(ii) Let $S_n = \frac{n}{n^2+1^2} + \frac{n}{n^2+2^2} \ldots + \frac{n}{2n^2}$

$$= \sum_{r=1}^{n} \frac{n}{n^2+r^2} = \sum_{r=1}^{n} \frac{1}{n\left(1+\frac{r^2}{n^2}\right)}$$

Hence, $S = \lim_{n\to\infty} S_n = \lim_{n\to\infty} \frac{1}{n}\sum_{r=1}^{n} \frac{1}{1+r^2/n^2}$

$$\int_0^1 \frac{dx}{1+x^2} = [\tan^{-1} x]_0^1 = \tan^{-1} 1 - \tan^{-1} 0 = \frac{\pi}{4}$$

(iii) $\lim_{n\to\infty} \left(\frac{1}{2n+1} + \frac{1}{2n+2} + \ldots + \frac{1}{6n}\right)$

$$= \lim_{n\to\infty} \frac{1}{n}\sum_{r=1}^{4n} \frac{1}{2+r/n}$$

$$= \int_0^4 \frac{dx}{2+x} = [\log(2+x)]_0^4 = \log 6 - \log 2 = \log 3$$

(iv) $\lim_{n\to\infty} \sum_{r=0}^{n-1} \frac{1}{\sqrt{4n^2-r^2}} = \lim_{n\to\infty} \frac{1}{n}\sum_{r=0}^{n-1} \frac{1}{\sqrt{4-\left(\frac{r}{n}\right)^2}}$

$$= \int_0^1 \frac{dx}{\sqrt{4-x^2}} = \left[\sin^{-1}\frac{x}{2}\right]_0^1 = \sin^{-1}\frac{1}{2} = \frac{\pi}{6}$$

(v) Let $S_n = \frac{1^p + 2^p + \ldots + n^p}{n^{p+1}}, p > 0$

$$= \sum_{r=1}^{n} \frac{r^p}{n^{p+1}}$$

$$= \sum_{r=1}^{n} \frac{1}{n}\left(\frac{r}{n}\right)^p$$

$$\therefore \quad S = \lim_{n\to\infty} S_n = \lim_{n\to\infty} \frac{1}{n}\sum_{r=1}^{n}\left(\frac{r}{n}\right)^p$$

$$= \int_0^1 x^p\,dx = \left[\frac{x^{p+1}}{p+1}\right] = \frac{1}{p+1}$$

4. (i) We know that,

$$|x| = \begin{cases} x, & x \ge 0 \\ -x, & x < 0 \end{cases}$$

$$\int_{-1}^{1} e^{|x|}\,dx = \int_{-1}^{0} e^{-x}\,dx + \int_0^1 e^x\,dx$$

$$= -[e^{-x}]_{-1}^{0} + [e^x]_0^1$$

$$= (-1 + e^1) + (e^1 - 1)$$

$$= 2e - 2$$

(ii) Clearly, $f(x) = x^3 - x = x(x-1)(x+1)$

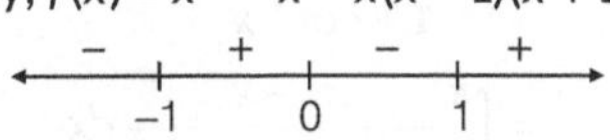

$$\text{So, } f(x) = \begin{cases} (x^3 - x), & -1 < x < 0 \\ -(x^3 - x), & 0 < x < 1 \\ (x^3 - x), & 1 < x < \infty \end{cases}$$

So, $\int_{-1}^{2} |x^3 - x|\,dx$

$$= \int_{-1}^{0} (x^3 - x)\,dx + \int_0^1 [-(x^3 - x)]\,dx + \int_1^2 (x^3 - x)\,dx$$

$$\Rightarrow \quad I = \left[\frac{x^4}{4} - \frac{x^2}{2}\right]_{-1}^{0} - \left[\frac{x^4}{4} - \frac{x^2}{2}\right]_0^1 + \left[\frac{x^4}{4} - \frac{x^2}{2}\right]_1^2$$

$$= -\left(\frac{1}{4} - \frac{1}{2}\right) - \left(\frac{1}{4} - \frac{1}{2}\right) + \left[(4-2) - \left(\frac{1}{4} - \frac{1}{2}\right)\right]$$

$$= \frac{3}{4} + (4 - 2)$$

$$\therefore \quad I = \frac{11}{4}$$

(iii) Let $I = \int_0^4 [x]\,dx$

$$= \int_0^1 [x]\,dx + \int_1^2 [x]\,dx + \int_2^3 [x]\,dx + \int_3^4 [x]\,dx$$

$$= 0 + \int_1^2 1\,dx + \int_2^3 2\,dx + \int_3^4 3\,dx$$

$$= [x]_1^2 + 2[x]_2^3 + 3[x]_3^4$$

$$= (2-1) + 2(3-2) + 3(4-3) = 1 + 2 + 3 = 6$$

(iv) We have, $\int_1^2 [3x]\,dx$

$$= \int_1^{4/3} [3x]\,dx + \int_{4/3}^{5/3} [3x]\,dx + \int_{5/3}^{2} [3x]\,dx$$

$$= 3\int_1^{4/3} dx + 4\int_{4/3}^{5/3} dx + 5\int_{5/3}^{2} dx.$$

$$= 3\left[\frac{4}{3} - 1\right] + 4\left[\frac{5}{3} - \frac{4}{3}\right] + 5\left[2 - \frac{5}{3}\right]$$

$$= 3\cdot\frac{1}{3} + 4\cdot\frac{1}{3} + 5\cdot\frac{1}{3} = 4$$

(v) Let $I = \int_{-2}^{1} \left[\left(1 + \cos\left(\frac{\pi x}{2}\right)\right)\right] dx$

$$= \int_{-2}^{-1} 1\,dx + \int_{-1}^{0} 0\,dx + \int_0^1 1\,dx = 1 + 0 + 1 = 2$$

(vi) We have, $\int_0^{100} [\tan^{-1} x]\,dx$

$$= \int_0^{\tan 1} [\tan^{-1} x]\,dx + \int_{\tan 1}^{100} [\tan^{-1} x]\,dx$$

$$= \int_0^{\tan 1} 0\,dx + \int_{\tan 1}^{100} 1\,dx$$

$$= 0 + 100 - (\tan 1) = 100 - \tan 1$$

5. (i) Let $I = \int_0^{\pi/2} \frac{\sin x}{\sin x + \cos x}\,dx$ ...(i)

[using $\int_0^a f(x)\,dx = \int_0^a f(a-x)\,dx$]

$$\Rightarrow \quad I = \int_0^{\pi/2} \frac{\sin\left(\frac{\pi}{2} - x\right)}{\sin\left(\frac{\pi}{2} - x\right) + \cos\left(\frac{\pi}{2} - x\right)}\,dx$$

$$\Rightarrow \quad I=\int_0^{\pi/2} \frac{\cos x}{\sin x+\cos x}\,dx \quad \text{...(ii)}$$

On adding Eqs. (i) and (ii), we get

$$2I=\int_0^{\pi/2} \frac{\sin x+\cos x}{\sin x+\cos x}\,dx$$

$$=\int_0^{\pi/2} dx = [x]_0^{\pi/2}$$

$$\Rightarrow \quad 2I=\frac{\pi}{2}$$

$$\therefore \quad I=\frac{\pi}{4}$$

(ii) Let $I=\int_0^{\pi/2} \log \tan x\,dx$ ...(i)

[using $\int_0^a f(x)\,dx=\int_0^a f(a-x)\,dx$]

$$\Rightarrow \quad I=\int_0^{\pi/2} \log \tan\left(\frac{\pi}{2}-x\right)dx$$

$$\Rightarrow \quad I=\int_0^{\pi/2} \log \cot x\,dx \quad \text{...(ii)}$$

On adding Eqs. (i) and (ii), we get

$$2I=\int_0^{\pi/2} (\log \tan x+\log \cot x)\,dx$$

$$=\int_0^{\pi/2} \log(\tan x \cdot \cot x)\,dx$$

$$\Rightarrow \quad 2I=\int_0^{\pi/2} 0\,dx$$

$$\therefore \quad I=0$$

(iii) Let $I=\int_0^1 x(1-x)^n\,dx$

[using $\int_0^a f(x)\,dx=\int_0^a f(a-x)\,dx$]

$$=\int_0^1 (1-x)x^n\,dx$$

$$=\int_0^1 (x^n-x^{n+1})\,dx$$

$$=\left[\frac{x^{n+1}}{n+1}-\frac{x^{n+2}}{n+2}\right]_0^1$$

$$=\frac{1}{n+1}-\frac{1}{n+2}$$

$$=\frac{n+2-n-1}{(n+1)(n+2)}$$

$$=\frac{1}{(n+1)(n+2)}$$

(iv) Let $I=\int_0^{\pi} \frac{x}{1+\sin x}\,dx$ ...(i)

[using $\int_0^a f(x)\,dx=\int_0^a f(a-x)\,dx$]

$$=\int_0^{\pi} \frac{\pi-x}{1+\sin(\pi-x)}\,dx$$

$$\Rightarrow \quad I=\int_0^{\pi} \frac{\pi-x}{1+\sin x}\,dx \quad \text{...(ii)}$$

On adding Eqs. (i) and (ii), we get

$$2I=\pi\int_0^{\pi} \frac{1}{1+\sin x}\,dx$$

$$=\pi\int_0^{\pi} \frac{1-\sin x}{\cos^2 x}\,dx$$

$$=\pi\int_0^{\pi} (\sec^2 x-\sec x\tan x)\,dx$$

$$=\pi[\tan x-\sec x]_0^{\pi}$$

$$=\pi[(0-(-1)-(0-1)]=2\pi$$

(v) Let $I=\int_0^{\pi/2} \frac{\sqrt{\tan x}}{(\tan x-1)}\,dx$ ...(i)

[using $\int_0^a f(x)\,dx=\int_0^a f(a-x)\,dx$]

$$=\int_0^{\pi/2} \frac{\sqrt{\cot x}}{(\cot x-1)}\,dx$$

$$\Rightarrow \quad I=-\int_0^{\pi/2} \frac{\sqrt{\tan x}}{\tan x-1}\,dx \quad \text{...(ii)}$$

On adding Eqs. (i) and (ii), we get

$$2I=0 \Rightarrow I=0$$

(vi) Let $I=\int_a^b \frac{f(x)}{f(a+b-x)+f(x)}\,dx$ ...(i)

[using $\int_a^b f(x)\,dx=\int_a^b f(a+b-x)\,dx$]

$$I=\int_a^b \frac{f(a+b-x)}{f(x)+f(a+b-x)}\,dx \quad \text{...(ii)}$$

On adding Eqs. (i) and (ii), we get

$$\Rightarrow \quad 2I=\int_a^b 1\,dx$$

$$2I=[x]_a^b$$

$$\therefore \quad I=\frac{b-a}{2}$$

6. (i) Let $I=\int_0^{2\pi} \cos^5 x\,dx$

Let $f(x)=\cos^5 x$

$$f(2\pi-x)=[\cos(2\pi-x)]^5=\cos^5 x$$

$$=\int_0^{2\pi} \cos^5 x\,dx=2\int_0^{\pi} \cos^5 x\,dx$$

[$\because \int_0^{2a} f(x)\,dx=2\int_0^a f(a-x)\,dx$, if $f(2a-x)=f(x)$]

Now, $f(\pi-x)=\cos(\pi-x)=-\cos x=-f(x)$

$$\therefore \quad 2\int_0^{\pi} \cos^5 x\,dx=0 \Rightarrow I=0$$

(ii) Let $I=\int_0^{\pi} \frac{x}{1+\cos^2 x}\,dx$ ...(i)

$$=\int_0^{\pi} \frac{\pi-x}{1+\cos^2(\pi-x)}\,dx$$

$$\Rightarrow \quad I=\int_0^{\pi} \frac{\pi-x}{1+\cos^2 x}\,dx \quad \text{...(ii)}$$

On adding Eqs. (i) and (ii), we get

$$2I = \pi \int_0^{\pi} \frac{1}{1+\cos^2 x} dx$$

$$\Rightarrow \quad I = \frac{\pi}{2} \int_0^{\pi} \frac{1}{1+\cos^2 x} dx$$

$$= \frac{\pi}{2} \cdot 2 \int_0^{\pi/2} \frac{1}{1+\cos^2 x} dx$$

$$\left[\because \int_0^{2a} f(x)\, dx = 2\int_0^{a} f(x)\, dx, \text{if } f(2a-x) = f(x)\right]$$

$$= \pi \int_0^{\pi/2} \frac{\sec^2 x}{\tan^2 x + 2} dx$$

$$= \frac{\pi}{\sqrt{2}} \left[\tan^{-1}\left(\frac{\tan x}{\sqrt{2}}\right)\right]_0^{\pi/2}$$

$$= \frac{\pi}{\sqrt{2}} \times \frac{\pi}{2} = \frac{\pi^2}{2\sqrt{2}}$$

(iii) Let $I = \int_0^{\pi} \frac{dx}{3\sin^2 x + 4\cos^2 x}$

$$= 2\int_0^{\pi/2} \frac{dx}{3\sin^2 x + 4\cos^2 x}$$

$$= 2\int_0^{\pi/2} \left(\frac{\sec^2 x\, dx}{3 + 4\tan^2 x}\right)$$

Put $t = \tan x$

$\Rightarrow dt = \sec^2 x\, dx$

$$= 2\int_0^{\infty} \frac{dt}{3 + 4t^2}$$

$$= \frac{1}{2}\int_0^{\infty} \frac{dt}{t^2 + \left(\frac{\sqrt{3}}{2}\right)^2}$$

x	π / 2	0
t	∞	0

$$= \frac{1}{2} \cdot \left[\frac{2}{\sqrt{3}} \tan^{-1}\left(\frac{2t}{\sqrt{3}}\right)\right]_0^{\infty}$$

$$= \frac{1}{2} \cdot \frac{2}{\sqrt{3}}\left(\frac{\pi}{2} - 0\right) = \frac{\pi}{2\sqrt{3}}$$

(iv) Let $I = \int_0^{\pi} x \log(\sin x)\, dx$ ...(i)

$$\Rightarrow \quad I = \int_0^{\pi} (\pi - x) \log[\sin(\pi - x)]\, dx$$

$$= \int_0^{\pi} (\pi - x) \log \sin x\, dx \quad \text{...(ii)}$$

On adding Eqs. (i) and (ii), we get

$$2I = \pi \int_0^{\pi} \log \sin x\, dx$$

$$\Rightarrow \quad I = \frac{\pi}{2} \int_0^{\pi} \log \sin x\, dx$$

$$= \frac{\pi}{2} \cdot 2\int_0^{\pi/2} \log \sin x\, dx$$

$$= \pi\left[\frac{\pi}{2} \log \frac{1}{2}\right]$$

$$\therefore \quad I = \frac{\pi^2}{2} \log \frac{1}{2}$$

7. (i) Let $I = \int_{-\pi/2}^{\pi/2} \sin^3 x\, dx$

Let $f(x) = \sin^3 x$

$$f(-x) = \sin^3(-x) = -\sin^3 x$$

$$f(-x) = -f(x)$$

$\therefore$ $f(x)$ is an odd function.

$$\left[\because \int_{-a}^{a} f(x)\, dx = 0 \text{ if } f(x) \text{ is odd function}\right]$$

$$\Rightarrow \quad \int_{-\pi/2}^{\pi/2} \sin^3 x\, dx = 0$$

(ii) Let $I = \int_{-\pi/2}^{\pi/2} \cos^2 x\, dx$

Let $f(x) = \cos^2 x$

$$f(-x) = \cos^2(-x) = \cos^2 x$$

$\therefore$ $f(x)$ is an even function

$$\Rightarrow \quad I = \int_{-\pi/2}^{\pi/2} \cos^2 x\, dx$$

$$= 2\int_0^{\pi/2} \cos^2 x\, dx$$

$$= 2\int_0^{\pi/2} \left[\frac{1+\cos 2x}{2}\right] dx$$

$$= \left[x + \frac{\sin 2x}{2}\right]_0^{\pi/2} = \left(\frac{\pi}{2} + 0\right) - 0 = \frac{\pi}{2}$$

(iii) Let $f(x) = e^{|x|}$

$$f(-x) = e^{|-x|} = e^{|x|} = f(x)$$

$\Rightarrow$ $f(x)$ is an even function.

$$I = \int_{-1}^{1} e^{|x|}\, dx = 2\int_0^{1} e^{|x|}\, dx$$

$$= 2\int_0^{1} e^{x}\, dx \quad [\because |x| = x, 0 < x < 1]$$

$$= 2[e^x]_0^1 = 2(e-1)$$

(iv) Let $f(x) = x^{10} \sin^7 x$

$$\Rightarrow f(-x) = (-x)^{10} \sin^7(-x) = -x^{10} \sin^7 x$$

$$\Rightarrow f(-x) = -f(x)$$

Thus, $f(x)$ is an odd function.

Hence, $\int_{-\pi}^{\pi} x^{10} \sin^7 x\, dx = 0$

$$\left[\because \int_{-a}^{a} f(x)\, dx = 0, \text{if } f(x) \text{ is an odd function}\right]$$

(v) Let $f(x)=\log\left(\frac{2-x}{2+x}\right)$

$$\Rightarrow f(-x)=\log\left(\frac{2+x}{2-x}\right)=\log\left(\frac{2-x}{2+x}\right)^{-1}=-\log\left(\frac{2-x}{2+x}\right)$$

$\because\ f(-x)=-f(x)$

So, $f(x)$ is an odd function.

Hence, $\int_{-1}^{1}\log\left(\frac{2-x}{2+x}\right)dx=0$

$[\because \int_{-a}^{a} f(x)\,dx=0$ if $f(x)$ is odd function$]$

8. (i) Let $y=\int_0^{x^2}\cos t\,dt$

$$\frac{dy}{dx}=\cos(x^2)\,(2x)=2x\cos x^2$$

(ii) Let $y=\int_{1/x}^{\sqrt{x}}\cos t^2\,dt$

$$\frac{dy}{dx}=\frac{\cos x}{2\sqrt{x}}+\cos\left(\frac{1}{x^2}\right).\frac{1}{x^2}$$

$$=\frac{1}{2\sqrt{x}}\cos x+\frac{1}{x^2}\cos\left(\frac{1}{x^2}\right)$$

(iii) Let $y=\int_{x^2}^{x^3}\frac{I}{\log t}\,dt$

$$\frac{dy}{dx}=\frac{1}{\log x^3}\cdot 3x^2-\frac{1}{\log x^2}\cdot 2x$$

$$=\frac{3x^2}{3\log x}-\frac{2x}{2\log x}=\frac{1}{\log x}(x^2-x)$$

(iv) Let $y=\int_0^{x^2}\frac{t^2-5t+4}{2+e^t}\,dt$

$$\frac{dy}{dx}=\frac{x^4-5x^2+4}{2+e^{x^2}}\cdot 2x$$

$$=\frac{2x\,(x-1)\,(x+1)\,(x-2)\,(x+2)}{2+e^{x^2}}$$

9. We have, $\int_{\pi/3}^{x}\sqrt{3-\sin^2 t}+\int_0^{y}\cos t\,dt=0$

On differentiating w.r.t. x, we get

$$\sqrt{3-\sin^2 x}+\cos y\frac{dy}{dx}=0\Rightarrow\frac{dy}{dx}=-\frac{\sqrt{3-\sin^2 x}}{\cos y}$$

10. We have, $f(t)=\int_1^{t^2}\frac{\sqrt{1+u^4}}{4}\,du$

$$\Rightarrow\quad f'(t)=\frac{\sqrt{1+t^8}}{4}\cdot\frac{d}{dt}(t^2)-\frac{\sqrt{1+1^4}}{4}\cdot\frac{d}{dt}(1)$$

$$\Rightarrow\quad f'(t)=\frac{\sqrt{1+t^8}}{4}\cdot(2t)-0=\frac{t}{2}\sqrt{1+t^8}$$

Also, $F(x)=\int_1^{x}f(t)\,dt\Rightarrow F'(x)=f(x)-0=f(x)$

$$\Rightarrow F''(x)=f'(x)=\frac{x}{2}\sqrt{1+x^8}$$

$$\Rightarrow\quad F''(2)=f'(2)=\frac{2}{2}\sqrt{1+2^8}=\sqrt{1+256}=\sqrt{257}$$

Level 2

1. (b) We have, $I=\int_0^1\frac{2^{2x+1}-5^{2x-1}}{10^x}\,dx$

$$=2\int_0^1\left(\frac{4}{10}\right)^x dx-\frac{1}{5}\int_0^1\left(\frac{25}{10}\right)^x dx$$

$$=2\int_0^1\left(\frac{2}{5}\right)^x dx-\frac{1}{5}\int_0^1\left(\frac{5}{2}\right)^x dx$$

$$=2\left[\frac{\left(\frac{2}{5}\right)^x}{\log\left(\frac{2}{5}\right)}\right]_0^1-\frac{1}{5}\cdot\left[\frac{\left(\frac{5}{2}\right)^x}{\log\left(\frac{5}{2}\right)}\right]_0^1$$

$$=\frac{2}{\log\left(\frac{2}{5}\right)}\left[\frac{2}{5}-1\right]-\frac{1}{5}\left[\frac{\frac{5}{2}-1}{\log\frac{5}{2}}\right]$$

$$=-\frac{3}{5}\left[\frac{2}{\log\left(\frac{2}{5}\right)}+\frac{1}{2\log\left(\frac{5}{2}\right)}\right]$$

2. (a) We have, $2\int_0^1\tan^{-1}x\,dx=\int_0^1\cot^{-1}(1-x+x^2)\,dx$

$$\Rightarrow\quad 2\int_0^1\tan^{-1}x\,dx=\int_0^1\left(\frac{\pi}{2}-\tan^{-1}(1-x+x^2)\right)dx$$

$$\Rightarrow\quad \int_0^1\tan^{-1}(1-x+x^2)\,dx=\frac{\pi}{2}-2\int_0^1\tan^{-1}x\,dx$$

$$\Rightarrow\int_0^1\tan^{-1}(1-x+x^2)\,dx$$

$$=\frac{\pi}{2}-2\left[x\tan^{-1}x-\frac{1}{2}\log(1+x^2)\right]_0^1$$

$$\Rightarrow\int_0^1\tan^{-1}(1-x+x^2)\,dx=\frac{\pi}{2}-2\left[\frac{\pi}{4}-\frac{1}{2}\log 2\right]=\log 2$$

3. (a) We have, $f(y)=e^y, g(y)=y, y>0$

$$F(t)=\int_0^t f(t-y)\,g(y)\,dy$$

$$=\int_0^t e^{t-y}\,y\,dy$$

$$=e^t\int_0^t e^{-y}\,y\,dy$$

$$=e^t\left([-ye^{-y}]_0^t+\int_0^t e^{-y}\,dy\right)$$

$$=e^t\left(-te^{-t}-[e^{-y}]_0^t\right)$$

$$=e^t\left(-te^{-t}-e^{-t}+1\right)$$

$$=e^t-(1+t)$$

4. (b) We have, $I(m,n)=\int_0^1 t^m(1-t)^n\,dt$

$$\Rightarrow\quad I(m+1,n-1)=\int_0^1 \underset{\text{I}}{t^{m+1}}\,\underset{\text{II}}{(1-t)^{n-1}}\,dt$$

$$\Rightarrow \quad I(m+1, n-1) = \left[-\frac{t^{m+1}(1-t)^n}{n}\right]_0^1 + \frac{m+1}{n} \int_0^1 t^m (1-t)^n \, dt$$

$$\Rightarrow \quad I(m+1, n-1) = 0 + \frac{m+1}{n} I(m, n)$$

$$\Rightarrow \quad I(m, n) = \frac{n}{m+1} I(m+1, n-1)$$

5. (d) Let $I_1 = \int_0^{\pi/2} (\sin x)^{\sqrt{3}+1} dx$

and $I_2 = \int_0^{\pi/2} (\sin x)^{\sqrt{3}-1} dx$

We know that,

$$\int (\sin x)^n dx = -\frac{1}{n}(\sin x)^{n-1} \cos x + \frac{n-1}{n} \int (\sin x)^{n-2} dx$$

$$\therefore \int_0^{\pi/2} (\sin x)^{\sqrt{3}+1} dx = \left[-\frac{1}{\sqrt{3}+1}(\sin x)^{\sqrt{3}} \cos x\right]_0^{\pi/2} + \frac{\sqrt{3}}{\sqrt{3}+1} \int_0^{\pi/2} (\sin x)^{\sqrt{3}-1} dx$$

$$\Rightarrow \quad I_1 = \left[-\frac{1}{\sqrt{3}+1} \cos x (\sin x)^{\sqrt{3}}\right]_0^{\pi/2} + \frac{\sqrt{3}}{\sqrt{3}+1} I_2$$

$$\therefore \quad I_1 = 0 + \frac{\sqrt{3}}{\sqrt{3}+1} I_2 \Rightarrow \frac{I_1}{I_2} = \frac{\sqrt{3}}{\sqrt{3}+1}$$

$$\therefore \quad \frac{\int_0^{\pi/2} (\sin x)^{\sqrt{3}+1} dx}{\int_0^{\pi/2} (\sin x)^{\sqrt{3}-1} dx} = \frac{\sqrt{3}}{\sqrt{3}+1}$$

6. (a) Let $I = \int_{\alpha}^{\alpha+1} \frac{dx}{(x+\alpha)(x+\alpha+1)}$

$$= \int_{\alpha}^{\alpha+1} \frac{(x+\alpha+1)-(x+\alpha)}{(x+\alpha)(x+\alpha+1)} dx$$

$$= \int_{\alpha}^{\alpha+1} \left(\frac{1}{x+\alpha} - \frac{1}{x+\alpha+1}\right) dx$$

$$= [\log_e (x+\alpha) - \log_e (x+\alpha+1)]_{\alpha}^{\alpha+1}$$

$$= \left[\log_e \left(\frac{x+\alpha}{x+\alpha+1}\right)\right]_{\alpha}^{\alpha+1}$$

$$= \log_e \frac{2\alpha+1}{2\alpha+2} - \log_e \frac{2\alpha}{2\alpha+1}$$

$$= \log_e \left(\frac{2\alpha+1}{2\alpha+2} \times \frac{2\alpha+1}{2\alpha}\right)$$

$$= \log_e \left(\frac{9}{8}\right) \quad \text{[given]}$$

$$\Rightarrow \frac{(2\alpha+1)^2}{4\alpha(\alpha+1)} = \frac{9}{8} \Rightarrow 8[4\alpha^2 + 4\alpha + 1] = 36(\alpha^2 + \alpha)$$

$$\Rightarrow \quad 8\alpha^2 + 8\alpha + 2 = 9\alpha^2 + 9\alpha$$

$$\Rightarrow \quad \alpha^2 + \alpha - 2 = 0$$

$$\Rightarrow \quad (\alpha+2)(\alpha-1) = 0 \Rightarrow \alpha = 1, -2$$

From the options, we get

$$\alpha = -2$$

7. (a) Let $I = \int_1^{e^{37}} \frac{\pi \sin(\pi \log x)}{x} dx$

Let $\pi \log x = t$

x	e^{37}	1
t	37π	0

$$\frac{\pi}{x} dx = dt$$

$$\therefore \quad I = \int_0^{37\pi} \sin t \, dt = -[\cos t]_0^{37\pi}$$

$$= -[\cos 37\pi - \cos 0]$$

$$= -[-1-1] = 2$$

8. (d) Let $I = \int_0^{\log_e 5} \frac{e^x \sqrt{e^x - 1}}{e^x + 3} dx$

Put $e^x - 1 = t^2$ [$\because e^x > 1$ for $x > 0$]

therefore $e^x dx = 2t \, dt$

Now, $I = \int_0^2 \frac{t}{t^2+4}(2t) \, dt$

$$= 2\int_0^2 \frac{t^2}{t^2+4} dt$$

$$= 2\int_0^2 \left(1 - \frac{4}{t^2+4}\right) dt$$

$$= 4 - \frac{8}{2}\left[\left(\tan^{-1}\frac{t}{2}\right)\right]_0^2$$

$$= 4 - 4\tan^{-1}(\tan^{-1} 1 - \tan^{-1} 0)$$

$$= 4 - 4\left(\frac{\pi}{4}\right) = 4 - \pi$$

9. (c) $f(x) = \int_1^x \frac{\ln t}{1+t} dt$ ...(i)

$$\Rightarrow \quad f\left(\frac{1}{x}\right) = \int_1^{1/x} \frac{\ln t}{1+t} dt$$

Put $t = \frac{1}{z}$

$\therefore dt = -\frac{1}{z^2} dz$

$$\therefore f\left(\frac{1}{x}\right) = \int_1^x -\frac{\ln z}{(z+1)/z}\left(-\frac{1}{z^2} dz\right)$$

$$= \int_1^x \frac{\ln z}{z(z+1)} dz$$

$$= \int_1^x \frac{\ln t}{t(t+1)} dt \quad \text{...(ii)}$$

So, $f(x)+f\left(\frac{1}{x}\right)=\int_1^x \frac{\ln t}{t+1}\left(1+\frac{1}{t}\right)dt$ [from Eqs. (i) and (ii)]

$$=\int_1^x \frac{\ln t}{t}\,dt$$

$$=\left[\frac{1}{2}(\ln t)^2\right]_1^x$$

$$=\frac{1}{2}(\ln x)^2$$

$$\therefore\ f(e)+f\left(\frac{1}{e}\right)=\frac{1}{2}(\ln e)^2=\frac{1}{2}$$

10. (d) $f(x)=x^5+5x-1$

$$f(1)=5 \Rightarrow f^{-1}(5)=1$$

$$f(2)=41 \Rightarrow f^{-1}(41)=2$$

Put $f^{-1}(x)=t$

$$\therefore\quad I=\int_5^{41}\frac{dx}{[f^{-1}(x)]^5+5f^{-1}(x)}$$

$$=\int_1^2\frac{f'(t)\,dt}{t^5+5t}=\int_1^2\frac{5t^4+5}{t^5+5t}\,dt$$

$$=[\log(t^5+5t)]_1^2$$

$$=\log 42-\log 6=\log 7$$

11. (b) The integral, $I=\int_0^{1/2}\frac{x^2}{(1-x^2)^{3/2}}\,dx$

Let $x=\sin\theta$, so $dx=\cos\theta d\theta$

x	$1/2$	0
θ	$\pi/6$	0

$$\therefore\quad I=\int_0^{\pi/6}\frac{\sin^2\theta}{(\cos^2\theta)^{3/2}}\cos\theta\,d\theta=\int_0^{\pi/6}\tan^2\theta\,d\theta$$

$$=\int_0^{\pi/6}(\sec^2\theta-1)d\theta=[\tan\theta-\theta]_0^{\pi/6}$$

$$=\frac{1}{\sqrt{3}}-\frac{\pi}{6}=\frac{k}{6}\qquad\text{[given]}$$

$$\therefore\ k=2\sqrt{3}-\pi$$

12. (b) Let $I=\int_0^x e^{zx}\,e^{-z^2}\,dz$

$$=\int_0^x e^{zx-z^2}\,dz=\int_0^x e^{\frac{x^2}{4}-\left(z-\frac{x}{2}\right)^2}\,dz$$

Put $z-\frac{x}{2}=t$

$\Rightarrow\quad dz=dt$

$$\Rightarrow\quad I=e^{x^2/4}\int_{-x/2}^{x/2}e^{-t^2}\,dt$$

$$=e^{\frac{x^2}{4}}(2)\int_0^{x/2}e^{-t^2}\,dt$$

$$=e^{\frac{x^2}{4}}\cdot 2\int_0^x e^{-\frac{u^2}{4}}\left(\frac{1}{2}\right)du,\qquad\left[\text{where } t=\frac{u}{2}\right]$$

$$=e^{\frac{x^2}{4}}\int_0^x e^{-\frac{u^2}{4}}\,du$$

$$=e^{\frac{x^2}{4}}\int_0^x e^{-\frac{z^2}{4}}\,dz$$

Therefore, $f(x)=e^{\frac{x^2}{4}}$

So, $\int e^x\left[\log(f(x))+\frac{x}{2}\right]dx$

$$=\int e^x\left(\frac{x^2}{4}+\frac{x}{2}\right)dx$$

[using Integrating by part rule]

$$=\frac{x^2e^x}{4}+C$$

13. (a) Given, $I=\int_0^\pi\sqrt{|\cos x|-|\cos^3 x|}\,dx$

$$\Rightarrow\quad I=\int_0^\pi\sqrt{|\cos x||1-\cos^2 x|}\,dx$$

$$=\int_0^\pi\sqrt{|\cos x||\sin^2 x|}\,dx$$

$$=\int_0^\pi\sqrt{|\cos x|}\,\sin x\,dx$$

Put $\cos x=t$

x	π	0
θ	-1	1

$$\Rightarrow\quad -\sin x\,dx=dt$$

$$\Rightarrow\quad I=-\int_1^{-1}\sqrt{|t|}\,dt$$

$$=\int_{-1}^0\sqrt{-t}\,dt+\int_0^1\sqrt{t}\,dt$$

$$\Rightarrow\quad I=-\frac{2}{3}[(-t)^{3/2}]_{-1}^0+\frac{2}{3}[t^{3/2}]_0^1$$

$$=-\frac{2}{3}(0-1)+\frac{2}{3}(1-0)$$

$$I=\frac{4}{3}$$

14. (c) Let $I=\int_{-\pi/4}^{3\pi/4}\frac{e^{\pi/4}\,dx}{(e^x+e^{\pi/4})(\sin x+\cos x)}$

$$=\int_{-\pi/4}^{3\pi/4}\frac{dx}{\sqrt{2}\,(e^{x-\pi/4}+1)\cos\left(x-\frac{\pi}{4}\right)}$$

Put $x-\frac{\pi}{4}=t$

$$=\frac{1}{\sqrt{2}}\int_{-\pi/2}^{\pi/2}\frac{dt}{(e^t+1)\cos t}\qquad\text{...(i)}$$

$$=\frac{1}{\sqrt{2}}\int_{-\pi/2}^{\pi/2}\frac{dt}{(e^{\pi/2-\pi/2-t}+1)\cos(\pi/2-\pi/2-t)}$$

$$=\frac{1}{\sqrt{2}}\int_{-\pi/2}^{\pi/2}\frac{dt}{(e^{-t}+1)\cos(-t)}$$

$$=\frac{1}{\sqrt{2}}\int_{-\pi/2}^{\pi/2}\frac{e^t\,dt}{(e^t+1)\cos t} \quad \text{...(ii)}$$

On adding Eqs. (i) and (ii), we get

$$2I=\frac{1}{\sqrt{2}}\int_{-\pi/2}^{\pi/2}\sec t\,dt$$

$$\Rightarrow \quad I=\frac{1}{2\sqrt{2}}\int_{-\pi/2}^{\pi/2}\sec t\,dt$$

$$\therefore \quad K=\frac{1}{2\sqrt{2}}$$

15. (c) We have, $\int_0^{\pi/3}\frac{\tan\theta}{\sqrt{2k\sec\theta}}d\theta=1-\frac{1}{\sqrt{2}}$, $(k>0)$

Let $I=\int_0^{\pi/3}\frac{\tan\theta}{\sqrt{2k\sec\theta}}d\theta$

$$=\frac{1}{\sqrt{2k}}\int_0^{\pi/3}\frac{\tan\theta}{\sqrt{\sec\theta}}d\theta$$

$$=\frac{1}{\sqrt{2k}}\int_0^{\pi/3}\frac{(\sin\theta)}{(\cos\theta)\sqrt{\frac{1}{\cos\theta}}}d\theta$$

$$=\frac{1}{\sqrt{2k}}\int_0^{\pi/3}\frac{\sin\theta}{\sqrt{\cos\theta}}d\theta$$

Let $\cos\theta=t\Rightarrow-\sin\theta\,d\theta=dt\Rightarrow\sin\theta\,d\theta=-dt$

For lower limit, $\theta=0\Rightarrow t=\cos 0=1$

For upper limit, $\theta=\frac{\pi}{3}\Rightarrow t=\cos\frac{\pi}{3}=\frac{1}{2}$

$$\Rightarrow \quad I=\frac{1}{\sqrt{2k}}\int_1^{1/2}\frac{-dt}{\sqrt{t}}=\frac{-1}{\sqrt{2k}}\int_1^{1/2}t^{-\frac{1}{2}}dt$$

$$=-\frac{1}{\sqrt{2k}}\left(\frac{t^{-\frac{1}{2}+1}}{-\frac{1}{2}+1}\right)_1^{\frac{1}{2}}=-\frac{1}{\sqrt{2k}}[2\sqrt{t}]_1^{\frac{1}{2}}$$

$$=-\frac{2}{\sqrt{2k}}\left[\sqrt{\frac{1}{2}}-\sqrt{1}\right]=\frac{2}{\sqrt{2k}}\left(1-\frac{1}{\sqrt{2}}\right)$$

$$\because \quad I=1-\frac{1}{\sqrt{2}} \quad \text{[given]}$$

$$\therefore \quad \frac{2}{\sqrt{2k}}\left(1-\frac{1}{\sqrt{2}}\right)=1-\frac{1}{\sqrt{2}}\Rightarrow\frac{2}{\sqrt{2k}}=1$$

$$\Rightarrow \quad 2=\sqrt{2k}\Rightarrow 2k=4\Rightarrow k=2$$

16. (a) $I=\int_0^{\pi/2}\sqrt{\tan x}\,dx$ ...(i)

$$\Rightarrow \quad I=\int_0^{\pi/2}\sqrt{\cot x}\,dx \quad \text{...(ii)}$$

On adding Eqs. (i) and (ii), we get

$$2I=\int_0^{\pi/2}(\sqrt{\tan x}+\sqrt{\cot x})\,dx$$

$$=\sqrt{2}\int_0^{\pi/2}\frac{\sin x+\cos x}{\sqrt{\sin 2x}}dx$$

$$=\sqrt{2}\int_0^{\pi/2}\frac{\sin x+\cos x}{\sqrt{1-(\sin x-\cos x)^2}}dx$$

Put $\sin x-\cos x=t\Rightarrow(\sin x+\cos x)\,dx=dt$

$$=\sqrt{2}\int_{-1}^{1}\frac{dt}{\sqrt{1-t^2}}=2\sqrt{2}\int_0^1\frac{dt}{\sqrt{1-t^2}}$$

$$=2\sqrt{2}[\sin^{-1}t]_0^1=2\sqrt{2}\cdot\frac{\pi}{2}=\sqrt{2}\pi$$

17. (c) $\text{In}=\int_0^{\pi/4}\tan^n x\sec^2 x\,dx$

Let $\tan x=t$

$\sec^2 x\,dx=dt$

$$=\int_0^1 t^n\,dt$$

$$=\left[\frac{t^{n+1}}{n+1}\right]_0^1$$

$$\text{In}=\frac{1}{n+1}$$

$$\therefore \quad I_1=\frac{1}{2},I_2=\frac{1}{3},I_3=\frac{1}{4}$$

Reciprocals of the terms $I_1,I_2,I_3\ldots$ are in AP therefore $I_1,I_2,I_3,\ldots$ are in HP.

18. (c) Given integral

$$I=\int_{\pi/6}^{\pi/3}\tan^3 x\cdot\sin^2 3x(2\sec^2 x\sin^2 3x+3\tan x\cdot\sin 6x)dx$$

$$=\int_{\pi/6}^{\pi/3}[2\tan^3 x\sec^2 x\sin^4 3x$$

$$+3\tan^4 x(2\sin^3 3x\cos 3x)]dx$$

$$=\int_{\pi/6}^{\pi/3}\left[\frac{d}{dx}\left(\frac{\tan^4 x\sin^4 3x}{2}\right)\right]dx$$

$$=\left[\frac{1}{2}\tan^4 x\sin^4 3x\right]_{\pi/6}^{\pi/3}=\frac{1}{2}\left[(9\times 0)-\left(\frac{1}{9}\times 1\right)\right]=-\frac{1}{18}$$

19. (a) Let $I=\int_0^1\cos\left(2\cot^{-1}\sqrt{\frac{1-x}{1+x}}\right)dx.$

Let $x=\cos\theta\Rightarrow dx=-\sin\theta\,d\theta$

$\theta=\cos^{-1}x$

At $x=0,\theta=\frac{\pi}{2}$ and at $x=1,\theta=0$

Now, $\sqrt{\frac{1-x}{1+x}}=\sqrt{\frac{1-\cos\theta}{1+\cos\theta}}=\tan\frac{\theta}{2}=\cot\left(\frac{\pi}{2}-\frac{\theta}{2}\right)$

$$I=\int_{\pi/2}^{0}\cos\left[2\left(\frac{\pi}{2}-\frac{\theta}{2}\right)\right](-\sin\theta\,d\theta)$$

$$= \int_{\pi/2}^{0} \sin\theta \cos d\theta$$

$$= \frac{1}{2}\int_{\pi/2}^{0} \sin 2\theta \, d\theta$$

$$= -\frac{1}{4}[\cos 2\theta]_{\pi/2}^{0} = -\frac{1}{4}(1+1) = -\frac{1}{2}$$

20. (b) We have, $I = \int_0^{\pi/4} \frac{\sin\theta + \cos\theta}{9 + 16\sin 2\theta} d\theta$

Let $\sin\theta - \cos\theta = t$

$(\sin\theta + \cos\theta)\, d\theta = dt$

$$= \int_{-1}^{0} \frac{dt}{9 + 16(1-t^2)} \quad [\because t^2 = 1 - 2\sin\theta\cos\theta]$$

$$= \int_{-1}^{0} \frac{dt}{25 - 16t^2}$$

$$= \frac{1}{16}\int_{-1}^{0} \frac{1}{\left(\frac{5}{4}\right)^2 - t^2} dt$$

$$= \frac{1}{16} \times \frac{1}{2 \times \frac{5}{4}} \left[\ln \left| \frac{\frac{5}{4} + t}{\frac{5}{4} - t} \right| \right]_{-1}^{0}$$

$$= \frac{1}{40}\left[\ln 1 - \ln\frac{1}{9}\right]$$

$$= \frac{1}{40}(\ln 9) = \frac{1}{20}\ln 3$$

21. (a) We have, $I = \int_0^{\pi/2} \frac{dx}{3 + 4\sin x}$

$$= \int_0^{\pi/2} \frac{dx}{3 + 8\sin\frac{x}{2}\cos\frac{x}{2}}$$

$$= \int_0^{\pi/2} \frac{\sec^2\frac{x}{2}\, dx}{3\left(1 + \tan^2\frac{x}{2}\right) + 8\tan\frac{x}{2}}$$

$$= \int_0^{\pi/2} \frac{\sec^2\frac{x}{2}\, dx}{3\tan^2\frac{x}{2} + 8\tan\frac{x}{2} + 3}$$

Put $\tan\frac{x}{2} = t$

$\frac{1}{2}\sec^2\frac{x}{2}\, dx = dt$

$$= \int_0^1 \frac{2\, dt}{3t^2 + 8t + 3}$$

$$= \frac{2}{3}\int_0^1 \frac{dt}{t^2 + \frac{8}{3}t + 1}$$

$$= \frac{2}{3}\int_0^1 \frac{dt}{\left(t + \frac{4}{3}\right)^2 - \left(\frac{\sqrt{7}}{3}\right)^2}$$

$$= \frac{2}{3} \cdot \frac{1}{2\left(\frac{\sqrt{7}}{3}\right)} \left[\log \frac{t + \frac{4}{3} - \left(\frac{\sqrt{7}}{3}\right)}{t + \frac{4}{3} + \frac{\sqrt{7}}{3}} \right]_0^1$$

$$= \frac{1}{\sqrt{7}} \left[\log \frac{1 + \frac{4}{3} - \frac{\sqrt{7}}{3}}{1 + \frac{4}{3} + \frac{\sqrt{7}}{3}} - \log \frac{\frac{4}{3} - \frac{\sqrt{7}}{3}}{\frac{4}{3} + \frac{\sqrt{7}}{3}} \right]$$

$$= \frac{1}{\sqrt{7}} \left[\log \frac{7 - \sqrt{7}}{7 + \sqrt{7}} - \log \frac{4 - \sqrt{7}}{4 + \sqrt{7}} \right]$$

$$= \frac{1}{\sqrt{7}} \left[\log \left(\frac{\sqrt{7} - 1}{\sqrt{7} + 1} \right) \left(\frac{4 + \sqrt{7}}{4 - \sqrt{7}} \right) \right]$$

$$= \frac{1}{\sqrt{7}} \left[\log \left(\frac{8 - 2\sqrt{7}}{6} \cdot \frac{23 + 8\sqrt{7}}{9} \right) \right]$$

$$= \frac{1}{\sqrt{7}} \log \left(\frac{36 + 9\sqrt{7}}{27} \right)$$

$$= \frac{1}{\sqrt{7}} \log \left(\frac{4 + \sqrt{7}}{3} \right)$$

22. (a) We have, $\int_0^{\pi/2} \frac{dx}{4\cos^2 x + 9\sin^2 x}$

$$= \int_0^{\pi/2} \frac{\sec^2 x}{4 + 9\tan^2 x} dx$$

Put $\tan x = t$

$\sec^2 x\, dx = dt$

$$= \int_0^{\infty} \frac{dt}{4 + 9t^2}$$

$$= \frac{1}{9}\int_0^{\infty} \frac{dt}{\left(\frac{2}{3}\right)^2 + t^2}$$

$$= \frac{1}{9} \cdot \frac{1}{2/3} \left[\tan^{-1}\left(\frac{3t}{2}\right) \right]_0^{\infty}$$

$$= \frac{1}{6}[\tan^{-1}\infty - \tan^{-1} 0]$$

$$= \frac{1}{6}\left(\frac{\pi}{2}\right)$$

$$= \frac{\pi}{12}$$

23. (a) Let $I=\int_0^1 \frac{1-x}{1+x}\frac{dx}{\sqrt{x+x^2+x^3}}$

$$=\int_0^1 \frac{(1-x^2)\,dx}{(x^2+2x+1)\sqrt{x+x^2+x^3}}$$

$$=\int_0^1 \frac{\left(\frac{1}{x^2}-1\right)dx}{\left(x+\frac{1}{x}+2\right)\sqrt{x+\frac{1}{x}+1}}$$

Put $x+\frac{1}{x}+1=t^2$

$$\left(1-\frac{1}{x^2}\right)dx=2t\,dt$$

$$\Rightarrow \quad I=\int_\infty^{\sqrt{3}}\frac{-2t\,dt}{(t^2+1)t}$$

$$=2\int_{\sqrt{3}}^{\infty}\frac{dt}{t^2+1}$$

$$=2[\tan^{-1}t]_{\sqrt{3}}^{\infty}=2\left(\frac{\pi}{2}-\frac{\pi}{3}\right)$$

$$=2\left(\frac{\pi}{6}\right)=\frac{\pi}{3}$$

24. (d) Given,

$$I=\int_0^1\frac{dx}{(1+x)(2+x)\sqrt{x(1-x)}}$$

Substitute $x=\sin^2\theta$

$$I=\int_0^{\pi/2}\frac{2\sin\theta\cos\theta}{(1+\sin^2\theta)(2+\sin^2\theta)\sqrt{\sin^2\theta(1-\sin^2\theta)}}\cdot d\theta$$

$$=\int_0^{\pi/2}\frac{2\sec^4\theta}{(\sec^2\theta+\tan^2\theta)(2\sec^2\theta+\tan^2\theta)}d\theta$$

Put $\tan\theta=t$

$\Rightarrow \quad \sec^2\theta\,d\theta=dt$

When $\theta=0$, then $t=0$ and $\theta=\frac{\pi}{2}$, then $t=\infty$

$$=2\int_0^{\infty}\frac{(1+t^2)\,dt}{(1+2t^2)(2+3t^2)}$$

$$=2\int_0^{\infty}\left[\frac{1}{1+2t^2}-\frac{1}{2+3t^2}\right]dt$$

$$\Rightarrow \quad I=2\left[\frac{1}{\sqrt{2}}\tan^{-1}(\sqrt{2}t)-\frac{1}{\sqrt{2}\sqrt{3}}\tan^{-1}\left(\frac{\sqrt{3}t}{\sqrt{2}}\right)\right]_0^{\infty}$$

$$\Rightarrow \quad I=\frac{\pi}{\sqrt{2}}\left(1-\frac{1}{\sqrt{3}}\right)$$

$$=\frac{\pi(\sqrt{3}-1)}{\sqrt{6}}$$

25. (a) We have, $\int_{\log 2}^{x}\frac{dy}{\sqrt{e^y-1}}=\frac{\pi}{6}$

Put $t=e^y-1$

$\Rightarrow \quad y=\log(t+1)$

$\Rightarrow \quad dy=\frac{dt}{t+1}$

$$\Rightarrow \quad 2\int_1^{e^x-1}\frac{dt}{(t+1)2\sqrt{t}}=\frac{\pi}{6}$$

Let $\sqrt{t}=p$

$$\frac{1}{2\sqrt{t}}dt=dp$$

$$\Rightarrow \quad 2\int_1^{\sqrt{e^x-1}}\frac{dp}{p^2+1}=\frac{\pi}{6}$$

$$\Rightarrow \quad 2[\tan^{-1}p]_1^{\sqrt{e^x-1}}=\frac{\pi}{6}$$

$$\Rightarrow \quad \tan^{-1}\sqrt{e^x-1}-\tan^{-1}1=\frac{\pi}{12}$$

$$\Rightarrow \quad \tan^{-1}\sqrt{e^x-1}-\frac{\pi}{4}=\frac{\pi}{12}$$

$$\Rightarrow \quad \tan^{-1}\sqrt{e^x-1}=\frac{\pi}{12}+\frac{\pi}{4}$$

$$\Rightarrow \quad \tan^{-1}\sqrt{e^x-1}=\frac{\pi}{3}$$

$$\Rightarrow \quad \sqrt{e^x-1}=\tan\frac{\pi}{3}$$

$$\Rightarrow \quad \sqrt{e^x-1}=\sqrt{3}$$

$$\Rightarrow \quad e^x-1=3\Rightarrow e^x=4$$

$$\Rightarrow \quad x=\log_e 4$$

26. (a) Let $I=\int_0^1\frac{\sqrt{x}\,dx}{(1+x)(1+3x)(3+x)}$

Let $x=t^2\Rightarrow dx=2t\,dt$

$$I=\int_0^1\frac{t(2t)}{(t^2+1)(1+3t^2)(3+t^2)}dt$$

$$=\int_0^1\frac{(3t^2+1)-(t^2+1)}{(3t^2+1)(t^2+1)(t^2+3)}dt$$

$$I=\int_0^1\frac{dt}{(t^2+1)(3+t^2)}-\int_0^1\frac{dt}{(1+3t^2)(3+t^2)}$$

$$=\frac{1}{2}\int_0^1\frac{(3+t^2)-(t^2+1)}{(t^2+1)(3+t^2)}dt$$

$$+\frac{1}{8}\int_0^1\frac{(1+3t^2)-3(3+t^2)}{(1+3t^2)(3+t^2)}dt$$

$$=\frac{1}{2}\int_0^1\frac{dt}{1+t^2}-\frac{1}{2}\int_0^1\frac{dt}{t^2+3}$$

$$+\frac{1}{8}\int_0^1\frac{dt}{3+t^2}-\frac{3}{8}\int_0^1\frac{dt}{1+3t^2}$$

$$= \frac{1}{2}(\tan^{-1} t)_0^1 - \frac{3}{8\sqrt{3}}\left(\tan^{-1}\frac{t}{\sqrt{3}}\right)_0^1$$

$$- \frac{3}{8\sqrt{3}}(\tan^{-1}(\sqrt{3}t))_0^1$$

$$= \frac{1}{2}\left(\frac{\pi}{4}\right) - \frac{\sqrt{3}}{8}\left(\frac{\pi}{6}\right) - \frac{\sqrt{3}}{8}\left(\frac{\pi}{3}\right)$$

$$= \frac{\pi}{8} - \frac{\sqrt{3}}{16}\pi = \frac{\pi}{8}\left(1 - \frac{\sqrt{3}}{2}\right)$$

27. (a) We have, $\lim_{n\to\infty} \frac{1}{\sqrt{n}}\left(1 + \frac{1}{\sqrt{2}} + \frac{1}{\sqrt{3}} + \ldots + \frac{1}{\sqrt{n}}\right)$

$$= \lim_{n\to\infty} \frac{1}{\sqrt{n}} \sum_{r=1}^{n} \left(\frac{1}{\sqrt{r}}\right)$$

$$= \lim_{n\to\infty} \frac{1}{n} \sum_{r=1}^{n} \left(\frac{n}{\sqrt{n}\sqrt{r}}\right)$$

$$= \lim_{n\to\infty} \frac{1}{n} \sum_{r=1}^{n} \left(\frac{1}{\sqrt{r/n}}\right)$$

$$= \int_0^1 \frac{dx}{\sqrt{x}}$$

$$= 2[\sqrt{x}]_0^1 = 2(1-0) = 2$$

28. (b) Let $A = \lim_{n\to\infty}\left(\frac{(n!)^{1/n}}{n}\right)$

$$\log A = \lim_{n\to\infty} \log\left(\frac{1\cdot 2\cdot 3\ldots n}{n^n}\right)^{1/n}$$

$$= \lim_{n\to\infty} \log\left(\frac{1}{n}\cdot\frac{2}{n}\ldots\frac{n}{n}\right)^{1/n}$$

$$\Rightarrow \log A = \lim_{n\to\infty} \frac{1}{n}\sum_{r=1}^{n} \log\left(\frac{r}{n}\right)$$

$$\Rightarrow \log A = \int_0^1 \log x\, dx = [x\log x - x]_0^1$$

$\therefore\ \log A = -1$

$\therefore\quad A = e^{-1}$

29. (c) Let $A = \lim_{n\to\infty} \frac{1}{n}[(n+1)(n+2)\ldots(n+n)]^{1/n}$

$$A = \lim_{n\to\infty}\left[\left(1+\frac{1}{n}\right)\left(1+\frac{2}{n}\right)\left(1+\frac{3}{n}\right)\ldots\left(1+\frac{n}{n}\right)\right]^{1/n}$$

Therefore, $\log A = \lim_{n\to\infty} \frac{1}{n}\left[\log\left(1+\frac{1}{n}\right) + \log\left(1+\frac{2}{n}\right) + \ldots + \log\left(1+\frac{n}{n}\right)\right]$

Now, rth term $= \frac{1}{n}\log\left(1+\frac{r}{n}\right)$

Put $n = \frac{1}{h}$ so that

rth term $= h\log(1+x)$, where $r = nx$ and $\frac{1}{n} = dx$

therefore

$$\log A = \int_0^1 \log(1+x)\, dx$$

$$= [x\log(1+x)]_0^1 - \int_0^1 \frac{x}{1+x}\, dx$$

$$= \log 2 - \int_0^1\left(1 - \frac{1}{1+x}\right) dx$$

$$= \log 2 - (1 - \log 2)$$

$$\Rightarrow \log A = 2\log 2 - 1$$

$$\therefore\quad A = e^{2\log 2 - 1} = \frac{4}{e}$$

30. (a) We have,

$$\lim_{n\to\infty} S_n = \lim_{n\to\infty}\left[\frac{1}{n} + \frac{n^2}{(n+1)^3} + \frac{n^2}{(n+2)^3} + \ldots + \frac{1}{8^n}\right]$$

$$= \lim_{n\to\infty} \frac{1}{n}\sum_{r=0}^{n} \frac{1}{\left(1+\frac{r}{n}\right)^3}$$

Therefore, the required limit is

$$\int_0^1 \frac{dx}{(1+x)^3} = -\frac{1}{2}\left[\frac{1}{(1+x)^2}\right]_0^1$$

$$= -\frac{1}{2}\left[\frac{1}{4} - 1\right] = -\frac{1}{2}\left(-\frac{3}{4}\right) = \frac{3}{8}$$

31. (c) Let $p = \lim_{n\to\infty}\left(\frac{(n+1)^{1/3}}{n^{4/3}} + \frac{(n+2)^{1/3}}{n^{4/3}} + \ldots + \frac{(2n)^{1/3}}{n^{4/3}}\right)$

$$= \lim_{n\to\infty} \sum_{r=1}^{n} \frac{(n+r)^{1/3}}{n^{4/3}}$$

$$= \lim_{n\to\infty} \sum_{r=1}^{n} \frac{\left(1+\frac{r}{n}\right)^{1/3} n^{1/3}}{n^{4/3}}$$

$$= \lim_{n\to\infty} \frac{1}{n}\sum_{r=1}^{n}\left(1+\frac{r}{n}\right)^{1/3}$$

Now, as per integration as limit of sum.

Let $\frac{r}{n} = x$ and $\frac{1}{n} = dx$ $\quad [\because n \to \infty]$

Then, upper limit of integral is 1 and lower limit of integral is 0.

So, $p = \int_0^1 (1+x)^{1/3}\, dx \quad \left[\because \lim_{n\to\infty} \frac{1}{n}\sum_{r=1}^{n} f\left(\frac{r}{n}\right) = \int_0^1 f(x)\, dx\right]$

$$= \left[\frac{3}{4}(1+x)^{4/3}\right]_0^1 = \frac{3}{4}(2^{4/3} - 1) = \frac{3}{4}(2)^{4/3} - \frac{3}{4}$$

32. (a) We have, $\lim_{n\to\infty}\sum_{r=1}^{n}\left(\frac{\sqrt{n}}{\sqrt{r}\,(\sqrt{r}+4\sqrt{n})^2}\right)$

$$= \lim_{n\to\infty}\frac{1}{n}\left[\sum_{r=1}^{n}\frac{n\sqrt{n}}{\sqrt{r}\,(\sqrt{r}+4\sqrt{n})^2}\right]$$

$$= \lim_{n\to\infty}\frac{1}{n}\left(\sum_{r=1}^{n}\left(\frac{1}{\sqrt{r/n}(\sqrt{r/n}+4)^2}\right)\right)$$

$$= \int_0^1 \frac{dx}{\sqrt{x}(\sqrt{x}+4)^2}$$

Let $\sqrt{x}+4=t \Rightarrow \frac{dx}{2\sqrt{x}}=dt$

$$= \int_4^5\left(\frac{2\,dt}{t^2}\right) = -\left(\frac{2}{t}\right)_4^5$$

$$= -2\left(\frac{1}{5}-\frac{1}{4}\right) = -2\times\left(-\frac{1}{20}\right) = \frac{1}{10}$$

33. (d) We have, $\lim_{n\to\infty}\frac{1}{n}\sum_{i=1}^{n}\frac{(2i-1)+8n}{(2i-1)+4n}$

$$\lim_{n\to\infty}\frac{1}{n}\sum_{i=1}^{n}\frac{\left(\frac{2i}{n}-\frac{1}{n}+8\right)}{\left(\frac{2i}{n}-\frac{1}{n}+4\right)}$$

$$\Rightarrow \int_0^1 \frac{2x+8}{2x+4}\,dx = \int_0^1 dx + \int_0^1 \frac{4}{2x+4}\,dx$$

$$= 1 + 4\cdot\left[\frac{1}{2}\ln(2x+4)\right]_0^1$$

$$= 1 + 2(\ln 6 - \ln 4)$$

$$= 1 + 2\ln\left(\frac{6}{4}\right) = 1 + 2\ln\left(\frac{3}{2}\right)$$

34. (b) Given, $\int_0^\infty \frac{\sin x}{x}\,dx = \frac{\pi}{2}$

Now, $\sin^3 x = \frac{3}{4}\sin x - \frac{1}{4}\sin 3x$

$$\therefore \int_0^\infty \frac{\sin^3 x}{x}\,dx = \frac{3}{4}\int_0^\infty \frac{\sin x}{x}\,dx - \frac{1}{4}\int_0^\infty \frac{\sin 3x}{x}\,dx$$

$$= \frac{3}{4}\int_0^\infty \frac{\sin x}{x}\,dx - \frac{1}{4}\int_0^\infty \frac{\sin u}{u}\,du$$

[let $3x = u \Rightarrow 3dx = du$]

$$= \frac{3}{4}\left(\frac{\pi}{2}\right) - \frac{1}{4}\left(\frac{\pi}{2}\right) = \frac{\pi}{4}$$

35. (a) We have,

$$\int_0^2 [x^2]\,dx = \int_0^1 [x^2]\,dx + \int_1^{\sqrt{2}} [x^2]\,dx + \int_{\sqrt{2}}^{\sqrt{3}} [x^2]\,dx + \int_{\sqrt{3}}^{2} [x^2]\,dx$$

$$\int_0^2 [x^2]\,dx = \int_0^1 0\cdot dx + \int_1^{\sqrt{2}} 1\,dx + \int_{\sqrt{2}}^{\sqrt{3}} 2\,dx + \int_{\sqrt{3}}^{2} 3\,dx$$

$$= 0 + (\sqrt{2}-1) + 2(\sqrt{3}-\sqrt{2}) + 3(2-\sqrt{3})$$

$$= 5 - \sqrt{3} - \sqrt{2}$$

36. (c) The graph of $y = \cos\left(\frac{\pi x}{2}\right)$ is

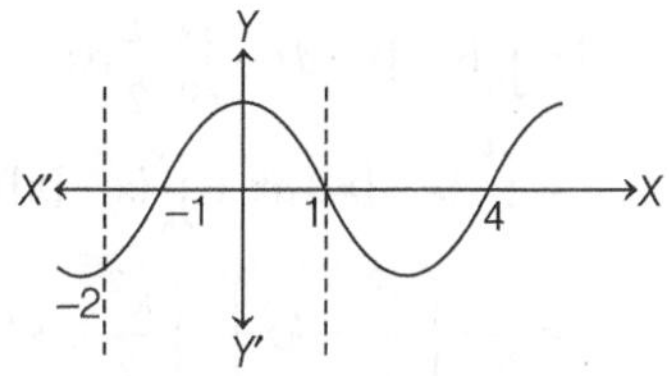

From graph,

$$\int_{-2}^{1}\left[x\left[1+\cos\frac{\pi x}{2}\right]+1\right]dx$$

$$= \int_{-2}^{-1}[x[1+(-1)]+1]\,dx + \int_{-1}^{1}[x(1+0)+1]\,dx$$

$$= [x]_{-2}^{-1} + \int_{-1}^{1}[x+1]\,dx$$

$$= (-1-(-2)) + \int_{-1}^{0} 0\,dx + \int_0^1 1\,dx = 1 + 1 = 2$$

37. (a) $\sum_{r=1}^{n}\int_0^1 f(r-1+x)\,dx$

$$= \int_0^1 f(x)\,dx + \int_0^1 f(1+x)\,dx + \int_0^1 f(2+x)\,dx + \ldots + \int_0^1 f(n-1+x)\,dx$$

$$= \int_0^1 f(x)\,dx + \int_1^2 f(x)\,dx + \int_2^3 f(x)\,dx + \ldots + \int_{n-1}^{n} f(x)\,dx$$

$$= \int_0^n f(x)\,dx$$

38. (c) Given, $4\alpha\int_{-1}^{2} e^{-\alpha|x|}dx = 5$

$$\Rightarrow 4\alpha\left[\int_{-1}^{0} e^{\alpha x}dx + \int_0^2 e^{-\alpha x}dx\right] = 5 \qquad \left[\because |x| = \begin{bmatrix} x, & x\ge 0 \\ -x, & x<0 \end{bmatrix}\right]$$

$$\Rightarrow 4\alpha\left[\left(\frac{e^{\alpha x}}{\alpha}\right)_{-1}^{0} + \left(\frac{e^{-\alpha x}}{-\alpha}\right)_0^2\right] = 5$$

$$\Rightarrow 4[(e^0 - e^{-\alpha}) - (e^{-2\alpha} - e^0)] = 5$$

$$\Rightarrow 4[1 - e^{-\alpha} - e^{-2\alpha} + 1] = 5$$

$$\Rightarrow 4[2 - e^{-\alpha} - e^{-2\alpha}] = 5$$

$$\Rightarrow 8e^{2\alpha} - 4e^{\alpha} - 4 = 5e^{2\alpha}$$

$$\Rightarrow 3e^{2\alpha} - 4e^{\alpha} - 4 = 0$$

$$\Rightarrow e^{\alpha} = \frac{4+\sqrt{16+48}}{6} \qquad [\because e^{\alpha} > 0, \forall\, \alpha \in R]$$

$$\Rightarrow e^{\alpha} = \frac{4+8}{6} = 2$$

$$\Rightarrow \alpha = \log_e 2$$

39. (b) We have, $\{x\} = x - [x]$

Let $k \le x < k+1$, where k is any integer. Then,

$$I = \int_0^x \left(\{x\} - \frac{1}{2}\right) dx = \int_0^x \left(x - [x] - \frac{1}{2}\right) dx$$

$$I = \int_0^x (x - [x])\, dx - \int_0^x \frac{1}{2}\, dx$$

$$= \int_0^k (x - [x])\, dx + \int_k^x (x - [x])\, dx - \frac{x}{2}$$

$$= \frac{k}{2} + \left(\frac{x^2}{2} - kx\right) - \left(\frac{k^2}{2} - k^2\right) - \frac{x}{2}$$

$$\Rightarrow \quad I = \frac{x^2}{2} - kx + \frac{k(k+1)}{2} - \frac{x}{2}$$

$$\Rightarrow \quad I = \frac{1}{2}(x^2 - 2kx + k^2) + \frac{1}{2}(k - x)$$

$$\Rightarrow \quad I = \frac{1}{2}(x - k)^2 - \frac{1}{2}(x - k)$$

$$= \frac{1}{2}(x - [x])^2 - \frac{1}{2}(x - [x])$$

$$= \frac{1}{2}\{x\}^2 - \frac{1}{2}\{x\} = \frac{1}{2}\{x\}(\{x\} - 1)$$

40. (b) Let $\quad I = \int_6^{16} \frac{\log x^2}{\log x^2 + \log(x^2 - 44x + 484)} dx$

$$\Rightarrow \quad I = \int_6^{16} \frac{\log x^2}{\log x^2 + \log(x - 22)^2} dx \quad \text{...(i)}$$

Using $\int_a^b f(x)\, dx = \int_a^b f(a + b - x)\, dx$

$$\Rightarrow \quad I = \int_6^{16} \frac{\log(22 - x)^2}{\log(22 - x)^2 + \log(22 - x - 22)^2} dx$$

$$\Rightarrow \quad I = \int_6^{16} \frac{\log(22 - x)^2}{\log x^2 + \log(22 - x)^2} dx \quad \text{...(ii)}$$

On adding Eqs. (i) and (ii), we get

$$2I = \int_6^{16} 1\, dx = [x]_6^{16}$$

$$\Rightarrow \quad 2I = 10$$

$$\therefore \quad I = 5$$

41. (d) The given functions are

$$g(x) = \log_e x,\ x > 0 \text{ and } f(x) = \frac{2 - x\cos x}{2 + x\cos x}$$

Let $\quad I = \int_{-\pi/4}^{\pi/4} g(f(x))dx$

Then, $\quad I = \int_{-\pi/4}^{\pi/4} \log_e \left(\frac{2 - x\cos x}{2 + x\cos x}\right) dx \quad \text{...(i)}$

Now, by using the property

$$\int_a^b f(x)dx = \int_a^b f(a + b - x)dx, \text{ we get}$$

$$I = \int_{-\pi/4}^{\pi/4} \log_e \left(\frac{2 + x\cos x}{2 - x\cos x}\right) dx \quad \text{...(ii)}$$

On adding Eqs. (i) and (ii), we get

$$2I = \int_{-\pi/4}^{\pi/4} \left[\log_e\left(\frac{2 - x\cos x}{2 + x\cos x}\right) + \log_e\left(\frac{2 + \cos x}{2 - x\cos x}\right)\right] dx$$

$$= \int_{-\pi/4}^{\pi/4} \log_e \left(\frac{2 - x\cos x}{2 + x\cos x} \times \frac{2 + x\cos x}{2 - x\cos x}\right) dx$$

$$[\because \log_e A + \log_e B = \log_e AB]$$

$$\Rightarrow \quad 2I = \int_{-\pi/4}^{\pi/4} \log_e (1) dx = 0$$

$$\Rightarrow \quad I = 0 = \log_e (1)$$

42. (a) Let $I = \int_{\sqrt{\ln 2}}^{\sqrt{\ln 3}} \left[\frac{x \sin(x^2)}{\sin x^2 + \sin \ln(6 - x^2)}\right] dx$

Let $x^2 = t \Rightarrow x\, dx = \frac{1}{2} dt$

$$= \frac{1}{2} \int_{\ln 2}^{\ln 3} \left(\frac{\sin t}{\sin t + \sin \ln(6 - t)}\right) dt \quad \text{...(i)}$$

$$= \frac{1}{2} \int_{\ln 2}^{\ln 3} \left(\frac{\sin(\ln 6 - t)}{\sin(\ln 6 - t) + \sin(t)}\right) dt \quad \text{...(ii)}$$

Adding Eqs. (i) and (ii), we get

$$2I = \frac{1}{2} \int_{\ln 2}^{\ln 3} \left(\frac{\sin(t) + \sin(\ln 6 - t)}{\sin(\ln 6 - t) + \sin(t)}\right) dt$$

$$= \frac{1}{2} \int_{\ln 2}^{\ln 3} dt$$

$$= \frac{1}{2}(\ln 3 - \ln 2)$$

$$= \frac{1}{2}\left[\ln\left(\frac{3}{2}\right)\right] \Rightarrow I = \frac{1}{4}\left[\ln\left(\frac{3}{2}\right)\right]$$

43. (d) Let $I = \int_0^{\pi} \frac{1}{e^{\cos x} + 1} dx \quad \text{...(i)}$

$$\Rightarrow \quad I = \int_0^{\pi} \frac{1}{e^{\cos(\pi - x)} + 1} dx$$

$$\Rightarrow \quad I = \int_0^{\pi} \frac{1}{e^{-\cos x} + 1} dx$$

$$\Rightarrow \quad I = \int_0^{\pi} \frac{e^{\cos x}}{1 + e^{\cos x}} dx \quad \text{...(ii)}$$

On adding Eqs. (i) and (ii), we get

$$2I = \int_0^{\pi} 1 \cdot dx$$

$$\Rightarrow \quad 2I = \pi$$

$$\therefore \quad I = \frac{\pi}{2}$$

44. (c) Let the integral

$$I = \frac{1}{a + b} \int_a^b x(f(x) + f(x + 1))dx \quad \text{...(i)}$$

On applying property

$\int_a^b f(x)dx = \int_a^b f(a + b - x)dx$, we get

$$I=\frac{1}{a+b}\int_a^b(a+b-x)(f(a+b-x)+f(a+b-x+1))dx$$

$$=\frac{1}{a+b}\int_a^b(a+b-x)(f(x+1)+f(x))dx \quad ...(ii)$$

$$\left[\begin{array}{l}\because f(a+b+1-x)=f(x);\\ \therefore f(a+b-x)=f(x+1)\end{array}\right]$$

On adding integrals Eqs. (i) and (ii), we get

$$2I=\frac{a+b}{a+b}\int_a^b(f(x)+f(x+1))dx$$

$$\Rightarrow \quad I=\frac{1}{2}\left[\int_a^b f(x)dx+\int_a^b f(x+1)dx\right]$$

$$=\frac{1}{2}\left[\int_a^b f(x)dx+\int_a^b f(a+b-x+1)dx\right]$$

$$=\frac{1}{2}\left[\int_a^b f(x)dx+\int_a^b f(x)dx\right]$$

$$I=\int_a^b f(x)dx$$

Put $x=t+1$, then upper limit $t=b-1$ and lower limit $t=a-1$ and $dx=dt$, so

$$I=\int_{a-1}^{b-1}f(t+1)dt=\int_{a-1}^{b-1}f(x+1)dx$$

45. (a) We have, $I=\int_0^\infty \frac{x\,dx}{(1+x)(1+x^2)}$

Put $\quad x=\tan\theta$

$dx=\sec^2\theta\,d\theta$

$$\Rightarrow \quad I=\int_0^{\pi/2}\frac{\tan\theta\cdot\sec^2\theta\cdot d\theta}{(1+\tan\theta)(1+\tan^2\theta)}$$

$$=\int_0^{\pi/2}\frac{\tan\theta\cdot\sec^2\theta}{(1+\tan\theta)(\sec^2\theta)}d\theta$$

$$\Rightarrow \quad I=\int_0^{\pi/2}\frac{\sin\theta}{\sin\theta+\cos\theta}d\theta \quad ...(i)$$

Using $\int_0^a f(x)\,dx=\int_0^a f(a-x)\,dx$

$$\Rightarrow \quad I=\int_0^{\pi/2}\frac{\cos\theta}{\sin\theta+\cos\theta}d\theta \quad ...(ii)$$

On adding Eqs. (i) and (ii), we get

$$2I=\int_0^{\pi/2}1\cdot d\theta$$

$$\Rightarrow \quad 2I=[\theta]_0^{\pi/2}$$

$$\therefore \quad I=\frac{\pi}{4}$$

46. (c) Let $I=\int_0^a\frac{g(x)}{f(x)+f(a-x)}dx \quad ...(i)$

Then, $I=\int_0^a\frac{g(a-x)}{f(a-x)+f(x)}dx \quad ...(ii)$

On adding Eqs. (i) and (ii), we get

$$2I=\int_0^a\frac{g(x)+g(a-x)}{f(x)+f(a-x)}dx$$

$$\Rightarrow \quad I=\frac{1}{2}\int_0^a\frac{g(x)+g(a-x)}{f(x)+f(a-x)}dx$$

Clearly, I vanishes, if

$$g(x)+g(a-x)=0\Rightarrow g(x)=-g(a-x)$$

47. (c) It is given that

$$I_1=\int_0^1(1-x^{50})^{100}dx$$

and $\quad I_2=\int_0^1(1-x^{50})^{101}dx=\int_0^1(1-x^{50})(1-x^{50})^{100}dx$

$$=\int_0^1(1-x^{50})^{100}dx-\int_0^1 x\cdot x^{49}(1-x^{50})^{100}dx$$

Now, let $I=\int_0^1 x\cdot x^{49}(1-x^{50})^{100}dx$

$$=[x\int x^{49}(1-x^{50})^{100}dx-\int(1\int x^{49}(1-x^{50})^{100}dx]_0^1$$

$$=\left[t^{1/50}\int\frac{(1-t)^{100}}{50}dt-\int\left(\int\frac{(1-t)^{100}}{50}dt\right)dx\right]_0^1$$

[where, $x^{50}=t$]

$$I=0-\frac{I_2}{50\times101}$$

$$\therefore \quad I_2=I_1-\frac{I_2}{5050}\Rightarrow\frac{5051}{5050}I_2=I_1$$

$$\because \quad I_2=\alpha I_1$$

$$\therefore \quad \alpha=\frac{5050}{5051}$$

48. (b) Let $I=\int_0^{2\pi}\frac{x\sin^8x}{\sin^8x+\cos^8x}dx \quad ...(i)$

$$\Rightarrow \quad I=\int_0^{2\pi}\frac{(2\pi-x)\sin^8(2\pi-x)}{\sin^8(2\pi-x)+\cos^8(2\pi-x)}dx$$

$$\Rightarrow \quad I=\int_0^{2\pi}\frac{(2\pi-x)\sin^8x}{\sin^8x+\cos^8x}dx \quad ...(ii)$$

On adding Eqs. (i) and (ii), we get

$$2I=\int_0^{2\pi}\frac{2\pi\sin^8x}{\sin^8x+\cos^8x}dx$$

$$\Rightarrow \quad 2I=2\times2\pi\int_0^{\pi}\frac{\sin^8x}{\sin^8x+\cos^8x}dx$$

$$\Rightarrow \quad I=2\pi\int_0^{\pi}\frac{\sin^8x}{\sin^8x+\cos^8x}dx$$

$$=4\pi\int_0^{\pi/2}\frac{\sin^8x}{\sin^8x+\cos^8x}dx \quad ...(iii)$$

$$=4\pi\int_0^{\pi/2}\frac{\sin^8\left(\frac{\pi}{2}-x\right)}{\sin^8\left(\frac{\pi}{2}-x\right)+\cos^8\left(\frac{\pi}{2}-x\right)}dx$$

$$\Rightarrow \quad I = 4\pi \int_0^{\pi/2} \frac{\cos^8 x}{\sin^8 x + \cos^8 x} dx \quad \text{...(iv)}$$

On adding Eqs. (iii) and (iv), we get

$$I = 2\pi \int_0^{\pi/2} \frac{\sin^8 x + \cos^8 x}{\sin^8 x + \cos^8 x} dx = 2\pi \left(\frac{\pi}{2}\right)$$

$$\therefore \quad I = \pi^2$$

49. (b) Let $I = \int_0^{\pi} \frac{x \sin x}{1 + \cos^2 x} dx$ ...(i)

Then, $I = \int_0^{\pi} \frac{(\pi - x) \sin (\pi - x)}{1 + \cos^2 (\pi - x)} dx$

$$\Rightarrow \quad I = \int_0^{\pi} \frac{(\pi - x) \sin x}{1 + \cos^2 x} dx \quad \text{...(ii)}$$

On adding Eqs. (i) and (ii), we get

$$2I = \pi \int_0^{\pi} \frac{\sin x}{1 + \cos^2 x} dx$$

Let $\cos x = t$

$\Rightarrow$ $-\sin x dx = dt$ when $x = 0, t = 1$

and when $x = \pi$ then $t = -1$

$$\Rightarrow \quad 2I = -\pi \int_1^{-1} \frac{1}{1+t^2} dt$$

$$\Rightarrow \quad 2I = -\pi[\tan^{-1} t]_1^{-1} = -\pi\left[-\frac{\pi}{4} - \frac{\pi}{4}\right]$$

$$\Rightarrow \quad 2I = \frac{\pi^2}{2}$$

$$\therefore \quad I = \frac{\pi^2}{4}$$

50. (d) Let $P = \int_0^{\pi} \frac{x^2 \cos x}{(1 + \sin x)^2} dx$

$$= \int_0^{\pi} x^2 \cdot \frac{\cos x}{(1 + \sin x)^2} dx$$

Integrate by parts

$$= \left[x^2 \left\{-\frac{1}{1 + \sin x}\right\}\right]_0^{\pi} + 2\int_0^{\pi} x \cdot \frac{1}{1 + \sin x} dx$$

$$= -\pi^2 + 2Q \quad \text{...(i)}$$

$$\left\{\text{where } Q = \int_0^{\pi} x \frac{1}{1 + \sin x} dx\right\}$$

Now, $Q = \int_0^{\pi} x \cdot \frac{1}{1 + \sin x} dx$

$$Q = \int_0^{\pi} \frac{(\pi - x)}{1 + \sin x} dx$$

$$\therefore \quad 2Q = \pi \int_0^{\pi} \frac{1}{1 + \sin x} dx$$

$$\Rightarrow \quad 2Q = 2\pi \int_0^{\pi/2} \frac{1}{1 + \sin x} dx$$

$$\Rightarrow \quad 2Q = 2\pi I$$

Put in Eq. (i),

$$P = -\pi^2 + 2\pi I$$

51. (d) Let $I = \int_{-\frac{\pi}{2}}^{\frac{\pi}{2}} \frac{dx}{[x] + [\sin x] + 4}$

$$= \int_{-\frac{\pi}{2}}^{-1} \frac{dx}{[x] + [\sin x] + 4} + \int_{-1}^{0} \frac{dx}{[x] + [\sin x] + 4}$$

$$+ \int_0^1 \frac{dx}{[x] + [\sin x] + 4} + \int_1^{\frac{\pi}{2}} \frac{dx}{[x] + [\sin x] + 4}$$

$$\because \quad [x] = \begin{cases} -2, & -\pi/2 < x < -1 \\ -1, & -1 \le x < 0 \\ 0, & 0 \le x < 1 \\ 1, & 1 \le x < \pi/2 \end{cases}$$

$$\text{and} \quad [\sin x] = \begin{cases} -1, -\pi/2 < x < -1 \\ -1, \ -1 < x < 0 \\ 0, \quad 0 < x < 1 \\ 0, \quad 1 < x < \pi/2 \end{cases}$$

$[\because$ For $x < 0, -1 \le \sin x < 0$ and for $x > 0, 0 < \sin x \le 1]$

$$\text{So, } I = \int_{-\frac{\pi}{2}}^{-1} \frac{dx}{-2 - 1 + 4} + \int_{-1}^{0} \frac{dx}{-1 - 1 + 4} + \int_0^1 \frac{dx}{0 + 0 + 4}$$

$$+ \int_1^{\frac{\pi}{2}} \frac{dx}{1 + 0 + 4}$$

$$= \int_{-\frac{\pi}{2}}^{-1} \frac{dx}{1} + \int_{-1}^{0} \frac{dx}{2} + \int_0^1 \frac{dx}{4} + \int_1^{\frac{\pi}{2}} \frac{dx}{5}$$

$$= \left(-1 + \frac{\pi}{2}\right) + \frac{1}{2}(0 + 1) + \frac{1}{4}(1 - 0) + \frac{1}{5}\left(\frac{\pi}{2} - 1\right)$$

$$= \left(-1 + \frac{1}{2} + \frac{1}{4} - \frac{1}{5}\right) + \left(\frac{\pi}{2} + \frac{\pi}{10}\right)$$

$$= \frac{-20 + 10 + 5 - 4}{20} + \frac{5\pi + \pi}{10}$$

$$= -\frac{9}{20} + \frac{3\pi}{5} = \frac{3}{20}(4\pi - 3)$$

52. (a) Let $I = \int_{-\pi}^{\pi} \left(\frac{2x(1 + \sin x)}{1 + \cos^2 x}\right) dx$

$$= \int_{-\pi}^{\pi} \frac{2x}{1 + \cos^2 x} dx + \int_{-\pi}^{\pi} \frac{2x \sin x}{1 + \cos^2 x} dx$$

$$= 0 + \int_{-\pi}^{\pi} \frac{2x \sin x}{1 + \cos^2 x} dx$$

$$\Rightarrow \quad I = 4\int_0^{\pi} \frac{x \sin x}{1 + \cos^2 x} dx$$

$$= 4\int_0^{\pi} \frac{(\pi - x) \sin (\pi - x)}{1 + \cos^2 (\pi - x)} dx$$

$$= 4\int_0^{\pi} \frac{(\pi - x) \sin x}{1 + \cos^2 x} dx$$

$$= 4\int_0^{\pi} \frac{\pi \sin x}{1+\cos^2 x} dx - 4\int_0^{\pi} \frac{x \sin x}{1+\cos^2 x} dx$$

$$= 4\int_0^{\pi} \frac{\pi \sin x}{1+\cos^2 x} dx - I$$

$$\Rightarrow \quad 2I = 4\int_0^{\pi} \frac{\pi \sin x}{1+\cos^2 x} dx$$

$$\Rightarrow \quad I = 2\int_0^{\pi} \frac{\pi \sin x}{1+\cos^2 x} dx$$

$$\Rightarrow \quad I = 4\pi \int_0^{\pi/2} \frac{\sin x}{1+\cos^2 x} dx$$

$$= -4\pi [\tan^{-1}(\cos x)]_0^{\pi/2}$$

$$= -4\pi \left(0 - \frac{\pi}{4}\right) = \pi^2$$

53. (b) We have,

$$I = \int_{-1/\sqrt{2}}^{1/\sqrt{2}} \left(\left(\frac{x+1}{x-1}\right)^2 + \left(\frac{x-1}{x+1}\right)^2 - 2\right)^{1/2} dx$$

$$= \int_{-1/\sqrt{2}}^{1/\sqrt{2}} \left(\left(\left(\frac{x+1}{x-1}\right) - \left(\frac{x-1}{x+1}\right)\right)^2\right)^{1/2} dx$$

$$= \int_{-1/\sqrt{2}}^{1/\sqrt{2}} \left|\frac{4x}{x^2-1}\right| dx$$

[using $\int_{-a}^{a} f(x)\,dx = 2\int_0^a f(x)\,dx$, if $f(x)$ is an even function]

$$= 2\int_0^{1/\sqrt{2}} \left|\frac{4x}{x^2-1}\right| dx$$

$$= -2\int_0^{1/\sqrt{2}} \frac{4x}{x^2-1} dx$$

$$= -4\int_0^{1/\sqrt{2}} \frac{2x}{x^2-1} dx$$

$$= -4[\ln|x^2-1|]_0^{1/\sqrt{2}}$$

$$\Rightarrow \quad I = 4\ln 2$$

$$\therefore \quad I = \ln 16$$

54. (c) Given integral $\int_{-\pi}^{\pi} |\pi - |x||\, dx$

$$= 2\int_0^{\pi} |\pi - |x||\, dx = 2\int_0^{\pi} |\pi - x|\, dx$$

$$= 2\int_0^{\pi} (\pi - x)\, dx \qquad [\because x \in (0, \pi) \Rightarrow x > 0]$$

$$= 2\left[\pi x - \frac{x^2}{2}\right]_0^{\pi}$$

$$= 2\left[\pi^2 - \frac{\pi^2}{2}\right]$$

$$= 2\left(\frac{\pi^2}{2}\right) = \pi^2$$

55. (c) Let $I = \int_{-\pi}^{\pi} \left(\frac{\cos^2 x}{1+a^x}\right) dx \qquad \text{...(i)}$

$$= \int_{-\pi}^{\pi} \left(\frac{\cos^2 x}{1+a^{-x}}\right) dx$$

$$= \int_{-\pi}^{\pi} \left(\frac{a^x \cos^2 x}{1+a^x}\right) dx \qquad \text{...(ii)}$$

Adding Eqs. (i) and (ii), we get

$$2I = \int_{-\pi}^{\pi} \left[\frac{(1+a^x)\cos^2 x}{(1+a^x)}\right] dx$$

$$= \int_{-\pi}^{\pi} (\cos^2 x)\, dx = 2\int_0^{\pi} (\cos^2 x)\, dx$$

$$= 2\int_0^{\pi} \left(\frac{1+\cos 2x}{2}\right) dx$$

$$\Rightarrow \quad I = \left[\frac{1}{2}\left(x + \frac{\sin 2x}{2}\right)\right]_0^{\pi} = \frac{\pi}{2}$$

56. (a) Since $f(x) = \cos 2x \cdot \log\left(\frac{1+x}{1-x}\right)$ is odd function.

$$\therefore \quad \int_{-1/2}^{1/2} \cos 2x \cdot \log\left(\frac{1+x}{1-x}\right) dx = 0$$

$$\left[\because \int_{-a}^{a} f(x)\,dx = \begin{cases} 0, \text{if } f(x) \text{ is odd function} \\ 2\int_0^a f(x)\,dx, \text{if } f(x) \text{ is even function} \end{cases}\right]$$

Hence, $\dfrac{\int_{-1/2}^{1/2} \cos 2x \cdot \log\left(\frac{1+x}{1-x}\right) dx}{\int_0^{1/2} \cos 2x \cdot \log\left(\frac{1+x}{1-x}\right) dx} = 0$

57. (a) Given integral, $I = \int_{-\pi/2}^{\pi/2} \frac{1}{1+e^{\sin x}} dx \qquad \text{...(i)}$

As we know that, $\int_a^b f(x)dx = \int_a^b f(a+b-x)dx$, so

$$I = \int_{-\pi/2}^{\pi/2} \frac{1}{1+e^{-\sin x}} dx = \int_{-\pi/2}^{\pi/2} \frac{e^{\sin x}}{e^{\sin x}+1} dx \qquad \text{...(ii)}$$

On adding Eqs. (i) and (ii), we get

$$2I = \int_{-\pi/2}^{\pi/2} \frac{1+e^{\sin x}}{1+e^{\sin x}} dx = \int_{-\pi/2}^{\pi/2} dx = \pi \Rightarrow I = \frac{\pi}{2}$$

58. (b) $I = \int_{-3\pi/2}^{-\pi/2} \{(x+\pi)^3 + \cos^2(x+3\pi)\}\, dx$

Putting $\pi + x = t$, we get

$$I = \int_{-\pi/2}^{\pi/2} \{t^3 + \cos^2(2\pi + t)\}\, dt$$

$$\Rightarrow \quad I = \int_{-\pi/2}^{\pi/2} t^3\, dt + \int_{-\pi/2}^{\pi/2} \cos^2 t\, dt$$

$$\left[\because \int_{-a}^{a} f(x)dx = \begin{cases} 2\int_0^a f(x)\,dx, & \text{if } f(x) \text{ is even} \\ 0, & \text{if } f(x) \text{ is odd} \end{cases}\right.$$

$$= 0 + 2\int_0^{\pi/2} \cos^2 t\,dt$$

$$= 2 \times \frac{\pi}{4} = \frac{\pi}{2}$$

59. (d) Let $I = \int_{-2}^{2} \frac{\sin^2 x}{\frac{1}{2} + \left[\frac{x}{\pi}\right]} dx$

Also, let $f(x) = \frac{\sin^2 x}{\frac{1}{2} + \left[\frac{x}{\pi}\right]}$

Then, $f(-x) = \frac{\sin^2(-x)}{\frac{1}{2} + \left[-\frac{x}{\pi}\right]}$ (replacing x by $-x$)

$$= \frac{\sin^2 x}{\frac{1}{2} + \left(-1 - \left[\frac{x}{\pi}\right]\right)}$$

$$\left[\because [-x] = \begin{cases} -[x], & \text{if } x \in I \\ -1 - [x], & \text{if } x \notin I \end{cases}\right]$$

$$\Rightarrow \quad f(-x) = -\frac{\sin^2 x}{\frac{1}{2} + \left[\frac{x}{\pi}\right]} = -f(x)$$

i.e. $f(x)$ is odd function.

$\therefore \quad I = 0$

$$\left[\because \int_{-a}^{a} f(x)\,dx = \begin{cases} 0, \text{if } f(x) \text{ is odd function} \\ 2\int_0^a f(x)dx, \text{if } f(x) \text{ is even function} \end{cases}\right]$$

60. (a) We have,

$$f(x) + f(x+1) = 2 \qquad \text{...(i)}$$

by $x \to x+1$

$$f(x+1) + f(x+2) = 2 \qquad \text{...(ii)}$$

From Eqs. (i) and (ii), we get

$$f(x) - f(x+2) = 0$$

$\therefore \quad f(x)$ is periodic with $T = 2$.

$$I_1 = \int_0^{2\times 4} f(x)\,dx = 4\int_0^2 f(x)\,dx \qquad \text{...(iii)}$$

$$I_2 = \int_{-1}^{3} f(x)\,dx = \int_0^4 f(x+1)\,dx$$

$$I_2 = \int_0^4 [2 - f(x)]\,dx$$

$$I_2 = 8 - 2\int_0^2 f(x)\,dx \qquad \text{...(iv)}$$

From Eqs. (iii) and (iv), we get

$$I_2 = 8 - \frac{I_1}{2}$$

$$\Rightarrow \quad I_1 + 2I_2 = 16$$

61. (d) We have, $\int_0^5 \frac{\tan^{-1}(x - [x])}{1 + (x - [x])^2} dx$

$$= 5\int_0^1 \frac{\tan^{-1}(x - [x])}{1 + (x - [x])^2} dx$$

$$= 5\int_0^1 \frac{\tan^{-1}(x - 0)}{1 + (x - 0)^2} dx$$

$$= 5\int_0^1 \frac{\tan^{-1} x}{1 + x^2} dx$$

Put $\tan^{-1} x = t$

$$\Rightarrow \quad \frac{1}{1+x^2}dx = dt$$

$$= 5\int_0^{\pi/4} t\,dt = 5\left[\frac{t^2}{2}\right]_0^{\pi/4}$$

$$= \frac{5}{2}\left[\frac{\pi^2}{16} - 0\right] = \frac{5\pi^2}{32}$$

62. (a) Let $I = \int_{-\pi/2}^{\pi/2} \frac{x^2 \cos x}{1 + e^x} dx$

We know that,

$$\int_{-a}^{a} f(x)\,dx = \int_0^a \{f(x) + f(-x)\}\,dx$$

$$\Rightarrow \quad I = \int_0^{\pi/2} \left\{\frac{x^2 \cos x}{1 + e^x} + \frac{(-x^2)\cos(-x)}{1 + e^{-x}}\right\} dx$$

$$\Rightarrow \quad I = \int_0^{\pi/2} x^2 \cos x\,dx$$

$$= [x^2 \sin x - 2(-x\cos x + \sin x)]_0^{\pi/2}$$

$$\therefore \quad I = \frac{\pi^2}{4} - 2$$

63. (a) We have, $I = \int_0^1 \frac{dx}{1 + x^{\pi/2}}$

$$x^2 < x^{\pi/2} < x, \forall\, x \in (0,1)$$

$$\Rightarrow \quad \frac{1}{1+x} < \frac{1}{1+x^{\pi/2}} < \frac{1}{1+x^2}$$

$$\Rightarrow \int_0^1 \frac{1}{1+x}dx < \int_0^1 \frac{dx}{1+x^{\pi/2}} < \int_0^1 \frac{dx}{1+x^2}$$

$$\Rightarrow \quad [\log_e x]_0^1 < I < [\tan^{-1} x]_0^1 \Rightarrow \log_e 2 < I < \frac{\pi}{4}$$

64. (b) Let $I = \int_{10}^{19} \frac{\sin x}{1 + x^a} dx < \int_{10}^{19} \frac{dx}{1 + x^a}$

$$\left[\because \sin x < 1 \Rightarrow \frac{\sin x}{1+x^a} < \frac{1}{1+x^a}\right]$$

$$\therefore \quad I < \int_{10}^{19} \frac{dx}{1+x^a} < \int_{10}^{19} \frac{dx}{1 + 10^a}$$

$$[\because 10 < x < 19 \Rightarrow 10^a + 1 < 1 + x^a < 19^a + 1]$$

$$\Rightarrow \quad I < \frac{9}{1+10^a}$$

$$\therefore \quad \frac{9}{1+10^a} < \frac{1}{9} \Rightarrow 1+10^a > 81$$

$$= 10^a > 80\ a = 2, 3, 4 \ldots$$

$\therefore$ Minimum odd value of 'a' is 3.

65. (c) Given integral

$$I = \int_1^2 \frac{dx}{\sqrt{2x^3 - 9x^2 + 12x + 4}}$$

Let a function $f(x) = \frac{1}{\sqrt{2x^3 - 9x^2 + 12x + 4}}$

Then, $f'(x) = \frac{-1(6x^2 - 18x + 12)}{2(2x^3 - 9x^2 + 12x + 4)^{3/2}}$

$$= -\frac{3(x^2 - 3x + 2)}{(2x^3 - 9x^2 + 12x + 4)^{3/2}}$$

$$= -3\frac{(x-1)(x-2)}{(2x^3 - 9x^2 + 12x + 4)^{3/2}}$$

$\because f'(1)$ and $f'(2)$ are zero and $f'(x)$ is positive $\forall\ x \in (1, 2)$, so $f(x)$ is an increasing function $\forall\ x \in (1, 2)$, so $f(1) < I < f(2)$

$$\Rightarrow \quad \frac{1}{\sqrt{2-9+12+4}} < I < \frac{1}{\sqrt{16-36+24+4}}$$

$$\Rightarrow \quad \frac{1}{3} < I < \frac{1}{2\sqrt{2}}$$

$$\Rightarrow \quad \frac{1}{9} < I^2 < \frac{1}{8}$$

66. (d) We have,

$$\int_0^x f(t)\, dt = \int_x^1 t^2 f(t)\, dt + \frac{x^{16}}{8} + \frac{x^6}{3} + a \quad \ldots(i)$$

For $x = 1$, $\int_0^1 f(t)\, dt = 0 + \frac{1}{8} + \frac{1}{3} + a = \frac{11}{24} + a \quad \ldots(ii)$

Differentiating both sides of Eq. (i) w.r.t. x', we get

$$f(x) = 0 - x^2 f(x) + 2x^{15} + 2x^5$$

$$\Rightarrow \quad f(x) = \frac{2(x^{15} + x^5)}{1+x^2}$$

$$\Rightarrow \quad 2\int_0^1 \frac{x^{15} + x^5}{1+x^2} dx = \frac{11}{24} + a \quad [\because \text{from Eq. (ii)}]$$

$$\Rightarrow \quad 2\int_0^1 (x^{13} - x^{11} + x^9 - x^7 + x^5)\, dx = \frac{11}{24} + a$$

$$\Rightarrow \quad 2\left(\frac{1}{14} - \frac{1}{12} + \frac{1}{10} - \frac{1}{8} + \frac{1}{6}\right) = \frac{11}{24} + a$$

$$\Rightarrow \quad a = -\frac{167}{840}$$

67. (a) We have, $\int_0^{t^2} x f(x)\, dx = \frac{2}{5} t^5$

Differentiating both sides w.r.t. 't', we get

$$t^2 \cdot f(t^2) \times 2t = \frac{2}{5} \times 5t^4$$

$$\Rightarrow \quad f(t^2) = t$$

Put $t = \frac{2}{5}$, then $f\left(\frac{4}{25}\right) = \frac{2}{5}$

68. (a) Let $f(x) = \int_0^x \frac{t^2}{1+t^4} dt = 2x - 1$, clearly $f(x)$ is continuous and $f'(x) = \frac{x^2}{1+x^4} - 2 = \frac{1}{\frac{1}{x^2} + x^2} - 2$

$$= \frac{1}{\left(x - \frac{1}{x}\right)^2 + 2} - 2 < 0$$

So, $f(x)$ is strictly decreasing function on $[0, 1]$ also $f(0) = 1 > 0$

and $f(1) = \int_0^1 \frac{t^2}{1+t^4} dt - 1 < 0 \quad \left[\because \int_0^1 \frac{t^2}{1+t^4} dt \le \int_0^1 \frac{1}{2} dt = \frac{1}{2}\right]$

So, $f(x) = 0$ has exactly one solution in $[0, 1]$.

69. (d) Given, $f(x) = \int_0^x g(t) dt$

On replacing x by $(-x)$, we get

$$f(-x) = \int_0^{-x} g(t) dt$$

Now, put $t = -u$, so

$$f(-x) = -\int_0^x g(-u) du = -\int_0^x g(u) du = -f(x)$$

$[\because g$ is an even function$]$

$\Rightarrow \quad f(-x) = -f(x) \Rightarrow f$ is an odd function.

Now, it is given that $f(x+5) = g(x)$

$\therefore \quad f(5-x) = g(-x) = g(x) = f(x+5)$

$[\because g$ is an even function$]$

$$\Rightarrow \quad f(5-x) = f(x+5) \quad \ldots(i)$$

Let $I = \int_0^x f(t) dt$

Put $t = u + 5 \Rightarrow t - 5 = u \Rightarrow dt = du$

$$\therefore \quad I = \int_{-5}^{x-5} f(u+5) du = \int_{-5}^{x-5} g(u) du$$

Put $\quad u = -t \Rightarrow du = -dt$, we get

$$I = -\int_5^{5-x} g(-t) dt = \int_{5-x}^{5} g(t) dt$$

$$\left[\because -\int_a^b f(x) dx = \int_b^a f(x) dx \text{ and } g \text{ is an even function}\right]$$

$$I = \int_{5-x}^{5} f'(t)dt \quad \text{[by Leibnitz rule } f'(x) = g(x)]$$

$$= f(5) - f(5-x) = f(5) - f(5+x) \quad \text{[from Eq. (i)]}$$

$$= \int_{5+x}^{5} f'(t)dt = \int_{5+x}^{5} g(t)dt$$

70. (b) We have, $\int_{-1}^{x} f(t)\,dt + f'''(3)\int_{x}^{0} dt$

$$= \int_{1}^{x} t^3\,dt - f'(1)\int_{0}^{x} t^2\,dt + f''(2)\int_{x}^{3} t\,dt$$

$$\Rightarrow \int_{-1}^{x} f(t)\,dt - xf'''(3) = \left(\frac{x^4}{4} - \frac{1}{4}\right) - f'(1)\frac{x^3}{3} + f''(2)\left(\frac{9}{2} - \frac{x^2}{2}\right)$$

Differentiating w.r.t. 'x', we get

$$f'(x) = 3x^2 - 2x\,f'(1) - f''(2)$$

$$\Rightarrow \quad f'(4) = 48 - 8f'(1) - f''(2)$$

71. (b) We have, $g(x) = \int_{2}^{x} \frac{t\,dt}{1+t^4}$

$$g'(x) = \frac{x}{1+x^4} \text{ or } g'(2) = \frac{2}{17}$$

Hence, $f(x) = e^{g(x)}$

$$f'(x) = e^{g(x)} \cdot g'(x) \Rightarrow f'(2) = e^{g(2)} \cdot g'(2)$$

$$\therefore \quad f'(2) = e^0 \times \frac{2}{17} \text{ [as } g(2) = 0]$$

$$\Rightarrow \quad f'(2) = \frac{2}{17}$$

72. (d) We have, $g(x) = \int_{x}^{a} \frac{f(t)}{t}\,dt$ ...(i)

Differentiating both sides w.r.t. 'x', we get

$$g'(x) = -\frac{f(x)}{x} \Rightarrow f(x) = -\,xg'(x)$$

Now, $\int_0^a f(x)\,dx = -\int_0^a xg'(x)\,dx$

$$= -\,xg(x)|_0^a + \int_0^a g(x)\,dx$$

$$= -\,a\,g(a) + \int_0^a g(x)\,dx$$

$$= \int_0^a g(x)\,dx \quad \text{[from Eq. (i), } g(a) = 0]$$

73. (a) Given, $\int_0^x f(t)\,dt = x^2 + \int_x^1 t^2 f(t)dt$

On differentiating both sides w.r.t. 'x', we get

$$f(x) = 2x + 0 - x^2 f(x)$$

$$\left[\because \frac{d}{dx}\left[\int_{\phi(x)}^{\psi(x)} f(t)dt\right] = f(\psi(x))\frac{d}{dx}\psi(x) - f(\phi(x))\frac{d}{dx}\phi(x)\right]$$

$$\Rightarrow \quad (1+x^2)f(x) = 2x$$

$$\Rightarrow \quad f(x) = \frac{2x}{1+x^2}$$

On differentiating w.r.t. 'x', we get

$$f'(x) = \frac{(1+x^2)(2) - (2x)(0+2x)}{(1+x^2)^2}$$

$$= \frac{2+2x^2-4x^2}{(1+x^2)^2} = \frac{2-2x^2}{(1+x^2)^2}$$

$$\therefore f'\left(\frac{1}{2}\right) = \frac{2-2\left(\frac{1}{2}\right)^2}{\left(1+\left(\frac{1}{2}\right)^2\right)^2} = \frac{2-2\left(\frac{1}{4}\right)}{\left(1+\frac{1}{4}\right)^2} = \frac{2-\frac{1}{2}}{\left(\frac{5}{4}\right)^2} = \frac{\frac{3}{2}}{\frac{25}{16}} = \frac{24}{25}$$

74. (3) $f(x) = \begin{vmatrix} \sin x + \sin 2x + \sin 3x & \sin 2x & \sin 3x \\ 3 + 4\sin x & 3 & 4\sin x \\ 1 + \sin x & \sin x & 1 \end{vmatrix}$

On applying $C_1 \to C_1 - (C_2 + C_3)$,

$$\begin{vmatrix} \sin x & \sin 2x & \sin 3x \\ 0 & 3 & 4\sin x \\ 0 & \sin x & 1 \end{vmatrix}$$

$$= \sin x(3 - 4\sin^2 x)$$

$$= 3\sin x - 4\sin^3 x$$

$$f(x) = \sin 3x$$

Now, $\int_0^{\pi/2} \sin 3x\,dx = \left[-\frac{\cos 3x}{3}\right]_0^{\pi/2}$

$$= 0 + \frac{1}{3} = \frac{1}{3}$$

So, $K = 3$

75. (2) Given, $I = \int_0^1 \frac{dx}{\sqrt{x} + \sqrt{2-x}}$

Substituting $x = 2\sin^2 t$

$$\Rightarrow \quad dx = 2\sin 2t\,dt$$

$$\Rightarrow \quad I = \int_0^{\pi/4} \frac{1}{\sqrt{2}\sin t + \sqrt{2}\cos t}\,dt$$

$$= \frac{1}{\sqrt{2}}\int_0^{\pi/4} \frac{1}{\sin t + \cos t}\,dt$$

Let $I_1 = \int \frac{1}{\sin t + \cos t}\,dt$

Put $u = \tan\frac{t}{2} \Rightarrow du = \frac{1}{2}\sec^2\frac{t}{2}\,dt$

$$\sin t = \frac{2u}{u^2+1}, \cos t = \frac{1-u^2}{1+u^2}, dt = \frac{2\,du}{u^2+1}$$

$$I_1 = 2\int \frac{1}{-u^2+2u+1}\,du$$

$$= 2\int \frac{1}{2-(u-1)^2}\,du$$

$$= 2\left(\frac{1}{2\sqrt{2}}\log\left|\frac{\sqrt{2}+u-1}{\sqrt{2}-u+1}\right|\right)$$

$$I_1 = \frac{1}{\sqrt{2}} \log \left(\frac{\sqrt{2} + \tan\frac{t}{2} - 1}{\sqrt{2} - \tan\frac{t}{2} + 1} \right)$$

Therefore, $I = \frac{1}{2}\left[\log \left| \frac{\sqrt{2} + \tan\frac{t}{2} - 1}{\sqrt{2} - \tan\frac{t}{2} + 1} \right| \right]_0^{\pi/4}$

$$= \sqrt{2} - \log(\sqrt{2} - 1)$$

Hence, $[I + \log(\sqrt{2} - 1)]^2 = (\sqrt{2})^2 = 2$

76. (5051) Let $I_n = \int_0^1 (1 - x^{50})^n \, dx$

Integrating by parts, we get

$$I_n = (1 - x^{50})^n \cdot x|_0^1 + \int_0^1 x \cdot 50n \cdot (1 - x^{50})^{n-1} \cdot x^{49} \cdot dx$$

$$= 0 + 50n \int_0^1 x^{50} \cdot (1 - x^{50})^{n-1} \cdot dx$$

$$= -50n \int_0^1 ((1 - x^{50}) - 1) \cdot (1 - x^{50})^{n-1} \cdot dx$$

$$= -50n \left[\int_0^1 (1 - x^{50})^n \cdot dx - \int_0^1 (1 - x^{50})^{n-1} \cdot dx \right]$$

$$= -50n (I_n - I_{n-1})$$

$$= -50n I_n + 50n I_{n-1}$$

$\Rightarrow \quad I_n = -50n I_n + 50n I_{n-1}$

$\Rightarrow (1 + 50n) I_n = 50n I_{n-1}$

$\Rightarrow \quad \frac{I_n}{I_{n-1}} = \frac{50n}{(1 + 50n)}$

Putting $n = 101$, then we get

$$\frac{I_{101}}{I_{100}} = \frac{50 \times 101}{(1 + 50 \times 101)} = \frac{5050}{5051}$$

So, $\quad 5050 \times \frac{I_{100}}{I_{101}} = 5051$

77. (7) We have,

$$\int_0^{1.5} x[x^2]\,dx = \int_0^1 x[x^2]\,dx + \int_1^{\sqrt{2}} x[x^2]\,dx + \int_{\sqrt{2}}^{1.5} x[x^2]\,dx$$

$$= \int_0^1 (0 \times x)\,dx + \int_1^{\sqrt{2}} 1(x)\,dx + \int_{\sqrt{2}}^{1.5} 2x\,dx$$

$$= \left[\frac{x^2}{2} \right]_1^{\sqrt{2}} + [x^2]_{\sqrt{2}}^{1.5}$$

$$= \left(1 - \frac{1}{2}\right) + \left(\frac{9}{4} - 2\right)$$

$$= \frac{1}{2} + \frac{1}{4} = \frac{3}{4}$$

Hence, $p + q = 3 + 4 = 7$

78. (1) We have,

$$L = \lim_{n \to \infty} \frac{2}{n}\left(f\left(\frac{1}{n}\right) + f\left(\frac{2}{n}\right) + \ldots + f(1) \right)$$

$$L = 2 \lim_{n \to \infty} \sum_{r-1}^{n} \ln\left(1 + \tan\frac{\pi x}{4}\right)$$

$$L = \frac{2}{\ln 2} \int_0^1 \ln\left(1 + \tan\frac{\pi x}{4}\right) dx \quad \text{...(i)}$$

$$L = \frac{2}{\ln 2} \int_0^1 \ln\left(1 + \tan\frac{\pi}{4}(1 - x)\right) dx$$

$$= \frac{2}{\ln 2} \int_0^1 \ln\left(1 + \tan\left(\frac{\pi}{4} - \frac{\pi}{4}x\right)\right) dx$$

$$= \frac{2}{\ln 2} \int_0^1 \ln\left(1 + \frac{1 - \tan\frac{\pi}{4}x}{1 + \tan\frac{\pi}{4}x}\right) dx$$

$$L = \frac{2}{\ln 2} \int_0^1 \ln\left(\frac{2}{1 + \tan\frac{\pi}{4}x}\right) dx$$

$$= \frac{2}{\ln 2} \int_0^1 \left[\ln 2 - \ln\left(1 + \tan\frac{\pi x}{4}\right)\right] dx \quad \text{...(ii)}$$

Adding Eqs. (i) and (ii), we get

$$2L = \frac{2}{\ln 2} \int_0^1 \ln 2\,dx$$

$\Rightarrow \quad 2L = \frac{2}{\ln 2} \cdot \ln 2 \Rightarrow L = 1$

79. (1) The integral, $I = \int_1^2 |2x - [3x]|\,dx$

$$= \int_1^{4/3} |2x - 3|\,dx + \int_{4/3}^{5/3} |2x - 4|\,dx + \int_{5/3}^2 |2x - 5|\,dx$$

$$= \int_1^{4/3} (3 - 2x)\,dx + \int_{4/3}^{5/3} (4 - 2x)\,dx + \int_{5/3}^2 (5 - 2x)\,dx$$

$$= [3x - x^2]_1^{4/3} + [4x - x^2]_{4/3}^{5/3} + [5x - x^2]_{5/3}^2$$

$$= \left(4 - \frac{16}{9} - 3 + 1\right) + \left(\frac{20}{3} - \frac{25}{9} - \frac{16}{3} + \frac{16}{9}\right) + \left(10 - 4 - \frac{25}{3} + \frac{25}{9}\right)$$

$$= (2 + 6) + \left(\frac{20}{3} - \frac{16}{3} - \frac{25}{3}\right) + \left(-\frac{16}{9} - \frac{25}{9} + \frac{16}{9} + \frac{25}{9}\right)$$

$$= 8 - \frac{21}{3} = 8 - 7 = 1$$

80. (2) Let $I = \int_0^1 \cot^{-1}(1 + x^2 - x)\,dx$

$\Rightarrow \quad I = \int_0^1 \cot^{-1}(1 - x(1 - x))\,dx$

$\Rightarrow \quad I = \int_0^1 \tan^{-1}\left(\frac{1}{1 - x(1 - x)}\right) dx$

$$= \int_0^1 \tan^{-1}\left(\frac{x+(1-x)}{1-x(1-x)}\right) dx$$

$$\Rightarrow \quad I = \int_0^1 (\tan^{-1} x + \tan^{-1}(1-x))\, dx$$

$$\Rightarrow \quad I = \int_0^1 \tan^{-1} x\, dx + \int_0^1 \tan^{-1}(1-x)\, dx$$

$$= \int_0^1 \tan^{-1} x\, dx + \int_0^1 \tan^{-1}(1-(1-x))\, dx$$

$$\left\{\because \int_0^a f(x)\, dx = \int_0^a f(a-x)\, dx\right\}$$

$$= \int_0^1 \tan^{-1} x\, dx + \int_0^1 \tan^{-1} x\, dx$$

$$= 2\int_0^1 \tan^{-1} x\, dx$$

$$= 2\left[x\tan^{-1} x - \frac{1}{2}\log|1+x^2|\right]_0^1$$

$$= 2\left(\frac{\pi}{4} - \frac{1}{2}\log 2\right)$$

$$\Rightarrow \quad I = 2\left(\frac{\pi}{4} - \log\sqrt{2}\right)$$

Comparing with $K\left(\frac{\pi}{4} - \log\sqrt{2}\right)$, we get

$$K = 2$$

81. (4) We have, $f(x) = \begin{cases} 1-x & : \quad 0 \le x < 1 \\ x-1 & : \quad 1 \le x < 2 \end{cases}$

and f is periodic with period 2.
Also, since $[-x] = 1 - [x]$ for non-integral values of x,
So, f is an even function.

Thus, $\int_{-10}^{10} f(x)\cos(\pi x)\, dx$

$$= 2\int_0^{10} f(x)\cos(\pi x)\, dx$$

$$= 2\cdot 5\int_0^2 f(x)\cos(\pi x)\, dx$$

$$= 10\int_0^2 f(x)\cos(\pi x)\, dx$$

$$= 10(I+J)$$

where $\quad, I = \int_0^1 f(x)\cos(\pi x)\, dx$

and $\quad J = \int_1^2 f(x)\cos(\pi x)\, dx$

Now, $\quad I = \int_0^1 f(x)\cos(\pi x)\, dx$

$$= \int_0^1 (1-x)\cos(\pi x)\, dx$$

$$= \int_0^1 [1-(1-x)\cos\pi(1-x)]\, dx$$

$$= -\int_0^1 x\cos(\pi x)\, dx$$

Also, $J = \int_1^2 (x-1)\cos(\pi x)\, dx$

$$= \int_0^1 t\cos(\pi(t+1))\, dt \text{ (let } x-1=t)$$

$$= -\int_0^1 t\cos(\pi t)\, dt$$

$$= -\int_0^1 x\cos(\pi x)\, dx = I$$

Thus, $\quad I = J$

$$= -\int_0^1 x\cos(\pi x)\, dx$$

$$= \left[-\frac{x}{\pi}\sin(\pi x)\right]_0^1 + \frac{1}{\pi}\int_0^1 \sin(\pi x)\, dx$$

$$= -\frac{1}{\pi^2}(\cos \pi x)_0^1 = \frac{2}{\pi^2}$$

Therefore, $\int_{-10}^{10} f(x)\cos(\pi x)\, dx$

$$= 10(I+J)$$

$$= 20I = 20 \times \frac{2}{\pi^2}$$

Now, $\frac{\pi^2}{10} \times \int_{-10}^{10} f(x)\cos(\pi x)\, dx$

$$= \frac{\pi^2}{10} \times \frac{40}{\pi^2} = 4$$

82. (1.5) Given integral $\int_0^2 ||x-1|-x|\, dx$

$$= \int_0^1 |-(x-1)-x|\, dx + \int_1^2 |x-1-x|\, dx$$

$$= \int_0^1 |1-2x|\, dx + \int_1^2 dx$$

$$= \int_0^{1/2} (1-2x)\, dx + \int_{1/2}^1 (2x-1)\, dx + 1$$

$$= [x-x^2]_0^{1/2} + [x^2-x]_{1/2}^1 + 1$$

$$= \frac{1}{2} - \frac{1}{4} + 1 - 1 - \frac{1}{4} + \frac{1}{2} + 1$$

$$= \frac{1}{2} + 1 = \frac{3}{2} = 1.5$$

83. (11) Let $\quad I = \int_{-2}^{2} \frac{3x^5 + 4x^3 + 2x^2 + x + 20}{x^2+4}\, dx$

$$= \int_{-2}^{2} \frac{3x^5 + 4x^3 + x}{x^2+4}\, dx + \int_{-2}^{2} \frac{2x^2+20}{x^2+4}\, dx$$

$\left[\because \int_{-a}^{a} f(x)\, dx = 0, \text{ if } f(x) \text{ is an odd function } 2\int_0^a f(x)\, dx, \text{ if } f(x) \text{ is an even function}\right]$

$$= 0 + 2\int_0^2 \frac{2x^2+20}{x^2+4} = dx$$

$$= 4\int_0^2 \frac{x^2+10}{x^2+4}\, dx$$

$$= 4\left[\int_0^2 \frac{x^2+4}{x^2+4}\, dx + \int_0^2 \frac{6}{x^2+4}\, dx\right]$$

$$= 4\left[[x]_0^2 + 6\cdot\frac{1}{2}\left[\tan^{-1}\frac{x}{2}\right]_0^2\right]$$

$$= 4\left[2+\frac{6}{2}(\tan^{-1} 1)\right]$$

$$= 8 + 12\left(\frac{\pi}{4}\right)$$

$$= 8 + 3\pi$$

So, $A + B = 8 + 3 = 11$

84. (2) We have, $f(x) = \int_1^x \sqrt{2-t^2}\, dt$

By Leibnitz rule, $f'(x) = \sqrt{2-x^2}$

Now, $x^2 - f'(x) = 0$

$$\Rightarrow \quad x^4 = 2 - x^2$$

$$\Rightarrow \quad x^4 + x^2 - 2 = 0$$

$$\Rightarrow \quad (x^2+2)(x^2-1) = 0$$

$\therefore$ $x = \pm 1$ are the real value of x.

Therefore, the number of real roots of the equation $x^2 - f'(x) = 0$ is 2.

85. (2) $f^2(x) = \int_0^x f(t) \dfrac{\cos t}{2+\sin t}\, dt$

Differentiating w.r.t. x using Leibnitz rule,

$$2f(x)\cdot f'(x) = f(x)\frac{\cos x}{2+\sin x}$$

$$2f'(x) = \frac{\cos x}{2+\sin x}$$

$$\Rightarrow \quad 2\int f'(x)\, dx = \int \frac{\cos x}{2+\sin x} dx$$

$$2f(x) = \log(2+\sin x) + \log c$$

Put $x = 0$, then

$$2f(0) = \log 2 + \log c \qquad [\because f(0) = 0]$$

$$\Rightarrow \quad \log c = -\log 2$$

$\therefore$ $2f(x) = \log(2+\sin x) - \log 2$

$$2f(x) = \log\left(\frac{2+\sin x}{2}\right)$$

$$\Rightarrow \quad f(x) = \frac{1}{2}\log\left(\frac{2+\sin x}{2}\right)$$

Hence, $K = 2$

86. (21) Since, $\int_0^n \{x\}dx = n\int_0^1 x dx$ $\qquad$ [as $n \in N, n > 1$]

$$= n\left[\frac{x^2}{2}\right]_0^1 = \frac{n}{2}$$

and $\int_0^n [x]dx = \int_0^1 0\, dx + \int_1^2 1 dx + \int_2^3 2 dx + \ldots + \int_{n-1}^n (n-1)dx$

$$= 0 + 1(2-1) + 2(3-2) + \ldots + (n-1)(n-(n-1))$$

$$= 1 + 2 + 3 + \ldots + (n-1) = \frac{n(n-1)}{2}$$

$$\because \quad \frac{n}{2}, \frac{n(n-1)}{2}, 10n(n-1) \text{ are in GP}$$

$$\Rightarrow \frac{1}{2}, \frac{n-1}{2}, 10(n-1) \text{ are also in GP} \qquad [\text{as } n > 1]$$

$$\Rightarrow \left(\frac{n-1}{2}\right)^2 = 5(n-1) \Rightarrow n - 1 = 20 \qquad [\text{as } n > 1]$$

$$\Rightarrow \quad n = 21.$$

87. (0) Given,

$$x\int_0^x (1-t)\sin f(t)\, dt = 2\int_0^x t\sin(f(t))\, dt$$

Differentiating w.r.t. 'x', we get

$$\int_0^x (1-t)\sin[f(t)]\, dt + x(1-x)\sin[f(x)] = 2x\sin(f(x))$$

$$\Rightarrow \quad \int_0^x (1-t)\sin(f(t))\, dt = x^2\sin(f(x)) + x\sin f(x)$$

Again, differentiating, w.r.t. 'x', we get

$$(1-x)\sin(f(x)) = 2x\sin(f(x)) + x^2\cos(f(x))f'(x) + \sin(f(x)) + x\cos(f(x))f'(x)$$

$$\Rightarrow \quad -3x\sin(f(x)) = (x+x^2)\cos(f(x))\cdot f'(x)$$

$$\Rightarrow \quad \frac{-3x}{x(1+x)} = \cot(f(x))\cdot f'(x)$$

$$\Rightarrow f'(x)\cot(f(x)) + \frac{3}{1+x} = 0$$

Level 3

1. (a) Let $n \le x < n+1$, where $n \in I$

$$I = \int_0^x \frac{2^t}{2^{[t]}} dt$$

$$= \int_0^n 2^{\{t\}} dt + \int_n^x 2^{\{t\}} dt$$

$$= n\int_0^1 2^{\{t\}} dt + \int_n^x 2^{\{t\}} dt$$

$$= n\int_0^1 2^t dt + \int_n^x 2^{t-n} dt$$

$$= n\left[\frac{2^t}{\ln 2}\right]_0^1 + \frac{1}{2^n}\left[\frac{2^t}{\ln 2}\right]_n^x$$

$$= \frac{n}{\ln 2}(2-1) + \frac{1}{2^n \ln 2}(2^x - 2^n)$$

$$= \frac{n}{\ln 2} + \frac{1}{\ln 2}(2^{x-n} - 1)$$

$$= \frac{[x] + 2^{\{x\}} - 1}{\ln 2}$$

2. (b) Let $I = \int_0^\pi \dfrac{x\sin^2 x}{1+2|\cos x|\sin x} dx$ $\qquad$ ...(i)

$\therefore$ $\quad I = \int_0^\pi \dfrac{(\pi - x)\sin^2 x}{1+2|\cos x|\sin x} dx$ $\qquad$ ...(ii)

On adding Eqs. (i) and (ii) we get,

$$2I = \pi \int_0^{\pi} \frac{\sin^2 x}{1 + 2|\cos x| \sin x} dx$$

$$\Rightarrow \quad I = \frac{\pi}{2} 2 \int_0^{\frac{\pi}{2}} \frac{\sin^2 x}{1 + 2\cos x \sin x} dx$$

$$= \pi \int_0^{\frac{\pi}{2}} \frac{\sin^2 x}{1 + \sin 2x} dx \quad \text{...(iii)}$$

$$\therefore \quad I = \pi \int_0^{\frac{\pi}{2}} \frac{\cos^2 x}{1 + \sin 2x} dx \quad \text{...(iv)}$$

On adding Eqs. (iii) and (iv), we get

$$2I = \pi \int_0^{\frac{\pi}{2}} \frac{dx}{(\cos x + \sin x)^2}$$

$$= \pi \int_0^{\frac{\pi}{2}} \frac{\sec^2 x}{(1 + \tan x)^2} dx$$

Putting $1 + \tan x = t$

$\sec^2 x \, dx = dt$

$$= \pi \int_1^{\infty} \frac{dt}{t^2}$$

$$= -\pi \left(\frac{1}{t}\right)_1^{\infty}$$

$$= -\pi(-1)$$

$$2I = \pi \Rightarrow I = \frac{\pi}{2}$$

3. (c) It is given that $f(x)$ is an even function. Therefore, $f(-x) = f(x)$ for all $x \in R$

$\Rightarrow \quad ax^2 - bx + c = ax^2 + bx + c$ for all $x \in R$

$$\Rightarrow \quad b = 0 \Rightarrow \frac{-b}{a} = 0$$

$$\Rightarrow \quad \alpha + \beta = 0 \Rightarrow \beta = -\alpha$$

Also, $\dfrac{f(-x)}{-x-\alpha} = \dfrac{f(x)}{-(x+\alpha)} = -\dfrac{f(x)}{(x-\beta)}$

and $\dfrac{f(-x)}{-x-\beta} = \dfrac{f(x)}{-x+\alpha} = -\dfrac{f(x)}{x-\alpha}$

$$\therefore f\left(\frac{f(-x)}{-x-\alpha}\right) = f\left(-\frac{f(x)}{x-\beta}\right) = f\left(\frac{f(x)}{x-\beta}\right)$$

and $f\left(\dfrac{f(-x)}{-x-\beta}\right) = f\left(\dfrac{f(x)}{x-\alpha}\right) = f\left(\dfrac{f(x)}{x-\alpha}\right)$

Let $I = \displaystyle\int_\alpha^\beta \frac{e^{f\left(\frac{f(x)}{x-\alpha}\right)}}{e^{f\left(\frac{f(x)}{x-\alpha}\right)} + e^{f\left(\frac{f(x)}{x-\beta}\right)}} dx$ ...(i)

$$I = \int_\alpha^\beta \frac{e^{f\left(\frac{f(-x)}{-x-\alpha}\right)}}{e^{f\left(\frac{f(-x)}{-x-\alpha}\right)} + e^{f\left(\frac{f(-x)}{-x-\beta}\right)}} dx$$

{Using $\int_a^b f(x)\,dx = \int_a^b f(a+b-x)\,dx$ and $\alpha + \beta = 0$}

$$I = \int_\alpha^\beta \frac{e^{f\left(\frac{f(x)}{x-\beta}\right)}}{e^{f\left(\frac{f(x)}{x-\beta}\right)} + e^{f\left(\frac{f(x)}{x-\alpha}\right)}} dx \quad \text{...(ii)}$$

From Eqs. (i) and (ii)

$$2I = \int_\alpha^\beta dx$$

$$2I = \beta - \alpha$$

$$I = \frac{\beta - \alpha}{2}$$

$$I = \frac{1}{2}\sqrt{(\alpha+\beta)^2 - 4\alpha\beta}$$

$$I = \frac{\sqrt{b^2 - 4ac}}{2|a|}$$

4. (d) We have,

$$f(x) = x + \int_0^{\frac{\pi}{2}} \sin x \cos y \, f(y)\, dy$$

$$f(x) = x + \sin x \int_0^{\frac{\pi}{2}} \cos y \, f(y)\, dy$$

Let $K = \int_0^{\frac{\pi}{2}} \cos y \, f(y)\, dy$

$\Rightarrow \quad f(x) = x + K \sin x$

and $\quad f(y) = y + K \sin y$

Now, $K = \int_0^{\frac{\pi}{2}} \cos y\, (y + K \sin y)\, dy$

$$K = \int_0^{\frac{\pi}{2}} y \cos y\, dy + K \int_0^{\frac{\pi}{2}} \cos y \sin y\, dy$$

[Put $\sin y = t$, $\cos y dy = dt$]

$$K = (y \sin y)_0^{\pi/2} - \int_0^{\frac{\pi}{2}} \sin y\, dy + K \int_0^1 t\, dt$$

$$K = \frac{\pi}{2} - 1 + K\left(\frac{1}{2}\right)$$

$$\Rightarrow \quad K = \pi - 2$$

So, $f(x) = x + (\pi - 2)\sin x$

5. (a) Let $I = \int_{\pi/4}^{\pi/2} (2\,\text{cosec}\, x)^{17} dx$ and

let $e^u + e^{-u} = 2\,\text{cosec}\, x$, then

$$x = \frac{\pi}{4}$$

$$\Rightarrow \quad e^u + e^{-u} = 2\sqrt{2}$$

$$\Rightarrow (e^u)^2 - 2\sqrt{2}\, e^u + 1 = 0$$

$$\Rightarrow \quad e^u = \sqrt{2} + 1$$

$$\Rightarrow \quad u = \log(1 + \sqrt{2})$$

and $x = \dfrac{\pi}{2} \Rightarrow e^u + e^{-u} = 2$

$$\Rightarrow \quad (e^u - 1)^2 = 0$$

$$\Rightarrow \quad e^u = 1 \Rightarrow u = 0$$

Now, $e^u + e^{-u} = 2\operatorname{cosec} x$

$\Rightarrow \quad (e^u)^2 - (2\operatorname{cosec} x)\, e^u + 1 = 0$

$$\Rightarrow \quad e^u = \operatorname{cosec} x + \cot x$$

$$e^{-u} = \frac{1}{\operatorname{cosec} x + \cot x}$$

$$= \operatorname{cosec} x - \cot x$$

$$\therefore \quad \cot x = \frac{e^u - e^{-u}}{2}$$

Again $e^u + e^{-u} = 2\operatorname{cosec} x$

$$\Rightarrow \quad (e^u - e^{-u})\, du = -2\operatorname{cosec} x \cot x\, dx$$

$$\Rightarrow \quad dx = -\frac{e^u - e^{-u}}{2\operatorname{cosec} x \cot x}\, du$$

$$\Rightarrow \quad dx = -\frac{e^u - e^{-u}}{(e^u + e^{-u})\left(\frac{e^u - e^{-u}}{2}\right)} \cdot du$$

$$\Rightarrow \quad dx = -\frac{2}{e^u + e^{-u}}\, du$$

$$\therefore \quad I = \int_{\pi/4}^{\pi/2} (2\operatorname{cosec} x)^{17}\, dx$$

$$I = \int_{\log(1+\sqrt{2})}^{0} (e^u + e^{-u})^{17} \frac{-2}{e^u + e^{-u}}\, du$$

$$I = \int_0^{\log(1+\sqrt{2})} 2(e^u + e^{-u})^{16}\, du$$

6. (b) Let $I = \pi^2 \int_0^2 \left(\sin \frac{\pi x}{2}\right)(x - [x])^{[x]} dx$

$$= \pi^2 \int_0^1 \sin\left(\frac{\pi x}{2}\right) x^0\, dx + \pi^2 \int_1^2 \sin\left(\frac{\pi x}{2}\right) \cdot (x-1)\, dx$$

$$= \pi^2 \left[\frac{-2}{\pi} \cos \frac{\pi x}{2}\right]_0^1 + \pi^2 \left[(x-1)\frac{2}{\pi}\left(-\cos \frac{\pi x}{2}\right)\right]_1^2 + \pi^2 \int_1^2 \frac{2}{\pi} \cos \frac{\pi x}{2}\, dx$$

$$= \pi^2 \left(\frac{2}{\pi}\right) + \frac{2\pi^2}{\pi}(1-0) + 2\pi \cdot \frac{2}{\pi}\left[\sin \frac{\pi x}{2}\right]_1^2$$

$$= 2\pi + 2\pi + 4(0-1)$$

$$= 4\pi - 4 = 4(\pi - 1)$$

7. (b) We have, $f'(x) = f(x) + \int_0^1 f(x)\, dx$

On differentiating both sides, we get

$$f''(x) = f'(x)$$

$$\Rightarrow \quad \int \frac{d(f'(x))}{f'(x)} = \int dx$$

$$\Rightarrow \quad \ln(f'(x)) = x + C$$

$$\Rightarrow \quad f'(x) = Ae^x \quad \text{...(i)}$$

$$\Rightarrow \quad \int f'(x)\, dx = \int Ae^x\, dx$$

$$\Rightarrow \quad f(x) = \int f'(x)\, dx = Ae^x + B \quad \text{...(ii)}$$

Now, $f(0) = 1 \Rightarrow A + B = 1 \Rightarrow B = 1 - A$

$$\therefore \quad f'(x) = f(x) + \int_0^1 (Ae^x + 1 - A)\, dx$$

$$\Rightarrow \quad Ae^x = (Ae^x + 1 - A) + [Ae^x + (1-A)x]_0^1$$

$$\Rightarrow \quad Ae^x = (Ae^x + 1 - A) + (Ae + (1-A) - A)$$

$$\Rightarrow 1 - A + (Ae + 1 - A - A) = 0$$

$$\Rightarrow \quad A(e-3) = -2$$

$$\Rightarrow \quad A = \frac{2}{3-e} \text{ and } B = 1 - \frac{2}{3-e} = \frac{1-e}{3-e}$$

or

$$f(\log 2) = \frac{4}{3-e} + \frac{1-e}{3-e} = \frac{5-e}{3-e}$$

8. (c) Let $g(x) = \int_0^{x^3} f(t)dt$

Now, $g(2) = \int_0^8 f(t)\, dt, g(0) = 0$

$$\int_0^8 f(t)\, dt = g(2) - g(1) + g(1) - g(0)$$

$$= \left[\frac{g(2) - g(1)}{2-1}\right] + \left[\frac{g(1) - g(0)}{1-0}\right]$$

$$= g'(\alpha) + g'(\beta) \quad \text{...(i)}$$

Now, $g(x) = \int_0^{x^3} f(t)\, dt$

By Leibnitz rule

$$g'(x) = 3x^2 f(x^3)$$

So,

$$g'(\alpha) = 3\alpha^2 f(\alpha^3)$$

$$g'(\beta) = 3\beta^2 f(\beta^3)$$

Put in Eq. (i)

$$\int_0^8 f(t)\, dt = 3\alpha^2 f(\alpha^3) + 3\beta^2 f(\beta^3)$$

$$= 3(\alpha^2 f(\alpha^3) + \beta^2 f(\beta^3))$$

9. (b) We have,

$$f(x) = x^2 + 3\int_0^{x^{\frac{1}{3}}} e^{-t^3} t^2 f(x - t^3)dt \quad \text{...(i)}$$

Put $\quad x - t^3 = z$

$$\therefore \quad f(x) = x^2 + \int_0^x e^{z-x} f(z)\, dz$$

$$= x^2 + e^{-x} \int_0^x e^z f(z)\, dz$$

On differentiating w.r.t. 'x', we get

$$f'(x) = 2x + e^{-x} e^x f(x) - e^{-x} \int_0^x e^z f(z)\, dz$$

$$\therefore \quad f'(x) = 2x + x^2$$

$$\therefore \quad f(x) = \frac{x^3}{3} + x^2 + C$$

From Eq. (i), $f(0) = 0$

So, $C = 0$

$$\therefore f(x) = \frac{x^3}{3} + x^2$$

10. (d) We have, $x\int_1^x y(t)\,dt=(x+1)\int_1^x ty(t)\,dt$

Differentiating w.r.t. 'x', we get

$$\int_1^x y(t)\,dt+xy(x)=\int_1^x ty\,(t)\,dt+(x+1)\,xy(x)$$

$$\Rightarrow \quad \int_1^x y(t)\,dt=\int_1^x ty\,(t)\,dt+x^2\,y\,(x)$$

On differentiating w.r.t. 'x', we get

$$y\,(x)=xy\,(x)+x^2\frac{dy}{dx}(x)+2xy\,(x)$$

$$\Rightarrow \quad (1-3x)\,y\,(x)=x^2\frac{d}{dx}(y\,(x))$$

$$\Rightarrow \quad \left(\frac{1}{x^2}-\frac{3}{x}\right)dx=\frac{d}{dx}(y\,(x))$$

Integrating, we get

$$-\frac{1}{x}-3\log x=\log y\,(x)+\log K$$

$$\Rightarrow \quad -\frac{1}{x}=\log(Kx^3y\,(x))$$

$$\Rightarrow \quad Kx^3\,y(x)=e^{-\frac{1}{x}}$$

$$\Rightarrow \quad y(x)=\frac{e^{-\frac{1}{x}}}{Kx^3}=\frac{C}{x^3}e^{-\frac{1}{x}}$$

11. (c) Let $L=\lim\limits_{n\to\infty}\left[\left(1+\frac{1}{n^2}\right)^{\frac{2}{n^2}}\left(1+\frac{2^2}{n^2}\right)^{\frac{4}{n^2}}\dots\left(1+\frac{n^2}{n^2}\right)^{\frac{2n}{n^2}}\right]$

$$=\lim_{n\to\infty}\prod_{r=1}^{n}\left(1+\frac{r^2}{n^2}\right)^{2r/n^2}$$

Taking log both sides, we get

$$\log L=\lim_{n\to\infty}\sum_{r=1}^{n}\frac{2r}{n^2}\log\left(1+\frac{r^2}{n^2}\right)$$

$$=\lim_{n\to\infty}\frac{1}{n}\sum_{r=1}^{n}\frac{2r}{n}\log\left(1+\frac{r^2}{n^2}\right)$$

$$=2\int_0^1 x\log(1+x^2)dx$$

Put $1+x^2=t\Rightarrow 2x\,dx=dt$

When $x=0, t=1$ and $x=1, t=2$

$$\therefore\ \log L=\int_1^2\log t\,dt=[t\log t-t]_1^2$$

$$\log L=(2\log 2-2)-(0-1)=\log 4-\log e=\log 4e^{-1}$$

$$L=4e^{-1}$$

12. (b) We have,

$$L=\lim_{n\to\infty}\prod_{r=1}^{n}(1+(rp)^2)^{1/r}$$

$$=\lim_{n\to\infty}\prod_{r=1}^{n}\left(1+\left(\frac{r}{n}\right)^2\right)^{1/r}\qquad\left[\because P=\frac{1}{n}\right]$$

$$\log L=\lim_{n\to\infty}\sum_{r=1}^{n}\frac{1}{r}\log\left(1+\left(\frac{r}{n}\right)^2\right)$$

$$\log L=\lim_{n\to 0}\frac{1}{n}\sum_{r=1}^{n}\frac{1}{r/n}\log\left(1+\left(\frac{r}{n}\right)^2\right)$$

$$\log L=\int_0^1\frac{\log(1+x^2)}{x}dx$$

Put $x^2=t\Rightarrow 2x\,dx=dt$

When $x\to 0\Rightarrow t\to 0$ and $x\to 1\Rightarrow t\to 1$

$$\therefore\ \log L=\frac{1}{2}\int_0^1\frac{\log(1+t)}{t}dt$$

$$=\frac{1}{2}\int_0^1\left(\frac{t-\frac{t^2}{2}+\frac{t^3}{3}-\frac{t^4}{4}\dots}{t}\right)dt$$

$$=\frac{1}{2}\int_0^1\left(1-\frac{t}{2}+\frac{t^2}{3}-\frac{t^3}{4}+\dots\right)dt$$

$$=\frac{1}{2}\left[t-\frac{t^2}{2^2}+\frac{t^3}{3^2}-\frac{t^4}{4^2}\dots\right]_0^1$$

$$=\frac{1}{2}\left[1-\frac{1}{2^2}+\frac{1}{3^2}-\frac{1}{4^2}+\dots\infty\right]$$

$$\log L=\frac{1}{2}\frac{\pi^2}{12}$$

$$\log L=\frac{\pi^2}{24}$$

13. (b, d) $v=\int_0^\infty\frac{x^2\,dx}{x^4+7x^2+1}$

Put $x=\frac{1}{t}\quad\Rightarrow dx=-\frac{1}{t^2}dt$

$$v=-\int_\infty^0\frac{\frac{1}{t^2}\cdot\frac{1}{t^2}dt}{\frac{1}{t^4}+\frac{7}{t^2}+1}=-\int_\infty^0\frac{dt}{t^4+7t^2+1}$$

$$v=\int_0^\infty\frac{dt}{t^4+7t^2+1}=\int_0^\infty\frac{dx}{x^4+7x^2+1}=u$$

$$\Rightarrow\qquad v=u$$

Hence, $\quad 2u=\int_0^\infty\frac{x^2+1}{x^4+7x^2+1}dx$

$$= \int_0^\infty \frac{1+\frac{1}{x^2}}{x^2+\frac{1}{x^2}+7}\,dx$$

$$= \int_0^\infty \frac{\left(1+\frac{1}{x^2}\right)}{\left(x-\frac{1}{x}\right)^2+3^2}dx$$

Put $x-\frac{1}{x}=t$

$\Rightarrow \left(1+\frac{1}{x^2}\right)dx = dt$

$$\int_{-\infty}^{\infty}\frac{dt}{t^2+(3)^2} = \frac{1}{3}\left[\tan^{-1}\left(\frac{t}{3}\right)\right]_{-\infty}^{\infty} = \frac{1}{3}\left[\frac{\pi}{2}-\left(-\frac{\pi}{2}\right)\right]$$

$$2u = \frac{2\pi}{6}$$

$\Rightarrow \qquad 2u = \frac{\pi}{3} \Rightarrow u = \frac{\pi}{6}$

So, $u = v = \frac{\pi}{6}$ and $6v = \pi$

$u+v = \frac{\pi}{3}$ and $2u+3v = \frac{5\pi}{6}$

14. (a,b) Let

$$I = \int_0^\alpha \frac{dx}{1-\cos\alpha\cos x}$$

$$= \int_0^\alpha \frac{dx}{\left(\cos^2\frac{x}{2}+\sin^2\frac{x}{2}\right)-\cos\alpha\left(\cos^2\frac{x}{2}-\sin^2\frac{x}{2}\right)}$$

$$= \int_0^\alpha \frac{dx}{(1-\cos\alpha)\cos^2\frac{x}{2}+(1+\cos\alpha)\sin^2\frac{x}{2}}$$

$$= \frac{1}{2}\int_0^\alpha \frac{\sec^2\frac{\alpha}{2}\sec^2\frac{x}{2}\,dx}{\tan^2\frac{\alpha}{2}+\tan^2\frac{x}{2}}$$

Put $\tan\frac{x}{2} = t \Rightarrow \frac{1}{2}\sec^2\frac{x}{2}\,dx = dt$

$$I = \int_0^{\tan\frac{\alpha}{2}} \frac{\sec^2\frac{\alpha}{2}}{t^2+\tan^2\frac{\alpha}{2}}dt$$

$$= \sec^2\frac{\alpha}{2}\cdot\cot\frac{\alpha}{2}\left[\tan^{-1}\left(\frac{t}{\tan\frac{\alpha}{2}}\right)\right]_0^{\tan\frac{\alpha}{2}}$$

$$= \frac{2}{\sin\alpha}\cdot\frac{\pi}{4} = \frac{\pi}{2\sin\alpha}$$

Hence, $A = \frac{\pi}{2}, B = 0$ and $A = \frac{\pi}{4}, B = \frac{\pi}{4\sin\alpha}$

15. (a, d) $\qquad I = \int_0^1 \frac{2x^2+3x+3}{(x+1)(x^2+2x+2)}dx$

$$= \int_0^1 \frac{2(x^2+2x+2)-(x+1)}{(x+1)(x^2+2x+2)}dx$$

$$= \int_0^1 \left(\frac{2}{x+1}-\frac{1}{x^2+2x+2}\right)dx$$

$$= [2\log(x+1)-\tan^{-1}(x+1)]_0^1$$

$$= 2\log 2-\tan^{-1}2+\tan^{-1}1$$

$$= 2\log 2-\tan^{-1}2+\frac{\pi}{4}$$

$$= \log 4-\left(\frac{\pi}{2}-\cot^{-1}2\right)+\frac{\pi}{4}$$

$$= -\frac{\pi}{4}+\log 4+\cot^{-1}2$$

16. (a, b, c) Let $I_n = \int_{-\pi}^{\pi}\left(\frac{\sin nx}{(1+\pi^x)\sin x}\right)dx$

$$= \int_{-\pi}^{\pi}\left(\frac{\sin nx}{(1+\pi^x)\sin x}\right)dx \qquad ...(i)$$

$$= \int_{-\pi}^{\pi}\left(\frac{\sin n(-x)}{(1+\pi^{-x})\sin(-x)}\right)dx$$

$$= \int_{-\pi}^{\pi}\left(\frac{\sin n(x)}{(1+\pi^{-x})\sin(x)}\right)dx$$

$$= \int_{-\pi}^{\pi}\left(\frac{\pi^x\sin n(x)}{(1+\pi^x)\sin(x)}\right)dx \qquad ...(ii)$$

Adding Eqs. (i) and (ii), we get

$$2I_n = \int_{-\pi}^{\pi}\left(\frac{(1+\pi^x)\sin n(x)}{(1+\pi^x)\sin(x)}\right)dx$$

$$= \int_{-\pi}^{\pi}\left(\frac{\sin n(x)}{\sin(x)}\right)dx$$

$$= 2\int_0^{\pi}\left(\frac{\sin n(x)}{\sin(x)}\right)dx$$

$\Rightarrow \qquad I_n = \int_0^{\pi}\left(\frac{\sin n(x)}{\sin(x)}\right)dx$

$\therefore \quad I_{n+2}-I_n = \int_0^{\pi}\left(\frac{\sin(n+2)x-\sin nx}{\sin x}\right)dx$

$$= \int_0^{\pi}\left(\frac{2\cos(n+1)x\times\sin x}{\sin x}\right)dx$$

$$= \int_0^{\pi} 2\cos(n+1)x\,dx$$

$$= \left(\frac{2\sin(n+1)x}{n+1}\right)_0^{\pi} = 0$$

$\Rightarrow \qquad I_{n+2} = I_n$

Now, $I_0 = 0$ and $I_1 = \int_0^{\pi}\left(\frac{\sin x}{\sin x}\right)dx = \pi$

Since $I_{n+2} = I_n$, So $I_1 = I_2 = I_5 = \ldots = I_{(2n-1)} = \pi$

and $I_0 = I_2 = I_4 = \ldots = I_{2n} = 0$

Thus, $\sum_{m=1}^{10} I_{2m+1} = I_3 + I_5 + \ldots + I_{21} = 10\pi$

and $\sum_{m=1}^{10} I_{2m} = I_2 + I_4 + \ldots I_{20} = 0$

17. (c, d) $g(x) = \int_{\sin x}^{\sin 2x} \sin^{-1}(t)\, dt$

$$g'(x) = 2(\cos 2x)\sin^{-1}(\sin 2x) - \cos x \sin^{-1}(\sin x)$$

So, $g'\left(\frac{\pi}{2}\right) = 2(-1)(0) - (0)\cdot(1) = 0$

and $g'\left(-\frac{\pi}{2}\right) = 0$

18. (b, d) For option (a)

Let $g(x) = e^x - \int_0^x f(t)\sin t\, dt$

$\therefore g'(x) = e^x - (f(x)\cdot \sin x) > 0, \forall x \in (0,1)$

$[\because f(x) \in (0,1) \text{ and } e^x > 0 \text{ for } x \in (0,1)]$

So, $g(x)$ is strictly increasing function.

$\Rightarrow g(x) > 1, \forall x \in (0,1)$

Thus, option (a) is not possible.

For option (b)

Let $h(x) = x^9 - f(x)$

Now, $h(0) = -f(0) < 0$ $\{\because f \in (0,1)\}$

Also, $h(1) = 1 - f(1) > 0$ $\{\because f \in (0,1)\}$

$\Rightarrow h(0)\cdot h(1) < 0$

So, option (b) is correct.

For option (c)

Let $G(x) = f(x) + \int_0^{\frac{\pi}{2}} f(t)\sin t\, dt$

$\Rightarrow G(x) > 0, \forall x \in (0,1)$ $[\because f \in (0,1)]$

So, option (c) is not possible.

For option (d)

Let $H(x) = x - \int_0^{\frac{\pi}{2}-x} f(t)\cos t\, dt$

$\therefore H(0) = 0 - \int_0^{\frac{\pi}{2}} f(t)\cos t\, dt < 0$

and $H(1) = 1 - \int_0^{\frac{\pi}{2}-1} f(t)\cos t\, dt > 0$

$\Rightarrow H(0)H(1) < 0$

So, option (d) is correct (using intermediate value theorem)

19. (a, c) For continuity at $x = a$

$$\lim_{x \to a^-} g(x) = 0$$

$$g(a) = \int_a^a f(t)\, dt = 0$$

$$\lim_{x \to a^+} g(x) = \lim_{x \to a^+} \int_a^x f(t)\, dt = 0 = \int_\alpha^a f(t)\, dt = 0$$

Hence, $g(x)$ is continuous at $x = a$.

For continuity at $x = b$.

$$\lim_{x \to b^-} g(x) = \lim_{x \to b^-} \int_a^x f(t)\, dt = \int_a^b f(t)\, dt$$

$$= \lim_{x \to b^+} g(x) = g(b)$$

Thus, $g(x)$ is continuous at $x = b$.

$$g'(x) = \begin{cases} 0, & x < a \\ f(x), & a < x < b \\ 0, & x > b \end{cases}$$

Since, $f(x) \geq 1$ for $x \in [a,b]$, $g(x)$ is non-differentiable at $x = a$ and $x = b$.

20. (a, c) $\int_0^\alpha f(x)\, dx$

Put $x = \alpha t$

$\therefore \int_0^\alpha f(x)\, dx = \alpha \int_0^1 f(\alpha t)\, dt \geq \alpha \int_0^1 f(t)\, dt$ as $\alpha \in (0,1)$

Also, $\alpha \int_0^1 f(t)\, dt \geq \alpha^2 \int_0^1 f(t)\, dt$

$\therefore \int_0^\alpha f(x)\, dx \geq \alpha^2 \int_0^1 f(x)\, dx$

21. (a, b) Let $I = \int_0^x \left\{\int_0^u f(t)\, dt\right\} du$

Integrating by parts, choose 1 as the second function

$$= \left\{u \int_0^u f(t)\, dt\right\}_0^x - \int_0^x f(u)\, u\, du$$

$$= x\int_0^x f(t)\, dt - \int_0^x f(u)\, u\, du$$

$$= x\int_0^x f(u)\, du - \int_0^x f(u)\, u\, du$$

$$I = \int_0^x f(u)(x-u)\, du$$

$$\left\{\text{using } \int_0^a f(x)\, dx = \int_0^a f(a-x)\, dx\right\}$$

$$I = \int_0^x u f(x-u)\, du$$

Sol. (Q.Nos. 22-24)

22. (a) $g\left(\frac{1}{2}\right) = \lim_{h \to 0^+} \int_h^{1-h} t^{-1/2}(1-t)^{-1/2}\, dt$

$$= \int_0^1 \frac{dt}{\sqrt{t - t^2}}$$

$$= \int_0^1 \frac{dt}{\sqrt{\frac{1}{4} - \left(t - \frac{1}{2}\right)^2}}$$

$$= \left[\sin^{-1}\left(\frac{t - \frac{1}{2}}{\frac{1}{2}}\right)\right]_0^1$$

$$= \sin^{-1} 1 - \sin^{-1}(-1) = \pi$$

23. (a) $g(a) = \lim_{h \to 0^+} \int_h^{1-h} t^{-a}(1-t)^{a-1} dt$

$$\Rightarrow g(1-a) = \lim_{h \to 0^+} \int_h^{1-h} t^{-(1-a)}(1-t)^{(1-a)-1} dt$$

$$= \lim_{h \to 0^+} \int_h^{1-h} t^{a-1}(1-t)^{-a} dt$$

$$t = \lim_{h \to 0^+} \int_h^{1-h} (1-t)^{a-1}(1-(1-t))^{-a} dt$$

$$= \lim_{h \to 0^+} \int_h^{1-h} (1-t)^{a-1} t^{-a} dt$$

$$g(1-a) = g(a)$$

$$\Rightarrow \quad g(a) - g(1-a) = 0$$

24. (d) $g'(a) = -g'(1-a)$

$$\Rightarrow \quad g'\left(\frac{1}{2}\right) = -g'\left(1-\frac{1}{2}\right) \Rightarrow g'\left(\frac{1}{2}\right) = 0$$

Sol. (Q.Nos. 25-27)

25. (b) Since f and g are continuous for $x \geq 0$ it follows that $\int_0^a f(x)\,dx, \int_0^a g(x)\,dx$ and $\int_0^a f(x)\,g(x)\,dx$ exists.

Now, $\int_0^a f(x)\,g(x)\,dx = \int_0^a f(a-x)\,g(a-x)\,dx$

$$= \int_0^a f(x)\,[2 - g(x)]\,dx$$

$$\{\because f(a-x) = f(x) \text{ and } g(x) + g(a-x) = 2\}$$

Therefore, $2\int_0^a f(x) \cdot g(x)\,dx = 2\int_0^a f(x)\,dx$

$$\Rightarrow \quad \int_0^a f(x)\,g(x)\,dx = \int_0^a f(x)\,dx$$

26. (d) Let $f(x) = e^{|\cos x|}\left(2\sin\left(\frac{1}{2}\cos x\right)\right)$

Since f is continuous for all real x, it follows that $\int_0^x f(x)\,dx$ exists. Also,

$$f(\pi - x) = e^{|\cos(\pi - x)|}\left(2\sin\left(\frac{1}{2}\right)\cos(\pi - x)\right)$$

$$= e^{|\cos x|}\left(2\sin\left(-\frac{1}{2}\cos x\right)\right)$$

$$= -e^{|\cos x|}\left(2\sin\left(\frac{1}{2}\cos x\right)\right)$$

$$f(\pi - x) = -f(x)$$

Therefore, $\int_0^\pi f(x)\,dx = \int_0^\pi f(\pi - x)\,dx = -\int_0^\pi f(x)\,dx$

So, $\quad \int_0^\pi f(x)\,dx = 0$

27. (a) Let $\quad I = \int_{\pi/4}^{3\pi/4} \frac{x}{1+\sin x}\,dx \quad$ ...(i)

$$I = \int_{\pi/4}^{3\pi/4} \frac{\pi - x}{1+\sin(\pi - x)}\,dx \quad \text{...(ii)}$$

On adding Eqs. (i) and (ii),

$$2I = \pi \int_{\pi/4}^{3\pi/4} \frac{dx}{1+\sin x}$$

$$= \pi \int_{\pi/4}^{3\pi/4} \frac{1-\sin x}{\cos^2 x}\,dx$$

$$= \pi \int_{\pi/4}^{3\pi/4} (\sec^2 x - \sec x \tan x)\,dx$$

$$= \frac{\pi}{2}\{[\tan x]_{\pi/4}^{3\pi/4} - [\sec x]_{\pi/4}^{3\pi/4}\}$$

$$= \frac{\pi}{2}[(-1-1) - (-\sqrt{2} - \sqrt{2})]$$

$$= \pi(\sqrt{2} - 1)$$

28. (i) $\to$ (A), (ii) $\to$ (D), (iii) $\to$ (B), (iv) $\to$ (C)

(i) $f(x+1) = f(x+3)$ or $f(x) = f(x+2)$

Thus, $f(x)$ is periodic with period 2.

Then, $\int_a^{a+b} f(x)\,dx$ is independent of a, for which b is multiple of 2. Thus, $b = 2, 4, 6, \ldots$

(ii) Let $I = 25\int_0^{\pi/4} (\tan^6(x - [x]) + \tan^4(x - [x])\,dx$

$$\{\because 0 < \pi \leq \frac{\pi}{4} \Rightarrow [x] = 0\}$$

$$= 25\int_0^{\pi/4} (\tan^6 x + \tan^4 x)\,dx$$

$$= 25\int_0^{\pi/4} \tan^4 x(1 + \tan^2 x)\,dx$$

$$= 25\int_0^{\pi/4} \tan^4 x \sec^2 x\,dx$$

$$= 25\left(\frac{\tan^5 x}{5}\right)_0^{\pi/4}$$

$$= 25 \times \frac{1}{5} = 5$$

(iii) Let $I = \int_1^4 \frac{\tan^{-1}[x^2]}{\tan^{-1}[x^2] + \tan^{-1}[25 + x^2 - 10x]}\,dx \quad$ ...(i)

On applying $\int_a^b f(x)\,dx = \int_a^b f(a+b-x)\,dx$, we get

$$I = \int_1^4 \frac{\tan^{-1}[(5-x)^2]}{\tan^{-1}[(5-x)^2] + \tan^{-1}[x^2]}\,dx \quad \text{...(ii)}$$

Adding Eqs. (i) and (ii), we get

$$2I = \int_1^4 1\,dx$$

$$2I = 3$$

$$I = \frac{3}{2}$$

$$[I] = \left[\frac{3}{2}\right] = 1$$

(iv) Let $y = \sqrt{x + \sqrt{x + \sqrt{x + \ldots}}}$

$$\Rightarrow y = \sqrt{x + y} \Rightarrow y^2 - y - x = 0$$

$$\Rightarrow \quad y = \frac{1 \pm \sqrt{1+4x}}{2}$$

$$y = \frac{1 + \sqrt{1+4x}}{2} \quad [\because y > 1]$$

$\therefore \quad I=\int_0^2 \frac{1+\sqrt{1+4x}}{2}\,dx$

$$=\left[\frac{x}{2}+\frac{(1+4x)^{3/2}}{\frac{3}{2}\cdot 2\times 4}\right]_0^2$$

$$=\left(1+\frac{27}{12}\right)-\left(0+\frac{1}{12}\right)$$

$$=1+\frac{26}{12}=\frac{19}{6}$$

$\therefore \quad I=3$

29. (i) → (C), (ii) → (A), (iii) → (D), (iv) → (B)

(i) We have, $I=\int_{1/e}^{e^2}\left|\frac{\log x}{x}\right|dx$

$$=\int_{1/e}^{1}\frac{-\log x}{x}\,dx+\int_1^{e^2}\frac{\log x}{x}\,dx$$

$$=-\frac{1}{2}[(\log x)^2]_{1/e}^{1}+\frac{1}{2}[(\log x)^2]_1^{e^2}$$

$$=-\frac{1}{2}\left[0-\left(\log\frac{1}{e}\right)^2\right]+\frac{1}{2}[(\log e^2)^2-0]$$

$$=\frac{1}{2}+\frac{1}{2}(2^2)=\frac{1}{2}+2=\frac{5}{2}$$

(ii) We have, $\int_0^9 \{\sqrt{x}\}\,dx$

$$=\int_0^9 \{\sqrt{x}-[\sqrt{x}]\}\,dx \; \{\because [t] \text{ is the greatest integer } f^n\}$$

$$=\int_0^9 \sqrt{x}\,dx-\left(\int_0^1 [\sqrt{x}]\,dx+\int_1^4 [\sqrt{x}]dx+\int_4^9 [\sqrt{x}]dx\right)$$

$$=\frac{2}{3}[x^{3/2}]_0^9-\left(0+\int_1^4 1\,dx+\int_4^9 2dx\right)$$

$$=\frac{2}{3}(9)^{3/2}-([x]_1^4+[2x]_4^9)$$

$$=\frac{2}{3}(3)^3-[3+10]=18-13=5$$

(iii) We have, $I=\int_0^{\pi/2}\frac{f(x)}{f(x)+f\left(\frac{\pi}{2}-x\right)}dx$ ...(i)

$$I=\int_0^{\pi/2}\frac{f\left(\frac{\pi}{2}-x\right)}{f\left(\frac{\pi}{2}-x\right)+f(x)}dx \quad \text{...(ii)}$$

On adding Eqs. (i) and (ii), we get

$$2I=\int_0^{\pi/2} 1\,dx$$

$$2I=\frac{\pi}{2}\Rightarrow I=\frac{\pi}{4}$$

Hence $k=4$

(iv) We have, $\lim_{n\to\infty}\left(\frac{1}{1+n^2}+\frac{2}{2^2+n^2}+\ldots\frac{n}{n^2+n^2}\right)$

$$=\lim_{n\to\infty}\sum_{k=1}^{n}\frac{k}{n^2+k^2}$$

$$=\lim_{n\to\infty}\sum_{k=1}^{n}\frac{1}{n^2}\times\frac{k}{1+\left(\frac{k}{n}\right)^2}$$

$$=\lim_{n\to\infty}\sum_{k=1}^{n}\frac{1}{n}\cdot\frac{\left(\frac{k}{n}\right)}{1+\left(\frac{k}{n}\right)^2}$$

$$=\int_0^1\frac{x}{1+x^2}\,dx$$

$$=\frac{1}{2}[\log(1+x^2)]_0^1$$

$$=\frac{1}{2}\log 2$$

Hence, $k=2$

30. (i) → (B) (ii) → (A), (iii) → (D), (iv) → (C)

(i) Let $f(x)=ax^2+bx$

Given that, $\int_0^1 f(x)\,dx=1$

$\Rightarrow \quad 2a+3b=6$

$\Rightarrow (a,b)\equiv(0,2)$ and $(3,0)$

Hence required number of polynomial is 2.

(ii) Let $I=\int_{-2}^{2}\frac{3x^2}{(1+e^x)}dx$ ...(i)

$$=\int_{-2}^{2}\frac{3(-x)^2}{1+e^{-x}}dx$$

$$\therefore \quad I=\int_{-2}^{2}\frac{e^x(3x^2)}{e^x+1} \quad \text{...(ii)}$$

On adding Eqs. (i) and (ii),

$$2I=\int_{-2}^{2}\frac{3x^2}{1+e^x}dx+\int_{-2}^{2}\frac{e^x(3x^2)}{e^x+1}dx$$

$$=\int_{-2}^{2}3x^2\,dx$$

$$=2\int_0^2 3x^2\,dx$$

$$\left\{\because \int_{-a}^{a} f(x)\,dx=2\int_0^a f(x)\,dx \text{ if } f(x) \text{ is an even } f^n\right\}$$

$\Rightarrow \quad 2I=16$

$I=8$

(iii) We have, $I=\dfrac{\int_{-1/2}^{1/2}\cos 2x\log\left(\frac{1+x}{1-x}\right)dx}{\int_0^{1/2}\cos 2x\log\left(\frac{1+x}{1-x}\right)dx}$

Let $\quad f(x)=\cos 2x\log\left(\frac{1+x}{1-x}\right)$

$\therefore \quad f(-x) = \cos 2(-x) \ln\left(\frac{1-x}{1+x}\right)$

$$= -\cos 2x \left(\ln\left(\frac{1+x}{1-x}\right)\right)$$

$f(-x) = -f(x)$

Thus, $f(x)$ is an odd function.

$\Rightarrow \quad I = 0$

(iv) $\text{sgn}\ (x - [x]) = \begin{cases} 1, & \text{if } x \text{ is not an integer} \\ 0, & x \text{ is an integer} \end{cases}$

Hence, $\int_{-1}^{3} \text{sgn}\,(x - [x]) = 4(1-0) = 4$

31. (i) $\rightarrow$ (D), (ii) $\rightarrow$ (C), (iii) $\rightarrow$ (A), (iv) $\rightarrow$ (B)

(i) We have, $T_r = r\text{th term} = \dfrac{n}{(n+r)\sqrt{r(2n+r)}}$

Put $\quad h = \dfrac{1}{n}$

So, $\quad T_r = \dfrac{\sqrt{h}}{(1+rh)\sqrt{r(2+rh)}}$

$$= \frac{h}{(1+rh)\sqrt{rh(2+rh)}}$$

Therefore, required sum

$$= \int_0^1 \frac{dx}{(1+x)\sqrt{x(2+x)}}$$

$$= \int_0^1 \frac{dx}{(1+x)\sqrt{(1+x)^2 - 1}}$$

$$= [\sec^{-1}(1+x)]_0^1$$

$$= \sec^{-1} 2 - \sec^{-1} 1$$

$$= \frac{\pi}{3} - 0 = \frac{\pi}{3}$$

(ii) We have,

$r\text{th term} = \dfrac{1}{\sqrt{2rn - r^2}}$

$$= \frac{\sqrt{h}}{\sqrt{2r - r^2 h}}, \text{ where } h = \frac{1}{n}$$

$$= \frac{h}{\sqrt{2rh - r^2h^2}}$$

Therefore required sum is

$$\int_0^1 \frac{dx}{\sqrt{2x - x^2}}$$

$$= \int_0^1 \frac{dx}{\sqrt{1-(1-x)^2}}$$

$$= -[\sin^{-1}(1-x)]_0^1$$

$$= -\sin^{-1} 0 + \sin^{-1} 1$$

$$= \frac{\pi}{2}$$

(iii) We have, $I = \int_{\log 1/2}^{\log 2} \sin\left[\dfrac{e^x - 1}{e^x + 1}\right] dx$

$$= \int_{-\log 2}^{\log 2} \sin\left[\frac{e^x - 1}{e^x + 1}\right] dx$$

Let $\quad f(x) = \sin\left[\dfrac{e^x - 1}{e^x + 1}\right]$

$$f(-x) = \sin\left(\frac{1 - e^x}{1 + e^x}\right)$$

$$= -\sin\left(\frac{e^x - 1}{e^x + 1}\right)$$

$f(-x) = -f(x)$

Hence, $f(x)$ is an odd function.

So, $\quad I = 0$

(iv) Given that, $\int_0^\infty \dfrac{\sin x}{x}\, dx = \dfrac{\pi}{2}$

$$\therefore \int_0^\infty \frac{\sin^3 x}{x}\, dx = \int_0^\infty \left(\frac{\frac{3}{4}\sin x - \frac{1}{4}\sin 3x}{x}\right) dx$$

$$= \frac{3}{4}\int_0^\infty \frac{\sin x}{x}\, dx - \frac{1}{4}\int_0^\infty \frac{\sin 3x}{x}\, dx$$

$$= \frac{3}{4}\int_0^\infty \frac{\sin x}{x}\, dx - \frac{1}{4}\int_0^\infty \frac{\sin u}{u}\, du \ (\text{let } u = 3x)$$

$$= \frac{3}{4}\cdot\frac{\pi}{2} - \frac{1}{4}\cdot\frac{\pi}{2} = \frac{\pi}{4}$$

Sol. (Q.Nos. 32-34)

32. (2) $\quad S_1 = \int_{\frac{\pi}{8}}^{\frac{3\pi}{8}} \sin^2 x\, dx$

$$= \int_{\frac{\pi}{8}}^{\frac{3\pi}{8}} \left(\frac{1 - \cos 2x}{2}\right) dx$$

$$= \left(\frac{\pi}{2} - \frac{\sin 2x}{4}\right)_{\frac{\pi}{8}}^{\frac{3\pi}{8}} = \frac{\pi}{8}$$

So, $\quad \dfrac{16 S_1}{\pi} = \dfrac{16}{\pi} \times \dfrac{\pi}{8} = 2$

33. (1.5) $\quad S_2 = \int_{\frac{\pi}{8}}^{\frac{3\pi}{8}} \sin^2 x\,|4x - \pi|\, dx \quad$ ...(i)

$$S_2 = \int_{\frac{\pi}{8}}^{\frac{3\pi}{8}} \sin^2\left(\frac{3\pi}{8} + \frac{\pi}{8} - x\right)\left|4\left(\frac{3\pi}{8} + \frac{\pi}{8} - \pi\right)\right| dx$$

$S_2 = \int_{\frac{\pi}{8}}^{\frac{3\pi}{8}} \cos^2 x\,|\pi - 4x|\, dx \quad$ ...(ii)

From Eqs. (i) and (ii), we get

$$2S_2 = \int_{\frac{\pi}{8}}^{\frac{3\pi}{8}} |4x - \pi|\, dx$$

$$= -\int_{\frac{\pi}{8}}^{\frac{\pi}{4}} (4x-\pi)\,dx + \int_{\frac{\pi}{4}}^{\frac{3\pi}{8}} (4x-\pi)\,dx$$

$$= -[2x^2 - \pi x]_{\pi/8}^{\pi/4} + [2x^2 - \pi x]_{\pi/4}^{3\pi/8}$$

$$= -\left(\frac{\pi^2}{8} - \frac{\pi^2}{4} - \frac{\pi^2}{32} + \frac{\pi^2}{8}\right) + \left[\frac{9\pi^2}{32} - \frac{3\pi^2}{8} - \frac{\pi^2}{8} + \frac{\pi^2}{4}\right]$$

$$= \frac{\pi^2}{32} + \frac{\pi^2}{32}$$

$$2S_2 = \frac{2\pi^2}{32}$$

$$S_2 = \frac{\pi^2}{32}$$

So, $\frac{48S_2}{\pi} = \frac{48}{\pi^2} \times \frac{\pi^2}{32} = \frac{3}{2} = 1.5$

34. (4) Let $I = \int_0^{\pi/2} g_2(x)\,dx$

$$I = \int_0^{\pi/2} |4x - \pi|\,dx$$

$$I = \int_0^{\pi/4} (\pi - 4x)dx + \int_{\pi/4}^{\pi/2} (4x - \pi)\,dx$$

$$I = [\pi x - 2x^2]_0^{\pi/4} + [2x^2 - \pi x]_{\pi/4}^{\pi/2}$$

$$I = \left(\frac{\pi^2}{4} - \frac{\pi^2}{8}\right) + \left[\left(\frac{\pi^2}{2} - \frac{\pi^2}{2}\right) - \left(\frac{\pi^2}{8} - \frac{\pi^2}{4}\right)\right]$$

$$I = \frac{\pi^2}{4} - \frac{\pi^2}{8} - \frac{\pi^2}{8} + \frac{\pi^2}{4}$$

$$I = \frac{\pi^2}{2} - \frac{\pi^2}{4}$$

$$= \frac{\pi^2}{4}$$

$\Rightarrow \quad k = 4$

Sol. (Q.Nos. 35-37)

We have, $g(x) = x^C \cdot e^{Cx}$

$$g'(x) = C \cdot x^{C-1} \cdot e^{Cx} + x^C \cdot e^{Cx} \cdot C$$

$$= \frac{C}{x} x^C e^{Cx} + x^C \cdot e^{Cx} \cdot C$$

$$= C \cdot x^C e^{Cx} \left(\frac{1}{x} + 1\right)$$

Now, $f(x) = \int_0^x t \cdot e^{2t} (1 + 3t^2)^{1/2}\,dt$

Differentiate using Leibnitz's rule

$$f'(x) = x \cdot e^{2x} (1 + 3x^2)^{1/2}$$

$$L = \lim_{x \to \infty} \frac{f'(x)}{g'(x)}$$

$$= \lim_{x \to \infty} \frac{x \cdot e^{2x} (1 + 3x^2)^{1/2}}{C \cdot x^c e^{cx} \left[\frac{1}{x} + 1\right]}$$

$$= \lim_{x \to \infty} \frac{x^2 \cdot e^{2x} \left(\frac{1}{x^2} + 3\right)^{1/2}}{C \cdot x^C \cdot e^{Cx} \left(\frac{1}{x} + 1\right)}$$

$$= \lim_{x \to \infty} \frac{x^{(2-C)} \cdot e^{(2-C)x} \left(\frac{1}{x^2} + 3\right)^{1/2}}{C\left(\frac{1}{x} + 1\right)}$$

35. (2) For finite limit

$$2 - c = 0$$

$\therefore \quad c = 2$

36. (0.86) $L = \lim_{x \to \infty} \frac{x^0 e^0 \left(\frac{1}{x^2} + 3\right)^{1/2}}{2\left(\frac{1}{x} + 1\right)}$

$$L = \frac{\sqrt{3}}{2} = 0.86$$

37. (2) $g(x) = x \Rightarrow x^2 e^{2x} - x = 0$

$$x(xe^{2x} - 1) = 0$$

$$x = 0 \text{ or } e^{2x} = \frac{1}{x}$$

e^{2x} and $\frac{1}{x}$ intersect at one point.

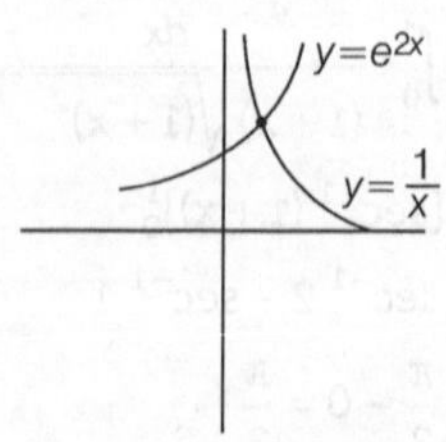

$\therefore$ Number of real roots are 2.

38. (2) $I = \int_0^{1/2} \frac{(1+\sqrt{3})}{((x+1)^2 (1-x)^6)^{1/4}}\,dx$

$$= \int_0^{1/2} \frac{(1+\sqrt{3})\,dx}{(x+1)^2 \left(\frac{(1-x)^6}{(1+x)^6}\right)^{1/4}}$$

Put $\left(\frac{(1-x)}{1+x}\right) = t$

$\Rightarrow \quad \frac{-2dx}{(1+x)^2} = dt$

$\therefore \quad I = \int_1^{1/3} \frac{(1+\sqrt{3})\,dt}{-2t^{3/2}}$

$$= \frac{-(1+\sqrt{3})}{2} \left[\frac{-2}{\sqrt{t}}\right]_1^{1/3}$$

$$= (1+\sqrt{3})(\sqrt{3} - 1) = 2$$

39. (192) $I=\int_0^{\pi/4}(\pi x-4x^2)\ln(1+\tan x)\,dx$ ...(i)

$I=\int_0^{\pi/4}(\pi x-4x^2)[\ln 2-\ln(1+\tan x)]\,dx$ ...(ii)

$\left\{\text{Using } \int_0^a f(x)\,dx=\int_0^a f(a-x)dx\right\}$

On adding Eqs. (i) and (ii),

$$2I=\ln 2\int_0^{\pi/4}(\pi x-4x^2)\,dx$$

$$I=\frac{\ln 2}{2}\left[\frac{\pi x^2}{2}-\frac{4}{3}x^3\right]_0^{\pi/4}$$

$$=\frac{\ln 2}{2}\left[\frac{\pi^3}{32}-\frac{\pi^3}{48}\right]$$

$$I=\frac{\pi^3}{192}\ln 2$$

Hence, $k=192$

40. (9) We have $\alpha=\int_0^1 e^{9x+3\tan^{-1}x}\left(\frac{12+9x^2}{1+x^2}\right)dx$

Put $9x+3\tan^{-1}x=t$

$$\alpha=\int_0^{9+\frac{3\pi}{4}}e^t\,dt$$

$$\Rightarrow \quad \alpha=e^{9+\frac{3\pi}{4}}-1\Rightarrow 1+\alpha=e^{9+\frac{3\pi}{4}}$$

So, $\log|1+\alpha|-\frac{3\pi}{4}=\log\left(e^{9+\frac{3\pi}{4}}\right)-\frac{3\pi}{4}=9$

41. (2) We have, $f(x)=\frac{e^x}{1+e^x}$

$$f(-x)=\frac{e^{-x}}{1+e^{-x}}=\frac{1}{1+e^x}$$

$$f(x)+f(-x)=\frac{e^x}{1+e^x}+\frac{1}{1+e^x}$$

$$f(a)+f(-a)=1$$

Now, $I_1=\int_{f(-a)}^{f(a)}xg\{x(1-x)\}\,dx$ ...(i)

Using $\int_a^b f(a+b-x)\,dx=\int_a^b f(x)\,dx$ and $f(a)+f(-a)=1$, we get

$$I_1=\int_{f(-a)}^{f(a)}(1-x)\,g\{(1-x)(1-(1-x)\}\,dx$$

$\Rightarrow \quad I_1=\int_{f(-a)}^{f(a)}(1-x)\,g\,\{x(1-x)\}\,dx$ ...(ii)

On adding Eqs. (i) and (ii), we get

$$2I_1=\int_{f(-a)}^{f(a)}g\{x(1-x)\}\,dx$$

$$2I_1=\int_{f(-a)}^{f(a)}g\{x(1-x)\}\,dx$$

$$\Rightarrow \quad 2I_1=I_2\Rightarrow\frac{I_2}{I_1}=2$$

42. (2) Given, $f(x)=x^3-\frac{3x^2}{2}+x+\frac{1}{4}$

$$=\frac{1}{4}(4x^3-6x^2+4x+1)$$

$$=\frac{1}{4}(4x^3-6x^2+4x-1+2)$$

$$f(x)=\frac{1}{4}[x^4-(1-x)^4]+\frac{2}{4}$$

$$\therefore \quad f(1-x)=\frac{1}{4}[(1-x)^4-x^4]+\frac{2}{4}$$

$\therefore f(x)+f(1-x)=\frac{2}{4}+\frac{2}{4}=1$ ...(i)

Replacing x by $f(x)$, we get

$$f(f(x))+f(1-f(x))=1$$

$$\therefore \quad f(f(x))=1-f(1-f(x))$$

Again from Eq. (i),

$$1-f(x)=f(1-x)=1-f(1-x)$$

Again $\quad 1-f(1-x)=f(x)$

$\therefore \quad I=\int_{1/4}^{3/4}f(f(x))\,dx=\int_{1/4}^{3/4}f(x)\,dx$ ...(ii)

$I=\int_{1/4}^{3/4}f(1-x)\,dx$ ...(iii)

On adding Eqs. (ii) and (iii),

$$2I=\int_{1/4}^{3/4}(f(x)+f(1-x))\,dx$$

$$=\int_{1/4}^{3/4}1\,dx$$

$$2I=\left[\frac{3}{4}-\frac{1}{4}\right]$$

$$\therefore \quad I=\frac{1}{4}$$

So, $\quad P=\frac{1}{4}$

$$8P=8\times\frac{1}{4}=2$$

43. (2) $\int_0^1 4x^3\left\{\frac{d^2}{dx^2}(1-x^2)^5\right\}dx$

Integrate using by parts, we get

$$\left\{4x^3\frac{d}{dx}(1-x^2)^5\right\}_0^1-\int_0^1 12x^2\frac{d}{dx}(1-x^2)^5\,dx$$

$$=[4x^3\times 5(1-x^2)^4(-2x)]_0^1$$

$$-12\left[[x^2(1-x^2)^5]_0^1-\int_0^1 2x(1-x^2)^5dx\right]$$

$$=0-0-12(0-0)+12\int_0^1 2x(1-x^2)^5dx$$

$$=12\left[-\frac{(1-x^2)^6}{6}\right]_0^1$$

$$=12\left(0+\frac{1}{6}\right)=2$$

44. (2) $g(x)=\int_x^{\pi/2}(f'(t)\operatorname{cosec}t-f(t)\operatorname{cosec}t\cot t)\,dt$

$$=\int_x^{\pi/2}\frac{d}{dt}(f(t)\operatorname{cosec}t)\,dt$$

$$=f\left(\frac{\pi}{2}\right)\operatorname{cosec}\left(\frac{\pi}{2}\right)-f(x)\operatorname{cosec}x$$

$$=3(1)-\frac{f(x)}{\sin x}=3-\frac{f(x)}{\sin x}$$

$$\therefore\quad \lim_{x\to 0}g(x)=\lim_{x\to 0}\left(3-\frac{f(x)}{\sin x}\right)$$

$$=3-\lim_{x\to 0}\frac{f(x)}{\sin x}$$

$$=3-\lim_{x\to 0}\frac{f'(x)}{\cos x}\ \{\text{using LH Rule}\}$$

$$=3-\frac{f'(0)}{\cos(0)}=3-1=2$$

45. (2) By Leibnitz Rule,

$$\frac{dy}{dx}=-\sin x\int_{\pi^2/16}^{x^2}\frac{\cos\sqrt{\theta}}{1+\sin^2\sqrt{\theta}}\,d\theta+\cos x\left(\frac{\cos x}{1+\sin^2 x}\times 2x\right)$$

Therefore, $\left(\frac{dy}{dx}\right)_{x=\pi}=0+(-1)\left(\frac{-1}{1}\times 2\pi\right)$

$$=2\pi$$

So, $k=2$

46. (0) $I_2=\int_1^{\operatorname{cosec}\theta}\frac{1}{x(x^2+1)}dx$

Put $\frac{1}{x}=t$

$\Rightarrow\quad -\frac{1}{x^2}dx=dt$

If $x=1,t=1$

$x=\operatorname{cosec}\theta$ and $t=\sin\theta$

$$=-\int_1^{\sin\theta}\frac{t}{1+t^2}dt$$

$$I_2=-\int_1^{\sin\theta}\frac{x}{1+x^2}dx$$

$$I_2=-I_1$$

$$\therefore\begin{vmatrix} I_1 & I_1^2 & I_2\\ e^{I_1+I_2} & I_2^2 & -1\\ 1 & I_1^2+I_2^2 & -1\end{vmatrix}=\begin{vmatrix} I_1 & I_1^2 & -I_1\\ 1 & I_1^2 & -1\\ 1 & 2I_1^2 & -1\end{vmatrix}$$

Applying, $R_3\to R_3-R_2$

$$=\begin{vmatrix} I_1 & I_1^2 & -I_1\\ 1 & I_1^2 & -1\\ 0 & I_1^2 & 0\end{vmatrix}$$

$$=-I_1^2(-I_1+I_1)$$

$$=0$$

CHAPTER

11

AREA UNDER CURVES

LEVEL 01 \ Starter Level

11.1 Area under Simple Curve

1 Find the area of region bounded by the line $2y = 5x + 7$, X-axis and the lines $x = 2$ and $x = 8$.

2 Find the area of region bounded by $y = -1$, $y = 2$, $x = y^3$ and $x = 0$.

3 Find the area above X-axis bounded by the line $x = 4$ and the curve $y = f(x)$, where
$$f(x) = \begin{cases} x^2, & 0 \le x \le 1 \\ \sqrt{x}, & x > 1 \end{cases}$$

4 If $y = 2\sqrt{1-x^2}$, $x \in [0, 1]$, evaluate the area enclosed between the curve and X-axis.

5 Find the area in first quadrant bounded by the parabola $y = 4x^2$ and the lines $x = 0$, $y = 1$ and $y = 4$.

6 Find the area of the region for which $0 < y < 3 - 2x - x^2$ and $x > 0$.

7 Find the area bounded by the curve $x = at^2$ and $y = 2at$ between the ordinates corresponding to $t = 1$ and $t = 2$.

8 Find the area bounded by the curve $y = |x - 2|$, $x = 1$, $x = 3$ and the X-axis.

9 Find the area of region bounded by the curve
$$y = 1 + |x + 1|, x = -2, x = 3 \text{ and } y = 0$$

10 Find the area bounded by $y = \log|x|$ and $y = 0$.

11 Find the area bounded by the curve $y = \cos x$, X-axis and the ordinates $x = 0$ and $x = 2\pi$.

12 Find the area inclosed by the curve $y = \cos^2 x$ and line $x = 0$, $x = \pi$ and X-axis.

13 Find the area of the smaller region of the circle $x^2 + y^2 = a^2$ cut-off by the line $x = \frac{a}{\sqrt{2}}$.

14 If $x = a$ line bisects the area under the curve $y = \frac{1}{x^2}$, $1 \le x \le 9$, then a is equal to.

11.2 Area between Curves

15 Find the area bounded by the curve $y = x$ and $y = x^3$.

16 Find the area enclosed between the curves $y^2 = x$ and $y = |x|$.

17 Find the area of smaller region bounded by the ellipse $\frac{x^2}{a^2} + \frac{y^2}{b^2} = 1$ and the straight line $\frac{x}{a} + \frac{y}{b} = 1$.

18 Find the area of region bounded by parabola $y = x^2$ and $y = |x|$.

19 Find the area of region bounded by the curve $y = \frac{3x^2}{4}$ and the line $3x - 2y + 12 = 0$.

20 Find the area of the region between the circles $x^2 + y^2 = 4$ and $(x-2)^2 + y^2 = 4$.

21 Find the area of region bounded by the curve $y = x^2 + 2$, $y = x$, $x = 0$ and $x = 3$.

22 Find the area bounded by the curves $y = (x-1)^2$, $y = (x+1)^2$ and $y = \frac{1}{4}$.

23 Using integration, find the area of ΔABC whose vertices have coordinates A (2, 5), B (4, 7) and C (6, 2).

24 Find the area of triangular region bounded by the lines $x + 2y = 2$, $y - x = 1$ and $2x + y = 7$.

LEVEL 02 \ JEE Main Level

Single Choice Correct Questions

11.1 Area Under Simple Curve

1 The area bounded by the coordinate axes and the curve $\sqrt{x} + \sqrt{y} = 1$ is

(a) 1 sq unit (b) $\frac{1}{2}$ sq unit (c) $\frac{1}{3}$ sq unit (d) $\frac{1}{6}$ sq unit

2 The area bounded by the curve $y = \sin^{-1} x$ and the line $x = 0$, $|y| = \frac{\pi}{2}$ is

(a) 2 sq units (b) 3 sq units (c) 1 sq unit (d) 4 sq units

3 The area (in sq units) of smaller portion of the circle $x^2 + y^2 = 4$ cut-off by the line $x = 1$ is

(a) $\frac{4\pi - 3\sqrt{3}}{3}$ (b) $\frac{4\pi + 3\sqrt{3}}{3}$ (c) $\frac{4\pi + 3\sqrt{3}}{2}$ (d) $\frac{4\pi - 3\sqrt{3}}{2}$

4 The area bounded by $y = x|\sin x|$ and X-axis between $x = 0$ and $x = 2\pi$ is

(a) π sq units (b) 2π sq units (c) 4π sq units (d) 3π sq units

5 The area bounded by the $y = |\sin x|$, X-axis and the line $|x| = \pi$ is (in sq units)

(a) 2 sq units (b) 1 sq unit (c) 4 sq units (d) 3 sq units

6 If $f(x) = \cos^{-1}(\cos x) - \sin^{-1}(\sin x)$ in $[0, 2\pi]$, then the area bounded by $f(x)$ and the X-axis is

(a) $\frac{\pi^2}{2}$ sq units (b) $\frac{\pi^2}{4}$ sq units (c) π^2 sq units (d) None of these

7 The area (in sq units) bounded by the curve $f(x) = ce^x$, $(c > 0)$, the X-axis and the two ordinates $x = p$ and $x = q$ is proportional to

(a) $f(p) \cdot f(q)$ (b) $|f(p) - f(q)|$ (c) $f(p) + f(q)$ (d) $\sqrt{f(p) \cdot f(q)}$

8 If A_1 is the area of the parabola $y^2 = 4ax$ lying between vertex and the latusrectum and A_2 is the area between the latusrectum and the double ordinate $x = 2a$, then $\frac{A_1}{A_2}$ is equal to

(a) $2\sqrt{2} - 1$ (b) $\frac{1}{7}(2\sqrt{2} + 1)$ (c) $\frac{1}{7}(2\sqrt{2} - 1)$ (d) $2\sqrt{2} + 1$

9 Let $S(\alpha) = \{(x, y) : y^2 \le x, 0 \le x \le \alpha\}$ and $A(\alpha)$ is area of the region $S(\alpha)$. If for λ, $0 < \lambda < 4$, $A(\lambda) : A(4) = 2 : 5$, then λ equals **[JEE Main 2019]**

(a) $2\left(\frac{4}{25}\right)^{\frac{1}{3}}$ (b) $4\left(\frac{2}{5}\right)^{\frac{1}{3}}$ (c) $4\left(\frac{4}{25}\right)^{\frac{1}{3}}$ (d) $2\left(\frac{2}{5}\right)^{\frac{1}{3}}$

10 The area bounded by the curve $x = \begin{cases} -2-y, & y < -1 \\ y^3, & -1 \le y \le 1 \\ 2-y, & y > 1 \end{cases}$ and $x = 0$ is

(a) $\frac{5}{2}$ sq units (b) $\frac{3}{2}$ sq units (c) $\frac{1}{3}$ sq units (d) $\frac{7}{2}$ sq units

11 The area of region bounded by the curve $y = x(x-1)^2$, $y = 2$ and $x = 0$ is

(a) $\frac{5}{2}$ sq units (b) $\frac{10}{3}$ sq units (c) $\frac{5}{3}$ sq units (d) $\frac{7}{3}$ sq units

11.2 Area between Two Curves

12 What is the area bounded by the curve $y = e^x$, $y = e^{-x}$ and the straight line $x = 1$?

(a) $\left(e + \frac{1}{e}\right)$ sq units (b) $\left(e - \frac{1}{e}\right)$ sq units

(c) $\left(e + \frac{1}{e} - 2\right)$ sq units (d) $\left(e - \frac{1}{e} - 2\right)$ sq units

13 The area of region bounded by the curves $y = ex\log x$ and $y = \frac{\log x}{ex}$ is

(a) $\left(\frac{e^2 - 5}{4e}\right)$ sq units (b) $\left(e - \frac{5}{4}\right)$ sq units (c) $\left(\frac{e}{4} - 5\right)$ sq units (d) $\left(\frac{e}{4} - \frac{1}{4e}\right)$ sq units

14 The area enclosed by the curves $y = |x - 1|$ and $y = -|x - 1| + 1$ is

(a) 1 sq unit (b) $\frac{1}{2}$ sq unit (c) 21 sq units (d) None of these

15 The region represented by $|x - y| \le 2$ and $|x + y| \le 2$ is bounded by a **[JEE Main 2019]**

(a) rhombus of side length 2 units (b) rhombus of area $8\sqrt{2}$ sq units

(c) square of side length $2\sqrt{2}$ units (d) square of area 16 sq units

16 The area bounded between the curve $y = \tan x$ and tangent drawn to it at $x = \frac{\pi}{4}$ and $y \ge 0$ is

(a) $\frac{1}{4}(\log 4 - 1)$ sq units (b) $\frac{1}{2}(\log 4 - 1)$ sq units

(c) $\frac{1}{4}(\log 4 + 1)$ sq units (d) $\frac{1}{2}(\log 4 + 1)$ sq units

17 The area of the region enclosed by the curves $y = x$, $x = e$, $y = \frac{1}{x}$ and positive X-axis is

(a) 1 sq units (b) $\frac{3}{2}$ sq units (c) $\frac{5}{2}$ sq units (d) $\frac{1}{2}$ sq units

18 The area (in sq units) of the region

$$\left\{(x, y): 0 \le y \le x^2 + 1, 0 \le y \le x + 1, \frac{1}{2} \le x \le 2\right\} \text{ is}$$

[JEE Main 2020]

(a) $\frac{23}{16}$ (b) $\frac{79}{24}$ (c) $\frac{79}{16}$ (d) $\frac{23}{6}$

19 The area (in sq units) of the region, given by the set $\{(x, y) \in R \times R \mid x \ge 0, 2x^2 \le y \le 4 - 2x\}$, is

(a) $\frac{8}{3}$ (b) $\frac{17}{3}$ (c) $\frac{13}{3}$ (d) $\frac{7}{3}$

20 Given, $f(x) = \begin{cases} x, & 0 \le x < \frac{1}{2} \\ 1 - x, & \frac{1}{2} \le x \le 1 \end{cases}$ and $g(x) = \left(x - \frac{1}{2}\right)^2$, $x \in R$.

Then, area bounded by the curve $y = f(x)$ and $y = g(x)$ between the lines $2x = 1$ and $2x = \sqrt{3}$ is

[JEE Main 2020]

(a) $\frac{1}{2} + \frac{\sqrt{3}}{4}$ sq units (b) $\frac{1}{2} - \frac{\sqrt{3}}{4}$ sq units

(c) $\frac{1}{3} + \frac{\sqrt{3}}{4}$ sq units (d) $\frac{\sqrt{3}}{4} - \frac{1}{3}$ sq units

21 Consider a region $R = \{x, y) \in R : x^2 \le y \le 2x\}$. If a line $y = \alpha$ divides the area of region R into two equal parts, then which of the following is true?

[JEE Main 2020]

(a) $3\alpha^2 - 8\alpha + 8 = 0$ (b) $\alpha^3 - 6\alpha^{3/2} - 16 = 0$

(c) $3\alpha^2 - 8\alpha^{3/2} + 8 = 0$ (d) $\alpha^3 - 6\alpha^2 + 16 = 0$

22 The area in (sq units) of the part of the circle $x^2 + y^2 = 36$, which is outside the parabola $y^2 = 9x$, is

[JEE Main 2021]

(a) $24\pi + 3\sqrt{3}$ (b) $12\pi + 3\sqrt{3}$

(c) $12\pi - 3\sqrt{3}$ (d) $24\pi - 3\sqrt{3}$

23 The area bounded by the curves $y = \sqrt{5 - x^2}$ and $y = |x - 1|$ is

(a) $\frac{5\pi}{4} - 2$ sq units (b) $\frac{5\pi - 2}{4}$ sq units (c) $\frac{5\pi - 2}{2}$ sq units (d) $\frac{\pi}{2} - 5$ sq units

24 The area of the region bounded by $y = x + 1$ and the curve $y = x^3 - x^2 - x + 1$ is

(a) $\frac{52}{12}$ sq units (b) $\frac{37}{6}$ sq units (c) $\frac{37}{12}$ sq units (d) $\frac{52}{6}$ sq units

25 The area enclosed between the curve $y = \sin x$ and $y = \cos x$ for $\frac{\pi}{4} \le x \le \frac{5\pi}{4}$ is

(a) $\sqrt{2}$ sq units (b) $2\sqrt{2}$ sq units (c) 2 sq units (d) 1 sq unit

26 The smallest area bounded by the curves $y = x - \sin x$ and $y = x + \cos x$ is

(a) $\sqrt{2}$ sq units (b) $2\sqrt{2}$ sq units (c) 2 sq units (d) None of these

27 The area enclosed by the curve $y = \sin x + \cos x$ and $y = |\cos x - \sin x|$ over the interval $\left[0, \frac{\pi}{2}\right]$ is

(a) $4(\sqrt{2} - 1)$ sq units (b) $2\sqrt{2}(\sqrt{2} - 1)$ sq units

(c) $2(\sqrt{2} + 1)$ sq units (d) $2\sqrt{2}(\sqrt{2} + 1)$ sq units

28 The area (in sq units) bounded by the curves

$$y = \cos^{-1}|\cos x| \text{ and } y = (\cos^{-1}|\cos x|)^2,\ x \in [0, \pi] \text{ is}$$

(a) $\frac{4}{3} + \frac{\pi^2}{4}\left(\frac{\pi}{3} - 1\right)$ (b) $\frac{4}{3} + \frac{\pi^2}{4}\left(\frac{\pi}{3} + 1\right)$

(c) $\frac{2}{3} + \frac{\pi^2}{4}\left(\frac{\pi}{3} - 1\right)$ (d) $\frac{2}{3} + \frac{\pi^2}{4}\left(\frac{\pi}{3} + 1\right)$

29 The positive values of the parabola k for which the area of the figure bounded by the curve $y = \sin(kx)$, $x = \frac{2\pi}{3k}$, $x = \frac{5\pi}{3k}$ and X-axis is less than 2 can be

(a) $\frac{1}{8} < k < \frac{3}{8}$ (b) $0 < k < \frac{1}{8}$ (c) $1 < k < 2$ (d) $\frac{3}{8} < k < \frac{5}{8}$

30 The area of the region, enclosed by the circle $x^2 + y^2 = 2$ which is not common to the region bounded by the parabola $y^2 = x$ and the straight line $y = x$, is **[JEE Main 2020]**

(a) $\frac{1}{3}(12\pi - 1)$ sq units (b) $\frac{1}{6}(12\pi - 1)$ sq units

(c) $\frac{1}{6}(24\pi - 1)$ sq units (d) $\frac{1}{3}(6\pi - 1)$ sq units

31 Area of the region bounded by the curves $y = \sqrt{x}$, $x \in [0, 1]$, $y = x^2$, $x \in [1, 2]$, $y = -x^2 + 2x + 4$, $x \in [0, 2]$ and Y-axis is

(a) $\frac{10}{7}$ sq units (b) $\frac{3}{5}$ sq units (c) $\frac{19}{3}$ sq units (d) $\frac{4}{3}$ sq units

32 The area of region bounded by the curves $x + 2y^2 = 0$ and $x + 3y^2 = 1$ is equal to

(a) $\frac{2}{3}$ sq units (b) $\frac{4}{3}$ sq units (c) $\frac{5}{3}$ sq units (d) $\frac{1}{3}$ sq units

33 The area (in sq units) of the region

$$A = \{(x, y) \in R \times R \mid 0 \le x \le 3,\ 0 \le y \le 4, y \le x^2 + 3x\} \text{ is}$$ **[JEE Main 2019]**

(a) $\frac{53}{6}$ (b) 8 (c) $\frac{59}{6}$ (d) $\frac{26}{3}$

34 The ratio of the area's bounded by the curves $y^2 = 12x$ and $x^2 = 12y$ is divided by the line $x = 3$, is

(a) 15 : 49 (b) 9 : 15 (c) 7 : 15 (d) 7 : 5

35 The area of region described by $A = \{(x, y) : x^2 + y^2 \le 1 \text{ and } y^2 \le 1 - x\}$ is

(a) $\left(\frac{\pi}{2} + \frac{4}{3}\right)$ sq units (b) $\left(\frac{\pi}{2} - \frac{4}{3}\right)$ sq units (c) $\left(\frac{\pi}{2} - \frac{2}{3}\right)$ sq units (d) $\left(\frac{\pi}{2} + \frac{2}{3}\right)$ sq units

36 The area (in sq units) of the region

$$\{(x, y) : x \ge 0,\ x + y \le 3,\ x^2 \le 4y \text{ and } y \le 1 + \sqrt{x}\} \text{ is}$$

(a) $\frac{59}{12}$ (b) $\frac{3}{2}$ (c) $\frac{7}{3}$ (d) $\frac{5}{2}$

37 The area (in sq units) of region bounded by $\{(x, y) : x^2 + y^2 \le 2ax,\ y^2 \ge ax,\ x \ge 0,\ y \ge 0\}$ is

(a) $\left(\frac{\pi}{4} + \frac{2}{3}\right)a^2$ (b) $\left(\frac{2}{3} - \frac{\pi}{2}\right)a^2$ (c) $\left(\frac{\pi}{4} - \frac{2}{3}\right)a^2$ (d) $\left(\frac{2}{3} + \frac{\pi}{2}\right)a^2$

38 Area (in sq units) of the region outside $\frac{|x|}{2}+\frac{|y|}{3}=1$ and inside the ellipse $\frac{x^2}{4}+\frac{y^2}{9}=1$ is **[JEE Main 2020]**

(a) $6(\pi-2)$ (b) $3(\pi-2)$ (c) $3(4-\pi)$ (d) $6(4-\pi)$

39 Let f be a differentiable function satisfying the condition $f\left(\frac{x}{y}\right)=\frac{f(x)}{f(y)}, y\neq 0, f(y)\neq 0\ \forall\ x, y\in R$ and $f'(1)=2$. If the smaller area enclosed by $y=f(x)$ and $x^2+y^2=2$ is A, then the value of A is

(a) $\left(\frac{1}{6}+\frac{\pi}{3}\right)$ sq units (b) $\left(\frac{1}{6}-\frac{\pi}{3}\right)$ sq units (c) $\left(\frac{1}{3}-\frac{\pi}{2}\right)$ sq units (d) $\left(\frac{1}{3}+\frac{\pi}{2}\right)$ sq units

40 The area (in sq units) of the region $A=\{(x, y):|x|+|y|\leq 1, 2y^2\geq |x|\}$ is **[JEE Main 2020]**

(a) $\frac{1}{6}$ (b) $\frac{5}{6}$ (c) $\frac{1}{3}$ (d) $\frac{7}{6}$

41 The area bounded by the curve $4y^2=x^2(4-x)(x-2)$ is equal to

(a) $\frac{\pi}{8}$ sq units (b) $\frac{3\pi}{8}$ sq units (c) $\frac{3\pi}{2}$ sq units (d) $\frac{\pi}{16}$ sq units

42 The area (in sq units) of the region bounded by the curves $y=2^x$ and $y=|x+1|$, in the first quadrant is **[JEE Main 2019]**

(a) $\frac{3}{2}$ (b) $\log_e 2+\frac{3}{2}$ (c) $\frac{1}{2}$ (d) $\frac{3}{2}-\frac{1}{\log_e 2}$

43 The area enclosed by curve $y=x^2+x+1$ and its tangent at (1, 3) and between the ordinates $x=-1$ and $x=1$ is

(a) $\frac{4}{3}$ (b) $\frac{1}{3}$ (c) $\frac{8}{3}$ (d) $\frac{16}{3}$

44 The area (in sq units) bounded by the parabola $y=x^2-1$, the tangent at the point (2, 3) to it and the Y-axis is **[JEE Main 2019]**

(a) $\frac{8}{3}$ (b) $\frac{56}{3}$ (c) $\frac{32}{3}$ (d) $\frac{14}{3}$

Numerical Value Type Questions

45 The area bounded by the curve $|y|=\sin(2x)$, where $0<x<2\pi$, is ......

46 The area enclosed between the curve $y=\ln(x+e)$, $x=\ln\left(\frac{1}{y}\right)$ and X-axis is ......

47 The area bounded by the curves $y=3-|x|$ and $y=|x-1|$ is ......

48 The area bounded by the curves $y=\ln|x|$ and $y=1-|x|$ is ......

49 The area of the region bounded by the X-axis and the curves defined by

$$\begin{cases} y=\tan x, & -\frac{\pi}{3}\leq x\leq\frac{\pi}{3} \\ y=\cot x, & \frac{\pi}{6}\leq x\leq\frac{\pi}{2}\end{cases}$$ is $\log k$, then the value of k is ......

50 If the line $y=mx$ bisects the area enclosed by the lines $x=0$, $y=0$, $x=\frac{3}{2}$ and the curve $y=1+4x-x^2$, then $12m$ is equal to ......

51 The area of the region $A=\{(x, y): 0\leq y\leq x|x|+1$ and $-1\leq x\leq 1\}$ is ......

52 Let a and b respectively be the points of local maxima and local minima of the function $f(x)=2x^3-3x^2-12x$. If A is the total area of the region bounded by $y=f(x)$, the X-axis and the lines $x=a$ and $x=b$, then $4A$ is equal to ......

53 The area of region bounded by $x=-1$, $x=2$, $y=x^2+1$ and $y=2x-2$ is ......

54 Let T be the tangent to the ellipse $E: x^2+4y^2=5$ at the point $P(1, 1)$. If the area of the region bounded by the tangent T, ellipse E and line $x=1$ and $x=\sqrt{5}$ is $\alpha\sqrt{5}+\beta+\gamma\cos^{-1}\left(\frac{1}{\sqrt{5}}\right)$, then $|\alpha+\beta+\gamma|$ is equal to ......

55 The area of region bounded by the parabola $(y-2)^2=(x-1)$, the tangent to it at the point, whose ordinate is 3 and the X-axis, is ......

56 The set of point (x, y) in the plane satisfying $x^{2/5}+|y|=1$ form a curve enclosing a region of area $\frac{p}{q}$ sq units, when p and q are relatively prime positive integers, then $p-q$ is ......

LEVEL 03 \ JEE Advanced Level

Single Choice Correct Questions

1 Let $f:[0,\infty)\to R$ be a continuous and strictly increasing function such that $f^3(x)=\int_0^x t f^2(t)dt\ \forall x>0$. The area enclosed by $y=f(x)$, the X-axis and the ordinate at $x=3$ is

(a) $\frac{1}{2}$ sq unit (b) 1 sq unit (c) 2 sq unit (d) $\frac{3}{2}$ sq unit

2 Given, $f(x)=\int_0^x e^t(\log_e \sec t-\sec^2 t)dt$, $g(x)=-2e^x\tan x$, then the area (in sq units) bounded by the curves $y=f(x)$ and $y=g(x)$ between the ordinates $x=0$ and $x=\frac{\pi}{3}$, is

(a) $\frac{1}{2}e^{\pi/3}\log_e 2$ (b) $e^{\pi/3}\log_e 2$

(c) $\frac{1}{4}e^{\pi/3}\log_e 2$ (d) $e^{\pi/3}\log_e 3$

3 The parabola $y^2=4x$ and $x^2=4y$ divide the square region bounded by the lines $x=4$, $y=4$ and the coordinate axis. If S_1, S_2 and S_3 are respectively the areas of these parts numbered from top to bottom, then $S_1:S_2:S_3$ is

(a) 1 : 2 : 1 (b) 1 : 2 : 3 (c) 2 : 1 : 2 (d) 1 : 1 : 1

4 The area of region $\{(x, y): 0\le x\le \frac{9}{4}, 0\le y\le 1, x\ge 3y, x+y\ge 2\}$ is

(a) $\frac{11}{32}$ sq unit (b) $\frac{35}{96}$ sq unit (c) $\frac{37}{96}$ sq unit (d) $\frac{13}{32}$ sq unit

5 The area of the region $\{(x, y): x\ge 0, x+y\le 3, x^2\le 4y, y\ge 1+\sqrt{x}\}$ is

(a) $\frac{5}{2}$ sq unit (b) $\frac{59}{12}$ sq unit (c) $\frac{3}{2}$ sq unit (d) $\frac{7}{3}$ sq unit

6 The area of the region $\{(x, y)\in R^2: y\ge\sqrt{|x+3|}, 5y\le x+9\le 15\}$ is equal to

(a) $\frac{1}{6}$ sq unit (b) $\frac{4}{3}$ sq unit (c) $\frac{3}{2}$ sq unit (d) $\frac{5}{3}$ sq unit

7 The area (in sq units) of the region $\{(x, y): xy \le 8,\ 1 \le y \le x^2\}$ is

(a) $16\log 2 - 6$ (b) $8\log 2 - \frac{7}{3}$ (c) $16\log 2 - \frac{14}{3}$ (d) $8\log 2 - \frac{14}{3}$

8 The area (in sq units) of region bounded by the curves $y = x^2$, $y = |2 - x^2|$ and $y = 2$, which lies to the right of the line $x = 1$, is

(a) $\left(\frac{12 - 20\sqrt{3}}{2}\right)$ (b) $\frac{20 - \sqrt{2}}{3}$ (c) $\frac{20 - 12\sqrt{2}}{3}$ (d) $\left(\frac{12 - 20\sqrt{2}}{3}\right)$

9 If $g(x) = 2x + 1$, $h(x) = 4x^2 + 4x + 5$ and $h(x) = (fog)(x)$, then the area (in sq units) enclosed by the graph of the function $y = f(x)$ and the pair of tangents drawn to it from the origin is

(a) $\frac{8}{3}$ (b) $\frac{32}{3}$ (c) $\frac{16}{3}$ (d) None of these

10 The smaller area enclosed by $y = f(x)$, when $f(x)$ is polynomial of least degree satisfying $\lim_{x \to 0}\left[1 + \frac{f(x)}{x^3}\right]^{1/x} = e$ and the circle $x^2 + y^2 = 2$ above the X-axis, is

(a) $\frac{\pi}{2}$ sq units (b) $\frac{3}{5}$ sq units (c) $\frac{\pi}{2} - \frac{3}{5}$ sq units (d) $\frac{\pi}{2} + \frac{3}{5}$ sq units

More than One Correct

11 The curve $y = a\sqrt{x} + bx$ passes through the point (1, 2) and the area enclosed by the curve, the X-axis and the line $x = 4$ is 8 sq units, then

(a) $a + 2b = 1$ (b) $2a + b = 1$ (c) $a = -1$ and $b = 3$ (d) $a = 3$ and $b = -1$

12 The area of the region bounded by the curve $y = e^x$ and lines $x = 0$, $y = e$ is

(a) $e - 1$ (b) $\int_1^e \ln(e + 1 - y)\,dy$ (c) $e - \int_0^1 e^x\,dx$ (d) $\int_1^e \ln y\,dy$

13 The area of the region bounded by the curve $y = e^{x^2}$ and the line $y = e$ is

(a) $2\int_1^e \sqrt{\log_e y}\,dy$ (b) $2e - \int_{-1}^{1} e^{x^2}\,dx$

(c) $\int_{-1}^{1} (e - e^{x^2})\,dx$ (d) $2\int_0^1 \sqrt{x}\,e^{x^2}\,dx$

14 If the line $x = \alpha$ divides the area of the region $R = \{(x, y) \in R^2 : x^3 \le y \le x,\ 0 \le x \le 1\}$ into two equal parts, then

(a) $0 < \alpha \le \frac{1}{2}$ (b) $\frac{1}{2} < \alpha < 1$

(c) $2\alpha^4 - 4\alpha^2 + 1 = 0$ (d) $\alpha^4 + 4\alpha^2 - 1 = 0$

15 The area bounded by curves $y = x^3$ and $y = (x)^{1/3}$ is/are

(a) lies in I quadrant (b) lies in III quadrant

(c) is equal to 1 sq unit (d) is equal to 4 sq units

16 Area enclosed between the curves $|y| = 1 - x^2$ and $x^2 + y^2 = 1$ is

(a) enclosed area is symmetrical with respect to coordinate axis

(b) $\left(\pi - \frac{8}{3}\right)$ sq units

(c) $\left(2\pi - \frac{8}{3}\right)$ sq units

(d) enclosed area is symmetrical about $y = 2$ line

17 Which of the following have the same bounded area?

(a) $f(x) = \sin x$ and $g(x) = \sin^2 x$, where $0 \le x \le 10\pi$

(b) $f(x) = \sin x$ and $g(x) = |\sin x|$, where $0 \le x \le 20\pi$

(c) $f(x) = |\sin x|$ and $g(x) = \sin^3 x$, where $0 \le x \le 10\pi$

(d) $f(x) = \sin x$ and $g(x) = \sin^4 x$, where $0 \le x \le 10\pi$

18 For what values of m, the area of the region bounded by the curves $y = x - x^2$ and the line $y = mx$ equals $\frac{9}{2}$?

(a) -4 (b) -2 (c) 2 (d) 4

Linked Comprehension Type

Paragraph I Directions (Q.Nos. 19-21)

Consider the area $S_0, S_1, S_2 \dots$ bounded by the X-axis and halfwaves of the curve $y = e^{-x} \sin x$, where $x \ge 0$.

19 The value of S_0 is

(a) $\frac{1}{2}(1 + e^{\pi})$ sq units (b) $\frac{1}{2}(1 + e^{-\pi})$ sq units

(c) $\frac{1}{2}(1 - e^{-\pi})$ sq units (d) $\frac{1}{2}(e^{\pi} - 1)$ sq units

20 The sequence $S_0, S_1, S_2 \dots$ form a GP with common ratio

(a) $\frac{e^{\pi}}{2}$ (b) $e^{-\pi}$ (c) e (d) $\frac{e^{-\pi}}{2}$

21 $\sum_{n=0}^{\infty} S_n$ is equal to

(a) $\frac{1 + e^{n}}{1 - e^{-n}}$ (b) $\frac{\frac{1}{2}(1 + e^{\pi})}{1 - e^{-\pi}}$

(c) $\frac{1}{2(1 - e^{-\pi})}$ (d) None of these

Paragraph II Directions (Q.Nos. 22-24)

Consider an ellipse $\frac{x^2}{a^2} + \frac{y^2}{b^2} = 1$

22 The area of the greatest rectangle that can be inscribed in the ellipse is

(a) ab sq units (b) $2ab$ sq units (c) $\frac{ab}{2}$ sq units (d) $\sqrt{ab}$ sq units

23 The area included between the ellipse and the greatest rectangle inscribed in the ellipse is

(a) $ab(\pi - 1)$ sq units (b) $2ab(\pi - 1)$ sq units

(c) $ab(\pi - 2)$ sq units (d) None of these

24 The area (in sq units) bounded by the ellipse and its latusrectum, where $a > b$ and e is the eccentricity of ellipse is

(a) $\frac{ab}{2}[e\sqrt{1 - e^2} + \sin^{-1} e]$ (b) $ab[e\sqrt{1 - e^2} + \sin^{-1} e]$

(c) $ab[e\sqrt{1 - e^2} - \sin^{-1} e]$ (d) $\frac{ab}{2}[e\sqrt{1 - e^2} - \sin^{-1} e]$

Matrix Match Type Questions

25 Match the following.

	Column I		Column II
(i)	Area enclosed by $y=\|x\|, \|x\|=1$ and $y=0$ is	(A)	4
(ii)	Area enclosed by the curve $y=\sin x$ $x=0, x=2\pi$ and $y=0$ is	(B)	1
(iii)	If the area of region bounded by $x^2 \le y$ and $y \le x+2$ is $k/4$, then k is equal to	(C)	27/4
(iv)	Area of the quadrilateral formed by tangents at the ends of latusrectum of ellipse $\frac{x^2}{9}+\frac{y^2}{5}=1$ is	(D)	18

26 Match the following.

	Column I		Column II
(i)	Area of region formed by points (x, y) satisfying $[x]^2=[y]^2$ for $0 \le x \le 4$ is equal to (where [.] denotes greatest integer function)	(A)	48
(ii)	The area of region formed by points (x, y) satisfying $x+y \le 6$, $x^2+y^2 \le 6y$ and $y^2 \le 8x$ is $\frac{k\pi-2}{12}$, then k is equal to	(B)	27
(iii)	The area in the first quadrant bounded by curve $y=\sin x$ and the line $\frac{2y-1}{\sqrt{2}-1}=\frac{2}{\pi}(6x-\pi)$ is $\left[\frac{\sqrt{3}-\sqrt{2}}{2}-\frac{(\sqrt{2}+1)\pi}{k}\right]$, then k is equal to	(C)	7
(iv)	If the area bounded by the graph of $y=xe^{-ax}$, $(a>0)$ and the abscissa axis is $\frac{1}{9}$, then the value of 'a' is equal to	(D)	3

Numerical Value Type Questions

Question Stem Directions (Q.Nos. 27-29) The answer to each question is a **NUMERICAL VALUE.**

Suppose that $f(\alpha)$ denotes the area of the region bounded by $x=0$, $x=2$, $y^2=4x$ and $y=|\alpha x-1|+|\alpha x-2|+\alpha x$, where $\alpha \in [0, 1]$.

27 The value of $f(\alpha)+\frac{8\sqrt{2}}{3}$, when $\alpha=0$, is ....

28 The value of $f(\alpha)+\frac{8\sqrt{2}}{3}$, when $\alpha=1$, is ....

29 The area of region bounded by parabola $y^2=4x$ and a normal drawn it with gradient -1 is $\frac{k}{3}$, then k is equal to ....

Question Stem Directions (Q.Nos. 30-32) The answer to each question is a **NUMERICAL VALUE.**

Computing area with parametrically represented boundaries. If the boundary of a figure is represented by parametric equation i.e. $x = x(t)$, $y = y(t)$, then the area of the figure is evaluated by one of the three formulas

$$S = -\int_{\alpha}^{\beta} y(t)\, x'(t)\,dt,\ S = \int_{\alpha}^{\beta} x(t) y'(t)\,dt \text{ and } S = \frac{1}{2}\int_{\alpha}^{\beta} (xy' - yx')\,dt,$$

where α and β are the values of the parameter t corresponding respectively to beginning and the end of the transversal of the curve corresponding to increasing t.

30 The area of the region bounded by an arc of the cycloid $x = a(t - \sin t)$, $y = a(1 - \cos t)$ and the X-axis is $k\pi a^2$ sq units, then k is equal to ....

31 The area of the loop described as $x = \frac{t}{3}(6 - t)$ and $y = \frac{t^2}{8}(6 - t)$ is $\frac{k'}{5}$ sq units, then k' is equal to ....

32 If the curve given by parametric equation $x = t - t^3$ and $y = 1 - t^4$ forms a loop for all values of $t \in [-1, 1]$, then the area of the loops is $\frac{a}{b}$ sq units, then $b - a$ is equal to ....

33 The area of region defined by $||x| + |y|| \geq 1$ and $x^2 + y^2 \leq 1$ is $\pi - k$, then the value of k is ......

34 The graph of $f(x) = x^2$ and $g(x) = Cx^3$ $(C > 0)$ intersect at two points. If the area of the region bounded between $f(x)$ and $g(x)$ over the interval $\left[0, \frac{1}{C}\right]$ is equal to $\frac{2}{3}$, then the value of $\frac{1}{C} + \frac{1}{C^2}$ is ......

35 If the area of the region bounded between the parabolas $y = x - bx^2$ and $y = \frac{x^2}{b}$ is maximum, then the value of b is ......

36 If the area of the region of the plane bounded by $(|x|, |y|) \leq 1$ and $xy \leq \frac{1}{2}$ is $a + \ln b$, then $a + b$ is ......

37 Let $f : [-3, 1] \to R$ be given as $f(x) = \begin{cases} \min\{(x + 6), x^2\} & -3 \leq x \leq 0 \\ \max\{\sqrt{x}, x^2\}, & 0 \leq x \leq 1 \end{cases}$

If the area bounded by $y = f(x)$ is A, then the value of $6A$ is equal to ......

38 Let $f(x) = \text{Max}\{x^2, (1 - x)^2, 2x(1 - x)\}$, where $0 \leq x \leq 1$, if the area of region bounded by the curves $y = f(x)$, X-axis, $x = 0$ and $x = 1$ is $\frac{p}{q}$, then $p + q$ is ......

39 A farmer F_1 has a land in the shape of a triangle with vertices at $P(0, 0)$, $Q(1, 1)$ and $R(2, 0)$. From this land a neighbouring farmer F_2 takes away the region which lies between the side PQ and a curve of the form $y = x^n$ $(n > 0)$, if the area of the region taken away by the farmer F_2 is exactly 30% of the area of ΔPQR, then the value of n is ......

40 For a point P in the plane, let $d_1(P)$ and $d_2(P)$ be the distance of the point P from the lines $x - y = 0$ and $x + y = 0$, respectively. The area of the region R consisting of all points P lying in the first quadrant of the plane and satisfying $2 \leq d_1(P) + d_2(P) \leq 4$ is ...

41 The area (in sq. units) of the region bounded by the curves $x^2 + 2y - 1 = 0$, $y^2 + 4x - 4 = 0$ and $y^2 - 4x - 4 = 0$ in the upper half plane, is ......

ANSWER KEY

Level 1

1. 96 sq units 2. $\frac{17}{4}$ sq units 3. 5 sq units 4. $\frac{\pi}{2}$ sq units 5. $\frac{7}{3}$ sq units 6. $\frac{5}{3}$ sq units

7. $\frac{56a^2}{3}$ sq units 8. 1 sq unit 9. $\frac{27}{2}$ sq units 10. 2 sq units 11. 4 sq units 12. $\frac{\pi}{2}$ sq units

13. $\left(\frac{\pi a^2}{4} - \frac{a^2}{2}\right)$ sq u nits 14. $\frac{9}{5}$ 15. $\frac{1}{2}$ sq units 16. $\frac{1}{6}$ units 17. $\frac{ab}{2}\left(\frac{\pi}{2} - 1\right)$ sq units

18. $\frac{1}{3}$ sq units 19. 27 sq units 20. $\left(\frac{8\pi}{3} - 2\sqrt{3}\right)$ sq units

21. $\frac{21}{2}$ sq units 22. $\frac{1}{3}$ sq unit 23. 7 sq units 24. 6 sq units

Level 2

1. (d)	**2.** (a)	**3.** (a)	**4.** (c)	**5.** (c)	**6.** (c)	**7.** (b)
8. (b)	**9.** (c)	**10.** (b)	**11.** (b)	**12.** (c)	**13.** (a)	**14.** (b)
15. (c)	**16.** (a)	**17.** (b)	**18.** (b)	**19.** (d)	**20.** (d)	**21.** (c)
22. (d)	**23.** (b)	**24.** (c)	**25.** (b)	**26.** (b)	**27.** (b)	**28.** (a)
29. (c)	**30.** (b)	**31.** (c)	**32.** (b)	**33.** (c)	**34.** (a)	**35.** (a)
36. (d)	**37.** (c)	**38.** (a)	**39.** (d)	**40.** (b)	**41.** (c)	**42.** (d)
43. (c)	**44.** (a)	**45.** (4)	**46.** (2)	**47.** (4)	**48.** (3)	**49.** (4)
50. (26)	**51.** (2)	**52.** (114)	**53.** (9)	**54.** (1.25)	**55.** (9)	**56.** (1)

Level 3

1. (d)	**2.** (b)	**3.** (d)	**4.** (a)	**5.** (a)	**6.** (c)	**7.** (c)
8. (c)	**9.** (c)	**10.** (d)	**11.** (a, d)	**12.** (b, c, d)	**13.** (a, b, c)	**14.** (b, c)
15. (a, b, c)	**16.** (a, b)	**17.** (a, c, d)	**18.** (b, d)	**19.** (b)	**20.** (b)	**21.** (b)
22. (b)	**23.** (c)	**24.** (b)				

25. (i)–B, (ii)–A, (iii)–D, (iv)–C

26. (i)–C, (ii)–B, (iii)–A, (iv)–D

27. (6)	**28.** (5)	**29.** (64)	**30.** (3)	**31.** (27)	**32.** (19)	**33.** (2)
34. (6)	**35.** (1)	**36.** (5)	**37.** (41)	**38.** (44)	**39.** (4)	**40.** (6)
41. (2)						

SOLUTIONS

Level I

1. Given, lines are $2y = 5x + 7$, $x = 2$ and $x = 8$

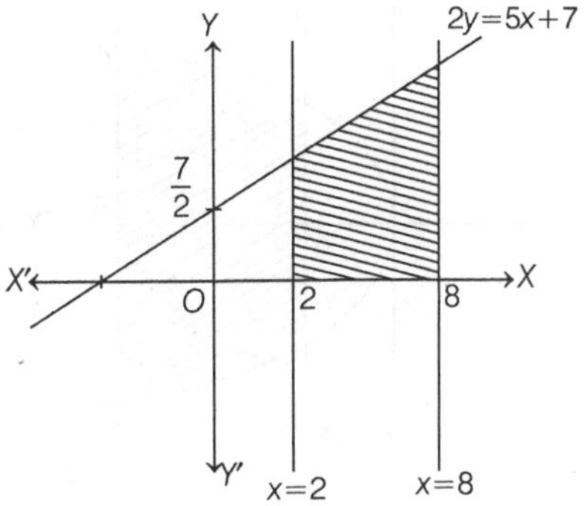

$\therefore$ Required area $= \int_2^8 y\,dx$

$$= \frac{1}{2}\int_2^8 (5x + 7)\,dx$$

$$= \frac{1}{2}\left[\frac{5x^2}{2} + 7x\right]_2^8$$

$$= 96 \text{ sq units}$$

2. Given, curve are $x = y^3, y = -1, y = 2$ and $x = 0$

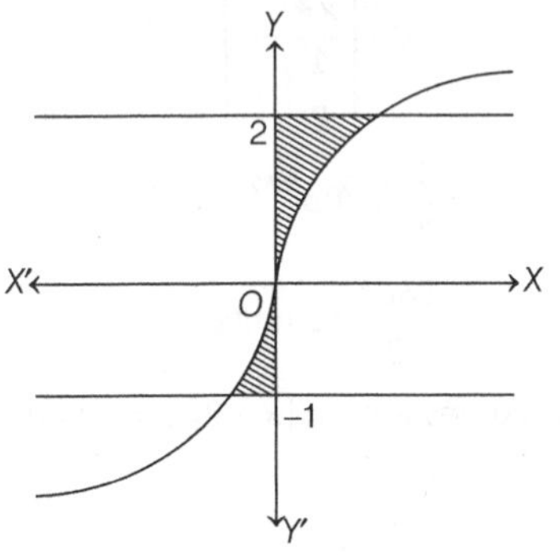

$\therefore$ Required area,

$$A = \left|\int_{-1}^0 y^3 dy\right| + \int_0^2 y^3\,dy$$

$$= \left|\left[\frac{y^4}{4}\right]_{-1}^0\right| + \left[\frac{y^4}{4}\right]_0^2$$

$$= \left|-\frac{1}{4}\right| + 4$$

$$= \frac{1}{4} + 4$$

$$= \frac{17}{4} \text{ sq units}$$

3. Given, curve $f(x) = \begin{cases} x^2, & 0 \le x \le 1 \\ \sqrt{x}, & x > 1 \end{cases}$

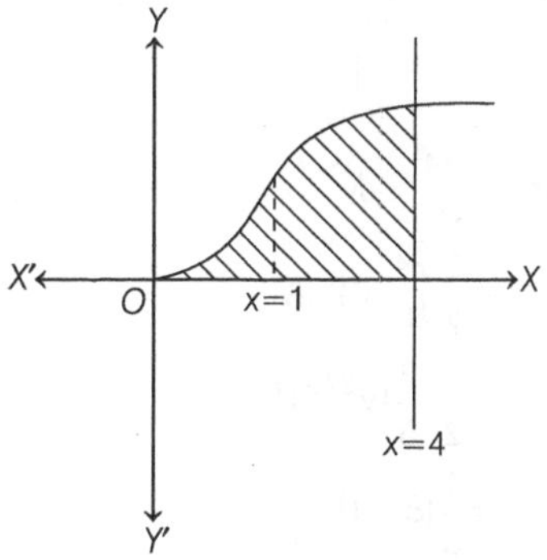

$\therefore$ Required area $= \int_0^1 x^2\,dx + \int_1^4 \sqrt{x}\,dx$

$$= \left[\frac{x^3}{3}\right]_0^1 + \frac{2}{3}[x^{3/2}]_1^4$$

$$= \frac{1}{3} + \frac{2}{3}(8 - 1)$$

$$= \frac{15}{3}$$

$$= 5 \text{ sq units}$$

4. Given, equation of curve is

$$y = 2\sqrt{1 - x^2}$$

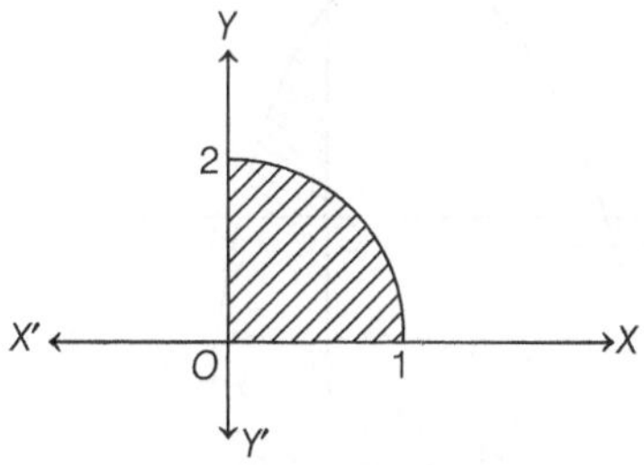

$\therefore$ The area of shaded region

$$= \int_0^1 y\,dx$$

$$= 2\int_0^1 \sqrt{1 - x^2}\,dx$$

$$= 2\left[\frac{x}{2}\sqrt{1 - x^2} + \frac{1}{2}\sin^{-1} x\right]_0^1$$

$$= 2\left[0 + \frac{1}{2}\frac{\pi}{2} - 0 - 0\right]$$

$$= \frac{\pi}{2} \text{ sq units}$$

5. Given, curve $y = 4x^2$ and lines $x = 0,\ y = 1$ and $y = 4$

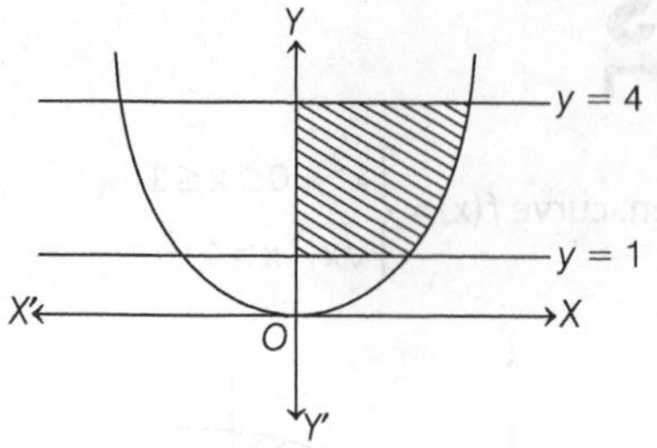

$\therefore$ Required area $= \int_1^4 x\, dy$

$= \frac{1}{2}\int_1^4 \sqrt{y}\, dy$

$= \frac{1}{2}\cdot\frac{2}{3}[y^{3/2}]_1^4$

$= \frac{1}{3}[8-1]$

$= \frac{7}{3}$ sq units

6. Given, $0 < y < 3 - 2x - x^2$ and $x > 0$

Consider parabola

$$y = 3 - 2x - x^2$$

$$\Rightarrow \quad (x+1)^2 = -(y-4)$$

Vertex of this parabola is $(-1, 4)$.

The required region in graph of $(x+1)^2 = -(y-4)$ and $x, y > 0$ is the region of parabola in I quadrant.

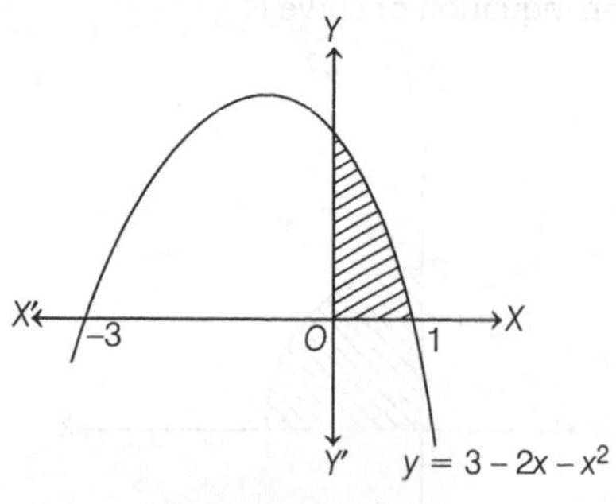

$\therefore$ Required area $= \int_0^1 (3 - 2x - x^2)\, dx$

$= \left[3x - x^2 - \frac{x^3}{3}\right]_0^1$

$= \frac{5}{3}$ sq units

7. The given equations of curve are

$$x = at^2 \text{ and } y = 2at \Rightarrow t = \frac{y}{2a}$$

So, $\quad x = a\left(\frac{y}{2a}\right)^2$

$$\Rightarrow \quad x = \frac{ay^2}{4a^2}$$

$$\Rightarrow \quad y^2 = 4ax$$

When $t = 1 \Rightarrow x = a$

When $t = 2 \Rightarrow x = 4a$

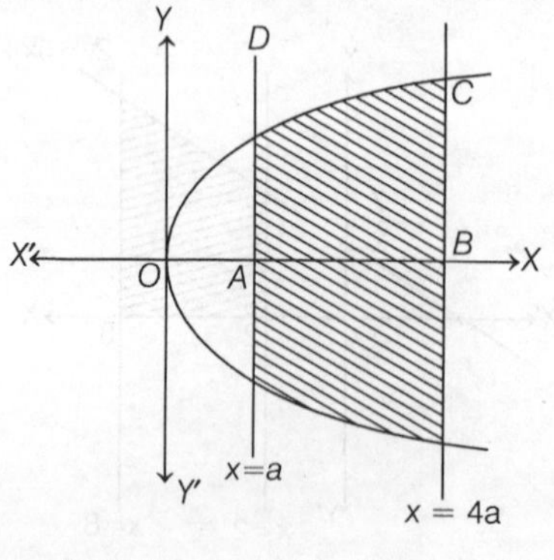

The given parabola is symmetric about X-axis.

So, area of shaded region

$= 2 \times$ area of $ABCD$

$= 2\int_a^{4a} y\, dx$

$= 2\int_a^{4a} 2\sqrt{ax}\, dx$

$= 4\sqrt{a}\int_a^{4a} x^{1/2}\, dx$

$= 4\sqrt{a}\left[\dfrac{x^{\frac{1}{2}+1}}{\frac{1}{2}+1}\right]_a^{4a}$

$= \frac{8}{3}\sqrt{a}\,(8a^{3/2} - a^{3/2})$

$= \frac{56a^2}{3}$ sq units

8. Given, curves are $y = |x - 2|$, $x = 1$ and $x = 3$

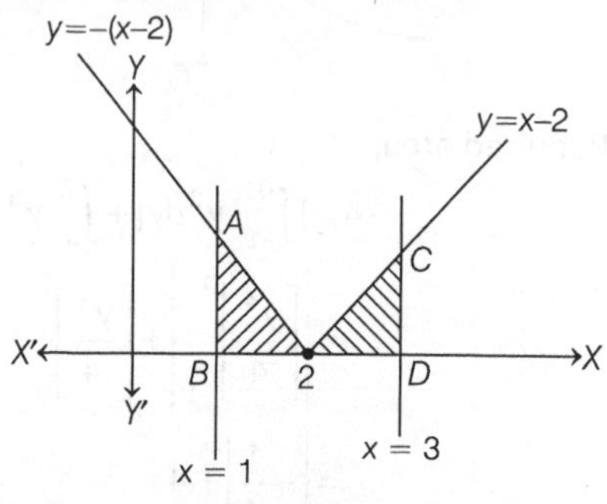

Line $x = 1$ and $y = -x + 2$ intersects at (1, 1)

and $x = 3$ and $y = x - 2$ intersects at (3, 1)

So, area of shaded region $=$ ar (ΔAOB) + ar (ΔCOD)

$= \int_1^2 (-x+2)\, dx + \int_2^3 (x-2)\, dx$

$$=\left[-\frac{x^2}{2}+2x\right]_1^2+\left[\frac{x^2}{2}-2x\right]_2^3$$

$$=\left[(-2+4)-\left(-\frac{1}{2}+2\right)\right]+\left[\left(\frac{9}{2}-6\right)-(2-4)\right]$$

$$=\frac{1}{2}+\frac{1}{2}=1 \text{ sq unit}$$

9. We have, $y=1+|x+1|$

$$y=\begin{cases}1-x-1, & x<-1\\ 1+x+1, & x>-1\end{cases}$$

$$y=\begin{cases}-x, & x<-1\\ x+2, & x>-1\end{cases}$$

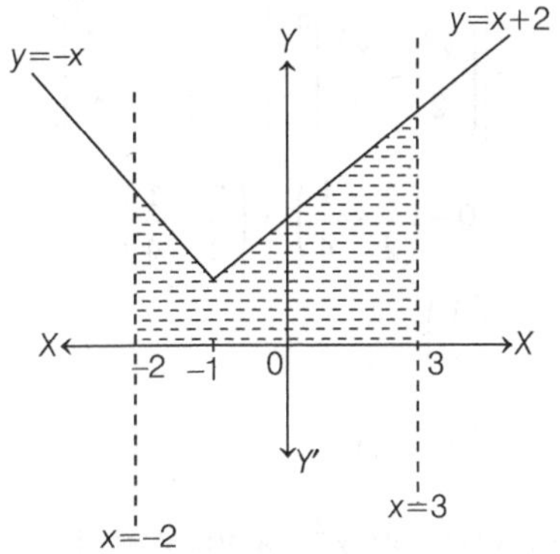

$\therefore$ Area of shaded region

$$=\int_{-2}^{-1}-x\,dx+\int_{-1}^{3}(x+2)\,dx$$

$$=-\left[\frac{x^2}{2}\right]_{-2}^{-1}+\left[\frac{x^2}{2}+2x\right]_{-1}^{3}$$

$$=-\left(\frac{1}{2}-2\right)+\left[\left(\frac{9}{2}+6\right)-\left(\frac{1}{2}-2\right)\right]$$

$$=\frac{27}{2}\text{ sq units}$$

10. We have, $y=\log|x|$ and $y=0$

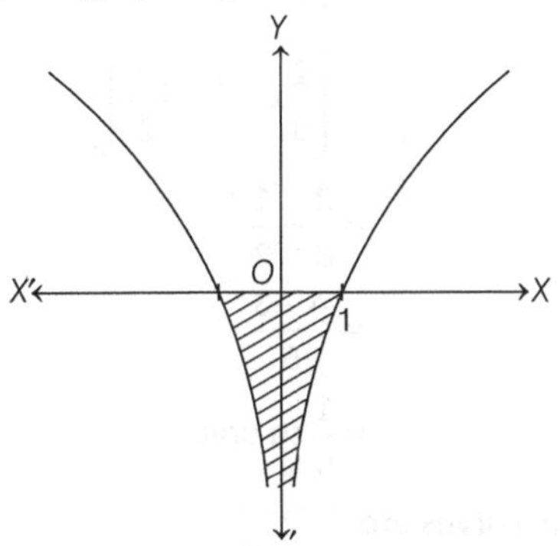

$\therefore$ Required area $=2\left|\int_0^1 \log x\,dx\right|$

$$=2\left|[x(\log x-1)]_0^1\right|$$

$$=2(1)=2\text{ sq units}$$

11. The graph of $y=\cos x$ is given as

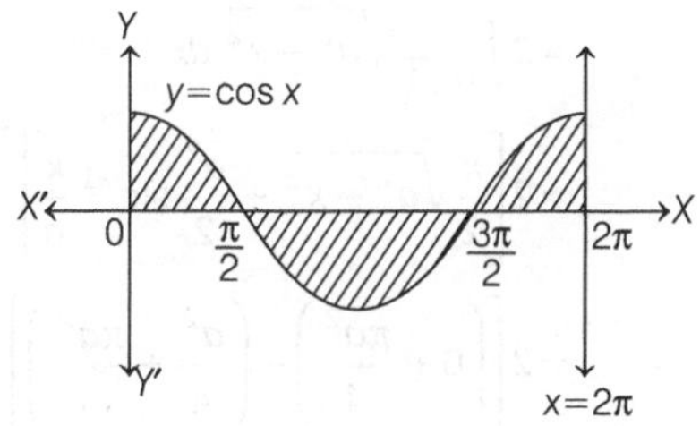

$\therefore$ Required area

$$=\int_0^{\pi/2}\cos x\,dx+\left|\int_{\pi/2}^{3\pi/2}\cos x\,dx\right|+\int_{3\pi/2}^{2\pi}\cos x\,dx$$

$$=[\sin x]_0^{\pi/2}+\left|[\sin x]_{\pi/2}^{3\pi/2}\right|+[\sin x]_{3\pi/2}^{2\pi}$$

$$=[1-0]+|-1-1|+[0-(-1)]$$

$$=1+2+1$$

$$=4\text{ sq units}$$

12. Graph of given curve is

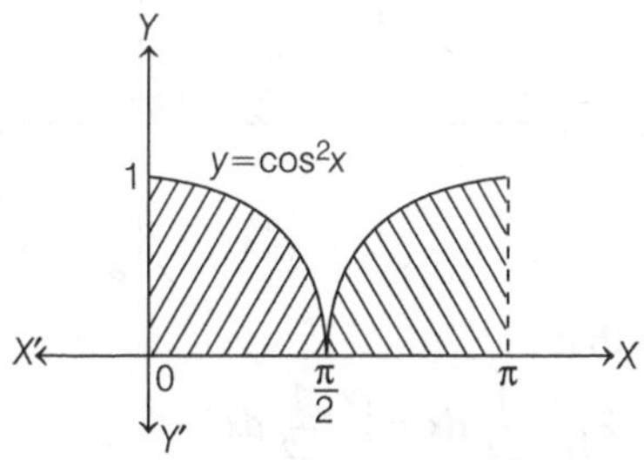

$\therefore$ The required area $=\int_0^{\pi}\cos^2 x\,dx$

$$=\int_0^{\pi}\left[\frac{1+\cos 2x}{2}\right]dx$$

$$=\frac{1}{2}\int_0^{\pi}(1+\cos 2x)\,dx$$

$$=\frac{1}{2}\left[x+\frac{\sin 2x}{2}\right]_0^{\pi}$$

$$=\frac{1}{2}(\pi)=\frac{\pi}{2}\text{ sq units}$$

13. We have, circle $x^2+y^2=a^2$ and line $x=\frac{a}{\sqrt{2}}$

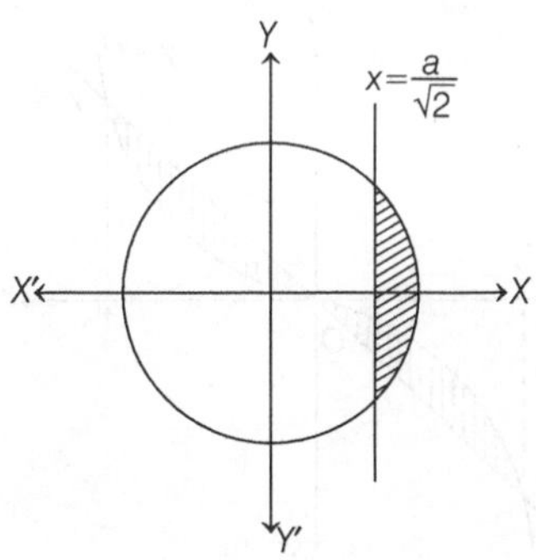

$\therefore$ Area of required region

$$= 2\int_{a/\sqrt{2}}^{a} \sqrt{a^2 - x^2}\, dx$$

$$= 2\left[\frac{x}{2}\sqrt{a^2 - x^2} + \frac{a^2}{2}\sin^{-1}\frac{x}{a}\right]_{a/\sqrt{2}}^{a}$$

$$= 2\left[\left(0 + \frac{\pi a^2}{4}\right) - \left(\frac{a^2}{4} + \frac{\pi a^2}{8}\right)\right]$$

$$= 2\left[\frac{\pi a^2}{8} - \frac{a^2}{4}\right]$$

$$= \left(\frac{\pi a^2}{4} - \frac{a^2}{2}\right) \text{ sq units}$$

14. Given, curve is

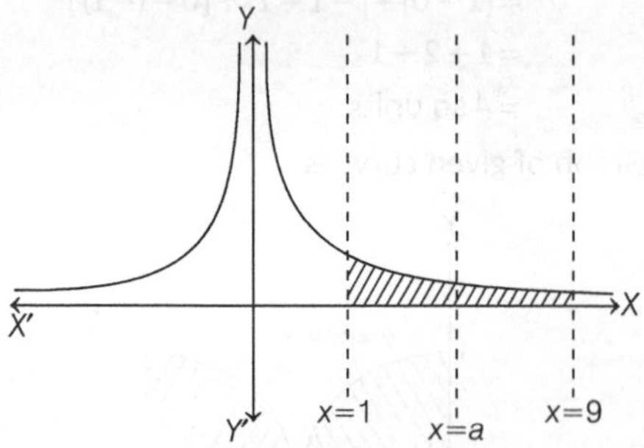

It is given that

$$2\int_1^a \frac{1}{x^2}\, dx = \int_1^9 \frac{1}{x^2}\, dx$$

$$\Rightarrow \quad 2\left[-\frac{1}{x}\right]_1^a = \left[-\frac{1}{x}\right]_1^9$$

$$\Rightarrow \quad 2\left(-\frac{1}{a} + 1\right) = \left(-\frac{1}{9} + 1\right)$$

$$\Rightarrow \quad -\frac{2}{a} + 2 = -\frac{1}{9} + 1$$

$$\Rightarrow \quad -\frac{2}{a} = -\frac{10}{9}$$

$$\Rightarrow \quad a = \frac{9}{5}$$

15. The given equations of curve are $y = x$ and $y = x^3$.

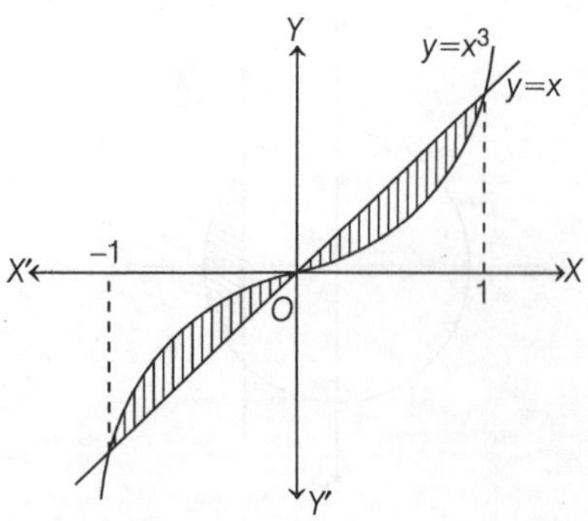

For find the point of intersection of the two curves

$$y = x \text{ and } y = x^3$$

$$\Rightarrow \quad x^3 = x$$

$$\Rightarrow \quad x^3 - x = 0$$

$$\Rightarrow \quad x(x^2 - 1) = 0$$

$$\Rightarrow \quad x(x-1)(x+1) = 0$$

So, $\quad x = 0, 1, -1$

The area of shaded region

$$= \left|\int_{-1}^{0} (y_{\text{line}} - y_{\text{curve}})\, dx\right| + \int_0^1 (y_{\text{line}} - y_{\text{curve}})\, dx$$

$$= \left|\int_{-1}^{0} (x - x^3)\, dx\right| + \int_0^1 (x - x^3)\, dx$$

$$= \left|\left[\frac{x^2}{2} - \frac{x^4}{4}\right]_{-1}^{0}\right| + \left[\frac{x^2}{2} - \frac{x^4}{4}\right]_0^1$$

$$= \left|0 - \left(\frac{1}{2} - \frac{1}{4}\right)\right| + \left(\frac{1}{2} - \frac{1}{4}\right)$$

$$= \left(\frac{1}{4}\right) + \frac{1}{4}$$

$$= \frac{1}{2} \text{ sq unit}$$

16. Given, curves are $y^2 = x$ and $y = |x|$

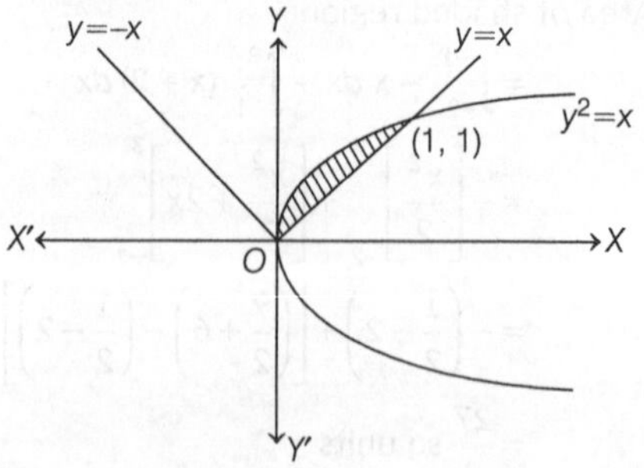

$\therefore$ Required area $= \int_0^1 (\sqrt{x} - x)\, dx$

$$= \left[\frac{2}{3}x^{3/2} - \frac{x^2}{2}\right]_0^1$$

$$= \frac{2}{3} - \frac{1}{2}$$

$$= \frac{4-3}{6}$$

$$= \frac{1}{6} \text{ sq units}$$

17. Given, curves are

$$\frac{x^2}{a^2} + \frac{y^2}{b^2} = 1 \text{ and } \frac{x}{a} + \frac{y}{b} = 1$$

$$\Rightarrow \quad y = \frac{b}{a}(a - x)$$

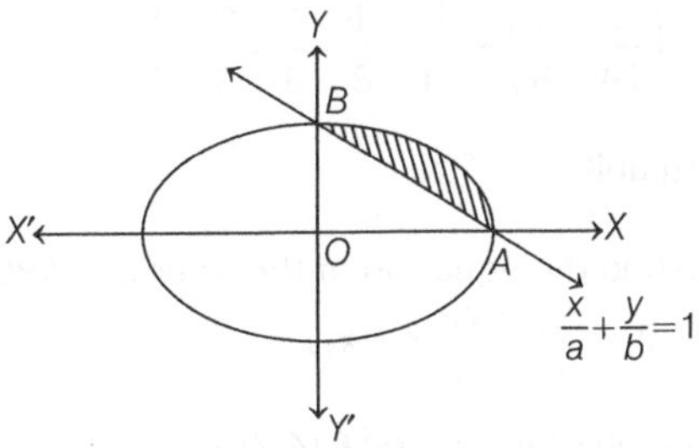

Hence, the required area

$$= \frac{\pi ab}{4} - \text{ar}(\Delta OAB)$$

$$= \frac{\pi ab}{4} - \frac{b}{a}\int_0^a (a-x)\,dx$$

$$= \frac{\pi ab}{4} - \frac{b}{a}\left[ax - \frac{x^2}{2}\right]_0^a$$

$$= \frac{\pi ab}{4} - \frac{b}{a}\left[a^2 - \frac{a^2}{2}\right]$$

$$= \frac{\pi ab}{4} - \frac{b}{a}\left(\frac{a^2}{2}\right)$$

$$= \frac{\pi ab}{4} - \frac{ab}{2} = \frac{ab}{2}\left(\frac{\pi}{2} - 1\right) \text{ sq units}$$

18. Given, $y = x^2$ and $y = |x|$

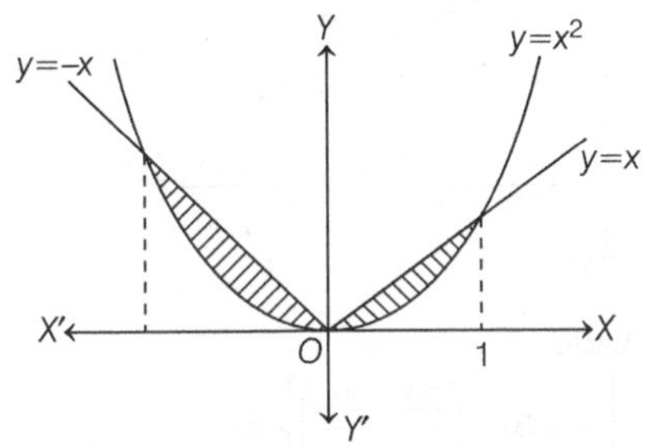

The shaded area is symmetrical about Y-axis.

∴ Area of shaded region

$$= 2\int_0^1 (x - x^2)\,dx$$

$$= 2\left[\frac{x^2}{2} - \frac{x^3}{3}\right]_0^1$$

$$= 2\left(\frac{1}{2} - \frac{1}{3}\right)$$

$$= 2\cdot\frac{1}{6} = \frac{1}{3} \text{ sq unit}$$

19. We have, $y = \frac{3x^2}{4}$ and line $3x - 2y + 12 = 0$

$$\Rightarrow \quad y = \frac{3x+12}{2}$$

So, $\frac{3x+12}{2} = \frac{3x^2}{4}$

$$\Rightarrow \quad 3x^2 - 6x - 24 = 0$$

$$\Rightarrow \quad x^2 - 2x - 8 = 0$$

$$\Rightarrow \quad x^2 - 4x + 2x - 8 = 0$$

$$\Rightarrow \quad x(x-4) + 2(x-4) = 0$$

$$\Rightarrow \quad (x-4)(x+2) = 0$$

So, $x = -2, \ 4$

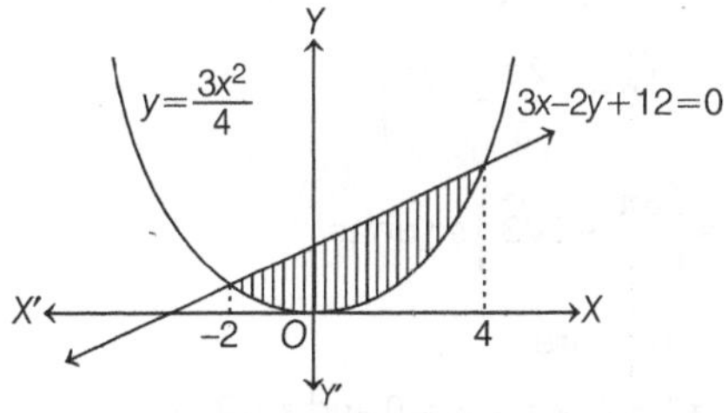

Hence, the required area

$$= \int_{-2}^{4}\left[\left(\frac{3x+12}{2}\right) - \frac{3x^2}{4}\right]dx$$

$$= \frac{1}{4}\int_{-2}^{4} (6x + 24 - 3x^2)\,dx$$

$$= \frac{3}{4}\int_{-2}^{4} (2x + 8 - x^2)\,dx$$

$$= \frac{3}{4}\left[x^2 + 8x - \frac{x^3}{3}\right]_{-2}^{4}$$

$$= \frac{3}{4}\left[\left(16 + 32 - \frac{64}{3}\right) - \left(4 - 16 + \frac{8}{3}\right)\right]$$

$$= \frac{3}{4}(36) = 27 \text{ sq units}$$

20. Given, curves are

$$x^2 + y^2 = 4 \text{ and } (x-2)^2 + y^2 = 4$$

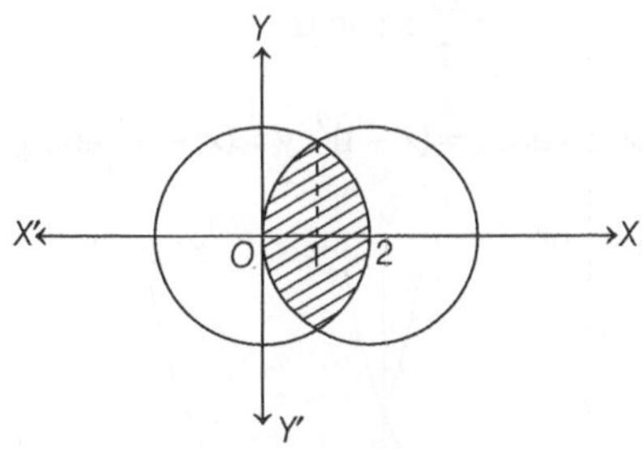

Both circles intersect at $x = 1$ and shaded area is symmetric about X-axis.

∴ Required area,

$$A = 2\left[\int_0^1 \sqrt{4-(x-2)^2}\,dx + \int_1^2 \sqrt{4-x^2}\,dx\right]$$

$$= 2\left[\frac{(x-2)}{2}\sqrt{4-(x-2)^2} + \frac{4}{2}\sin^{-1}\left(\frac{x-2}{2}\right)\right]_0^1$$

$$+ 2\left[\frac{x}{2}\sqrt{4-x^2} + \frac{4}{2}\sin^{-1}\left(\frac{x}{2}\right)\right]_1^2$$

$$= 2\left[-\frac{\sqrt{3}}{2} + 2\sin^{-1}\left(-\frac{1}{2}\right) - 2\sin^{-1}(-1)\right]$$

$$+ 2\left[2\sin^{-1}(1) - \frac{\sqrt{3}}{2} - 2\sin^{-1}\left(\frac{1}{2}\right)\right]$$

$$= 2\left(-\frac{\sqrt{3}}{2} - \frac{\pi}{3} + \pi + \pi - \frac{\sqrt{3}}{2} - \frac{\pi}{3}\right)$$

$$= 2\left(2\pi - \frac{2\pi}{3} - \sqrt{3}\right)$$

$$= \left(\frac{8\pi}{3} - 2\sqrt{3}\right) \text{ sq units}$$

21. Given, curves are

$y = x^2 + 2,\ y = x, x = 0$ and $x = 3$

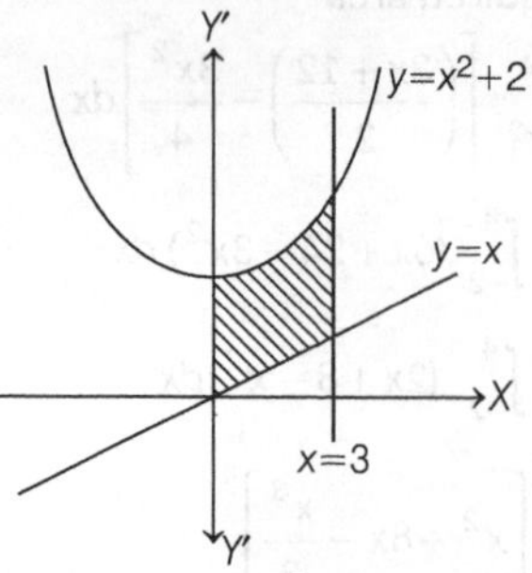

$\therefore$ Required area $= \int_0^3 [(x^2 + 2) - x]\, dx$

$$= \left[\frac{x^3}{3} + 2x - \frac{x^2}{2}\right]_0^3$$

$$= \left(9 + 6 - \frac{9}{2}\right)$$

$$= \frac{21}{2} \text{ sq units}$$

22. Given, curves are $y = (x-1)^2, y = (x+1)^2$ and $y = \frac{1}{4}$

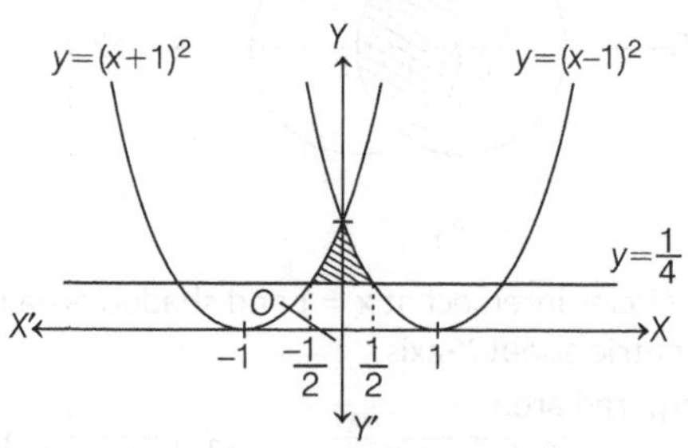

Hence, the area

$$= \int_{-1/2}^{0} \left[(x+1)^2 - \frac{1}{4}\right] dx + \int_0^{1/2} \left[(x-1)^2 - \frac{1}{4}\right] dx$$

$$= \left[\frac{(x+1)^3}{3} - \frac{x}{4}\right]_{-1/2}^{0} + \left[\frac{(x-1)^3}{3} - \frac{x}{4}\right]_0^{1/2}$$

$$= \frac{1}{3} - \left(\frac{1}{24} + \frac{1}{8}\right) - \frac{1}{24} - \frac{1}{8} + \frac{1}{3} = \frac{2}{3} - \frac{1}{3}$$

$$= \frac{1}{3} \text{ sq unit}$$

23. First, we find the equation of the sides of ΔABC by using $y - y_1 = \frac{y_2 - y_1}{x_2 - x_1}(x - x_1)$

The equation of $A(2, 5)$ and $B(4, 7)$ is

$$y - 5 = \frac{7-5}{4-2}(x - 2)$$

$$\Rightarrow \quad x - y + 3 = 0$$

The equation of BC is

$$y - 7 = \frac{2-7}{6-4}(x - 4)$$

$$\Rightarrow \quad 5x + 2y - 34 = 0$$

The equation of AC is

$$y - 5 = \frac{2-5}{6-2}(x - 2)$$

$$\Rightarrow \quad 3x + 4y - 26 = 0$$

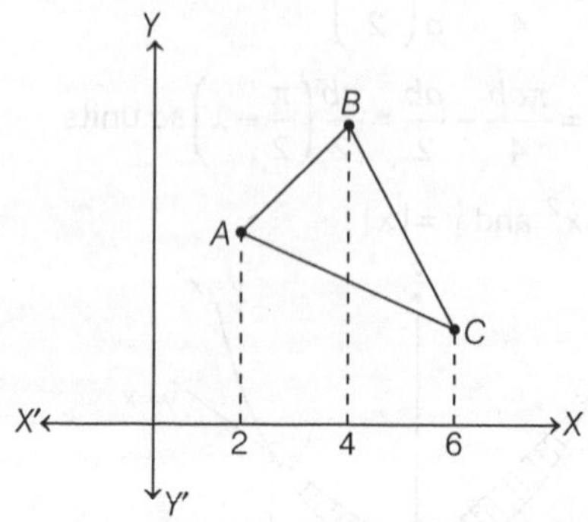

$\therefore$ Area of ΔABC

$$= \int_2^4 \left\{(x+3) - \left(\frac{26-3x}{4}\right)\right\} dx + \int_4^6 \left\{\left(\frac{34-5x}{2}\right) - \left(\frac{26-3x}{4}\right)\right\} dx$$

$$= \frac{1}{4}\int_2^4 (7x - 14)\, dx + \frac{1}{4}\int_4^6 (42 - 7x)\, dx$$

$$= \frac{1}{4}\left[\left(\frac{7x^2}{2} - 14x\right)_2^4 + \left(42x - \frac{7x^2}{2}\right)_4^6\right]$$

$$= \frac{1}{4}[\{(56 - 56) - (14 - 28)\} + \{(252 - 126) - (168 - 56)\}]$$

$$= 7 \text{ sq units}$$

24. The equations of given lines are

$$x + 2y = 2 \quad \text{...(i)}$$

$$y - x = 1 \quad \text{...(ii)}$$

and $\quad 2x + y = 7 \quad$ ...(iii)

Solving Eqs. (i), (ii) and (iii) pair wise, we get

$A(0,1), B(2,3)$ and $C(4,-1)$

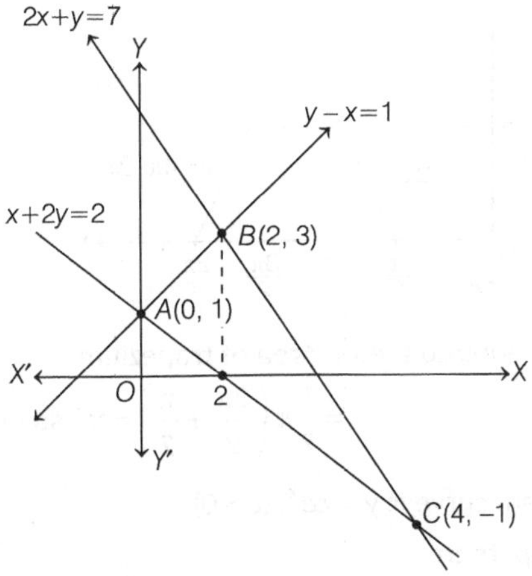

Area of ΔABC

$$=\int_0^2 \left\{(x+1)-\left(\frac{2-x}{2}\right)\right\} dx + \int_2^4 \left\{(7-2x)-\left(\frac{2-x}{2}\right)\right\} dx$$

$$=\int_0^2 \frac{3x}{2} dx + \int_2^4 \left(6-\frac{3x}{2}\right) dx$$

$$=\left[\frac{3x^2}{4}\right]_0^2 + \left[6x-\frac{3x^2}{4}\right]_2^4$$

$=6$ sq units

Level 2

1. (d) The given curve is

$$\sqrt{x}+\sqrt{y}=1$$

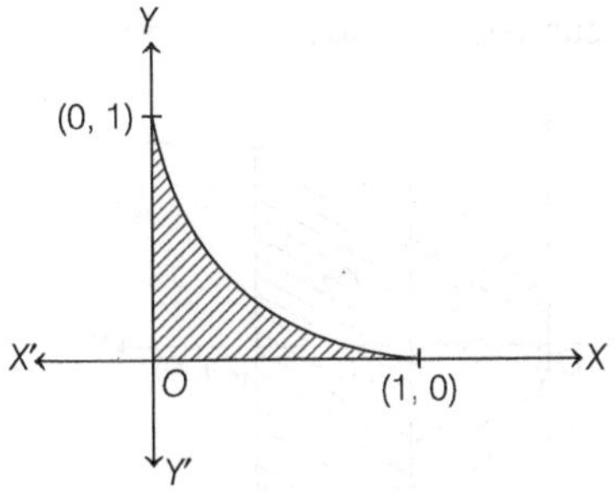

$\therefore$ The rquired area

$$=\int_0^1 (1-\sqrt{x})^2 dx$$

$$=\int_0^1 (1+x-2\sqrt{x}) dx$$

$$=\left[x+\frac{x^2}{2}-2\cdot\frac{2}{3}x^{3/2}\right]_0^1$$

$$=1+\frac{1}{2}-2\cdot\frac{2}{3}(1)$$

$$=\frac{1}{6} \text{ sq unit}$$

2. (a) Given, curve $y=\sin^{-1} x$

$\Rightarrow \quad x=\sin y$

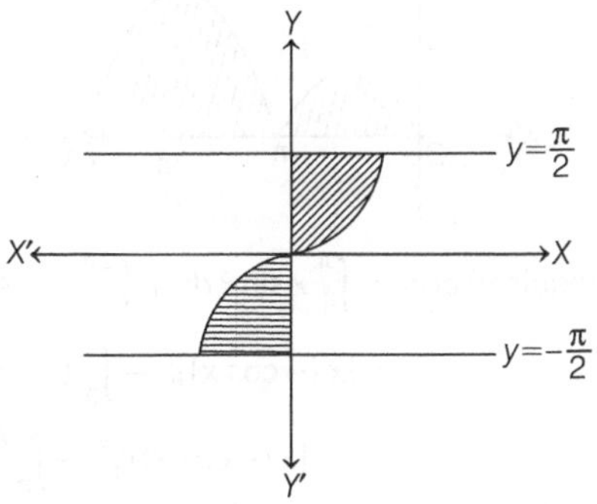

The shaded region is symmetrical about X-axis.

So, required area $=2\int_0^{\pi/2} x\, dy$

$$=2\int_0^{\pi/2} \sin y\, dy$$

$$=-2[\cos y]_0^{\pi/2}$$

$$=-2(0-1)$$

$$=2 \text{ sq units}$$

3. (a) Given, the equation of the circle is $x^2+y^2=4$ and line $x=1$

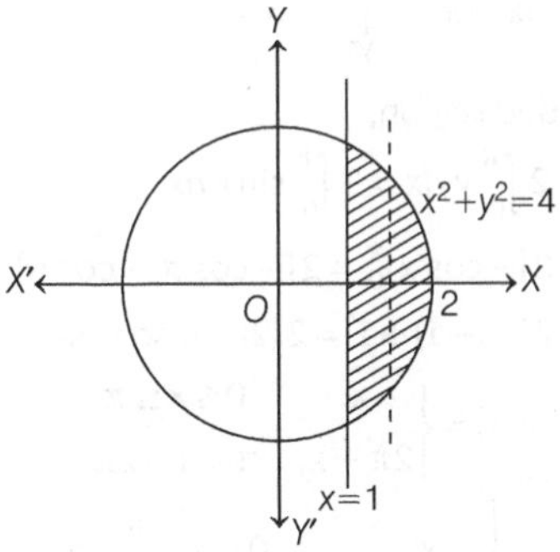

$\therefore$ The required area

$$=2\int_1^2 y\, dx$$

$$=2\int_1^2 \sqrt{4-x^2}\, dx$$

$$=2\left[\frac{x}{2}\sqrt{4-x^2}+\frac{2^2}{2}\sin^{-1}\frac{x}{2}\right]_1^2$$

$$=0+4\sin^{-1}(1)-2\left\{\frac{\sqrt{3}}{2}+2\sin^{-1}\left(\frac{1}{2}\right)\right\}$$

$$=\frac{4\pi-3\sqrt{3}}{3} \text{ sq units}$$

4. (c) We have, $y = x|\sin x|$

$$y = \begin{cases} x\sin x, & \text{if } \sin x \geq 0 \text{ i.e. } 0 \leq x \leq \pi \\ -x\sin x, & \text{if } \sin x < 0 \text{ i.e. } \pi < x \leq 2\pi \end{cases}$$

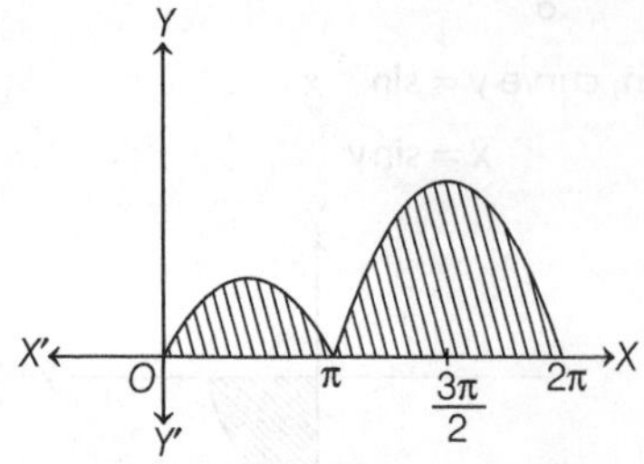

The required area $= \int_0^{\pi} x\sin x\, dx + \int_{\pi}^{2\pi} (-x\sin x)\, dx$

$$= [x(-\cos x)]_0^{\pi} - \int_0^{\pi} (-\cos x)\, dx$$

$$- [x(-\cos x)]_{\pi}^{2\pi} + \int_{\pi}^{2\pi} (-\cos x)\, dx$$

$$= \pi + [\sin x]_0^{\pi} + (2\pi + \pi) - [\sin x]_{\pi}^{2\pi}$$

$$= \pi + 3\pi = 4\pi \text{ sq units}$$

5. (c) The bounded area in the graph of $y = |\sin x|, |x| = \pi$ and X-axis is

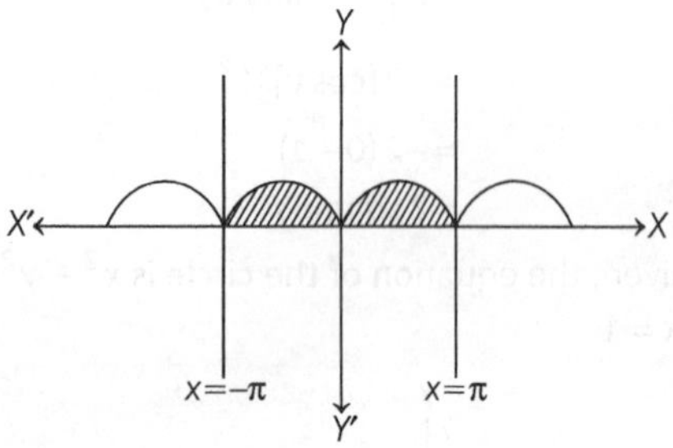

$\therefore$ Area of shaded region,

$$A = 2\int_0^{\pi} y\, dx = 2\int_0^{\pi} \sin x\, dx$$

$$= 2[-\cos x]_0^{\pi} = 2[-\cos \pi + \cos 0]$$

$$= 2[-(-1) + 1] = 2(2) = 4 \text{ sq units}$$

6. (c) Let $\cos^{-1}(\cos x) = \begin{cases} x, & 0 \leq x \leq \pi \\ 2\pi - x, & \pi < x \leq 2\pi \end{cases}$

and $\sin^{-1}(\sin x) = \begin{cases} x, & 0 \leq x \leq \frac{\pi}{2} \\ \pi - x, & \frac{\pi}{2} < x \leq \frac{3\pi}{2} \\ x - 2\pi, & \frac{3\pi}{2} < x \leq 2\pi \end{cases}$

Hence, $f(x) = \begin{cases} 0, & \text{if } x \in \left(0, \frac{\pi}{2}\right) \\ 2x - \pi, & \text{if } x \in \left(\frac{\pi}{2}, \pi\right) \\ \pi, & \text{if } x \in \left(\pi, \frac{3\pi}{2}\right) \\ 4\pi - 2x, & \text{if } x \in \left(\frac{3\pi}{2}, 2\pi\right) \end{cases}$

So, the graph of $f(x)$ is

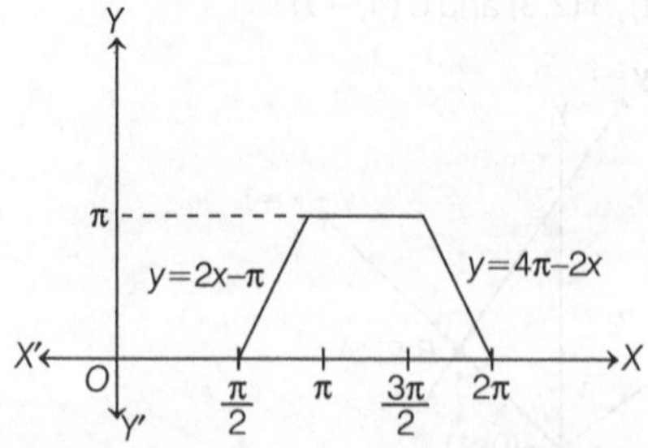

$\therefore$ The required area = Area of trapezium

$$= \frac{1}{2}\pi\left(\frac{3\pi}{2} + \frac{\pi}{2}\right) = \pi^2 \text{ sq units}$$

7. (b) Given, curve is $y = ce^x$, $(c > 0)$

Let $q > p$, then

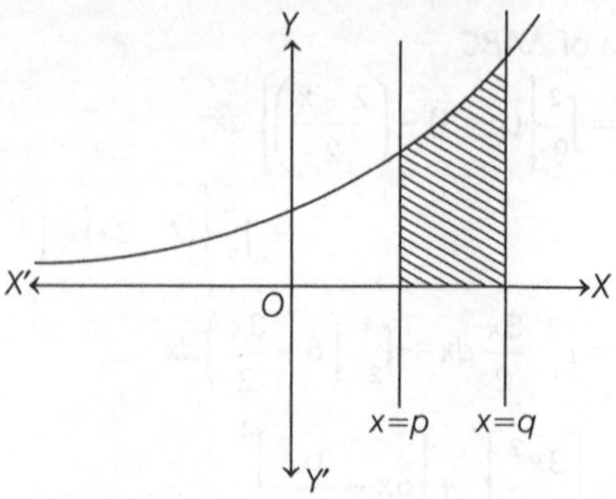

$\therefore$ Required area

$$= \int_p^q ce^x dx = c[e^x]_p^q$$

$$= c(e^q - e^p)$$

$$= ce^q - ce^p$$

$$= f(q) - f(p)$$

$$= |f(p) - f(q)|$$

8. (b) Given, curve is $y^2 = 4ax$

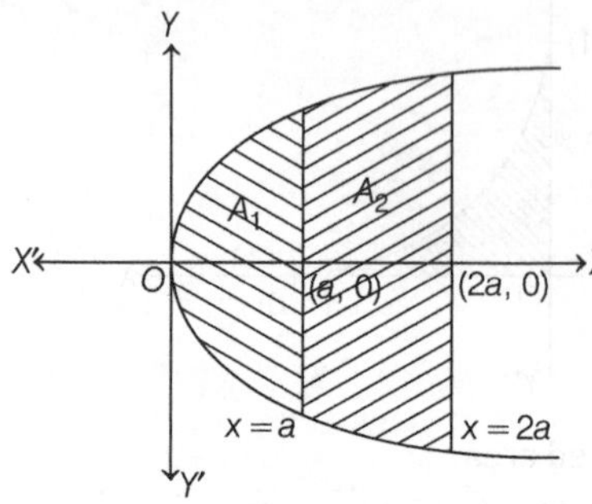

$\therefore \quad A_1 = 2\int_0^a \sqrt{4ax}\, dx$

$$= 4\sqrt{a}\int_0^a \sqrt{x}\, dx$$

$$= 4\sqrt{a}\left[\frac{2}{3}x^{3/2}\right]_0^a$$

$$=\frac{8}{3}\sqrt{a}(a^{3/2})$$

$$=\frac{8}{3}a^2 \text{ sq units}$$

$$\therefore \quad A_2=2\left[\int_a^{2a}\sqrt{4ax}\,dx\right]$$

$$=2\cdot 2\sqrt{a}\left[\frac{2}{3}x^{3/2}\right]_a^{2a}$$

$$=\frac{8}{3}\sqrt{a}[(2a)^{3/2}-(a)^{3/2}]$$

$$A_2=\frac{8}{3}a^2[2\sqrt{2}-1] \text{ sq units}$$

Now, $$\frac{A_1}{A_2}=\frac{\frac{8}{3}a^2}{\frac{8}{3}a^2(2\sqrt{2}-1)}$$

$$=\frac{1}{2\sqrt{2}-1}\times\frac{2\sqrt{2}+1}{2\sqrt{2}+1}$$

$$=\frac{1}{7}(2\sqrt{2}+1)$$

9. (c) Given,

$$S(\alpha)=\{(x,y): y^2\le x, 0\le x\le \alpha\} \text{ and}$$

$A(\alpha)$ is area of the region $S(\alpha)$

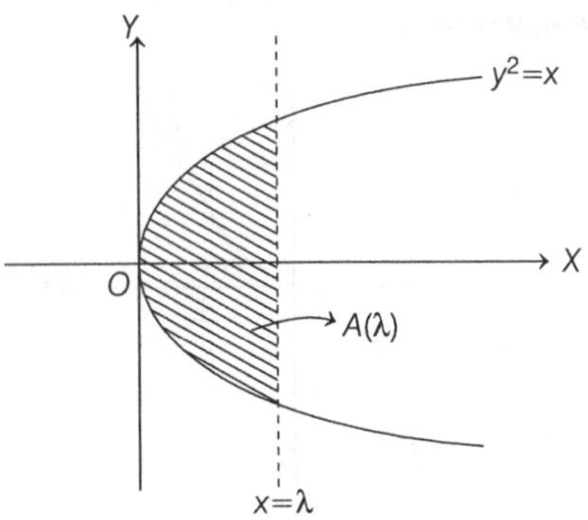

Clearly, $$A(\lambda)=2\int_0^{\lambda}\sqrt{x}\,dx=2\left[\frac{x^{3/2}}{3/2}\right]_0^{\lambda}=\frac{4}{3}\lambda^{3/2}$$

Since, $$\frac{A(\lambda)}{A(4)}=\frac{2}{5}, (0<\lambda<4)$$

$$\Rightarrow \quad \frac{\lambda^{3/2}}{4^{3/2}}=\frac{2}{5}$$

$$\Rightarrow \quad \left(\frac{\lambda}{4}\right)^3=\left(\frac{2}{5}\right)^2$$

$$\Rightarrow \quad \frac{\lambda}{4}=\left(\frac{4}{25}\right)^{1/3}$$

$$\Rightarrow \quad \lambda=4\left(\frac{4}{25}\right)^{1/3}$$

10. (b) Required area is same as area bounded by curve

$$y=\begin{cases}-2-x, & x<-1\\ x^3, & -1\le x\le 1 \text{ and } y=0\\ 2-x, & x>1\end{cases}$$

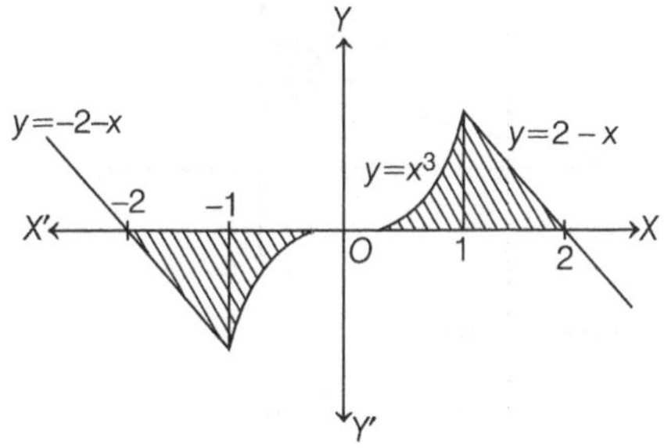

$$\therefore \text{ Required area}=2\int_0^1 x^3dx+2\int_1^2(2-x)\,dx$$

$$=2\left[\frac{x^4}{4}\right]_0^1+2\left[2x-\frac{x^2}{2}\right]_1^2$$

$$=2\left[\frac{1}{4}\right]+2\left[4-\frac{4}{2}-2+\frac{1}{2}\right]=\frac{3}{2} \text{ sq units}$$

11. (b) Given, curve is $y=x(x-1)^2$

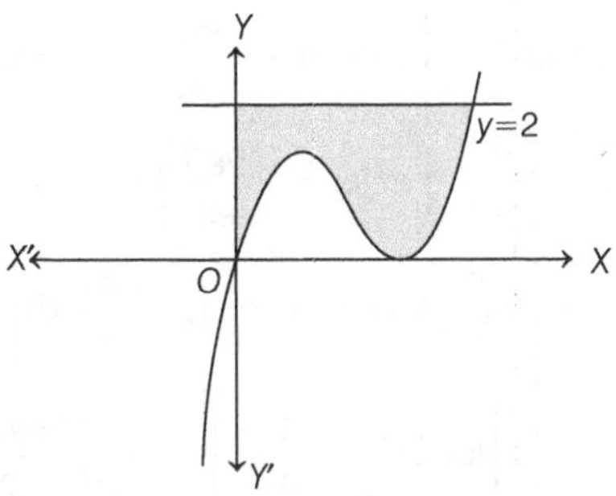

Hence, the required area

$$=\int_0^2 (2-x(x-1)^2)\,dx$$

$$=\int_0^2 [2-x(x^2-2x+1)]\,dx$$

$$=\int_0^2 [2-(x^3-2x^2+x)]\,dx$$

$$=\left(2x-\frac{x^4}{4}+\frac{2}{3}x^3-\frac{x^2}{2}\right)_0^2$$

$$=\left(4-4+\frac{16}{3}-2\right)$$

$$=\frac{16-6}{3}=\frac{10}{3} \text{ sq units}$$

12. (c) The equation of curves are

$$y=e^x \text{ and } y=e^{-x}$$

$$\therefore \quad e^x=\frac{1}{e^x}$$

$\Rightarrow \quad e^{2x} = e^0$

$\Rightarrow \quad x = 0$

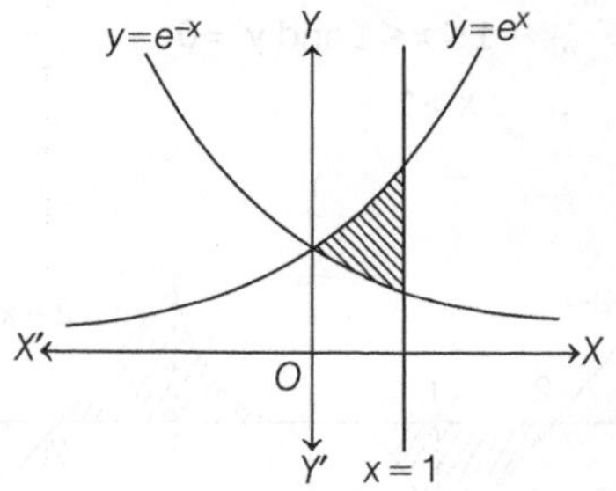

$\therefore$ Required area

$$= \int_0^1 (e^x - e^{-x})\,dx$$
$$= [e^x + e^{-x}]_0^1$$
$$= e + e^{-1} - e^0 - e^0$$
$$= e + \frac{1}{e} - 1 - 1$$
$$= \left(e + \frac{1}{e} - 2\right) \text{ sq units}$$

13. (a) The intersection of curves $y = ex\log x$ and $y = \dfrac{\log x}{ex}$ is $x = 1$ and $\dfrac{1}{e}$ $\left[\because ex\log x = \dfrac{\log x}{ex} \Rightarrow x = 1/e \text{ and } x = 1\right]$

So, area $= \left|\int_{1/e}^1 \left(ex\log x - \dfrac{\log x}{ex}\right)dx\right|$

$$= \left|e\int_{1/e}^1 x\log x\,dx - \frac{1}{e}\int_{1/e}^1 \frac{\log x}{x}\,dx\right|$$
$$= \left|e\left[\log x\cdot\frac{x^2}{2} - \frac{1}{4}x^2\right]_{1/e}^1 - \frac{1}{e}\left[\frac{(\log x)^2}{2}\right]_{1/e}^1\right|$$
$$= \left|e\left[-\frac{1}{4} - \left(\frac{(-1)e^{-2}}{2} - \frac{1}{4}e^{-2}\right)\right] - \frac{1}{e}\left(0 - \frac{1}{2}\right)\right|$$
$$= \left|-\frac{e}{4} + \left(\frac{3}{4}e^{-1}\right) + \frac{2}{4e}\right|$$
$$= \left|\frac{-e}{4} + \frac{5}{4e}\right| = \left(\frac{e^2 - 5}{4e}\right) \text{ sq units}$$

14. (b) Given, curves are $y = |x - 1|$ and $y = -|x - 1| + 1$

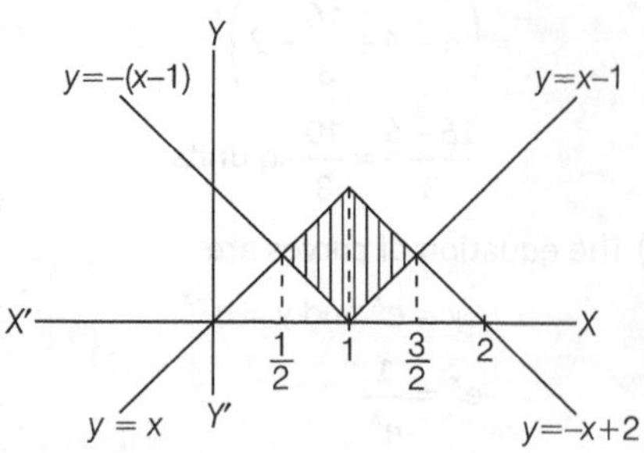

$\therefore$ Required area

$$= \int_{1/2}^1 [x - (-x + 1)]\,dx + \int_1^{3/2} [(-x + 2) - (x - 1)]\,dx$$
$$= \int_{1/2}^1 (2x - 1)\,dx + \int_1^{3/2} (-2x + 3)\,dx$$
$$= [x^2 - x]_{1/2}^1 + (-x^2 + 3x)_1^{3/2}$$
$$= \frac{1}{4} + \frac{1}{4} = \frac{1}{2} \text{ sq unit}$$

15. (c) The given inequalities are $|x - y| \le 2$ and $|x + y| \le 2$. On drawing, the above inequalities, we get a square

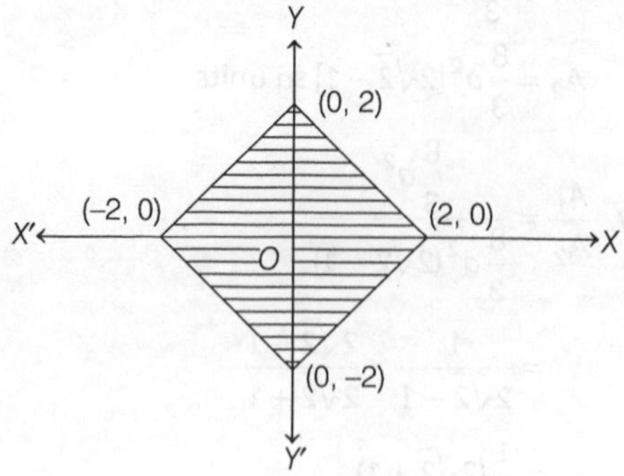

Now, the area of shaded region is equal to the area of a square having side length $\sqrt{(2-0)^2 + (0-2)^2}$

$$= 2\sqrt{2} \text{ units}$$

16. (a) Given, $y = \tan x$

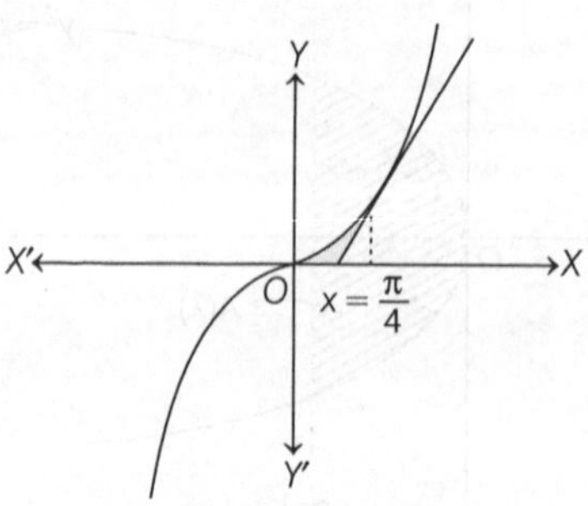

Slope at $x = \dfrac{\pi}{4}$ is $\left.\dfrac{dy}{dx}\right|_{x=\frac{\pi}{4}} = \sec^2\left(\dfrac{\pi}{4}\right) = 2$

and if $x = \dfrac{\pi}{4}$, then $y = 1$

Then, equation of tangent

$$(y - 1) = 2\left(x - \frac{\pi}{4}\right)$$

$$\Rightarrow \quad y = 2x + 1 - \frac{\pi}{2}$$

Now, at $y = 0$, $x = \dfrac{\pi}{4} - \dfrac{1}{2}$

So, $A = \displaystyle\int_0^{\pi/4} \tan x\,dx - \int_{\pi/4-1/2}^{\pi/4} \left(2x + 1 - \frac{\pi}{2}\right)dx$

$$= [\ln \sec x]_0^{\pi/4} - \left[x^2 + x - \frac{\pi x}{2}\right]_{\pi/4-1/2}^{\pi/4}$$

$$A = \left(\frac{\ln 4 - 1}{4}\right) \text{ sq units}$$

17. (b) The graph of given curves

$y = x, x = e$ and $y = \frac{1}{x}$ is

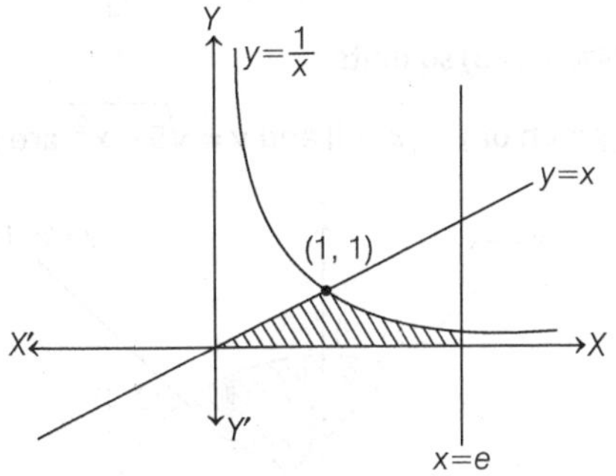

$\therefore$ Area of required region

$$= \int_0^1 x\,dx + \int_1^e \frac{1}{x}\,dx$$

$$= \left[\frac{x^2}{2}\right]_0^1 + [1 \log x]_1^e$$

$$= \frac{1}{2} + [1 - 0] = \frac{3}{2} \text{ sq units}$$

18. (b) Given region

$\{(x, y) : 0 \le y \le x^2 + 1, 0 \le y \le x + 1, \frac{1}{2} \le x \le 2\}$

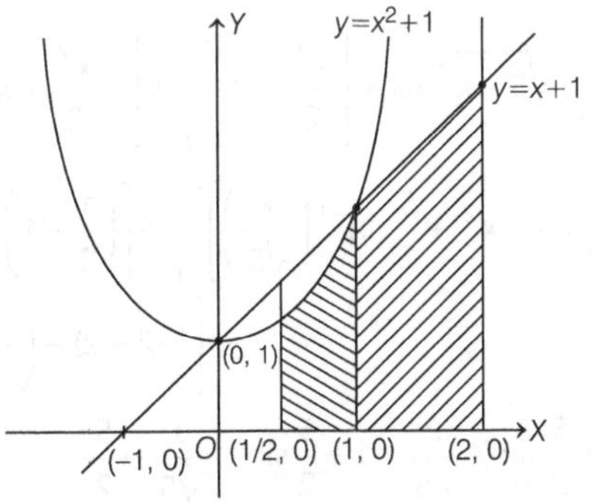

= Shaded region in the figure

$\therefore$ Area of required region is

$$= \int_{1/2}^1 (x^2 + 1)\,dx + \int_1^2 (x + 1)dx$$

$$= \left[\frac{x^3}{3} + x\right]_{1/2}^1 + \left[\frac{x^2}{2} + x\right]_1^2$$

$$= \frac{1}{3} + 1 - \frac{1}{24} - \frac{1}{2} + \frac{4}{2} + 2 - \frac{1}{2} - 1$$

$$= 3 + \frac{1}{3} - \frac{1}{24} = \frac{72 + 8 - 1}{24} = \frac{79}{24} \text{ sq unit}$$

19. (d) Given that

$\{(x, y) \in R \times R \mid x \ge 0, 2x^2 \le y \le 4 - 2x\}$

Let curve $y = 2x^2$ and line $y = 4 - 2x$

For point of intersection,

$$4 - 2x = 2x^2$$

$\Rightarrow \quad 2x^2 + 2x - 4 = 0 \Rightarrow (x - 1)(x + 2) = 0$

So, $\quad x = 1$ and $x = -2$

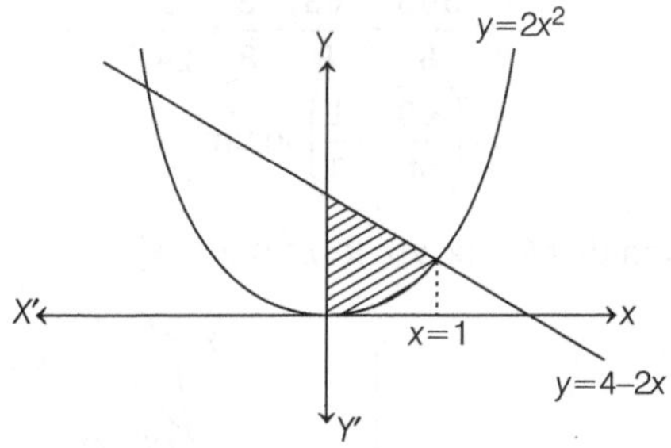

$\therefore$ Required area

$$= \int_0^1 [(4 - 2x) - (2x^2)]\,dx$$

$$= \left[4x - x^2 - \frac{2x^3}{3}\right]_0^1$$

$$= 4 - 1 - \frac{2}{3} = \frac{7}{3} \text{ sq units}$$

20. (d) We have,

$$f(x) = \begin{cases} x, & 0 \le x < \frac{1}{2} \\ 1 - x, & \frac{1}{2} \le x \le 1 \end{cases}$$

and $\quad g(x) = \left(x - \frac{1}{2}\right)^2 \; x \in R$

So, graph of $f(x)$ and $g(x)$ is

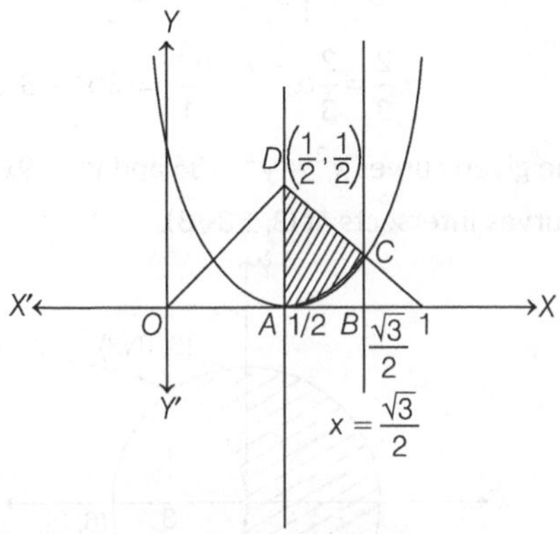

The point of intersection of the curve $f(x) = (1 - x)$, $\frac{1}{2} \le x \le 1$ and $g(x) = \left(x - \frac{1}{2}\right)^2$ is $C\left(\frac{\sqrt{3}}{2}, \left(1 - \frac{\sqrt{3}}{2}\right)\right)$.

$$\therefore \text{ Required area} = \int_{1/2}^{\sqrt{3}/2} \left[(1-x) - \left(x - \frac{1}{2}\right)^2\right] dx$$

$$= \int_{1/2}^{\sqrt{3}/2} \left(\frac{3}{4} - x^2\right) dx$$

$$= \left[\frac{3}{4}x - \frac{x^3}{3}\right]_{1/2}^{\sqrt{3}/2}$$

$$= \frac{3\sqrt{3}}{8} - \frac{\sqrt{3}}{8} - \frac{3}{8} + \frac{1}{24}$$

$$= \left(\frac{\sqrt{3}}{4} - \frac{1}{3}\right) \text{ sq units}$$

21. (c) Given that $R = \{x, y) \in R : x^2 \le y \le 2x\}$

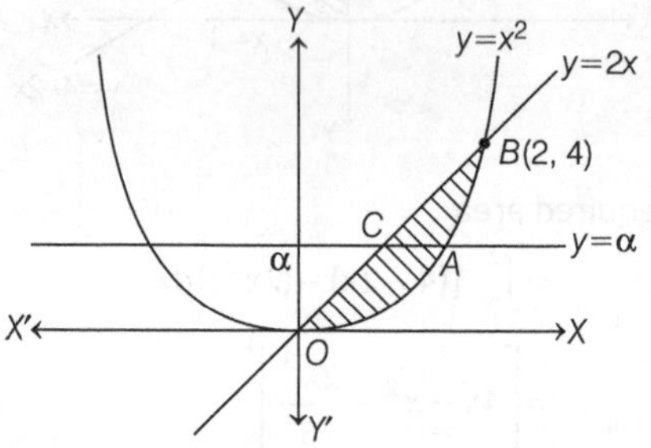

According to the question,
Area of $OABCO = 2$ area of $OACO$

$$\int_0^4 \left(\sqrt{y} - \frac{y}{2}\right) dy = 2\int_0^{\alpha} \left(\sqrt{y} - \frac{y}{2}\right) dy$$

$$\Rightarrow \left[\frac{2}{3}y^{3/2} - \frac{y^2}{4}\right]_0^4 = 2\left[\frac{2}{3}y^{3/2} - \frac{y^2}{4}\right]_0^{\alpha}$$

$$\Rightarrow \frac{16}{3} - 4 = 2\left[\frac{2}{3}(\alpha)^{3/2} - \frac{\alpha^2}{4}\right]$$

$$\Rightarrow \frac{4}{3} = 2\left[\frac{2}{3}\alpha^{3/2} - \frac{\alpha^2}{4}\right]$$

$$\Rightarrow \frac{2}{3} = \frac{2}{3}\alpha^{3/2} - \frac{\alpha^2}{4} \Rightarrow 3\alpha^2 - 8\alpha^{3/2} + 8 = 0$$

22. (d) The given curves $x^2 + y^2 = 36$ and $y^2 = 9x$
The curves intersects at $(3, \pm 3\sqrt{3})$.

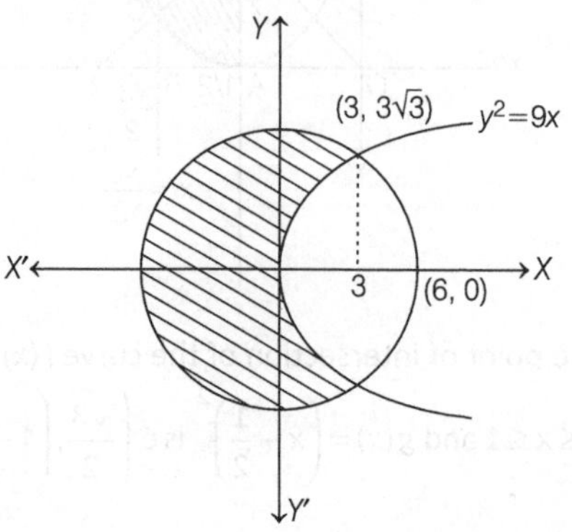

So, required area

$$= \pi r^2 - 2\left[\int_0^3 \sqrt{9x}\, dx + \int_3^6 \sqrt{36 - x^2}\, dx\right]$$

$$= 36\pi - 12\sqrt{3} - 2\left[\frac{x}{2}\sqrt{36 - x^2} + 18 \sin^{-1}\left(\frac{x}{6}\right)\right]_3^6$$

$$= 36\pi - 12\sqrt{3} - 2\left[9\pi - \left(\frac{9\sqrt{3}}{2} + 3\pi\right)\right]$$

$$= (24\pi - 3\sqrt{3}) \text{ sq units}$$

23. (b) The graph of $y = |x - 1|$ and $y = \sqrt{5 - x^2}$ are shown

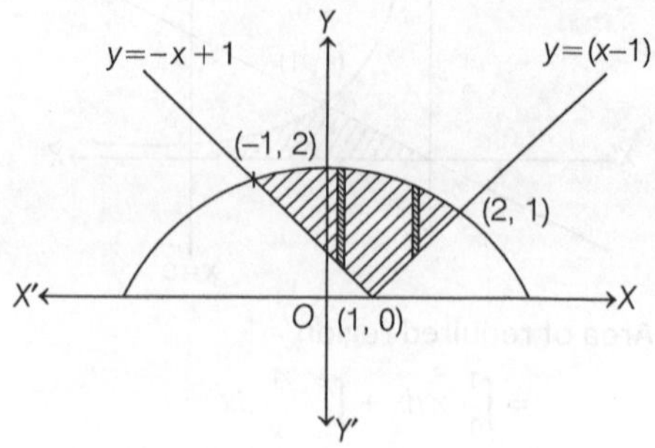

Let A be the area bounded by the given curves, then

$$A = \int_{-1}^{1} [\sqrt{5 - x^2} - (-x + 1)]\, dx + \int_1^2 [\sqrt{5 - x^2} - (x - 1)]\, dx$$

$$A = \int_{-1}^{1} \sqrt{5 - x^2}\, dx + \int_1^2 \sqrt{5 - x^2}\, dx + \int_{-1}^{1} (x - 1)\, dx + \int_1^2 (-x + 1)\, dx$$

$$= \int_{-1}^{2} \sqrt{5 - x^2}\, dx + \left[\frac{x^2}{2} - x\right]_{-1}^{1} + \left[-\frac{x^2}{2} + x\right]_1^2$$

$$= \left[\frac{x}{2}\sqrt{5 - x^2} + \frac{5}{2}\sin^{-1}\left(\frac{x}{\sqrt{5}}\right)\right]_{-1}^{2} + \left[\left(\frac{1}{2} - 1\right) - \left(\frac{1}{2} + 1\right)\right] + \left[(-2 + 2) - \left(-\frac{1}{2} + 1\right)\right]$$

$$= 1 + \frac{5}{2}\sin^{-1}\frac{2}{\sqrt{5}} + 1 + \frac{5}{2}\sin^{-1}\frac{1}{\sqrt{5}} - \frac{5}{2}$$

$$= -\frac{1}{2} + \frac{5}{2}\left[\sin^{-1}\left(\frac{2}{\sqrt{5}}\sqrt{1 - \frac{1}{5}} + \frac{1}{\sqrt{5}}\sqrt{1 - \frac{4}{5}}\right)\right]$$

$$= -\frac{1}{2} + \frac{5}{2}\sin^{-1}(1)$$

$$= -\frac{1}{2} + \frac{5}{2}\left(\frac{\pi}{2}\right)$$

$$= \frac{5\pi}{4} - \frac{1}{2}$$

$$= \left(\frac{5\pi - 2}{4}\right) \text{ sq units}$$

24. (c) We have,

$$y = x^3 - x^2 - x + 1$$

$$y = (x-1)^2(x+1)$$

i.e. two coincident roots are $x = 1, 1$ and other root is $x = -1$

$$\frac{dy}{dx} = 3x^2 - 2x - 1$$

$$\Rightarrow \quad (x-1)(3x+1) = 0$$

$$\Rightarrow \quad x = 1 \text{ and } x = -\frac{1}{3}$$

and solving $y = x + 1$ and $y = x^3 - x^2 - x + 1$

$$\Rightarrow \quad x = 0, 2, -1$$

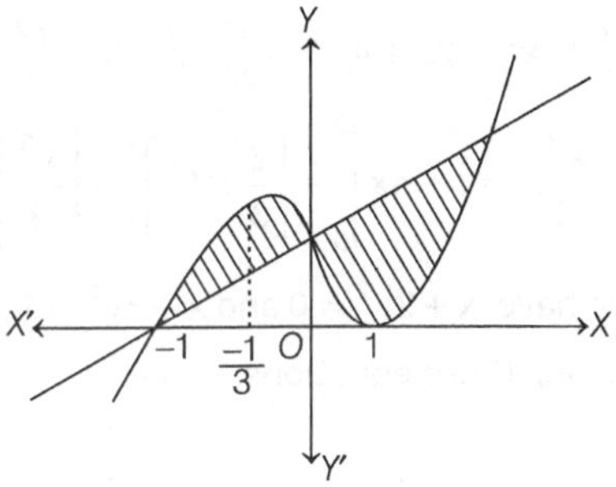

$\therefore$ Area, $A = \int_{-1}^{0} [(x^3 - x^2 - x + 1) - (x+1)]\,dx$

$$+ \int_0^2 [(x+1) - (x^3 - x^2 - x + 1)]\,dx$$

$$= \left[\frac{x^4}{4} - \frac{x^3}{3} - x^2\right]_{-1}^{0} + \left[x^2 + \frac{x^3}{3} - \frac{x^4}{4}\right]_0^2$$

$$= \left(1 - \frac{7}{12}\right) + \frac{8}{3}$$

$$= \frac{11}{3} - \frac{7}{12} = \frac{37}{12} \text{ sq units}$$

25. (b) The graph of $y = \sin x$ and $y = \cos x$

$$\cos x \le \sin x \text{ for } x \in \left[\frac{\pi}{4}, \frac{5\pi}{4}\right]$$

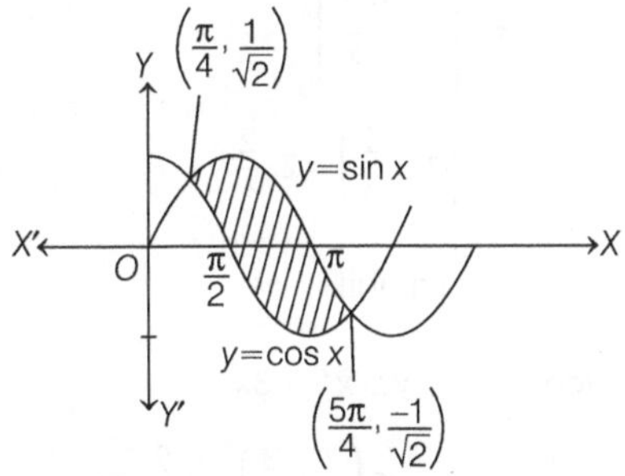

$\therefore$ The required area is

$$A = \int_{\pi/4}^{5\pi/4} (\sin x - \cos x)\,dx$$

$$= [-\cos x - \sin x]_{\pi/4}^{5\pi/4}$$

$$= -\left(\cos\frac{5\pi}{4} + \sin\frac{5\pi}{4}\right) + \left(\cos\left(\frac{\pi}{4}\right) + \sin\frac{\pi}{4}\right)$$

$$= -\left(-\frac{1}{\sqrt{2}} - \frac{1}{\sqrt{2}}\right) + \left(\frac{1}{\sqrt{2}} + \frac{1}{\sqrt{2}}\right)$$

$$= \sqrt{2} + \sqrt{2} = 2\sqrt{2} \text{ sq units}$$

26. (b) Solving given curves, we have

$$x - \sin x = x + \cos x$$

$$\therefore \quad \tan x = -1$$

For smallest bounded area, we must consider two consecutive solutions

i.e. $x = \frac{-\pi}{4}, \frac{3\pi}{4}$

$\therefore$ Required area

$$= \int_{-\pi/4}^{3\pi/4} [(x + \cos x) - (x - \sin x)\,dx$$

$$= \int_{-\pi/4}^{3\pi/4} [\cos x + \sin x]\,dx$$

$$= [\sin x - \cos x]_{-\pi/4}^{3\pi/4}$$

$$= \left(\frac{1}{\sqrt{2}} + \frac{1}{\sqrt{2}}\right) - \left(-\frac{1}{\sqrt{2}} - \frac{1}{\sqrt{2}}\right)$$

$$= 2\sqrt{2} \text{ sq units}$$

27. (b) Since, $\sin x$ and $\cos x > 0 \,\forall\, x \in \left[0, \frac{\pi}{2}\right]$, the graph of $y = \sin x + \cos x$ always lies above the graph of $y = |\cos x - \sin x|$.

Also, $\cos x > \sin x \,\forall\, x \in \left[0, \frac{\pi}{4}\right]$

and $\sin x > \cos x \,\forall\, x \in \left[\frac{\pi}{4}, \frac{\pi}{2}\right]$

So, area $= \int_0^{\pi/4} [(\sin x + \cos x) - (\cos x - \sin x)]\,dx$

$$+ \int_{\pi/4}^{\pi/2} [(\sin x + \cos x) - (\sin x - \cos x)]\,dx$$

$$= (4 - 2\sqrt{2}) \text{ sq units}$$

28. (a) Consider,

$$y = \cos^{-1}|\cos x|;\ x \in (0, \pi)$$

$$y = \cos^{-1}(\cos x)$$

$$y = x \quad \text{...(i)}$$

and $\quad y = (\cos^{-1}|\cos x|)^2$

$$y = x^2 \quad \text{...(ii)}$$

$\therefore$ Area of required region

$$= 4\int_0^1 (x - x^2)\,dx + 4\int_1^{\pi/2} (x^2 - x)\,dx$$

$$= 4\left[\frac{x^2}{2} - \frac{x^3}{3}\right]_0^1 + 4\left[\frac{x^3}{3} - \frac{x^2}{2}\right]_1^{\pi/2}$$

$$= \frac{2}{3} + \frac{\pi^2}{6}\left(\frac{\pi-3}{2}\right) + \frac{2}{3}$$

$$= \frac{4}{3} + \frac{\pi^2}{4}\left(\frac{\pi}{3} - 1\right) \text{ sq units}$$

29. (c) We have, $y = \sin kx$

Required area

$$A = \int_{2\pi/3k}^{\pi/k} (\sin kx)dx - \int_{\pi/k}^{5\pi/3k} (\sin kx)\,dx$$

$$= -\left[\frac{\cos kx}{k}\right]_{2\pi/3k}^{\pi/k} + \left[\frac{\cos kx}{k}\right]_{\pi/k}^{5\pi/3k}$$

$$= -\frac{1}{k}\left[-1 - \cos\frac{2\pi}{3}\right] + \frac{1}{k}\left[\cos\frac{5\pi}{3} - \cos\pi\right]$$

$$= -\frac{1}{k}\left[-1 + \frac{1}{2}\right] + \frac{1}{k}\left[\frac{1}{2} + 1\right]$$

$$= \frac{1}{2k} + \frac{3}{2k} = \frac{4}{2k} = \frac{2}{k}$$

But given that $\frac{2}{k} < 2$

$\Rightarrow \quad k > 1$

30. (b) Let the area of the region, enclosed by the parabola $y^2 = x$ and straight line $y = x$ is

$$A = \int_0^1 (\sqrt{x} - x)dx$$

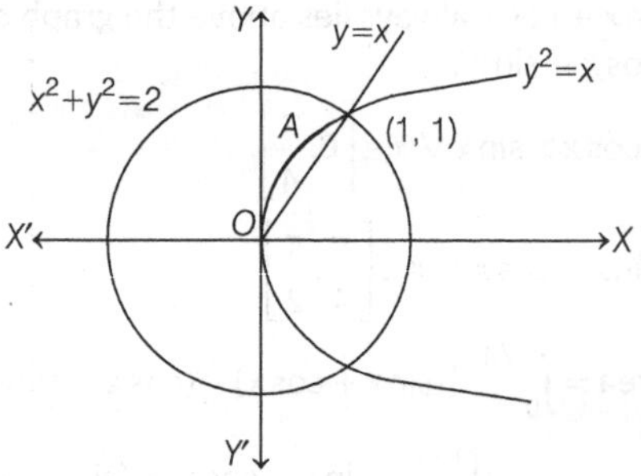

$$= \left[\frac{x^{3/2}}{3/2} - \frac{x^2}{2}\right]_0^1$$

$$= \left(\frac{1}{3/2} - \frac{1}{2}\right)$$

$$= \left(\frac{2}{3} - \frac{1}{2}\right)$$

$$= \frac{4-3}{6} = \frac{1}{6}$$

$\because$ Area of circle having radius $r = \sqrt{2}$ units is $\pi r^2 = 2\pi$

$\therefore$ The area of the region, enclosed by the circle $x^2 + y^2 = 2$, which is not common to the region bounded by the parabola $y^2 = x$ and the straight line $y = x$, is $2\pi - \frac{1}{6} = \frac{1}{6}(12\pi - 1)$ sq units

31. (c) The graph of curves is given below

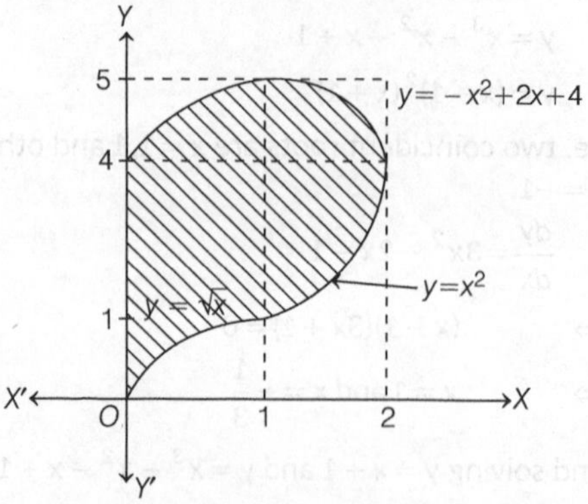

So, the area of shaded region is

$$A = \int_0^2 (-x^2 + 2x + 4)\,dx - \int_0^1 \sqrt{x}\,dx - \int_1^2 x^2\,dx$$

$$= \left[-\frac{x^3}{3} + x^2 + 4x\right]_0^2 - \left[\frac{2}{3}x^{3/2}\right]_0^1 - \left[\frac{x^3}{3}\right]_1^2 = \frac{19}{3} \text{ sq units}$$

32. (b) We have, $x + 2y^2 = 0$ and $x + 3y^2 = 1$

On solving these equations,

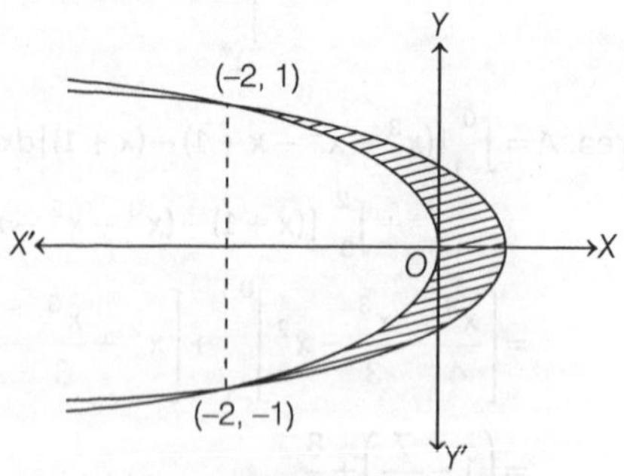

we get points of intersection $(-2, 1)$ and $(-2, -1)$.

$\therefore$ The required area

$$= 2\int_0^1 [(1 - 3y^2) - (-2y^2)]\,dy$$

$$= 2\int_0^1 (1 - y^2)\,dy$$

$$= 2\left[y - \frac{y^3}{3}\right]_0^1$$

$$= 2\left(1 - \frac{1}{3}\right) = 2 \cdot \frac{2}{3}$$

$$= \frac{4}{3} \text{ sq units}$$

33. (c) Given, $\quad y \le x^2 + 3x$

$$\Rightarrow \quad y \le \left(x + \frac{3}{2}\right)^2 - \frac{9}{4}$$

$$\Rightarrow \quad \left(x + \frac{3}{2}\right)^2 \ge \left(y + \frac{9}{4}\right)$$

Since, $0 \le y \le 4$ and $0 \le x \le 3$

$\therefore$ The diagram for the given inequalities is

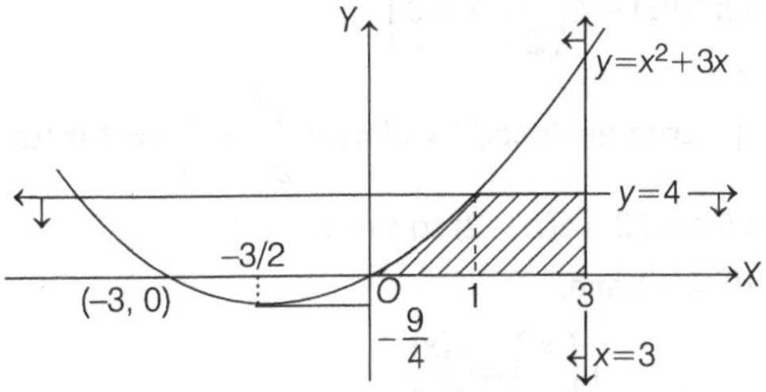

and points of intersection of curves $y = x^2 + 3x$ and $y = 4$ are (1, 4) and (–4, 4)

Now, required area $= \int_0^1 (x^2 + 3x)dx + \int_1^3 4\,dx$

$$= \left[\frac{x^3}{3} + \frac{3x^2}{2}\right]_0^1 + [4x]_1^3$$

$$= \frac{1}{3} + \frac{3}{2} + 4(3-1)$$

$$= \frac{2+9}{6} + 8 = \frac{11}{6} + 8$$

$$= \frac{59}{6} \text{ sq units}$$

34. (a) Given, curves are $y^2 = 12x$ and $x^2 = 12y$

Both curves intersect at $x = 12$ and $x = 0$

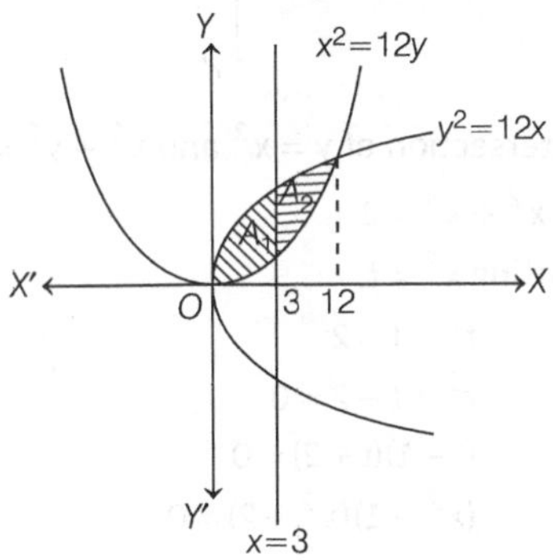

So, $A_1 = \int_0^3 \left(2\sqrt{3}\sqrt{x} - \frac{x^2}{12}\right)dx$

$$= \left[\frac{4\sqrt{3}}{3}x^{3/2} - \frac{x^3}{36}\right]_0^3$$

$$= \frac{45}{4} \text{ sq units} \quad \text{...(i)}$$

and $A_2 = \int_3^{12} \left(2\sqrt{3}\sqrt{x} - \frac{x^2}{12}\right)dx$

$$= \left[\frac{4\sqrt{3}}{3}x^{3/2} - \frac{x^3}{36}\right]_3^{12}$$

$$= \frac{147}{4} \text{ sq units} \quad \text{...(ii)}$$

Hence, from Eqs. (i) and (ii), we get

$$\frac{A_1}{A_2} = \frac{45/4}{147/4} = \frac{15}{49}$$

35. (a) $x^2 + y^2 \le 1$ represents the region inside the circle $x^2 + y^2 = 1$

and $y^2 \le 1 - x$ represents the region inside parabola $y^2 = 1 - x$ towards focus.

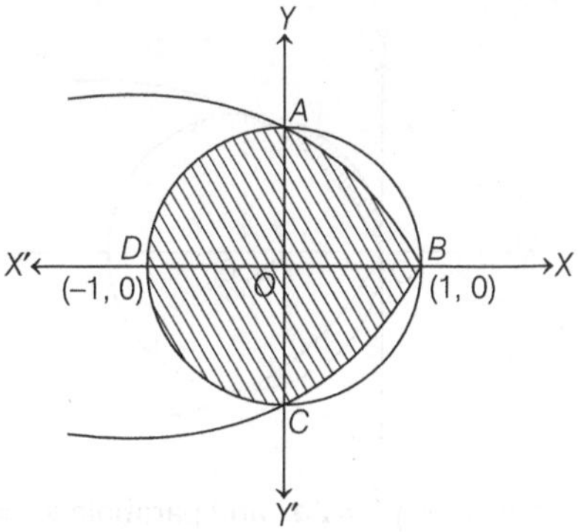

$\therefore$ The area of shaded region

= Area of semi-circle + Area of shaded region *OABCO*

$$= \frac{1}{2}\pi(1)^2 + 2\int_0^1 (1 - y^2)\,dy$$

$$= \frac{\pi}{2} + 2\left[y - \frac{y^3}{3}\right]_0^1$$

$$= \frac{\pi}{2} + 2\left[1 - \frac{1}{3}\right] = \left(\frac{\pi}{2} + \frac{4}{3}\right) \text{ sq units}$$

36. (d) $\therefore$ Required area

$$= \int_0^1 (1 + \sqrt{x})dx + \int_1^2 (3 - x)dx - \int_0^2 \frac{x^2}{4}dx$$

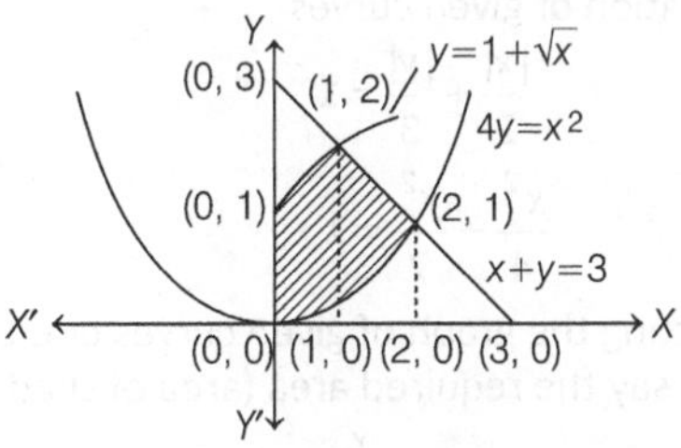

$$= \left[x + \frac{x^{3/2}}{3/2}\right]_0^1 + \left[3x - \frac{x^2}{2}\right]_1^2 - \left[\frac{x^3}{12}\right]_0^2$$

$$= \left(1 + \frac{2}{3}\right) + \left(6 - 2 - 3 + \frac{1}{2}\right) - \left(\frac{8}{12}\right)$$

$$= \frac{5}{3} + \frac{3}{2} - \frac{2}{3}$$

$$= 1 + \frac{3}{2}$$

$$= \frac{5}{2} \text{ sq units}$$

37. (c) Given, curves are

$$\{(x, y): x^2 + y^2 \le 2ax,\ y^2 \ge ax,\ x \ge 0,\ y \ge 0\}$$

Let $\qquad x^2 + y^2 = 2ax$

$\Rightarrow \qquad x^2 - 2ax + a^2 + y^2 = a^2$

$\Rightarrow \qquad (x-a)^2 + (y)^2 = a^2$

It is a circle with radius a and centre $(a, 0)$ and $y^2 = ax$ is a parabola.

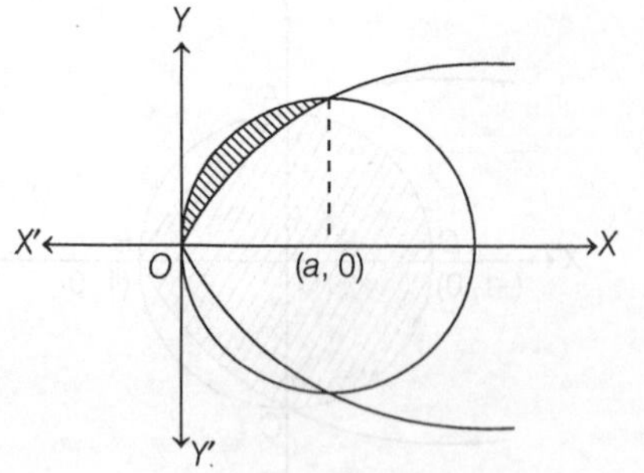

So, circle $x^2 + y^2 = 2ax$ and parabola $y^2 = ax$ intersect at $(\pm a, 0)$.

Hence, required area $= \int_0^a (y_2 - y_1)\,dx$

$$= \int_0^a \left(\sqrt{a^2 - (x-a)^2} - \sqrt{a}\sqrt{x}\right) dx$$

$$= \left(\frac{x-a}{2}\sqrt{a^2 - (x-a)^2} + \frac{a^2}{2}\sin^{-1}\left(\frac{x-a}{a}\right) - \frac{2\sqrt{a}}{3}x^{3/2}\right)_0^a$$

$$= -\frac{2\sqrt{a}}{3}(a)^{3/2} - \frac{a^2}{2}\sin^{-1}(-1)$$

$$= \left(-\frac{2}{3}a^2 + \frac{\pi a^2}{4}\right) = \left(\frac{\pi}{4} - \frac{2}{3}\right)a^2 \text{ sq unit}$$

38. (a) Equation of given curves

$$\frac{|x|}{2} + \frac{|y|}{3} = 1 \qquad \dots \text{(i)}$$

and, $$\frac{x^2}{4} + \frac{y^2}{9} = 1 \qquad \dots \text{(ii)}$$

On plotting the graph of given curves due to symmetry, we can say the required area (area of shaded region)

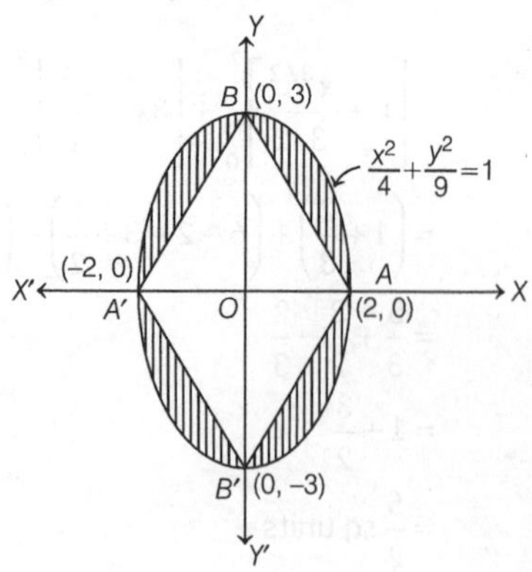

= Area enclosed by ellipse – 4 (Area of ΔAOB)

$$= \pi(2)(3) - 4\left(\frac{1}{2} \times 2 \times 3\right)$$

[$\because$ area enclosed by ellipse $\frac{x^2}{a^2} + \frac{y^2}{b^2} = 1$ is πab sq units]

$$= 6\pi - 12 = 6(\pi - 2) \text{ sq units}$$

39. (d) We have,

$$f\left(\frac{x}{y}\right) = \frac{f(x)}{f(y)}$$

Let $\qquad f(x) = x^n$

$\qquad f'(x) = nx^{n-1}$

Given that $f'(1) = 2$

$\Rightarrow \qquad n = 2$

So, $\qquad f(x) = x^2$

Now, graph of the curves $f(x) = x^2$ and $x^2 + y^2 = 2$ is

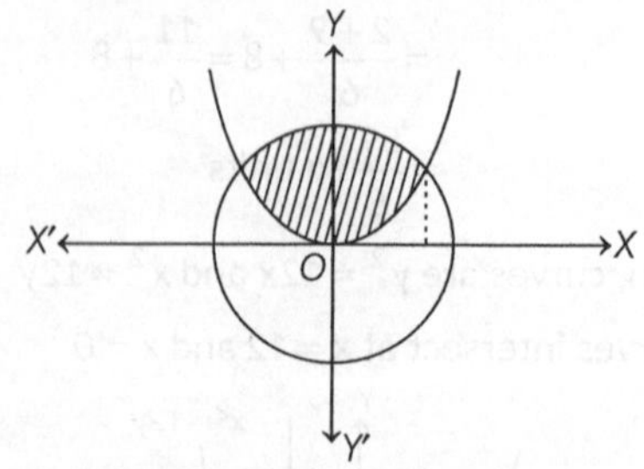

For intersection of $y = x^2$ and $x^2 + y^2 = 2$,

$$x^2 + x^4 = 2$$

On putting $x^2 = t$,

$$t^2 + t = 2$$

$\Rightarrow \qquad t^2 + t - 2 = 0$

$\Rightarrow \qquad (t-1)(t+2) = 0$

$\Rightarrow \qquad (x^2 - 1)(x^2 + 2) = 0$

$\Rightarrow \qquad x = \pm 1$

Now, required area

$$A = 2\int_0^1 [\sqrt{2 - x^2} - x^2]dx$$

$$= 2\left[\frac{x}{2}\sqrt{2 - x^2} + \frac{2}{2}\sin^{-1}\left(\frac{x}{\sqrt{2}}\right) - \frac{x^3}{3}\right]_0^1$$

$$= 2\left[\frac{1}{2} + \frac{\pi}{4} - \frac{1}{3}\right]$$

$$= 2\left(\frac{1}{6} + \frac{\pi}{4}\right)$$

$$\therefore A = \left(\frac{1}{3} + \frac{\pi}{2}\right) \text{ sq units}$$

40. (b) Given that the area of region

$A=\{(x,y):|x|+|y|\le 1, 2y^2\ge |x|\}$

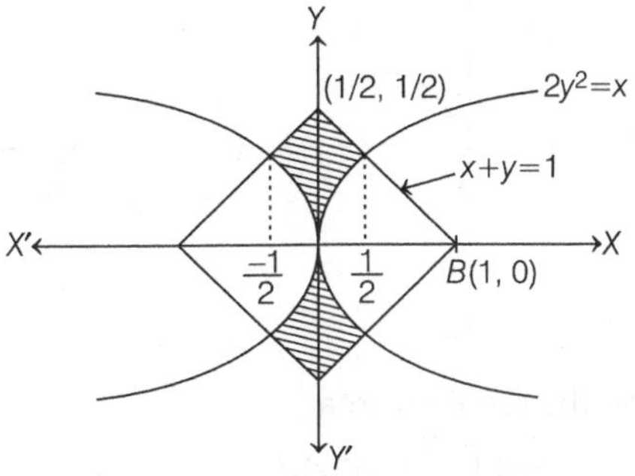

For point of intersection of $x+y=1$ and $2y^2=x$,

$\Rightarrow \quad 2y^2=1-y$

$\Rightarrow \quad 2y^2+y-1=0$

$\Rightarrow \quad (2y-1)(y+1)=0$

$\Rightarrow \quad y=\frac{1}{2}$ or -1

When $y=\frac{1}{2}$, then $x=\frac{1}{2}$

So, point of intersection is $\left(\frac{1}{2},\frac{1}{2}\right)$

Hence, required area

$$=4\int_0^{1/2}\left[(1-x)-\left(\sqrt{\frac{x}{2}}\right)\right]dx$$

$$=4\int_0^{1/2}\left(1-x-\frac{\sqrt{x}}{\sqrt{2}}\right)dx$$

$$=4\left[x-\frac{x^2}{2}-\frac{1}{\sqrt{2}}\cdot\frac{2}{3}-x^{3/2}\right]_0^{1/2}$$

$$=4\left[\frac{1}{2}-\frac{1}{8}-\frac{2}{3\sqrt{2}}\left(\frac{1}{2\sqrt{2}}\right)\right]$$

$$=4\left[\frac{1}{2}-\frac{1}{8}-\frac{1}{6}\right]=4\left(\frac{5}{24}\right)=\frac{5}{6}\text{ sq units}$$

41. (c) We have, $4y^2=x^2(4-x)(x-2)$

$$|y|=\frac{|x|}{2}\sqrt{(4-x)(x-2)}$$

$\Rightarrow \quad y_1=\frac{x}{2}\sqrt{(4-x)(x-2)}$

and $\quad y_2=-\frac{x}{2}\sqrt{(4-x)(x-2)}$

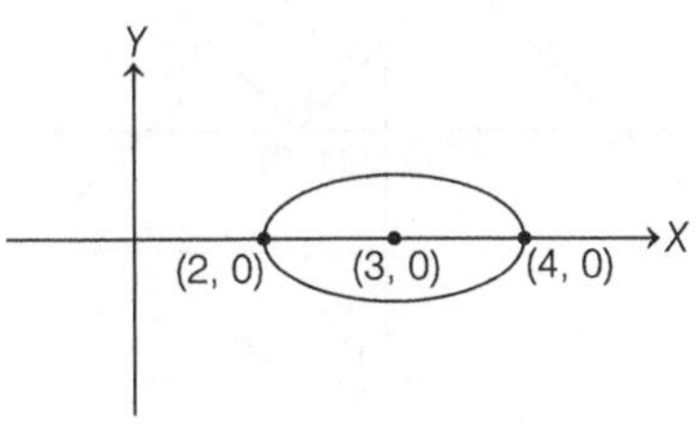

$D: x\in[2,4]$

$\therefore$ Required area

$$A=\int_2^4 (y_1-y_2)\,dx=\int_2^4 x\sqrt{(4-x)(x-2)\,dx} \quad \text{...(i)}$$

Applying $\int_a^b f(x)\,dx=\int_a^b f(a+b-x)\,dx$

$$A=\int_2^4 (6-x)\sqrt{(4-x)(x-2)}\,dx \quad \text{...(ii)}$$

Adding Eqs. (i) and (ii), we get

$$2A=6\int_2^4 \sqrt{(4-x)(x-2)}\,dx$$

$$A=3\int_2^4 \sqrt{1-(x-3)^2}\,dx$$

$$A=3\cdot\frac{\pi}{2}\cdot(1)^2=\frac{3\pi}{2}\text{ sq units}$$

42. (d) Given, equations of curves

$$y=2^x \text{ and } y=|x+1|=\begin{cases}x+1 & ,x\ge -1\\ -x-1 & ,x<-1\end{cases}$$

$\because$ The figure of above given curves is

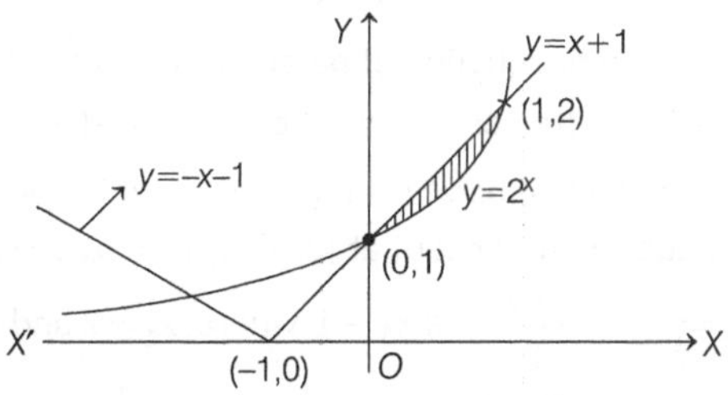

In first quadrant, the above given curves intersect each other at (1, 2).

So, the required area $=\int_0^1((x+1)-2^x)\,dx$

$$=\left[\frac{x^2}{2}+x-\frac{2^x}{\log_e 2}\right]_0^1$$

$$\left[\because \int a^x dx=\frac{a^x}{\log_e a}+C\right]$$

$$=\left[\frac{1}{2}+1-\frac{2}{\log_e 2}+\frac{1}{\log_e 2}\right]$$

$$=\left(\frac{3}{2}-\frac{1}{\log_e 2}\right)\text{ sq units}$$

43. (c) We have, $y=x^2+x+1$

$$\frac{dy}{dx}=2x+1$$

The slope of tangent is

$$\frac{dy}{dx}=3 \text{ at } x=1$$

The equation of tangent is

$$y-3=3(x-1)$$

$\Rightarrow \quad y=3x$

The required area ... (continued)

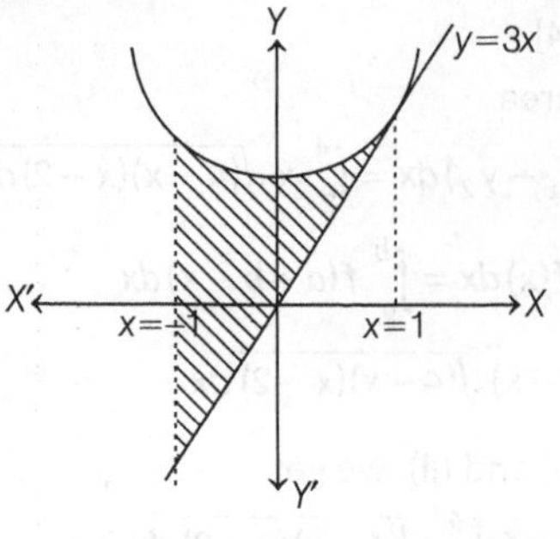

$\therefore$ The required area

$$= \int_{-1}^{1} (x^2 + x + 1 - 3x)\,dx$$

$$= \int_{-1}^{1} (x^2 - 2x + 1)\,dx$$

$$= \left[\frac{x^3}{3} - x^2 + x\right]_{-1}^{1}$$

$$= \left(\frac{1}{3} - 1 + 1\right) - \left(-\frac{1}{3} - 1 - 1\right) = \frac{2}{3} + 2 = \frac{8}{3} \text{ sq units}$$

44. (a) Given, equation of parabola is $y = x^2 - 1$, which can be rewritten as $x^2 = y + 1$ or $x^2 = (y - (-1))$.

$\Rightarrow$ Vertex of parabola is (0, – 1) and it is open upward.

Equation of tangent at (2, 3) is given by $T = 0$

$$\Rightarrow \quad \frac{y + y_1}{2} = x\,x_1 - 1, \text{ where, } x_1 = 2 \text{ and } y_1 = 3.$$

$$\Rightarrow \quad \frac{y + 3}{2} = 2x - 1$$

$$\Rightarrow \quad y = 4x - 5$$

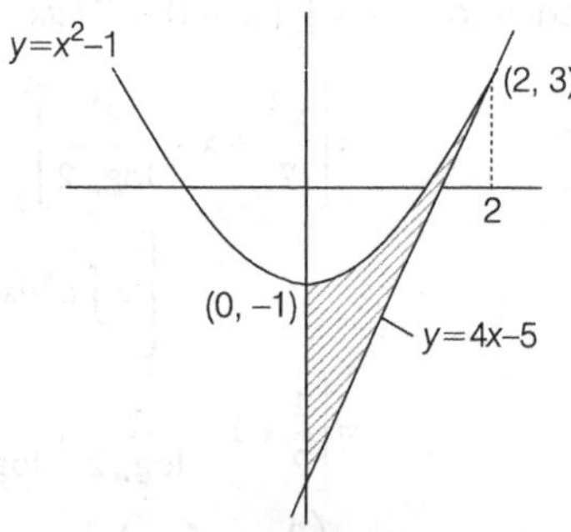

Now, required area = area of shaded region

$$= \int_0^2 (y(\text{parabola}) - y(\text{tangent}))\,dx$$

$$= \int_0^2 [(x^2 - 1) - (4x - 5)]\,dx$$

$$= \int_0^2 (x^2 - 4x + 4)\,dx = \int_0^2 (x - 2)^2\,dx$$

$$= \left|\frac{(x-2)^3}{3}\right|_0^2 = \frac{(2-2)^3}{3} - \frac{(0-2)^3}{3}$$

$$= \frac{8}{3} \text{ sq units}$$

45. (4) Given, curve is $|y| = \sin(2x)$, where $0 < x < 2\pi$

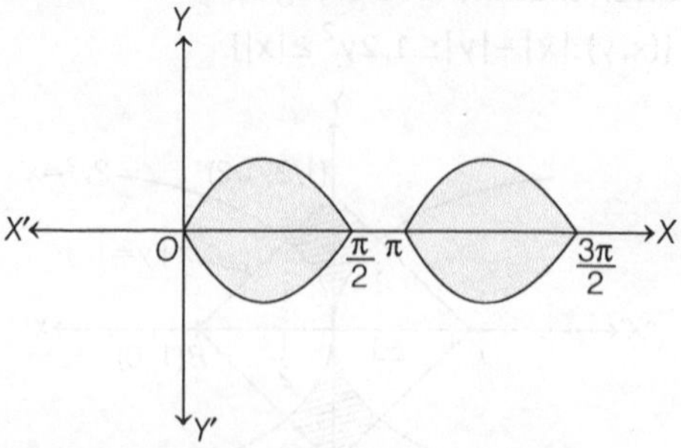

Hence, the required area

$$= 4\int_0^{\pi/2} (\sin 2x)\,dx$$

$$= 4\left(-\frac{\cos 2x}{2}\right)_0^{\pi/2}$$

$$= -2(-1 - 1)$$

$$= 4 \text{ sq units}$$

46. (2) The given curve are $y = \ln(x + e)$ and $x = \ln\left(\frac{1}{y}\right)$

$$\Rightarrow \quad \frac{1}{y} = e^x \Rightarrow y = e^{-x}$$

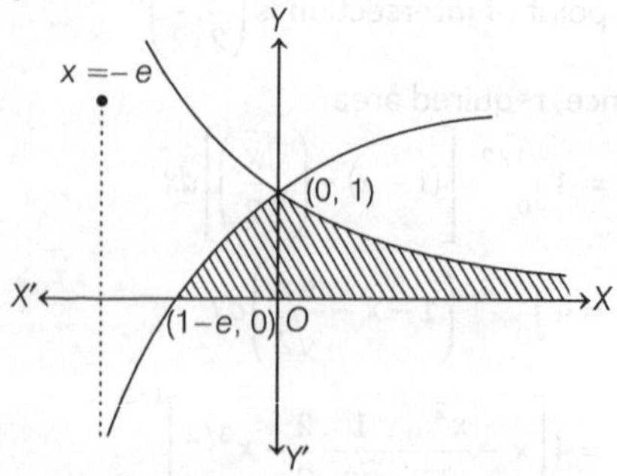

Hence, the required area

$$= \int_{1-e}^{0} \ln(x + e)\,dx + \int_0^{\infty} e^{-x}\,dx$$

$$= \int_1^e \ln t\,dt + \int_0^{\infty} e^{-x}\,dx \qquad [\because \text{putting } x + e = t]$$

$$= [t \ln t - t]_1^e - [e^{-x}]_0^{\infty}$$

$$= (e - e) - (0 - 1) - [0 - 1]$$

$$= 1 + 1 = 2$$

47. (4) Given, curves are $y = 3 - |x|$ and $y = |x - 1|$

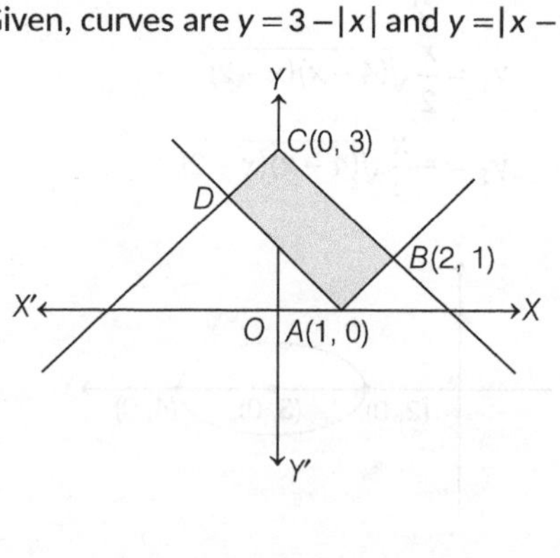

Here, $AB=\sqrt{1^2+1^2}=\sqrt{2}$

and $\quad BC=\sqrt{2^2+2^2}=\sqrt{8}=2\sqrt{2}$

Hence, the required area $=AB\times BC$

$$=\sqrt{2}\times 2\sqrt{2}=4 \text{ sq units}$$

48. (3) Given, curves are

$$y=\ln|x| \text{ and } y=1-|x|$$

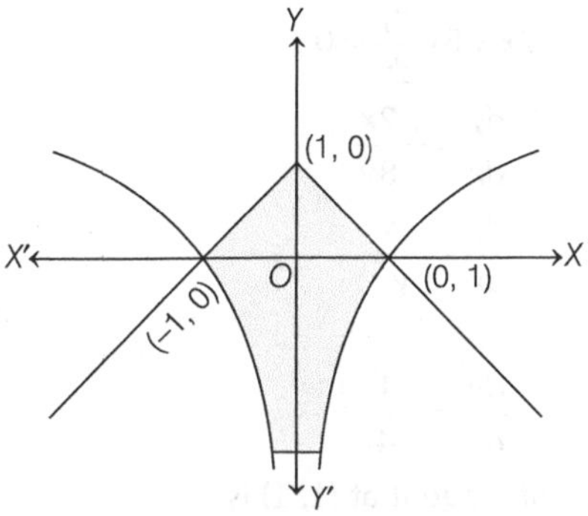

Hence, the required area

$$=2\left[\frac{1}{2}\times 1\times 1+\int_{-\infty}^{0} e^y\,dy\right]$$

$$=2\left(\frac{1}{2}+1\right)=2\times\frac{3}{2}=3 \text{ sq units}$$

49. (4) Given, curves are

$$\begin{cases} y=\tan x, & -\dfrac{\pi}{3}\le x\le\dfrac{\pi}{3} \\ y=\cot x, & \dfrac{\pi}{6}\le x\le\dfrac{\pi}{2} \end{cases}$$

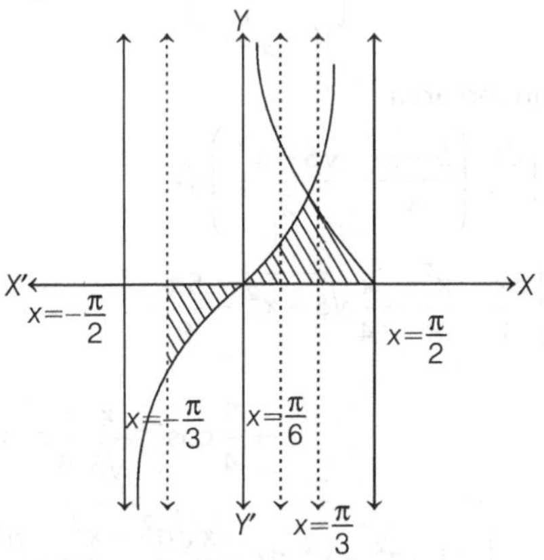

Hence, the required area

$$=\left|\int_{-\pi/3}^{0}\tan x\,dx\right|+\int_{0}^{\pi/4}\tan x\,dx+\int_{\pi/4}^{\pi/2}\cot x\,dx$$

$$=|[\log\sec x]_{-\pi/3}^{0}|+[\log\sec x]_0^{\pi/4}+[\log(\sin x)]_{\pi/4}^{\pi/2}$$

$$=\log(2)+\log(\sqrt{2})-\log\left(\frac{1}{\sqrt{2}}\right)$$

$$=\log 2+2\log(\sqrt{2})$$

$$=\log 2+\log 2=\log 4 \text{ sq units}$$

$$\Rightarrow \qquad k=4$$

50. (26) Graph of given curves

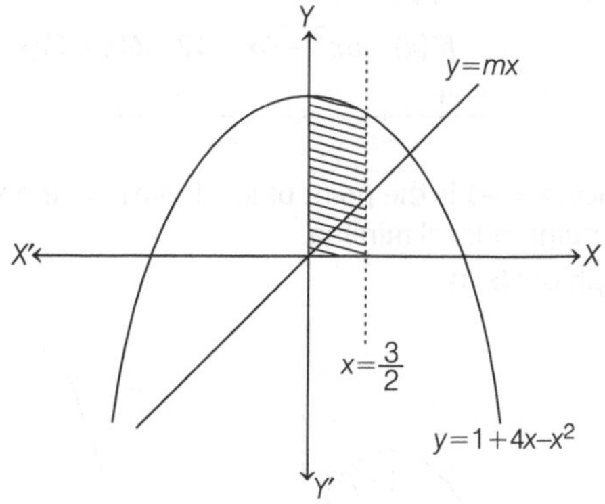

According to questions

$$\frac{1}{2}\int_0^{3/2}(1+4x-x^2)dx=\int_0^{3/2}mx\,dx$$

$$\Rightarrow \frac{1}{2}\left[\left(x+2x^2-\frac{x^3}{3}\right)\right]_0^{3/2}=\frac{m}{2}[x^2]_0^{3/2}$$

$$\Rightarrow \frac{3}{2}+\frac{9}{2}-\frac{9}{8}=\frac{9}{4}m \Rightarrow \frac{48-9}{8}=\frac{9}{4}m$$

$$\Rightarrow m=\frac{39}{18}=\frac{13}{6}$$

Hence, $12\,m=12\times\dfrac{13}{6}=26$

51. (2) Let $y=x|x|+1$

$$=\begin{cases}-x^2+1, & x<0 \\ x^2+1, & x\ge 0\end{cases}$$

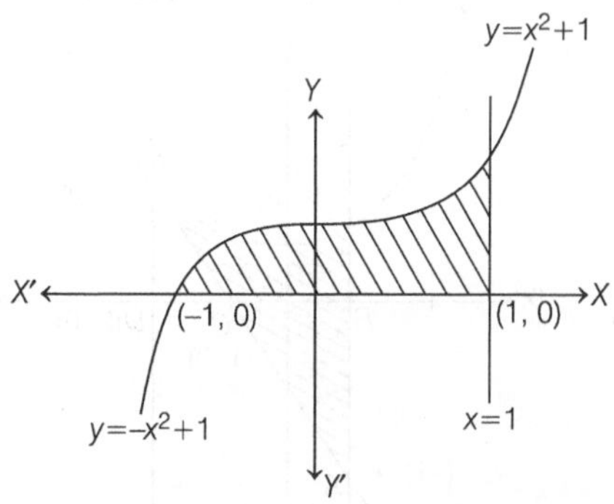

$\therefore$ Required area

$$=\int_{-1}^{0}(-x^2+1)\,dx+\int_0^1(x^2+1)\,dx$$

$$=\left[-\frac{x^3}{3}+x\right]_{-1}^{0}+\left[\frac{x^3}{3}+x\right]_0^1$$

$$=0-\left(\frac{1}{3}+(-1)\right)+\left(\frac{1}{3}+1\right)$$

$$=-\frac{1}{3}+1+\frac{1}{3}+1$$

$$=2 \text{ sq units}$$

52. (114) We have, $f(x) = 2x^3 - 3x^2 - 12x$

$$f'(x) = 6x^2 - 6x - 12 = 6(x-2)(x+1)$$

$$\underset{}{+}\quad \underset{-1}{|} \quad - \quad \underset{2}{|} \quad +$$

Hence, $x = -1$ is the point of local maxima and $x = 2$ is the point of local minima.

Graph of $f(x)$ is

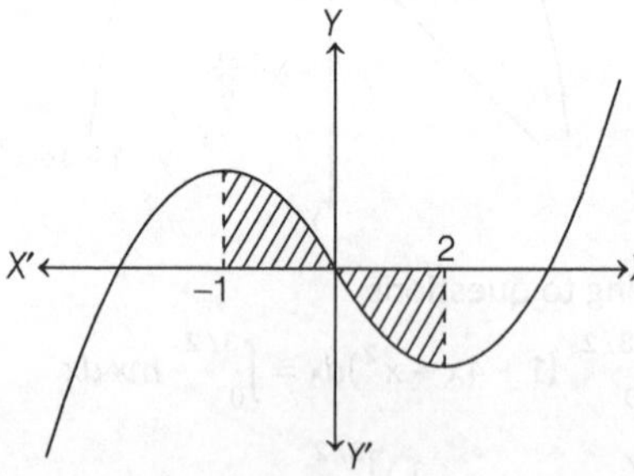

Area, $A = \int_{-1}^{0} (2x^3 - 3x^2 - 12x)\,dx + \left| \int_0^2 (2x^3 - 3x^2 - 12x)\,dx \right|$

$$= \left[\frac{x^4}{2} - x^3 - 6x^2 \right]_{-1}^{0} + \left| \left[\frac{x^4}{2} - x^3 - 6x^2 \right]_0^2 \right|$$

$$= \frac{9}{2} + |-24| = \frac{9}{2} + 24 = \frac{57}{2}$$

Hence, $4A = 4 \times \frac{57}{2} = 114$

53. (9) The shaded region for given equations is

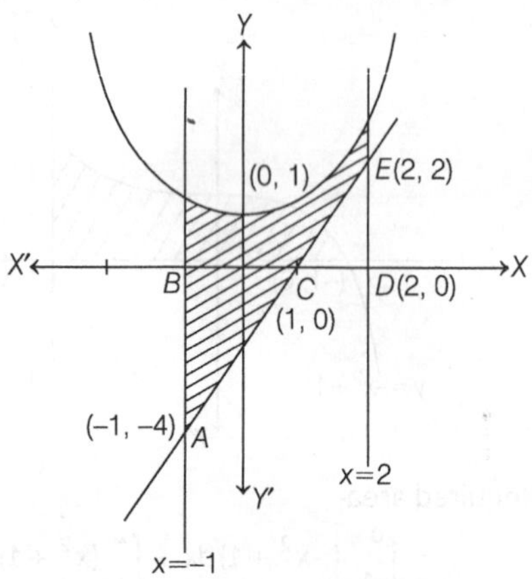

Area, $A_1 = \text{ar}(\Delta ABC) = \frac{1}{2}(2)(4) = 4$ sq units

Area, $A_2 = \text{ar}(CDE)$

$$A_2 = \frac{1}{2} \times 1 \times 2 = 1 \text{ sq unit}$$

and area of region between parabola and X-axis

$$A_3 = \int_{-1}^{2} (x^2 + 1)\,dx = \left[\frac{x^3}{3} + x \right]_{-1}^{2}$$

$$= \left(\frac{8}{3} + 2 \right) - \left(\frac{-1}{3} - 1 \right) = 6$$

$\therefore$ Required area $= 6 + 4 - 1 = 9$ sq units

54. (1.25) Given, ellipse

$$E: x^2 + 4y^2 = 5$$

Differentiating w.r.t. x, we get

$$2x + 8y\frac{dy}{dx} = 0$$

$$\frac{dy}{dx} = -\frac{2x}{8y}$$

$$\frac{dy}{dx} = \frac{-x}{y}$$

At (1, 1),

$$\frac{dy}{dx} = -\frac{1}{4}$$

Equation of tangent at (1, 1) is

$$y - 1 = -\frac{1}{4}(x-1)$$

$$\Rightarrow \quad x + 4y = 5$$

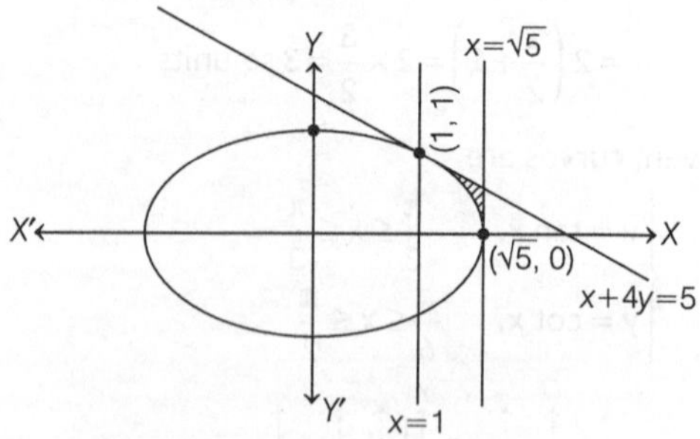

$\therefore$ Required area

$$= \int_1^{\sqrt{5}} \left(\frac{5-x}{4} - \frac{\sqrt{5-x^2}}{2} \right) dx$$

$$= \left[\frac{5x}{4} - \frac{x^2}{8} - \frac{x}{4}\sqrt{5-x^2} - \frac{5\pi}{8} + \frac{5}{4}\cos^{-1}\frac{x}{\sqrt{5}}\frac{5}{4}\cos^{-1}\left(\frac{x}{\sqrt{5}}\right) \right]_1^{\sqrt{5}}$$

$$\left[\because \int \sqrt{a^2 - x^2}\,dx = \frac{x\sqrt{a^2-x^2}}{2} + \frac{a^2}{2}\sin^{-1}\left(\frac{x}{a}\right) \right]$$

$$= \frac{5}{4}\sqrt{5} - \frac{5}{4} - \frac{5}{4}\cos^{-1}\left(\frac{1}{\sqrt{5}}\right)$$

By comparing, $\alpha = \frac{5}{4}$, $\beta = -\frac{5}{4}$ and $\gamma = -\frac{5}{4}$

$$\therefore \quad |\alpha + \beta + \gamma| = \left| \frac{5}{4} - \frac{5}{4} - \frac{5}{4} \right|$$

$$= \left| -\frac{5}{4} \right| = 1.25$$

55. (9) We have, $(y-2)^2=(x-1)$ ...(i)

When $y=3 \Rightarrow x=2$

Then, point of contact is (2, 3)

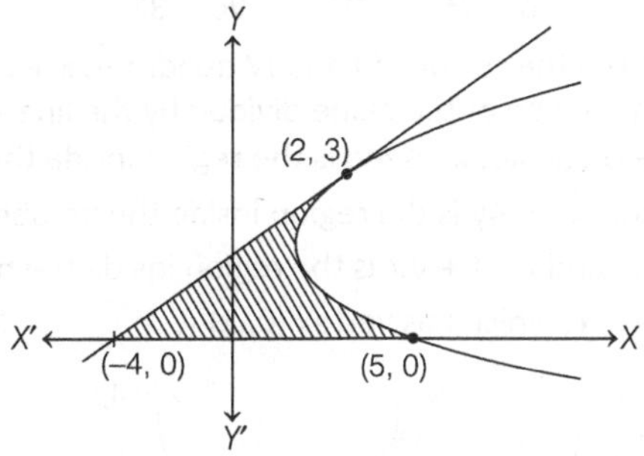

Differentiating Eq. (i),

$$2(y-2)y'=1$$

$$\Rightarrow \quad y'=\frac{1}{2(y-2)}$$

At (2, 3),

$$y'=\frac{1}{2}$$

Equation of tangent is

$$(y-3)=\frac{1}{2}(x-2)$$

$$\Rightarrow \quad x-2y+4=0$$

Now, area of required region

$$=\int_0^3 [((y-2)^2+1)-(2y-4)]dy$$

$$=\int_0^3 (y^2-6y+9)dy$$

$$=\left[\frac{y^3}{3}-3y^2+9y\right]_0^3$$

$$=9-27+27=9 \text{ sq units}$$

56. (1) We have, $x^{2/5}+|y|=1$

Then, curves are $x^{2/5}+y=1$ and $x^{2/5}-y=1$

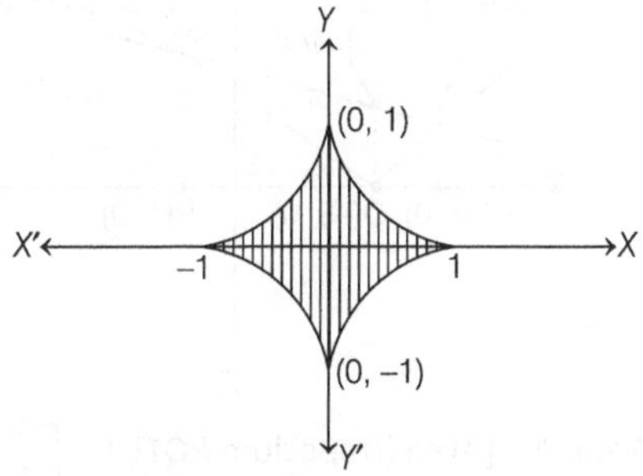

$\therefore$ Required area $=4\int_0^1(1-x^{2/5})dx$

$$=4\left[x-\frac{5}{7}x^{7/5}\right]_0^1$$

$$=4\left[1-\frac{5}{7}\right]$$

$$=4\left(\frac{2}{7}\right)=\frac{8}{7}$$

Hence, $p-q=8-7=1$

LEVEL 3

1. (d) We have, $f^3(x)=\int_0^x t f^2(t)\,dt$

On differentiating w.r.t. x, we get

$$3f^2(x)f'(x)=xf^2(x)$$

$$\Rightarrow \quad f^2(x)[3f'(x)-x]=0$$

As, $\quad f(x)\neq 0$

$$\Rightarrow \quad f'(x)=\frac{x}{3}$$

$$\Rightarrow \quad f(x)=\frac{x^2}{6}+C$$

But $\quad f(0)=0$

$$\Rightarrow \quad C=0$$

$$\Rightarrow \quad f(x)=\frac{x^2}{6}$$

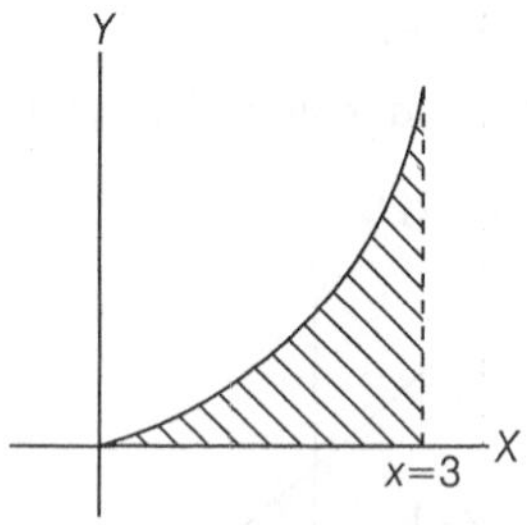

$\therefore$ Area, $A=\frac{1}{6}\int_0^3 x^2dx=\left[\frac{x^3}{18}\right]_0^3=\frac{3}{2}$ sq units

2. (b) We have, $f(x)=\int_0^x e^t(\log_e \sec t-\sec^2 t)\,dt$

$$=\int_0^x e^t[(\log_e \sec t-\tan t)+(\tan t-\sec^2 t)]dt$$

$$=[e^t(\log_e \sec t-\tan t)]_0^x$$

$$=e^x(\log_e \sec x-\tan x)$$

$\therefore$ Required area

$$A=\int_0^{\pi/3} [e^x(\log_e \sec x-\tan x)-(-2e^x\tan x)]dx$$

$$=\int_0^{\pi/3} e^x(\log_e \sec x+\tan x)dx$$

$$=[e^x\log_e \sec x]_0^{\pi/3}$$

$$=e^{\pi/3}\log_e \sec\frac{\pi}{3}$$

$$=(e^{\pi/3}\log_e 2) \text{ sq units}$$

3. (d) The graph of $y^2 = 4x$, $x^2 = 4y$ and lines $x = 4$, $y = 4$ is

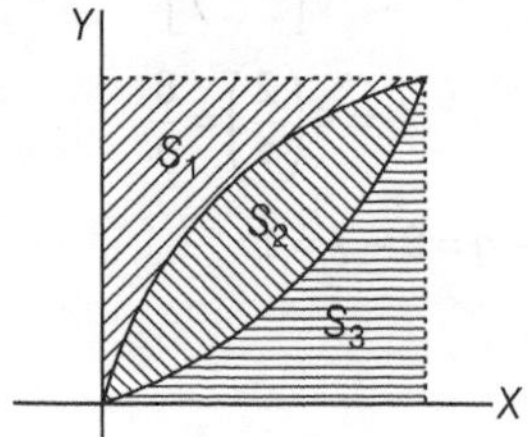

$$\therefore \quad S_1 = \int_0^4 \frac{y^2}{4} dy = \frac{1}{4}\left[\frac{y^3}{3}\right]_0^4 = \frac{64}{4 \times 3} = \frac{16}{3} \text{ sq units}$$

$$S_2 = \int_0^4 \left(\sqrt{4x} - \frac{x^2}{4}\right) dx$$

$$= 2 \times \frac{2}{3}[x^{3/2}]_0^4 - \frac{1}{12}[x^3]_0^4$$

$$= \frac{4}{3} \times (4)^{3/2} - \frac{1}{12} \times 64$$

$$= \frac{4}{3} \times 8 - \frac{16}{3} = \frac{16}{3} \text{ sq units}$$

and $S_3 = \int_0^4 \frac{x^2}{4} dx = \frac{1}{12}(64) = \frac{16}{3}$ sq units

Therefore, $S_1 : S_2 : S_3 = 1 : 1 : 1$

4. (a) The bounded region of $\{(x, y) : 0 \le x \le \frac{9}{4}, 0 \le y \le 1, x \ge 3y, x + y \ge 2\}$ is

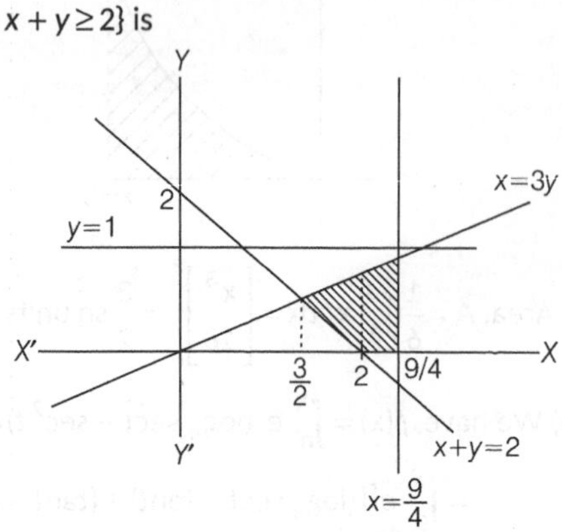

$$\therefore \quad \text{Area} = \int_{3/2}^{2} \left[\frac{x}{3} - (2 - x)\right] dx + \int_2^{9/4} \frac{x}{3} dx$$

$$= \int_{3/2}^{2} \left[\frac{4x}{3} - 2\right] dx + \int_2^{9/4} \frac{x}{3} dx$$

$$= \left[\frac{2x^2}{3} - 2x\right]_{3/2}^{2} + \frac{1}{6}[x^2]_2^{9/4}$$

$$= \left[\left(\frac{8}{3} - 4\right) - \left(\frac{3}{2} - 3\right)\right] + \frac{1}{6}\left[\frac{81}{16} - 4\right]$$

$$= \left[\frac{8}{3} - \frac{3}{2} - 1\right] + \frac{1}{6}\left(\frac{81 - 64}{16}\right)$$

$$= \left(\frac{16 - 9 - 6}{6}\right) + \frac{1}{6}\left(\frac{17}{16}\right)$$

$$= \frac{1}{6} + \frac{17}{96} = \frac{16 + 17}{96} = \frac{33}{96} = \frac{11}{32} \text{ sq units}$$

5. (a) $x \ge 0$ is the region of I and IV quadrant. $x + y \le 3$ is half the region of the plane divided by the line $x + y = 3$, where origin lies $x^2 \le 4y$, is the region inside the parabola $x^2 = 4y$ is the region inside the parabola $x^2 = 4y$ and $y \ge 1 + \sqrt{x}$ is the region inside the parabola $y = 1 + \sqrt{x}$, which has vertex (0, 1)

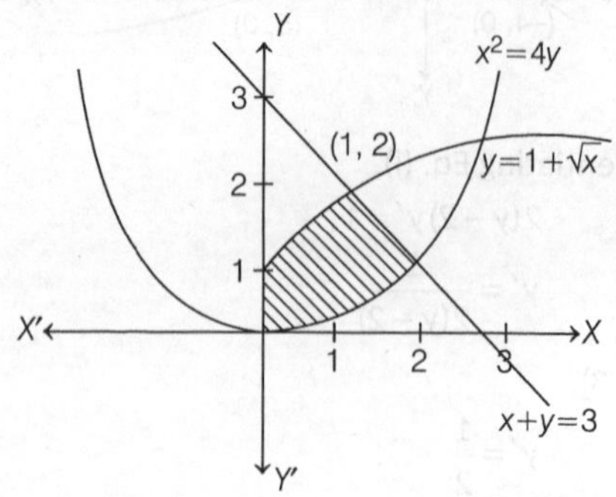

$\therefore$ Area of shaded region

$$= \int_0^1 (1 + \sqrt{x}) dx + \int_1^2 (3 - x) dx - \int_0^2 \frac{x^2}{4} dx$$

$$= \left[x + \frac{2}{3}x^{3/2}\right]_0^1 + \left[3x - \frac{x^2}{2}\right]_1^2 - \left(\frac{x^3}{12}\right)_0^2$$

$$= \left(1 + \frac{2}{3}\right) + \left\{(6 - 2) - \left(3 - \frac{1}{2}\right)\right\} - \frac{8}{12}$$

$$= \frac{5}{3} + \frac{3}{2} - \frac{2}{3} = 1 + \frac{3}{2} = \frac{5}{2} = \frac{5}{2} \text{ sq units}$$

6. (c) We have, $5y \le x + 9 \le 15$ and $y \ge \sqrt{|x + 3|}$

$$y^2 \ge \begin{cases} x + 3, & \text{if } x \ge -3 \\ -x - 3, & \text{if } x < -3 \end{cases}$$

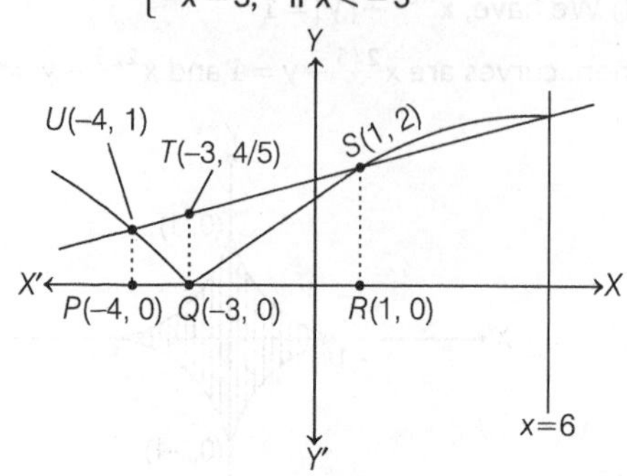

$\therefore$ Area, A = [Area (trapezium $PQTU$) $- \int_{-4}^{-3} \sqrt{-x - 3}\, dx$]

+ [Area (trapezium $QRST$ $- \int_{-3}^{1} \sqrt{x + 3}\, dx$]

Area A = (Area of trapezium $PRSU$)

$$- \int_4^{-3} \sqrt{-x - 3}\, dx - \int_{-3}^{1} \sqrt{x + 3}\, dx$$

$$= \frac{1}{2}(1+2)\cdot 5 - \int_{-4}^{-3}\sqrt{-x-3} - \int_{-3}^{1}\sqrt{x+3}$$

$$= \frac{15}{2} - \frac{2}{3} - \frac{16}{3} = \frac{15}{2} - \frac{18}{3} = \frac{3}{2} \text{ sq units}$$

7. (c) The graph of $\{(x, y): xy \le 8,\ 1 \le y \le x^2\}$ is

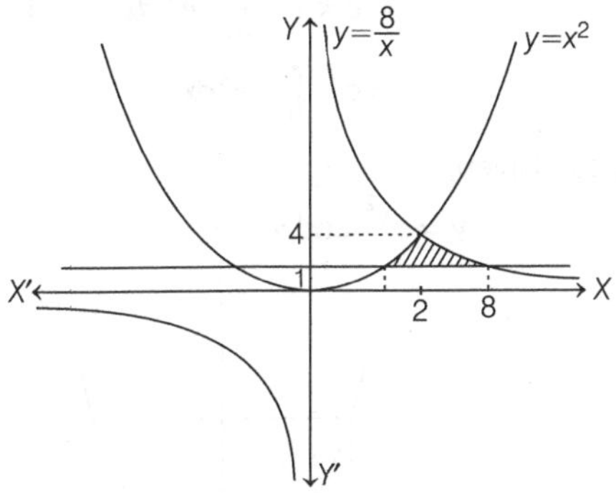

$$\therefore \text{ Required area} = \int_1^4 \left(\frac{8}{y} - \sqrt{y}\right) dy$$

$$= \left[8\ln y - \frac{2}{3}y^{3/2}\right]_1^4$$

$$= \left[8\ln 4 - \frac{2}{3}4^{3/2}\right] - \left(0 - \frac{2}{3}\right)$$

$$= 16\ln 2 - \frac{16}{3} + \frac{2}{3}$$

$$= \left(16\ln 2 - \frac{14}{3}\right) \text{ sq units}$$

8. (c) Given, curves are $y = x^2$, $y = |2 - x^2|$ and $y = 2$

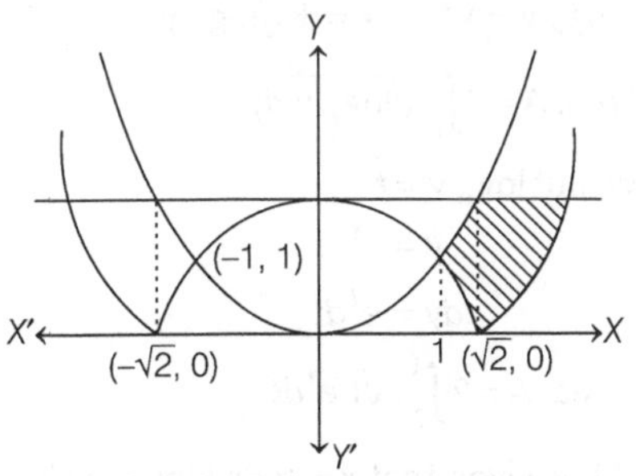

$$\therefore \text{ Required area} = \int_1^{\sqrt{2}} [x^2 - (2 - x^2)]\,dx + \int_{\sqrt{2}}^{2} [2 - (x^2 - 2)]\,dx$$

$$= \int_1^{\sqrt{2}} (2x^2 - 2)\,dx + \int_{\sqrt{2}}^{2} (4 - x^2)\,dx$$

$$= \left[\frac{2}{3}x^3 - 2x\right]_1^{\sqrt{2}} + \left[4x - \frac{x^3}{3}\right]_{\sqrt{2}}^{2}$$

$$= \left(\frac{20 - 12\sqrt{2}}{3}\right) \text{ sq units}$$

9. (c) Given, $g(x) = (2x + 1)$

and $h(x) = 4x^2 + 4x + 5$

$\Rightarrow \quad h(x) = (2x + 1)^2 + 4$

Here, $h(x) = f[g(x)] = (2x + 1)^2 + 4 = f(2x + 1)$

Let $(2x + 1) = t \Rightarrow f(t) = t^2 + 4$

$\therefore \quad f(x) = x^2 + 4$

Let equation of tangent, which passes through origin, is

$$y = mx$$

On solving $y = mx$ and $y = x^2 + 4$, we get

$$x^2 - mx + 4 = 0$$

For tangency, put $D = 0$

$$m^2 = 16 \Rightarrow m = \pm 4$$

So, the tangents are $y = 4x$ and $y = -4x$

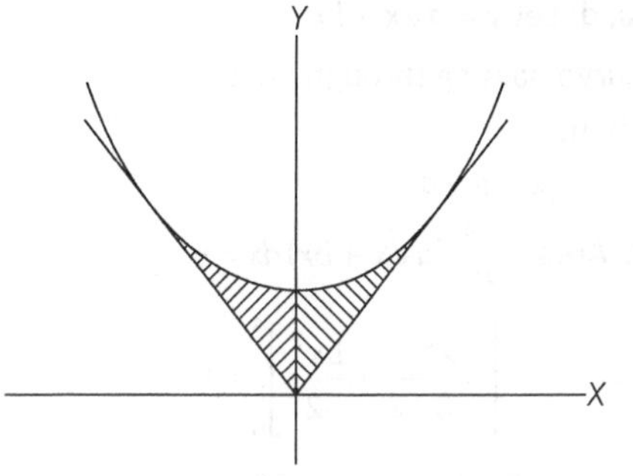

$$\therefore \text{ The required area} = 2\int_0^2 (x^2 + 4) - 4x]\,dx$$

$$= 2\int_0^2 (x - 2)^2\,dx$$

$$= \frac{2}{3}[(x - 2)^3]_0^2 = \frac{2}{3}(8) = \frac{16}{3} \text{ sq units}$$

10. (d) We have, $\lim_{x \to 0}\left[1 + \frac{f(x)}{x^3}\right]^{1/x}$ exists

So, $\lim_{x \to 0} \frac{f(x)}{x^3} = 0$

It means $f(x) = a_4x^4 + a_5x^5 + \ldots + a_nx^n,\ a_n \ne 0,\ n \ge 4$

$f(x)$ is of least degree $f(x) = a_4x^4$

$$\lim_{x \to 0}\left[1 + \frac{f(x)}{x^3}\right]^{1/x} = e,\ a_4 = 1,\ f(x) = x^4$$

The graph of $y = x^4$ and $x^2 + y^2 = 2$

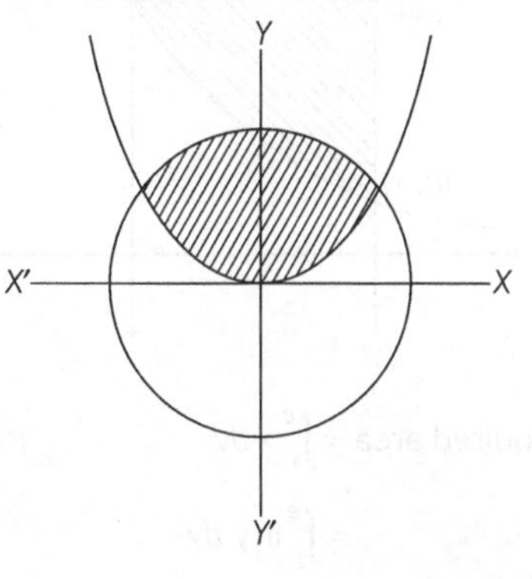

$$\therefore \text{ Area} = 2\int_0^1 (\sqrt{2-x^2} - x^4)\,dx$$

$$\text{Area} = 2\left[\frac{x\sqrt{2-x^2}}{2} + \frac{2}{2}\sin^{-1}\left(\frac{x}{\sqrt{2}}\right) - \frac{x^5}{5}\right]_0^1$$

$$= 2\left[\left(\frac{1}{2} + \frac{\pi}{4} - \frac{1}{5}\right) - 0\right]$$

$$= 1 + \frac{\pi}{2} - \frac{2}{5}$$

$$= \left(\frac{\pi}{2} + \frac{3}{5}\right) \text{ sq units}$$

11. (a, d) Let $y = a\sqrt{x} + bx$

Curve passing through (1, 2).

Then,

$$2 = a + b \quad \text{...(i)}$$

$$\therefore \text{ Area} = \int_0^4 (a\sqrt{x} + bx)\,dx = 8$$

$$\Rightarrow \left[a\frac{x^{3/2}}{3/2} + \frac{bx^2}{2}\right]_0^4 = 8$$

$$\Rightarrow \left[\frac{2}{3}ax^{3/2} + \frac{b}{2}x^2\right]_0^4 = 8$$

$$\Rightarrow \frac{16}{3}(a) + 8b = 8$$

$$\Rightarrow \frac{2a}{3} + b = 1$$

$$\Rightarrow 2a + 3b = 3 \quad \text{...(ii)}$$

From Eqs. (i) and (ii), we get

$$a = 3 \text{ and } b = -1$$

Hence, $a + 2b = 3 + 2(-1)$

$$= 1$$

12. (b, c, d) Given, curve $y = e^x$

$$\Rightarrow x = \ln y$$

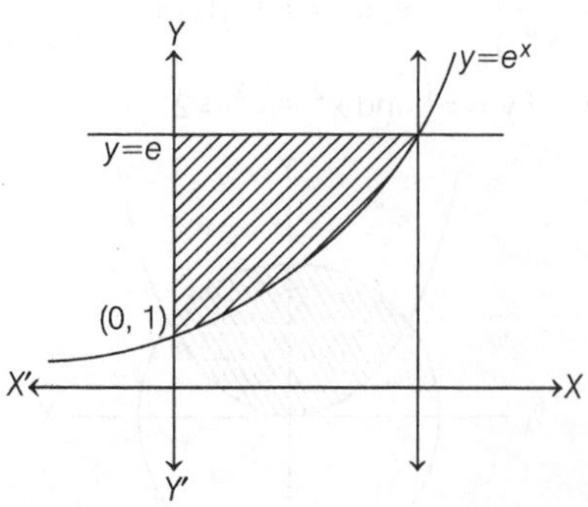

$$\therefore \text{ Required area} = \int_1^e x\,dy$$

$$= \int_1^e \ln y\,dy$$

Using $\int_a^b f(x)\,dx = \int_a^b f(a+b-x)\,dx$,

$$= \int_1^e \ln(1+e-y)\,dy$$

Further, required area

$$= e \times 1 - \int_0^1 e^x\,dx \quad \left[\because \int_0^1 e^x\,dx = e-1\right]$$

$$= e - \int_0^1 e^x dx$$

13. (a,b,c) Given,

$$y = e^{x^2} \text{ and } y = e$$

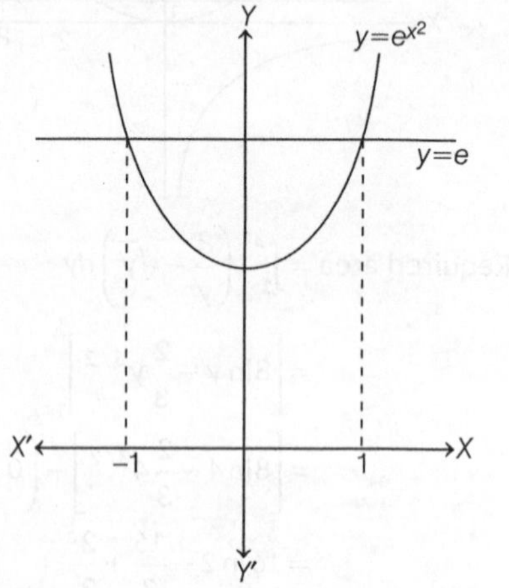

Area, $A = \int_{-1}^1 (e - e^{x^2})dx = 2e - \int_{-1}^1 e^{x^2}\,dx$

$$y = e^{x^2}$$

$\therefore \quad \log_e y = x^2 \Rightarrow x = \pm\sqrt{\log_e y}$

$\therefore$ Area, $A = 2\int_1^e \sqrt{\log_e y}\,dy$

Now, put $\log_e y = t$

$\therefore \quad y = e^t$

$\therefore \quad dy = e^t dt$

So, area, $A = 2\int_0^1 \sqrt{t}\,e^t dt$

14. (b,c) It is given that $x = \alpha$ divides the shaded region into two equal parts.

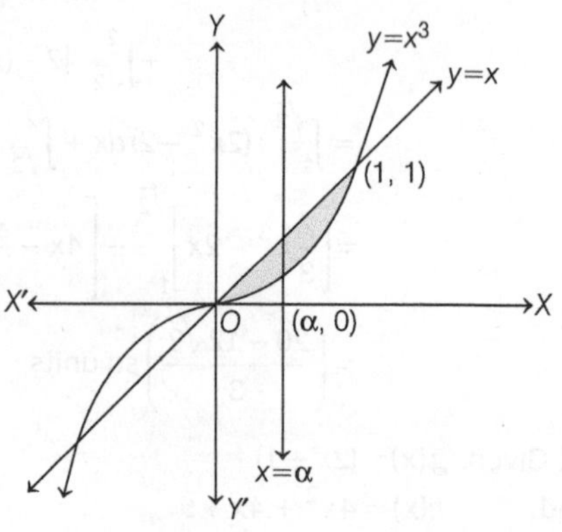

$\therefore \quad 2\int_0^{\alpha}(x-x^3)dx=\int_0^1(x-x^3)dx$

$\Rightarrow \quad 2\left[\frac{x^2}{2}-\frac{x^4}{4}\right]_0^{\alpha}=\left[\frac{x^2}{2}-\frac{x^4}{4}\right]_0^1$

$\Rightarrow \qquad \alpha^2-\frac{\alpha^4}{2}=\frac{1}{4} \Rightarrow 2\alpha^4-4\alpha^2+1=0$

Let $f(\alpha)=2\alpha^4-4\alpha^2+1$.

Then, $f(1)=-1<0$ and $f\left(\frac{1}{2}\right)=\frac{1}{8}>0 \Rightarrow \frac{1}{2}<\alpha<1$

15. (a,b,c) Given, curve $y=x^3$ and $y=x^{1/3}$ is mirror image about $y=x$ line

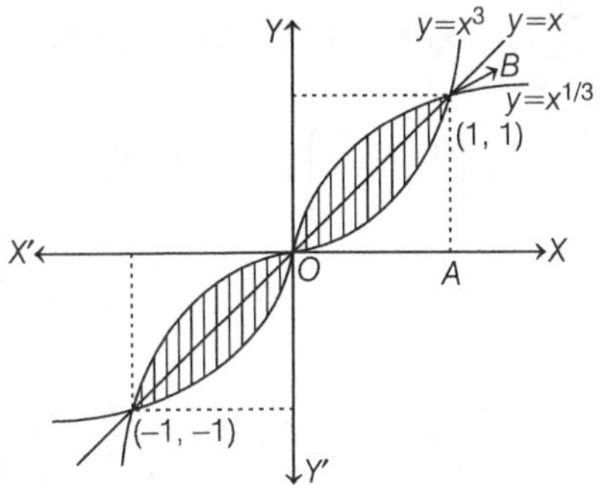

$\therefore$ Area $=4$ (area of $\Delta OAB-\int_0^1 x^3dx$)

$=4\left(\frac{1}{2}\times1\times1-\left[\frac{x^4}{4}\right]_0^1\right)$

$=4\left(\frac{1}{2}-\frac{1}{4}\right)=4\left(\frac{1}{4}\right)=1$ sq unit

16. (a,b) Given, curves are $|y|=1-x^2$ and $x^2+y^2=1$

Enclosed area is symmetrical about both axes.

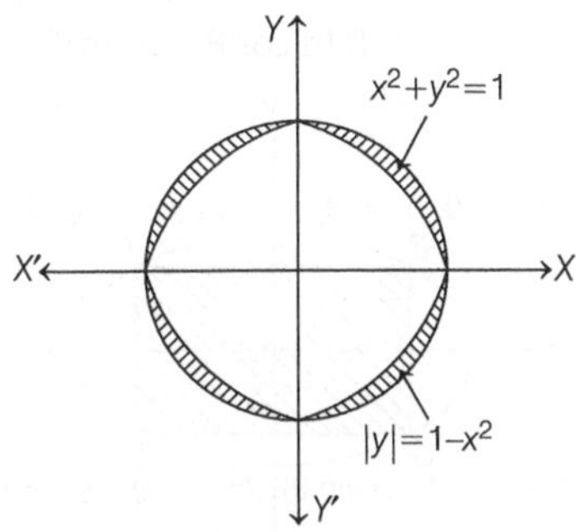

$\therefore$ Area = Area of circle $-4\int_0^1(1-x^2)\,dx$

$=\pi(1)^2-4\left[x-\frac{x^3}{3}\right]_0^1$

$=\pi-4\left(1-\frac{1}{3}\right)$

$=\left(\pi-\frac{8}{3}\right)$ sq units

17. (a,c,d)

(a)

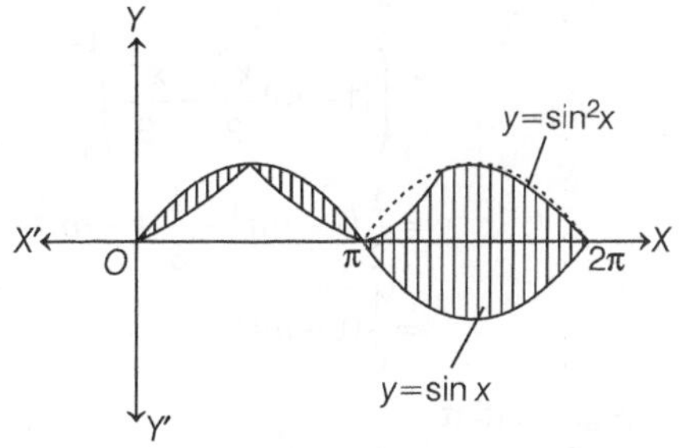

We know that area bounded by $y=\sin x$ and X-axis for $x\in[0,\pi]$ is 2 sq. units. Then, area bounded by $y=\sin x$ and $y=\sin^2 x$ is 4 sq. units for $x\in[0,2\pi]$.

Then, for $x\in[0,10\pi]$, the area bounded is 20 sq units.

(b)

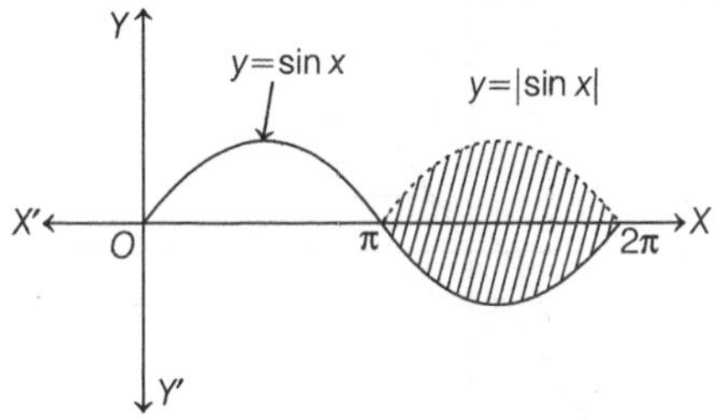

The area bounded by $y=|\sin x|$ and $y=|\sin x|$ for $x\in[0,2\pi]$ is 4 sq units.

Then, for $x\in[0,20\pi]$, the area bounded is 40 sq units

(c)

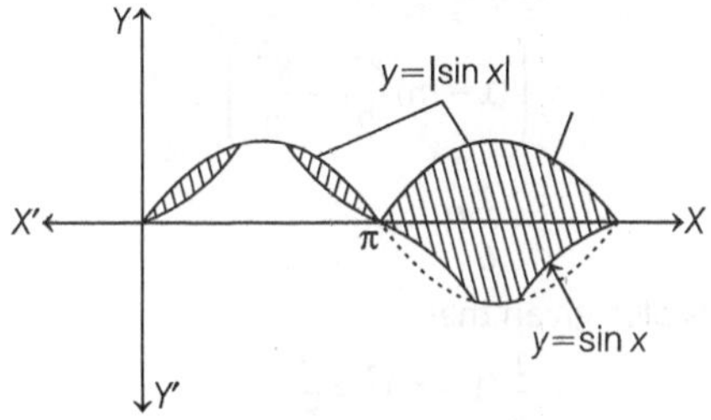

The area bounded by $y=\sin x$ and $y=\sin^3 x$ for $x\in[0,2\pi]$ is 4 sq units. Then, for $x\in[0,10\pi]$, the area bounded is 20 sq units.

(d) Similarly, the area bounded by $y=\sin x$ and $y=\sin^4 x$ for $x\in[0,10\pi]$ is 20 sq units.

18. (b,d)

Case-I : When $m<0$

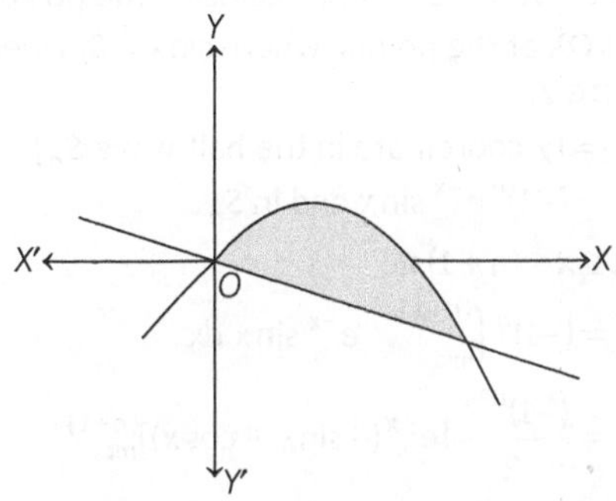

$\therefore$ Area of the region $= \int_0^{1-m} [(x - x^2) - mx]\,dx$

$$= \left((1-m)\frac{x^2}{2} - \frac{x^3}{3}\right)_0^{1-m}$$

$$= \frac{1}{2}(1-m)^3 - \frac{1}{3}(1-m)^3$$

$$= \frac{1}{6}(1-m)^3$$

Also, it is given that

$$\frac{1}{6}(1-m)^3 = \frac{9}{2}$$

$$\Rightarrow \quad (1-m)^3 = 27$$

$$\Rightarrow \quad (1-m) = 3 \Rightarrow m = -2$$

Case II : When $m > 0$

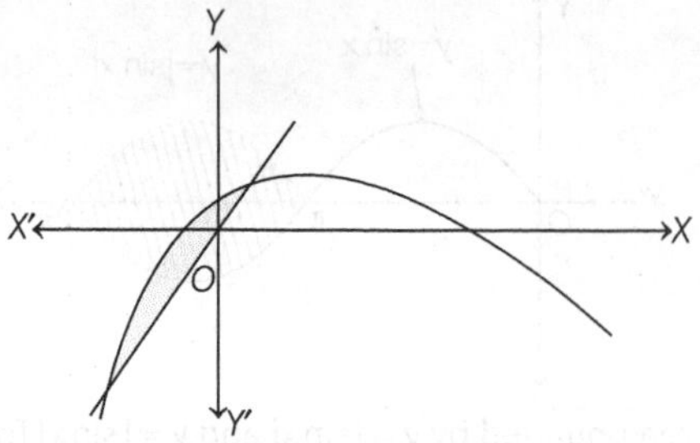

$\therefore$ Area of the region

$$= \int_{1-m}^{0} (x - x^2 - mx)\,dx$$

$$= \left((1-m)\frac{x^2}{2} - \frac{x^3}{3}\right)_{1-m}^{0}$$

$$= -\frac{1}{6}(1-m)^3$$

It is also given that

$$-\frac{1}{6}(1-m)^3 = \frac{9}{2}$$

$$\Rightarrow \quad (1-m)^3 = -27$$

$$\Rightarrow \quad 1 - m = -3$$

$$\Rightarrow \quad m = 4$$

Sol. (Q.Nos. 19-21)

19. (b) Since, $-1 \le \sin x \le 1$, the curve $y = e^{-x}\sin x$ is bounded by the curves $y = e^{-x}$ and $y = e^{-x}$.

Also, the curve $y = e^{-x}\sin x$ intersects the positive semi-axis OX at the points, where $\sin x = 0$, where $x_n = n\pi, n \in Z$.

Also, $[y_n] = [y\text{-coordinate in the half-wave } S_n]$

$= (-1)^n e^{-x}\sin x$ and in S_n

$n\pi \le x \le (n+1)\pi$

$$\therefore \quad S_n = (-1)^n \int_{n\pi}^{(n+1)\pi} e^{-x}\sin x\,dx$$

$$= \frac{(-1)^{n+1}}{2}[e^{-x}(-\sin x + \cos x)]_{n\pi}^{(n+1)\pi}$$

$$= \frac{(-1)^{n+1}}{2}[e^{-(n+1)\pi}(-1)^{n+1} - e^{-n\pi}(-1)^n]$$

$$= \frac{e^{-n\pi}}{2}(1 + e^{-\pi}) \Rightarrow \frac{S_{n+1}}{S_n} = e^{-\pi}$$

and $S_0 = \frac{1}{2}(1 + e^{-\pi})$ sq units

20. (b) Therefore, the sequence $S_0, S_1, S_2, \ldots$ forms an infinite GP with common ratio $e^{-\pi}$, we have

21. (b) $\sum_{n=0}^{\infty} S_n = \dfrac{\frac{1}{2}(1+e^{\pi})}{1 - e^{-\pi}}$

$$\because \quad \sum_{k=0}^{n} S_k = S_0 + S_1 + S_2 + S_3 + \ldots + S_n$$

$$= \left(\frac{1+e^{-\pi}}{2}\right) + e^{-\pi}\frac{(1+e^{-\pi})}{2} + e^{-2\pi}\frac{(1+e^{-\pi})}{2} + \ldots + \frac{e^{-n\pi}(1+e^{-\pi})}{2}$$

$$\sum_{k=0}^{\infty} S_k = \lim_{n\to\infty}\left(\sum_{k=0}^{n} S_n\right) = \left(\frac{1+e^{-\pi}}{2}\right)\cdot\frac{1}{1-e^{-\pi}}$$

[common ratio $= e^{-\pi}$]

$$\therefore \quad \sum_{n=0}^{\infty} S_n = \frac{(1+e^{-\pi/2})}{1-e^{-\pi}}$$

Sol. (Q.Nos. 22-24)

22. (b) Let $A(a\cos\theta, b\sin\theta)$ be any point on an ellipse in I quadrant.

So, coordinates of $B(-a\cos\theta, b\sin\theta)$

$C(-a\cos\theta, -b\sin\theta)$

$D(a\cos\theta, -b\sin\theta)$

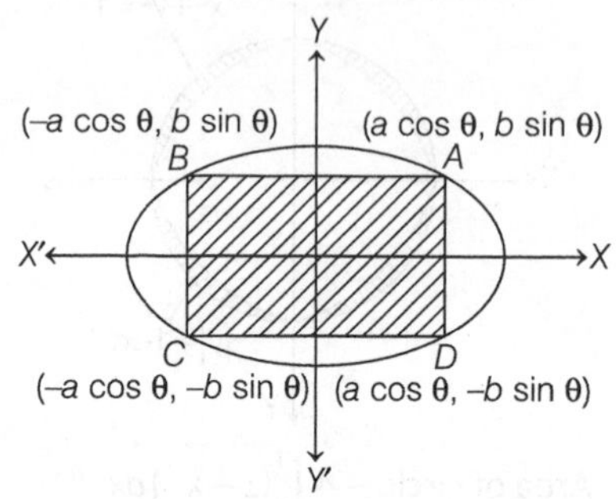

Now, area of rectangle $ABCD = CD \times AB$

$$= (2a\cos\theta)(2b\sin\theta)$$

$$= 2ab\sin 2\theta$$

Area of rectangle will be greatest, when $\sin 2\theta$ have its maximum value i.e. $\sin 2\theta = 1$

$\therefore$ Area $= 2ab$ sq units

23. (c) We know that area of ellipse $= \pi ab$

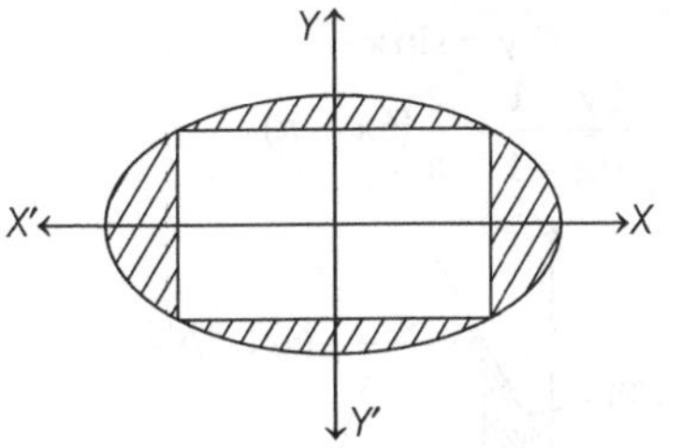

$\therefore$ Required area of shaded region

$= \text{Area of ellipse} - \text{Area of rectangle}$

$= \pi ab - 2ab = ab(\pi - 2)$ sq units

24. (b) Given ellipse $\dfrac{x^2}{a^2} + \dfrac{y^2}{b^2} = 1$

Equation of latusrectum $= x = ae$

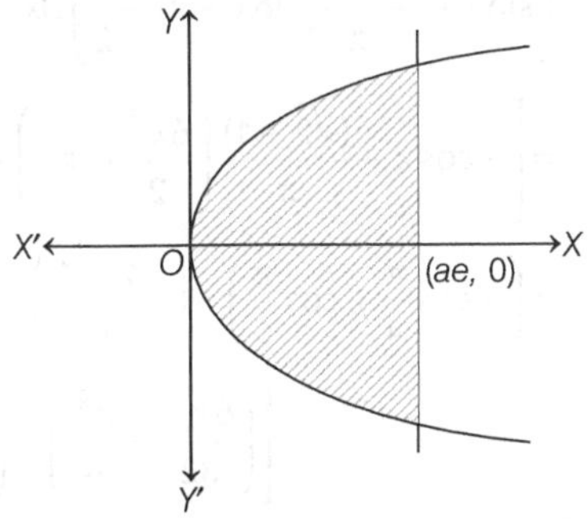

$\therefore$ Area of shaded region

$$2 \cdot \int_0^{ae} \frac{b}{a} \sqrt{a^2 - x^2}\, dx$$

$$= \frac{2b}{a} \left[\frac{x}{2} \sqrt{a^2 - x^2} + \frac{a^2}{2} \sin^{-1} \frac{x}{a} \right]_0^{ae}$$

$$= \frac{2b}{2a} \left[ae \sqrt{a^2 - a^2 e^2} + a^2 \sin^{-1} \frac{ae}{a} \right]$$

$$= ab\,(e\sqrt{1 - e^2} + \sin^{-1} e) \text{ sq units}$$

25. (i) $\to$ (B), (ii) $\to$ (A), (iii) $\to$ (D), (iv) $\to$ (C)

(i) We have, $y = |x|$, $|x| = 1$ and $y = 0$

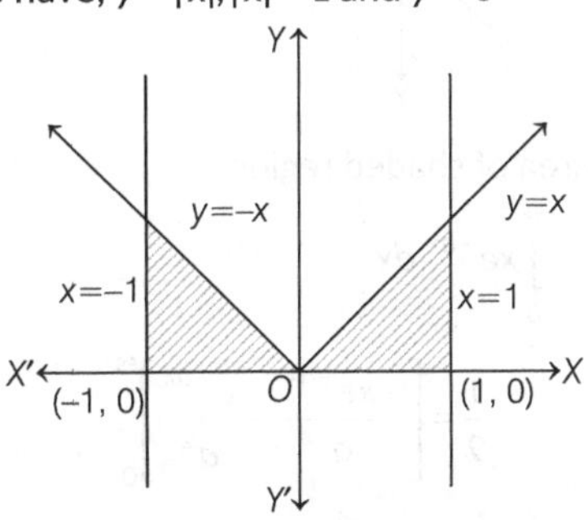

$\therefore$ Required area $= 2\int_0^1 x\, dx$

$$= 2\left[\frac{x^2}{2}\right]_0^1 = 2 \times \frac{1}{2} = 1 \text{ sq unit}$$

(ii) We have, $y = \sin x$, $x = 0$, $x = 2\pi$ and $y = 0$

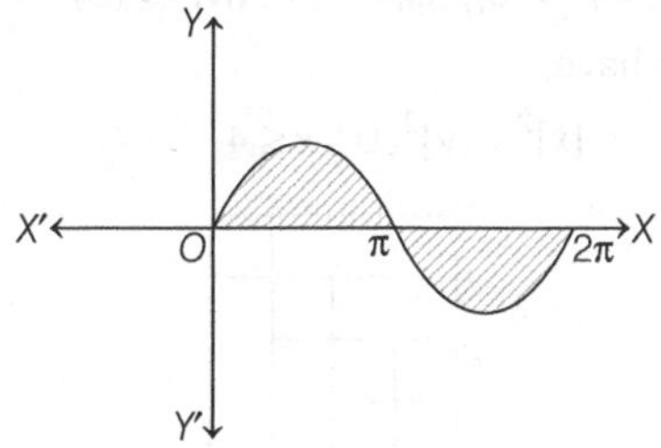

$\therefore$ Area of shaded region

$$2\int_0^{\pi} \sin x\, dx = 2[-\cos x]_0^{\pi}$$

$$= 2[-\cos \pi + \cos 0] = 2(1 + 1) = 4 \text{ sq units}$$

(iii) We have, $x^2 = y$ and $y = x + 2$

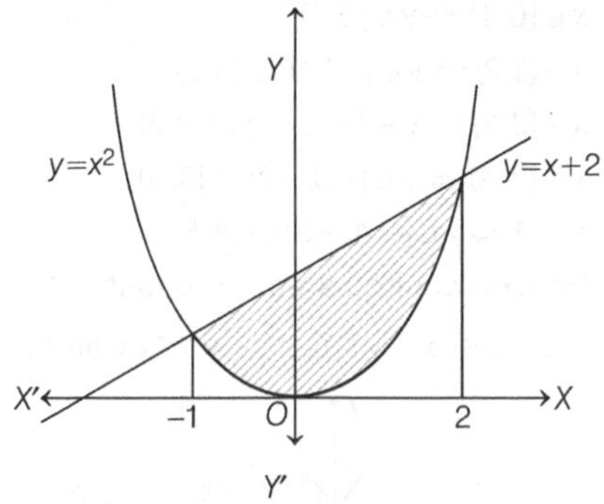

$\therefore$ Area of shaded region

$$\int_{-1}^{2} (x + 2 - x^2)\, dx = \left[\frac{x^2}{2} + 2x - \frac{x^3}{3}\right]_{-1}^{2}$$

$$= \left(2 + 4 - \frac{8}{3}\right) - \left(\frac{1}{2} - 2 + \frac{1}{3}\right) = \frac{9}{2} \text{ sq units}$$

$$\therefore \frac{k}{4} = \frac{9}{2} \Rightarrow k = 18$$

(iv) We have, $\dfrac{x^2}{9} + \dfrac{y^2}{5} = 1$

Here, $a^2 = 9$ and $b^2 = 5$

$(ae)^2 = 9 - 5 = 4 \Rightarrow ae = 2$

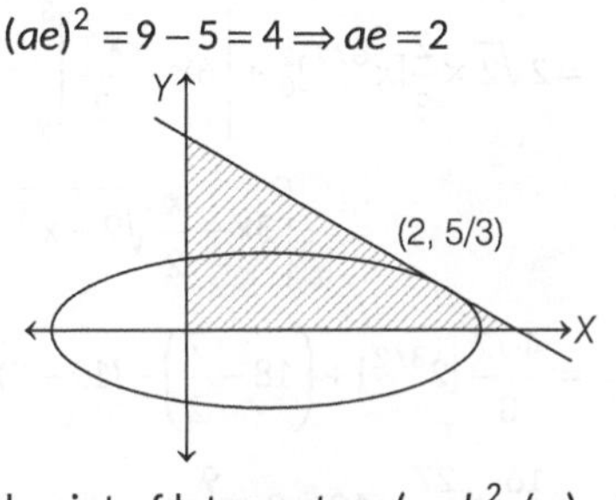

End point of latusrectum $(ae, b^2/a) = (2, 5/3)$

$\therefore$ Equation of tangent at $\left(2, \dfrac{5}{3}\right)$ is $\dfrac{2x}{9} + \dfrac{y}{3} = 1$

$\therefore$ x-intercept $= \dfrac{9}{2}$ and y-intercept $= 3$

$\therefore$ Area of shaded region $= \dfrac{1}{2} \times \dfrac{9}{2} \times 3 = \dfrac{27}{4}$ sq units

26. (i) →(C), (ii) → (B), (iii) → (A), (iv) → (D)

(i) We have,

$$[x]^2 = [y]^2, 0 \le x \le 4$$

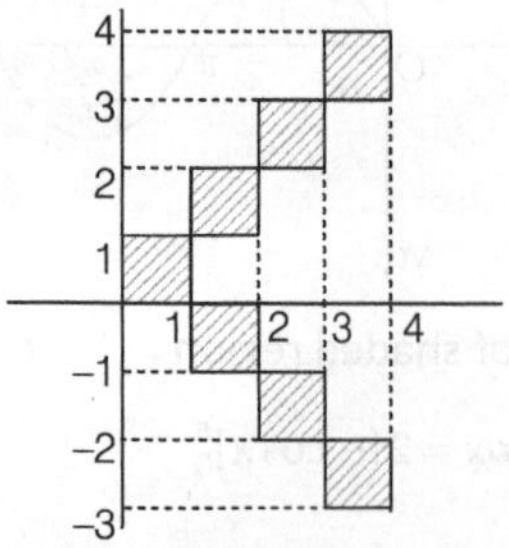

$$[y] = \pm [x]$$

$x \in [0,1) \Rightarrow y \in [0,1)$

$x \in [1,2) \Rightarrow y \in [-1,0) \cup [1,2)$

$x \in [2,3) \Rightarrow y \in [-2,-1) \cup [2,3)$

$x \in [3,4) \Rightarrow y \in [-3,-2) \cup [3,4)$

$x = 4 \Rightarrow y \in [-4,-3) \cup [4,5)$

Hence, required area = 7 sq units

(ii) We have, $x + y \le 6$, $x^2 + y^2 \le 6y$ and $y^2 \le 8x$

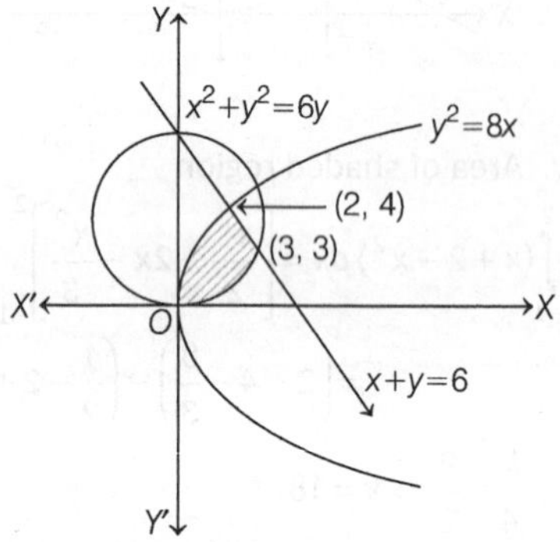

∴ Area of shaded region

$$\int_0^2 \sqrt{8x}\, dx + \int_2^3 (6-x)\, dx - \int_0^3 (3 - \sqrt{9-x^2})\, dx$$

$$= 2\sqrt{2} \times \frac{2}{3}[x^{3/2}]_0^2 + \left[6x - \frac{x^2}{2}\right]_2^3 - \left[3x - \frac{x}{2}\sqrt{9-x^2} - \frac{9}{2}\sin^{-1}\frac{x}{3}\right]_0^3$$

$$= \frac{4\sqrt{2}}{3}[2^{3/2}] + \left(18 - \frac{9}{2}\right) - (12-2) - \left[9 - \frac{9}{2}\frac{\pi}{2}\right]$$

$$= \frac{16}{3} + \frac{27}{2} - 10 - 9 + \frac{9}{4}\pi$$

$$= \frac{9}{4}\pi - \frac{1}{6}$$

$$= \frac{27\pi - 2}{12}$$

∴ $k = 27$

(iii) We have,

$$y = \sin x$$

and $$\frac{2y-1}{\sqrt{2}-1} = \frac{2}{\pi}(6x - \pi)$$

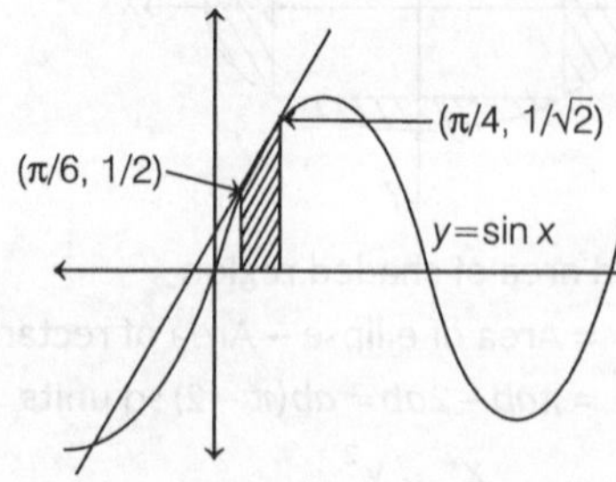

Area of region

$$\int_{\pi/6}^{\pi/4} \left(\sin x - \frac{(\sqrt{2}-1)}{\pi}(6x - \pi) - \frac{1}{2} \right) dx$$

$$= \left[-\cos x - \frac{(\sqrt{2}-1)}{\pi}\left(\frac{6x^2}{2} - \pi x\right) - \frac{1}{2}x \right]_{\pi/6}^{\pi/4}$$

$$= \left(-\cos\frac{\pi}{4} + \cos\frac{\pi}{6}\right) - \left(\frac{\sqrt{2}-1}{\pi}\right)\left[\left(\frac{6\pi^2}{32} - \frac{\pi^2}{4}\right) - \left(\frac{6\pi^2}{72} - \frac{\pi^2}{6}\right)\right] - \frac{1}{2}\left(\frac{\pi}{4} - \frac{\pi}{6}\right)$$

$$= \frac{\sqrt{3}-\sqrt{2}}{2} - \frac{(\sqrt{2}+1)\pi}{48}$$

∴ $k = 48$

(iv) We have, $y = xe^{-ax}$ and $y = 0$

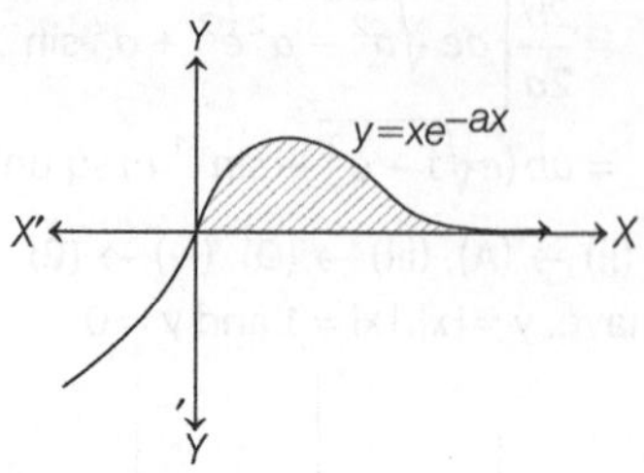

∴ Area of shaded region

$$\int_0^\infty xe^{-ax}\, dx$$

$$\Rightarrow \frac{1}{9} = \left[\frac{-xe^{-ax}}{a} - \frac{e^{-ax}}{a^2}\right]_0^\infty$$

$$\Rightarrow \frac{1}{9} = 0 + \frac{1}{a^2}$$

$$\Rightarrow a^2 = 9$$

$$\Rightarrow a = 3 \quad [\because a > 0]$$

Sol. (Q.Nos. 27-29)

27. (6) When $\alpha = 0$, then $f(0)$ denotes that area of the shaded region bounded by $x = 0$, $y = 3$, $x = 2$ and $y^2 = 4x$

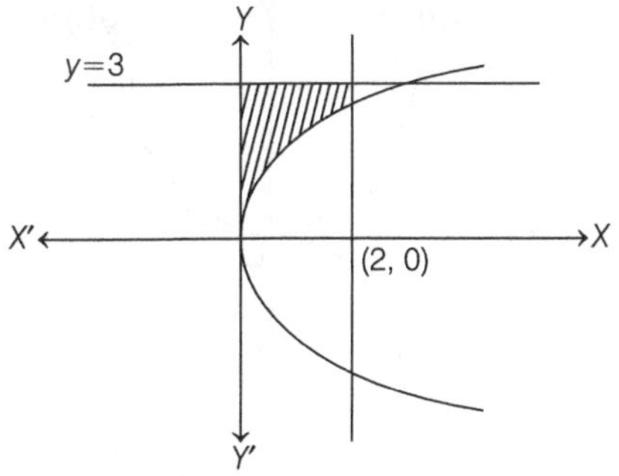

$$\therefore \quad f(0) = \int_0^2 (3 - y)\,dx$$

$$= \int_0^2 (3 - 2\sqrt{x})\,dx$$

$$= \left[3x - \frac{4x^{3/2}}{3}\right]_0^2 = 6 - \frac{8\sqrt{2}}{3}$$

So, $f(0) + \frac{8\sqrt{2}}{3} = 6$

28. (5) When $\alpha = 1$, then $f(1)$ is the area of shaded region bounded by $x = 0, x = 2, y^2 = 4x$

and $\quad y = |x - 1| + |x - 2| + x$

$$y = \begin{cases} -x + 3 & , \quad x < 1 \\ x + 1 & , \quad 1 \le x < 2 \\ 3x - 3 & , \quad x \ge 2 \end{cases}$$

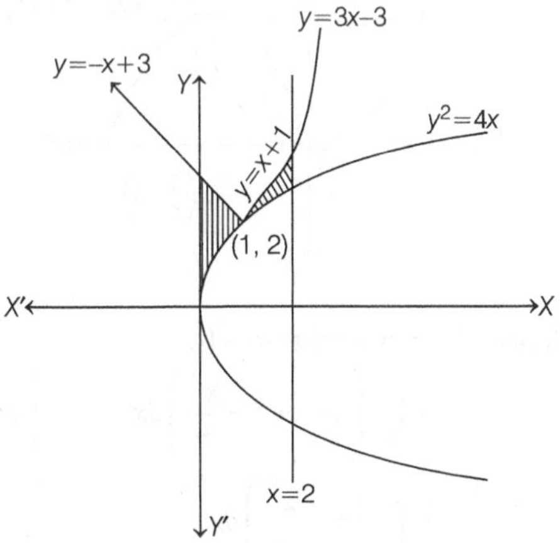

$$f(1) = \int_0^1 (-x + 3 - 2\sqrt{x})\,dx + \int_1^2 (x + 1 - 2\sqrt{x})\,dx$$

$$= \left[-\frac{x^2}{2} + 3x - \frac{4}{3}x^{3/2}\right]_0^1 + \left[\frac{x^2}{2} + x - \frac{4}{3}x^{3/2}\right]_1^2$$

$$= \left(-\frac{1}{2} + 3 - \frac{4}{3}\right) + \left(2 + 2 - \frac{8\sqrt{2}}{3} - \frac{1}{2} - 1 + \frac{4}{3}\right)$$

$$f(1) = 5 - \frac{8}{3}\sqrt{2} \Rightarrow f(1) + \frac{8}{3}\sqrt{2} = 5$$

29. (64) Given parabola $y^2 = 4x$

Slope of normal $= -1$

$\therefore$ Equation of normal is $y = 3 - x$

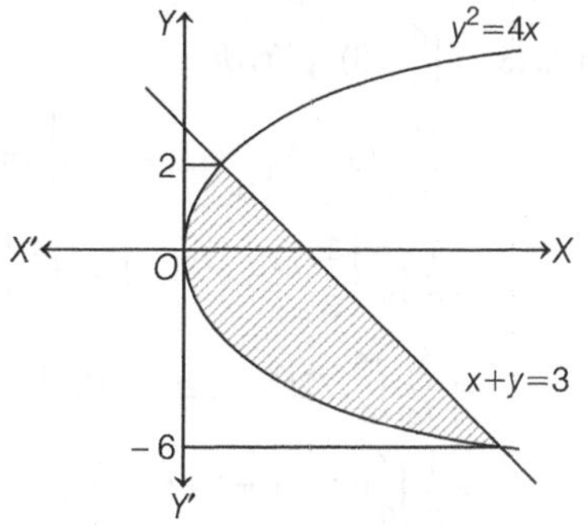

$\therefore$ Area of shaded region

$$\int_{-6}^2 \left(3 - y - \frac{y^2}{4}\right) dy$$

$$= \left[3y - \frac{y^2}{2} - \frac{y^3}{12}\right]_{-6}^2$$

$$= \frac{64}{3} \text{ sq units}$$

$$\Rightarrow \quad k = 64$$

Sol. (Q.Nos. 30-32)

Given cycloid,

$$x = a(t - \sin t)$$

and $\quad y = a(1 - \cos t)$

Thus, area of the region bounded by an arc is

$$S = \left| -\int_\alpha^\beta y(t)\,x'(t)\,dt \right|$$

$$\alpha = 0, \beta = 2\pi$$

$$= \left| -\int_0^{2\pi} a(1 - \cos t)\,a(1 - \cos t)\,dt \right|$$

30. (3) $\quad S = \left| -\int_0^{2\pi} a(1 - \cos t)a(1 - \cos t)\,dt \right|$

$$= \left| -a^2 \int_0^{2\pi} (1 - 2\cos t + \cos^2 t)\,dt \right|$$

$$= \left| -a^2 \int_0^{2\pi} 1 - 2\cos t + \left(\frac{1 + \cos 2t}{2}\right) dt \right|$$

$$= \left| -\frac{a^2}{2} \int_0^{2\pi} (3 - 4\cos t + \cos 2t)\,dt \right|$$

$$= \left| -\frac{a^2}{2}\left[3t - 4\sin t + \frac{\sin 2t}{2}\right]_0^{2\pi} \right|$$

$$= |-3\pi a^2|$$

$$= 3\pi a^2 \text{ sq units}$$

$$\Rightarrow \quad k = 3$$

31. (27) Let $x = \frac{t}{3}(6-t), y = \frac{t^2}{8}(6-t)$

$$\Rightarrow \quad x = \frac{1}{3}(6t - t^2), y = \frac{1}{8}(6t^2 - t^3)$$

Now, Area $(S) = \int_\alpha^\beta x(t) \cdot y'(t)\, dt$

$$= \int_0^6 \frac{t}{3}(6-t)\left\{\frac{1}{8}(12t - 3t^2)\right\} dt$$

$$= \frac{1}{8}\int_0^6 \left\{\frac{1}{3}(6t - t^2)(12t - 3t^2)\right\} dt$$

$$= \frac{1}{24}\int_0^6 [72t^2 - 12t^3 - 18t^3 + 3t^4]\, dt$$

$$= \frac{1}{8}\int_0^6 [t^4 - 10t^3 + 24t^2]\, dt$$

$$= \frac{1}{8}\left[\frac{t^5}{5} - \frac{10t^4}{4} + \frac{24t^3}{3}\right]_0^6$$

$$= \frac{1}{8}\left[\frac{6^5}{5} - \frac{10(6)^4}{4} + 8(6)^3\right]$$

$$= \frac{1}{8}\left[\frac{7776}{5} - 3240 + 1728\right]$$

$$= \frac{215}{5 \times 8} = \frac{27}{5} \text{ sq units}$$

$$\Rightarrow \quad k' = 27$$

32. (19) $\frac{dx}{dt} = 1 - 3t^2$ and $\frac{dy}{dt} = 1 - 4t^3$

So, $x\frac{dy}{dt} - y\frac{dx}{dt} = (t - t^3)(1 - 4t^3) - (1 - t^4)(1 - 3t^2)$

$$= t^6 - 3t^4 - t^3 + 3t^2 + t - 1$$

$\therefore$ Required area $= \frac{1}{2}\int_{-1}^{1} (t^6 - 3t^4 - t^3 + 3t^2 + t - 1)\, dt$

$$= \frac{16}{35} \text{ sq units}$$

$$\Rightarrow \quad b - a = 35 - 16 = 19$$

33. (2) Given, curve $||x| + |y|| \geq 1$ and $x^2 + y^2 \leq 1$

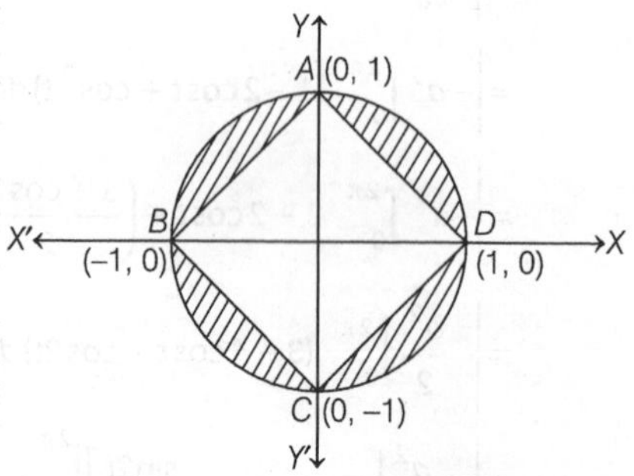

$\therefore$ Area of shaded region = area of circle

$$-2 \times \text{area of } \Delta ABD = \pi(1)^2 - 2\left(\frac{1}{2} \times 2 \times 1\right)$$

Area $= \pi - 2$

Hence, $k = 2$

34. (6) The two curves $f(x) = x^2$ and $g(x) = cx^3$ intersect at $A\left(\frac{1}{c}, \frac{1}{c^2}\right)$ and $O(0, 0)$

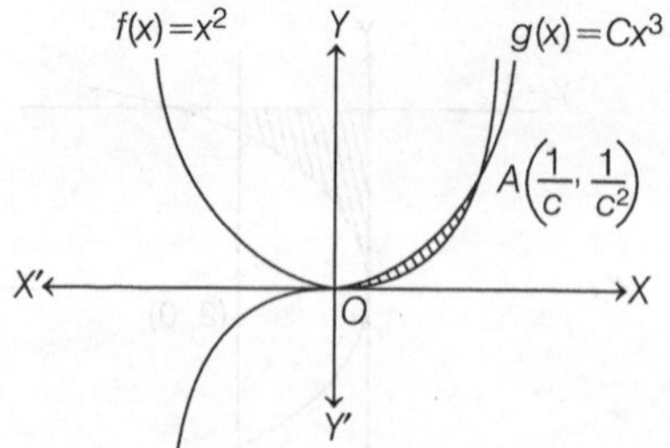

$$\therefore \quad \int_0^{1/c} (y_1 - y_2)\, dx = \frac{2}{3}$$

$$\Rightarrow \quad \int_0^{1/c} (x^2 - Cx^3)\, dx = \frac{2}{3}$$

$$\Rightarrow \quad \left[\frac{x^3}{3} - \frac{Cx^4}{4}\right]_0^{1/c} = \frac{2}{3}$$

$$\Rightarrow \quad \frac{1}{12c^3} = \frac{2}{3} \Rightarrow c = \frac{1}{2}$$

Hence, $\frac{1}{c} + \frac{1}{c^2} = 6$

35. (1) Given, curves are $y = x - bx^2$ and $y = \frac{x^2}{b}$

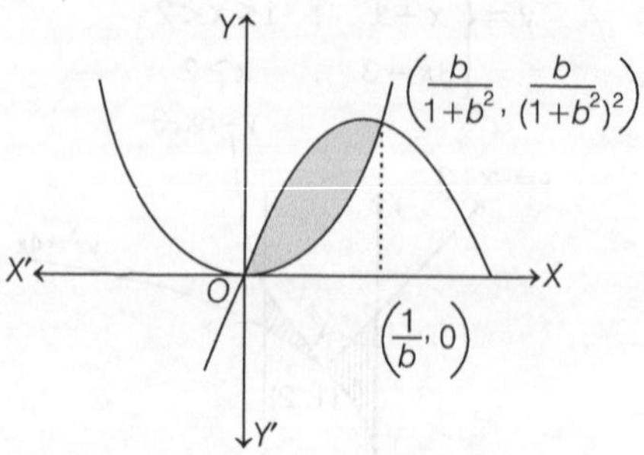

Hence, the required area = (A)

$$= \int_0^t \left(x - bx^2 - \frac{x^2}{b}\right) dx \qquad \left[\text{let } t = \frac{b}{b^2 + 1}\right]$$

$$= \int_0^t \left(x - \frac{x^2}{t}\right) dx$$

$$= \left(\frac{x^2}{2} - \frac{x^3}{3t}\right)_0^t = \left(\frac{t^2}{2} - \frac{t^2}{3}\right)$$

$$= \frac{3t^2 - 2t^2}{6} = \frac{t^2}{6} = \frac{1}{6}\left(\frac{b}{b^2 + 1}\right)^2$$

Taking log both sides, we get

$$\log A = 2\log b - 2\log(b^2 + 1) - \log 6$$

$$\Rightarrow \quad \frac{1}{A}\frac{dA}{db} = \frac{2}{b} - \frac{4b}{(b^2+1)}$$

$$\Rightarrow \quad \frac{dA}{db} = A\left(\frac{2(b^2-2b+1)}{b(b^2+1)}\right)$$

For maximum or minimum,

$$\frac{dA}{db} = 0 \text{ gives } b = 1$$

Since, $b > 0$, so $b = 1$

Thus, A is maximum, when $b = 1$

36. (5) Given that $(|x|, |y|) \le 1$ and $xy \le \frac{1}{2}$

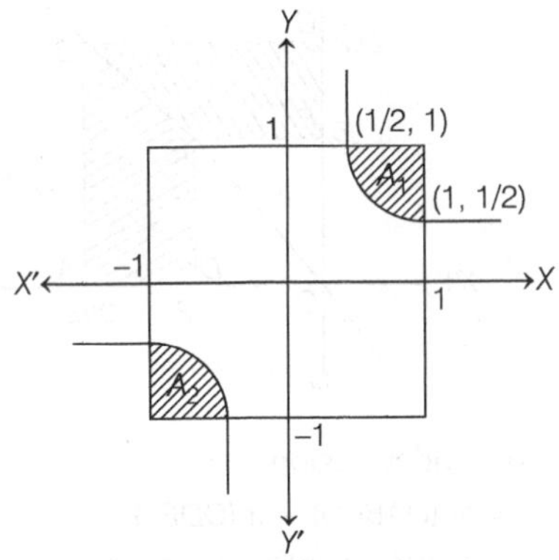

$$\therefore \quad A_1 = A_2 = \int_{1/2}^{1}\left(1 - \frac{1}{2x}\right)dx$$

$$= \left(x - \frac{1}{2}\ln x\right)_{1/2}^{1} = \frac{1}{2} - \frac{1}{2}\ln 2$$

Required area = Area of square $-(A_1 + A_2)$

$$= 4 - 2\left(\frac{1}{2} - \frac{1}{2}\ln 2\right)$$

$$= 4 - (1 - \ln 2)$$

$$= 3 + \ln 2$$

So, $a = 3$ and $b = 2$

Therefore, $a + b = 5$

37. (41) Let $f : [-3, 1] \to R$ be given as

$$f(x) = \begin{cases} \min\{(x+6), x^2\} & -3 \le x \le 0 \\ \max\{\sqrt{x}, x^2\}, & 0 \le x \le 1 \end{cases}$$

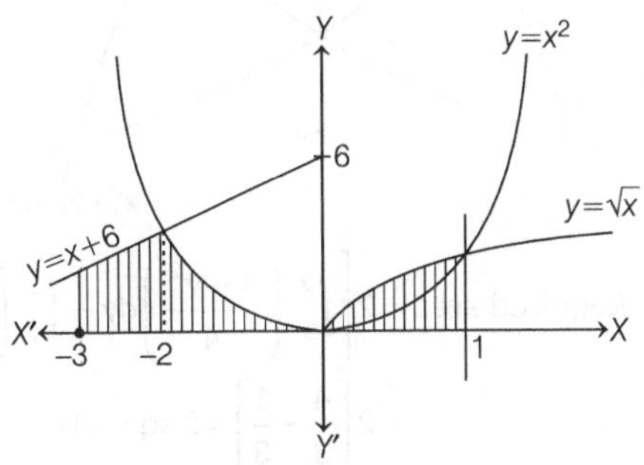

$\therefore$ Area bounded by the $f(x)$ and X-axis is A

$$= \int_{-3}^{-2}(x+6)\,dx + \int_{-2}^{0} x^2 dx + \int_{0}^{1}\sqrt{x}\,dx$$

$$= \left[\frac{x^2}{2} + 6x\right]_{-3}^{-2} + \left[\frac{x^3}{3}\right]_{-2}^{0} + \left[\frac{2}{3}x^{3/2}\right]_{0}^{1}$$

$$A = \frac{41}{6}$$

$$6A = 41$$

38. (44) $f(x) = \max\{x^2, (1-x)^2, 2x(1-x)\}, 0 \le x \le 1$

We draw the graph of

$$y = x^2 \quad \text{...(i)}$$

$$y = (1-x)^2 \quad \text{...(ii)}$$

$$y = 2x(1-x) \quad \text{...(iii)}$$

On solving Eqs. (i) and (iii), we get

$$x^2 = 2x(1-x)$$

$$\Rightarrow \quad 3x^2 = 2x$$

$$\Rightarrow \quad x = 0 \text{ or } x = \frac{2}{3}$$

On solving Eqs. (ii) and (iii), we get

$$(1-x)^2 = 2x(1-x)$$

$$\Rightarrow \quad x = \frac{1}{3} \text{ and } x = 1$$

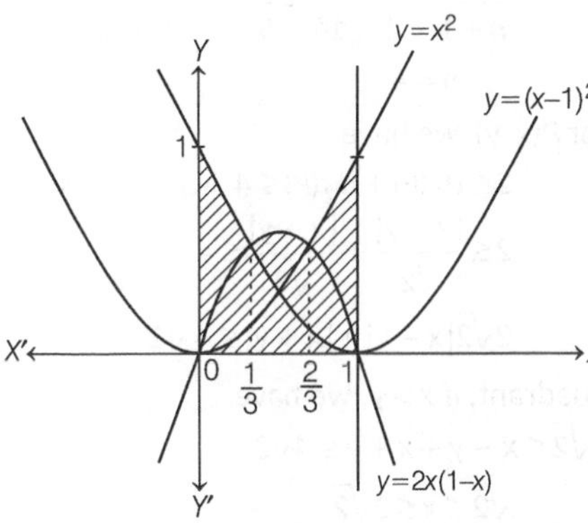

From the figure it is clear that

$$f(x) = \begin{cases} (1-x)^2, & 0 \le x \le \frac{1}{3} \\ 2x(1-x), & \frac{1}{3} < x \le \frac{2}{3} \\ x^2, & \frac{2}{3} < x \le 1 \end{cases}$$

$\therefore$ The required area

$$A = \int_0^1 f(x)\,dx$$

$$= \int_0^{1/3}(1-x)^2 dx + \int_{1/3}^{2/3} 2x(1-x)dx + \int_{2/3}^{1} x^2 dx$$

$$= \left[-\frac{1}{3}(1-x)^3\right]_0^{1/3} + \left[\left(x^2 - \frac{2x^3}{3}\right)\right]_{1/3}^{2/3} + \left[\frac{x^3}{3}\right]_{2/3}^{1}$$

$$=-\frac{1}{3}\left(\frac{2}{3}\right)^3+\frac{1}{3}+\left(\frac{2}{3}\right)^2-\frac{2}{3}\left(\frac{2}{3}\right)^3$$
$$-\left(\frac{1}{3}\right)^2+\frac{2}{3}\left(\frac{1}{3}\right)^3+\frac{1}{3}-\frac{1}{3}\left(\frac{2}{3}\right)^3=\frac{17}{27}$$

So, $p+q=17+27=44$

39. (4)

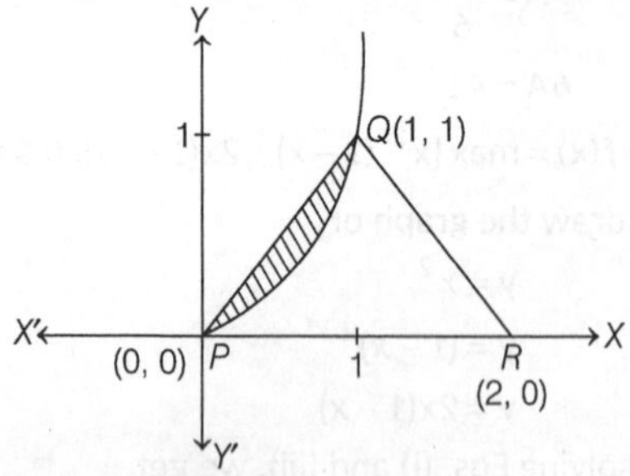

According to question,

$$\int_0^1 (x-x^n)dx=\frac{3}{10}\left(\frac{1}{2}\times 2\times 1\right)$$
$$\Rightarrow \quad \left[\frac{x^2}{2}-\frac{x^{n+1}}{n+1}\right]_0^1=\frac{3}{10}$$
$$\Rightarrow \quad \frac{1}{2}-\frac{1}{n+1}=\frac{3}{10}$$
$$\Rightarrow \quad \frac{1}{n+1}=\frac{1}{2}-\frac{3}{10}=\frac{1}{5}$$
$$\Rightarrow \quad n=4$$

40. (6) For $P(x, y)$, we have

$$2\le d_1(P)+d_2(P)\le 4$$
$$\Rightarrow \quad 2\le\frac{|x-y|}{\sqrt{2}}+\frac{|x+y|}{\sqrt{2}}\le 4$$
$$\Rightarrow \quad 2\sqrt{2}|x-y|+|x+y|\le 4\sqrt{2}$$

In I quadrant, if $x>y$, we have

$$2\sqrt{2}\le x-y+x+y\le 4\sqrt{2}$$
$$\Rightarrow \quad \sqrt{2}\le x\le 2\sqrt{2}$$

The region of points satisfying these inequalities is

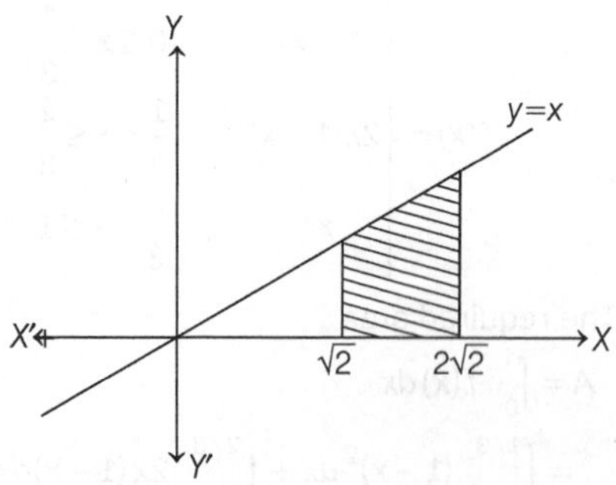

In I quadrant, if $x<y$, we have

$$2\sqrt{2}\le y-x+x+y\le 4\sqrt{2}$$
$$\sqrt{2}\le y\le 2\sqrt{2}$$

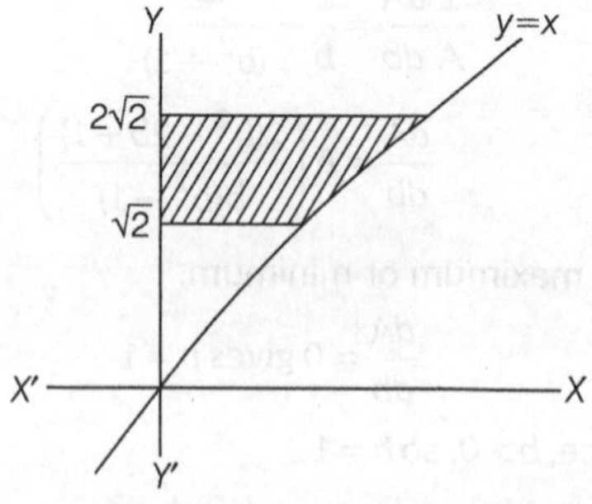

Combining above two regions, we have

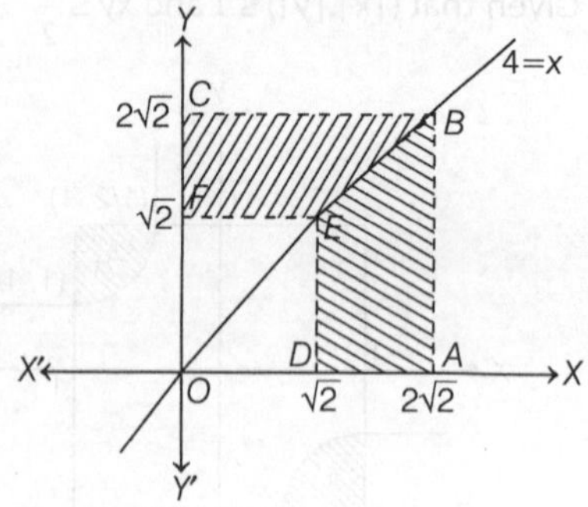

The area of shaded region

$$= \text{ar}(OABCO)-\text{ar}(ODEF)$$
$$=2\sqrt{2})^2-(\sqrt{2})^2=8-2=6 \text{ sq. units}$$

41. (2) Given, curves are

$$x^2+2y-1=0\Rightarrow x^2=-2\left(y-\frac{1}{2}\right)$$

and $\quad y^2+4x-4=0\Rightarrow y^2=-4(x-1)$

and $\quad y^2-4x-4=0\Rightarrow y^2=4(x+1)$

So, the graph of given curves are

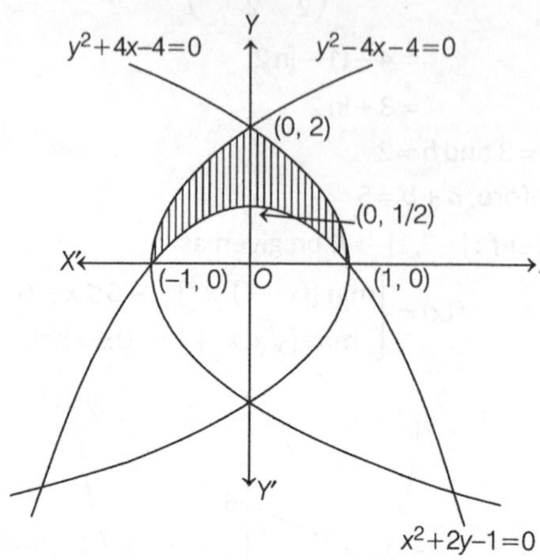

$$\therefore \text{ Required area}=2\left[\int_0^2\left(\frac{4-y^2}{4}\right)dy-\int_0^1\left(\frac{1-x^2}{2}\right)dx\right]$$
$$=2\left[\frac{4}{3}-\frac{1}{3}\right]=2 \text{ sq units}$$

CHAPTER

12

DIFFERENTIAL EQUATION

LEVEL 01 \ Starter Level

12.1 Order and Degree of Differential Equation

1 Find the order and degree (if defined) of the following differential equations.

(i) $\left(\frac{dy}{dx}\right)^2 + \frac{dy}{dx} - x\sin y = 0$

(ii) $\frac{dy}{dx} + y = \frac{1}{\frac{dy}{dx}}$

(iii) $\cos^{-1}\left(\frac{dy}{dx}\right) = 2x + y$

(iv) $\sqrt{1-x^2}\,dx + \sqrt{1-y^2}\,dy = 0$

(v) $\frac{d^4y}{dx^4} = \left\{c + \left(\frac{dy}{dx}\right)^2\right\}^{3/2}$

(vi) $2\log\left(\frac{dy}{dx}\right) = 2x + 3y$

(vii) $\frac{d^2y}{dx^2} - 2\sin\left(\frac{dy}{dx}\right) + y = 0$

(viii) $3\frac{d^2y}{dx^2} = 2x\log\left(\frac{dy}{dx}\right)$

(ix) $y^2 + 2\frac{dy}{dx} = \frac{1}{2}\int y\,dx$

12.2 Formation of Differential Equation

2 Find form the differential equation of the family of curves represented by the following equations, where *a* and *b* are arbitrary constants.

(i) $y^2 = 4ax$

(ii) $\frac{x}{a} + \frac{y}{b} = 1$

(iii) $y = a\cos(x + b)$

(iv) $\frac{x^2}{a^2} + \frac{y^2}{b^2} = 1$

(v) $y = ae^{3x} + be^{-2x}$

3 Find form the differential equation of family of circles which passes through origin and whose centre lies on X-axis.

4 Find form the differential equation of the family of parabolas having their axis of symmetry coincident with the X-axis.

12.3 Solution of Differential Equation Using Inspection Method

5 Solve the following differential equations.

(i) $x\,dy + y\,dx + x^2 dx = 0$

(ii) $x^3 \dfrac{dy}{dx} + 4x^2 \tan y = \sec y$

(iii) $\dfrac{x\,dy}{x^2+y^2} = \left(\dfrac{y}{x^2+y^2} - 1\right)dx$

(iv) $ye^{x/y} dx = (xe^{x/y} + y^2)dy \ (y \neq 0)$

12.4 Solution of Differential Equation Using Variable Separation

6 Solve the following differential equations.

(i) $(x+1)\dfrac{dy}{dx} = 3xy$

(ii) $\sec^2 x \tan y\,dx + \sec^2 y \tan x\,dy = 0$

(iii) $\dfrac{dy}{dx} = \dfrac{1+y^2}{1+x^2}$

(iv) $\log\left(\dfrac{dy}{dx}\right) = ax + by$

(v) $\dfrac{dy}{dx} = \dfrac{(1+y^2)}{xy(1+x^2)}$

(vi) $\dfrac{dy}{dx} = 2e^x y^3,\ y(0) = \dfrac{1}{2}$

(vii) $x\,dy + y\,dx = xy\,dx,\ \ y(1) = 1$

7 Solve the following differential equations.

(i) $\dfrac{dy}{dx} = \cos(x+y)$

(ii) $\dfrac{dy}{dx} = (x+y+1)^2$

(iii) $\dfrac{dy}{dx} = \dfrac{x-y+3}{2(x-y)+5}$

(iv) $\dfrac{dy}{dx} = \dfrac{6x-2y-7}{3x-y+4}$

12.5 Homogeneous Differential Equation

8 Solve the following differential equations.

(i) $x^2 y\,dx - (x^3 + y^3)dy = 0$

(ii) $x\,dy - y\,dx = \sqrt{x^2+y^2}\,dx$

(iii) $(1+e^{x/y})dx + e^{x/y}\left(1 - \dfrac{x}{y}\right)dy = 0$

(iv) $(xe^{y/x} + y)dx = x\,dy,\ y(1) = 1$

(v) $\dfrac{dy}{dx} = \dfrac{x+2y+3}{2x+3y+4}$

12.6 Linear Differential Equation

9 Solve the following differential equations.

(i) $\dfrac{dy}{dx} - \dfrac{y}{x} = 2x^2,\ \ x > 0$

(ii) $y\,dx + (x - y^3)dy = 0$

(iii) $(1+x^2)\dfrac{dy}{dx} + 2xy - 4x^2 = 0,\ y(0) = 0$

(iv) $(x^2+1)\dfrac{dy}{dx} - 2xy = (x^4 + 2x^2 + 1)\cos x,\ y(0) = 0$

(v) $(1+y^2)dx = (\tan^{-1} y - x)dy,\ y(0) = 0$

(vi) $\dfrac{dy}{dx} - x^3 y^3 + xy = 0$

(vii) $\dfrac{dy}{dx} + \dfrac{y}{x}\log y = \dfrac{y}{x^2}(\log y)^2$

(viii) $(1-x^2)\dfrac{dy}{dx} + xy = xy^2$

12.7 Application of Differential Equation

10 The slope of the curve passing through (3, 4) at any point is the reciprocal of twice the ordinate of that point, then find the curve.

11 Find the equation of the curve passing through (5, 1), which has constant subtangent.

12 Find the orthogonal trajectory of $y^2 = 4ax$.

13 Find the curve that passes through the point (2, 3) and has the property that the segment of any tangent to it lying between the coordinate axis is bisected by the point of contact.

LEVEL 02 \ JEE Main Level

Single Choice Correct Questions

12.1 Degree and Order of Differential Equation

1 The degree of differential equation $\left(\frac{d^2y}{dx^2}\right)^2 = \left(1+\frac{dy}{dx}\right)^{1/2}$ is

(a) 2 (b) 3 (c) 4 (d) None of these

2 The degree of differential equation

$\left(\frac{d^3y}{dx^3}\right)^{2/3} + 4 - 3\left(\frac{d^2y}{dx^2}\right) + 5\left(\frac{dy}{dx}\right) = 0$ is

(a) 3 (b) 2 (c) 2/3 (d) not define

3 The differential equation of all parabolas, whose axis of symmetry is parallel to X-axis of order

(a) 2 (b) 3 (c) 1 (d) 4

4 The general solution of differential equation is $y = ae^{bx+c}$, where a, b and c are arbitrary constants. The order of differential equation is

(a) 3 (b) 2 (c) 1 (d) None of these

5 The order and degree of the differential equation $a = \frac{\left[1+\left(\frac{dy}{dx}\right)^2\right]^{3/2}}{\frac{d^2y}{dx^2}}$, where a is a constant,

are m and n respectively, then (m, n) is

(a) (2, 3) (b) (3, 2) (c) (2, 2) (d) None of these

6 The order and degree of the differential equation of all tangent lines to the parabola $y = x^2$ are

(a) 1, 2 (b) 2, 3 (c) 2, 1 (d) 1, 1

12.2 Formation of Differential Equation

7 The differential equation which respresents the family of curves $y = ae^{bx}$, where a and b are arbitrary constants, is

(a) $y' = y^2$ (b) $y'' = y' \cdot y$

(c) $y \cdot y'' = y'$ (d) $y \cdot y'' = (y')^2$

8 The differential equation of the family of the circles with fixed radius 5 units and centre on the line $y = 2$ is

(a) $(y-2)\left(\frac{dy}{dx}\right)^2 = 25-(y-2)^2$
(b) $(y-2)^2\left(\frac{dy}{dx}\right)^2 = 25-(y-2)^2$
(c) $(x-2)^2\left(\frac{dy}{dx}\right)^2 = 25-(y-2)^2$
(d) $(x-2)\left(\frac{dy}{dx}\right)^2 = 25-(y-2)^2$

9 A differential equation representing the family of parabolas with axis parallel to Y-axis and whose length of latusrectum is the distance of the point $(2, -3)$ from the line $3x+4y=5$, is given by

(a) $10\frac{d^2y}{dx^2} = 11$
(b) $11\frac{d^2x}{dy^2} = 10$
(c) $10\frac{d^2x}{dy^2} = 11$
(d) $11\frac{d^2y}{dx^2} = 10$

10 The differential equation of the family of curves, $x^2 = 4b(y+b), b \in R$, is **[JEE Main 2020]**

(a) $xy'' = y'$
(b) $x(y')^2 = x+2yy'$
(c) $x(y')^2 = x-2yy'$
(d) $x(y')^2 = 2yy'-x$

11 The differential equation of the family of parabolas having vertex at the origin and the Y-axis as the axis of symmetry, is

(a) $2y_1 - y = 0$
(b) $2y_1 + xy = 0$
(c) $xy_1 - 2y = 0$
(d) $yy_1 - 2x = 0$

12 Form the differential equation of the family of curves having equation $y = (\sin^{-1} x)^2 + A\cos^{-1} x + B$, is

(a) $(1-x^2)\frac{d^2y}{dx^2} - x\frac{dy}{dx} = 2$
(b) $(1+x^2)\frac{d^2y}{dx} - x\frac{dy}{dx} = 0$
(c) $x\frac{dy}{dx} = (1+x^2)\frac{d^3y}{dx^3}$
(d) $(1+x^2)\frac{d^3y}{dx^3} = \frac{dy}{dx}$

12.3 Solution of Differential Equation Using Inspection Method

13 The solution of the differential equation $(x+y)(dx-dy) = dx+dy$ is

(a) $x-y = Ke^{x-y}$
(b) $x+y = Ke^{x+y}$
(c) $x+y = K(x+y)$
(d) $x+y = Ke^{x-y}$

14 The solution of the differential equation $y\,dx+(x+x^2y)dy = 0$ is

(a) $-\frac{1}{xy}+C$
(b) $-\frac{1}{xy}+\log y = C$
(c) $\frac{1}{xy}+\frac{1}{\log y} = C$
(d) $\log y = Cx$

15 The solution of the differential equation $\frac{dy}{dx} = \frac{yf'(x)-y^2}{f(x)}$ is

(a) $f(x) = y+C$
(b) $f(x) = y(x+C)$
(c) $f(x) = x+C$
(d) None of these

16 If a curve $y = f(x)$ passes through the point $(1, -1)$ and satisfies the differential equation $y(1+xy)dx = x\,dy$, then $f\left(-\frac{1}{2}\right)$ is equal to

(a) $-\frac{2}{5}$
(b) $-\frac{4}{5}$
(c) $\frac{2}{5}$
(d) $\frac{4}{5}$

12.4 Solution of Differential Equation by Variable Separable

17 The solution of $e^x \tan y\, dx + (1-e^x)\sec^2 y dy = 0$ is

(a) $\tan y = C(1-e^x)$
(b) $\sec y = C(1-e^x)$
(c) $\tan y(1-e^x) = C$
(d) $\sec y = (1-e^x)$

18 The solution of $\frac{dy}{dx} = \frac{ax+h}{by+k}$ represents a parabola, if

(a) $a = 2$ and $b = 1$
(b) $a = 2$ and $b = 2$
(c) $a = 0$ and $b = 2$
(d) $a = 0$ and $b = 0$

19 If $x^3 dy + xy\, dx = x^2 dy + 2y\, dx$; $y(2) = e$ and $x > 1$, then $y(4)$ is equal to **[JEE Main 2020]**

(a) $\frac{\sqrt{e}}{2}$
(b) $\frac{3}{2} + \sqrt{e}$
(c) $\frac{3}{2}\sqrt{e}$
(d) $\frac{1}{2} + \sqrt{e}$

20 The solution of $\frac{dy}{dx} + \frac{x}{y^2}\frac{(1+y^3)}{(1+x^2)} = 0$ is

(a) $(1-x^2)+(1+y^2) = C$
(b) $(1+x^2)+(1+y^3) = C$
(c) $(1+x^2)^3(1+y^3)^2 = C$
(d) $(1+x^2)(1+y^3)^2 = Cxy^2$

21 If $y = y(x)$ is the solution of the differential equation $\frac{5+e^x}{2+y}\cdot\frac{dy}{dx} + e^x = 0$ satisfying $y(0) = 1$, then the value of $y(\log_e 13)$ is **[JEE Main 2020]**

(a) 0
(b) -1
(c) 1
(d) 2

22 If $\frac{dy}{dx} = \frac{2^x y + 2^y \cdot 2^x}{2^x + 2^{x+y}\log 2}$, $y(0) = 0$, then for $y = 1$, the value of x lies in the interval

(a) (1, 2)
(b) $\left(\frac{1}{2}, 1\right)$
(c) (2, 3)
(d) $\left(0, \frac{1}{2}\right]$

23 If $y = y(x)$ and $\frac{2+\sin x}{y+1}\left(\frac{dy}{dx}\right) = -\cos x$, $y(0) = 1$, then $y\left(\frac{\pi}{2}\right)$ is equal to

(a) $\frac{1}{3}$
(b) $\frac{2}{3}$
(c) $-\frac{1}{3}$
(d) 1

24 The equation of the curve through the point (1, 0), whose slope is $\frac{y-1}{x^2+x}$, is

(a) $(y-1)(x+1)+2x = 0$
(b) $2x(y-1)+(x+1) = 0$
(c) $x(y-1)(x+1)+2 = 0$
(d) $x(y+1)+y(x-1) = 0$

25 If $y = y(x)$ is the solution of the differential equation, $e^y\left(\frac{dy}{dx} - 1\right) = e^x$ such that $y(0) = 0$, then $y(1)$ is equal to **[JEE Main 2020]**

(a) $2 + \log_e 2$
(b) $2e$
(c) $1 + \log_e 2$
(d) $\log_e 2$

26 The equation of the curve passing through the point $\left(0, \frac{\pi}{3}\right)$ satisfying the differential equation $\sin x \cos y dx + \cos x \sin y\, dy = 0$, is

(a) $\cos x \cos y = \frac{\sqrt{3}}{2}$
(b) $\sin x \sin y = \frac{\sqrt{3}}{2}$
(c) $\sin x \sin y = \frac{3}{2}$
(d) $\cos x \cos y = \frac{1}{2}$

27 Solution of the differential equation $\left(\frac{x+y-1}{x+y-2}\right)\frac{dy}{dx}=\left(\frac{x+y+1}{x+y+2}\right)$, given that $y=1$, when $x=1$, is

(a) $\log\left|\frac{(x-y)^2-2}{2}\right|=2(x+y)$
(b) $\log\left|\frac{(x-y)^2+2}{2}\right|=2(x-y)$
(c) $\log\left|\frac{(x+y)^2+2}{2}\right|=2(x-y)$
(d) None of these

28. The solution of the differential equation

$$\frac{dy}{dx}-\frac{y+3x}{\log_e(y+3x)}+3=0 \text{ is (where } C \text{ is a constant of integration)}$$ **[JEE Main 2020]**

(a) $x-\log_e(y+3x)=C$
(b) $y+3x-\frac{1}{2}(\log_e x)^2=C$
(c) $x-2\log_e(y+3x)=C$
(d) $x-\frac{1}{2}\{\log_e(y+3x)\}^2=C$

29. The solution of the differential equation $(x+2y)(dx-dy)=dx+dy$ is

(a) $2(y-x)+4\log|3x+6y-1|=C$
(b) $2(x-y)+2\log|3x+6y-1|=C$
(c) $2(x+2y)+\log|x+2y-1|=C$
(d) None of these

30. The solution of the differential equation, $\frac{dy}{dx}=(x-y)^2$, when $y(1)=1$, is **[JEE Main 2019]**

(a) $\log_e\left|\frac{2-y}{2-x}\right|=2(y-1)$
(b) $-\log_e\left|\frac{1+x-y}{1-x+y}\right|=x+y-2$
(c) $\log_e\left|\frac{2-x}{2-y}\right|=x-y$
(d) $-\log_e\left|\frac{1-x+y}{1+x-y}\right|=2(x-1)$

12.5 Homogeneous Differential Equation

31 The solution of $(x^2+xy)dy=(x^2+y^2)dx$ is

(a) $\log x=\log(x-y)+\frac{y}{x}+K$
(b) $\log x=2\log(x-y)+\frac{y}{x}+K$
(c) $\log x=\log(x-y)+\frac{y}{x}+K$
(d) None of these

32 If $\frac{dy}{dx}=\frac{xy}{x^2+y^2}$; $y(1)=1$; then the value of x satisfying $y(x)=e$, is **[JEE Main 2020]**

(a) $\frac{1}{2}\sqrt{3}e$
(b) $\sqrt{3}e$
(c) $\sqrt{2}\,e$
(d) $\frac{e}{\sqrt{2}}$

33 The solution of the differential equation

$$\left\{x\cos\left(\frac{y}{x}\right)+y\sin\left(\frac{y}{x}\right)\right\}y\,dx=\left\{y\sin\left(\frac{y}{x}\right)-x\cos\left(\frac{y}{x}\right)\right\}x\,dy \text{ is}$$

(a) $xy\cos\left(\frac{y}{x}\right)=K$
(b) $\frac{x}{y}\cos\left(\frac{y}{x}\right)=K$
(c) $\frac{1}{xy}\cos\left(\frac{y}{x}\right)=K$
(d) None of these

34 If $x\,dy = \left(y + x\dfrac{f\left(\frac{y}{x}\right)}{f'\left(\frac{y}{x}\right)}\right)dx$, then $f\left(\frac{y}{x}\right)$ is equal to

(a) Cx (b) $\frac{Cx^2}{2}$ (c) $\frac{C}{x}$ (d) None of these

35 If $x\frac{dy}{dx} = y(\log y - \log x + 1)$, then the solution of the differential equation is

(a) $\log\frac{x}{y} = Cy$ (b) $\log\frac{y}{x} = Cy$ (c) $\log\frac{y}{x} = Cx$ (d) None of these

36 The solution curve of $\frac{dy}{dx} = \frac{y^2 - 2xy - x^2}{y^2 + 2xy - x^2}$, $y(-1) = 1$, is

(a) a straight line (b) a circle (c) a parabola (d) ellipse

37 If $y\frac{dy}{dx} = x\left[\frac{y^2}{x^2} + \dfrac{\phi\left(\frac{y^2}{x^2}\right)}{\phi'\left(\frac{y^2}{x^2}\right)}\right]$, $x > 0, \phi > 0$ and $y(1) = -1$, then $\phi\left(\frac{y^2}{4}\right)$ is equal to

(a) $4\phi(2)$ (b) $4\phi(1)$ (c) $2\phi(1)$ (d) $\phi(1)$

38 The curve amongst the family of curves represented by the differential equation $(x^2 - y^2)dx + 2xy\,dy = 0$, which passes through (1, 1), is **[JEE Main 2019]**

(a) a circle with centre on the Y-axis (b) a circle with centre on the X-axis
(c) an ellipse with major axis along the Y-axis (d) a hyperbola with transverse axis along the X-axis

39 If a curve $y = f(x)$, passing through the point (1, 2), is the solution of the differential equation, $2x^2 dy = (2xy + y^2)dx$, then $f\left(\frac{1}{2}\right)$ is equal to **[JEE Main 2020]**

(a) $\frac{1}{1 + \log_e 2}$ (b) $\frac{1}{1 - \log_e 2}$ (c) $1 + \log_e 2$ (d) $\frac{-1}{1 + \log_e 2}$

40 A curve passes through the point $\left(1, \frac{\pi}{6}\right)$, let the slope of the curve at each point (x, y) be $\frac{y}{x} + \sec\frac{y}{x}$, $x > 0$, then the equation of the curve is

(a) $\sin\left(\frac{y}{x}\right) = \log x + \frac{1}{2}$ (b) $\text{cosec}\, x\left(\frac{y}{x}\right) = \log x + 2$ (c) $\sec\left(\frac{2y}{x}\right) = \log x + 2$ (d) $\cos\left(\frac{2y}{x}\right) = \log x + \frac{1}{2}$

12.6 Linear Differential Equation

41 The solution of differential equation $\cos x\, dy = y(\sin x - y)dx$, $0 < x < \frac{\pi}{2}$, is

(a) $\sec x = (\tan x + C)y$ (b) $y\sec x = \tan x + C$ (c) $y\tan x = \sec x + C$ (d) $\tan x = (\sec x + C)y$

42 If $y = y(x)$ is the solution curve of the differential equation $x^2 dy + \left(y - \frac{1}{x}\right)dx = 0$, $x > 0$ and $y(1) = 1$, then $y\left(\frac{1}{2}\right)$ is equal to **[JEE Main 2019]**

(a) $\frac{3}{2} - \frac{1}{\sqrt{e}}$ (b) $3 + \frac{1}{\sqrt{e}}$ (c) $3 + e$ (d) $3 - e$

43 If $y(t)$ is solution of $(t+1)\frac{dy}{dt}-ty=1$, $y(0)=-1$, at $t=1$ the solution is

(a) $e+\frac{1}{2}$ (b) $-\frac{1}{2}$ (c) $\frac{1}{2}$ (d) $e-\frac{1}{2}$

44 The solution of the differential equation $x\frac{dy}{dx}+2y=x^2 (x\neq 0)$ with $y(1)=1$, is [JEE Main 2019]

(a) $y=\frac{x^2}{4}+\frac{3}{4x^2}$ (b) $y=\frac{x^3}{5}+\frac{1}{5x^2}$

(c) $y=\frac{3}{4}x^2+\frac{1}{4x^2}$ (d) $y=\frac{4}{5}x^3+\frac{1}{5x^2}$

45 The solution of differential equation $\frac{dy}{dx}+\frac{1}{2}y\sec x=\frac{\tan x}{2y}$, where $0\leq x<\frac{\pi}{2}$ and $y(0)=1$ is given by

(a) $y^2=1-\frac{x}{\sec x+\tan x}$ (b) $y^2=1+\frac{x}{\sec x+\tan x}$

(c) $y=1+\frac{x}{\sec x+\tan x}$ (d) $y=1-\frac{x}{\sec x+\tan x}$

46 The solution of the differential equation $(xy^4+y)dx-x\,dy=0$ is

(a) $4x^4y^3+3x^3=Cy^3$ (b) $3x^3y^4+4y^3=Cx^3$

(c) $3x^4y^3+4x^3=Cy^3$ (d) None of these

47 Let $y=y(x)$ be the solution of the differential equation $\cos x\frac{dy}{dx}+2y\sin x=\sin 2x$, $x\in\left(0,\frac{\pi}{2}\right)$. If $y\left(\frac{\pi}{3}\right)=0$, then $y\left(\frac{\pi}{4}\right)$ is equal to [JEE Main 2020]

(a) $2-\sqrt{2}$ (b) $2+\sqrt{2}$ (c) $\sqrt{2}-2$ (d) $\frac{1}{\sqrt{2}}-1$

48 Solution of differential equation $e^x(x+1)dx+(ye^y-xe^x)dy=0$, $f(0)=0$, is

(a) $xe^xe^{-y}+y^2=0$ (b) $2xe^xe^{-y}+y^2=0$

(c) $2xe^xe^{-y}-y^2=0$ (d) $xe^xe^{-y}-y^2=0$

49 Let $y=y(x)$ be the solution curve of the differential equation, $(y^2-x)\frac{dy}{dx}=1$, satisfying $y(0)=1$. This curve intersects the X-axis at a point whose abscissa is [JEE Main 2020]

(a) 2 (b) $2-e$ (c) $-e$ (d) $2+e$

50 Let $y=y(x)$ be the solution of the differential equation,

$xy'-y=x^2(x\cos x+\sin x), x>0$. If $y(\pi)=\pi$, then $y''\left(\frac{\pi}{2}\right)+y\left(\frac{\pi}{2}\right)$ is equal to [JEE Main 2020]

(a) $2+\frac{\pi}{2}+\frac{\pi^2}{4}$ (b) $1+\frac{\pi}{2}$ (c) $1+\frac{\pi}{2}+\frac{\pi^2}{4}$ (d) $2+\frac{\pi}{2}$

51 If $y(x)$ is the solution of the differential equation

$$\frac{dy}{dx}+\left(\frac{2x+1}{x}\right)y=e^{-2x}, x>0, \text{ where } y(1)=\frac{1}{2}e^{-2}, \text{ then}$$ [JEE Main 2019]

(a) $y(x)$ is decreasing in $\left(\frac{1}{2},1\right)$ (b) $y(x)$ is decreasing in (0, 1)

(c) $y(\log_e 2)=\log_e 4$ (d) $y(\log_e 2)=\frac{\log_e 2}{4}$

52 A function $y = f(x)$ satisfies $(x+1) f'(x) - 2(x^2+x) f(x) = \frac{e^{x^2}}{(x+1)}$, $\forall x > -1$, if $f(0) = 5$, then $f(x)$ is

(a) $\left(\frac{3x+5}{x+1}\right)e^{x^2}$ (b) $\left(\frac{6x+5}{x+1}\right)e^{x^2}$ (c) $\left(\frac{6x+5}{(x+1)^2}\right)e^{x^2}$ (d) $\left(\frac{5-6x}{x+1}\right)e^{x^2}$

12.7 Application of Differential Equation

53 The equation of the curve in which the subnormal varies as the square of the ordinate is

(a) $y = Ce^{-Kx}$ (b) $y = C\log Kx$
(c) $y = Ce^{Kx}$ (d) None of these

54 Orthogonal trajectories of the curve $x^{2/3} + y^{2/3} = a^{2/3}$, where a is any arbitrary constant, is

(a) $x^{2/3} - y^{2/3} = C$ (b) $x^{4/3} - y^{4/3} = C$
(c) $x^{4/3} + y^{4/3} = C$ (d) $x^{1/3} - y^{1/3} = C$

55 The orthogonal trajectories to the curve $y^2 = 4a(x+a)$ is

(a) $(\sqrt{x^2+y^2}) = 2x + C$ (b) $2\sqrt{x^2+y^2} = x + C$
(c) $x^2 + y^2 = 2x + C$ (d) None of these

56 The orthogonal trajectories of the family of curves $a^{n-1}y = x^n$ are given by (a is the arbitrary constant)

(a) $x^n + n^2 y = \text{constant}$ (b) $ny^2 + x^2 = \text{constant}$
(c) $n^2 x + y^n = \text{constant}$ (d) $n^2 x - y^n = \text{constant}$

57 The curve such that the square of the intercept cut by any tangent of the Y-axis is equal to the product of the coordinates of the point of tangency, is

(a) $\sqrt{\frac{y}{x}} + \log x = C$ (b) $\sqrt{\frac{y}{x}} - \log x = C$
(c) $2\sqrt{\frac{y}{x}} + \log x = C$ (d) $2\sqrt{\frac{y}{x}} - \log x = C$

58 A tangent to the curve $y = f(x)$ cuts the line $y = x$ at a point, which is at a distance of one unit from Y-axis , the equation of the curve is

(a) $\frac{x+1}{y+1} = C$ (b) $\frac{x-1}{y-1} = C$
(c) $(x-1)(y-1) = C$ (d) None of these

59 Given that the slope of the tangent to a curve $y = y(x)$ at any point (x, y) is $\frac{2y}{x^2}$. If the curve passes through the centre of the circle $x^2 + y^2 - 2x - 2y = 0$, then its equation is

[JEE Main 2019]

(a) $x^2 \log_e |y| = -2(x-1)$ (b) $x \log_e |y| = x - 1$
(c) $x \log_e |y| = 2(x-1)$ (d) $x \log_e |y| = -2(x-1)$

60 The equation of a curve passing through (0, 1) and having gradient $\frac{-(y+y^3)}{1+x+xy^2}$ at (x, y) is

(a) $\frac{1}{2}xy + \tan^{-1} y = \frac{\pi}{4}$ (b) $2xy - \tan^{-1} y = \frac{\pi}{4}$
(c) $xy + \tan^{-1} y = \frac{\pi}{4}$ (d) $xy + \frac{1}{2}\tan^{-1} y = \frac{\pi}{4}$

61 A right circular cone with radius (R) and height (H) contains a liquid, which evaporates at a rate proportional to its surface area in contact with air (proportionality constant $= k > 0$), the time after which the cone will be empty, is

(a) $\frac{R}{k}$ (b) $\frac{k}{R}$ (c) $\frac{H}{R}$ (D) $\frac{H}{k}$

Numerical Value Type Questions

62 If $y = y(x)$ is the solution of the differential equation $\frac{5+e^x}{2+y} \cdot \frac{dy}{dx} + e^x = 0$ satisfying $y(0) = 1$, then the value of $|y(\log_e 13)|$ is ......

63 Let $y = y(x)$ be solution of the differential equation $\log\left(\frac{dy}{dx}\right) = 3x + 4y$, with $y(0) = 0$. If $y\left(-\frac{2}{3}\log 2\right) = -\frac{1}{k}\log 2$, then the value of k is ......

64 If $y = y(x)$ satisfies the differential equation $8\sqrt{x}(\sqrt{9+\sqrt{x}})dy = (\sqrt{4+\sqrt{9+\sqrt{x}}})^{-1}\, dx$, $x > 0$ and $y(0) = \sqrt{7}$, then the value of $y(256)$ is ....

65 Solution of differential equation
$\frac{dy}{dx} = \frac{x^2+y^2+3x+3y+2xy+1}{x^2+y^2-3x-3y+2xy+1}$ is $y - \frac{m}{n}\log|(x+y)^2+1| = x + C$, then the value of $(m-n)$ is ......

66 The equation of curve whose slope at (x, y) is $-\frac{(x+y)}{x}$ and which passes through the point (2, 1) is $x^2 + lxy = k$, then the value of $(k + l)$ is ......

67 If for $x \geq 0$, $y = y(x)$ is the solution of the differential equation,
$(x+1)dy = \{(x+1)^2 + y - 3\}\, dx$, $y(2) = 0$, then $y(3)$ is equal to .... **[JEE Main 2020]**

68 If the curve $y = y(x)$ represented by the solution of the differential equation $(2xy^2 - y)dx + x\,dy = 0$ passes through the intersection of the lines $2x - 3y = 1$ and $3x + 2y = 8$, then $|y(1)|$ is equal to ......

69 Let $f : [1, \infty] \to [2, \infty]$ be a differentiable function such that $f(1) = 2$ if $6\int_1^x f(t)\, dt = 3xf(x) - x^3 - 5$ for all $x \geq 1$, then the value of $f(2)$ is ......

70 Let f be a real valued differentiable function on R such that $f(1) = 1$. If the y-intercept of the tangent at any point $P(x, y)$ on the curve $y = f(x)$ is equal to the cube of the abcissa of P, then the value of $f(-3)$ is equal to ......

71 The rate of growth of bacteria culture is proportional to the number of bacteria present and the bacteria count is 1000 at initial time. The number of bacteria is increased by 20% in 2 h. If the population of bacteria is 2000 after $\frac{k}{\log\left(\frac{6}{5}\right)}$ h, then $\left(\frac{k}{\log 2}\right)^2$ is equal to ......

72 The equation of the curve passing through the origin if the middle point of the segment of its normal from any point of the curve to the X-axis lies on the parabola $2y^2 = x$ is $y^2 - kx = 1 - e^{kx}$, then the value of k is ......

LEVEL 03 \ JEE Advanced Level

Single Choice Correct Questions

1. If $f(x)$ is a function such that $x\int_0^x (1-t)f(t)dt = \int_0^x t\,f(t)dt$, $f(1)=1$, then $f(x)$ is equal to

(a) $y = x^3 e^{\left(\frac{x-1}{x}\right)}$ (b) $yx^3 = e^{\left(\frac{x-1}{x}\right)}$ (c) $y = \frac{1}{x^3}e^{\left(\frac{1-x}{x}\right)}$ (d) $y = x^3 e^{\frac{1}{x}}$

2 Solution of the given differential equation $y\cos\left(\frac{y}{x}\right)(x\,dy - y\,dx) + x\sin\left(\frac{y}{x}\right)(x\,dy + y\,dx) = 0$, $y(1) = \frac{\pi}{2}$, is

(a) $xy\sin\left(\frac{x}{y}\right) = \frac{\pi}{4}$ (b) $xy\sin\left(\frac{y}{x}\right) = \frac{\pi}{2}$ (c) $xy\sin\left(\frac{y}{x}\right) = \pi$ (d) $2xy\sin\left(\frac{y}{x}\right) = 0$

3 Let C_1 be the curve obtained by the solution of differential equation $2xy\frac{dy}{dx} = y^2 - x^2$, $x > 0$. Let the curve C_2 be the solution of $\frac{2xy}{x^2-y^2} = \frac{dy}{dx}$. If both the curves passes through (1, 1), then the area enclosed by the curves C_1 and C_2 is equal to

(a) $\frac{\pi}{4}+1$ (b) $\pi+1$ (c) $\pi-1$ (d) $\frac{\pi}{2}-1$

4 Let $y = y(x)$ be a solution curve of the differential equation $(y+1)\tan^2 x\,dx + \tan x\,dy + y\,dx = 0$ $x \in \left(0, \frac{\pi}{2}\right)$. If $\lim_{x\to 0^+} xy(x) = 1$, then the value of $y\left(\frac{\pi}{4}\right)$ is

(a) $-\frac{\pi}{4}$ (b) $\frac{\pi}{4}-1$ (c) $\frac{\pi}{4}+1$ (d) $\frac{\pi}{4}$

5 The integral curve of the differential equation $x(1-x\ln y)\frac{dy}{dx} + y = 0$, which passes through the point $\left(e, \frac{1}{e}\right)$ is

(a) $\{(\ln y - 1) + ey\}x = 1$ (b) $\{(\ln y + 1) + ey\}x = 1$
(c) $\{(\ln y + 1) + y\}x = 2$ (d) $\{(\ln y + 1) + y\}x = 1$

6 A differentiable function f satisfies the relation $f(xy) = xf(y) + yf(x)$ for every x, y in R. If $f'(1) = 1$, then the function f is

(a) $y = x\log|x| + 3$ (b) $y = x\log|x|$ (c) $2y = x\log x$ (d) $2y = x\log x - 3$

7 Let $f(x)$ be differentiable on the interval $(0, \infty)$, such that $f(1) = 1$ and $\lim_{t\to x}\frac{t^2 f(x) - x^2 f(t)}{t-x} = 1$ for each $x > 0$, then $f(x)$ is

(a) $\frac{1}{3x} + \frac{2x^2}{3}$ (b) $-\frac{1}{3x} + \frac{4x^2}{3}$ (c) $-\frac{1}{x} + \frac{2}{x^2}$ (d) $\frac{1}{x}$

8 The function $y = f(x)$ is the solution of the differential equation $\frac{dy}{dx} + \frac{xy}{x^2-1} = \frac{x^4+2x}{\sqrt{1-x^2}}$ in $(-1, 1)$ satisfying $f(0) = 0$, then $\int_{-\sqrt{3}/2}^{\sqrt{3}/2} f(x)dx$ is equal to

(a) $\frac{\pi}{3} - \frac{\sqrt{3}}{2}$ (b) $\frac{\pi}{3} - \frac{\sqrt{3}}{4}$ (c) $\frac{\pi}{6} - \frac{\sqrt{3}}{4}$ (d) $\frac{\pi}{6} - \frac{\sqrt{3}}{2}$

9 A curve passes through (2, 0) and the slope of tangent at point $P(x, y)$ equals $\frac{(x+1)^2+y-3}{(x+1)}$.

The area enclosed by the curve and the X-axis in fourth quadrant is

(a) $\frac{2}{3}$ (b) $\frac{1}{3}$ (c) $\frac{4}{3}$ (d) None of these

More than One Correct

10 The general solution of differential equation $x(1+y^2)dx+y(1+x^2)dy=0$ is/are

(a) $\frac{1}{2}\ln(1+x^2)(1+y^2)=C$ (b) $(1+y^4)=C(1+x^2)$

(c) $(1+x^2)(1+y^2)=K$ (d) $\ln\left(\frac{(1+x^2)}{(1+y^2)}\right)=K$

11 Let $f:R\to R$ and $g:R\to R$ be two non-constant differentiable functions. If $f'(x)=(e^{(f(x)-g(x))})g'(x)$ for all $x\in R$ and $f(1)=g(2)=1$, then which of the following statement is (are) true?

(a) $f(2)<1-\log 2$ (b) $f(2)>1-\log 2$ (c) $g(1)>1-\log 2$ (d) $g(1)<1-\log 2$

12 A solution curve of the differential equation $(x^2+xy+4x+2y+4)\frac{dy}{dx}-y^2=0, x>0$ passes through the point (1, 3). The solution curve

(a) intersects $y=x+2$ exactly at one point (b) intersects $y=x+2$ exactly at two points

(c) intersects $y=(x+2)^2$ (d) does not intersect $y=(x+3)^2$

13 If $y(x)$ satisfies the differential equation $\frac{dy}{dx}-y\tan x=2x\sec x$ and $y(0)=0$, then

(a) $y\left(\frac{\pi}{4}\right)=\frac{\pi^2}{8\sqrt{2}}$ (b) $y'\left(\frac{\pi}{4}\right)=\frac{\pi^2}{18}$ (c) $y\left(\frac{\pi}{3}\right)=\frac{\pi^2}{9}$ (d) $y'\left(\frac{\pi}{3}\right)=\frac{4\pi}{3}+\frac{2\pi^2}{3\sqrt{3}}$

14 The solution of the differential equation $x\cos x\left(\frac{dy}{dx}\right)+y(x\sin x+\cos x)=1$ is

(a) $xy=\sin x+C\cos x$ (b) $xy\sec x=\tan x+C$

(c) $xy+\sin x+C\cos x=0$ (d) 0

15 Given a function g which has a derivative $g'(x)$ for every real x and satisfies $g'(0)=2$ and $g(x+y)=e^y g(x)+e^x g(y)$ for all x and y, then

(a) $g(x)$ is increasing for all $x\in[-1,\infty)$ (b) range of $g(x)$ is $\left[-\frac{2}{e},\infty\right)$

(c) $g''(x)>0, \forall x<0$ (d) $\lim_{x\to 0}\frac{g(x)}{x}=1$

16 A curve is such that any point P on it is as far from the origin as from the point in which the normal at P meets the X-axis. Then, the curve is a

(a) circle (b) parabola (c) ellipse (d) hyperbola

17 Let the function $y=f(x)$ satisfies the differential equation

$x^2\frac{dy}{dx}=y^2e^{1/x}$, $(x\neq 0)$ and $\lim_{x\to 0^-}f(x)=1$. Identify the correct statement(s).

(a) Range of $f(x)$ is $(0, 1)-\left\{\frac{1}{2}\right\}$ (b) $f(x)$ is bounded

(c) $\lim_{x\to 0^+}f(x)=1$ (d) $\int_0^e f(x)\,dx>\int_0^1 f(x)\,dx$

18 Let $f:[0,\infty)\to R$ be a continuous function such that $f(x)=1-2x+\int_0^x e^{x-t}f(t)\,dt$ for all $x\in[0,\infty)$. Then, which of the following statement(s) is (are) true? **[JEE Advanced 2018]**

(a) The curve $y=f(x)$ passes through the point $(1, 2)$

(b) The curve $y=f(x)$ passes through the point $(2, -1)$

(c) The area of the region $\{(x,y)\in[0,1]\times R: f(x)\le y\le\sqrt{1-x^2}\}$ is $\frac{\pi-2}{4}$

(d) The area of the region $\{(x,y)\in[0,1]\times R: f(x)\le y\le\sqrt{1-x^2}\}$ is $\frac{\pi-1}{4}$

Linked Comprehension Type

Paragraph I Directions (Q.Nos. 19-21)

Let $y=f(x)$ satisfies the equation $f(x)=(e^x+e^{-x})\cos x-2x-\int_0^x (x-t)f'(t)dt$

19 $y=f(x)$ satisfies the differential equation

(a) $\frac{dy}{dx}+y=e^x(\cos x-\sin x)-e^{-x}(\cos x+\sin x)$

(b) $\frac{dy}{dx}-y=e^x(\cos x-\sin x)+e^{-x}(\cos x+\sin x)$

(c) $\frac{dy}{dx}+y=e^x(\cos x+\sin x)-e^{-x}(\cos x-\sin x)$

(d) $\frac{dy}{dx}-y=e^x(\cos x-\sin x)+e^{-x}(\cos x-\sin x)$

20 The value of $f(0)+f'(0)$ is

(a) -1 (b) 0 (c) 1 (d) 2

21 $f(x)$ is equal to, $f(0)=2$

(a) $e^{-x}(\cos x-\sin x)+\frac{e^x}{5}(3\cos x+\sin x)+\frac{2}{5}e^{-x}$

(b) $e^{-x}(\cos x+\sin x)+\frac{e^x}{5}(3\cos x-\sin x)-\frac{2}{5}e^{-x}$

(c) $e^{-x}(\cos x-\sin x)+\frac{e^x}{5}(3\cos x-\sin x)+\frac{2}{5}e^{-x}$

(d) $e^{-x}(\cos x+\sin x)+\frac{e^x}{5}(3\cos x-\sin x)-\frac{2}{5}e^{-x}$

Matrix Match Type Questions

22 Match the following.

	Column I		Column II
(i)	$\frac{dx}{dy}+x=e^{-y}\sec^2 y$	(A)	$e^y=e^x+\frac{x^3}{3}+C$
(ii)	$(y\log x-1)ydx=x\,dy$	(B)	$y(Cx+\log ex)=1$
(iii)	$\frac{dy}{dx}=e^{x+y}$	(C)	$xe^y=\tan y+C$
(iv)	$\frac{dy}{dx}=e^{x-y}+x^2e^{-y}$	(D)	$e^x+e^{-y}=C$

23 Match the following.

	Column I		Column II
(i)	If the function $y = e^{4x} + 2e^{-x}$ is a solution of the differential equation $\dfrac{\dfrac{d^3y}{dx^3} - 13\dfrac{dy}{dx}}{y} = k$, then the value of $k/3$ is	(A)	2
(ii)	Let $y(x)$ be the solution of the differential equation $(x \log x)\dfrac{dy}{dx} + y = 2x \log x \ (x \geq 1)$, then $y(e)$ is	(B)	0
(iii)	Let $x^k + y^k = a^k, (a, k > 0)$ and $\dfrac{dy}{dx} + \left(\dfrac{y}{x}\right)^{1/3} = 0$, then $k = \dfrac{p}{q}$. The value of $(p+q)$ is	(C)	4
(iv)	If the solution of differential equation $x^2 \dfrac{d^2y}{dx^2} + 2x\dfrac{dy}{dx} = 12y$ is $y = Ax^m + Bx^{-n}$, then possible value of $\lvert m+n \rvert$ is	(D)	5

24 Match the following.

	Column I		Column II
(i)	If $y(x)$ is a solution of $\dfrac{dy}{dx} - y = 1 - e^{-x}$ and $y(0) = y_0$ has a finite value when $x \to +\infty$, then $\left\lvert \dfrac{1}{y_0} \right\rvert$ is equal to	(A)	6
(ii)	If $y(x)$ is a solution of $\dfrac{dy}{dx} - y \ln 2 = 2^{\sin x}(\cos x - 1)\ln 2$, y being bounded when $x \to +\infty$, then $y\left(\dfrac{\pi}{2}\right) + y^2\left(\dfrac{\pi}{2}\right)$ is	(B)	8
(iii)	A function $y = f(x)$ satisfies $xf'(x) - 2f(x) = x^4 f^2(x), \forall x > 0$ and $f(1) = -6$, then $f'(3^{1/5})$ is equal to	(C)	9
(iv)	Let f be a continuous function satisfying the equation $\int_0^x f(t)\,dt + \int_0^x tf(x-t)\,dt = e^{-x} - 1$, then the value of $e^{10}f(10)$ is	(D)	2

Numerical Value Type Questions

Question Stem Directions (Q.Nos. 25-27) The answer to each question is a **NUMERICAL VALUE.**

General solution of the differential equation $f(x)dx + g(y)dy = 0$ is $\int f(x)dx + f(y)\,dy = 0$.

25 General solution of the equation $\dfrac{dy}{dx} = \sin(x+y) + \cos(x+y)$ is $1 + \tan\left(\dfrac{x+y}{k}\right) = Ce^x$, then k is equal to ....

26 Solution of $\frac{dy}{dx}=\frac{x(2\log x+1)}{\sin y+y\cos y}$ is $y\sin y=x^k\log x+C$, then k is equal to ....

27 Solution of $\frac{dy}{dx}+\frac{y}{x}=\frac{1}{(1+\log x+\log y)^2}$ is $xy\,[1+\{\log(xy)\}^2]=\frac{x^k}{k}+C$, then k is equal to ....

Question Stem Directions (Q.Nos. 28-30) The answer to each question is a **NUMERICAL VALUE.**

A differential function $y=g(x)$ satisfies $\int_0^x (x-t+1)g(t)dt=x^4+x^2$ for all $x\geq 0$.

28 $y=g(x)$ satisfies the differential equation $\frac{dy}{dx}+y=\alpha x^2+\beta$, then $\alpha-\beta$ is equal to ....

29 The value of $g(0)$ is equal to ....

30 The value of $\int_0^1 \frac{12}{g'(x)+g(x)+10}dx$ is equal to $\frac{\pi}{k}$, then k is equal to ....

31 The curve satisfying the equation $\frac{dy}{dx}=\frac{y(x+y^3)}{x(y^3-x)}$ and passing through the point $(4,-2)$ is $y^3=kx$, then the value of $|k|$ is ......

32 Let $y=y(x)$ be the solution of the differential equation $dy=e^{\alpha x+y}dx$, $\alpha\in N$. If $y(\log 2)=\log 2$ and $y(0)=\log\left(\frac{1}{2}\right)$, then the value of α is ......

33 Let $f:R\to R$ be a differentiable function with $f(0)=0$ if $y=y(x)$ satisfies the differential equation $\frac{dy}{dx}=(2+5y)(5y-2)$, then the value of $\lim_{x\to-\infty} f_{(x)}$ is ...... **[JEE Advanced 2018]**

34 The number of straight lines which satisfies the differential equation $\frac{dy}{dx}+x\left(\frac{dy}{dx}\right)^2=y$ is m, then the value of $m+2$ is ......

35 The ordinate and the normal at any point P on the curve meet the X-axis at points A and B respectively, the equation of the family of curves satisfying the condition $AB=$ arithmetic mean of abscissa and ordinate of P is $\left(\frac{y}{x}-A\right)\left(\sqrt{B\left(\frac{y}{x}\right)+1}\right)=\sqrt{\left(\frac{C}{x^3}\right)}$, then $A+B$ is equal to ......

36 A function $y=f(x)$ satisfies the differential equation $f(x)\sin 2x-\cos x+(1+\sin^2 x)f'(x)=0$ with initial condition $y(0)=0$. The value of $f\left(\frac{\pi}{6}\right)$ is $\frac{A}{B}$, then $|A-B|$ is equal to ......

37 Let $f:[3,5]\to R$ be a twice differentiable function on $(3,5)$ such that $f(x)=e^{-x}\int_3^x(3t^2+2t+4f'(t))\,dt$. If $f'(4)=\frac{\alpha e^{\beta}-224}{(e^{\beta}-4)^2}$, then $\alpha+\beta$ is equal to ......

38 If the solution of the differential equation $\frac{dy}{dx}=\frac{1}{x\cos y+\sin 2y}$ is $x=Ce^{\sin y}-k(1+\sin y)$, then the value of k is ......

39 Let $y'(x)+y(x)g'(x)=g(x)g'(x)$, $y(0)=0$, $x\in R$, where $g'(x)$ denotes $\frac{d}{dx}\{g(x)\}$ and $g(x)$ is a given non-constant differentiable function on R with $g(0)=g(2)=0$, then $y(2)$ is equal to ......

ANSWER KEY

Level 1

1. (i) order = 1, degree= 2 (ii) order = 1, degree = 2
(iii) order = 1, degree = 1 (iv) order = 1, degree = 1
(v) order = 4, degree = 2 (vi) order = 1, degree = 2
(vii) order = 2, degree = not defined (viii) order = 2, degree = not defined
(ix) order = 2, degree = 1

2. (i) $y^2 = 2xy\frac{dy}{dx}$ (ii) $\frac{d^2y}{dx^2} = 0$ (iii) $\frac{d^2y}{dx^2} + y = 0$
(iv) $xy\frac{d^2y}{dx^2} + x\left(\frac{dy}{dx}\right)^2 - y\left(\frac{dy}{dx}\right) = 0$ (v) $\frac{d^2y}{dx^2} - \frac{dy}{dx} - 6y = 0$

3. $2x - 2r + 2y\frac{dy}{dx} = 0$ 4. $y\frac{d^2y}{dx^2} + \left(\frac{dy}{dx}\right)^2 = 0$

5. (i) $xy + \frac{1}{3}x^3 = C$ (ii) $x^4 \sin y = \frac{x^2}{2} + C$ (iii) $\tan^{-1}\left(\frac{y}{x}\right) = -x + C$ (iv) $e^{x/y} = y + C$

6. (i) $\log y = 3x - 3\log(x+1) + C$ (ii) $\tan x \tan y = C$
(iii) $y - x = C(1 + xy)$ (vi) $\frac{-e^{-by}}{b} = \frac{e^{ax}}{a} + C$
(v) $(1 + y^2)(1 + x^2) = x^2C^2$ (vi) $4y^2(2 - e^x) = 1$
(vii) $xy = e^{x-1}$

7. (i) $\tan\left(\frac{x+y}{2}\right) = x + C$ (ii) $\tan^{-1}(x + y + 1) = x + C$
(ii) $2(x - y) + \log(x - y + 2) = x + C$ (iv) $(2x - y) - 15\log|3x - y + 19| = C$

8. (i) $\frac{-x^3}{3y^3} + \log|y| = C$ (ii) $(y + \sqrt{x^2 + y^2})^2 = C^2x^4$
(iii) $x + ye^{x/y} = C$ (iv) $e^{-y/x} = \frac{1}{e} - \log|x|$
(v) $\frac{1}{\sqrt{3}}\log\left|\frac{(x-1)+\sqrt{3}(y+2)}{(x-1)-\sqrt{3}(y+2)}\right| - \frac{1}{2}\log|x^2 - 3y^2 - 2x\ 12y - 11| = \log C$

9. (i) $y = x^3 + Cx$ (ii) $xy = \frac{y^4}{4} + C$
(iii) $y = \frac{4x^3}{3(1 + x^2)}$ (iv) $y = (x^2 + 1)\sin x$
(v) $e^{\tan^{-1} y}(x - \tan^{-1} y + 1) = 1$
(iv) $\frac{e^{-x^2}}{y^2} = (x^2 + 1)e^{-x^2} + C$
(vii) $\frac{1}{x\log y} - \frac{1}{2x^2} = C$
(viii) $\frac{1}{y} = 1 + \frac{C}{\sqrt{1 - x^2}}$

10. $y^2 = x + 13$
11. $K\log y = x - 5$
12. $2x^2 + y^2 = 2C$
13. $xy = 6$

Level 2

1. (c)	**2.** (b)	**3.** (b)	**4.** (b)	**5.** (c)	**6.** (a)	**7.** (d)
8. (b)	**9.** (d)	**10.** (b)	**11.** (c)	**12.** (a)	**13.** (d)	**14.** (b)
15. (b)	**16.** (d)	**17.** (a)	**18.** (c)	**19.** (c)	**20.** (c)	**21.** (b)
22. (a)	**23.** (a)	**24.** (a)	**25.** (c)	**26.** (d)	**27.** (d)	**28.** (d)
29. (a)	**30.** (d)	**31.** (b)	**32.** (b)	**33.** (a)	**34.** (a)	**35.** (c)
36. (a)	**37.** (b)	**38.** (b)	**39.** (a)	**40.** (a)	**41.** (a)	**42.** (d)
43. (b)	**44.** (a)	**45.** (a)	**46.** (c)	**47.** (c)	**48.** (b)	**49.** (b)
50. (d)	**51.** (a)	**52.** (b)	**53.** (c)	**54.** (b)	**55.** (a)	**56.** (b)
57. (c)	**58.** (b)	**59.** (c)	**60.** (c)	**61.** (d)	**62.** (1)	**63.** (4)
64. (3)	**65.** (1)	**66.** (10)	**67.** (3)	**68.** (1)	**69.** (6)	**70.** (9)
71. (4)	**72.** (2)					

Level 3

1. (a)	**2.** (b)	**3.** (d)	**4.** (d)	**5.** (b)	**6.** (b)	**7.** (a)
8. (b)	**9.** (c)	**10.** (a, c)	**11.** (b, c)	**12.** (a, d)	**13.** (a, d)	**14.** (a, b)
15. (a, b)	**16.** (a, d)	**17.** (a, b, d)	**18.** (b, c)	**19.** (a)	**20.** (b)	**21.** (c)

22. (i)–C, (ii)–B, (iii)–D, (iv)–A

23. (i)–C, (ii)–A, (iii)–D, (iv)–B

24. (i)–D, (ii)–A, (iii)–B, (iv)–C

25. (2)	**26.** (2)	**27.** (2)	**28.** (10)	**29.** (0)	**30.** (4)	**31.** (2)
32. (2)	**33.** (0.4)	**34.** (4)	**35.** (3)	**36.** (3)	**37.** (16)	**38.** (2)
39. (0)						

SOLUTIONS

Level I

1. (i) We have, $\left(\frac{dy}{dx}\right)^2+\frac{dy}{dx}-x\sin y=0$

Hence, order = 1
and degree = 2

(ii) We have, $\frac{dy}{dx}+y=\frac{1}{\frac{dy}{dx}}$

$$\Rightarrow \left(\frac{dy}{dx}\right)^2+y\left(\frac{dy}{dx}\right)=1$$

Hence, order = 1
and degree = 2

(iii) We have, $\cos^{-1}\left(\frac{dy}{dx}\right)=2x+y$

$$\Rightarrow \frac{dy}{dx}=\cos(2x+y)$$

Hence, order = 1
and degree = 1

(iv) We have, $\sqrt{1-x^2}\,dx+\sqrt{1-y^2}\,dy=0$

$$\Rightarrow \frac{dy}{dx}=-\frac{\sqrt{1-x^2}}{\sqrt{1-y^2}}$$

Hence, order = 1
and degree = 1

(v) We have, $\frac{d^4y}{dx^4}=\left\{C+\left(\frac{dy}{dx}\right)^2\right\}^{3/2}$

On squaring both sides,

$$\left(\frac{d^4y}{dx^4}\right)^2=\left\{C+\left(\frac{dy}{dx}\right)^2\right\}^3$$

Hence, order = 4
and degree = 2

(vi) We have, $2\log\left(\frac{dy}{dx}\right)=2x+3y$

$$\Rightarrow \log\left(\frac{dy}{dx}\right)^2=2x+3y$$

$$\Rightarrow \left(\frac{dy}{dx}\right)^2=e^{2x+3y}$$

Hence, order = 1 and degree = 2

(vii) We have, $\frac{d^2y}{dx^2}-2\sin\left(\frac{dy}{dx}\right)+y=0$

Hence, order = 2
and degree = not defined

(viii) We have, $3\frac{d^2y}{dx^2}=2x\log\left(\frac{dy}{dx}\right)$

Hence, order = 2
and degree = not defined

(ix) We have, $y^2+2\frac{dy}{dx}=\frac{1}{2}\int y\,dx$

Differentiating both sides w.r.t. x, we get

$$2y\frac{dy}{dx}+2\frac{d^2y}{dx^2}=\frac{y}{2}$$

Hence, order = 2
and degree = 1

2. (i) We have, $y^2=4ax$ ...(i)

Differentiating both sides w.r.t. x, we get

$$2y\frac{dy}{dx}=4a$$

On putting in Eq. (i),

$$y^2=\left(2y\frac{dy}{dx}\right)x$$

$$\Rightarrow y^2=2xy\frac{dy}{dx}$$

which is the required differential equation.

(ii) We have, $\frac{x}{a}+\frac{y}{b}=1$

Differentiating both sides w.r.t. x, we get

$$\frac{1}{a}+\frac{1}{b}\frac{dy}{dx}=0$$

Again, differentiating both sides w.r.t x, we get

$$\frac{1}{b}\frac{d^2y}{dx^2}=0$$

$$\Rightarrow \frac{d^2y}{dx^2}=0$$

which is the required differential equation.

(iii) We have,

$$y=a\cos(x+b) \quad ...(i)$$

Differentiating both sides w.r.t. x, we get

$$\frac{dy}{dx}=-a\sin(x+b)$$

Again, differentiating w.r.t. x, we get

$$\frac{d^2y}{dx^2}=-a\cos(x+b)$$

From Eq. (i),

$$\frac{d^2y}{dx^2}=-y\Rightarrow\frac{d^2y}{dx^2}+y=0$$

which is the required differential equation.

(iv) We have, $\frac{x^2}{a^2}+\frac{y^2}{b^2}=1$

Differentiating both sides w.r.t. x, we get

$$\frac{2x}{a^2}+\frac{2y}{b^2}\frac{dy}{dx}=0$$

$$\Rightarrow \quad \frac{2y}{b^2}\frac{dy}{dx}=-\frac{2x}{a^2}$$

$$\Rightarrow \quad \frac{y}{x}\frac{dy}{dx}=-\frac{b^2}{a^2}$$

Again, differentiating both sides w.r.t. x, we get

$$\Rightarrow \quad \frac{y}{x}\cdot\frac{d^2y}{dx^2}+\left(\frac{x\frac{dy}{dx}-y\cdot(1)}{x^2}\right)\frac{dy}{dx}=0$$

$$\Rightarrow \quad xy\frac{d^2y}{dx^2}+x\left(\frac{dy}{dx}\right)^2-y\frac{dy}{dx}=0$$

(v) We have, $y=ae^{3x}+be^{-2x}$

Differentiating both sides w.r.t. x, we get

$$\frac{dy}{dx}=3ae^{3x}-2be^{-2x} \quad \text{...(i)}$$

Again, differentiating both sides w.r.t. x, we get

$$\frac{d^2y}{dx^2}=9ae^{3x}+4be^{-2x} \quad \text{...(ii)}$$

Subtracting Eq. (i) from Eq. (ii),

$$\frac{d^2y}{dx^2}-\frac{dy}{dx}=6ae^{3x}+6be^{-2x}$$

$$=6(ae^{3x}+be^{-2x})$$

$$\Rightarrow \quad \frac{d^2y}{dx^2}-\frac{dy}{dx}=6y$$

$$\Rightarrow \quad \frac{d^2y}{dx^2}-\frac{dy}{dx}-6y=0$$

3. The equation of circle which passes through origin, whose centre lies on X-axis, is

$$(x-r)^2+y^2=r^2$$

$$\Rightarrow \quad x^2-2xr+r^2+y^2=r^2$$

$$\Rightarrow \quad x^2-2xr+y^2=0$$

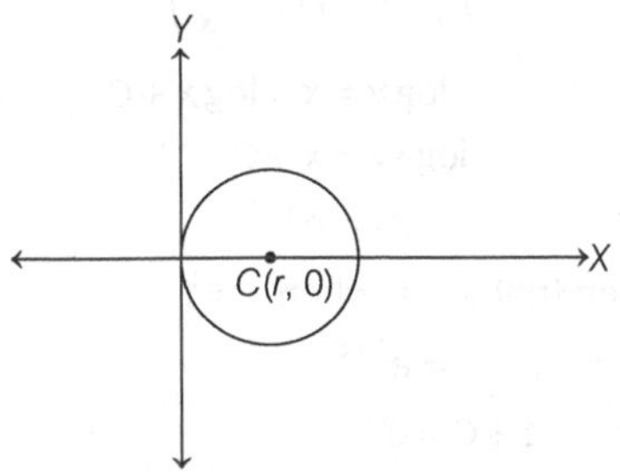

Differentiating both sides w.r.t. x, we get

$$2x-2r+2y\frac{dy}{dx}=0$$

4. The equation of family of parabolas having their axis of symmetry X-axis, is

$$y^2=4a(x-h) \quad \text{...(i)}$$

This equation contains two arbitrary constants a and b.
Differentiating both sides of Eq. (i) w.r.t. x, we get

$$2y\frac{dy}{dx}=4a$$

$$\Rightarrow \quad y\frac{dy}{dx}=2a \quad \text{...(ii)}$$

Again, differentiating both sides w.r.t. x, we get

$$y\frac{d^2y}{dx^2}+\left(\frac{dy}{dx}\right)^2=0$$

5. (i) We have, $x\,dy+ydx+x^2dx=0$

$$\Rightarrow \quad d(xy)+\frac{1}{3}(3x^2dx)=0$$

$$\Rightarrow \quad d(xy)+\frac{1}{3}d(x^3)=0$$

On integrating both sides, we get

$$xy+\frac{1}{3}x^3=C$$

which is the required solution of the given differential equation.

(ii) We have, $x^3\frac{dy}{dx}+4x^2\tan y=\sec y$

$$\Rightarrow \quad x^3\cos y\frac{dy}{dx}+4x^2\sin y=1$$

$$\Rightarrow \quad x^4\cos y\frac{dy}{dx}+4x^3\sin y=x$$

$$\Rightarrow \quad d(x^4\sin y)=x\,dx$$

On integrating both sides, we get

$$x^4\sin y=\frac{x^2}{2}+C$$

(iii) We have, $\frac{x\,dy}{x^2+y^2}=\left(\frac{y}{x^2+y^2}-1\right)dx$

$$\Rightarrow \quad \frac{x\,dy-y\,dx}{x^2+y^2}=-dx$$

$$\Rightarrow \frac{d}{dx}\left\{\tan^{-1}\left(\frac{y}{x}\right)\right\}=-dx$$

On integrating both sides, we obtain

$$\tan^{-1}\frac{y}{x}=-x+C$$

(iv) We have, $ye^{x/y}dx-xe^{x/y}dy=y^2dy$

$$\Rightarrow \quad e^{x/y}\left[\frac{y\,dx-x\,dy}{y^2}\right]=dy$$

$$\Rightarrow \quad d(e^{x/y})=dy$$

On integrating both sides, we get

$$e^{x/y}=y+C$$

6. (i) We have, $(x+1)\frac{dy}{dx}=3xy$

$$\frac{1}{y}dy=\frac{3x}{x+1}dx$$

On integrating both sides, we get

$$\int\frac{1}{y}dy=3\int\frac{x}{x+1}dx$$

$$\int\frac{1}{y}dy=3\int\frac{x+1-1}{x+1}dx$$

$$\log y=3[x-\log(x+1)]+C$$

$$\log y=3x-3\log(x+1)+C$$

(ii) We have,

$$\sec^2 x\tan y\,dx+\sec^2 y\tan x\,dy=0$$

$$\Rightarrow \sec^2 y\tan x\,dy=-\sec^2 x\tan y\,dx$$

$$\Rightarrow \frac{\sec^2 y}{\tan y}dy=-\frac{\sec^2 x}{\tan x}dx$$

On integrating both sides, we get

$$\int\frac{\sec^2 y}{\tan y}dy=-\int\frac{\sec^2 x}{\tan x}dx$$

$$\Rightarrow \log\tan y=-\log\tan x+\log C$$

$$\Rightarrow \log(\tan y\tan x)=\log C$$

$$\Rightarrow \tan x\tan y=C$$

which is the required solution.

(iii) We have, $\frac{dy}{dx}=\frac{1+y^2}{1+x^2}$

$$\Rightarrow \frac{1}{1+y^2}dy=\frac{1}{1+x^2}dx$$

On integrating both sides, we get

$$\int\frac{1}{1+y^2}dy=\int\frac{1}{1+x^2}dx$$

$$\Rightarrow \tan^{-1}y=\tan^{-1}x+\tan^{-1}C$$

$$\Rightarrow \tan^{-1}y-\tan^{-1}x=\tan^{-1}C$$

$$\Rightarrow \tan^{-1}\left(\frac{y-x}{1+xy}\right)=\tan^{-1}C$$

$$\Rightarrow \frac{y-x}{1+xy}=C$$

$$\Rightarrow y-x=C(1+xy),$$

which is the required solution.

(iv) We have, $\log\left(\frac{dy}{dx}\right)=ax+by$

$$\Rightarrow \frac{dy}{dx}=e^{ax+by}$$

$$\Rightarrow \frac{dy}{dx}=e^{ax}\cdot e^{by}$$

$$\Rightarrow e^{-by}dy=e^{ax}dx$$

On integrating both sides, we get

$$\int e^{-by}dy=\int e^{ax}dx$$

$$\Rightarrow -\frac{e^{-by}}{b}=\frac{e^{ax}}{a}+C$$

(v) The given differential equation is

$$\frac{dy}{dx}=\frac{(1+y^2)}{xy(1+x^2)}$$

$$\Rightarrow \frac{y\,dy}{1+y^2}=\frac{dx}{x(1+x^2)}$$

$$\Rightarrow \frac{y\,dy}{1+y^2}=\left(\frac{1}{x}-\frac{x}{1+x^2}\right)dx$$

On integrating both sides, we get

$$\log|1+y^2|=2\log|x|-\log|1+x^2|+2\log C$$

$$\Rightarrow (1+y^2)(1+x^2)=x^2C^2$$

(vi) We have, $\frac{dy}{dx}=2e^xy^3$

$$y^{-3}dy=2e^xdx$$

On integrating both sides, we get

$$\int y^{-3}dy=2\int e^x\,dx$$

$$-\frac{1}{2y^2}=2e^x+C \quad \text{...(i)}$$

When $x=0$, then $y=\frac{1}{2}$

$$-2=2+C$$

$$\Rightarrow C=-4$$

Putting in Eq. (i),

$$-\frac{1}{2y^2}=2e^x-4$$

$$\Rightarrow 4y^2(e^x-2)=-1$$

$$4y^2(2-e^x)=1$$

(vii) We have, $x\,dy+y\,dx=xy\,dx$

$$x\,dy=(x-1)y\,dx$$

$$\Rightarrow \frac{1}{y}dy=\left(1-\frac{1}{x}\right)dx$$

On integrating both sides, we get

$$\int\frac{1}{y}dy=\int\left(1-\frac{1}{x}\right)dx$$

$$\Rightarrow \log y=x-\log x+C$$

$$\Rightarrow \log xy=x+C$$

$$\Rightarrow xy=e^{x+c} \quad \text{...(i)}$$

Given that $y=1$, when $x=1$

$$1=e^{1+C}$$

So, $1+C=0$

$$\Rightarrow C=-1$$

Putting $C=-1$ in Eq. (i), we get

$$xy=e^{x-1}$$

7. (i) We have, $\frac{dy}{dx} = \cos(x+y)$

Let $x + y = v$

Differentiating both sides w.r.t. x, we get

$\Rightarrow \quad 1 + \frac{dy}{dx} = \frac{dv}{dx}$

$\Rightarrow \quad \frac{dy}{dx} = \frac{dv}{dx} - 1$

So, $\frac{dv}{dx} - 1 = \cos v$

$\Rightarrow \quad \frac{dv}{dx} = 1 + \cos v$

$\Rightarrow \quad \frac{1}{1+\cos v} dv = dx$

$\Rightarrow \quad \frac{1}{2}\sec^2 \frac{v}{2} dv = dx$

On integrating both sides, we get

$$\int \frac{1}{2}\sec^2 \frac{v}{2} dv = \int dx$$

$$\tan\frac{v}{2} = x + C$$

$\Rightarrow \quad \tan\left(\frac{x+y}{2}\right) = x + C$ [putting $v = x + y$]

(ii) We have, $\frac{dy}{dx} = (x+y+1)^2$

Let $x + y + 1 = v$

Differentiating both sides w.r.t. x, we get

$1 + \frac{dy}{dx} = \frac{dv}{dx}$

$\Rightarrow \quad \frac{dy}{dx} = \frac{dv}{dx} - 1$

$\therefore \quad \frac{dv}{dx} - 1 = v^2$

$\Rightarrow \quad \frac{dv}{dx} = v^2 + 1$

$\Rightarrow \quad \frac{1}{v^2+1} dv = dx$

On integrating both sides, we get

$$\int \frac{1}{v^2+1} dv = \int dx$$

$\Rightarrow \quad \tan^{-1} v = x + C$

$\Rightarrow \quad \tan^{-1}(x+y+1) = x + C$

(iii) We have, $\frac{dy}{dx} = \frac{x-y+3}{2(x-y)+5}$

Let $x - y = v$

Differentiating both sides w.r.t. x, we get

$1 - \frac{dy}{dx} = \frac{dv}{dx}$

$\Rightarrow \quad \frac{dy}{dx} = 1 - \frac{dv}{dx}$

$\therefore \quad 1 - \frac{dv}{dx} = \frac{v+3}{2v+5}$

$\Rightarrow \quad \frac{dv}{dx} = 1 - \frac{v+3}{2v+5}$

$= \frac{2v+5-v-3}{2v+5}$

$\Rightarrow \quad \frac{dv}{dx} = \frac{v+2}{2v+5}$

$\Rightarrow \quad \frac{2v+5}{v+2} dv = dx$

$\Rightarrow \quad \left(2 + \frac{1}{v+2}\right) dv = dx$

On integrating both sides, we get

$$\int \left(2 + \frac{1}{v+2}\right) dv = \int dx$$

$\Rightarrow \quad 2v + \log(v+2) = x + C$

$\Rightarrow \quad 2(x-y) + \log(x-y+2) = x + C$

(iv) We have, $\frac{dy}{dx} = \frac{6x-2y-7}{3x-y+4}$

$\Rightarrow \quad \frac{dy}{dx} = \frac{2(3x-y)-7}{(3x-y)+4}$

Let $3x - y = v$,

Differentiating both sides w.r.t. x, we get

$3 - \frac{dy}{dx} = \frac{dv}{dx}$

$\Rightarrow \quad \frac{dy}{dx} = 3 - \frac{dv}{dx}$

$\Rightarrow \quad 3 - \frac{dv}{dx} = \frac{2v-7}{v+4}$

$\Rightarrow \quad \frac{dv}{dx} = 3 - \frac{2v-7}{v+4} = \frac{3v+12-2v+7}{v+4}$

$\Rightarrow \quad \frac{dv}{dx} = \frac{v+19}{v+4}$

$\Rightarrow \quad \frac{v+4}{v+19} dv = dx$

On integrating both sides, we get

$$\int \frac{v+19-15}{v+19} dv = \int dx$$

$\Rightarrow \quad v - 15\log|v+19| = x + C$

$\Rightarrow \quad (3x-y) - 15\log|3x-y+19| = x + C$

$\Rightarrow \quad (2x-y) - 15\log|3x-y+19| = C$

8. (i) The given differential equation is

$x^2y\,dx - (x^3+y^3)dy = 0$

$\Rightarrow \quad \frac{dy}{dx} = \frac{x^2y}{x^3+y^3}$

which is a homogeneous differential equation.

Putting $y = vx$, we get

$$\frac{dy}{dx} = v + x\frac{dv}{dx}$$

$$\Rightarrow \quad v + x\frac{dv}{dx} = \frac{v}{1+v^3}$$

$$\Rightarrow \quad x\frac{dv}{dx} = \frac{v}{1+v^3} - v$$

$$\Rightarrow \quad x\frac{dv}{dx} = -\frac{v^4}{1+v^3}$$

$$\Rightarrow \quad \frac{1+v^3}{v^4}dv = -\frac{dx}{x}$$

On integrating both sides, we get

$$\int\left(\frac{1}{v^4} + \frac{1}{v}\right)dv = -\int\frac{1}{x}dx$$

$$\Rightarrow \quad -\frac{1}{3v^3} + \log|v| = -\log|x| + C$$

$$\Rightarrow \quad -\frac{1}{3v^3} + \log|vx| = C$$

$$\Rightarrow -\frac{x^3}{3y^3} + \log\left|\frac{y}{x}\cdot x\right| = C$$

$$\Rightarrow \quad -\frac{x^3}{3y^3} + \log|y| = C$$

which is the required solution.

(ii) We have,

$$x\,dy - y\,dx = \sqrt{x^2+y^2}\,dx$$

$$\frac{dy}{dx} = \frac{\sqrt{x^2+y^2}+y}{x}, x \neq 0$$

It is a homogeneous differential equation.

Putting $y = vx$, we get

$$\frac{dy}{dx} = v + x\frac{dv}{dx}$$

$$v + x\frac{dv}{dx} = \sqrt{1+v^2} + v$$

$$x\frac{dv}{dx} = \sqrt{1+v^2}$$

$$\frac{dv}{\sqrt{1+v^2}} = \frac{dx}{x}$$

On integrating both sides, we get

$$\int\frac{dv}{\sqrt{1+v^2}} = \int\frac{dx}{x}$$

$$\Rightarrow \quad \log|v + \sqrt{1+v^2}| = \log x + \log C$$

$$\Rightarrow \quad |v + \sqrt{1+v^2}| = |Cx|$$

$$\Rightarrow \quad \left|\frac{y}{x} + \sqrt{1+\frac{y^2}{x^2}}\right| = |Cx|$$

$$\Rightarrow \quad (y + \sqrt{x^2+y^2})^2 = (Cx^2)^2$$

$$\Rightarrow \quad (y + \sqrt{x^2+y^2})^2 = C^2x^4$$

which is the required solution.

(iii) We have, $(1+e^{x/y})dx + e^{x/y}\left(1-\frac{x}{y}\right)dy = 0$

$$\Rightarrow \quad \frac{dx}{dy} = \frac{-e^{x/y}\left(1-\frac{x}{y}\right)}{(1+e^{x/y})}$$

It is a homogeneous differential equation.

So, putting $x = vy$, we have

$$\frac{dx}{dy} = v + y\frac{dv}{dy}$$

$$\Rightarrow \quad v + y\frac{dv}{dy} = \frac{-e^v(1-v)}{1+e^v}$$

$$\Rightarrow \quad y\frac{dv}{dy} = \frac{-e^v + ve^v - v - ve^v}{1+e^v}$$

$$\Rightarrow \quad y\frac{dv}{dy} = \frac{-e^v - v}{1+e^v}$$

$$\Rightarrow \frac{1+e^v}{e^v+v}dv = -\frac{1}{y}dy$$

On integrating both sides, we get

$$\int\frac{1+e^v}{e^v+v}dv = -\int\frac{1}{y}dy$$

$$\Rightarrow \quad \log|v+e^v| = -\log|y| + \log C$$

$$\log|(v+e^v)\cdot y| = \log C$$

$$\Rightarrow \quad \left(\frac{x}{y} + e^{x/y}\right)y = C$$

$$\Rightarrow \quad x + ye^{x/y} = C$$

(iv) We have, $(xe^{y/x} + y)dx = x\,dy$

$$\frac{dy}{dx} = e^{y/x} + \frac{y}{x}$$

This is a homogeneous differential equation.

Putting $y = vx$, we get

$$\frac{dy}{dx} = v + x\frac{dv}{dx}$$

$$\Rightarrow \quad v + x\frac{dv}{dx} = e^v + v$$

$$\Rightarrow \quad e^{-v}dv = \frac{dx}{x}$$

On integrating both sides, we get

$$\int e^{-v}dv = \int\frac{dx}{x}$$

$$\Rightarrow \quad -e^{-v} = \log|x| + C$$

$$\Rightarrow \quad -e^{-y/x} = \log|x| + C$$

When $y = 1$, then $x = 1$

$$C = -e^{-1}$$

So, $-e^{-y/x} = \log|x| - \frac{1}{e}$

$$\Rightarrow \quad e^{-y/x} = \frac{1}{e} - \log|x|$$

(v) The given differential equation is

$$\frac{dy}{dx}=\frac{x+2y+3}{2x+3y+4}$$

Let $x=X+\alpha$ and $y=Y+\beta$

Thus, $\frac{dy}{dx}=\frac{(X+\alpha)+2(Y+\beta)+3}{2(X+\alpha)+3(Y+\beta)+4}$

$$\frac{dy}{dx}=\frac{(X+2Y)+(\alpha+2\beta+3)}{2X+3Y+(2\alpha+3\beta+4)} \quad ...(i)$$

Let us choose α and β in such a way that

$\alpha+2\beta+3=0$ and $2\alpha+3\beta+4=0$

On solving, we get

$\alpha=1$ and $\beta=-2$

Eq. (i) reduces to

$$\frac{dy}{dx}=\frac{X+2Y}{2X+3Y} \quad ...(ii)$$

which is a homogeneous equation.

Putting $Y = vX$, we get

$$\frac{dY}{dX}=v+X\frac{dv}{dX}$$

So, $v+X\frac{dv}{dX}=\frac{1+2v}{2+3v}$

$$\Rightarrow \frac{2+3v}{1-3v^2}dv=\frac{dX}{X}$$

On integrating both sides, we get

$$\int\frac{2+3v}{1-3v^2}dv=\int\frac{dX}{X}$$

$$\Rightarrow \frac{1}{\sqrt{3}}\log\left|\frac{1+\sqrt{3}v}{1-\sqrt{3}v}\right|-\frac{1}{2}\log|(1-3v^2)| =\log X+\log C$$

$$\Rightarrow \frac{1}{\sqrt{3}}\log\left(\frac{X+\sqrt{3}Y}{X-\sqrt{3}Y}\right)-\frac{1}{2}\log|(X^2-3Y^2)|=\log C$$

$$\Rightarrow \frac{1}{\sqrt{3}}\log\left|\frac{(x-1)+\sqrt{3}(y+2)}{(x-1)-\sqrt{3}(y+2)}\right| -\frac{1}{2}\log|x^2-3y^2-2x-12y-11|=\log C$$

9. (i) We have,

$$\frac{dy}{dx}-\frac{y}{x}=2x^2$$

which is a linear differential equation of the form of

$$\frac{dy}{dx}+Py=Q,$$

where $P=-\frac{1}{x}$ and $Q=2x^2$

$$\therefore \text{ IF } =e^{\int Pdx}=e^{-\int\frac{1}{x}dx}=e^{-\log x}=\frac{1}{x}$$

$\therefore$ The required solution is

$$y\cdot\left(\frac{1}{x}\right)=\int(2x^2)\frac{1}{x}dx+C$$

$$\Rightarrow \frac{y}{x}=2\int x\,dx+C$$

$$\Rightarrow \frac{y}{x}=\frac{2x^2}{2}+C$$

$$\Rightarrow y=x^3+Cx,$$

which is the required solution.

(ii) The given differential equation is

$$y\,dx+(x-y^3)dy=0$$

$$\Rightarrow \frac{dx}{dy}+\frac{x}{y}=y^2$$

which is a linear differential equation of the form of

$$\frac{dx}{dy}+Px=Q,$$

where $P=\frac{1}{y}$ and $Q=y^2$

$$\therefore \text{ IF}=e^{\int P\,dy}=e^{\int\frac{1}{y}dy}=e^{\log y}=y$$

So, the required solution is

$$x\cdot y=\int y^2\cdot y\,dy+C$$

$$xy=\int y^3dy+C$$

$$xy=\frac{y^4}{4}+C$$

(iii) The given differential equation can be written as

$$\frac{dy}{dx}+\frac{2x}{1+x^2}y=\frac{4x^2}{1+x^2}$$

which is the linear differential equation of the form of $\frac{dy}{dx}+Py=Q,$

where $P=\frac{2x}{1+x^2}$ and $Q=\frac{4x^2}{1+x^2}$

$$\therefore \text{ IF}=e^{\int P\,dx}=e^{\int\frac{2x}{1+x^2}dx}=e^{\log(1+x^2)}=1+x^2$$

So, solution is

$$y(1+x^2)=\int\frac{4x^2}{1+x^2}(1+x^2)dx+C$$

$$\Rightarrow y(1+x^2)=\frac{4x^3}{3}+C$$

It is given that $y=0$, when $x=0$, we get

$$C=0$$

$$\therefore y=\frac{4x^3}{3(1+x^2)}$$

which is the required solution.

(iv) We have,

$$(x^2+1)\frac{dy}{dx}-2xy=(x^4+2x^2+1)\cos x$$

$$\Rightarrow \frac{dy}{dx}-\frac{2x}{x^2+1}y=(x^2+1)\cos x$$

which is a linear differential equation with

$$P=-\frac{2x}{x^2+1} \text{ and } Q=(x^2+1)\cos x$$

$$\therefore \text{ IF}=e^{\int P\,dx}=e^{-\int\frac{2x}{x^2+1}dx}=e^{-\log(x^2+1)}=(x^2+1)^{-1}$$

$$\text{IF}=\frac{1}{x^2+1}$$

So, the required solution is

$$y\cdot\frac{1}{x^2+1}=\int(x^2+1)\cos x\cdot\frac{1}{x^2+1}dx+C$$

$$\Rightarrow \frac{y}{x^2+1}=\int\cos x\,dx+C$$

$$\Rightarrow \frac{y}{x^2+1}=\sin x+C \qquad \text{...(i)}$$

It is given that $y(0)=0$

$$0=C$$

Putting in Eq. (i), we get

$$\frac{y}{x^2+1}=\sin x$$

$$\Rightarrow y=(x^2+1)\sin x$$

(v) We have,

$$(1+y^2)dx=(\tan^{-1}y-x)dy$$

$$\Rightarrow \frac{dx}{dy}+\frac{x}{1+y^2}=\frac{\tan^{-1}y}{1+y^2}$$

which is a linear differential equation with

$$P=\frac{1}{1+y^2}$$

and $$Q=\frac{\tan^{-1}y}{1+y^2}$$

$$\therefore \text{ IF}=e^{\int P\,dy}$$

$$=e^{\int\frac{1}{1+y^2}dy}=e^{\tan^{-1}y}$$

So, the required solution is

$$x\cdot e^{\tan^{-1}y}=\int e^{\tan^{-1}y}\cdot\frac{\tan^{-1}y}{1+y^2}dy+C$$

Let $\tan^{-1}y=t$

$$\Rightarrow \frac{1}{1+y^2}dy=dt$$

$$\therefore x\cdot e^{\tan^{-1}y}=\int t\cdot e^t dt$$

$$\Rightarrow x\cdot e^{\tan^{-1}y}=e^t(t-1)+C$$

$$\Rightarrow xe^{\tan^{-1}y}=e^{\tan^{-1}y}(\tan^{-1}y-1)+C$$

It is given that $y(0)=0$

$$C=1$$

$$\therefore xe^{\tan^{-1}y}=e^{\tan^{-1}y}(\tan^{-1}y-1)+1$$

$$\Rightarrow e^{\tan^{-1}y}(x-\tan^{-1}y+1)=1$$

(vi) We have,

$$\frac{dy}{dx}-x^3y^3+xy=0$$

$$\Rightarrow \frac{dy}{dx}+xy=x^3y^3$$

Dividing by y^3 on both sides,

$$\frac{1}{y^3}\frac{dy}{dx}+xy^{-2}=x^3 \qquad \text{...(i)}$$

Let $$y^{-2}=v\Rightarrow\frac{1}{y^3}\frac{dy}{dx}=-\frac{1}{2}\frac{dv}{dx}$$

From Eq. (i),

$$-\frac{1}{2}\frac{dv}{dx}+vx=x^3\Rightarrow\frac{dv}{dx}-2xv=-2x^3$$

$$\therefore \text{ IF}=e^{-2\int x\,dx}=e^{-x^2}$$

So, solution is

$$v\cdot e^{-x^2}=\int(-2x^3)e^{-x^2}dx+C$$

Let $$-x^2=z\Rightarrow-2xdz=dz$$

$$\therefore v\cdot e^{-x^2}=\int(-z\cdot e^z)\,dz+C$$

$$\Rightarrow =-\int(z\cdot e^z)\,dz+C$$

$$=-(z-1)\,e^z+C$$

$$=-(-x^2-1)\,e^{-x^2}+C$$

$$\Rightarrow v\cdot e^{-x^2}=(x^2+1)e^{-x^2}+C$$

$$\Rightarrow \frac{e^{-x^2}}{y^2}=(x^2+1)e^{-x^2}+C$$

(vii) The given differential equation is

$$\frac{dy}{dx}+\frac{y}{x}\log y=\frac{y}{x^2}(\log y)^2$$

Dividing on both sides by $y(\log y)^2$,

$$\frac{1}{y(\log y)^2}\cdot\frac{dy}{dx}+\frac{1}{(\log y)}\cdot\frac{1}{x}=\frac{1}{x^2}$$

Let $$\frac{1}{\log y}=v\Rightarrow\frac{1}{y(\log y)^2}\frac{dy}{dx}=-\frac{dv}{dx}$$

$$\therefore -\frac{dv}{dx}+\frac{1}{x}\cdot v=\frac{1}{x^2}$$

$$\Rightarrow \frac{dv}{dx}-\frac{v}{x}=-\frac{1}{x^2},$$

which is a linear differential equation.

Thus, $$\text{IF}=e^{-\int\frac{1}{x}dx}=e^{-\log x}=\frac{1}{x}$$

Hence, the solution is $$v\left(\frac{1}{x}\right)=\int-\frac{1}{x^2}\left(\frac{1}{x}\right)dx+C$$

$$\Rightarrow \frac{1}{x\log y}=\frac{1}{2x^2}+C$$

$$\Rightarrow \frac{1}{x\log y}-\frac{1}{2x^2}=C$$

(viii) The given differential equation is

$$(1-x^2)\frac{dy}{dx}+xy=xy^2$$

$$\Rightarrow \quad \frac{dy}{dx}+\frac{xy}{1-x^2}=\frac{xy^2}{1-x^2}$$

$$\Rightarrow \quad \frac{1}{y^2}\cdot\frac{dy}{dx}+\frac{x}{y(1-x^2)}=\frac{x}{1-x^2}$$

Let $\frac{1}{y}=v \Rightarrow \frac{1}{y^2}\frac{dy}{dx}=-\frac{dv}{dx}$

$$\therefore \quad \frac{dv}{dx}-\frac{vx}{1-x^2}=\frac{-x}{1-x^2},$$

which is a linear differential equation.

$$\therefore \quad \text{IF}=e^{-\int\frac{x}{1-x^2}dx}=e^{\frac{1}{2}\log|1-x^2|}=\sqrt{1-x^2}$$

So, solution of given differential equation is

$$v\sqrt{1-x^2}=-\int\frac{x\sqrt{1-x^2}}{1-x^2}dx$$

$$=-\int\frac{x}{\sqrt{1-x^2}}dx$$

$$\Rightarrow \quad \frac{1}{y}\sqrt{1-x^2}=\sqrt{1-x^2}+C$$

$$\Rightarrow \quad \frac{1}{y}=1+\frac{C}{\sqrt{1-x^2}}$$

10. It is given that

$$\frac{dy}{dx}=\frac{1}{2y}$$

$$\Rightarrow \quad 2y\,dy=dx$$

On integrating both sides, we get

$$y^2=x+C$$

Now, when $x=3$, then $y=4$

which gives $C=13$

So, required equation is

$$y^2=x+13$$

which is a parabola.

11. Given that

Subtangent $=\frac{y}{\frac{dy}{dx}}=K$ (constant)

$$\Rightarrow \quad K\frac{dy}{dx}=y$$

$$\Rightarrow \quad K\frac{1}{y}dy=dx$$

On integrating both sides, we get

$$K\log y=x+C$$

$\because$ Curve passes through (5, 1)

$$\Rightarrow \quad C=-5$$

Hence, the required curve is

$$K\log y=x-5$$

12. We have, $\quad y^2=4ax$ ...(i)

$$2y\frac{dy}{dx}=4a \quad \text{...(ii)}$$

Eliminating a from Eqs. (i) and (ii), we get

$$y^2=2y\frac{dy}{dx}x$$

Replacing $\frac{dy}{dx}$ by $-\frac{dx}{dy}$, we get

$$y=2\left(-\frac{dx}{dy}\right)x$$

$$2x\,dx+y\,dy=0$$

Integrating both sides, we get

$$x^2+\frac{y^2}{2}=C$$

$$\Rightarrow \quad 2x^2+y^2=2C$$

which is required orthogonal trajectory.

13. Let $P(x, y)$ be any point on the curve, the equation of tangent of P is

$$Y-y=\frac{dy}{dx}(X-x)$$

Thus, cuts coordinate axes at $A\left(x-y\frac{dx}{dy},0\right)$ and $B\left(0,y-x\frac{dy}{dx}\right)$

It is given that $P(x, y)$ is mid-point of AB.

$$\therefore \quad 2x=x-y\frac{dx}{dy} \text{ and } 2y=y-x\frac{dy}{dx}$$

$$\Rightarrow \quad \frac{dy}{dx}=-\frac{y}{x}$$

$$\Rightarrow \quad x\,dy+y\,dx=0$$

$$\Rightarrow \quad d(xy)=0$$

$$\Rightarrow \quad xy=C$$

The curve passes through (2, 3)

$$\therefore \quad 2\times 3=C$$

$$\Rightarrow \quad C=6$$

The equation of curve is

$$xy=6$$

Level 2

1. (c) The given differential equation is

$$\left(\frac{d^2y}{dx^2}\right)^2 = \left(1+\frac{dy}{dx}\right)^{1/2}$$

Squaring both sides,

$$\left(\frac{d^2y}{dx^2}\right)^4 = \left(1+\frac{dy}{dx}\right)$$

Hence, degree = 4

2. (b) Given, differential equation is

$$\left(\frac{d^3y}{dx^3}\right)^{2/3} + 4 - 3\left(\frac{d^2y}{dx^2}\right) + 5\left(\frac{dy}{dx}\right) = 0$$

$$\Rightarrow \quad \left(\frac{d^3y}{dx^3}\right)^{2/3} = 3\frac{d^2y}{dx^2} - 5\frac{dy}{dx} - 4$$

$$\Rightarrow \quad \left(\frac{d^3y}{dx^3}\right)^{2} = \left(3\frac{d^2y}{dx^2} - 5\frac{dy}{dx} - 4\right)^3$$

∴ Required degree = 2

3. (b) The general equation of all parabolas, whose axis of symmetry is parallel to X-axis, is

$$x = ay^2 + by + c$$

where a, b and c are arbitrary constants.

There are three arbitrary constants.

So, the required differential equation is of order 3.

4. (b) We have, $\quad y = ae^{(bx+c)}$ ...(i)

Differentiating both sides w.r.t. x, we get

$$\frac{dy}{dx} = ae^{bx+c}(b)$$

$$\Rightarrow \quad \frac{dy}{dx} = abe^{bx+c} \quad \text{...(ii)}$$

Again, differentiating both sides w.r.t. x, we get

$$\frac{d^2y}{dx^2} = ab^2 \cdot e^{bx+c}$$

$$\Rightarrow \quad \frac{d^2y}{dx^2} = b^2y \quad \text{[from Eq. (i)] ...(iii)}$$

From Eqs. (i) and (ii), $b = \frac{1}{y}\cdot\frac{dy}{dx}$

Putting in Eq. (iii),

$$\frac{d^2y}{dx^2} = y\cdot\frac{1}{y^2}\left(\frac{dy}{dx}\right)^2$$

$$\frac{d^2y}{dx^2} = \frac{1}{y}\left(\frac{dy}{dx}\right)^2$$

∴ Order = 2

5. (c) The given equation when expressed as a polynomial is derivative as

$$a^2\left(\frac{d^2y}{dx^2}\right)^2 - \left[1+\left(\frac{dy}{dx}\right)^2\right]^3 \quad [a \text{ is constant} = 0]$$

Hence, order = 2

and degree = 2

6. (a) The parametric form of the given equation is

$$x = t \text{ and } y = t^2$$

The equation of any tangent at t is

$$2xt = y + t^2$$

Differentiating both sides w.r.t. x, we get

$$2t = y' \quad \left[\because y' = \frac{dy}{dx}\right]$$

Putting the value in the equation of the tangent, we get

$$\frac{2xy'}{2} = y + \left(\frac{y'}{2}\right)^2$$

$$\Rightarrow \quad 4xy' = 4y + (y')^2$$

The order of this equation is 1 and degree is 2.

7. (d) The equation of the family of curve is

$$y = ae^{bx}$$

$$\Rightarrow \quad \log y = \log a + bx$$

Differentiating both sides w.r.t. x, we get

$$\frac{1}{y}\frac{dy}{dx} = b$$

Again, differentiating both sides w.r.t. x, we get

$$\frac{1}{y}\frac{d^2y}{dx^2} - \frac{1}{y^2}\left(\frac{dy}{dx}\right)^2 = 0$$

$$\Rightarrow \quad y\frac{d^2y}{dx^2} = \left(\frac{dy}{dx}\right)^2$$

$$\Rightarrow \quad yy'' = (y')^2$$

8. (b) Let equation of circle with radius 5 units and centre is on $y = 2$, is

$$(x-h)^2 + (y-2)^2 = 5^2 \quad \text{...(i)}$$

$$\Rightarrow \quad 2(x-h) + 2(y-2)\cdot y' = 0$$

$$\Rightarrow \quad 2(y-2)y' = -2(x-h)$$

Squaring both sides,

$$\Rightarrow \quad (y-2)^2\cdot(y')^2 = (x-h)^2$$

From Eq. (i),

$$\Rightarrow \quad (y-2)^2(y')^2 = 25-(y-2)^2$$

$$\Rightarrow \quad (y-2)^2\left(\frac{dy}{dx}\right)^2 = 25-(y-2)^2$$

9. (d) Given that length of latusrectum is the distance of the point $(2, -3)$ from the line $3x + 4y = 5$

$\therefore$ Length of latus rectum $= \left|\frac{3(2)+4(-3)-5}{\sqrt{3^2+4^2}}\right| = \frac{11}{5}$

Let equation of parabola with axis parallel to Y-axis is

$$(x-h)^2 = \frac{11}{5}(y-k)$$

Differentiating both sides w.r.t. x, we get

$$2(x-h) = \frac{11}{5}\frac{dy}{dx}$$

Again, differentiating both sides w.r.t. x, we get

$$2 = \frac{11}{5}\frac{d^2y}{dx^2}$$

$$11\frac{d^2y}{dx^2} = 10$$

10. (b) Given equation of family of curves

$$x^2 = 4b(y+b), b \in R$$

$$\Rightarrow \quad x^2 = 4by + 4b^2 \quad \text{...(i)}$$

On differentiating Eq. (i) w.r.t. x, we get

$$2x = 4b\frac{dy}{dx}$$

$$\Rightarrow \quad 4b = \frac{2x}{y'} \quad \left[\text{where } \frac{dy}{dx} = y'\right] \quad \text{...(ii)}$$

From Eqs. (i) and (ii), on eliminating b, we get

$$x^2 = \frac{2x}{y'}y + 4\left(\frac{x}{2y'}\right)^2$$

$$\Rightarrow \quad x^2(y')^2 = 2xyy' + x^2$$

$$\Rightarrow \quad x(y')^2 = x + 2yy'$$

11. (c) The equation of the family of parabolas having vertex at the origin and the Y-axis as the axis of symmetry, is

$$y = ax^2$$

$$\Rightarrow \quad \frac{dy}{dx} = 2ax$$

$$\Rightarrow \quad \frac{dy}{dx} = \frac{2y}{x} \quad [\because y = ax^2 \Rightarrow a = y/x^2]$$

$$\Rightarrow \quad xy_1 - 2y = 0$$

which is the required differential equation.

12. (a) We have, $y = (\sin^{-1}x)^2 + A\cos^{-1}x + B$

$$= (\sin^{-1}x)^2 - A\sin^{-1}x + \frac{\pi A}{2} + B \quad \text{...(i)}$$

Differentiating both sides w.r.t. x, we get

$$\frac{dy}{dx} = \frac{2\sin^{-1}x}{\sqrt{1-x^2}} - \frac{A}{\sqrt{1-x^2}}$$

$$\Rightarrow \quad (1-x^2)\left(\frac{dy}{dx}\right)^2 = 4(\sin^{-1}x)^2 - 4A\sin^{-1}x + A^2$$

$$= 4y - 2\pi A - 4B + A^2$$

[from Eq. (i)]

Again, differentiating both sides w.r.t. x, we get

$$2(1-x^2)\left(\frac{dy}{dx}\right)\left(\frac{d^2y}{dx^2}\right) - 2x\left(\frac{dy}{dx}\right)^2 = 4\frac{dy}{dx}$$

$$\Rightarrow \quad (1-x^2)\frac{d^2y}{dx^2} - x\frac{dy}{dx} = 2,$$

which is the required differential equation.

13. (d) We have, $(x+y)(dx-dy) = dx + dy$

$$\Rightarrow \quad dx - dy = \frac{dx+dy}{x+y}$$

$$\Rightarrow \quad d(x-y) = \frac{d(x+y)}{x+y}$$

On integrating both sides, we get

$$\Rightarrow \quad x - y = \log(x+y) + \log C$$

$$\Rightarrow \quad C(x+y) = e^{x-y}$$

$$\Rightarrow \quad x + y = Ke^{x-y}, \text{ where } K = \frac{1}{C}$$

14. (b) We have,

$$y\,dx + (x + x^2y)\,dy = 0$$

$$\Rightarrow \quad y\,dx + x\,dy + x^2y\,dy = 0$$

$$\Rightarrow \quad y\,dx + x\,dy = -x^2y\,dy$$

Dividing on both sides by x^2y^2, we get

$$\Rightarrow \quad \frac{y\,dx + x\,dy}{x^2y^2} = -\frac{1}{y}dy \Rightarrow \frac{d(xy)}{x^2y^2} = -\frac{dy}{y}$$

On integrating both sides, we get

$$-\frac{1}{xy} = -\log y + C$$

$$\Rightarrow \quad \log y - \frac{1}{xy} = C$$

15. (b) We have, $\frac{dy}{dx} = \frac{yf'(x) - y^2}{f(x)}$

$$\Rightarrow \quad yf'(x)dx - f(x)\,dy = y^2dx$$

$$\Rightarrow \quad \frac{yf'(x)dx - f(x)dy}{y^2} = dx$$

$$\Rightarrow \quad d\left\{\frac{f(x)}{y}\right\} = dx$$

On integrating both sides, we get

$$\frac{f(x)}{y} = x + C$$

$$\Rightarrow \quad f(x) = y(x+C)$$

16. (d) The differential equation is

$$y(1+xy)dx = x\,dy$$

$$\Rightarrow \quad y dx + xy^2 dx = x\,dy$$

$$\Rightarrow \quad \frac{x\,dy - y\,dx}{y^2} = x\,dx$$

$$\Rightarrow \quad \frac{-(y\,dx - xdy)}{y^2} = x\,dx$$

$$\Rightarrow \quad -d\left(\frac{x}{y}\right) = x\,dx$$

On integrating both sides, we obtain

$$\frac{-x}{y} = \frac{x^2}{2} + C \quad \text{...(i)}$$

Given curve passes through (1, – 1)

$$\therefore \quad 1 = \frac{1}{2} + C \Rightarrow C = \frac{1}{2}$$

Puttiing $C = \frac{1}{2}$ in Eq. (i),

$$y(x^2+1) + 2x = 0$$

Putting $x = -\frac{1}{2}$, we get

$$y = \frac{4}{5}$$

Hence, $f\left(-\frac{1}{2}\right) = \frac{4}{5}$

17. (a) We have,

$$e^x \tan y dx + (1-e^x)\sec^2 y\,dy = 0$$

$$\Rightarrow \quad e^x \tan y\,dx = -(1-e^x)\sec^2 y\,dy$$

$$\Rightarrow \quad -\frac{e^x}{1-e^x}dx = \frac{\sec^2 y\,dy}{\tan y}$$

On integrating both sides, we get

$$-\int \frac{e^x}{1-e^x}dx = \int \frac{\sec^2 y}{\tan y}dy$$

Putting $1 - e^x = t$ and $\tan y = P$

$$-e^x dx = dt \text{ and } \sec^2 y\,dy = dP$$

$$\Rightarrow \quad \int \frac{1}{t}dt = \int \frac{1}{P}dP$$

$$\Rightarrow \quad \log t = \log P + \log C$$

$$\Rightarrow \quad \log t = \log PC$$

$$\Rightarrow \quad t = PC$$

$$\Rightarrow \quad (1-e^x) = (\tan y)C$$

$$\Rightarrow \quad \tan y = \frac{(1-e^x)}{C} \Rightarrow \tan y = C_1(1-e^x)$$

where $C_1 = \frac{1}{C}$

18. (c) We have, $\frac{dy}{dx} = \frac{ax+h}{by+k}$

$$\Rightarrow \quad \int (by+k)\,dy = \int (ax+h)\,dx$$

$$\Rightarrow \quad \frac{by^2}{2} + ky = \frac{ax^2}{2} + hx + C$$

which represents a parabola only when

$$a = 0, b = 2 \text{ or } b = 0, a = 2$$

19. (c) Given differential equation,

$$x^3 dy + xydx = x^2 dy + 2ydx$$

$$\Rightarrow \quad (x^3 - x^2)dy = y(2-x)dx$$

$$\Rightarrow \quad \int \frac{dy}{y} = \int \frac{2-x}{x^2(x-1)}dx$$

Making partial fractions for RHS, we get

$$\frac{2-x}{x^2(x-1)} = \frac{A}{x} + \frac{B}{x^2} + \frac{C}{x-1}$$

$$\Rightarrow \quad 2 - x = Ax(x-1) + B(x-1) + Cx^2$$

On comparing $A + C = 0, B - A = -1$ and $B = -2$

$$\therefore \quad A = -1 \text{ and } C = 1$$

$$\therefore \quad \int \frac{dy}{y} = -\int \frac{dx}{x} - 2\int \frac{dx}{x^2} + \int \frac{dx}{x-1}$$

$$\Rightarrow \quad \log_e y = -\log_e x + \frac{2}{x} + \log_e(x-1) + C$$

$$\because \quad y(2) = e$$

$$\Rightarrow \quad 1 = -\log_e 2 + 1 + 0 + C \Rightarrow C = \log_e 2$$

$\therefore$ At $x = 4$,

$$\log_e y = -\log_e(4) + \frac{1}{2} + \log_e(3) + \log_e 2$$

$$\Rightarrow \quad \log_e\left(\frac{2y}{3}\right) = \frac{1}{2} \Rightarrow y = \frac{3}{2}\sqrt{e}$$

$$\Rightarrow \quad y(4) = \frac{3}{2}\sqrt{e}$$

20. (c) We have,

$$\frac{dy}{dx} + \frac{x(1+y^3)}{y^2(1+x^2)} = 0$$

$$\Rightarrow \quad \frac{dy}{dx} = \frac{-x(1+y^3)}{y^2(1+x^2)}$$

$$\Rightarrow \quad \frac{y^2}{1+y^3}dy = \frac{-x}{1+x^2}dx$$

Integrating both sides, we get

$$\int \frac{y^2}{1+y^3}dy = -\int \frac{x}{1+x^2}dx$$

Put $1 + y^3 = t$ and $1 + x^2 = P$

$$\Rightarrow \quad 3y^2 dy = dt \Rightarrow 2x\,dx = dP$$

$\therefore \quad \frac{1}{3}\int \frac{1}{t}dt = -\frac{1}{2}\int \frac{dP}{P}$

$\Rightarrow \quad \frac{1}{3}\log t = -\frac{1}{2}\log P + \log C$

$\Rightarrow \quad \frac{1}{3}\log(1+y^3) + \frac{1}{2}\log(1+x^2) = \log C$

$\Rightarrow \quad 2\log(1+y^3) + 3\log(1+x^2) = \log C$

$\Rightarrow \quad (1+y^3)^2(1+x^2)^3 = C$

21. (b) Given differential equation,

$$\frac{5+e^x}{2+y}\cdot\frac{dy}{dx} + e^x = 0$$

$\Rightarrow \quad \int \frac{dy}{2+y} + \int \frac{e^x}{5+e^x}dx = 0$

$\Rightarrow \log_e|2+y| + \log_e|5+e^x| = \log_e C$

$\Rightarrow \quad |2+y||5+e^x| = C$

$\because \quad y(0) = 1 \Rightarrow C = 18$

$\therefore$ At $x = \log_e 13$,

$|2+y||5+13| = 18$

$\Rightarrow \quad |2+y| = 1$

$\Rightarrow \quad y+2 = \pm 1 \Rightarrow y = -1$ or -3

22. (a) We have, $\frac{dy}{dx} = \frac{2^x(y+2^y)}{2^x(1+2^y\ln 2)}$

$\Rightarrow \quad \int \frac{(1+2^y\ln 2)}{(y+2^y)}dy = \int dx$

$\Rightarrow \quad \ln|y+2^y| = x + C$

If $x = 0$, then

$y = 0 \Rightarrow C = 0$

$\Rightarrow \quad x = \ln|y+2^y|$

At $y = 1$, $x = \ln 3 \in (1, 2)$

23. (a) We have,

$$\frac{2+\sin x}{y+1}\frac{dy}{dx} = -\cos x$$

$\Rightarrow \quad \frac{1}{y+1}dy = -\frac{\cos x}{2+\sin x}dx$

Integrating both sides, we get

$$\int \frac{1}{y+1}dy = -\int \frac{\cos x}{2+\sin x}dx$$

$\Rightarrow \quad \log(y+1) = -\log(2+\sin x) + \log K$

$\Rightarrow \quad \log(y+1) + \log(2+\sin x) = \log K$

$\Rightarrow \quad (y+1)(2+\sin x) = K$

Now, when $x = 0$, then

$y = 1$

$2(2) = K$

$\Rightarrow \quad K = 4$

So, $(y+1)(2+\sin x) = 4$

Now, at $x = \frac{\pi}{2}$

$(y+1)(2+1) = 4$

$\Rightarrow \quad y+1 = \frac{4}{3} \Rightarrow y = \frac{1}{3}$

24. (a) Given that

$$\frac{dy}{dx} = \frac{y-1}{x^2+x}$$

$\Rightarrow \quad \frac{1}{y-1}dy = \frac{1}{x^2+x}dx$

On integrating both sides, we get

$$\int \frac{1}{y-1}dy = \int \frac{1}{x(x+1)}dx$$

$$= \int\left(\frac{1}{x} - \frac{1}{x+1}\right)dx$$

$\Rightarrow \quad \log(y-1) = \log x - \log(x+1) + \log C$

$\Rightarrow \quad \log\left(\frac{(y-1)(x+1)}{x}\right) = \log C$

$\Rightarrow \quad (y-1)(x+1) = xC$

It passes through (1, 0)

$C = -2$

So, $\quad (y-1)(x+1) = -2x$

$\Rightarrow \quad (y-1)(x+1) + 2x = 0$

25. (c) The given differential equation,

$$e^y\left(\frac{dy}{dx} - 1\right) = e^x$$

$\Rightarrow \quad \frac{dy}{dx} - 1 = e^{x-y}$

Let $x - y = t \Rightarrow 1 - \frac{dy}{dx} = \frac{dt}{dx}$

$\therefore \quad -\frac{dt}{dx} = e^t$

$\Rightarrow \quad -\int \frac{dt}{e^t} = \int dx$

$\Rightarrow \quad e^{-t} = x + C$

$\Rightarrow \quad e^{-(x-y)} = x + C$

$\Rightarrow \quad e^{y-x} = x + C$

$\because$ $y(0) = 0$, so $e^{0-0} = 0 + C \Rightarrow C = 1$

$\therefore \quad e^{y-x} = x + 1$

$\Rightarrow \quad y = x + \log_e(x+1)$

So, $\quad y(1) = 1 + \log_e 2$

26. (d) Given differential equation is

$$\sin x\cos y\,dx + \cos x\sin y\,dy = 0$$

$\Rightarrow \quad \sin x\cos y\,dx = -\cos x\sin y\,dy$

$\Rightarrow \quad \tan x\,dx = -\tan y\,dy$

$\Rightarrow \quad \tan x\,dx + \tan y\,dy = 0$

On integrating both sides, we get

$$\log \sec x + \log \sec y = \log C$$

$\Rightarrow \quad \log(\sec x \sec y) = \log C$

$\Rightarrow \quad \sec x \sec y = C$

$\Rightarrow \quad \cos x \cos y = \dfrac{1}{C}$

$\Rightarrow \quad \cos x \cos y = C_1 \quad$ ...(i)

It passes through the point $\left(0, \dfrac{\pi}{3}\right)$, then

$$\cos 0 \cos \frac{\pi}{3} = C_1$$

$$C_1 = \frac{1}{2}$$

From Eq. (i),

$$\cos x \cos y = \frac{1}{2}$$

27. (d) We have,

$$\left(\frac{x+y-1}{x+y-2}\right)\frac{dy}{dx} = \left(\frac{x+y+1}{x+y+2}\right)$$

Let $x + y = u \Rightarrow 1 + \dfrac{dy}{dx} = \dfrac{du}{dx}$

$\therefore \quad \left(\dfrac{u-1}{u-2}\right)\left(\dfrac{du}{dx} - 1\right) = \left(\dfrac{u+1}{u+2}\right)$

$\Rightarrow \quad \dfrac{du}{dx} - 1 = \left(\dfrac{u+1}{u+2}\right)\left(\dfrac{u-2}{u-1}\right)$

$\Rightarrow \quad \dfrac{du}{dx} = \dfrac{u^2 - u - 2}{u^2 + u - 2} + 1$

$\Rightarrow \quad \dfrac{du}{dx} = \dfrac{2u^2 - 4}{u^2 + u - 2}$

$\Rightarrow \quad \dfrac{u^2 + u - 2}{u^2 - 2} du = 2dx$

$\Rightarrow \quad \left(1 + \dfrac{u}{u^2 - 2}\right) du = 2dx$

On integrating both sides, we get

$$u + \frac{1}{2}\log|u^2 - 2| = 2x + C$$

$\Rightarrow \quad (x+y) + \dfrac{1}{2}\log|(x+y)^2 - 2| = 2x + C$

$\Rightarrow \quad 2(y-x) + \log|(x+y)^2 - 2| = 2C \quad$ ...(i)

When $x = 1$, then $y = 1$

$\therefore \quad \log 2 = 2C$

Substituting the value in Eq. (i), we get

$$2(y-x) + \log|(x+y)^2 - 2| = \log 2$$

$\Rightarrow \quad 2(y-x) + \log\left|\dfrac{(x+y)^2 - 2}{2}\right| = 0$

28. (d) Given differential equation is

$$\frac{dy}{dx} - \frac{y+3x}{\log_e(y+3x)} + 3 = 0$$

Put $\quad y + 3x = t \Rightarrow \dfrac{dy}{dx} + 3 = \dfrac{dt}{dx}$

$\therefore \quad \dfrac{dt}{dx} - \dfrac{t}{\log_e(t)} = 0$

$\Rightarrow \quad \displaystyle\int \frac{\log(t)}{t} dt = \int dx$

$\Rightarrow \quad \dfrac{1}{2}(\log_e t)^2 = x + C'$

$\Rightarrow \quad x - \dfrac{1}{2}\{\log_e(y+3x)\}^2 = C$, where $C' = -C$.

29. (a) The given differential equation is

$$(x+2y)(dx - dy) = dx + dy$$

$\Rightarrow \quad (x+2y-1)dx = (x+2y+1)dy$

$\Rightarrow \quad \dfrac{dy}{dx} = \dfrac{(x+2y-1)}{x+2y+1)} \quad$ ...(i)

Let $x + 2y = v \Rightarrow 1 + 2\dfrac{dy}{dx} = \dfrac{dv}{dx}$

$\therefore \quad \dfrac{dy}{dx} = \dfrac{1}{2}\left(\dfrac{dv}{dx} - 1\right)$

$\Rightarrow \quad \dfrac{1}{2}\left(\dfrac{dv}{dx} - 1\right) = \dfrac{v-1}{v+1}$

$\Rightarrow \quad \dfrac{dv}{dx} = \dfrac{2v-2}{v+1} + 1$

$\Rightarrow \quad \dfrac{dv}{dx} = \dfrac{3v-1}{v+1}$

$\Rightarrow \quad \left(\dfrac{v+1}{3v-1}\right) dv = dx$

$\Rightarrow \quad \left(\dfrac{3v-1+4}{3v-1}\right) dv = 3dx$

$\Rightarrow \quad \left(1 + \dfrac{4}{3v-1}\right) dv = 3dx$

On integrating both sides, we get

$\Rightarrow \quad v + 4\log|3v-1| = 3x + C$

$\Rightarrow \quad (x+2y) + 4\log|3x+6y-1| = 3x + C$

$\Rightarrow \quad 2(y-x) + 4\log|3x+6y-1| = C$

30. (d) We have, $\dfrac{dy}{dx} = (x-y)^2$

which is a differential equation of the form of

$$\frac{dy}{dx} = f(ax + by + c)$$

Put $x - y = t$

$\Rightarrow \quad 1 - \dfrac{dy}{dx} = \dfrac{dt}{dx} \Rightarrow \dfrac{dy}{dx} = 1 - \dfrac{dt}{dx}$

$\therefore \quad 1 - \dfrac{dt}{dx} = t^2 \quad \left[\because \dfrac{dy}{dx} = (x-y)^2\right]$

$$\Rightarrow \quad \frac{dt}{dx}=1-t^2 \Rightarrow \int \frac{dt}{1-t^2}=\int dx$$

[separating the variables]

$$\Rightarrow \quad \frac{1}{2}\log_e\left(\frac{1+t}{1-t}\right)=x+C$$

$$\left[\int \frac{dx}{a^2-x^2}=\frac{1}{2a}\log_e\left|\frac{a+x}{a-x}\right|+C\right]$$

$$\Rightarrow \quad \frac{1}{2}\log_e\left(\frac{1+x-y}{1-x+y}\right)=x+C \qquad [\because t=x-y]$$

Since, $y=1$ when $x=1$, therefore

$$\frac{1}{2}\log_e\left(\frac{1+0}{1+0}\right)=1+C$$

$$\Rightarrow \quad C=-1 \qquad [\because \log 1=0]$$

$$\therefore \quad \frac{1}{2}\log_e\left(\frac{1+x-y}{1-x+y}\right)=x-1$$

$$\Rightarrow \quad -\log_e\left|\frac{1-x+y}{1+x-y}\right|=2(x-1)$$

$$\left[\because \log\frac{1}{x}=\log x^{-1}=-\log x\right]$$

31. (b) We have,

$$(x^2+xy)\,dy=(x^2+y^2)dx$$

$$\Rightarrow \quad \frac{dy}{dx}=\frac{x^2+y^2}{x^2+xy}$$

Let $y=vx \Rightarrow \dfrac{dy}{dx}=v+x\dfrac{dv}{dx}$

So, $$v+x\frac{dv}{dx}=\frac{1+v^2}{1+v}$$

$$x\frac{dv}{dx}=\frac{1+v^2-v-v^2}{1+v}$$

$$\Rightarrow \quad x\frac{dv}{dx}=\frac{1-v}{1+v}$$

$$\Rightarrow \quad \int\frac{1+v}{1-v}dv=\int\frac{1}{x}dx \Rightarrow \int-\left(1-\frac{2}{1-v}\right)dv=\int\frac{1}{x}dx$$

$$\Rightarrow \quad -v-2\log(1-v)=\log x+\log C$$

or $$-\frac{y}{x}-2\log\left(\frac{x-y}{x}\right)=\log x+\log C$$

$$\Rightarrow \quad \log x=2\log(x-y)+\frac{y}{x}+K,$$

where $K=\log C$

32. (b) Given differential equation,

$$\frac{dy}{dx}=\frac{xy}{x^2+y^2}$$

Put $y=vx \Rightarrow \dfrac{dy}{dx}=v+x\dfrac{dv}{dx}$

$$\therefore \quad v+x\frac{dv}{dx}=\frac{v}{1+v^2}$$

$$\Rightarrow \quad x\frac{dv}{dx}=\frac{v-v-v^3}{1+v^2}$$

$$\Rightarrow \quad \left(\frac{1}{v^3}+\frac{1}{v}\right)dv=-\frac{dx}{x}$$

$$\Rightarrow \quad -\frac{1}{2v^2}+\log_e|v|=-\log_e|x|+C$$

[on integrating both sides]

$$\Rightarrow \quad -\frac{x^2}{2y^2}+\log_e|y|=C \qquad \left[\because v=\frac{y}{x}\right]$$

$$\because \quad y(1)=1$$

$$\therefore \quad C=-\frac{1}{2}$$

Now, as $y=e$, then

$$-\frac{x^2}{2e^2}+1=-\frac{1}{2}$$

$$\Rightarrow \quad \frac{3}{2}=\frac{x^2}{2e^2}$$

$$\Rightarrow \quad x^2=3e^2$$

$$\Rightarrow \quad x=\sqrt{3}e$$

33. (a) The given differential equation is

$$\left\{x\cos\left(\frac{y}{x}\right)+y\sin\left(\frac{y}{x}\right)\right\}y\,dx=\left\{y\sin\left(\frac{y}{x}\right)-x\cos\left(\frac{y}{x}\right)\right\}x\,dy$$

$$\therefore \quad \frac{dy}{dx}=\frac{\left\{\cos\left(\frac{y}{x}\right)+\frac{y}{x}\sin\left(\frac{y}{x}\right)\right\}}{\left\{\frac{y}{x}\sin\left(\frac{y}{x}\right)-\cos\left(\frac{y}{x}\right)\right\}}\cdot\left(\frac{y}{x}\right)$$

This is a homogeneous equation.

Let $y=vx \Rightarrow \dfrac{dy}{dx}=v+x\dfrac{dv}{dx}$

$$\therefore \quad v+x\frac{dv}{dx}=\frac{v\cos v+v^2\sin v}{v\sin v-\cos v}$$

$$\Rightarrow \quad x\frac{dv}{dx}=\frac{v\cos v+v^2\sin v-v^2\sin v+v\cos v}{v\sin v-\cos v}$$

$$\Rightarrow \quad x\frac{dv}{dx}=\frac{2v\cos v}{v\sin v-\cos v}$$

$$\Rightarrow \quad \int\frac{v\sin v-\cos v}{v\cos v}dv=\int\frac{2\,dx}{x}$$

$$\Rightarrow \quad -\log(v\cos v)=2\log x+\log C$$

$$\Rightarrow \quad \frac{1}{v\cos v}=Cx^2$$

$$\Rightarrow \quad \cos\left(\frac{y}{x}\right)=\frac{1}{Cxy}$$

$$\Rightarrow \quad xy\cos\left(\frac{y}{x}\right)=K \qquad \left[\text{where } \frac{1}{C}=K\right]$$

34. (a) We have,

$$x\,dy = \left(y + x\frac{f\left(\frac{y}{x}\right)}{f'\left(\frac{y}{x}\right)}\right)dx$$

$$\Rightarrow \quad \frac{dy}{dx} = \frac{y}{x} + \frac{f\left(\frac{y}{x}\right)}{f'\left(\frac{y}{x}\right)}$$

Putting $y = vx$, $\quad \frac{dy}{dx} = v + x\frac{dv}{dx}$

$$\therefore \quad v + x\frac{dv}{dx} = v + \frac{f(v)}{f'(v)}$$

$$\Rightarrow \quad \int \frac{f'(v)}{f(v)}dv = \int \frac{1}{x}dx$$

$$\Rightarrow \quad \log f(v) = \log x + \log C$$

$$\Rightarrow \quad f(v) = xC$$

$$f\left(\frac{y}{x}\right) = xC$$

35. (c) We have,

$$x\frac{dy}{dx} = y(\log y - \log x + 1)$$

$$\Rightarrow \quad \frac{dy}{dx} = \frac{y}{x}\left(\log\frac{y}{x} + 1\right)$$

Putting $y = vx$, $\frac{dy}{dx} = v + x\frac{dv}{dx}$

$$\therefore \quad v + x\frac{dv}{dx} = v(\log v + 1)$$

$$\Rightarrow \quad \frac{dv}{v\log v} = \frac{dx}{x}$$

Integrating both sides, we get

$$\int \frac{dv}{v\log v} = \int \frac{dx}{x}$$

$$\Rightarrow \quad \log(\log v) = \log x + \log C$$

$$\Rightarrow \quad \log v = xC$$

$$\Rightarrow \quad \log\frac{y}{x} = xC$$

36. (a) We have, $\quad \frac{dy}{dx} = \frac{y^2 - 2xy - x^2}{y^2 + 2xy - x^2}$

Putting $y = vx$, $\frac{dy}{dx} = v + x\frac{dv}{dx}$

$$\therefore \quad v + x\frac{dv}{dx} = \frac{v^2 - 2v - 1}{v^2 + 2v - 1}$$

$$\Rightarrow \quad x\frac{dv}{dx} = \frac{(v^2 - 1) - 2v}{(v^2 - 1) + 2v} - v$$

$$\Rightarrow \quad = \frac{(v^2 - 1)(1 - v) - 2v(1 + v)}{(v^2 - 1) + 2v}$$

$$\Rightarrow \quad x\frac{dv}{dx} = \frac{(v+1)[-(v-1)^2 - 2v]}{v^2 - 1 + 2v}$$

$$\Rightarrow \quad x\frac{dv}{dx} = \frac{-(v^2+1)(v+1)}{v^2 - 1 + 2v}$$

$$\Rightarrow \quad \frac{(v^2 - 1) + 2v}{(v^2+1)(v+1)}dv = -\frac{1}{x}dx$$

$$\Rightarrow \quad \frac{2v(v+1) - 1(v^2+1)}{(v^2+1)(v+1)}dv = -\frac{1}{x}dx$$

$$\Rightarrow \quad \left(\frac{2v}{v^2+1} - \frac{1}{v+1}\right)dv = -\frac{1}{x}dx$$

On integrating both sides, we get

$$\log(v^2 + 1) - \log(v + 1) = -\log x + \log C$$

$$\Rightarrow \quad \log\frac{v^2+1}{v+1} = \log\frac{C}{x}$$

So, $\quad \frac{v^2+1}{v+1} = \frac{C}{x}$

$$\frac{y^2 + x^2}{y + x} = C$$

When $x = -1$, $y = 1$, then $C = \infty$

So, $\quad \frac{y^2 + x^2}{y + x} = \infty$

$$x + y = 0$$

which represents a straight line.

37. (b) We have, $y\frac{dy}{dx} = x\left[\frac{y^2}{x^2} + \frac{\phi\left(\frac{y^2}{x^2}\right)}{\phi'\left(\frac{y^2}{x^2}\right)}\right]$

Let $y = tx \Rightarrow \frac{dy}{dx} = t + x\frac{dt}{dx}$

$$\therefore \quad tx\left(t + x\frac{dt}{dx}\right) = x\left[t^2 + \frac{\phi(t^2)}{\phi'(t^2)}\right]$$

$$\Rightarrow \quad t^2 + xt\frac{dt}{dx} = t^2 + \frac{\phi(t^2)}{\phi'(t^2)}$$

$$\int \frac{t\phi'(t^2)}{\phi(t^2)}dt = \int \frac{dx}{x}$$

Let $\phi(t^2) = P \Rightarrow \phi'(t^2)2t\,dt = dP$

$$\therefore \quad \frac{1}{2}\int \frac{1}{P}dP = \int \frac{dx}{x}$$

$$\Rightarrow \quad \frac{1}{2}\ln P = \ln x + \ln C$$

$$\Rightarrow \quad \ln P = \ln(xC)^2$$

$$\Rightarrow \quad \phi(t^2) = x^2 K \quad [\text{where}, K = C^2]$$

$$\Rightarrow \quad \phi\left(\frac{y^2}{x^2}\right) = x^2 K$$

Now, $y(1) = -1$

$$\Rightarrow \quad \phi(1) = K,$$

Then, $\phi\left(\frac{y^2}{4}\right) = 4K$

$$\Rightarrow \quad \phi\left(\frac{y^2}{4}\right) = 4\phi(1)$$

38. (b) Given differential equation is

$(x^2 - y^2)dx + 2xy\,dy = 0$, which can be written as

$$\frac{dy}{dx} = \frac{y^2 - x^2}{2xy}$$

Put $y = vx$ [$\because$ it is in homogeneous form]

$$\Rightarrow \quad \frac{dy}{dx} = v + x\frac{dv}{dx}$$

Now, differential equation becomes

$$v + x\frac{dv}{dx} = \frac{v^2x^2 - x^2}{2x(vx)}$$

$$\Rightarrow \quad v + x\frac{dv}{dx} = \frac{(v^2 - 1)x^2}{2vx^2}$$

$$\Rightarrow \quad x\frac{dv}{dx} = \frac{v^2 - 1}{2v} - v$$

$$= \frac{v^2 - 1 - 2v^2}{2v}$$

$$\Rightarrow \quad x\frac{dv}{dx} = -\frac{1 + v^2}{2v}$$

$$\Rightarrow \quad \int \frac{2v\,dv}{1 + v^2} = -\int \frac{dx}{x}$$

$$\Rightarrow \quad \ln(1 + v^2) = -\ln x - \ln C$$

$$\left[\because \int \frac{f'(x)}{f(x)}dx \Rightarrow \ln|f(x)| + C\right]$$

$$\Rightarrow \quad \ln|(1 + v^2)Cx| = 0 \quad [\because \ln A + \ln B = \ln AB]$$

$$\Rightarrow \quad (1 + v^2)Cx = 1 \quad [\log_e x = 0 \Rightarrow x = e^0 = 1]$$

Now, putting $v = \frac{y}{x}$, we get

$$\left(1 + \frac{y^2}{x^2}\right)Cx = 1 \Rightarrow C(x^2 + y^2) = x$$

$\because$ The curve passes through (1, 1), so

$$C(1 + 1) = 1 \Rightarrow C = \frac{1}{2}$$

Thus, required curve is $x^2 + y^2 - 2x = 0$, which represents a circle having centre (1, 0).

$\therefore$ The solution of given differential equation represents a circle with centre on the X-axis.

39. (a) Given differential equation is

$$2x^2 dy = (2xy + y^2)\,dx$$

$$\Rightarrow \quad \frac{dy}{dx} = \frac{2xy + y^2}{2x^2}$$

which is a hmogeneous differential equation

Let $\quad y = vx \Rightarrow \frac{dy}{dx} = v + x\frac{dv}{dx}$

$\therefore$ The differential equation becomes

$$v + x\frac{dv}{dx} = \frac{2vx^2 + v^2x^2}{2x^2}$$

$$\Rightarrow \quad x\frac{dv}{dx} = \frac{2v + v^2 - 2v}{2}$$

$$\Rightarrow \quad x\frac{dv}{dx} = \frac{v^2}{2} \Rightarrow 2\frac{dv}{v^2} = \frac{dx}{x}$$

On integrating both sides, we get

$$\Rightarrow \quad 2\left(\frac{-1}{v}\right) = \log_e x + C$$

$$\Rightarrow \quad \log_e x + \frac{2x}{y} + C = 0 \quad \text{...(i)}$$

$\because$ The curve (i) passes through the point (1, 2).

So, $\quad C = -1$

$$\therefore \quad \log_e x + \frac{2x}{y} = 1$$

Now, at $x = \frac{1}{2}$

$$\log_e\left(\frac{1}{2}\right) + \frac{1}{y} = 1$$

$$\Rightarrow \quad \frac{1}{y} = 1 + \log_e 2$$

$$\Rightarrow \quad y = f\left(\frac{1}{2}\right) = \frac{1}{1 + \log_e 2}$$

40. (a) It is given that

$$\frac{dy}{dx} = \frac{y}{x} + \sec\left(\frac{y}{x}\right), x > 0 \quad \text{...(i)}$$

Let $y = vx \Rightarrow \frac{dy}{dx} = v + x\frac{dv}{dx}$

Substituting these values in Eq. (i), we obtain

$$v + x\frac{dv}{dx} = v + \sec v$$

$$x\frac{dv}{dx} = \sec v$$

$$\Rightarrow \quad \cos v\,dv = \frac{1}{x}dx$$

Integrating both sides, we get

$$\sin v = \log x + C$$

$$\Rightarrow \quad \sin\left(\frac{y}{x}\right) = \log x + C \quad \text{...(ii)}$$

It passes through the point $\left(1, \frac{\pi}{6}\right)$.

$\therefore \quad \sin\left(\frac{\pi}{6}\right) = \log 1 + C \Rightarrow C = \frac{1}{2}$

Putting $C = \frac{1}{2}$ in Eq. (ii), we obtain

$$\sin\left(\frac{y}{x}\right) = \log x + \frac{1}{2}$$

41. (a) We have, $\cos x\, dy = y(\sin x - y)\, dx$

$\Rightarrow \quad \frac{dy}{dx} = \frac{y\sin x - y^2}{\cos x}$

$\Rightarrow \quad \frac{dy}{dx} = y\tan x - y^2 \sec x$

$\Rightarrow \quad \frac{1}{y^2}\frac{dy}{dx} - \frac{1}{y}\tan x = -\sec x$

$\Rightarrow \quad -\frac{1}{y^2}\frac{dy}{dx} + \frac{1}{y}\tan x = \sec x$

Let $\frac{1}{y} = t \Rightarrow -\frac{1}{y^2}\frac{dy}{dx} = \frac{dt}{dx}$

$\therefore \quad \frac{dt}{dx} + (\tan x)t = \sec x$

$\therefore$ IF $= e^{\int \tan x\, dx} = e^{\log \sec x} = \sec x$

Hence, the required solution is

$$t \cdot \sec x = \int \sec x \cdot \sec x\, dx + C$$

$\Rightarrow \quad \frac{1}{y}\sec x = \tan x + C \qquad [\because \int \sec^2 x\, dx = \tan x]$

$\Rightarrow \quad \sec x = (\tan x + C)y$

42. (d) We have, $x^2 dy + \left(y - \frac{1}{x}\right)dx = 0, x > 0, y(1) = 1$

$\Rightarrow \quad x^2 dy + \frac{(xy-1)}{x}dx = 0$

$\Rightarrow \quad x^2 dy = -\left(\frac{xy-1}{x}\right)dx$

$\Rightarrow \quad \frac{dy}{dx} = \frac{1-xy}{x^3}$

$\Rightarrow \quad \frac{dy}{dx} = \frac{1}{x^3} - \frac{y}{x^2}$

$\Rightarrow \quad \frac{dy}{dx} + \frac{1}{x^2}y = \frac{1}{x^3}$

$\therefore$ IF $= e^{\int \frac{1}{x^2}dx} = e^{-1/x}$

So, $\quad ye^{-1/x} = \int \frac{1}{x^3}e^{-1/x}dx$

$\Rightarrow \quad ye^{-1/x} = e^{-1/x}\left(1 + \frac{1}{x}\right) + C \qquad$ ...(i)

If $y(1) = 1$

$$1 \cdot e^{-1} = e^{-1}(2) + C$$

$\Rightarrow \quad C = -e^{-1} = -\frac{1}{e}$

From Eq. (i), $y \cdot e^{-1/x} = e^{-1/x}\left(1 + \frac{1}{x}\right) - \frac{1}{e}$

Now, $\quad y\left(\frac{1}{2}\right) = 3 - \frac{1}{e} \times e^2$

$\Rightarrow \quad y\left(\frac{1}{2}\right) = 3 - e$

43. (b) We have, $(t+1)\frac{dy}{dt} - ty = 1$

$\Rightarrow \quad \frac{dy}{dt} - \frac{t}{t+1}y = \frac{1}{t+1}$, where $P = -\frac{t}{t+1}$

$\therefore \quad$ IF $= e^{\int P dx} = e^{\int -\frac{t}{t+1}dt}$

$= e^{\int\left(-1+\frac{1}{t+1}\right)dt}$

$= e^{-t+\log(t+1)}$

IF $= e^{-t}(t+1)$

Solution of given linear differential equation is

$$y \cdot e^{-t}(t+1) = \int e^{-t}(t+1)\frac{1}{(t+1)}dt + C$$

$\Rightarrow \quad ye^{-t}(t+1) = -e^{-t} + C$

At $t = 0, y = -1$

$$-1 = -1 + C \Rightarrow C = 0$$

So, $y(t+1) = -1$

At $\quad t = 1, y = -\frac{1}{2}$

44. (a) Given differential equation is

$$x\frac{dy}{dx} + 2y = x^2, (x \neq 0)$$

$\Rightarrow \quad \frac{dy}{dx} + \left(\frac{2}{x}\right)y = x,$

which is a linear differential equation of the form

$$\frac{dy}{dx} + Py = Q$$

Here, $P = \frac{2}{x}$ and $Q = x$

$\therefore \quad$ IF $= e^{\int \frac{2}{x}dx} = e^{2\log x} = x^2$

Since, solution of the given differential equation is

$$y \times \text{IF} = \int (Q \times \text{IF})\, dx + C$$

$\therefore \quad y(x^2) = \int (x \cdot x^2)\, dx + C \Rightarrow yx^2 = \frac{x^4}{4} + C$

$\because \quad y(1) = 1$, so $1 = \frac{1}{4} + C \Rightarrow C = \frac{3}{4}$

$\therefore \quad yx^2 = \frac{x^4}{4} + \frac{3}{4}$

$\Rightarrow \quad y = \frac{x^2}{4} + \frac{3}{4x^2}$

45. (a) We have, $\frac{dy}{dx} + \frac{1}{2}(\sec x)y = \frac{\tan x}{2y}$

$\Rightarrow \quad 2y\frac{dy}{dx} + y^2 \sec x = \tan x$

Putting $u = y^2$, $\frac{du}{dx} = 2y\frac{dy}{dx}$

So, $\frac{du}{dx} + (\sec x)u = \tan x$ ...(i)

which is a linear differential equation.

$\therefore$ IF $= e^{\int \sec x\, dx} = e^{\log(\sec x + \tan x)} = \sec x + \tan x$

Solution of Eq. (i) is given by

$u(\sec x + \tan x) = \int \tan x(\sec x + \tan x)dx + C$

$\Rightarrow \quad y^2(\sec x + \tan x) = \sec x + \tan x - x + C$ ...(ii)

It is given $y = 1$, when $x = 0$, we get $C = 0$

Putting in Eq. (ii), we get

$y^2(\sec x + \tan x) = \sec x + \tan x - x$

$\Rightarrow \quad y^2 = 1 - \frac{x}{\sec x + \tan x}$

46. (c) We have, $(xy^4 + y)\,dx - x\,dy = 0$

$\Rightarrow \quad \frac{dy}{dx} = \frac{xy^4 + y}{x}$

$\Rightarrow \quad \frac{dy}{dx} - \frac{y}{x} = y^4$

$\Rightarrow \quad \frac{1}{y^4}\frac{dy}{dx} + \left(-\frac{1}{y^3}\right)\frac{1}{x} = 1$

Let $-y^{-3} = u$

$\Rightarrow \quad 3y^{-4}\frac{dy}{dx} = \frac{du}{dx}$

$\therefore \quad \frac{1}{3}\frac{du}{dx} + \frac{u}{x} = 1$

$\Rightarrow \quad \frac{du}{dx} + \frac{3}{x}u = 3$

which is a linear differential equation.

$\therefore$ IF $= e^{\int \frac{3}{x}dx} = e^{3\log x} = x^3$

So, solution of given linear equation is

$ux^3 = \int 3x^3 dx + C_1$

$\Rightarrow \quad -\frac{x^3}{y^3} = \frac{3x^4}{4} + C_1$

$\Rightarrow \quad -4x^3 = 3x^4y^3 + 4y^3C_1$

$\Rightarrow \quad 3x^4y^3 + 4x^3 = Cy^3$ [where $C = -4C_1$]

47. (c) Given differential equation,

$(\cos x)\frac{dy}{dx} + 2y\sin x = \sin 2x, x \in \left(0, \frac{\pi}{2}\right)$

$\Rightarrow \frac{dy}{dx} + 2(\tan x)y = 2\sin x$, is in the form of linear differential equation, so

IF $= e^{2\int \tan x dx} = \sec^2 x$

$\therefore$ Solution is $y(\sec^2 x) = \int 2\tan x \sec x dx + C = 2\sec x + C$

$\because \quad y\left(\frac{\pi}{3}\right) = 0 \Rightarrow C = -4$

So, the required equation is

$y(\sec^2 x) = 2\sec x - 4$

At $x = \frac{\pi}{4}$, we get

$2y = 2\sqrt{2} - 4$

$\therefore \quad y\left(\frac{\pi}{4}\right) = \sqrt{2} - 2$

48. (b) We have,

$e^x(x+1)\,dx + (ye^y - xe^x)dy = 0$

$\Rightarrow \quad e^x(x+1)\frac{dx}{dy} - xe^x = -ye^y$ ...(i)

Putting $xe^x = t$, $e^x(x+1)\frac{dx}{dy} = \frac{dt}{dy}$

$\therefore$ Eq. (i) reduces to

$\frac{dt}{dy} - t = -ye^y$

which is a linear differential equation.

$\therefore$ IF $= e^{-\int dy} = e^{-y}$

$\therefore$ Solution is $t \cdot e^{-y} = -\int ye^y e^{-y} dy + C$

$\Rightarrow \quad x \cdot e^x \cdot e^{-y} = -\frac{y^2}{2} + C$

Since, $f(0) = 0 \Rightarrow C = 0$

$\therefore \quad xe^x e^{-y} = -\frac{y^2}{2}$

$\Rightarrow \quad 2xe^x e^{-y} + y^2 = 0,$

which is the required solution.

49. (b) Given differential equation,

$(y^2 - x)\frac{dy}{dx} = 1$

$\Rightarrow \quad \frac{dx}{dy} + x = y^2$ is a linear differential equation.

$\therefore \quad$ IF $= e^{\int 1 dy} = e^y$

$\therefore$ Solution of given differential equation is

$xe^y = \int y^2 e^y dy + C$

$= y^2 e^y - 2\int ye^y dy + C$

$$= y^2e^y - 2\{ye^y - \int e^y dy\} + C$$

$\Rightarrow \quad xe^y = y^2e^y - 2ye^y + 2e^y + C$

$\because \quad y(0) = 1$

$\Rightarrow \quad 0e = 1e - 2e + 2e + C \Rightarrow C = -e$

$\therefore$ Particular solution of given differential equation is

$$xe^y = y^2e^y - 2ye^y + 2e^y - e \quad \text{...(i)}$$

$\because$ Curve (i) intersects the X-axis, then $y = 0$

$\therefore \quad xe^{\circ} = 2 - e$

$\Rightarrow \quad x = 2 - e$

Therefore, abscissa of point of intersection of curve (i) and X-axis is $(2 - e)$.

50. (d) Given differential equation,

$$xy' - y = x^2(x\cos x + \sin x), x > 0$$

$\Rightarrow \frac{dy}{dx} - \frac{1}{x}y = x(x\cos x + \sin x)$, which is the form of linear differential equation of the form

$$\frac{dy}{dx} + Py = Q$$

So, $\quad \text{IF} = e^{-\int \frac{dx}{x}} = \frac{1}{x}$

$\therefore$ Solution of given differential equation

$y(\text{IF}) = \int Q(\text{IF})dx + C$, where

$Q = x(x\cos x + \sin x)$

$\Rightarrow \quad \frac{y}{x} = \int (x\cos x + \sin x)dx + C$

$= x\sin x + C \quad [\because \int \{xf'(x) + f(x)\}\,dx = xf(x) + C]$

$\because \quad y(\pi) = \pi \Rightarrow C = 1$

$\therefore \quad y = x^2 \sin x + x$

$\therefore \quad y' = 2x\sin x + x^2\cos x + 1$

$\Rightarrow \quad y'' = 2x\cos x + 2\sin x + 2x\cos x - x^2 \sin x$

So, $y''\left(\frac{\pi}{2}\right) + y\left(\frac{\pi}{2}\right) = \left(2 - \frac{\pi^2}{4}\right) + \left(\frac{\pi^2}{4} + \frac{\pi}{2}\right) = 2 + \frac{\pi}{2}$

51. (a) We have, $\frac{dy}{dx} + \left(\frac{2x+1}{x}\right)y = e^{-2x}, (x > 0)$

which is of the form $\frac{dy}{dx} + Py = Q$, where

$$P = \frac{2x+1}{x} \text{ and } Q = e^{-2x}$$

Now, $\quad \text{IF} = e^{\int P dx} = e^{\int \left(\frac{1+2x}{x}\right)dx} = e^{\int \left(\frac{1}{x}+2\right)dx}$

$$= e^{\ln x + 2x} = e^{\ln x} \cdot e^{2x} = x \cdot e^{2x}$$

So, the solution of the given equation is

$$y \cdot (\text{IF} = \int (\text{IF})Q\,dx + C$$

$\Rightarrow \quad y(xe^{2x}) = \int (x\,e^{2x} \cdot e^{-2x})\,dx + C$

$$= \int x\,dx + C = \frac{x^2}{2} + C \quad \text{... (i)}$$

Since, $y = \frac{1}{2}e^{-2}$, when $x = 1$

$\therefore \quad \frac{1}{2}e^{-2}.e^2 = \frac{1}{2} + C \Rightarrow C = 0$

$\therefore \quad y\,(xe^{2x}) = \frac{x^2}{2} \quad$ [using Eq. (i)]

$\Rightarrow \quad y = \frac{x}{2}e^{-2x}$

Now, $\frac{dy}{dx} = \frac{1}{2}e^{-2x} + \frac{x}{2}e^{-2x}(-2)$

[by using product rule of derivative]

$$= e^{-2x}\left\{\frac{1}{2} - x\right\} < 0,$$

if $\frac{1}{2} < x < 1$

Thus, $y(x)$ is decreasing in $\left(\frac{1}{2}, 1\right)$.

and $y(\log_e 2) = \frac{\log_e 2}{2}e^{-2\log_e 2}$

$$= \frac{1}{2}(\log_e 2)e^{\log_e 2^{-2}}$$

$$= \frac{1}{2}(\log_e 2)\cdot 2^{-2} = \frac{1}{8}\log_e 2$$

52. (b) We have,

$$f'(x) - \frac{2x(x+1)}{x+1}f(x) = \frac{e^{x^2}}{(x+1)^2}$$

which is linear differential equation.

$\therefore \quad \text{IF} = e^{\int -2x\,dx} = e^{-x^2}$

Thus, solution is

$$f(x) \cdot e^{-x^2} = \int \frac{dx}{(x+1)^2} + C$$

$$f(x) \cdot e^{-x^2} = -\frac{1}{x+1} + C$$

Given that $f(0) = 5$, thus $C = 6$

$\therefore \quad f(x) = \left(\frac{6x+5}{x+1}\right)e^{x^2}$

53. (c) Since, sub-normal is $y\frac{dy}{dx}$.

We have, $\quad y\frac{dy}{dx} = Ky^2 \quad$ [K is a constant]

$\Rightarrow \quad \frac{dy}{y} = K dx$

Integrating both sides, we get

$$\log y = Kx + \log C$$

$\Rightarrow \quad y = Ce^{Kx}$

54. (b) We have,

$$x^{2/3}+y^{2/3}=a^{2/3}$$

$$\Rightarrow \quad \frac{2}{3}x^{-1/3}+\frac{2}{3}y^{-1/3}\frac{dy}{dx}=0$$

$$\Rightarrow \quad \frac{dy}{dx}=-\frac{x^{-1/3}}{y^{-1/3}}$$

Replacing $\frac{dy}{dx}$ by $-\frac{dx}{dy}$,

$$\Rightarrow \quad \frac{dx}{dy}=\frac{x^{-1/3}}{y^{-1/3}}$$

$$\Rightarrow \quad \int x^{1/3}dx=\int y^{1/3}dy$$

$$\Rightarrow \quad x^{4/3}-y^{4/3}=C$$

55. (a) The given curve is

$$y^2=4a(x+a) \qquad \text{...(i)}$$

$$\Rightarrow \quad 2y\frac{dy}{dx}=4a$$

$$\Rightarrow \quad a=\frac{y}{2}\frac{dy}{dx} \qquad \text{...(ii)}$$

Eliminating a from Eqs. (i) and (ii),

$$y^2=2y\frac{dy}{dx}\left(x+\frac{y}{2}\frac{dy}{dx}\right)$$

$$\Rightarrow \quad y^2=2xy\frac{dy}{dx}+y^2\left(\frac{dy}{dx}\right)^2$$

Replacing $\frac{dy}{dx}$ by $-\frac{dx}{dy}$, we get

$$y^2=2xy\left(-\frac{dx}{dy}\right)+y^2\left(\frac{dx}{dy}\right)^2$$

$$\Rightarrow \quad y^2\left(\frac{dy}{dx}\right)^2+2xy\left(\frac{dy}{dx}\right)-y^2=0$$

Let $\frac{dy}{dx}=P$

$$\therefore \quad y^2P^2+2xyP-y^2=0$$

$$\Rightarrow \quad (yP+x)^2=x^2+y^2$$

$$\Rightarrow \quad yP+x=\sqrt{x^2+y^2}$$

$$\Rightarrow \quad \frac{x\,dx+y\,dy}{\sqrt{x^2+y^2}}=dx$$

$$\Rightarrow \quad \frac{2x\,dx+2y\,dy}{\sqrt{x^2+y^2}}=2dx$$

$$\Rightarrow \quad d(\sqrt{x^2+y^2})=2dx$$

Integrating both sides, we get

$$(\sqrt{x^2+y^2})=2x+C$$

which is the required orthogonal trajectory.

56. (b) The equation of the given family of curves is

$$a^{n-1}y=x^n \qquad \text{...(i)}$$

$$\Rightarrow \quad (n-1)\log a+\log y=n\log x$$

On differentiating both sides w.r.t. x, we get

$$\frac{1}{y}\frac{dy}{dx}=\frac{n}{x} \qquad \text{...(ii)}$$

This is the differential equation of the family of curves given in Eq. (i).

The differential equation of the orthogonal trajectories of Eq. (i) is obtained by replacing $\frac{dy}{dx}$ by $-\frac{dx}{dy}$ in Eq. (ii).

Replacing $\frac{dy}{dx}$ by $-\frac{dx}{dy}$ in Eq. (ii), we get

$$\frac{1}{y}\times\left(-\frac{dx}{dy}\right)=\frac{n}{x}$$

$$\Rightarrow \quad x\,dx+ny\,dy=0$$

On integrating both sides, we get

$$\frac{x^2}{2}+n\frac{y^2}{2}=C$$

$$\Rightarrow \quad x^2+ny^2=2C$$

57. (c) The equation of tangent to the curve at (x, y), is

$$Y-y=\frac{dy}{dx}(X-x)$$

Putting $X=0$, we get

$$Y=y-x\frac{dy}{dx}$$

It is given that

$$\left(y-x\frac{dy}{dx}\right)^2=xy$$

$$\Rightarrow \quad \left(y-x\frac{dy}{dx}\right)=\sqrt{xy}$$

$$\Rightarrow \quad x\frac{dy}{dx}+\sqrt{xy}=y$$

$$\Rightarrow \quad \frac{dy}{dx}+\sqrt{\frac{y}{x}}=\frac{y}{x}$$

Putting $y=vx$,

$$\frac{dy}{dx}=v+x\frac{dv}{dx}$$

$$\therefore \quad v+x\frac{dv}{dx}+\sqrt{v}=v$$

$$\Rightarrow \quad x\frac{dv}{dx}=-\sqrt{v}$$

$$\Rightarrow \quad \frac{1}{\sqrt{v}}dv=-\frac{1}{x}dx$$

Integrating both sides, we get

$$2\sqrt{v}=-\log|x|+C$$

$$\Rightarrow \quad 2\sqrt{\frac{y}{x}}+\log|x|=C$$

58. (b) The equation of the tangent at $P(x, y)$ is

$$Y - y = \frac{dy}{dx}(X - x)$$

It cuts the line $y = x$, so the point of intersection

$$= \left(\frac{y - x\frac{dy}{dx}}{1 - \frac{dy}{dx}} \right)$$

It is given that

$$\frac{y - x\frac{dy}{dx}}{1 - \frac{dy}{dx}} = 1$$

$$\Rightarrow \quad y - x\frac{dy}{dx} = 1 - \frac{dy}{dx}$$

$$\Rightarrow \quad (1 - x)\frac{dy}{dx} = (1 - y)$$

$$\Rightarrow \quad \frac{dy}{(y-1)} = \frac{dx}{(x-1)}$$

Integrating both sides, we get

$$\log|x - 1| = \log|y - 1| + \log C$$

$$\Rightarrow \quad \log\left|\frac{x-1}{y-1}\right| = \log C$$

$$\Rightarrow \quad \left(\frac{x-1}{y-1}\right) = C$$

59. (c) Given, $\frac{dy}{dx} = \frac{2y}{x^2}$

$$\Rightarrow \quad \int \frac{dy}{y} = \int \frac{2}{x^2} dx \quad \text{[integrating both sides]}$$

$$\Rightarrow \quad \log_e|y| = -\frac{2}{x} + C \quad \text{...(i)}$$

Since, curve (i) passes through centre (1, 1) of the circle

$$x^2 + y^2 - 2x - 2y = 0$$

$$\therefore \quad \log_e(1) = -\frac{2}{1} + C \Rightarrow C = 2$$

$\therefore$ Equation of required curve is

$$\log_e|y| = -\frac{2}{x} + 2 \quad \text{[putting } C = 2 \text{ in Eq. (i)]}$$

$$\Rightarrow \quad x\log_e|y| = 2(x - 1)$$

60. (c) We have, $\frac{dy}{dx} = \frac{-(y + y^3)}{1 + x + xy^2}$

$$\Rightarrow \quad \frac{dx}{dy} = -\frac{1 + x(1 + y^2)}{y + y^3}$$

$$\Rightarrow \quad \frac{dx}{dy} = -\left[\frac{1}{y(y^2+1)} + \frac{x(1+y^2)}{y(1+y^2)}\right]$$

$$\therefore \quad \frac{dx}{dy} + \frac{x}{y} = \frac{-1}{y(1+y^2)}$$

$$\text{IF} = e^{\int \frac{dy}{y}} = e^{\log y} = y$$

Hence, the solution is

$$x \cdot y = -\int \frac{1}{y(1+y^2)} y\,dy + C$$

$$= -\int \frac{dy}{1+y^2} + C$$

$$xy = -\tan^{-1} y + C$$

This passes through (0, 1), $C = \frac{\pi}{4}$

Hence, $xy + \tan^{-1} y = \frac{\pi}{4}$

61. (d) Let at time t, r and h be the radius and height of cone of water, respectively. Thus, at time t, surface area of liquid in contact with air $= \pi r^2$.

Now, according to question,

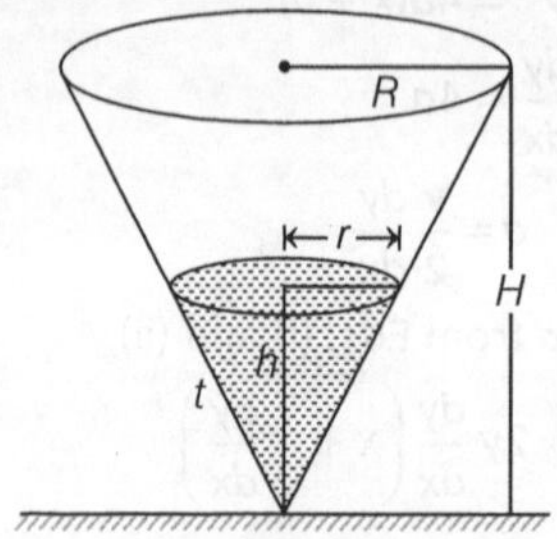

$$-\frac{dV}{dt} \propto \pi r^2$$

[$\because$ – sign shows that V decreases with time]

$$\text{or} \quad \frac{dV}{dt} = -k\pi r^2 \quad \text{...(i)}$$

But from the figure, we get [similar triangles]

$$\frac{r}{h} = \frac{R}{H} \text{ or } h = \frac{rH}{R}$$

$$\text{or} \quad \frac{1}{3}\frac{d}{dt}\left[r^2 \frac{rH}{R}\right] = -kr^2 \quad \text{[from Eq. (i)]}$$

$$\text{or} \quad \frac{r^2H}{R}\frac{dr}{dt} = -kr^2$$

$$\text{or} \quad \frac{dr}{dt} = -\frac{kR}{H}$$

$$\text{or} \quad r = \frac{-kR}{H}t + C \quad \text{[integrating both sides]}$$

Now, at $t = 0$, $r = R$

$$\therefore \quad R = 0 + C \text{ or } C = R$$

$$\therefore \quad r = \frac{-kRt}{H} + R$$

Now, let the time at which cone is empty be T. Then, at T, $r = 0$ (no liquid is left)

$$\therefore \quad 0 = \frac{-kRT}{H} + R \text{ or } T = \frac{H}{k}$$

62. (1) We have,

$$\frac{5+e^x}{2+y}\cdot\frac{dy}{dx}+e^x=0 \Rightarrow \frac{5+e^x}{2+y}\cdot\frac{dy}{dx}=-e^x$$

Integrating both sides, we get

$$\int\frac{dy}{2+y}=\int\frac{-e^x}{e^x+5}dx$$

$$\Rightarrow \ln(y+2)=-\ln(e^x+5)+K$$

$$\Rightarrow (y+2)\cdot(e^x+5)=C \quad [\because K=\log C]$$

$$\because \quad y(0)=1 \Rightarrow C=18$$

$$\therefore \quad y+2=\frac{18}{e^x+5}$$

At $x=\log 13$,

$$y+2=\frac{18}{13+5}=1$$

$$\Rightarrow y=-1$$

$$|y(\log 13)|=1$$

63. (4) Given that

$$\log\left(\frac{dy}{dx}\right)=3x+4y$$

$$\Rightarrow \frac{dy}{dx}=e^{3x+4y}$$

$$\Rightarrow \frac{dy}{dx}=e^{3x}\cdot e^{4y}$$

$$[\because e^{a+b}=e^a\cdot e^b, \forall\, a,b\in R]$$

$$\Rightarrow \int e^{-4y}dy=\int e^{3x}dx$$

$$\Rightarrow \frac{e^{-4y}}{-4}=\frac{e^{3x}}{3}+C$$

$y=0$, when $x=0$

$$\Rightarrow -\frac{1}{4}=\frac{1}{3}+C \Rightarrow C=\frac{-7}{12}$$

So, $$\frac{e^{-4y}}{-4}=\frac{e^{3x}}{3}-\frac{7}{12}$$

$$\Rightarrow e^{-4y}=\frac{4e^{3x}-7}{-3}$$

$$\Rightarrow 4y=\ln\left(\frac{3}{7-4e^{3x}}\right)$$

when $x=-\frac{2}{3}\ln 2$, then

$$4y=\ln\left(\frac{3}{6}\right) \Rightarrow y=\frac{1}{4}\ln\left(\frac{1}{2}\right) \Rightarrow y=-\frac{1}{4}\ln 2$$

Hence, $k=4$

64. (3) The given differential equation is

$$8\sqrt{x}(\sqrt{9+\sqrt{x}})dy=(\sqrt{4+\sqrt{9+\sqrt{x}}})^{-1}dx \quad [x>0]$$

$$\Rightarrow dy=\frac{1}{\{8\sqrt{x}(\sqrt{9+\sqrt{x}})\}(\sqrt{4+\sqrt{9+\sqrt{x}}})}dx$$

Let $4+\sqrt{9+\sqrt{x}}=t$

$$\Rightarrow \frac{1}{2\sqrt{9+\sqrt{x}}}\times\frac{1}{2\sqrt{x}}dx=dt$$

$$\therefore \quad dy=\frac{1}{2\sqrt{t}}dt$$

On integrating both sides, we get

$$y=\sqrt{t}+C \Rightarrow y=\sqrt{4+\sqrt{9+\sqrt{x}}}+C$$

It is given that $y=\sqrt{7}$ when $x=0$.

Then, $C=0$

So, $y=\sqrt{4+\sqrt{9+\sqrt{x}}}$

When $x=256$, then

$$y=\sqrt{4+\sqrt{9+\sqrt{256}}}=3 \text{ [since, } x>0 \text{ given so, } y>0]$$

Hence, $y(256)=3$

65. (1) The given differential equation is

$$\frac{dy}{dx}=\frac{x^2+y^2+3x+3y+2xy+1}{x^2+y^2-3x-3y+2xy+1}$$

$$\frac{dy}{dx}=\frac{(x+y)^2+3(x+y)+1}{(x+y)^2-3(x+y)+1}$$

Let $x+y=v \Rightarrow 1+\frac{dy}{dx}=\frac{dv}{dx}$

$$\therefore \quad \frac{dv}{dx}-1=\frac{v^2+3v+1}{v^2-3v+1}$$

$$\Rightarrow \frac{dv}{dx}=\frac{2v^2+2}{v^2-3v+1}$$

$$\Rightarrow \frac{v^2-3v+1}{v^2+1}dv=2dx$$

$$\Rightarrow \left(1-\frac{3}{2}\left(\frac{2v}{v^2+1}\right)\right)dv=2dx$$

Integrating both sides, we get

$$v-\frac{3}{2}\log|v^2+1|=2x+C$$

$$\Rightarrow y-\frac{3}{2}\log|(x+y)^2+1|=x+C$$

Hence, $m-n=3-2=1$

66. (10) It is given that

$$\frac{dy}{dx}=-\left(\frac{x+y}{x}\right)$$

$$\Rightarrow \frac{dy}{dx}=-1-\frac{y}{x}$$

Let $y=vx \Rightarrow \frac{dy}{dx}=v+x\frac{dv}{dx}$

$$\therefore \quad v+x\frac{dv}{dx}=-1-v$$

$$\Rightarrow x\frac{dv}{dx}=-1-2v$$

$$\Rightarrow \quad \frac{dv}{2v+1} = -\frac{dx}{x}$$

Integrating both sides, we get

$$\log|2v+1| + \log x^2 = \log C$$

$$\Rightarrow \quad (2v+1)\cdot x^2 = C$$

$$\Rightarrow \quad \left(\frac{2y}{x}+1\right)\cdot x^2 = C$$

$$\Rightarrow \quad x^2 + 2xy = C$$

which is passes through (2, 1), we get $C = 8$

$$x^2 + 2xy = 8$$

Hence, $k + l = 8 + 2 = 10$

67. (3) Given differential equation for $x \geq 0$ is

$$(x+1)\,dy = \{(x+1)^2 + y - 3\}\,dx$$

$$\Rightarrow \quad \frac{dy}{dx} - \frac{1}{x+1}y = (x+1) - \frac{3}{x+1}$$

$\because$ The above differential equation is in the form of linear, so

$$\text{IF} = e^{-\int \frac{dx}{1+x}} = \frac{1}{1+x}$$

$\therefore$ Solution of the differential equation is

$$\frac{y}{1+x} = \int \left[(x+1) - \frac{3}{x+1}\right]\frac{dx}{1+x}$$

$$\Rightarrow \quad \frac{y}{1+x} = \int \left(1 - \frac{3}{(x+1)^2}\right)dx$$

$$\Rightarrow \quad \frac{y}{1+x} = x + \frac{3}{x+1} + C \qquad \text{...(i)}$$

$\because \quad y(2) = 0$

$\therefore \quad 0 = 2 + 1 + C$

$\Rightarrow \quad C = -3$

So for $y(3)$, on putting $x = 3$ and $C = -3$ in Eq. (i), we get

$$\frac{y}{4} = 3 + \frac{3}{4} - 3$$

$$\Rightarrow \quad y = 3$$

68. (1) Given, $(2xy^2 - y)\,dx + x\,dy = 0$

$$\Rightarrow \quad \frac{dy}{dx} + 2y^2 - \frac{y}{x} = 0$$

$$\Rightarrow \quad -\frac{1}{y^2}\frac{dy}{dx} + \frac{1}{y}\left(\frac{1}{x}\right) = 2$$

Let $\frac{1}{y} = t \Rightarrow -\frac{1}{y^2}\frac{dy}{dx} = \frac{dt}{dx}$

$$\therefore \quad \frac{dt}{dx} + \frac{1}{x}t = 2$$

$$\text{IF} = e^{\int \frac{1}{x}dx} = x$$

So, the solution is

$$t\cdot x = \int 2x\,dx + C$$

$$\frac{x}{y} = x^2 + C \qquad \text{...(i)}$$

It passes through the point of intersection of lines $2x - 3y = 1$ and $3x + 2y = 8$ i.e. (2, 1).

$$\therefore \quad \frac{2}{1} = 4 + C \Rightarrow C = -2$$

So, $\frac{x}{y} = x^2 - 2$

Put $x = 1$, then $\frac{1}{y} = 1 - 2 = -1$

$$\Rightarrow \quad y(1) = -1$$

$$|y(1)| = 1$$

69. (6) Given, $6\int_1^x f(t)\,dt = 3x f(x) - x^3 - 5,\ x \geq 1$

On differentiating both sides, we get

$$6f(x) = 3f(x) + 3xf'(x) - 3x^2$$

$$f'(x) - \frac{1}{x}f(x) = x$$

Putting $y = f(x)$,

$$\frac{dy}{dx} - \frac{1}{x}y = x$$

which is a linear differential equation.

So, $\text{IF} = e^{\int -\frac{1}{x}dx} = e^{-\log x} = \frac{1}{x}$

Solution of given differential equation is

$$y\cdot\frac{1}{x} = \int \frac{1}{x}\cdot x\,dx + C$$

$$\Rightarrow \quad y\frac{1}{x} = x + C$$

Put $x = 1, y = 2 \Rightarrow C = 1$

$$\Rightarrow \quad \frac{y}{x} = x + 1 \Rightarrow y = x^2 + x$$

So, $f(x) = x^2 + x$

$$f(2) = (2)^2 + 2 = 6$$

70. (9) Let equation of tangent at $P(x, y)$ is

$$Y - y = m(X - x)$$

For y-intercept, $X = 0$

$$Y = y - mx$$

Given that

$$y - mx = x^3$$

$$\Rightarrow \quad x\frac{dy}{dx} - y = -x^3 \qquad \left[\because m = \frac{dy}{dx}\right]$$

$$\Rightarrow \quad \frac{dy}{dx} - \frac{1}{x}y = -x^2$$

$$\text{IF} = e^{-\int \frac{1}{x}dx} = e^{-\log x} = \frac{1}{x}$$

So, solution of given linear differential equation is

$$y\frac{1}{x}=\int \frac{1}{x}(-x^2)\,dx$$

$$\Rightarrow \quad y=-\frac{x^3}{2}+Cx$$

Given that $f(1)=1 \Rightarrow C=\frac{3}{2}$

Hence, $y=-\frac{x^3}{2}+\frac{3}{2}x$

$$\Rightarrow \quad f(-3)=9$$

71. (4) Given that $\frac{dx}{dt} \propto x$

$$\Rightarrow \quad \frac{dx}{dt}=\lambda x \qquad [\lambda = \text{constant}]$$

$$\Rightarrow \quad \frac{dx}{x}=\lambda\, dt$$

$$\Rightarrow \quad \int_{1000}^{x}\frac{dx}{x}=\int_0^t \lambda\, dt$$

$$\Rightarrow \quad \log x-\log 1000=\lambda t$$

$$\Rightarrow \quad \log\left(\frac{x}{1000}\right)=\lambda t$$

Put $t=2$, $x=1200$

$$\log\left(\frac{12}{10}\right)=2\lambda$$

$$\lambda=\frac{1}{2}\ln\left(\frac{6}{5}\right)$$

Now, $\log\left(\frac{x}{1000}\right)=\frac{t}{2}\log\left(\frac{6}{5}\right)$

$$x=1000\,e^{t/2\ln\left(\frac{6}{5}\right)}$$

when $x=2000$ at $t=\frac{k}{\ln\left(\frac{6}{5}\right)}$

$$\Rightarrow \quad 2000=1000\,e^{\left(\frac{k}{2\ln(6/5)}\times \ln\left(\frac{6}{5}\right)\right)}$$

$$\Rightarrow \quad 2=e^{k/2}$$

$$\Rightarrow \quad \frac{k}{2}=\ln 2 \Rightarrow k=2\ln 2$$

$$\Rightarrow \quad \left(\frac{k}{\ln 2}\right)=2$$

Hence, $\left(\frac{k}{\ln 2}\right)^2=4$

72. (2) Equation of normal at any point $P(x, y)$ is

$$\frac{dy}{dx}(Y-y)+(X-x)=0$$

This meets X-axis at $A\left(x+y\frac{dy}{dx}, 0\right)$

Mid-point of AP is $\left(x+\frac{1}{2}y\frac{dy}{dx}, \frac{y}{2}\right)$, which lies on the parabola $2y^2=x$. Therefore,

$$2\cdot\frac{y^2}{4}=x+\frac{1}{2}y\frac{dy}{dx}$$

$$\Rightarrow \quad y^2=2x+y\frac{dy}{dx}$$

Putting $y^2=t$, so that $2y\frac{dy}{dx}=\frac{dt}{dx}$, we get

$$\frac{dt}{dx}-2t=-4x$$

which is a linear differential equation.

$$\text{IF}=e^{-2\int dx}=e^{-2x}$$

Therefore, solution is given by

$$t\cdot e^{-2x}=-4\int xe^{-2x}dx+C$$

$$=-4\left[-\frac{1}{2}xe^{-2x}+\int\frac{1}{2}e^{-2x}dx\right]+C$$

$$\Rightarrow \quad y^2e^{-2x}=2xe^{-2x}+e^{-2x}+C$$

Since, curve passing through (0, 0), we get

$$C=-1$$

Therefore,

$$y^2e^{-2x}=2xe^{-2x}+e^{-2x}-1$$

$$\Rightarrow \quad y^2=2x+1-e^{2x}$$

$$\Rightarrow \quad y^2-2x=1-e^{2x}$$

On comparing,

$$k=2$$

LEVEL 3

1. (a) Given, $x\int_0^x(1-t)\,f(t)\,dt=\int_0^x t\,f(t)\,dt$, $f(1)=1$

Applying Newton and Leibnitz formula, we get

$$\Rightarrow \quad x(1-x)\,f(x)+\int_0^x(1-t)\,f(t)\,dt=x\,f(x)$$

$$\Rightarrow \quad x\,f(x)-x^2\,f(x)-x\,f(x)+\int_0^x(1-t)\,f(t)\,dt=0$$

$$\Rightarrow \quad -x^2\,f(x)+\int_0^x(1-t)\,f(t)\,dt=0$$

$$\Rightarrow \quad -2x\,f(x)-x^2f'(x)+(1-x)\,f(x)=0$$

$$\Rightarrow \quad -x^2f'(x)-3x\,f(x)+f(x)=0$$

$$\Rightarrow \quad x^2f'(x)+(3x-1)\,f(x)=0$$

$$\Rightarrow \quad f'(x)+\left(\frac{3x-1}{x^2}\right)y=0 \qquad [\because y=f(x)]$$

$$\Rightarrow \quad \frac{dy}{y}+\left(\frac{3x-1}{x^2}\right)dx=0$$

Integrating on both sides, we get

$$\Rightarrow \quad \int \frac{dy}{y} + \int \left[\frac{3}{x} - \frac{1}{x^2}\right] dx = C$$

$$\Rightarrow \quad \log|y| + 3\log|x| + \frac{1}{x} = C$$

$$\Rightarrow \quad \log|y \cdot x^3| + \frac{1}{x} = C$$

When $x = 1, y = 1$, then $C = 1$

Hence, the solution is

$$\log|y \cdot x^3| + \frac{1}{x} = 1$$

$$\Rightarrow \quad \log|y \cdot x^3| = \left(1 - \frac{1}{x}\right) = \left(\frac{x-1}{x}\right)$$

$$\Rightarrow \quad y \cdot x^3 = e^{\left(\frac{x-1}{x}\right)}$$

2. (b) The given differential equation is

$$y\cos\left(\frac{y}{x}\right)(x\,dy - y\,dx) + x\sin\left(\frac{y}{x}\right)(x\,dy + y\,dx) = 0$$

$$\Rightarrow \quad \frac{y}{x}\cos\left(\frac{y}{x}\right)\left(\frac{dy}{dx} - \frac{y}{x}\right) + \sin\left(\frac{y}{x}\right)\left(\frac{dy}{dx} + \frac{y}{x}\right) = 0$$

$$\Rightarrow \quad v\cos v\left(v + x\frac{dv}{dx} - v\right) + \sin v\left(v + x\frac{dv}{dx} + v\right) = 0$$

$$\left[\text{let } \frac{y}{x} = v\right]$$

$$\Rightarrow \quad xv\cos v\frac{dv}{dx} + \sin v\left(2v + x\frac{dv}{dx}\right) = 0$$

$$\Rightarrow \quad x(v\cos v + \sin v)\frac{dv}{dx} = -2v\sin v$$

$$\Rightarrow \quad x\frac{dv}{dx} = -\frac{2v\sin v}{(v\cos v + \sin v)}$$

$$\Rightarrow \quad \frac{(v\cos v + \sin v)}{v\sin v}dv = -\frac{2dx}{x}$$

$$\Rightarrow \quad \left(\cot v + \frac{1}{v}\right)dv = -\frac{2dx}{x}$$

Integrating both sides, we get

$$\log|\sin v| + \log|v| = \log C - 2\log x$$

$$\Rightarrow \quad v\sin v = \frac{C}{x^2}$$

$$\Rightarrow \quad \left(\frac{y}{x}\right)\sin\left(\frac{y}{x}\right) = \frac{C}{x^2}$$

$$\Rightarrow \quad (xy)\sin\left(\frac{y}{x}\right) = C$$

When $x = 1,\ y = \frac{\pi}{2}$, then $C = \frac{\pi}{2}$

Hence, the solution is $(xy)\sin\left(\frac{y}{x}\right) = \frac{\pi}{2}$

3. (d) we have,

$$2xy\frac{dy}{dx} = y^2 - x^2$$

$$\frac{dy}{dx} = \frac{y^2 - x^2}{2xy}$$

Put $y = vx$

$$\Rightarrow \quad \frac{dy}{dx} = v + x\frac{dv}{dx}$$

$$\therefore \quad v + x\frac{dv}{dx} = \frac{v^2 - 1}{2v}$$

$$\Rightarrow \quad x\frac{dv}{dx} = \frac{v^2 - 1 - 2v^2}{2v}$$

$$= \frac{-(v^2 + 1)}{2v}$$

$$\Rightarrow \quad \frac{2v}{v^2 + 1}dv = -\frac{dx}{x}$$

Integrating both sides, we get

$$\ln(v^2 + 1) = -\ln x + \ln C$$

$$\Rightarrow \quad v^2 + 1 = \frac{C}{x}$$

$$\Rightarrow \quad \frac{y^2}{x^2} + 1 = \frac{C}{x}$$

$$\Rightarrow \quad x^2 + y^2 = Cx$$

It passes through (1, 1).

$$\therefore \quad x^2 + y^2 - 2x = 0$$

Similarly for second differential equation

$$\frac{dy}{dx} = \frac{2xy}{x^2 - y^2}$$

Equation of curve is

$$x^2 + y^2 - 2y = 0$$

Now, required area is

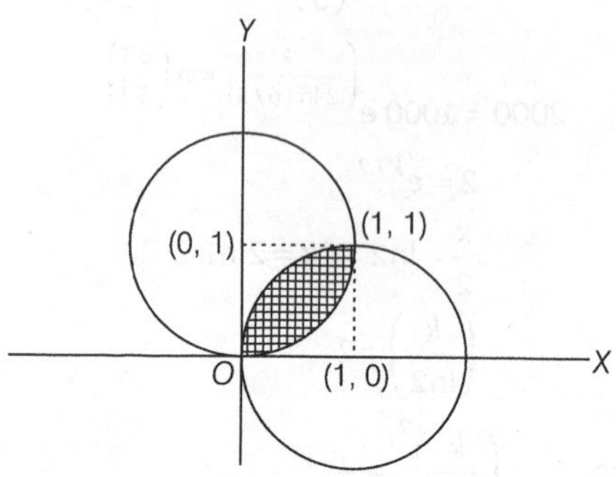

$$= 2\int_0^1 (\sqrt{2x - x^2} - x)\,dx$$

$$= \left(\frac{\pi}{2} - 1\right) \text{sq units}$$

4. (d) We have, $(y+1)\tan^2 x\,dx + \tan x\,dy + y\,dx = 0$

$$\Rightarrow \quad \frac{dy}{dx} + \frac{\sec^2 x}{\tan x} y = -\tan x$$

$$\text{IF} = e^{\int \frac{\sec^2 x}{\tan x} dx} = \tan x$$

So, $\quad y \cdot \tan x = -\int \tan^2 x\,dx$

$$\Rightarrow \quad y \tan x = -\tan x + x + C$$

$$\Rightarrow \quad y = -1 + \frac{x}{\tan x} + \frac{C}{\tan x}$$

$$\Rightarrow \quad \lim_{x\to 0} xy = -x + \frac{x^2}{\tan x} + \frac{Cx}{\tan x} = 1$$

$$\Rightarrow \quad C = 1$$

So, $y(x) = \cot x + x\cot x - 1$

$$y\left(\frac{\pi}{4}\right) = \frac{\pi}{4}$$

5. (b) We have, $x(1 - x\ln y)\frac{dy}{dx} + y = 0$

Let $\ln y = z$

$$\Rightarrow \quad \frac{1}{y}\frac{dy}{dx} = \frac{dz}{dx}$$

$$\Rightarrow \quad y\frac{dx}{dy} = \frac{dx}{dz}$$

$$\therefore \quad \frac{dx}{dz} + x = x^2 z$$

$$\Rightarrow \quad \frac{1}{x^2}\frac{dx}{dz} + \frac{1}{x} = z$$

which is a linear differential equation.

Let $\frac{1}{x} = v \Rightarrow -\frac{1}{x^2}\frac{dx}{dz} = \frac{dv}{dz}$

$$\therefore \quad -\frac{dv}{dz} + v = z \Rightarrow \frac{dv}{dz} - v = -z$$

which is a linear differential equation.

$$\therefore \quad \text{IF} = e^{-\int dz} = e^{-z}$$

Hence, the solution is

$$v \cdot e^{-z} = \int -ze^{-z} dz + C$$

$$\Rightarrow \quad v \cdot e^{-z} = (z+1)e^{-z} + C$$

$$\Rightarrow \quad v = (z+1) + Ce^{z}$$

$$\Rightarrow \quad \frac{1}{x} = (\ln y + 1) + Ce^{\ln y}$$

$$\Rightarrow \quad \frac{1}{x} = (\ln y + 1) + Cy$$

which is passing through $\left(e, \frac{1}{e}\right)$, so $C = e$

Hence, the required integral curve is

$$\frac{1}{x} = (\ln y + 1) + ey$$

$$\Rightarrow \quad \{(\ln y + 1) + ey\}x = 1$$

6. (b) Put $x = 1, y = 1$, then $f(1) = 0$

Put $y = 1 + h$

Now, $f\{x(1+h)\} = xf(1+h) + (1+h)f(x)$

$$\Rightarrow \quad f(x+xh) - f(x) = xf(1+h) + hf(x)$$

$$\Rightarrow \quad f(x+xh) - f(x) = x\{f(1+h) - f(1)\} + hf(x)$$

$$\Rightarrow \quad \left(\frac{f(xh+x) - f(x)}{xh}\right) \cdot x = \frac{x(f(1+h) - f(1))}{h} + \frac{hf(x)}{h}$$

$$\Rightarrow \quad \lim_{h\to 0}\left(\frac{f(xh+x) - f(x)}{xh}\right)x = \lim_{h\to 0}\left(\frac{f(1+h) - f(1)}{h}\right)x + f(x)$$

$$\Rightarrow \quad xf'(x) = xf'(1) + f(x)$$

$$\Rightarrow \quad xf'(x) = x + f(x)$$

$$\Rightarrow \quad x\frac{dy}{dx} = x + y$$

$$\Rightarrow \quad \frac{dy}{dx} = 1 + \frac{y}{x}$$

$$\Rightarrow \quad \frac{dy}{dx} + \left(-\frac{1}{x}\right)y = 1$$

which is a linear differential equation.

$$\therefore \quad \text{IF} = e^{-\int \frac{dx}{x}} = e^{-\log x} = \frac{1}{x}$$

Hence, the solution is

$$\frac{y}{x} = \int \frac{dx}{x} + C$$

$$\Rightarrow \quad \frac{y}{x} = \log|x| + C$$

When $x = 1, y = 0$, then $c = 0$

Thus, the equation of the given curve is

$$y = x\log|x|$$

7. (a) We have, $\lim_{t\to x} \frac{t^2 f(x) - x^2 f(t)}{t - x} = 1$

Using L-Hospital rule,

$$\lim_{t\to x} \frac{2t\,f(x) - x^2 f'(t)}{1} = 1$$

$$\Rightarrow \quad 2x\,f(x) - x^2 f'(x) = 1$$

If $y = f(x)$, then $f'(x) = \frac{dy}{dx}$

$$\therefore \quad x^2\frac{dy}{dx} - 2xy = -1$$

$$\Rightarrow \quad \frac{dy}{dx} - \frac{2}{x}y = -\frac{1}{x^2}$$

It is a linear differential equation.

So, $\text{IF} = e^{\int P dx} = e^{\int -\frac{2}{x} dx}$

$$= e^{-2\log x}$$

$$= e^{\log \frac{1}{x^2}} = \frac{1}{x^2}$$

Solution of given differential equation is

$$y\cdot\frac{1}{x^2}=\int -\frac{1}{x^2}\cdot\frac{1}{x^2}dx+C$$

$$\Rightarrow \quad \frac{y}{x^2}=\frac{1}{3x^3}+C$$

$$\Rightarrow \quad y=\frac{1}{3x}+Cx^2$$

Given, $x=1, y=f(1)=1$

$$\therefore \quad 1=\frac{1}{3}+C$$

$$\Rightarrow \quad C=\frac{2}{3}$$

Hence, $y=\frac{1}{3x}+\frac{2}{3}x^2$

8. (b) We have,

$$\frac{dy}{dx}+\frac{xy}{x^2-1}=\frac{x^4+2x}{\sqrt{1-x^2}}$$

This is a linear differential equation.

$$\therefore \quad \text{IF}=e^{\int\frac{x}{x^2-1}dx}=e^{\frac{1}{2}\log|x^2-1|}=\sqrt{1-x^2}$$

Solution of given equation is

$$y\sqrt{1-x^2}=\int (x^4+2x)dx+C$$

$$\Rightarrow \quad y\sqrt{1-x^2}=\frac{x^5}{5}+x^2+C$$

Putting $x=0$ and $y=f(0)=0$, we get

$$C=0$$

So, $y\sqrt{1-x^2}=\frac{x^5}{5}+x^2$

$$f(x)=y=\frac{x^5}{5\sqrt{1-x^2}}+\frac{x^2}{\sqrt{1-x^2}}$$

$$\therefore \quad \int_{-\sqrt{3}/2}^{\sqrt{3}/2} f(x)dx=\int_{-\sqrt{3}/2}^{\sqrt{3}/2}\frac{x^5}{5\sqrt{1-x^2}}dx+\int_{-\sqrt{3}/2}^{\sqrt{3}/2}\frac{x^2}{\sqrt{1-x^2}}dx$$

$$=2\int_0^{\sqrt{3}/2}\frac{x^2}{\sqrt{1-x^2}}dx$$

$\left[\int_{-a}^{a} f(x)\,dx=0, \text{ if } f(x) \text{ is odd} =2\int_0^a f(x)\,dx, \text{ if } f(x) \text{ is even}\right]$

$$=2\int_0^{\pi/3}\sin^2\theta\, d\theta, \text{ where } x=\sin\theta$$

$$=\int_0^{\pi/3}(1-\cos 2\theta)\,d\theta$$

$$=\left[\theta-\frac{1}{2}\sin 2\theta\right]_0^{\pi/3}$$

$$=\frac{\pi}{3}-\frac{\sqrt{3}}{4}$$

9. (c) Given, slope of the tangent at point (x, y) is

$$\frac{dy}{dx}=\frac{(x+1)^2+y-3}{x+1} \quad \text{...(i)}$$

Put $x+1=X,\ y-3=Y$

$$\therefore \quad \frac{dy}{dx}=\frac{dY}{dX}$$

From Eq. (i), $\frac{dY}{dX}=X+\frac{Y}{X}$

$$\Rightarrow \quad \frac{dY}{dX}-\frac{Y}{X}=X$$

Above is a linear differential equation.

$$\therefore \quad \text{IF}=e^{\int\frac{-1}{X}dX}=\frac{1}{X}$$

Solution is

$$Y\cdot\frac{1}{X}=\int X\cdot\frac{1}{X}dX+C$$

$$\Rightarrow \quad \frac{Y}{X}=X+C$$

So, $\frac{y-3}{x+1}=x+1+C$

The curve passes through the point (2, 0).

$$\therefore \quad -\frac{3}{3}=3+C$$

$$\Rightarrow \quad C=-4$$

Hence, required equation is

$$y-3=(x+1)^2-4(x+1)$$

$$y=x^2-2x$$

$$\Rightarrow \quad y+1=(x-1)^2$$

This equation represents a parabola with vertex at (1, – 1) which meets the X-axis at (0, 0) and (2, 0).

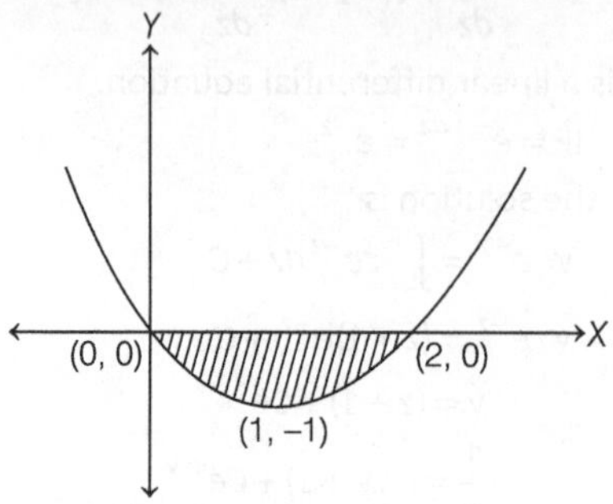

∴ Required area in fourth quadrant is

$$\left|\int_0^2 y\,dx\right|=\left|\int_0^2 (x^2-2x)\,dx\right|$$

$$=\left|\left[\frac{x^3}{3}-x^2\right]_0^2\right|$$

$$=\frac{4}{3}\text{ sq units}$$

10. (a,c) Given equation is

$$x(1+y^2)dx + y(1+x^2)dy = 0$$

$$\Rightarrow \quad x(1+y^2)dx = -y(1+x^2)dy$$

Integrating both sides,

$$\int \frac{x}{1+x^2}dx = -\int \frac{y}{1+y^2}dy$$

Put $1+x^2 = t \Rightarrow 2x\,dx = dt$

and $1+y^2 = u \Rightarrow 2y\,dy = du$

$$\therefore \quad \frac{1}{2}\int \frac{dt}{t} = -\frac{1}{2}\int \frac{du}{u}$$

$$\Rightarrow \quad \frac{1}{2}\ln(1+x^2) = -\frac{1}{2}\ln(1+y^2) + C_1$$

$$\Rightarrow \quad \frac{1}{2}\ln(1+x^2) + \frac{1}{2}\ln(1+y^2) = C_1$$

$$\Rightarrow \quad \frac{1}{2}\ln(1+x^2)(1+y^2) = C \qquad [\text{where } C = C_1]$$

$$\Rightarrow \quad (1+x^2)(1+y^2) = e^{2C}$$

$$\Rightarrow \quad (1+x^2)(1+y^2) = K \qquad [\text{where } e^{2C} = K]$$

11. (b,c) We have,

$$f'(x) = (e^{(f(x)-g(x))})g'(x),\ f(1) = g(2) = 1$$

$$\therefore \quad e^{-f(x)}f'(x) = e^{-g(x)}g'(x)$$

Integrating both sides, we get

$$e^{-f(x)} = e^{-g(x)} + C \qquad \text{...(i)}$$

Putting $x = 1$, $e^{-f(1)} = e^{-g(1)} + C$

$$\therefore \quad e^{-1} = e^{-g(1)} + C \qquad [as\ f(1) = 1] \text{...(ii)}$$

Putting $x = 2$ in Eq. (i), we get

$$e^{-f(2)} = e^{-g(2)} + C$$

$$\therefore \quad e^{-f(2)} = e^{-1} + C \qquad [\text{as } g(2) = 1] \text{...(iii)}$$

Subtracting Eq. (iii) from Eq. (ii), we get

$$e^{-1} - e^{-f(2)} = e^{-g(1)} - e^{-1}$$

$$\Rightarrow \quad e^{-f(2)} = 2e^{-1} - e^{-g(1)} < 2e^{-1} \qquad [\text{as } e^{-g(1)} > 0]$$

$$\Rightarrow \quad -f(2) < \log(2e^{-1})$$

$$\Rightarrow \quad f(2) > 1 - \log 2$$

Similarly, $g(1) > 1 - \log 2$

12. (a,d) The given differential equation is

$$(x^2 + xy + 4x + 2y + 4)\frac{dy}{dx} - y^2 = 0,\ x > 0$$

$$\Rightarrow \quad \{(x^2 + 4x + 4) + (xy + 2y)\}\frac{dy}{dx} - y^2 = 0$$

$$\Rightarrow \quad \{(x+2)^2 + y(x+2)\}\frac{dy}{dx} - y^2 = 0$$

$$\Rightarrow \quad (x+2)(x+y+2)\frac{dy}{dx} = y^2$$

$$\Rightarrow \quad \frac{dy}{dx} = \frac{y^2}{(x+2)(x+2+y)} \qquad \text{...(i)}$$

Let $x + 2 = X$ and $y = Y$

Then, $\frac{dy}{dx} = \frac{dY}{dX}$

Put in Eq. (i), $\quad \frac{dY}{dX} = \frac{Y^2}{X(X+Y)}$

$$\Rightarrow \quad (X^2 + XY)dY = Y^2dX$$

$$\Rightarrow \quad -\frac{1}{Y}dY = \frac{X\,dY - Y\,dX}{X^2}$$

$$\Rightarrow \quad -\frac{1}{Y}dY = d\left(\frac{Y}{X}\right)$$

On integrating both sides, we get

$$-\log|Y| = \frac{Y}{X} + C$$

$$\Rightarrow \quad -\log|y| = \frac{y}{x+2} + C \qquad \text{...(ii)}$$

The given curve passes through the point (1, 3).

$$\therefore \quad -\log 3 = 1 + C$$

$$\Rightarrow \quad C = -\log 3 - 1$$

Substituting the value of C in Eq. (ii), we obtain

$$-\log|y| = \frac{y}{x+2} - \log 3 - 1 \qquad \text{...(iii)}$$

For intersection with $y = x + 2$, put in Eq. (iii),

$$-\log(x+2) = -\log 3$$

$$\Rightarrow \quad x + 2 = \pm 3$$

$$\Rightarrow \quad x = 1 \qquad [\because x > 0]$$

Put $x = 1$ in $y = x + 2$,

$$y = 3$$

So, the curve intersects $y = x + 2$ at exactly one point.

Similarly, for $y = (x+2)^2$

$$-2\log|x+2| = (x+2) - \log 3 - 1$$

$$\Rightarrow \quad x + 1 + 2\log(x+2) = \log 3$$

We find $f(x) = x + 1 + 2\log(x+3)$ is strictly increasing function for $x > 0$ and its minimum value is

$$f(0) = 1 + 2\log 2 > \log 3$$

So, the above equation has no solution.

So, $y = (x+2)^2$ does not intersect, the solution curve again for intersection with $y = (x+3)^2$.

Put $y = (x+3)$ in Eq. (iii),

$$-\log(x+3)^2 = \frac{(x+3)^2}{x+2} - \log 3 - 1$$

$$\Rightarrow \quad \log\frac{(x+3)^2}{3} + \frac{(x+3)^2}{x+2} - 1 = 0$$

For $x > 0$, $\log\frac{(x+3)^2}{3} + \frac{(x+3)^2}{x+2} - 1 > 0$

So, the above equation has no solution.

Hence, $y = (x+3)^2$ does not intersect the solution curve.

13. (a,d) We have,

$$\frac{dy}{dx} - y\tan x = 2x\sec x$$

Given equation is linear differential equation.

So, $\quad IF = e^{\int P dx} = e^{-\int \tan x dx}$

$$= e^{\log \cos x} = \cos x$$

$\therefore$ Solution is

$$y \cdot \cos x = \int 2x \cdot \sec x \cdot \cos x \, dx + C$$

$$y \cos x = \int 2x \, dx + C$$

$$y \cos x = x^2 + C$$

$$y(0) = 0 \Rightarrow C = 0$$

$\therefore \quad y = x^2 \sec x$

Hence, $y\left(\frac{\pi}{4}\right) = \frac{\pi^2}{8\sqrt{2}}, y\left(\frac{\pi}{3}\right) = \frac{2\pi^2}{9}$

$$y'(x) = 2x\sec x + x^2 \sec x \tan x$$

$$y'\left(\frac{\pi}{4}\right) = \frac{\pi}{\sqrt{2}} + \frac{\pi^2}{16}\sqrt{2}$$

$$= \frac{\pi}{\sqrt{2}} + \frac{\pi^2}{8\sqrt{2}} = \frac{9\pi^2}{8\sqrt{2}}$$

$$y'\left(\frac{\pi}{3}\right) = \frac{2\pi}{3} \times 2 + \frac{\pi^2}{9} \times 2\sqrt{3}$$

$$= \frac{4\pi}{3} + \frac{2\sqrt{3}\pi^2}{9}$$

$$y'\left(\frac{\pi}{3}\right) = \frac{4\pi}{3} + \frac{2\pi^2}{3\sqrt{3}}$$

14. (a, b) We have,

$$x\cos x\left(\frac{dy}{dx}\right) + y(x\sin x + \cos x) = 1$$

$$\Rightarrow \quad \frac{dy}{dx} + y\left(\tan x + \frac{1}{x}\right) = \frac{1}{x\cos x}$$

This is a linear differential equation.

$\therefore \quad IF = e^{\int\left(\tan x + \frac{1}{x}\right)dx}$

$$= e^{-\ln \cos x + \ln x}$$

$$= e^{\ln\left(\frac{x}{\cos x}\right)}$$

$$= \frac{x}{\cos x}$$

Solution of given differential equation is

$$y \cdot \frac{x}{\cos x} = \int \frac{1}{x\cos x} \frac{x}{\cos x} dx$$

$$\Rightarrow \quad \frac{xy}{\cos x} = \int \sec^2 x \, dx$$

$$\Rightarrow \quad \frac{xy}{\cos x} = \tan x + C$$

$$\Rightarrow \quad xy\sec x = \tan x + C$$

and $\quad \frac{xy}{\cos x} = \frac{\sin x}{\cos x} + C$

$$xy = \sin x + C\cos x$$

15. (a,b)

We have,

$$g(x+y) = e^y g(x) + e^x g(y)$$

Put $x = y = 0$, then $g(0) = 0$

Now, $g(x+h) = e^h g(x) + e^x g(h)$

$$\Rightarrow \quad \frac{g(x+h) - g(x)}{h} = \frac{e^h - 1}{h} g(x) + e^x \frac{g(h) - g(0)}{h}$$

$$\Rightarrow \lim_{h \to 0} \frac{g(x+h) - g(x)}{h} = \lim_{h \to 0} \frac{e^h - 1}{h} g(x) + e^x \lim_{h \to 0} \frac{g(h) - g(0)}{h}$$

$$\Rightarrow \quad g'(x) = g(x) + e^x g'(0)$$

$$\Rightarrow \quad g'(x) = g(x) + 2e^x \quad [\because g'(0) = 2]$$

$$\Rightarrow \quad g'(x) - g(x) = 2e^x$$

This is a linear differential equation.

$\therefore$ IF $= e^{\int -1 \, dx} = e^{-x}$

So, solution is

$$g(x)e^{-x} = \int e^{-x} 2e^x dx$$

$$g(x)e^{-x} = 2x + C$$

Now, $\quad g(0) = 0$

$$g(0) = C$$

$$\Rightarrow \quad C = 0$$

$$g(x)e^{-x} = 2x$$

$$g(x) = 2xe^x$$

Now, $\quad g'(x) = 2e^x(1+x)$

$-$ $\quad$ $+$
-1

$g(x)$ is increasing $x \in [-1, \infty)$.

So, minimum value of $g(x)$ at $x = -1$

$$g(-1) = 2(-1)e^{-1} = -\frac{2}{e}$$

So, range $= \left[-\frac{2}{e}, \infty\right)$.

Now, $g''(x) = 2e^x(x+2) > 0$ when $x > -2$

Now, $\quad \lim_{x \to 0} \frac{g(x)}{x} = \lim_{x \to 0} \frac{2xe^x}{x}$

$$= \lim_{x \to 0} 2e^x = 2$$

16. (a,d) In ΔPMG

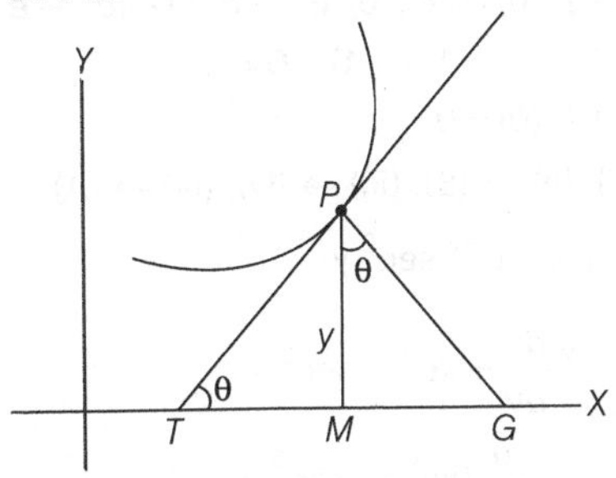

$$\frac{y}{PG} = \cos\theta$$

$\Rightarrow \quad PG = y\sec\theta = r = \sqrt{x^2 + y^2}$ [given]

$\therefore \quad y^2 \sec^2\theta = x^2 + y^2$

$\Rightarrow \quad y^2(1 + \tan^2\theta) = x^2 + y^2$

$\Rightarrow \quad y^2 + y^2\left(\frac{dy}{dx}\right)^2 = x^2 + y^2$

$\Rightarrow \quad y\,dy \pm x\,dx = 0$

or $\quad y^2 \pm x^2 = k^2$

This represents a circle or equilateral hyperbola.

17. (a, b, d) We have,

$$x^2\frac{dy}{dx} = y^2 e^{1/x}$$

$\Rightarrow \quad \frac{1}{y^2}dy = \frac{e^{1/x}}{x^2}dx$

On integrating both sides, we get

$$-\frac{1}{y} = -e^{1/x} + C$$

$\therefore \quad \lim_{x \to 0^-} f(x) = 1 \Rightarrow -1 = 0 + C$

$C = -1$

$\therefore \quad -\frac{1}{y} = -e^{1/x} - 1$

$\Rightarrow \quad y = \frac{1}{e^{1/x} + 1}$ ...(i)

Thus, from Eq. (i),

$$0 < y < 1$$

So, $y = f(x)$ is bounded.

$\Rightarrow \quad \frac{dy}{dx} = \frac{-1}{(1 + e^{1/x})^2} \times e^{1/x}\left(-\frac{1}{x^2}\right)$

$$= \frac{e^{1/x}}{x^2(1 + e^{1/x})^2}$$

$\therefore \quad \frac{dy}{dx} > 0, \forall\, x \in R - \{0\}$

$$\lim_{x \to \infty} \frac{1}{1 + e^{1/x}} = \frac{1}{2}$$

Thus, 1/2 does not belongs to range of $f(x)$.

So, range $f(x)$ is $(0, 1) - \left\{\frac{1}{2}\right\}$.

and $\lim_{x \to 0^+} \frac{1}{1 + e^{1/x}} = 0$

$\therefore$ Graph of the function is

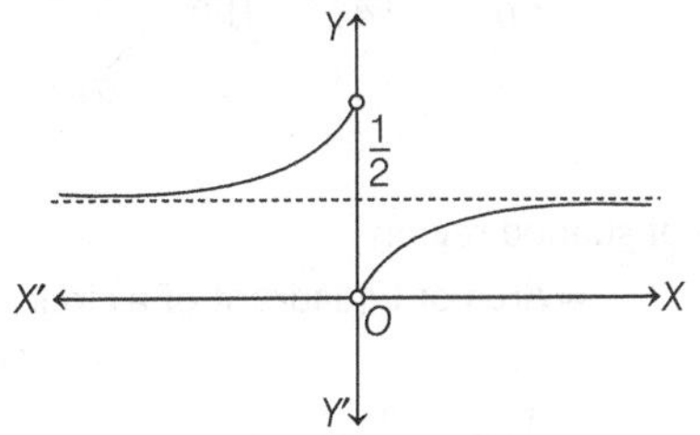

From the graph option (a), (b) and (d) are correct.

18. (b, c) We have,

$$f(x) = 1 - 2x + \int_0^x e^{x-t} f(t)\,dt$$

On multiplying e^{-x} both sides, we get

$$e^{-x}f(x) = e^{-x} - 2xe^{-x} + \int_0^x e^{-t}f(t)dt$$

On differentiating both side w.r.t. x, we get

$$e^{-x}f'(x) - e^{-x}f(x) = -e^{-x} - 2e^{-x} + 2xe^{-x} + e^{-x}f(x)$$

$\Rightarrow \quad f'(x) - 2f(x) = 2x - 3$ [dividing both sides by e^{-x}]

Let $\quad f(x) = y$

$\Rightarrow \quad f'(x) = \frac{dy}{dx}$

$\therefore \quad \frac{dy}{dx} - 2y = 2x - 3$

which is linear differential equation of the form

$$\frac{dy}{dx} + Py = Q.$$

Here, $P = -2$ and $Q = 2x - 3$.

Now, $\quad IF = e^{\int P\,dx} = e^{\int -2\,dx} = e^{-2x}$

$\therefore$ Solution of the given differential equation is

$$y \cdot e^{-2x} = \int \underset{\text{I}}{(2x - 3)}\, \underset{\text{II}}{e^{-2x}}\,dx + C$$

$$y \cdot e^{-2x} = \frac{-(2x-3)\cdot e^{-2x}}{2} + 2\int \frac{e^{-2x}}{2}dx + C$$

[by using integration by parts]

$\Rightarrow \quad y \cdot e^{-2x} = \frac{-(2x-3)\,e^{-2x}}{2} - \frac{e^{-2x}}{2} + C$

$\Rightarrow \quad y = (1 - x) + Ce^{2x}$

On putting $x = 0$ and $y = 1$, we get

$$1 = 1 + C \Rightarrow C = 0$$

$\therefore \quad y = 1 - x$

$y = 1 - x$ passes through $(2, -1)$

Now, area of region bounded by curve

$y = \sqrt{1 - x^2}$ and $y = 1 - x$ is shows as

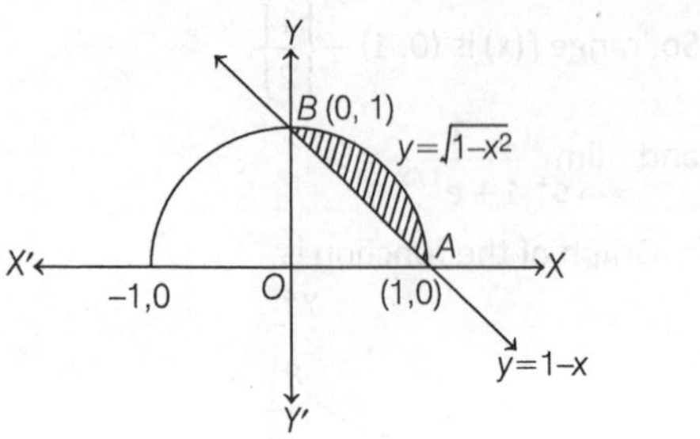

∴ Area of shaded region

$$= \text{Area of I quadrant of a circle} - \text{Area of } \Delta OAB$$

$$= \frac{\pi}{4}(1)^2 - \frac{1}{2} \times 1 \times 1$$

$$= \frac{\pi}{4} - \frac{1}{2} = \frac{\pi - 2}{4}$$

Sol. (Q.Nos. 19-21)

19. (a) **20.** (b) **21.** (c)

We have, $f(x) = (e^{-x} + e^{x})\cos x - 2x - \int_0^x (x - t)f'(t)dt$

$$f(x) = (e^{x} + e^{-x})\cos x - 2x - [x\int_0^x f'(t)dt - \int_0^x tf'(t)dt]$$

$$f(x) = (e^{x} + e^{-x})\cos x - 2x - [xf(x) - xf(0) - \{[tf(t)]_0^x - \int_0^x f(t)dt\}]$$

$$f(x) = (e^{x} + e^{-x})\cos x - 2x - xf(x) + 2x + [xf(x) - \int_0^x f(t)dt]$$

[as $f(0) = 2$]

$$\Rightarrow \quad f(x) = (e^{x} + e^{-x})\cos x - \int_0^x f(t)\,dt \quad \text{...(i)}$$

On differentiating both sides w.r.t. x,

$$f'(x) + f(x) = \cos x(e^{x} - e^{-x}) - (e^{x} + e^{-x})\sin x \quad \text{...(ii)}$$

$$\Rightarrow \frac{dy}{dx} + y = e^{x}(\cos x - \sin x) - e^{-x}(\cos x + \sin x) \quad \text{...(iii)}$$

[$\because y = f(x)$]

This is linear differential equation.

$\text{IF} = e^{x}$

∴ Solution is

$$y \cdot e^{x} = \int e^{2x}(\cos x - \sin x)dx - \int(\cos x + \sin x)dx$$

$$\Rightarrow y \cdot e^{x} = \int e^{2x}(\cos x - \sin x)dx - (\sin x - \cos x) + C$$

$$\Rightarrow y \cdot e^{x} = e^{2x}\left(\frac{3}{5}\cos x - \frac{1}{5}\sin x\right) - (\sin x - \cos x) + C$$

When $x = 0$, $y = 2$

$$2 = \frac{3}{5} + 1 + C$$

$$C = \frac{2}{5}$$

$$y = e^{x}\left(\frac{3}{5}\cos x - \frac{1}{5}\sin x\right) - (\sin x - \cos x)e^{-x} + \frac{2}{5}e^{-x}$$

$$f(0) + f'(0) = \cos(0)(e^{0} - e^{-0}) - (e^{0} + e^{-0})\sin(0)$$

$$= 1(1 - 1) - 0 = 0$$

$$\therefore \quad f(0) + f'(0) = 0$$

22. (i) → (C), (ii) → (B), (iii) → (D), (iv) → (A)

(i) $\frac{dx}{dy} + x = e^{-y}\sec^2 y$

$$\Rightarrow e^{y}\frac{dx}{dy} + xe^{y} = \sec^2 y$$

$$\Rightarrow \frac{d}{dy}(xe^{y}) = \sec^2 y$$

$$\Rightarrow xe^{y} = \tan y + C$$

(ii) We have,

$$(y\log x - 1)y\,dx = x\,dy$$

$$\Rightarrow \frac{dy}{dx} = (y\log x - 1)\frac{y}{x}$$

$$\Rightarrow \frac{dy}{dx} + \frac{y}{x} = y^2\left(\frac{\log x}{x}\right)$$

$$\Rightarrow \frac{1}{y^2}\frac{dy}{dx} + \frac{1}{xy} = \frac{\log x}{x}$$

Let $\frac{1}{y} = z \Rightarrow -\frac{1}{y^2}\frac{dy}{dx} = \frac{dz}{dx}$

$$\therefore \quad -\frac{dz}{dx} + \frac{z}{x} = \frac{\log x}{x}$$

$$\Rightarrow \frac{dz}{dx} - \frac{z}{x} = \frac{-\log x}{x}$$

Now, $\text{IF} = e^{\int -\frac{1}{x}dx} = e^{\log\frac{1}{x}} = \frac{1}{x}$

Solution is $z\left(\frac{1}{x}\right) = \int -\frac{\log x}{x} \cdot \frac{1}{x}dx + C$

$$= -\int \frac{\log x}{x^2}dx$$

$$= -\left[-\frac{1}{x}\log x - \int -\frac{1}{x} \cdot \frac{1}{x}dx\right] + C$$

$$= \frac{1}{x}\log x + \frac{1}{x} + C$$

Therefore, $z = \log x + 1 + Cx$

$$z = \log(ex) + Cx$$

$$\Rightarrow \frac{1}{y} = \log(ex) + Cx$$

$$\Rightarrow 1 = y\{\log(ex) + Cx\}$$

(iii) We have,

$$\frac{dy}{dx} = e^{x+y}$$

$$\Rightarrow \frac{dy}{dx} = e^{x} \cdot e^{y}$$

$$\Rightarrow \int e^{-y}dy = \int e^{x}dx$$

$$\Rightarrow -e^{-y} = e^{x} + C$$

$$\Rightarrow e^{x} + e^{-y} = C$$

(iv) We have,

$$\frac{dy}{dx} = e^{x-y} + x^2 e^{-y}$$

$$\Rightarrow \quad e^y dy = (e^x + x^2)dx$$

$$\Rightarrow \quad \int e^y dy = \int (e^x + x^2)dx + C$$

$$\Rightarrow \quad e^y = e^x + \frac{x^3}{3} + C$$

23. (i) → (C), (ii) → (A), (iii) → (D), (iv) → (B)

(i) We have, $y = e^{4x} + 2e^{-x}$

$$\Rightarrow \frac{dy}{dx} = 4e^{4x} - 2e^{-x}$$

$$\Rightarrow \frac{d^2y}{dx^2} = 16e^{4x} + 2e^{-x}$$

$$\Rightarrow \frac{d^3y}{dx^3} = 64e^{4x} - 2e^{-x}$$

Now, $\frac{d^3y}{dx^3} - 13\frac{dy}{dx} = (64e^{4x} - 2e^{-x})$

$$-13\,(4e^{4x} - 2e^{-x})$$

$$= 12e^{4x} + 24e^{-x}$$

$$\Rightarrow \frac{d^3y}{dx^3} - \frac{13dy}{dx} = 12(e^{4x} + 2e^{-x})$$

$$\Rightarrow \quad \frac{d^3y}{dx^3} - 13\frac{dy}{dx} = 12y$$

$\therefore \quad k = 12$

and $\quad \frac{k}{3} = 4$

(ii) Given, $\frac{dy}{dx} + \left(\frac{1}{x\log x}\right)y = 2$

This is linear differential equation.

$$\text{IF} = e^{\int \frac{1}{x\log x}dx} = e^{\log(\log x)} = \log x$$

So, $y \cdot \log x = \int 2\log x\, dx + C$

$\Rightarrow y \cdot \log x = 2[x\log x - x] + C$

Put $x = 1$, then $C = 2$

Hence, $y\log x = 2[x\log x - x] + 2$

At $x = e$

$y\log e = 2[e\log e - e] + 2$

$y = 2$

(iii) We have, $\quad x^k + y^k = a^k$

$$\Rightarrow kx^{k-1} + ky^{k-1}\frac{dy}{dx} = 0$$

$$\Rightarrow \quad \frac{dy}{dx} + \left(\frac{x}{y}\right)^{k-1} = 0$$

and given that

$$\frac{dy}{dx} + \left(\frac{y}{x}\right)^{1/3} = 0 \quad \text{...(ii)}$$

On comparing Eqs. (i) and (ii), we get

$$k - 1 = \frac{-1}{3} \Rightarrow k = \frac{2}{3}$$

So, $p + q = 2 + 3 = 5$

(iv) We have, $y = Ax^m + Bx^{-n}$

$$\therefore \quad \frac{dy}{dx} = Amx^{m-1} - nBx^{-n-1}$$

$$\therefore \frac{d^2y}{dx^2} = Am(m-1)x^{m-2} + n(n+1)x^{-n-2}$$

Putting values in $\frac{x^2d^2y}{dx^2} + 2x\frac{dy}{dx} = 12y$, we get

$m(m+1)Ax^m + n(n-1)Bx^{-n} = 12(Ax^m + Bx^{-n})$

i.e. $m(m+1) = 12$ and $n(n-1) = 12$

i.e. $m = 3, -4$ and $n = 4, -3$

24. (i) → (D), (ii) → (A), (iii) → (B), (iv) → (C)

(i) We have, $\frac{dy}{dx} - y = 1 - e^{-x}$

$$\Rightarrow \quad e^{-x}\frac{dy}{dx} - ye^{-x} = e^{-x} - e^{-2x}$$

$$\Rightarrow e^{-x}dy - ye^{-x}\,dx = (e^{-x} - e^{-2x})\,dx$$

$$\Rightarrow \quad \int d(ye^{-x}) = \int (e^{-x} - e^{-2x})\,dx$$

$$\Rightarrow \quad ye^{-x} = -e^{-x} + \frac{e^{-2x}}{2} + C$$

When $x = 0$, $y = y_0$

$$\therefore \quad y_0 = -1 + \frac{1}{2} + C \Rightarrow C = y_0 + \frac{1}{2}$$

$$\therefore\; ye^{-x} = -e^{-x} + \frac{e^{-2x}}{2} + y_0 + \frac{1}{2}$$

When $x \to +\infty$

$$\therefore \quad 0 = 0 + 0 + y_0 + \frac{1}{2} = \frac{1}{y_0} = -2$$

$$\Rightarrow \left|\frac{1}{y_0}\right| = |-2| = 2$$

(ii) We have, $\frac{dy}{dx} - y\ln 2 = 2^{\sin x}(\cos x - 1)\ln 2$

$$\text{IF} = e^{\int -\ln 2\,dx} = e^{-x\ln 2} = e^{\ln 2^{-x}} = 2^{-x}$$

$\therefore$ Solution of differential equation is

$$y2^{-x} = \int 2^{\sin x - x}(\cos x - 1)\ln 2\,dx$$

$$\Rightarrow \quad y2^{-x} = 2^{\sin x - x} + C$$

$$\Rightarrow \quad y = 2^{\sin x} + C \cdot 2^x$$

$x \to \infty$, y is bounded

$\therefore \quad C = 0$

$y = 2^{\sin x}$

When $x = \frac{\pi}{2}$, $y = 2^{\sin(\pi/2)} = 2$ and $y^2 = (2)^2 = 4$

$$\therefore\; y\left(\frac{\pi}{2}\right) + y^2\left(\frac{\pi}{2}\right) = 2 + 4 = 6$$

(iii) We have, $xf'(x)-2f(x)=x^4f^2(x)$ $[x>0]$

$$\Rightarrow \quad x\frac{dy}{dx}-2y=x^4y^2$$

$$\Rightarrow \quad \frac{1}{y^2}\frac{dy}{dx}-\frac{2}{xy}=x^3$$

Put $-\frac{1}{y}=z \Rightarrow \frac{1}{y^2}\frac{dy}{dx}=\frac{dz}{dx}$

$\therefore$ Equation becomes

$$\frac{dz}{dx}+\frac{2z}{x}=x^3$$

$$\text{IF}=e^{\int\frac{2}{x}dx}=e^{2\log x}=x^2$$

Solution of differential equation is

$$zx^2=\int x^5\,dx=\frac{x^6}{6}+C$$

$$\Rightarrow \quad -\frac{1}{y}=\frac{x^4}{6}+Cx^{-2}$$

When $x=1, y=-6$

$$\therefore \quad \frac{1}{6}=\frac{1}{6}+C \Rightarrow C=0$$

$$\therefore \quad -\frac{1}{y}=\frac{x^4}{6}\Rightarrow y=-6x^{-4}$$

$$\frac{dy}{dx}=24x^{-5}$$

$$\Rightarrow \quad f'(x)=24x^{-5}$$

$$f'(3^{1/5})=24(3)^{5\times\left(-\frac{1}{5}\right)}=\frac{24}{3}=8$$

(iv) We have, $\int_0^x f(t)\,dt+\int_0^x tf(x-t)\,dt=e^{-x}-1$

$$=\int_0^x f(t)\,dt+\int_0^x (x-t)\,f(t)\,dt=e^{-x}-1$$

$$=x\int_0^x f(t)\,dt+\int_0^x (1-t)\,f(t)\,dt=e^{-x}-1$$

Differentiating both sides w.r.t. x, we get

$$\int_0^x f(t)\,dt+xf(x)+(1-x)\,f(x)=-e^{-x}$$

Again, differentiating both sides w.r.t. x, we get

$$f(x)+\{f(x)+xf'(x)\}+\{f'(x)-xf'(x)\}-f(x)=e^{-x}$$

$$\Rightarrow \quad f'(x)+f(x)=e^{-x}$$

$$\text{IF}=e^x$$

Solution is

$$\therefore \quad y\cdot e^x=\int 1\,dx$$

$$\Rightarrow \quad f(x)e^x=x+C,\ f(0)=-1$$

$$\Rightarrow \quad f(x)e^x=x-1$$

When $x=10$

$$f(10)\,e^{10}=10-1=9$$

Sol. (Q.Nos. 25-27)

25. (2) The given equation is

$$\frac{dy}{dx}=\sin(x+y)+\cos(x+y)$$

Put $x+y=u$

$$\frac{du}{dx}=1+\sin u+\cos u$$

$$=(1+\cos u)+\sin u$$

$$=2\cos^2\frac{u}{2}+2\sin\frac{u}{2}\cos\frac{u}{2}$$

Therefore, $\dfrac{1}{2\cos^2\frac{u}{2}+2\sin\frac{u}{2}\cos\frac{u}{2}}du=dx$

$$\Rightarrow \quad \frac{\frac{1}{2}\sec^2\frac{u}{2}}{1+\tan\frac{u}{2}}du=dx$$

Integrating both sides, we get

$$\int\frac{\frac{1}{2}\sec^2\frac{u}{2}}{1+\tan\frac{u}{2}}du=x+C_1$$

$$\Rightarrow \quad \log\left(1+\tan\frac{u}{2}\right)=x+C_1$$

$$\Rightarrow \quad 1+\tan\left(\frac{x+y}{2}\right)=e^{x+C_1}$$

$$\Rightarrow \quad 1+\tan\left(\frac{x+y}{2}\right)=Ce^x \qquad [C=e^{C_1}]$$

$$\Rightarrow \quad k=2$$

26. (2) The given equation is

$$\frac{dy}{dx}=\frac{x(2\log x+1)}{\sin y+y\cos y}$$

$$\Rightarrow \quad (\sin y+y\cos y)dy=(2x\log x+x)dx$$

Integrating both sides, we get

$$\int(\sin y+y\cos y)dy=\int(2x\log x+x)dx+C$$

$$\Rightarrow \quad -\cos y+\cos y+y\sin y$$

$$=x^2\log x-\int x^2\cdot\frac{1}{x}dx+\int x\,dx+C$$

$$\Rightarrow \quad y\sin y=x^2\log x+C$$

$$\Rightarrow \quad k=2$$

27. (2) We have,

$$\frac{dy}{dx}+\frac{y}{x}=\frac{1}{(1+\log x+\log y)^2}$$

$$=\frac{1}{(1+\log xy)^2}$$

Put $xy=u$, so that

$$\frac{du}{dx}=\frac{x}{(1+\log u)^2}$$

Therefore, $(1+\log u)^2\, du = x\, dx$

$$\Rightarrow \quad \int (1+\log u)^2\, du = \int x\, dx + C$$

$$\Rightarrow \quad u(1+\log u)^2 - \int \frac{2(1+\log u)}{u} \cdot u\, du = \frac{x^2}{2} + C$$

$$\Rightarrow \quad u(1+\log u)^2 - 2u - 2\left[u\log u - \int 1\, du\right] = \frac{x^2}{2} + C$$

$$\Rightarrow \quad u[1+2\log u + (\log u)^2] - 2u\log u = \frac{x^2}{2} + C$$

$$\Rightarrow \quad u[(1+(\log u)^2] = \frac{x^2}{2} + C$$

$$\Rightarrow \quad xy[1+\{\log(xy)\}^2] = \frac{x^2}{2} + C$$

$$\Rightarrow \quad k = 2$$

Sol. (Q.Nos. 28-30)

28. (10) We have,

$$\int_0^x (x-t+1)g(t)dt = x^4 + x^2, \forall\, x \geq 0$$

$$\Rightarrow \quad x\int_0^x g(t)dt - \int_0^x tg(t)dt + \int_0^x g(t)dt = x^4 + x^2$$

On differentiating both sides, we get

$$\int_0^x g(t)dt + xg(x) - xg(x) + g(x) = 4x^3 + 2x$$

$$\Rightarrow \quad \int_0^x g(t)dt + g(x) = 4x^3 + 2x \qquad \text{...(i)}$$

Again, differentiating both sides, we get

$$g(x) + g'(x) = 12x^2 + 2$$

$$\Rightarrow \quad y + \frac{dy}{dx} = 12x^2 + 2$$

$$\Rightarrow \quad y + \frac{dy}{dx} = 12x^2 + 2$$

$$\Rightarrow \quad \alpha = 12, \beta = 2$$

$$\therefore \quad \alpha - \beta = 12 - 2 = 10$$

29. (0) From Eq. (i),

$$\int_0^x g(t)dt + g(x) = 4x^3 + 2x$$

When $x = 0$

$$0 + g(0) = 0 \Rightarrow g(0) = 0$$

30. (4) Let $I = \int_0^1 \frac{12}{g'(x) + g(x) + 10}\, dx$

$$I = \int_0^1 \frac{12}{12x^2 + 2 + 10}\, dx$$

$$I = \int_0^1 \frac{dx}{x^2+1}$$

$$= [\tan^{-1} x]_0^1$$

$$= \frac{\pi}{4}$$

$\Rightarrow k = 4$

31. (2) We have, $\frac{dy}{dx} = \frac{y(x+y^3)}{x(y^3-x)}$

$$\Rightarrow \quad (xy^3 - x^2)dy - (xy + y^4)dx = 0$$

$$\Rightarrow \quad y^3(xdy - y\, dx) - x(x\, dy + y\, dx) = 0$$

$$\Rightarrow \quad x^2y^3\frac{(x\, dy - y\, dx)}{x^2} - x(x\, dy + y\, dx) = 0$$

$$\Rightarrow \quad x^2y^3 d\left(\frac{y}{x}\right) - xd\,(xy) = 0$$

Dividing by x^3y^2,

$$\frac{y}{x} d\left(\frac{y}{x}\right) - \frac{d(xy)}{x^2y^2} = 0$$

Integrating, we get

$$\frac{1}{2}\left(\frac{y}{x}\right)^2 + \frac{1}{xy} = C$$

It passes through the point $(4, -2)$, then

$$\frac{1}{8} - \frac{1}{8} = C \Rightarrow C = 0$$

$$\therefore \quad y^3 = -2x$$

Hence, $|k| = |-2| = 2$

32. (2) We have, $dy = e^{\alpha x + y} dx$

$$\int e^{-y} dy = \int e^{\alpha x} dx$$

$$\Rightarrow \quad -e^{-y} = \frac{e^{\alpha x}}{\alpha} + C \qquad \text{...(i)}$$

Put $y(\log 2) = \log 2$, we get

$$-\frac{1}{2} = \frac{2^\alpha}{\alpha} + C \qquad \text{...(ii)}$$

Put $y(0) = \log\frac{1}{2}$ in Eq. (i),

$$-2 = \frac{1}{\alpha} + C \qquad \text{...(iii)}$$

Subtracting Eq. (ii) from Eq. (iii),

$$\frac{2^\alpha - 1}{\alpha} = \frac{3}{2} \Rightarrow \alpha = 2 \qquad [\text{as } \alpha \in N]$$

33. (0.4) Given that

$$\frac{dy}{dx} = (5y+2)(5y-2)$$

$$\Rightarrow \quad \frac{dy}{dx} = 25y^2 - 4$$

$$\Rightarrow \quad \frac{1}{25}\int \frac{dy}{y^2 - \left(\frac{2}{5}\right)^2} = \int dx$$

$$\Rightarrow \quad \frac{1}{25} \cdot \frac{5}{4} \log\left|\frac{y - \frac{2}{5}}{y + \frac{2}{5}}\right| = x + C$$

$$\Rightarrow \quad \frac{1}{20}\log\left|\frac{5y-2}{5y+2}\right| = x + C$$

Given that $f(0) = y(0) = 0$

$$\therefore \quad C = 0$$

Hence, $\left|\frac{2-5y}{2+5y}\right| = e^{20x}$

$$\therefore \quad \lim_{x \to -\infty}\left|\frac{2-5y}{2+5y}\right| = \lim_{x \to -\infty} e^{20x}$$

$$\Rightarrow \quad \lim_{x \to -\infty}\left|\frac{2-5y}{2+5y}\right| = 0$$

$$\Rightarrow \quad 2 - 5\lim_{x \to -\infty} y = 0$$

$$\Rightarrow \quad \lim_{x \to -\infty} y = \frac{2}{5} = 0.4$$

34. (4) Let equation of straight line

$$y = mx + c \qquad \text{...(i)}$$

where, $m = \frac{dy}{dx}$

It satisfies

$$\frac{dy}{dx} + x\left(\frac{dy}{dx}\right)^2 = y$$

$$\Rightarrow \quad m + x(m^2) = y$$

From Eq. (i),

$$m + m^2 x = mx + C$$

$$\Rightarrow \quad m^2 x - mx + m - C = 0$$

$$\Rightarrow \quad x(m^2 - m) + (m - c) = 0$$

It is possible, when

$$m^2 - m = 0 \text{ and } m = C$$

$$\Rightarrow \quad m(m-1) = 0$$

So, $m = 0$ or $m = 1$ and $m = C$

If $m = 0, C = 0,$

$$y = 0$$

and if $m = 1, C = 1$

Then, $y = x + 1$

Hence, the number of straight line is 2.

Thus, $m = 2$, because the number of straight lines which satisfied the given differential equation is m.

So, the value of $m + 2 = (2 + 2) = 4$.

35. (3) The equation of the normal at $P(x, y)$ is

$$Y - y = -\frac{dx}{dy}(X - x)$$

Putting $Y = 0$, we get

$$-y = -\frac{dx}{dy}(X - x)$$

$$\Rightarrow \quad X = y\frac{dy}{dx} + x$$

Thus, $B = \left(y\frac{dy}{dx} + x, 0\right)$ and $A = (x, 0)$

So, it is given that

$$x + y\frac{dy}{dx} - x = \frac{x+y}{2}$$

$$\Rightarrow \quad y\frac{dy}{dx} = \frac{x+y}{2}$$

$$\Rightarrow \quad \frac{dy}{dx} = \frac{x+y}{2y}$$

Let $y = vx \Rightarrow \frac{dy}{dx} = v + x\frac{dv}{dx}$

$$\therefore \quad v + x\frac{dv}{dx} = \frac{x + vx}{2vx} = \frac{1+v}{2v}$$

$$\Rightarrow \quad x\frac{dv}{dx} = \frac{1+v}{2v} - v = \frac{1+v-2v^2}{2v}$$

$$\Rightarrow \quad \int\frac{2v\,dv}{2v^2 - v - 1} = -\int\frac{dx}{x}$$

$$\Rightarrow \quad \int\frac{2v\,dv}{(2v+1)(v-1)} = -\int\frac{dx}{x}$$

$$\Rightarrow \quad \int\frac{\{(2v+1)-1\}\,dv}{(2v+1)(v-1)} = -\int\frac{dx}{x}$$

$$\Rightarrow \int\frac{dv}{v-1} - \int\frac{dv}{(2v+1)(v-1)} = -\int\frac{dx}{x}$$

$$\Rightarrow \int\frac{dv}{v-1} - \frac{1}{3}\int\left[\frac{1}{v-1} - \frac{2}{2v+1}\right]dv = -\int\frac{dx}{x}$$

$$\Rightarrow \quad \frac{2}{3}\int\frac{dv}{v-1} + \frac{2}{3}\int\frac{dv}{2v+1} = -\int\frac{dx}{x}$$

$$\Rightarrow \quad 2\int\frac{dv}{v-1} + 2\int\frac{dv}{2v+1} = -3\int\frac{dx}{x}$$

$$\Rightarrow \quad 2\log|v-1| + \frac{2}{2}\log|2v+1| = -3\log|x| + \log C$$

$$\Rightarrow \quad \log|v-1| + \frac{1}{2}\log|2v+1| = \frac{1}{2}\log\left(\frac{C}{x^3}\right)$$

$$\Rightarrow \quad (v-1)\sqrt{(2v+1)} = \sqrt{\left(\frac{C}{x^3}\right)}$$

$$\Rightarrow \quad \left(\left(\frac{y}{x}\right) - 1\sqrt{\left(2\left(\frac{y}{x}\right) + 1\right)}\right) = \sqrt{\left(\frac{C}{x^3}\right)}$$

$$\left[\text{since, } v = \frac{y}{x}\right]$$

By comparing, we get

$$A = 1, B = 2$$

$$\Rightarrow \quad A + B = 3$$

36. (3) $y \cdot \sin 2x - \cos x + (1+\sin^2 x)\frac{dy}{dx} = 0$

$$\left[\text{let } f(x) = y, f'(x) = \frac{dy}{dx}\right]$$

$$\Rightarrow \quad (1+\sin^2 x)\frac{dy}{dx} + (\sin 2x)y = \cos x$$

$$\Rightarrow \quad \frac{dy}{dx} + \left(\frac{\sin 2x}{1+\sin^2 x}\right)y = \frac{\cos x}{1+\sin^2 x}$$

This is a linear differential equation.

$$\therefore \quad \text{IF} = e^{\int P dx} = e^{\int \frac{\sin 2x}{1+\sin^2 x} dx}$$

Let $1+\sin^2 x = t \Rightarrow \sin 2x dx = dt$

$$\text{IF} = e^{\int \frac{dt}{t}} = e^{\ln t} = t$$

$$\text{IF} = (1+\sin^2 x)$$

Here solution is

$$y \cdot (1+\sin^2 x) = \int \frac{\cos x}{(1+\sin^2 x)}(1+\sin^2 x)dx$$

$$\Rightarrow \quad y(1+\sin^2 x) = \sin x + C \qquad \text{...(i)}$$

Given that $y(0) = 0$

$$\Rightarrow \quad C = 0$$

Substituting in Eq. (i),

$$y = \frac{\sin x}{1+\sin^2 x}$$

Now, when $x = \frac{\pi}{6}$, then $y = \frac{\sin\frac{\pi}{3}}{1+\sin^2\frac{\pi}{3}} = \frac{2}{5}$

Hence, $|A-B| = |2-5| = 3$

37. (16) Given that

$$f(x) = e^{-x}\int_3^x \{3t^2 + 2t + 4f'(t)\}dt$$

$\therefore f(3) = 0$

Now, $\quad e^x f(x) = \int_3^x \{3t^2 + 2t + 4f'(t)\}dt$

$$\Rightarrow \quad e^x f(x) + e^x f'(x) = 3x^2 + 2x + 4f'(x)$$

$$\Rightarrow \quad (e^x - 4)\frac{dy}{dx} + e^x y = (3x^2 + 2x)$$

This is a linear differential equation.

So, $\text{IF} = e^{\int P dx} = e^{\int \frac{e^x}{e^x - 4}dx} = e^{\ln(e^x - 4)} = e^x - 4$

Solution is $y \cdot (e^x - 4) = \int\left(\frac{3x^2+2x}{(e^x-4)}\cdot(e^x-4)\right)dx$

$$\Rightarrow \quad y(e^x - 4) = x^3 + x^2 + C$$

Put $x = 3$

Then, $y = 0 \Rightarrow C = -36$

$$y = \frac{x^3 + x^2 - 36}{e^x - 4}$$

$$f'(x) = \frac{(3x^2+2x)(e^x-4) - (x^3+x^2-36)e^x}{(e^x-4)^2}$$

Put $x = 4$, then $f'(4) = \frac{12e^4 - 224}{(e^4-4)^2}$

$$\therefore \quad \alpha = 12 \text{ and } \beta = 4$$

Hence, $\alpha + \beta = 12 + 4 = 16$

38. (2) We have,

$$\frac{dy}{dx} = \frac{1}{x\cos y + \sin 2y}$$

$$\Rightarrow \quad \frac{dy}{dx} = \frac{1}{x\cos y + 2\sin y\cos y}$$

$$\therefore \quad \frac{dx}{dy} = x\cos y + 2\sin y\cos y$$

$$\Rightarrow \quad \frac{dx}{dy} + (-\cos y)x = 2\sin y\cos y$$

This is a linear differential equation.

$\text{IF} = e^{\int -\cos y dy} = e^{-\sin y}$

Thus, solution is

$$x \cdot e^{-\sin y} = 2\int e^{-\sin y}\cdot \sin y\cos y\, dy$$

$$= -2\sin y e^{-\sin y} + 2\int e^{-\sin y}\cos y dy$$

$$\Rightarrow \quad xe^{-\sin y} = -2\sin y e^{-\sin y} - 2e^{-\sin y} + C$$

$$\Rightarrow \quad x = -2\sin y - 2 + Ce^{\sin y}$$

$$x = Ce^{\sin y} - 2(1+\sin y)$$

$$\therefore \quad k = 2$$

39. (0) Given that

$$y'(x) + y(x)g'(x) = g(x)g'(x),\ y(0) = 0,\ x \in R$$

$$\frac{d}{dx}\{y(x)\} + g'(x)y(x) = g(x)g'(x)$$

$\text{IF} = e^{\int g'(x)dx} = e^{g(x)}$

So, $y(x)\cdot e^{g(x)} = \int e^{g(x)}\cdot g(x)\cdot g'(x)dx + C$

Let $g(x) = t$

$$\Rightarrow \quad g'(x)dx = dt$$

$$y(x)e^{g(x)} = \int t\cdot e^t dt + C$$

$$= t\cdot e^t - e^t + C$$

$$\Rightarrow \quad y(x)e^{g(x)} = e^{g(x)}\{g(x) - 1\} + C$$

$$\Rightarrow \quad y(x) = (g(x) - 1) + Ce^{-g(x)}$$

Put $x = 0$, then

$$y(0) = (g(0) - 1) + Ce^{-g(0)}$$

$$\Rightarrow \quad C = 1$$

$$\Rightarrow \quad y(x) = (g(x) - 1) + e^{-g(x)}$$

When $x = 2$, then

$$y(2) = (g(2) - 1) + e^{-g(2)}$$

$$\Rightarrow \quad y(2) = (0 - 1) + e^{-(0)} = -1 + 1 = 0$$